T0133457

The Handbook of Mobile Middleware

The Handbook
of Mobile
Middleware

Edited by
Paolo Bellavista • Antonio Corradi

Auerbach Publications
Taylor & Francis Group
Boca Raton   New York

Auerbach Publications is an imprint of the
Taylor & Francis Group, an informa business

Auerbach Publications
Taylor & Francis Group
6000 Broken Sound Parkway NW, Suite 300
Boca Raton, FL 33487-2742

© 2007 by Taylor & Francis Group, LLC
Auerbach is an imprint of Taylor & Francis Group, an Informa business

International Standard Book Number-10: 0-8493-3833-6 (Hardcover)
International Standard Book Number-13: 978-0-8493-3833-5 (Hardcover)

Library of Congress Cataloging-in-Publication Data

Bellavista, Paolo.
   The handbook of mobile middleware / Paolo Bellavista and Antonio Corradi.
     p. cm.
   Includes bibliographical references and index.
   ISBN 0-8493-3833-6 (978-0-8493-3833-5 : alk. paper)
    1. Mobile computing--Handbooks, manuals, etc. 2. Middleware--Handbooks, manuals, etc. 3. Ubiquitous computing--Handbooks, manuals, etc. I. Corradi, A. (Antonio) II. Title.

QA76.59B42 2006
004.165--dc22
                                                     2006005314

Visit the Taylor & Francis Web site at
http://www.taylorandfrancis.com

and the Auerbach Web site at
http://www.auerbach-publications.com

Preface

Because device miniaturization and wireless communications are making it more feasible to use mobility-enhanced services to exploit all potential and all the opportunities of mobile computing, the ultimate goal of mobility scenarios is becoming the realization of ubiquitous, pervasive, and eventually disappearing computing — in other words, the seamless and transparent collaboration of wireless devices with most human activities without the need for explicit user or administration intervention. This vision, however, introduces novel challenges for the support infrastructure that require novel middleware solutions capable of addressing connectivity-level, location-dependent, and context-dependent support issues, all of which are crucial for advanced adaptive services for mobile environments.

The purpose of this advanced reference book is to provide an exhaustive overview of the work done in recent years in very different fields related to software support of mobile computing. The goal is to present some relevant results obtained and lessons learned in a single compendium that adopts the original and comprehensive perspective of developing and deploying mobile middleware. The book begins by presenting mobile middleware motivations, requirements, and technologies; it then proposes a taxonomy of solutions found in the literature and organized on the basis of their increasingly complex goals (mobility and disconnection handling, location-based support, and context-based support). In addition, the book pays particular attention to the variety of application domains in which mobile middleware is demonstrating its feasibility and effectiveness and shows by example the pros, cons, and tradeoffs of the emerging mobile middleware solutions. Some of the goals that have inspired our work include:

- ■ Clarifying and organizing timely reference information about a hot topic in a way that we believe will enhance its relevance
- ■ Assembling the authoritative opinions of many recognized experts in the field to produce a significant collection of important but different perspectives
- ■ Producing a compendium useful for experts in the field as a reference volume or comparison stimulus as well as for people who wish to improve their knowledge via an easy-to-follow guide that also provides deep technical details
- ■ Providing a tool that organizes our knowledge in the field of mobile middleware and spreading this knowledge throughout our research community

We approached the design of this book by considering what might be suitable for graduate students and Ph.D. candidates, as well as for researchers and practitioners in the areas of both mobile computing and application-level environments for the provisioning of mobility-enhanced services. This book could serve as a reference in this area and as the basis for an advanced mobile computing seminar class.

We begin in Section 1 by establishing a common background regarding mobile computing. Some innovative technologies that are emerging and providing relevant solutions in the development of mobile middleware are covered in Section 2, which has the objective not of presenting single solutions but of surveying the main developments contributing to such innovations. Section 3 addresses the primary requirements of support infrastructures and stresses the crucial requirement of maintaining session information independently of user, terminal, and resource mobility during service provisioning.

The next three sections share the goal of presenting a complete overview of real results deriving from investigations and discussions in the field of mobile middleware at large. We have classified mobile middleware into three categories. Section 4 reports on mobile middleware that supports seamless connectivity during the access of traditional services; such middleware is typically designed and deployed for wired networks via wireless terminals. This section also presents solutions for transparent network connectivity, data consistency, and resource and service replication. Section 5 presents mobile middleware that addresses the further challenge of increasing complexity — that is, the support of novel location-dependent services that must deal with client location or client–server positions at provisioning time. Section 6 addresses mobile middleware for novel and more complex application scenarios where service contents must be adapted on the fly to the currently applicable context; doing so allows the deployment of services that also depend on the profile of client terminals, on user preferences, on server congestion, and, in general, on any relevant property describing the distributed resources involved in mobility-enhanced service provisioning.

Section 7 is devoted to presenting the primary application areas where the need for mobile middleware solutions has already emerged in a clear and obvious way. The goal of this section is to report lessons learned from practical experience by presenting real cases of mobile middleware use in several domains as well as an overview of the state of the art and future directions of mobile applications, from wearable computing to ubiquitous entertainment and from context-dependent content distribution networks to collaborative applications in war or disaster scenarios.

Many people have contributed much competence and expertise to this book. All of the authors have enthusiastically supported the project and devoted so much of their (little) available time for such a demanding duty. We sincerely thank all of our other colleagues who supported us during the initial book design phase and reviewed some of the chapters relevant to their fields of expertise. Also, we express out appreciation to the editorial staff (Sartaj Sahni for inviting us to consider the idea of a mobile middleware book; John Wyzalek, Karen Schober, and Kari Budyk for their continuous organizational support), without whose deep involvement and motivation this book could not have come to fruition. Their suggestions, advice, and support have been invaluable and fundamental. Finally, we must mention our families and friends, who have supported us throughout the project and demonstrated great patience.

Paolo Bellavista
Antonio Corradi

The Editors

 **Paolo Bellavista** is an associate professor of computer engineering within the Department of Electronics, Computer Science, and Systems (DEIS) of the University of Bologna, Italy, where he teaches several courses on computer basics and operating systems. He received the Laurea degree in Electronics Engineering and his Ph.D. in computer engineering from the University of Bologna. His research activities range from mobile computing to mobile agent-based middleware, from pervasive wireless computing to location-/context-aware services, from replication in mobile *ad hoc* networks to adaptive multimedia. He is a member of several professional associations, including IEEE, ACM, and AICA (Italian Association for Computing). He is an associate technical editor of the *IEEE Communication Magazine*. Readers can find more information about him at his Web page (http://lia.deis.unibo.it/Staff/PaoloBellavista/).

 **Antonio Corradi** is a full professor of computer engineering within the Department of Electronics, Computer Science, and Systems (DEIS) of the University of Bologna, Italy, where he teaches several courses on basic computer networks and advanced distributed systems and infrastructures. He received the Laurea degree in electronics engineering from the University of Bologna and his M.S. degree in electrical engineering from Cornell University. His research interests focus on parallel and distributed systems, support infrastructures for mobile and dynamic services, advanced middleware for pervasive and mobile computing, and context- and location-aware provisioning. He is a member of several professional associations, including IEEE, ACM, and AICA (Italian Association for Computing). Readers can find more information about him at his Web page (http://lia.deis.unibo.it/Staff/AntonioCorradi/).

Contributors

Sachin Agarwal received his Ph.D. in computer engineering at Boston University in 2005, his master's degree in computer engineering from Boston University in 2002, and his bachelor's degree in electronics and communication engineering from the Regional Engineering College, Warangal, India, in 2000. He began work at Deutsche Telekom Laboratories, Berlin, Germany, in 2005 as a postdoctoral research scientist.

Alessandra Agostini earned a M.Sc. in information science from the University of Milan, Italy. She is an assistant professor in the DICo Department of the University of Milano and has participated in various European Union funded research projects; in particular, she was project manager of the Esprit LTR Campiello project, which studied ubiquitous technologies to support cultural exchanges among communities worldwide.

Nadeem Akhtar is a research fellow at the Centre for Communication Systems Research (CCSR) at the University of Surrey. He received his M.E. degree in Telecommunications from the Indian Institute of Science, Bangalore, India, in 2000 and is pursuing a Ph.D. in routing and quality of service in mobile *ad hoc* networks. He was involved with the IST EVOLUTE project and has been a member of the Mobile Research Group since 2001. He has also worked on the Mobile Virtual Centre of Excellence in Mobile and Personal Communications (Mobile VCE) project.

Nancy Alonistioti has earned B.Sc. and Ph.D. degrees in informatics and telecommunications from the University of Athens, Greece. She had been working as an expert at the National Telecommunications Commission and is a project manager and senior researcher in the Communication Networks Laboratory at the University of Athens. She has participated in several national and European projects; she was technical manager of the IST projects MOBIVAS and ANWIRE. She is involved in the IST projects E2R and LIAISON, focusing on reconfigurable mobile environments and on location-based services for working environments, respectively.

Stefan Arbanowski is director of the Competence Centre Smart Environments at Fraunhofer Institute for Open Communication Systems (FOKUS) in Berlin, Germany, where his work is focused on the provision of telecommunications services in a variety of environments. He received his Ph.D. and M.Sc. in computer science at the Technical University Berlin. He has had over 60 papers published in journals and conference proceedings in the area of mobile service provisioning and I-centric communications. He is active in the Wireless World Research Forum and was chairman elect of the WWRF Service Platform Working Group for 2004–2005.

Jean Bacon is a reader in distributed systems and a fellow of Jesus College Cambridge. She leads the Opera research group at the University of Cambridge Computer Laboratory and is editor-in-chief of *Distributed Systems Online,* the IEEE's first online-only magazine. She is also a member of the governing body of the IEEE Computer Society.

Dineshbalu Balakrishnan is working on his Ph.D. in computer science at the School of Information Technology and Engineering (SITE), University of Ottawa, Canada, where he received his master's degree in 2004. He earned his B.E. in computer science from the University of Madras, India, in 2001.

Guruduth Banavar is the senior manager of the Pervasive Computing Infrastructure department at the IBM T.J. Watson Research Center in Hawthorne, New York. His group works on several aspects of pervasive computing, such as context-based and notification systems, wearable and embedded systems, programming tools, and security technologies for pervasive applications. He received his Ph.D. in computer science from the University of Utah.

Nalini Belaramani received her master's degree in computer science from the University of Hong Kong in 2002. During her study, she helped design the Facet programming model for the Sparkle project. She is now a Ph.D. student at the University of Texas at Austin.

Claudio Bettini is a professor of computer science at the DICo Department of the University of Milan. He is also a research professor at the Center for Secure Information Systems of George Mason University, Fairfax, Virginia. He received his Ph.D. in computer science from the University of Milan in 1993. He is a member of ACM Sigmod.

Andrzej Bieszczad is an assistant professor of computer science at California State University, Channel Islands. His research has focused on building computer programs that mimic high-level processing functions of the human brain and on the study of intelligent methods for applications, primarily in computer networking. He has worked in the R&D departments of leading computer and telecommunications companies, including Alcatel, Bell-Northern Research (Nortel Networks), and Bell Laboratories (Lucent). He earned his Ph.D. in electrical engineering and master's degree in computer science at Carleton University, Canada, as well as a master's degree in informatics from the Jagiellonian University, Poland. He is the author of numerous published papers.

Gordon Blair graduated from Strathclyde University with a first-class honours degree in computer science in 1980 and earned a Ph.D. in the same subject in 1983. Since then, he has worked at Lancaster University and holds a chair in distributed systems at this institution. He is also an adjunct professor at Tromsø University and a Visiting Researcher at the Simula Research Laboratory (both in Norway). He is Chair of the Steering Committee for the ACM/IFIP/Usenix Middleware conference and has been on the program committees of many conferences in his field. He is the author of more than 200 published papers.

Raouf Boutaba is an associate professor in the School of Computer Science of the University of Waterloo, Ontario, Canada, and was previously with the Department of Electrical and Computer Engineering at the University of Toronto. Before joining academia, he founded and was director of the telecommunications and distributed systems division of the Computer Science Research Institute of Montreal (CRIM). He received his B.S. in computer engineering from the University of Annaba, Algeria, in 1988. He received his M.Sc. and Ph.D. degrees in computer science from the University of Paris VI, France, in 1990 and 1994, respectively.

Dario Bruneo is a research associate at the Engineering Faculty of the University of Messina. He received his degree in computer engineering from the Engineering Faculty of University of Palermo, Italy, in 2000 and his Ph.D. degree in advanced technologies for information engineering at the University of Messina, Italy, in 2005.

Giacomo Cabri is a research associate in computer science at the University of Modena and Reggio Emilia. He received the Laurea degree in electronic engineering from the University of Bologna in 1995, and his Ph.D. in computer science from the University of Modena and Reggio Emilia in 2000.

Vinny Cahill is an associate professor of computer science at Trinity College, Dublin. He earned his B.A., M.Sc., and Ph.D. degrees in computer science from the University of Dublin. He is a co-founder and member of the editorial board of *IEEE Pervasive Computing* and is a member of the ACM and IEEE Computer Society. He has published numerous peer-reviewed papers in the general area of distributed systems.

Andrew T. Campbell is an associate professor of electrical engineering at Columbia University, New York, and a member of the COMET Group. He received his Ph.D in computer science in 1996 and was a recipient of the National Science Foundation CAREER Award in 1999 for his research in programmable mobile networking. Prior to joining academia, he spent 10 years working on transport and operating systems issues in industry. He spent a sabbatical year at the Computer Lab, Cambridge University, as an Engineering and Physical Sciences Research Council (EPSRC) visiting fellow.

Guanling Chen is an assistant professor of computer science at the University of Massachusetts, Lowell. He received his Ph.D in Computer Science from Dartmouth College in 2004.

Ling-Jyh Chen received his B.Ed. degree in information and computer education from National Taiwan Normal University in 1998 and his M.S. and Ph.D. in computer science from the University of California, Los Angeles, in 2002 and 2005, respectively. He joined the Institute of Information Science as assistant research fellow in 2005.

Yih-Farn (Robin) Chen received a Ph.D. degree in computer science from University of California, Berkeley, an M.S. in computer science from the University of Wisconsin, Madison, and a B.S. in electrical engineering from National Taiwan University. He is on the staff of AT&T Labs–Research and a member of the IETF OPES Working Group. He served as a program co-chair of WWW2003 and was a guest co-editor of a special issue of *IEEE Internet Computing* on mobile applications.

Marco Chiani is a full professor in the Department of Electronics, Computer Science, and Systems (DEIS) at the University of Bologna. During the summer of 2001 he was a Visiting Scientist at AT&T Research Laboratories in Middletown, NJ. He is a frequent visitor at the Massachusetts Institute of Technology, where he holds a Research Affiliate appointment. He is the past chair (2002–2004) of the Radio Communications Committee of the IEEE Communication Society and the current *Wireless Communications* editor for *IEEE Transactions on Communications*.

Marco Conti is a senior researcher at IIT–CNR. He has served as the technical program committee chair of several IFIP-TC6 conferences and is the author of more than 150 published papers in the area of computer-network architectures and protocols. He is co-editor of the book *Mobile Ad Hoc Networking* (2004), is an associate editor of *Pervasive and Mobile Computing Journal*, and serves on the editorial board of *IEEE Transactions on Mobile Computing*, *Ad Hoc Networks Journal*, and *ACM Mobile Computing and Communications Review*.

Gianpaolo Cugola received his Dr.Eng. degree in electronic engineering from Politecnico di Milano, where he spent most of his professional life. In 1998, he received the Dimitri N. Chorafas Foundation prize for engineering and technology for his Ph.D. thesis on software development environments. He is an associate professor at Politecnico di Milano and also a guest professor at the University of Lugano. He collaborates as Information Director with the ACM Software Engineering Interest

Group (SIGSoft). He has been involved in several projects financed by the European Union commission and by the Italian governor. He is the co-author of several scientific papers published in international journals and conference proceedings.

Sajal K. Das is a professor of computer science and engineering at the University of Texas at Arlington. He was the founding director of the Center for Research in Wireless Mobility and Networking (CReWMaN) at the university, where he received their College of Engineering Research Excellence Award in 2003 and Outstanding Research Award in 2005. He is frequently invited as a keynote speaker at international conferences and symposia. He has had over 250 research papers pubilshed, has directed numerous industry- and government-funded projects, and holds four U.S. patents in wireless mobile networks. He is the editor-in-chief of the *Pervasive and Mobile Computing Journal*, is on the editorial boards of numerous journals, and is a co-founder of many conferences, including IEEE PerCom.

Shirshanka Das received his B.Tech. degree in computer science from the Indian Institute of Technology, Delhi, in 2001, and his M.S. and Ph.D. in computer science from the University of California, Los Angeles, in 2003 and 2005, respectively. He is an application architect at PayPal, Inc.

Franca Delmastro received her Laurea degree in computer engineering from the University of Pisa in 2002. She is working toward her Ph.D. in information engineering at the Institute for Informatics and Telematics of the Italian National Research Council.

Ovidiu Drugan is a Ph.D. student at DMMS, Department of Informatics, University of Oslo.

Margaret H. Dunham received her B.A. and M.S. degrees in mathematics from Miami University, Oxford, Ohio, and her Ph.D. degree in computer science from Southern Methodist University. She served as editor of the *ACM SIGMOD Record* from 1986 to 1988 and has served on the program and organizing committees for several ACM and IEEE conferences. She was a guest editor for a special section of *IEEE Transactions on Knowledge and Data Engineering* devoted to main memory databases, as well as a special issue of the *ACM SIGMOD Record* devoted to mobile computing in databases. She was general chair of the ACM SIGMOD 2000 conference. She is an associate editor for *IEEE Transactions on Knowledge Engineering* and is author of a recently published book, *Data Mining: Introductory and Advanced Topics* (2002).

Ashutosh Dutta is a senior research scientist in Telcordia Technology's Internet Network Research Laboratory. Before joining Telcordia, he was the director of computing facilities at Columbia University's Computer Science Department. He has received Telcordia Technologies CEO awards, the Science Applications International Corporation (SAIC) best paper award, and the IEEE EIT 2005 best paper award. He earned his B.S. in electrical engineering from India, his master's in computer science from NJIT, and a professional engineering degree in electrical engineering from Columbia University, where he is pursuing his Ph.D.

Geir Egeland holds a bachelor's degree in engineering from the University of Bristol. For the last ten years, he has worked as a research scientist in the field of mobile network and is a research scientist with Telenor R&D. He was formerly with the Norwegian Defence Research Establishment (NDRE), where as a research scientist he worked on the design and analysis of MAC and routing protocols for mobile *ad hoc* networks.

Markus Endler obtained his Dr.rer.nat. in computer science from the Technical University in Berlin in 1992 and the title Professor Livre-docente from the University of São Paulo in 2001. He worked as a researcher at the GMD Forschungstelle Karlsruhe (Germany) and as an assistant professor at the Institute of Mathematics and Statistics of the University of São Paulo. Since 2001, he has been with the Department of Informatics of the Pontifícia Universidade Católica in Rio de Janeiro. He is member of the ACM, the Brazilian Computer Society (SBC), IFIP WG6.1, and the Steering Committee of the ACM/IFIP Middleware Conference.

Paal E. Engelstad completed his Ph.D. in resource discovery in mobile *ad hoc* and personal area networks in 2005. He has also earned bachelor's and master's degrees in applied physics from NTNU, Norway, and a bachelor's degree in computer science from the University of Oslo, Norway. After working five years in industry, he joined Telenor R&D. He has published numerous refereed papers and holds three patents (two pending).

Jieyan Fan received his bachelor's and master's degrees in electrical engineering from Shanghai Jiaotong University, Shanghai, China, in 2001 and 2004, respectively. He is pursuing his Ph.D. in electrical and computer engineering from the University of Florida, Gainesville.

Luca Ferrari is a Ph.D. student in computer science at the University of Modena and Reggio Emilia, where he received his Laurea degree in computer science engineering in 2002.

Paulo Ferreira is an associate professor of computer and information systems at the Technical University of Lisbon, Portugal. In 1996, he received his Ph.D. degree in computer science from Université Pierre et Marie Curie. Since 1986, he has been a researcher at INESC–ID, where he leads the Distributed Systems Group. He is the author or co-author of more than 50 peer-reviewed scientific communications and has served on the program committees of several international journals, conferences, and workshops in the area of distributed systems.

Tim Finin is a professor of computer science and electrical engineering at the University of Maryland, Baltimore County (UMBC). He has over 30 years of experience in the applications of artificial intelligence to problems in information systems. He holds degrees from MIT and the University of Illinois. Prior to joining the UMBC, he held positions at Unisys, the University of Pennsylvania, and the MIT AI Laboratory. He is the author of over 225 refereed publications and has received research grants and contracts from a variety of sources. He is a former AAAI councilor and serves on the board of directors of the Computing Research Association.

Chien-Liang Fok received his bachelor's degree in computer science and computer engineering in 2002 from Washington University, St. Louis, where he is a Ph.D. candidate in the Department of Computer Science and Engineering and is a member of the Mobile Computing Laboratory.

Xia Gao is a senior research engineer at DoCoMo Communications Laboratories USA, Inc. (DoCoMo USA Labs). He received a B.E. (1992) and M.E. (1995) from Huazhong University of Science and Technology, China, and a Ph.D. degree (2001) from the University of Illinois at Urbana–Champaign. He joined DoCoMo USA Labs in 2001 to conduct research on 4G mobile networks.He has authored numerous peer-reviewed papers and served on the technical program committee of WCNC 2003, ICC 2003, and Globecom 2003.

Michael Georgiades is a Research Fellow at the Centre for Communication Systems Research (CCSR) at the University of Surrey. He received his B.Eng. degree in communications and radio engineering from King's College London in 2000 and his M.Sc. degree in telecommunications from University College London in 2001. In 2002, he worked as a systems development engineer for INSIG Ltd. (U.K.) on wireless Internet solutions. Since then, he has been working at the Mobile Group of CCSR and has been involved in the EU-funded IST Ambient Networks and IST EVOLUTE projects. He is also studying for a Ph.D. on context aware mobility management in all-IP networks.

Mario Gerla received a graduate degree in engineering from the Politecnico di Milano in 1966 and M.S. and Ph.D. in engineering from the University of California, Los Angeles, in 1970 and 1973, respectively. After working for Network Analysis Corporation, he joined the faculty of the Computer Science Department at UCLA, where he is now a professor. He has worked on the design, implementation, and testing of various wireless *ad hoc* network protocols (channel access, clustering, routing, and transport) within the DARPA WAMIS and GloMo projects and, most recently, the ONR MINUTEMAN project. He is also conducting research on QoS routing, multicasting protocols, and TCP transport for the next-generation Internet.

Majid Ghaderi received his B.S. and M.S. degrees in computer engineering from Sharif University of Technology, Iran, in 1999 and 2001, respectively. He is a Ph.D. student in computer science at the University of Waterloo, Canada.

Arif Ghafoor is a professor in the Electrical and Computer Engineering Department and Director of the Distributed Multimedia Systems Laboratory at Purdue University, W. Lafayette, Indiana. He earned his Ph.D. in electrical engineering from Columbia University, New York, and is a fellow of the IEEE.

Vittorio Ghini is an assistant professor of computer science at the University of Bologna, Italy. He received his Laurea degree (1997) and Ph.D. (2002) in computer science from the University of Bologna.

Vera Goebel is a professor in the Department of Informatics of the University of Oslo, where she works in the Distributed Multimedia Systems Group.

Paul Grace is a research associate in the Computing Department at Lancaster University. He earned his Ph.D. (2004) and M.Sc. in distributed systems (2000) from the same institution and graduated from the University of York with a B.Sc. in computer science. He was the primary architect and developer of the ReMMoC framework for tackling middleware heterogeneity.

Mads Haahr is a multidisciplinarian who lectures computer science at Trinity College, Dublin. He earned B.Sc. and M.Sc. degrees from the University of Copenhagen and a Ph.D. from the University of Dublin. He edits the multidisciplinary academic journal *Crossings: Electronic Journal of Art and Technology* and gives away true random numbers on the Internet (www.random.org).

Oliver Haase received his diploma degree in computer science from Karlsruhe University, Germany, in 1993 and his Ph.D. from Siegen University, Germany, in 1997. After working on Internet telephony and mobile service platforms at the NEC Computer & Communication Research Labs in Heidelberg, Germany, he then joined the High-Speed Data Networking Research Department at Bell Labs Research in Holmdel, New Jersey. Since 2005, he has been a professor of software engineering and distributed system at Constance University of Applied Sciences, Germany.

Qi Han is pursuing a Ph.D. from the Bren School of Information and Computer Science at the University of California, Irvine. She earned her M.S. in computer science from the Huazhong University of Science and Technology, China. She is developing adaptive middleware techniques for collecting various dynamic context data in heterogeneous environments to support context-aware applications and is a student member of the IEEE.

Hamid Harroud is working on his Ph.D. in computer science at the School of Information Technology and Engineering (SITE), University of Ottawa, Canada. He received his engineering degree from the Mohammadia School of Engineering, Morocco, in 1997.

Qi He earned his B.S. degree in mathematics from Tsinghua University, Beijing, China, and his M.S. degree in computer science from the University of Maryland,

College Park, in 1997. He is a project scientist at Carnegie Mellon University. His research has focused on leveraging cryptographic methodology to construct an agent-based security infrastructure to address security issues in ubiquitous computing.

Dazhi Huang is a Ph.D. student in the Department of Computer Science and Engineering at Arizona State University, Tempe. He received his B.S. degree in computer science from Tsinghua University in China.

Yun Huang is a Ph.D. candidate studying in the Bren School of Information and Computer Science at the University of California, Irvine. She is devising efficient resource discovery algorithms and data placement strategies for providing mobile users with multimedia services by leveraging heterogeneous and intermittently available grid resources. She earned an M.S. in computer science from the University of California, Irvine, and received her B.S. degree in computer science from Tsinghua University, China.

Nayeem Islam is vice president of the mobile software labs at NTT DoCoMo. He earned his Ph.D. at the University of Illinois, Urbana–Champaign; his M.S. from Stanford; and a B.S.E. from Princeton University, all in computer science.

Rittwik Jana received his B.E. degree in electrical engineering from the University of Adelaide, Australia, in 1994, and his Ph.D. degree from the Australian National University, Canberra, in 1999. He was an engineer with the Defense Science and Technology Organization (DSTO), Australia, from 1996 to 1999 and since then has been a member of the technical staff at AT&T Labs–Research. His research work has been in the area of mobile and wireless communications, from physical-layer modem design to application-layer software development.

Carl-Gustav Jansson has been a professor in the Department of Computer and Systems Sciences of the Royal Institute of Technology (KTH) since 1998.

Martin Jonsson is an M.Sc.E.E. and is pursuing a Ph.D. at the Department of Computer and Systems Sciences of the Royal Institute of Technology (KTH).

Anupam Joshi is an associate professor of computer science and electrical engineering at the University of Maryland, Baltimore County (UMBC). He obtained a B.Tech. degree in electrical engineering from IIT Delhi in 1989, and his master's degree (1991) and Ph.D. (1993) in computer science from Purdue University, W. Lafayette. He is the author of over 100 technical papers and has obtained research support from NSF, NASA, DARPA, DoD, IBM, AetherSystens, HP, AT&T, and Intel. He has presented tutorials in conferences, served as guest editor for special issues of *IEEE Personal Communication* and *Communications of the ACM*, and served as an associate editor of *IEEE Transactions of Fuzzy Systems*.

Theo G. Kanter earned his technical doctorate in computer communications from the Royal Institute of Technology (KTH) in Stockholm, Sweden. He is a senior researcher at Ericsson Research in the area of service-layer technologies and also a guest researcher at the Center of Wireless Systems of the Royal Institute of Technology (KTH) in Stockholm. He holds a number of patents in the area of architectures and middleware for context-aware mobile services.

Ahmed Karmouch earned his M.S. and Ph.D. degrees in computer science at the University of Paul Sabatier, Toulouse, France. He has served as a research engineer at the Institut National de Recherche en Informatique et en Automatique (INRIA) Paris, France; as a senior manager for Bull SA, Paris; and as director of research at the Ottawa Medical Communications Research Group, University of Ottawa. Since 1991, he has been a professor of electrical and computer engineering and computer science at the School of Information Technology and Engineering, University of Ottawa. He is involved in several projects with the Telecommunications Research

Institute of Ontario, Nortel Networks, Bell Canada, Mitel, National Research Council Canada, Centre National de Recherche Scientique, March Networks, CANARIE, Communications & Information Technology Ontario (Cito), and the TeleLearning National Center of Excellence. He has published over 180 papers and served as guest editor for *IEEE Communications, Computer Communications*, and *Multimedia Tools and Applications Journal*.

John Keeney is employed as a postdoctoral researcher in the Computer Science Department of Trinity College, Dublin. He holds a B.A.I. degree in computer engineering and a Ph.D. in computer science from the University of Dublin. His current research is focused on semantic-based autonomic management of networks and ubiquitous computing spaces.

Sean Kelley received his M.S. degree in computer science from Southern Methodist University, Dallas, Texas, in 2005. He received his B.S. Business Administration–MIS degree from the University of Texas at Austin in 2002. He has extensive experience as a data warehouse architect and consultant.

Pradeep Khosla is the Philip and Marsha Dowd Professor at the College of Engineering and School of Computer Science at Carnegie Mellon University, Pittsburgh, Pennsylvania. He is also the head of the Electrical and Computer Engineering Department and the Information Networking Institute, as well as founding director of the CyLab at Carnegie Mellon. He served as a DARPA program manager in the Software and Intelligent Systems Technology Office (SISTO), Defense Sciences Office (DSO), and Tactical Technology Office (TTO), where he managed advanced research and development programs in information technology and intelligent systems.

Fredrik Kilander holds a postdoctorate position at the Department of Computer and Systems Sciences of the Royal Institute of Technology (KTH) in Stockholm.

David Kotz is a professor of computer science, director of the Center for Mobile Computing, and executive director of the Institute for Security Technology Studies at Dartmouth College, Hanover, New Hampshire. He completed his Ph.D in computer science from Duke University, Durham, North Carolina, in 1991.

Michael E. Kounavis, a senior research scientist with Intel Research and Development, is working on cryptography and data integrity. He has co-authored numerous technical papers and seven U.S. patent applications. In 2004, he obtained his Ph.D degree in programming network architectures from Columbia University, New York, and his M.Sc. (1998) and B.Sc. (1996) degrees from Columbia and the National Technical University of Athens, Greece, respectively.

Mohan Kumar is a professor of computer science and engineering at the University of Texas at Arlington. He received his Ph.D. from the Indian Institute of Science in 1992. He is the author of more than 100 articles published in refereed journals and conference proceedings and is an editor of the *Pervasive and Mobile Computing* journal and the *Computer Journal*. He is a co-founder of IEEE PerCom and served as program chair for PerCom 2003 and general chair for PerCom 2005.

Vivien W.M. Kwan received her master's degree in computer science from the University of Hong Kong in 2002. She helped develop the intelligent proxy system for the Facet-based software architecture of the Sparkle Project.

Francis C.M. Lau received his Ph.D. in Computer Science from the University of Waterloo, Ontario, Canada, in 1986. He is a professor in Computer Science at the University of Hong Kong.

Letizia Leonardi is a full professor in computer science at the University of Modena and Reggio Emilia, where she teaches basic and advanced computer science courses. She received her Laurea degree in electronic engineering in 1982 and Ph.D. in computer science in 1989, both from the University of Bologna.

Francesco Lilli received his degree in electronic engineering in 1995 from Polytechnic of Turin, Italy. Also in 1995, he earned certification as a professional engineer at Padua University, Italy. He then performed postdoctoral work and research at the FIAT Research Centre in Orbassano, Turin, Italy, where he studied and developed on-board telematic systems. After serving as a researcher at CRF, he worked on the development of safety systems for the European Community research project IN-ARTE. In 2001, he became a project coordinator for the GALLANT and GALILEI projects and became involved in the IST ACTMAP project. In 2004, he became the head of the Telematic Technologies Department at the FIAT Research Centre.

Wei Li is pursuing a Ph.D. in the Department of Computer and Systems Sciences of the Royal Institute of Technology (KTH) in Stockholm.

Peter Lönnqvist holds an M.Sc. in Psychology and studies people's interaction with artifacts in new worlds at the Department of Computer and Systems Sciences of the Royal Institute of Technology (KTH) in Stockholm.

Chenyang Lu is an assistant professor in the Department of Computer Science and Engineering at Washington University in St. Louis, Missouri. He has published numerous refereed research papers and was the recipient of the National Science Foundation CAREER Award in 2005. He received his Ph.D. from the University of Virginia in 2001, his M.S. degree from the Chinese Academy of Sciences in 1997, and his B.S. degree from the University of Science and Technology of China in 1995, all in computer science.

Zakaria Maamar is an associate professor of computer sciences at Zayed University, Dubai, United Arab Emirates.

Thomas Magedanz, Ph.D., is head of the 3Gbeyond division at the Fraunhofer Institute for Open Communication Systems (FOKUS), Germany, which also provides the national Open 3Gb test bed. In addition, he is a full professor at the Technical University of Berlin in the field of next-generation telecommunication infrastructures. He is an editorial board member of several journals and the author of more than 120 technical papers. He is the author of two books and is a regularly invited speaker at major international telecom events and conferences.

Gerald Q. Maguire, Jr., Ph.D., has been a professor at the Royal Institute of Technology (KTH) since 1994.

Marco Mamei has been a contract researcher at the University of Modena and Reggio Emilia since 2004. He obtained his Laurea degree in computer science in 2001 and Ph.D. in computer science in 2004, both from the University of Modena and Reggio Emilia.

Kazuhiro Minami is a postdoctoral researcher in computer science at Dartmouth College, Hanover, New Hampshire. He earned his Ph.D. in computer science from Dartmouth College in 2006.

Shivajit Mohapatra received his Ph.D. from the Donald Bren School of Information and Computer Science at the University of California, Irvine, in 2005. He is a senior research scientist at Motorola Labs, where his research focus is in the area of *ad hoc* and mobile computing. He received his master's degree from UCI and his bachelor's degree from the Birla Institute of Technology and Science, Pilani.

Soraya Kouadri Mostéfaoui is a research assistant in the Mobile Information Systems Laboratory of the Computer Science Department of the University of Applied Sciences of Western Switzerland, Fribourg, Switzerland.

Ellen Munthe-Kaas is an associate professor at the Distributed Multimedia Systems Group, Department of Informatics, University of Oslo.

Amy L. Murphy is an assistant professor in the Department of Informatics at the University of Lugano, Switzerland. She received a B.S. degree in computer science from the University of Tulsa in 1995, and her M.S. (1997) and D.Sc. (2000) degrees from Washington University, St. Louis, Missouri. She served as an assistant professor at the University of Rochester, New York, and a visiting researcher at Politecnico di Milano, Italy, before joining the department in Lugano.

Faïza Najjar received her Ph.D. in computer science from the University of Tunis, ElManar, in 1999. Since 2000, she has been an assistant professor at the National School of Computer Science and Engineering in Manouba, Tunisia.

Alok Nandan received his B.Tech. degree in computer science from the Indian Institute of Technology, Kharagpur, in 2001 and his M.S. and Ph.D. in computer science from the University of California, Los Angeles, in 2003 and 2005, respectively. He joined Microsoft as a research program manager in 2005.

Chandra Narayanaswami manages a group of researchers at IBM Research in Hawthorne, New York, that is exploring several aspects of mobile computing, including form factors, novel applications, power management, user interfaces, and device symbiosis. He has received 19 IBM Invention Achievement Awards and an Outstanding Technical Innovation Award. He has published extensively and is a holder of several patents. He has served on the program committees for several leading conferences in mobile computing and was the general chair for the IEEE Symposium on Wearable Computers in 2003. He obtained a Ph.D. in computer and systems engineering from Rensselaer Polytechnic Institute, Troy, New York, and a B.S. in electrical engineering from the Indian Institute of Technology, Bombay.

Nanjangud C. Narendra is a research staff member at IBM India Research Lab, Bangalore, India.

Spyros Panagiotakis received his B.Sc. in physics and his M.Sc. in electronic automation from the Department of Physics of the University of Athens, Greece, in 1997 and 1999, respectively. Since 2000, he has been a member of the Communication Networks Laboratory (University of Athens) and has participated in several national and European projects. He is working as a researcher for the FP6 IST project LIAISON, focusing on location-based services for working environments, while pursuing his Ph.D. at the Department of Informatics and Telecommunications of the University of Athens.

Fabio Panzieri is a professor of computer science at the Faculty of Science of the University of Bologna, Italy. He obtained his Laurea degree in 1978 in information science from the University of Pisa, Italy, and his Ph.D. degree in computer science in 1985 from the University of Newcastle upon Tyne, U.K., where he was a research associate in the Computing Laboratory.

Jim Parker received his B.S. degree in computer science from James Madison University, Harrisonburg, Virginia, in 1985 and his M.S. degree in computer science from the University of Maryland, Baltimore County (UMBC) in 1998. He is a Ph.D. candidate in computer science at UMBC and is a member of the eBiquity research group at UMBC.

Anand Patwardhan received his B.E. degree in computer engineering from the University of Pune, India, in 2000 and his M.S. degree in computer science and engineering from Oregon Health and Science University, Portland, Oregon, in 2002. He is a Ph.D. candidate in the Computer Science and Electrical Engineering Department at the University of Maryland, Baltimore County (UMBC).

Raymond Paul has been a professional electronics engineer, software architect, developer, tester, and evaluator for the past 26 years. He serves as the technical director for Command and Control (C2) Policy, Office of the Secretary of Defense, Networked Information Infrastructure; in this position, he supervises command-and-control systems engineering development for objective, quantitative, and qualitative measurements concerning the status of software/systems engineering resources and evaluating project outcomes to support major investment decisions. He holds a doctorate in software engineering and is an active fellow of the IEEE Computer Society. He has published more than 67 articles on software engineering in various technical journals and symposia proceedings and has authored chapters in four technical books concerning software engineering.

Filip Perich is a senior research engineer at Cougaar Software, in McLean, Virginia, and an adjunct assistant professor at the University of Maryland, Baltimore County (UMBC). He received his Ph.D. degree in computer science from UMBC in 2004 and his M.S. degree in computer science in 2002; he earned his B.A. degree in mathematics from Washington College, Maryland, in 1999. He is a member of the eBiquity research group at UMBC, has authored over 20 referred publications, and has served as a conference organization and committee member for multiple conferences and workshops.

Gian Pietro Picco is an associate professor at the Department of Electronics and Information of Politecnico di Milano, Italy. He received his M.Sc. degree in electronic engineering from Politecnico di Milano in 1993 and his Ph.D. in computer science from Politecnico di Torino in 1998. He visited Washington University in St. Louis, Missouri, as a research assistant and then as a visiting assistant professor. He has been with Politecnico di Milano since 1999.

Evaggelia Pitoura received her B.Sc. from the University of Patras, Greece, in 1990 and her M.Sc. (1993) and Ph.D. (1995) in computer science from Purdue University, W. Lafayette, Indiana. Since 1995, she is has been on the faculty of the Department of Computer Science of the University of Ioannina, Greece, where she leads the distributed data management group. Her publications include more than 80 articles in international journals and conferences and a book on mobile computing. She has also co-authored two tutorials on mobile

computing for IEEE ICDE 2000 and 2003. She has received the IEEE ICDE 1999 best paper award and two recognition of service awards from ACM.

Thomas Plagemann is a professor at the Department of Informatics of the University of Oslo, where he heads the Distributed Multimedia Systems Group.

Christos Politis received his engineering degree from Technological University of Athens, Greece, in 1996, his M.Sc. in mobile and satellite communications from the University of Surrey (UniS) in 1999, and his Ph.D. in mobile networking from the Centre for Communication Systems Research (CCSR) at the same university in 2004. Past positions include telecommunications engineer with INTRACOM SA, Athens; IT engineer at AMSAT; wireless communications engineer at Hellenic Air Force; and general staff and senior researcher with CCSR, where he was involved in several EU-funded IST projects. He is the R&D manager with OFCOM. He is a patent holder and the author of more than 35 papers published in international journals and conference proceedings.

Radu Popescu-Zeletin is a professor at the Technical University Berlin and Director of the Fraunhofer Institute for Open Communication Systems (FOKUS). He led the R&D department of the BERKOM project of the German Telekom pilot project for the development of new applications in the broadband ISDN environment. He has been active in standardization committees (DIN, ISO, EURESCOM) and has contributed to the development of telecommunication standards. He is chairman-elect of the Wireless World Research Forum (WWRF) Working Group 2. He earned his Ph.D. from the University of Bremen, Germany, and certification from the Technical University Berlin. He is a doctor *honoris causa* of the Polytechnical Institute, Bucharest, and professor *honoris causa* of the Catholic University of Campinas, Brazil. He is a bearer of the Public Service Medal of the Republic of Romania.

Antonio Puliafito is a full professor of computer engineering at the University of Messina, where he is the coordinator of the Ph.D. course in advanced technologies for information engineering. He has contributed to the development of the software tools WebSPN, MAP, and DAVID, which are widely used today, and is in charge of the ICT and e-learning initiatives for the University of Messina.

Matija Puzar is a Ph.D. student at DMMS, Department of Informatics, University of Oslo.

Ilja Radusch received his M.Sc. in computer science from the Technical University of Berlin. Since then, he has been a researcher with the Open Communication Systems department of the Technical University of Berlin, where he is responsible for the AVM (Autonomous Distributed Microsystems) project, which is funded by the German Ministry of Education and Research.

David Reich is a senior software engineer and tools architect in IBM's Application Integration Middleware division. He has been with IBM for 18 years. He has held positions in programming and technical leadership, spent several years in development management, and took a side trip into corporate IT. He holds several patents (with many more pending) and has authored three books and numerous trade journal articles. He is also a requested speaker at industry conferences worldwide. He has B.S. and M.S. degrees in computer science from the State University of New York at Albany.

Daniele Riboni received his M.Sc. degree in computer science from the University of Milan in 2002 and is a Ph.D. student at the DICo Department of the same university.

Marco Roccetti is a professor of computer science at the Faculty of Science of the University of Bologna, Italy. He received his Laurea degree in electronics engineering from the University of Bologna. He has served as the general co-chair for SCS International Conferences on Simulation and Multimedia in Engineering

Education (2002, 2003) and for IEEE Workshops on Networking Issues in Multimedia Entertainment (2004, 2005). He has authored and co-authored more than 100 technical refereed papers published in the proceedings of international conferences and journals.

Ricardo Rocha Ricardo Rocha is a Ph.D. candidate at the Department of Informatics, Pontifícia Universidade Católica do Rio de Janeiro. He received his M.Sc. in computer science from IME–USP, Brazil, in 2001. He is member of the ACM and the Brazilian Computer Society (SBC).

Gruia-Catalin Roman is a professor and chairman of the Department of Computer Science and Engineering at Washington University in St. Louis, Missouri. He was a Fulbright Scholar at the University of Pennsylvania, where he received a B.S. degree (1973), an M.S. degree (1974), and a Ph.D. (1976), all in computer science. He was an associate editor for ACM TOSEM and served as the general chair of ICSE 2005.

Manuel Roman is a member of the Mobile Software Lab at DoCoMo Communications Laboratories USA, Inc. He received his Ph.D. from the University of Illinois at Urbana–Champaign.

Hana Rubinsztejn is a Ph.D. candidate at the Department of Informatics, Pontifícia Universidade Católica do Rio de Janeiro. She received her M.Sc. in computer science from the Universidade Estadual de Campinas (UNICAMP), Brazil, in 2001.

Vagner Sacramento is a Ph.D. candidate at the Department of Informatics, Pontifícia Universidade Católica do Rio de Janeiro. He received his M.Sc. in computer science from the Universidade Federal do Rio Grande do Norte (UFRN), Brazil, in 2002.

George Samaras received a Ph.D. in computer science from Rensselaer Polytechnic Institute, Troy, New York, in 1989. He is an associate professor at the University of Cyprus, Greece. He was previously at IBM Research in Triangle Park, North Carolina, where he served as the lead architect of

IBM's distributed commit architecture. He has co-authored a book on data management for mobile computing and holds a number of patents relating to transaction processing technology. He received the best paper award of the 1999 IEEE ICDE. He has also served as program co-chair and program committee member on a number of conferences.

Norun Sanderson is a Ph.D. student at DMMS, Department of Informatics, University of Oslo.

Roberto Saracco has been a researcher for over 30 years in the telecommunications area of the Telecom Italia Research Lab. He has been director of the Future Centre, has worked on a World Bank project in Latin America to foster the application of innovation to business, and has led the technology trajectory and disruptions group within the EU program FISTERA.

Marco Scarpa received his degree in computer engineering in 1994 from the University of Catania, Italy, and the Ph.D. in computer science in 2000 from the University of Turin, Italy. From 2000 to 2001, he was an assistant professor of computer science at the Faculty of Engineering of Catania University. He is an associate professor in operating systems at the Faculty of Engineering of Messina University.

Henning Schulzrinne received his Ph.D. from the University of Massachusetts in Amherst. He was a member of the technical staff at AT&T Bell Laboratories, Murray Hill, New Jersey, and an associate department head at the Fraunhofer Institute for Open Communication Systems (FOKUS) in Berlin before joining the Computer Science and Electrical Engineering departments at Columbia University, New York. He is chair of the Department of Computer Science. Protocols he has helped develop, such as RTP, RTSP, and SIP, are now Internet standards used by almost all Internet telephony and multimedia applications.

Basit Shafiq received a B.S. degree in electronics engineering from the GIK Institute of Engineering Sciences and Technology, Pakistan, in 1998 and an M.S. degree in electrical engineering from Purdue University, W. Lafayette, Indiana, in 2001. He is working toward a Ph.D. degree in the School of Electrical and Computer Engineering at Purdue University.

Waseem Sheikh received a B.S. degree in electronics engineering from the GIK Institute of Engineering Sciences and Technology, Pakistan, in 2000 and an M.S. degree in electrical engineering from Purdue University, West Lafayette, Indiana, in 2002. He is working toward a Ph.D. degree in the School of Electrical and Computer Engineering at Purdue University.

Muhammad Sher is a Ph.D. research fellow at the Technical University of Berlin and the Fraunhofer Institute for Open Communication Systems (FOKUS), Berlin, Germany. In addition, he is assistant professor for the Faculty of Applied Sciences, International Islamic University, Islamabad, Pakistan, in the field of computer networks and network security. He is an author of more than 25 research papers and one book. He received the National NCR IT Excellence Award in the field of research and development in 2000 and the Dr. Razi-ud-Din Saddiqui Award from IIU for his book in 2004.

Pauline P.L. Siu received her M.Sc. in computer science from the University of Hong Kong in 2004. She helped develop the context-aware state management system (CASM) for the Sparkle project.

Katrine S. Skjelsvik is a Ph.D. student at DMMS, Department of Informatics, University of Oslo.

Stephan Steglich obtained his Ph.D. degree in Computer Science from the Technical University Berlin in 2003. He has worked intensively in the research area of intelligent mobile agents and since 1999 has been involved in research activities in the area of user-centric communication. He has worked on a number of projects related to human–machine interaction, UMTS/VHE, personalization, and user profiling. He manages international-

and national-level research activities and has been an organizer and a member of program committees of several international conferences. He has participated in standardization activities within the Object Management Group and gives lectures at the Technical University Berlin.

Rahim Tafazolli is the head of the Mobile Communications Research Group in CCSR, School of Electronics and Physical Sciences, the University of Surrey. He is the author of more than 300 research papers published in refereed journals and international conference proceedings. He holds more than 15 patents in the field of mobile communications. He is an advisor and consultant to a number of mobile companies and founder and past chairman of the International Conference on 3G Mobile Technologies. He is also a member of the IEEE Committee on U.K. Regulations on Information Technology and Telecommunications, a member of the Wireless World Research Forum (WWRF) Vision Committee, past chairman of the New Technologies group of the WWRF, and academic coordinator of the U.K. Mobile Virtual Centre of Excellence.

Javid Taheri received his bachelor's degree in electrical engineering in 1998 and his master's degree in electrical engineering in 2000, both from Sharif University of Technology, Tehran, Iran. He is a Ph.D. student in the field of mobile computing at the School of Information Technologies, University of Sydney, Australia. Before beginning his Ph.D. program, he worked as a professional software developer for the largest car manufacturer in Iran, IKCO, where he developed sophisticated programs to achieve machine vision in highly industrial environments such as automobile assembly lines and body shops.

Giovanni Turi has been a junior researcher at the CNR Institute for Informatics e Telematics (IIT) in Pisa, Italy, since 2003. He received the Laurea degree in computer science from the University of Pisa in 2000, after which he spent two years

as consultant and software engineer for IBM and Agilent Technologies, working on software related to license and remote management services. Since 2002, he has been pursuing a Ph.D. in information engineering at the University of Pisa.

Can Türker heads the Data Integration Group of the Functional Genomics Center Zurich (FGCZ). Before joining FGCZ in 2005, he was a postdoctorate fellow in the Database Group at ETH Zurich. He earned his Ph.D. degree in computer science from the University of Magdeburg in 1999. He is the author or co-author of the lecture books *Object Databases*, *SQL:1999 and SQL:2003*, and *Mobile and Wireless Information Systems* (all in German). He has served as the chair for a number of program committees of database-related conferences.

Luís Veiga received his B.Sc. (1998) and M.Sc. (2001) degrees in computer engineering from the Technical University of Lisbon (Instituto Superior Técnico), Portugal, where he is a lecturer and Ph.D. candidate in the Computer and Information Systems Department. He has been a researcher at INESC–ID (Distributed Systems Group) since 1999 and has participated in projects such as Mnemosyne, MobileTrans, OBIWAN, DGC-Rotor, and UbiRep. He has authored or co-authored numerous peer-reviewed scientific papers published in workshop and conference proceedings and journals, and he has served as a reviewer for international conferences.

Nalini Venkatasubramanian is an associate professor at the Bren School of Information and Computer Science, University of California, Irvine. When she was a member of the technical staff at Hewlett-Packard Laboratories in Palo Alto, California, she worked on large-scale distributed systems and interactive multimedia applications. She has also worked on various database management systems and on programming languages/compilers for high-performance machines. She earned her M.S. degree and Ph.D. in computer science from the University of Illinois, Urbana–Champaign.

Cho-Li Wang received his B.S. degree in computer science and information engineering from National Taiwan University in 1985. He earned his M.S. and Ph.D. degrees in computer engineering from the University of Southern California in 1990 and 1995, respectively. He is an associate professor of the Department of Computer Science at the University of Hong Kong.

Tony White is an associate professor of computer science at Carleton University, Ontario, Canada, where he is conducting research into problems in autonomic computing, telecommunications, and peer-to-peer computing. He has written over 60 published papers and is the co-author of six patents (two pending). He earned a master's degree in theoretical physics from Cambridge University, England, and a Ph.D. in electrical engineering from Carleton University.

Wai-Kwong Wing received his B.S. degree in computer science and information systems in 2001 from the University of Hong Kong, where he is a Ph.D. candidate.

K. Daniel Wong received his B.S.E. degree in electrical engineering from Princeton University, New Jersey, and M.S. and Ph.D. degrees in electrical engineering from Stanford University, California. He has been a senior research scientist at Telcordia since 1998. Since 2003, he has also taught at the Malaysia University of Science and Technology (MUST). He is a member of the editorial board of *IEEE Communications Surveys and Tutorials* and the author of *Wireless Internet Telecommunications* (2004). He received the G. David Forney, Jr., Prize from Princeton University in 1992, the Telcordia Technologies CEO Award in 2002, and the best paper award at IEEE EIT 2005.

Dapeng Wu received his bachelor's degree in electrical engineering in 1990 from the Huazhong University of Science and Technology, Wuhan, China, his master's degree in electrical engineering in 1997 from Beijing University of Posts and Telecommunications, Beijing, China, and his Ph.D. in electrical and computer engineering in 2003 from Carnegie Mellon University, Pittsburgh, Pennsylvania. Since 2003, he has been an assistant professor in the Electrical and Computer Engineering Department at the University of Florida, Gainesville. He is an associate editor for the *IEEE Transactions on Vehicular Technology* and received the *IEEE Circuits and Systems for Video Technology (CSVT) Transactions* best paper award in 2001.

Stephen S. Yau is a professor in the Department of Computer Science and Engineering at Arizona State University, Tempe. He served as chair of the department from 1994 to 2001. He was previously with the University of Florida, Gainesville, and Northwestern University, Evanston, Illinois. He served as the president of the IEEE Computer Society and editor-in-chief of *IEEE Computer* magazine. He received a B.S. degree from National Taiwan University, Taipei, and M.S. and Ph.D. from the University of Illinois, Urbana–Champaign, all in electrical engineering. He is a life fellow of the IEEE and a fellow of American Association for the Advancement of Science.

Eiko Yoneki is a Ph.D. candidate in the Computer Laboratory at the University of Cambridge, England. She received a postgraduate diploma in computer science from the University of Cambridge in 2002. Previously, she spent several years working for IBM on various networking products.

Franco Zambonelli has been an associate professor in computer science at the University of Modena and Reggio Emilia since 2001. He obtained his Laurea degree in electronic engineering in 1992 and his Ph.D. in computer science in 1997, both from the University of Bologna. He was a founding member of the Autonomic Communication Forum.

Dong Zhou is a research engineer at Mobile Software Lab at DoCoMo Communications Laboratories USA, Inc. (DoCoMo USA Labs). He earned a Ph.D. in computer science from the Georgia Institute of Technology, Atlanta, and his B.S. and M.S. degrees in computer science from Nanjing University, China.

Albert Y. Zomaya is head of the school and CISCO Systems chair professor of Internetworking in the School of Information Technologies, University of Sydney, Australia. Previously, he was a full professor in the Electrical and Electronic Engineering Department at the University of Western Australia, during which time he also led the Parallel Computing Research Laboratory and held visiting positions at Waterloo University and the University of Missouri–Rolla. He is the author or coauthor of six books and more than 200 articles in technical journals and conference proceedings and is the editor of numerous books. He is an associate editor for 14 journals, the founding editor of the Wiley Book Series on Parallel and Distributed Computing, and a founding coeditor of the Wiley Book Series on Bioinformatics. He was chair of the IEEE Technical Committee on Parallel Processing and serves on its executive committee. He received the 1997 Edgeworth David Medal from the Royal Society of New South Wales for outstanding contributions to Australian Science and an IEEE Computer Society's Meritorious Service Award in 2000.

Contents

Section 1

FUNDAMENTALS

Chapter 1

Toward a Software Infrastructure for Ubiquitous Disappearing Computing

Roberto Saracco

CONTENTS

Introduction

In the next ten years, we will witness a continuous evolution of technology much like we have observed over the last 50 years. We have come to expect it; it has become a part of our life. No one is surprised to find less expensive and more powerful PCs, larger television screens, higher resolution digital cameras, and so on. Hence, should it not be easy to predict what is going to happen, from a technological perspective? Yes and no: "Yes," because, indeed, we know a lot about technology; "no," because many factors must be considered that we know very little about.

Fifteen years ago people were saying that it was no longer an issue of *technology push*; instead, what we were beginning to see was a *market pull*. Technology evolution could make something possible but was not sufficient to steer the market. The market was deciding which of the available technologies to use, so it was indeed the market that was pulling technology. Now we are on the brink of a major change that is going to have a deep and possibly disturbing effect on research and its relation with the evolution of technology. We are on the dawn of a *market push*.

Researchers within big enterprises (small enterprises can no longer afford to do research in technology areas) and universities are having a tough time securing funding for their work. To obtain funding, they must tell a story that investors can understand; for example, they might explain that conducting research in a particular area will generate a significant market response and a healthy return on their investment. Researchers must point out what investors probably already know as a result of analyzing a particular market — that the current market is promising so let's exploit it by injecting new technology.

This is happening everywhere. The venture capitalists and investor angels in the Silicon Valley are no longer investing money in 20 different start-ups without really understanding what they are up to and assuming that if just one hits the bull's eye then they will recoup their investment and then some. Today, such investors want solid data on return expectations, and researchers can offer such information only by talking about today, not about tomorrow.

Investing in a certain technology is likely to push its evolution; hence, it is the market today that steers the evolution of technology. It is likely that we are going to experience, and suffer through, a *market push* in the near future. This can be very bad for those countries and enterprises that lack the vision and courage to invest. They will cultivate only a linear evolution and will be displaced by those who are more aggressive.

I have had several opportunities to talk to some key people in a number of industries about the potential evolution of sensors, e-tags, multiple radio access infrastructures, and potentially unlimited radio band-width. In most cases, the reaction I have gotten is of the "let's wait and see" variety. These are not issues to be dealt with tomorrow morning so why care? Individual enterprises often are not able to make a dent in the overall evolution of a technology, particularly when this evolution requires the significant build-up of infrastructures: physical (networks), logical (platforms), or relational (regulations and standards).

In South Korea, the government has launched a comprehensive "8–3–9" strategy for which they have identified eight broad classes of services, three basic infrastructures, and nine "growth engines." The three infrastructures are the Broadband Convergent Networks, Ubiquitous Service Network, and IPv6. (I will be discussing the first two later in this chapter.) The point I want to make here, in the introduction, is that, unless some entity such as a government or a national or supranational initiative funds and steers effort toward building a comprehensive infrastructure, it will not happen as result of individual enterprise nor demand from the market.

The title of this chapter is "Toward a Software *Infrastructure* for Ubiquitous Disappearing Computing." Achieving such an infrastructure is not going to be the result of accidents or the uncoordinated efforts of individual enterprises. It will require much more. The United States, Europe, and East Asia are all competing in the evolution of the technology business and are trying to maintain or acquire a dominant position. China, because of the sheer volume of its market; Japan, because of its well-coordinated industry sector; and South Korea, because of its government-led initiatives, are all very strong threats to current U.S. dominance and European hopes.

A discussion of the "2G–3G–4G" evolution may be instructive here. Europe won the 2G battle because it set a universal standard, the GSM. In the United States, 3G is based on a market-oriented paradigm of letting the market decide what is best; Europe is losing ground to U.S. companies holding patents. Because 4G is still on the starting block, it may be premature to draw any conclusion, but a very real risk is that 4G technology will be steered and dominated by terminal equipment manufacturers from Korea and Japan.

This is a book about technology, but it is also about the future. I believe we should try to put our technical expectations into the wider perspective of a global vision, and that is what I have tried to do in this opening chapter.

Technology Evolution

Let's consider some of the newer technologies driving the business incentive to create a software infrastructure for ubiquitous disappearing computing: storage, processing, communications, and data capturing. In considering each of them, I will make reference to the role the technology is likely to play in supporting ubiquitous disappearing computing and in establishing a software infrastructure.

Storage

Several technologies are available for storing information, including silicon-based, polymer, magnetic, holographic, and nanostorage. All may play a role in the topic at hand, although the two most important are the silicon-based and magnetic technologies. Holographic storage will likely be used to support specific areas where searching through huge amounts of data is important (such as in e-citizen applications managed by institutions). Nanostorage is still a little bit into the future and may be considered, from a conceptual point of view and in terms of applications, as a 1000-fold evolution of silicon-based memory. It may be available somewhere toward the end of the next decade. Polymer memory offers potential as read-only memory; it is capable of storing terabytes of information in very-low-cost packaging. Memory that is credit-card sized will be able to store 2000 movies and could eliminate the need to send movies over a network.

Silicon and magnetic memories will play a major role in many areas where data must be stored, retrieved, and processed. Magnetic storage has slower transfer times but it is much less expensive than silicon-based storage. In 2005, top-of-the-line hard drives could store 1 TB at a cost of about $1000. Even more interesting is a tiny hard disk with a capacity of 1.5 GB available since 2004 at a cost of about $10. By 2010, we can reasonably expect to find 1-TB hard drives in any medium-class consumer PC being sold, and the tiny hard drives will be found in every cellphone and portable device with a capacity on the order of 10 GB or so.

Silicon-based memory will exceed 1 GB per chip, and compact flash memories will top 100 GB early in the next decade. Much smaller storage will be embedded in sensors, directly in the sensor chip. System-on-a-chip (SoC) will include a significant amount of memory, not just for processing purposes, as is the case today, but also to serve as storage. More and more objects will have local storage capability to track the history of the object itself. This can be in the form of rewritable radio-frequency identification (RFID) tags. In the next decade, any object, including consumables, is quite likely to have an RFID tag, and many will have rewritable tags. By the end of the next decade, a sheet of paper could have an embedded rewritable tag to supplement the information written on the paper. The aspect of synchronization will become very important, and a number of service will be offering to support synchronization.

Processing

Processing has been steadily evolving over the last 40 years at an astonishing rate (overall, storage has evolved faster than processing but in more sporadic bursts), and it should continue to do so for the next seven years, at least; however, ten years ago it had already reached a threshold for certain types of chips for which the processing capacity offered had only marginal demand, thus leading to a drop in price. This allowed the insertion of processing capabilities at basically no cost in a number of objects. Think about checking into a hotel and the desk clerk giving you a key card to enter your room. In either the key card or the room lock is a processor performing some operations. The same goes for sprinklers or for greeting cards that play a tune when touched. Within the next four years the same will happen to microprocessors found in today's PCs. This will provide tremendous processing power for a variety of objects.

The evolution of processing will continue in several directions. Increased processing speed will no longer be the measuring stick of evolution; rather, we are going to see other characteristics becoming more relevant, such as low energy consumption (and low dissipation), communications capabilities embedded on a chip, programmable wiring to adapt the architecture to the task at hand (which may turn out to be a way to further decrease power consumption), the coexistence of several microprocessors on the same sliver of silicon, highly specialized architectures, SoC, and direct coupling with optoelectronic interfaces. Additionally, we will see the emergence of alternatives to silicon, such as molecular computers and hybrid structures (bioelectronics) that are particularly suitable to the area of sensors. Quantum computing is still far away, and it is impossible to know when such processors will be available, if ever. For the sake of this chapter, we will disregard quantum computers, but molecular computers, particularly their application to sensors, are considered. Energy and communications capabilities (often tightly coupled) are possibly the most important factors to be addressed when discussing the factors enabling ubiquitous disappearing computing.

From the point of view of business, the key issue is the trend toward embedding processors in any objects and the shift toward SoC. In principle, processing may enter into the design of any object, but most of the companies producing these objects today are not prepared to manage this technology. It is possible, from a purely technological point of view, to place a processor in any table. From the service point of view, however, the manufacturer would ask why customers would want to have processors in their tables. One might also ask if any customer would be willing to pay the price to get a processing-enabled table, but this kind of question will no longer apply as the manufacturing costs for embedded processing decline. The other question — why customers would want a table with an embedded chip — remains to be answered, though, and that is exactly the type of question most companies are not prepared to answer. When a company does find a convincing answer to this question, it will immediately gain a competitive advantage in the market, displacing other competitors. Catching up will not be easy because of the changes required in the production process. A carpenter is not likely to be prepared to manage the embedding of chips in his tables when he discovers that a market exists for such a thing. It is not simply the assembly that matters, though; it is also the configuration, the general architecture, ... the list goes on and on.

When SoC becomes the normal way to provide functionality, chips will no longer be a commodity. The cost of assembling together several chips, as is being done today to provide certain functions, is not going to decrease, thus this approach is less desirable compared to producing similar products using SoC (the single chip is going to follow Moore's law, thus price will decrease over time). Ubiquitous disappearing computing requires a significantly different production approach.

Communications

The field of communications has progressed in a completely different way compared to storage and processing. Transmission capacity has basically remained stable for many years, providing more capacity by deploying extra cable. The advent of digital transmissions in the 1960s created the first discontinuity, which introduced a multiplication factor of approximately 30. A second, greater, discontinuity occurred at the turn of this century, with the massive deployment of optical fibers and the introduction of dense wavelength-division multiplexing (D-WDM) and coarse wavelength-division multiplexing (C-WDM; still in progress). The capacity of networks multiplied thousands of times in the blink of an eye, completely disrupting the market. The economic bottleneck is no longer the transmission of data but the conversion of data from analog to digital and the other way around. The technological bottlenecks are found in the various technologies comprising today's networks. Backbones have overabundant capacity. The advent of Asymmetrical Digital Subscriber Line (ADSL) and the like has provided the capacity to satisfy demand as it is voiced (and the customer is prepared to pay for it), with some notable exceptions in some areas. Metropolitan area networks (MANs) have both technological and economic bottlenecks that force operators to make some tough decision over the medium term.

In the longer term, signs clearly point to an all-optical network (AON), where capacity is no longer an issue; however, no consensus has been reached with regard to the architecture of this network. It is likely that this network will include a phase of interconnected optical islands, none of which requires signal amplification. Signal amplification may take place (with associated costs) in the links connecting the various islands. Distribution and some metropolitan rings may be based on C-WDM, and links and backbones will utilize D-WDM.

As important as these issues are for an operator, they are not the most crucial, as all of these technical options are being played out on their turf and are under their control. Much more difficult is understanding the strategic choices to be made regarding a completely new phenomenon that has just emerged: alternative access infrastructures potentially owned by myriad players and in many cases completely outside the operator's control.

This new development can be traced back to the emergence of competition in the wireless market, where for the first time operators have had to compete and cooperate to ensure that their customers can talk to one another regardless of the access network being used. However, additional issues are just emerging due to the proliferation of WiFi access and the transparency of the network technology to the service. On the one hand, these developments are highly desirable for all parties, but on the other hand they are eliminating distinctive values of the infrastructure.

Hotels that have traditionally charged high fees for international calls are seeing their revenues drop because travelers are connecting via Skype and paying less for an intercontinental call than a local call. The same situation will arise for fixed-line and mobile operators alike, possibly within the next five years (or sooner, in some areas).

Progress in propagation technology, along with the already-mentioned evolution in processing that allows communication on every chip, is creating a completely different scenario that was simply inconceivable ten years ago. Low-power, pulse-based communications (ultrawideband, or UWB) and other varieties of WiFi (such as 802.11n, which provides 540 Mbps capacity per micro cell) are increasing radio access capacity to the point where supply exceeds demand, thus removing the need for regulating the spectrum.

By the next decade, the progress expected in processing and energy storage may well lead to a negotiation of interference by terminals, thus effectively making available unlimited bandwidth for all practical purposes. That, together with the existence of a variety of infrastructures (including the one provided by the terminal itself), will deal the final blow to the concept of the "value of the infrastructure." Operators will not be unprepared. The value of the infrastructure will not simply vanish into thin air; instead, it will shift to services and other types of infrastructures, such as one enabling services, which I am discussing in this chapter.

Data Capturing

Data capturing technologies will evolve significantly. Under this banner we have a variety of technologies, such as ones that capture images of three-dimensional objects, that locate objects, that spot or track people and goods, that capture various sources of data such as music, sounds, biometric parameters, voice, emotions, gestures … the list is very long. However, for the sake of this discussion, the relevant data capture technologies are the ones that sense the environment and are embedded in objects. Tagging technologies are well suited for identifying, locating, and providing the relative positions of objects in a given environment which can be derived from calculating the identification signal propagation time.

The evolution of sensor technology will provide ways to create an awareness of the surrounding environment — to a limited extent in the sensor itself and to a greater extent within the network of a sensor. Complete awareness is probably beyond the capability of these networks and may require external computation and the integration of other information.

Particularly interesting is the increased flexibility of individual sensors that can change their sensing strategy and what they are sensing (to a certain extent) according to a cooperative defined strategy. This requires evolution in another direction that is also central to any discussion of software infrastructures for ubiquitous disappearing computers: the emergence of autonomous systems. Autonomous systems form islands of independent processing, creating an inner environment clearly separated from the outer environment. Decisions are made based on the outer environment as perceived at the edges and obviously on the goal of the autonomous network and the transitions required to maintain internal stability. (This is exactly what occurs in a living being, where the biological processes endeavor to maintain the *status quo* by taking into account the goals of the being and the external environment as perceived at the boundary.) Autonomous systems do not communicate with the external environment; they are only aware of the situation at the boundary. This is quite a departure from the way engineers are used to designing systems. They tend to sit together for lengthy periods of time to agree on communications standards and who is doing what, and then they have to negotiate for feature interactions.

Quite simply, such an approach does not work for complex systems. The time required to agree on an interface would be greater than the time required for that interface to evolve, thus making any agreement irrelevant. This is what is already beginning to happen in the communications area due to the emergence of a variety of access infrastructures, and this has already happened with regard to interactions between applications. The concepts of plug-ins, client–server architectures, and peer-to-peer communications are testimony of a completely different approach to communications.

Sensor networks, designed to reduce energy consumption, will likely decide on a case-by-case basis the appropriate communications protocol. The software radio, which will begin being applied in 2007 and beyond, is another instance. Technology is evolving not just in performance but also in the direction of creating new paradigms that in turn are creating new challenges to the value chains and to the market.

Paradigm Evolution

What it is meant by "ubiquitous computing"? This term can be interpreted in at least three ways. It can refer to (1) the pervasiveness of systems providing local processing power, (2) the presence of processing capacity in many objects within any environment, or (3) the possibility of accessing processing power at any point to satisfy any processing need. In the first meaning, the focus is basically on the progressive dissemination of PCs, laptops, and personal digital assistants (PDAs). In the second meaning, the focus is on smart objects — everyday objects that are modified by embedding the processing power necessary to interact with the environment and to process information locally. This application is sometimes referred to as "pervasive computing." In the third meaning, the focus is on connectivity, on a network able to connect any processing request with processing capabilities residing in some other place. In this meaning, the attention shifts to grid structures. All of these three meanings are of interest for the topic at hand.

One for Many, One Each, Many for One, Many for Everybody

> In the beginning, man created the mainframe. And this was so expensive and so complex that only few elected could access it and share its resources.

Hence, *one for many!* At one time, it was difficult to imagine that many companies, not to mention many people, would need to access processing resources. Then the PC arrived and the paradigm shifted to *one each*. Now people have laptops, PDAs, cellphones, and organizers, and the paradigm has shifted to *many for one*. Let's look at this last paradigm and consider that "many" also includes the pocket calculator, the car navigator, the remote control — basically any object that in one way or another is processing today information. The evolution of these systems has basically been in line with Moore's law. The crucial point here it is not to determine when Moore's law will stop working because of physical limitations; rather, it is determining whether we will reach a point where any further increase of processing power will no longer be required by the market.

Over the years, demand has always exceeded the supply of processing power, in both the business and consumer markets (in the latter, due in no small part to video games and digital video and film processing). Is this situation going to change? In fact, we are already beginning to see the first signs of change. Intel and AMD are no longer trumpeting the increased speed of new processors, although the fact that a chip hit the 1-GHz milestone and then 2 GHz was known to everybody because of advertisements and articles in the media. Beginning in 2004, few headlines celebrated hitting the 3-GHz or more recently the 4-GHz mark. People do not really care anymore. The rate of replacing PCs is decreasing, and their life-cycles are growing longer. As a consequence, we might expect a significant decrease in the price of microprocessors that in turn will make possible their broader application, as is the case today, where the microchips of the 1980s have found use in remote controls, keys, and locks. Notice that what goes down is the chip price, not necessarily the price of the PC. The PC box contains much more than the microprocessor; for example, it contains wires and has packaging costs that are not likely to decrease significantly.

Personal computer sales in the world are now stable at around 130 million pieces. Currently, PDA sales are around 11 million units, and cellphone sales are around 400 million. In Italy, the 2003 figure was 13 million PCs, 6 of which were bought for family use, a market having reached a penetration of about 28 percent. In the United States, the penetration is stabilizing at around 50 percent of homes, and we are witnessing a lengthening of the life of PCs there. Features offered by new models are not sufficient to push consumers toward replacing what they already have.

Personal computers have given rise to a strong discontinuity in the market. It is interesting to note that the factor leading to this discontinuity is not technological progress measured in terms of MIPS (million instructions per second, or the processing power); rather, it is the size of the market. One implication is that PDAs, not tablet PCs, are likely to create further discontinuity. A transformation of the business is likely to derive from the penetration of microprocessors into everyday objects, as this would create a market size at least one order of magnitude larger than the current one. Smart objects and grids are the next revolution. Ubiquitous disappearing computers along with the supporting communications and software infrastructures discussed here are likely to be the "brave new world." This creates a new paradigm. When computers are embedded in any object and they have become invisible, they are no longer "your" computers; instead, they are shared with everybody. Does the market really need this kind of paradigm shift, from *many for one* to *many for everybody?*

Storage Infrastructures

Local storage technology has increased enormously and has become a way of life. People have 100 MB dangling from their key holders. Rather than continuing to embed ever greater storage capacity into devices, today we see storage devices (e.g., USB memory sticks) embedding digital cameras, MP3 players, and WiFi gateways. Some telecommunications companies have begun to offer storage services to the consumer market. Customers take pictures with their cellphones, and their providers offer to save the pictures on their networks so the customers can share them with friends and family. Google is offering 1 GB of memory space in their network for e-mail. Customers will never have to delete e-mails to free space; they can essentially record their social lives there and access the space at any time. Nokia is working on the idea that everything passing through a customer's cellphone can be stored in the network for later use. A lot of information passes through cellphones: calls, messages, agendas, and pictures. In the future, this information will be supplemented by metadata, such as time of day, location, and nearby cellphones. The network becomes a place to store information to make better use of it. It is possible to actually create a network based on information.

Only ten years ago distributed storage was developed as a response to the need to replicate information for availability and reliability purposes. We are now seeing two completely new approaches to distributed storage. One is typified by the OceanStore project, which proposes to use the myriad servers connected to the Internet as repositories of fragments of information. The information is split into little pieces, which are duplicated a thousand times and distributed on the network. Only the owner of the information knows the retrieval algorithm and therefore can reassemble the split fragments into coherent information. A failure in any server is not going to have any affect because that same fragment is present on a thousand other servers. Any hacker attacking a server will only retrieve nonsensical data. This approach is not particularly new. It is the one used by animals' brains to store memories and experience. That is why our memories are so long lasting, in spite of the continuous death of our brain cells (one a second, according to some studies, or 100.000 a day!).

The second approach for distributed storage is DataGrid. The objective of DataGrid is to make it possible to share huge quantities of data and allow parallel analyses, the results of which would be information that is not the sum of what is available but something brand new and whose value is greater than the sum of the parts. Each data storage is basically influencing every other one in a dynamic way but without requiring any change in any of them. In this case, we are in the autonomous system environment.

Oracle has made its grid software development kit available to interface its databases and produce a grid infrastructure. The company clearly hopes that its database will become the storage medium for any grid application. This is interesting given the number of enterprises using Oracle who may now find a seamless path to becoming part of a business grid. Oracle's vision is one of complete transparency. The client should no longer care where the data is stored nor where the data will be processed. Electronic Arts has created a virtual-reality game based on these Oracle interfaces and exploiting the grid architecture to support parallel gaming of up to 100,000 players. A project in the Adriatic Basin is working along these same lines to create a shared ubiquitous infrastructure for business to enhance service provision.

In 2002, Oxford University announced the use of the grid for their eDiamond project, aimed at sharing medical information on breast cancer. The name of the project underscores the many facets that are addressed, including the aim of using the same data for both advanced research and everyday care of patients. The project is part of a wider initiative aimed at making processing power, information, and applications available to the scientific community. eDiamond does more than simply provide access to information for institutions, researchers, medical doctors, and patients alike. It also guarantees the privacy of sensitive information and provides applications for analyses, cure identification, and progress monitoring. Researchers can study and compare hundreds of thousands of cases and analyze the effect of the different therapies. It is interesting to note that the basic applications are commercially available; they have not been created specifically for the project. The value added is the transparency, dissemination, and accessibility provided by the software infrastructure.

Profiling

"Profiling" is a name that has been used extensively in marketing to identify characteristics of market segments and to associate potential clients with a specific segment. It is going to take a completely different twist in the coming years to radically change the concept of service itself, from both the client's and the provider's points of view. Beyond that, profiling will be applied to any entity engaging in any activity on a network. Profiling will become a sort of characterization of user expectations merged with the experiences of specific users, their current needs, their locations, and the environments within their reach. It is this web of connections reaching into the environment that is of particular interest for the purposes of this chapter. The presence of a software infrastructure connecting a variety of ubiquitous invisible computers can significantly change the way we access and make use of services. Again, we are confronted with the concept of autonomous systems. The capability and willingness to disclose these environments and negotiate services with providers will change from situation to situation. Service selection and adaptation may be done only within the current environment or may result from negotiations with providers.

The creation of a profile is likely to become a very sophisticated activity, something that itself is a service provided by trusted parties. Intelligent agent technologies are particularly attractive for monitoring experiences and creating and updating profiles. They may also be used in the negotiation of services.

A profile may result from the synchronization of information captured at different times by different devices. This synchronization should take place automatically and seamlessly, but at the same time it should be trusted and under the control of the person or object owning the profile. The sharing of profiling information should be at the owner's discretion, which opens up a can of worms. To get the maximum from profiling, the owner must share information, but at the same time it can be difficult to maintain control over the information. The software infrastructure should provide for some anonymity features to allow information to be exchanged without revealing identities.

Information Searching and Sharing

At the turn of the last century, Pointcast proposed a mechanism to push information to people. Users would subscribe to an information channel from their PCs, and every time they used their PCs information would be pushed on the screen for their perusal. The service did not create sufficient market interest (in the sense of people willing to pay for it) and was discontinued.

In the meantime, the amount of information available has continued to grow; it has doubled in the last three years, whereas the previous doubling took over 30 years. The expectation is for continuous increases, with a doubling every two to three years for the next 15 years. Additionally, more and more of this information is digital and is stored somewhere, thus it is potentially within reach.

Having too much information is not that different from having very little. Suppose you have just ten pointers as a result from a query on the Internet. Would that be so different from receiving 200,000 results? No. Most people look only at the first four to six result lines returned by Google. Almost no one ever continues to look beyond the third page of Google results. The market is ripe for some sort of information push and for different ways of information sharing.

As was already pointed out when discussing the DataGrid, the value of information resides more in relationships than in a single bit of information. The overall information of an environment may be more valuable than the information contained in one of the entities of that environment. It may be more interesting to grasp the know-how of a research center than to determine what each researcher there knows. Knowledge management is likely to significantly shift direction by taking a much more holistic view.

Today, knowledge is quite segmented, and the more outstanding an organization is the more likely it is that there is no single point to focus on. Actually, the knowledge society is exactly what the name implies: a web of interactions that are the knowledge. The model where the value of a library can be found in its individual books is rapidly fading away, to be replaced by the value of an interconnected world. This is going to have a tremendous impact on the market. A seemingly far-fetched example: When an association of engineers in Italy commissioned a study in 2002 on the possible future of engineers, they found that the future will see the aggregation of engineers into enterprises, each containing a variety of engineers providing different skills. Engineers will no longer be hired as individuals but as a collective force to address ever more complicated issues.

The software infrastructures will be crucial enablers in this transformation from information to relations, from data to meaning. The so-called platforms and middleware, further addressed in this book, are necessary components in this transformation.

Terminals as Network Nodes

Terminals, be they cellphones, PDAs, digital cameras, or WiFi enabled, are embedding more and more processing power, storage capacity, and a growing communications capability. This trend is going to continue in the future. Already today some terminals can act as gateways for other devices. A cellphone may be used as such by a PDA, which will communicate via Bluetooth® with the cellphone, which in turn will relay information anywhere by providing network connectivity.

By the middle of the next decade we can expect terminals to become network nodes capable of selecting an appropriate communications channel based on the user profile and the specific service being requested. A significant portion of this decision making will be completely transparent to the end user. Already today, when using a 3G cellphone, the user is not aware of whether the communication is carried out on a 3G or 2G network, and he does not care. When we are roaming abroad we usually do not notice whose network we are using. The same will happen with multi-standard cellphones where connectivity may be provided by either WiFi or cellular networks.

We might not even give much thought to the fact that the video call we are receiving on a window in the television on which we were watching a football game was actually routed there by our cellphone, which was aware of the existence of such a television and that we are right in front of it. It will seem entirely ordinary to have the video call pop up on the screen. Such a scenario requires a significant effort on the technology, infrastructure, and standardization side to make it happen in a seamless way. The usual rule applies: The easier a service is for a user, the more complex its implementation. The main issue here, however, is not a technical one but one of shifting power among different players. The situation at the beginning of 2005 is still pretty uncertain. Who is going to play the role of the integrator, the orchestra director, of the PC, entertainment system, applications, services, and those who are behind these?

In Korea, this concept of integration is considered fundamental to tomorrow's business. Most importantly, the home environment will seamlessly extend into and integrate with all other environments. The software infrastructure supporting the seamless use of any appliance and service within the home will have to reach out and accommodate services, information, and applications residing on other platforms — at schools, offices, and healthcare facilities, for example.

The network of today will extend its reach into any environment and into any device, and these devices will no longer be seen as terminals but as network nodes. They will be designed and produced in ways that are beyond the control of operators but they will find themselves forced to interact with them. Actually, successful operators will learn to leverage them to provide more services financed by their customers.

Terminals as Networks

The next step is almost natural. When terminals have all these capabilities they may also begin communicating with others in the vicinity. For terminals that are not in the vicinity, other terminals will be used to establish a connection path to distant ones. Moving in this direction requires some significant progress in propagation studies and *ad hoc* and mesh networks, all items on researchers' agendas that are unlikely to lead to marketable products within this decade, as some significant hurdles must still be overcome. Notice, however, that an infrastructure based on terminals with basically no external infrastructure support is in keeping with the concept of ubiquitous disappearing computers and requires significant evolution in the software infrastructure. This is the idea behind generating an infrastructure out of the physical connection capabilities offered by terminals. We are likely to see this kind of network-based infrastructures in some market niches and in some application sectors.

The Market View

To close this chapter it is appropriate to examine our previous discussion from a market point of view. The market is going to be affected more by the paradigm shifts presented in the previous section than by the technology evolution, which remains in the background as a crucial enabler but nothing more. Clearly, the story would be different if we were to consider a manufacturing business where adopting one technology over another may tip the scales of success toward one manufacturer over another one. Here we are looking at the end-user market, the kinds of services this market is likely to need, and the implications on the value chains.

Technology Evolution Versus Market

Technology is going to evolve in the future just as it has done in the past, but even more so. A variety of researchers, enterprises, and businesses will be inventing, perfecting, and deploying technology at a faster pace. The era of standardization and painful and lengthy discussions among international committees is, to a certain extent, over. I do not mean to suggest that we will no longer have standardization committees, forums, and international organizations trying to come to agreements. Actually, these are essential. What I am saying is that technology innovation and deployment are almost impossible to plan.

Some studies performed within the FISTERA group, in cooperation with the Polytechnic of Torino, Italy, seem to indicate that the expected evolution, as forecasted by scientists, will not behave according to a completely random set but it will not behave as a planned set, either. It will be somewhere in between. The expected evolution would seem to conform to the theories of "small worlds"; that is, evolution happens in an unstructured way but some attractors over time steer the direction, creating a sort of structure. FISTERA has identified some of these attractors within the technology set studied: batteries, storage, embedded systems, microkernel and *ad hoc* protocols, bandwidth, information semantics, and radio propagation. It is interesting to note that most of these relate to the topic of this chapter and book — namely, embedded systems, microkernel and *ad hoc* protocols, information semantics, and radio propagation. Batteries and storage may be considered as instrumental and therefore loosely related to the topic.

These attractors, however, only underscore the obvious fact that evolution in these technology areas is likely to accelerate evolution in many other connected areas. The interesting attractors should be studied at the market level, as it is the market that will push technology evolution in certain directions. The South Korean identification of eight services to be promoted, three infrastructures, and nine information technology growth engines is a clear attempt to stimulate the overall scenario by forcing the creation of attractors. Software infrastructures, platforms, and middleware can be viewed as being instrumental in enabling some attractors to initiate a virtual circle of evolution.

The market today has plenty of technology from which to choose. Technology in a way is not the stumbling block. The iPod® miracle, as many call it, is filled to the brim with technology, but it was not technology that created the miracle. Plenty of MP3 players were on the market with very similar characteristics, but there was only one iPod. And the iPod has become an attractor. Look at ipodder.com, which provides a way to receive audio information in a push mode, and podcasting, which is a new way for people to communicate. iPod, as an attractor, is not simply making business for the company that owns it; rather, it is generating additional business for potentially anybody. It has basically created an infrastructure, a *de facto* standard, that no one had to agree upon.

The embedding of processing capacity in appliances along with communications is creating yet another attractor. Why would an appliance producer begin to create processing-enabled appliances? Ariston, an Italian white goods company, has incorporated processing into its product lines. Margherita is a washing machine that can connect to the Internet and communicate with the repair center when it malfunctions. In principle, it can also talk to the refrigerator. Why should it do that? A dialogue among the various appliances can decrease power consumption and can ensure that it remains within the agreed-upon target usage. If the refrigerator knows that the washing machine is heating water, it will wait to run its compressor. The same goes for the microwave oven. When a user turns on the microwave oven, it immediately asks the washing machine to suspend heating water for the next 20 seconds to avoid overloading the power line. The Internet connection used by the washing machine for proactive maintenance control can, in the future, be used to obtain information on how to wash a certain dress, the identity of which is embedded in an RFID tag woven in the fabric. And, of course, all of this is performed automatically. The pervasive computing environment takes care of that.

How much does it cost to embed a computer in an appliance? Ubicom offers a microprocessor for this kind of application at less than $10. This microprocessor also embeds communications capability. Emware offers software supporting the remote monitoring of these appliances.

Microsoft's Smart Personal Object Technology (SPOT) software foresees a variety of everyday objects embedding processing and communications capability. A simple key chain may receive traffic information via FM signals and tell us which route to take. Of course, a car navigator can access the same information and prompt us on the quickest way to get home. Assume, though, that one night we are not going directly home; in fact, earlier in the day we called a restaurant to make a reservation. The cellphone is aware of this call, and it can inform the car navigator of our destination as we step into the car. Or, a digital frame in our living room may select and display photographs depending on who is in the room, recognizing what that person likes and learning from that, or it can show the faces of people who have left messages on the answering machine. Again, these are trivial examples of a seamless intelligent environment with ubiquitous pervasive computers.

Clearly, in moving from technology to market, two basic questions must be answered: What is a person willing to pay and does a sufficient market exist? Most of the examples we can think of in the area of ubiquitous pervasive computing and intelligent environments are fairly trivial, including the ones I have already mentioned. The point is that our life is made up, 90 percent of the time, by such rather trivial things but they become integral parts of our life. Once I checked into a low-cost hotel in the United States with my younger son. In the room he looked for the remote control to switch on the television. It was an old model and had no remote control. When I explained that to him he could not believe that there was a time when televisions did not have remote controls. "Did you really have to walk from the couch to the television set to change channels? Unbelievable!" Before remote controls were the norm, how much would we have been prepared to pay for a remote control? Probably nothing, as most of the people would not have seen any need for a remote control. It was such a straightforward thing to manually change the channel. This is how habits are entrenched in our culture, and that is why the intelligent environment we are imagining today looks rather irrelevant and not worth the cost; however, once we get there we will never look back. Simplicity is a must for any technology that hopes to win the market, and the grid is a step in that direction.

From Information to Applications: The Grid

The Internet was designed to provide ubiquitous computing, but it does more than that in that today it serves as a data communications network, a way to share information. More importantly, though, is the fact that it does not serve as a way to share processing resources, basically because these got so inexpensive over time that the need for additional resources was felt only in some niches. By the middle of the 1990s, many of the PCs connected to the Internet were idle most of the time. The idea was put forward to take advantage of this idle time, and SETI@home was born. This occurred in July 1996, a date we can probably mark as the birth of massively distributed processing. In a very short time, hundreds of thousands of computers were participating in the effort, resulting in an overall processing capacity exceeding that of supercomputers.

A few years before that, however, in 1992, many telecommunications operators decided to initiate a consortium to create a distributed environment to facilitate the creation of complex applications and services, building upon the variety of applications already existing and ones that would come. The initiative was known as TINA (Telecommunications Information Networking Architecture). In the end, though, the goals (or dreams) were not achieved, as it was probably ahead of its time; however, the idea of sharing applications as basic blocks in a distributed environment that can be used as building blocks to create other applications is still valid today and still remains a challenge for the future, a challenge taken up by the application grid, one of the many forms the grid concept has taken. I have already mentioned the grid in terms of processing to harvest processing capacity through the use of many processors interconnected in a distributed environment. The DataGrid is a way to share massively distributed information to derive meta information of higher value.

The application grid is a further step. The basic idea is to harvest application resources available in a distributed environment. This objective is achieved through the use of three types of interfaces: the connection protocol, the resource protocol, and the collective protocol. The connection protocol provides a set of interfaces that allows one application to establish a connection with another application to provide authentication services and negotiation support while establishing and carrying out the communication. The resource protocol allows an application to discover the existence of other applications that may be available in that particular distributed environment. At the same time, it establishes a mechanism to declare what an application is doing, thus allowing others to understand if it is of interest for their purpose. The collective protocol supports the aggregation of several distributed applications, managing them, from the point of view of the application calling upon them, as if they were a single application providing a higher level service (this is, as a matter of fact, the result of the services provided by the individual applications, appropriately synchronized). The interactions that the collective protocol is required to support are very complicated and as of today represent the biggest obstacle for effective use of an application grid. Still lacking is what the HTML and browser have used to tie together information created in a distributed environment. Current implementations of the application grid work only in very specific areas where the people who have designed the individual applications have followed agreed-upon ways to describe them.

It is not possible to predict the kind of evolution the application grid will have in future years. Surely, if an effective solution is developed, it will have a significance similar to that of HTML and Mosaic on the Internet. In this case, however, the impact will not be felt directly by the consumer, as has been the case for the Web, but by the businesses offering complex services, wherever they are in the world. This would really change the market and value chains. A company in a developing country in Africa may come up with a service that exploits the capabilities of applications embedded in appliances in Bill Gates' home to provide him with outstanding services. That company could use the same service to provide enhanced facilities to a client in Asia making use of the devices the client has installed in his home which are likely to be different from those of Bill Gates. The existence of a software infrastructure tying together various computers present in any environment will allow this kind of service creation and delivery. Notice the importance here of the concept of "creation," which takes place at the delivery point.

Tagging

At the end of 2004, China had produced 5 billion RFID tags. This number may exceed 1000 billion in the next decade. The impact on business will be astounding, but it is not going to be, as some people are speculating, due to changes in the production or supply chain enabled by this technology; rather, these changes will be significant because they will make the production and supply processes more efficient and less costly. Of major importance will be the creation of a direct link between production and the end client because of the transformation of products into services. This is really going to change the value chain.

For this to happen, in addition to reducing the cost to manufacture and embed e-tags, a pervasive communications infrastructure is necessary, as is overcoming privacy concerns in the developed world, at least in the deployment phase. My personal opinion is that, while we are discussing privacy violation concerns that might result from the use of tags, some countries, such as China, are moving forward and creating a *de facto* environment that will force the remainder of the world to go along with it. I do not want to take a stance on complex ethical issues such as stem cell research and cloning, but I do want to emphasize that while many countries do not condone research in these areas others are continuing with it. If this research will lead to some drugs that prove effective in curing cancer, I doubt that we will not take advantage of them because of ethical considerations. At that point, the value will shift to those who invested and brought forward the innovative products and solutions. It may be better to remain a part of the game, keeping a close eye on all issues, including ethical ones, rather than peering in through the window.

E-tags are nothing more than a technology, and not a particularly sophisticated one, but the context they create is a very sophisticated one that enables ambient intelligence and acts as an attractor for a variety of infrastructures, terminals, and services. Software infrastructures for ubiquitous disappearing computing are going to be affected by the evolution of tags and their usage. At the same time, they can take advantage of tags to create a richer fabric for a variety of applications. We may expect to see tags systematically pervading any environment by the end of this decade, although it is likely that most of the impact will be felt in the next decade.

Sensor Networks

Sensor networks are fairly similar to tags with regard to their global effect on business, although the impact is likely to be limited to some vertical markets. They may become part of a pervasive infrastructure and managed as such. Today, sensor networks focus on very specific tasks and are deployed to respond to specific needs of a company, institution, or government. As sensors become more flexible and the range of parameters they can capture becomes broader, we may see growing interest among infrastructure-related enterprises, such as telecom operators, utilities, and municipalities, to provide sensor networks by piggybacking on their infrastructure and selling the information captured as just another service.

A typical home may contain a variety of sensors designed to detect malfunctions, fires, and intruders, but some also contain monitoring systems that track the well-being of the inhabitants. Cars, too, have sensors, which gather information that may be used by others (e.g., to monitor pollution level). The environment will be progressively populated by sensors. More and more objects will embed sensors (most already do) and will be capable of sharing the information captured (which today is not the case).

Cellphones may contain sensors and are in an ideal position to share the information captured. The very presence of a cellphone is a sensor by itself; for example, one could monitor the presence of people in certain areas to determine the potential for congestion. Of course, marketing data can easily be derived from such information. Assuming that the identities of the people being tracked are not disclosed, one could create a wealth of information for advertisers. Similarly, the control of epidemics may be aided by this kind of knowledge.

Today we are basing many decisions on a few very costly sensors that must be carefully tuned to ensure their appropriate functioning. Tomorrow, a vast amount of data will be derived from hundreds of thousands of sensors, and averaging of such information will greatly offset possible local inaccuracies. Moreover, today we have to plan in advance the type of information we want to capture and deploy the appropriate sensors in the most effective places well in advance. In the future, sensors will be everywhere, and it will be simply a matter of deciding what we want to know and receiving an answer almost immediately.

Telecom operators must decide whether to enter into this market by deploying the required infrastructures (obviously leveraging on what they already have, which is a lot) or to just sit and wait for someone else to do it and hope to get a share of the pie in terms of traffic rights.

Impacts on the Value Chains

The advent of an intelligent environment composed of a software infrastructure that supports communications locally and acts as a gateway with the outer world and the variety of applications independently developed by a variety of actors and present on appliances of any type will deeply transform the way we live and the way business is being carried out. Today's assumption is one of a value chain that ends at the retail point.

Additionally, each link in the value chain effectively separates an actor from others further down (or up) the value chain. Telecom operators, as an example, have had a very clear termination point for their network (the NT, or network terminator) and whatever is beyond that is no longer their business. ADSL has begun to change that because ASDL is viewed as connectivity by the operator but as a service by the client. The result has been a nightmare on both sides. When customers call the support center to complain that the service is not working, the engineers there perform remote checks and come back to them saying that on their side, up to the NT, everything looks fine and it is not a network problem. Customers do not care where the problem is; they just want it fixed. Most of the time, fixing the problem would require getting inside the customer's PC, the set-top box, or the appliance to look at the software and the drivers of newly activated peripherals — something that the operator is not technically prepared to do and, more than that, is not prepared culturally to do.

This very same situation is going to be faced by product manufacturers. As the product operates and cooperates in an intelligent environment it can talk back to the producer. As it talks, it morphs itself into a service. This is a great opportunity for establishing a connection with the end customer but it is a double-edged sword. Companies selling services can continuously monitor the ways their customers are using their services and can come back to the customer with new services, thus further developing their business. They can focus advertisements on specific customers and deliver them at the most effective moment.

At the same time, a customer buying a product basically loses all the leverage over the seller once he has paid for it. He can obviously complain if something is wrong, but it is very difficult to get any money back. A customer buying a service pays for it as he is using it. The very moment he stops using it, he stops paying for it. Making money from a service requires continuously satisfying customers; making money on a product means convincing a customer once. The target market of a car manufacturer is people who have yet to buy cars, not those who already have bought cars.

As the world shifts toward services, competition will truly become global as services can be delivered from anywhere in the world to anywhere in the world. Products are likely to become a way to deliver services, and in many cases products will be given away for free in the hope of making money on the services. This is basically what has already happened in the ink-jet market, where the printer is either free with the purchase of a computer or is sold at a very low price, and revenue is made on cartridges. Another example would be gaming consoles, which are sold at a loss in return for the potential profits to be made on the games.

Such a development can be a threat for many local businesses who risk losing their location advantage, but at the same time it can open up new opportunities. Tourists visiting a nice spot in Italy may subscribe to a service to enjoy virtual visits once they are back home. Or, when they are back home these same tourists can use their entertainment systems to show friends and family what they saw. Think about developing an album of pictures that here and there contains links to the places visited that allow a real-time view of what is going on there. The world is shrinking, and lots of opportunities and challenges await.

Getting There

In closing this chapter, it might be worth considering for a moment the problems facing us as we attempt to make ubiquitous computing a reality and to also consider the problems that will result from such an achievement. The rapid growth of applications and processing power, combined with reduced costs in production processes that have enabled the embedding of microprocessors in any object at a marginal cost, is leading to the creation of environments having a local intelligence. The evolution of telecommunications, specifically on the access side, is making it possible to interconnect these local environments and potentially control them from a central location. These two opposing forces, one centrifugal (local processing) and the other centripetal (telecommunications as the core), are not balanced and in my opinion the centrifugal one will take the upper hand. In a way, this should be good news for those who fear the advent of "Big Brother." The localization of processing power brings along the issue of management. New architectures and new structures are required, and the evolution toward the autonomic systems is a step in this direction. Looking further into the future, the existence of highly distributed and communicating processing structures may radically change what we consider a telecommunications network to be and in particular where its boundaries are.

Acknowledgments

The ideas presented here derive from work performed in the FISTERA project (http://fistera.jrs.es), funded by the European Community, and from discussion with many researchers within TILAB and in other organizations in Europe and abroad. They do not necessarily represent the positions of Telecom Italia Lab nor of Telecom Italia.

Chapter 2

Mobile Computing

Radu Popescu-Zeletin, Stefan Arbanowski,
Stephan Steglich, and Ilja Radusch

CONTENTS

Mobility

To begin, *mobility* can be swiftly explained by stating that things are moving from A to B; therefore, explaining mobile computing and appropriate middleware systems simply requires us to state *what* exactly and *how* exactly things are moving. This is, as you will discover, more difficult than one might anticipate.

Apparently, from a network-technology-centric point of view the first — and only — thing moving is a *network terminal device*. Research in mobile computing should therefore focus on allowing mobile devices to easily disconnect and reconnect to different networks, but this seems rather insufficient. Because most terminal devices are not able to move autonomously, one could argue that research should rather have the *user* moving the device to be addressed. Additionally, such user mobility can be extended to cases where users move from one device to another, demanding support for the same services. This can be expressed with the more general term of "personal mobility." Most users, however, do not care about the inner workings of their devices; they just want, among other things, the appropriate *services* to be available at all times, which we can denote as "service mobility." Because a service is expressed through *program code*, we have now identified the following four key types of mobility:

- Terminal mobility
- Personal mobility
- Service mobility
- Code mobility

Terminal mobility, then, subsumes in general all changes in the network topology. These changes can be due to the physical disconnection of devices (and reconnection to other networks) or moving wireless network devices in and out of the radio footprint of their neighbors. Examples are laptops connected via Ethernet or wireless LAN or mobile phones moving from one base station to another. Coping with terminal mobility is a well-known and well-addressed problem, as the network reference model implied that providing network access was the first (and to some people only) problem to solve; however, research in mobile computing showed that this approach must be extended for most practical applications of mobile users.

Personal mobility differs from device mobility as users do not need to carry terminal devices with them but instead use different terminal devices in their vicinity. So, a user accessing his mailbox from different locations would not carry a specialized terminal device but would contact his central mailbox server directly from available terminals. Another example is the follow-me phone, where all new calls addressed to the user are routed to the fixed-line telephone nearest the current location of the user. Additionally, personal mobility is often divided into *user mobility*, covering the description above, and *user session mobility* (often also referred to as *session mobility*). In the latter, users can carry their current session data to various terminal devices. Extending the concept of the follow-me phone, user session mobility allows active conversations to be transferred to the new location. Furthermore, the counterparts of personal and session mobility are actor and role session mobility. With *actor mobility* a specific user is replaced by a group of people belonging to a certain role; for example, all employees currently working at the reception would be able to answer service calls. *Role session mobility* implies the reactivation of actor sessions at the new terminals.

This more general paradigm of user-centric mobility emerged with the advance of "ubiquitous computing," a concept coined by Mark Weiser in 1991 [54]. The traditional view of explicitly used computers and terminal devices is superseded by smart and autonomous computing technology embedded in every device. Instead of relying on specialized devices that must be carried and maintained by the user, such as mobile phones, the focus is now on services provided for the user, such as reachability for phone calls, as mentioned above. In this regard, service mobility means that each device or each group of devices able to record and render audio could be made to function like a traditional phone for the user. Additionally, this scenario also foretells two additional concepts important for personal and service mobility: dynamic *adaptation* and *personalization* of devices and services. We will explain these further in the following sections.

Challenges in Mobile Computing

In the previous section, we established what types of mobility are relevant for mobile computing. The key challenges can be divided into physical, connectivity, performance, and terminal challenges.

Physical challenges reflect the fact that mobile devices are often more fragile and more vulnerable to damage or loss, thus rendering service provision to the mobile user impossible. Furthermore, mobile terminals today rely on limited energy sources. While this restraint can be relieved somewhat with true service mobility, applications and middleware for mobile computing must be careful about energy consumption. Although it would seem that few choices are available to compensate for the loss of equipment or final discharge of batteries, research into mobile computing systems can nonetheless seek solutions by utilizing, for example, graceful degradation (i.e., shut down less useful services first in case of a power shortage).

Connectivity challenges are obvious in mobile systems. Due to mobility, the connectivity of devices can be unstable in both performance and reliability; for example, users or devices may not be connected to the network at all over extended periods of time or the bit rate available to the application might vary over time and location.

The first two challenges generally imply the third: *performance challenges*. These are twofold. First, mobile devices are usually less powerful compared to their static counterparts. This is mostly due to constraints in size, weight, and ergonomics for mobile devices. Second, mobility — or, in general, all changes to a given system — greatly inhibits traditional optimization. Furthermore, given the connectivity challenges outlined above, services for the user must be developed with reduced bandwidth in mind (compared to static setups); thus, the performance of mobile services is often perceived as less satisfying by the user compared to similar services in static networks.

Terminal challenges arise from the vast variety of devices available. Providing consistent service quality for a wide range of devices and environments requires maintenance of several input sources or a general framework to dynamically generate output according to the terminal capabilities. This includes the need for adapting service content as well as the need for supporting various input methods, ranging from keyboards to mobile phones, as well as single-value input devices such as switches or sensors and varying screen sizes and color resolutions.

Based on these challenges, we explain in this chapter the fundamental functional and nonfunctional requirements for mobile computing. One approach to tackle terminal, connectivity, and performance challenges is to bring parts of the service logic to the terminal device. Application code (e.g., an applet) is transferred to the terminal and must utilize the available resources as best as possible. The next section describes this code mobility in further detail.

Code Mobility

Traditional approaches to achieve personal and service mobility have often implied *code mobility*, whereby program code is either explicitly or implicitly migrated to another device. This approach stems from the fact that this is a simple way of describing distributed service logic to a new device. Code mobility, as described in Fuggetta et al. [53], discriminates computational environments (CEs) (i.e., originating and target systems) hosting one or several execution units (EUs) along with resources. Examples of EUs are sequential flows of computation or different threads of a multi-threaded process.

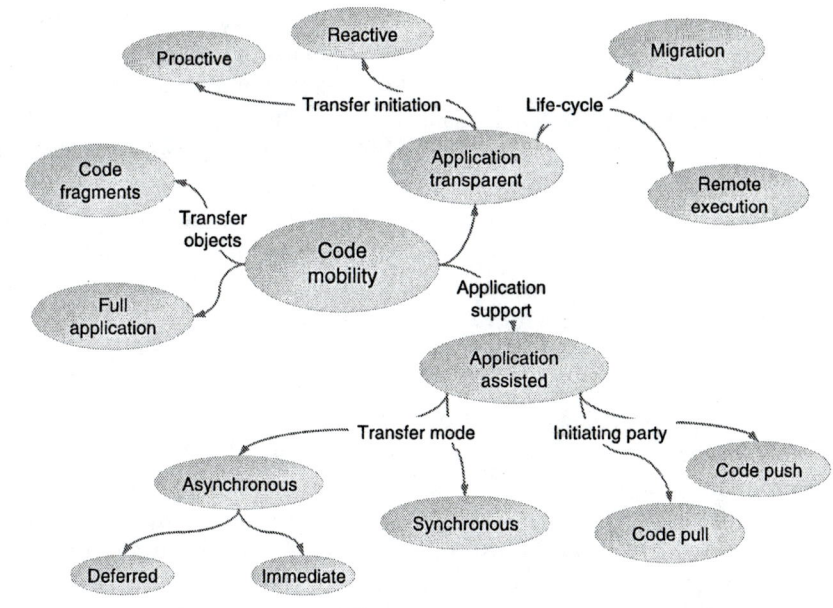

Figure 2.1 Classification of code mobility.

In general, moving a running application from one (originating) CE to another (target) CE requires three basic steps. First is the *transfer of program code* (i.e., passing executable code to the target system). Second, one must *restore the execution state* at the new system; of course, this implies that the execution state was safely backed up at the originating system. Third, all *resources or data* associated with the old system must be propagated to the target system. Research in mobile computing has shown that these three basic steps can be implemented in various ways. A detailed classification of code mobility is given in Figure 2.1.

With regard to the implementation of code mobility in mobile middleware systems, we distinguish two key issues: application support (i.e., how much developer involvement is demanded) and which objects are to be transferred during code mobility. The former denotes whether code mobility is hidden from the developer (application transparent) through the framework or relies on special application assistance. The second key issue characterizes whether full applications are transferred or mere code fragments. The notion of code fragments also includes application-specific scripting code. These issues have great impact on the middleware systems implementing code mobility and their ability to solve the challenges of mobile computing outlined above.

Regarding code mobility that is transparent to the application (also referred to as *strong mobility*), we can further distinguish it according to the circumstances under which the code transfer is initiated: proactive or reactive. In the case of a mobile user, this translates to either before the user changes his location (proactive) or after he has changed his location (reactive). This is an important tradeoff with regard to the performance challenges mentioned earlier, as proactive systems will almost certainly appear more responsive in user experience than reactive systems. Additionally, we distinguish whether applications on the originating system are removed completely (migration) or are duplicated on the target system (remote execution). The latter is especially useful for displaying locally rendered content on devices near to the user or for load balancing by distributing heavy computation to one or several different systems.

On the other hand, application-assisted code mobility (often denoted as *weak mobility*) is further distinguished according to the transfer mode: asynchronous or synchronous, depending on whether or not the originating EU is suspended during transfer. Furthermore, asynchronous code can be executed immediately on the target CE or be deferred until a given condition (e.g., the first invocation request or an external event occurred) is satisfied. Additionally, weak mobility can be characterized by the transfer direction. Program code can be pushed to the target CE or pulled from the originating CE.

Within application-transparent implementations, execution state is either automatically or semi-automatically transferred through special serialization and deserialization functions. Within application-assisted systems, the logical execution state is usually encoded in the code transfer call or must be restored by the application itself. Likewise, associating the execution state with a resource can help in developing application-assisted code mobility, because the transfer task is then assigned to the resource mobility subsystem. Figure 2.2 depicts the classification of resource mobility as a complement to code mobility.

Resource mobility is characterized by three aspects: resource binding, transfer method, and transfer constraints. Resources can include almost anything in a mobile computing environment, ranging from status variables of code objects to files on a network share to printers connected to a CE. Resources can be bound by reference, value, or type. Type is most applicable for resources bound to a specific CE, such as printers, which must and can be replaced by similar resources available in the target CE. Furthermore, resources are also characterized by their transfer constraints at the time of the resource transfer. They are freely transferable, bound to the system, or only uniquely transferable. Please note that these constraints can be highly volatile; for example, a huge file that, considering the current bandwidth available, is bound to the originating CE can become

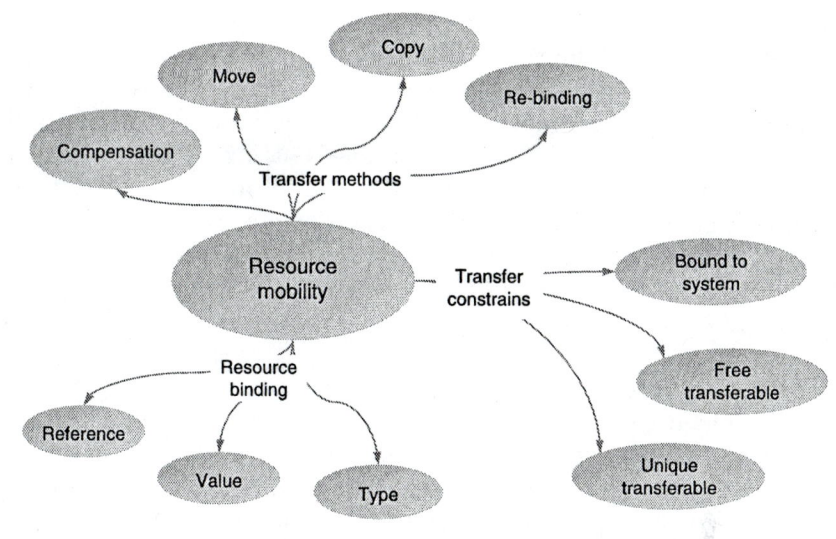

Figure 2.2   Classification of resource mobility.

a freely transferable resource when the bandwidth is improved. The former would imply that this resource is addressed by reference at the target CE. Uniquely transferable resources describe resources that cannot or must not be duplicated (e.g., resources describing tokens). These must either be addressed by reference or be moved between CEs. Resource transfers can occur by moving or copying resource values, rebinding to resource references, or compensating with similar resource types.

Personal and Service Mobility

Traditional services architectures of telecommunications and information systems have been designed and implemented from the bottom up in an independent way, from network to end system, as have the corresponding services offered to consumers: public switched telephone networks (PSTNs) for telephony, cable distribution networks for television, radio networks for mobile telephony, etc. The results of this bottom–up process are a continuous search for killer applications for the expensive infrastructure and a cumbersome and costly integration of services over different communication systems; however, the development of the Internet, pervasive computing, and sensor networks technologies has provided the required technological basis to devise future communication architectures via a top–down approach.

At one time, the communication space of humans was limited to their actual physical surroundings (village, home, or office) due to the limited spatial range of human senses. Morse's telegraph system led to the development of an ever-expanding communication space. Thanks to the telegraph, people were for the first time confronted with communication content (or "news") about people and locations of public interest that they had not directly experienced. The introduction of telephony expanded the average communication range as well as communication content as fast as the telephone network grew and the price for phone calls fell. By now, people were able to communicate with relatives across continents about topics purely relevant to the sender and recipient. This differed greatly from the communication habits of a couple of hundred years ago, when either the communication content had to be important enough to justify the cost of someone carrying the message or the content had to be durable, such that less expensive but more time-consuming message propagation methods did not render the message obsolete.

Eventually with the introduction of mobile phones, it became possible not only to reach locations very far away but also to address and communicate with people regardless of their location. Later, with asynchronous services such as electronic mail and short message service (SMS), the dimension of time was expanded. Today, people can send e-mails and do not have to be concerned about whether or not the addressees are ready to receive the messages. Over time, technology has eliminated distances in time and space, or at least has made the boundaries almost imperceptible. Of course, allowing people to interact with each other over unlimited space and time implies an exponential increase of messages and communication channels for each user. Reducing the number of messages addressed to a user to a comfortable level is essential to future communication and one task of *personalization*. Likewise, making communication channels interchangeable and thereby reducing their number for the user is the task of *adaptation*.

From the perspective of someone utilizing future communication services, one may draw some initial conclusions:

■ Individuals are interested in the content, not the presentation, of a specific service. Further, in various situations or locations, the service presentation must differ to suit the current situation; for example, someone who is driving a car requires a different service presentation than a user who is sitting in a chair in an office.

■ A human being has a limited and individualized communication space (the user does not know everybody in the world and is not interested in everything); hence, services must adapt to and personalize each individual communication space.

- Presentation of a service has to adapt to the quality of each individual's senses, life stage, and environmental situation.

Out of these requirements, we observe that modeling the individual communication space for future communication systems is the starting point for developing an all-embracing service architecture, which is further extended to a full reference model for future I-centric services. I-centric services refer generally to all services that are able to adapt to and personalize service behavior according to the needs and preferences of the user.

The *individual communication space* is defined by a set of personalized and adaptable I-centric services offered to the user (the individual) by various objects within the user's range. Objects are, in general, logical representations of substantial (e.g., hardware, lights, terminals) or insubstantial (e.g., software, money, preferences, user context) entities, providing well-defined services and encapsulating specific semantics. See Figure 2.3 for examples of objects in the individual communication space.

Individual communication spaces grow and shrink over time based on the individual's life stage, personal interests, and working and living environments, as well as the availability of new kinds of telecommunication services and devices. Furthermore, objects may and will pertain to different communication spaces. They can be controlled by individuals or other objects. More importantly, all communication between individuals

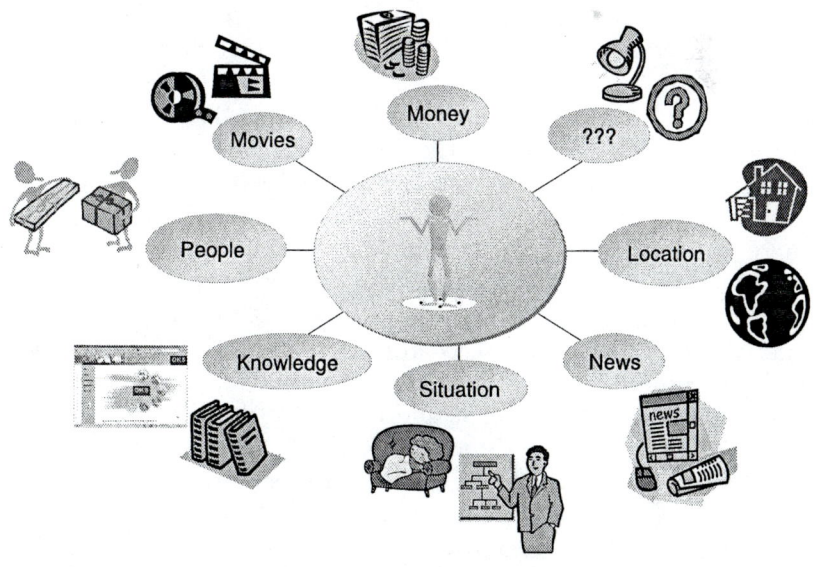

Figure 2.3 The individual communication space.

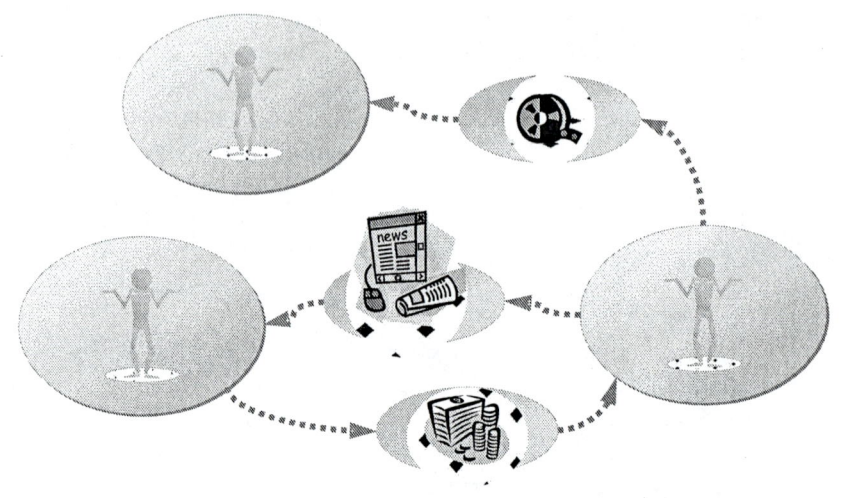

Figure 2.4   Communication through sharing common objects.

takes place by sharing objects of their respective communication spaces (see Figure 2.4); for example, a text message is first created in the communication space of the sender and then passed to the recipient. This message is likely to be transformed along the way to the recipient due to different networks or encoding schemes. Future communication systems may extend this transformation process up to the presentation layer, adapting messages according to the best suited end device available to the recipient.

Each individual communication space must provide a set of objects, the services of which an individual can use to achieve his goals. Individuals always communicate with objects in their environment according to a certain context. Orchestrating these objects in context provides the definition of relationships and causalities between different objects of an individual communication space. A context represents a "universe of discourse" in an individual communication space. It defines relationships and causalities of an individual to and between particular objects of the individual communication spaces currently providing I-centric services.

Being surrounded with objects and interacting through and with these in a specific context is quite natural for human beings; however, computer systems so far have no understanding and (more importantly) almost no access to these objects. Therefore, service mobility requires software representations of these objects. They must be able to cooperate within a service platform providing universal access to, as well as ways for basic semantic understanding of, all objects involved in human communication.

The aim of the service platform is to provide a seamless environment for these objects. This service platform must be open, distributed, and scalable, integrating heterogeneous devices ranging from tiny actuators to large computers. It must combine architectures, operating systems, middleware, programming models, and tools to support location and context sensitivity, personalization, and real-time adaptation. An object represented within this service platform may be as simple as a sensor or as complex as a portable device, a car, or a building.

To enable *ad hoc* interaction of previously unrelated objects, the service platform must also provide an interaction model between objects. This interaction model describes the dynamic cooperation of these objects to perform a specific task. Together with an organizational model, which describes relations between objects such as ownership issues, such types of interactions can be used to stimulate the social behavior of objects, such as multi-agent systems do [1,7,39].

Furthermore, objects may and will pertain to different contexts. Generic contexts are defined independently from a specific environment; however, an individual acting in his own communication space is always in a specific environment that must be captured by the communication system; therefore, an active context has to be modeled in the service platform. An *active context* defines the relationship of an individual to and between particular resources and people at a certain moment in time in a specific environment. Selecting and activating a context involves:

- Identification of objects required in the context and in a specific ambient environment
- Evaluation of the relationships and causalities between these objects
- Orchestration of these objects to perform the required I-centric service

The difference between *context* and *active context* is characterized by the entities, which are considered in relations and causalities. *Context* refers only to objects as an abstract model of what kind of objects must be taken into account in a certain context, whereas an *active context* refers to the selected resources that have been identified during the activation process. Active contexts have a dynamic nature reflecting the current environment in which an individual resides. A context is active when it is adapted to a certain environment at a certain moment in time. It defines the relationships and causalities of an individual to a particular number of objects at certain a moment in time, in a certain environment.

I-centric services can define, manage, and activate or deactivate contexts in an individual communication space, taking the preferences of

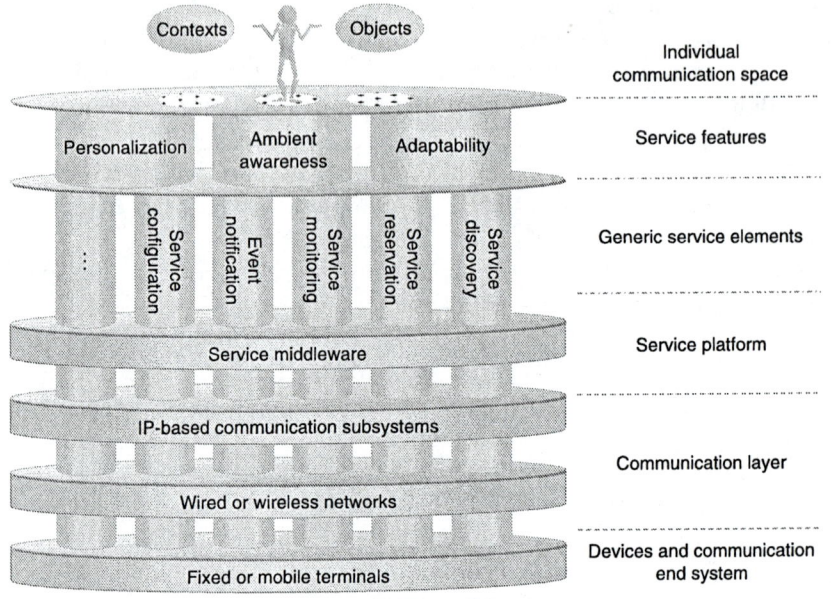

Figure 2.5 I-centric reference model.

individuals and ambient information into account. They support an individual (I-centric), adaptive, personalized, and ambient-aware way to interact with objects in individual communication spaces. Based on the evaluation of personal preferences, service capabilities, and sensed information about the actual environment, the individual can be provided with I-centric services for his actual demands. A reference model for this I-centric service architecture that provides ubiquitous service mobility based on the above rationales has been defined by the World Wireless Research Forum (WWRF) and is further explained below.

Reference Model for the I-Centric Service Architecture

Figure 2.5 illustrates the reference model for the I-centric service architecture. The reference model is divided into four horizontal basic layers (terminals, networks, Internet Protocol [IP]-based communication subsystem, and service platform) and several vertical supporting generic service elements, as well as service features, which are all explained further in the following text.

Communication Layer

The *IP-based communication subsystem* is responsible for providing the linkage between different objects in the communication spaces. These links have to be maintained and managed even when they are subject to change because of roaming between different network topologies or access networks. Non-IP-based communication networks might exist underneath the IP-based communication subsystem. They have to be wrapped by bridging facilities to include them in I-centric communication systems. IP communication is seen as the common denominator to harmonize heterogeneous network infrastructures. The IP-based communication subsystem consists of three layers:

- *Service support layer,* which provides well-defined application programming interfaces (APIs) for the service platform to access the IP-based communication subsystem
- *Network control and management layer,* which combines the traditional concepts of network management with required real-time aspects needed for systemwide control functions
- *IP transport layer,* which basically represents OSI layer four

The *wired or wireless network layer* implements all aspects of the physical connections between different objects. Due to the hierarchical structure of the reference model, a connection in the IP-based communication subsystem might use multiple connections in the underlying network. *Devices and communication end systems* provide the physical infrastructure that hosts all other layers. They can serve as switches responsible for connecting different networks or even as multimodal terminals able to interact with a certain individual.

Service Platform Layer

The service platform layer is responsible for shaping the communication system, based on individual communication spaces, contexts, preferences, and ambient information. Additionally, it activates or deactivates objects (as advised by I-centric services), identifies causalities between them based on sensed environmental data, controls the services offered by these objects, and converts data structures and operations for interactions between services. The equipment is configured dynamically, its state is profiled, distributed objects are controlled, service creation and deployment are supervised, and the interaction among domains is enabled by the platform. The service platform is an infrastructure that supports the development and operation of I-centric services by providing a set of service features:

- Execution environment for services and objects
- Generic abstraction as well as semantic description of objects and services
- Deployment of services
- Discovery of services and objects
- Generic access interfaces for services on objects
- Interworking of services and objects

The service platform is divided as follows:

- *Application support API* — Provides well-defined APIs to applications, services, and objects. It offers universal access to generic service elements that can be used by developers of these entities to ease and fasten the process of design, implementation, deployment, and management.
- *Service middleware* — Provides the actual runtime environment of applications, services, and objects. It supports their secure, QoS-aware, and managed execution.

Moreover, the service platform provides functional blocks that directly support the I-centric approach. These functional blocks manage ambient information, preferences, and adaptability to be offered to I-centric services. To fulfill the functionalities requested by I-centric communications, I-centric service platforms impose requirements on the underlying communication subsystem. Furthermore, the service platform provides abstract software representations of objects in the individual communication space, referred to as cooperative objects (COs). These cooperative objects utilize the following properties:

- *Autonomy* — COs are autonomous entities. Each of them can act fully independently from the others. They should interact in a peer-to-peer manner without any dependencies from specialized servers (client–server paradigm). Whenever two COs are connected in a network, they should be able to use the services provided by the other, without any central instance.
- *Ad hoc communication* — A characteristic of COs is their potential mobility. It should be taken into consideration that COs can appear, disappear, and move along the network rapidly. COs have to be able to communicate, to use core functions of the system, and to negotiate service usage. Furthermore, they must provide a generic

and semantic description of their services in order to enable *ad hoc* usage.

- *Diversity of processing capabilities* — COs are entities with all conceivable intermediate stages of the possible spectrum of computing capabilities, from highly capable computers and PDAs to plain light switches with almost no computing power. All COs provide a standardized interface for access and understanding their capabilities.
- *Communication technology independence* — Every device or software, irrespective of the network communication technology it supports, may use benefits of the service platform and, in turn, offer its services to other COs. The prerequisite for participation is conformance to the CO interface and CO description standards.
- *Scalability* — The number of objects in the individual communication space can grow rapidly, is very dynamic, and changes over the time; therefore, interfaces and communication protocols must consider this characteristic and include appropriate concepts.

The characteristics of a CO are further defined by its properties and the services it offers. A CO is dynamically characterized by its status, which is nonambiguous for every point in time. The status of a CO can be modified solely through the invocation of its services. Services define the capabilities of a CO. A service is represented by a group of operations that can be executed.

Although some COs can act autonomously, others must make use of the services of other COs before they can complete their own services; for example, consider a service, wrapped by a CO, that sustains a constant brightness level in each room the user enters. This service must find a CO offering a brightness-measuring service as well as one or several COs able to adjust the brightness level (via, for example, lights, dimmers, window blinds) to build and execute its own service.

Because cooperative objects wrap objects of the individual communication space, a CO usually wraps a device able to interoperate with the user. Thus, the CO maps the device capabilities to the operations of its service interface and makes these available to all COs in the service platform. However, as seen in the constant brightness example, a CO can also wrap certain service functionality without wrapping any devices.

Service Interfaces

A CO offers two different types of interfaces: operational interfaces and management interfaces. Operational interfaces represent CO-specific services that can be invoked to perform actions on a CO or to request CO-specific information. Complementing the operational interfaces are the management interfaces. They allow the discovery of CO services, service control, monitoring, and configuration of a CO. Most notable for management interfaces are the monitoring, configuration, reservation, and discovery interfaces:

- *Monitoring* allows subscription to CO resources.
- *Configuration* allows configuration of the CO resource data.
- *Reservation* allows reservation of CO utilization.
- *Discovery* allows a CO to advertise its capabilities in the CO system.

These necessary management interfaces are explained further in the following sections.

Generic Service Elements

The main features of I-centric communication (ambient awareness, personalization, and adaptability) affect all layers; therefore, supporting functions must be provided as a vertical solution. The reference model introduces the concept of *generic service elements*, which implement common functionalities on all layers. I-centric communication systems will have to cope with such issues as numerous service providers, always connected individuals, automatic service adaptation, and ambient awareness. Aspects such as dynamic service discovery and service provisioning in unknown environments and personalized services usage requires new mechanisms to support I-centric communication systems.

To simplify the definition and realization of I-centric services and applications, a set of reusable software components is necessary to support functionalities common to the different services and applications. These components are referred to as *generic service elements* to emphasis their general applicability for all kinds of services. Generic service elements can be seen as a toolbox with which complex services can be assembled and executed dynamically. The vertical approach allows I-centricity on all layers (e.g., for establishing I-centric private virtual networks). Notable generic service elements are as follows:

- *Service creation* covers the building and composition of generic services.
- *Service deployment* allows the distribution of services even in unknown and distributed environments.
- *Service discovery* is a mechanism for discovering service features that are provided within a certain environment or by a certain physical resource.

- *Service configuration* is the process for configuring the resources needed for a specific service.
- *Service reservation* manages the exclusive usage of objects.
- *Event notification* publishes and subscribes the interfaces for event distribution.

Each generic service element exposes a well-defined collection of interface specifications designed for its specific domain. The idea is to equip the same types of objects with standardized interfaces for functional (usage) and nonfunctional (management) interfaces. From the aspect of telecommunications, open service APIs such as OSA/Parlay build the basis for such interfaces [55].

The following sections present the functions that a CO system architecture should offer to facilitate CO communication and cooperation. This scenario can be used as example of CO interaction in the system: Several COs that control electrical appliances in an office build a CO environment. A new CO joins this environment. The newcomer offers a service that turns off the office appliances (e.g., lighting, air conditioning system) at 7 p.m. The existing COs in the environment should supply some mandatory functions that allow the newcomer to become integrated in the environment so it will be able to communicate and cooperate with the other COs. A first collection of these mandatory functions is listed below.

Service Discovery

A CO may require other COs to accomplish its services. When a CO changes its location and joins a new CO environment, it must discover the services offered by other COs that it requires before its services can be executed. Because it knows what services are necessary, the CO should be able to initiate a search for COs that provide these services. The discovery facility is required by the system to support CO cooperation. At the same time, a newcomer CO in the system should have the capability to announce services it can offer to other COs in the system.

The discovery mechanism specifies procedures that enable publishing of CO capabilities in the network/group and a dynamic search for COs based on certain criteria. This chapter describes mechanisms that could be leveraged in CO networks/groups to allow discovery of COs and their services. It also discusses the data that should be exchanged among COs to facilitate the discovery process. Generally, it is possible to obtain data regarding CO availability in the system in two ways:

- Announcements
- Search requests

An announcement is a message containing information on some topic that should be published. The announcement messages are spread across the system parts. In the case of COs, the object properties and offered services can be announced. The search procedure allows retrieval of the locations of specific COs or services based on well-defined search criteria. In a CO system or group, searching for COs with special properties or offering desired services would be possible.

To enhance the efficiency of the discovery mechanism, a hybrid approach embracing both CO announcements and search requests has been proposed. The discovery interface includes operations for both the announcement of COs and searching. CO announcements are a mechanism that allows COs to spread information about their capabilities throughout the CO network. A CO announcement is a brief message specifying the CO properties and services. When joining a network/group, the new CO can send announcements to notify the other COs of its presence and to supply information about itself. When it leaves the network/group, the CO so informs the other COs so they can update their information on currently available COs and services. Because many COs will very likely leave the network unexpectedly, limiting the validity time of announcements is being considered.

A CO announcement contains basic information specifying the CO and the services it offers. It should also include data required to contact the CO and a description of services offered. When a CO joins a network/ group, it spreads announcements across the system using the discovery interface. COs that are interested in using the services of this particular CO can save the announcement and later connect to the CO and request further data or service descriptions.

When a CO has just joined a network and must use the services of other participants immediately, without waiting for their announcements, it must fall back on the second type of discovery by issuing a search request. The search mechanism allows the active search of all COs registered at the service platform based on specific criteria. When a CO joins a network, it can immediately begin searching for the services it requires. The search is performed by broadcasting search messages across the network.

The CO announcement and search mechanisms are assumed to operate in a decentralized system. In this case, when publishing services by an announcement or searching for the services of other COs, the announcement and search messages should be broadcast to all COs present in the network.

In contrast, in systems with centralized discovery services, all the data concerning the COs currently available in the system is collected in the service; consequently, it is not necessary to broadcast announcements to all the COs, as an announcement sent only to this central entity suffices.

The search for a CO or service can be performed in a similar way, by sending a search request to the discovery service. Because the discovery service possesses information on all CO available in the network or group, it is possible to retrieve information on all CO possessing certain properties from it.

Service Reservation

For certain applications, COs should be able to reserve the utilization of other CO resources. This requirement has arisen due to the need to get exclusive access to services. Operations enabling the reservation of CO utilization are collected in the reservation interface. The three modes for CO reservation are *pessimistic, optimistic,* and *unlimited access.* The *unlimited access* mode indicates that the CO is not currently reserved. In this mode, reservation operations can be accepted. In the *optimistic access* mode, all COs except for the reserving CO can only access the resource data of that CO in the read-only mode. Other COs can inquire about the services that the CO offers and the current state of its parameters, but no service invocations or configuration operations are allowed. State notification subscriptions through the monitoring interface are permitted for everyone, as well. A CO reserved in the *pessimistic* mode acts as if it has disappeared from the network. For the duration of the reservation, it does not announce its presence in the network and does not react to search requests sent by objects in the network; it is visible only to the reserving CO. Further, only the reserving CO can execute its services, subscribe to the state notifications of the CO, and monitor and configure its resource data.

Service Monitoring

Cooperative objects also provide interface functions enabling monitoring of their services' state. Services can require monitoring capabilities to control the service state. The CO state can be actively monitored either by polling or by notification of state changes. Thereby, monitoring by polling consists of periodically checking the state of the CO by the observer. The monitoring of service state by a polling mechanism can be disadvantageous, as it is difficult to estimate the most appropriate polling frequency. If too high of a frequency is chosen, it can lead to unnecessary network workload; if it is too low, the service client would get the information about the new service state long time after it has changed. For this reason, event notifications can be used to complement the polling operations. Event notifications are sent across the network to inform other objects about CO state changes. The COs that would like to monitor the state of certain COs (called further *notification subscribers,* or *subscribers* for short) subscribe the state change notifications at a CO of interest. Event notifications are sent to all subscribers when the CO state changes.

Event Notification

A mechanism for the propagation of state information is useful to avoid resource-consuming polling. When a CO state changes, other COs can subscribe to receive immediate notification of this state change. With the introduction of a subscription concept, COs may subscribe to receive notification of any events they are interested in. Event notifications that are sent to subscribers can be predefined. When subscribers are informed of a state change immediately, as soon as it occurs, this type of notification is referred to as *on-change notification.* Another option is to allow subscribers to predefine how often they would like to get information about the CO state; if the subscriber chooses this option, the notifications are sent to it once during a specified time interval, even if the state has changed since the last notification. This option is referred to as *interval notification.* Other options concerning when event notifications are issued can be conceived, but in this chapter only these two options are considered.

Service Configuration

Cooperative object interfaces also provide functions that allow configuration of the resource data and service parameters of COs. This function can be used to configure CO location information, services, notification intervals, etc. The configuration interface offers operations enabling the configuration of CO resource data. Because every CO possesses more or less unique collections of resource data, information about the parameters to configure should be requested before the first operation is invoked. All resources that can be configured in a CO belong to its resource data and are specified in the CO configuration. It is also possible to set the current CO configuration to one of the predefined configurations specified in the profile.

Service Features

Personalization

Personalization is considered the key factor for I-centric communication. Information and services must become increasingly tailored to individual preferences to make the usage of services easier and the perception of the individual communication space richer. Personalization integrates and relates aspects such as user preferences, user roles, and user tasks. An extended personalization concept allows value networks (e.g., value chains) of content providers, network providers, and service providers to offer personalized services to mobile users in a way that suits their needs at a specific place and time. For I-centric communication, this means that objects available in an individual communication space must adapt to the preferences of individuals. The personalization service feature models each individual in the I-centric service platform by managing its preferences and providing these preferences to I-centric services. Furthermore, personalization provides the information for modeling preferences for an individual communication space in the I-centric system [56]. Personalization gathers profile information (containing preferences) and incorporates dynamic behavior to enrich the stored and gathered information and enable proactive I-centric services. This personalization leads to an overall profiling infrastructure managing the individual preferences.

Ambient Awareness

Ambient awareness is the functionality provided by an I-centric system to sense and exchange information about the individual's current environment [56]. In future communication systems, services will be tailored to the contexts of the individual's communication space, and the services will adapt themselves to changes in the environment. The services must be able to deal with the changing environments of nomadic individuals, and these adaptations to current situations (in a certain context) must be hidden from the individual. In addition, the environment itself can be influenced by the presence and activities of an individual and adapt itself accordingly.

The ambient-awareness service feature gathers ambient information from the network, application, individual, terminal, and contexts. The gathering of ambient information from various sources, depending on the individual's mobility and roaming, is an integral part of ambient awareness. Sensor networks embedded in mobile equipment, communication networks, and living and working environments will sense who the user is, where he is, what he is doing, and what the environmental conditions are to provide ambient information to I-centric services.

In general, *ambient information* is information that can be collected, gathered, or sensed from the environment using the objects of the individual communication space of a certain individual. Ambient information includes temporal and spatial characteristics such as user input, temperature, noise level, light intensity, and the presence of other people, to give just a few examples. Ambient information can also include geographical information (e.g., location), environmental information (e.g., temperature), and life conditions (e.g., blood pressure).

I-centric services require ambient information in order to adapt to the environment. Temporal and spatial characteristics are only two examples of information that may affect the service behavior. Note that a particular environment can restrict the functionality requested in a certain context. Interacting in a "TV context" while driving a car may reduce the available functionality to "record the movie for later viewing" or listening just to the audio part.

Adaptability

Adaptability provides the functionality to adapt I-centric services to personal preferences and environmental conditions; therefore, adaptability can be seen as a function that activates a context based on whatever information is provided by ambient awareness and personalization. In general, I-centric adaptability translates the wishes of individuals (which are usually inaccurate, incomplete, and sometimes even contradictory) into a set of rules precise enough to be automated with sufficient reliability. It has implications in the structure of the services to allow adaptability and is the engine that activates a context at a certain moment in time in a certain environment [56]. Adaptation typically results in a substantial change in the connectivity characteristics, entering into a new service domain, or changing terminal devices in the service session. Adaptability requires the adaptation of media, content, and service behavior. Over the past several years, several concepts for adaptation have been developed [24,46]:

- Communication streams are altered during transmission (e.g., bitrate adaptation).
- Media types are changed (e.g., text-to-speech conversion).
- The type of presentation is adapted (e.g., downscaling an image to fit a PDA screen).
- The content of a message is altered (e.g., adding or stripping information).
- The service behavior is modified (e.g., by customer service control functions).

Adaptability is not only reactive. When the battery of a mobile device dies or the connectivity is broken, many actions become impossible; however, something could have been done beforehand to prevent these situations. Adaptation, therefore, has to be proactive, as well, which in turn requires predictability of the near future.

Summary

In this chapter, we discussed various types of mobility. Beginning with traditional code mobility, we developed a reference model for personal and service mobility for I-centric services. This reference model combines an IP-based communication layer with a universal service platform, which is used by generic service elements to provide the necessary infrastructure for more elaborate service features, such as personalization, ambient awareness, and adaptability.

We identified these three service features as basic building blocks for I-centric services. These I-centric services are aware of the user context and the environment, including the objects surrounding the user. The service platform provides the means to interact with these objects in a generic way. We also presented a general vision of the I-centric service architecture for personal service mobility. In summary, personal service mobility consists of:

- A coherent adaptation framework
- Personalized user interaction
- Ambient-aware user interaction
- Device-independent service invocation

These general requirements reflect the relevant areas of concern and suggest the starting points for development of an approach to enable I-centric services.

References

[1] Arbanowski, S., Breugst, M., Busse, I., and Magedanz, T., Impact of standard mobile agent technology on telecommunications, in *Proc. of the 5th Conf. on Computer Communications (AFRICOM–CCDC'98)*, Tunis, Tunisia, October 20–22, 1998, pp. 189–203.

[2] Abowd, G.D., Dey, A.K. et al., Towards a better understanding of context and context awareness, in *Proc. of the First Int. Symp. on Handheld and Ubiquitous Computing (HUC'99)*, Karlsruhe, Germany, September 27–29, 1999.

[3] Barbir, K., Bennett, N., Penno, R., Pham, H.T. et al., *A Framework for Service Personalization*, Internet draft, 2002, http://quimby.gnus.org/internet-drafts/draft-barbir-opes-fsp-00.txt.

[4] van Bekkum, M., Bijlsma, M., van Kranenburg, H., and Lankhorst, M., *Personal Service Environment: Analysis and Research Issues*, Telematica Instituut, Enschede, The Netherlands, 2000.

[5] Bunt, H., Ahn R., Beun R. et al., *Cooperative Multimodal Communication in the DenK Project*, Institute for Language Technology and Artificial Intelligence (ITK), Tilburg University, Tilburg, The Netherlands; Institute for Perception Research (IPO), Eindhoven, The Netherlands; Faculty of Mathematics and Computing Science, Eindhoven University of Technology, Eindhoven, The Netherlands.

[6] Butler, M. H., *Current Technologies for Device Independence*, HP Laboratories, Bristol, 2001.

[7] Breugst, M. and Magedanz, M., On the usage of standard mobile agent platforms in telecommunication environments, in *Proc. of the 5th ACTS IS&N Conf.*, Antwerp, Belgium, May 25–28, 1998.

[8] Carroll, L., *Alice's Adventures in Wonderland/Through the Looking-Glass*, Bloomsbury, London, 2001.

[9] OMG, *The Common Object Request Broker: Architecture and Specification*, Revision 2.2, Object Management Group, Needham, MA, 1998.

[10] DEC, *Universal Messaging* [white paper], Digital Equipment Corporation, Maynard, MA, 1997.

[11] Dey, A. and Abowd, G., Towards a better understanding of context and context awareness, in *Proc. of the Computer–Human Interaction 2000 (CHI 2000) Workshop on the What, Who, Where, When, and How of Context Awareness*, The Hague, The Netherlands, April 1–6, 2000.

[12] Eckardt, T., Magedanz, T., and Pfeifer, T., On the convergence of distributed computing and telecommunications in the field of personal communications, in *Proc. of Kommunikation in Verteilten Systemen (KiVS'95)*, Franke, K. et al., Eds., Springer, Berlin, 1995, pp. 46–60.

[13] Eckardt, T., Magedanz, T., Ulbricht, C., and Popescu-Zeletin, R., Generic personal communications support for open service environments, in *Proc. of the IFIP World Conf. on Mobile Communications*, Canberra, Australia, September, 1996.

[14] Eckardt, T., Ed., Deutsche Telekom Project: Personal Communications Support in TINA, Report No. 1, *GMD Fokus*, June, 1996.

[15] ETSI, *Universal Mobile Telecommunication Systems (UMTS): Service Aspects, Service Principles*, Technical Specification TS 22.01v3.1.0, European Telecommunications Standards Institute, Sophia Antipolis, France, 1997.

[16] Faroogui, K. and Logrippo, L., *Introduction to ODP Computational Model*, Department of Computer Science, University of Ottawa, Canada.

[17] Fink, J., Koenemann, J., Noller, S., and Schwab, I., Putting personalization into practice, *Comm. ACM*, 45(5), 41–42, 2002.

[18] International Telecommunication Union, Telecommunication Standardization Sector (ITU-T), *Memorandum of Understanding on Global Mobile Personal Communications by Satellite*, World Telecommunications Policy Forum, October 1996/February 1997.

[19] Göbel, S., Buchholz, S., Ziegert, T., and Schill, A., Device independent representation of web-based dialogs and contents, in *Proc. of the IEEE Youth Forum in Computer Science and Engineering (YUFORIC'01)*, Valencia, Spain, November 29–30, 2001.

[20] Göbel, S., Buchholz, S., Ziegert, T., and Schill, A., *Software Architecture for the Adaptation of Dialogs and Contents to Different Devices*, Department of Computer Science, Technische Universität Dresden, Germany, 2002.

[21] Guntermann, M. et al., Integration of advanced communication services in the personal services communication space: a realisation study, *Mobile Kommunikationssysteme*, March, 127–131, 1994.

[22] van der Meer, S., Arbanowski, S., Steglich, S., and Popescu-Zeletin, R., The human communication space: toward I-centric communications, in *Proc. of the Computer–Human Interaction 2000 (CHI 2000) Workshop on the What, Who, Where, When, and How of Context Awareness*, The Hague, The Netherlands, April 1–6, 2000.

[23] Arbanowski, S., van der Meer, S., Steglich, S., and Popescu-Zeletin, R., *I-Centric Communications*, Springer-Verlag, Berlin, 2001.

[24] van der Meer, S., Arbanowski, S., and Steglich, S., Flexible control of media gateways for service adaptation, in *Proc. of the 5th IEEE Intelligent Network Workshop*, Cape Town, South Africa, May 7–11, 2000.

[25] Arbanowski, S., van der Meer, S., and Popescu-Zeletin, R., I-centric services in the area of telecommunication: the I-talk service, in *Proc. of the 6th Ifip Tc6/Wg6.7 Conf. on Intelligence in Networks (SmartNet 2000)*, Vienna, Austria, September 18–22, 2000, pp. 499–508.

[26] Arbanowski, S., van der Meer, S., Steglich, S., and Popescu-Zeletin, R., The human communication space: towards I-centric communications, *J. Pers. Ubiquitous Comput.*, 5(1), 34–37, 2000.

[27] Steglich, S. and Popescu-Zeletin, R., Towards I-centric user interaction, in *Proc. of the ICME 2001 Int. Conf. on Multimedia and Expo*, Tokyo, Japan, August 22–25, 2001.

[28] van der Meer, S., Arbanowski, S., and Steglich, S., User-centric communications, in *Proc. of the IEEE Int. Conf. on Telecommunications (IEEE ICT 2001)*, Bucharest, Romania, June 4–7, 2001, pp. 452–444.

[29] Arbanowski, S. and Steglich, S., Profiling contextual information, in *Proc. of the IEEE Int. Conf. on Parallel Architectures and Compilation Techniques (PACT'2001), Workshop on Ubiquitous Computing and Communication*, Barcelona, Spain, September 10–12, 2001.

[30] Arbanowski, S. and Steglich, S., Profile information based service creation, in *Proc. of the 4th Asia–Pacific Symp. on Information and Telecommunication Technologies*, Tribhuvan University, Kathmandu, Nepal, 2001.

[31] Arbanowski, S. and Steglich, S., Service architectures for 3G and beyond, in *Proc. of the Sixth SICE Annual Conf.*, Fukui, Japan, August 4–6, 2003.

[32] Radusch, I., Arbanowski, S., Steglich, S., and Popescu-Zeletin, R., I-centric services based on super distributed objects, in *Proc. of the Med–Hoc NET'2003 Workshop*, Mahdia, Tunesia, March 26–27, 2003.

[33] Arbanowski, S., Steglich, S., and Popescu-Zeletin, R., Super distributed objects: an execution environment for I-centric services, in *Proc. of the 9th IEEE Int. Workshop on Object-Oriented Real-Time Dependable Systems (WORDS 2003F)*, Capri Island, Italy, October 1–3, 2003.

[34] van der Meer, S. and Arbanowski, S., Service interoperability through advanced media gateways, in *Proc. of the 6th Ifip Tc6/Wg6.7 Conf. on Intelligence in Networks (SmartNet 2000)*, Vienna, Austria, September 18–22, 2000, pp. 583–595.

[35] van der Meer, S. and Arbanowski, S., Flexible media and content adaptation for communication systems, in *Proc. of the IEEE Conf. on Protocols for Multimedia Systems (PROMS 2000)*, Krakow, Poland, October 22–25, 2000, pp. 461–477.

[36] Ducatel, K., Bogdanowicz, M., Scapolo, F., Leijten, J., and Burgelman, J-C., *Scenarios for Ambient Intelligence in 2010*, ISTAG Report, European Commission, Institute for Prospective Technological Studies, Seville, 2001.

[37] Manber, U., Patel, A., and Robison, J., Experience with personalization on Yahoo!, *Comm. ACM*, 43(8), 35–39, 2000.

[38] Mandyam, S., Vedati, K., Kuo, C., and Wang, W., *User Interface Adaptations: Indispensable for Single Authoring*, position paper of W3C Workshop on Device Independent Authoring Techniques, SAP University, S. Leon-Rot, Germany, September 25–26, 2002.

[39] Martin, D., The open agent architecture: a framework for building distributed software systems, *Appl. Artificial Intell.*, 13(1/2), 91–128, 1999.

[40] Menkhaus, G., Architecture for client-independent web-based applications, in *IEEE Proc. of the TOOLS–Europe Conf.*, Zürich, Switzerland, March 12–14, 2001.

[41] Mohamad, Y. et al., *Supporting Device Independent Accessible Authoring by a Next Generation Web Publishing Framework*, position paper of W3C Workshop on Device Independent Authoring Techniques, SAP University, S. Leon-Rot, Germany, September 25–26, 2002.

[42] Müller, A., Forbrig, P., and Cap, C., *Model-Based User Interface Design Using Markup Concepts*, Department of Computer Science, Rostock University, Rostock, Germany, 2001.

[43] Nanneman, D., Unified messaging: a progress report, *Telecomm. Mag.*, March, 1997.

[44] IPTC, *NITF 3.0: News Industry Text Format*, International Press Telecommunications Council, Windsor, U.K., 2001.

[45] W3C, *The Platform for Privacy Preferences 1.0 (P3P1.0) Specification*, World Wide Web Consortium Recommendation, 16 April 2002.

[46] Pfeifer, T., Automatic Conversion of Communication Media, Ph.D. dissertation, Institute for Open Communications Systems (OKS), Technical University of Berlin, 1999.

[47] The Parlay Technical Team, *Parlay APIs 2.1: Mobility Data Definitions*, The Parlay Group, Inc., San Ramon, CA, 2000.

[48] Rakotonirainy, A., Wai Loke, S., and Fitzpatrick, G., *Context Awareness for the Mobile Environment*, proposal for Computer–Human Interaction 2000 (CHI 2000) Workshop, April, 2000.

[49] Raymond, K., Reference model of open distributed processing (RM-ODP): introduction, in *Proc. of the Int. Conf. on Open Distributed Processing (ICODP'95)*, Brisbane, Australia, February 20–24, 1995.

[50] Lassila, O. and Swick, R., *Resource Description Framework (RDF) Model and Syntax Specification*, World Wide Web Consortium (W3C) Recommendation, February 22, 1999.

[51] Rossi, G., Schwabe, D., and Guimarães, R., Designing personalized web applications, in *Proc. of the Tenth Int. World Wide Web (WWW10) Conf.*, Hong Kong, May 1–5, 2001, pp. 275–284.

[52] Sandkuhl, K., Ein Referenzmodell für informationslogistische Anwendungen, in *Report Informationslogistik*, Vol. 1, *Auflage*, Deiters, W. and Lienemann, C., Eds., Symposion Publishing, Düsseldorf, Germany, 2001.

[53] Fuggetta, A., Picco, G.P., and Vigna, G., Understanding code mobility, *Trans. Software Eng.*, 24(5), 342–361, 1998.

[54] Weiser, M., The computer for the 21st century, *Sci. Am.*, 265(3), 94–104, 1991.

[55] OSA/Parlay Group, http://www.parlay.org/.

[56] WWRF, *WG2: Service Infrastructure of the Wireless World*, white paper on service personalization, ambient awareness, and service adaptability, Wireless World Research Forum, http://www.wireless-world-research.org/.

Chapter 3

Wireless Technologies

Marco Chiani

CONTENTS

Introduction

This chapter provides an overview of communication technologies for wireless networks. The main characteristics of wireless technologies are presented, with special emphasis on the emerging techniques that are changing the mobile wireless communications capability from voice applications to broadband multimedia, toward the goal of "multimedia wireless communications anytime, anywhere." The topic is so vast and changing so rapidly that obviously this brief chapter cannot provide exhaustive coverage. For this reason, the reader who wants to know more is invited to refer to specific journals covering wireless communications, mainly those by the Institute of Electrical and Electronics Engineers (IEEE), some of which are listed at the end of this chapter, which provide up-to-date research results or tutorial overviews of the latest developments [1–3,5]. In this chapter, we first review some basic facts about wireless communications, including technologies such as multiple antennas and multicarrier modulation. We then describe the main characteristics of current wireless systems, with a discussion on possible evolution toward cognitive radio.

Technical Challenges in Wireless Communications

Data Rates, Mobility, and Area Coverage

Wireless communications have significant peculiarities that must be clearly identified, and it is important not to apply to wireless communication systems concepts or solutions that are valid only for cable or fiber systems. Mobile communication systems aim to provide mobile users the same services as those provided by wired networks; however, channel interference, the scarceness of radio resources, and mobility itself impose severe limitations on the quality of service in terms of data rates and area coverage. The tradeoff between data rates and user mobility results in specific solutions for different application scenarios (see Figure 3.1). It is worthwhile to note the strong relationships among data rates, area coverage, and mobility. In mobile cellular communication systems, small cells are preferable to achieve high data rates. At the same time, small cells imply efficient handover mechanisms and implementations that are more and more demanding as a user's speed increases.

The traditional way of categorizing digital wireless networks operating worldwide is based on the distinction between cellular networks (primarily carrying voice calls and with extensive area coverage) and local and personal area networks (wireless local area networks [WLANs] and wireless personal area networks [WPANs]). Recently, however, the evolution of the latest generation of cellular networks with the potential to provide

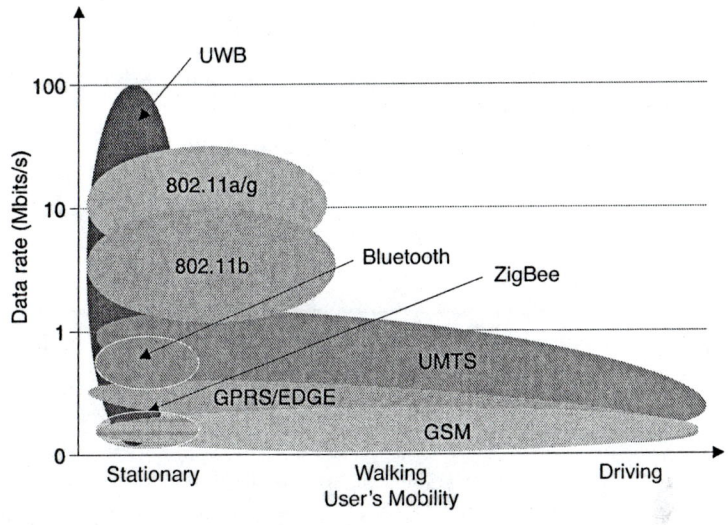

Figure 3.1 Comparison of wireless communication systems.

multimedia content to mobile users, area coverage being extended through the use of several WLAN spots, and the possibility of Voice-over-IP (VoIP) are making the distinction somewhat less clear. A more clear taxonomy could be based on usage of the radio spectrum and the use of licensed wireless applications such as the cellular Global System for Mobile Communications (GSM) and Universal Mobile Telecommunications System (UMTS), as opposed to systems using unlicensed bands such as WLANs and WPANs.

Wireless Channel Characteristics

Mobile radio propagation is the reason for the pronounced difference between wireless and wired communication systems. In wireless systems, we can identify such complicating factors as thermal noise (related to the physics of circuits and apparatus, hence unavoidable), signal power attenuation (related to the distance between transmitting and receiving antennas), multipath propagation (due to more rays arriving at the receiving antenna after reflection, diffraction, and scattering), and interfering signals (related to the scarcity of the radio spectrum and the consequent need for several users to use the same spectrum band). Moreover, for mobile radio systems where users are moving, the radio channel changes in an unpredictable way. In this subsection, we address path loss and multipath propagation for a single-transmitter/single-receiver scenario; interference issues are discussed later in this chapter.

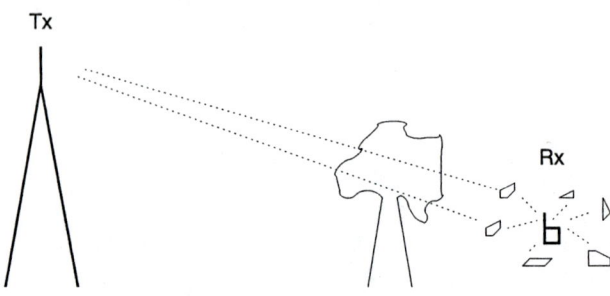

Figure 3.2 Example of wireless communication, including path loss, shadowing, and multipath.

Path loss is due to dissipation of the power radiated by the transmitting antenna. Consider an ideal free-space environment with no obstructions between the transmitter and the receiver, where the signal propagates along a straight line. This scenario is referred to as a *line-of-sight* (LOS) channel. In this case, the received signal power (P_r) is related to the transmitted signal power (P_t) and to the link distance (d) by $P_r = \alpha P_t d^{-2}$, where the constant α depends on the carrier wavelength and on the antenna directional gains. Unfortunately, in many cases, we cannot use this simple LOS model for the radio channel; in fact, the radiated electromagnetic field is diffracted, reflected, and scattered by a multiplicity of obstacles, such as trees and walls, buildings, and vehicles, before reaching the mobile wireless receiver. The presence of objects and obstacles in the environment produces at the receiving antenna several copies of the transmitted signals. Another important phenomenon that can be observed for high-speed users is the frequency shift of the received signal due to the Doppler effect.

Without going into the details of propagation, an instructive example is depicted in Figure 3.2. Here we can distinguish two different phenomena. The first is related to the power loss due to the presence of obstructing objects and is usually modeled by means of a deterministic path loss (related to distance d and to a quite broad classification of the environment) with, superimposed, a random variation of power, called *shadowing*. This random fluctuation is caused by variations in the obstructions (e.g., terrain obstruction such as hills, trees, manmade obstructions such as buildings) such that the received signal power can vary considerably at different locations, even when they are at the same radial distance from the transmitter. This effect is often referred to as the *large-scale propagation effect*, as it describes variations that can be observed by moving the receiver over a length of the order of the dimension of the obstacles

Table 3.1 Frequencies and Corresponding Wavelengths

Carrier Frequency	Wavelength (cm)
900 MHz	33
1800 MHz	16.7
2400 MHz	12.5
5 GHz	6
10 GHz	3

obstructing the propagation (10 to 100 m outdoors, less for indoors). This slow variation is well described by a log-normal distribution.

A second effect is related to the presence of many objects surrounding the antennas that act as scatterers. Instead of one path, we now have a multipath channel. Signals arriving from different paths can add constructively or destructively, depending on their relative phases as determined by reflections and by the delays associated with each path. Now, denote the light speed by $c = 3·108$ m/s and assume we are transmitting an unmodulated carrier with frequency f_0. Let us first focus on a particular path. Observe that we have a phase variation of π radians when we move the receiver position by half a wavelength ($\lambda/2$) along the direction of the wave. If we assume more paths are coming from different directions, we can understand why, in the multipath scenario, a displacement of the order of $\lambda = c/f_0$ in the receiver antenna position or in the position of the surrounding objects causes different changes in the phases of the paths that can result in a dramatic variation in the overall received power. Just to give an idea of the relevance of this phenomenon, Table 3.1 provides the wavelengths for some carrier frequencies of interest. As can be seen in this table, a change in the position of the user or of surrounding objects of only a few centimeters can cause a large fluctuation in the received power; thus, this effect is often referred to as the *small-scale propagation effect*.

Figure 3.3 illustrates the typical behavior of the received power for a carrier frequency of 2.4 GHz, where variations of tenths of decibels can be observed due to the presence of multipath. The same figure also shows the average of the received power over a window of a few wavelengths. By averaging, we can remove the small-scale propagation effect; thus, the behavior of this average power can be described in terms of shadowing on top of the power as predicted by path-loss models.

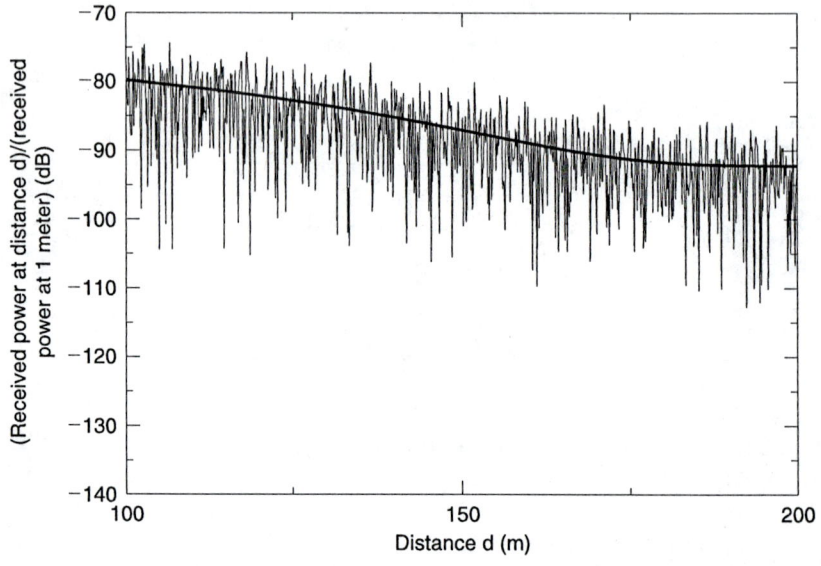

Figure 3.3   Example of received power (normalized, in dB), including fast fading, shadowing, and path loss for a carrier frequency of 2.4 GHz. The smoother curve is the large-scale propagation effect due to path loss and shadowing, obtained by averaging over few wavelengths.

Multiple Antenna Systems: Diversity, Interference Mitigation, and MIMO

We have just shown that a major problem in wireless communications is the received power fluctuation due to multiple paths. To overcome this problem, diversity techniques can be employed. Diversity systems combine different copies of the same information (copies possibly subject to independent fading) so as to minimize the probability of a reception failure. Diversity can be achieved by exploiting time (e.g., by error-correcting codes and interleaving), frequency (e.g., by frequency hopping and error-correcting codes), or space (with multiple antennas).

Over the last several decades multiple antennas have been used to combat fast fading. When multiple antennas are used to counteract fast fading, the advantage is an increased robustness with respect to the deleterious effects of multipath [1,2,7–9]. As an example, with one transmitting antenna we can use two receiving antennas at the receiver (single-input/multiple-output, or SIMO) (Figure 3.4) and choose at each instant the output of the antenna with the strongest signal power. If the receiving antenna elements are sufficiently spaced apart, the fading can be assumed to be independent on the two antennas. Hence, if $p < 1$ is the probability

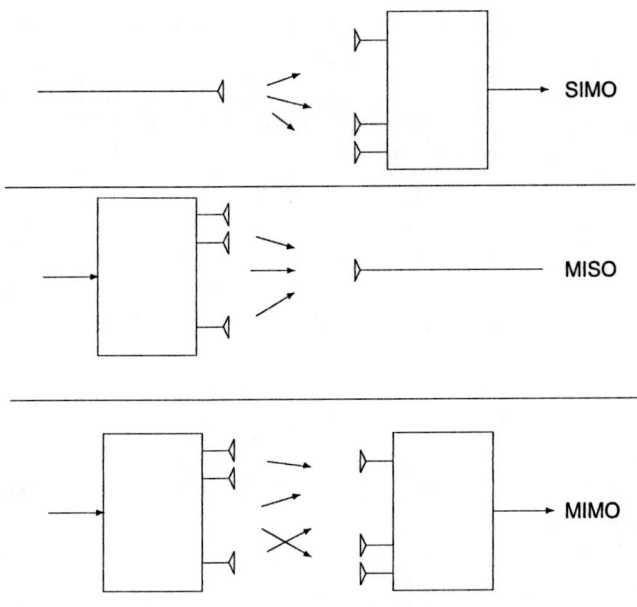

Figure 3.4 Single-input/multiple-output, multiple-input/single-output, and multiple-input/multiple-output systems.

that one antenna is experiencing a deep fade, then the probability that both antennas are in a deep fade and therefore that the communication is degraded is $p^2 < p$. In this case, we are exploiting the spatial dimension to achieve a diversity gain.

Receiver diversity is well known and has been employed for quite some time to improve wireless links; however, there is also an interest in determining whether or not it is possible to achieve diversity with multiple transmitting antennas and possibly one receiving antenna (multiple-input/single-output, or MISO) (see Figure 3.4). This scenario arises, for example, with regard to the downlink (DL) in mobile cellular systems, the link between the radio base station and the mobile user. Putting more antenna elements on the base station is simple enough, but doing the same on the user's terminal is not as easy because of space limitations. Only recently techniques have been developed to provide diversity with multiple transmitting antennas (transmitter diversity). Transmitter diversity is one of the novel techniques introduced in cellular mobile communications third-generation standards [4]. Another well-known technique is to use smart antennas to mitigate the effect of co-channel interference. Simple approaches for interference reduction include sectored antennas and multibeam antenna systems. To maximize the desired output signal power and reduce interfering signals as much as possible, a more advanced technique consists of weighting and adding the signals from multiple antenna elements at the receiver (e.g., to modify the radiation pattern if signals have a clear direction of arrival, although with dense multipaths where rays come from many directions this geometric interpretation is not useful). The use of smart antennas to reduce co-channel interference has a strong impact on the system capacity [8,14].

In the last few years, it has been also recognized that the capacity (in terms of bps/Hz) of wireless communication links can be increased by using multiple antennas both at the transmitter and at the receiver (multiple-input/multiple-output, or MIMO) (see Figure 3.4), thus exploiting the spatial dimension to construct virtual parallel channels [7,9,10, 15,17]. Motivated by theoretical capacity analysis, the increasing demand for higher capacity has given rise to the proposal of practical transmission schemes based on MIMO, where different signals are simultaneously transmitted to achieve high spectral efficiencies. These schemes are known as high-spectral-efficiency MIMO systems. Toward achieving these capacities, a promising transmission system, called *D-BLAST* (Diagonal Bell Laboratories Layered Space-Time), has been proposed [10]. This scheme is able to provide a high spectral efficiency in a rich and quasi-static scattering environment. Due to the large computational complexity required for this scheme, a simplified version, called *V-BLAST* (Vertical BLAST), has also been proposed [11]. The large spectral efficiency of transmission systems based on MIMO is due to their capacity to exploit the spatial dimension in environments characterized by rich scattering, thus allowing high spectral efficiencies with an important multiplexing advantage [9,16,18].

Modulation and Error Control Techniques

The radio resource is so limited and precious that it must be used with the maximum possible efficiency. In this regard, one important parameter is the number of bits per second (bps) per frequency units we are able to transmit — that is, the spectral efficiency in terms of bps/Hz. From basic communications theory, we recall that a modulation format with L points in the constellation can transmit $\log_2 L$ bps/Hz. For example, the theoretical spectral efficiency of binary phase-shift keying (BPSK) is 1 bps/Hz, and for quadrature phase-shift keying (QPSK) it is 2 bps/Hz. From this perspective, it seems convenient to use higher order modulations such as 64 quadrature amplitude modulation (QAM), giving us 6 bps/Hz. Unfortunately, the requirements in terms of link budget are more strict as the modulation order increases, and the wireless channel impairments are so severe that the difficulties in demodulating these high-order constellation signals increase with the data rate. Indeed, by increasing the data rate the signal band increases and so increases the distortion due to the multipaths. One possible solution to counteract channel distortion due to multipaths consists of subdividing the available band into several sub-bands over which the channel is nearly nondistorting. Over each sub-band a low data rate signal can be transmitted with the maximum possible constellation size using BPSK, QPSK, or 16-QAM, depending on the channel quality for that subchannel. By multiplexing all subchannels a high data rate is achieved. This is the idea behind multicarrier modulation techniques, such as orthogonal frequency-division multiplexing (OFDM), which is one of the most important recent advances in wideband wireless communication systems.

Moreover, the presence of severe channel impairments requires the adoption of powerful error-correcting codes (channel codes) to recover errors introduced by the wireless channel. So, the actual spectral efficiency must include the redundancy added for error correction. The most important error-correcting codes in wireless applications are convolutional codes, turbo codes, and low-density parity check codes (LDPCCs). Spectral efficiency is further reduced due to the redundancy introduced by the error-correcting code. So, for example, a rate 1/2 channel code with QPSK gives only 1 bps/Hz. If the target would be, for example, 1 Gbps, this means that a frequency bandwidth of 1 GHz would be needed! If we realize that the radio spectrum ranges from few hundreds of KHz to few GHz in total (for all applications), it is apparent that to target wireless Gbps systems we must resort to higher spectral efficiencies.

Indeed, the solution to the high-data-rate problem in wireless systems is multiple antenna systems (MIMO), as discussed in the previous section. With MIMO it is possible to achieve a very high spectral efficiency, taking advantage of the scattering to obtain as many virtual parallel channels as possible between the number of transmitting and receiving antennas. For example, a 3×3 MIMO (three transmitting and three receiving antennas) with QPSK can achieve a spectral efficiency of 6 bps/Hz. MIMO technologies are thus of extreme importance for high-data-rate wireless systems [15,16,18].

Multiple Access and Resource Allocation

When the channel used to communicate is radio, users in a given area must share the common radio resource to keep interference at tolerable levels. The capacity of the system, then, in terms of served users per area is strictly related to the capability to cope with co-channel interference. The three basic methods to provide multiple access for cellular mobile systems are frequency-division multiple access (FDMA), time-division multiple access (TDMA), and code-division multiple access (CDMA).

Consider first FDMA. In its simplest form, it divides a given frequency band into sub-bands and allocates each sub-band to an active user. In TDMA, time is repetitively subdivided into frames and each frame into a fixed number of time slots. Each active user is assigned a specific time slot per frame. CDMA, the third type of multiple-access technique for cellular mobile radio systems, is based on spread spectrum (SS). Spreading can be obtained by frequency hopping (FH), by time hopping (TH), or by multiplication in the time domain of the data with a higher rate sequence (direct sequence SS, or DSSS). More precisely, in DSSS we can roughly assume that for each information bit (0 or 1) a sequence of bits (called *chips*) or its complement is transmitted. The code sequence of $N > 1$ chips is called the *spreading sequence*, and the resulting SS transmitted signal has a bandwidth much larger than the data rate. In direct sequence CDMA (DS-CDMA), each user is assigned a spreading sequence, and sequences are chosen to be nearly orthogonal, so, even if users transmit at the same time and in the same frequency band, it is still possible to distinguish the various information bits from the different users at the receiver end.

In circuit-switched systems, for all three cases a logical channel (a carrier for FDMA, a time slot for TDMA, or a code sequence for CDMA) can be allocated to a user as long as needed (i.e., until a call is completed). In contrast, in packet-switched systems the data is bundled into blocks of bits (packets) that are individually transmitted through the network; in this case, the channel (carrier, time slot, or code) is only assigned to a given packet for the time required to transmit that packet.

Other access techniques include random access protocols such as ALOHA and its evolutions, in some cases jointly utilized with collision avoidance mechanisms, such as the well-known Carrier Sense Multiple Access/Collision Avoidance (CSMA/CA) used in WLANs. Here, a station wishing to send a message listens to the channel. If the channel is free, the station waits a prescribed time and then transmits. This time interval is introduced to reduce the chance of collision with a message that has been transmitted by another station and has not yet been sensed. If the channel is busy, access is still deferred until the medium is sensed free, then the station waits for a prescribed time extended in a random manner (random backoff), and finally transmits if no transmission on the medium is sensed during this time. In this mechanism, random backoff is introduced to reduce the probability that more waiting stations will begin to transmit at the same time after the channel is released; however, collisions may still occur, as other stations might have begun transmitting at the same time. So, after transmission, the transmitting station monitors the channel; if a collision is detected, it stops transmitting and defers retransmission for a random time interval to reduce the possibility of users again colliding.

In cellular systems, the available radio resources are used in a cellular environment with reuse constraints built in to keep interference at a tolerable level. The capacity of the system (e.g., in terms of served users per unit area) depends on how often in space the same resource is reused. To increase system capacity, one possibility is to reduce the cell size from larger macrocells to microcells and picocells. An immediate consequence of reducing cell sizes is that handoffs become more frequent. Another possibility is to use smart antennas to mitigate co-channel interference. Other methods to improve system capacity include dynamic channel allocation (DCA) strategies to reduce call blocking, as well as power control for reducing interference.

In fixed channel allocation (FCA), channels are assigned permanently to each cell following a prescribed reuse patterns. In contrast, DCA refers to techniques where channels are dynamically assigned to cells according to traffic demands to control co-channel interference. DCA techniques range from one where no permanent assignments of channels to cells are made and all radio resources are kept in a pool to techniques where channels are nominally assigned to cells but it is possible for a cell to borrow unused channels from other cells when necessary.

In any case, communication is possible if the signal-to-interference ratio (SIR) is above a minimum level that depends on the radio interface (e.g., coding, modulation, receiver structure). A particular case is that of CDMA; here, because all signals are superimposed in time and frequency, a strong received power for one signal can prevent the reception of other users' signals. In the uplink (from users to the base station, or BS), because users are generally at different distances, if the transmitted power levels were the same for all users then the corresponding received power levels at the BS would be very different, obscuring some users. It is therefore necessary to implement techniques to change the transmitted power to produce received power levels at the BS that are as constant as possible for all users. This is the basic concept of power control techniques. Changing the transmitted power level can also be used to cope with multipath power fluctuations. For this reason, cellular systems based on CDMA have adopted the so-called *fast power control* with a high adaptation rate to both control interference and reduce fast fading. Another possibility is to adaptively change coding and modulation format [27,28].

Current Wireless Systems and Beyond

Cellular Systems

Table 3.2 summarizes the main characteristics of second-generation (2G) digital cellular phone standards. These systems were principally focused on providing voice communication services over a wide area for users with

Table 3.2 2G Cellular System Characteristics

Characteristic	GSM	IS-136	IS-95
Multiple access	TDMA	TDMA	CDMA
Modulation	GMSK	$\pi/4$ DQPSK	BPSK/QPSK
Uplink frequencies (MHz)	890–915, 1715–1785	824–849	824–849
Downlink frequencies (MHz)	935–960, 1810–1880	869–894	869–894
Carrier separation (KHz)	200	30	1250
Compressed speech rate (Kbps)	13/6.5	7.95	1.2–9.6 (variable)

Note: BPSK, binary phase-shift keying; CDMA, code-division multiple access; DQPSK, differential quadrature phase-shift keying; GMSK, Gaussian minimum-shift keying; QPSK, quadrature phase-shift keying; TDMA, time-division multiple access.

high mobility. For speech transmission, the data rate is on the order of 10 kilobits per second (Kbps), and the resulting equivalent spectrum occupancy per active user is around 25 to 30 KHz (note that, in GSM, each carrier has eight time slots and each can carry a speech channel). The objective of wide coverage has led to the well-known concept of subdividing the area in the cells (cellular space division), with base stations taking care of users within each cell. When a communicating user moves from one cell to another, a procedure known as *handover* must take place to ensure continuity in the service during the transition to the new base station.

In the late 1990s, the 2G systems became 2.5G with some modifications introduced to support data services in addition to voice. In particular, the GSM evolution has included high-speed circuit-switched data (HSCSD), General Packet Radio Service (GPRS), and Enhanced Data Rates for GSM Evolution (EDGE). For HSCSD, up to four time slots can be assigned to a single user, with an overall data rate of up to 57.6 Kbps. In GPRS, a major enhancement was introduced with packet-switched data in addition to circuit-switched voice. The maximum data rate for GPRS is 171.2 Kbps when all eight time slots of a GSM carrier are assigned to a single user. With EDGE the GSM is enhanced by the introduction of adaptive modulation and coding, with data rates of up to 384 Kbps.

Further innovations led to third-generation (3G) mobile cellular standards, the main characteristics of which are summarized in Table 3.3. All of these approaches are aimed at supporting high mobility in conjunction with advanced services such as high-data-rate Internet access and video

Table 3.3 3G Cellular System Characteristics

Characteristic	CDMA2000	W-CDMA
Subclass	1xEV-DO Rev. A	UMTS
Channel bandwidth (MHz)	1.25	5
Peak data rate (Mbps)	3.1 (DL), 1.8 (UL)	2 (14.4 with HSDPA)
Modulation	QPSK/8-PSK/16-QAM (DL), QPSK/8-PSK (UL)	QPSK/16-QAM (DL), QPSK (UL)
Power control	600 Hz	1500 Hz

Note: DL, downlink; HSDPA, high-speed downlink packet access; PSK, phase-shift keying; QAM, quadrature amplitude modulation; QPSK, quadrature phase-shift keying; UL, uplink; UMTS, Universal Mobile Telecommunications System.

communication, and they are based on the CDMA technique. Data rates range from 384 Kbps up to 2.4 Mbps, depending on the level of interference and on channel impairments. Both fast power control and transmitter diversity techniques are adopted in 3G systems.

Wireless Local Area Networks

Table 3.4 provides a description of current wireless local area network (WLAN) standards. Note that in LANs the original focus was on using wireless communications in place of cables for data communication. Because the primary application was originally considered to be "local" (indoor), these systems operate in unlicensed radio bands and support of user mobility is quite limited. The baseline standard, IEEE 802.11, has allocated 83.5 MHz of bandwidth in the unlicensed 2.4-GHz radio band for this purpose. Because this is an unlicensed band, it is necessary to transmit at a low power for conventional modulation systems or to use some form of spectrum spreading for higher power levels. The 802.11 standard includes both frequency-hopping spread spectrum (FHSS) and direct sequence spread spectrum (DSSS). The latter uses Barker sequences for spreading, allowing data rates up to 2 Mbps with a channel bandwidth of 22 MHz. An evolution of the baseline standard is IEEE 802.11b, where complementary code keying (CCK) is used instead of Barker sequences, and a data rate of up to 11 Mbps can be achieved. The maximum link range is on the order of 100 m.

A further evolution is represented by the IEEE 802.11a standard, which allows operating in an unlicensed band around 5 GHz (in some parts of the world, 300 MHz are allocated from 5.2 to 5.825 GHz, but not in Europe) and data rates of up to 54 Mbps. This standard incorporates orthogonal

Table 3.4   IEEE 802.11 Wireless LAN Link Layer Standards

	IEEE Standard			
	802.11	802.11a	802.11b	802.11g
Bandwidth (MHz)	83.5	300	83.5	83.5
Frequency range (GHz)	2.4–2.4835	5.15–5.25 (lower) 5.25–5.35 (middle) 5.725–5.825 (upper)	2.4–2.4835	2.4–2.4835
Number of channels	3	12 (4 per subband)	3	3
Modulation	BPSK, QPSK	OFDM	BPSK, QPSK	OFDM
Spectrum spreading	FH, DS (Barker)	None	DS (CCK)	None
Channel coding	—	Conv. (rate 1/2, 2/3, 3/4)	—	Conv. (rate 1/2, 2/3, 3/4)
Maximum data rate (Mbps)	2	54	11	54
Range (m)	—	27–30 (lower band)	75–100	30
Random access	CSMA/CA	CSMA/CA	CSMA/CA	CSMA/CA

Note: BPSK, binary phase-shift keying; CCK, complementary code keying; CSMA/CA, Carrier Sense Multiple Access/Collision Avoidance; DS, direct sequence; FH, frequency hopping; OFDM, orthogonal frequency-division multiplexing; QPSK, quadrature phase-shift keying.

frequency-division multiplexing (OFDM) to cope with the distortion due to large signal bandwidth and multipath propagation. Here, different transmitter power levels are specified, ranging from 50 mW to 1 W, to permit both indoor and outdoor applications. Because the 5-GHz band is not available worldwide, the IEEE 802.11g standard was introduced which has the same characteristics of IEEE 802.11a but allows operating in the 2.4-GHz unlicensed band. In principle, under 802.11g data rates could reach 54 Mbps; however, it is important to note that all systems working in the unlicensed band are limited primarily by the number of active users and, more generally, by the unpredictable amount of interference. This interference depends both on the number of active transmitters in the area using the same band (not necessarily adopting the same standard) and on the presence of manmade interference such as that due to microwave ovens. For this reason, the data rate that can be practically achieved could be considerably lower than the maximum data rate noted above. Multiple antenna systems have been recently introduced in WLAN systems to cope with multipath propagation and to increase the data rate by utilizing the high-spectral-efficiency MIMO techniques described earlier.

Wireless Personal and Body Area Networks

In this section, we summarize the characteristics of some standards and emerging proposals for short-distance wireless networks. The most important standards are in the IEEE 802.15 family, where already-issued standards include IEEE 802.15.4 (related to ZigBee™) and IEEE 802.15.1 (related to Bluetooth®). Ultrawideband (UWB) is emerging as the technology that could bring about further evolution. ZigBee generally targets wireless networks with a large number of nodes and low energy consumption, is able to operate for many months with a single battery charge, and has relatively low data rates (less than 250 Kbps). Bluetooth targets applications with fewer nodes, requires higher data rates (up to 1 Mbps), and features low-latency voice channels.

Ultrawideband is an emerging technology characterized by signals with a large occupied bandwidth (larger than 500 MHz). UWB can be obtained by utilizing very short pulses in the time domain (so-called *impulse radio*) or by using OFDM. To prevent disturbing primary users, UWB signals must have a very low power spectral density; as a consequence, the overall transmit power must be kept low, and the distance range is on the order of some tenths of meters. UWB has some advantages over traditional systems, including its potential for very high data rates, its ability to cope with multipath propagation, its inherent localization capability, and its ease of implementation (for impulse radio UWB). UWB will most probably be the technology to support both low-data-rate wireless sensor networks and high-data-rate systems for short-range wireless multimedia applications (see Table 3.5).

Finally, we should mention IEEE 802.16 (and the related WiMax), which was designed to provide wireless access to buildings as an alternative to fiberoptic or coaxial links. The standard, which is intended more for fixed wireless stations than for mobile users, addresses frequencies from 10 to 66 GHz, with data rates of up to 70 Mbps over large distances. For users with moderate mobility, the standard adopts frequencies ranging from 2 to 11 GHz to cope with situations where there is no line-of-sight propagation.

Table 3.5 Principal Wireless Short-Range Network Standards

	ZigBee (802.15.4)	Bluetooth (802.15.1)	UWB (Under Study)
Frequency range (GHz)	2.4–2.4835	2.4–2.4835	3.1–10.6
Bandwidth (MHz)	83.5	83.5	7500
Modulation	BPSK, OQPSK	GFSK	BPSK, QPSK, PPM, or PAM
Spectrum spreading	DS	FH	FH/OFDM or TH, DS impulse radio
Maximum data rate (Mbps)	0.25	1	>100
Range (m)	30	10	10
Power consumption (mW)	5–20	40–100	80–150
Access	CSMA/CA	TDMA	Undefined
Networking	Mesh/Star/Tree	Subnet clusters (8 nodes)	Undefined

Note: BPSK, binary phase-shift keying; CSMA/CA, Carrier Sense Multiple Access/Collision Avoidance; DS, direct sequence; FH, frequency hopping; GFSK, Gaussian frequency-shift keying; OFDM, orthogonal frequency-division multiplexing; OQPSK, offset quadrature phase-shift keying; PAM, pulse-amplitude modulation; PPM, pulse position modulation; QPSK, quadrature phase-shift keying, TDMA, time-division multiple access, TH, time hopping.

Licensed Versus Unlicensed Spectrum: Spectrum Regulation and Cognitive Radio

Traditionally, we divide the radio spectrum into licensed and unlicensed bands. In unlicensed bands, users can freely transmit, but with some limitations on the power spectral density to minimize the amount of interference for other users operating in the same frequency band. If too many users are operating in the same unlicensed band in the same area, however, the resulting interference can be sufficiently high to prevent communications. For this reason, we cannot have any guaranteed quality of service for systems in the unlicensed band; in other words, we must consider these systems as being not very efficient in terms of served users per area. Licensed services, on the other hand, are designed to efficiently utilize the spectrum.

An emerging concept in spectrum usage is known as *cognitive radio* (CR) [22–26]. Indeed, a key observation is that the radio spectrum is often not used very efficiently. For example, in some applications, the licensed spectrum is not used in some geographical areas, or, in some emergency applications, the use of the spectrum has a very low duty cycle (e.g., services that use the spectrum only occasionally but with high priority). By investigating radio spectrum usage in some revenue-rich urban areas, it was observed that some frequency bands are largely unoccupied most of the time, that some other frequency bands are only partially occupied, and that the remaining frequency bands are heavily used [23–26]. Based on this research, a spectrum hole has been defined as occurring when a band of frequencies assigned to a primary user is not utilized by that user at a particular time and in a specific geographic location [23]. To improve spectrum utilization, these spectrum holes could be utilized by secondary users at the appropriate time and location.

The means for achieving such an efficient utilization of the spectrum is cognitive radio. The basic idea in cognitive radio is that a device can sense its environment and location and then alter its power, frequency, modulation, and other parameters so as to dynamically reuse available spectrum. This could in theory allow multidimensional reuse of the spectrum over space, frequency, and time, thus eliminating the spectrum and bandwidth limitations that have slowed wireless services development.

This new technology is the natural evolution of software-defined radio (SDR). With SDR, the software embedded in a radio terminal, for example, would define the parameters under which the phone should operate in real time as its user moves from place to place. Cognitive radio is even smarter than SDR, as the aim is to have a radio that can sense and is aware of its environment and that can learn from its environment for the best spectrum and resources usage. Some of the most important tasks that a cognitive radio should encompass are related to radio environment analysis (for the detection of spectrum holes and to estimate interference effects), channel identification (for estimating channel state information and available channel capacity), dynamic spectrum management, and cooperation and competition among users (related to game theory).

An early example of altering the traditional spectrum subdivision was the release by the Federal Communications Commission (FCC) of up to 7.5 GHz of bandwidth for ultrawideband signaling in the region between 3.1 and 10.6 GHz. In this band where other licensed applications are operating, UWB transmission is allowed with some limits in the transmitted power to ensure a negligible impact in terms of interference. Although the power levels allowed for UWB are extremely low (–41 dBm/MHz), with UWB the FCC has allowed for the first time unlicensed use across otherwise licensed bands. Cognitive radio, then, could represent a more complete solution as it actively looks for unused spectrum and begins to transmit inside those bands, stopping when necessary if the primary users show up.

Concluding Remarks

We have reviewed the main characteristics of wireless communication systems, from channel characteristics to multiple antennas and orthogonal frequency-division multiplexing. With regard to wireless physical-layer technology, various solutions for different data rates and mobilities are currently available; however, the future will certainly witness the integration of radio interfaces to provide users a single, wireless terminal that can operate according to the various standards depending on both available resources and user profiles. In particular, the emerging concept of cognitive radio and dynamical usage of the radio spectrum could represent a major breakthrough for the evolution of wireless systems.

Acknowledgments

The author wishes to thank Dr. A. Giorgetti and Dr. G. Liva for their careful reading of the manuscript. This research was supported in part by the University of Bologna (Progetto Pluriennale d'Ateneo).

References

[1] Stuber, G.L., *Principles of Mobile Communication*, 2nd ed., Kluwer Academic, Norwell, MA, 2001.
[2] Rappaport, T.S., *Wireless Communications*, Prentice Hall, Upper Saddle River, NJ, 1996.
[3] Parsons, J.D., *The Mobile Radio Propagation Channel*, John Wiley & Sons, New York, 1992.
[4] Holma, H. and Toskala, A., *WCDMA for UMTS: Radio Access for Third Generation Mobile Communications*, rev. ed., John Wiley & Sons, New York, 2002.
[5] Goldsmith, A., *Wireless Communications*, Cambridge University Press, Cambridge, U.K., 2005.
[6] IEEE, *Part 11: Wireless LAN Medium Access Control (MAC) and Physical Layer (PHY) Specifications: High-Speed Physical Layer in the 5 GHz Band*, IEEE Std. 802:11TM, Institute of Electrical and Electronics Engineers, New York, 1999 (http://standards.ieee.org/).
[7] Winters, J.H., Salz, J., and Gitlin, R.D., The impact of antenna diversity on the capacity of wireless communication systems, *IEEE Trans. Commun.*, 42(2/3/4), 1740–1751, 1994.

[8] Winters, J.H., Smart antennas for wireless systems, *IEEE Pers. Commun. Mag.*, 5(1), 23–27, 1998.

[9] Winters, J.H., On the capacity of radio communication systems with diversity in a radio fading environment, *IEEE J. Selected Areas Commun.*, 5(5), 871–878, 1987.

[10] Foschini, G.J., Layered space-time architecture for wireless communication in a fading environment using multiple antennas, *Bell Labs Tech. J.*, 1(2), 41–59, 1996.

[11] Foschini, G.J., Golden, G.D., and Valenzuela, A., Simplified processing for high special efficiency wireless communication employing multi-element arrays, *IEEE J. Select. Areas Commun.*, 17(11), 1841–1851, 1999.

[12] Telatar, E., Capacity of multi-antenna Gaussian channels, *Eur. Trans. Telecommun.*, 10, 585–595, 1999.

[13] Foschini, G.J. and Gans, M.J., On limits of wireless communications in a fading environment when using multiple antennas, *Wireless Pers. Commun.*, 6, 311–335, 1998.

[14] Chiani, M., Win, M.Z., and Zanella, A., Error probability for optimum combining of M-ary PSK signals in the presence of interference and noise, *IEEE Trans. Commun.*, 51(11), 1949–1957, 2003.

[15] Chiani, M., Win, M.Z., and Zanella, A., On the capacity of spatially correlated MIMO Rayleigh fading channels, *IEEE Trans. Inform. Theory*, 49(10), 2363–2371, 2003.

[16] Zanella, A., Chiani, M., and Win, M.Z., MMSE reception and successive interference cancellation for MIMO systems with high spectral efficiency, *IEEE Trans. Wireless Commun.*, 4(3), 1244–1253, 2005.

[17] Giorgetti, A., Smith, P.J., Shafi, M., and Chiani, M., Mimo capacity, level crossing rates and fades: the impact of spatial/temporal channel correlation, *KICS/IEEE Int. J. Commun. Networks*, 5(2), 104–115, 2003 (special issue on coding and signal processing for MIMO systems).

[18] Paulaj, A.J., Gore, D.A., Nabar, R.H., and Bolcskei, H., An overview of MIMO communications: a key to gigabit wireless, *Proc. IEEE*, 92(2), 198–218, 2004.

[19] Proakis, J.G., *Digital Communications*, 4th ed., McGraw-Hill, New York, 2001.

[20] Tarokh, V., Seshadri, N., and Calderbank, A.R., Space-time codes for high data rate wireless communication: performance criterion and code construction, *IEEE Trans. Inform. Theory*, 44(2), 744–765, 1998.

[21] Vucetic, B. and Yuan, J., *Space–Time Coding*, John Wiley & Sons, New York, 2003.

[22] FCC, *Spectrum Policy Task Force*, ET Docket No. 02-135, Federal Communications Commission, Washington, D.C., 2002.

[23] Kolodzy, P. et al., Next generation communications: kickoff meeting, in *Proc. Defense Advanced Research Projects Agency (DARPA) Workshop*, October 17, 2001.

[24] McHenry, M., Frequency agile spectrum access technologies, in *Proc. of FCC Cognitive Radio Workshop*, May 19, 2003.

[25] Staple, G. and Werbach, K., The end of spectrum scarcity, *IEEE Spectrum*, 41(3), 48–52, 2004.

[26] Haykin, S., Cognitive radio: brain-empowered wireless communications, *IEEE J. Selected Areas Commun.*, 23(2), 201–220, 2005.

[27] Nanda, S., Balachandran, K., and Kumar, S., Adaptation techniques in wireless packet data services, *IEEE Commun. Mag.*, 38(1), 54–64, 2000.

[28] Conti, A., Win, M.Z., and Chiani, M., On the performance of slow adaptive M-QAM with antenna subset diversity in fading channels, in *Proc. of IEEE Global Telecomm. Conf.*, Dallas, TX, December, 2004, pp. 3373–3378.

Chapter 4

Mobile *Ad Hoc* Communication Issues

Hamid Harroud, Dineshbalu Balakrishnan,
and Ahmed Karmouch

CONTENTS

Introduction

The availability of portable computing devices and advances in wireless networking technologies have contributed to the growing acceptance of mobile computing applications and opened the door for the possibility of seamless and pervasive services in mobile environments. However, due to the restraints of limited device capabilities, network connectivity, transmission range, and frequent changes caused by user or device mobility, a considerable burden is placed on applications to be deployed in an environment where mobile devices must connect to each other through automatic configuration and communicate with each other over wireless links.

Mobile devices (also referred to as *mobile nodes*), including phones, personal digital assistants (PDAs), laptops, and sensors, dynamically cooperate with each other to form and set up a mobile *ad hoc* network (MANET) [1] by wirelessly communicating with other nearby mobile devices without the support of a fixed infrastructure or centralized controlling system. *Ad hoc* communication is a type of spontaneous communication wherein software and devices communicate directly with other nodes within wireless transmission range and indirectly with other nodes by relying on some nodes that act as routers. To exchange information with another node, a dynamically determined multi-hop route may be required, depending on various parameters such as the distance between nodes, directions, and the mobile *ad hoc* network topology, with nodes joining, leaving, and moving at any time.

Ad hoc communication and mobile *ad hoc* network environments can be used in smaller areas, such as conferences, or in larger areas, such as for battlefield communications or disaster recovery. Table 4.1 lists some of the common applications of *ad hoc* networks that benefit from *ad hoc* communication [1,2].

Among the services required by the various *ad hoc* applications are service discovery and location service. Joining an *ad hoc* environment, mobile nodes must explore and locate the available services in the environment, and these exploration and location activities must be carried out in a context-aware manner, using the current position of the node, proximity, available resources, and additional context information. Mobile agents have also been proposed to serve as a mechanism to support

TABLE 4.1 *Ad Hoc* Network Applications

Field	Application
Telecooperation	Delay-sensitive applications, interactive television
Collaborative groupware	Scheduling
Hypermedia	Web-based transactions
Home and enterprise networking	Personal area networks (PANs)
Mobile agent models	Digital personal assistants (PDAs)
Sensor networks	Mobile wireless local area networks (WLANs)
Context-aware systems	Location-dependent applications
Emergency services	Disaster recovery, military communications

transient data sharing between nodes within communication range to highly simplify the development and deployment of various *ad hoc* applications.

Ad Hoc Communication Approaches

Ad hoc networks have gained great importance in a variety of domains, including home and sensor networks, the fields of education and entertainment, and other industries. Providing efficient *ad hoc* communication network environments requires the use of appropriate approaches for solving challenging issues related to an *ad hoc* network. These issues can be related to the interconnectivity of the mobile devices, the routing protocols (the *ad hoc* network topology frequently changes and multihop communication is required), and the applications and services to be provided to mobile users. The main characteristics of *ad hoc* network environments, their classification, and the technology required in such environments are discussed in the following subsections.

Characteristics and Classifications

The main characteristics of an *ad hoc* environment include [3]:

- Autonomicity
- System and device heterogeneity
- Flexibility and scalability
- Self-configuration
- Dynamic network topology

The main constraints, from a mobile *ad hoc* environment point of view, include [3]:

- Wireless medium constraints
- Resource constraints
- Connectivity (bandwidth) constraints
- Security and privacy issues

Ad hoc communications can be classified based on the environments in which they are to be used, such as:

- Ubiquitous computing environments
- Pervasive real-time environments
- Agent-based computing environments
- Ambient computing environments

They can also be classified based on their configuration:

- Flat *ad hoc* networks
- Hierarchical networks
- Proactive
- Reactive
- Sensor-based
- Semantic- and collaboration-based

A broad technical classification of the use of *ad hoc* communications could be based on the utilization of such communications in networks and applications (i.e., *ad hoc* networks and *ad-hoc*-communications-based applications). Because the literature of *ad hoc* networks is too extensive to be analyzed in detail here, this chapter focuses on *ad hoc* applications and provides a detailed introduction to service discovery.

Ad Hoc Networking

An *ad hoc* network [4,35] is formed dynamically by mobile devices and is managed by the nodes that enter and leave this network; for example, mobile agents themselves could organize and administer *ad hoc* wireless networks. Thus, an *ad hoc* network may have any network topology and may or may not have a gateway to any particular fixed network. The users might be able to bring their mobile devices and get connected to the network without any prior configuration. The *ad hoc* network can be visualized as a wireless LAN where the users bring in various types of

Figure 4.1   A mobile *ad hoc* network example.

wireless devices such as PDAs or notebooks. When the devices are turned on, they are spontaneously connected and can communicate with other connected devices or use the services that are in this network. Due to this spontaneous and dynamic nature of *ad hoc* networks, they are proving to be an interesting and challenging problem for researchers.

The operation of mobile *ad hoc* networks does not rely on fixed infrastructures. As they are autonomously formed by wireless associations between mobile terminals (and users), they are highly flexible, infrastructure independent, and convenient. But, the features do have a price, such as *ad hoc* wireless networking constraints. One of the major concerns of such a network is the security of the nodes, as they are more prone to attacks such as eavesdropping or spoofing. The attacks largely arise due to dynamic reconfigurability and the absence of a single centralized controller. An example of a MANET is illustrated in Figure 4.1.

The main concepts that challenge researchers from an *ad hoc* networking point of view include routing, connectivity, and service and resource discovery [3]. In MANETs, efficient routing protocols and standards are required to establish communication between involved networks. These routing standards should take into consideration the constraints of a mobile *ad hoc* environment. Among the several proposed routing solutions, the hybrid routing protocols, which proactively route nearby nodes and reactively route distant nodes, are promising [3]. Other routing approaches that could be employed in a mobile *ad hoc* environment include location-based routing, time-based routing, and clustering.

Various predictive architectures could be used to reduce the impact of wireless constraints in *ad hoc* wireless networks. Another important concern of *ad hoc* wireless networks is security. Several frameworks proposed for the security of *ad hoc* wireless networks are still in the development phase [5].

Due to the popularity of Internet Protocol (IP) standards, one has to employ IP addresses even though they face several constraints in mobile *ad hoc* networks due to node mobility and overhead [3]; hence, novel solutions are required for Internet connectivity and addressing. Among the several connectivity models that have been proposed, the main possible approaches include: (1) the use of Mobile IP for connectivity, (2) the use of subnets for addressing, and (3) utilization of network address translation (NAT) for Internet connectivity [3].

Mobile *Ad Hoc* Applications

The introduction of terminologies such as the Bluetooth®, 802.11, and Hyperlan greatly facilitated the deployment of *ad hoc* technology. In contrast to applications specific to the military domain, several new *ad hoc* networking applications appeared. In this section, we focus on addressing *ad hoc* networking challenges by using *ad hoc* approaches via applications and tools such as Session Initiation Protocol (SIP), agents, mobile agents, context awareness, XML, and conferencing tools. The scalability and flexibility characteristics of mobile *ad hoc* networks make this technology attractive for several applicative scenarios such as personal area networks (PANs).

Context, Mobile Agents, and Policies

Context plays an important role in managing *ad hoc* environments. Because the environment is dynamic and reconfigurable, knowledge of the related entities in the *ad hoc* environment becomes essential. Capturing the context and analyzing and matching context data from various related resources allows collaboration among various devices. Context can also be used to address the security issues in an *ad hoc* space in that it can act as a firewall against entities possessing irrelevant or questionable context. The use of mobile agents in *ad hoc* communications is also an important alternative and is further discussed later.

Policies are rules or conditions set by the user in order to govern the behavior of entities with a specific domain (for example, an *ad hoc* environment). Policies [7] are generally applied in security (for restricting access), management (to assign rules for participating entities), and conversational policies (to structure and carry out conversations between entities). Several policy languages aim at formalizing the specifications of policies so they can be represented and interpreted by machines [7]. Future applications could learn from the current system behavior and create policies at run-time, an asset of MANETs. The behavior of an agent system can thus be modified without influencing the system architecture.

Virtual Conferencing

Unlike a traditional conference wherein people must be physically present, virtual conference participants can be physically separated along networks but virtually joined in a meeting place where they can participate in live interaction and information exchange. Virtual conference applications include audio and video conferences, common multimedia conferences, and *ad hoc* conferences. An *ad hoc* conference is a dynamic meeting based on *ad hoc* networks where users randomly join or leave the network.

Session Initiation Protocol

The Session Initiation Protocol (SIP) is an application-layer-controlling protocol that can establish, modify, and terminate multimedia sessions or calls. It is a standardized signaling protocol for establishing real-time calls and conferences over the Internet [8]; its basic architecture is client–server based in nature. SIP is most commonly utilized in applications such as multimedia conferences, distance learning, and Internet telephony. Although the text-based SIP protocol is similar in both syntax and semantics to the Hypertext Transfer Protocol (HTTP) [9], it can be easily extended, unlike HTTP. This extensibility feature aids in the provision of various services such as instant messaging, call transfer, and call control. Because SIP is a general-purpose protocol, it is independent of packet layers and supports both the User Datagram Protocol (UDP) and the Transmission Control Protocol (TCP). SIP cooperates with other protocols for multimedia communication and control; for example, it works with the Session Description Protocol (SDP) for multimedia session description during SIP session establishment [10], and it works with the Real-Time Transport Protocol (RTP) [11] for real-time data transportation after SIP session establishment.

The main entities in SIP are the user agent, SIP proxy server, SIP redirect server, and registrar [28]. The SIP user agent works at the client end and frequently updates users' contact information to the SIP registrar. Every SIP entity has a unique SIP address for the purposes of identification. The SIP address is presented in the form of an SIP universal resource locator (URL) as follows: "sip: username@domain." The six request methods defined in the SIP by which entities exchange SIP messages are [28]:

- INVITE is used by the SIP user agent to initiate a session.
- BYE terminates a session between two users.
- ACK confirms session establishment.
- CANCEL terminates call processing.
- REGISTER registers a user's SIP address with the SIP registrar server.
- OPTIONS queries server capabilities without setting up a call.

Agent Technology and Platforms

A common interest among researchers is providing global and efficient technologies, standards, execution environments, and security solutions for mobile middleware [12]. A related interest is the employment of agent technology to develop these requirements. An agent is an entity that represents a person, an organization, or an application and which independently or by interacting with other agents executes a task or set of tasks. Agents can be created, moved, cloned, or destroyed dynamically which adds to the flexibility of their usage. For example, in the case of intelligent agents, the itinerary may change dynamically depending on the agents' status at a particular terminal or node in a network. The agents are autonomous and usually carry out specific sets of tasks they are programmed to do. The most interesting feature of these agents is that they are mobile, which means they can be created at one location and executed in another. Agents can also be viewed as components of a software application in certain cases [13].

Agent models, unlike client–server models, promise an entirely new approach to problem solving. Instead of the client sending a request to the server, a representation of the client (its agent) moves itself to the server, executes its task, and brings back the results using the Agent Communication Language (ACL). The efficiency of applications related to information technology could be improved by utilizing agent technology [14]. The main attributes of agent technology that would benefit *ad hoc* communications include:

- Agents will become more prominent as the Internet and Web technologies continue to grow [15]. They are well suited for mobile applications due to their small size, limited bandwidth requirements, and properties such as adaptability, scalability, and mobility.
- Agent-based computing can be considered as a natural extension to object-oriented programming [15]. The distributed nature of agents makes them extensible and simple.

- Voice recognition, mobility, location sensitivity, and personalized intelligent filters can be enabled in software applications using agents [16,17]. Information monitors and filters considerably reduce the mobile user's small-size, low-resolution screen problem.

The main drawbacks of agent technology include:

- Trust, privacy, and security issues
- Unreliability of nascent agent platforms and execution environments for mobile devices
- Time required to implement agents learning from people, people learning from agents, and agents improving the creative performance of people

The third constraint arises because agents should have access to all data that may be relevant, the accessed data should be scriptable and recordable, and every user's behavior is different.

For an agent to move from one node or network to another and interact with the foreign agents, the platform at the target must be able to recognize this agent and understand its language; therefore, it is necessary to have a common platform that recognizes the agents and its semantics. An agent platform is the execution environment wherein agents are created, moved, cloned, and destroyed. So far, several agent platforms have been built as applications on the operating system; these agent platforms have general specifications given by the Foundation for Intelligent Physical Agents (FIPA) [16]. Current prevalent agent platforms include JADE (http://jade.tilab.com) and FIPA-OS [18]; lightweight versions for mobile wireless devices are the Lightweight Extensible Agent Platform (LEAP) (http://leap.crm-paris.com) and MicroFIPA-OS [18], respectively.

Personal Assistant

New trends and emerging technologies focus on the performance of user tasks with the least user intervention. The personal assistant application framework was developed using agent technology [29]. It is an effort taken to develop a new application for mobiles that semi-autonomously assists users in utilizing various Internet applications on their mobile devices with minimal involvement; that is, the software agent-based application simplifies employing the Internet in mobile devices. Assistance is provided based on the particular user's personal preferences, so each user could have a customized personal assistant [19].

The personal assistant framework is not restricted to providing assistance with specific Internet applications for mobile device users but can also be used for various other applications. The components of the framework keep track of the actions performed by the user and communicate with each other to perform tasks; for example, a mobile user who wants to schedule a meeting with his colleagues can instruct his personal assistant to do so by specifying his case-specific preferences. If necessary, the personal assistant will communicate with external agents in remote platforms; that is, inter-platform agent communication is facilitated through integration with other agent-based systems. Common tasks performed by personal assistants include:

- Access and store most of the frequently used information and quickly search and locate documents of interest and importance.
- Organize information hierarchically according to the user's personal settings and preferences.
- Eliminate the mobile user's dependence on using touch-screen keypads and small-sized buttons by including icons and other visual aids in the graphical user interface (GUI).
- Assist user-oriented services such as e-mail retrieval, file or media transfers, and reporting weather conditions and news.
- By integrating with the mobile-agent-based *ad hoc* communication system, assist services such as printing, PDF writing, MP3 playback, and conferencing.

Although the personal assistant is intended for mobile devices, an agent-based system could also execute in devices such as laptops and tablet PCs. Related contributions of the personal assistant framework include: (1) design and implementation of the proxy agent that resolves problems in mobile devices such as unstable execution environments, overloading, and insufficient resources; and (2) design and implementation of the adaptation agent that resolves compatibility problems in mobile devices. The latter is achieved through a decrease in the hardware and software requirements of mobile devices utilizing personal assistants. Other aspects include support for unrecognized file formats without additional software requirements and the need for an active wireless interface only when sending and receiving data. The personal assistant can result in adaptability problems [20], primarily due to constraints in their execution environments and current mobile device limitations.

Mobile Agents

Inter-platform agent communication can be effectively performed by using the concept of agent mobility [21]. Mobile agents (i.e., agents possessing the agent mobility feature) dynamically move from one location (i.e., agent environment) to another under their own control to perform tasks. They are capable of performing various roles in an *ad hoc* network, such as routing [22], network management, and security. Network connectivity and communication are less reliable in an *ad hoc* network, but mobile nodes can hand their tasks over to an agent and wait for it to return with the results. Due to this capability of mobile agents, network traffic is optimized because the agent carries out its tasks when the connection is reliable and enough bandwidth is available to execute the task. Mobile agents require approximately four times less bandwidth to complete tasks involving intense remote communication compared to the client–server approach [21]. Moreover, mobile agents provide security at a higher level, on top of the network layer, thereby reducing security threats to the mobile nodes. Agents provide authentication of requests and maintain the confidentiality of private information, attributes that are highly sought after in *ad hoc* networks. Because mobile nodes primarily run on batteries, saving power becomes another important criterion. The agents play an important role in reducing power usage by carrying out tasks on behalf of the nodes even after the mobile nodes have left the network. Thus, from both the user point of view and the network point of view, mobile agents can be employed for a wide variety of operations in the *ad hoc* network.

Agent systems could produce performance bottlenecks due to a lack of resources or overloading [23]; for example, the personal assistant model could suffer from this problem. Although this problem was predicted and handled in Balakrishnan and Karmouch [29], who used proxy and adaptation agents, other common solutions include the migration of agents to foreign hosts via the agent mobility concept [23]. An alternative approach to agent mobility is agent cloning [23]. The cloning approach solves both the insufficient resources issue and the agent overloading problems. This approach features both agent mobility and task transfer. An agent clone could be created on a foreign host, and tasks could be transferred from an overloaded agent in the local host to the cloned agent. This cloned agent could finally die after performing the required tasks. If additional resources are available at the local host, then agents could first be cloned locally and later migrate to a foreign host [23].

Sample *Ad Hoc* Application Scenario

In this section, we provide an example of a mobile-agent-based *ad hoc* communications project to aid in understanding *ad hoc* communications (Figure 4.2). The central idea of this mobile *ad hoc* communications project is to bring various types of users and services together in a network where they can collaborate with one another and share services in the network.

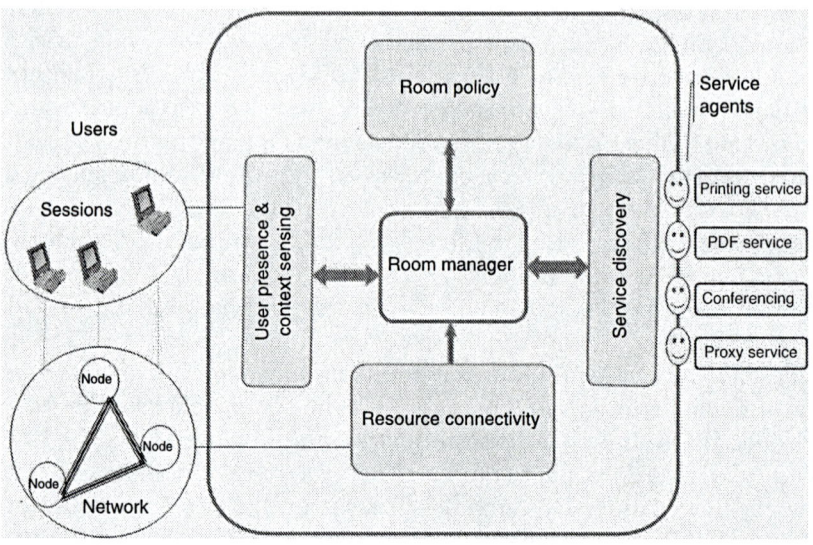

Figure 4.2 Overview of the mobile-agent-based *ad hoc* communications project.

In Figure 4.2, the room manager is the central controller. The main purpose of this project is to able collaboration among different entities (users and services). The users and services enter and leave the room in a dynamic fashion. The users may or may not carry personal devices such as laptops or PDAs. The presence of these entities is identified spontaneously, and appropriate tools for collaborating with the environment are supplied to the client devices in case they do not possess the required tools. The underlying network can be wired or wireless, depending on the devices that are connecting to it. Every privileged user is made aware of the services and other users in the room after passing policy tests. The users are allowed to share their own services and use the available authorized services when they need to, and this is handled by an exclusive service discovery module. The service discovery module provides suitable service matching of requests from clients in the same room or from other similar rooms. When the target devices are identified by their capabilities, a communication session is formed among them and monitored constantly via the SIP. This session is managed by using the context information of the room and that of the participating entities. The resource connectivity module is responsible for creating and managing the communication sessions between the entities. It provides session management for the user–service or user–user interactions.

Every user, after connecting to the *ad hoc* network, will be provided with a GUI that displays other users and services in the network with which the user can collaborate. These services and user lists are displayed after performing policy checks for the users and services. If the user wants to use these services, he simply selects the service and clicks on "submit." The request goes to the appropriate service agent, and the agent migrates to the user's terminal to perform further operations.

Overall Environment

The prototype of this project was implemented in a resource rich environment with wired and wireless LANs. It was tested in Java because it is system independent and ensures adaptability when it comes to mobile devices. The agents (e.g., service agents) were created and deployed using the FIPA-OS agent platform. For hand-held devices, the compact MicroFIPA-OS version was used along with the Jeode JAVA runtime environment (http://www.insignia.com). Because the agents use a common platform, all the agents must follow the FIPA standards if they must communicate with one another or move from one platform to another platform. Necessary information is cached by the proxy agent as XML files, and the created files are named after the conversation ID ACL field to ease the information-retrieval process.

The main components — the room manager, CMS server, proxy agent, and service agents (e.g., printing service agent) — ran on Pentium IV 2.4-GHz systems with 512 MB RAM. The radiofrequency (RF) sensor (MANTIS kit) used to identify the physical presence of entities in a room was from RF Code. The reader was equipped with the 802.11b wireless protocol and connected to the wired LAN through an ORINOCO access point. Two portable devices (a Compaq IPAQ 3800 Series Pocket PC and a Pentium III 1.1-GHz laptop) were used as terminal devices for mobile users. Both the mobile devices featured the 802.11b wireless capability. The personal assistant was implemented on the Pocket PC and the laptop was loaded with a PDF writer, which the user of the laptop could opt to share with the environment. In addition, four Pentium IV Windows XP desktops connected to the wired LAN were used as terminals for users not possessing a personal device.

Tools and Techniques

This section introduces the different tools and techniques employed in the *ad hoc* communications project prototype in addition to the already discussed techniques. These discussions will help the reader better understand the application scenario described in the previous section.

MANTIS Kit

The MANTIS kit (http://www.rfcode.com) was used for tag sensing. There is an error of approximately ±5 feet (maximum) in calculating the distance of a tag. This directly affects the precision in finding a tag if it is inside or outside the room; therefore, we set a time limit of 20 seconds between beacon emissions. If the beacon was not heard for more than 20 seconds, then the tag was considered to be out of range. Major issues that arise in such a scenario are the interface and communication among various types of services, because no one common language or protocol is understood by all services. X10 (http://www.x10.com) is a communication language that allows electrical appliances to talk with each other. The prototype implementation basically monitors, manages, and stores contextual parameters. The entities considered are services and people in a physical room space. There can be more than one room space, and the users and services in these rooms can collaborate with each other.

XML

eXtensible Markup Language (XML) was used as the encoding language for entity profiles because it provides portability and flexibility. The profiles of the entities are represented in XML, and object backup is performed after converting them to XML and storing them as XML files. Also, the contents of any ACL message (including the fields performative, sender address, receiver address, content, language, encoding, ontology, protocol, conversation ID, reply with, and in reply to) are stored in an XML file. An XML file containing a particular ACL message is either stored in the mobile device or in the workstation, or even both, depending on the contents and significance of the particular ACL message. The Xerces Java parser [25] was employed to retrieve the cached XML files.

Audio Conference Tool

The Robust Audio Tool (RAT) [30] is a tool primarily developed for multiparty audio conferencing over the Internet. It can be started from the command line as follows: Prompt> rat [options] <IP address/port>. For multicasting, the IP address must be in the range of 224.2.0.0 to 224.2.255.255 (except while using admin scope). The port number must be an even number and at least 1024. The IP address and the port number indicate the address where a multicast conference could be started. All participants must start RAT at the same IP address and port number to join a particular multicast conference.

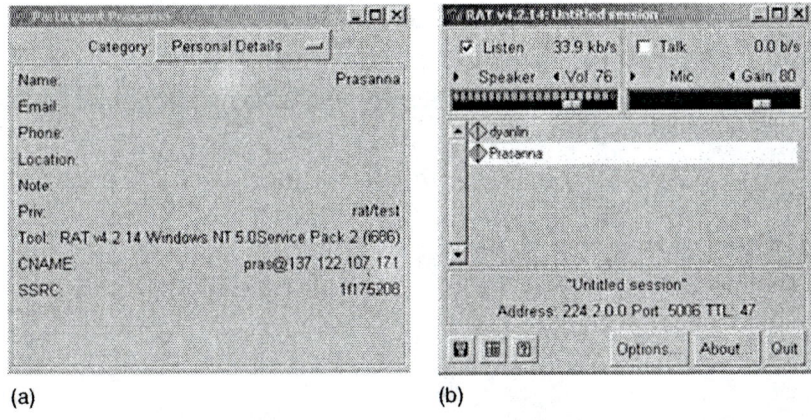

(a)   (b)

Figure 4.3   RAT windows: (a) user profile window, and (b) main window.

As shown in Figure 4.3b, RAT has a main window that displays all conference participants. The current speaker is highlighted, and clicking on any participant displays the user profile window, which includes the participant's name, e-mail, transmission status, etc. (Figure 4.3a). In this example, two users are participating in the ongoing conference, and the highlighted user "Prasanna" is speaking. More information about RAT can be obtained from http://www-mice.cs.ucl.ac.uk/multimedia/software/rat.

Video Conference Tool

A video conferencing application, VIC [31] is a tool for multiparty video conferencing over the Internet and was developed by the Network Research Group at Lawrence Berkeley National Laboratory and the University of California, Berkeley. VIC, similar to RAT, can also be started from the command line, and the even port number must be at least 5002. Again, all participants must start the tool at the same IP address and port number to take part in the same video conference. As illustrated in Figure 4.4, VIC has a main window that streams all of the conference participants' videos. In addition to video streaming, the participants' names, e-mails, etc. are also displayed. Clicking the button labeled "Info" will lead to a window displaying a user's detailed profile, including RTP status. The "Menu" button at the bottom of the main window opens the "Menu" window. By checking "Transmit" displayed at the top of the "Menu" window, the corresponding user's image will be transmitted to other conference participants. Clicking "Release" halts all ongoing transmissions. Any participant's "thumb" image can be

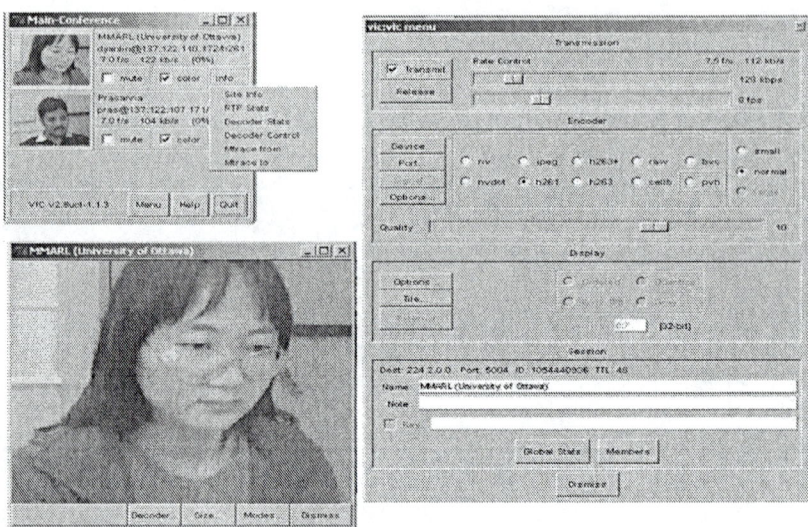

Figure 4.4 VIC user interface windows.

enlarged in a new window. More information about VIC can be found at http://www-mice.cs.ucl.ac.uk/multimedia/software/vic.

Service Discovery

Service discovery has become a highly desirable feature in today's networks. It allows for the automatic and spontaneous discovery and configuration of shared network services [36]. Because the network topology is constantly changing in an *ad hoc* environment, service discovery becomes even more important but more difficult. Possible services or resources include printing, writing PDFs, storage, access to databases or files, and Internet access. A variety of research is being conducted in the field of service discovery for *ad hoc* networks, but here we discuss an entirely agent-based approach to take advantage of the benefits introduced by agent technology [12,23].

Although it is critical to ensure that the network is aware of its active services at all times, an important goal is to create as little work for the user side as possible; hence, a push-based scheme would provide the user side with all the available services regardless of their current requirements. Such a scheme would also reduce delays for the user when initiating the services. Using a common centralized approach, the required agents include service agents (SAs) to represent the services, personal agents (PAs) to represent the users, and a fixed central service discovery agent

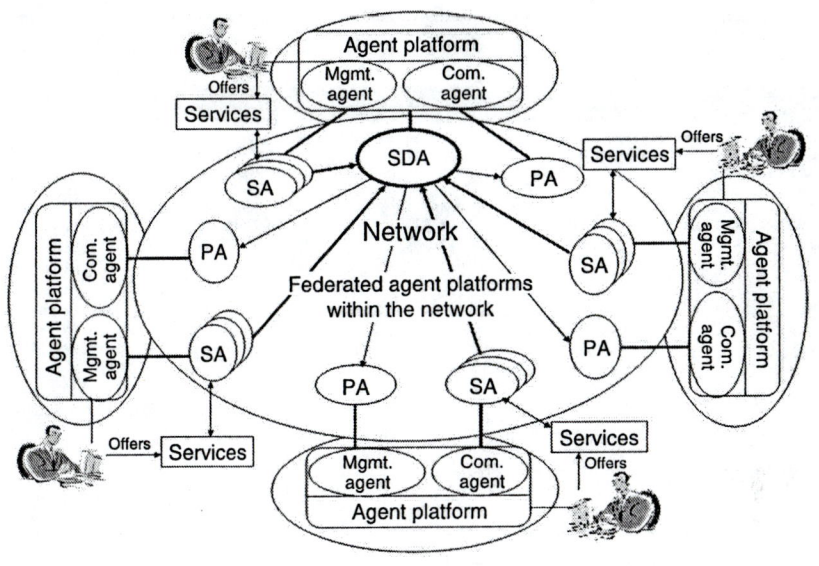

Figure 4.5 Service discovery using a central discovery agent.

(SDA) that manages services and satisfies the needs of both SAs and PAs (see Figure 4.5). Because services can come and go without notice in an *ad hoc* network, the SAs must be dynamically generated and destroyed.

Service agents must register and deregister with the SDA. PAs are responsible for listening to service announcements and making search requests. As service information changes, announcements are made by the SDA to those PAs authorized to access the affected services; therefore, mobile users always have the most current list of services in the network they are allowed to use, thus reducing their search effort to a minimum.

A user who enters the conference room with a mobile device is detected (via tag sensors) by the context-awareness system of the *ad hoc* environment (in this case, by the MANTIS kit). The presence of a user triggers the verification of the corresponding sensed tag ID. If the tag has a profile associated with it, then that profile is parsed and a presence agent is created. A tag ID could also be associated with a service, and similarly a service agent would be created after a service becomes available. The presence/service agent is set to monitor the activities of the user or service and transmit the readings to the application. Any further conversations with the entity will pass through this agent. The agent will also be responsible for changes in the state values of the context variables. Some software services may not be associated with any tags. In this case, their availability is determined by their registration to the platform when devices on which they execute are connected.

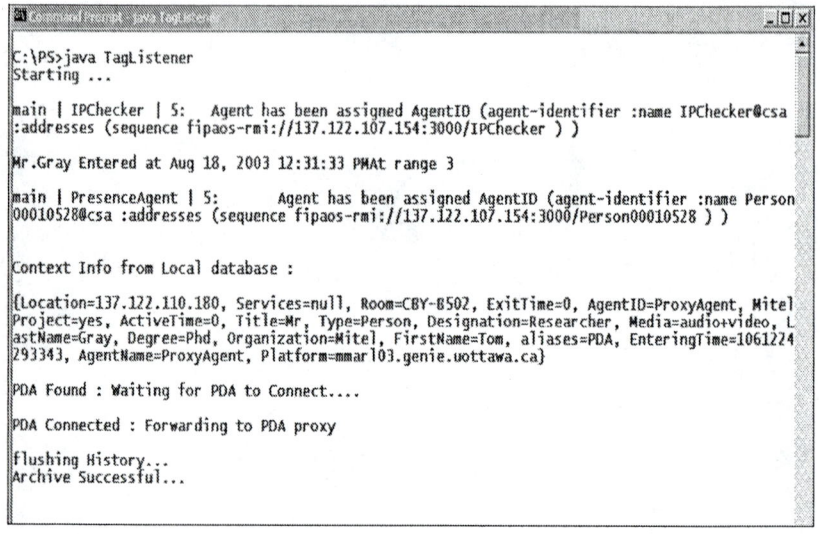

Figure 4.6 Person entering with PDA.

Samples

This section further explains the mobile-agent-based *ad hoc* communications model by illustrating the user interface snapshots, command prompt snapshots, and code snippets. Figure 4.6 and Figure 4.7 show screenshots of the log window that displays the activities that take place when a user

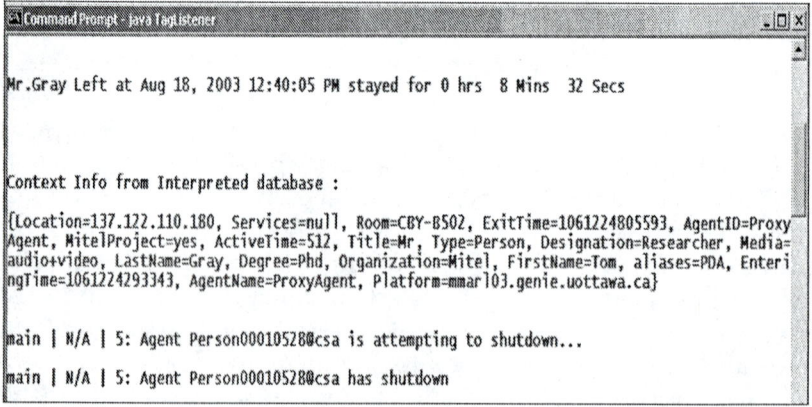

Figure 4.7 Person leaving with PDA.

IndexNo	Date	Services	EnterTime	ExitTime	ActiveTime	Status	Service Access	Access Time
29	1.0553495E+12	null	1.0553495305E+12	1.0553495703006E+12	14	0	HP LaserJet	1.0553493681E+12
36	1.0554467E+12	null	1.05544570837E+12	1.0554478303693E+12	1125	1	HP LaserJet	1.05544756E+12
38	1.0833682E+12	null	1.058368113718E+12	1.05836816021E+12	48	0		
39	1.0833683E+12	null	1.058368296343E+12	1.058368930539E+12	46	1		
40	1.0833684E+12	null	1.058368330552E+12	1.058368937875E+12	49	1		
41	1.0833684E+12	null	1.058368401812E+12	1.0583684493653E+12	48	0		
42	1.0833688E+12	null	1.058368740359E+12	1.0583687904653E+12	46	2		
43	1.0833689E+12	null	1.058368916306E+12	1.0583696864046E+12	48	0		
44	1.0833689E+12	null	1.058368987328E+12	1.0583695464414E+12	46	3		
45	1.0833690E+12	null	1.058368969375E+12	1.0583690108206E+12	48	0		
46	1.0833691E+12	null	1.058368932378E+12	1.0583690781094E+12	46	2		

Figure 4.8   User archive.

enters the room with a PDA and leaves after some time. The first line of the screenshot displays the time and distance at which the user was spotted via the MANTIS kit. The name of the user is obtained from the local database after mapping the tag ID of the user with the profiles. The context information from the local database is displayed as a hash table containing the basic and cached attributes. Similar procedures will be followed if a service enters and leaves the room.

The next two snapshots deal with caching of necessary information. Figure 4.8 shows a screenshot of the user's archive over a period of time. The enter time and exit time of the user are archived along with the status of the user at a particular time. The status value can be interpreted as follows:

0 — The user is available for interaction.
1 — The user is busy.
2 — The user is not at his desk.
3 — The user will be there for a few minutes.

As discussed previously, the ACL messages are cached in XML format. This is illustrated in Figure 4.9.

An integral part of the personal assistant GUI, the ACL message creation and sending screen is shown in Figure 4.10. This frame allows mobile users to create ACL messages in their PDAs by filling in mandatory and optional ACL fields. Users can select values from combo boxes for mandatory ACL fields. The mandatory ACL elements are distinguished from optional ACL elements by the button components placed between them. The user can

```
<?xml version="1.0" ?>
<Document>
  <Performative value="request" />
  <Sender value="UIAgent@pda01.genie.uottawa.ca" />
  <Receiver value="RM@main.genie.uottawa.ca" />
  <Content value="PrintGUI (Receive,Ready)" />
  <Language value="Prolog" />
  <Encoding />
  <Ontology value="RequestService_RM" />
  <Protocol value="FIPA-SL" />
  <ConversationID value="CID_20030624_RM" />
  <ReplyWith value="Service002" />
  <InReplyTo />
  <ReplyTo />
  <ReplyBy value="03-06-25" />
</Document>
```

Figure 4.9   Cached XML file composed of ACL elements.

then send the created ACL message to the intended receiver by clicking on the appropriate button. This figure represents a user's request for the printing service via the personal assistant model.

Alternatively, services can be selected from the available list via the integrated mobile-agent-based *ad hoc* communications model. Based on the user's policy profiles, the list of available services and users (for

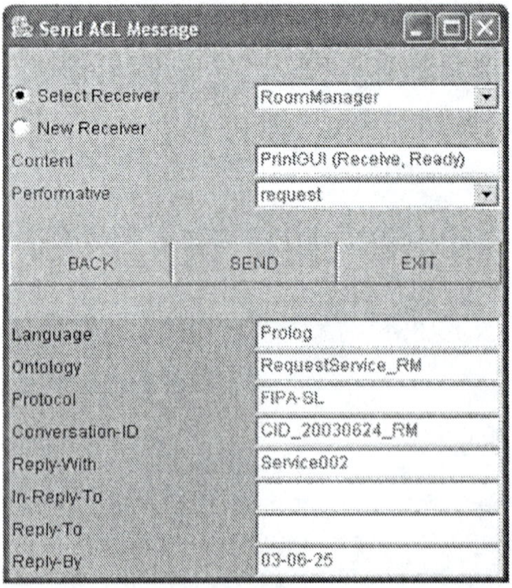

Figure 4.10   User interface snapshot — create and send ACL message screen.

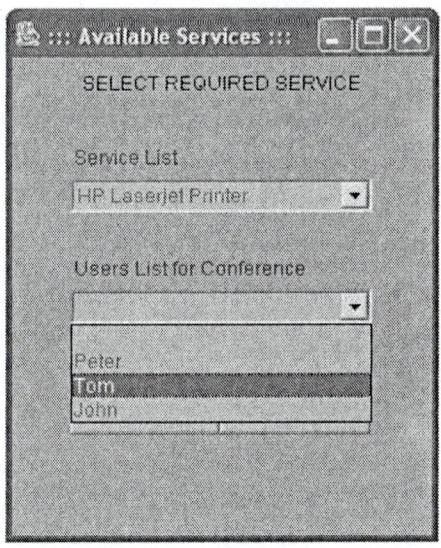

Figure 4.11   Snapshot displaying available users.

conferencing) varies. They are visible in the available services GUI via the room manager when the user successfully logs on. A sample snapshot is shown in Figure 4.11. For example, if the printing service is selected, the printing service GUI (Figure 4.12) is moved to the user's device via the room manager (and proxy agent). Figure 4.13, Figure 4.14, and Figure 4.15 illustrate use of the SIP to provide conference and sidebar services.

Discussion

Some of the significant points regarding this sample scenario include:

- Because the personal assistant application is integrated with the mobile-agent-based *ad hoc* communications system, authorized users of the application can also utilize various services offered by the mobile-agent-based *ad hoc* communications system.
- The GUI listing the available services and users and the GUI of the requested service are dynamically moved to the personal assistant interface via the proxy agent. Instead of actually making the agents mobile, we can clone them via the proxy service based on the requirements.

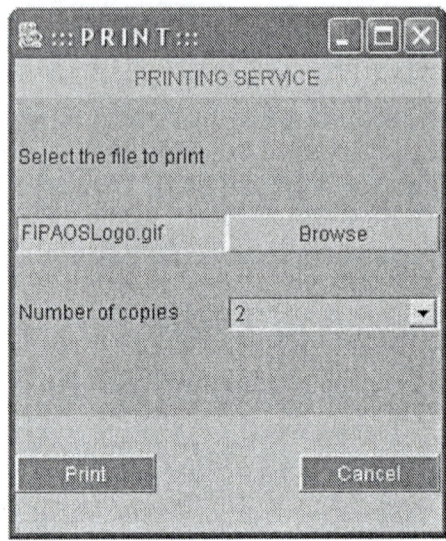

Figure 4.12   The printing service snapshot.

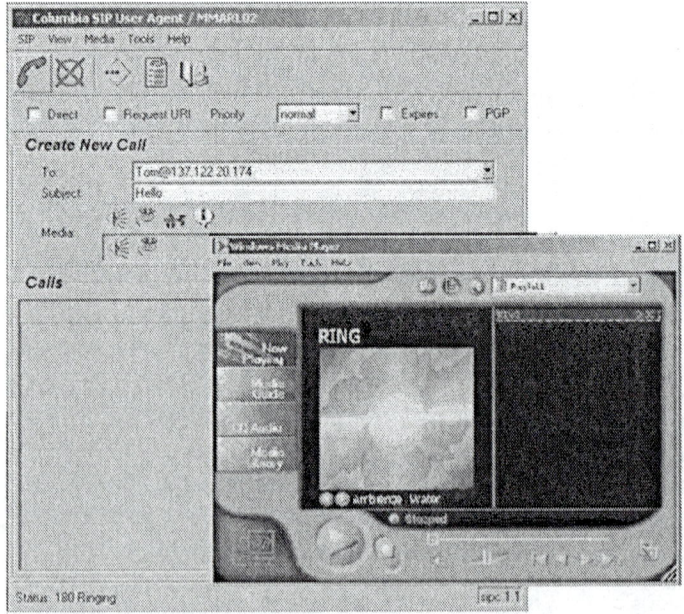

Figure 4.13   How a SIP call is initiated via the SIP user agent.

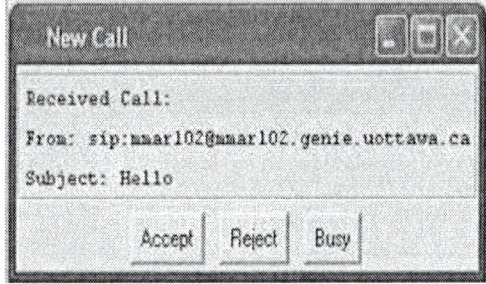

Figure 4.14   SIP INVITE interface.

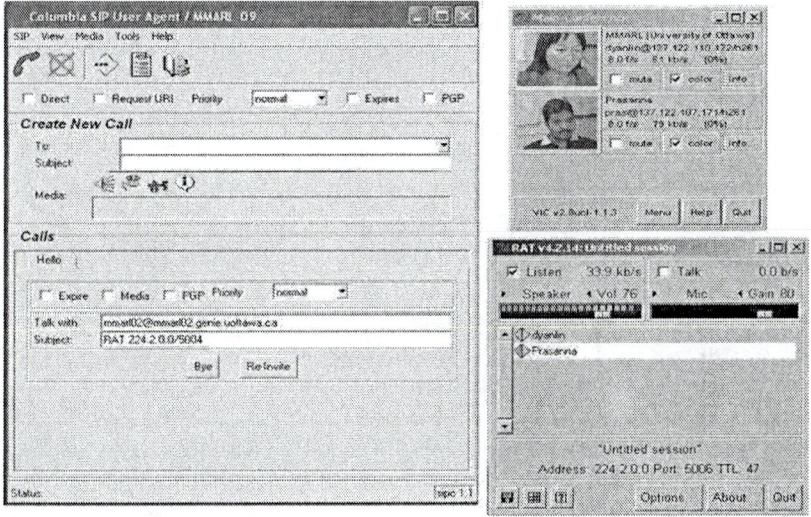

Figure 4.15   Users communicating in the main conference.

■ The mobile device user can then execute the requested service using the service GUI and forward necessary information to the proxy agent. Any data passed between the room manager and personal assistant is intercepted by the proxy agent. The intercepted data is then cached, forwarded, and adapted accordingly. The adaptation is performed by the adaptation agent integrated with the proxy agent.

■ The GUI listing the available services automatically gets updated whenever the status (present or absent) of a service changes.

- The GUI also lists the users currently available to join the public conference or initiate a private conference. Service requests are sent to the room manager, which in turn moves the GUI of the requested service to the appropriate user's device.
- A private conference (i.e., a sidebar association) is a conference controlled by the user. It is more like a scheduled conference in that users may define policies for sidebars, such as scheduled conference times, maximum number of participants allowed, sidebar media types, etc.

Conclusion

This chapter has provided an overview of the great potential of *ad hoc* networks for providing new communication models in mobile environments and enhancing the existing infrastructure-based communication protocols. In fact, mobile *ad hoc* networks are expected to become an important part of fourth-generation wireless communication networks [3], as they can be used to extend base station coverage and address current deficiencies of the infrastructure-based network. The increasing use of wireless devices with the emergence of potential mobile applications may further expand the use of mobile *ad hoc* networks in future pervasive computing environments. A list of such potential applications and a sample usage scenario of an *ad hoc* communications network have been provided in this chapter. Sensor networks, as components of *ad hoc* communications networks, can be utilized for military applications (e.g., to detect chemical or biological weapons), as environmental sensing networks, and as traffic sensors to monitor traffic congestion. Even though the concept of mobile *ad hoc* networks has been around for a while, many challenging issues remain to be solved, such as the power consumption of the devices and the proactive and reactive routing protocols required (as an *ad hoc* network frequently changes and multi-hop communication is required). The scope of research on *ad hoc* networks is too vast to be covered here; as a result, several aspects of *ad hoc* networks have not been discussed, including routing protocols, security, and interlayer interactions, to name a few.

References

[1] Basagni S., Myers, A.D., and Syrotiuk, V.R. Mobility-independent flooding for real-time, multimedia applications in *ad hoc* networks, in *Proc. of the 1999 IEEE Emerging Technologies Symp. on Wireless Communications and Systems*, Richardson, TX, April 12–13, 1999.

[2] Kaminsky, A., *Infrastructure for Distributed Applications in Ad Hoc Networks of Small Mobile Wireless Devices*, IT Lab Technical Report, May 22, 2001 (http://www.cs.rit.edu/~anhinga/publications/AnhingaPaper200105 22.pdf).

[3] Hoebeke, J., Moerman, I., Dhoedt, B., and Demeester, P., *An Overview of Mobile Ad Hoc Networking: Applications and Challenges*, Department of Information Technology, Ghent University, Belgium, 2004.

[4] Johnson, D.B. and Maltz, D.A., Dynamic source routing in *ad hoc* wireless networks, in *Mobile Computing*, Imielinski, T. and Korth, H., Eds., Kluwer Academic, Norwell, MA, 1996, pp. 153–181.

[5] Hughes, B. and Cahill, V., *Towards Real-Time Event-Based Communication in Mobile Ad Hoc Wireless Networks*, Tech. Report TCD-CS-2003-25, Distributed Systems Group, Department of Computer Science, Trinity College, Dublin, Ireland, 2003.

[6] Corson, S. and Macker, J., *Mobile Ad Hoc Networking (MANET): Routing Protocol Performance Issues and Evaluation Considerations* [memo], Request for Comments 2501, The Internet Society, Reston, VA, January, 1999 (http://www.ietf.org/rfc/rfc2501.txt).

[7] Kagal, L., Finin, T., and Joshi, A., A Policy Language for a Pervasive Computing Environment, in *Proc. of the IEEE 4th International Workshop on Policies for Distributed Systems and Networks (POLICY 2003)*, Lake Como, Italy, June 4–6, 2003, pp. 63–76.

[8] Session Initiation Protocol (SIP), Computer Science Department, Columbia University (http://www.cs.columbia.edu/sip/).

[9] Fielding, R., Gettys, J., Mogul, J., Frystyk, H., and Berners-Lee, T., *Hypertext Transfer Protocol (HTTP/1.1)* [memo], Request for Comments 2068, The Internet Society, Reston, VA, January, 1997 (http://www.ietf.org/rfc/rfc2616.txt).

[10] Handley, M. and Jacobson, V., *SDP: Session Description Protocol* [memo], Request for Comments 2327, The Internet Society, Reston, VA, April, 1998 (http://www.ietf.org/rfc/rfc2327.txt).

[11] Schulzrinne, H., Casner, S., Frederick, R., and Jacobson, V., *RTP: A Transport Protocol for Real-Time Applications* [memo], Request for Comments 1889, The Internet Society, Reston, VA, January, 1996 (http://www.freesoft.org/CIE/RFC/1889/index.htm).

[12] Poslad, S. and Calisti, M., Towards improved trust and security in FIPA agent platforms, in *Proc. of the Autonomous Agents 2000 Workshop on Deception, Fraud and Trust in Agent Societies (AGENTS 2000)*, Barcelona, Spain, June 3–7, 2000, pp. 87–90.

[13] Millier, M., Software agents, in *Proc. of the Conf. on Human Factors in Computing Systems (CHI 97)*, Atlanta, GA, April 18–23, 1997 (http://www.acm.org/sigs/sigchi/chi97/proceedings/tutorial/mm.htm).

[14] Nwana, H.S., Software agents: an overview, *Knowl. Eng. Rev.*, 11(3), 1–40, 1996.

[15] AgentLand, http://www.agentland.com.

[16] Foundation for Intelligent Physical Agents (FIPA), http://www.fipa.org.

[17] *CRUMPET (Creation of User-Friendly Mobile Services Personalised for Tourism)*, Information Society Technologies, London, U.K. (http://www.ist-crumpet.org).

[18] *FIPA-OS and MicroFIPA-OS Agent Platforms*, Emorphia, Ltd., Harlow, U.K. (http://fipa-os.sourceforge.net).

[19] Englmeier, K. and Mothe, J., Trustworthy personal assistance: a design objective for interactive agents, tech. meta-track, in *Proc. of the 2001 Americas Conference on Information Systems*, Boston, MA, August 3–5, 2001 (CD-ROM).

[20] Raatikainen, K. et al., Monads: adaptation agents for nomadic users, in *Proc. of World Telecom '99* (http://www.cs.helsinki.fi/research/monads).

[21] D'Agents Tutorials, Computer Science Department, Dartmouth College, Hanover, NH (http://agent.cs.dartmouth.edu/tutorials).

[22] Marwaha, S., Tham, C.K., and Srinavasan, D., Mobile agents based routing protocol for mobile *ad hoc* networks, in *Proc. of the 2002 IEEE Global Telecommunications Conference (GLOBECOM'02)*, Taipei, Taiwan, November 17–21, 2002.

[23] Shehory, O. et al., Agent cloning: an approach to agent mobility and resource allocation, *IEEE Comm.*, 36(7), 58–67, 1998.

[25] *Xerces Java Parser, v. 1.2.1*, The Apache XML Project (http://xml.apache.org/xerces-j).

[27] *PersonalJava Application Environment*, Sun Microsystems, Santa Clara, CA (http://java.sun.com/products/personaljava).

[28] Rosenberg, J. et al., *SIP: Session Initiation Protocol*, IETF Internet Draft, Internet Engineering Task Force (http://www.ietf.org/internet-drafts/draft-ietf-sip-rfc2543bis-09.pdf).

[29] Balakrishnan, D. and Karmouch, A., A personal assistant for mobile device users using agents, in *Proc. of the 22nd Biennial Symp. on Communications*, Kingston, Ontario, Canada, May 31–June 3, 2004, pp. 405–407.

[30] *Robust Audio Tool (RAT)*, University College, London (http://www-mice.cs.ucl.ac.uk/multimedia/software/rat).

[31] *Videoconferencing Tool (VIC)*, Network Research Group, Lawrence Berkeley National Laboratory, and University of California, Berkeley (http://www-mice.cs.ucl.ac.uk/multimedia/software/vic).

[32] Gerla, M. and Tsai, J.T., Multicluster, mobile, multimedia radio network, *J. Wireless Networks*, 1(3), 255–265, 1995.

[33] Plagemann, T., Goebel, V., Griwodz, C., and Halvorsen, P., Towards middleware services for mobile *ad hoc* network applications, in *Proc. of the Ninth IEEE Workshop on Future Trend of Distributed Computing Systems (FTDCS)* San Juan, Puerto Rico, May 27–30, 2003, pp. 249–255.

[34] El-Sayed, A. and Roca, V., A survey of proposals for an alternative group communication service, *IEEE Network*, 17(1), 46–51, 2003.

[35] Perkins, C.E. and Royer, E.M., *Ad hoc* on-demand distance vector routing, in *Proc. of the Second IEEE Workshop on Mobile Computing Systems and Applications*, New Orleans, LA, February 25–26, 1999, pp. 90–100.

[36] Lu, Y., Karmouch, A., Ahmed, M., and Impey, R., Agent-based service discovery in *ad hoc* networks, in *Proc. of the 22nd Biennial Symp. on Communications*, Queen's University, Kingston, Ontario, Canada, May 31–June 3, 2004.

Chapter 5

Infrastructure Versus *Ad Hoc* Wireless Networks: Mobility Issues and Solutions

Ling-Jyh Chen, Shirshanka Das,
Mario Gerla, and Alok Nandan

CONTENTS

Introduction

Over the past two decades we have witnessed a major shift from fixed to mobile phones all over the world. The number of mobile phones has surpassed that of fixed phones in most countries. This is no surprise, as telephony is an inherently mobile application. The same shift has been happening in personal computing platforms, from desktops to laptops and personal digital assistants (PDAs). The proliferation of mobile computing platforms has coincided with the emergence of mobile data communications, namely, wireless local area network (WLAN) 2.5G and 3G cellular data services. This phenomenon has caused a major paradigm shift in the way we access the Internet, from fixed to wireless and mobile. Already, wireless Internet access has exceeded wired access; in fact, one expects that within a few years the great majority of clients will be not only wireless and portable but also equipped with multiple wireless interfaces. Today, the majority of Internet applications are still stationary in nature; that is, we e-mail, browse the Web, download files, and play Internet games from our homes or offices. We exploit the wireless interface primarily to avoid cables; however, we are witnessing an emergence of truly mobile access scenarios (from cars or public transport vehicles or while walking in shopping malls). In parallel is the emergence of new mobile applications and services, such as location-based services, car navigation services, dynamic workgroups, pervasive computing, and interaction with the environment.

Clearly, several concurrent factors are favoring the mobilization of computing and communications, including the miniaturization of devices (which have become portable); the availability of long-life, light-load batteries; the availability of efficient wireless access media (cellular and wireless LANs, personal wireless media, and *ad hoc* networks); and the emergence of mobile, nomadic applications. As these mobile scenarios are emerging, another important layer of services must be implemented to make them possible: mobile middleware. Mobile middleware is essential for both mobilizing legacy infrastructure applications (e.g., e-mail, Web browsing) and enabling applications that are purely mobile (e.g., car navigation safety).

In this chapter, we focus on various wireless media access schemes and on the mobile middleware requirements of each. Three major wireless access and network techniques exist today: cellular, wireless (infrastructure) LAN, and *ad hoc* wireless networks. Because cellular and wireless LAN technologies are well understood and have been extensively covered in the open literature, here we focus mainly on the emerging *ad hoc* wireless and personal networks. In the process, we identify key mobile applications and discuss emerging mobile middleware functions required to support them.

In the following section, we study mobility management in last-hop wireless networks and discuss various handoff solutions that manage client movements. The next section provides a review of the MANET architecture and study its evolution from battlefield to commercial applications, which is followed by a discussion of scalable routing in presence of mobility. Then, we examine the need of P2P overlays in MANETs, with specific application to commercial scenarios. Finally, we study an emerging commercial application (file sharing in the vehicular network), and we propose P2P swarming middleware for this application.

Mobility Management in Last-Hop Wireless Networks: Horizontal Handoff and Mobile IP

We begin with an overview of mobile services required in last-hop wireless networks. The scenario of interest is potentially extremely broad. It includes cellular networks (e.g., GSM, CDMA, 2.5G, 3G); indoor wireless LANs (e.g., 802.11abg for basic Internet access; 802.11n for high-speed access; 802.11e for QoS-oriented multimedia access); outdoor, urban wireless access networks (e.g., 802.11s for urban Mesh Network access; 802.16 for urban high-speed distribution); and short-range, low-data-rate access networks (e.g., Bluetooth® and ZigBee™ for personal and pervasive access). All of these scenarios are examples of infrastructure-type networks. The physical environment is partitioned into cells. Each cell is controlled by an access point that acts as a gateway to the wired Internet. (Some of the schemes, such as 802.11s and 802.16, do have a wireless fixed backbone between the access point and the Internet gateway, but the following considerations still apply to these cases.) A user may move from one cell to another, either within the same technology (e.g., UMTS) or across technologies. In fact, a mobile client is often equipped with multiple radio interfaces and can roam across technologies. When this happens, the user must "re-register" with the new access point or base station. This registration may happen at one or more layers of the protocol stack. Often, it happens at the middleware layer, thus requiring mobile middleware services. This registration procedure goes under the name of *handoff*. The next section describes various handoff modes.

Handoffs

Handoff occurs when the user switches between different network access points. Handoff techniques have been well studied and deployed in the domain of cellular systems and are gaining a great deal of momentum in wireless computer networks as Internet Protocol (IP)-based wireless networking increases in popularity. Differing in the number of network interfaces involved during the process, handoff can be characterized as either *vertical* or *horizontal* [1], as depicted in Figure 5.1. A vertical handoff involves two different network interfaces, which usually represent different technologies; for example, when a mobile device moves out of an 802.11b network and into a 1xRTT network, a vertical handoff occurs. A horizontal handoff occurs between two network access points that use the same technology and device interface; for example, when a mobile device moves between 802.11b network domains, the handoff event would be considered horizontal because the connection is disrupted by the change of 802.11b domain (i.e., different frequency channel) but not by the change of wireless technology.

A handoff is *seamless* if it maintains the connectivity of all applications running on the mobile device. Seamless handoffs are designed to provide continuous end-to-end data service in the face of any link outages that might occur during switchover. Low latency and minimal packet loss are the two critical design goals. Low latency requires that the switch from one path to the other be completed almost instantaneously; service interruptions must be minimized.

Various seamless handoff techniques have been proposed [2–5]. These proposals can be classified into two categories: *network layer approaches* and *upper layer approaches*. Network layer approaches are based on IP

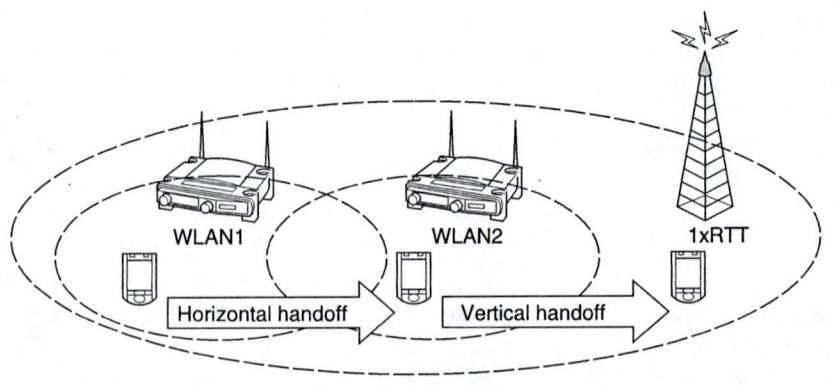

Figure 5.1   Horizontal and vertical handoff.

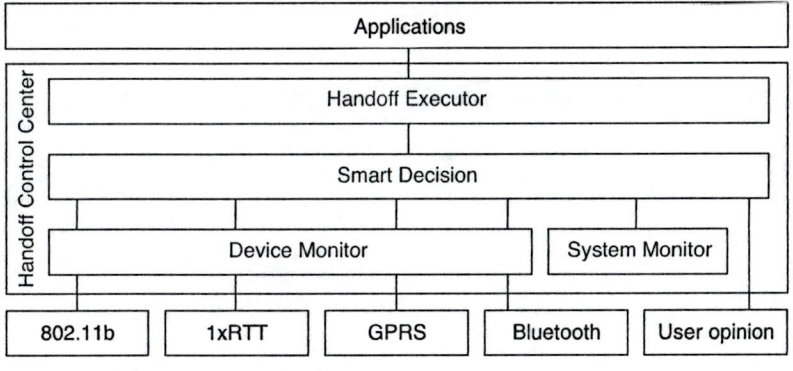

Figure 5.2 Smart Decision Model.

address "indirection" through a home agent and a foreign agent. They can be accomplished using IPv6 [6] or Mobile IPv4 [7] standards. These network layer approaches, however, are costly to implement. They require the deployment of several agents on the Internet for relaying or redirecting the data to the moving host (MH). For these reasons, the upper layer approaches are becoming increasingly popular. These approaches implement a session layer (in fact, a mobile middleware layer) above the transport layer that hides any connection changes at the underlying layers and makes them transparent to the application [8–12]. Some upper layer approaches implement mobility support at the transport layer, thus requiring the development of new transport layer protocols such as the Stream Control Transmission Protocol (SCTP) [13] and the Transmission Control Protocol/Multi-Home (TCP-MH) [14].

Smart Vertical Handoff

As mentioned earlier, several devices today have multiple radio interfaces. The opportunity then arises to select the best radio option, leading to a user-initiated (as opposed to infrastructure-driven) vertical handoff. Basically, smart handoff has all the ingredients of soft handoff; in addition, it includes mobile middleware software to select the best alternative. In this section, we present an example of smart handoff, the *Smart Decision Model* [15], to support flexible configuration in executing vertical handoffs. Figure 5.2 illustrates the Smart Decision Model. In this figure, a Handoff Control Center (HCC) provides the connection between the network interfaces and the upper-layer applications. The HCC is composed of four components: Device Monitor (DM), System Monitor (SM), Smart Decision

```
Smart Decision Process
Priority Phase:
    1.  Add all available interfaces into candidate
        list.
    2.  Remove user specified devices from the
        candidate list.
    3.  If candidate list is empty, add back removed
        devices from step 1.
    4.  Continue with Normal Phase.
Normal Phase:
    1.  Collect information on every wireless
        interface in the candidate list from the DM
        component.
    2.  Collect current system from SM component.
    3.  Use the score function to obtain the score of
        every wireless interface in the candidate list.
    4.  Handoff all current transmissions to the
        interface with the highest score if different
        from current device.
```

Figure 5.3 Algorithm for making smart decisions on HCC.

(SD), and Handoff Executor (HE). The DM is responsible for monitoring and reporting the status of each network interface (i.e., the signal strength, link capacity, and power consumption of each interface). The SM monitors and reports system information (e.g., current remaining battery). The SD integrates user preferences (obtained from user set default values) and all other available information provided by the DM and SM to make a "smart decision," or identify the best network interface to use at that moment. The HE then performs the device handoff if the current network interface is different from the best network interface. The HCC is implemented in the vertical handoff testbed to perform automatic handoffs to the best network interface. In our design, the SD has two phases: *priority* phase and *normal* phase. The SD algorithm is illustrated in Figure 5.3.

Priority and normal phases are necessary in the SD to accommodate user-specific preferences regarding the usage of network interfaces; for example, a user may decide not to use a device when the device could cause undesirable interferences with other devices (e.g., 802.11b and 2.4GHz cordless phones). With priority and normal phases in place, the SD module provides flexibility in controlling the desired network interface to the user. Additionally, the SD deploys a *score function* to calculate a score for every wireless interface; the handoff target device is the network interface with the highest score. More specifically, we must consider k factors in calculating the score. The final score of interface i will be the sum of k weighted functions. The *score function* used is the following:

$$S_i = \sum_{j=1}^{k} w_j f_{j,i} \quad 0 < S_i < 1, \sum_{j=1}^{k} w_j = 1 \tag{5.1}$$

In the equation, w_j stands for the weight of factor k, and $f_{j,i}$ represents the normalized score of interface i of factor j. The best target connection interface at any given moment, then, is the one that achieves the highest score among all candidate interfaces. We further break down the score function into three components, accounting for usage expense (E), link capacity (C), and power consumption (P); therefore, Equation 5.1 becomes:

$$S_i = w_e f_{e,i} + w_c f_{c,i} + w_p f_{p,i} \tag{5.2}$$

Additionally, a corresponding function exists for each term $f_{e,i}$, $f_{c,i}$, and $f_{p,i}$, and the ranges of the functions are bounded from 0 to 1. The functions are illustrated below:

$$f_{e,i} = \frac{1}{e^{\alpha_i}}, f_{c,i} = \frac{e^{\beta_i}}{e^M}, f_{p,i} = \frac{1}{e^{\gamma_i}}, \quad \text{where } \alpha_i \geq 0, M \geq \beta_i \geq \gamma_i \geq 0 \tag{5.3}$$

The coefficients α_i, β_i, and γ_i can be obtained via a lookup table or a well-tuned function. In Equation 5.3, we used the inversed exponential equation for $f_{e,i}$ and $f_{p,i}$ to bound the result to between 0 and 1 (i.e., these functions are normalized) and properly model user preferences. For $f_{c,i}$, a new term (M) is introduced as the denominator to normalize the function, where M is the maximum bandwidth requirement demanded by the user. When not specified by the user, the default value of M is defined as the maximum link capacity among all available interfaces. Note that the properties of bandwidth and those of usage cost and power consumption are opposite (i.e., the more bandwidth the better, whereas lower cost and power consumption are preferred).

Managing Server Rate and Content During Handoff: CapProbe

So far we have considered the client side of the handoff; that is, the client selects the best option. Suppose now that the client is a thin client — say, a smart phone. It is receiving a soccer game video stream from the server. The client moves from indoor 802.11 at 5 Mbps to outdoor 1xRTT at 100 Kbps. A smooth handoff guarantees that the connection is maintained, but it is obvious that chaos will result unless the server (or

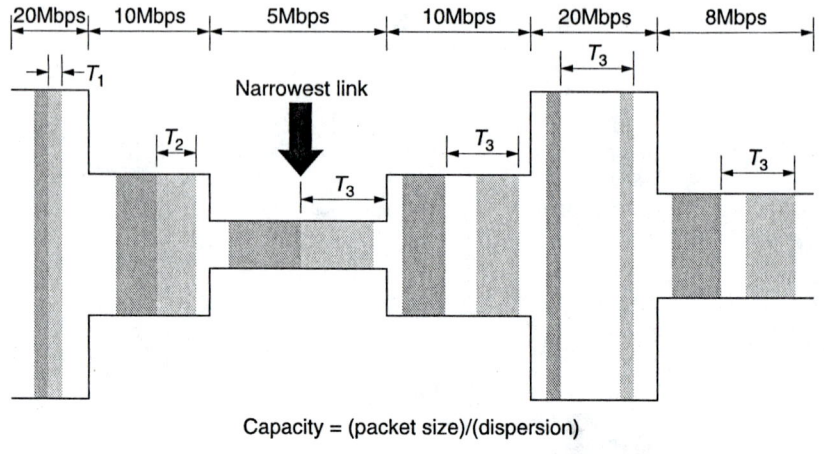

Capacity = (packet size)/(dispersion)

Figure 5.4   CapProbe: a simple and fast capacity estimation tool.

the transcoding-capable proxy that caches the server stream) detects the change in client Internet access capacity and adjusts its rate and content accordingly (from full-motion video to highly compressed MPEG4 video or even still frames).

A new mobile middleware software (server adaptation middleware) is required to make the server (or proxy) immediately aware of client changes and to select the best server delivery strategy. We have recently implemented such a middleware service using a basic capacity-estimation technique known as *CapProbe* [17]. CapProbe is a packet-pair method that measures the capacity of the narrow link on the path (in our case, invariably the last wireless hop) with extreme speed (order of seconds). The concept is illustrated in Figure 5.4. A packet pair is launched by the source. The packets get separated along the path due to varying link capacities. The ratio of packet size to the inter-packet interval at the destination yields the narrow link capacity. Details on the actual CapProbe tool can be found in Chen et al. [16]. Referring to Figure 5.5, we see an Internet path ending with an 802.11 link. This was the actual setup of an experiment carried out at UCLA in the Network Research Lab [18]. The 802.11 client is exposed to interference from a Bluetooth user operating in the same frequency. Interference notwithstanding, CapProbe manages to evaluate the exact capacity of the 802.11 channel.

The server mobile middleware embeds periodic packet pairs in the multimedia stream (by transmitting some of the video packets back to back) and is constantly informed (by client feedback) of the last-hop capacity. It can then dynamically adjust the rate and content to client

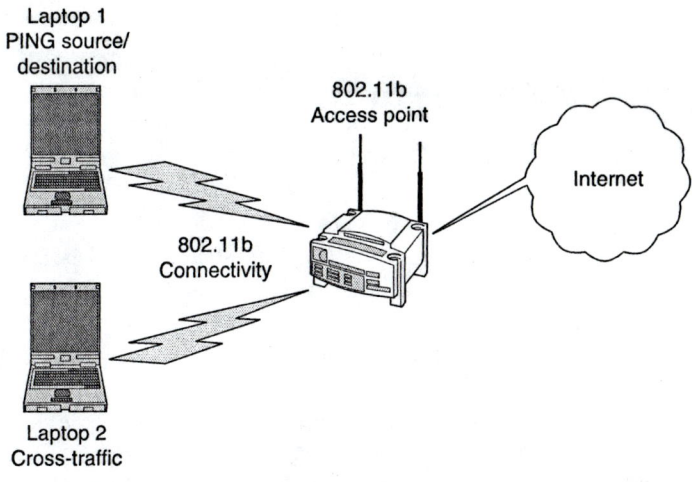

Figure 5.5   CapProbe testbed with last-hop wireless link.

capacity. Using the same principle of client feedback, the server middleware can also adjust to changes in device form and type (e.g., the user switches from a laptop to smart phone when stepping out of a car, yet maintains UMTS connectivity).

The Mobile *Ad Hoc* Network (MANET)

A wireless mobile *ad hoc* network (MANET) is a network established for a special, often extemporaneous, service customized to particular applications. An *ad hoc* network is typically set up for a limited period of time, in an environment that may change from application to application. As opposed to the Internet, where the TCP/IP protocol suite supports a vast range of applications, in a MANET the protocols are tuned to a specific customer and application (e.g., send a video stream across a battlefield, detect a fire in the forest, establish a videoconference among several teams engaged in a rescue effort). The customers move and the environments may change dynamically and unpredictably. For the MANET to retain its efficiency, the *ad hoc* protocols at various layers may have to self-tune to adjust to environment, traffic, and mission changes. From these properties emerges a vision of the MANET as being an extremely flexible, malleable, yet robust and formidable network architecture. Indeed, it is an architecture that can be deployed to monitor the habits of birds in their natural habitat, organized to coordinate rescue crews after a tsunami disaster, or structured to launch deadly attacks onto unsuspecting enemies.

MANETs are set apart from conventional wired or wireless infrastructure type networks by a number of unique attributes and requirements. Perhaps the two most critical attributes are self-configurability and mobility. A third important requirement (which is critically impacted by the first two) is scalability. A review of these attributes follows:

- *Self-configurability* — The MANET is deployed and managed independently of any preexisting infrastructure. This is the most important prerequisite to qualify a wireless network as *ad hoc*; consequently, the network must autonomously determine its own configuration parameters including addressing, routing, clustering, position identification, and power control. In some large networks, special nodes (e.g., mobile backbone nodes) coordinate their position and motion to provide coverage of disconnected islands. In this way, an infrastructure can be created within the *ad hoc* network itself.

- *Mobility* — The fact that nodes move is probably the most important attribute of MANETs. Mobility differentiates MANETs from their close cousins, the sensor networks. Mobility dictates network- and application-level protocols. For example, rapid deployment in unexplored areas with no infrastructure may require that some of the nodes form scouting teams or swarms. These, in turn, coordinate among themselves to create a task force or a mission. Mobility may be in some cases a challenge for the designer and may become part of the solution in other cases. We can have several types of mobility models: individual random mobility, group mobility, motion along preplanned routes, etc. The mobility model can have a major impact on the selection of a routing scheme and thus can influence performance.

- *Scalability* — In both military and civilian applications (e.g., large battlefield deployments, urban vehicle grids) the *ad hoc* network can grow to several thousands of nodes. For wireless infrastructure-type networks (e.g., urban mesh networks), scalability is simply handled by a hierarchical construction. Mobility appears to be the discriminator between easy and difficult scaling. A hierarchical model is very scalable in static networks (as demonstrated by the Internet). Limited mobility in an infrastructure can be easily handled using Mobile IP or other handoff and redirection techniques. Pure *ad hoc* networks, due to their self-configuring nature and consequent unrestricted mobility, do not tolerate a classic hierarchy structure and Mobile IP approach. Thus, mobility on a large scale is one of the most critical challenges in *ad hoc* designs.

The Evolution of MANETs: From Battlefield to Campus Networks and Urban Grids

In the early 1970s, MANETs were born on the heels of the success of the Advanced Research Projects Agency (ARPANet), when the Defense Advanced Research Projects Agency (DARPA) recognized the strategic importance of the packet-switching technology in the automated battlefield. Since then, the military has been the major sponsor of MANET research and development in industry and academia. A few years ago, the National Science Foundation (NSF) also joined in the support of MANET research to explore the transfer of this technology to civilian and possibly commercial applications. Support of MANETs by the industry, however, has been minimal (as compared to other areas of networking), in part due to the fact that commercial applications have been very slow in materializing. Because of the source of the funding, it is no surprise that most of the MANET problems addressed today by researchers are directed toward large-scale, specialized scenarios, such as battlefields, civilian defense, and disaster recovery. These are typically self-configured networks, totally decoupled from any commercial network infrastructure. One may say that even the network scenarios addressed by the MANET Internet Engineering Task Force (IETF) working group are better fit to military and civilian disaster recovery applications than to commercial ones.

Recent new technology developments might give rise to new alternatives in the MANET area and help the transition to commercial MANET applications. The first emerging technology is the personal area network (PAN), spearheaded by Bluetooth (802.15.1) and by the recently introduced ZigBee and 802.15.4 standards. It would make sense to interconnect a few Bluetooth piconets in a small-scale MANET (called a *scatternet*) to facilitate work-group communications (e.g., the exchange of business cards, files, and images) and to have a more efficient connection to the Internet (e.g., 802.11, UMTS). The second technology is the wireless LAN (802.11). The 802.11 technology and its derivatives dominate in the home, in university and industrial campuses, in public areas (e.g., malls, airport lounges, coffee shops), and in urban mesh networks. The single-hop wireless LAN, however, has range limitations. Two- or three-hop MANETs can be used to opportunistically extend the range of the wireless LAN. The third technology is digital short-range communications (DSRC). This technology addresses car- to-car and car-to-Internet communications for navigation safety purposes. The DSRC technology will pave the way for the "urban communications grid" concept, where car-to-car communications between any two vehicles will be made possible in a MANET, without using the fixed Internet. Although navigation safety is the top DSRC priority, the urban grid will eventually enable the support of a broad range of new mobile applications.

This brief overview shows that many different MANET scenarios are possible, each enabling different types of peer-to-peer (P2P) and overlay applications. In the next section, we present some emerging P2P examples for different MANET scenarios.

Handling Large Scale and Mobility in the Battlefield

Future battlefield operations will be characterized by the massive deployment of autonomous agents such as unmanned ground vehicles (UGVs) and unmanned air vehicles (UAVs). These autonomous agents will be sent to the front lines for intelligence, surveillance, strikes, enemy anti-aircraft suppression, damage assessment, search and rescue, and other tactical operations. These agents will interact with and support ground and airborne manned assets (e.g., tank battalions and jet fighter and helicopter squadrons). They will also communicate with ground sensors. One can easily imagine how this scenario could involve thousands of mobile nodes (some manned, some unmanned) and several more thousands of smaller, fixed nodes. Similar large-scale mobile networks could be formed to facilitate recovery from extensive civilian disasters, such as earthquakes, tsunamis, chemical spills, and urban terrorist attacks.

A critical problem in *ad hoc* networks is routing. If the *ad hoc* network is stationary, then hierarchical routing proves to be a very scalable solution. When the network is mobile, however, the hierarchical routing solution introduces excessive overhead because the hierarchical addresses must be continuously updated to reflect the dynamically changing topology. Mobility causes problems also with other protocol layers besides routing (e.g., Media Access Control [MAC] layer or TCP). In particular, one of the major challenges in *ad hoc* TCP design is dealing with path disruptions caused by mobility. In large-scale routing, however, mobility can also be an asset, in that it can be exploited to improve performance. In this section, we show that group mobility can be harnessed via *landmarking* to lead to more scalable routing. Moreover, if mobile backbone nodes are deployed in the *ad hoc* network, connectivity can be enhanced.

Landmark Routing for Group Mobility

Typically, when wireless network size and mobility increase (beyond certain thresholds), current flat proactive routing schemes (i.e., distance vector and link state) become altogether unfeasible because of line and processing overhead. Chen and Gerla [21] introduced a novel table-driven routing protocol for wireless *ad hoc* networks named Landmark *Ad Hoc* Routing (LANMAR), which combines the features of Fisheye State Routing (FSR) [19] and landmark routing [21,22]. The novel idea behind LANMAR is the notion of keeping track of logical subnets in which the members have a commonality of interests and are likely to move as a group (e.g., brigade in the battlefield, colleagues in the same organization, a group of students from the same class). Moreover, a landmark node is elected in each subnet. LANMAR improves scalability by reducing the routing table size and update overhead. More precisely, it resolves the routing table scalability problem by using an approach similar to the landmark hierarchical routing proposed for wired networks [21,22]. In the original landmark scheme, the hierarchical address of each node reflects its position within the hierarchy and helps find a route to it. Each node has full knowledge of all the nodes within the immediate vicinity. At the same time, each node keeps track of the next hop on the shortest path to various landmarks at different hierarchical levels. Routing is consistent with the landmark hierarchy, and the path is gradually refined from a top-level hierarchy to low levels as a packet approaches its destination.

Kleinrock and Stevens [20] applied the wired network landmark concept to FSR to reduce the routing update overhead for nodes that are far away. Each logical subnet has one node serving as a landmark. Beyond the fisheye scope, the update frequency of the landmark nodes remains unaltered, but the update frequency of the regular nodes is reduced to zero. As a result, each node maintains accurate routing information about the immediate neighborhood and as well as the landmark nodes. When a node must relay a packet, if the destination is within its neighbor's scope (as indicated in the routing table), the packet will be forwarded directly; otherwise, the packet will be routed toward the landmark corresponding to the destination logical subnet. The packet does not have to go all the way to the landmark; instead, when the packet moves within the scope of the destination, it is routed to it directly.

At the beginning of the execution, no landmark exists. The LANMAR protocol uses only the FSR functionality. As the FSR computation progresses, one of the nodes will learn (from the FSR table) that more than a certain number of group members (say, N) are in the FSR scope. It then proclaims itself as a landmark for this group. The landmark information will be broadcast to the neighbors jointly with the topology update packets. In case of a tie, the lowest ID breaks the tie. The competing nodes defer. When a landmark dies, its neighbors will detect the silence after a given timeout. A new round of landmark election then begins anew for the group in question.

In conclusion, LANMAR is an excellent example of a routing protocol that exploits group mobility by "summarizing" routes and reducing table storage and line overhead. Simulation results have shown that a LANMAR-empowered network can easily scale to thousands of nodes.

MANETs and P2P Mobile Middleware

In the wired Internet, a P2P network is basically an overlay network justified by the need for specialized functions that are not possible (or not cost effective) in the IP layer. These functions must be performed at the middleware or application layer. Classic examples of Internet overlay networks include real-time multicast overlays, which overcome the lack of multicast support in the IP routers, and P2P distributed index systems such as Gnutella, BitTorrent, and Pastry. These P2P indices are typically implemented as overlays that permit efficient content routing based on distributed hash tables (DHTs), for example. Content-based routing is not possible in the IP layer.

MANETs give rise to an even stronger need for P2P overlays for the following reasons: (1) mobile *ad hoc* applications require sophisticated routing functions (e.g., location awareness, content addressing) that are well beyond what is available from standard routing protocols, and (2) the unpredictability of the radio channel combined with user mobility poses major challenges to routing and to connectivity. The preferred strategy to overcome these problems is to implement customization functions in the upper layers and P2P networking overlays while keeping the basic routing and transport protocols simple.

As an example of a MANET overlay, consider a delay-tolerant file-sharing application that includes hosts partly on the Internet and partly on *ad hoc* "opportunistic" network extensions. Wireless nomadic users can rapidly change their connectivity to the Internet from Kbps (e.g., GPRS) to Mbps (e.g., 802.11). Temporarily, the users may also become disconnected. The use of the standard network routing protocols may lead to inefficiencies, violation of delay constraints, and possibly retransmission of large portions of the file. A P2P overlay network instead can keep track of connectivity among the various hosts. The overlay network can extend to wired, wireless, and *ad hoc* network segments. It can predict disconnection and reconnection dynamics and can exploit them to deliver files efficiently and within constraints (by using, for example, intermediate proxy nodes for bundle storing and forwarding).

Another promising environment for the emergence of opportunistic *ad hoc* networking and P2P mobile middleware is the vehicle communication grid. Future cars will come equipped with radios (for safe navigation) and with plenty of on-board storage and processing power. Car-to-car communications will be enabled by a standard architecture derived from DSRC and promoted by the Institute of Electrical and Electronics Engineers (IEEE) and the Department of Transportation. Most importantly, cars will have a captive audience — the passengers — with plenty of time to burn. In the following section, we describe CarTorrent, a hypothetical application for the vehicular grid. CarTorrent, inspired by the Internet-based BitTorrent distributed file-sharing system, allows cars to partially download multimedia files from highway WiFi access points and to cooperatively complete file assembly using a unique P2P mobile middleware solution.

CarTorrent: Mobile Middleware for Vehicle Networks

CarTorrent is a cooperative strategy for content delivery and sharing in future vehicular networks [23]. CarTorrent represents an interesting example of mobile middleware in a scenario that oscillates spatially from being infrastructure supported to being completely infrastructure independent. CarTorrent targets the problem of downloading files to a moving car from the Internet. CarTorrent aims to utilize efficiently the unused bandwidth between hot spots on the freeway. Without it, cars would have to park at a hot spot (kiosk) and wait to get served. This section discusses the issues involved in using such a strategy from the standpoint of vehicular *ad hoc* networks (VANETs). VANET applications will include onboard active safety systems leveraging vehicle–vehicle or roadside–vehicle networking. These systems may assist drivers in avoiding collisions. Non-safety applications include real-time traffic congestion and routing information, high-speed tolling, mobile "infotainment," and content delivery (as discussed here), among many others.

Content Delivery Techniques for Vehicular Networks

Future vehicular networks are expected to deploy short-range communication technology for inter-vehicle communications. In addition to vehicle–vehicle communication, users will be able to access the multimedia-rich Internet from within the vehicular network. Kids sitting in the back seat of the car could play online games with their friends sitting at home, while Mom in the front seat might want to check out www.cnn.com and www.sigalert.com for the latest breaking news and traffic alerts on all the major freeways. Within a limited radius, access to the Internet would be in the form of infostations or WiFi hot spots. We are focusing here on the content-delivery application, where Internet content must be delivered to the user (upon request) within a certain time constraint.

Content can be obtained directly from the hot spot, but also from peers. Referring to the latter mode, *swarming* is a peer-to-peer content delivery mechanism that utilizes parallel download among a mesh of cooperating peers. Scalability is achieved because the system capacity increases with the number of peers participating in the system. The primary purpose of the protocols is twofold: First, from the conventional *server perspective*, reduce the load of the origin server or the content publisher. Second, from the *client perspective*, reduce the download time.

On the Internet, the above file content download and sharing procedure is embodied in BitTorrent, a popular file-sharing tool that accounts for a significant proportion of Internet traffic. BitTorrent is a swarming P2P file-sharing solution. Simply put, BitTorrent allows a single source to disseminate a single file to many users by having each user share what they just downloaded. It can be used to share any type of file of nearly any size, with minimal bandwidth investment by the original distributors.

BitTorrent requires a few things to run: a client, a torrent, and a tracker. The client opens a .torrent file, chooses a location to save the file, and connects to the tracker. The tracker keeps track of how much each user is downloading and uploading and what parts they have and gives information to the client about where to get the next piece of the file. Note that BitTorrent downloads are in a primarily random order, although it prefers to get pieces that the fewest people have so even if no one person has the entire file then every piece will be available. Also, the tracker watches the users' "karma." A user's download speed is tied to his upload speed, so a user who is not uploading much is likely to have a low download speed. Thus, BitTorrent builds its overlays by randomly selecting peers, a fact that can be potentially wasteful in a MANET environment.

A Swarming Protocol for Vehicular Networks

Consider a VANET with short-range communication technology. Given an average speed of 50 miles per hour and a gateway radio range of 500 meters, a simple calculation gives a car a transmission window to and from a fixed Internet access point on the order of a minute at the most. Taking into account competition from other cars, it is possible that the available bandwidth is not sufficient to allow each user to download e-mail and songs, as well as browse multimedia-rich Web sites in the short time they are connected to the gateway. Another practical issue is that, on inter-city highways, the gateways would be hosted by gas stations and food concessions and thus would be located farther apart — say, every 5 to 10 miles. Thus, the vehicle would be connected for about a minute to the Internet before being disconnected for around 5 minutes. As we shall see, the high mobility of nodes in VANETs coupled with the intermittent connectivity to the Internet provides an incentive for individual nodes to *cooperate* while accessing the Internet to achieve some level of seamless connectivity.

For these reasons, an interesting problem is the design of cooperative protocols to improve client-perceived performance of the vehicular network as a whole. The key contribution of CarTorrent is the development of P2P mobile middleware that includes the following features:

- A gossip mechanism to propagate content availability information
- A proximity-driven content selection and delivery strategy
- Leveraging the broadcast nature of wireless networks to reduce redundant message transmission

Before presenting the protocol, we define the network model and introduce some definitions. The network consists of a set of N nodes with the same computation and transmission capabilities, communicating through bidirectional wireless links between each other. This is the infrastructureless *ad hoc* mode of operation. Wireless gateways at regular intervals provide access to the rest of the Internet using infrastructure support (either wired or multi-hop wireless). The data unit for the swarming protocol is a *chunk*; that is, the content is broken up into equal-sized chunks, each with its own unique identity. These chunks are shared and transferred among the peers. We assume that each node is reachable from every other node.

CarTorrent has the same generic structure of any swarming protocol. Peers downloading a file form a mesh and exchange pieces of the file among themselves. However, the wireless setting of VANETs, characterized by limited capacity, intermittent connectivity, and a high degree of churn in nodes (cars), requires adaptation in specific ways. Figure 5.6 illustrates the basic operation of the CarTorrent protocol. Components of the CarTorrent protocol include *peer discovery, peer and content selection*, and *content discovery and selection*. For the sake of brevity, we provide here just a simple, intuitive version of the protocol; readers should refer to Nandan et al. [23] for a more detailed discussion of the protocol and of the various options.

When a new car enters the vehicular network (e.g., enters a freeway or a section of freeway with access points), it requests the gateway for the particular file. If the gateway has the file in its cache, it begins uploading a chunk to the node. The node begins downloading chunks from the gateway while it is in range. The gateway also bootstraps it with a list of the last-known peers (cars) requesting the same file and when. Thus, the car has an idea of how popular the file is and how likely it is to benefit from cooperative strategies.

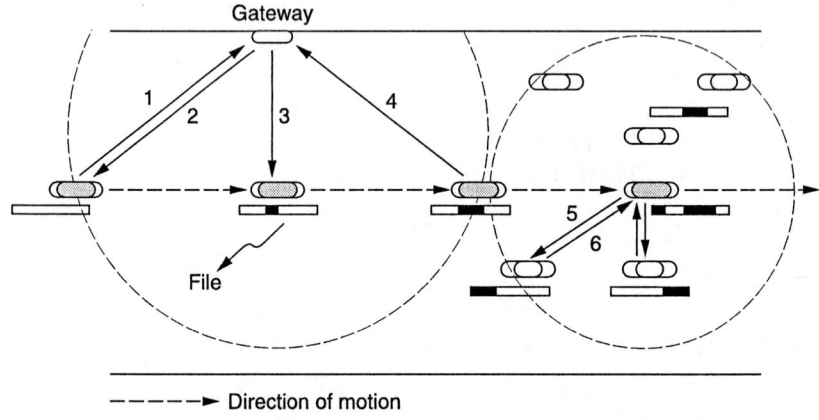

Figure 5.6 The basic operation of the CarTorrent protocol: A node (car) enters the radio range of a gateway (1), initiates the connection (2), and begins downloading (3) pieces of the file. When it goes out of range (4), it begins to gossip (5) and discovers other peers (nodes) with same content and exchanges pieces of the file (6).

Peers generate gossip messages from time to time to advertise their presence and current content. A naïve gossiping scheme has the potential of generating a large number of gossip messages in addition to being subject to the problem of messages ping-ponging when two peers keep exchanging stale data. This scenario uses a gossip scheme utilizing methods that minimize redundant forwarding, such as minimum connected set forwarding, passive clustering, or multipoint relay. Only the essential set of neighbors forwards the data/control packet for a specified number of hops. Forwarding nodes detect and suppress duplicates.

In the simplified swarming protocol, the newcomer (e.g., node A) forwards upstream (in the direction of traffic) a gossip control packet with the list of chunks it requires. Selected intermediate nodes (the forwarding nodes for this file) turn on the forwarding flag (e.g., according to the passive clustering scheme). The nearest peer (e.g., node B), a few hops away, upon receiving the gossip packet will respond with the first requested chunk. It also piggybacks its own current list. The forwarding nodes broadcast the chunk, which is thus propagated back to node A. When node A receives the first chunk it requested, it responds by transmitting in turn the first chunk that node B requested (if any) and so on until node B has received all the chunks it can possibly get from node A. Basically, this is a send-and-wait protocol between nodes A and B that is concluded when node B has received all it required from node A. From this point, the transfer is simply downstream, from node B to node A, until nodes A and B have the same content. Typically, if the file is popular and the peer population dense, the transfer will be mostly downstream, from node B to node A. The reader will appreciate the fact that this swarming scheme requires chunk transmissions only between neighboring peers; thus, the download overhead is independent of network size and peer population, and the scheme scales to any network size. Because the basic scheme employs User Datagram Protocol (UDP) transport and broadcast MAC, a potential congestion is a concern. To avoid congestion, rate control can be used. We refer the interested reader to the details in Nandan et al. [23].

The Future of VANETs

Research on vehicular networks has made tremendous strides over the past decade. Prominent players such as BMW, Daimler–Chrysler, and Toyota are looking very closely at this area to determine the appropriate mix of ingredients to make life easier for the driver without reducing control or sacrificing privacy. Infotainment within the vehicle is one of those gray areas where it is difficult to determine when entertainment becomes distraction. We envision the day when a driver is zipping down the highway, listening to his favorite radio station, and hears a really good song. He hits the download button on his player and, when passing a gateway, initiates a CarTorrent download of the file. After crossing the gateway, the player begins gossiping with neighboring cars advertising the driver's interest in the file. Other cars are advertising some of the pieces and the player begins downloading pieces from them. In about 5 to 10 minutes, all the pieces of the file have been assembled by downloading through the gateway and exchanging pieces with the neighboring cars. From then on, the driver can keep playing that song until he gets it out of his head. Until that day, research on vehicular networks will continue to strive toward getting information to cars in a better and faster way.

Conclusions

In this chapter, we have reviewed two types of wireless networks (infrastructure and *ad hoc*) and have evaluated the impact of mobility. The two systems indeed present very different mobility models and problems. For the infrastructure, the key issue is handoff; we have examined the model of the nomadic client that can connect to the infrastructure with multiple wireless interfaces (e.g., GPRS, UMTS, 802.11) and must select the most convenient one to switch to. For the *ad hoc* environment, one of the key issues is the design of routing algorithms that can scale and are also robust to mobility. We identified two different *ad hoc* scenarios and studied the routing problems associated with each. First, we focused on the large-scale automated battlefield scenario, where mobile middleware allows recognition and exploitation of group motion, creating a robust hierarchical routing solution based on landmarks. Then, we shifted our attention to commercial applications and studied the vehicular network scenario by examining the file-sharing application CarTorrent. We found that even in this case the routing solution is highly dependent on coordinated car motion. Here, again, mobile middleware is required to build a routing overlay that supports swarming among cars. In summary, mobility impacts last-hop wireless (i.e., infrastructure) applications in different ways than *ad hoc* networks. In both cases, however, mobile middleware is required to efficiently manage mobility.

References

[1] Stemm, M. and Katz, R.H., Vertical handoffs in wireless overlay networks, *Mobile Networks Appl.*, 3(4), 335–350, 1998.

[2] Dommety, G. et al., *Fast Handovers for Mobile IPv6*, Internet Engineering Task Force (IETF) Internet Draft, March, 2002 (http://www3.ietf.org/proceedings/02jul/I-D/draft-ietf-mobileip-fast-mipv6-04.txt).

[3] Hsieh, R., Zhou, Z.G., and Seneviratne, A., S-MIP: a seamless handoff architecture for Mobile IP, in *Proc. of IEEE INFOCOM 2003*, San Francisco, CA, April 1–3, 2003.

[4] Johnson, D.B., Perkins, C., and Arkko, J., *Mobility Support in IPv6*, Internet Engineering Task Force (IETF) Internet Draft, May, 2002 (http://tools.ietf.org/wg/mip6/draft-ietf-mobileip-ipv6/draft-ietf-mobileip-ipv6-17.txt).

[5] El Malki, K. et al., *Low Latency Handoffs in Mobile IPv4*, draft-ietf-monileip-lowlatency-handoffs-v4-03.txt, Internet Engineering Task Force (IETF) Internet draft, November, 2001.

[6] Deering, S. and Hinden, R., *Internet Protocol, Version 6 (IPv6)* [memo], Request for Comments 2460, The Internet Society, Reston, VA, December, 1998 (http://www.ietf.org/rfc/rfc2460.txt).

[7] Perkins, C., Ed., *IP Mobility Support for IPv4* [memo], Request for Comments 3344, The Internet Society, Reston, VA, August, 2002 (http://mailman.rfc-editor.org/pipermail/rfc-dist/2002-September/000078.html).

[8] Ghini, V., Pau, G., Salomoni, P., Roccetti, M., and Gerla, M., Smart download on the go: a wireless internet application for music distribution over heterogeneous networks, in *Proc. of the IEEE Int. Conf. on Communications (ICC 2004)*, Paris, June 20–24, 2004.

[9] Handley, M., Schulzrinne, H., Schooler, E., and Rosenberg, J., *SIP: Session Initiation Protocol* [memo], Request for Comments 2543, The Internet Society, Reston, VA, March, 1999 (http://www.ietf.org/rfc/rfc2543.txt).

[10] Maltz, D. and Bhagwat, P., MSOCKS: an architecture for transport layer mobility, in *Proc. of IEEE INFOCOM 1998*, San Francisco, CA, March 29–April 2, 1998, pp. 1037–1045.

[11] Schlaeger, M., Rathke, B., Bodenstein, S., and Wolisz, A., Advocating a remote socket architecture for Internet access using wireless LANs, *Mobile Networks Appl.*, 6(1), 23–42, 2001.

[12] Snnoeren, A.C., A Session-Based Approach to Internet Mobility, Ph.D. thesis, Massachusetts Institute of Technology, Cambridge, MA, 2002.

[13] Stewart, R. et al., *Stream Control Transmission Protocol* [memo], Request for Comments 2960, The Internet Society, Reston, VA, October, 2000 (http://www.ietf.org/rfc/rfc2960.txt).

[14] Matsumoto, A., Kozuka, M., Fujikawa, K., and Okabe, Y., *TCP Multi-Home Options*, Internet Engineering Task Force (IETF) Internet Draft, October, 2003 (http://www.potaroo.net/ietf/all-ids/draft-arifumi-tcp-mh-00.txt).

[15] Chen, L.-J., Sun, T., Chen, B., Rajendran, V., and Gerla, M., A smart decision model for vertical handoff, in *Proc. of the 4th ANWIRE International Workshop on Wireless Internet and Reconfigurability (ANWIRE 2004)*, Athens, Greece, May 14, 2004.

[16] Chen, L.-J., Sun, T., Yang, G., Sanadidi, M.Y., and Gerla, M., *Ad Hoc Probe: Path Capacity Probing in Wireless Ad Hoc Networks*, Technical Report TR050005, Computer Science Department, University of California, Los Angeles, 2005.

[17] Kapoor, R., Chen, L.-J., Lao, L., Gerla, M., and Sanadidi, M.Y., CapProbe: A simple and accurate capacity estimation technique, in *Proc. of ACM Special Interest Group on Data Communication (SIGCOMM 2004)*, Portland, OR, August 30–September 3, 2004.

[18] University of California, Los Angeles, Network Research Laboratory, http://www.cs.ucla.edu/NRL/.

[19] Pei, G., Gerla, M., and Chen, T.-W., Fisheye state routing in mobile *ad hoc* networks, in *Proc. of the 20th Int. Conf. on Distributed Computing Systems (ICDCS) Workshop on Wireless Networks and Mobile Computing (WWNMC 2000)*, Taipei, Taiwan, April 10–13, 2000.

[20] Kleinrock, L. and Stevens, K., *Fisheye: A Lenslike Computer Display Transformation*, Technical Report, Computer Science Department, University of California, Los Angeles, 1971.

[21] Chen, T.-W. and Gerla, M., Global state routing: a new routing scheme for *ad hoc* wireless networks, in *Proc. of IEEE Int. Conf. on Communications (ICC 98)*, Atlanta, GA, June 7–11, 1998, pp. 171–175.

[22] Xu, K., Hong, X., and Gerla, M., Landmark routing in *ad hoc* networks with mobile backbones, *J. Parallel Distributed Comput.*, February (special issue), 110–123, 2003.

[23] Nandan, A. et al., Cooperative downloading in vehicular *ad hoc* networks, in *Proc. of Second Annual Wireless On-Demand Network Systems and Services (WONS 2005)*, St. Moritz, Switzerland, January 19–21, 2005.

Chapter 6

Evolution of Application Models for Pervasive Computing

Guruduth Banavar

CONTENTS

Introduction

Pervasive computing fundamentally takes computing off the desktop and into the spaces where we live and work in every day. It is about enabling access to relevant applications and data at any location and on any device in a manner that is customized to the user and the task at hand. Mark Weiser [19] called it "invisible" computing, and much work has been done in support of that vision.

One of the key requirements of pervasive applications is mobility. Because mobile devices come with many capabilities, mobile applications must run on a wide variety of devices, including the devices embedded in various environments and devices carried by users. Applications must also support varying levels of network connectivity. Ideally, an application is hosted on the network and is able to execute on any device with multiple levels of connectivity.

Another key requirement is that applications must adapt themselves to the dynamics of the environment; for example, applications must customize themselves to interact with a user in a manner appropriate to the user's current context (such as location and activity), exploiting locally available devices and services without distracting the user from the task at hand. This implies that the application must identify and bind to data sources that provide the correct information, compose the information from these sources to create information that is useful for an application, and, finally, use that information in meaningful ways within the application itself.

Consider a simple example of a pervasive computing application: a pervasive calendar application [3]. First, the application will be able to run on multiple device platforms — from a networked phone (with a limited user interface and limited bandwidth but always connected) to a smart personal digital assistant (PDA) (with a richer user interface and higher bandwidth but not always connected) to a conference room computer (with a very rich user interface and very high bandwidth and always connected). Furthermore, users should be able to interact with this application using multiple user interface modalities, such as a graphical interface, a voice interface, or a combination of the two. Second, the application will be sensitive to the environment in which it is running; for example, if a user brings up a calendar at home, the application might bring up a family calendar by default, and, if the user brings up a calendar in an office when running late for a meeting, the application might bring up a work calendar with information about the meeting highlighted.

In this chapter, we discuss the evolution of the underlying programming models that allow application developers to build such applications. For the purposes of this discussion, we consider two classes of applications: *interactive* applications, which involve a user, and *sense-and-respond* applications, which combine data from a variety of sources and react to them in significant ways. These two classes are not mutually exclusive; in fact, the above example application combines elements of both classes.

Application models were originally designed for mobile applications written for highly resource-constrained devices. More recently, mobile devices are becoming highly capable in terms of processor and memory. This allows the device to host a significant amount of runtime middleware services capable of richer presentation and data management. The device programming model is evolving in significant ways to support richer user experiences. One development is the notion of extreme componentization, where applications are specified in terms of small reusable components that are composed dynamically to create specific instantiations. These components can be created in radically different ways, yet they have to come together to support the application function. This chapter discusses this trend.

Context-aware applications such as the one described earlier have been discussed in the literature for awhile now; however, a significant trend in pervasive applications is the evolution of a number of sensor-based applications. These applications capture large amounts of data from a variety of heterogeneous sensor data sources, combine that data in various ways to determine trends and boundary conditions, and activate business logic appropriately. An example is when a radiofrequency identification (RFID)-based, supply-chain application has detected that the trend during a holiday season will likely require many more items of a certain type to be transported from a warehouse to retail locations in a certain area. The programming model challenges regarding the support of these kinds of applications are many, and this chapter also discusses this evolving area.

Banavar et al. [3] identified some of the current approaches being used to address the complexity of building these kinds of pervasive applications. One technique is to capture the basic user interaction structures and control flow in a manner that can be reused across multiple devices and modalities. Another technique is to encapsulate business logic and data in a manner that can be reused regardless of which host on which a component is instantiated. Yet another technique is to specify the required context data for an application and allow the infrastructure to manage the specific data formats, locations, and combinations of physical data sources to provide the actual data. We summarize these techniques at the end of this chapter.

Interactive Pervasive Computing Applications

This section considers the evolution of application models for interactive pervasive computing applications and articulates the challenges in the context of this evolution. Interactive pervasive computing applications are of two types: *mobile personal applications* and *smart-space applications*. Mobile personal applications are the classic pervasive applications commonly found on mobile devices such as phones and PDAs. Smart-space applications are those that run on a collection of devices within integrated and highly interactive environments.

To discuss the programming models for interactive pervasive computing applications, it is useful to consider the classic model–view–controller application structure [14], where the view represents the presentation, and the controller represents the application flow, including navigation, validation, error handling, and event handling. The view and the controller together deal with the user interaction of the application. The model component includes the application logic as well as the data underlying the application logic.

Mobile Personal Applications

In the early days of mobile devices, the predominant applications were native, standalone applications on devices such as the Palm and Windows CE. Because the applications were standalone, the application model was quite simple and straightforward, even primitive relative to their counterparts on desktop computers. The libraries of view components were relatively simple and straightforward, with standard controls and not very sophisticated interaction possibilities. The controller consisted of event handlers and low-level navigation mechanisms. The model consisted of basic storage and data-handling capabilities available natively on the device. The programming abstractions were not very high level, and the programmer was responsible for memory management, multitasking management, etc.

A key requirement of the data model of these early applications was the ability to share and back up the data underlying applications; for example, the data underlying personal information management (PIM) applications, such as calendar and address book, had to be shared with other devices that a user had, such as a PC. This need was supported by placing devices in their cradles or docking stations. Programming model mechanisms evolved for transferring data from a handheld device on a docking station to the PC storage (e.g., the *Conduit* programming model for Palm devices). This programming model, which is an early model for supporting disconnected operation, supports the function of maintaining consistency between a data store on the handheld device and a data store on the PC. Typically, this includes the ability to mark data structures as "dirty" while operating in a disconnected mode and reconciling the data structures once connected. Irreconcilable updates are flagged to the user, who is expected to be able to resolve them one way or another.

Thin-Client Application Models

True networked PDA applications began to emerge around the time that the World Wide Web was expanding its reach. The earliest of these models were met with the requirement of accommodating and extending the Web application programming model. The early Web model, also referred to as a *thin-client application model*, was built around the fact that a browser is located on the client device that is capable of rendering a presentation described by Hypertext Markup Language (HTML). The server side of an application implemented: (1) the module that generated the presentation markup (view), (2) the control flow and event handling (controller), and (3) the data model and business logic underlying that application. This application model evolved to use a view programming model such as Java Server Pages (JSPs) [13], a controller programming model such as Struts [17], and a data/business logic programming model such as Enterprise JavaBeans (EJB) [9]. From the user experience point of view, Web applications typically contained text, limited forms of media, and form-based applications.

Web browsers were being implemented for a variety of mobile devices. Web applications began to be implemented to generate markup language that could be appropriately rendered on mobile devices; however, a problem began to emerge in that the number of devices with different capabilities was expanding beyond anyone's expectations. The differences in capabilities included the markup language (e.g., WML, CHTML), as well as the user interface of the devices, such as the size and capabilities of the display, input mechanisms, color and media capabilities, network capability, and so forth. Every Web application that had to be accessed via multiple devices had to be customized for each of the devices. The complexity of developing and maintaining those applications began to increase dramatically.

The initial solution approach for customizing Web content to devices was *transcoding* [20]. A transcoder is an intermediary agent between a server that generates markup and a device that consumes the markup. The transcoder is responsible for understanding the capabilities of the device and to suitably modify the markup while it is on its way from the server to the device. The transcoder interprets rules for how to transform its input to its output. Unfortunately, transcoding solutions only had limited success, as they were too complex for third parties to write the rules and policies by which the transcoding module could understand and modify the content. Moreover, the original authors of the content typically want full control over the content and do not want intermediaries to modify the content without their approval.

As a result, authoring-oriented approaches for multi-device Web applications emerged. One example of this approach is Multi-Device Authoring

Technology (MDAT) [4]. This technology allows application developers to specify a generic form of an application (the view and the controller) describing the overall function. The author is then able to "specialize" the application to particular targets by specifying the deltas from the generic application to individual target devices or classes of target devices. The notion of specialization is akin to subclassing in object-oriented programming, where the generic application is akin to the superclass containing common behavior and the specialized versions are akin to subclasses. Given this specification, the MDAT tool generates the concrete device-specific versions of the Web applications for each target device, including the view (JSPs) and the controller (Struts). Furthermore, the application developer is able to test the generated application and modify it, if necessary. The tool supports the ability to save the changes to the generated application so the developer does not lose the changes if the generic version of the application is modified and the device-specific version regenerated.

This description of Web application models is not limited to graphical user interfaces (GUIs) but also includes voice applications. In an application that uses voice rather than GUIs as the user interaction paradigm, the view and controller components are implemented in a radically different way, as voice interaction is quite complex to implement. First of all, voice input requires special information for proper functioning — for example, a grammar that describes the vocabulary that can be used for input. In more complex cases, acoustic models may also be associated with voice recognition. Once we have achieved that, consider the simple case of presenting a list of choices. If the list is long, it is not practical to have a single arbitrarily ordered list; instead, the list should be appropriately prioritized (e.g., based on the likelihood of selection) and split into sublists that are easier to present. Then, we have the case of exception handling. In a voice interface, the user is always presented with the option of escaping out of a menu, or selection list, or any dialog. This presents a different method for flow control for applications compared with the form-based interaction presented in Web-based GUI applications. The view and controller parts of a voice application are implemented by markup languages such as VoiceXML [18], and the model part of the application is implemented by the standard component models such as EJB. Specialized tooling for specifying VoiceXML-based application flows at a high level have evolved to support this application model.

From a mobile user interaction point of view, GUI-only or voice-only applications are less than optimal. A GUI provides a high-bandwidth way for users to consume information, but data input is not as easy due to the limited nature of input mechanisms on mobile devices. On the other hand, a voice interface provides a high-bandwidth way for users to input information, but data output is not as easy to consume. The idea of *multimodal* applications is to combine multiple modalities, such as GUI and voice, to combine the benefits of each of these modalities. Markup languages such as XHTML+Voice (X+V) [21] have evolved to support this type of bimodal application. The application model for creating bimodal applications is also based on the model–view–controller application structure, where the view and controller are encapsulated within the X+V specification. The tooling for X+V is an interesting combination of GUI and VoiceXML tools, where the GUI specification forms the basis and the voice interactions take the form of annotations.

An early application model that emerged to support offline operation was exemplified by AvantGo. The idea was that a user could specify the Web universal resource locators (URLs) that should be downloaded into the device for offline use. When the device was in the cradle, the connected PC downloaded the contents of these URLs and stored them on the device. This model allowed users to browse such content as news, weather, stocks, and the like while offline; however, the obvious disadvantage was that any new content that was previously not planned for could not be accessed. Web-based transactions were queued while offline and released when the device was docked.

A more recent development in the support of offline Web applications is the more powerful form-based application models such as XForms [22]. XForms supports a clean separation of form presentation from form data. An XForms document encapsulates the specification of user interaction elements as well as control-flow within that set of elements. Data that is presented to the user or collected from the user through these user interaction elements is represented via an XML data structure. The XML data structure is downloaded to the device when the XForms document is downloaded to the device, and the updated XML data structure is uploaded back to the server. The XML data on the server can be received and processed by standard Web componentry such as Web services.

Rich-Client Application Models

The application models presented above support a form of user interaction that is quite limited in the following sense. First, the presentation structures are form-based controls, such as selection lists and text inputs, rather than the larger and more sophisticated set of user interaction elements found in highly interactive computing applications. Second, the application is partitioned in such a way that frequent round-trip communication with the server is required, resulting in a lack of instant response. This results in a user experience that is far from what can be expected from the processing and storage capabilities of the device hardware on which an application is running. For the purposes of this discussion, let us consider two key elements of good user experience: *responsiveness* and *expressiveness.*

Rich-client application models aim to achieve the responsiveness and expressiveness of interactive computing commensurate with the capabilities of the device hardware platform. Let us consider the two factors in turn. To achieve good responsiveness, we need to support a set of capabilities directly on the device platform, so as to minimize the impact of server round-trip communication. This also has the benefit of supporting disconnected or weakly connected operation. And, incidentally, this must be done to minimize the impact on the programming model; that is, a device-local implementation of application services should not impose a new programming model but should instead use as much of the existing server programming model as is feasible. To achieve good expressiveness, we need to support a broader set of functions on the platform, including interaction, graphics, storage, and media. Not every device can support the same rich set of features, so a subset of this set of features will have to be supported according to the capabilities of each device platform.

Let us first consider local implementation of application services. Picking up from the Web application programming model, the first approach that comes to mind for supporting better responsiveness is to create a device-local (or embedded) implementation of Web application services. For example, an embedded Web application server and an embedded database on a device can support the downloading of Web applications to the device so as to significantly reduce round-tripping to the server. This also allows us to download enough data and computation to the device so application interaction can continue in the absence of a connection to the server (which is an important requirement in itself). This then introduces the necessity to synchronize the data between the device-local services and the network-based services. These functions have been implemented in existing prototypes, with minimal impact to the Web server programming model.

To support better expressiveness, the first thing that comes to mind is to support richer graphical interaction models. This includes the ability for arbitrary two-dimensional and three-dimensional graphics, animation, and rich media support; however, better expressiveness goes beyond rich graphics. It includes notions of the kinds of operations one can perform with storage, such as save at will, browse, reorganize, and protect. It includes support for common interactive desktop operations, such as copy-and-paste, undo, and restructure. It includes rich context-sensitive help and automated editing support. And, very importantly, it includes the notion of multitasking: viewing the tasks being performed and managing them. Implementing all of these features in a programming model goes far beyond the Web programming model discussed earlier and requires a device-local implementation of a number of significant new services.

Implementing this set of services into a device platform poses a new challenge. Although devices are increasing in capabilities, the range of devices is large, and many of them still have quite limited capabilities in terms of processing and storage. The solution to this problem is to implement the services as components that can be managed intelligently. This is a serious engineering challenge in the sense that the system should be broken down into manageable components that can be loaded and unloaded from the network as necessary. New component models that support this kind of function are emerging.

Ultimately, an application can be viewed as a composition of a collection of distributed components, some of which will reside on the user's device and others on various nodes on a network. The component programming model will vary depending on the function being provided and the class of platforms it is intended to run. The composition programming model, on the other hand, should be uniform so as to allow putting together any set of components with compatible interfaces to support application functions. Once composed, the problem will be to partition the components into the various nodes of the network, based on the available resources, the required responsiveness and throughput, and other factors such as cost and security. This is ultimately the vision for Web services.

In this line of evolution, future application models for mobile interactive applications will support a seamless, heterogeneous, managed, component model. Components from multiple vendors that internally use different programming models can be combined together seamlessly and managed in a uniform manner. This type of application model will support the best kinds of innovation in interaction models and allow independent functions to come together seamlessly in support of user needs.

Smart-Space Applications

Smart or active spaces have been an early and consistent topic of ubiquitous computing research. The basic concept of smart spaces is that people will be surrounded by visible and invisible technology that can sense and act, communicate, reason, and interact to make their environment a better place to live and work. A smart space is thus an indoor or outdoor environment with computing elements that can perform the functions mentioned earlier in a robust, self-managing, and scaleable way.

Smart spaces offer services that are composed from both the devices embedded in the environment and portable devices worn or carried by users into the spaces. The goal is for the combination of imported and native devices to support the information and collaboration needs of the users in that space. Smart spaces may:

- Perceive and identify users, their actions, and even their intent.
- Facilitate interaction with information rich sources.
- Support local and distributed collaboration.
- Anticipate and support user needs during task performance.
- Provide enriched records and summaries for later use.

The earliest and most well-known smart-space experiment was the Xerox PARC [19], where Weiser and his colleagues developed what they called *tabs*, *pads*, and *boards*, which are inch-scale devices like active sticky notes, foot-scale devices like note pads, and yard-scale displays, respectively. They placed many dozens of tabs, several pads, and a few boards in indoor spaces and allowed people to accomplish their tasks by using these devices. In his classic article [19], Weiser also discusses a smart outdoor space where a car driver is able to look into a "foreview" mirror to check for traffic in his projected path and to easily find a parking space.

More recent experiments have integrated commercial off-the-shelf devices and technologies to achieve some of the above objectives. The Gaia project at the University of Illinois in Urbana–Champaign [11] is a good example for considering the programming model for active spaces. This project has built distributed-operating-system-like functions, such as events, signals, shared file systems, and security, and has extended them with concepts such as context awareness and device transparency. Applications are built using an application framework that includes standard services and requires others to be specified by the developer.

The high-level programming model [10] for such a smart space includes entities such as users, services, applications, devices, and other physical objects. Application developers refer to these entities at an abstract level. The application framework maps these virtual references to actual physical objects in a space using the constraints specified by the developer, the available resources in the space, the policies specified for the space, and the current context of the space. The framework uses an ontological representation to capture knowledge about the various entities and their relationships and optimizes the discovery and binding of virtual references to the best physical resources. Programmers can also manage the activities in the space at a high level by specifying high-level commands such as starting, stopping, and moving components and taking action when users or devices enter or exit spaces. These high-level abstractions make it considerably easier to program smart-space applications than was possible previously.

Sense-and-Respond Pervasive Computing Applications

Context awareness is the ability of computing applications to be aware of the environment in which the computation is taking place and potentially to adapt accordingly. The attributes of the environment, such as location, destination, other people in the vicinity, and activity being performed, are referred to as the *context* of the application. The need for context-aware applications arises because pervasive computing makes applications available in contexts other than a computer workstation with a keyboard, mouse, and screen. The user of a pervasive-computing application will typically be focused upon some task other than the use of a computing device and may even be unaware that he or she is using a computing device. Applications must customize themselves to interact with a user in a manner appropriate to the user's current context and activities, exploiting locally available devices, without distracting the user from the task at hand.

A context-aware application has a sensing aspect and a responding aspect. The sensing aspect binds to data sources, collects data, analyzes the data, and ensures that the data is relevant to the application. If so, it notifies the responding aspect, which takes an action appropriate to the sensed data. The complexity of these kinds of applications comes from the fact the data sources are (1) heterogeneous (e.g., indoor location can come from 802.11 triangulation or active badges or other means), (b) dynamic (i.e., sources may come on and go off depending on many factors such as the location of a person), and (c) low level (e.g., a person's location, phone usage, and a calendar entry noting that the person is in a meeting are all low-level data that together signify the fact that the person cannot be interrupted).

The programming model for context-aware applications [7] consists of several parts. At the most basic level, an application should be able to specify a data source at a high level so it can be independent of the physical source of the data. It would be the responsibility of the underlying infrastructure to discover and bind to the appropriate data source based on the needs and the availability. The types of data and their relationships can be captured in a data type of hierarchy. Also, the infrastructure should be able to accommodate different types of data sources. Some data sources, such as request–response Web services, are passive or pull based. Other data sources, such as sensors that trigger alarms, are active or push based. A flexible infrastructure is capable of discovering both kinds of data sources. An application can then pull the current value from a passive data source or subscribe to be notified each time an active data source generates a new value. Finally, the programming model should support the flexible composition of data from multiple sources. The language for specifying the composition typically resembles a rule language and supports notions of aggregation, filtering, and correlation.

Context-aware applications are instances of sensor-based applications. The general structure of sensor-based applications consists of a collection of data sources, a hierarchy of entities for collecting and composing the data from these sources, and some decision logic to take action based on the data collected. For example, an RFID supply-chain application has a multitude of RFID-tagged objects moving from manufacturer to consumer, a number of data aggregation points (e.g., at the manufacturer's dock, at a warehouses, at retailers), and a back-end information technology system that gets the information from the aggregators to make decisions about the demand for particular items and how best to manufacture and inventory those items. Many aspects of the programming model for context-aware systems are applicable in this broader context as well.

Even more broadly speaking, sensor applications are an instance of sense-and-respond applications. As mentioned earlier, the responding part of context-aware applications supports the adaptation of an application to a user's environment, or the responding part of an RFID supply-chain application supports the throttling of the supply chain to support demand. In many cases, sensor applications have a physical actuator that affects the real world; for example, in an industrial setting, a sensor that detects an overheated motor could result in the motor being slowed down or shut off, or the filters of an oil well pumping a high level of impurities could be adjusted. In a sense, these applications behave like control systems. From a programming model point of view, the challenge is whether these low-level control systems can be integrated effectively with the higher level information technology decision systems to produce a seamless application model that supports end-to-end sense-and-respond applications.

The elements of such an end-to-end, sense-and-respond programming model include event discovery and binding, event propagation, event correlation/aggregation, event storage, decision logic, and actuation. Discovery and binding support a dynamic and heterogeneous set of data sources. Event propagation supports the gathering of event data from pull-based and push-based data sources. Event correlation and aggregation support the combination of multiple event streams to create higher level events that are of interest to applications. Event storage helps retain historical events so they can be reconstituted for supporting applications that require long-range sensed data. The decision logic in applications uses events at all levels to arrive at a decision about how to react to the available sensed data. And, finally, actuation takes action based on the decisions made. These elements must come together in a comprehensive way for future systems to support the full range of sense-and-respond applications.

Summary of Current Programming Model Approaches

In this section, we summarize the current state of pervasive programming models [3]. As mentioned before, a key issue that programming models are trying to address is application development *complexity* to adequately deal with heterogeneous devices, varying degrees of connectivity, and dynamic data sources. *Reuse* of application components is the fundamental means of addressing this complexity. Four basic approaches to enhancing reuse have been applied, based on the well-known model–view–controller application structure (briefly described earlier), and these are described in the following subsections. These approaches have reached different levels of maturity in research projects and commercial offerings. Several challenges remain before these approaches can become widely useful.

Device-Independent Views

Device-independent views allow an application to capture the basic user interaction structures that should be reused across multiple devices and modalities. This device-independent representation describes the intent behind the user interaction within a view component (such as a page), rather than the actual physical representation of a user-interface control. For example, the fact that an application requires users to input their ages is represented by a generic INPUT element with a range constraint. An adaptation engine determines, based on the target device characteristics, usability considerations, user preferences, and whether the INPUT element should be realized as a text field, a selection list, or even voice input. Several device-independent view representations have evolved over the years, including UIML [1], AUIML/Druid [15], XForms [8], and Microsoft's Mobile Controls [16].

Automatic runtime adaptation is the common technique used to convert the device-independent representation to a device-specific representation. The runtime adaptation engine retrieves the device identifier via the request header of a Web application (specifically, the *user agent* field) and maps that to a database record containing detailed device information. The information in this database record guides the adaptation of the device-independent representation to device-specific representations. Microsoft, Oracle, and Volantis have commercial products using some variation of runtime adaptation.

One of the pitfalls of this approach is its reliance on automatic runtime adaptation of the device-independent representation. Fully automatic adaptation can work in certain cases, when the content is simple or when the device variations are not too great; however, experience shows that it is extremely difficult for fully automatic adaptation to produce highly customized and usable interfaces that are comparable to handcrafted user interfaces. This is especially true in modern, highly interactive applications. As a result, most successful systems that use this technique provide a way for developers to provide additional information, or metadata, to guide or augment the runtime adaptation process.

Design-time adaptation is a technique that converts the device-independent representation to device-specific representations before the application is deployed to the runtime. The result of design-time adaptation is a set of target-specific artifacts that can be viewed and manipulated by the developer. At the end of this process, the developer ends up with a set of target-specific view components, similar to the components that a developer would have built by hand [5]. This approach has two major advantages: One, the developer has full control over the adaptation process and the generated artifacts. If the developer is not satisfied with the output, the process can be rerun with different parameters until the result is satisfactory. The generated artifacts can also be manipulated to add device-specific capabilities for particular devices. Two, there is no runtime performance overhead for translating applications, because the translations have occurred at design time.

Design-time adaptation only supports devices that are known at design time. If new devices must be supported after an application has been deployed, it may not be reasonable to depend on the application provider to target those devices via the design-time tool. Also, for dynamic content (again, that will be unknown at design time), it is necessary to have some level of runtime adaptation. For these reasons, some systems, such as MDAT [4], support a hybrid of design-time and runtime adaptation. Design-time adaptation results in one or more device-specific application versions that can be deployed to a Web application server.

Platform-Independent Controllers

As described earlier, the controller of an application represents the control-flow, including data validation and error handling, typically via event handlers. A platform-independent controller allows an application to specify the overall control-flow across multiple execution platforms but still allows an application to have different control-flow structures for different devices and uses. The reasons why the controller of an application must be targeted to multiple devices include the following:

■ Different devices may have different types of input hardware, ranging from a keyboard, tracking device, and microphone on a personal computer to a pair of buttons and a scrolling wheel on a wristwatch.

- The flow of an application may be different on different devices; for example, an application that contains a secure transaction may not support this transaction on a device that does not have the appropriate level of security infrastructure. Similarly, an application that supports rich content may choose to skip those pages on devices that are not capable of presenting rich content.
- When a device-independent page is adapted and rendered on multiple devices, the page may be split into multiple device-specific pages for any device that is too small to contain the entire page.
- The controller execution framework may be different for different device platforms. Recall that a device platform is the end-to-end distributed platform that supports the execution of all components of the application. One device platform may support a Java-based Struts framework, whereas another may support a different framework, such as the base Servlet framework, or a different language altogether, such as PHP or C#.

As a result, a complete solution for targeting multiple devices must include the application controller. One approach [4] is to represent the controller in a declarative way using a *generic* graph representation, where the nodes are device-independent pages and the arcs are control-flow transitions from one page to another. This representation addresses the three requirements above as follows:

- Developers can modify the flow of the application for particular target devices. These are represented as incremental changes to the generic controller.
- When a device-independent page is split into multiple pages, the appropriate controller elements for navigating among those pages are also automatically generated.
- The concrete controller code for specific controller platforms (e.g., Struts) is automatically generated from the declarative controller representation. The specific controller framework can be changed as necessary.

Host-Independent Models

Networked mobile applications vary in the distribution of logic and data between the mobile device and the server. In a thin-client application, views are generated on the server and then rendered on the client device by a component such as a Web browser. Controller logic, model logic, and model data all reside on the server, so disconnected operation is impossible. At the other end of the spectrum, a rich-client application resides entirely on the client device. It maintains its own fully functional model, which may be synchronized from time to time with replicas of the model on a server.

We need a programming model that allows the model components of an application (such as the view and controller components) to be shared by multiple versions of a disconnectable application — that is, by connected and disconnected versions of the application. In the ideal scenario, the logic and the data for the model component are specified once, and the tools and infrastructure supporting the programming model extract the appropriate subset of logic and data for the disconnected mode on each supported device. In reality, this extraction process will likely have to be guided extensively by the developer. The developer will likely specify the model, view, and controller in a generic way (view and controller as described in previous sections), and the tools will enable the developer to incrementally refine this generic representation for particular target environments. This is an ongoing area of work, and significant issues remain to be resolved. Ultimately, host-independent models allow an application to encapsulate the business logic and data in a manner that can be reused regardless of which host a component is instantiated on.

Source-Independent Context Data

An application obtaining data from heterogeneous sources with inconsistent availability and quality of service should not name a specific source of data such as a sensor, a Web service, or a database; rather, it should describe the kind of data that is required so the underlying infrastructure can discover an appropriate source for the data. This approach, known as descriptive, data-centric, or intentional naming [2,6,12], has a number of advantages. It allows the system to select the best available source of data, based on current conditions. If the selected source should fail, the infrastructure can rebind to another source satisfying the same description, thus making the application more robust. New data sources satisfying a description can be introduced or old data sources removed without modifying the application; likewise, the application can be ported to an environment having a different set of sources for the described data.

The basic idea of this approach is for an application to specify the desired context data without specifying the exact location and data type of the source or whether it is coming from multiple sources. These are considerations that will be handled transparently by the infrastructure. In some cases, the infrastructure may discover a data source, such as a device or a Web service, that directly provides the described data; for example, suppose an application specifies that it is interested in a Boolean value for "Is Jane at lunch?" The infrastructure may discover a data source that directly reports whether Jane's location is the cafeteria. Alternatively, the infrastructure may discover a programmed component, referred to as a *composer* in Cohen et al. [7], that computes the described data from other data. In our example, some combination of Jane's calendar, office status, and computer status might be combined by a composer to determine with a certain degree of certainty whether she is at lunch. A composer may be reusable across multiple applications and may itself be built on top of other composers that handle lower-level, more generic, data. For example, the query "Is X at lunch?" could be answered using the answer to a query of the form "Is X located at Y?" and queries of that form might themselves be answered by consulting multiple sources of location data (e.g., active badge, 802.11, or cell tower) with different resolutions and inferring a composite location with a certain degree of confidence.

Some data sources, such as request–response Web services, are passive or pull based. Other data sources, such as sensors that trigger alarms, are active or push based. A flexible infrastructure is capable of discovering both kinds of data sources. An application can then pull the current value from a passive data source or subscribe to be notified each time an active data source generates a new value. The kind of source-independent data specification described here allows an application to specify the intended context data to be supplied by reusable infrastructure components, which in turn are concerned with the specific data formats, locations, and combinations of physical data sources that provide the actual data.

Conclusions

This chapter has given a retrospective on the evolution of programming models for pervasive computing applications, the two major classes of which are interactive applications and sense-and-respond applications. Interactive applications consist of mobile personal applications and smart-space applications. Mobile personal applications evolved from the early standalone applications to cradle-based applications to Web-based applications. Web applications evolved from the early browser-based applications to multiple-device application models to richer XForms-based applications. Voice and multimodal applications converged with the Web application programming model. All of these thin-client application models, however, were limited in their user experience.

To support a better user experience, including responsiveness and expressiveness, richer forms of the programming model are evolving. Responsiveness is typically supported by having device-local services, which incidentally also support disconnected and weakly connected operations. Due to resource constraints, device-local services are managed via a strong component model that supports composition and lifecycle management. The ultimate vision here is to have a program-ming model that supports the composition and management of heter-ogeneous components to provide a seamless user experience.

Smart-space applications support a different kind of interactive expe-rience, one where a number of computing devices embedded in a space helps support the task that a user is trying to accomplish. In the early experiments in this domain, the programming model was low level and extremely complex. In more recent days, a higher level programming model has evolved for such an environment which exposes the resources, services, and their relations at a high level and allows the developer to specify the application at the abstract level with the infrastructure mapping it to the physical resources.

Context-aware applications support the adaptation of applications to the environment of a user. These applications have evolved from stand-alone applications using one or a few data sources to a common generic infrastructure that supports a number of applications using a variety of data sources. The programming model for these applications can support more general sense-and-respond applications, such as RFID and industrial control applications. The eventual goal in this domain is to have a comprehensive end-to-end sense and respond application model that supports event discovery, binding, storage, propagation, correlation/aggre-gation, and actuation.

Acknowledgments

This article is a compendium of many ideas that have evolved from projects and discussions with many individuals in the pervasive computing group at IBM. The author is grateful to many researchers, and especially to Norman Cohen and Danny Soroker, for the thoughts behind this article.

References

[1] Abrams, M., Phanouriou, C., Batongbacal, A.L., Williams, S.M., and Shuster, J.E., UIML: an appliance-independent XML user interface language, *WWW8/Computer Networks*, 31(11–16), 1695–1708, 1999.
[2] Adjie-Winoto, W., Schwartz, E., Balakrishnan, H., and Lilley, J., The design and implementation of an intentional naming system, in *Proc. of the 17th ACM Symp. on Operating Systems Principles (SOSP '99)*, Kiawah Island Resort, SC, December 12–15, 1999 [published in *Operating Syst. Rev.*, 33(5), 186–201, 1999].
[3] Banavar, G., Cohen, N., and Soroker, D., Pervasive application development: approaches and pitfalls, in *Mobile Computing Handbook*, Ilyas, M. and Mahgoub, I., Eds., Auerbach, New York, 2004.

[4] Banavar, G. et al., An authoring technology for multi-device web applications, *IEEE Pervasive Comput.*, 3(3), 83–93, 2004.

[5] Bergman, L.D., Banavar, G., Soroker, D., and Sussman, J., Combining handcrafting and automatic generation of user-interfaces for pervasive devices, in *Proc. of the Fourth Int. Conf. on Computer-Aided Design of User Interfaces (CADUI 2002)*, Valenciennes, France, May 15–17, 2002, pp. 155–166.

[6] Bowman, M., Debray, S.K., and Peterson, L.L., Reasoning about naming systems, *ACM Trans. Programming Languages Syst.*, 15(5), 795–825, 1993.

[7] Cohen, N.H., Purakayastha, A., Wong, L., and Yeh, D.L., iQueue: a pervasive data-composition framework, in *Proc. of the 3rd Int. Conf. on Data Management*, Singapore, January 8–11, 2002, pp. 146–153.

[8] Dubinko, M., Klotz, Jr., L.L., Merrick, R., and Raman, T.V., Eds., *XForms 1.0*, World Wide Web Consortium (W3C) recommendation, October 14, 2003 (http://www.w3.org/TR/xforms/).

[9] Enterprise JavaBeans Technology, http://java.sun.com/products/ejb/.

[10] Ranganathan, A., Chetan, S., Al-Muhtadi, J., Campbell, R.H., and Mickunas, M.D., Olympus: a high-level programming model for pervasive computing environments, in *Proc. of the IEEE Int. Conf. on Pervasive Computing and Communications (PerCom 2005)*, Kauai Island, HI, March 8–12, 2005.

[11] Román, M., Hess, C.K., Cerqueira, R., Ranganathan, A., Campbell, R.H., and Nahrstedt, K., Gaia: a middleware infrastructure to enable active spaces, *IEEE Pervasive Comput.*, 1, 74–83, 2002.

[12] Intanagonwiwat, C., Govindan, R., and Estrin, D., Directed diffusion: a scalable and robust communication paradigm for sensor networks, in *Proc. of the Sixth Annual Int. Conf. on Mobile Computing and Networking (MobiCom 2000)*, Boston, MA, August 6–11, 2000, pp. 56–67.

[13] Java Server Pages Technology, http://java.sun.com/products/jsp/.

[14] Krasner, G. and Pope, S., A cookbook for using the model–view–controller user interface paradigm in smalltalk-80, *J. Object-Oriented Programming*, 1(3), 26–49, 1988.

[15] Merrick, R.A., Defining user interfaces in XML, in *Proc. of Petrochemical Open Standards Consortium (POSC) Annual Meeting*, London, September 28–30, 1999 (http://www.posc.org/notes/sep99/sep99_rm.pdf).

[16] Microsoft, *Mobile Web Development with ASP.NET*, Microsoft Corporation, Redmond, WA, 2003 (http://msdn.microsoft.com/vstudio/device/mobilecontrols/default.aspx).

[17] Apache Struts Web Application Framework, http://struts.apache.org/.

[18] VoiceXML Forum, http://www.voicexml.org/.

[19] Weiser, M., The computer for the twenty-first century, *Sci. Am.*, 265(3), 94–104, 1991.

[20] IBM WebSphere Transcoding Publisher, http://www.ibm.com/software/pervasive/transcoding_publisher/.

[21] XHTML + Voice Profile 1.0, http://www.w3.org/TR/xhtml+voice/.

[22] *XForms: The Next Generation of Web Forms*, http://www.w3.org/MarkUp/Forms/.

Chapter 7

Mobile Middleware: Definition and Motivations

Dario Bruneo, Antonio Puliafito,
and Marco Scarpa

CONTENTS

Basic Concepts

One of the main reasons for the wide and rapid spread of computer networks is the concept of *transparency*, traditionally embodied in the layered protocol stack [41]. Due to this simple but extremely powerful approach, users accessing a computer network are not aware (and do not want to be) of the technical details, protocols, and management issues. Usually, users want to run an application and get results without any knowledge of all the operations involved. If the application is a distributed application, very likely a connection must be established, followed by negotiation with regard to the most appropriate communication parameters and other complex activities. All of these activities are hidden to most network users.

Distributed Systems

With the increasing use of computer networks, the software architecture has changed radically. From a relatively simple architecture (*centralized system*) where all the software components are executed on a single machine, new software architectures (*distributed systems*) have been developed where software is organized into different modules distributed to different *elaboration nodes*, and data is exchanged by means of a communication network. In a centralized system, an application runs on a single node and constitutes a single process; the workstation is the only *active* component of the system because it hosts the application itself. Terminals share the resources of the workstation in such a way that different users can use the application. Terminals can even exist without a CPU, being equipped instead with a communication interface only for sending commands to the running application on the workstation. To summarize, in such a system the communication network is used to interconnect *stupid* nodes (*terminals*) with the *elaboration* node, as depicted in Figure 7.1.

In a distributed system an application is composed of more processes running on different nodes of the network. All the processes are cooperating closely, and they execute in parallel. So, a characteristic of distributed systems is that the processes do not share memory but instead rely on message exchange, thus introducing delay during the computation. The reference architecture changes, as depicted in Figure 7.2.

Because the network communication infrastructure has become very inexpensive over time, the computational power achieved by a distributed system is less expensive than that of an equivalent workstation. Moreover, the entire infrastructure is more manageable, as it is relatively simpler to increase the resources, to balance the workload, and to run parallel applications. In a distributed system, the reference programming paradigm

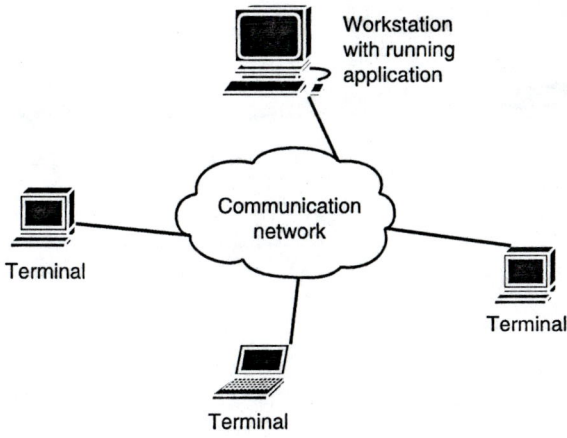

Figure 7.1 Example of a centralized system.

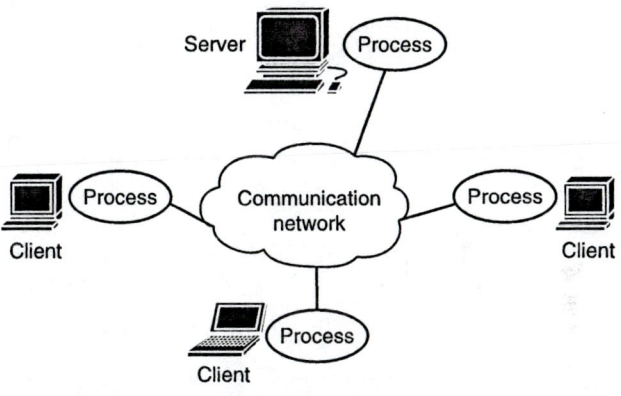

Figure 7.2 Example of a distributed system.

usually adopted is the *client–server* model. *Client* and *server* are distinct processes running on different nodes characterized by a well-defined interface. The server usually makes available some procedures for handling the data that are designed for responding to criteria of general effectiveness. The actual data processing is left to the server, where *ad hoc* procedures for the desired processing can be executed. The typical functional scheme is (Figure 7.3):

- The client asks the server for a service.
- The server performs the requested elaboration and sends the results back to the client.

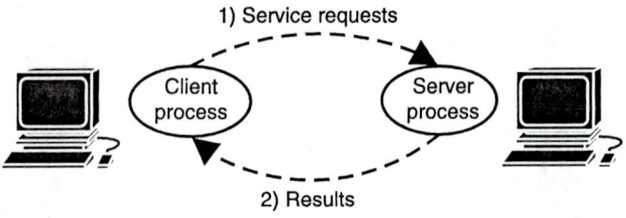

Figure 7.3 Client–server interaction.

Note that the distinction between *client* and *server* is based purely on function; in other words, a server could be a client of another server.

The Middleware Layer

Programming in a distributed system environment is a very tricky activity for developers, who must manage all the details related to the communication (e.g., addressing, error handling, data representation). This is due to the fact that developers use the low-level abstraction provided by the network operating system. To free developers from the expensive and time-consuming activity of resolving all the problems related to the network management, greater abstraction must be introduced. This is exactly the role of the *middleware* layer.

Middleware is a software layer between the operating system and the applications that provides a higher degree of abstraction in distributed programming. Using middleware, a programmer can develop an improved-quality distributed software by using the most appropriate, correct, and efficient solutions embedded in the middleware. In other words, utilizing middleware to build distributed systems frees developers from the implementation of low-level details related to the network, such as concurrency control, transaction management, and network communication, in such a way that they can focus on application requirements. Now consider the International Organization for Standardization (ISO)/Open Source Initiative (OSI) network reference model. Because middleware allows programmers to develop distributed systems as integrated computing facilities, it must address shortcomings of the network operating system; therefore, it implements the session and presentation layers of the ISO/OSI reference model [41], as depicted in Figure 7.4. In this environment, developers are able to request parameterized services from remote components and they can execute them without worrying about implementation of the session and presentation layers. Some examples of middleware successfully used thus far include OMG's CORBA™ [36], Microsoft's COM [8], SUN's Java Remote

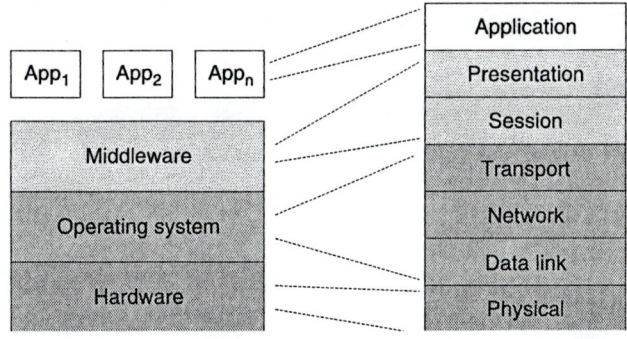

Figure 7.4 Software models and their relation with network models.

Method Invocation (RMI) [40], IBM's MQSeries™ [23], and remote procedure calls (RPCs), which were introduced by SUN in the 1980s [7].

As an example of middleware, we can take a look at RPC behavior. RPC middleware was created to give developers some kind of mechanism for accessing remote resources using just a local procedure call to hide all the details on the connection setup, marshal all the parameters, and hide all the problems related to the heterogeneity of different platforms. When the client performs a call, it is intercepted by the middleware, which runs the code for gathering the parameters, opening a socket, and so on (see Figure 7.5). The user is completely unaware of what happens. The middleware generates an appropriate message containing all the data related to that call and transmits it to the server, which is able to understand the request sent from the client and execute the procedure. When the

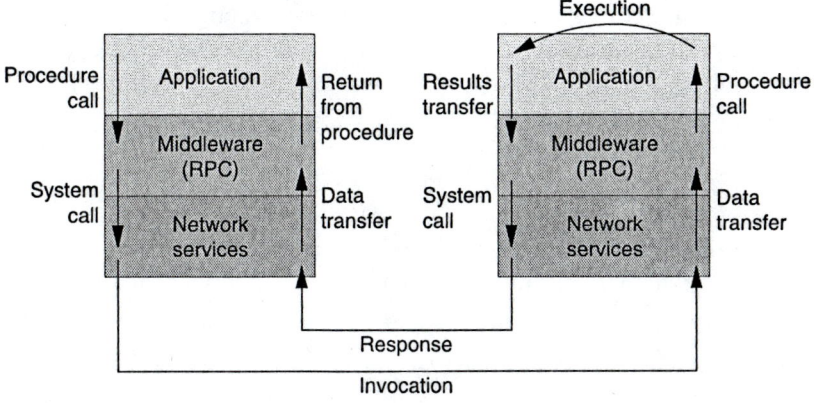

Figure 7.5 Remote procedure call (RPC).

results are ready, they are sent back to the client middleware, which returns to the application. The user perceives the process as being nothing more than a traditional local call.

Middleware Requirements for Fixed Distributed Systems

Middleware is a software layer that hides all the implementation details of the communication infrastructure to developers in such a way that they can overlook all the problems arising due to directly using network operating system primitives. To create this kind of abstraction, some requirements must be satisfied. It should be pointed out that some of these requirements must be altered when the middleware is to be used for mobile systems:

- *Communication* — First of all, the middleware has to guarantee the exchange of data among the nodes of a distributed system in such a way that the system appears to the user to be an integrated system. When two entities communicate, they exchange data as parameters of some kind of service; due to the heterogeneity of the nodes in the distributed system, the internal representation of data could be different, and this problem must be anticipated by the middleware. Moreover, all the parameters must be appropriately composed to be correctly exchanged; the middleware has to implement marshaling–unmarshaling operations.
- *Coordination* — The evolution of processes in distributed systems has to be controlled to achieve maximum cooperation among the different activities. Usually, most threads execute on the same host concurrently, and all of them must be synchronized with the remote one. Another issue is the fact that a distributed system could be viewed as a big repository of services; each service runs on a host, and it is available for invocation from other components of the system. In general, it is unknown when an invocation might come, and it is a waste of resources for the service to constantly execute; for these reasons, *activation* and *deactivation* must be provided by the middleware in order to start and stop the services when needed.
- *Reliability* — The network layers below the middleware ensure that communication errors and faults of the network are transparently recovered, but other kinds of errors can arise in the usual activities of a distributed application. As an example, nothing can be said about correct execution of a service or the order of the requests. This kind of reliability is implemented

directly by the middleware by using the *best effort, at most once, at least once*, and *exactly once* services [17]. Reliability may also be increased by replicating the services on multiple hosts to make them more readily available, even if a host is unavailable for either internal or external reasons.

■ *Scalability* — In this context, scalability is the ability to manage a growing load in the future. In fact, in centralized systems the load is limited by the load the node can support, but in distributed systems a new load request can be directed toward different servers according to necessity, while the user must be unaware of the real used resource. The typical mechanism used to accomplish this issue is *transparency* (i.e., location, migration, access, and replication transparency) [17,34]. Particularly important is *location transparency*, which demands that components do not know the physical location of the components with which they interact. A detailed discussion on transparency can be found in Emmerich [17]. For efficient location transparency, a load-balancing mechanism must be provided to either reduce or increase load on different hosts when a service is started or stopped somewhere in the distributed system.

■ *Security* — Because the Internet is an important component of the actual communication network, it is not possible to protect connections from third parties. Sometimes, the data involved in distributed computations is private data, and users want to protect it from unauthorized access. The middleware itself should be able to provide cryptography mechanisms to the users.

Another issue is verifying the real identity of users to protect the server from unauthorized people and to ensure the high quality of a given service for the client. At least four security services can be incorporated:

■ Protection of data against reading by unauthorized users
■ Forbidding the creation and deletion of messages to unauthorized users
■ Identity checking
■ Possibility of using electronic signed documents

How Mobility Affects Middleware Design

The rapid growth of wireless technologies and the development of smaller and smaller devices have led to the widespread use of mobile computing. Each user, equipped with a portable device, is able to access miscellaneous services in any way at any time and anywhere thanks to the connectivity powered by modern network technologies. Mobile access to distributed applications and services raises many new issues. A major problem is the wireless technology itself, as the bandwidth provided is orders of magnitude lower than in the wired networks, signal loss is very frequent, and the noise level is influenced by external conditions. A second aspect is related to the mobile devices, which are characterized by scarce resources in terms of CPU, RAM, display, and storage; in particular, they are equipped with smart batteries, which limit the autonomy of the device (in terms of power consumption) and affect both wireless transmission and access to services that require a high computational load. Finally, a third aspect that must be considered is user mobility, which causes problems related to signal loss during movement to a new cell (handoff), as well as problems with address management caused by users traversing through different administrative domains and the need to adapt services to the position of the user.

Traditional middleware solutions are not able to adequately manage these issues. Originally designed for use in a static context, such middleware systems hide low-level network details to provide a high level of transparency to applications. In mobile environments, though, the context is extremely dynamic and cannot be managed by *a priori* assumptions, so it is necessary to implement reconfiguration techniques that can react to changes in the operating context and develop powerful mechanisms to propagate such changes until the application level is reached. It is necessary, therefore, to develop new middleware for mobile systems in the early stages of the design phase that will provide mobility support.

Mobile Middleware Requirements

To highlight the issues related to the mobile middleware design we present a typical mobile computing scenario, focusing on the main aspects that must be taken into account. Consider the case of a user equipped with a personal digital assistant (PDA) who wishes to receive information about movies playing at the nearest cinema. The user will select a movie after viewing some video clips. To meet the user's requirements, the system must:

- Determine the user's actual position, search for the requested services, and select the ones more appropriate for the user.
- Be aware of the user's habits (e.g., the user's favorite movie genre), so it can select the results to be presented to the user.
- Establish and rate the quality of service (QoS) level to be used.
- Format the information to be presented according to the hardware and software features of the device used.

- Manage the user's mobility by allocating the resources in advance for using the service (e.g., the bandwidth required for streaming) and managing the changes of address.
- Manage the load of the wired network so as to optimize the resources by changing the server from which the streaming will be retrieved.

It may happen that the user goes from a WiFi area to a General Packet Radio Service (GPRS) area with a consequent decrease in the bandwidth available, or the user might change devices (e.g., the user may switch to a PC upon arriving home). In these cases, the system should be able to reconfigure the parameters to resume the view from the point where it was stopped and to adapt it to the new device. Some operations of clip transcoding, reallocation of resources, and QoS management will therefore be necessary. When the movie has been chosen, the user may decide to buy tickets online, in which case the system should be able to manage the transaction with regard to issues of both security and the possible disconnection and reconnection of the user.

Using our example of the moviegoer, some interesting issues involved in the management of advanced services in mobile environments can be identified, such as service discovery, QoS, service adaptation, and load balancing. Service discovery plays a primary role, because it allows the user to access the list of available services. The techniques currently used include the use of distributed directory services, which are more scalable than the centralized approach [30]. The user position and the user profile will have to be managed to filter the services available by choosing, initially, only those corresponding to the user preferences and habits and located near the user's actual position.

The streaming of multimedia flows over wireless channels requires very strict QoS requirements. In fact, any variation in the quality of the transmission can degrade the application requested and make it inaccessible; for example, when a user is viewing a movie clip, the bit rate cannot be reduced below a specific threshold without affecting the viewing quality. The QoS management is therefore very important.

A major problem is the limited bandwidth available in wireless channels. This involves the investigation of new techniques for bandwidth reservation to anticipate the user's handoff and reallocate resources in the new access point that will be visited by the user [11]. The system cannot disregard the user's actual position, as it has to propose services and resources available near the user's position. This is necessary because of restrictions in terms of QoS and because a crucial factor is the distance (latency) between the server and the mobile client.

Furthermore, the system will have to be able to recognize features of the device to adapt the service requested to the type of terminal used; for example, if the same movie clip is to be viewed on a notebook with a 16.8-million-color display and on a PDA with a 256-color display, different resolutions will be required. Considering the bandwidth savings required in wireless environments, the need to transcode the same video in several formats to tailor it according to the client device is evident; for example, an MPEG-2 video could be encoded in MPEG-4 when switching from a wired to a wireless station. Information can be distributed on the wired network, which allows quicker and more effective access to the resources and makes available more copies of the same video, even in different formats.

Load-balancing operations are necessary to better exploit the use of storage and computing resources. Wireless channels are characterized by frequent disconnections, which cause QoS degradation and possible failures due to data loss during a transaction; therefore, the system should be able to manage such sudden disconnections by providing mechanisms for information replication and transaction recovery. Finally, all of these operations must be carried out taking into consideration the limited resources of mobile devices, particularly power consumption. Some power-saving techniques must be applied to reduce the wireless transmission load during both downloads and uploads [3].

In light of the many issues that must be addressed when developing mobile middleware, it can be observed that the use of traditional middleware originally conceived for fixed distributed systems is not always feasible in such complex and heterogeneous environments. This is due primarily to the remarkable differences between the two operating environments (see Table 7.1). These differences call for new design strategies that take into account features of the mobile environment to overcome,

Table 7.1   Main Differences Between Distributed and Mobile Environments

	Distributed Environments	Mobile Environments
Bandwidth	High	Low
Context	Static	Dynamic
Connection type	Stable	Unstable
Mobility	No	Yes
Communication	Synchronous	Asynchronous
Resource availability	High	Low

in an efficient way, all of the issues discussed here [16,31]. Such design strategies can be classified into three main types: context management, connection management, and resource management.

Context Management

The main assumption of middleware for fixed distributed systems (i.e., static representation of the context that is not transparent to upper layers) is too restrictive in mobile environments, as such environments are characterized by frequent context changes. Mobile computing applications must adapt their behavior to these changes to overcome issues related to high-level service provisioning. Context transparency, in fact, makes the development of complex applications easier, but it does not allow the service level to make decisions about the environments in which such applications must run [14]. This approach is powerful in systems where the operating conditions are static or where changes can be considered as exceptional and predictable events.

Mobile environments do not satisfy these requirements. Disconnections can occur either voluntarily (due to power-saving policies) or suddenly (due to signal loss), wireless technologies differ greatly in terms of performances and reliability, and portable devices have a high degree of heterogeneity. To enable applications to adapt to such context evolutions, parameter reconfiguration at provision time is necessary [4,13]. Mobile middleware systems cannot hide the context at the upper layers but instead must both represent it and announce changes until the service layer is reached. It is at this layer, in fact, that is possible to make decisions about the best way to react to context changes. For example, in a multimedia streaming session, in response to a drastic reduction in bandwidth, the system could activate the movie transcoding tasks (by reducing the bit rate or the color depth) or could decide to transmit only the audio data, dropping all the video packets (possible cases might include football matches, video conferences, video clips). Clearly, such decisions can be made only when the service typology is known and not on the basis of low-level information. It is necessary to implement middleware systems in such a way as to achieve a trade-off between transparency and awareness [12].

The operating context in a mobile computing scenario can be divided into three main aspects: user context, device context, and network context. The user context is composed of information related to the user's position, to user preferences, and to the QoS level requested. The device context includes details on the status of the available resources (e.g., CPU, batteries, display) and on the relative position of the device in the network — for example, in terms of latency between the device and a service

Table 7.2   Context Representation

Context Type	Description
User context:	
Location	Real position of the user in the wireless environment
Profile	User preferences and habits
QoS level	QoS level requested by the user accessing the services
Device context:	
Profile	Device features (e.g., CPU type, display)
Resource status	Updated usage level of the device resource
Location	Latency from other devices and from the service providers
Network context:	
Technology	Adopted wireless technology (e.g., 802.11, Bluetooth, GPRS)
Noise level	Quality level of the wireless signal
Activity	Wireless activity in terms of throughput
Bandwidth available	Bandwidth available in the wireless environment
Addressing	Address protocol used by the wireless infrastructure

provider or in terms of distance from other hardware components (e.g., mobile devices, printers). The network context contains all the information about the available bandwidth, noise level, wireless technologies adopted (e.g., 802.11, Bluetooth®), and addressing. Such context representation is shown in Table 7.2.

One of the main design considerations of context-aware middleware is the study of a representation of the operating context to capture its features and to make them available to the upper layers. Such a representation has to be flexible and powerful to allow applications to easily react at provision time to the frequent context changes. A common technique adopted for context representation is the definition of metadata; in particular, profiles and policies comprise metadata, which describes with a high level of abstraction context features and actions to carry out in case of changes. The metadata, represented by a meta-language (e.g., XML [1]), has to be separated from the implementation details to simplify the management operations. Suitable binding techniques must be impslemented to enable applications to change their execution at runtime according to the established policies. Such requirements have made computational reflection, introduced in Smith [39], an attractive technique to adopt in the mobile middleware design. Reflection is the ability of a software system to monitor its computation and to change, if necessary, the way it is executed. The two phases of monitoring and adapting are generally referred to as *intro-spection* and *interception*. Discussions of context-aware middleware based on the concept of reflection can be found in the literature [2,29].

A context feature that has aroused quite a bit of interest in recent years is location management [25]. Location-aware middleware that is able to provide services according to the user's position and to manage the user's movements has been developed for such scenarios as e-health [37], e-learning [24], and cultural heritage [28].

Connection Management

User mobility, intermittent signals, and resource management policies give rise to the frequent disconnection and reconnection of mobile devices. Such behavior, which is not encountered in traditional distributed systems, makes unsuitable the adoption of a synchronous communication system that is based on the assumption that the sender and receiver are continuously connected during the communication phases. Mobile middleware has to provide an asynchronous communication system able to carry out all the tasks, notwithstanding the intermittent link between sender and receiver. To this end, solutions to decouple the sender and the receiver are required. Decoupled middleware systems have to manage the issues related to data synchronization by implementing data replication techniques. One of the most widely adopted solutions is the use of tuple space systems, which provide shared memory areas where both the sender and the receiver can put their data in an asynchronous way [33]. When a message has been sent (that is, after a *write* operation), the sender can continue its tasks without waiting for the receiver to carry out the *read* operation; thus, a mobile user can make a query, disconnect from the network, and, when reconnected, retrieve the results of that query. Examples of tuple-space-based middleware include TSpaces™ [43] and JavaSpaces™ [22].

Another technique adopted for transaction management in mobile environments is the use of data subsets downloaded in the mobile device to create a local representation of information scattered over the wired network; offline transactions can be carried out by mobile users using these local data subsets, and the actual operations can be carried out when the user goes online [18,38]. For example, these subsets can be adopted in an e-commerce scenario to allow users to download a part of the product list and to create, offline, a local shopping cart with the selected products. When an online connection is established, the system will retrieve the local shopping cart to complete the order.

A drawback of such solutions is the data synchronization that requires the use of advanced techniques; in our offline shopping scenario, we have to deal, for example, with issues related to potential price updates on the official product list that are not indicated on the older, local product list, or we might have to take into account problems related to product availability. The system should be able to disseminate these updates and to carry out effective comparisons among data, verifying the correctness of the transactions. Such operations depend greatly on the manner in which the data is structured.

Another issue related to connection management is the provision of services based on the concept of session (e.g., multimedia streaming). A temporary disconnection or change of address could cause the loss of the session and the end of service provisioning. Such issues can be solved by adopting proxies capable of decoupling the client and server and hiding these disconnections from the service layer [5,9]. Proxies have to interact with the specific protocol involved in service provisioning, and then their development is strictly related to the particular typology of the service that we want to use.

Resource Management

The design of mobile middleware and the management of the discussed techniques are strictly related to the hardware resources to be used for their execution, which are usually quite scarce, thus introducing another constraint on the design of such systems: Mobile middleware has to be lightweight [44]. Mobile middleware, on the one hand, has to implement techniques capable of guaranteeing powerful usage of available resources by reducing, for example, wireless transmissions and by adapting service typology to the real features of the client devices. On the other hand, mobile middleware has to be designed to be efficient to avoid overloading the device itself.

First, we must take into account the use of the sensors required to accomplish the goals of the middleware; an appropriate context representation, in fact, foresees the use of several sensors (e.g., of position) for the collection of the data that must be monitored to manage the context evolutions. The use of sensors must to be restricted as much as possible because such components are quite greedy in terms of resource consumption; for example, if location management is needed, triangulation techniques could be implemented rather than using global position system (GPS) modules on the mobile devices [26].

A second aspect is related to the computational load of middleware; in addition to being light in terms of memory usage, mobile middleware

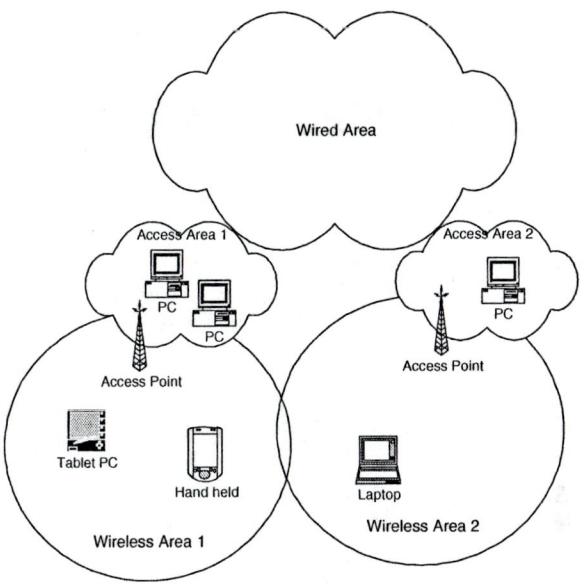

Figure 7.6 A nomadic computing scenario.

must reduce the amount of data to process so the limited computing resources of mobile devices are not overloaded. To this end, it is important to design highly modular middleware systems that are capable of activating only the modules absolutely necessary for the required operations (deactivating at runtime the unnecessary ones) [4] and delegating to other parts of the system (e.g., the wired infrastructure) those operations that require a high computational load, such as multimedia content tailoring [9].

Middleware for Nomadic Systems

Nomadic systems are characterized by a fixed infrastructure that provides a wireless access to mobile devices through access points. Thanks to this wireless link, mobile users are able to access services offered in the wired network. Designers of mobile middleware for nomadic systems must take into consideration many issues that involve both the services offered (in terms of their heterogeneity and complexity) and the main features (performance and dependability) of the processing and storage equipments. A typical nomadic computing scenario is shown in Figure 7.6. Such a setup includes the following areas:

- Wireless
- Access
- Wired

The wireless area is the coverage area of an access point, where one or more mobile devices can be found. The access area is the contact area between the wireless area and the wired area. It consists of access points that allow mobile devices to access services available in the wired area. The wired area is the core network infrastructure.

Managing the issues related to service provisioning in wireless environments calls for a middleware solution that covers all the areas presented in this scenario [6]. In such a way, interactions between the wired and the wireless components of the system can occur. Moreover, the distribution of this middleware over the three described areas will allow carrying out expensive tasks whenever more convenient resources are available. By responding to the resource management issues, mobile middleware will never overload mobile devices.

It is in the wired area where all the tasks requiring a high computational load must be carried out. Such tasks are related to service adaptation, data storage, and load balancing. Powerful management of these tasks is mandatory to make effective the QoS strategies provided by the other areas of the scenario and to improve service provisioning [45]. Let us consider, for example, tailoring a multimedia format to the features of a client device. It can be easily observed that when we send high-resolution multimedia data to a smart device we will experience a high percentage of wasted resources, affecting other users present in the same cell and making useless any resource reservation policy.

The access area contains tasks related to resource reservation and context management. Resource reservation can be performed by admission-control techniques that restrict access to the wireless environment. To guarantee service provisioning during user movement, the middleware has to be able to reserve in advance resources in the new access point where the user is headed [10]. The advanced reservation strategies must know the user position and manage the bandwidth of each cell by leaving a portion of it for users coming from neighboring cells. QoS levels have to be created to allow bandwidth reconfiguration according to the users' needs and resource availability. The middleware, at provision time, can automatically adapt the bandwidth assigned to a user to accept new users from neighboring cells, thus reducing the call-blocking probability. Proxy solutions able to carry out tasks on behalf of the user have to be inserted in this area to overcome issues related to signal loss [5,11].

The wireless area hosts middleware components related to mobile devices. Due to the limited resources of such devices, these middleware components must be reduced to only an interface with the system. In this area, we can find mobility management modules (only in the case of GPS systems), user and device profile management modules, the graphical interface for user–device interactions, and a communication mechanism that exchanges data with other middleware components present in the scenario.

Middleware for Ad Hoc Systems

Ad hoc systems are communication infrastructures where users can communicate notwithstanding previous agreements and can connect, disconnect, or move around in the surrounding space. This means that the nodes that form the network cannot rely on a fixed infrastructure nor on a central coordinating entity, because they all have the same computational potential and the same probability of disconnecting or migrating. To ensure the exchange of information among users, some routing protocols must be created. Such protocols route the packets in multi-hop paths, which consider a changing network topology.

The development of a mobile middleware for *ad hoc* systems is heavily influenced by the features of such environments. In fact, the lack of a fixed infrastructure limits most design choices. Any type of centralization has to be removed, as the presence of a static entity capable of carrying out tasks such as discovery of service, QoS management, and so on is not allowed. All of these tasks have to be accomplished using distributed approaches. Also, the mobile middleware implementation has to take into account the high degree of mobility typical of such environments and must produce an exhaustive context representation. The lack of a wired infrastructure restricts the execution of such middleware to mobile devices, as any distribution over different areas of the scenario, such as in the case of nomadic systems, is not possible. For this reason, the design of middleware for *ad hoc* systems has the main goal of achieving a high level of simplicity [44]. The most resource-expensive tasks cannot be delegated to other network components scattered in the system; instead, it is necessary to implement techniques that can accomplish the middleware operations using only the limited resources provided by mobile devices.

Although the development of mobile middleware for *ad hoc* systems is still in the early stages, some solutions and some guidelines are presented in the literature. Some authors have tried to modify middleware for nomadic systems to operate in such complex environments [20]; others have designed entirely new paradigms adapted to the features of *ad hoc* networks [35]. Peer-to-peer (P2P) is one of the most used paradigms; it is based on the concept of information dissemination between nodes and on the use of advanced techniques searching and provisioning services in accordance with the network topology [27].

To overcome the resource consumption issues, cooperation techniques between nodes have to be developed such that it will be possible to utilize the least-frequently used devices for accomplishing complex tasks. These operations can be provided by designing middleware systems with a high degree of modularity which are able to manage loading and unloading operations at runtime and execute the component middleware in a parallel way.

Available Technologies for Mobile Middleware

Mobile middleware design calls for new network technologies able to manage the continuous changes of the environment and to provide cooperative mechanisms that overcome the lack of resources of portable devices. In this section, we show how the use of the mobile agent technology and of the grid computing paradigm provides an effective strategy in the deployment of an overall architecture that achieves the goals of middleware.

Mobile Agent Technology

A software agent is a kind of software package that is smart enough to act as an assistant to accomplish some tasks on behalf of human beings [19]. The most salient feature that distinguishes agents and ordinary code is autonomy. Agents can cooperate with other agents to carry out more complex tasks than they themselves could handle. One special kind of agent, the mobile agent, can move from one system to another to access remote resources or even to meet other agents. The big success of mobile agents can be attributed to their ability to combine the typical features of software agents (e.g., autonomy, delegation) with the opportunity to migrate by moving from one position to another [42]. On the one hand, this feature allows the operations to be decentralized; on the other, interaction is possible with the environment around the agent. The scalability of the system can be increased, and some context- and location-aware mechanisms can be used.

Mobile agents seem to be a natural choice for dealing with issues related to providing advanced services in mobile environments. Agent programming technologies have emerged as a flexible and complementary way to manage the resources of distributed systems due to the increased flexibility in adapting to the dynamically changing requirements of such systems [15]. Such technology is considered to be both promising and challenging with regard to addressing personal or terminal mobility issues. Mobile agents are considered to be an enabling technology for automated, flexible, and customized service provisioning in a highly distributed way, as network nodes become active and take part in the computation of applications and provisioning of customized services. In addition to the clear separation of key functionality and aspects of deployment on the functional side, such technology offers potential technical advantages. Among them are reduced communication cost, reduced bandwidth usage, the possibility of using remote interfaces, and support for offline computation.

Mobile agents enable both temporal (i.e., over time) and spatial (i.e., over different nodes of the network) distributions of the service logic. These distributions add another technical advantage (namely, scalability), while at the same time such bottlenecks of centralized approaches as reduced network availability and malfunctioning are avoided. What makes this approach so appealing is how the previously discussed benefits of mobile agents address the typical issues and restrictions of wireless communication (e.g., low-bandwidth, high-latency networks; high bit error rate; low processing power; small area available for the user interface).

Grid Computing Paradigm

When developing a powerful infrastructure that must provide services of a guaranteed quality while maintaining the user's profile and addressing characteristics of mobile devices and the limited availability of resources, it becomes apparent that the wired and the wireless components are not as different as they might appear to be at first. In fact, strong coordination between these two environments is necessary. The grid computing paradigm [21] is a valid solution to implement distributed management strategies in the wired part of the system, which must strongly interact with the mechanisms available in the wireless part to provide ever more sophisticated services with a high level of QoS. It is extremely important to develop an infrastructure that provides effective QoS management and allows mobile users to benefit from the service requested, regardless of the device used and the users' moves.

Grid refers to a new distributed computational infrastructure that provides an innovative method for accessing and distributing data and resources. The idea on which the grid is based is allowing people to share transparently and on a wide scale computational data and resources with members of communities working toward the same purposes. Interest in the grid technology has been growing, primarily among scientific communities. The grid technology arose from the wide use of Internet computing. The grid makes use of the idle resources available on the Internet to perform distributed computational operations.

The grid is intended to provide a more rational use of the resources distributed on the network. This rational use includes many operations such as load balancing, QoS management, and secure access. Incorporation of the grid in the design of mobile middleware systems would make possible such operations as:

- *More effective management of distributed data in the network* (e.g., databases, online libraries, video-clips) — This offers the opportunity of moving data to bring it closer to the user or to manage overload and fault tolerance situations.
- *Management of services according to the user's device* — For the same user to access services through different environments, the terminal equipment may have to be different; for example, a terminal with a high-resolution screen may be desirable at home, but a handheld terminal with a low-resolution screen may be the cost of mobility. Clearly, as the environment changes, the content to be delivered to the user also changes. A video conference, for example, may be delivered as video and voice communications at a fixed terminal or as voice and text only at a mobile terminal.
- *Service discovery* — It would be possible to offer this feature through the use of distributed strategies based on data replication.

Grid computing would seem to be a powerful strategy for the development of middleware for *ad hoc* environments. In fact, the use of cooperative and resource-sharing techniques between nodes is mandatory in environments not equipped with any wired infrastructure.

The so-called *wireless grid* [32] is a new research field aimed at introducing grid computing concepts to systems composed of resource-constrained devices to carry out complex tasks in a parallel way by sharing the resources of idle devices. The peer-to-peer paradigm is used to discover the services and the resources provided by the devices, and component-based programming techniques are adopted to bind at runtime the software modules shared by nodes. At the moment, the use of a mobile grid in the design of mobile middleware for *ad hoc* environments appears to be one of the most interesting directions to pursue.

References

[1] eXtensible Markup Language (XML), http://www.w3.org/XML/.
[2] OpenORB Project, http://openorb.sourceforge.net.
[3] Anastasi, G., Conti, M., Gregori, E., and Passarella, A., A performance study of power-saving policies for Wi-Fi hotspots, *Computer Networks*, 45, 295–318, 2004.

[4] Bellavista, P., Corradi, A., Montanari, R., and Stefanelli, C., Context-aware middleware for resource management in the wireless Internet, *IEEE Trans. Software Eng.*, 29(12), 1086–1099, 2003.

[5] Bellavista, P., Corradi, A., and Stefanelli, C., Mobile agent middleware for mobile computing, *IEEE Comput.*, 34(3), 73–81, 2001.

[6] Bellavista, P., Corradi, A., and Stefanelli, C., Application-level QoS control for video-on-demand, *IEEE Internet Comput.*, 7(6), 16–24, 2003.

[7] Birrell, A.D. and Nelson, B.J., Implementing remote procedure calls, *ACM Trans. Comput. Syst.*, 2, 39–59, 1984.

[8] Box, D., *Essential COM*, Addison-Wesley, Boston, MA, 1998.

[9] Bruneo, D., Villari, M., Zaia, A., and Puliafito, A., VoD services for mobile wireless devices, in *Proc. of the IEEE Symp. on Computers and Communications (ISCC'2003)*, Antalya, Turkey, June 30–July 3, 2003, pp. 602–207,

[10] Bruneo, D., Paladina, L., Paone, M., and Puliafito, A., Resource reservation in mobile wireless networks, in *Proc. of the IEEE Symp. on Computers and Communications (ISCC'2004)*, Alexandria, Egypt, June 28–July 1, 2004, pp. 460–465.

[11] Bruneo, D., Villari, M., Zaia, A., and Puliafito, A., QoS management for MPEG-4 flows in wireless environment, *Microprocessors Microsyst.*, 27(2), 85–92, 2003.

[12] Capra, L., Emmerich, W., and Mascolo, C., Middleware for mobile computing: awareness vs. transparency, in *Proc. of the 8th Workshop on Hot Topics in Operating Systems (HotOS)*, Schloss Elmau, Germany, May, 2001, pp. 164–169.

[13] Capra, L., Emmerich, W., and Mascolo, C., CARISMA: Context-aware reflective middleware system for mobile applications, *IEEE Trans. Software Eng.*, 29(10), 929–945, 2003.

[14] Chan, A. and Chuang, S.-N., MobiPADS: a reflective middleware for context-aware mobile computing. *IEEE Trans. Software Eng.*, 29(12), 1072–1085, 2003.

[15] La Corte, A., Puliafito, A., and Tomarchio, O., An agent-based framework for mobile users, in *Proc. of the European Research Seminar on Advances in Distributed Systems (ERSADS'99)*, Madeira, Portugal, April 23–28, 1999.

[16] Eliassen, F. et al., Next generation middleware: requirements, architecture and prototypes, in *Proc. of the 7th IEEE Workshop on Future Trends of Distributed Computing Systems (FTDCS'99)*, Capetown, South Africa, December 20–22, 1999, pp. 60–65, 1999.

[17] Emmerich, W., *Engineering Distributed Objects*, John Wiley & Sons, New York, 2000.

[18] Mascolo, C. et al., Xmiddle: a data-sharing middleware for mobile computing, *Wireless Pers. Commun. Int. J.*, 21(1), 77–103, 2002.

[19] Etzioni, O. and Weld, D.S., Intelligent agents on the Internet: fact, fiction, and forecast, *IEEE Expert*, 10(3), 44–49, 1995.

[20] Fok, C., Roman, G., and Hackmann, G., A lightweight coordination middleware for mobile computing, in *Proc. of the Sixth Int. Conf. on Coordination Models and Languages*, Pisa, Italy, February 24–27, 2004, pp. 135–151.

[21] Foster, I., Kesselman, C., and Tuecke, S., The anatomy of the grid: enabling scalable virtual organizations, *Int. J. Supercomputer Appl.*, 15(3), 200–222, 2001.

[22] Freeman, E., Hupfer, S., and Arnold, K., *JavaSpaces: Principles, Patterns, and Practice*, Addison-Wesley, Boston, MA, 1999.

[23] Gilman, L. and Schreiber, R., *Distributed Computing with IBM MQSeries*, John Wiley & Sons, New York, 1996.

[24] Griswold, W.G. et al., ActiveCampus: experiments in community-oriented ubiquitous computing, *IEEE Comput.*, 37, 73–81, 2004.

[25] Hazas, M., Scott, J., and Krumm, J., Location-aware computing comes of age, *IEEE Comput.*, 37, 95–97, 2004.

[26] Hightower, J. and Borriello, G., Location systems for ubiquitous computing, *IEEE Comput.*, 34(8), 57–66, 2001.

[27] Hsieh, H. and Sivakumar, R., On using peer-to-peer communication in cellular wireless data networks, *IEEE Trans. Mobile Comput.*, 3(1), 57–72, 2004.

[28] Krosche, J., Baldzer, J., and Boll, S., MobiDENK: mobile multimedia in monument conservation, *IEEE Multimedia*, 11, 72–77, 2004.

[29] Ledoux, T., OpenCorba: a reflective open brocker, in *Proc. of the Second Int. Conf. on Meta-Level Architectures and Reflection*, Vol. 1616, Lecture Notes in Computer Science, Springer, Berlin, 1999, pp. 197–214.

[30] Lee, D.L., Xu, J., Zheng, B., and Lee, W., Data management in location-dependent information services, *IEEE Pervasive Comput.*, 1(3), 65–72, 2002.

[31] Mascolo, C., Capra, L., and Emmerich, W., Middleware for mobile computing, in *Networking 2002 Tutorial Papers*, Vol. 2497, Lecture Notes on Computer Science, Springer, Berlin, 2002, pp. 20–58.

[32] McKnight, L.W., Howison, J., and Bradner, S., Wireless grids: distributed resource sharing by mobile, nomadic, and fixed devices, *IEEE Internet Comput.*, 8(4), 24–31, 2004.

[33] Nitzberg, B. and Lo, V., Distributed shared memory: a survey of issues and algorithms, *IEEE Comput.*, 24, 52–60, 1991.

[34] ISO, *Open Distributed Processing: Reference Model*, Technical Report ISO 10746-1, International Organization for Standardization, Geneva, Switzerland, 1998.

[35] Plagemann, T., Goebel, V., Griwodz, C., and Halvorsen, P., Towards middleware services for mobile *ad hoc* network applications, in *Proc. of the 9th IEEE Workshop on Future Trends of Distributed Computing Systems (FTDCS'03)*, San Juan, Puerto Rico, May, 2003, pp. 249–255.

[36] Pope, A., *The CORBA Reference Guide: Understanding the Common Object Request Broker Architecture*, Addison-Wesley, Boston, MA, 1998.

[37] Rodriguez, M.D., Favela, J., Martinez, E.A., and Munoz, M.A., Location-aware access to hospital information and services, *IEEE Trans. Inform. Technol. Biomed.*, 8(4), 448–455, 2004.

[38] Satyanarayanan, M., Kistler, J.J., Kumar, P., Okasaki, M.E., Siegel, E.H., and Steere, D.C., CODA: a highly available file system for a distributed workstation environment, *IEEE Trans. Comput.*, 39(4), 447–459, 1990.

[39] Smith, B., Reflection and semantics in LISP, in *Proc. of the 11th Annual ACM Symp. on Principles of Programming Languages*, Salt Lake City, UT, January, 1984, pp. 23–35.

[40] *JavaSoft: Java Remote Method Invocation Specification*, Revision 1.5, jdk 1.2 edition, Sun Microsystems, Santa Clara, CA, 1992.

[41] Tanenbaum, A.S., *Computer Networks*, 4th ed., Prentice Hall, Upper Saddle River, NJ, 2002.

[42] Waldo, J., Mobile code, distributed computing, and agents, *IEEE Intelligent Syst.*, 16(2), 10–12, 2001.

[43] Wyckoff, P. et al., TSpaces, *IBM Syst. J.*, 37(3), 454–474, 1998.

[44] Yu, Y., Krishnamachari, B., and Prasanna, V.K., Issues in designing middleware for wireless sensor networks, *IEEE Network*, 18(1), 15–21, 2004.

[45] Zaia, A., Bruneo, D., and Puliafito, A., A scalable grid-based multimedia server, in *Proc. of Workshop on Emerging Technologies for Next Generation GRID (ETNGRID-2004)*, Modena, Italy, June 14–16, 2004, pp. 337–342.

Section 2

EMERGING TECHNOLOGIES FOR MOBILE MIDDLEWARE

Chapter 8

Name Resolution and Service Discovery on the Internet and in *Ad Hoc* Networks

Paal E. Engelstad and Geir Egeland

CONTENTS

Name Resolution

As human beings, we prefer to remember the name of a computer. Computers, on the other hand, prefer to address each other by numbers, which on the Internet are 32 bits or 128 bits long, depending on the Internet Protocol (IP) used (IPv4 or IPv6). This is one reason why we need a naming service that handles mapping between computer names that we humans find convenient to remember and the network addresses (i.e., numbers) that computers deal with. Another reason is that, according to the Internet model, an IP address does not identify a host, such as a Web server, but a network interface. Although the host makes changes to its network interface or network attachment, it is convenient for the users and applications to allow the name of the host to remain unchanged. As such, keeping different identifiers at different layers helps keep the protocol layers more independent and also reduces problems associated with layering violations. From a middleware perspective, naming services are a question of keeping higher layer names of entities independent of their lower layer identfiers and their actual locations. Here, the naming service not only helps the users of applications but is also just as much an aid for the software developer. This section familiarizes the reader with the most common naming services used on the Internet and how names can be resolved in wireless mobile *ad hoc* networks (MANETs).

An Architecture for Naming Services

A fundamental facility in any computer network is the naming service, which is the means by which names are associated with network addresses. Network addresses are found based on their names; for example, to use an electronic mail system the user must provide the name of the recipient to whom the mail is being sent. To access a Web site, the user must provide the universal resource locator (URL) of the site, which again is translated into the network address of the computer hosting the Web site. To give an example, the Domain Name System (DNS) [1,2] maps the host name of the University of Oslo's public Web server, which is www.uio.no, to the IP address 129.240.4.44. Another example is a Voice-over-IP (VoIP) system that maps a Session Initiation Protocol (SIP) identifier to an E.164 number; for example, the URL sip:dave@my_telecom.com:5060 is translated to +47 904 30 495. The association between a name and the lower layer identifier of a network entity is called a *binding*. Some naming services, such as DNS, can do reverse mapping (i.e., map an IP address to a corresponding higher layer name) and mapping from one higher layer name to another. In the following section, we will describe a generic model for naming services, which will serve as a reference model for the remaining part of the chapter.

A Generic Model

The process of looking up a name in a computer network normally consists of the following steps: First, the binding between higher layer names and lower layer addresses must be registered in the network. This procedure can be referred to as *registration*. The registration is normally done only once, and the binding is provided through some administrator authority. The bindings are normally registered with a server that holds a binding cache. An example of such an entity is a DNS server. With a strict authentication regime, the registration can be done automatically, as illustrated in message 1 of Figure 8.1. Second, the network entities that desire to resolve a name must be informed of the corresponding binding that is registered in the network. This procedure can be referred to as *name resolution.* The two main approaches to name resolution are:

- *Push approach* — The bindings are proactively broadcast to all network entities that might have to use the bindings for name resolution some time in the future.
- *Pull approach* — A network entity that desires to resolve a name to a network entity issues a request on demand at the time the binding is needed.

Due to scalability issues, the *pull* model has been chosen for use on the Internet, and this model is the focus of the following sections.

The generic model involves four networking entities for the registration and name resolution procedures:

- A *user entity* (UE) represents the network entity issuing a request to resolve a name.
- A *named entity* (NE) represents the network entity that a name points to.
- A *naming authority* (NA) is authorized to create a binding between names and addresses of named entities (NEs).
- A *Caching Coordinator Entity* (CCE) provides intermediate storage of bindings.

User Entity

The role of the user entity is to resolve the mapping between a name and the lower layer identifiers of an NE by retrieving a binding from the naming service. The UE may be software components or actual end users who want to look up a specific name. In most cases, the UE will offer a low-level functionality directed toward the system components. In the Domain Name System, requests are issued by the *resolver* in the computer's operating system. The initiative to activate the resolver can come from an end user typing a Web address in a browser or from an application requiring access to a binding. The request will end up at the entity caching the binding for the name requested, and the binding containing the resolved identifiers will be returned to the resolver. This is illustrated with messages 2 and 3 in Figure 8.1.

Named Entity

The named entity is the network entity (e.g., host or computer) identified by a name. For communication, it uses the network interfaces referred to by the lower layer identifiers of a binding.

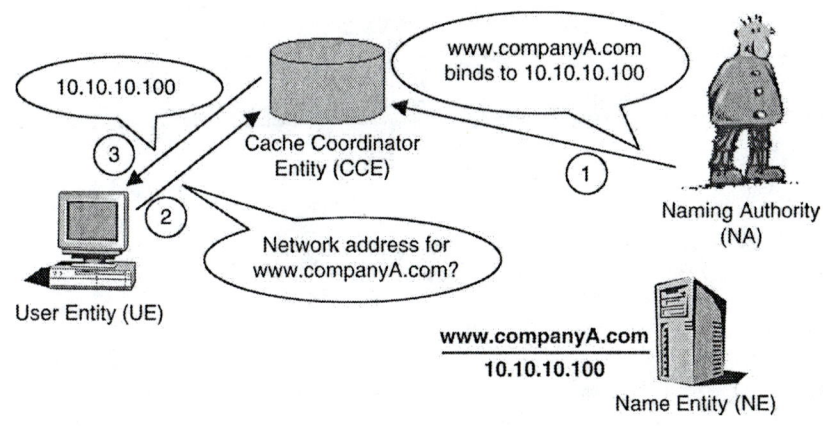

Figure 8.1 A Cache Coordinator Entity (e.g., the enterprise DNS server) is updated with a new binding for its named entity (e.g., a public Web server).

Naming Authority

The naming authority is the authority or system of authorities permitted to assign names to named entities and bindings between the names and the lower layer identifiers of the NEs. This is normally only configured once but might have to be updated if parameters of the network configuration changes. For example, an Internet Service Provider (ISP) can perform the administrative task of configuring their DNS server to map the network address of a customer's public Web server to the network address assigned to the customer, or the network administrator of an enterprise network can configure the company's local DNS server to map a computer's name to a fixed network address.

Solutions exist that enable a named entity to update its own binding directly with a Caching Coordinator Entity. This requires the CCE to be able to authenticate the NE. With no authentication, it would be possible for someone to insert false information into the caching entity and, in a worst-case scenario, impersonate another network entity by hijacking its binding. The NE must also be authorized to automatically update its binding, thus the node also plays a role in the naming authority.

Caching Coordinator Entity

The primary task of the Caching Coordinator Entity is to act as a cache for name bindings. The CCEs are important for the efficiency of the naming service in the presence of a large number of user entities and named entities. Normally, the NE knows the location of the CCE with which it

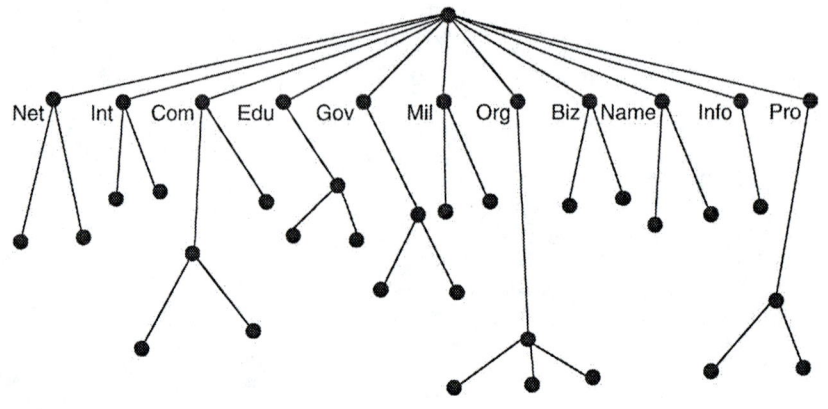

Figure 8.2   Structure of the DNS name space.

is supposed to register its bindings. The UE, which somehow has to retrieve this information, normally does not know the location of the CCE where the binding is located. For the DNS, the location of the local CCE (i.e., the local DNS server) is normally provided dynamically to the UE by, for example, mechanisms such as the Dynamic Host Configuration Protocol (DHCP). In this case, the UE normally uses the local CCE to locate and retrieve a binding stored at another CCE on the Internet.

The Domain Name System

The distributed database of the DNS is indexed by domain names. Each domain is basically just a path in an inverted tree referred to as the *domain name space*. Figure 8.2 illustrates the structure of the domain name space. The practical operation of the DNS system consists of three modules:

- The *DNS resolver*, which generates DNS requests on behalf of software programs
- The *recursive DNS server*, which searches through the DNS in response to queries from resolvers and returns answers to those resolvers
- The *authoritative DNS server*, which responds to queries from recursive DNS servers

The DNS resolver acts as the user entity described earlier, and the DNS servers act as CCEs. The registration is normally a manual process, and the named entities normally do not take part in the name resolution process (unless they use DNS Secure Dynamic Updates [5]).

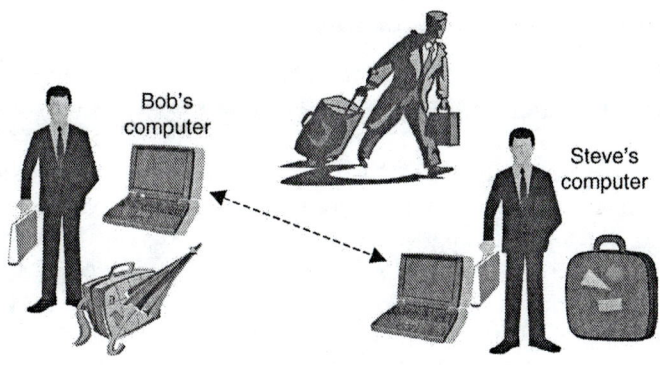

Figure 8.3 Computers communicating without a DNS server.

Resolving Names Without the Use of DNS Servers

In some situations, it is not feasible to make use of the Domain Name System to resolve name bindings, and in some cases the DNS might not even be available. For example, consider a setting where people who happen to meet in an airport lounge want to make use of the wireless local area network (WLAN) feature of their laptops to connect to each other to exchange music or other information they may find interesting (Figure 8.3). Without any DNS service, they would have to identify themselves using the network address, which for human beings is not very appealing. If it was somehow possible to define a separate name space in addition to the DNS, and if some mechanism could advertise and resolve these names in such a spontaneous setting, users could search and identify names in a more human-friendly way. Multicast DNS [6] and Link Local Multicast Name Resolution [7] are two competing solutions addressing this scenario.

Multicast DNS Name Resolution

Multicast DNS (mDNS) utilizes familiar DNS programming interfaces, packet formats, and operating semantics in a small network where no conventional DNS server has been installed [6]. In short, it enables a node to search for the network address of a computer named X by sending a multicast DNS message asking: "Does anyone know the network address of node X?" If a node with the name X is present on the network, it will respond by sending back a DNS response containing information about its network address. Multicast DNS is a part of the Mac OS® X operating system, where its implementation is called *Rendezvous*.

Link Local Multicast Name Resolution

Link Local Multicast Name Resolution (LLMNR) is a peer-to-peer name resolution protocol focused on enabling resolution of names on the local link [7]. LLMNR utilizes the DNS packet format and supports all DNS formats, types, and classes. LLMNR is not intended as a replacement for DNS, and as a result it is only used when a DNS server is either not available or is not providing an answer to a query. LLMNR differs from mDNS in many ways. First, LLMNR is an Internet Engineering Task Force (IETF) standards track specification, while the mDNS that is used in Apple Rendezvous is not. LLMNR is designed for use only on the local link, but mDNS also offers sitewide usage. Furthermore, mDNS sends multicast responses as well as multicast queries.

Name Resolution in Ad Hoc Networks

Mobile *ad hoc* networking was developed from military research on packet radio networks. In the late 1990s, however, the topic was included as a working group item of the IETF. The goal of the IETF was "to develop a peer-to-peer mobile routing capability in a purely mobile, wireless domain. This capability will exist beyond the fixed network (as supported by traditional IP networking) and beyond the one-hop fringe of the fixed network" [8].

Characteristics of Ad Hoc Networks

A mobile *ad hoc* network consists of mobile routers, often simply referred to as *nodes*. They are free to move about arbitrarily, and wireless technology is used for direct communication between the nodes. Due to the dynamic nature of the wireless media and the arbitrary mobility of the nodes, the network forms a random, multi-hop graph that changes with time. The network is an autonomous system that may operate in isolation, or it may optionally have gateways that connect it as a "stub" network to a fixed network infrastructure. Because a node is not necessarily in direct radio range with any other node in the network, the nodes must participate in the routing process and be willing to forward packets on behalf of other nodes in the network.

An *ad hoc* network is a network that is created spontaneously, without support from the existing fixed Internet infrastructure. The network might be formed when people equipped with portable PCs come together at conferences, or such a network can be used to combine a user's personal wireless devices into a personal area network (PAN). *Ad hoc* networks may also be formed during emergency situations when legacy network infrastructures are unavailable or damaged. Yet another application of *ad hoc* networking is on a military battlefield where fixed network infrastructures are unavailable or not feasible to use.

The salient characteristics of *ad hoc* networks do not include simply the dynamics of the network topology; in addition, the links are bandwidth constrained and of ever-changing capacity. Furthermore, the nodes often rely on energy-constrained batteries to move about freely, so energy conservation is an important design goal for many *ad hoc* networking technologies. Due to the dynamic networking topology and the fact that nodes might enter and leave the network frequently, it is often also assumed that an *ad hoc* network is without any preexisting infrastructure and that it is difficult to maintain an infrastructure in such a dynamic environment. Because of the lack of a preexisting infrastructure, it is anticipated that direct peer-to-peer communication between nodes will be popular on *ad hoc* networks. This means that any node may in principle operate as a server (e.g., Web server or SIP server) and be contacted directly by other MANET nodes. Any node may also operate as a client and contact other servers available in the network.

A mobile *ad hoc* network that is equipped with IP-based routing is normally referred to as a MANET. The two primary approaches to routing in MANETs are *reactive routing* and *proactive routing*. These two approaches are detailed below, with an emphasis on reactive routing.

Reactive and Proactive Routing Protocols

The routing protocols in mobile *ad hoc* networks can be reactive or proactive. A reactive routing protocol has no prior knowledge of the network topology but finds a route to a given destination on demand. A proactive routing protocol, on the other hand, tries to always have a complete updated picture of the network topology. Reactive routing protocols are normally preferred when nodes are highly mobile, when only a subset of nodes are communicating at any one time, and when communication sessions last for relatively long times. Proactive routing protocols, on the other hand, are preferred for lower levels of mobility and when communication is random and sporadic. Reactive routing is not well known to most people, probably because routing on the fixed Internet is proactive in nature.

A number of reactive routing protocols have been proposed over the years. The most widely studied and popular proposals include the *Ad Hoc* On-Demand Distance Vector (AODV) routing protocol [9] and the Dynamic Source Routing (DSR) routing protocol [10]. Reactive protocols allow source nodes to discover routes to an IP address on demand. Most proposals, including AODV and DSR, work as follows: When a source router requires a route to a destination IP address for which it does not

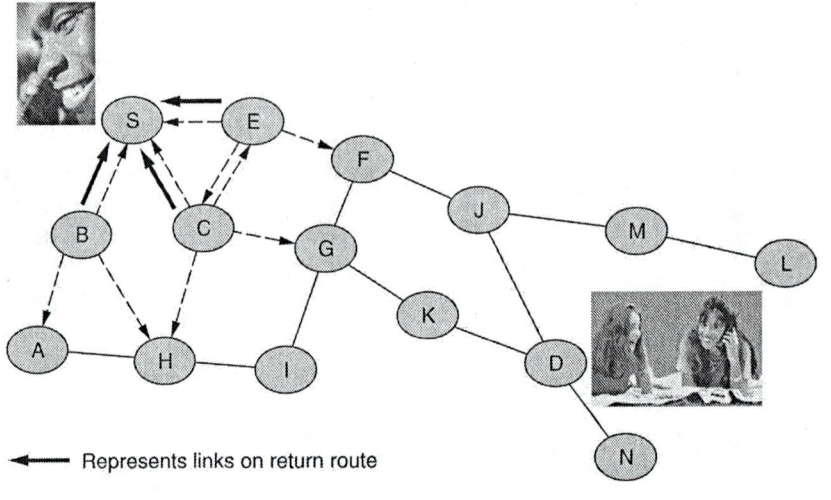

Figure 8.4   Route requests in AODV.

already have a route, it issues a route request (RREQ) packet. The packet is broadcast by controlled flooding throughout the network, and it sets up a return route to the source (Figure 8.4). If a router receiving the RREQ is either the destination or has a valid route to the destination IP address, it unicasts a route reply (RREP) back to the source along the reverse route. The RREP normally sets up a forward route from source to destination; thus, the pair of RREQ and RREP messages sets up a bidirectional unicast route between source and destination. When the source router receives the RREP, it may begin to forward data packets to the destination. (The acronyms RREQ and RREP are borrowed from AODV.)

Most protocols let routes that are inactive eventually time out. If a link becomes unavailable while the route is active, the routing protocol normally implements an algorithm to repair the route. Often the router upstream of the link breakage would send an error message upstream toward the source. The *Ad Hoc* On-Demand Distance Vector is a protocol that stores state information in the network. Routers that receive RREQs set up the return routes in the route tables as backward pointers to the source router, and RREPs that are propagated back to the source along the reverse route leave forward pointers to the destination in the route tables. The Dynamic Source Routing protocol, on the other hand, does not rely on routing state in the network; instead, DSR uses source routing. The RREQ collects the IP addresses of all the nodes that it has passed on the way to the destination. The destination subsequently sends a route reply by source routing back to the source of the request, providing it with the source route to the destination.

The Importance of Name Resolution in MANETs

Name resolution is an important feature in an *ad hoc* network, as addresses may change relatively frequently due to the network dynamics (nodes entering and leaving the network). Furthermore, it is often not possible to use the address as a well-known identifier, because IP addresses for nodes on the MANET will normally be autoconfigured at random, and nodes may also have to change addresses due to addressing conflicts. Devices and resources should instead be identified by stable and unique higher layer names (e.g., fully qualified domain names).

If a MANET is connected to the Internet, a MANET node may use the existing mechanism for name resolution on the Internet (namely, DNS) to look up the IP address of another MANET node; however, in most scenarios the MANET will not always be permanently connected to a fixed infrastructure, and the DNS infrastructure on the Internet might be unavailable. Relying entirely on the DNS on the Internet would not be a robust solution to name resolution in the MANET.

One option would be to introduce a DNS infrastructure into the MANET; however, DNS is designed with a fixed network in mind and has a relatively static, centralized, and hierarchical architecture that does not fit well with MANETs. Without a name resolution method in place, MANET users cannot easily use the applications that are developed for fixed networks for local communication on the MANET. The following text explores name resolutions in *ad hoc* networks. The focus is primarily on name resolution in reactive MANETs, because name resolution in proactive MANETs might be a less challenging task.

Architectures for Resolving Host and Service Names in Ad Hoc Networks

For name resolution in *ad hoc* networks, a MANET node may play the same role as discussed earlier for fixed networks. A node may act as a user entity that wants to resolve a name, as a named entity that wants to make its services available to other MANET nodes, or as a Caching Coordinator Entity that holds a central repository for cached bindings and assists other UEs and NEs with name resolution. A binding maps a name to an IP address and possibly a port number that the UE may subsequently use to contact the NE. Three name resolution architectures must be considered for *ad hoc* networks:

- Distributed architecture
- Coordinator-based architecture
- Hybrid architecture

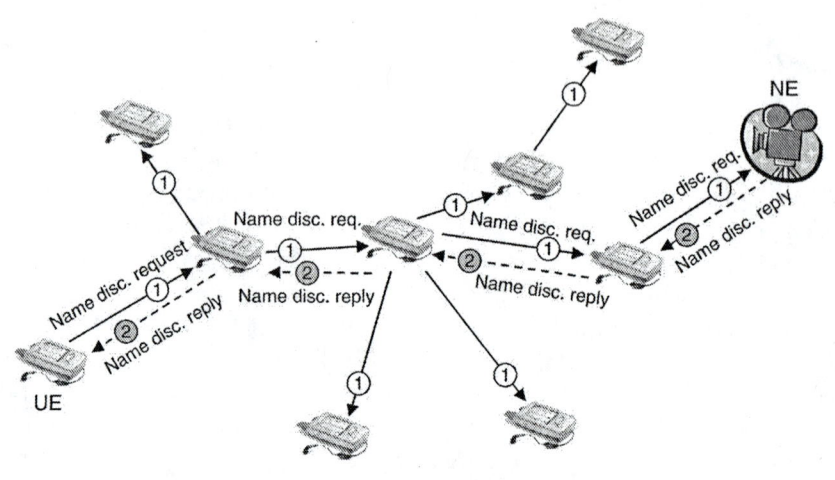

Figure 8.5   Distributed architecture with user entities (UEs) and named entities (NEs).

Distributed Architecture

As shown in Figure 8.5, this architecture contains no CCE. Instead, a user entity floods the name resolution request (NREQ) throughout its surroundings in the network. The flooding can be limited by a *flooding scope* parameter. Each named entity responds to an NREQ for its own name (i.e., no name caching is allowed) with a unicast name resolution reply (NREP).

Coordinator-Based Architecture

Certain nodes in the MANET are chosen to be Caching Coordinator Entities, a role quite similar to a DNS server. The interactions among user entities, named entities, and Caching Coordinators are illustrated in Figure 8.6. The CCEs announce their presences to the network by periodically flooding CCE announcement messages. The flooding can be limited to a certain number of hops, as determined by the *Coordinator announcement scope* parameter. Due to the dynamics of *ad hoc* networks, the NEs must be allowed to register their bindings automatically with the CCE; hence, an NE that receives CCE announcements unicasts name registration messages to register its bindings (i.e., names and associated IP addresses) with CCEs in its surroundings. A UE that has received CCE announcement messages may unicast an NREQ to a selected CCE to discover desired services. The CCE finally responds with a unicast NREP. The selected CCE is often referred to as an *affiliated* CCE.

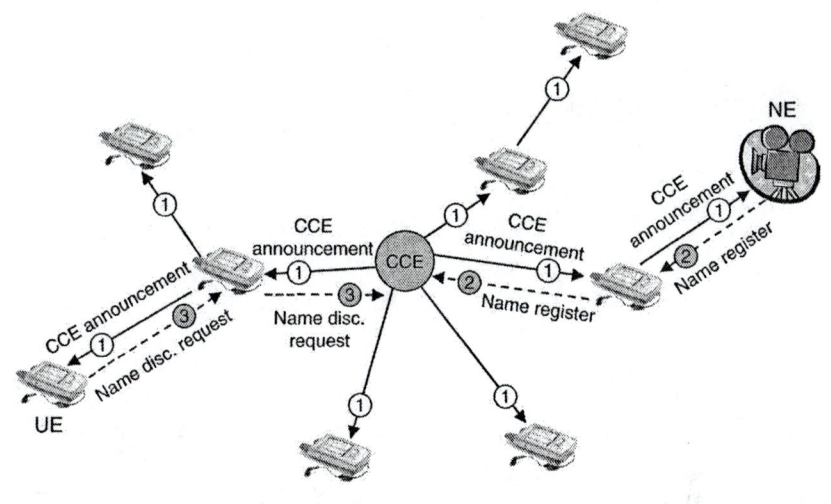

Figure 8.6 Coordinator-based architecture with user entities (UEs), Caching Coordinator Entities (CCEs), and named entities (NEs).

Hybrid Architecture

This architecture combines the two architectures described previously. User entities within the Coordinator announcement scope of one or more Caching Coordinator Entities will register their bindings with them; however, they must also be ready to respond to flooded NREQs. When a UE unicasts a NREQ to its affiliated CCE in line with the Coordinator-based architecture (Figure 8.6), the CCE responds with a positive or negative NREP; however, if there is no CCE in the surroundings of the UE or if the affiliated CCE returned a negative NREP, the UE will simply fall back to the distributed architecture (Figure 8.5). Both CCEs and NEs may respond to a flooded NREQ with a positive NREP that matches the requested service.

Intermediate Node Caching

An additional alternative (or a supplement) to the three architectures is to use intermediate node caching. The name resolution may, for example, follow the distributed architecture, but intermediate nodes are allowed to cache bindings found in NREPs that they are forwarding. Later, when receiving a NREP for a cached binding, the intermediate node resolves the name on behalf of the named entity according to the cached binding. The bindings should contain a lifetime value that controls for how long a binding should be kept valid in a cache.

Emerging Principles for Name Resolution in Reactive Ad Hoc Networks

Many *ad hoc* routing protocols are designed to conserve the scarce networking resources by reducing the need for and negative impact of systemwide flooded broadcasts. Flooded broadcasts exhaust the available bandwidth on the network and reduce the scalability in terms of number of nodes accommodated on the network. Broadcasts also consume the battery power of all networked devices. Reactive routing protocols, such as AODV, are designed to reduce to the greatest extent possible the need for systemwide flooded broadcasts associated with route discovery. Although route discovery is efficient in terms of reducing the number of flooded broadcasts from two to one, name resolution that is not optimized with respect to the route discovery would not work efficiently with reactive routing. The process of contacting a node on the MANET would require two or three broadcasts, as illustrated in the left side of Figure 8.7. Two broadcasts are necessary for the initial name resolution, because the user entities first have to flood a name resolution request. The reply returned by a node that can resolve the name also requires flooding, because the node does not have a route to the node that issued the request. Finally, the UE will have to flood a regular RREQ to find a route to the resolved IP address.

Alternatively, if the node resolving the name to an IP address is the named entity of the name (and not a node that has cached the name binding), the specification might mandate that the reply be returned by unicast to the user entity. Then, before replying, the node must first flood a RREQ to discover and set up a unicast route to the UE and send the name resolution reply by unicast along this route. It would then be possible to reduce the number of flooded broadcasts from three to two, because the UE already has a route to the resolved IP address as a result of the name resolution process when it contacts the NE. Further details are provided in Engelstad et al. [11,12].

It is possible to reduce the number of flooded broadcast to one, as illustrated in the right side of Figure 8.7. The solution is to use routing messages as carriers for name resolution. First, the user entity floods the name resolution request. By piggybacking the NREQ on a route request packet, a return route to the UE is formed as part of this flooding. By also piggybacking a name resolution reply on a route request packet, the NREP is sent by unicast along the return route to the UE. The RREP also ensures that a forward route is formed as part of this transmission. When the UE finally contacts the service at the IP address of the resolved name, the service request is unicast along the forward route that was put in place by the RREP. In summary, only one flooded broadcast is required in total. The idea of using routing messages as carriers has been proposed for name resolution in Engelstad et al. [11]. In fact, the same mechanism can also

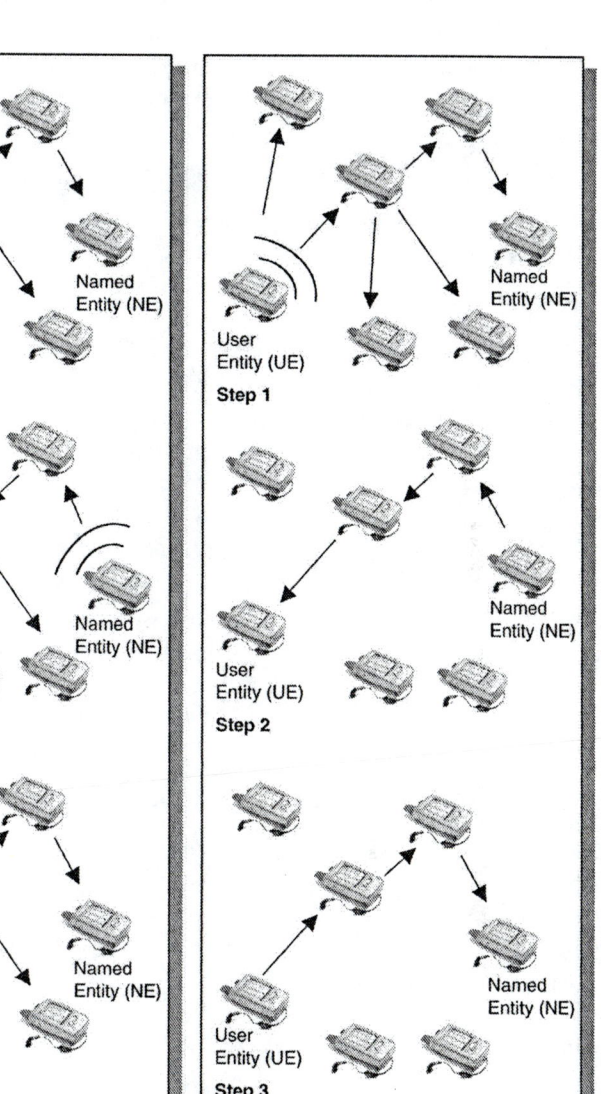

Figure 8.7   Name resolution without optimization (left) and optimized (right).

be used for service resolution, as we will see in the discussion on service discovery below.

Needless to say, this broadcast issue is a smaller problem in proactive *ad hoc* networks, because all unicast communication (including the NREP) can be sent along unicast routes established beforehand by the routing protocol. The broadcasting of the NREQ might benefit from reusing the efficient flooding capabilities (e.g., using multi-point relays) that are built-in features of many proactive protocols, including the Optimized Link State Routing (OLSR) protocol [13].

A Proposal for Name Resolution in Reactive Ad Hoc Networks

Overview

A mechanism for name resolution in MANET is proposed in Engelstad et al. [14]. It is mainly targeted at users that can supply their MANET node with a fully qualified domain name (FQDN) from the globally unique DNS name space. The user may have control over some part of the DNS name space or may have received the FQDN from an organization to which the user belongs or subscribes. The proposed name resolution scheme shares characteristics of the Link Local Multicast Name Resolution protocol and the multicast DNS protocol for local-link name resolution, presented earlier. The mechanism proposed for *ad hoc* networks specifies compressed message formats that allow for bandwidth-efficient name lookups. As an option, it also specifies message formats that reuse the format of DNS messages, which allows for name lookups that are fully compatible with DNS.

Name Resolution Requests and Replies

The proposed scheme uses the distributed architecture presented earlier, with no intermediate node caching. No Caching Coordinators are allowed; instead, only user entities and named entities are present on the MANET. When a name resolution request is broadcast by flooding throughout the MANET, each node with an NE processes the request. By carrying the NREQ as an extension to a route request (Figure 8.8), the number of broadcasts required for name resolution is reduced, as explained earlier in this chapter; thus, a return unicast route to the UE of the request is already in place for a node that wants to respond to the NREQ.

The destination IP address contained in the RREQ, which indicates the address to which a route is sought, is set to a predefined value. This can be a zero address, a broadcast address, or a preassigned multicast address to which no node can cache a route. Intermediate nodes without a valid address mapping for the requested name will not respond to the RREQ part of the message. The NREP is carried as an extension to an RREP message (Figure 8.8). The user entity sending the NREP will normally include its own IP address as the destination IP address in the RREP message to ensure that a forward route is formed. By carrying the response in an RREP message, a responder that is identified by the name that is sought can supply the UE with the resolved IP address in addition to a

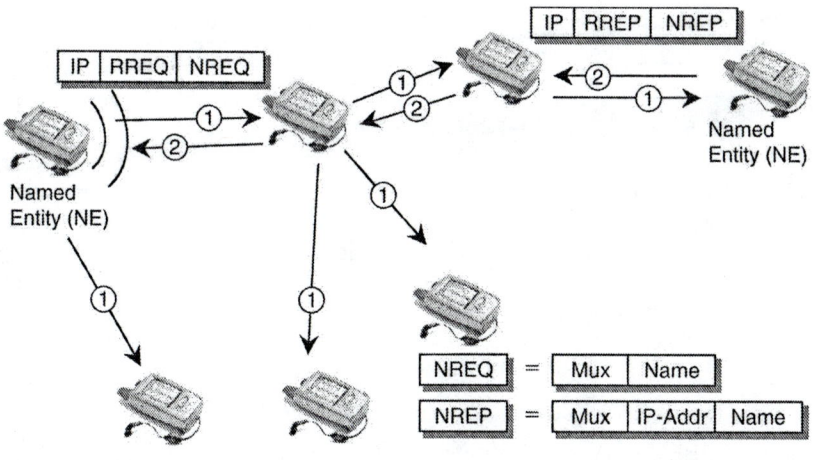

Figure 8.8 A name resolver (NR) floods a name resolution request (NREQ), carried by a route request (RREQ) header, throughout the network (1). A name server (NS) process with the requested name-to-address mapping unicasts a name resolution reply (NREP), carried by a route reply (RREP) header, back along the reverse route formed by the RREQ (2).

unicast route to that IP address. The UE does not have to issue an additional broadcast to discover a route to the resolved address when it subsequently tries to contact that address.

Interaction with External Networks

Mobile *ad hoc* networks might be connected to external networks through Internet gateways (IGWs). An IGW is a MANET router that also is a host or a router on an external network (with Internet connectivity). The IGW may have access to a conventional DNS server over the external network, and it may also provide other MANET nodes with access to the external network. The scheme proposes to use each IGW as a DNS proxy, as shown in Figure 8.9. The main advantage of using the IGW as a DNS proxy is that there is only one name resolution scheme used on the MANET and that nodes can resolve names in one single process.

Response Selection

A flooded name resolution request might result in reception of multiple name resolution replies. If the named entity present in the MANET has registered its name in the DNS, both the NE and each Internet gateway present in the network may return an NREP. Furthermore, if the NREP

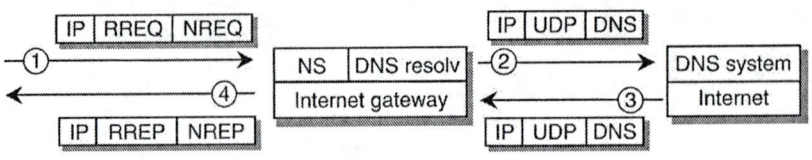

Figure 8.9 An Internet gateway (IGW) that receives a name resolution request (NREQ) on the MANET (1) may try to resolve the requested name using the Domain Name System (DNS) on the Internet (2). A successful response (3) may be injected as a name resolution reply (NREP).

contains a DNS SRV resource record to resolve a non-unique service name (as explained later in this chapter), many NEs present in the MANET may respond to the same NREP. To deal with the possibility of multiple responses, the user entity should wait for some milliseconds to collect responses that might arrive. The proposal includes response selection rules that ensure that a response from an NE present locally on the MANET will always have preference over responses arriving from other nodes, such as IGWs, because a local response might be more reliable and up to date. Furthermore, a direct route through the MANET should normally have preference compared to a route that goes through external networks. If the UE has multiple addresses from which to select after applying this selection rule, it should select (as a secondary selection rule) an IP address to which it has a route that is preferred by the routing protocol. That means that it will normally select an IP address to which it has valid routes and select the IP address that is the fewest hops away from the UE.

Service Discovery

The only information two endpoints communicating over the Internet must know, besides their own configuration, is the network address of the end point with which they are communicating. Any of the two end points might offer a wide range of services (e.g., e-mail, FTP, HTTP), but no defined way exists for the other end point to find out which of these services are being offered. Instead, the users themselves have to remember the name of the computer offering the services and to know in advance the Transmission Control Protocol (TCP) or User Datagram Protocol (UDP) port numbers associated with the set of services desired or to rely on the well-known port numbers. Would it not be easier if services and the associated IP addresses and port numbers could be searched for and discovered dynamically? This was indeed the case for Macintosh computers that used AppleTalk® networking. The Mac user did not require any assistance from a network administrator or even a complicated manual to locate services. Service discovery worked automatically and was operated by using a simple interface.

This section introduces the reader to various service discovery mechanisms that enable the same type of functionality that the AppleTalk protocol offered [15]. Further, the reader is introduced to service discovery mechanisms for an *ad hoc* network where traditional service discovery mechanisms might be deficient and where service discovery is particularly important because the availability of services is dependent on the network dynamics.

A Generic Model

Users typically want to accomplish a certain task, not query a list of devices to find out what services are running. It makes far more sense for a client to ask a single question ("What print services are available?") than to query each available device with a question ("What services are you running?") and sift through the results looking for printers. This latter approach, which is referred to as *device-centric*, not only is time consuming but also generates a tremendous amount of network traffic, most of it useless. On the other hand, a *service-centric* approach sends a single query that generates only relevant replies. This is what service discovery is all about. The process of discovering services on a computer network is similar to the process of looking up a name described in the previous chapter, but instead of asking "Who has this name?" service discovery asks the question: "Who offers these services?" Normally three entities, or *agents*, are necessary in a service discovery architecture for a computer network:

- A user agent (UA) that represents the network entity issuing a request to find a service
- A service agent (SA) that represents the service offered
- A directory agent (DA) for intermediate storage of information about available services

User Agent

The user agent represents the client part in the process of discovering services. The UA may be a software component or an end user who wants to locate a specific service. In most cases, the UA will offer a low-level functionality directed toward system components.

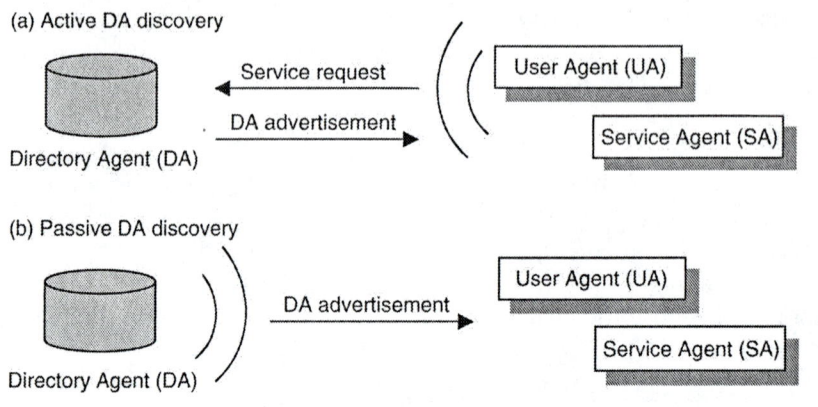

Figure 8.10 Active and passive directory agent (DA) discovery.

Service Agent

The service agent represents the service in the architecture. This can be the actual service or some entity representing it; such an entity is called a *proxy-SA*. An SA will advertise the services it offers by either broadcast or multicast service messages. If a directory agent exists, the SA will try to register with it.

Directory Agent

A directory agent acts as a cache and will merely collect information from the service agents and forward it on demand to user agents. The UAs and SAs can use either passive or active DA discovery to find a DA (Figure 8.10). The service model can be divided into three architectures: *distributed, centralized,* and *hybrid.* In the distributed architecture, no DAs are present. The UAs will issue a multicast message to find the SAs. The reply from the SAs can be unicast, or multicast if the UAs have the ability to use and manage a local cache of available SAs. In the centralized architecture, one or more DAs are present.

A variation of the centralized architecture is shown in Figure 8.11, where the UA is retrieving service discovery bindings from the DA according to the *pull* model. In this example, UAs are also proactively caching announced bindings according to the *push* model, as illustrated in the left part of the figure. (The *push* and *pull* approaches were described in the previous section on name resolution.) In the hybrid approach, a UA will first try to contact a DA according to the centralized architecture and fall back to the distributed approach if no DA can be located.

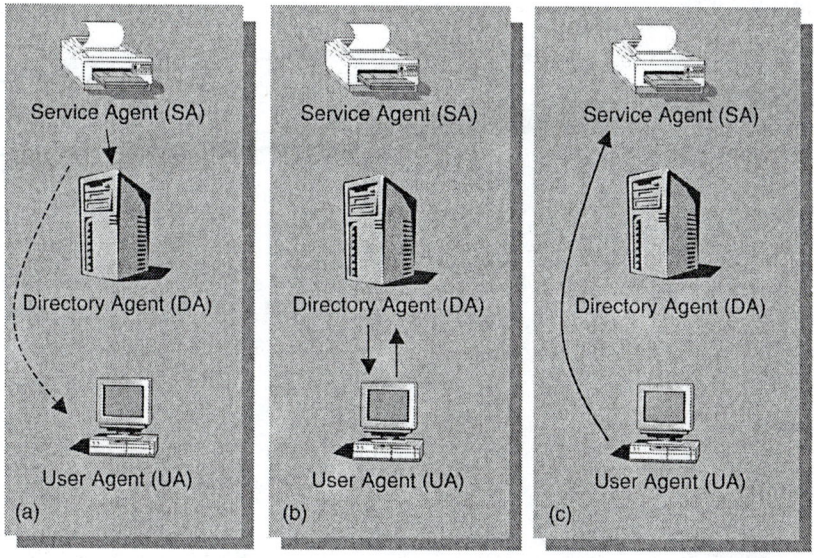

Figure 8.11 **A service agent announces its services, and announcements can be cached in both the directory agent (DA) and the user agent (UA) (a). If the UA is searching for a service it has not heard being announced, it can broadcast a service request or contact a DA directly. The DA will respond with information about where the service can be located (b). The UA can access a service it has learned about via a DA or from the SA directly (c).**

Service Discovery on the Internet

Current Practice for Service Discovery

When an application on computer A wants to connect with an application on computer B, computer A requires the network address of computer B and the port number of the service. The network address is necessary to route A's request to B. Because B may offer a multitude of services, a port number is used to distinguish between the different services offered at the same network address. Existing practice in IP networks is to go through a three-step process to obtain the network address and the port number of the services. The three steps consists of:

- Mapping the service name to the name of the computer offering the service
- Mapping the port number of the service to a service name
- Resolving the computer name to a network address using DNS or a local name service

Today's IP network has no widely used mechanism to undertake the first step of this process. A common method of advertising services is to map the network address of the computer offering services and the service name with the DNS service. Examples of such services can be a public Web server, which is given the name "www" (e.g., www.some_domain. com) or a public file transfer server, which is given the name "ftp" (e.g., ftp.some_domain.com).

The procedure for mapping port numbers to a service name is pretty simple, and a one-to-one relationship exists between the service name and the port number. The port number is assigned by the Internet Assigned Number Authority (IANA) and is normally maintained in a database on the local computer. The mapping between a service name and a port is typically:

<name> <port number>/<transport protocol> <aliases>

An example of mapping between a port and a service name is:

http → 80/tcp → www → www-http

The mapping of a network address to the computer offering the service is normally done using address resolution through DNS or through a cache at the local computer, if no access to any name server is available. As can be seen, the procedure for discovering services on the Internet is cumbersome and not very efficient.

Service Location Protocol

The Service Location Protocol (SLP) is an emerging Internet standard provided by the IETF for automatic service discovery on the Internet [16]. SLP provides a framework to allow networking applications to discover the existence, location, and configuration of networked services in enterprise networks. Traditionally, to locate services on the network, users of network applications have been required to supply the host name or network address of the machine that provides a desired service. Ensuring that users and applications are supplied with the correct information has, in many cases, become an administrative nightmare. SLP was inspired by the AppleTalk® protocol [15], a mechanism created by Apple that proved to be a huge success due to the simplicity and benefits of the solution. The only drawback was that it did not scale very well. The main focus of SLP is to be a mechanism that acts as an enabler for plug-and-play functionality in IP networks with automatic and dynamic bindings between services and service users. The SLP protocol introduces three major components into the network:

- *User agent (UA)* — The SLP user agent is a software entity that is looking for the location of one or more services. This search is usually implemented (at least partially) as a library to which client applications link, and it provides client applications with a simple interface for accessing SLP registered service information.
- *Service agent (SA)* — The SLP service agent is a software entity that advertises the location of one or more services. SLP advertisement is designed to be both scalable and effective, minimizing the use of network bandwidth through the use of targeted multicast messages and unicast responses to queries.
- *Directory agent (DA)* — The SLP directory agent is a software entity that acts as a centralized repository for service location information. In a large network with many UAs and SAs, the amount of multicast traffic involved in service discovery can become so large that network performance degrades. By deploying one or more DAs, both SAs and UAs make it a priority to discover available DAs, as the use of a DA minimizes the amount of multicast messages sent by the protocol on the network. The only SLP-registered multicast in a network with DAs is for active and passive DA discovery.

The Service Location Protocol introduces dynamic naming services without the need for any centralized name server or other agents. Because SLP uses IP multicast for this purpose, it requires the cooperation of IP routers that implement IP multicast. IP multicast is used for such features as IP-based audio and video broadcasting and video conferencing, but IP multicasting may not be completely implemented across some intranets. In the absence of IP multicasting, SLP name lookups will only work within the subnet on which they are performed or within the groups of subnets over which IP multicast is supported.

The Service Location Protocol suffers from a lack of implementation support, because individual companies, such as Apple and Microsoft, are pushing different service discovery technologies. Generally, standardization is a good thing but not very useful if no one provides implementation support for the standards. The SLP has proved to be useful, especially for UNIX® variants, and an open-source project exists to support service discovery on operating systems such as Linux™ and BSD. (An implementation of SLP can be downloaded from www.openslp.org.)

DNS Service Resource Records

An alternative to building up a new SLP infrastructure on the Internet is to reuse the existing DNS infrastructure and allow for service discovery as an extension to DNS. Extensions to DNS are enabled through the use of DNS resource records (DNS RRs). The most common resource record for service location is the DNS SRV resource record [17]. DNS SRV was originally designed to locate services on the global Internet. As an example, let us assume that company_A has implemented the use of SRV in its DNS server. Entries for the Hypertext Transfer Protocol (HTTP) and the Simple Mail Transfer Protocol (SMTP) would look like this in the zone file for company_A:

```
$ORIGIN company_A.com.
@                       SOA server.company_A.com.
                            root.company_A.com. (
1995032001 3600 3600 604800 86400 )
                        NS server.company_A.com.
http.tcp.www            SRV 0 0 80 server.company_A.com.
smtp.tcp                SRV 0 0 25 mail.company_A.com.
server                  A172.30.79.10
mail                    A172.30.79.11
```

For example, to locate an HTTP server that supports TCP and provides Web service, it does a lookup for:

```
_http._tcp.www.company_A.com.
```

If the use of SRV had been widely deployed, a DNS server would have answered with a list of Web servers that satisfied the searching criteria.

With no existing directory agent, this mechanism depends solely on the DNS system. Critics claim that the deployment of it puts an extra burden on an already overloaded DNS system. Furthermore, DNS SRV only allows for simple service name resolution and has little support for the type of service attribute negotiation that is accommodated by SLP.

XML Web Services/UDDI

An eXtensible Markup Language (XML) Web service is a service that accommodates direct interaction using XML-based messaging (such as Simple Object Access Protocol, or SOAP [18]) over Internet-based protocols, such as HTTP. The interfaces and bindings of the Web service are defined, described, and discovered by XML [19]. In addition to being able to describe and invoke a Web service, publication of and discovery of Web services should also be accommodated. The Universal Description, Discovery, and Integration (UDDI) specification [20] is commonly accepted to be the standard mechanism to handle this. UDDI registries provide a publishing interface to allow for creation and deletion of entries in the registry and an inquiring interface to search for entries in the registry by different search criteria. The interfaces are invoked by SOAP messages, and as such UDDI itself can be thought of as an XML Web service. Each entry in the UDDI registry contains three parts:

- The white pages contain business information.
- The yellow pages contain the service a business provides.
- The green pages contain the specific services provided and technical information sufficient for a programmer to write an application that makes use of the service.

Web services are described by XML using the Web Services Description Language (WSDL) specification [21].

Other Service Discovery Protocols

The Salutation protocol [22] released by the Salutation Consortium in 1996 predates SLP. It introduces the same concept as a directory agent, referred to as the Salutation Manager (SLM). In the Salutation protocol, user agents and service agents are referred to as clients and servers, respectively. The Salutation Manager Protocol (SMP) and Cisco's Transport Manager™ are used for the actual communication, utilizing remote procedure calls (RPCs). SLM will also assist in establishing a session pipe over which a service access between the client and the server can occur.

Jini™ [23] is another technology for service discovery that runs on top of Java. It allows clients to join a Jini lookup service, which maintains dynamic information about services in the network. The client can use it for simple service discovery by requesting information about a particular device. An attractive feature of Jini is that it also allows clients to download Java code from the lookup service which is used to access the service; however, it requires that the server must have already uploaded the Java proxy that the client downloads from the lookup service. Jini also supports the concepts of federations, where groups of devices may register with each other to make their services available within the group.

Service Discovery on Link Local Networks

Simple Service Discovery Protocol

The Simple Service Discovery Protocol (SSDP) [24] is a part of Microsoft's Universal Plug and Play (UPnP™) [25] and provides a mechanism that network clients can use to discover network services. UPnP supports self-configuration networks by enabling the ability to automatically acquire an

IP address, announce a name, learn about the existence and capabilities of other elements in the network, and inform others about their own capabilities.

The UPnP protocols are based on open Internet-based communications standards. UPnP is based on IP, TCP/User Datagram Protocol (UDP), HTTP, and XML. The SSDP protocol specifies the use of multicast of UDP/HTTP for announcements of services. The content of the service announcements are described using XML. Hypertext Transfer Protocol Unicast (HTTPU) and Hypertext Transfer Protocol Multicast (HTTMU) are used by SSDP to generate requests over unicast and multicast. The SSDP architecture introduces three entities into the network:

- *SSDP service* — The SSDP service is a service agent and represents the individual resources in an SSDP-enabled network. The agent is defined in two versions, depending on whether or not an SSDP proxy is available in the network. An SSDP service without proxy support is a simple service where all messages are sent on an SSDP reserved multicast group.
- *SSDP client* — The SSDP client is a user agent. Initially, the SSDP client will search for a proxy, followed by a search for other relevant resources. If an SSDP proxy is available, all requests are done using unicast. If it is not, the SSDP searches for services using multicast. The SSDP client will cache all information about services and uses a time stamp to manage the accuracy of the cache.
- *SSDP proxy* — The SSDP proxy is a directory agent that gathers information about available resources in the network. The proxy can be viewed as a regular resource that caches and manages all service information. An SSDP proxy is not a mandatory element in an SSDP-enabled network, but it does improve the scalability when deployed in large networks.

Multicast DNS

Domain Name System service discovery is a way of using standard DNS programming interfaces, servers, and packet formats to browse the network for services. As shown earlier, multicast DNS (mDNS) [26] can be used to resolve names without the use of any DNS server. The same multicast mechanism can be used to search for services, by requesting a binding for the type of service wanted instead of requesting a binding for a name. This is illustrated in Figure 8.12.

When a mDNS query is sent out for a given service type and domain, any matching services reply with their names. The result is a list of available services from which to choose. For example, an application that is searching

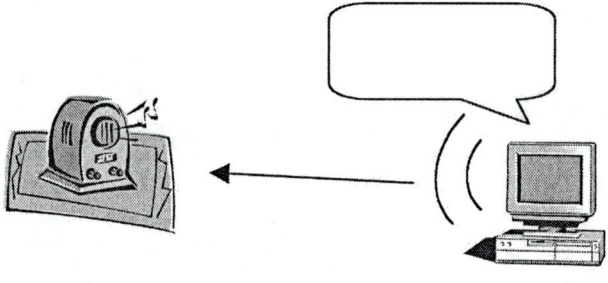

Figure 8.12   The principle of multicast DNS.

for a printer that supports TCP and is located in the company_A domain would issue a query for:

```
_lpr._tcp.company_A.com
```

Then, every printer attached to the LAN will answer with information about its services, such as color and pages per minute. Using a mechanism for automatic service discovery greatly simplifies the job of connecting PCs, terminals, wireless units, and consumer electronics.

The caching of multicast packets can prevent hosts from requesting information that has already been requested; for example, when one host requests, say, a list of LPR print spoolers, the list of printers comes back via multicast so all local hosts can see it. The next time a host requires a list of print spoolers, it already has the list in its cache and does not have to reissue the query.

Service Discovery in Ad Hoc Networks

Service Discovery Mechanism for MANETs

Discovery of services and other named resources is an important feature for the usability of mobile *ad hoc* networks. The characteristics of *ad hoc* networks were described earlier. We recall that a MANET is anticipated to be without any preexisting infrastructure and that nodes may enter or leave the network at any time. This makes efficient and timely service discovery a challenging task. In a MANET, any node may in principle operate as a server and provide its services to other MANET nodes. Any node may also operate as a client and use the service discovery protocol to detect available services in the network. The service attributes include service characteristics and service binding information, such as IP addresses, port numbers and protocols, that allow the client to initiate the selected service on the

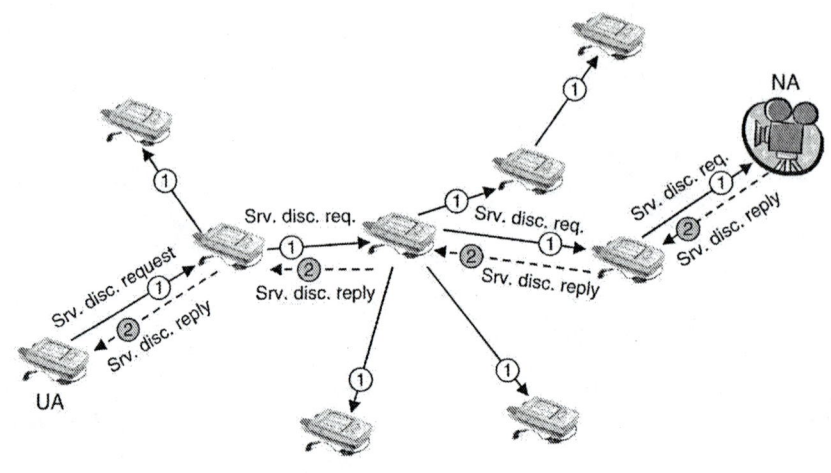

Figure 8.13 Distributed architecture.

appropriate server. Existing service discovery mechanisms, described in previous sections, are designed with a fixed network in mind and might not fit well with MANETs. Before a service discovery mechanism for *ad hoc* networks can be designed, we must determine the necessary principles for service discovery in *ad hoc* networks and evaluate which service discovery architecture that is most suitable. The reader will observe that name resolution and service discovery have many similar features in terms of both discovery principles and possible architectures.

Service Location Architectures for Service Discovery on MANETs

The architectures available for name resolution, as described for name resolution above, are also available for service discovery. In the context of service discovery, the Caching Coordinator Entity is normally referred to as a service coordinator (SC), the user entity is referred to as a user agent (UA), and the named entity (NE) is referred to as a service agent (SA). Furthermore, the architectures are often referred to as distributed, service-coordinator-based, and hybrid service location architectures. The distributed architecture for MANETs is illustrated in Figure 8.13, and the service-coordinator-based architecture is shown in Figure 8.14. The hybrid architecture is a combination of the two: The UA tries to discover services according to the service-coordinator-based architecture but falls back to the distributed approach if the selected SC does not have the desired binding or if no SC can be found.

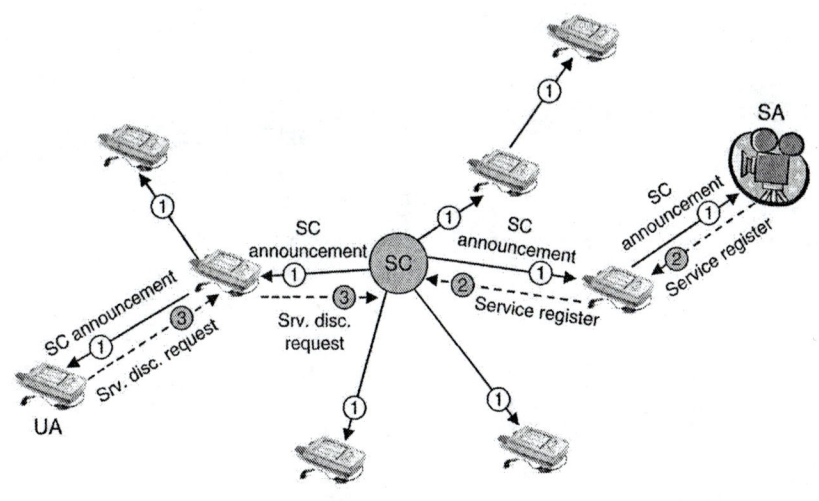

Figure 8.14   Service-coordinator-based architecture.

Emerging Principles for Service Discovery on Reactively Routed MANETs

The emerging principles for name resolution in reactive MANETs, in which resolution requests and replies are carried by routing messages (outlined earlier), are also useful for service discovery. In the context of service discovery, service discovery requests (SREQs) and service discovery replies (SREPs) are piggybacked on RREQ and RREP packets, respectively (Figure 8.15). The advantages of piggybacking service discovery on routing messages include:

- Reverse routes to the user agent (i.e., client) are established along with the SREQ so no additional route discovery is necessary to relay the SREP back to the requestor.
- Forward routes to the SC are established along with the SC announcements so SREQs and service registrations can be unicast to the SC.
- A forward route is established along with the SREP so no additional route discovery is necessary for further communication with the node issuing the reply.

Figure 8.15 shows how service discovery can be streamlined with the reactive routing protocol. Service Discovery Requests are piggybacked on routing request packets, and service discovery replies are piggybacked on routing reply packets. In addition, for the hybrid architecture, the SC

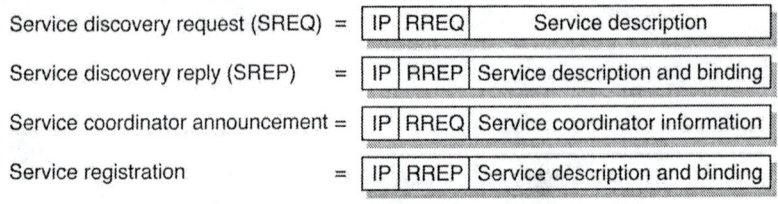

Figure 8.15   **Routing packets carry service discovery messages.**

announcements are piggybacked on RREQ packets, and service registrations are piggybacked on RREP packets. Thus, both the SC-based, hybrid and distributed architectures can take advantage of this procedure.

Proposed Solution for Service Discovery in Reactive Ad Hoc Networks

A solution for service discovery in reactive *ad hoc* networks has been proposed in Koodli and Perkins [27]. It uses basically the same mechanism as was presented for name resolution earlier in the chapter. It is based on the distributed architecture without the use of any service coordinators, but here the intermediate nodes are allowed to cache service bindings and respond immediately if a valid binding is found. It also uses the same technique to carry the discovery messages by the routing packets to allow both services and the routes to the nodes providing these services to be discovered in one round-trip. Koodli and Perkins [27] defined a *service binding* as a mapping of a service name to an IP address. Different encoding schemes, such as a service port request or service URL, can be used to request a binding for an IP address. A service discovery request for a service URL contains a service-type string and a service request predicate of formats that are defined by SLP. The format of the URL and the authentication block contained in the corresponding service discovery reply are also defined by SLP; hence, not only are formats of SLP reused but the authorization block also ensures that the service authorization features of SLP are maintained. The use of SREQs for service ports assumes that the user agents know in advance the well-defined (TCP or UDP) port number associated with the requested service; hence, the SREQ only has to contain the port number associated with the service application requested. The proposed service discovery protocol considers the case where the UA has neither a service binding nor an active route to a node providing the desired service. It also considers the case where the route is active but the service binding has expired (or is absent) and the case where the service binding is active but the route has expired.

Evaluation of Service Location Architectures in Ad Hoc Networks

As a slight simplification, one could say that all of the service discovery protocols presented earlier are based on two baseline mechanisms for the management of service discovery information:

- Information about services offered on the network is stored on one or a few centralized nodes, referred to as service coordinators (SCs) in this chapter.
- Information about each service is stored on each node that is offering the service.

In previous sections, we defined the service discovery architectures according to the two mechanisms above. A solution that is based only on the first mechanism is a service-coordinator-based architecture, and a solution based only on the second mechanism is a distributed architecture. Finally, a solution based on a mixture of both the first and the second mechanisms is a hybrid architecture. In the next section, we evaluate the performance of the hybrid and distributed architectures in a reactively routed MANET. The architectures have been presented in detail earlier in the chapter.

Architecture Evaluation

The evaluation of the distributed and hybrid architectures is based on results from Engelstad et al. [28]. The architectures can be configured by different settings of the following two parameters:.

- *Flooding scope* — This parameter determines the maximum number of hops a flooded service discovery request is allowed to traverse in the network (e.g., the flooding scope is four hops in Figure 8.13 and two hops in Figure 8.14).
- *SC announcement scope* — This parameter determines the maximum number of hops a flooded SC announcement is allowed to traverse in the network (e.g., Figure 8.14 illustrates a situation with an SC announcement scope of two hops). This parameter is used only in the hybrid architecture. Alternatively, the distributed architecture can be considered as a special case of a hybrid architecture where the SC announcement scope is set to zero.

The objective is to optimize the benefits of additional service availability provided by the use of service coordinators against the cost of additional overhead and possibly higher delay. We will consider the performance measured by delay, by the service availability, and by the message overhead.

Engelstad et al. [28] observed that the differences in delays between the two architectures are only on the order of a few milliseconds. Because service discovery is normally part of the service initiation, users would normally accept an initial delay (e.g., when retrieving a Web page on the Internet or for setting up an IP telephony call); hence, it was concluded that the small observed differences in delay between the two architectures should be considered negligible in this context. With delay out of the picture, the key question is reduced to whether the increased service availability is worth the increase in message overhead. The service availability can be defined as [29]:

$$\text{Service availability} = \frac{\text{Number of positive service replies}}{\text{Total number of service requests generated}}$$

A positive service discovery reply indicates successful contact with this server via the given access information (i.e., a route to the resolved server can be found).

It is observed in Figure 8.16 that the service availability is indeed higher with the hybrid approach. Figure 8.16 also shows how the presence of service coordinators (i.e., for the hybrid architecture) influences the service availability. When comparing architectures that use the same flooding scope, we find that the hybrid architecture improves the service availability as compared to the distributed query-based architecture. The main reason why SCs improve the service availability is that in some cases the SC will be positioned in between the user agent and service agent. We may return to Figure 8.14 for an example of such a situation. Here, the UA and SA are four hops apart, but both are only two hops away from the SC. The SC announcement and flooding scopes are both of two hops; hence, the SA is able to register its services with the SC, and the UA is able to discover the server by means of the SC. However, because the UA is four hops away from the SA, the UA would not able to discover the SA if the distributed architecture with a flooding scope of two hops had been used.

From Figure 8.16 we observe that, with SC announcement scopes of one or two hops, the service availability is improved by 8.7 or 20.9 percent, respectively, at a server density of 5 percent. Because the introduction of SCs improves the service availability, it comes as no surprise that the service availability increases with an increasing SC announcement scope.

Although the service coordinators introduced in the hybrid architecture yield higher service availability, they also result in extra message overhead, as observed in Figure 8.17. The SCs introduce two proactive elements to the network — namely, SC announcements and service registrations. These messages will take up a fixed bandwidth regardless of whether or not there are clients doing service discoveries. In addition, these two types of

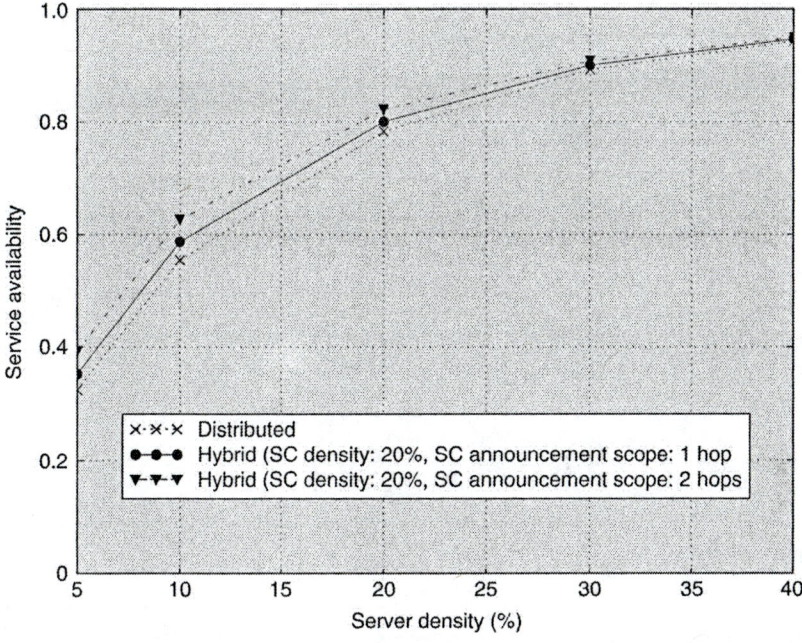

Figure 8.16   The introduction of SCs improves the service availability.

messages will also trigger pure route discovery messages when a reactive routing protocol is being used. By comparing Figure 8.16 and Figure 8.17, we observe that the additional cost of using SC in terms of percentage increase in message overhead is much higher than the additional benefits provided in terms of percentage increase in service availability.

A more rigorous analysis that compares the two architectures has been undertaken in Engelstad et al. [28]. It takes into consideration a large range of control parameters, such as server density, service coordinator density, flooding scopes, SC announcement scopes, reasonable request frequencies, number of different types of services, level of mobility, and so forth. It is also argued that the conclusion is valid independent of the lengths of the service discovery messages.

In Engelstad et al. [28], it is generally observed that for any hybrid configuration with a given service coordinator announcement scope and flooding scope it is always possible to find a distributed configuration (with some flooding scope) that outperforms the hybrid configuration in terms of both higher service availability and lower messaging overhead. As the opposite is not the case, it is concluded that the distributed architecture outperforms the hybrid architecture. Service discovery protocols that are using SCs (or functionality similar to directory agents) do not

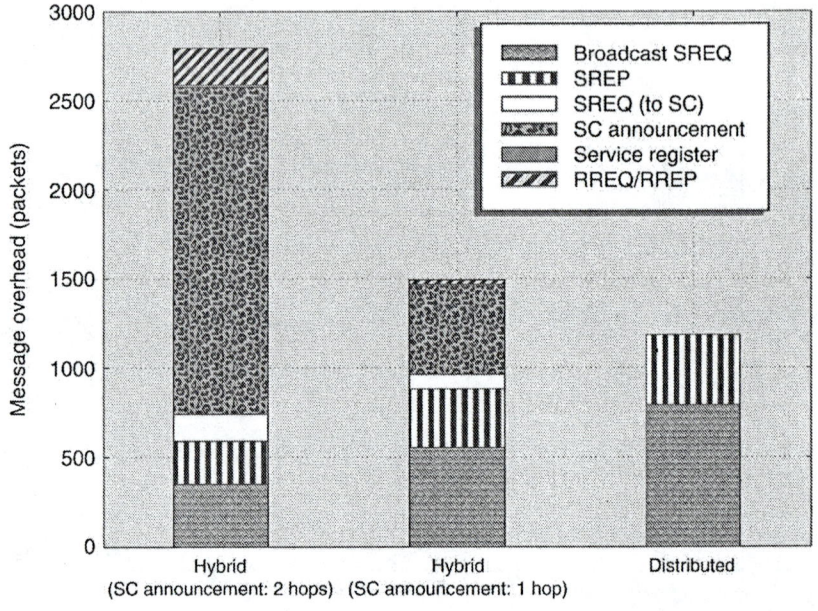

Figure 8.17 Detailed comparison of message overhead by message type (server density at 20 percent, flooding scope of two hops).

work well in *ad hoc* networks with reactive routing. The main reason is that the increase in service availability by adding SCs is negligible compared to the extra message overhead it causes.

In addition to the analyses presented in Engelstad et al. [28], several other arguments would suggest against introducing service coordinators in MANETs at large. First, the distributed architecture is considerably less complex than the hybrid architecture. Furthermore, in a dynamic topology with network entries and departures, the service coordinators of the hybrid architecture have the disadvantage of sometimes providing the user agents with "false positives" (i.e., with outdated bindings of servers that have already left the network). Moreover, the hybrid approach may call for a separate complicated mechanism for electing SCs, which might require a substantial amount of network resources.

References

[1] Mockapetris, P., *Domain Names: Concepts and Facilities*, Request for Comments 1034, Internet Engineering Task Force (IETF), November, 1987 (http://www.ietf.org/rfc/rfc1034.txt).

[2] Mockapetris, P., *Domain Names: Implementation and Specification*, Request for Comments 1035, Internet Engineering Task Force (IETF), November, 1987 (http://www.ietf.org/rfc/rfc1035.txt).

[3] Microsoft TechNet, *Windows Internet Naming Service (WINS)*, Microsoft Corporation, Redmond, WA (http://www.microsoft.com/technet/archive/windows2000serv/evaluate/featfunc/nt5wins.mspx).

[4] Sun Microsystems, *The Network Information Service (NIS)/Yellow Pages*, Sun Microsystems, Santa Clara, CA (http://www.sun.com).

[5] Wellington, B., *Secure Domain Name System (DNS) Dynamic Update*, Request for Comments 3007, Internet Engineering Task Force (IETF), November, 2000 (http://www.ietf.org/rfc/rfc3007.txt).

[6] Chesire, S. and Krochmal, M., *Multicast DNS*, IETF Internet Draft, draft-cheshire-dnsext-multicastdns-04.txt, February, 2004 (http://files.dns-sd.org/draft-cheshire-dnsext-nbp.txt).

[7] Aboba, B., Thaler, D., and Esibov, L., *Linklocal Multicast Name Resolution (LLMNR)*, draft-ietf-dnsext-mdns-39.txt, March, 2005 (http://tools.ietf.org/wg/dnsext/draft-ietf-dnsext-mdns/draft-ietf-dnsext-mdns-39.txt).

[8] Macker, J. and Corson, S., *Mobile Ad Hoc Networking (MANET): Routing Protocol Performance Issues and Evaluation Considerations*, Request for Comments 2501, Internet Engineering Task Force (IETF), January, 1999 (http://www.ietf.org/rfc/rfc2501.txt).

[9] Perkins, C.E., Royer, E.M. and Das, S.R., *Ad Hoc On-Demand Distance Vector (AODV) Routing*, Request for Comments 3561, Internet Engineering Task Force (IETF), July, 2003 (http://www.ietf.org/rfc/rfc3561.txt).

[10] Johnson, D.B., Maltz, D.A. and Hu, J.-C., *The Dynamic Source Routing Protocol*, IETF Internet Draft, draft-ietf-manet-dsr-10.txt, July, 2004 (http://www.ietf.org/internet-drafts/draft-ietf-manet-dsr-10.txt).

[11] Engelstad, P.E., Do, T.V. and Egeland, G., Name resolution in on-demand MANETs and over external IP networks, in *Proc. of the IEEE Int. Conf. on Communications 2003 (ICC 2003)*, Seattle, WA, May 28–30, 2003 (http://www.unik.no/~paalee/publications/NR-paper-for-ICC2003.pdf).

[12] Engelstad, P.E., Do, T.V. and Jønvik, T.E., Name resolution in mobile *ad hoc* networks, in *Proc. of the 10th Int. Conf. on Telecom 2003 (ICT 2003)*, Tahiti, February 23–March 1, 2003 (http://www.unik.no/~paalee/publications/NR-paper-for-ICT2003.pdf).

[13] Clausen, T. and Jacquet, P., Eds., *Optimized Link State Routing Protocol (OLSR)*, Request for Comments 3626, Internet Engineering Task Force (IETF), October 2003 (http://www.ietf.org/rfc/rfc3626.txt).

[14] Engelstad, P.E., Egeland, G., Koodli, R., and Perkins, C.E., *Name Resolution in On-Demand MANETs and over External IP Networks*, IETF Internet Draft, draft-engelstad-manet-name-resolution-01.txt, February, 2004 (http://www.unik.no/personer/paalee/publications/draft-engelstad-manet-name-resolution-01.txt).

[15] Apple Computers, *Inside AppleTalk*, 2006, http://www.developer.apple.com/MacOs/opentransport/docs/dev/Inside_AppleTalk.pdf.

[16] Guttman, E., Perkins, C., Veizades, J., and Day, M., *Service Location Protocol, Version 2*, Request for Comments 2608, Internet Engineering Task Force (IETF), June, 1999 (http://www.ietf.org/rfc/rfc2608.txt).

[17] Gulbrandsen, A., Vixie, P., and Esibov, L., *A DNS RR for Specifying the Location of Services (DNS SRV)*, Request for Comments 2782, Internet Engineering Task Force (IETF), February, 2000 (http://www.ietf.org/rfc/rfc2782.txt).

[18] Gudgin, M. et al., *SOAP Version 1.2, Part 1: Messaging Framework*, World Wide Web Consortium (W3C) Recommendation, June, 2003 (http://www.w3.org/TR/soap12-part1/).

[19] Bray, T. et al., *Extensible Markup Language (XML) 1.0 (Second Edition)*, World Wide Web Consortium (W3C) Recommendation, October, 2000 (http://www.w3.org/TR/2000/REC-xml-20001006).

[20] OASIS UDDI, *Universal Description, Discovery, and Integration (UDDI) 2.0*, Organization for the Advancement of Structured Information Standards, Boston, MA (http://www.uddi.org).

[21] Christensen, E., Curbera, F., Meredith, G., and Weerawarana, S., *Web Services Description Language (WSDL) 1.1*, World Wide Web Consortium (W3C) Note, March, 2001 (http://www.w3.org/TR/2001/NOTE-wsdl-20010315).

[22] The Salutation Consortium, *Salutation Architecture Specification Version 2.1*, 1999.

[23] Sun Microsystems, *Jini Network Technology*, Sun Microsystems, Santa Clara, CA (http://www.jini.org).

[24] Goland, Y. et al., *Simple Service Discovery Protocol 1.0*, IETF Internet Draft, draft-cai-ssdp-v1-03.txt, October, 1999 (http://www.ietf.org/internet-drafts/draft-cai-ssdp-v1-03.txt).

[25] Microsoft, *Universal Plug-and-Play (UPnP) Forum*, Microsoft Corporation, Redwood, WA (http://www.upnp.org).

[26] Cheshire, S. and Krochmal, M., *Multicast DNS*, IETF Internet Draft, draft-cheshire-dnsext-multicastdns-04.txt, February, 2004 (http://mirrors.isc.org/pub/www.watersprings.org/pub/id/draft-cheshire-dnsext-multicastdns-04.txt).

[27] Koodli, R. and Perkins, C.E., *Service Discovery in On-Demand Ad Hoc Networks*, IETF Internet Draft, draft-koodli-manet-servicediscovery-00.txt, October, 2002 (http://mirrors.isc.org/pub/www.watersprings.org/pub/id/draft-koodli-manet-servicediscovery-00.txt).

[28] Engelstad, P.E., Zheng, Y., Koodli, R., and Perkins, C.E., Service discovery architectures for on-demand *ad hoc* networks, *Int. J. Ad Hoc Sensor Wireless Networks*, 2(1), 27–58, 2006.

[29] Engelstad, P.E. and Zheng, Y., Evaluation of service discovery architectures for mobile *ad hoc* networks, in *Proc. of the 2nd Annual Conf. on Wireless On-Demand Networks and Services (WONS 2005)*, St. Moritz, Switzerland, January 19–21, 2005.

Chapter 9

Data Synchronization

Sachin Agarwal

CONTENTS

Introduction

Distributed applications are modeled around replicating copies of the same data on many hosts in a network for a variety of reasons. For one, system designers can alleviate the single-server implosion problem and instead distribute client requests for data across many hosting servers. Second, making data locally available on a host speeds up applications because the applications do not block for network input/output as data is transmitted; in some networks, this latency can be on the order of hundreds of milliseconds. Also, a desirable side effect is data robustness due to the existence of redundant copies on geographically separated hosts.

Replicating data on mobile devices allows users to cut the chord and carry data around without the need to always be connected to the data server. Newer mobile devices are increasingly equipped with WiFi [1], Bluetooth® [2], and cellular-based data links to the Internet. Thus, the amount of dynamic content stored on these devices has increased as compared to earlier devices that only offered very limited or no connectivity; consequently, stored data was rarely updated. The ability to disconnect with the network, make local changes, and then synchronize these changes back into the system makes the mobile device a vital extension of modern databases and collaborative tools.

Data synchronization allows users to work with offline data and make independent edits (additions, deletions, and modifications) to the data while disconnected from the network. These edits are later synchronized with a server or possibly other mobile devices. Thus, data synchronization is an enabling mechanism that removes the stringent requirement of maintaining constant connectivity and allows users to run applications while being disconnected from the network.

Data synchronization middleware has been an important component of mobile device software offerings. For example, Microsoft provides the ActiveSync® [21] application on its Pocket PC platform to enable synchronization between users' PCs and Pocket PCs. In particular, ActiveSync handles Pocket PC synchronization with address books, calendars, and other personal information management (PIM) software installed on users' computers. Similarly, the Palm® platform offers HotSync® [20], which allows similar communication and synchronization between Palm personal digital assistants (PDAs) and user computers. As another example, in the cellular handset arena some Siemens and Sony Ericsson handsets use XTNDConnectPC [28].

In this chapter, we explore the data synchronization process in mobile devices that have limited power resources, high latency connections, and limited persistent storage memory. We first discuss why efficient data synchronization is vital to many distributed applications, in both the wired and wireless mobile world. We discuss the important problem of conflicts and some well known strategies to resolve these conflicts. We then survey a number of commonly used synchronization techniques by examining their high-level algorithms and compare their characteristics, highlighting their strengths and weaknesses.

In this context, we divide our data synchronization survey into two categories. The first category deals with the "atomic" process of two-device synchronization — that is, the low-level algorithms that two hosts (say, a mobile Pocket PC and a computer) use to synchronize their data. The second category in our survey is that of multiple mobile devices synchronizing with each other. This category includes enterprisewide mobility solutions that permit offline modes of operation on multiple devices when they are outside network coverage. These solutions seek to provide seamless data synchronization when devices reconnect to the network.

Our discussion includes the various tradeoffs among network latency, computational complexity, and the amount of memory used to store synchronization metadata. These tradeoffs can serve as guides to system designers when they choose one data synchronization algorithm over another based on the requirements of their application and characteristics of the underlying network and devices. In our discussion, we consider distributed data to be a *database* that is replicated on many hosts and must be synchronized periodically. Usually, this database is a set of key–value pairs representing information in a distributed application. The value in the key–value pair is a *data item*. We use the terms *hosts*, *devices*, and *mobile devices* interchangeably, depending on the context.

Need for Efficient Data Synchronization

The usability of distributed systems often deteriorates as hosts edit their databases independently, making the data more and more dissimilar, a condition referred to as data *inconsistency* among distributed copies of the database. *Consistency* is defined as the (desirable) condition in which a read operation on a data item on any of the hosts participating in the distributed system yields the same result at a given time. Moreover, the result of a read operation should correspond to the value written to the data item during the last write operation (on that data item) regardless of the hosts on which the read and write operations are performed. This type of strict data consistency is difficult to achieve on most networks. Instead, the conditions are usually relaxed and a certain latency between application of an edit to a data item and its propagation to other hosts in the network is tolerated. This tolerance value is an application-dependent

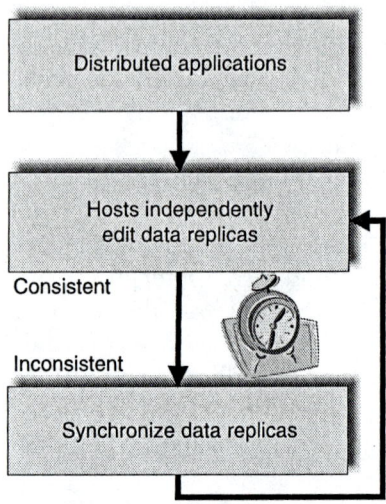

Figure 9.1 Distributed applications repeatedly call data synchronization routines during their lifetimes.

design parameter, although hard constraints, such as underlying network delays, set the minimum bounds on when an update will be propagated to all other hosts.

Data synchronization has been a vital part of almost all distributed systems. Figure 9.1 shows a typical distributed application algorithm with respect to replication and synchronization. The important observation is that data synchronization is a repetitive and frequent process during the life of the application; hence, each run of the data synchronization protocol must be highly efficient because it is invoked multiple times on multiple hosts participating in the distributed application.

Distributed databases and file systems such as Coda [3] must synchronize data on a periodic basis to be useful to users, as we discuss in a later section. Numerous Internet-based services, such as OSPF [4], link state updates, PGP key services [5,6], the Internet Domain Name System (DNS) [7], document versioning and backup systems, and server mirroring; also, Akamai®-type [8] load-balancing applications routinely use data replication and synchronization in one form or another. Open-source utilities such as rsync [11] and Unison [12] are used to synchronize files with a minimum transfer of redundant information. These utilities actually synchronize file data instead of key–value-pair databases and successfully preserve the structure of the text while synchronizing file edits efficiently. Document and source code versioning systems such as Concurrent

Versions System (CVS) [30] and Subversion (SVN) [31] use variants of replication and file synchronization algorithms to keep consistent copies of edited files.

In the mobile world, many PDAs and cellphones offer data synchronization facilities with users' computers. These facilities support personal information management (PIM) tools, such as address books and calendars, that can be made consistent across many mobile devices and also on Internet-based PIM. For example, Bayou [9] proposed to enable collaboration among mobile users who intermittently connect to access distributed calendars, e-mails, documents, databases, etc. Real-time collaborative and database applications (e.g., inventory control in warehouses that use PDAs and tablets as inventory database front ends) also require data synchronization on a continual, repetitive, and often real-time basis.

Mobile devices are especially dependent on data synchronization middleware to maintain consistency with other user-owned devices such as laptops, PCs, and Internet-based PIM solutions because the data synchronization middleware that connects them to PCs is often the only communication conduit to the outside world. In addition, the fact that mobile devices almost always find the most utility in disconnected situations makes data synchronization an indispensable part of their functionality. Many enterprisewide mobile device deployments routinely use some form of data synchronization to keep mobile devices updated. For example, mobile extensions of database servers such as Oracle® Lite™ [32] and Sybase® iAnywhere™ [33] have elaborate mechanisms to resynchronize and merge mobile client data as they return to connected states.

Conflicts While Synchronizing Data

When two hosts synchronize data, a problematic condition known as a *conflict* can arise. A conflict occurs when the synchronization algorithm is unable to make an unambiguous decision about which copy of a synchronizing host's particular data item is to be used and which copies are to be discarded (i.e., each host's database has a different data value corresponding to the same key). An example of this type of situation is illustrated in Figure 9.2. Initially, the home computer and the office computer are both synchronized and their databases are identical, but then the users of both independently modify the data item corresponding to the key "Wed," thus creating a conflict. When the two computers synchronize, it will not be possible to determine which data item to overwrite and which data item to keep for the key "Wed" ("Walk" or "Weep").

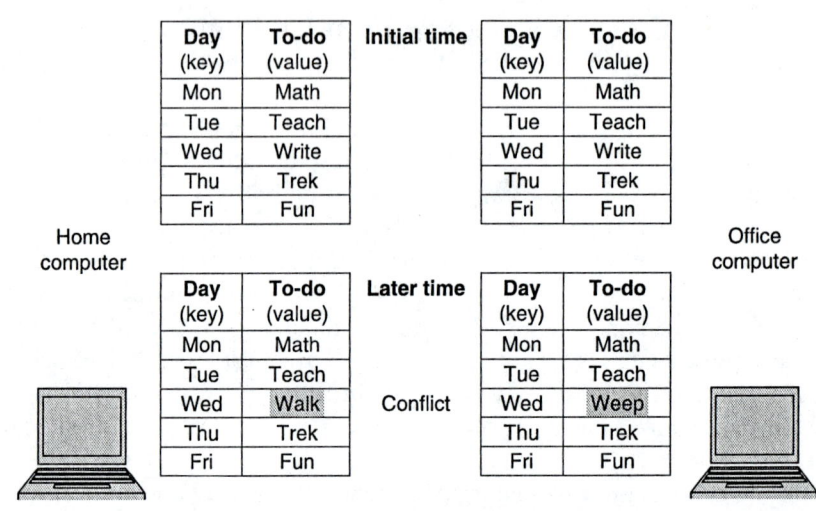

Home
computer

Office
computer

Day (key)	To-do (value)	Initial time	Day (key)	To-do (value)
Mon	Math		Mon	Math
Tue	Teach		Tue	Teach
Wed	Write		Wed	Write
Thu	Trek		Thu	Trek
Fri	Fun		Fri	Fun

Day (key)	To-do (value)	Later time	Day (key)	To-do (value)
Mon	Math		Mon	Math
Tue	Teach		Tue	Teach
Wed	Walk	Conflict	Wed	Weep
Thu	Trek		Thu	Trek
Fri	Fun		Fri	Fun

Figure 9.2   A conflict occurs when both hosts modify the same data item and then try to synchronize at a later time.

Hosts usually make a reference copy of the synchronized database at the end of synchronization that they can use to differentiate between new edits and the original values so they can decide which data items have been modified since the last data synchronization, but this modification information is not useful in our above example because both hosts have modified the same data item. Techniques such as storing and comparing the time of the edits is not universally applicable, because there are no guarantees on clock synchronization, particularly in heterogeneous mobile networks.

Conflicts are usually resolved either by human intervention or through arbitrary rules; for example, a rule could be that the office computer is always right. In this case, the data item corresponding to the key "Wed" will become "Weep" if this rule is applied to our example. In simple settings such as PC–PDA synchronization, the middleware orchestrating the synchronization (e.g., ActiveSync on Pocket PCs, HotSync on Palm PDAs) may prompt the human user to resolve the conflict by choosing one data item over the other.

So far we have discussed conflicts from the perspective of a key–value-pair database, but analogous situations may occur in text files — for example, when a character at a particular location in the text is changed independently on the synchronizing hosts. We briefly revisit the problem of conflicts in our discussion on SyncML®. In the rest of the chapter, however, we skirt this important issue of conflicts and instead direct the user toward some excellent references in database literature [10].

Characterization of Data-Synchronization Applications

Based on What Is Exchanged During Synchronization

Mobile and fixed hosts can synchronize databases in two possible ways:

- They can exchange data items that have changed since the last synchronization.
- They can exchange a list of operations performed on the database since the last synchronization was completed. Each recipient host can then apply the received list of operations to the reference copy (stored at the end of the last synchronization) of the database in order to obtain the sending host's current database. A comparison yields the differences and the next synchronization state can be computed.

Both approaches are used in mobile devices, although the former is probably more useful when the operations are very complex. The second approach might be useful when updates are being propagated from a limited-bandwidth mobile client to a server because it may result in significant communication savings as compared to sending the data items. For example, if the data item in question is a large image and some image processing was performed between synchronizations on the mobile client, then sending the list of image processing operations instead of sending the image can significantly reduce the communication.

Based on Application Tolerance to Inconsistency

Data synchronization services a wide variety of applications that have varying tolerance requirements with regard to consistency. We can classify data synchronization on the basis of the time between two successive data synchronizations. This time varies depending on the applications employing the data synchronization protocols. We show these requirements for some representative applications in Figure 9.3. Collaborative tools and databases usually synchronize in real time and provide a high degree of consistency. Other systems such as PGP key-servers or the Internet DNS servers may synchronize among each other every few hours. Mobile device users synchronize their PIM information with their personal computers and cellular phones every few days or even weeks. It is important to ascertain this frequency of synchronization for a distributed application in the design phase itself so an appropriate data synchronization algorithm can be adopted.

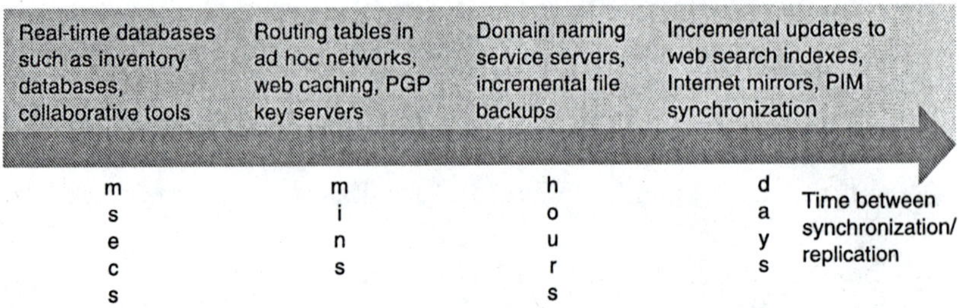

Real-time databases such as inventory databases, collaborative tools	Routing tables in ad hoc networks, web caching, PGP key servers	Domain naming service servers, incremental file backups	Incremental updates to web search indexes, Internet mirrors, PIM synchronization
m s e c s	m i n s	h o u r s	d a y s

Time between synchronization/ replication

Figure 9.3 These representative applications have varying requirements of time between two successive data synchronizations.

Mobile devices impose stricter restrictions on resources such as battery power, network bandwidth, processing capabilities, and available memory, and these factors also have to be taken into account when designing an appropriate synchronization protocol. In general, communication-intensive data synchronization protocols have a direct bearing on the power consumed by the mobile device because signaling is a power-hungry process. Although computationally intensive protocols also deplete battery power by way of running the central processing unit (CPU) for extended periods, designers first seek to reduce communication complexity because it is more power hungry in general.

Server Push Versus Client Pull

The popularity of the RIM BlackBerry™ [34] mobile push-based, e-mail device highlights the notion of server push-based services for mobile devices. In the server push model, the server periodically initiates and synchronizes with mobile clients. This model usually results in more consistency because the synchronization step is initiated by the server at the appropriate times; moreover, the server administrator can guarantee that a certain update will reach *all* clients by some specified time. This type of guarantee is very important in many corporate settings, such as for inventory management, processing financial data, etc. The obvious downside to this model, of course, is the additional burden the server assumes (i.e., keeping track of all clients and their data states in order to determine which client to synchronize with next as data gets updated on the server).

Push-based synchronization is very useful in the one-way multicast synchronization model, where a server simply multicasts updates to all clients at periodic intervals. There are no updates in the reverse direction (i.e., from the clients to the server). This model may be important in keeping all clients in the same synchronized state as determined by the server; for example, in a mobile network composed of one server and many clients, the server could multicast an updated version of a program or database to each client device.

In the client pull model, a client specifically requests synchronization from a server. A good example is a Palm PDA user pushing the HotSync button on the PDA that initiates synchronization between the HotSync application running on the PC and the PDA. This model is more prone to inconsistencies between the copies stored on various clients because the decision of when to synchronize is left to the clients. On the positive side, client-based synchronization architectures are more scalable because clients only synchronize when they have to.

Host-to-Host Data Synchronization Techniques

In this section, we discuss some common approaches to data synchronization in a two-host setting. None of these approaches is a one-stop solution to the data synchronization problem. By exposing some of the important strengths and weakness of these approaches we hope to make the reader aware of the tradeoffs involved in each approach when designing the data synchronization component of a system. Agarwal et al. [13] have elucidated some of the data-synchronization techniques mentioned here and performed trials to substantiate their findings about the overhead of these techniques. In this section, we describe five proposed approaches to host-to-host data synchronization: time stamps, version vectors, copy sync, CPI sync, and SyncML.

Time Stamps

A "mark-dirty" flag is associated with each key–value pair in a database. The flag indicates if the data item (value) was edited since the last time the database was synchronized with the other host. The algorithm works by examining these flags to find those data items in a host's database that have changed since the last time the host synchronized with the other host. The scheme is computationally fast; a single linear pass through the database yields all the items that have been modified and must be transmitted to the other host for synchronization.

An immediate problem with this scheme is that it only works well when the same two devices synchronize; otherwise, a separate set of flags (one set per additional host in the network) is required for each host in the network. Each device, then, has to maintain metadata regarding the sets of flags, the number of which grows linearly with the number of devices as well as with the number of records in the database.

Time stamps also suffer from a serious problem that causes inefficient bandwidth usage. We illustrate this problem in Figure 9.4. Here we have hosts A, B, and C that hold databases with items {a}, {x, y, z}, and {c}, respectively. Hosts A/C synchronize during Sync 1, then hosts A/B and B/C synchronize during Sync 2. During Sync 2, the items x, y, and z that newly populate databases at hosts A and C are marked as "new" (i.e., the mark-dirty flag is set with respect to each other). So, when hosts A/C synchronize again during Sync 4, items x, y, and z are exchanged again even though both hosts already have these data items. The scenario discussed in this example is common when many mobile devices synchronize with each other in an *ad hoc* manner.

Time-stamp-based synchronization is popular in mobile devices because it can theoretically achieve the minimum communication complexity possible: $O(d)$, where d differences exist among the synchronizing databases. Part of its popularity stems from the the fact that most mobile devices do not synchronize with many other hosts in today's architectures. Moreover, a popular architecture in multiple device synchronization is that of only the one central server synchronizing individually with each other mobile client. The problem of Figure 9.4 is avoided in this model, and only one set of mark-dirty flags has to be maintained on each mobile client (i.e., with respect to the central server). A variant of time-stamp synchronization, called *fastsync*, is implemented for Palm HotSync as well as Pocket PC ActiveSync, the former supporting one set of mark-dirty flags and the latter keeping two sets of mark-dirty flags.

Version Vectors

Version vectors can be thought of as a more refined implementation of time stamps. They are also referred to as *version time stamps*. A version vector is a array of counters stored on each synchronizing host. There is one version vector for each data record in the database, with the length of the vector being equal to the number of synchronizing hosts (k) in the network. For simplicity of discussion, let us assume that the hosts have IDs 1, 2, 3, ..., k. Then, each host (x) stores a k element version vector (v_x) for each data record (j) in the database. To clarify further, we consider all version vectors associated with a certain fixed record j among the n records in the database in the subsequent discussion. Each element (counter) of the version vector is set to 0 initially. Whenever host x edits record j, it increments the counter-associated vector $v_x[x]$ by 1. When a host synchronizes record j with another host (y), it obtains v_y from host y and compares its own version vector v_x with v_y. Then,

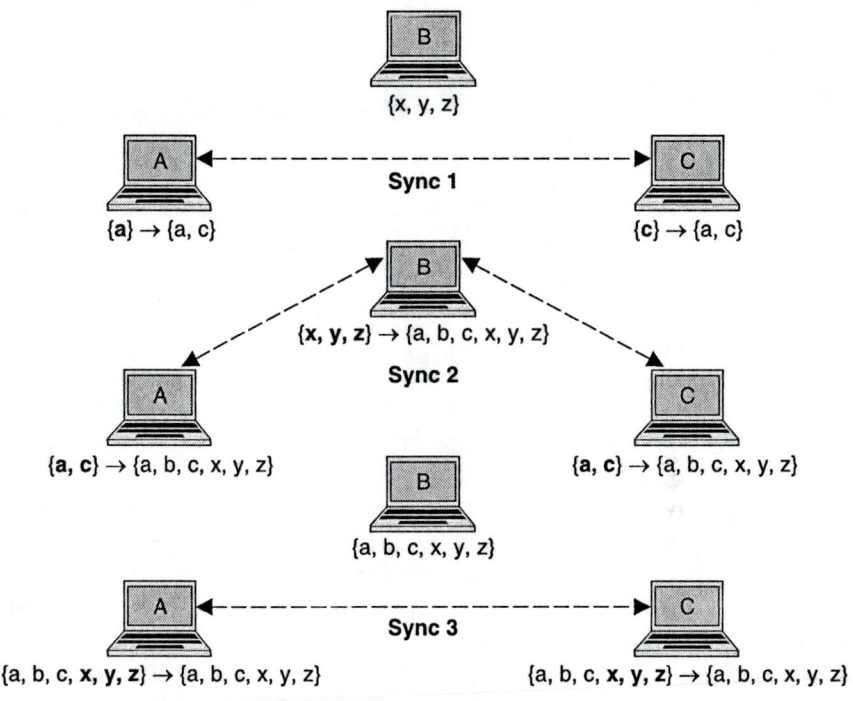

Figure 9.4   Time-stamps are inefficient. Data items highlighted in bold are exchanged during a synchronization operation. During Sync 4 hosts A and C end up exchanging x, y, and z even though these are already available on both the hosts.

- If $v_x[i] \geq v_y[i]$ for all i = 1, 2, 3, ..., k, then the value of j stored on host x is the more current value, and this is written to the next synchronized state; v_x is left unchanged.

- If $v_x[i] \leq v_y[i]$ for all i = 1, 2, 3, ..., k, then the value of j stored on host y is the more current value, and this value is retrieved from host y and written to the next synchronized state; v_x is set to v_y.

Otherwise, a conflict is signaled. An obvious flaw of version vectors is the amount of metadata necessary for synchronization. Each host has to store n such k-ary vectors. Another drawback is the recurring cost of exchanging version vectors *before* any real data is exchanged.

Copy Sync

This form of synchronization involves comparing two databases on an item-by-item basis to determine the edits (additions, deletions, or modifications) on the constituent data items. One of the hosts can volunteer to compare the two databases and then send back any changes required on the other host's database to achieve a synchronized state. The approach is wasteful in communication bandwidth and suffers from high latency because the entire database has to be first copied onto one host from another before comparisons are made; however, certain benefits can justify copy sync under some circumstances. If we assume that an n-item database is organized in an indexed random access list, then the computational complexity is simply $O(n)$. This is also the lower bound on any comparison algorithm (for sorted data) as each data item has to be at least read to factor it into the comparison (hence, we need at least n reads).

The algorithm is simple to implement and has the desirable quality of being *asymmetric*. By asymmetric, we mean that the bulk of the computation (the comparison) can be done on the more computationally capable host. In many mobile devices, computational resources come at a premium; in fact, some mobile devices run a single thread of execution. It suits such devices to leave the synchronization routine to, say, a PC if it is synchronizing with one. If synchronization is to be run over high-speed links such as a Universal Serial Bus (USB) [14] or Ethernet local area network (LAN) (e.g., in PC–PDA synchronization) then communication is not a bottleneck for moderate-size databases. Mobile devices and databases do not have to keep track of any synchronization metadata, and it is not necessary to upgrade currently deployed databases to hold additional synchronization metadata such as the mark-dirty flags of time stamps or the version vector arrays of version vectors.

Copy sync has significant shortcomings, some of which are particularly relevant in mobile device networks. Communication bandwidth and latency are major bottlenecks in mobile networks. Common mobile connection technologies such as General Packet Radio Service (GPRS) [15] do not yet offer high-bandwidth, low-latency links. In fact, copy sync is sometimes referred to as "slow sync" in the literature due to the high communication cost (time) of the algorithm. Although WiFi hotspots are increasingly available and metropolitan mesh networks [16] are making significant inroads, the bandwidth offered is shared bandwidth. If many mobile devices frequently run communication-intensive tasks such as copy sync algorithms, then the per-device bandwidth would suffer greatly. Many revenue models for wireless network access charge users on the basis of the amounts of data downloaded and uploaded; thus, copy sync is not viable for many users.

CPISync

Characteristic Polynomial Interpolation Synchronization (CPISync) is a set synchronization algorithm that was first proposed by Minsky et al. [17]. It was adopted in the mobile device scenario by implementing it on a Palm PC–PDA synchronization [18], as well as a Linux™-based PDA-to-PDA synchronization setting. CPISync achieves the theoretical information lower bound (this lower bound simply states that the minimum communication required is at least the size of the difference between the synchronizing databases) on the communication required to synchronize two remotely held databases to within a small constant. If hosts A and B have databases $S_A = \{a_1, a_2, ..., a_n\}$ and $S_B = \{b_1, b_2, ..., b_k\}$, respectively, that differ in no more than m differences, then the algorithm is as follows. Hosts A and B compute their characteristic polynomials $P_A(z)$ and $P_B(z)$:

$$P_A(z) = (z - a_1)(z - a_2)...(z - a_n)$$
$$P_B(z) = (z - b_1)(z - b_2)...(z - b_k)$$

They evaluate $P_A(z)$ and $P_B(z)$ at m sample points $\{s_1, s_2, ..., s_m\}$. These are known *a priori* on both hosts, and none of the sample points is a set element of either S_A or S_B. Hosts A and B now have evaluation sets E_A and E_B, respectively:

$$E_A = \{P_A(s_1), P_A(s_2), ..., P_A(s_m)\}$$
$$E_B = \{P_B(s_1), P_B(s_2), ..., P_B(s_m)\}$$

Host A sends E_A to host B. Note that the size of the communication is $O(m)$ bits. Host B can now obtain m evaluations of a rational function:

$$R = \left\{ \left(s_1, \frac{P_A(s_1)}{P_B(s_1)} \right), \left(s_2, \frac{P_A(s_2)}{P_B(s_2)} \right), ..., \left(s_m, \frac{P_A(s_m)}{P_B(s_m)} \right) \right\}$$

Host B now interpolates these tuple points to obtain $R(z)$. Note that the degree of the rational function (i.e., the sum of the degrees of the numerator and the denominator) cannot exceed m because $R(z) = [P_A(z)]/[P_B(z)]$, and sets A and B differ in no more than m elements. The numerator and denominator of $R(z)$ can be factored separately to obtain sets $S_A - S_B$ and $S_B - S_A$.

Interpolation and factorization are computationally expensive, with expected time $O(m^3)$. The authors of the original CPISync paper described a simple mechanism to guess m so it is no more than $O(d)$, where d is the number of differing entries in the set; thus, the communication is $O(d)$ and the computation is $O(d^3)$.

The communication complexity of CPISync is clearly better than copy sync because it requires no wholesale transfer of the database prior to synchronization. In addition, it is not necessary to maintain metadata such as the mark-dirty flags of time stamps; however, the computational requirements of CPISync are more intensive compared to copy sync and time stamps. Recent results [26] have reduced the computational complexity to a linear level (instead of cubic), although these results rely on multiple rounds of communication between the synchronizing hosts. In many cases, a hybrid of some or all of the approaches discussed above yields the best compromise; for example, the Palm OS mobile platform [20] uses a single set of mark-dirty flags on their databases, so a PDA synchronizing with the same host results in a time-stamp-based synchronization. However, if a user alternates regularly between work and home computers, then the set of mark-dirty flags is always ignored and a (slow) copy sync is run to complete the synchronization. Microsoft's ActiveSync [21] supports two sets of mark-dirty flags on its Pocket PC platform, but beyond two hosts the protocol will revert to copy sync. Hybrid protocols can be surprisingly effective and practical, although they are heuristic solutions that do not scale well with the network size for more general applications.

Open Mobile Alliance (SyncML®)

We have discussed some of the common approaches to synchronizing two hosts and have emphasized that each algorithm has its benefits and weaknesses. A major problem in the real world is that heterogeneous devices, networks, and applications implement proprietary synchronization protocols that seldom cooperate. A large number of (mostly) incompatible synchronization protocols for mobile devices is available today; for example, the HotSync protocol [19] that synchronizes Palm-OS-based PDAs and smart phones does not integrate with Microsoft's ActiveSync on the Windows® mobile Pocket PC and smart phone platforms. Similarly, platforms such as Symbian™ OS [22] and ARM-based Linux distribution PIM [23] are isolated from each other in the sense of not being able to synchronize data seamlessly. This produces a significant cost overhead when deploying heterogeneous multiple-vendor mobile devices in an enterprise setting because of the need to install third-party adaptors, or, worse, end users may have to compromise on their platforms of choice to accommodate synchronizing to other devices.

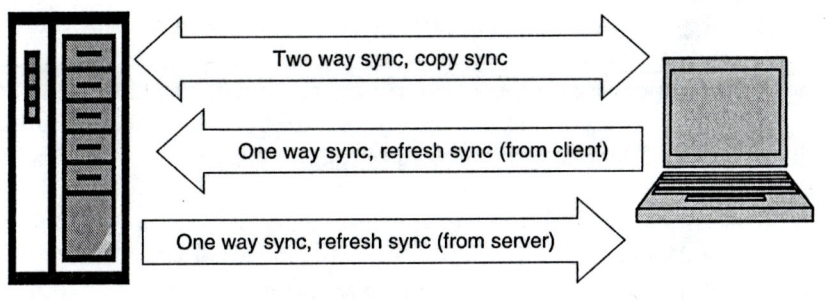

Figure 9.5 Some of the common flavors of synchronization supported in the SyncML interface.

In 2000, many players in the mobile devices industry came together and devised the SyncML [24] initiative to alleviate these incompatibility problems. SyncML was recently consolidated into the Open Mobile Alliance (OMA) [25]. All devices that conform to the SyncML standard must provide an implementation of the standard SyncML interface, thus guaranteeing a degree of synchronization interoperability between heterogeneous devices. Although the specifics of the underlying implementation can vary, the common SyncML interface allows devices to synchronize among each other seamlessly.

A SyncML-compliant protocol is based on a client–server architecture where one device acts as a client and initiates the synchronization protocol. The SyncML protocol is formally divided into two parts. The SyncML Representation protocol [24] deals with the semantics and representation of SyncML messages, whereas the SyncML Synchronization protocol [24] defines the steps undertaken on a client and server when they synchronize data.

A client synchronizes with a server by initiating contact and authenticating itself. The many possible modes of data synchronization in SyncML are shown in Figure 9.5. Two-way sync is a normal time-stamp-based or other bandwidth-efficient way to synchronize. The copy sync protocol is usually initiated under conditions when two-way sync is not feasible — for example, when the client and server synchronize for the first time or the synchronization metadata cannot be used for some other reason. A one-way sync, on the other hand, results in the initiating host getting all modifications from the request's recipient. SyncML provides devices with additional functionality that might not be considered data synchronization in the traditional sense; for example, "refresh sync" essentially replicates information from one host to the other. The designers of the protocol thought it necessary to include such functionality to make the protocol practical; for example, "refresh sync" simply copies over the data from the initiating host to the other, overwriting all data on the recipient host.

The synchronizing devices can decide on the particular synchronization method based on *synchronization anchors* that are exchanged when a client contacts the server. A mismatch of these synchronization anchors (set on the client and the server at the end of the previous synchronization) indicates that the database has been *reset* since the last synchronization. A reset means that the flags associated with the records of the synchronizing database cannot be trusted, and this warrants a slow sync where all data is replicated from the server to the client (or the other way). In the case when the anchors match up, only the modifications have to be transmitted in both directions.

Conflicts in SyncML are usually resolved at the server using pre-established rules (not part of the formal specification but user implemented), although a provision is made for client-side conflict resolution. In the latter case, the server only returns notifications of conflicts to the clients, and it is up to the synchronization engine of the client to actually resolve the conflicts and apply these resolved conflicts on its copy of the data. This approach may be useful, for example, when a user of the client is required to make choices between data items to resolve the conflict.

Conflicts may be classified based on the reason for their occurrence. Update conflicts, which are by far the most common conflicts, occur when two devices modify the same data item (i.e., corresponding to the same key). In light of the fact that SyncML is designed for mobile clients, it differentiates between soft deletions and hard deletions. Soft deletions are usually done on the mobile client to conserve storage space and are not intended to delete the data on the server. Hard deletes, on the other hand, signal that the data has to be deleted on the server also. A hard–soft conflict occurs when the mobile client does a soft delete and the server does a hard delete. SyncML does not provide any mechanisms to resolve these conflicts, but it does include mechanisms to notify the client when they occur. SyncML gives developers a universal interface that can truly make interoperable heterogeneous device synchronization possible. Just how pervasive this standard will become will depend on whether some of the major vendors are willing to adopt SyncML as their standard of choice rather than holding on to their propriety synchronization protocols.

Multiple Device Synchronization

In a multi-host distributed application setting, a single host-to-host data synchronization is just one of many synchronization operations that must take place to bring all the hosts to an identical synchronized state. In many distributed systems, this "identical" state is never achieved because data synchronization and the introduction of inconsistencies are ongoing

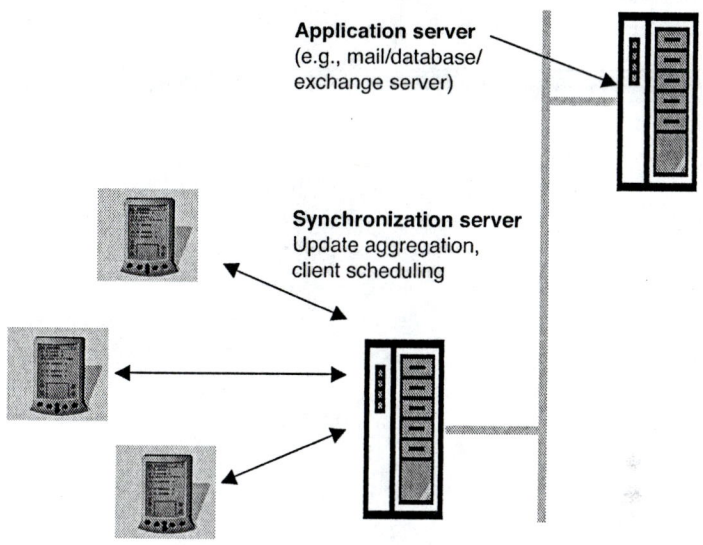

Figure 9.6 A synchronization server acts as an update aggregator and proxy for the application server.

processes; however, a well-designed distributed system will not allow more than a tolerable number of inconsistencies between any two hosts in the system, the tolerance being an application-dependent parameter.

Most corporate mobile solutions seek to provide field personnel with mobile devices on which they can enter or access enterprise data. The scale of the deployment can be very large, often spanning hundreds or even thousands of mobile devices. At this scale, problems such as conflict management and server implosion take on an entirely new urgency, and solutions that merely throw more hardware and computational power at the server return diminishing benefits because of the sheer size of the network and the number of updates being independently generated in the field.

To illustrate these problems, we consider a standard model for multiple mobile device synchronization whose variants are used in many commercial systems, as shown in Figure 9.6. To handle a large number of mobile clients, a separate synchronization server is set up as a proxy between the application server and the mobile clients. The synchronization server collects and aggregates updates received from the various clients and periodically sends these updates to the application server. The application server applies the updates and returns a new synchronized database copy to the synchronization server. This state becomes the next current state of the database and is synchronized with the clients connecting to the system.

The approach trades real-time updating of client data with scalability; that is, even though client updates are not applied to the data of the application server in real time, this approach scales well up to hundreds and even thousands of clients. The non-real-time limitation in effect keeps the application server from continually processing client updates and becoming a scalability bottleneck. The shortcoming of the approach is that mobile clients do not necessarily get the updates recently committed by other clients because the available synchronized state stored on the synchronization server that is downloaded to the client might not have been updated. Good implementations usually let clients keep their own updates rather than have them roll back (their own recent updates) to the current synchronized state downloaded from the synchronization server; thus, users at least do not see the slow synchronization problem overtly.

The Coda File System

The Coda file system developed at Carnegie Mellon University is an example of a multiple-client distributed file system. A prototype implementation is available [27] and has been critically reviewed and studied for the many features it offers. The Coda file system is useful when users wish to share files or other information in a consistent manner. A good example of its application would be when many users collaborate, share, and work on the same document. The Coda file system has been ported to mobile devices, and it offers disconnected operation for mobile clients under less than ideal conditions of server failures and network partitioning due to partial network disconnections. It has been demonstrated to work well with the mobile computing model of intermittent connectivity and repeated synchronization when the mobile client re-enters the connected world.

The system abstracts out the concept of a file system regardless of the connectivity status. Under normal circumstances, users will not detect that they have been disconnected from the file system. The underlying file system caches user updates in a way that can be synchronized with the server on reconnection. The Coda file system also caches popular files (those accessed frequently by the user) when it is connected to the server, a process known as *hoarding*. When users are disconnected, these files can be served to the users if they so desire. Of course, there is no guarantee that a file read requested by a user will succeed (the file may not exist in the cache), but by making the cache sufficiently large there is a good possibility that a user would find the file in the local cache during disconnected operation. This push-based caching provides a transparent illusion of connectivity to most casual users. The file system provides transparent resynchronization of the file when the user reconnects to the system. Conflicts are possible due to concurrent updates of the same data file. As in other systems discussed previously, these can be resolved either through automatic update rules or by manual user intervention.

When a client is connected to a server, the Coda file system enforces serial write privileges on files; that is, only one client is allowed to write to a file, and other clients are only allowed to read the file. This ensures that no conflicts arise while the clients are connected to the server. Of course, this serialized write mechanism cannot be implemented on disconnected clients, but the file synchronization (automatic or manual) is designed to handle these situations. The Coda file system also provides the important functionality of server replication based on an implementation of version vectors.

Conclusions

Seamless and efficient synchronization across heterogeneous networks is an important goal for the mobile wireless networked world. A wide variety of synchronization issues can come up in any mobile device system design. In this chapter, we have examined some of the possible host-to-host as well as centralized synchronization approaches and protocols and highlighted their strengths and weaknesses. Although the choice of the synchronization approach depends on the application as well as underlying hardware and connectivity constraints, system designers do have a variety of options from which to choose.

Time-stamp-based approaches provide fast synchronization but do not scale well with network size. A more sophisticated version of time stamps, called version vector synchronization, stores a large amount of metadata, although it is highly efficient in locating conflicts. Copy sync, although easy to implement, is communication intensive. CPISync is computationally intensive and useful only when the number of differences between the synchronizing hosts is small. Standards such as SyncML guarantee a standard interface that enables interoperability between the various data synchronization protocols and algorithms.

Multiple device synchronization remains a key challenge with regard to scalability and the number of mobile clients it can support. This challenge has taken on a new urgency with the emergence of mobile and sensor devices that repeatedly replicate, update, and synchronize data throughout their lifetime. The Coda file system is a good example of a multi-device mobile file system that supports disconnected operation, synchronizing updates, and handling data consistency gracefully.

Acknowledgments

The author is grateful to the National Science Foundation, which funded his research in the area of data synchronization at Boston University, and Ari Trachtenberg, who introduced him to the subject at the Laboratory of Networking and Information Systems in the Electrical and Computer Engineering department at Boston University. He also wishes to thank David Starobinski for his input.

References

[1] IEEE standards wireless zone, http://standards.ieee.org/wireless/index.html.
[2] Bluetooth®, https://www.bluetooth.org/.
[3] Coda File System, http://www.coda.cs.cmu.edu/.
[4] Moy, R., *OSPF Version 2*, Request for Comments 1328, Network Working Group, April, 1998 (http://www.faqs.org/rfcs/rfc2328.html).
[5] International PGP Home Page: http://www.pgpi.org/pgpi/.
[6] *SKS: Synchronizing Key Server*, http://www.nongnu.org/sks/.
[7] DNS protocol-related documents, http://www.faqs.org/rfcs/dns-rfcs.html.
[8] Akamai®, http://www.akamai.com.
[9] Xerox Palo Alto Research Center (PARC)'s Bayou project, http://www2.parc.com/csl/projects/bayou/.
[10] Silberschartz, A., Korth, H.F., and Sudarshan, S., *Database System Concepts*, 3rd ed., McGraw-Hill, New York, 1997.
[11] Tridgell, A., Efficient Algorithms for Sorting and Synchronization, Ph.D. thesis, The Australian National University, Acton, 2000.
[12] Unison file synchronization, http://www.cis.upenn.edu/~bcpierce/unison/.
[13] Agarwal, S., Starobinski, D., and Trachtenberg, A., On the scalability of data synchronization protocols for PDAs and mobile devices, *IEEE Network*, 16(4), 22–28, 2002.
[14] Universal Serial Bus, http://www.usb.org/home.
[15] GSM World, http://www.gsmworld.com/technology/gprs/index.shtml.
[16] Akyildiz, I.F., Wang, X. and Wang, W., *Wireless Mesh Networks: A Survey*, http://www.ece.gatech.edu/research/labs/bwn/mesh.pdf&e=9799.
[17] Minsky, Y., Trachtenberg, A., and Zippel, R., Set reconciliation with nearly optimal communication complexity, in *Proc. of the IEEE Int. Symp. on Information Theory*, Washington, D.C., June 24–29, p. 232.
[18] Trachtenberg, A., Starobinski, D., and Agarwal, S., Fast PDA synchronization using characteristic polynomial interpolation, in *Proc. of IEEE INFOCOM 2002*, New York, June 23–27, 2002, pp. 1510–1519.
[19] Rhodes, N. and McKeehan, J., *Palm Programming: The Developer's Guide*, O'Reilly, Sebastopol, CA, 1999.
[20] Palm Source™, http://www.palmsource.com/.
[21] Microsoft ActiveSync®, http://www.microsoft.com/windowsmobile/pocketpc/.

[22] Symbian™ OS, http://www.symbian.com/.

[23] *Familiar Linux for Pocket PC*, http://www.handhelds.org/.

[24] SyncML, http://www.openmobilealliance.org/tech/affiliates/syncml/syncm-lindex.html.

[25] Open Mobile Alliance, http://www.openmobilealliance.org/.

[26] Minsky, Y. and Trachtenberg, A., Scalable set reconciliation, technical, in *Proc. of the 40th Annual Allerton Conf. on Communication, Control, and Computing*, Monticello, IL, October 3–5, 2002.

[27] Coda File System, http://www.coda.cs.cmu.edu/.

[28] Extended Systems, http://www.extendedsystems.com/.

[29] Terry, D.B., Demers, A.J., Petersen, K., Spreitzer, M.J., Theimer, M.M., and Welch, B.B., Session guarantees for weakly consistent replicated data, in *Proc. of the Third Int. Conf. on Parallel and Distributed Information Systems*, Austin, TX, September 28–30, 1994, pp. 140–149.

[30] Concurrent Versions System (CVS), http://www.gnu.org/software/cvs/.

[31] Subversion, http://subversion.tigris.org/.

[32] Oracle® Database Lite 10g, http://www.oracle.com/technology/products/lite/index.html.

[33] Sybase® iAnywhere™, http://www.ianywhere.com/.

[34] RIM BlackBerry™, http://www.blackberry.com/na/index.shtml.

Chapter 10

Uncoupling Coordination: Tuple-Based Models for Mobility

Giacomo Cabri, Luca Ferrari, Letizia Leonardi, Marco Mamei, and Franco Zambonelli

CONTENTS

Introduction

Computing is becoming intrinsically mobile and ubiquitous [4,11]. Computer-based systems are going to be embedded in all of our everyday objects and environments. These systems will typically be communication enabled and capable of coordinating with each other within the context of complex distributed applications to, for example, support our cooperative activities, monitor and control our environments [3], and improve our interactions with the physical world [19]. Also, because most of the embeddings will be intrinsically mobile, such as a car or a human, distributed active components will have to effectively interact with each other and effectively coordinate their activities in a context-aware way, despite the network and environmental dynamics induced by mobility. From now on, we adopt the term *agents* to indicate the active components of a distributed application. Identifying proper coordination models and the associated middleware services to effectively rule and control agents' activities is a key research issue.

    Among several proposals, tuple-based models rooted in the Linda coordination language [13] appear very suitable for supporting coordination activities in mobile computing settings. By promoting indirect coordination via a sort of shared dataspace, tuple-based coordination uncouples interacting agents and relieves them from the need of knowing each other *a priori* and of knowing their respective positions, information that would be otherwise costly to obtain in dynamic and mobile computing scenarios. Also, shared dataspace coordination models, such as tuple-based ones, naturally support context-aware coordination models.

Beginning with the basic suitability of tuple-based coordination to mobile computing, a number of diverse solutions can be conceived and have been proposed to actually promote tuple-based coordination in the form of middleware-level services for mobile computing. The aim of this chapter is to provide an overview of the range of such possible solutions and to survey the most relevant proposals for tuple-based coordination in mobile computing systems. Specific attention is given to the software engineering implications — that is, to the analysis of the support that the surveyed models and middleware give to the software architect or programmer, in term of abstractions, tools, and application programming interfaces (APIs).

Tuple-Based Coordination

Tuple-based coordination was first introduced in the late 1980s in the form of the Linda coordination language for concurrent and parallel programming [13], and it consisted of a limited set of primitives, the *coordination primitives*, to access a *tuple space*. Later, in the 1990s, the model gained widespread recognition as a general-purpose coordination paradigm for distributed programming. The atomic units of interaction in tuple-based coordination are *tuples*. A tuple is a structured set of typed data items. Coordination activities between application agents (including synchronization) can take place indirectly via the exchange of tuples through a shared *tuple space*, a sort of shared dataspace that acts simply as a tuple container. The coordination primitives provided to agents grant access to a shared tuple space.

A tuple can be written in the tuple space by an agent performing the out output primitive. As an example, out ("amount", 10, a) writes a tuple with three fields: the string amount, the integer 10, and the contents of the program variable a. Two input primitives (rd and in), provided to associatively retrieve data from the tuple space, read or extract, respectively, a tuple from the tuple space.

A *matching rule* governs tuple selection to retrieve tuples from a tuple space in an associative way: input operations take a *template* as their argument, and the returned tuple is one *matching* the template. To match, the template and the tuple must be of the same length, the field types must be the same, and the values of constant fields must be identical. For example, the operation in("amount", ?b, a) looks for a tuple containing the string amount as its first field, followed by a value of the same type as the program variable b, and the value of the variable a. The notation ?b indicates that the matching value is to be bound to variable b after retrieval. If the above tuple ("amount", 10, a) has been inserted in the tuple space, then performing the previous in operation activates the matching mechanism that associates the value 10 to the program variable b. Input operations are *blocking*; that is, they return only when a matching tuple is found, thus providing a mechanism for two agents to indirectly synchronize based on the occurrences of the tuples. When multiple tuples match a template, one is selected nondeterministically.

Other Linda operations include inp and rdp, which are the *predicative, nonblocking* versions of in and rd and which return true if a matching tuple has been found and false otherwise. Another operation is eval, which is intended to create an active tuple — a tuple for which one or more fields do not have a definite value but which must be computed by function calls. When such a tuple is emitted, a new process is created for each function call to be computed. Eventually, when all these processes have performed their computations, the active tuple is replaced by a regular (passive) tuple, the function calls of which are replaced by the corresponding computed values. However, the eval primitive has never been widely adopted, because process creation and the lifecycle must always be dealt with (in actually implemented systems) outside the tuple-based coordination system.

Based on the basic simple model presented above, different models can be developed for tuples (e.g., based on objects, records, logic predicates) and for pattern-matching mechanisms (e.g., object matching, data matching, logic unification). The differences in these models are of little relevance to our discussion in this chapter, where we primarily focus on architectural and distributed software engineering issues.

Tuple-Based Coordination and Mobility

Let us now review the main challenges that have to be addressed when developing distributed mobile applications and show how tuple-based models can effectively deal with these challenges. To ground the discussion, an example mobile computing application is introduced.

Mobile Computing Core Challenges

The core problems related to mobile computing derive from the fact that applications will be embedded in complex, open, dynamic, and ever-changing environments [10]. In particular:

- Agents are connected to each other via *dynamic wireless networks*. Apart from technological issues, what really matters is that these networks will be ever changing. Components will be dynamically added or removed from them. Their topology will change because of node mobility. An agent, executing in such a network, would perceive an ever-changing environment running in different places and different contexts. Available communication partners can become unreachable in a matter of a few seconds and new ones can show up.
- Besides being dynamic, these networks will be extremely *heterogeneous* and *huge*. Consider a scenario a few years hence in which a large city such as Boston might have several wireless base stations in every building — a number of nodes on the order of 10^7. If most of the electrical devices in the buildings and those carried by people are also wirelessly networked, then the total number of nodes could be as high as 10^{10}. If these nodes communicate peer-to-peer with nearby devices, then one could envision the entire city as being connected into a mobile *ad hoc* network approximately 10^3 hops in diameter.
- Because pervasive and mobile computing systems will be everywhere and will have an impact on every moment of our life, characteristics such as *security* and *robustness* will become even more important. Hackers could gain entry to our cell/smart phones and viruses, for example, could prevent our cars from braking.
- Even worse, these systems are inherently *difficult to test and debug*. Emergent unexpected situations can arise only when the system is actually deployed, and offline simulations can lead to wrong solutions. Moreover, in a dynamic system where components are mobile and wirelessly interacting, debugging is extremely difficult [10]: Who is talking with whom? What happened in the past?

Other than mere technological issues, the above are mostly modeling and conceptual issues that impact the software engineering principles behind mobile application development [31].

Why Tuple-Based Coordination Models

Keeping in mind the above issues, the reasons why tuple-based coordination models (although originally conceived for parallel and concurrent systems) have been found to be suitable for developing open, distributed, and mobile applications can be summarized as follows [5]:

- *Uncoupling* — The use of a tuple space as the coordination medium uncouples the coordinating components both in space and time. An agent can perform an out operation independently of the presence or even the existence of the retrieving agent and can terminate its execution before such a tuple is actually retrieved. Moreover, because agents do not have to be in the same place to interact, the tuple space helps to minimize locality issues. In a scenario such as mobile computing, where agents can come and go at any time and can be at any location in a possibly large network, the uncoupling feature is dramatically important.

- *Associative addressing* — The template used to retrieve a tuple specifies what *kind* of tuple is requested, rather than *which* tuple. This well suits mobile agent scenarios. In a wide and dynamic environment, a complete and updated knowledge of all execution environments and of other application agents may be difficult or even impossible to acquire. As agents would somehow require pattern-matching mechanisms to deal with uncertainty, dynamicity, and heterogeneity (as intrinsically exhibited by mobile computing scenarios), it is worthwhile integrating these mechanisms directly in the coordination model to simplify agent programming and reduce application complexity.

- *Context awareness* — A tuple space can act as a natural repository of contextual information that allows agents to access information about what is happening in the surrounding operational environment.

- *Security and robustness* — A tuple space can be put in charge of controlling all interactions performed via tuples, independently of the identity of the involved agents. This fact, together with the simplicity of the model, increases the degree of robustness of ant systems based on such coordination models, particularly mobile computing systems.

- *Separation of concerns* — Coordination languages focus on the issue of coordination only; they are not influenced by characteristics of the host programming language or of the involved hardware architecture. This leads to a clear coordination model, simplifies programming, and intrinsically suits open and dynamic scenarios.

Summarizing, the Linda coordination model grants the flexibility and the adaptability required in developing applications in mobile computing scenarios.

A Case Study Application

At this point, it may be helpful to provide an application case study. Because it involves a wide range of mobile computing applications, let us consider a system to support visitors, each assumed to be carrying a mobile device, to a large museum. The devices carried by the users can be exploited to help the visitors achieve such goals as retrieving information about art pieces, effectively orientating themselves in the museum, and meeting up with each other (in the case of organized groups). Two problems that might arise would be (a) gathering and exploiting information related to art pieces the visitors want to see; and (2) planning and coordinating their movements with other, possible unknown visitors (e.g., to avoid crowds or queues or to meet together at a desired location).

To this end, we can assume that: (a) the visitors are provided with a software agent running on some wireless handheld device, such as a computer or a cellphone, that gives the visitor information on the art pieces and suggestions on where and when to move; (b) the museum has an adequate embedded network infrastructure based on tuple-based coordination models; and (c) both the devices and the handheld infrastructures have localization mechanisms to determine where they actually are located in the museum. With regard to the infrastructure embedded in the museum walls (associated with either each piece of art or each museum room), the museum must have a wired network of computer hosts, each capable of communicating with each other and with the mobile devices located in their proximity via the use of a short-range wireless link. Within such an infrastructure, a multiplicity of tuple spaces (e.g., one per museum room, plus any other ones that may be necessary for administrative reasons) can be made available to agents to interact with each other and to retrieve museum information.

In spite of the rather simplified description, this kind of system is a case study that captures in a powerful way features and constraints of mobile computing system:

- It represents a very dynamic scenario. The system has to cope with various museum floor plans and a variable number of visitors entering and exiting the museum at different times and who quite possibly will ignore or misunderstand the advice given by their handheld devices. The uncoupling of tuple-based coordination models minimizes such issues.

- Inside large museums can be thousands of embedded electronic devices and people with mobile devices. Multiple systems can be running concurrently within the museum computer infrastructure (e.g., lighting and heating systems) and other systems connected to these other services. The associative mechanism of tuple spaces helps in managing the appropriate information without imposing a rigid schema on application agents and on the infrastructure.
- Agents (in this case, visitors) have the primary goal of discovering what the museum holds (i.e., to achieve context awareness). Tuple spaces can be assumed to be a digital representation of a museum room from which to obtain information about the context.
- The system should be secure and robust, as malicious or badly programmed agents could try to penetrate the system. Embedded hosts can break down, wireless networks can have glitches, and any kind of unexpected situation could arise. The system can cope with these anomalies by controlling the requests posted in the form of tuples for security's sake and redirecting the requests to other tuple spaces in a flexible way when parts of the systems are not available.
- By managing all interactions via tuple spaces, security and monitoring rules can be defined and enforced separately from the logic of the museum services.

The above scenario and the associated coordination problems are of a very general nature and are pertinent to such widely varying scenarios as traffic management and forklift activity in a warehouse, where navigator-equipped vehicles provide guidance to their operators. Or, consider software agents exploring the Web, where mobile software agents coordinate distributed research on various Web sites.

Middleware Taxonomy

Given the above-mentioned advantages of tuple-based models for mobile computing scenarios, it is not surprising that several middleware infrastructures and services relying on tuple-based coordination models have been recently proposed. Although based on the same general concepts, these systems tend to focus on different aspects of the aforementioned problems and consequently adopt very different architectural solutions. To study and compare such different systems, it is very important to: (1) focus on a specific comparable subset of the services offered by different middleware systems, and (2) to produce an effective taxonomy on which to ground the comparison.

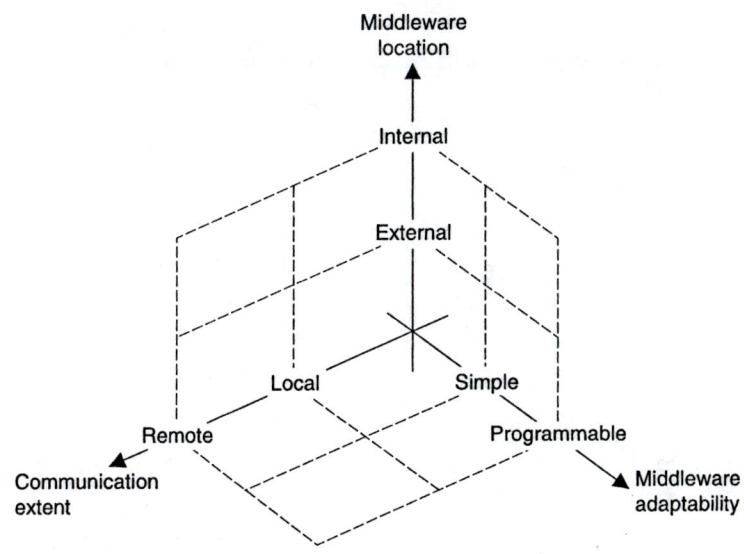

Figure 10.1 The middleware 3D taxonomy schema.

With regard to the former point, we will focus on those services supporting the coordination of agents from a software engineering perspective. From such a perspective, we can identify two fundamental building blocks of every coordination activity that has to be supported: *interaction mechanisms* and *context awareness*. On the one hand, it is obvious that coordination requires some form of interaction. Agents need to communicate in some way to decide, plan, and synchronize their actions. On the other hand, the very nature of coordination requires context awareness. An agent can meaningfully work together and combine efforts with other agents only if it is somehow aware of what is around it (i.e., its context). Of course, these two building blocks are tightly interwoven in that contextual information can be communicated only via the available interaction mechanisms. As already noted, tuple-based models are particularly effective in supporting both of these activities; in fact, tuple spaces provide both an uncoupled communication mechanism and a repository for contextual information. Still, the specifics of different architectural solutions carry on different advantages and drawbacks.

Thus, with regard to the latter point, our proposal is to classify middleware infrastructures along three main axes (see Figure 10.1):

- Middleware location
- Communication extent
- Middleware adaptability

On this basis, we can analyze how the positioning of a specific proposal along each of these axes may impact the interaction mechanisms and context awareness of the agents.

Middleware Location

We can classify various middleware infrastructures by considering the following question: "Is the middleware something *external*, to which agents connect, or is it something *internal*, such as an agent subsystem (i.e., API)?" With regard to this question, we have developed two landmark categories of middleware (*landmark* in the sense that intermediate–borderline systems can also be conceived):

- *External middleware* — The middleware is something like an external service accessed by the agents; for example, a middleware server offering a shared data space to which agents can post and retrieve data would belong to this category.
- *Internal middleware* — The middleware is strictly local, and each agent is provided with its own private instance of the middleware; for example, a middleware API that wraps and enriches the agent communication features would belong to this category.

External Middleware: Impact on Interaction Mechanisms and Context Awareness

The middleware locality strongly influences how and with whom agents interact and, consequently, how contextual information is exchanged. Let us focus on the external middleware case by considering the case study application. In our museum example, we can suppose that the museum building is provided with a network of middleware servers, installed in every room and providing suitable services to enable the interactions of agents. Agents connect to the closest middleware server to interact by, for example, posting and retrieving messages or subscribing to events (Figure 10.2a).

The first obvious point we can make is that, by definition, this kind of middleware requires a deployed infrastructure, which means that this kind of middleware is not suitable for *ad hoc* situations where coordinating the agents (interaction plus context awareness) has to be achieved in environments that are not instrumented. On the other hand, the availability of an infrastructure greatly reduces the agent load (thus saving device batteries), in that agents can demand of the infrastructure many operations; for example, the pattern-matching operations of the tuples can be performed on the infrastructure without consuming PDA resources.

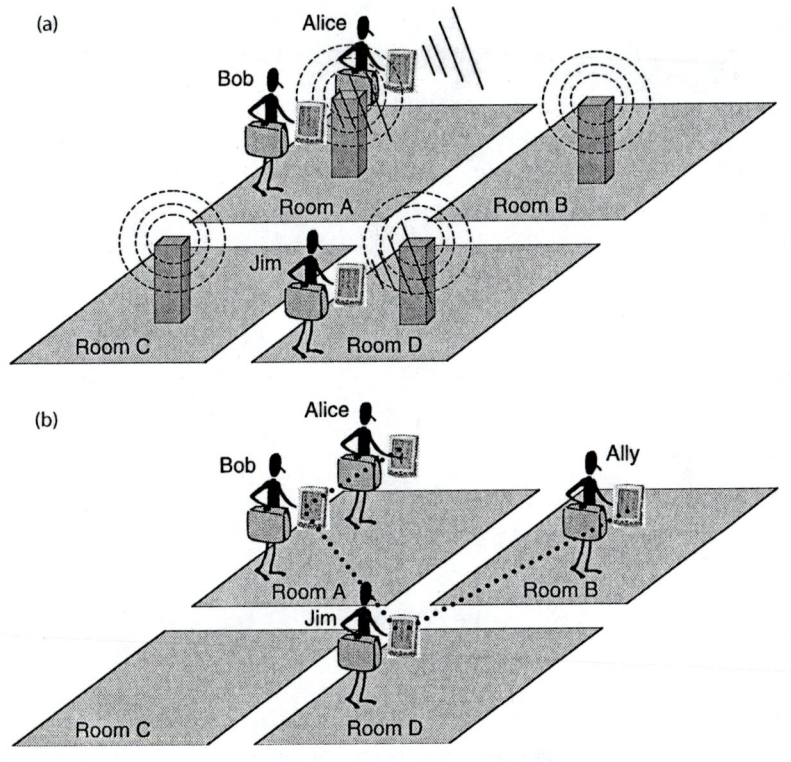

Figure 10.2 (a) Museum with a set of external middleware servers accessed by agents on a location basis. (b) The users' agents are networked (e.g., via a mobile *ad hoc* network) and they interact directly with each other by means of an internal middleware.

Another important consideration is that an external middleware, possibly accessed by a vast number of agents, can be the source of scalability and robustness problems, representing a candidate bottleneck and a single point of failure. If the middleware server in a room breaks down, communication in that room would be disrupted; however, the use of external middleware (especially when relying on a fixed network infrastructure) avoids most of the resource constraints of mobile and pervasive devices and thus is subject to fewer hardware problems (e.g., exhausted batteries or wireless network glitches). Moreover, from the middleware developer point of view, an external middleware is inherently simpler than an internal one. External middleware, being detached from the agents, does not require managing its own mobility and consequent reconfigurations and dynamics.

Internal Middleware: Impact on Interaction Mechanisms and Context Awareness

Turning our attention to the internal middleware case, we can suppose that the users' PDAs connect in a mobile *ad hoc* network (MANET) that allows agents to communicate (see Figure 10.2b). In this scenario, most of the points made earlier are turned upside down: Internal middleware is suitable for coordinating the activities of agents in environments that are not instrumented. It is thus less expensive to implement, because it does not require installation costs. It is inherently scalable and robust in that it eliminates bottlenecks and single points of failure. It is inherently more complex, however, in that the middleware (internal to the agents) must react to the movements of agents and to consequent reconfigurations.

Communication Extent

We now want to classify various middleware infrastructures by considering the following question: "Given a set of middleware instances (whether internal or external), how are they connected?" In this case, we can identify two types of infrastructures:

- *Local middleware* — In this kind of middleware, the various instances are only locally connected or are not connected at all. This means that a middleware instance can either communicate only with other neighboring middleware instances or communicate only with connected agents.
- *Remote middleware* — This kind of middleware enables long-range, multi-hop communications between its instances.

Local Communication: Impact on Interaction Mechanisms and Context Awareness

Let us consider again Figure 10.2a, where it is clear that Bob and Alice can interact by means of the services (e.g., shared space facility) offered by the middleware server installed in room A. But, how can Bob and Jim interact? It is clear that, if the middleware allows only local communication across its instances, Bob and Jim cannot interact with the abstractions promoted by the middleware (because they access separate, or disjoint, shared data spaces). If they want to communicate, they must meet in a specific room and use the middleware in that room to interact. Note that communicating entities still must meet in the same room even when the middleware is internal, but it enables a single-hop communication. Consider the internal middleware scenario of Figure 10.2b and communication between Alice and Ally. If a barrier between Room A and Room B prevents transmission of a communication signal, the only way Alice can communicate with Ally is to exploit the bridge offered by the already established communication between Bob (with whom Alice can communicate, as she is in the same room) and Jim (who can communicate with Ally). But, this will require middleware able to establish multi-hop communication, particularly three hops (see Figure 10.2b). For simple internal middleware that can only communicate by a single hop, this is not possible, and the only way to achieve communication in this case is to have the communicating parties within the same room.

Of course, the same locality scope applies to context awareness. In fact, each agent can only know information about what is happening in its immediate neighborhood, and it is hoped that these will be the most relevant for its actual execution. Although this strict locality requirement for the interaction and perception of agents can be a severe limitation, it is not necessarily bad. The locality scope reduces the problem of information overload and allows the system to better scale with increasing size.

In the museum application, for example, having local communication middleware would mean that agents could receive information related only to art pieces in the room where they are. Most of the time, this is not a problem as it is likely that visitors will request information about art pieces they are actually viewing. Moreover, this enables the system to scale better, in that an agent is not bothered with unnecessary information related to irrelevant, faraway items. On the other hand, strict local communications can represent a major obstacle for the motion coordination task. If, for example, two visitors located on opposite sides of the museum want to meet somewhere, it will be difficult to coordinate such a meeting. Because they cannot interact, their only choices are to wander randomly or to exploit information previously published within their locality scope by other agents; for example, one of the two visitors moving through the museum can store the tuple "I will be in Room A at 10 a.m." in all the middleware instances it connects with, and the other visitor can use this information to meet up with that person at 10 a.m.

Remote Communication: Impact on Interaction Mechanisms and Context Awareness

In this kind of middleware, all the middleware instances can interact with each other. Figure 10.2a presents the case where, for example, all the servers are networked and data entering one server is automatically replicated in all the others. This, of course, would permit long-range interactions. In such middleware, the locality scope for agent interactions is considerably weakened, but this approach increases the system flexibility as an agent can be informed of relevant information happening far away. On the other hand, it can create scalability problems and information overload. To this end, further methods to filter and reduce accessible information should be implemented. In the museum application, for example, multi-hop communication would enable visitors to access information about every art piece in the museum from wherever they are. Moreover, it would allow motion coordination even between faraway agents, which would be able to exchange messages, regardless of their actual position, to decide a common motion strategy. The problem with this approach relates to information overload and overconsumption of network bandwidth, so visitors must be able to filter only relevant information and high-level constraints must be enforced to limit bandwidth usage.

Middleware Adaptability

We can classify various middleware infrastructures by considering the following question: "Is the middleware capable of supporting the computational activities of agents by means of programmable behaviors?" Once again, we propose two landmark classifications of middleware:

- *Simple middleware* — In this case, the middleware is not able to support any computational activity; all the computations are left to the agents. This kind of middleware provides a predefined set of capabilities implemented in a fixed way and does not allow the middleware itself or an agent to change or customize the middleware features.
- *Programmable middleware* — This type of middleware is a system that is able to dynamically download, store, and execute foreign code. Agents can thus program the middleware, not only by reshaping its predefined set of features but also by implanting new programs and services. These new implanted services can be associated with some triggering conditions to let the middleware execute those procedures whenever the proper conditions are met.

Simple Middleware: Impact on Interaction Mechanisms and Context Awareness

Simple middleware cannot adapt (or be adapted) to changing situations and provides agents only with a fixed set of unchangeable tools. Typically, simple middleware enables direct communication between agents, such that agents can exchange string-like messages or method invocations. In our museum example, it might be desirable to adapt visitor information to the PDA displays and user settings; however, simple middleware is not flexible enough to adapt services to the different types of visitor PDAs. Such an operation could be done only in a static way, as the middleware is only able to manage a predefined set of device profiles.

Moreover, the fact that the middleware supports only string-based communication or method invocation can be a constraint in some applications; for example, representing contextual information by means of plain strings might not be expressive enough and can force agents to execute complex algorithms to understand the information and decide what to do. In our museum example, although knowledge of the coordinates of all the agents in the museum would be complete contextual information for motion coordination tasks, it would still be difficult for an agent to decide what to do (i.e., where to go) on the basis of such rough information. Despite these drawbacks, simple middleware is quite easy to implement and can be successfully applied in those scenarios that do not require complex features. Moreover, the simplicity of the middleware is likely to lead to generally better performance.

Programmable Middleware: Impact on Interaction Mechanisms and Context Awareness

Programmable middleware, which is able to store and execute foreign code, can perform any kind of adaptation. Not only can its mechanisms be adapted to changing situations (e.g., the middleware could be programmed to automatically compress specified pieces of data, depending on the available bandwidth), but (taking the approach to the extreme) virus-like communication based on mobile code can also be enacted. Here, the information is exchanged by mobile code; thus, a message would be able to autonomously specify its routing, automatically adapt and change its own content, and execute any kind of required action. Of course, the flexibility of this kind of middleware comes at the expense of security issues, which naturally arise when possibly malicious code is allowed to run in the middleware.

Programmable middleware offers great flexibility. Agents can program the middleware to let it filter and aggregate relevant contextual information [9], and context information sources can describe their information not only by simple messages but also by complex programs. These programs can contain the algorithms on how to parse or interpret the contextual information or the routing mechanism, thus making possible information fusion and aggregation [16].

In our museum application, agents could flexibly program the middleware to receive information suitable for their display capabilities. They could embed programs in the middleware to let it react to special events in ways possibly not foreseen when the middleware was first deployed; for example, the middleware could be programmed to block communication between a group of students' PDAs when their teacher has posted a question. Finally, with regard to the motion coordination problem, we can imagine an agent being able to send via the middleware something like a program (e.g., a routing algorithm) that allows other agents to reach a specific destination by executing the received algorithm.

Current Middleware Infrastructures

In this section, we are going to survey various middleware infrastructures according to the classification schema provided earlier. Because the number of proposed models and types of middleware is overwhelming and is still growing very rapidly, we have tried here to provide a classification system based on an exploration of the introduced middleware range. For each type of middleware, some relevant implementations are presented. In Figure 10.3, we depict the middleware properly considered to be located in the taxonomy range. We present the models and types of middleware populating this space by means of an imaginary walk along each one of the three axes. If a proposal represents several dimensions, its placement in the taxonomy schema is determined by the axis of the most relevant dimension.

A Walk Along the Communication Extent Axis

For the communication extent axis, we can see that the taxonomy space is divided into two regions (see Figure 10.4). In the rightmost region, local communication middleware defines boundaries for agent communication and context awareness. This can turn out to be a problem, if an agent has to acquire a global picture of the application state, but it can also implicitly alleviate the problem of information overload and bandwidth overexploitation. In the leftmost region, remote communication middleware does not impose boundaries in agent communication and thus increases the flexibility of the system; however, if not properly controlled, this type of system can lead to information overloading and exhaustion of network bandwidth.

In particular, we can consider those approaches that are nearer to the origin of the axis — those that are *local*, *external*, and *simple* (star 1 in Figure 10.3). Here we would find the JavaSpaces™ [12] proposal from Sun Microsystems, which includes the specification of a framework for distributed network services that is implemented using the Java language.

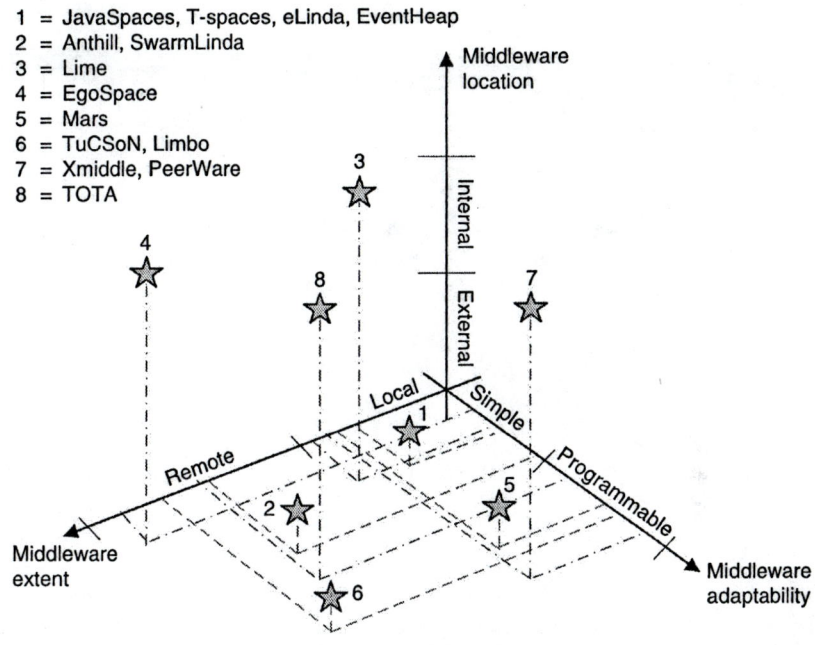

1 = JavaSpaces, T-spaces, eLinda, EventHeap
2 = Anthill, SwarmLinda
3 = Lime
4 = EgoSpace
5 = Mars
6 = TuCSoN, Limbo
7 = Xmiddle, PeerWare
8 = TOTA

Figure 10.3 Surveyed middleware in the taxonomy schema.

The JavaSpaces specification defines a tuple space where special tuples, instances of the Entry class, can be stored and retrieved through the Java serialization mechanism. JavaSpaces can be accessed directly addressing the tuple space instance and exploiting its interface services, including a simple pattern-matching mechanism. Although simple, JavaSpaces provides a few interesting features, such as the already mentioned tuple pattern-matching mechanism, being able to define a "leasing time" on a tuple (i.e., a time to live), support for tuple inserting notification, and the capability to extend the base tuple class with user-defined data and services. Interesting implementations of the JavaSpaces specification include GigaSpaces [14], which also enables access to the tuple space by means of Simple Object Access Protocol (SOAP), and AutevoSpaces [15], which relies on a distributed tuple space implementation.

TSpaces™ [30] is the IBM answer to JavaSpaces and provides a Java implementation of a tuple space enhanced with nonblocking tuple access services, rendezvous-specific services, indexing, and support for tuple activity notification. Similar to the previously discussed tuple spaces, TSpaces can be directly addressed in order to exploit the provided services, even if several TSpaces can be aggregated in order to build a single space of spaces.

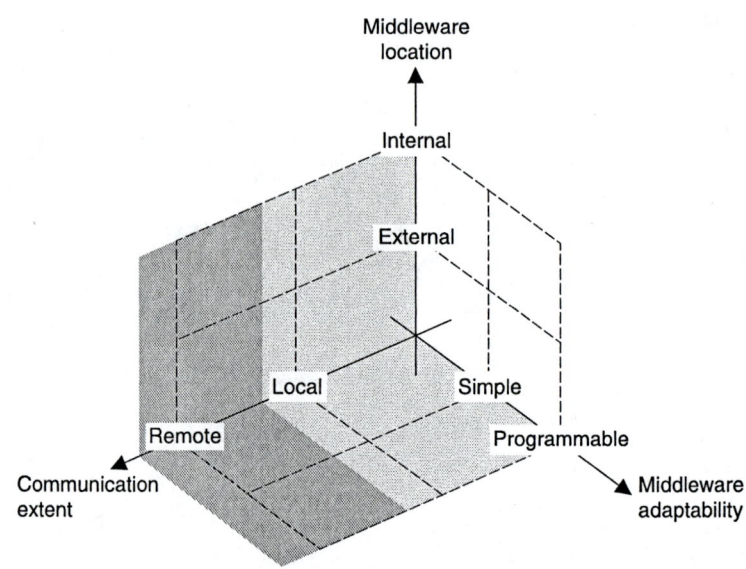

Figure 10.4   Different zones along the communication extent axis.

Another interesting approach is Linda [29], which provides a Java implementation of a tuple space with a programmable matching engine. The latter can provide specialized matching logics, such as, for example, the search for a minimum value in the tuple space. Nevertheless, the matching engine cannot be dynamically programmed, which means the matching features must be known *a priori*. EventHeap [17] allows tuples (called *events*) to be made of fields whose values are not known *a priori* (post fields) or are known but are not relevant (virtual fields); thus, the exact behavior of the tuple space will depend on the value assigned later to those fields in the tuples. EventHeap has been implemented on top of TSpaces, and has been rewritten in Java starting from scratch.

Moving along the communication axis of the taxonomy, we find the approaches that are catalogued as *remote*, *external*, and *simple* (star 2 in Figure 10.3). These systems can interact with each other, which means that different instances can exchange data and information. Often, this is achieved through a peer-to-peer network, where different hosts run an instance of the middleware and such instances connect to other instances running on other hosts. An interesting approach in this direction is represented by SwarmLinda [6], a Java implementation that exploits eXtensible Markup Language (XML) documents to describe tuples. SwarmLinda is based on the concepts of swarm intelligence and multi-agent systems modeling the tuple space as a set of nodes and provides services (e.g., inserting, retrieving) performed by ants that travel across the nodes and

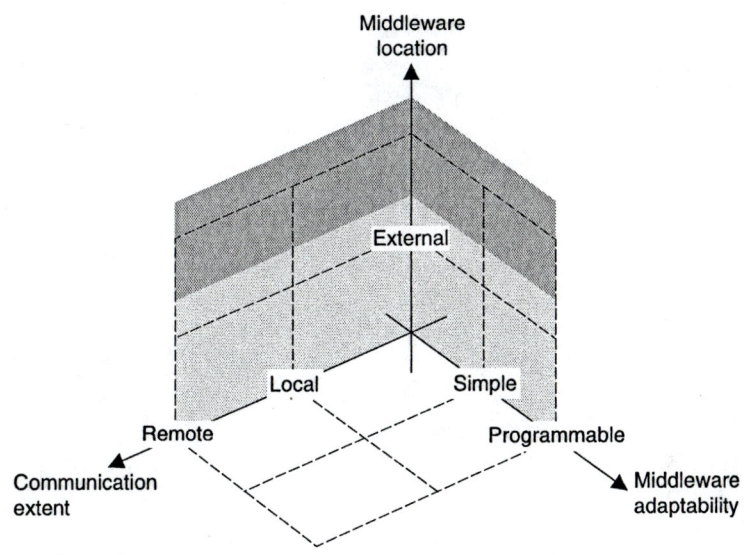

Figure 10.5 Different zones along the middleware location axis.

search for one or more tuples. The interesting feature of SwarmLinda is its tuple aggregation based on pattern criteria, so similar tuples will be close together and kept (possibly) in the same node space. This implies that, although the entire system can be seen as a composition of distributed tuple spaces, it is really a single tuple space with clients connected to different instances but who perceive the system as being unique.

Another interesting approach, quite similar to SwarmLinda, is Anthill [1], a middleware that relies on the Java implementation of JXTA™ [18] and provides a self-organizing network of interconnected units (called *nests*) visited by ants, with agents assigned to one or more tasks (e.g., tuple inserting, tuple retrieving). It is important to note that ants cannot communicate together directly but must leave information that can be exploited by other ants; this kind of indirect communication is called *stigmergy*.

A Walk Along the Location Axis

Focusing our attention on the location axis, we can find two other regions (see Figure 10.5). The bottom one is where external middleware, which requires underlying common infrastructures, resides. These infrastructures can have problems in a MANET scenario. Moreover, these infrastructures can induce scalability and robustness problems because the middleware can, in principle, be accessed by a large number of agents, thus producing bottlenecks or single points of failure. The top region, characterized by internal middleware, can be painlessly applied in a MANET scenario, because such models and middleware do not require a common infrastructure accessible by different agents. Also, internal middleware infrastructures tend to scale with the size of the system, as they are replicated in every agent; for the same reason they are robust they tend not to cause single points of failure.

Linda in a Mobile Environment (LIME) [24] is an *internal* middleware that is characterized also as *simple* and *local* (star 3 in Figure 10.3). The key idea of LIME is that each mobile entity, either a software agent or a physical device, is associated with a personal tuple space, accessed through an interface tuple space (ITS). When mobile entities meet together, their ITSs are transparently merged to allow coordination. In other words, each mobile entity performs a tuple operation over its personal tuple space, which is updated with other personal tuple space information when possible. It is important to note that LIME allows the definition of private tuple spaces, which will not be exchanged with other mobile entities; moreover, it supports reactivity — that is, the capability of performing a particular operation when a specific tuple is found in the tuple space.

EgoSpace [28] is an *internal, remote,* and *simple* middleware (star 4 in Figure 10.3) that connects each entity belonging to the network and running a middleware instance. In this way, a distributed and collaborative architecture can be built and information (i.e., tuples) can be shared among the instances. An important feature of EgoSpace is the capability of expressing the interest level for information belonging to specific geographical areas; thus, it is possible to define boundaries for searching for and retrieving tuples, in addition to saving bandwidth.

A Walk Along the Adaptability Axis

The final dimension we follow is the one defining the middleware adaptability (see Figure 10.6). Here, we can define a leftmost region characterized by simple (i.e., not programmable) middleware infrastructures that are easier to use because they have only a fixed set of capabilities. The drawback of this simplicity is that they are best suited only to relatively static scenarios, because their fixed set of capabilities cannot be customized to varying dynamics and unexpected situations. In the rightmost region, however, programmable middleware infrastructures are extremely flexible and well suited to dynamic application scenarios. The drawback of all this flexibility is that programmable middleware tends to be more complex to use and can introduce security concerns by offering the possibility of installing a foreign, possibly malicious, code.

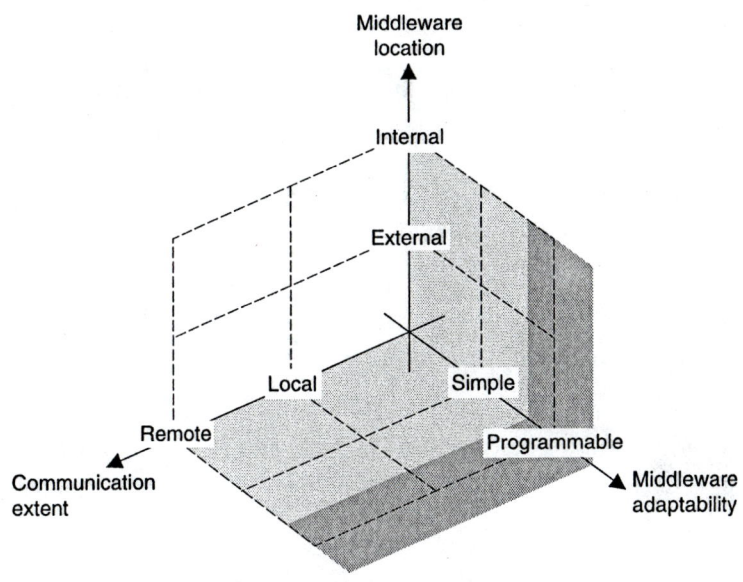

Figure 10.6   Different zones along the adaptability axis.

Focusing on this latter type of middleware, we can consider those models that are *local, programmable,* and *external* (star 5 in Figure 10.3). An interesting approach is Mobile Agent Reactive Spaces (MARS) [5], which extends JavaSpaces™. MARS defines a set of independent spaces, each one tied to a host (i.e., a local execution environment) that agents can access through the MARS interface to perform tuple operations. Each MARS instance can support reactivity, which is the capability of performing operations when a tuple operation (e.g., inserting, reading, extracting) is performed. To do this, MARS exploits a metatuple space (metaspace) that stores as tuples reactions to be performed. Each time a tuple operation is performed, the metaspace is searched for a reaction tuple and, if the latter is found, the reaction is executed.

Now it is time to take a look at middleware that is *programmable* and *external* but *remote,* thus allowing their instances to communicate (star 6 in Figure 10.3). A representative example is the Tuple Centres Spread over Networks (TuCSoN) system [22], which focuses on the coordination of mobile agents. TuCSoN defines the concept of a tuple center, an instance running on an Internet host that can be connected to other tuple centers. Tuple centers are uniquely named across the Internet, so agents can directly interact with any of them that are network aware or can interact with the local tuple space that will interact transparently with the others, if agents are not network aware. It is important to note that a tuple center is not simply a tuple space, as it can support specification tuples that define the reaction logic for communicative actions in the tuple space. A different approach is that followed by Limbo [8], which defines a set of tuple spaces interacting each other but where the programmable feature is reached by subtyping an existing tuple space. In other words, when a client wants to specify the behavior of the tuple space, it has to place a few tuples describing that behavior in the tuple space and then place a special "create tuple space" tuple that will be handled by the tuple space itself and will cross into a new tuple space with the specified behavior.

Other Mixed Approaches

For yet another middleware location, we can survey *programmable, internal,* and *local* middleware (star 7 in Figure 10.3). An interesting project, in this field, is represented by XMIDDLE [21], which proposes tuple spaces based on XML trees, where each mobile entity carries on its own tree, which is merged with other trees when agents meet together. Thanks to its exploitation of the XML language, XMiddle can share tailored portion of data, depending on the differences between the spaces that are synchronizing. Most important, the use of XML allows XMiddle to handle structured data, thus it is able to associate a piece of information with the exchanged data, allowing the implementation of different synchronization and reconciliation protocols.

Another interesting approach is PeerWare [7], where again each node runs a middleware instance handling its local data and merging the data each time it joins other peers. It is important to note that this approach explicitly recognizes the local space and the global space (the one made by all the local spaces available online), defining different primitives for both the spaces. Furthermore, a special set of primitives is available for both the spaces, in order to define the programmability of the space.

The last region of our middleware taxonomy is *programmable, internal,* and *remote* middleware; here we can find the TOTA [19] approach. The key idea behind TOTA is that a tuple must be propagated in the surrounding environment following a specific rule; in other words, the information is composed by the tuple itself and the rules to be applied on the tuple. When a tuple is outputted, it is distributed among each node (running the TOTA middleware), meaning that the tuple can be copied as it is or it can be modified in order to reflect changes in the environment.

Open Issues and Research Directions

A number of apparently very diverse research areas (e.g., peer-to-peer overlay networks, swarm intelligent systems, pervasive and ubiquitous computing systems) share some common issues with tuple-based coordination models and middleware. Still, the relations and synergies between these areas are mostly unexplored, thus representing fertile ground for further investigation.

Overlay Networks and Overlay Data Structures

Overlay networks can be defined as routing distributed data structures providing agents with a suitable application-specific view of the network (i.e., they allow agents to perceive a specific overlay topology of the network) [25,26]. These structures are typically created by deploying across the network suitable routing information and are at the basis of a number of mobile computing scenarios. In many applications, in fact, the utility of a network of mobile computing devices derives primarily from the data and information it holds. The identity of the individual nodes storing the data tends to be less relevant. Consider, for example, sensor network applications or file-sharing applications between mobile devices. Suitable interaction models and communication abstractions would have to be flexible and tailored to the application, not tied to the identities of the individual components. In this context, overlay networks can offer several application-specific mechanisms to route information in a dynamic network:

- *Location-based routing*, where an agent takes advantage of location information to access resources within suitable locality constraints (e.g., "find all the printers on my floor" or "find the closest gas station")
- *Content-based routing*, where data is searched and accessed on the basis of its content rather than on the basis of the network addresses or location of the nodes

An example of the latter case, in a mobile sensor network scenario [26], would be an agent interested in the occurrence of the data named "truck sightings"; the network must provide the means to effectively access such data wherever it might be.

Tuple-based models are a particularly fertile ground to investigate to develop much more advanced kinds of overlay networks. For example, a set of tuple spaces networked with each other can naturally lead to a semantically enriched overlay network, from which to retrieve data in a more meaningful way than in current approaches. Moreover, programmable tuple spaces such as MARS [5] and TOTA [19] can naturally support the reconfiguration, updating, and maintenance of such overlay networks.

Overlay data structures are not focused only on the network topology; instead, they can generalize overlay networks by encoding and providing agents with several pictures, possibly locally confined, of specific aspects of the operational environments of the agents. Agents can access overlay data structures to achieve awareness and possibly modify specific contextual aspects. The strength of these overlay data structures is that they can be accessed piecewise as the application agents visit different places of the distributed environment. This allows the agents to access the appropriate information at the correct location. Again, tuple spaces are a perfect abstraction on which to build such overlays.

Stigmergy and Swarm Intelligence

Overlay data structures realized on top of tuple-based models can naturally accommodate stigmergic interaction patterns. These patterns are at the core of swarm intelligent systems [2], which are systems where a large number of simple agents coordinate (often mimicking natural and biological systems) in an indirect way, via sensing digital pheromones in a virtual environment, to achieve — in an adaptive and self-organizing way — tasks that far exceed their capabilities as single individuals [23].

Tuple-based models are naturally suited to supporting these interaction patterns, in that tuple spaces can be used to store the pheromones at the basis of stigmergy interactions. Moreover, active or, better, reactive tuple spaces can naturally provide functionalities to allow pheromones (implemented by means of tuples) to change as needed (e.g., to diffuse across the network, to be aggregated and combined, to evaporate if not used and reinforced). It is easy to see that such innovative scenarios give rise to endless directions for research regarding tuple-based coordination models. How can tuple-based models be employed to control and govern self-organizing stigmergic coordination activities? How can tuple spaces fill the semantic gap between heterogeneous agents that coordinate by means of tuples and "ant-like" agents that simply coordinate by reacting upon pheromone sensing? What applications can be enabled by such rich coordination models? We do not have any answers to these questions yet.

Pervasive Spaces and Tuple Spaces

Recent research in the area of pervasive and ubiquitous computing suggest that, to enable spontaneous interactions among a set of computer-based devices embedded in an environment (e.g., smart rooms, smart furniture, smart objects) and also to support advanced interaction models between users and the surrounding (computer-enriched) environment, the environment itself should be rendered in some sort of digital abstraction. These considerations have led to several proposals for middleware infrastructure based on the concept of *active spaces* [27]. Active spaces are a sort of digital representation of a physical environment, where each resource in the environment (a user, a computer, as well as any computer-based device and any computer-based object) has a digital representation and is provided with mechanisms to interact with the other resources. Thus, active spaces act as a type of shared data spaces, where high-level interaction patterns can be promoted and from which high-level contextual information can be obtained. Remaining to be investigated are how and to what extent tuple-based coordination models and active-spaces-based approaches can be made to coexist and converge into a single coordination model that supports messaging, synchronization, and the exchange of raw data and of semantic high-level information. We must still investigate how and to which extent emerging pervasive computing technologies can be used to conceive new architectural solutions for tuple-based middleware models. For example, one could think of exploiting the stable memory of radio-frequency identification (RFID) tags to deploy, in a massively distributed way, tuples and tuple spaces in any physical environment [20].

Conclusions

In this chapter we have discussed how tuple-based models and middleware can indeed offer valuable tools to support the coordination and context awareness of uncoupled agents in mobile computing scenarios. We hope that the taxonomy introduced here and the critical survey of existing systems have helped readers to reach a better understanding of the software engineering issues involved and that they steer developers toward the adoption of specific tuple-based middleware systems suitable to their purposes. Despite the suitability of tuple-based coordination models for mobile computing, developers and researchers should be aware that a number of fascinating research issues are yet worth investigating.

Acknowledgments

This work was supported by the Italian MIUR and CNR within the project "IS-MANET, Infrastructures for Mobile *Ad Hoc* Networks."

References

[1] Babaoglu, O., Meling, H., and Montresor, A., Anthill: a framework for the design and the analysis of peer-to-peer systems, in *Proc. of the 4th European Research Seminar on Advances in Distributed Systems (ERSADS '01)*, Bertinoro, Italy, May 14–16, 2001.

[2] Bonabeau, E., Dorigo, M., and Theraulaz, G., *Swarm Intelligence*, Oxford University Press, London, 1999.

[3] Borcea, C. et al., Cooperative computing for distributed embedded systems, in *Proc. of the 22nd Int. Conf. on Distributed Computing Systems (ICDC'02)*, Vienna, Austria, July, 2002.

[4] Cabri, G., Leonardi, L., Mamei, M., and Zambonelli, F., Location-dependent services for mobile users, *IEEE Trans. Systems, Man, Cybernetics, Part A: Systems Humans*, 33(6), 667–681, 2003.

[5] Cabri, G., Leonardi, L., and Zambonelli, F., MARS: a programmable coordination architecture for mobile agents, *IEEE Internet Comput.*, 4(4), 26–35, 2000.

[6] Charles, A., Menezes, R., and Tolksdorf, R., On the implementation of SwarmLinda, in *Proc. of the 42nd Annual ACM Southeastern Conf.*, Huntsville, AL, April 2–3, 2004.

[7] Cugola, G. and Picco, G.P., *PeerWare: Core Middleware Support for Peer-to-Peer and Mobile Systems*, Technical Report, http://peerware.sourceforge.net/.

[8] Davies, N., Wade, S., Friday, A., and Blair, G., Limbo: a tuple space based platform for adaptive mobile applications, in *Proc. of the Int. Conf. on Open Distributed Processing/Distributed Platforms (ICODP/ICDP '97)*, Toronto, Canada, May 27–30, 1997.

[9] Dey, A. and Abowd, G., The context toolkit: aiding the development of context-aware applications, in *Proc. of the Conf. on Human Factors in Computing Systems (CHI '99)*, ACM Press, New York, 1999, pp. 434–441.

[10] Edwards, K. and Grinter, R., At home with ubiquitous computing: seven challenges, in *Proc. of the 2001 ACM Handheld and Ubiquitous Computing Conf. (UbiComp)*, Atlanta, GA, October, 2001.

[11] Estrin, D., Culler, D., Pister, K., and Sukjatme, G., Connecting the physical world with pervasive networks, *IEEE Pervasive Comput.*, 1(1), 59–69, 2002.

[12] Freeman, E., Hupfer, S., and Arnold, K., *JavaSpaces Principles, Patterns, and Practice*, Addison–Wesley, Boston, MA, 1999.

[13] Gelernter, D. and Carriero, N., Coordination languages and their significance, *Comm. ACM*, 35(2), 96–107, 1992.

[14] GigaSpaces Technologies, Ltd., http://www.gigaspaces.com/index.html.

[15] Intamission, Ltd., *AutevoSpaces Product Overview*, http://www.intramission.com/downloads/datasheets/AutevoSpaces-Overview.pdf.

[16] Intanagonwiwat, C., Govindan, R., and Estrin, D., Directed diffusion: a scalable and robust communication paradigm for sensor networks, in *Proc. of the Sixth ACM MOBICOM Conf.*, Boston, MA, August, 2000.

[17] Johanson, B. and Fox, A., The event heap: a coordination infrastructure for interactive workspaces, in *Proc. of the 4th IEEE Workshop on Mobile Computing Systems and Applications (WMCSA 2002)*, Callicoon, NY, June 21–22, 2002.

[18] The JXTA Project, http://www.jxta.org.

[19] Mamei, M. and Zambonelli, F., Programming pervasive and mobile computing applications with the TOTA middleware, in *Proc. of the 2nd IEEE Int. Conf. on Pervasive Computing and Communication (PerCom 2004)*, Orlando, FL, March 14–17, 2004.

[20] Mamei, M. and Zambonelli, F., Spreading pheromones in everyday environment through RFID technology, in *Proc. of the 2nd IEEE Symp. on Swarm Intelligence*, Pasadena, CA, June, 2005.

[21] Mascolo, C., Capra, L., and Emmerich, W., An XML-based middleware for peer-to-peer computing, in *Proc. of the 1st IEEE Int. Conf. on Peer-to-Peer Computing*, Linkoping, Sweden, August 25–27, 2001.

[22] Omicini, A. and Zambonelli, F., Coordination for Internet application development, *J. Autonomous Agents Multi-Agent Syst.*, 2(3), 251–269, 1999.

[23] Parunak, V., Brueckner, S., and J. Sauter, J., Digital pheromones for coordination of unmanned vehicles, in *Proc. of the Workshop on Environments for Multi-Agent Systems (E4MAS)*, New York, July 19, 2004.

[24] Picco, G.P., Murphy, A.L., and Roman, G.C., LIME: a middleware for logical and physical mobility, in *Proc. of the 21st Int. Conf. on Distributed Computing Systems (ICDCS-21)*, Phoenix, AZ, April 2001.

[25] Rao, A., Papadimitriou, C., Ratnasamy, S., Shenker, S., and Stoica, I., Geographic routing without location information, in *Proc. of the Ninth ACM MOBICOM Conf.*, San Diego, CA, September 14–19, 2003.

[26] Ratsanamy, S. et al., GHT: a geographic hash table for data-centric storage, in *Proc. of the 2002 Int. Workshop on Wireless Sensor Networks and Applications*, Atlanta, GA, September 28, 2002.

[27] Roman, M. et al., Gaia: a middleware infrastructure for active spaces, *IEEE Pervasive Comput.*, 1(4), 74–83, 2002.

[28] Roman, G.C., Julien, C., and Huang, Q., Network abstractions for context-aware mobile computing, in *Proc. of the Int. Conf. on Software Engineering (ICSE 2002)*, Orlando, FL, May 19–25, 2002.

[29] Wells, G.C., New and improved: Linda in Java, in *Proc. of the Third Int. Conf. on Principles and Practice of Programming Java (PPPJ)*, Las Vegas, NV, June 16–18, 2004.

[30] Wyckoff, P., McLaughry, S.W., Lehman, T.J., and Ford, D.A., T spaces, *IBM Syst. J.*, 37(3), 454–474, 1998.

[31] Zambonelli, F., Gleizes, M.P., Mamei, M., and Tolksdorf, R., Spray computers: explorations in self organization, *J. Pervasive Mobile Comput.*, 1(1), 1–20, 2005.

Chapter 11

Content-Based Publish–Subscribe in a Mobile Environment

Gianpaolo Cugola, Amy L. Murphy,
and Gian Pietro Picco

CONTENTS

Introduction

Modern distributed computing demands not only scalability, as witnessed by the Internet, but also an unprecedented degree of adaptability to dynamic conditions. Mobile computing is evidence of this trend. The mobility of network nodes undermines many of the traditional assumptions of distributed systems: The topology becomes fluid as hosts move and yet retain the ability to communicate wirelessly; communication occurs over a shared medium that is not only unreliable but also largely unpredictable, as it strongly depends on the characteristics of the local environment; and hosts and therefore applications frequently experience disconnection, which is no longer just a network accident but is often induced deliberately for long periods of time to save power. Other modern distributed scenarios raise similar issues in terms of dynamicity; peer-to-peer networks and sensor networks come to mind.

Coping with these demands is a challenging task. In recent years, the *publish–subscribe* paradigm has emerged as a promising and effective way to tackle many of these issues. The implicit and asynchronous communication paradigm that characterizes publish–subscribe supports a high degree of decoupling among the components of a distributed application. In principle, it is possible to add or remove one component without affecting the others — only the dispatcher, the element in charge of collecting subscriptions and routing messages, has to be aware of the change. Clearly, this form of decoupling would be desirable in a scenario where the set of available components undergoes continuous change, as in the mobile one. Nevertheless, much of the potential of the publish–subscribe *model* still remains to be unleashed by publish–subscribe *systems*. Indeed, many of the available distributed publish–subscribe middleware exploit a dispatching network arranged in a tree overlay for increased scalability, but their designs usually do not tolerate any form of topological reconfiguration. Paradoxically, therefore, these systems cannot be exploited precisely in those application scenarios where decoupling would be most beneficial.

In this chapter, we discuss challenges of and solutions for content-based publish–subscribe in a mobile scenario. Although we focus on our own research in the field [1,13–16,18–20,27,33,36], we also provide the reader with a discussion of related and alternative approaches, thus covering the entire spectrum of the state of the art.

Publish–Subscribe: An Overview

Distributed applications exploiting publish–subscribe middleware are organized as a collection of autonomous components (*clients*) which interact by *publishing* messages and by *subscribing* to the classes of messages they are interested in. The core component of the middleware, the *dispatcher*, is responsible for collecting subscriptions and forwarding messages from publishers to subscribers. This scheme results in a high degree of decoupling among the communicating parties. These ideas have been recently popularized by a wealth of systems, each interpreting the publish–subscribe paradigm in a different way. (For more detailed comparisons, see Carzaniga et al. [8], Cugola et al. [17], Eugster et al. [22], and Rosenblum and Wolf [40].)

A first point of differentiation is the expressiveness of the subscription language, drawing a line between *subject-based* and *content-based* systems. In the first case, subscriptions contain only the name of a class of messages — usually called *subject*, *channel*, or *topic* — chosen among a set of predefined classes. In content-based systems, the selection of a message is determined entirely by the client, which uses expressions (often called *filters*) that allow sophisticated matching on the message content.

The second point of differentiation is the architecture of the dispatcher, which can be either centralized or distributed. In this chapter, we focus on the latter type. In this middleware, a set of *brokers* (see Figure 11.1) is interconnected in an overlay network; they cooperatively route subscriptions and messages sent by clients connected to them, thus increasing the scalability of the system. In this context, the main design decisions concern the topology of interconnection and the routing strategy. Although the first approaches based on a graph topology are beginning to appear [1,15], most of the available systems are based on a tree topology, as this simplifies routing (e.g., by avoiding the possibility of routing loops) and provides a high degree of scalability.

Several tree-based routing strategies can be found in the literature [5,8,17], with the most basic ones shown and compared in Figure 11.1. The simplest approach is *message forwarding*, in which a published message is forwarded by a broker to all the others along the dispatching tree. Subscriptions are never propagated beyond the broker receiving them. This broker stores these subscriptions in a *subscription table* that is used to determine which clients, if any, should receive incoming messages.

Message forwarding may generate high overhead because messages are sent to all brokers regardless of the interests of the clients attached to them. An alternative strategy, called *subscription forwarding*, limits this overhead by spreading knowledge about subscriptions throughout the

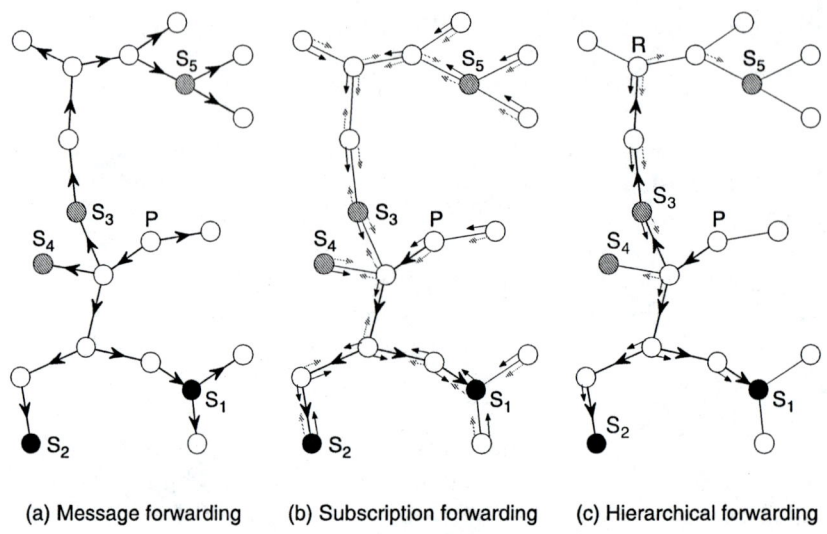

(a) Message forwarding   (b) Subscription forwarding   (c) Hierarchical forwarding

Figure 11.1   **Publish–subscribe routing strategies.**

system. When a broker receives a subscription from one of its clients, not only does it store the associated filter in its subscription table as in message forwarding, but it also forwards it to all the neighboring brokers. During this propagation, each dispatcher behaves as a subscriber with respect to its neighbors; consequently, each of them records the filter associated with the subscription in its own subscription table and forwards it again to all its neighboring dispatchers except the one that sent it. This process effectively sets up routes for messages through the reverse path followed by subscriptions. (Note that this scheme is optimized to avoid forwarding the same message filter in the same direction; moreover, some systems perform even more aggressive optimizations by exploiting "coverage" relations among filters [8,9,26,45].)

Finally, *hierarchical forwarding* strikes a balance between the two aforementioned strategies by assuming a rooted tree topology. Subscriptions are forwarded toward the root to establish the routes that published messages will follow downstream toward subscribers. Messages, in fact, are always propagated upstream up to the root and flow downstream along the tree only if a matching subscription has been received from the corresponding subtree.

Figure 11.1 provides a graphical representation of the three strategies. Brokers S_1 and S_2 subscribed (through their clients; not shown in the figure) to the same black filter, and S_3, S_4, and S_5 subscribed to gray. The small arrows represent the content of the subscription tables for the corresponding filters. Broker P published a message matching the black filter but not the gray one. The path followed by this message is indicated by the large arrows. In hierarchical forwarding, broker R is the root of the dispatching tree.

Mobility and Publish–Subscribe: The Issues

The overview in the previous section showed how several approaches enable distributed content-based publish–subscribe; however, most of the research effort has focused either on how to provide efficient pattern matching and message forwarding or on efficient routing strategies for pushing scalability. Research essentially aims at improving the *performance* of content-based publish–subscribe in large-scale settings, implicitly assuming a static dispatching infrastructure. This is unfortunate because, as mentioned in the introduction, the characteristics of the publish–subscribe model and, more specifically, the high degree of decoupling it enables make it amenable to highly dynamic scenarios such as those defined by mobility. Nevertheless, this scenario is possible only if the systems embodying the model are expressly designed to take into account the assumptions and challenges posed by the target dynamic scenario.

Mobility poses several challenges to the design of publish–subscribe middleware. The most evident is that the topology of the system, usually assumed static by existing systems, now becomes dynamic and undergoes continuous reconfiguration as the mobile nodes move. Depending on the mobility scenario, these factors may have different impacts.

In many cases, mobility is relegated to the periphery of the system; for example, this is true for the *nomadic* scenarios many of us experience while traveling or even when moving from office to home. The user detaches from one network (e.g., office) and reconnects to a different one (e.g., home or a hotel room). The entry point to the network has changed, yet, thanks to dedicated protocols (e.g., Dynamic Host Configuration protocol [DHCP] and virtual private networks [VPNs]), the user retains access to the basic networking services. Similar considerations hold for those scenarios where users change their network entry points while in movement and protocols such as Mobile IP [34] transparently maintain connectivity at the network level. Notably, in the first case wireless communication is a nice but unnecessary feature, whereas in the second one it becomes key to enabling unconstrained and continuous movement. In both these scenarios, however, only the end nodes are mobile; the networking infrastructure, which handles routing and other functions, is assumed to be stable.

The same concepts can be applied to the typical architecture of a publish–subscribe system by observing that clients play the role of end nodes, as they do not provide network functionality, and brokers assume the roles of routers and switches. The impact of mobility on publish–subscribe is similar to its impact on networking — namely, modifications that shield clients from the complexity of dealing with mobility while leaving the behavior of the infrastructure largely unaffected. Interestingly, the same applies when the system exhibits logical mobility of code or agents [28] (e.g., because the publish–subscribe clients are mobile agents, which detach and reattach to the closest broker during migration).

At the other extreme, *mobile ad hoc networks* (MANETs) [35,44] define the most radical mobility scenario, where no assumption is made about the dynamic topology of the system and the networking infrastructure itself is assumed to be mobile. The impact of mobility in this case is disruptive and no longer limited to the clients dwelling at the fringes of the system, as the intermediate nodes in charge of routing and other network functions are now assumed to be mobile. Moreover, most application scenarios for MANETs actually blur the distinction between end nodes and intermediate nodes by assuming that *all* the network nodes implement the functionality required to enable routing. As a consequence, networking protocols must be reconsidered from the ground up to accommodate the new deployment assumptions, as witnessed by the appearance of entirely new routing protocols (e.g., those described by Perkins [35]).

Again, publish–subscribe systems face similar problems in that they demand significant and radical changes in the behavior of the dispatching infrastructure. For example, subscription information can no longer be associated permanently with the link from which it came, because the subscriber can move and become connected through a route involving a different set of links. Moreover, as in networking scenarios, the distinction between infrastructure and application nodes becomes blurred, effectively introducing a different application model where all client hosts are also brokers [30].

Interestingly, analogous considerations hold for scenarios where the communication topology is dynamic but not caused by mobility and wireless communication. For example, in peer-to-peer (P2P) networks, the hosts and physical communication links are fixed, but the *logical* topology of the overlay network along which file searches are disseminated undergoes continuous change as peers join and leave. Exploiting a publish–subscribe system in this scenario raises challenges similar to those discussed thus far in that the architecture of the P2P network (e.g., based on a hierarchical supernode infrastructure or totally decentralized) determines the level of dynamicity required in the dispatching infrastructure.

Other challenges are peculiar to mobility, such as those related to the physical communication media. Wireless communication removes the need for cables and therefore is a key enabler of mobility; however, the price paid for this freedom is lower performance and reduced reliability, which must often be taken into account at the higher application layers. For example, unicast communication is the fundamental building block for many distributed applications in fixed environments, where it enjoys an efficient network implementation. In mobile scenarios, multi-hop unicast can be expensive, as it often requires several local broadcasts (and corresponding replies) to find a suitable route [35]; therefore, it should be used sparingly in the development of middleware for these scenarios.

Similarly, the communication links of a conventional distributed system are often thought of as being fairly reliable. For example, the fault model assumed by many systems and protocols — notably, the Transmission Control protocol (TCP) — is one where communication failures are rare and transient; that is, the communication target is assumed to become reachable again. In mobile scenarios, disconnections are frequent, not only because the communication medium is more sensitive (e.g., to fluctuations in the propagation of radiowaves induced by the environment) but also because disconnection is no longer an accident; rather, it is often deliberately induced by the user or application to, for example, save battery power. (Although power management is another relevant issue in mobility that affects not only the host but also network communication, we do not have the opportunity in this chapter to touch upon these issues further.) Failures are not guaranteed to be transient; for example, cars moving on a highway in opposite directions may never meet again.

Reliability is usually taken into account at the network level; however, in the field of mobility, it is often useful to reduce the size of the network stack by blurring the distinctions among levels for the sake of reducing the system footprint and enabling optimizations (e.g., reduce the use of unicast). In the specific case of publish–subscribe, another challenge to reliability comes directly from the application level, where the messages being routed on the overlay network may get lost along stale routes due to the topological reconfiguration induced by mobility. The net result of these considerations is that the design of publish–subscribe middleware must often deal directly with reliability.

Finally, in addition to posing challenges to the implementation of the core communication layer, mobility makes it necessary to take a new approach to the development of distributed applications — namely, one that is *context aware*. By definition, mobile hosts change their location in the physical space and in doing so experience a different *context* in terms of the physical (e.g., temperature, light, reachable hosts) or logical (e.g., application services) constituents of the environment. Devising programming abstractions to properly capture, disseminate, and exploit context is an open research problem. Publish–subscribe and in particular content-based systems appear to provide a sound foundation for many approaches, thanks to their decoupling and reactive paradigm of interaction. The rest of this chapter analyzes many of these issues in more detail.

Dealing with Mobile Clients

The first and simplest form of mobility that should be supported by publish–subscribe middleware tailored to mobile scenarios is that of clients, by offering them the possibility to disconnect from the dispatching infrastructure and reconnect from a different place at a later time. This facility is fundamental to effectively supporting scenarios of mobility, such as nomadic computing, that involve only the "leaves" of the system (i.e., the clients). In these situations, the publish–subscribe middleware must offer appropriate mechanisms to make mobility transparent to the other components by reconfiguring routing and storing messages addressed to the moving clients until they reconnect.

In the presence of a centralized dispatcher, supporting mobile clients is just a matter of buffering messages addressed to disconnected clients until they reconnect. The problem becomes much more complex in the presence of a distributed dispatcher. In this case, a client must be allowed to disconnect from the broker currently acting as its entry point to the dispatching network and to later reconnect to a broker potentially different from the previous one, which is usually chosen as the closest one to the new location of the client. To support this form of mobility, not only must the publish–subscribe middleware buffer messages addressed to the client while it is disconnected, but it must also be able to change the brokers' subscription tables when the client reconnects. This requires a distributed protocol that coordinates the brokers involved and avoids losing (or duplicating) the messages sent while the reconnection process is running.

JEDI [17] was the first distributed publish–subscribe middleware to offer this form of mobility. It adopts a hierarchical forwarding routing strategy and expects clients to proactively inform the middleware when moving away from or arriving at a broker. Figure 11.2 illustrates the procedure that takes place when client C detaches from broker B_1, moves, and reattaches to B_2. Upon disconnection, B_1 begins buffering the messages addressed to C. When C reconnects at B_2, the latter initiates and coordinates the distributed protocol to rearrange the routing information and retrieve the messages buffered at B_1. This protocol consists of the following steps: First, B_2 repropagates the subscriptions held by C to set up the new routes that will steer messages toward the new location of C. Any message received

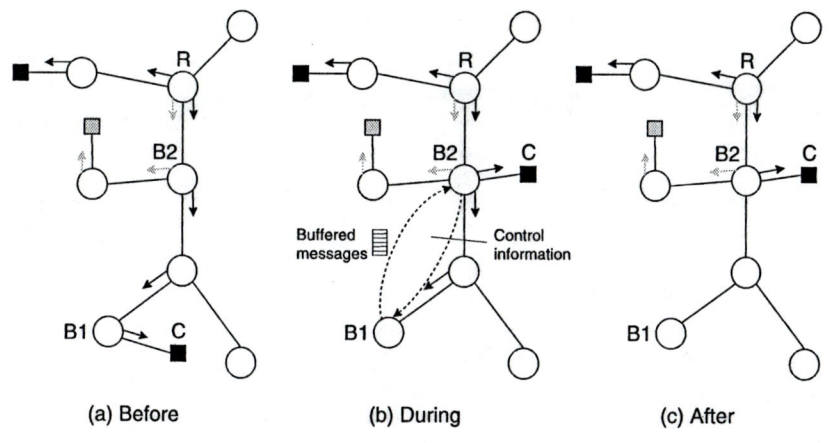

Figure 11.2 Dealing with client mobility in Jedi: the situation before, during, and after migration of client C.

as a consequence of these subscriptions is buffered at B_2 until the reconnection process ends. After the new routes are in place, B_2 asks B_1 to stop buffering messages, to remove C's subscriptions, and to forward the buffered messages. These messages, together with those buffered at B_2, represent the entire set of messages circulated in the system during the migration of C. Some duplicates may be present in this set because the old routes and the new ones coexisted for a short time, as shown in Figure 11.2b; however, these duplicates are easily detected and discarded at B_2. The filtered set of messages is finally sent to C, ending the reconnection process.

Similar distributed protocols, albeit in the context of a subscription-forwarding routing strategy, are adopted by the extended version of Siena in Caporuscio et al. [7], Elvin in Sutton et al. [43], REBECA in Fiege et al. [24], and the system described in Podner and Lovrek [39].

Dealing with Mobile Brokers: An Integrated Approach

As mentioned earlier, dealing with mobility scenarios that make no assumptions about the stability of the infrastructure requires an entirely different approach from the one discussed in the previous section. The topological reconfiguration induced by mobility disrupts the very dispatching infrastructure, so new solutions are required to preserve its operation and yet support its dynamicity.

The reconfiguration problem we address in this section can be defined informally as *to adapt the dispatching infrastructure of a distributed publish–subscribe system to changes in the topology of the underlying physical network and to do so without interrupting the normal system operation.* In the following, we focus on content-based systems that adopt a subscription-forwarding strategy and an unrooted tree overlay, as these are assumed by the majority of existing systems. For these systems, the reconfiguration problem stated previously can be broken down into three subproblems, namely:

- Repairing the overlay dispatching network to retain connectivity among brokers without creating loops
- Reconciling the subscription information held by each broker to keep it consistent with the topological changes without interfering with the normal processing of subscriptions and unsubscriptions
- Recovering messages lost during reconfiguration

In this section, we present solutions to these problems, based on our own research on the topic [13,14,18,20,27,33,36]. To reduce the complexity, each problem is addressed separately by leveraging the fact that the three problems are orthogonal. When the techniques we describe here to solve each problem are combined in a single coherent system (e.g., the REDS system we describe later), they provide an integrated solution to the overall problem of dealing with mobile brokers.

Repairing the Overlay

Given that the overlay network we consider is a tree, we have two options to consider for repairing: to allow cycles to form and remove them later or to disallow the formation of cycles.

We choose the second approach because it is most appropriate when considering updates to the subscription tables, as seen in the next section. We also consider two different types of failures: link and broker. From a theoretical perspective, link failure creates two trees with exactly the same nodes as before the link break. Repair, therefore, involves adding a link with endpoints in each of the two trees. Failure of a node with n neighbors results in n partitions, which require the addition of $(n - 1)$ new links.

The main challenge to address in repairing the overlay network is selecting these links to repair the tree. We have developed two approaches, the first specifically for mobile *ad hoc* networks and the second for dynamic networks in which connectivity exists between each pair of brokers (e.g., P2P networks). In the following, we consider these two scenarios, separately.

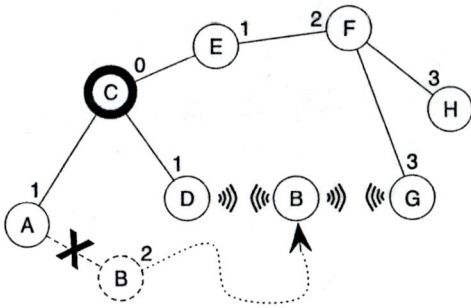

Figure 11.3   Overlay network in a mobile environment where B moves out of contact with A and into contact with D and G. All brokers are labeled with their numerical depth from leader C.

Mobile Networks

Our work to build and maintain an overlay network in a mobile environment is based on prior study of multicast in MANETs. In particular, we started from the Mobile *Ad Hoc* On-Demand Distance Vector (MAODV) protocol [41], because it focuses on building and maintaining a single tree containing the mobile nodes participating in a multicast group.

In MAODV, one node is identified as the leader, and all nodes know their distance from it in the tree. When a link breaks (or a node fails, causing several links to break), the nodes farthest from the leader initiate the reconnection process, searching for a link that will reconnect their subtree to the subtree of the parent through a node with a depth less than or equal to their own. This constraint guarantees that insertion of a link will not create a cycle. For example, in Figure 11.3, if the link fails between A and B, B will search for a new link between its subtree and the subtree of its parent A. D satisfies the depth constraint; consequently, the link (B,D) is added and the depths of all nodes are updated.

Identification of potential links is accomplished by broadcasting a route request (RREQ) message a small number of hops from B. Any node with a depth less than or equal to the depth of B responds with a route reply (RREP) message that follows the reverse path of the RREQ, identifying the path reconnecting the two trees. In MAODV, nodes not participating in the multicast group may also serve as routers in the multicast tree. Our approach [33] assumes that all nodes act as brokers in the publish–subscribe network; a similar assumption was made in Huang and Garcia-Molina [30]. With this assumption, we have designed more efficient mechanisms for reconnecting the tree, specifically changing the propagation rules of the RREQ message and altering the selection criteria for the new link.

In our approach, members of the tree are allowed to propagate the RREQ message, an act disallowed in MAODV to prevent the formation of loops. This forwarding of the RREQ extends the limits of the search for a broker with suitable depth and the identification of a path between it and the requesting broker. To prevent the introduction of loops, we prohibit the RREQ from being propagated across a non-tree link more than once; in other words, the RREQ cannot propagate from B to G (a non-tree link) and then to H (a second non-tree link). This disallows the addition of links (B,G) and (G,H), a situation that forms cycles in the overlay tree; however, G may forward the RREQ from B to F along the overlay tree, a situation explicitly prohibited in MAODV. In this case, F meets the depth criteria and responds to the RREQ message, indicating the possible addition of the link between B and G, an option for restoring the tree that MAODV does not identify.

Our second extension to MAODV is the selection criteria for the new link. In Figure 11.3, both D and F reply to the RREQ message with options for repairing the tree, and B must choose one. The main selection criteria in MAODV is the length of the path to the tree, which in MAODV may include several nodes that are actually not part of the multicast group. In our approach, where all nodes act as brokers, the distance to the tree is always one hop; therefore, we adopt a selection criteria that minimizes the effect of the reconfiguration on the subscription tables. Specifically, the reply with the shortest path between the endpoint of the old link on the tree (A) and the new endpoint (D or G) is selected. In the example, D's reply is selected because the path length from A to D is shorter than from A to G. A more detailed description of this approach, together with experimental results collected from our implementation, can be found in Mottola et al. [33].

Fixed Networks

As an alternative to networks where connectivity is determined by broker proximity, we have also devised a protocol [27] for a fixed network scenario, as found in P2P networks. In this scenario, dynamicity comes from the addition and removal of brokers, not links, and the presence of a fixed network enables the addition of a link between any pair of connected brokers. Our solution is again inspired by MAODV but adapted to the aforementioned scenario requirements. Furthermore, because we have greater control over connectivity, we also enforce a maximum degree on each broker, thus limiting its message forwarding burden.

In our approach, we exploit three types of repair procedures: local, global, and root-specific. In a local recovery, only brokers close to the recovering one are involved. Global recovery reaches brokers anywhere in the tree. Root-specific protocols come into play only when the root (i.e., the leader in MAODV terminology) broker fails. Local recovery exploits the fact that all brokers know the identities of their siblings as well as the identities of some of their direct ancestors (brokers on the path between itself and the root). When a broker fails, the tree can be reconnected by linking its former children to each other and at least one of these children to an ancestor. We have developed protocols that balance the broker degree, preventing all brokers from connecting to the same ancestor (creating a star network with high broker degree) and similarly preventing all but one child from connecting to one another (creating a line with low broker degree). The local recovery procedure also allows a broker to refuse a request if the addition of the broker as a child will increase its degree beyond a predefined limit. In this case, the request to find a parent is forwarded downstream from the refusing broker in hopes of finding a broker that has not yet reached its maximum degree. This technique is surprisingly effective, exploiting the trend that brokers farthest from the root have lower degrees.

Global recovery comes into play when local recovery fails because broker degree requirements cannot be met or because a new parent cannot be identified among the siblings and the ancestors, a case that arises when a cluster of brokers fails. In these situations, the broker seeks a new parent from a cache that it maintains of other brokers in the tree. This cache is populated, for example, by recording the source of messages propagated over the tree. To find a new parent, a broker is selected from the cache and is sent a request to allow the requesting broker to become a child. To avoid loops, we adopt an algorithm based on the notion of tree depth, similar to MAODV. If the broker has a lower depth than the requesting broker, it can accept to become the new parent; otherwise, it can forward the request upstream to find a broker with lower depth, forward it downstream to find a broker with lower broker degree, or simply reject the request.

It may still happen that a broker cannot identify a new broker to serve as its parent. In this case, it declares itself to be a new root, creating its own tree. Because our goal is to maintain a *single* overlay tree, we need a mechanism to merge trees when they discover each another. For this, we assign an identifier to each tree. After a broker has declared itself to be a root, it periodically contacts brokers in its global cache. If a broker is found with a different tree identity, the two trees are merged.

This notion of merging trees can also be exploited in the case of root failure. When the root fails, all its former children declare themselves to be roots of their subtrees. They then exploit their global cache to identify the other subtrees and re-merge the tree; unfortunately, this may take a long time, during which message routing on the tree is disrupted. We therefore defined a protocol specific to root failure, essentially electing a new root among the former children of the old root and allowing the remaining children to connect to this new root or to one another. By combining local, global, and root-specific protocols we can keep a tree connected despite brokers frequently being added and removed. A full evaluation of the effectiveness of these techniques is available in Frey and Murphy [27].

Reconciling Routing Information

After ensuring maintenance of the overlay tree, the next step is maintaining the subscription tables to allow messages to continue to reach the subscribers. Here we consider protocols that address link loss rather than broker loss, because the latter case can be addressed as a combination of several link repair actions.

When a link fails between a pair of brokers the overlay management protocols described in the previous section take on the responsibility of finding the replacement link. When this link has been found, the subscription tables must be updated so all messages that traversed the now-broken link are sent across the new link to reach the subscribers on the other subtree. We have developed a series of protocols to accomplish this, each with different requirements from the overlay management protocols and with different assumptions about the environment [18,20,36].

The first solution, which we refer to as a *strawman protocol*, is the only proposal previously suggested in the literature [8]. This approach utilizes only the usual publish–subscribe subscription and unsubscription messages. When a link disappears, a broker behaves as if it received unsubscription messages from the former neighbor, updating its subscription table and propagating the unsubscription message if necessary. This has the effect of stopping message forwarding across the broken link. When the new link is added, its endpoints send subscriptions to one another for all entries in their subscription table, allowing messages to flow across the new link.

While this approach successfully reconfigures the subscription tables, it may cause unnecessary overhead; for example, consider the scenario in Figure 11.4 in which only one broker in a subtree is a subscriber. When the link breaks between A and B, the unsubscription process removes all entries in the subscription tables of the brokers in B's subtree. When the subscription process begins across the new link (C,D), it reinserts most of these entries exactly as before, creating unnecessary overhead to remove many subscriptions that are immediately reinserted. To overcome this, we experimented with delaying the unsubscription process until the subscription process is complete. This reversal technique, that we refer to as

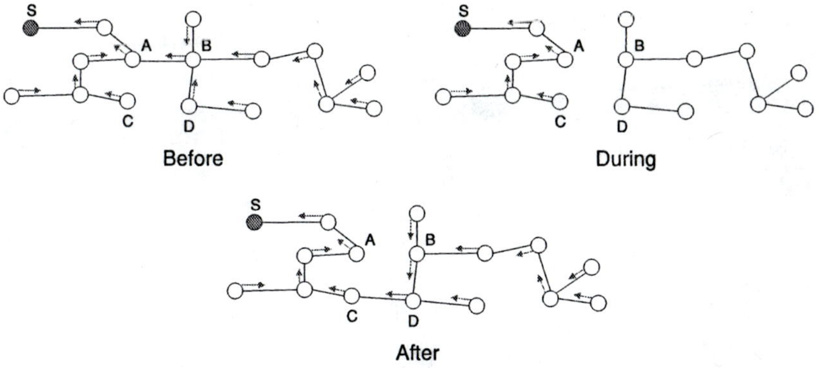

Figure 11.4 A dispatching tree before, during, and after a reconfiguration performed using Strawman. The shaded broker is a subscriber. Arrows indicate the propagation direction for messages.

deferred unsubscription, is effective in reducing the overhead of reconfiguration, up to 50 percent over the strawman protocol in simulation studies characterized by a large number of reconfigurations. In the simple example above, it prevents the removal and replacement of all subscriptions on B's subtree. Details about two different mechanisms for deferring subscriptions can be found in Picco et al. [36] and Cugola et al. [20].

Analysis of the publish–subscribe behavior reveals that reconfiguration is restricted to the brokers on the path between the endpoints of the old and new links, termed the *reconfiguration path* [18]. The subscription tables of all other brokers remain unchanged. In Figure 11.4, the reconfiguration path is composed of the brokers from A to C, across the new link from C to D, and from D to B. To exploit this property, we designed a protocol [18] that begins at one endpoint of the old link and moves along the reconfiguration path, updating the subscription tables as it progresses. One drawback of this protocol is the requirement that the path must remain intact during the entire reconfiguration. If a second link fails on the reconfiguration path, the reconfiguration messages stop propagating and the system is left with inconsistent subscription tables. A second drawback is the need to know the identity of the brokers on the reconfiguration path, an additional requirement for the overlay management protocol. Finally, this protocol is complex when considering the details to address the subscriptions and unsubscriptions that occur during the reconfiguration. On the other hand, this protocol can achieve overhead reductions up to 78 percent over the strawman protocol in scenarios where reconfigurations do not overlap.

To bridge between the resilient Deferred Unsubscription protocol and the efficient Reconfiguration Path protocol, we have designed a new protocol that exchanges information among the brokers on the old and new links which we refer to as the *Informed Link Activation protocol* [47]. Specifically, the endpoints on the old link send the contents of their subscription tables to the endpoints of the new links. By combining these with their own subscription tables, the endpoints of the new link calculate which subscriptions to send across the new link. Again, this is complicated by the insertion and removal of subscriptions during reconfiguration, but the protocol is not as complex as the Reconfiguration Path protocol. With these approaches that share information between the old and new link, we have shown that few brokers outside the reconfiguration path are affected by reconfiguration, thus resulting in an overhead reduction of up to 76 percent, similar to results for the reconfiguration path approach but in the presence of concurrent reconfigurations.

Each of these protocols operates with varying expectations from the overlay management protocol and tolerance for changes during tree repair. This leads to a number of tradeoffs that must be considered when selecting the protocol for a given system. For example, although the reconfiguration path approach has clear advantages with respect to reduction of overhead, it adds the burden to the overlay management protocol to identify all nodes on the path and requires the environment to keep the path stable during reconfiguration. The Deferred Unsubscription protocol makes no assumptions about either stability or knowledge passed from the overlay management protocol; however, its overhead reduction is not as significant. The Informed Link Activation protocol falls in between the reconfiguration path and Deferred Unsubscription protocols both in terms of overhead reduction and required knowledge. Notably, the endpoints of the old link must be informed of the identities of the endpoints of the new link in order to send information to aid reconfiguration. In summary, our suite of protocols provides many options to the system designer, who can select the most appropriate protocol based on the characteristics of the deployment environment.

Recovering Lost Messages

The last problem hampering content-based publish–subscribe on a dynamic topology is recovering lost messages. Even in the presence of reliable links, messages can be lost due to the reconfiguration of the dispatching network, as routing tables are changed while a message is in transit and therefore may cause its forwarding along stale routes. In this section, we describe a solution based on epidemic algorithms that does not make any assumptions about the cause of message loss and therefore enjoys general applicability.

The idea behind *epidemic* (or *gossip*) algorithms [6,21] is for each process to communicate periodically its partial knowledge about the system state to a random subset of other processes, thus contributing toward building a shared view of the global state. The interaction between hosts can exploit a push or pull style. In a *push* style, each process gossips periodically to disseminate its view of the system. In a *pull* style, each process requests the transmission of information from other processes. Usually, a push approach exploits gossip messages containing a *positive* digest, and a pull approach exploits a *negative* digest (i.e., containing the portion of the state known to be missing). Regardless of the scheme adopted, the probabilistic and decentralized nature of epidemic algorithms brings many desirable properties: a constant, equally distributed load on the processes in the system which improves scalability; resilience to changes in the system configuration, including topological ones; a simple implementation; and low computational overhead.

In our case, the state to be reconciled is the set of messages that have appeared in the system; nevertheless, the nature of content-based publish–subscribe systems adds to the complexity of the problem. Unlike subject-based publish–subscribe and IP multicast, not only are messages not bound to a subject or group determining their routing but they may also match multiple subscriptions instead of a single group. Together, these features greatly complicate the task of identifying the subset of brokers that may hold a missing message.

The solutions we describe share a common structure. Each broker periodically starts a new gossip round, during which it contacts other brokers potentially holding a copy of the lost messages. The broker playing this *gossiper* role builds a gossip message and sends it along the dispatching tree. The content of the gossip message and its routing by the other brokers along the tree vary according to the solutions we describe next. We assume that each broker caches the messages received and that a unicast mechanism is available for sending missing messages (e.g., using the reverse path of gossip messages along the tree or through an out-of-band transport protocol).

Push

Our first solution uses proactive gossip push with positive digests. At each gossip round, the gossiper chooses randomly a filter p from its subscription table, constructs a digest of the identifiers (the pair given by the source identifier and a monotonically increasing sequence number associated with the source is sufficient) of all the cached messages matching p, builds a gossip message containing the digest, and labels it with p. The gossip message is then propagated along the dispatching tree as if it were a normal message matching p. The only difference is that, to limit overhead, the gossip message is forwarded only to a random subset of the neighbors subscribed to p. To increase the chance of eventually finding all the brokers interested in the cached messages (thus speeding up convergence), p is selected from the entire subscription table instead of just the local subscriptions.

When a broker receives a gossip message labeled with p, it checks if it is subscribed to this filter and if all the identifiers contained in the digest correspond to previously received messages. The identifiers of the missed messages are included in a request message sent to the gossiper, which replies by sending a copy of the messages. Both messages are exchanged through the unicast channel mentioned above.

Pull

A pull approach implies the ability to detect lost messages. In subject-based systems, this is easily achieved by using a sequence number per source and per subject. In content-based systems this task is complicated by the absence of a notion of subject and by the fact that each broker receives only those messages whose content matches the filters it is subscribed to. As detailed in Costa et al. [14], this problem can be solved by tagging each message with (1) the identifier of its source, (2) information about all the filters matched by the message, and (3) a sequence number for each filter that is incremented at the source each time a message is published for that filter. This information is bound to each message at its source — an opportunity that arises due to subscription forwarding, where subscriptions are known to all brokers. Event loss is detected when a broker receives a message matching a filter p for which the sequence number, associated with p in the message identifier, is greater than the one expected for p from that message source.

Based on this detection technique, we have defined two approaches exploiting different routing strategies: one steers gossip messages toward the subscribers and the other steers them toward the publisher:

- *Subscriber-based pull* — Upon detecting a lost message, a broker inserts the corresponding information (i.e., source, matched filter, and sequence number associated with the filter and source) in a buffer (i.e., *Lost*). When the next gossip round begins, the broker (now a gossiper) picks a filter p from among those associated with local subscriptions, selects the messages in *Lost* matching p, and inserts the corresponding information in a digest attached to a new gossip message. (Unlike push, subscriptions are not drawn from the entire subscription table, as the goal here is to retrieve messages relevant to the gossiper rather than disseminating information about received messages.) Finally, the gossip message is labeled with p and routed as in the push solution. A broker receiving the gossip message checks its cache against the requested messages and, if any are found, sends them back to the gossiper. Note how the replying broker need not be subscribed to p. In fact, the broker could have received the gossip message because it sits on a route toward a subscriber for p and could have received (and cached) some of the messages missed by the gossiper because they also match a filter ($p' \neq p$) the broker is subscribed to.

■ *Publisher-based pull* — This scheme requires that published messages are cached not only by the brokers that received them but also by the source and that the address of each broker encountered on the route toward a subscriber is appended to the published message. Processing occurs similarly to the previous scheme, but gossip messages are routed toward publishers instead of subscribers.

These solutions are described in greater detail as well as formalized in Costa et al. [14]. Moreover, in Costa et al. [13] we evaluated their performance through simulation. The results confirmed that the approach is effective and provided insights about how to tune the parameters (most notably, the interval between two gossip rounds and the size of the message cache) to achieve the desired level of reliability. Interestingly, we discovered that neither of the pull solutions alone guarantees a satisfactory performance. Instead, the combination of the two, performed by randomly choosing subscriber- or publisher-based pull according to a given probability, performs similarly to push, albeit with lower overhead in the case of infrequent reconfiguration.

REDS: Mobile Publish–Subscribe in Practice

Developed at Politecnico di Milano, REDS (Reconfigurable Dispatching System) [19] puts the mechanisms and algorithms described in the previous sections into practice. REDS is publicly available at http://zeus.elet.polimi.it/reds as open source under the Lesser General Public License (LGPL) and is implemented entirely in Java. Its distinctive and innovative feature is its reconfigurability, a property made available on two different planes. The first concerns the configuration of the middleware architecture and allows the selection of different mechanisms (e.g., the format of messages and filters or the routing strategy) for different deployment scenarios. The second concerns the dynamic reconfiguration of the topology of the REDS distributed dispatcher and addresses the problems of

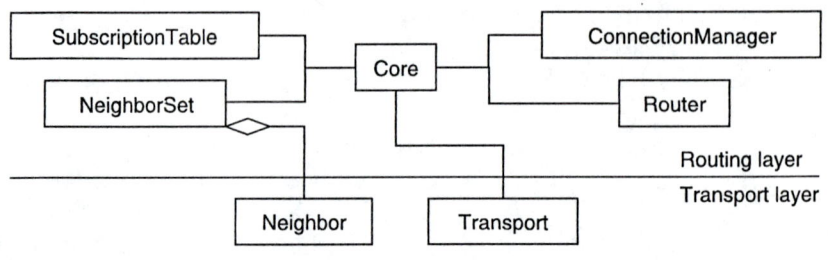

Figure 11.5   The architecture of a REDS broker.

maintaining the overlay network of REDS brokers in the face of topological changes, efficiently restoring stale subscription information and recovering messages lost during the reconfiguration.

To achieve these goals, REDS is conceived of as a framework (in the object-oriented sense) of Java classes that allows programmers to easily build a publish–subscribe middleware explicitly tailored to their application domain. In particular, REDS defines the architecture of a generic broker organized as a set of components implementing well-defined interfaces that represent several aspects of a publish–subscribe system. For example, with regard to messages and filters, REDS defines two interfaces encompassing the minimal set of methods required by a publish–subscribe broker to operate. By implementing these interfaces, developers are free to define their own message formats and, more importantly, their own filters, without having to change the rest of the system. Here, however, we are interested in the components used to build the brokers constituting the REDS distributed dispatcher. As shown in Figure 11.5, each REDS broker is organized as a set of components grouped into two layers, the *transport* and the *routing* layers:

- *Transport* — The transport layer encapsulates the mechanisms used to transport messages, (un)subscriptions, and any other kind of broker-specific control messages through the network. In performing this task, it hides the wire protocol adopted to move data around and the mechanisms used to address and access brokers and clients and to setup the dispatching network in case the dispatcher is distributed. It includes an instance of the `Transport` component and a set of `Neighbors`. The `Transport` component is in charge of receiving incoming requests from neighboring brokers or clients (e.g., connection requests, subscriptions), interpreting them, and calling the appropriate methods on the `Core` component that, as we describe later, is the pivot of the whole architecture. Similarly, `Neighbor` instances are used internally by the components of the routing layer as proxies to interact with the broker's neighbors, thus hiding the details concerned with accessing the underlying network layer. The current version of REDS provides two different implementations for the transport layer and consequently for the two components `Transport` and `Neighbor`: one using TCP links to connect each broker with its neighbors and one based on User Datagram protocol (UDP) datagrams.

■ *Routing* — The routing layer includes three components (`Core`, `Router`, and `ConnectionManager`) which share two data structures: `SubscriptionTable` and `NeighborSet`. The `SubscriptionTable` plays a critical role in recording the subscriptions received by the broker's neighbors. By encapsulating the algorithm used to efficiently match messages against a set of filters, its implementation has a great impact on the performance of the broker. The `NeighborSet` is a convenience data structure provided as a way to simplify the task of accessing, from within the routing components, information concerned with the broker's neighbors represented by `Neighbor` instances.

As its name suggests, the `Core` is the central element of a REDS broker. It holds the two aforementioned data structures and mediates communication among the other components, which therefore do not have direct visibility to each other. This design choice yields two benefits: It increases the decoupling among components, which need to be aware only of the existence of the `Core`, and it provides a central place through which all communication is funneled, therefore opening opportunities for transparently intercepting and modifying messages before redirecting them to the intended components. As an example, we are currently exploiting this possibility in the implementation of the epidemic algorithm described earlier.

The `Router` is in charge of implementing the specific routing strategy. Its methods are invoked by the `Transport`, through the `Core`, to notify it that a subscription, unsubscription, or message has been received from one of the broker's neighbors and must be routed according to the strategy it encapsulates.

The `ConnectionManager` is the key component of our architecture that provides support for publish–subscribe on a dynamic topology. It is in charge of (1) maintaining the overlay dispatching network connected and (2) efficiently rearranging the brokers' subscription tables when the topology of the network changes. It operates in a reactive way, being notified by the `Transport` (through the `Core`) when a new client or broker requests to connect or disconnect and, most importantly, when the transport cannot reach a neighbor that was formerly connected. The protocols described earlier are currently implemented in REDS as specializations of this component.

Related Approaches

In this section, we discuss other approaches to publish–subscribe in mobile scenarios by focusing on the few other publish–subscribe middleware systems capable of tolerating reconfigurations of the dispatching network and faulty links, as well as on middleware that provides solutions specific to content-based publish–subscribe on MANETs. Finally, we touch on how location — a fundamental aspect in mobile scenarios — can be introduced into the publish–subscribe communication model.

Reconfigurable and Fault-Tolerant Publish–Subscribe

The ability to deal with dynamic reconfiguration of the dispatching network topology is not common in content-based publish–subscribe middleware. The most relevant exception is Hermes [37,38], a scalable and reconfigurable publish–subscribe middleware that uses peer-to-peer techniques to build and maintain its overlay routing network. Hermes provides a slightly limited form of content-based routing, termed *type- and attribute-based routing* [23]. Type-based routing is comparable to subject-based routing but preserves inheritance among message types. On top of this routing mechanism, Hermes adds content-based filtering on message attributes. Each message type is associated with a rendezvous point, which takes on the same role as the core node in core-based tree multicast [2]. The Hermes P2P substrate associates a specific Hermes broker to any rendezvous point and helps in building the dispatching tree associated with the associated message type. The self-organization and stabilization features of this P2P substrate allow Hermes to handle dynamic addition, removal, and failure of brokers; however, Hermes does not address the problem of recovering messages lost during reconfiguration.

This latter problem has been addressed by the developers of JORAM [3] and Gryphon [5], which focus on fault tolerance and reliability by allowing a set of brokers to operate as a redundant cluster, but not specifically for mobility. A new feature in JORAM 4.2 allows a set of brokers to be grouped together and operate as a single, redundant cluster to transparently handle network handover and broker fail-over. The JORAM brokers that are part of the same cluster communicate and coordinate by using JGroups [4], a toolkit for reliable multicast communication developed at Cornell University. Similarly, an approach based on a redundant network of brokers to deal with link failures and broker crashes has been recently proposed for the Gryphon [46] system.

Publish–Subscribe on MANETs

Earlier we described a comprehensive set of approaches to dynamically reconfigure a tree of brokers and clients to efficiently support publish–subscribe interactions in different mobile scenarios exhibiting various degrees of mobility. In these approaches, however, message dispatching still relies on a tree-shaped overlay graph. In a MANET environment, and especially if hosts move frequently, the overhead of maintaining the broker tree may overcome the advantages offered by this topology. Very little literature has addressed this problem, however.

In Huang and Garcia-Molina [29], the spectrum of approaches (from centralized to distributed) to publish–subscribe is described, and some possible extensions to mobile environments and particularly MANETs are briefly analyzed. The paper does not provide any complete solutions to the problem but does offer a good starting point by eliciting the problems involved. Similarly, the preliminary work described in Skjelsvik et al. [42] analyzes the issues involved in designing a publish–subscribe middleware for MANETs, focusing specifically on the routing problem.

The work in Huang and Garcia-Molina [30] describes a distributed protocol to build optimized publish–subscribe trees in wireless networks. The authors make rather constraining assumptions; for example, they expect to have knowledge about the placement of publishers (which must become the root of the dispatching tree) and about the statistical distribution of messages with respect to subscriptions (to satisfy the optimality requirements of routing). Moreover, they consider only quasi-static scenarios where nodes move only occasionally and then settle down for a period on the order of minutes. Chen and Schwan [12] describe an alternate mechanism to reconfigure an overlay dispatching network depending on the changes in the physical topology of the MANET and on the current brokers' load; however, each broker must be provided with a global view of the network topology, the mechanism does not handle partitions, and the approach requires an underlying unicast protocol.

The problem of providing content-based publish–subscribe for highly mobile scenarios without relying on a tree overlay has recently been tackled by our research group. In Costa and Picco [15], the authors describe a semi-probabilistic routing algorithm that relies on an overlay network of brokers organized in an undirected connected graph. This topology is easier to maintain than a tree and it is intrinsically more tolerant to reconfigurations and faults, as it provides multiple routes between any two brokers. Routing is partially deterministic and partially probabilistic. Subscriptions are forwarded as in subscription forwarding, but only up to a given distance (i.e., number of hops) from the subscriber, thus providing accurate routing information within a certain horizon from the subscriber.

Along its route, a message is routed using this deterministic information, if available. If there is no such information to determine the next hop, the decision is made probabilistically by forwarding the message along a randomly selected subset of the available links. The simulations in Costa and Picco [15] confirm that proper tuning of the span of the subscription horizon and of the fraction of randomly selected links yields very good performance (in terms of event delivery and overhead), even in highly dynamic scenarios. In particular, the semi-probabilistic approach performs better than a purely probabilistic (or deterministic) approach.

Baldoni et al. [1] have suggested letting each broker autonomously decide about forwarding messages, based on its estimated distance from the closest subscriber, and performing forwarding by using the broadcast facility provided by wireless network cards. This allows other neighboring brokers to decide if they must forward the message or if they should cancel forwarding. In particular, in a MANET with quickly moving mobile nodes the distance between two such nodes can be estimated by measuring the time since they were most recently in communication range. Routing works as follows: Each broker listens for messages broadcast by neighboring brokers. When a broker receives such a message, it stores it and delays forwarding for a time interval proportional to the estimation it has made of its distance from the closest subscriber. During this time interval, forwarding is canceled if a message with the same identifier is forwarded by some neighboring broker (to avoid unnecessarily flooding the network). When the delay expires and if forwarding has not been canceled, the message is broadcast to the neighboring brokers, which reason similarly. The simulations in Baldoni et al. [1] confirmed that this is an efficient technique. It exploits the broadcast nature of wireless communication to send multiple copies of the same message via a single transmission; it avoids the burden of link breakage detection, and it provides an intrinsic resilience to the topological changes caused by the mobility of the nodes.

Location- and Context-Aware Publish–Subscribe

The very notion of mobility is tightly coupled with the notion of *location*. Indeed, in mobile scenarios, the ability to send messages only toward specific locations or that of subscribing to messages published by components located in specific areas could be beneficial to implementing interactions that take into account mobility. Unfortunately, commonly available publish–subscribe middleware does not offer location-based services as part of the API and only few systems address the problem.

In Cugola and Munoz de Cote [16], the authors provide a categorization of possible location-based publish–subscribe services and describe an algorithm to introduce them efficiently in a distributed publish–subscribe middleware system, using a subscription-forwarding routing strategy. In this scheme, each broker is provided with a *location table* used to route location-aware messages and subscriptions. Information about the actual location of publishers and subscribers is forwarded along the network of brokers to populate each broker's location table. This information is used both at subscription and publish time. If a component subscribes to messages coming from a specific area A, location tables are used to limit forwarding of the subscription only toward A. Similarly, if a component publishes a message toward area A, location tables are used (together with conventional subscription tables) to route the message only toward subscribers located in area A, if any. A similar approach is reported in Fiege et al. [24], where the authors describe an extension to the REBECA [26] middleware to implement location-based subscriptions. The main difference between this approach and the previous one is that the scheme proposed by Fiege et al. [24] does not take advantage of information about the actual location of clients to limit forwarding of location-based subscriptions. As a consequence, location-based subscriptions flood the entire network of the brokers in REBECA, potentially reaching areas that are not relevant. A different approach is pursued in Fiege et al. [25], where a general notion of *scope* is introduced to structure message availability and notification by restricting the visibility of published messages to a subset of subscribers in the system — those in the requested scope. In principle, scope can be defined using location or other forms of contextual information, thus obtaining a form of context awareness similar to those described above.

Scalable Timed Events and Mobility (STEAM) [32] is a publish–subscribe middleware designed for deployment over MANETs. STEAM targets application scenarios that include a large number of application components that communicate using wireless technology in an *ad hoc* scenario. In STEAM, messages are valid in a certain geographical area surrounding the publisher. In other words, STEAM provides a special form of location-based publishing service in which location is expressed relative to the publisher. The STEAM implementation is specifically tailored to MANETs and takes advantage of a proximity-based group communication service [31] that uses the number of hops traveled by messages at the Media Access Control (MAC) networking layer to approximate distance.

The work in Chen et al. [11] tackles the different, but related, problem of efficiently filtering a stream of messages representing the current location of clients against a set of spatial predicates. The goal is to determine the set of clients that could be interested in receiving some messages based on their position. The authors propose a middleware system based on a centralized spatial matching engine, which collects subscriptions and delivers them to clients. Clients are in charge of matching those subscriptions against their current position. The results of this process are given back to the engine.

Finally, Solar takes a complementary approach [10]. Solar is a distributed publish–subscribe system explicitly developed to disseminate location and contextual information to a set of distributed components. The emphasis, therefore, is not on constraining the propagation of messages and subscriptions based on location; instead, it is on using the publish–subscribe infrastructure to efficiently disseminate contextual data (e.g., gathered by sensors) that can be processed and used by the distributed application. Solar abstracts context information as messages and allows components to subscribe to the kind of information to be notified of when their context changes. Moreover, components may use Solar services to aggregate low-level context information into more expressive and easier to manage high-level ones.

Conclusions

The publish–subscribe model holds the potential to become of fundamental importance in mobile computing, but only if the technology supporting it embodies the mechanisms and algorithms necessary to cope with the dynamicity of this environment. In this chapter, we presented the challenges posed by the mobile environment, described our own solutions for bringing dynamicity in content-based publish–subscribe technology, and surveyed alternative state-of-the art proposals in the field.

References

[1] Baldoni, R., Beraldi, R., Cugola, G., Migliavacca, M., and Querzoni, L., Content-based routing in highly dynamic mobile networks, *Int. J. Pervasive Computers Comput. (JPCC)*, 1(4), 2006.

[2] Ballardie, T., Francis, P., and Crowcroft, J., Core based trees, in *Proc. ACM SIGCOMM'93*, San Francisco, CA, August, 1993.

[3] Balter, R., *JORAM: The Open Source Enterprise Service Bus*, Technical Report, ScalAgent Distributed Technologies, Cedex, France, 2004 (www.scalagent.com/pages/en/datasheet/040322-joram-whitepaper-en.pdf).

[4] Ban, B., *Design and Implementation of a Reliable Group Communication Toolkit for Java*, Technical Report, Cornell University, Ithaca, NY, 1998 (www.cs.cornell.edu/home/bba/).

[5] Banavar, G., Chandra, T., Mukherjee, B., Nagarajarao, J., Strom, R.E., and Sturman, D.C., An efficient multicast protocol for content-based publish–subscribe systems, in *Proc. IEEE Int. Conf. on Distributed Computing Systems (ICDCS'99)*, Austin, TX, May, 1999.

[6] Birman, K.P., Hayden, M., Ozkasap, O., Xiao, Z., Budiu, M., and Minsky, Y., Bimodal multicast, *ACM Trans. Comput. Syst.*, 17(2), 41–88, 1999.

[7] Caporuscio, M., Carzaniga, A., and Wolf, A.L., Design and evaluation of a support service for mobile, wireless publish/subscribe applications, *IEEE Trans. Software Eng.*, 29(12), 1059–1071, 2003.

[8] Carzaniga, A., Rosenblum, D.S., and Wolf, A.L., Design and evaluation of a wide-area event notification service, *ACM Trans. Comput. Syst.*, 19(3), 332–383, 2001.

[9] Chand, R., and Felber, P.A., A scalable protocol for content-based routing in overlay networks, in *Proc. of the 2nd IEEE Int. Symp. on Network Computing and Applications*, Cambridge, MA, April, 2003, p. 123.

[10] Chen, G. and Kotz, D., Solar: an open platform for context-aware mobile applications, in *Proc. of the 1st Int. Conf. on Pervasive Computing*, Zurich, Switzerland, June, 2002, pp. 41–47.

[11] Chen, X., Chen, Y., and Rao, F., An efficient spatial publish/subscribe system for intelligent location-based services, in *Proc. of Int. Workshop on Distributed Event-Based Systems (DEBS'03)*, San Diego, CA, June, 2003.

[12] Chen, Y. and Schwan, K., Opportunistic overlays: efficient content delivery in mobile *ad hoc* networks, in *Proc. of the 6th ACM/IFIP/USENIX Int. Middleware Conf.*, Vol. 3790, Lecture Notes in Computer Science, Springer, Berlin, 2005, pp. 354–374.

[13] Costa, P., Migliavacca, M., Picco, G.P., and Cugola, G., Epidemic algorithms for reliable content-based publish–subscribe: an evaluation, in *Proc. of the 24th Int. Conf. on Distributed Computing Systems (ICDCS'04)*, Tokyo, Japan, March, 2004, pp. 552–561.

[14] Costa, P., Migliavacca, M., Picco, G.P., and Cugola, G., Introducing reliability in content-based publish–subscribe through epidemic algorithms, in *Proc. of Int. Workshop on Distributed Event-Based Systems (DEBS'03)*, San Diego, CA, June, 2003.

[15] Costa, P. and Picco, G.P., Semi-probabilistic content-based publish–subscribe, in *Proc. of the 25th Int. Conf. on Distributed Computing Systems (ICDCS'05)*, Columbus, OH, June, 2005.

[16] Cugola, G. and Munoz de Cote, J.E., On introducing location awareness in publish–subscribe middleware, in *Proc. of Int. Workshop on Distributed Event-Based Systems (DEBS'05)*, Columbus, OH, June, 2005.

[17] Cugola, G., Di Nitto, E., and Fuggetta, A., The JEDI event-based infrastructure and its application to the development of the OPSS WFMS, *IEEE Trans. Software Eng.*, 27(9), 827–850, 2001.

[18] Cugola, G., Frey, D., Murphy, A.L., and Picco, G.P., Minimizing the reconfiguration overhead in content-based publish–subscribe, in *Proc. of the 19th ACM Symp. on Applied Computing (SAC'04)*, Nicosia, Cyprus, March, 2004, pp. 1134–1140.

[19] Cugola, G. and Picco, G.P., *REDS: A Reconfigurable Dispatching System*, Technical Report, Politecnico di Milano, Italy, 2005 (www.elet.polimi.it/upload/picco).

[20] Cugola, G., Picco, G.P., and Murphy, A.L., Towards dynamic reconfiguration of distributed publish–subscribe systems, in *Proc. of the 3rd Int. Workshop on Software Engineering and Middleware (SEM)*, Vol. 2596, Lecture Notes in Computer Science, Springer, Berlin, 2002, pp. 187–202.

[21] Demers, A., Greene, D., Hauser, C., Irish, W., Larson, J. et al., Epidemic algorithms for replicated database maintenance, *Operating Syst. Rev.*, 22(1), 8–32, 1988.

[22] Eugster, P., Felber, P., Guerraoui, R., and Kermarrec, A.-M., The many faces of publish/subscribe, *ACM Comput. Surv.*, 2(35), 114–131, 2003.

[23] Eugster, P.T., Guerraoui, R., and Damm, C.H., On objects and events, in *Proc. of ACM Conf. on Object-Oriented Programming Systems, Languages, and Applications (OOPSLA 2001)*, Tampa Bay, FL, October, 2001, pp. 254–269.

[24] Fiege, L., Gartner, F.C., Kasten, O., and Zeidler, A., Supporting mobility in content-based publish/subscribe middleware, in *Proc. of the 4th ACM/IFIP/ USENIX Int. Middleware Conf.*, Rio de Janeiro, Brazil, June, 2003.

[25] Fiege, L., Mezini, M., Muhl, G., and Buchmann, A.P., Engineering event-based systems with scopes, in *Proc. of the 16th European Conf. on Object-Oriented Programming (ECOOP02)*, Vol. 2374, Lecture Notes in Computer Science, Springer, Berlin, 2002, pp. 309–333.

[26] Fiege, L., Muhl, G., and Gartner, F.C., Modular event-based systems, *Knowledge Eng. Rev.*, 17(4), 359–388, 2002.

[27] Frey, D. and Murphy, A.L., *Maintaining Publish–Subscribe Overlay Tree in Large Scale Dynamic Networks*, Technical Report, Politecnico di Milano, Italy, 2005 (www.elet.polimi.it/upload/frey).

[28] Fuggetta, A., Picco, G.P., and Vigna, G., Understanding code mobility, *IEEE Trans. Software Eng.*, 24(5), 342–361, 1998.

[29] Huang, Y. and Garcia-Molina, H., Publish/subscribe in a mobile environment, in *Proc. of the 2nd ACM Int. Workshop on Data Engineering for Wireless and Mobile Access (MobiDe'01)*, Santa Barbara, CA, May, 2001, pp. 27–34.

[30] Huang, Y. and Garcia-Molina, H., Publish/subscribe tree construction in wireless *ad hoc* networks, in *Proc. ACM Int. Conf. on Mobile Data Management (MDM'03)*, Melbourne, Australia, January, 2003, pp. 122–140.

[31] Killijian, M., Cunningham, R., Meier, R., Mazare, L., and Cahill, V., Towards group communication for mobile participants, in *Proc. of ACM Workshop on Principles of Mobile Computing (POMC'2001)*, Newport, RI, August, 2001, pp. 75–82.

[32] Meier, R. and Cahill, V., STEAM: event-based middleware for wireless *ad hoc* networks, in *Proc. of Int. Workshop on Distributed Event-Based Systems (DEBS'02)*, Vienna, Austria, July, 2002.

[33] Mottola, L., Cugola, G., and Picco, G.P., *A Self-Repairing Tree Overlay Enabling Content-Based Routing on Mobile Ad Hoc Networks*, Technical Report, Politecnico di Milano, Italy, 2005 (www.elet.polimi.it/upload/picco).

[34] Perkins, C.E., *IP Mobility Support*, Request for Comments 2002, Internet Engineering Task Force (IETF), 1996 (http://www.ietf.org/rfc/rfc2002.txt).

[35] Perkins, C.E., Ed., *Ad Hoc Networking*, Addison-Wesley, Boston, MA, 2000.

[36] Picco, G.P., Cugola, G., and Murphy, A.L., Efficient content-based event dispatching in presence of topological reconfiguration, in *Proc. IEEE Conf. on Distributed Computing Systems (ICDCS'03)*, Providence, RI, May, 2003, pp. 234–243.

[37] Pietzuch, P.R. and Bacon, J.M., Hermes: a distributed event-based middle-ware architecture, in *Proc. of Int. Workshop on Distributed Event-Based Systems (DEBS'02)*, Vienna, Austria, July, 2002.

[38] Pietzuch, P.R. and Bacon, J.M., Peer-to-peer overlay broker networks in an event-based middleware, in *Proc. of the 2nd Int. Workshop on Distributed Event-Based Systems (DEBS'03)*, June 2003.

[39] Podnar, I. and Lovrek, I., Supporting mobility with persistent notifications in publish–subscribe systems, in *Proc. of Int. Workshop on Distributed Event-Based Systems (DEBS'04)*, Edinburgh, Scotland, May, 2004.

[40] Rosenblum, D.S. and Wolf, A.L., A design framework for Internet-scale event observation and notification, in *Proc. of the 6th European Software Engineering Conf. held jointly with the 5th Symp. on the Foundations of Software Engineering (ESEC/FSE97)*, Zurich, Switzerland, September, 1997.

[41] Royer, E.M. and Perkins, C.E., Multicast operation of the *ad hoc* on-demand distance vector routing protocol, in *Proc. of the 5th ACM/IEEE Int. Conf. on Mobile Computing and Networking (MOBICOM'99)*, Seattle, WA, August, 1999, pp. 207–218.

[42] Skjelsvik, K.S., Goebel, V., and Plagemann, T., Distributed event notification for mobile *ad hoc* networks, *IEEE Distributed Syst. Online*, 5(8), 2004.

[43] Sutton, P., Arkins, R., and Segall, B., Supporting disconnectedness: transparent information delivery for mobile and invisible computing, in *Proc. of the IEEE Int. Symp. on Cluster Computing and the Grid (CCGRID'01)*, Brisbane, Australia, May, 2001.

[44] Toh, C.-K., *Ad Hoc Mobile Wireless Networks*, Prentice Hall, Upper Saddle River, NJ, 2002.

[45] Triantafillou, P. and Economides, A., Subscription summarization: a new paradigm for efficient publish/subscribe systems, in *Proc. of the 24th Int. Conf. on Distributed Computing Systems (ICDCS'04)*, Tokyo, Japan, March, 2004.

[46] Zhao, Y., Sturman, D., and Bhola, S., Subscription propagation in highly available publish/subscribe middleware, in *Proc. of the 5th ACM/IFIP/USENIX Int. Middleware Conf.*, Toronto, Canada, October, 2004, pp. 274–293.

[47] Cugola, G., Frey, D., Murphy, A.L., and Picco, G.P., *Content-Based Routing for Publish-Subscribe on a Dynamic Topology: Concepts, Protocols, and Evaluation*, Technical Report, 2006, www.elet.polimi.it/upload/picco.

Chapter 12

Code Mobility and Mobile Agents

Andrzej Bieszczad and Tony White

CONTENTS

Introduction

In this chapter, we discuss the fundamentals of distributed systems based on mobile code. We begin by providing a brief historical perspective and a discussion of theoretical principles which will help the reader to understand the fundamentals of code mobility as well as its place in the toolbox of a software engineer. We continue with descriptions of several enabling technologies. Mobile agents constitute one of several mobile code paradigms that we present next. Then, we consider the numerous advantages that have been attributed to mobile code. Understanding the issues is a requirement of efficient use of any technology, so we scrutinize a number of them here, including the most controversial one: security. The chapter concludes with a discussion of mechanisms for building mobile code frameworks and a brief note on relevant standardization activities. We include references to selected publications that we used extensively to prepare the chapter. The reader should be aware that we have left out several aspects of mobile code (e.g., applications, patterns, more substance on standards) due to space limitations.

Code Mobility Principles

In the age of the Internet it is difficult to find a single piece of software that does not have to deal (at least to some degree) with the distributed nature of information and computing systems. The prevalence of distributed systems has yielded numerous technologies that are used to harness their complexity. A relatively recent addition to the software engineering toolbox, *mobile code*, has generated a lot of excitement in many research circles. In essence, code mobility is an evolution of established distributed systems in which data is transported to and from stationary computational units toward systems in which it is code that moves while the data may (it does not have to) stay in place.

The genesis of mobile code can be traced back to process migration techniques in distributed operating systems. As underlying computing systems were evolving from a single processing unit to multiple units distributed in space, traditional static approaches to code generation that bound execution entities to known locations began to break. To balance the load, operating systems had to relocate processes between participating executing environments. One way to do so assumes that both computing platforms (the source and the destination) are homogeneous. In this case, the process on the source machine can be encapsulated together with its state in a transport unit and shipped to the destination machine, where it is unwrapped and restarted as if it has just resumed operation locally from a suspended state. In general, the three types of code migration are:

- *Transparent involuntary migration* — A distributed operating system relocates a process as necessary to attain certain global goals.
- *On-demand migration* — Code is relocated to provide certain functionality in the target location.
- *Autonomous migration* — Code logic determines migration patterns.

Unfortunately, the association of code mobility with viruses plaguing computer systems has caused a black cloud to hang over this technology.

Taxonomy of Code Mobility

Relocating computational logic between two execution environments can be achieved in several distinct ways. Let us begin with *logical mobility*. Imagine that we have two execution environments that have code of a certain computational unit available from their respective local code repositories. Furthermore, suppose that a protocol is in place that allows for the transfer of execution states between execution environments. In such a case, the state of an executing unit on the source machine can be transported to the destination machine, where a replica of the code running on the first machine is loaded from a local repository and started with the transferred state as the initial state of the execution. The whole process can also be viewed as process cloning.

In the remainder of this chapter, we deal primarily with *physical mobility*, mobility that involves the physical relocation of program code; that is, the executable code that constitutes a computational unit active at one point in time in one execution environment is physically transported to another location and restarted in the new execution environment. If transported code is accompanied by the state at which the process was suspended at the source location and restarted in exactly the same state in the new environment, we say that it has *strong mobility*. Strong mobility requires homogeneous execution environments or a very sophisticated adaptation layer that allows for state recreation in the destination. If process code is transported without any memory of its former execution, then we refer to it as having *weak mobility*. In this case, the code is started in a new execution environment as if it was loaded from a local repository.

The behavior of a newly started process may depend on contacting some rendezvous point and finding out directives for the task at hand. This behavior can be set by default or, more commonly, by some managing entity that brought on transport of the code in the first place. In another approach, the managing entity may contact the newly started process just after its activation and provide details for further computation.

Enabling Technologies

The subject of the transport — mobile code — is simply a form of computer program, a computational unit. To be transferred into a process, a program requires a computer; in other words, it must have an execution environment. In the course of execution, a computational unit acquires access to a variety of resources that are subject to management policies that apply to the entire or part of the distributed data space.

A computer and an operating system are fundamental parts of an *execution environment.* They provide mechanisms that transfer a given computational unit into an active process. Distributed operating systems extend a notion of an execution environment to a network of computers. If management of multiple processes running in such a distributed environment is not the goal, then migration functionality does not have to be an integral part of the operating system. Nevertheless, a layer supporting code migration must be in place in any event if the system is to support any kind of code mobility. The closer that layer is to the operating system, the better code migration efficiency, security, and transparency that can be achieved; however, such improvement usually comes at the expense of flexibility.

An execution environment supporting migration requires several mechanisms to stop a process, acquire a version of its code that is suitable for transport (possibly its state), unbind any references to resources accessed by the process, package the code and the state in an envelope appropriate for transport, establish a communication link with the destination execution environment, and physically relocate the envelope to the destination. The role of the execution environment on the receiving end is complementary. It has to expose a communication port for establishing connectivity, then receive an envelope from the source, conduct thorough security checks, unpack the envelope, and retrieve a computational unit with (possibly) its preserved state, resolve resource references (as discussed in the following section), and recreate an active process from all available components. Some of the details in this scheme depend on a type of transfer — for example, presence of process state.

Execution environments come in a variety of sizes and shapes. For a process migrating as part of a load-balancing scheme, a sole distributed operating system suffices. In other cases, additional facilities at various levels of abstraction are required. The facilities might be incorporated into an operating system, be part of an operational platform (e.g., Java Virtual Machine), or be run as applications (e.g., Web browsers that execute applets).

A *computational unit* is a concept representing a unit of executable computer code. A computational unit can be initiated as an active process running in some execution environment. It is a static entity that is convenient for packaging and transfer. A computational unit is transformed into a running process in the course of process instantiation. A process can be transformed back into a computational unit — for example, when the unit is to be migrated.

A computational unit may take many forms. A program written in a high-level language may be considered a computational unit. *Source code* can be transported between computers, recompiled locally, and run as a process. Transporting code in its source format offers many advantages. Portability is guaranteed as long as a compiler for the encoding language is locally available in the destination. A local linker and loader provide code arrangement and binding of necessary resources, so the migration platform is relatively easy to implement. On the other hand, the availability of local resources necessary for compilation, linking, and execution is a problem.

Interpreted code can alleviate these problems while preserving many advantages of code written in a high-level compiled language. It does not require recompilation, relinking, and reloading in the destination, because these operations are not necessary in code interpretation. The price for the improvement is a need for an interpreter — an additional computational layer — that reads commands of the received code and undertakes actions that they stipulate. The process of interpretation is *de facto* an execution cycle for statements of interpreted code; however, resolving resource references is more difficult, because the interpreter has to perform that task on behalf of arriving code.

A common complaint against the use of interpreted languages that most designers agree upon is their relatively low efficiency. *Intermediate code* is a compromise between interpreted and machine code. Intermediate code is generated as output from a compiler, so it can take advantage of all optimization techniques exploited in modern compilers. On the other hand, the code is still executed by a necessary interpreter — called a *virtual machine* — that separates the program from the hosting machine. Due to the relative simplicity of constructs used in intermediate code, virtual machines provide much greater efficiency in code execution. Some code may actually execute in a native machine code, as just-in-time compilation technology might be used. At the same time, intermediate code tends to be more compact than its source version. All of this does not invalidate the advantages attributed to interpreted code with the controlled execution environment that provides the basis for comprehensive security management.

No other form of code can surpass *native machine code* with regard to execution efficiency. Unfortunately, a number of serious issues make native code a bad candidate for mobility. Machine code is extremely difficult to analyze, so it often poses an intolerable, undetectable security threat. The domain for code mobility has to be homogeneous. The size of computational units of machine code might also be an issue, because if special care is not taken native code tends to be extensive.

The process of tying a computational unit to a resource is called *binding*. Binding can be static or dynamic. In *static binding*, a resource is allocated during program instantiation. If a resource is acquired as a result of code execution, we have a *dynamic binding*.

From a perspective of code mobility, resources can be transferable or nontransferable. A *transferable resource* is a resource that can be relocated to another execution environment together with the migrating code. A *nontransferable resource* cannot be moved to another location. In some cases, migrating resources may be possible (making them transferable), but the migration might be undesirable (for example, due to their size), so the resources can be tagged appropriately to prevent their transport.

Numerous types of resources are available in execution environments; for example, a resource can be disk space, a space in memory, a printer, a file handle, an object, etc. A binding process allocates a resource to a computational unit, which obtains a resource identifier. The *identifier* is a handle allowing the process to access a resource. The *value* of a resource depends on its type. The value may be an address in memory, a number, a string, a socket number, a file descriptor, and so on. Some values allocated to a process are static, while others can change to reflect the dynamics of computation.

As we said earlier, in strong mobility the process state moves with the mobile code. *State* is a collection of resources with their values. Not all of them are handled in the same way by the migration mechanism; for example, some resources have transient values that do not have to be replicated in the new location.

Mobile Code Paradigms

Having a cellular telephone in a pocket, pouch, or purse has become as pervasive as wearing a watch on one's wrist. In fact, the two technologies are beginning to converge, as both are considered indispensable in our busy lives. As evidenced in other chapters of this book, many modern mobile telephones employ technologies that have transferred them into powerful computing devices — in the jargon used in this chapter, execution environments for computational units. We will use mobile telephones and mobile networking infrastructure to illustrate the mobile code paradigms.

Traditional distributed systems utilize a *client–server paradigm* to transfer data in the course of a computation process. Although the two communicating entities have well-defined roles in the paradigm, such a client–server relationship can be established dynamically in response to a need to exchange data. Running processes can even be servers and clients at the same time, serving data to others on the one hand and obtaining data from others on the other.

The essence of a client–server paradigm is that one process, a client, requires something that it cannot do or get on its own, so it asks another process usually (but not necessarily) running in another execution environment — the server — for help. The server can help because it has access to certain logic or certain data that the client does not.

Let us consider a scenario in which you, a mobile telephone user, try to use your device to buy a good bottle of wine for your spouse's birthday. You have a telephone that exploits cutting-edge technologies, so you can connect to the Internet. You access a Web site that specializes in matching customer preferences with offers from numerous wine vendors. Your telephone, acting as a client, connects to the server, which conducts an interview with the goal of obtaining specifics of the search (it can be as simple as a Hypertext Markup Language [HTML] form to fill out or a more complex conversational interaction). After you provide the detailed nuances of your spouse's tasting buds and preferred *Appellation d'Origin Controlee*, the server's logic performs its magic and you are presented with the best offers matching your request. You select some wines with a bouquet that should please your spouse and originate an order. Many client–server interactions among your telephone, the server, and other parties (servers of wine vendors, payment institutions, and shipping companies, to name a few) will be required to complete the purchase.

It is important to notice in this context that, after the service has been performed, the client has the data that it asked for, but not the logic and not all available data. To purchase another bottle for your parents' silver anniversary, you have to send a corresponding request to the server again. If you decide to go to a store to purchase the wine, you will not know what logic the server used to match wine with your spouse's taste when you were buying online. That very logic is what makes money for the broker, so it wants to protect it! The service is useless, however, if you have access to a vendor's catalog and not the server. Sending descriptions of all wines available in the store to the server that knows how to select one would be extremely tedious. If there is some time limit on the purchase — for example, you are attending an auction — such a solution is just impossible. And, what do you do if your wireless connection to the network is dead because you are out of range?

As you can see, although the client–server paradigm is extremely useful — and therefore virtually ubiquitous in our networked world — it does have certain important limitations. In the context of our scenarios, the client–server paradigm may be both inefficient and unreliable, because the task at hand is attained through a series of requests and responses that must travel back and forth between the client and the server. The server may not be reachable, the latency may be intolerable, or the amount of data to transport could be excessive.

The way to overcome the problems that may plague the client–server model under some circumstances is to use a different computational model instead. Imagine, for example, that the business model of another provider of Web services is different from the one we discussed in the preceding section. In this new model, it is selling a wine-recommendation program that makes money. Such a program can be downloaded to your mobile phone, so you can run it whenever you are trying desperately to figure out the difference between a *vin de pays* and a *vin de table*. The new model, which we call *code on demand*, overcomes problems with transporting large databases and with network latency.

Another way to overcome the problems is to apply the so-called *remote code evaluation* paradigm instead. In this model, a program containing some logic is also transported, but this time in the opposite direction. For example, after trying numerous available online wine-advisory services, you conclude that none of them can even come close to your mastery of *terroir* and *cepages*. As a technically savvy individual, you may write a program for performing a very specialized *vin* query. You can install the program on your mobile phone, but you need a database to act upon. You have two choices. You can restrict the use of the program to cases when you are at a store, because you or a personal area network (PAN) agent can act as an intermediary between your program and the store database. Alternatively, you can send the program to the execution environment hosting an online wine database and request that the program be run locally where access to the database is not a problem. Your smart program travels to the remote location and — after instantiation as a process — it performs the search and returns a selection of wines from the remote database. The program can annihilate itself, it can be deleted manually after its task is carried out, or it may stay resident, either permanently or at least for a while in a cache so you may perform similar queries if you need to buy wine again in the near or distant future.

There has always been some confusion about the relation of mobile code to *mobile agents*. Many just equate both — quite incorrectly, we have to say. Code is just one characteristic of a mobile agent, and it is not the one that determines *agenthood*. Although not explicit in the name, implicitly, intelligence is an attribute of a mobile agent; therefore, mobile agents tend to be considered close cousins of *intelligent agents*. The following is *a* rather than *the* definition of an intelligent agent. We believe that most researchers in the area will agree that a software agent is a software entity that includes at least the following characteristics:

- It acts on behalf of others.
- It performs a task delegated by another entity (be it human or machine) autonomously, without further supervision from the delegating authority.
- It is proactive, so it attempts to achieve the objective, the goal, set by the delegating entity without further prompts.
- It is reactive, so it is capable of responding to changes in the environment by modifying its behavior.

Additionally, an intelligent agent may exhibit a certain degree of capability to:

- Learn, so it improves its capabilities for the future
- Cooperate, so it exhibits social behavior in order to improve its chance to accomplish the goal
- Move, so it can relocate to a different execution environment if carrying out the task at hand requires it to do so

The roots of intelligent agents lie in *artificial intelligence* (AI) and in *distributed systems*. In a nutshell, AI is concerned with discovering methodologies and technologies that address so-called *hard problems* — that is, problems that are too difficult to resolve using traditional, analytical means or whose solution would require a prohibitive length of time to obtain or to execute. AI does not strive to always provide an exact solution to a given problem. A solution that is good enough suffices. How do we make that judgment? We need an evaluation function that provides a measure of *goodness*. With such a function at hand, we can apply the technique until the function yields a value that is within an acceptable margin of error. Heuristics such as this have helped to solve many otherwise NP-complete (i.e., tractable, albeit nonscalable) problems.

Distributed AI (DAI) deals with systems built out of a number of agents that interact following a variety of patterns. Communication with other agents is a fundamental capability of an intelligent agent, because collaboration is a critical component of agent-based computation. The ability to tackle problems with the entire system through comprehensive and coherent communication constitutes what is called *weak intelligence*. In contrast, *strong intelligence* refers to methodologies, techniques, algorithms, and mechanisms derived from artificial intelligence.

Advantages of Mobile Code

Numerous advantages have been attributed to the use of mobile code. One thing must be very clear here, however. Many problems that can be solved with mobile code can also be treated by other means, but very often a mobile-code-based implementation brings something unique and desirable. To withstand the test of time, systems have to adapt to the changing environment in which they operate. *Adaptation* implies reconfiguration triggered and guided by the changes. If we tried to define what the characteristics of adaptability are, we would certainly include flexibility, scalability, and customizability among the major ones. Modularity and mobility are fundamental characteristics of mobile code to provide system adaptability. We need to note that mobile code is not adaptive *per se*; instead, it can be used to design adaptive systems in which modules are installed or replaced in response to contextual transformations.

We already analyzed how critical mobile code is in reorganizing systems to optimize their use of resources. Although *load balancing* in distributed operating systems has many constraints, such as location and homogeneity, distributed systems that employ mobile code platforms overcome many of them. Mobile code can be sent to any geographical location as long as the location has an execution environment for accepting and running it. Because a mobile platform usually separates the code from the hardware, heterogeneous resources can be pulled together into a uniform computing environment.

Data proximity is very important in environments with high network latency and voluminous databases. Imagine that we need to control a Martian explorer in some unknown environment. A control signal sent from Earth will get to Mars in eight or so minutes, and anything that the robot senses will be seen by Earth observers in the same amount of time. Therefore, if a reaction to an event were to come from the Earth control center, it would have a 16-minute *round-trip* delay; remote real-time control is impossible. Instead, a controller specialized for the new environment can be sent to the robot and executed locally.

In another scenario, imagine that we need to search a very large database (recall our earlier wine purchasing examples). If our connection to the database is composed of only super-fast link segments, then we can transport data back and forth for a remote analysis. Of course, we may have a security issue, but that can be addressed by a secure connection; however, if we cannot get any guarantees on the *quality of service* (QoS) provided by the connection — a common occurrence on the Internet — then transporting large volumes of data is impractical. Mobile code provides an alternative, because code can be sent to the database to analyze the data locally. Usually, only a small amount of the results of that analysis must be sent back over the network.

The most common and undisputable characteristic of current-day computer systems is their complexity; consequently, in spite of improving software engineering methodologies, it is difficult to predict all possible problems and remove all potential issues. In fact, fault management is now a standard part of system development — a fact that acknowledges that errors are inevitable. As we mentioned earlier, systems must be able to adapt to emerging novelties in the environment, but this adaptation has to be accompanied by a capability to deal with arising problems. The objective, then, is to build *fault tolerance* into systems that allows them to deal graciously with inevitable faults.

There are many sources of faults in computer systems: software bugs, hardware malfunctions, human errors, and collaboration problems that may be the consequences of communication link failures, interface problems, third-party problems that propagate into the system, etc. Each type of error requires a specific approach to fix or prevent it, so it is difficult to generalize remedies. Nevertheless, in the following we attempt to suggest some uses of mobile code in this context.

Modularity that allows the designers to deal with complexities is not unique to mobile code, but code mobility provides another dimension in the struggle with system problems; for example, faulty modules can usually be exchanged without bringing down the whole system. Mobile computational units can be moved to another location if their current hosting environment is error prone. Code cloning and parallel execution in multiple execution environments improve chances for critical parts of the system being executed under any condition. If one copy fails, the others may carry out the task at hand.

Self-management functionality is one of the requirements of modern computer systems. Employing elements of artificial intelligence together with code mobility may bring immune-system-like functionality; for example, *self-repair* based on mobile code is a step in the direction of designing robust, dependable systems that provide the necessary confidence — the lack of which many application areas just cannot afford.

A client–server model requires the presence of a connection between the communicating endpoints. Without a connection, neither data nor control signals can be exchanged. Still, in spite of incredible technological advances, communication links are not resistant to failures, nor can they survive dramatic changes in the underlying infrastructure; for example, moving a device to another location very often results in a lost connection. In addition, in spite of huge investments made by telecommunications service providers, the networking infrastructure is still not ubiquitous, so from time to time we end up in a dead zone with no services. Mobile code provides a system designer with a tool for building solutions to the client–server dilemma in such environments. Software can be designed in such a way that computation can be continued with *intermittent connectivity*. When communication is possible, either the server or the client can accept code for performing certain tasks that can be carried out without constant exchanges of data, as is the case in a regular client–server model.

To be reliable, a system has to be built out of nonorthogonal components. There must be multiple ways to carry out each task. Full *redundancy* is commonly used to ensure that a backup exists for every function of the system. Modularity and mobility allow for fine granularity to be built into redundancy; for example, a system may be engineered in such a way that each task has (at least) a dual nature. Instead of one process fulfilling the needs, two (or more!) processes are running in parallel. Ideally, they perform the same computation, if possible; however, even a simple watchdog can be used to ensure that the job is done. If a watched process dies or stalls, the watchdog can restart the process in the same location if possible. If that is not possible, the task can be completed at another location. Due to code mobility, a replacement can be uploaded to an alternate execution environment and started there.

One can also imagine a system in which every request is routed to a proxy that sends mobile code units implementing the feature to numerous locations. All installed units may be run in parallel or a single one may be elected to execute while all the others are waiting in standby and watchdog modes. If the executing process fails, one of its stand-by shadows is activated.

Mobile Code Issues

Like every technology that brings abundant benefits, mobile code has its own share of issues. Even in its simplest form, from an execution environment perspective (i.e., native machine code), mobile code requires the presence of numerous mechanisms. Unbinding, rebinding, code transport, verification, and security all require a systemwide infrastructure. The additional layer of complexity not only poses implementation hurdles but also affects its functionality that is not related to mobile code. The overhead may be considerable in mobile code systems that use high-level languages as the basis for mobility; for example, a language interpreter (e.g., for Tool Control Language, or Tcl) or an intermediate code virtual machine (e.g., for Java, C#) is required.

One obvious consequence of the overhead required for code mobility is code execution *efficiency*. Earlier in this chapter, we discussed the convenience of using interpreted languages versus languages that produce intermediate code. Neither technology can beat execution of native machine code. Interpreting involves two levels of code execution: One is the hardware, and another is the virtual machine or interpreter. As we also discussed, compiler optimization is much more difficult for interpreted code, although a lot of progress has been made in this area. Although a compiler optimizes machine code offline, any optimization of interpreted code has to be done dynamically online, or partially offline and online in the case of intermediate code.

In addition, code transport introduces a certain *latency*, as the code must be packaged, transmitted, and then securely unpacked. The underlying networking infrastructure may aggravate the latency problem. For example, transmission on a slow dial-up link will take more time than transmission over a fast fiber connection. The problem is that the transport mechanism cannot make any assumptions, because multiple segments might be involved in one connection. It is impossible for the application or transport layer to find out what technology is used on each of the segments.

Mobile code requires special care from its designers. In some cases, mobile code must be constructed in a way that ensures its *serialization* — that is, its resilience to transmission over communication links. In strong mobility, the process state has to be transferred along with the code; however, some of the elements of the state might be *transient* (temporary). A software designer has to deal with such a problem; for example, by using language facilities in Java, one can exclude certain data members of an object from being serialized. Object orientation and linked references introduce another problem: chains of state elements. The links may constitute multi-level hierarchies, so a decision must be made whether a *deep* or a *shallow* snapshot image of the process state should be taken.

The efficiency issues that we discussed in the previous section constitute another challenge to a software designer. They force program architects to contemplate the non-algorithmic aspects of system design; for example, designers have to incorporate considerations of transport latency into the algorithm so the behavior of the program is not affected by *transmission delays*. They also must take into account that mobile code will be executing in an environment with a variety of capabilities and inconsistent access to resources.

Without any doubt, *security* is the most challenging issue for systems incorporating mobile code. It is also one of the most controversial issues in mobile code research circles. This should not be a surprise to all who have heard about worms and viruses exposing computer system users to multimillion dollar losses. Hackers employ mobile code to do their mischief, so it is no wonder that the very concept of mobile code generates bad feelings. The many security threats in distributed systems might be classified in three wide categories: threats related to disclosure of information, threats that can affect the information, and threats that disable system functionality.

Knowledge means power, so attempting to obtain information from any possible source is a common practice of many endeavors. Physical enclosure, commonly used to protect information in the predigital world, is no longer an option. Data is stored in interconnected networks, so it is exposed to attempted unauthorized exploration and exploitation. Data in transit poses another risk. Complete databases rarely have to be relocated, but serving data chunks is a fundamental part of the functionality of almost every system. Each time a piece of data is put in a transport medium, the threat of eavesdropping becomes reality. Both passive and active intruding techniques can be applied; for example, beaming data from an unprotected wireless station might lead to a passive attack if somebody else accidentally or intentionally intercepts the transmission. Breaching the security of a firewall on a wireless router is an example of an active intrusion technique.

An attempt to alter stored information is a much more serious threat than just eavesdropping. While eavesdropping can be passive, an attempt to change data requires the proactive behavior of an intruder and might be fatal to the users of the modified data. Every execution platform and the entire network can be compromised and affected.

Even systems that secure their resources appropriately so they can be neither disclosed nor maliciously modified are still at risk. Intruders applying denial-of-service attacks may overwhelm a system with too many tasks to handle. A classic example is an attempt to establish a large number of Transmission Control Protocol (TCP) connections at the same time that may virtually disable the networking capability of a server. In the context of mobile code, migrating large volumes of code to one location with the purpose of choking it can achieve a similar effect. Dumping too many migration requests on a network may also be fatal, because of the danger of network congestion.

Numerous vulnerabilities of mobile code systems can be identified. The addition of a networking dimension to the computing infrastructure exposes it in numerous ways. It is not only mobile code that can harm, although it has been the most notorious way of inflicting damage. Any malicious party can use the network to tamper with any resource connected to that network.

Mobile code in the form of a worm or a virus has already proven its destructive power. Intrusion by code executed locally is easier, because one of the barriers — the network — is not present. Systems can be compromised in a variety of ways, from sending joyful messages (the first self-replicating virus was created to display *Merry Christmas!*) to wiping out complete functionality. The excessive use of resources by a single agent or an uncontrollable influx of migrating agents can be the roots of denial-of-service attacks against a host without exploring its functionality. Mobile code is also vulnerable to attacks from hosting execution environments in numerous ways. Data that it carries may be compromised, its code may be altered, its migration pattern might be modified, its algorithms might be reverse engineered, and the code could be cloned for unauthorized reuse in replay attacks.

Mobile code platforms may host mobile code coming in from many sources and acting on behalf of numerous entities. Malicious mobile code may not only try to compromise the hosting platform but also target other users of the platform. Both local applications and other visiting mobile code may be targets of attacks. Administrators of computing resources usually provide comprehensive protection to hosts; therefore, intruders may target mobile code platforms that run as applications on hosts, because they may perceive them as less sturdy. Malicious mobile agents may target middleware rather than the host that it is running on, because they have relatively direct access to it. Remote intruders may also attempt to harm a platform if the security for the platform is not as strong as the one protecting the entire host.

If a compromised platform stops working, then the problem becomes visible and can be fixed. Hidden alterations may be more damaging in the long run, because they may expose other vulnerabilities in some subtle ways. Detecting such intrusions may be difficult, because the mobile code framework may still be working; for example, the infected platform may be reconfigured to redirect code of a search agent to the intruder's site.

Neither mobility nor remote communication would be possible without a networking infrastructure, so it is one of the critical resources that must be protected. Misuse of mobile code or its erroneous behavior can be dangerous or even fatal to the network. Uncontrollable increase in traffic is the root cause of network congestion. Flooding the network with a lot of migration requests and transporting large amounts of code jeopardize the fundamental functionality of the network in the same way as excessive amounts of data.

Intruders can use numerous types of security attacks against a mobile code system. We already mentioned many of them. The following is a more structured list:

- Masquerading (pretending to be someone else)
- Denial of service (e.g., flooding a platform or flooding the network)
- Unauthorized access
- Repudiation (denying past acts)
- Eavesdropping (data or code) (e.g., reverse engineering)
- Alteration (data or code)
- Unauthorized cloning (data or code for replay attacks)

A mobile code framework has to incorporate an apparatus that addresses the vulnerabilities that we just discussed and that prevents all forms of attacks. The security mechanism must have the capacity to protect:

- Middleware
- Hosts
- Data
- Code
- The network

We describe a number of mechanisms in our further discussion of security that focuses on mobile code platforms.

Mobile Code Frameworks

The mechanisms necessary to support code migration fall into three categories: migration, collaboration, and security. *Migration* mechanisms deal with the relocation of mobile code between execution environments. Relocation is a multistage process, not just the physical transport of code between two execution environments. Before initiating any action in response to an external request (e.g., from an agent), the framework has to verify that the new execution environment is compatible with it. That may require verifying execution capabilities, resource availability, and platform version, as well as security capability checks (discussed further on). If a targeted environment does not guarantee that the code will be able to run successfully after relocation, then the party requesting the migration must be consulted.

Before any code can be transferred, the decision has to be made about all resources bound to the running process. Earlier in this chapter, we discussed the types of relationships between resources and running processes. That relationship determines what action must be taken upon code migration. Some of the resources will have to be moved or copied; others will be accessed remotely or recreated in the new environment.

After taking care of the resources, the framework is ready for code migration, which requires engaging the underlying networking infrastructure. Network transport provides both reliable and unreliable services. In reliable services, a successful transport is assured. If the framework employs unreliable services (in which case flexibility is an advantage), then the framework itself must ensure that the transport of code and data is successful. Of course, we need to keep in mind code and data vulnerabilities, as security is of paramount importance. Again, we postpone that discussion for a little longer.

Upon accepting a computational unit in a new environment, the platform at the new location must instantiate it in an execution environment and rebind all resources. If all of that is successful, the new migrated code is activated.

Collaboration mechanisms support collaboration between computational units. They may be local or remote and direct or indirect. *Local communication* is contained within a single execution environment; that is, the communicated information does not leave the execution environment. Of course, any process that is local may gain access to local information given appropriate admission rights. In *remote communication*, a sender and a receiver are running in two execution environments, so a message is transmitted outside of the underlying local platform. A network infrastructure must be in place to support remote communication. A mobile code platform may leave communication totally up to the agents; however, as we discussed earlier, that exposes the platform to a number of security attacks. Therefore, platforms should provide communication mechanisms that both simplify the use of the networking infrastructure and provide security. In *direct communication*, agents establish a communication link that they use to exchange data. A mobile code platform may assist active agents in establishing such a link, but they do not assist the agents in transporting data.

Link and session management is the responsibility of a communicating agent. Message queuing, error detection and recovery, synchronization of multi-party conferences, etc., put a lot of pressure on agent code, so platforms commonly provide indirect means of communication. In *indirect communication*, a mobile code platform is an intermediary between two or more communicating agents. Two communication schemes can be put in place. The first one is e-mail-like message forwarding. The second scheme is based on shared spaces, or (to use a term from AI) blackboard systems.

A *message-forwarding system* usually mandates the use of envelopes for delivering messages. An envelope template includes provisions for delivery and return addresses along with possibly numerous parameters that spell out details of the request. For example, a sender may ask that a message be formatted in a special way, encrypted, sent through a specific transportation means, acknowledged, etc.

A *shared space*, or *blackboard*, is a shared memory area that can be written to and read from by many parties. The simplest possibility is just a data structure that contains message slots with general access rights. To gain a better understanding, let us analyze an analogy. If a team tries to solve a problem, it often resorts to a brainstorming session during which everybody presents opinions on a blackboard (or, more commonly today, a whiteboard), so every member of the team can see and analyze them. Numerous teams might be assembled to tackle different aspects of a problem or to deal with completely different problems. They would not use just one whiteboard, but many. Similarly, inter-agent communication over a single shared space with a flat organization would not be very efficient. A blackboard system may provide a more sophisticated means to communicate through the capability to create communities of interest. Agents can collaborate in many communities, but because they have a choice they are not overwhelmed with information that can now be delivered only to a targeted audience. The possibilities are limitless if the scheme allows for creating hierarchies.

Imagine staring at a whiteboard for hours while no new information is being written on it. It is much more efficient to check the board out only if somebody tells you that its content has changed. A blackboard system may provide analogous notification services. The notification scheme may allow for complex triggers that are based on communities of interest, sources of information, times of posting, etc.

If a mobile code platform allows for direct communication between executing agents, then it may be desirable to restrict who can talk to whom. This can be achieved in a variety of ways. In *direct communication*, an agent may take care of the issue completely. If connections are established with assistance from a platform, then an agent may ask the platform to restrict connectivity by others through names or passwords. In yet another scheme, only agents that enter a certain *cone of silence* are able to communicate with each other.

In *indirect communication*, if a message delivery system using envelopes provides an adequate level of security (as we discuss later on in this chapter), then the information contained in the payload of the envelope is secure. The sender of the message selects parties that should be given access rights to given data and the delivery system ensures that only those parties receive the data.

On the other hand, blackboard systems present a greater security risk, because their very functionality is based on shared access. Some parts of a blackboard may be public, and any party can read from them. Usually *access control lists* (ACLs) are used to verify rights to obtain certain data; parties trying to read the data have to present appropriate credentials. In a sense, a common-interest space protected by an ACL constitutes a cone of silence, because agents can communicate only if they enter the protected zone.

Security is a dominant issue in any system and software development in general, not only in relation to mobile code. The designer of a mobile code platform can use or at least borrow from plenty of security techniques, mechanisms, schemes, and methodologies. Mobile code security can be designed around language-based, language-independent, or operating system (OS)-based mechanisms; for example, a designer of a system implemented in Java can utilize the security sandbox that the Java Virtual Machine provides. The operating system may allow for interception of system calls, so such calls are monitored and potentially rejected if the system believes that they constitute a security breach. Unfortunately, most security provisions have to be designed by hand.

Intrusion detection systems protect platforms from intruders. They use numerous technique to detect an invasion. *Digital signatures* provide a mechanism for ensuring data integrity. To generate a digital signature, the source platform that is sending mobile code hashes the code and encrypts the result of the hashing using a *private key*. By using the source's *public key* to decrypt the signature, computing the hash, and comparing the two, the target platform verifies that the code was not tampered with.

Certification is a common way of verifying identities. Mobile code may carry a *certificate* that identifies its source by encrypting the source's public key. Mobile code arriving at a platform must present a certificate. To recover the source's public key, the target platform decrypts the certificate using a public key of the *certification authority* that published the certificate. The decrypted key can be used to verify the code signature as explained in the previous section. The target platform may determine a set of capabilities for the mobile code based on its source identity.

A mobile code platform may subject incoming mobile code to *static code analysis*. It is a procedure that examines code in an attempt to detect patterns that may cause problems. Such analysis can be applied to source or compiled code. A mobile code platform may implement a *code admission policy* based on code signatures, certificates, and results of code analysis. A platform may admit mobile code (that is instantiated as a running program) only if it comes from a trusted source and has appropriate credentials, and no problem is detected during the analysis.

In contrast to its static counterpart, *dynamic code analysis* is the process of examining code that has already been instantiated as a running process. Before any code instruction is executed, it is analyzed for potential security breaches. The platform may restrict code from executing actions that it believes are not safe. An analysis may be integrated with a capability control based, for example, on the source of the code.

A *security zone*, or a *sandbox*, is a mechanism for containing a computational unit in a restricted execution space. It is a basis for two levels of protection. First, a sandbox separates the code from other processes, both local and visiting. If the code causes a problem, then the problem is contained to one zone. Furthermore, a security zone facilitates controlled access to certain resources. The restrictions can be based on the identity of the source of the code.

Observing the state of an executing process may yield some warnings on its potentially destructive behavior; for example, a data structure may grow dynamically to a level that is dangerous to the functionality of the hosting system. An agent that adds a number of connections that goes beyond some reasonable limit is another example of a symptom of potential trouble.

One or more mobile code platforms may implement a *trust-building scheme*. Each mobile agent may be assigned an attribute that is a measure of trust and is part of the agent's credentials. Trust can be based on the past behavior of an agent or an agent owner. Trust can also be imported or exported if a propagation scheme is in place. A measure of trust may be the basis for a capability set allocated to a particular agent. A host might be less restrictive to agents with good trust credentials.

Some mobile code may carry with it the migration path that led to the current location. An analysis of such information may be useful in determining the probability of that agent being compromised and therefore dangerous; for example, an agent may have visited places that are known sources of security problems. This information may be a basis for lowering credentials of the agent.

Some researchers have proposed incorporating *formal methods* in verification of mobile code. Incoming code is accompanied by one or more *proofs* that can be executed on the destination host before the agent is allowed to execute. The proofs that are based on the logic of the original agent verify that the code still implements the same logic and therefore has not been modified during transport. Unfortunately, generating proofs automatically is difficult, and doing it by hand is very tedious and error prone.

Hardware-based security provisions are usually superior, because it is difficult to tamper with them; for example, the presence of a secure smartcard may be necessary for accessing certain system calls on a host. Inserting a card into the system may enable execution of associated mobile agents. Such a mechanism may be applied at a bank automated teller machine (ATM). The system may verify the authenticity, take a picture, record a video, etc., and then enable some customized operation implemented by the customer's agents.

Code can be considered data, so techniques similar to the ones used to protect data can be also used to protect code. *Encryption* techniques can be used to prevent third parties from stealing code in transit. Code is treated as data, so all data protection mechanisms are applicable in this context. *Obfuscation* makes code unintelligible to an eavesdropper. The process is performed on code to prevent reverse engineering. The code still has to be valid, because it must execute in the target environment; therefore, reverse engineering might be difficult but — unfortunately — still possible.

A new area of research is *encrypted functions*. Researchers have proposed techniques that involve transitions in functional space. A function implemented by mobile code can be encrypted beyond recognition to yield another function, which is sent and executed in a foreign environment. The result of the execution is an encrypted desired outcome that can be decrypted using a key belonging to the owner of the agent. The result can be sent to a trusted host (e.g., home) for decryption.

Another area of novel research is *dynamically generated code*. Instead of carrying all required code or fetching missing pieces from remote locations, mobile code can generate more chunks of code after being installed in the target environment. Because the code is generated on the spot, it is not exposed during transport. Reverse engineering may also be difficult if the host is malicious. As we discussed earlier, *tamper-resistant hardware* may protect a mobile code platform. It may also protect agents, because they will execute only in environments that are physically enabled by associated hardware (e.g., a smart card).

Some hosts may be interested in paths traveled by mobile code. As we saw earlier, this information may be part of a protection scheme. Analyzing trails might also be malicious; for example, a host may implement a targeted attack scheme that is driven by the past history of an incoming agent. Therefore, it might be necessary to *obscure a traveled trail*. In the simplest case, an agent may just forget all visited places. If a trail is necessary for some reason, then the data may be encrypted rather than carried in the open. *Tracing migration paths* can also be used to protect agents from wandering into undesired areas or, in more general terms, performing unexpected transitions. An agent may report visited locations to some verifier (e.g., its home platform) that may trigger a corrective action if necessary.

An external *verifier* may also be used for ensuring that a mobile agent executes only its original algorithm. An agent may report milestones during an execution, so the verifier can observe its behavior. The behavior is checked to detect any abnormalities; for example, if it cannot match (with some degree of accuracy) one of its normal behavioral patterns, then the agent might be declared invalid and a corrective action may be undertaken.

The normalcy of agent behavior can also be verified in another way. Instead of instructing one agent to perform a given task, two or more agents are dispatched for the same job. A verification scheme can be put in place that verifies that the execution outcomes of each of them are compatible. That can be achieved by implementing (as explained earlier) an external verifier (e.g., home platform), or the scheme may involve inter-agent communication and distributed verifications of behaviors.

To protect itself from accusations of any wrongdoing during a visit to a certain location, an agent may record its activity. This can be done locally in a data structure that the agent will carry or remotely by sending reports to an external observer (e.g., home platform). Each critical action to be undertaken by the agent must be digitally signed by the platform using the private key of the platform. An agent may refuse to execute anything that is not permitted by the hosting platform. With digital signatures on record, the platform cannot later claim illegality of any of the agent's actions.

Some mobile code may use or generate secret data that must be concealed. One way of dealing with the problem is to leave secrets in a secure place and retrieve them over secure connections only when they are needed. If secret data is generated dynamically, it can be sent home over a secure connection and the local copies discarded. Some data is not secret but cannot change. In this case, *message integration checks* can be applied; a hash function is applied to the data, and the result is signed using a private key of the sender. If the value decrypted with a matching public key does not match the recomputed hash value at the receiver, then the data has been compromised.

A common cause of flooding of networks is accidental or mischievous replication of data or code and dumping it on a network as quickly as possible. Wandering *mobile code junk*, mobile code that was injected into the network and never terminated by carelessness or design, is another problem. The networking infrastructure is usually well protected against potential attacks, but mobile code platforms may contribute to network security in several ways that we discuss next. The first method limits agent lifespan to prevent flooding with eternal generations of agents. Incorporating a time-to-live parameter within a migrating computational unit enables the control of multi-hop migration patterns. The value can be protected from tampering by encryption with a secret key shared by all platforms. When an agent is created (or rejuvenated), the time-to-live value is set to a certain threshold. At each stop, the value is decrypted, decreased, and checked against zero. If the number is zero, then the number of hops has exceeded the allowance and the agent is destroyed; otherwise, the decreased value is encrypted again and attached to the agent. Earlier, we discussed a mobile code admission policy. An analogous policy can be used to control agent migration from a platform. An agent's credentials can be used to make a decision on granting permission to leave.

Another security provision can be based on regulation of a host-leaving rate. Agents would have to queue their requests to migrate, and serving of the queue could be spaced in time, prioritized, or based on a volume of permits per time unit. Agent credentials can also be used in policies that regulate agent replication. Agents may not be allowed to replicate at all. Less restrictively, the number of replications can be controlled. That can be achieved by a mechanism constrained to a single platform that counts replication requests and matches the number with the maximum allowance of that platform. Another mechanism can be implemented systemwide with an agent carrying a number of replications. That number is handled on multiple platforms in the same way as the time-to-live mechanism described earlier. If the number of replications on any platform exceeds the threshold, the agent is not allowed to replicate anymore.

Implementing security mechanisms is rarely an easy task. Many details can very often escape attention and then haunt system administrators. Some problems cannot be addressed at all, because many security attacks explore system bugs that are not known before the attack actually takes place. Very often, fixing a bug occurs after damage has already been done. In systems that utilize authentication facilities, there is an inherent issue of a level of trust that can be put into certification authorities. Yet another issue relates to the scope of the authority. It is difficult to build one centralized control scheme, and distributing security management is not easy. To understand why, consider the use of a shared key to encrypt the time to live or an allowance to replicate numbers. How can one distribute secret keys without assurance that they are not compromised? This discussion represents just the tip of the iceberg. We are just warning potential designers and, it is hoped, making them more sensitive to the security issues.

Standards

Standardization efforts to set rules for interoperability between platforms for intelligent and for mobile agents were very vigorous at the peak of interest in the area but these efforts have lost their impetus in recent years; nevertheless, a lot of thought went into the documents that were accepted as standards. The Object Management Group (OMG) supported work on standardization of mobile code platforms known as the *Mobile Agent System Interoperability Facility* (MASIF). MASIF addresses issues of interoperability between heterogeneous mobile code platforms. It standardizes agent management, agent transfer between homogeneous (or very similar) platforms, agent and system naming conventions, system types, agent location syntax, and agent tracking.

The second standard was developed by the *Foundation for Intelligent Physical Agents* (FIPA). In 2005, FIPA's members voted to join the Institute of Electrical and Electronics Engineers (IEEE) Computer Society to become its eleventh standardization committee: the FIPA Standards Committee. FIPA does not deal with mobile code, but as we explained earlier in this chapter, mobile agents usually include some intelligence to perform their mission. The core of the standard is agent collaboration through exchanging messages; therefore, the standard addresses weak rather than strong intelligence. It may be a disappointment to some, but neither FIPA nor MASIF addresses any AI-based mechanisms that would deliver *intelligence*. Neither standard offers any solution recipes for designers seeking guidance in such matters.

The phrase *intelligent agents*, which has become the name of the field, is therefore misleading for those who do not understand the difference between weak and strong intelligence.

The FIPA standard provides specifications for intelligent communication between entities that will be able to implement some strong AI methods, thus allowing them to use that framework in some way to improve the chances for achieving a task at hand. Which AI methods and how the agents should use them are not subject matters of the FIPA standard. The large body of standardization documents released by FIPA dwarfs the MASIF standard. The standard covers numerous areas related to agent communication and attempts to address other aspects of *agent-based computing*.

Concluding Remarks

In this chapter, we explored models of computation based on mobile code. After scrutinizing characteristics of mobile code, we showed that it might be a tempting computing paradigm for software engineers including those working on middleware for mobility. We introduced several types of mobile code, including mobile agents. We discussed the issues that have so far prevented widespread use of mobile code. We explained that security concerns are the biggest obstacle in the progress of technologies utilizing mobile code. We suggested numerous methodologies and techniques that can be employed to ensure required levels of system, data, and code protection. We concluded with an overview of standards that apply to mobile and intelligent agents.

References

[1] Picco, G.P., Mobile agents: introduction, *J. Microproc. Microsyst.*, 25(2), 65–74, 2001.

[2] Fuggetta, A., Picco, G.P., and Vigna, G., Understanding code mobility, *IEEE Trans. Software Eng.*, 24(5), 352–361, 1998.

[3] Loureiro, S., Molva, R., and Roudier, Y., Mobile code security, in *Proc. of ISYPAR 2000 (4ème Ecole d'Informatique des Systèmes Parallèles et Répartis)*, Toulouse, France, February 1–3, 2000.

[4] Jansen, W. and Karygiannis, T., *Mobile Agent Security*, NIST Special Publ. No. 800-19, National Institute of Standards and Technology, Gaithersburg, MD, 1999.

[5] MASIF Standard, http://www.omg.org/cgi-bin/doc?orbos/97-10-05.

[6] FIPA Standards, http://www.fipa.org/.

Chapter 13

Proxy-Based Adaptation for Mobile Computing

Markus Endler, Hana Rubinsztejn,
Ricardo Rocha, and Vagner Sacramento

CONTENTS

Introduction

The use of proxies is commonplace in today's networks, where they are used for a huge variety of network services. A proxy is an intermediary placed in the path between a server and its clients. Proxies are used for saving network bandwidth, reducing access latency, and coping with network and device heterogeneity. In the specific case of mobile computing and wireless communication, proxies are mainly used to overcome the three major problems of these networks: throughput and latency differences between the wired and the wireless links, host mobility, and limited resources of the mobile hosts (MHs). Although proxies may be used also for implementing specific services in mobile *ad hoc* networks, usually they are used in infrastructured mobile networks, because their functions commonly place high demands on both processing and memory. Thus, in this chapter, we primarily discuss proxy-based architectures for infrastructured mobile networks.

In most cases, proxies act as protocol translators, caches, and content adapters for clients with network or device constraints and are placed on, or close to, the border between the wired and the wireless networks, such as at the wireless *access points* (APs), also referred to as *base stations* or *mobility support stations*. In addition to these canonical functions, however, proxies can perform a wide range of other complex tasks on behalf of the mobile clients, such as handover, session or consistency management, personalization, authentication, checkpointing, and service/resource discovery, among others. The major advantages of using a proxy-based architecture for serving mobile clients, when compared to an end-to-end approach, include the following:

- All mobility- and wireless-dependent transformations (translation, transcoding) can be assigned to the proxy and need not be handled by the servers, allowing legacy services to be easily adopted for mobile access.

- Any processing required for protocol and content transformations is distributed to other nodes and only where they are required, avoiding an overload at the servers.

- Placing a proxy at (or close to) a node with a wireless interface enables more agile and accurate monitoring of the wireless link quality, detection of MH disconnections, and better selection of the required adaptation.

- Transformations at any communication layer can be implemented and are more easily adapted or customized according to the specific capabilities of the wireless links.

As expected, there is a huge amount of work on proxy-based middleware for mobile and wireless computing, each effort solving the problems specific to some sort of service or application, such as Web access, multimedia streaming, and database access. Many authors use the terms *gateway*, *intermediary*, and *agent* instead of *proxy*. Although there might be some subtle differences in their meanings, we will use these terms interchangeably and adopt the general definition of a proxy as being *an entity that intercepts communication or performs some service on behalf of some mobile client.*

In spite of the huge diversity of proxy-centered architectures and proposals, we have identified two orthogonal forms of classifying and comparing all proxy-based approaches. The first dimension takes into account some general characteristics of the proxy-based architecture, and the second dimension focuses on the tasks (i.e., functionalities) assigned to the proxies. These two classifications are further detailed later in this chapter.

Obviously, other possible criteria could be used to classify proxy-based approaches. Dikaiakos [13] has written a very interesting survey about proxy-based infrastructures specifically for the Web. He proposes a classification of proxy approaches in three dimensions: *system architecture*, *functionality*, and *interactions*. Regarding system architecture, he distinguishes between centralized and distributed architectures, options for proxy placement, and proxy configurability and programmability. Concerning functionality, he proposes six broad categories, which are consistent with our task categorization. Finally, with regard to interactions, Dikaiakos considers whether the proxy supports synchronous or asynchronous communication. In addition, the article also compares eight proxy-based architectures and frameworks for the Web in deep detail; hence, we recommend it as complementary reading to the interested reader.

Architecture-Based Classification

In this section, we discuss a classification of proxy-based approaches that emphasizes general features of the software architecture and which is largely independent of the specific task assigned to the proxies. In particular, we have found that proxy-based architectures can be classified according to aspects such as *level, placement, single-/multi-protocol*, and *communication* and *extensibility*.

Level

Because proxies may be used for handling adaptation or customization at various software levels, we believe that this is a suitable classification criterion. In this respect, proxies can be used at three generic levels:

■ *Communication level* — At this level, proxies are in charge of handling all sorts of issues related to the communication protocols and abstractions. The main goal is to take device mobility and the use of wireless links transparent to the higher software layers. Typical adaptations at this level are wired–wireless protocol translation or optimization, buffering, handover management, etc. Examples are several proposals for Transmission Control Protocol (TCP) extensions for wireless networks [15] and wireless CORBA, an extension to the Common Object Request Broker Architecture (CORBA™) [5].

■ *Middleware level* — At the middleware level, proxies perform general tasks neither tailored to a specific type of application nor related to a specific communication protocol. Examples are some forms of content adaptation [19,34], consistency management of cached data [2,25], service or resource discovery [9], and security and authentication functions, among others.

■ *Application level* — Some proxy-based architectures are focused on a specific type of application, such as Web browsing [4,21,29], database access [2], and peer-to-peer (P2P) data sharing [44]. In this case, proxies execute tasks tailored to specific requirements and functions of an application class. For example, to compare caching in Web and database applications, the former handles heterogeneous objects and essentially aims at reducing response time, whereas the latter usually handles homogeneous data but requires management of cache consistency.

Placement and Distribution

Concerning the placement of proxies, we adopt the well-known classification suggested by Pitoura and Samaras [38] for proxy-based architectures, which defines the following main structures: a proxy executing only at a stationary node of the network (*server-side*); a proxy only at the mobile node (*client-side*); a pair of proxies, one executing at a stationary host and the other at the mobile host (also referred to as the *interceptor model*); and a proxy that can move between a stationary node and the mobile device (*migratory proxy* or *migratory agent*). Although most systems use either a server-side proxy or a proxy pair, examples of pure client-side proxies include those in the Coda file system [26]. As has been discussed elsewhere [38], server-side proxies are suitable for any kind of device, but client-side proxies normally require devices with more computing resources (i.e., thick clients). Migratory proxies have been suggested and implemented by several research groups as a means of transferring computing tasks from the MH to the network and of following the MH while it moves between networks [43]. Another aspect concerns the distribution of the proxy-specific adaptation and management functionality in the architecture: It may be *centralized* if all functionality is bundled into each proxy [21,25] or *decentralized* if it consists of several cooperating proxies, where each is responsible for some subset of the functions [3,34].

Single-/Multi-Protocol

Proxy architectures fall into two groups with respect to the number of communication protocols they support. Most systems handle a single protocol, such as TCP or the Hypertext Transfer Protocol (HTTP), and support specific adaptations of these protocols aiming to bridge the wired–wireless gap. Other proxy-based architectures, however, also adopt a multi-protocol approach, in which the proxy supports wired–wireless translation using several protocols (e.g., UDP, SMTP, SMS, WSP) and is able to dynamically switch between these protocols for delivering the data to the user independently of which wireless or cellular network the user is currently connected with. Examples of the latter group are iMobileEE [10], TACC [19], and eRACE [12].

Communication

This aspect characterizes proxy-based architectures with respect to the way a proxy communicates with the client, the server, and other proxies. Essentially, a proxy can communicate with both endpoints, the server and the client, in two modes: In the synchronous mode, the proxy performs the adaptation task and replies to the client in response to an explicit client request. In the asynchronous mode, the proxy does long-term work on behalf of the user (e.g., based on the user's preferences) and sends asynchronous notifications to the client. This asynchronous mode is common when proxies play the role of user agents, searching, collecting, and aggregating information on behalf of a user. Examples of architectures supporting both communication modes are the Wireless Application Protocol (WAP) [17], Web Intermediaries (WBI) [3], and MoCA's ProxyFramework [40,41]. Some architectures also support communication among proxies, usually for the purposes of session and handover management, checkpointing, and multicasting, among others. In this respect, communication can be direct or indirect. In the first mode, a proxy knows — perhaps through its client — which other proxy it needs to interact with [8,35]. In the second mode, the server (or another proxy) serves as a router of the messages exchanged among the peer proxies.

Extensibility/Programmability

Proxy extensibility (i.e., the ability to adapt and customize its functions) is also an important criterion to differentiate architectures. In most systems, the proxy has predefined adaptive behavior, usually determined by the current state of the execution environment. As a first step toward extensibility, some approaches provide a generic framework in which proxies can be easily tailored to the specific needs of an application or middleware at deployment time, such as in MoCA's ProxyFramework [40,41]. Yet another group of proxy infrastructures further support the dynamic loading of filters or new modules implementing specific functionality, such as those presented in Zenel [47].

Common Proxy Tasks

In this section, we present the other way to classify proxy-based approaches, which is by the main task, or function, executed by the proxies. One should note that this classification does not render disjoint categories, as several of the tasks discussed in this section are in fact somewhat intertwined. For example, protocol translation and optimization are key tasks in almost any proxy architecture; hence, several of the other tasks may also be regarded as a kind of translation or optimization. Moreover, the set of tasks discussed here is unavoidably incomplete, as several other application-specific functions could be assigned to proxies. Nevertheless, we believe we have selected the most common proxy tasks discussed in the literature.

Protocol Translation and Optimization

Because most conventional communication protocols for wired networks are usually not suited for wireless links because of their higher error rates, smaller throughput, higher cost and latency, mutual interference, intermittent connectivity, etc., one of the most common tasks of proxies is to deal with protocol translation, as well as optimizations of wireline protocols for wireless links. Wired–wireless protocol translation is required at many layers of the protocol stack, but in this section we focus on protocol issues of the transport layer and above, and lower-level transcodings are considered to be below the middleware level.

In addition to the plain translation between protocol formats (i.e., header transcoding, data alignment, and data encoding), proxies may also have to deal with an array of other communication-specific issues, such as flow control, error detection and recovery, and medium multiplexing, which essentially aim at optimizing data transfer over the wireless link and smoothing the wired–wireless gap. This is particularly true for connection-oriented protocols, such as TCP, whose mechanism for flow control does not react properly to disconnections, burst packet losses, or fluctuations in round-trip delay. This has motivated the development of several so-called *TCP split connection protocols* (e.g., MTCP, I-TCP, M-TCP, SRP) [15], where a proxy performs the mapping between the conventional TCP and an optimized transport protocol for the wireless link. Another example is the *Wireless-Profiled TCP*, adopted in the WAP 2.0 standard by the acronym WTCP [18] and used in i-Mode [14]. WTCP was developed for wireless metropolitan area networks (MANs) and wide area networks (WANs) and essentially uses the ratio of inter-packet separation as the primary metric for rate control, rather than packet loss and timeouts. Many other examples can be found of proxies being used for protocol translation at the session or application layers; for example, the WAP gateway is responsible for converting between wire-line session, presentation, and application-level protocols and the corresponding protocols of the WAP protocol stack [17].

A related task commonly assigned to proxies is that of optimizing data transfer of a conventional protocol over the wireless link. Protocol optimization essentially has two goals: to achieve higher bandwidth utilization and to provide smaller round-trip delay. The usual optimization techniques include caching of data, connection multiplexing, header and payload compression, adaptive flow control, and data volume reduction. HTTP and TCP are probably the most frequently cited protocols that have been optimized for wireless networks. Most optimizations done in the TCP split connection approach are based on the following general principles: using separate error and flow controls on each side of the connection (wireless/wire-line); performing faster recovery of wireless errors due to shorter round-trip times (RTTs); hiding transmission errors from the sender; and generating selective/spontaneous TCP acknowledgments (ACKs) to avoid window resizing.

Concerning HTTP, the main problems associated with communication over a wireless link include the following: human-readable and verbose headers; transfer of data objects without compression; huge RTT incurred by the use of a connection-oriented transport protocol and frequent Domain Name System (DNS) lookups; and separate HTTP requests for each inline image, such as buttons, icons, and bullet marks. One of the earliest works attempting to optimize HTTP over wireless links was Mowgli [29], which employs an HTTP proxy pair using asynchronous messages over long-lived transport-level connections with header and payload compression. IBM's WebExpress [21] also uses a proxy pair to optimize HTTP traffic through caching, differentiating (i.e., transmitting the delta between an HTTP result and a cached base Web object), HTTP header reduction, and multiplexing of several HTTP connections over a single TCP connection. More recently, Rodriguez et al. [39] have proposed a proxy-based architecture also aimed at reducing the RTT caused by DNS lookup. In fact, most current infrastructures for mobile Web access are proxy based and perform some of the above-mentioned HTTP optimizations [13].

Content Adaptation

Although protocol translation deals with protocol-specific adaptations and optimizations, content adaptation is largely protocol independent and aims at transforming the messsage payload for optimized transmission and presentation at the mobile device. The specific kind of adaptation used is determined primarily by the application requirements, and may take into account the following issues: the quality of the wireless link (broadband, cellular); characteristics of the device, such as its computational power (CPU, memory); output capabilities (screen size, gray-scale screens); and supported protocols (e.g., HTML, WML). A wide range of approaches for content adaptation for different kinds of data has been proposed, including techniques such as data distillation or refinement, summarization, intelligent filtering, and transcoding. Although no unique and widely accepted definitions of these terms exist, in the following we use the most common definitions found in the literature. Because the term *transcoding* is often used to denote any of the previous types of adaptation, we also use it to discuss general techniques and present architectures supporting a larger spectrum of content adaptations.

Distillation and Refinement

Distillation is a highly lossy, real-time, data-specific compression technique that attempts to eliminate redundant or unnecessary information while preserving most of the *semantic content* of the data. Distillation is thus a general term for several forms of data compression, which may or may not be based on coding standards and representations; for example, JPEG is a lossy compression method where compression rates can be controlled according to desired image quality.

An example of non-coding-based distillation could be a transformation where images are scaled down on each dimension to reduce their total size, thereby also reducing their binary representations. Yet another example of distillation is a reduction of color depth or color-map size. The resulting representation, though poorer in color and resolution than the original, is nonetheless still recognizable and therefore useful to the user.

Alternatively, the user may want to see the highly precise content of some part of the original data — for example, by zooming in on a section of a graphic or image or by rendering a particular PostScript page with figures without having to render the other pages. *Refinement* refers to the process of selecting some part of a document in its original quality. In fact, one can define a distillation–refinement space for each type of data (e.g., text, image, video), where distillation and refinement can be applied orthogonally to the data to reduce its binary size.

ActiveProxies [19], developed within the BARWAN project, was a pioneering piece of work focusing on data distillation and refinement. Active proxies are a means to perform on-the-fly content adaptation to support variations in the network, device characteristics, and software capabilities. The Transformation, Aggregation, Caching, and Customization (TACC) model provides mechanisms for the composition of TACC *workers*, where each worker handles the distillation or refinement for a specific Multipurpose Internet Mail Extensions (MIME) type. The project built several workers to deal with text, image, and video content, such as distillers for GIF and JPEG images, for HTML, and for MPEG video streams.

Summarization

Summarization is a sort of lossy compression where *specific* parts of the original data are selected for presentation, aiming at the least possible loss of information. The most common data types summarized for mobile and wireless devices are text and video. Text summarization techniques have been researched for quite a while, but the recent desire to display Web contents on small screens has given the field a new impetus. A video summary (or abstract) is defined as a sequence of still or moving pictures, with or without audio, that presents the content of a video file in such a way that the user is provided with concise information about the content while the essential message of the original is preserved. It may be a shorter version of a video file assembled by picking important segments from the original or a series of short clips containing the essence of a longer video file, without a break in the presentation medium [28]. For transmission over a low-throughput connection, video summarization is useful for providing users with a video digest so they can obtain the content quickly and comprehensively.

A canonical example of a system that applies video summarization is Mowser [4], a server-side proxy for dynamic context-based modification of HTTP streams that uses content negotiation as described in the HTTP/1.1 specification. It selects the best representation of a data resource based on the browser-supplied preferences for media type, network connection, available resources, languages, and encoding. Mowser allows the user to set viewing/presentation preferences, such as starting point; color capability; video resolution; sound capability; maximum allowed size for text, image, video, and audio files; and size restrictions for image files.

Intelligent Filtering

Intelligent filtering is usually defined as a mechanism to transform, drop, or delay data delivery by applying filters on a data path, according to network or target device conditions. Mobiware [34] is a quality of service (QoS)-aware middleware platform for mobile multimedia applications. Mobiware introduces the concept of *active filters*, which can be dynamically dispatched during handoff to strategic points in the network (e.g., base stations, mobile devices) to provide media scaling of audio and video streams when and where needed. Its goal is to support valued-added QoS with the best utilization of available bandwidth and seamless media delivery. The two styles of filters are *active media* (for audio/video flows) and *adaptive forward error correction* (FEC). In Mobiware, so-called *QoS adaptation proxy* (QAP) objects play a central role in allowing mobile devices to probe resource availability and to adapt to changes in the quality of the wireless link. Zenel [47] was one of the first to propose a framework for generic filtering. His architecture consists of a *proxy server*, composed of a high-level proxy and a low-level proxy, and a *filter control* (EventManager). The high-level proxy supports filters for application-layer protocols, and the low-level proxy supports filters for network and transport layers. These filters may drop, delay, or transform any sort of data being transferred to and from the mobile host, such as to improve the perceived quality of the network.

Transcoding

Transcoding is the general process of transforming the format and representation of content. Data may be filtered, transformed, converted, or reformatted to make it accessible by a variety of devices. Transcoding is commonly used for the conversion of video formats (i.e., QuickTime to MPEG) or the adjustment of HTML and graphics files to the constraints of mobile devices (e.g., HTML to WML transcoding). It is often used when device characteristics prevent the content from being presented in its original format. In one approach to transcoding, the transformation depends only on the type of content, and in a second approach the conversion is specified by an external annotation describing specific requirements of the device and the adaptations to be performed.

Far more proxy-based architectures employ the first approach, but we begin here by describing a system based on the latter approach. Annotation-based Web content transcoding [20] is an example of the external annotation approach. This system handles HTML documents and focuses on page fragmentation for small-screen devices. Upon receiving a request from a mobile device, the proxy server adapts the document to the capabilities of the particular client on the basis of associated annotations. An annotation specifies the transformations and contains information to help a transcoding proxy select from several alternative representations the one that best suits the client device. In the remainder of this section we summarize some well-known systems adopting the pure content-based transcoding approach.

AT&T Mobile Network (AMN™) [10] is a proxy-based mobile platform designed to deliver customized multimedia services to users of mobile devices. The server-side multi-protocol proxy is composed of *devlets, infolets,* and *applets.* Devlets are protocol adapters that provide protocol interfaces to different mobile devices, infolets are responsible for obtaining information from various data sources, and applets incorporate the application-specific logic. The *proxy engine* supports user and device profiles for customization, performs content transcoding and adaptation, and invokes the proper applets and infolets to answer requests from devlets. The transcoders transform content based on the MIME type specified in the service request.

IBM's Internet Transcoding for Universal Access [32] is a transcoding system that adapts video, images, audio, and text to the devices with diverse capabilities using a proxy that allows the content to be summarized, translated, and converted on the fly. The system handles composite multimedia documents and device constraints.

The Mowgli [29] infrastructure consists of two mediators that use the MowgliHTTP protocol to communicate with each other, thus reducing the number of round-trips between the client and the server. Mowgli reduces HTTP data transfer over the wireless link by employing three different techniques: data compression, caching, and intelligent filtering.

Caching and Consistency Management

Caching of data close to (or at) the mobile host is a very common task assigned to proxies. The common and main goals of caching are to reduce traffic to and from the source server, restrict the user-perceived latency, conserve wireless bandwidth and the battery power of the mobile device, and handle client disconnections (i.e., support some limited functionality of the client application at mobile hosts while disconnected). For the first two goals (reducing traffic and latency) it may be sufficient to cache data at a node on the edge of the wired network (i.e., to use a server-side proxy), but for the remaining goals caching at a client-side proxy on the mobile host is necessary.

In principle, server-side caching for mobile hosts does not significantly differ from conventional proxy-based caching for wired network access (e.g., Web proxies). The main difference, however, is that in mobile communications there is a wider range of possible networks and devices (e.g., laptops, palmtops, or cellphones) used by clients and which have much different capabilities. To cope with such diversity, it is now common to store content in different formats and fidelities. This practice has a serious implication on caching: Because each request is treated independently, popular items (e.g., Web objects) might be cached at the same time in different formats, thus wasting valuable storage at the proxy. To solve this problem, several proposals have been made to combine active transcoding with adaptive caching at the proxy so as to transcode contents into the various formats closer to the client.

Client-side caching, on the other hand, aims at enabling some limited form of data access by the user during the time in which the mobile host is disconnected. The main problem is to handle involuntary disconnections and to guarantee consistency of the cached objects (e.g., files, database records), particularly when cached objects can be modified by clients, when more than one client can cache the same data object, or when the original copy of the object at the server can be modified by other means. Several approaches for handling cache consistency in these networks have been proposed in the context of databases [2], but significant work has also come from other areas, such as distributed file systems and other data-sharing applications.

Due to the high probability of disconnections and the limited wireless bandwidth, neither a pure detection-based approach (client detects inconsistencies) nor a pure avoidance-based approach (server sends *invalidation reports* to the cache holder whenever the original object is modified) can be used for guaranteeing cache consistency. However, several other strategies for cache consistency have been proposed that are based on stateful, stateless, or hybrid servers or on incremental approaches [2,7]. In fact, many recent studies suggest that invalidation-report-based caching management is better suited for mobile networks, but a major problem with invalidation reports is that disconnected clients may miss some of these reports. To overcome this problem and avoid stateful servers that must track which clients have received (and acknowledged) which reports, Kahol et al. [25] have proposed the *asynchronous stateful* (AS) caching scheme, whereby server-side proxies, called *home location caches* (HLCs), buffer the invalidation reports from servers while the MH is disconnected, and deliver these reports to the MH when it reconnects to the network. Furthermore, each time a MH migrates, this buffer of invalidation reports is transferred to an HLC that is close to the next access point. More recently, other cache invalidation schemes based on intermediates that claim to be more efficient have been proposed [45].

To ensure operation in spite of intermittent connectivity, two main approaches have been explored. The first is supporting eager prefetching of data objects and performing conflict resolution on demand. The second considers each mobile host as being an autonomous entity and regards the disconnected mode, rather than the connected mode, as the norm and not the exception. Here, hosts synchronize their data objects upon sporadic connections. An example of the first approach is the well-known Coda file system [26], whose client-side proxy Venus does predictive caching (*hoarding*) of files being used while the host has network connectivity and supports reintegration of these files at host reconnection. Similarly, the OSMOSE mobility framework [16] aims at general-purpose support for service continuity in spite of disconnections. A well-known example of the other approach (asynchronous operation) is the Bayou [44] system for P2P file sharing, where an anti-entropy protocol is executed among replicated data managers (i.e., peer-side proxies) to resolve conflicts of potentially inconsistent files on peer hosts.

Session Management

Many applications use the notion of a *session*, which in general consists of a set, or sequence, of coherent actions performed by a user. Although the concept of a session may differ from one type of application or service to another, all of them have the notion of a *session state*. In a mobile and wireless computing environment, session management is thus concerned with maintaining the session state of a service in spite of disconnections and migrations of the user. Notice that, in this context, migration can have several meanings. In the simplest form, users keep their devices and simply reconnect to a different AP within the same network or a different network, an approach referred to as *network migration*. A more complex kind of migration occurs when the user switches devices but wishes to continue using the same service from the new device (*device migration*). In this case, the session state must not only be transferred to the new device but probably must also be adapted to the new communication or transport protocol (e.g., HTTP to WAP). Finally, in a yet more sophisticated kind of migration, the user switches between different, albeit related, applications or services (*application migration*); for example, users may switch from synchronous to asynchronous communication upon noticing that their devices are connected to a wireless network with higher latency and smaller throughput. In this case, the session initiated with the first service must be *transformed* into the session of the new service.

Session management essentially deals with how to represent, encapsulate, and adapt the session state; how to transfer and install the session state at the new device; and how to implement mechanisms for controlling online sessions. Gardner and Shahi [30] have proposed a middleware-level proxy architecture that maintains voice and Web data sessions and allows users to seamlessly transfer session states between different devices or to share them with other users. It consists of two parts. A server-side proxy intercepts application-level commands and handles user authentication and authorization, session storage, and synchronization. The client side features a graphical user interface (GUI) for session administration (e.g., the user may keep several ongoing sessions) and application plug-ins for capturing the state of the associated applications, managing the transfer, and synchronizing session states between multiple clients. Central to their work is the definition of a session schema for capturing state information of Web browsing sessions.

Handover Management

Among the several advantages offered by the wireless network, user mobility is perhaps the most appealing benefit, as it enables users to access information from different locations even while they are moving; however, to support this, mobile networks must provide support for mobility management (i.e., handover management). A handover, or handoff, occurs when a user previously connected to some network reconnects to the same or to a new network. Handover management is mainly responsible for two tasks: updating the location and address of the MH to ensure that it can be reached, and transferring the session state of the MH from the old to the new network. Thus, essentially handover management is concerned with offering mobility transparency to the applications.

Wireless CORBA [5,35] and Mobile IP [37] are two examples of proxy-based infrastructures that support handover management. Both define a very similar architecture composed of three basic elements: home location agents (HLAs), proxies (in Mobile IP terminology, *foreign agents*), and the MHs. The HLA contains records of which proxy is serving which MH, and the proxies manage the handover and act as intermediaries for all communications between the MH and servers in the wired network. Due to the similarity between the Wireless CORBA (wCORBA) and Mobile IP approaches, in the following we describe only Wireless CORBA in some more detail.

The wCORBA specification supports mobility transparency of objects through a mobile interoperable object reference (mobile IOR) and a General Inter-ORB Protocol (GIOP) tunneling protocol, which handles handovers between access bridges in a technology-independent way. The access bridge plays the role of an object proxy, through which clients on the wired network can request execution of an object's method on a MH, and *vice versa*. Among other tasks, access bridges are also in charge of synchronizing the session state transferred between them when the MH performs a handover. This is implemented through notification messages about the mobility events of the MH which are exchanged among the access bridges and the HLA of the MH.

To provide mobility transparency in CORBA applications, wCORBA's mobile IOR is used to hide the mobility of the device from clients invoking operations on target objects at the device. Instead of informing the concrete address of the target object, the host and port fields of an Internet Inter-ORB Protocol (IIOP) profile, a mobile IOR indicates the address of either the HLA of the MH hosting the target object or the access bridge currently associated with the MH. In this way, all information addressed to the MH is routed through its HLA, if it has one, which in turn forwards the received information to the current access bridge of the MH (through a LOCATION_FORWARD message), which forwards the information directly to the MH. If the MH is homeless, the information is sent directly to the current access bridge of the MH.

In contrast to wCORBA and Mobile IP, the home proxy (HP)-based wireless Internet framework [8] provides mobility support through an application-level proxy. In addition to supporting handover management, this framework aims to facilitate the integration of mobility support with QoS management mechanisms. Just as in Mobile IP, all packets addressed to the MH are first routed to its home network, then the home proxy intercepts the packets and redirects them to the current subnet of the MH. Unlike the home agent in Mobile IP, the HP uses a split-connection approach, based on session-layer mobility (SLM) [27], to relay packets to the MH. Using this approach, the HP creates two separate connections, one with the MH and the other with the peer host, so it can route packets between them. This work extends the SLM for recovering and managing TCP connection states when the MHs perform handovers.

Discovery and Autoconfiguration

The dynamicity of mobile computing environments imposes stronger requirements for service discovery mechanisms than traditional distributed environments; for example, service discovery should handle changes in the availability of devices or services and should be able to choose the most suitable service for each client, according to its current context. Proxy-based system architectures can help to overcome such challenges by hiding network heterogeneity and dynamicity, as well as by reducing the complexity of the control mechanisms to manage such dynamism. By accessing a service through a proxy, instead of interacting directly with a particular instance of the service, a client is exempted from such responsibilities as choosing the most suitable service, executing the service-specific protocol, and making the appropriate reconfiguration when the execution context changes. This approach is adopted by some Jini™-based middlewares [9,23].

Other proxy-based architectures focus on dynamic service reconfiguration. In WebPADS [11], clients and servers communicate with each other over wireless links using a WebPADS proxy instance that provides a service discovery mechanism. When some adaptation is required (e.g., bandwidth decrease), the proxy loads the suitable service from the network and installs it in the WebPADS proxy. A service is developed as a *mobilet*, using the WebPADS application programming interface (API), and it can migrate from the service directory to a proxy. Such reconfiguration is transparent to clients and servers.

Security

Services for secure mobile communications may also employ a proxy-based approach. In a mobile environment, proxies can be used to decentralize the authentication process and allow the application to use a public-key security model on the wired network that may require a high computational effort, while keeping the computed functions at the device as simple as possible. Furthermore, by acting as intermediaries between clients and servers, proxies may provide a natural and efficient means of implementing anonymity for the mobile application, hiding the real identity of the requester whenever necessary. In this case, the proxies are used to represent mobile clients and may be responsible for handling privacy, authorization, user authentication, or data encryption.

In the proxy-based security architecture proposed by Burnside et al. [6], the proxies implement a public-key security model to control access over shared resources (e.g., files, printers). For guaranteeing security and privacy, this work uses two separate protocols: a protocol for secure device-to-proxy communication and a protocol for secure proxy-to-proxy communication.

The protocol for device-to-proxy communication sets up a secure channel that encrypts and authenticates all messages on the wireless link using symmetric keys. On the other hand, the proxy-to-proxy protocol uses a public key infrastructure to implement the access control through access control lists (ACLs) on the public or protected resources. If a requested resource is protected by an ACL, the request must be accompanied by a proof of authenticity, which shows that it is authentic, and a proof of authorization, which shows that the requester is authorized to perform the particular request on the particular resource. The proof of authenticity is typically a signed request, and the proof of authorization is typically a chain of certificates.

With respect to user authentication, much work has been done on employing proxy-based signature generation. For example, Park and Lee [36] have proposed a nominative proxy signature scheme, a method in which the proxy generates a nominative signature and transmits it to the certification authority instead of sending it to the original client. The advantages are that it preserves user anonymity and decreases the mobile client's cost of computing the signature. A refinement of this approach has recently been proposed by Seo and Lee [42] which guarantees non-repudiation by the proxy and does not assume a secure channel between the client and the proxy.

Other Tasks

In this section, we briefly discuss some other proxy tasks commonly found in the literature. Many other application- or protocol-specific tasks could be mentioned, but due to space limitations we cannot discuss them all in depth here.

Personalization refers to the function of tailoring the information content in response to a request, learning the user's profile or current preferences, and possibly performing some complex task on behalf of the user. For example, when searching for some information on the Web, the proxy can select the most suitable universal resource locators (URLs) for a given user request or infer a semantic match between the user's query terms and the objects referred by the URLs. As an example, eRACE [12] supports personalization in the form of user-specific differentiated services (filtering, data aggregation, personalized dissemination), which are determined by eXtensible Markup Language (XML)-encoded eRACE profiles.

Content creation by a proxy is possible when the infrastructure supports the offloading and execution of application- and client-specific code at the proxy. On behalf of the client and based on user preferences, the proxy might be able to autonomously access network services, discover new data sources, and retrieve information from several sources to produce new content that is a composition, a summary, or a selection of the different pieces of retrieved data. IBM's WBI [3] and TACC [19] are examples of infrastructures supporting the dynamic instantiation of content aggregation modules at general-purpose proxies.

Checkpointing may be necessary for recovering the state of a distributed application after a failure, but, unfortunately, checkpointing algorithms usually incur high overhead in terms of control messages. To support checkpoint-based recovery in mobile networks, several studies have proposed the use of proxies as representatives of the MHs and in charge of collecting and managing the states of the mobile client. For example, the checkpointing algorithm proposed in Ni et al. [33] uses a proxy coordinator on the wired network that acts as a representative for processes executing on the mobile host to avoid sending checkpoint control messages over wireless links.

Proxy Frameworks

Because proxies have been used as a general approach for handling dynamic adaptation, several efforts have been made to develop generic proxy architectures, or *proxy frameworks*, that can be customized or extended to solve a particular problem. An example of such an effort is the Internet Engineering Task Force's *Open Pluggable Edge Services* (OPES) [46], which proposes a reference architecture for Web proxies and addresses such issues as security, distribution, and dynamic configuration.

In this section, we describe common mechanisms used in proxy frameworks and compare well-known systems, such as TACC, RAPIDware, Mobiware, MARCH, Web Intermediaries, and MoCA's ProxyFramework. The RAPIDware [31] project has proposed adaptive proxy services for multimedia streams. Mobiware [34] is a QoS-aware middleware platform for multimedia applications that also provides support for handoff control. WBI [22] was developed at IBM for HTTP-based adaptations, such as personalizing contents, transcoding, or caching. MARCH [1], TACC [19], and MoCA's ProxyFramework [40,41] are general-purpose content adaptation frameworks.

Most proxy frameworks provide general-purpose solutions for the following four main issues: (1) implementation and composition of adaptation modules, called *adapters*; (2) description of the conditions in which the adapters should be applied; (3) monitoring of the context, such as the profile of the mobile device, the state of the application, and the network bandwidth; and (4) loading of adapters. In the remainder of this section, we discuss these features in more detail. A complementary discussion about proxy frameworks can be found in Dikaiakos [13].

Adapter Development

The main customization point of a content-adaptation proxy framework is the *adapter*, a module responsible for implementing a transcoding function for a message or its content. (Note that some publications use different names for the adapter, including *filter* [31], *transcoder* [3], and *worker* [19].) A proxy may use several adapter instances for implementing specific adaptations required for different clients or contexts. Taking into account the client's current context, a proxy determines at runtime which adapter should be used for manipulating data content. In some cases, more than one adapter can be selected for transcoding a message; therefore, some frameworks support the definition of priorities, ordering, and composition of adapters.

Most proxy frameworks are designed on the basis of extensibility mechanisms and component-based approaches to support the development and composition of adapters, as well as their loading into a proxy. In some frameworks, such as WBI, RAPIDware, and MARCH, adapters can be developed as independent and composable components that are stored in adapter repositories or libraries and deployed in proxies. Some frameworks provide classes of special-purpose adapters; for example, RAPIDware provides some FEC filters to improve the ability of the audio/video stream to tolerate errors in a wireless environment. The TACC model supports adapters for transformation (content adaptation), aggregation (information collecting), caching, and customization.

Adapter Selection

The choice of adapters and when to use them is an extensible characteristic of proxy frameworks that can be defined in two ways: via programmable interfaces or via rule-based configuration. An example of the first approach can be found in Mobiware, where the application requirements and the adaptations to be applied must be programmed using an API provided by the framework. When a rule-based configuration is supported, the developer must define rules that contain trigger conditions, described in terms of the client and network states (i.e., context); the adaptations to be executed; and sometimes also a priority of the rule. Usually, the rules are described manually via an XML file. In MARCH, the selection process evaluates a set of rules during session setup and, as a result, produces a sequence of adapters to be applied. In MoCA's ProxyFramework and WBI, rules are evaluated just before each message is sent to the client. Rule-based systems are easily configured and less error prone than the ones based on programmable interfaces; also, it is not necessary to deal with intrinsic programming details of the framework. Furthermore, only the content provider can decide which adaptation is acceptable in different contexts. By using rules, he or she may define the sequence of adaptations to apply to data, thus better controlling the composition of the data, which is a very complex task to automate.

Context Monitoring

The monitoring and gathering of context information — the client profile and conditions of the execution environment, such as the available resources, load, and energy at the mobile host and the network — are part of the desired functionality of proxy frameworks. Probing the network state, such as available bandwidth or connectivity, is generally done via a monitoring function or service, as in TACC, MARCH, and MoCA's Proxy-Framework. Information related to the client device may be obtained at the startup connection request [1], via a customization database containing profiles [19], or through monitoring of the resources of the device [40]. In most frameworks, context changes are notified through asynchronous events, which must be interpreted and processed by the proxy to execute the appropriate action.

Adapter Loading and Execution

According to how adapters are loaded and activated, proxy frameworks can be classified as *configurable* or *dynamic* proxies. In a *configurable proxy*, adapters are defined statically at proxy deployment time. The developer can change the behavior of the proxy by using trigger rules that define the order and context in which an adapter should be executed. A *dynamic proxy* supports dynamic and on-demand loading of adapters from an adapter repository, according to the current context.

Two examples of dynamic proxies are RAPIDware and MARCH. RAPID-ware provides a composable proxy framework to support the dynamic composition of services by fetching adapters (or filters) from a repository and instantiating and reconfiguring them dynamically at the proxy in response to the changing needs of mobile clients. MARCH provides a dynamic execution environment for adapters that facilitates the uploading of proxies to the server or the mobile client on a per-session basis. In MARCH, the mobile-aware server (MAS) component is in charge of deciding which adapters, chosen from the proxy repositories, are to be used and where to execute them.

An example of frameworks for the configurable deployment of proxies is Web Intermediaries (WBI). At proxy startup, the registered adapters (or plug-ins) are instantiated with the corresponding firing rule conditions and an associated priority. WBI supports the aggregation of adapters, and the proxy can be placed either on the server or on the client side. Another example is MoCA's ProxyFramework, where the adapters are instantiated during proxy initialization according to trigger rules (described in an XML configuration file) specifying the context in which the adaptation (or set of adaptations) should be applied. This framework also supports the chaining of adapters, use of priorities, and mechanisms for specifying caching policies.

Comparing the two approaches, the dynamic loading of adapters gives more flexibility to a system; however, configurable proxies support verification of a consistent combination/configuration of adapters. In addition, dynamic loading of adapters from a repository can be time consuming; therefore, this approach is more suited for systems where context changes are not very frequent.

Table 13.1 presents the cited frameworks, summarizing their main characteristics according to the aspects discussed in this section and earlier. Comparing the systems presented in the table, it is worth mentioning that all of them support content adaptation, caching management appears as the second most frequent functionality, and handover management is provided only by Mobiware. Furthermore, an equal number of systems address adapter loading (dynamic- versus configuration-based) and the form of adaptation selection (programmable versus trigger-rule configuration), suggesting that all of these approaches have their advantages and disadvantages. Concerning communication capabilities, only MoCA's ProxyFramework and WBI support asynchronous (publish–subscribe) communication, which has been recognized as well suited for mobile computing. Context awareness is also supported by most of the frameworks (except WBI), but only MARCH and MoCA's ProxyFramework also consider the state of the client's devices. Although it is quite difficult to compare the frameworks, Mobiware seems to be one of the most complete systems in terms of supported functionality, extensibility, and architecture.

Conclusion

In this chapter, we presented two classifications of proxy-based architectures for mobile computing, identified and discussed broad categories of responsibilities assigned to these intermediaries, and presented the most representative examples of such systems. Despite the widespread adoption of proxy-based architectures for mobile computing, a number of open challenges remain to be addressed to make proxy-based systems more flexible, scalable, and shaped to the specific requirements of current and future mobile networks and applications.

More precisely, some justified concerns have been raised with regard to the scalability of the proxy-based approaches. As the number of mobile users connecting through wireless links increases, server-side proxies may not be able to cope with the increasing computational demands of the mobile clients they represent. This is particularly true if the adaptation and transcoding performed at the proxies is specific for each mobile client

Table 13.1 Comparison of Extensible Proxy Approaches

	RAPIDware	Mobiware	MARCH	TACC	MoCA ProxyFramework	WBI
Purpose	Multimedia	Multimedia, QoS	General	General	General	Web applications
Level	Middleware	Middleware	Application	Middleware	Middleware	Application
Proxy placement	Server-side	Client-side and server-side	Server-side and proxy-pair	Server-side	Server-side	Client-side and server-side
Dynamic adapter loading	Yes	Yes	Yes	No	No	No
Adaptation selection	Programmable	Programmable	Trigger-rules configuration	Programmable	Trigger-rules configuration	Trigger-rules configuration
Functionality	Content adaptation	Content adaptation, handover management	Content adaptation	Caching, content adaptation	Caching, content adaptation	Caching, content adaptation
Communication	Synchronous	Synchronous	Synchronous	Synchronous	Synchronous, asynchronous	Synchronous, asynchronous
Context awareness	Wireless link	Wireless link	Device and wireless link	Wireless link	Device and wireless link	—

(e.g., takes into account the particular device characteristics and limitations) and is resource hungry, such as, for example, the transcoding of multimedia streams.

The obvious alternative to the use of proxies is the adoption of an end-to-end approach where information servers pre-transcode contents to an array of different formats and resolutions and deliver content to each client in the most suitable form and fidelity, according to the capabilities of the corresponding device and the current quality of its wireless connection. Because disk storage is becoming increasingly less expensive, several content providers are pursuing such an end-to-end approach.

A pure end-to-end approach, however, does have some drawbacks: (1) It requires the use of servers that are capable of interpreting the information about client capabilities and current network conditions; (2) it is not scalable because servers would have to store all their content in different formats and fidelities so as to adequately serve their mobile and resource-limited clients, in addition to the stationary clients; (3) it does not support dynamic and seamless switching from one format or fidelity to another during a transmission, which may be necessary for adapting to the variable quality of wireless connections; and (4) tasks such as disconnection and handover management, caching, and protocol translation cannot be properly handled by servers, as these tasks solve specific problems related to the mobility and resource limitations of the client devices, as well as to the specific characteristics of the wireless technologies. Hence, for most adaptations required by mobile applications, a proxy-based approach turns out to be the best choice.

A major challenge remaining, though, is to design proxy-based architectures that are scalable. One possibility is to combine the end-to-end and proxy approaches, such as is discussed in Joshi [24]. Another promising approach is to develop infrastructures that support the deployment of distributed and cooperative networks of intermediaries that collectively perform adaptations for a huge variety of devices and protocols, such as in IETF's proposals of Open Pluggable Edge Services (OPES) [46].

As the number of applications for mobile networks increases, and their services become more complex and personalized, proxies will be used for an increasing number of specialized functions. Although each type of application will have specific demands for proxy-based functions, we have identified a common and recurrent set of functions, structures, and architectural patterns in proxy implementations that can be used as the basis for developing proxies for specific needs. In our opinion, the demand is increasing for flexible and extensible tools and frameworks for the rapid development and customization of proxy-based architectures, at both the application and middleware levels.

The other trend we envisage in proxy-based architectures is that of dynamic proxy configuration, which allows shaping the proxy's functionality according to dynamic demand by the clients, server load, or current mobile network conditions. Ideally, we should have a library of standardized modules for content adaptation, protocol translation, caching management, etc., that could be automatically loaded, instantiated, and interconnected in a proxy framework, according to the specific needs and network conditions.

References

[1] Ardon, S., Gunningberg, P., LandFeldt, B., Portmann, M., Ismailov, Y., and Seneviratne, A., March: a distributed content adaptation architecture, *Int. J. Commun. Syst.*, 16(1), 97–115, 2003 (special issue on wireless access to the global Internet: mobile radio networks and satellite systems).

[2] Barbara, D., Mobile computing and databases: a survey, *Trans. Knowledge Data Eng.*, 11(1), 108–117, 1999.

[3] Barrett, R. and Maglio, P.P., Intermediaries: an approach to manipulating information streams, *IBM Syst. J.*, 38, 629–641, 1999.

[4] Bharadvaj, H., Joshi, A., and Auephanwiriyakul, S., An active transcoding proxy to support mobile Web access, in *Proc. of the 17th IEEE Symp. on Reliable Distributed Systems (SRDS'98)*, W. Lafayette, IN, October, 1998.

[5] Black, K., Currey, J., Kangasharju, J., Länsiö, J., and Raatikainen, K., *Wireless Access and Terminal Mobility in CORBA*, White Paper, Department of Computer Science, University of Helsinki, Finland, 2001.

[6] Burnside, M., Clarke, D., Mills, T., Maywah, A., Devadas, S., and Rivest, R., Proxy-based security protocols in networked mobile devices, in *Proc. of the 17th ACM Symp. on Applied Computing (SAC'02)*, Madrid, Spain, March, 2002, pp. 265–272.

[7] Cai, J., Tan, K.-L., and Ooi, B.C., On incremental cache coherency schemes in mobile computing environments, in *Proc. of the 13th Int. Conf. on Data Engineering (ICDE'97)*, Birmingham, U.K., April, 1997, pp. 114–123.

[8] Chan, J., Landfeldt, B., Liu, R., and Seneviratne, A., A home-proxy based wireless internet framework in supporting mobility and roaming of real-time services, *IEICE Trans. Commun.*, E84-B(4), 873–884, 2001 (special issue on mobile multimedia communications).

[9] Chen, H., Joshi, A., and Finin, T.W., Dynamic service discovery for mobile computing: intelligent agents meet Jini in the aether, *Cluster Comput.*, 4(4), 343–354, 2001.

[10] Chen, Y.-F., Huang, H., Jana, R., Jim, T., Hiltunen, M. et al., iMobile EE: an enterprise mobile service platform, *Wireless Networks*, 9(4), 283–297, 2003.

[11] Chuang, S.-N., Chan, A.T.S., Cao, J., and Cheung, R., Dynamic service reconfiguration for wireless Web access, in *Proc. of the Twelfth Int. Conf. on the World Wide Web (WWW 2003)*, Budapest, Hungary, May, 2003, pp. 58–67.

[12] Dikaiakos, M. and Zeinalipour-Yazti, D., *A Distributed Middleware Infrastructure for Personalized Services*, Technical Report TR-2001-4, Department of Computer Science, University of Cyprus, 2001.

[13] Dikaiakos, M.D., Intermediary infrastructures for the World Wide Web, *Comput. Networks*, 45(4), 421–447, 2004.

[14] NTT DoCoMo, *i-Mode*, www.nttdocomo.com/corebiz/services/imode.

[15] Elaarg, H., Improving TCP performance over mobile networks, *ACM Comput. Surv.*, 34, 357–374, 2002.

[16] Conan, D. et al., *Wp2 Frameworks: Architecture and API of the Mobility Framework*, Technical Report WP2-040108-1, Information Technology for European Advancement (ITEA), INT, INRIA/OASIS, and Thales, 2004.

[17] WAP Forum, *Wireless Application Protocol Architecture Specification*, WAP-210-WAPArch-20010712-a, 2001, www.wapforum.org.

[18] WAP Forum, *Wireless Profiled TCP, Version 31*, WAP-225-TCP-20010331-a, 2001, www.wapforum.org.

[19] Fox, A., Gribble, S., Chawathe, Y., and Brewer, E., Adapting to network and client variation using active proxies: lessons and perspectives, *IEEE Pers. Commun.*, 5(4), 10–19, 1998.

[20] Hori, M., Kondoh, G., Ono, K., Hirose, S.I., and Singhal, S., Annotation-based Web content transcoding, *Computer Networks*, 33(1–6), 197–211, 2000.

[21] Housel, B.C. and Lindquist, D.B., Webexpress: a system for optimizing web browsing in a wireless environment, in *Proc. of the 2nd ACM/IEEE Int. Conf. on Mobile Computing and Networking (MOBICOM'96)*, White Plains, NY, November, 1996, pp. 108–116.

[22] Ihde, S.C., Maglio, P.P., Meyer, J., and Barrett, R., Intermediary-based transcoding framework, in *Proc. of the Ninth Int. Conf. on the World Wide Web (WWW 2000)*, Amsterdam, The Netherlands, May, 2000.

[23] PsiNaptic, Inc., *Jmatos*, http://www.psinaptic.com/.

[24] Joshi, A., On proxy agents, mobility, and Web access, *Mobile Networks Appl.*, 5(4), 233–241, 2000.

[25] Kahol, A., Khurana, S., Gupta, S.K.S., and Srimani, P.K., A strategy to manage cache consistency in a disconnected distributed environment, *IEEE Trans. Parallel Distrib. Syst.*, 12(7), 686–700, 2001.

[26] Kistler, J.J. and Satyanarayanan, M., Disconnected operation in the Coda file system, *ACM Trans. Comput. Syst.*, 10(1), 3–25, 1992.

[27] Landfeldt, B., Larsson, T., Ismailov, Y., and Seneviratne, A., SLM, a framework for session layer mobility management, in *Proc. of the 10th IEEE Int. Conf. on Computer Communications and Networks (ICCCN'99)*, Boston, MA, October, 1999.

[28] Lienhart, R., Pfeiffer, S., and Effelsberg, W., Video abstracting, *Commun. ACM*, 40(12), 55–62, 1997.

[29] Liljeberg, M., Alanko, T., Kojo, M., Laamanen, H., and Raatikainen, K., Optimizing World-Wide Web for weakly connected mobile workstations: an indirect approach, in *Proc. of 2nd Int. Workshop on Services in Distributed and Networked Environments (SDNE'95)*, Whistler, Canada, June, 1996.

[30] Gardner, M. and Shahi, A., *Mobile Web Sessions for Mobile Computing*, Chimera Working Paper 2004-01, University of Essex, Colchester, U.K., 2004.

[31] McKinley, P.K., Padmanabhan, U.I., Ancha, N., and Sadjadi, S.M., Composable proxy services to support collaboration on the mobile Internet, *IEEE Trans. Comput.*, 52(6), 713–726, 2003.

[32] Mohan, R., Smith, J.R., and Li, C.-S., Adapting multimedia internet content for universal access, *IEEE Trans. Multimedia*, 1(1), 104–114, 1999.

[33] Ni, W., Vrbsky, S.V., and Ray, S., Low-cost nonblocking coordinated checkpointing in mobile computing systems, in *Proc. of the 8th IEEE Int. Symp. on Computers and Communications (ISCC'03)*, Kemer-Antalya, Turkey, June, 2003, pp. 1427–1434.

[34] Angin, O., Campbell, A.T., Kounavis, M.E., and Liao, R.R.-F., The Mobiware Toolkit: programmable support for adaptive mobile networking, *IEEE Pers. Commun. Mag.*, 5(4), 32–43, 1998 (special issue on adapting to network and client variability).

[35] OMG, *Wireless Access and Terminal Mobility in CORBA*, Object Management Group, Needham, MA, 2002 (http://www.omg.org/technology/documents/formal/telecom_wireless.htm).

[36] Park, H.-U. and Lee, I.-Y., A digital nominative proxy signature scheme for mobile communication, in *Proc. of the Third Int. Conf. on Information and Communications Security (ICICS'01)*, Xian, China, November, 2001, pp. 451–455.

[37] Perkins, C.E. and Johnson, D.B., Mobility support in IPv6, in *Proc. of the 2nd ACM/IEEE Int. Conf. on Mobile Computing and Networking (MOBICOM'96)*, White Plains, NY, November, 1996, pp. 27–37, 1996.

[38] Pitoura, E. and Samaras, G., *Data Management for Mobile Computing*, Kluwer, Dordrecht, 1998.

[39] Rodriguez, P., Mukherjee, S., and Rangararajan, S., Session level techniques for improving Web browsing performance on wireless links, in *Proc. of the Thirteenth Int. Conf. on the World Wide Web (WWW 2004)*, New York, May, 2004.

[40] Rubinsztejn, H.K., Endler, M., and Rodrigues, N., A framework for building customized adaptation proxies for mobile computing, in *Proc. of the IFIP Conf. on Intelligence in Communication Systems (INTELLCOMM 2005)*, Montreal, October 1–11, 2005.

[41] Sacramento, V., Endler, M., Rubinsztejn, H.K., Lima, L.S., Gonçalves, K., and do Nascimento, F.N., MoCA: a middleware for developing collaborative applications for mobile users, *IEEE Distributed Syst. Online*, 5(10), 2004.

[42] Seo, S.-H. and Lee, S.-H., New nominative proxy signature scheme for mobile communications, in *Proc. of Int. Conf. on Security and Protection of Information (SPI 2003)*, Brno, Czech Republic, April, 2003, pp. 149–156.

[43] Spyrou, C., Samaras, G., Pitoura, E., and Evripidou, P., Mobile agents for wireless computing: the convergence of wireless computational models with mobile-agent technologies, *Mobile Networks Appl.*, 9(5), 517–528, 2004.

[44] Terry, D.B., Petersen, K., Spreitzer, M.J., and Theimer, M.M., The case for non-transparent replication: examples from Bayou, *IEEE Data Eng.*, 21(4), 12–20, 1998.

[45] Wang, Z., Das, S., Che, H., and Kumar, M., A scalable asynchronous cache consistency scheme (SACCS) for mobile environments, *IEEE Trans. Parallel Distributed Syst.*, 15(11), 983–995, 2004.

[46] OPES Working Group, *Open Pluggable Edge Services*, Technical Report, Internet Engineering Task Force (IETF), 2006 (www.ietf.org/html.charters/opes-charter.html).

[47] Zenel, B., A general purpose proxy filtering mechanism applied to the mobile environment, *Wireless Networks*, 5(5), 391–409, 1999.

Chapter 14

Reflective Middleware

Paul Grace and Gordon Blair

CONTENTS

Introduction

The environmental characteristics of the mobile computing environment present middleware engineers with challenging problems. Primarily, mobile middleware research has focused on addressing the key problems — namely, unpredictable network connections, poor network quality of service (QoS), *ad hoc* interaction, and limited end-system resources. In many cases, mobile middleware has extended traditional object-oriented middleware, including the Common Object Request Broker Architecture (CORBA™) and Java Remote Method Invocation (RMI), to mobile settings. Example middleware solutions of this type are the Architecture for Location-Independent Computing Environments (ALICE) [1] and MobileRMI [2]; these provide solutions to maintain the Internet Inter-ORB Protocol (IIOP) (ALICE) and Java RMI (MobileRMI) connections transparently in the face of disconnection and poor network QoS. On the other hand, wireless communication is more naturally served by decoupled and opportunistic communication paradigms (i.e., mobile devices do not have to be simultaneously connected) and exploit connectivity whenever it becomes available. Tuple spaces [3] and publish–subscribe middleware [4] are examples of these paradigm types. These key problems, however, are compounded further by the dynamism of the mobile environment. As a mobile device moves, it naturally encounters changes in its environment: change in context (such as the device location), change in network conditions, and change in available resources. Therefore, when developing mobile middleware, it is not enough to consider only the primary properties of wireless networks; rather, solutions must be able to adapt dynamically to deal with changes in the context of a mobile device.

We now introduce five examples of middleware adaptation that illustrate how adaptation of the middleware behavior, based on environmental change, benefits mobile applications (this is not an exhaustive list):

■ *Mobile device heterogeneity* — Mobile devices have varied characteristics in terms of size, screen dimensions, processing power, system memory, network connections, etc.; therefore, a "one-size fits all" approach to mobile middleware development is not feasible. Instead, the middleware must be configurable to operate effectively across multiple device types.

■ *Network context change* — Fluctuations in network QoS, caused, for example, by a change in the network type of the device (e.g., from a GPRS to a 802.11b network connection), requires the middleware to adapt its behavior to the resultant changes in bandwidth, latency, etc. to maintain the performance levels of a given interaction. In a video application, for example, the media content streamed between mobile nodes may be filtered to reduce or increase the amount of data transmitted to maintain the video output (with a corresponding change in picture quality).

- *Connectivity* — During periods of use, a mobile device may remain permanently connected, but at other times the connection may be intermittent; hence, mobile middleware must adapt its interaction mechanism to suit the current connection conditions and application requirements.
- *Resource fluctuation* — Available battery power and system memory are limited resources on mobile devices and must be carefully managed in the face of high- and low-level loads; therefore, the middleware must adapt its resource use (e.g., by utilizing less power and memory when fewer resources are available and *vice versa*).
- *Middleware heterogeneity* — Mobile applications are not implemented using identical middleware platforms. Many mobile middleware solutions are available to application developers, including distributed-object-based, publish–subscribe, agent-based, and tuple spaces. Dynamically adaptable middleware must cope with this property to ensure that interaction between heterogeneous middleware types can continue.

To manage environmental change and perform adaptations of the types previously listed, mobile middleware must be able to adapt itself based on reasoning about both its current behavior and the current environmental conditions. The authors believe *reflection* is well suited to the development of such adaptive middleware; the technique provides principled mechanisms to inspect the structure and behavior of the middleware and make changes to both at runtime. In this chapter, we introduce the reader to the fundamental techniques of reflection and reflective middleware and examine complementary case studies that have applied reflection to the domain of mobile computing middleware. We also examine the closely related technology of dynamic aspect-oriented programming, which offers powerful techniques for performing systemwide dynamic adaptation (offering both functional and nonfunctional extensions) of middleware implementations.

Reflection

Overview of Reflection

Reflection is a technique that first emerged in the language community to support the design of more open and extensible languages (e.g., see Kiczales et al. [5]). The approach is nicely summarized by the following quote from Brian Cantwell Smith, the originator of early work on reflection [6]:

> In as much as a computational process can be constructed to reason about an external world in virtue of comprising an ingredient process (interpreter) formally manipulating representations of that world, so too a computational process could be made to reason about itself in virtue of comprising an ingredient process (interpreter) formally manipulating representations of its own operations and structures.

Hence, reflection is the capability of a system to reason about itself and act upon this information. For this purpose, a reflective system maintains a representation of itself that is causally connected to the underlying system that it describes. This is known as *causally connected self-representation* (CCSR) [7]. CCSR is often referred to as the *meta level*, and the system itself the *base level*; hence, changes made at the meta level via this self-representation are reflected in the underlying base level, and *vice versa*. The process of making the base level tangible and accessible at the meta level is known as *reification*. Operations to introspect and make changes to the meta level are commonly referred to as the Meta Object Protocol (MOP).

As discussed earlier, reflection has been predominantly applied to language design, and a wide variety of reflective languages are now available. Examples include CLOS [5], Sun's Core Java Reflection library [8], Iguana [9], and OpenC++ [10]. In this section, we concentrate on the application of reflection to the design of middleware systems. We focus in particular on the general techniques involved, which are common to the mobile middleware solutions described later in the chapter.

Reflective Middleware

One of the goals of this chapter is to convince the reader as to why we should make mobile middleware reflective. In the introduction, we presented five examples that illustrated the need for change in middleware behavior. In these situations, a fixed, static middleware platform is unsuitable; for example, such a platform cannot be configured to operate on heterogeneous devices, nor can it alter its behavior to react to changing environmental contexts. Instead, solutions that promote *openness* in the development of middleware are essential. That is, the internal details about the middleware implementations must be made available to support both configuration and also dynamic reconfiguration decisions. Reflection provides principled (as opposed to *ad hoc*) mechanisms for introspection and adaptation and hence is well suited to developing open solutions that manage dynamic change. Furthermore, middleware is ideally placed as the arbitrator between applications and the network environment; the middleware can reason about itself and the current environment and make dynamic changes that will benefit the running applications. Notably, reflective mechanisms also offer an ideal solution to the problem of middleware heterogeneity within the mobile domain (as discussed further later in this chapter). Reflection does not assume or require any particular communication paradigm. Reflective middleware can provide any communication style (e.g., tuple space) or, better, can provide different communication paradigms (e.g., remote procedure calls, events, messages) from which applications can dynamically choose the one that best suits their current needs.

In middleware platforms, two (complementary) styles of reflection have emerged:

- *Structural reflection* is concerned with the underlying structure of objects or components (e.g., in terms of interfaces supported). This is similar, for example, to the introspection features found in Java 1.2 and associated technologies, such as JavaBeans™ [8]. More advanced features may also be offered, such as the ability to adapt the structure of an object (e.g., to add new behavior at runtime). Similarly, some systems provide architectural reflection, whereby the software architecture of the system can be reified and altered (e.g., in terms of components and connectors) [11,12]. This can be applied to the very structure of the middleware platform itself, allowing customization of the architecture for the current environmental conditions. Finally, metadata or context can be viewed as a form of structural reflection, providing additional (meta) information about the underlying system (e.g., physical location, current battery levels, performance of the network).
- *Behavioral reflection* is concerned with activity in the underlying system (e.g., in terms of the arrival and dispatching of invocations). Typical mechanisms provided include the use of interceptors that support the reification of the process of invocation and the subsequent insertion of pre- or post-actions. Other systems provide similar capabilities through dynamic proxies [8].

Fundamental Reflective Middleware Architectures

We now investigate two key reflective middleware approaches: the Lancaster approach and DynamicTAO. These architectures underpin the reflective middleware solutions applied to the mobile computing domain, as will be seen in the case studies that follow.

The Lancaster Approach

The Lancaster approach [11] to reflective middleware development is based on a design philosophy that promotes a marriage of reflection, component technologies, and component frameworks. Notably, the approach can be used to create families of reflective middleware in domains, including multimedia computing, mobile computing, and sensor-based computing, among others. Components are the fundamental building blocks of these middleware, where a component is "a unit of composition with contractually specified interfaces, which can be deployed independently and is subject to third-party creation" [13]. It is important to note that the use of component programming promotes the benefits of *configurability, reconfigurability*, and *reuse* at the middleware level. All reflective middleware implementations that follow this philosophy have been developed using the OpenCOM component model [14], which supports explicit dependencies (bindings) between components; that is, it maintains information about the connections between one component's provided interface and the required interface of another component. Reflection is then used to provide a principled mechanism to inspect and dynamically adapt the component structure (in terms of components and their bindings) of the middleware implementation. Finally, component frameworks constrain the design space and the scope for evolution, where a component framework (CF) is defined as a collection of rules and contracts that govern the interaction of a set of components [13]. Hence, component frameworks act to ensure that the middleware behavior is not compromised by malicious or invalid reconfigurations.

Figure 14.1 illustrates the metaspace model that forms the basis of the Lancaster reflective middleware design. Every OpenCOM component has access to an associated *metaspace*. Three distinct metamodels represent the metaspace: *interface, architecture*, and *interception*. The interface and architecture metamodels provide structural reflection, and the interception metamodel supports behavioral reflection:

- The *interface metamodel* supports inspection of the provided and required interfaces of a component. Typically, it is possible to examine the operations available on these interfaces or dynamically invoke one of the operations.
- The *architecture metamodel* accesses the software architecture of a component represented by two elements: a *component graph* and a set of *architectural constraints*. The component graph is represented by a set of connected components, where a connection maps between a required and a provided interface in the same address space; hence, the architecture metamodel can be used to both discover and make changes to this structure at runtime. The

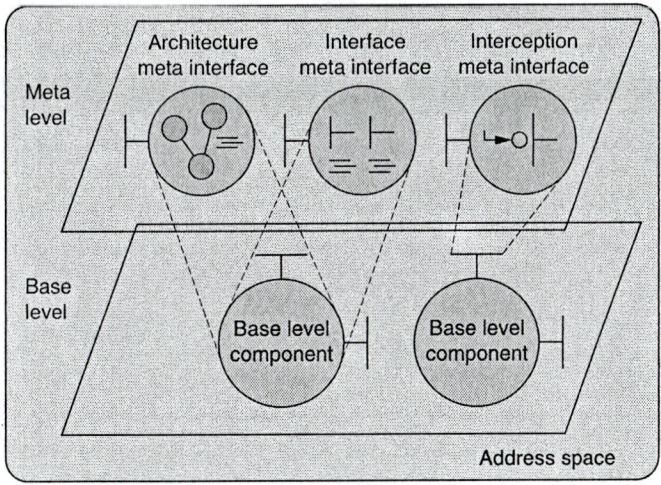

Figure 14.1   The metaspace structure of OpenCOM.

metamodel can also be used to discover the architectural constraints defined over the component graph; such constraints must be preserved during periods of adaptation and indeed are checked as changes occur, as discussed below.

■ The *interception metamodel* enables the dynamic insertion of *interceptors*, which support the insertion of pre- and post-behavior onto interfaces. These interceptors are executed before each operation invocation of an interface and after the operation has completed.

The architecture metamodel is fundamental when developing dynamic middleware solutions, as seen in the Reflective Middleware for Mobile Commuting (ReMMoC) framework described later; however, providing open access to the structure of the system and the ability to make runtime changes increases the likelihood of system failure and opens it up to third-party attack. To prevent invalid or malicious reconfigurations, OpenCOM supports the architecture metamodel by including a component framework model [15]. Here, a CF is a composite component (seen in Figure 14.2) that contains its own internal structure (a graph of components). Each CF supports the architecture MOP described above, which provides reflective operations to inspect and dynamically reconfigure the framework's local component architecture. Notably, the framework exports a health check mechanism (illustrated in Figure 14.2 as the required interface called IAccept); components providing rules about valid dynamic reconfigurations for this particular framework are then plugged into this interface. All reconfigurations are checked against these rules and valid changes are

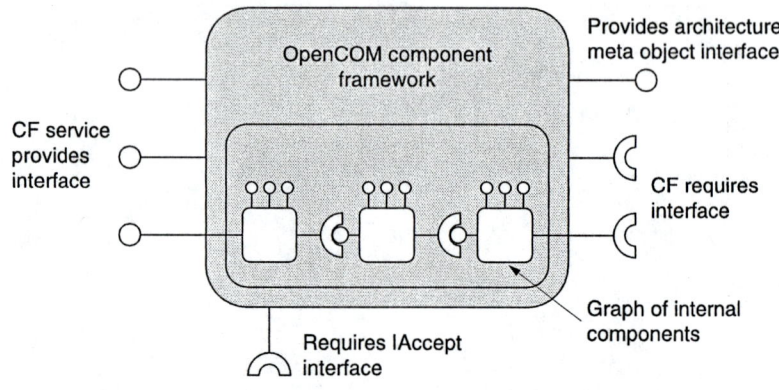

Figure 14.2 The OpenCOM framework model for maintaining system integrity.

accepted, but invalid attempts are prevented and the framework rolls back to its last safe state. This general architecture has been used to implement a small family of reflective middleware platforms, including OpenORB (a reflective CORBA ORB) [11] and ReMMoC [15].

DynamicTAO

DynamicTAO [16] is a reflective CORBA ORB built as an extension of the TAO middleware platform [17]. The ACE ORB (TAO) is a portable, flexible, extensible, and configurable ORB that conforms to the CORBA standard and utilizes the strategy design pattern to encapsulate different aspects of the ORB internal engine. In particular, TAO contains a configuration file that specifies the strategies the ORB uses to implement aspects such as concurrency, request demultiplexing, scheduling, and connection management. When the ORB is initiated, the configuration file is parsed and the selected strategies are loaded. DynamicTAO extends TAO to support runtime reconfiguration; this is achieved by keeping an explicit representation of the ORB internal components and of the dynamic interactions among them (this is identified as the metalevel). This reification allows the ORB to change its own specific strategies without having to restart its execution; this process is managed by elements known as *component configurators*, the role of which is to maintain the dependencies between a component and other system components. For example, each instance of the ORB contains a customized configurator called the *TAOConfigurator*, which contains hooks to which implementations of dynamicTAO strategies are attached. Example strategies are scheduling strategies, security strategies, and monitoring strategies; to change the current scheduling policy in place, the mounted strategy is removed and a new one is inserted.

The meta level of dynamicTAO presents a MOP that supports three key capabilities:

- Components can be transferred across the distributed system, so components not currently available on the local system can be fetched from remote repositories.
- Modules encapsulating different elements of the behavior of the ORB can be loaded and unloaded, which allows behavior to be added and removed from the middleware.
- The ORB configuration state can be inspected and modified dynamically to support dynamic adaptation of the internal ORB engine.

Four Complementary Case Studies

Overview

In the introduction, we described five examples of middleware adaptation that improves support for mobile application development and deployment. In this section, we examine mobile middleware solutions that have explicitly utilized reflective techniques to support these adaptation types. We first investigate the Universal Interoperable Core, which tackles device heterogeneity and middleware heterogeneity using configurable multi-ORB middleware personalities. Its successor, ExORB, then extends this solution to provide third-party software reconfiguration techniques in the face of failure and context changes. Third, ReMMoC offers improved techniques for overcoming device and middleware heterogeneity in dynamically changing environments; a wider range of middleware personalities is made available and the middleware supports dynamic self-reconfiguration. Finally, CARISMA (Context-Aware Reflective Middleware System for Mobile Applications) concentrates on responding to general context changes and resource fluctuations. Notably, these solutions are complementary in nature. We describe later combinations of these techniques for a more comprehensive approach to adaptation; for example, ReMMoC and CARISMA can usefully be combined to provide all five of the adaptation types defined in the introduction [18].

Universal Interoperable Core

The Universally Interoperable Core (UIC) [19] is a reflective middleware, the design of which is based on the previously described dynamicTAO architecture. The primary goal of UIC is to provide a tailorable middleware personality that can be changed from one service interaction type to

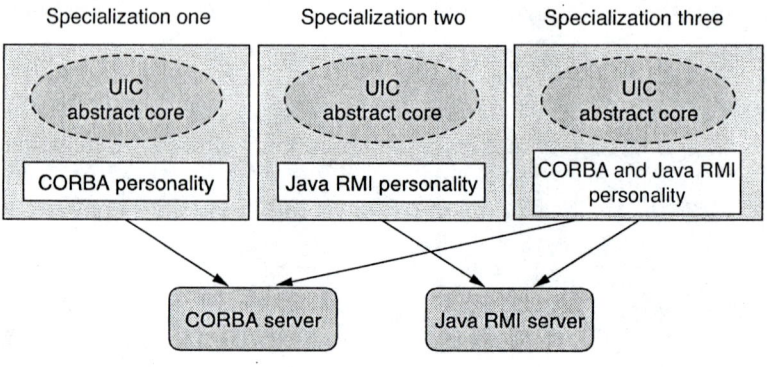

Figure 14.3   Specializing UIC personalities.

another. In one instance, the middleware can be configured to provide a CORBA ORB personality to interact with other CORBA-based applications, whereas an alternative configuration on another device may implement a Java RMI or Simple Object Access Protocol (SOAP)-based personality. Furthermore, to tackle device heterogeneity, the design of the platform is driven by the principle of "what you need is what you get" [19]. UIC determines that existing middleware platforms contain all possible functionality, even if the application only uses a subset; this is not suitable for devices with limited resources. UIC, therefore, provides more fine-grained configurability; for example, on devices with limited resources only a personality supporting CORBA client requests is configured, whereas on resource-rich devices a complete ORB implementation is configured.

At its core, UIC provides a skeleton of abstract components that form the base architecture. To enable the system to have the properties of particular middleware platforms (e.g., CORBA client, CORBA client and server, or SOAP server), components are dynamically added to specialize the abstract components. Hence, a UIC *personality* is a particular instance of the UIC obtained after this process of specialization, as illustrated in Figure 14.3. It can be seen from the diagram that personalities can be classified as single personality or multi-personality. A single personality interacts with a single middleware platform (e.g., the CORBA personality interacts only with a CORBA server), and a multi-personality UIC can interact with more than one platform at the same time (e.g., the CORBA and Java RMI personalities can interact with both CORBA and Java RMI servers).

The UIC personalities can be configured either statically or dynamically. In static configurations, personalities are built at compile time by statically assembling all the components together. The result is a single, fixed personality. In dynamic configurations, personalities are a collection of dynamically loadable libraries that can be fully reconfigured at run time.

The main benefit of the dynamic configuration is the ability to modify the architecture of the personalities dynamically without affecting the applications, but with added memory footprint.

ExORB

ExORB [20] extends the UIC approach further, examining in particular middleware for cellular telephones. Such devices require a middleware infrastructure to make it possible for device carriers to configure new software, upgrade existing software, and repair software bugs without manual intervention (which may otherwise involve the user returning the device). To address this, ExORB employs a new technique called *externalization* to explicitly describe the state, logic, and component architecture of the middleware platform at runtime; notably, this information is available remotely, thus allowing the software configuration to be controlled by a third party. The downside of such an open approach is that it must be designed to prevent malicious attacks on individual devices. ExORB is implemented using a technique referred to as Dynamically Programmable and Reconfigurable Software (DPRS), which gives rise to the following three abstractions:

- *Micro building blocks* (MBBs) are the smallest addressable functional unit in the system (i.e., a single operation or method). The local state of the MBB is held separately in a system-provided storage area. When an MBB is replaced, the state need not be transferred to the new unit; rather, it simply accesses the existing state.
- *Actions* specify the order in which MBBs execute; hence, they define the system logic. An action in DPRS is a deterministic directed graph, the nodes of which are the operational units and the edges the execution transitions.
- *Domains* aggregate collections of related MBBs. These store both the list of building blocks and the corresponding list of actions, plus the localized state of the domain; hence, collections of building blocks can be treated as single units (to be suspended, resumed, inserted, and removed).

The ExORB middleware demonstrates the capabilities of this abstraction to support *configurability*, *updateability*, and *upgradeability*. The middleware itself implements a multi-ORB personality (like UIC). In one typical configuration, the system offers an IIOP personality and an XML–RPC protocol personality; this configurations consists of 28 MBBs grouped into 11 domains that can be tailored and changed at runtime. In addition, the middleware can be customized to support client-side functionality only (domains supporting service-side behavior are not included), or an existing protocol implementation can be replaced by an optimized version. These operations are supported by the ability to externalize and adapt the domains within the implementation. More fine-grained reflection can be used to update or correct errors in the middleware. An example of this is that an MBB producing the IIOP header can be replaced if it begins to produce faulty messages or if an optimized operation is available. Such fine-grained reflection also supports the evolution of the software (where new behavior is added later). ExORB allows interceptor-like behavior to be introduced in the middleware; that is, additional operations can be invoked before and after sending the ORB requests (e.g., interceptors for encrypting and decrypting message buffers evolve the platform with new security features). The interception behavior is achieved by adding a new component implementing the additional behavior to the externalized structure and then simply updating the action logic to ensure that this operation is called before and after the ORB's "Send MBB."

ReMMoC

The ReMMoC framework [15] uses the Lancaster philosophy to overcome a specific problem in the mobile computing domain. When mobile devices move from location to location they do not know how the services and applications with which they will inevitably interact have been implemented. Mobile applications thus cannot easily be written assuming a single middleware standard because of the dynamic nature of interaction. Moreover, it is likely that many different middleware implementations will be encountered. This is often referred to as the *middleware heterogeneity problem* [21]. ReMMoC addresses this issue by presenting a reflective middleware to adapt dynamically between different middleware personalities at runtime; this ensures that a mobile application can continue operating in potentially many unknown locations.

As previously noted, UIC and ExORB initially identified the problem of middleware heterogeneity and promote a multi-personality ORB to address this problem. This approach, however, is limited in three respects: (1) an ORB-based middleware cannot cope with the diversity of interaction paradigms used in the mobile environment, including publish–subscribe, tuple spaces, and data-sharing; (2) a higher level abstraction to hide the application developer from middleware heterogeneity is not provided; and, most importantly, (3) third-party reconfiguration is not sufficient to handle changing interaction types as the device moves from location to location. The middleware must dynamically adapt itself based on information retrieved about the current environmental context; that is, it must

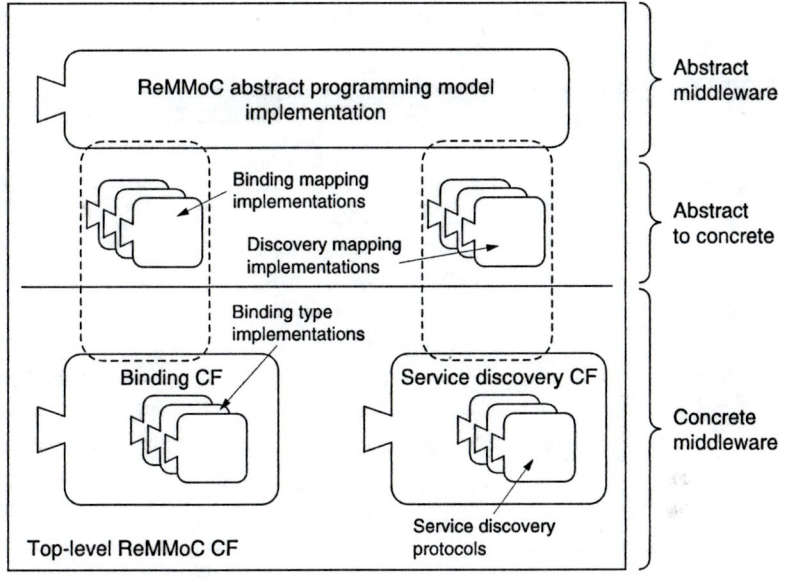

Figure 14.4   The ReMMoC framework for reconfigurable resource discovery and interaction.

find out which interaction type is required before reconfiguring to that type. To address these issues, ReMMoC is based on three fundamental principles: reconfigurable resource discovery, reconfigurable interaction, and a suitable common interaction abstraction (in this case, Web services).

The discovery framework illustrated in Figure 14.4 (as part of the overall ReMMoC architecture) is responsible for *reconfigurable resource discovery.* The role of the service discovery framework is to perform lookup operations across a set of different discovery protocols; for example, in one location, a tourist guide service advertised using Service Location Protocol (SLP) may be found, and in the next location the same service type may be found advertised using Universal Plug and Play (UPnP™). To meet this goal, the framework has two key characteristics:

■ The framework automatically mirrors the current environmental conditions (i.e., which discovery protocols are in use). If UPnP and SLP are both being used to advertise local resources, then both personalities are configured into the framework. Reconfiguration is based on a "cycle-and-see" approach; every location change forces the framework to discover local context information about discovery protocols in use, and a test for each known protocol determines whether or not its personality should be configured. In addition, the framework monitors the environment to detect for new protocols that have emerged, again forcing the protocol personality to be configured dynamically.

■ A single resource lookup abstraction allows lookup requests to be executed in parallel over all personalities configured in the framework; the found resources are then returned in a common format.

The principal function of the binding framework is to provide a configurable and dynamically *reconfigurable interaction* mechanism that allows mobile clients to bind and interoperate with application services implemented upon particular interaction paradigms (e.g., remote method invocation, publish–subscribe, asynchronous messaging). Reconfiguration of the binding framework is again controlled by the middleware itself. ReMMoC receives information from the service discovery framework to drive the correct configuration; that is, it finds a SOAP service and then reconfigures to SOAP. Fine-grained reconfiguration is also supported; for example, when a mobile device switches from an infrastructure-based wireless network to an *ad hoc* network, the lookup and interaction protocols can be reconfigured accordingly. Both SLP and the event subscriber personality utilize an Internet Protocol (IP) multicast component; however, this can be replaced by an epidemic-style multicast component (for example) that operates by intelligently flooding the *ad hoc* network.

Using dynamic reconfiguration to mirror protocols in the current environment does not protect the application developer from middleware heterogeneity. A programmer using this technology would need to explicitly program for each dynamic change; for example, when the discovered service is of type SOAP, a SOAP RPC invocation must be made, then when an event publisher is found the client must subscribe for events. ReMMoC promotes a *common interaction abstraction*, which has the following property: Applications invoke operations on abstract mobile services; that is, ReMMoC follows the Web services concept of separating the description of the behavior of a service from its interaction protocol. ReMMoC takes the information from an application programming interface (API) programming Web services and maps this onto the concrete binding and discovery protocols (as seen in Figure 14.4). Further details on this mapping process can be found in Grace [21].

CARISMA

The CARISMA platform [22], developed at University College London, is a reflective, policy-based framework for adapting the behavior and operation of an underlying middleware platform; it utilizes the XMIDDLE datasharing platform [23]. The work concentrates on the important issue of how context information (e.g., device context, such as power or memory; external context, such as network connection, bandwidth, or location) affects the performance of a mobile application and how middleware adaptation can be performed to maintain the best level of performance in the face of these changes.

In a particular context, an application may require the middleware to behave in a particular way; for example, an image-processing application may ask to display pictures in black and white rather than color when the battery power is low or compress the image before sending it across the network. Each application describes its adaptation requirements in an application profile that contains associations among the services that the middleware delivers, the policies that can be applied to deliver the services, and the context configurations that must hold in order for a policy to be applied.

Figure 14.5 illustrates the general layout of application policies and an example policy for a message sending service. For the previously described example, the middleware service "Messaging Service" has two policies to select from: the "PlainMessage" policy, with a context of greater than 40 percent capacity network bandwidth available, and a "CompressedMessage" policy, with a context of less than 40 percent network bandwidth available. Each time the application invokes the "Messaging Service," CARISMA consults the required profile and then selects the appropriate policy, based on the current context (bandwidth capacity).

Every application submits its policy to the middleware upon initialization; however, given the dynamic nature of mobile applications, it is expected that the policies themselves must be changed dynamically. CARISMA provides a reflective API that allows introspection and dynamic reconfiguration of this policy. CARISMA also manages the end-system resources of the mobile device being utilized by competing mobile applications. Different policies have different middleware requirements; for example, one policy may require increased throughput, but a competing policy may want to reduce battery power (these goals are in conflict with one another). Conflicts of this type are resolved by an auction protocol. Each application submits a bid for resource use citing nonfunctional concerns (e.g., security, performance, availability). The resource goes to the highest bidder. In a similar fashion, reflection allows the application to dynamically change the nonfunctional properties of its bid if its requirements dynamically change.

CARISMA promotes the use of higher level policies that control the behavior of middleware based on context metadata. Changes to context information alter the middleware behavior; similarly, dynamic changes to the policies themselves will affect the runtime behavior of the middleware. CARISMA cannot be singularly classified as reflective middleware; rather,

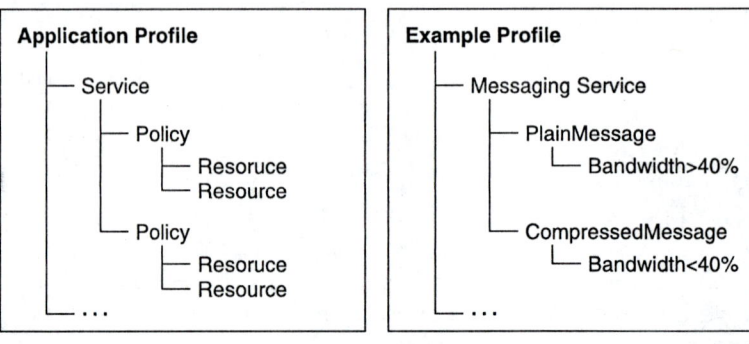

Figure 14.5 A CARISMA application profile. (From Capra, L. et al., *IEEE Trans. Software Eng.*, 29(10), 929–945, 2003. With permission.)

reflection is used to inspect and alter the system policies (this in turn alters the middleware behavior). It can effectively supplement existing reflective middleware solutions, adding support to dynamically alter the strategies for dynamic reconfiguration; for example, it can extend ReMMoC to support additional context-based reconfigurations, handle conflicting reconfiguration requests, and allow the middleware to evolve [18]. As an example, a policy could be added dynamically describing the following reconfiguration: When the context changes from a wireless infrastructure network type to an *ad hoc* interaction type, the binding framework replaces the IP multicast component with a network flooding component. Notably, the reflective properties of CARISMA ensure that the behavior of the middleware is extensible and evolvable to handle newly defined context-based reconfigurations.

In addition, a number of policy-based approaches to dynamic adaptation are not reflective in nature but could equally be utilized to manage dynamic reconfigurations in mobile environments or be integrated with adaptive middleware solutions. For example, Ponder [24] is a language for specifying policies for management and security in distributed systems. The Lancaster Context Architecture [25] provides mechanisms to resolve conflicts between multiple policies for mobile computing applications. Puppeteer [26] uses policies to manage resources such as battery power on mobile hosts; Puppeteer policies define how media presentation applications present smaller parts of the document or display them in lower resolutions when fewer resources are available. Odyssey [27] is an extension to the Coda file-sharing system, designed to support access to shared information from mobile hosts; the application specifies the policies to adapt the behavior of the platform in terms of utilization of system resources.

Dynamic Aspect-Oriented Programming

Introduction to Aspect-Oriented Programming

Aspect-oriented programming (AOP) [28] is a software engineering approach designed to tackle the problems of tangled code; that is, the basic functional implementation of a component becomes tangled with additional code for features such as security, persistence, logging, and monitoring. Developers often implement these features in an *ad hoc* manner across the system, which leads to increased system development, debugging, and evaluation time because of the increased system complexity. AOP supports the concept of *separation of concerns* to counter this problem in such a way that individual concerns such as security and monitoring code are not implemented within the base code; instead, these are each implemented as individual *aspects*, which are pieces of code that can then be woven into the base code at compile time. A single monitoring aspect, for example, can be woven into every base component of the system. Developers utilize *point cuts*, which identify positions in the code where these aspects should be attached.

Dynamic aspect-oriented programming promotes the same benefits as AOP, but the aspects are woven at *runtime* rather than compile time. This is an important technique for mobile computing middleware. In particular, AOP provides a series of techniques to enable the programmer to reason at a higher level about issues that cross-cut the structure of a system, with dynamic AOP allowing such concerns to be adapted to suit the current context. Security is a classic example of such a cross-cutting concern; aspects can be used to insert security policies and procedures at various points in the base-level code which can then be modified at runtime to reflect the current operating environment (e.g., type of network). This higher level view also promotes correctness of the resultant software and supports reasoning about interactions between aspects of the system structure. The approach is complementary to the aforementioned studies of reflection. Dynamic AOP supports higher level reasoning about software structure but this inevitably relies on underlying reflective mechanisms in the underlying implementation.

In the remainder of this section, we examine two techniques to dynamically insert aspects into middleware systems: *invasive* and *noninvasive*. Invasive dynamic AOP breaks the component architecture by weaving code within the base component implementation (i.e., behind the interface contracts), whereas noninvasive approaches utilize the component interfaces as point cuts, and these aspects are implemented as interceptors on the interfaces. The former approach tends to rely on code rewriting techniques, such as bye-code rewriting as supported by tools such as Javassist [29]. In contrast, noninvasive approaches tend to rely on behavioral reflection mechanisms, such as interception, to dynamically introduce or remove aspects.

MIDAS/PROSE: Invasive Dynamic AOP

MIDAS is a middleware layer developed at ETH Zurich for providing runtime extensions to mobile applications. Popovici et al. [30] observed that mobile applications must adapt and extend themselves to their current environment conditions; for example, a PDA may interoperate with application services from different mobile locations. In these different locations, however, different functionalities may be required to interact with the local services; for example, encryption layers must be added to allow interaction to happen at one location, whereas billing modules must be included to pay for services at another location. The applications cannot carry every piece of possible code around with them nor can the developer plan for every interoperation eventuality. It is the role of MIDAS, therefore, to add the functional extensions to the developer's basic code implementation (in this case, service interoperation) at runtime. When a function extension is required it is downloaded to the MIDAS middleware, which then dynamically weaves the code into the base application at runtime.

MIDAS is underpinned by PROSE (PROgrammable extenSions of sErvices), the purpose of which is to provide a dynamic AOP system; that is, PROSE provides the capability to do runtime weaving of functional extensions. PROSE and MIDAS are both implemented in Java, and PROSE relies on just-in-time (JIT) compilation (i.e., the original byte code is converted to native code at execution time to support efficient operation) to perform the dynamic insertion of aspects. Hence, PROSE instructs the JIT compiler to include the additional actions (the aspect code described as an extension) when converting from byte code to native code. In PROSE, potential point cuts are described in the base code by the developer. When the JIT compiler detects these point cuts it can add the new behavior in the native code. PROSE, then, promotes invasive dynamic AOP; that is, the internal implementation of a component is changed at runtime.

Lasagne: A Noninvasive Dynamic AOP

Lasagne [31] is an AOP framework that supports the dynamic customization of middleware platforms and distributed services. In Lasagne, aspects are introduced dynamically at system runtime in a noninvasive manner, and the selection of which aspects to compose is context sensitive. Core functionality can be woven with two categories of system extensions: new service functionality or nonfunctional services, such as authentication or persistence.

The aspect-oriented approach of Lasagne is based on *extensions*, where an extension encapsulates a slice of behavior that updates multiple components at the same time. For example, an authentication extension may cross-cut a number of components involved in a client–server request (potentially distributed). Only when the extension is applied across the complete system has the new nonfunctional service been dynamically added to the system. The extension is implemented by a set of *wrappers*, where a wrapper is the per-instance implementation of the aspect that is to be applied at each component. The dynamic insertion of wrappers is noninvasive; the point cut is outside the base component interface. Lasagne dynamically alters the message flow of the system to be directed to the wrappers before the component interface (using a technique similar to message interception).

Finally, Lasagne uses context information to decide when extensions should be dynamically employed. Contextual properties are defined and attached to specific service functionalities, and interceptors attached to the components of the middleware then inspect the values of these properties to decide which extensions should be executed. Furthermore, this decision is propagated with the message flow of the basic service to enable consistent, systemwide dispatching to all the wrappers of a selected extension.

Future Research Challenges

The case studies presented in the previous two sections have shown that reflection and the related technique of dynamic AOP offer promising solutions for providing principled adaptation of middleware in mobile computing environments; however, many questions still remain unanswered in this field. Open systems are vulnerable to many types of security attacks; therefore, what security measures should be taken to make reflective middleware secure? Reflection is also expensive in terms of system performance due to the storage of meta-information and the time involved in configuration and reconfiguration. How can reflection be made more efficient for devices with limited resources? Furthermore, we also see two fundamentally important, future directions where reflective middleware can be extended to support a richer set of mobile applications. We now discuss these topics in more detail.

Coordinated Adaptation

As can be seen, reflective middleware provides a set of underlying mechanisms that can then be supplemented by higher level statements of policy. In practice, the scope of such policies has been rather limited (e.g., focusing on the behavior of one node and also most likely to one layer or aspect of the software on that node); however, a large class of mobile applications involves collaboration between groups of mobile users. Examples include peer-to-peer (P2P) data sharing, shared workspaces, and multimedia conference applications. Here it is not enough to perform local adaptation. In our view, coordinated reconfiguration of middleware across entire systems is required to make adaptations that are beneficial to the operation of the application as a whole. We now present two examples where coordinated reconfiguration between collaborating nodes will be beneficial:

- *A multimedia conferencing application* — A common media filter may have to be agreed upon and applied so all members of the multicast can receive and view video frames if the sender changes the filter.
- *A P2P messaging application* — Suppose a set of mobile nodes is participating in a group messaging application based on a group multicast service. If a message sender changes from sending text messages to sending picture messages or video messages, then the members of the group must change to be able to receive the streaming data. The interaction type has changed from message based to streaming based. A local change at the sender only would not affect the remainder of the multicast group, which would be unable to receive the new messages.

As we see it, there are two important dimensions to coordinated reconfigurations: *horizontal* and *vertical*. The horizontal dimension refers to the various levels or layers that form the architecture of local instances of a middleware implementation. A reconfiguration in this dimension describes the changes that must be made across the local architecture to accommodate the new behavior; for example, a reconfiguration from event-based messaging to streaming media on the local host will affect different layers of the implementation, such as how components interact with a network transport component. Similarly, if we were to add new features to the middleware (e.g., security), then that would affect elements across the middleware implementation.

The vertical dimension refers to the coordinated agreement between peer-to-peer collaborators. This reconfiguration will affect the middleware service or interaction type in which the nodes themselves are participating; for example, this could involve a multicast stream of multimedia data, a virtually synchronous group communication service, a pool of shared resources, or advertisement and discovery of application services. Reconfiguration across the vertical dimension ensures that each member of the collaboration maintains the same consistent view of the middleware service (e.g., the same interaction method as described in the P2P messaging application above). We believe that reflective technologies can be employed to make reconfigurations across the vertical dimension. Rather than reifying the middleware on an individual node, information about the middleware service across nodes should be made available. Policies must be extended to support decision making across the group of nodes as a whole, and consensus mechanisms must be provided to commit dynamic reconfigurations across each local node of the participating group.

Autonomic Computing

Experience from the mobile community (and, indeed, other communities) has shown that the development of adaptive applications is complex. Although reflection does help by providing a strong element of separation of concerns, the underlying complexity of controlling adaptation does not go away. It is also not feasible for the middleware developer to program for every possible reconfiguration that may be required over the lifecycle of the middleware. Furthermore, third-party evolution (e.g., as provided by ExORB) is similarly often infeasible due to the complexity confronting the human developer, and in many mobile environments such update facilities are unavailable. This has led researchers to consider the extent to which platforms can be self-managed. More specifically, the field of *autonomic computing* has emerged, promoting a vision of software that is able to reconfigure itself, heal itself, and optimize itself [32]. It is an obvious and yet important fact that autonomic computing requires a level of openness; hence, it is very interesting to consider a potential marriage of autonomic techniques and reflective middleware platforms.

An example of autonomous middleware in the mobile environment is as follows. As mentioned above, ReMMoC supports different discovery protocols; hence, it can find services advertised by different mechanisms. In our current approach, the cycle-and-see philosophy would be used to discover the relevant protocol in use. This search, however, can be optimized. In particular, when a location has been visited, it is likely that on returning to that location the same service discovery protocol will be in use; therefore, the middleware can learn to optimize itself to ensure that as the user moves toward such a location the middleware can be reconfigured appropriately. In more general terms, middleware is ideally placed to learn about environmental conditions and can define reconfigurations that provide the best level of service in these conditions based on past experiences and not on fixed policies.

Summary

In this chapter, we have focused on the need to cope with dynamic change in mobile computing environments. These environments are inherently dynamic in nature, and fixed middleware implementations cannot cope with change. We have argued that dynamic, adaptive middleware is ideally placed to respond to the frequent changes in mobile environments and that reflection offers principled techniques to develop such middleware. To emphasize this point, reflective middleware solutions are becoming more prevalent in the mobile computing domain. To keep this discussion concise, we have presented four key individual, but complementary, reflection middleware solutions; these have illustrated how middleware adaptation can benefit mobile computing applications. The authors believe that future platforms in this domain will build upon the complementary and fundamental properties provided by these platforms to offer richer support to mobile application developers.

In addition, we have investigated the emerging role of dynamic AOP in the realm of mobile computing and mobile middleware. This technique, although not labeled as reflective, is complementary in nature, and we have identified the inherent similarities between dynamic AOP and behavioral reflection. Our discussions of two solutions for invasive and noninvasive dynamic AOP revealed that similar dynamic adaptations, as provided by their reflective counterparts, can be made (e.g., the addition of new functional implementations to existing base implementations). In addition, dynamic AOP offers improved software engineering techniques to maintain the separation of concerns across collaborating components (i.e., the complexity of reflective programming is hidden from the system developers).

Finally, we have identified coordinated reconfiguration and autonomous computing as two of the fundamental research challenges in the domain of adaptive mobile middleware. We believe these in turn will produce new techniques for applying reflection across distributed middleware instances and in turn produce more powerful and sophisticated middleware platforms.

References

[1] Haahr, M., Cunningham, R., and Cahill, V., Towards a generic architecture for mobile object-oriented applications, in *Proc. of the 2000 IEEE Workshop on Service Portability and Virtual Customer Environments*, San Francisco, CA, December, 2000, pp. 91–96.

[2] Campadello, S., Helin, H., Koskimies, O. and Raatikainen, K., Wireless Java RMI, in *Proc. of the 4th International Enterprise Distributed Object Computing Conf.*, Makuhari, Japan, September, 2000, pp. 114-123.

[3] Murphy, A., Picco, G., and Roman, G., LIME: a middleware for logical and physical mobility, in *Proc. of the 21st Int. Conf. on Distributed Computing Systems (ICDCS-21)*, Phoenix, AZ, April, 2001, pp. 524–533.

[4] Cugola, G., Di Nitto, E., and Fuggetta, A., The JEDI event-based infrastructure and its application to the development of the OPSS WFMS, *IEEE Trans. Software Eng.*, 9(27), 827–850, 2001.

[5] Kiczales, G., des Rivières, J., and Bobrow, D., *The Art of the Metaobject Protocol*, MIT Press, Cambridge, MA, 1991.

[6] Smith, B.C., Reflection and Semantics in a Procedural Programming Language, Ph.D. thesis, Massachusetts Institute of Technology, Cambridge, MA, 1982.

[7] Maes, P., Concepts and experiments in computational reflection, in *Proc. of the ACM Conf. on Object-Oriented Programming Systems, Languages and Applications (OOPSLA'87)*, Vol. 22, SIGPLAN Notices, ACM Press/Addison–Wesley, Boston, MA, 1987, pp. 147–155.

[8] Java Reflection API, http://java.sun.com/j2se/1.3/docs/guide/reflection/index.html.

[9] Gowing, B. and Cahill, V., Meta-object protocols for C++: the iguana approach, in *Proc. of Reflection'96*, San Francisco, CA, April, 1996, pp. 137–152.

[10] Chiba, S., A Metaobject protocol for C++, in *Proc. of the ACM Conf. on Object-Oriented Programming Systems, Languages, and Applications (OOPSLA'95)*, Austin, TX, October, 1995, pp. 285–299.

[11] Blair, G., Coulson, G., Andersen, A., Blair, L., Clarke, M. et al., The design and implementation of Open ORB 2, *IEEE Distributed Syst. Online*, 2(6), 2001.

[12] Cazzola, W., Savigni, W., Sosio, A., and Tisato, F., Rule-based strategic reflection: observing and modifying behaviour at the architectural level, in *Proc. of the 14th IEEE Int. Conf. on Automated Software Engineering (ASE'99)*, Cocoa Beach, FL, October, 1999, pp. 287–290.

[13] Szyperski, C., *Component Software: Beyond Object-Oriented Programming*, ACM Press/Addison-Wesley, Boston, MA, 1998.

[14] Clarke, M., Blair, G., Coulson, G., and Parlavantzas, N., An efficient component model for the construction of adaptive middleware, in *Proc. of Middleware 2001*, Heidelberg, Germany, November, 2001, pp. 160–178.

[15] Grace, P., Blair, G., and Samuel, S., A reflective framework for discovery and interaction in heterogeneous mobile environments, *ACM SIGMOBILE Mobile Comput. Comm. Rev.*, 9(1), 2–14, 2005 (special section on discovery and interaction of mobile services).

[16] Kon, F., Roman, M., Liu, P., Mao, J., Yamane, T. et al., Monitoring, security, and dynamic configuration with the dynamicTAO reflective ORB, in *Proc. of Middleware 2000*, New York, April, 2000, pp. 121–143.

[17] Schmidt, D. and Cleeland, C., Applying patterns to develop extensible ORB middleware, *IEEE Comm. Mag.*, 37(4), 54–63, 1999 (special issue on design patterns).

[18] Capra, L., Blair, G., Mascolo, C., Emmerich, W., and Grace, P., Exploiting reflection in mobile computing middleware, *ACM SIGMOBILE Mobile Comp. Comm. Rev.*, 6(4), 34–44, 2002.

[19] Roman, M., Kon, F., and Campbell, R., Reflective middleware: from your desk to your hand, *IEEE Distributed Syst. Online*, 2(5), 2001.

[20] Roman, M. and Islam, N., Dynamically programmable and reconfigurable middleware services, in *Proc. of the 5th ACM/IFIP/USENIX Int. Conf. on Middleware*, Toronto, Canada, November, 2004, pp. 372–396.

[21] Grace, P., Overcoming Middleware Heterogeneity in Mobile Computing Applications, Ph.D. thesis, Lancaster University, Lancaster, U.K., March, 2004.

[22] Capra, L., Emmerich, W., and Mascolo, C., CARISMA: context-aware reflective middleware system for mobile applications, *IEEE Trans. Software Eng.*, 29(10), 929–945, 2003.

[23] Mascolo, C., Capra, L., Zachariadis, S., and Emmerich, W., XMIDDLE: a data-sharing middleware for mobile computing, *Wireless Pers. Comm.*, 21(1), 77–103, 2002.

[24] Lymberopoulos, L., Lupu, E., and Sloman, M., An adaptive policy based framework for network services management, *J. Network Syst. Manage.*, 11(3), 277–303, 2003 (special issue on policy-based management).

[25] Efstratiou, C., Friday, A., Davies, N., and Cheverst, K., A platform supporting coordinated adaptation in mobile systems, in *Proc. of the 4th IEEE Workshop on Mobile Computing Systems and Applications*, Callicoon, NY, June, 2002, pp. 128–137.

[26] Flinn, J., de Lara, E., Satyanarayanan, M., Wallach, D., and Zwaenepoel, W., Reducing the energy usage of office applications, in *Proc. of Middleware 2001*, Heidelberg, Germany, November, 2001, pp. 252–272.

[27] Satyanarayanan, M., Mobile information access, *IEEE Pers. Comm.*, 3(1), 26–33, 1996.

[28] Kiczales, G., Lamping, J., Mendhekar, A., Maeda, C., Videira Lopes, C. et al., Aspect oriented programming, in *Proc. of the 11th European Conf. on Object-Oriented Programming (ECOOP'97)*, Jyväskylä, Finland, June, 1997, pp. 220–242.

[29] Chiba, S. and Nishizawa, M., An easy-to-use toolkit for efficient Java byte-code translators, in *Proc. of the 2nd Int. Conf. Generative Programming and Component Engineering (GPCE'03)*, Springer-Verlag, New York, 2003, pp. 364–376.

[30] Popovici, A., Frei, A., and Alonso, G., A proactive middleware platform for mobile computing, in *Proc. of the 4th ACM/IFIP/USENIX Int. Middleware Conf.*, Rio de Janeiro, Brazil, June, 2003, pp. 455–473.

[31] Truyen, E., Dynamic and Context-Sensitive Composition in Distributed Systems, Ph.D. thesis, Katholieke Universiteit Leuven, Belgium, 2004.

[32] Kephart, J. and Chess, D., The vision of autonomic computing, *IEEE Comp.*, 36(1), 41–50, 2003.

Chapter 15

Techniques for Dynamic Adaptation of Mobile Services

John Keeney, Vinny Cahill, and Mads Haahr

CONTENTS

Introduction

This chapter discusses the dynamic adaptation of software for mobile computing. The primary focus of this chapter is on techniques for adapting software as it runs and managing the application of these adaptations. In a mobile computing environment, the need for adaptation can often arise as a result of a spontaneous change in the context of the operating environment, ancillary software, or indeed the user. To exacerbate this problem, if that contextual change is in some way unanticipated, then the required adaptation may itself be unanticipated until the need for it arises. For this reason, this chapter is particularly concerned with supporting adaptations that are completely unanticipated [19]. This chapter discusses reflective and aspect-oriented techniques for dynamically adapting software for mobile computing. Policy-based management is then addressed as a mechanism to control such dynamic adaptation mechanisms. The chapter then introduces the Chisel dynamic adaptation framework, which supports completely unanticipated dynamic adaptation, and provides a case study where Chisel is used with the Architecture for Location-Independent Computing Environments (ALICE), a mobile middleware, to provide a flexible and adaptable middleware framework for mobile computing.

Issues in Dynamically Adaptable Mobile Applications and Middleware

The main difficulty with mobile computing is the inherent scarcity and variability of resources available for use by mobile computers as they move. The primary resource requirement of a mobile device when it is working as part of a distributed system is its network connection, often some form of wireless connection, which, when used by a device that is physically moving, suffers from unanticipated and possibly prolonged disconnections [14]. The reason why this issue is such a major problem for mobile computing is that the applications currently being developed are being built as distributed system applications that do not sufficiently account for these disconnections and reconnections [30].

Middleware for Mobile Computing

"Middleware can be viewed as a reusable, expandable set of services and functions that are commonly needed by many applications to function well in a networked environment" [1]. Traditional middleware systems provide abstractions and shelter applications from the complexities of the underlying environment, communication subsystems, and distribution mechanisms, thereby providing a single view of the underlying environment, as seen in traditional middleware systems such as COM+ [24], Java Remote Method Invocation (RMI) [39], and the Common Object Request Broker Architecture (CORBA™) [25]. A middleware system for mobile computing must be flexible to provide a homogeneous and stable programming model and interface for possibly erratic execution contexts. It is desirable that an adaptable middleware for mobile computing be open, allowing the application and the user to inspect the execution environment and manipulate the application and middleware in a mobile-aware manner, using application-specific and user-specific semantic knowledge.

Difficulties with Applications and Middleware for Mobile Computing

As environmental conditions change to unprecedented values unknown to the application designer, the middleware that provides abstractions for these environmental resources must dynamically adapt to support the applications that run on top of that middleware. As stated, one of the primary services provided by middleware is the ability to supply network communications services as these resources change. A key requirement for middleware for mobile computing is the ability to adapt to drastic changes in available resources, especially network connection availability [15]. The characteristics of the available connections can range from an inexpensive, very-high-bandwidth, low-latency connection, such as a high-speed wired local area network (LAN) connection, to a very expensive, low-bandwidth, high-latency connection such as a Global System for Mobile Communications (GSM) connection, where each communication protocol used may make use of different communication models and addressing modes.

Mobile computing applications should also be able to handle periods of disconnection, supported by the middleware underneath. The difficulties that are associated with such a range of connection characteristics are further compounded by the fact that these characteristics can change in an unanticipated manner. For example, these disconnections occur when the device moves out of range for wireless connections or an interface device is suddenly disconnected, as seen when a user suddenly disconnects the device from a synchronization cradle or removes a networking device currently in use.

A further issue with such a varied collection of communication technologies that can be leveraged for mobile computing is that the user may not wish to make full use of the available resources in an eager or greedy manner to maintain data connectivity; for example, even if a General Packet Radio Service (GPRS) connection is available, this connection is generally much more expensive than available wireless connections. A further example is the case where, although currently disconnected but with connections available, the user may be aware that a less expensive or more convenient connection resource will soon be available — something that cannot be anticipated in a generalized manner by the adaptable middleware platform. For these reasons, it is imperative that the added potential of the user's own resources, preferences, and intelligence is exploited.

Reflective Middleware

Principals and Key Ideas

A reflective computational system is one that reasons about its own computation. This is achieved by the system maintaining a representation (metadata) of itself that is causally connected to its own operation, so if the system changes its representation of itself then the system adapts [22]. With behavioral reflection in an object-oriented system, the reflective system reasons about and adapts its own behavior by associating meta objects with the objects in the application, where the meta objects control or adapt the behavior of the application objects [12]. In a reflective system, the communications between the meta objects and base objects takes place through a set of well-defined interfaces, referred to as that Meta Object Protocol (MOP) of the system [20].

Case Studies of Reflective Middleware

Although several reflective middleware frameworks have been discussed in detail in previous chapters, this section discusses two additional reflective systems that target middleware for dynamic adaptation. In addition, a number of systems described later in this chapter make use of reflective techniques but are discussed under a different category.

ACT

ACT [35,36] is a generic adaptation framework for CORBA™-compliant [25] ORBs that supports unanticipated dynamic adaptation. When the ORB is started, ACT is enabled by registering a specific portable request interceptor [25], intercepting every remote invocation request and handing them to a set of dynamically registered interceptors. These dynamically registered interceptors can be added in an unanticipated manner. Rule-based dynamic interceptors allow the request to be redirected to a different source or handed either to a number of local proxy components exporting the same interface as that of the destination server component [35] or to a generic local proxy component [36]. This generic proxy component can also be dynamically created in an unanticipated manner. This proxy in turn can request a rule-based decision-making component, which can incorporate an event service to either perform the invocation or change parameters and forward the request to its original destination or to a different destination. A prototype is described whereby the Quality Objects (QuO) framework [2], an aspect-oriented quality of Service (QoS) adaptation framework for CORBA ORBs, was used with a CORBA-compliant ORB to support completely unanticipated runtime aspect weaving in the ORB. A number of management interfaces were also provided to manage the runtime registration of new rule-based dynamic interceptors and the addition of new rules to these interceptors.

Correlate

Presented by the DistriNet research group at Katholieke Universiteit Leuven, Correlate [16,33,34,40], is a concurrent object-oriented language based on C++ (and later Java) to support mobile agents. It has a flexible runtime engine to support migration and location-independent inter-object communication. Each agent object has an associated meta object that can intercept creation, deletion, and all invocation messages for the object. This system allows nonfunctional aspects of the application to be separated from the application object. The nonfunctional behaviors are designed to be largely application independent; however, independent policy objects can be defined to contain application-specific information to assist in the operation of these meta-level nonfunctional behaviors. The meta-level system was initially used to implement nonfunctional concerns such as real-time operation, load balancing, security, and fault tolerance. Later, this system was used to customize ORBs, using application-specific requirements, as an adaptable graph of meta-level components that could be extended or adapted at runtime.

The application-independent nonfunctional behaviors are implemented as meta-object classes that can interact with the base program to adapt its operation using a message-based MOP. These meta-object classes define a set of possible property values in a policy template. Each application class has an associated singleton policy-class object, which is an instantiation of these templates and contains application-specific information. These singleton policy-class objects are consulted by the meta level before performing the nonfunctional behaviors of the application, allowing the operation to be customized in an application-specific manner.

This policy system, however, is limited because policy templates are imposed at the time the meta program is written. These templates, written in a declarative language, must fully define what possible customizations an application may require at a later stage. The policies, also written in the same declarative manner, select values for template properties according to the application classes with which they are associated. These templates cannot be changed, so adaptation in response to unanticipated requirements cannot be fully handled.

Policies are written before runtime by a system integrator, and these policies are then translated to code and compiled with the application and so cannot be changed at runtime. Unanticipated forms of dynamic adaptation cannot be achieved in this architecture as the meta-level programmer and template designer need complete *a priori* knowledge of the possible changes in context values that may occur; also, the set of customizations from which the meta level can choose is fixed at compile time.

Discussion

The use of reflective mechanisms for adaptable middleware is an old yet active research area. The main issue with reflection for the adaptation of middleware lies not with the use of reflection to adapt the structure, behavior, or architecture of middleware but with how the application of those adaptations is controlled and managed. This issue is of particular importance if the adaptation is required in response to an unanticipated change in the state, requirements, or context of the users, applications, or environment.

Aspect-Oriented Approaches to Dynamic Adaptation

Principals and Key Ideas

Aspect-oriented programming (AOP) [13,21] is a programming methodology that allows cross-cutting concerns to be declared as "aspects." A cross-cutting concern is a property or function of a system that cannot be cleanly declared in terms of individual components, because the application of the cross-cutting concern must be scattered or distributed across otherwise unrelated components. AspectJ [42], the *de facto* standard for AOP, introduced the concept of an aspect as a language construct used to specify a modular unit to encapsulate a cross-cutting concern, which is then woven into the application code at compile time. An aspect is defined in terms of *point cuts* (a collection of join-point locations within the application code where the aspect should be woven and conditional contextual values at those join points), *advice* (code executed before, after, or around a join point when it is reached), and *introductions* (Java code to be introduced into base classes) [42].

Aspect-oriented programming supports the production of these aspects in a manner that is separate from or *oblivious* to the application components [13] into which the aspects are later incorporated or woven at a specified or quantified set of join points. *Obliviousness*, one of the key components of AOP, refers to the degree of separation between the aspects of the system and how they can be developed independently without preparation, cooperation, or anticipation. Most AOP systems support weaving before runtime, but newer dynamic AOP systems (e.g., Wool and PROSE) described in this section allow aspects to be woven at load time or runtime, thereby allowing the incorporation of aspects into base programs to remain unanticipated until load time or runtime.

Case Studies of Dynamic Aspect-Oriented Systems

Wool

Wool [38] is a dynamic AOP framework that uses a hybrid aspect weaving approach utilizing both the Java Platform Debugger Architecture (JPDA) and the Java HotSwap mechanism [39]. Because JPDA supports remote activation of breakpoints at runtime, join-point hooks in the form of debugging breakpoints can be dynamically set from outside of the application. A point cut may be made up of a number of these hooks. Each aspect specifies a point cut and a set of advices to be executed when one of the point cut's join points (represented as breakpoints) is reached.

New aspects can be serialized and sent to the target the Java Virtual Machine (JVM) for weaving at any point cut. In one approach, when a join point is encountered, the inserted breakpoint redirects the operation to the Wool runtime component in a manner similar to a debugger, where advices are then executed.

The alternative approach allows the advice to be hotswapped into the application class, thereby improving performance if the join point is encountered repeatedly. This is achieved by using Javassist [7] to rewrite the class, without access to its source code, and having the adapted class replace the original application class using the Java HotSwap mechanism. This also removes the breakpoint, so calls to the debugger are removed; however, this mechanism means that all objects of the woven class will have the adaptation incorporated, so individual objects cannot be adapted. Currently, the aspect programmer must specify in the source code of the aspect whether the advice should be woven by the HotSwap mechanism or by the debug interface, so to achieve good performance the aspect writer should anticipate the access patterns of the point cut of the aspect. Wool does not support adding introductions, but a proposed solution is provided.

PROSE

PROSE (PROgrammable extenSions of sErvices) [26,29] is another dynamic AOP framework for Java that supports runtime aspect weaving. PROSE was originally intended as a framework for debugging or rapid prototyping of AOP systems which could later be completed using compile-time or load-time aspect weaving [29]. This was mainly due to its use of the Java Virtual Machine Debug Interface (JVMDI) [39], which resulted in a large performance penalty. A later version of PROSE [26] was implemented by modifying an open-source JVM, greatly improving its performance. In both versions, new aspects can be to dynamically woven, with support for these aspects to define new join points for which new interception hooks are created at weave time, thereby allowing PROSE to be used to support dynamic adaptation by weaving additional nonfunctional behaviors into the code at runtime.

A number of graphical user interfaces (GUIs) are included to manage the unanticipated weaving of new aspects at runtime; however, like Wool above, PROSE only supports weaving at a class level, so individual objects cannot be adapted individually. MIDAS [27], implemented as a spontaneous container [28], is middleware for the management of PROSE extensions that provides a distributed event-based system for the dissemination and management of aspects from a central server to mobile computers based on their location.

TRAP/J

TRAP/J [37] is a prototype unanticipated dynamic adaptation framework for Java. It combines compile-time aspect weaving using AspectJ [42] and unanticipated dynamic adaptation with wrapper classes and delegate classes. At compile time, the programmer selects a subset of application classes that will be adaptable. The TRAP/J system then automatically creates AspectJ code to replace all instantiations of the selected classes with wrapper class instantiations. Java code for each wrapper class and a meta-object class for that wrapper class are also automatically created. At runtime, each instantiated wrapper object has an instance of the original wrapped object and a meta object bound to it. These wrapper objects redirect all method calls to their meta objects, which in turn act as placeholders for a set of delegate objects that may handle the invocation of the method or adjust its parameters prior to execution by the original wrapped object. New, dynamically created delegates can be added or removed at runtime via a Remote Method Invocation (RMI) [39] interface using a management console. These delegates can be added on a per-object basis because the meta objects can supply a name for each instance and register it in an RMI registry.

This framework was used to demonstrate the dynamic adaptation of a network-enabled application by replacing instances of the `java.net.MulticastSocket` class with instances of an adaptable socket class, `MetaSocket` [18]. The TRAP/J framework, however, does not support completely unanticipated dynamic adaptation. The adaptation, its intelligent and controlled dynamic application, and the timing of its application all remain unanticipated until runtime, but the possible locations for the adaptations are specified in the application source code, as the version of AspectJ used requires access to the application source code. Despite improving the performance of the TRAP/J framework, this restriction greatly limits the nature of the unanticipated dynamic adaptations that can be applied. No information is provided about whether the generated meta-object class code can be modified prior to compilation and weaving.

In addition, TRAP/J seems to delegate the invocation of the method to only one delegate (the first one it finds implementing the method), but this ordering of delegates can be configured. This means that only one adaptation can be applied at a time because adaptation behaviors are not automatically composed. In addition, TRAP /J does not seem to allow the user to apply an easily recognizable name to the base object being adapted which may make it difficult for the user to identify the object to which adaptations should be dynamically applied. From the documentation, TRAP/J does not seem to support applying dynamic adaptations via new delegates on a structured class-wide or interface-wide basis because RMI registry look-ups are on a per-meta-object basis. Unlike Wool and PROSE, which support only the adaptation of classes, TRAP/J supports only the adaptation of individual objects at any one time.

Discussion

Dynamic AOP technologies would appear to be a promising area of research for dynamically adaptable middleware. Aspects can be used not only to implement nonfunctional concerns within the middleware but also to adapt or augment the functional behavior of the middleware [21]. This ability to dynamically adapt functionality or inject new functionality at clearly defined join points is of particular importance to middleware for mobile computing because dynamic and possibly unanticipated adaptation requirements are typical for mobile computing. The *separation of concerns* model of aspects reduces the difficulty of incorporating adaptations into complex middleware frameworks because the introduced cross-cutting concerns can be targeted correctly to the location requiring adaptation; however, current dynamic AOP methodologies such as Wool, PROSE, and TRAP/J are lacking a structured mechanism to dynamically specify these locations for dynamic adaptation and how these adaptations should be applied after the target software has begun execution in a manner that incorporates user, application, and environmental context at runtime. Despite this, this area of AOP-based dynamic adaptation of middleware is proving to be an active area of research and should quickly provide a number of solutions for this issue.

Policy-Based Management of Dynamic Adaptation

Principles and Key Ideas

Many traditional adaptable systems are composed of a single adaptation manager that is responsible for the entire adaptation process (i.e., monitoring, adaptation selection intelligence, and performing the actual adaptation). Because the intelligence to select appropriate adaptations and the mechanism to perform these adaptations are embedded directly within the adaptation manager, this type of system becomes inflexible and inappropriate for general use. By decoupling the adaptation mechanism from the adaptation manager and removing the intelligence mechanism that selects or triggers adaptations, the adaptation manager becomes more scalable and flexible. Policy specifications maintain a very clear separation of concerns between the adaptations available, the adaptation mechanism itself, and the decision process that determines when these adaptations are performed.

Policy specification documents are usually persistent text-based declarative representations of policy rules that ideally can be read, understood, and generated by users, programmers, and applications. A *policy rule* is defined as a rule governing the choices in behavior of a managed system [8]. Informally, a policy rule can be regarded as an instruction or authority for a manager to execute actions on a managed target to achieve an objective or execute a change.

An adaptation policy rule is usually made up of an event specification that triggers the rule, which is often fired as a result of a monitoring operation; an action to perform in response to the trigger; and a target object that is part of the managed system upon which that action is performed [8]. Many policies will also contain some restrictions or guards confining the rule action to appropriate occasions. This *event–condition–action* (ECA) format is standard for rule-based adaptation systems [4–6, 8,9,16,19,33–36,40], where an adaptation management system is responsible for monitoring these events, evaluating the conditions, and initiating the management action on the targeted managed object. In a policy-based dynamic adaptation system, it should be possible to edit the rule set and have it reinterpreted to support the dynamic addition of new rules or changes in policy.

Case Studies of Policy-Based Middleware

This section discusses two systems that employ policy-based management techniques to manage dynamic adaptation of middleware, but additionally the ACT, TRAP/J, and Correlate systems could also be described in terms of their use of policy-rule-based techniques. A number of mechanisms discussed in other chapters could also be discussed in terms of their use of rule-based management mechanisms.

RAM

Reflection for Adaptable Mobility (RAM) [4,9] from École des Mines de Nantes, takes the approach of completely separating functional and non-functional aspects of an application in a manner related to aspect-oriented programming. Using this separation of concerns approach, only the core application functionality is inserted into the application code, with all middleware services represented as nonfunctional concerns. Container meta objects wrap each application object and support the composition of other meta objects that implement these nonfunctional concerns. The wrapping of application objects with Containers occurs at either load time using Javassist [4,7] or at compile-time using AspectJ [9,42]. These meta objects provide the middleware services by selecting appropriate

`RoleProvider` objects for each service (i.e., the meta objects that provide the actual implementations of the services). Adaptation can occur by adding, removing, or reordering these `RoleProviders`.

The RAM approach also provides a resource manager, whereby the system maintains a tree of `MonitoredResource` objects, which describe a contextual resource or group of resources. These `MonitoredResource` objects are updated by `probe` objects that actively monitor the environment. `MonitoredResource` objects can be queried explicitly or alternatively by requesting change notifications to signal the adaptation engine when an interesting resource change occurs. The `Container` meta objects that wrap each application component can also expose the `Monitored-Resource` interface, supporting queries of application context as resources, thereby exploiting application-specific knowledge in the adaptation process.

The set of meta objects (aspects) to use in each `Container` is adapted at runtime by means of an adaptation engine that uses an application policy and a system policy, both written in a declarative Scheme-like language and both of which are passed to the adaptation engine when the application is started. The application policy defines point cuts (a dynamic set of join points, or `Container` objects) in the application and the named nonfunctional aspects to be used at these point cuts in an application-aware but resource-independent manner. The set of rules that determine which join points make up a point cut is also specified in the application policy, but these rules are dynamically evaluated, so this set of join points can change dynamically. The nonfunctional aspects woven at these point cuts are defined in the system policy in an adaptive condition–action model, where sets of application-independent but resource-aware conditions are dynamically evaluated to decide which meta objects will implement the nonfunctional aspect. When the conditions are dynamically evaluated, the bindings of meta objects can be changed, in a manner similar to dynamic aspect weaving; therefore, the set of join points that make up a point cut and the set of meta objects that implement an aspect can both be dynamically specified according to the rules in the policies.

The current system does not support dynamic changes to the policies and so cannot support unanticipated adaptation management logic; however, this capability is planned for future versions. In most cases where AspectJ is used, access to the source code of the application is also required. A version of RAM suggests using a configuration file to specify the set of join points that can be used and using AspectJ to create these join points at compile time rather than have `Containers` wrap every application object [11]. This means, however, that all possible locations for adaptation must be anticipated at compile time and access to the source code of the application is required. Preliminary designs for an adaptation framework extending RAM that would possibly support completely unanticipated adaptation by allowing dynamic specification of policies and dynamic selection of adaptation locations is presented in David and Ledoux [10], but this system has yet to be implemented.

CARISMA

Research carried out at University College London on the Context-Aware Reflective Middleware System for Mobile Applications (CARISMA) project [5,6] presents a design for peer-to-peer middleware based on service provision. Each node can export services and possible different behaviors or implementations for those services. Services can be selected according to user and application context information, as specified in an *application profile*, an eXtensible Markup Language (XML) policy document. Embedded in this application profile is the application-specific information that the middleware uses when binding to these services (e.g., which service behavior to use in response to changes in the execution context). The middleware is responsible for maintaining a view of the system environment by directly querying the underlying network-enabled operating system. Applications may request viewing and changing their profiles at runtime, thereby adapting the middleware as application-specific and user-specific requirements change dynamically.

This system also provides the ability for the application to be informed by the middleware of changes in specific execution conditions, supporting the development of resource-aware applications. This system is based on the provision of multiple implementations of the same service with different behaviors, in a manner similar to the strategy design pattern rather than adapting the service itself. The primary contribution of this work focuses on the identification and resolution of profile conflicts [6], not on the actual provision of an adaptable middleware implementation. No information is provided about how the services are implemented, if they can be dynamically loaded, how they implement their different strategies, or if these strategies can be expanded at runtime. It should be noted, however, that the application profile that controls how the system adapts and the mechanism for profile conflicts can both be adapted at runtime in an unanticipated manner. XMIDDLE [23], which appears to form the basis for CARISMA, is peer-to-peer data sharing middleware for mobile computing. In XMIDDLE, data is replicated as XML trees pending disconnections, with these trees being reconciled when possible in a policy-based manner according to application-specific conflict resolution data embedded in the shared data structures.

Benefits of Policy-Based Management of Dynamic Adaptations

An adaptable system that has its adaptation logic encoded directly into it cannot operate in a general-purpose manner or adapt in response to unanticipated changes, as often arises with an enabling technology such as middleware operating in an environment where the operating context changes erratically (as seen in a mobile computing environment). The use of a policy-based control model allows the clean decoupling of adaptation logic from the adaptation mechanism used by the adaptation framework.

The control logic to manage the dynamic application of an adaptation must be capable of specifying what adaptation should be applied, where and when it should be applied, and the conditions for restricting application of the adaptation, if necessary. Because many dynamic adaptations are necessarily required when some state, resource, or requirement has changed for the user, application, or execution environment, this dynamically specified control logic must also support the querying of this runtime context. Using dynamic loading and interpretation of policy directives can also be used to support the management of new unanticipated adaptations by allowing those new adaptations to be referred to dynamically, along with where they should be applied and what management logic should be used to control how and when those adaptations are applied.

Chisel and ALICE: Policy-Based Reflective Middleware for Mobile Computing

This section describes the Chisel dynamic adaptation framework and how it can be used with the ALICE middleware for mobile computing to create a dynamically adaptable middleware that can be used to adapt a standard network application in an unanticipated manner to operate in a mobile computing environment.

Chisel

The Chisel dynamic adaptation framework [19], developed in Trinity College Dublin, supports the application of arbitrary completely unanticipated dynamic adaptations to compiled Java software as it runs. An adaptation is completely unanticipated if the behavioral change contained in the adaptation, the location at which that adaptation is to be applied, the time when that adaptation will be applied, and the control logic that controls the application of the adaptation can all remain unanticipated until after the target software has begun execution [19].

The adaptations are achieved by dynamically associating Iguana/J meta-types [31,32] with any application object or class and so changing their behavior on the fly, without regard to the type of the object or class and indeed without access to its source code. The metatype of a class or object represents some coherent internal behavior change from its original source code behavior [31] (i.e., a behavioral change associated with the class or object). In Iguana/J, metatypes are implemented using custom MOPs to decide which parts of the object model to reify, writing a set of meta-object classes for these reifications to implement the new metatype behavior and then associating that metatype implementation with an object or class. In the Iguana literature, the terms *metatype association* and *MOP selection* are similar and refer to this association of MOP implementations with objects and classes. This association mechanism is performed using runtime behavioral reflection techniques, whereby selected parts of application objects and classes are reified and intercepted and the new metatype behavior inserted at this interception point. Iguana/J supplies the frame-work to instantiate these meta objects to reify the object model and correctly order metatypes if more than one is selected. Iguana/J provides a mech-anism to associate new metatypes with objects and classes at runtime, thereby changing the behavior of the system on the fly.

The execution of a new behavior embedded in the meta objects can then occur alongside or around the original behavior of the target object by wrapping the behavior of the target object and adapting or tailoring the intercepted operation or by introducing the new behavior before, after, or instead of the intercepted operation. New metatypes can be defined at any time and compiled offline using the Iguana/J metatype compiler, even as a target application is running. In this way, the adaptations to be applied can remain unprepared and unanticipated until needed. When a metatype is associated with a class, the behaviors that are changed are the static behaviors of the class, the behaviors of each current and future instance of the class, and the behavior of all subclasses and their current and future instances. Here, *static* refers to the behavior and data embedded in a class, instead of in each of its instances — for example, static methods, static data fields, and class initialization procedures, implemented using the `static` keyword in Java and C++.

The dynamic associations of these metatypes are driven by a dynamically specified and interpreted policy script. Using this policy script, the user can specify which classes or named objects should be adapted, either in a proactive manner or in a reactive event-based manner. The Chisel policy language, described in detail in Keeney [19], also supports the dynamic definition of new event types for use in reactive rules. In addition, the Chisel policy language allows events to be dynamically fired by other rules or in response to changes in dynamically specified contextual conditions.

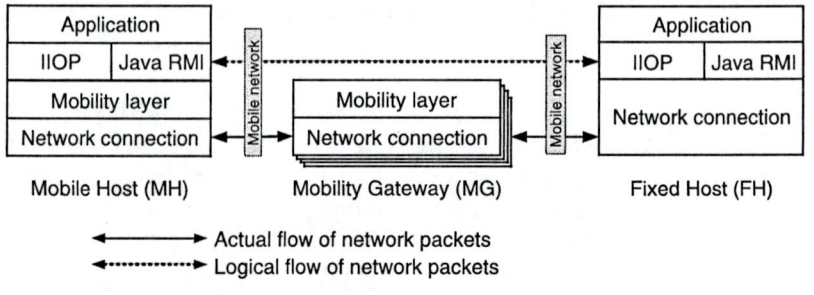

Figure 15.1 ALICE middleware framework.

In this manner, the timing and control logic for any dynamic metatype association can remain unspecified until during runtime and so remain unanticipated. By dynamically creating a new policy, specifying which class or object to adapt, and specifying which named metatype to associate, the location of the adaptation can also remain unanticipated until runtime. This use of runtime behavioral reflection and runtime specification and interpretation of adaptation policies allows the Chisel framework to support the completely unanticipated dynamic adaptation of any running Java application, without stopping it and without access to its source code.

ALICE

ALICE [3,15,41], also developed at Trinity College Dublin, is an architectural middleware framework that supports network connectivity in a mobile computing environment by providing a range of client–server protocols (Figure 15.1). In ALICE, *mobile hosts* are mobile devices that may interact with fixed computers, called *fixed hosts*. These connections are tunneled through *mobility gateways*, which are also fixed computers. The mobile host can become disconnected from a mobility gateway and later become reconnected to a different mobility gateway without interfering with the virtual connection to the fixed host. The ALICE mobility layer handles communications between devices by overriding socket functions while hiding which communication interface is being used for the connection. The mobility layer tracks available connections and picks one using a reconfigurable selection algorithm. When a disconnection occurs, the ALICE mobility layer will synchronously queue unsent data between the mobile host and the mobility gateway until a connection is re-established.

For this case study, a full Java implementation of the ALICE mobility layer was completed, based on work in Reference 41. It provides a class,

MASocket, that contains the ALICE connection behavior, which implements a socket interface similar to the standard Java socket class java.net.Socket. When the MASocket class is used instead of the standard Java socket, all messages from a mobile host to a fixed host are redirected via a mobility gateway. When the connection between the mobile host and the mobility gateway breaks, all network data is cached at the mobile host and the mobility gateway for later reconnection. This disconnection and reconnection occurs without the application being made aware of the disconnection.

Chisel and ALICE

To demonstrate the use of the Chisel dynamic adaptation framework, an off-the-shelf application, the Java Telnet Application/Applet [17], was adapted to operate in a mobile computing environment by dynamically adapting it to use the ALICE mobility layer, all without stopping the application and without changing or requiring access to the source code of the application in any way. The only initial assumption made about the internal programming of the application was that a standard Java socket, or a subclass of java.net.Socket, is used to connect the client and the telnet server, a reasonable assumption for any network enabled Java application.

A metatype, DoAliceConnection, was developed to intercept the creation of the socket connection to the telnet server and replace the socket in use with an instance of the ALICE MASocket. The metatype definition below specifies that the reified creation of objects should be intercepted and handled by the MetaObjectCreateALICEConn meta-object class:

```
protocol DoAliceConnection {
    reify Creation: MetaObjectCreateALICEConn();
}
```

This redirection behavior was embedded in the meta object class MetaObjectCreateALICEConn, as shown below. This redirection behavior is achieved by intercepting the creation of all socket objects, and if the connection is a not a local host connection or one used by ALICE then, by the use of the Java reflective application programming interface (API), the java.net.Socket constructor is replaced by the MASocket constructor. The application would be completely unaware of the change because the returned MASocket is extended from java.net.Socket and exposes the same interface.

```
class MetaObjectCreateALICEConn extends
        ie.tcd.iguana.MCreate {
  public Object create(Constructor cons, Object[]
            args) … {
    if(/*not a localhost connection, or a
            connection used by ALICE */){
      // Change the constructor, from
                java.net.Socket to MASocket
      cons = (Class.forName("MASocket")).getCon-
            structor( … );
    }
    Object result = proceed(cons, args);/*
            create the socket */
    return result;// result is either a normal
            socket or an MASocket
  }
};
```

This adaptation was then applied to the telnet application in a number of ways using the Chisel policy language [19]. One method was to apply this adaptation in a context-aware manner — that is, only perform the metatype association if the application was being used in a mobile computing environment, where the network connection was known to be intermittent. In the adaptation policy rules seen below, the DoAlice-Connection metatype is only associated with the java.net.Socket class if the UsingDodgyNet event fires. When the connection moves to a stable network connection, the UsingGoodNet event is fired, thereby re-enabling the use of standard Java sockets.

```
ON UsingDodgyNet java.net.Socket.DoAliceConnection
ON UsingGoodNet java.net.Socket.NullProtocol
```

The event UsingDodgyNet could be fired automatically by the Chisel event manager using an automatic rule definition and trigger rule, by the Chisel context manager when a wireless connection was detected, by the user using another event manipulation policy rule, etc. Similarly, the UsingGoodNet event could be fired when the network connection is deemed stable, by another policy rule, by some network monitoring code, or by the context manager. In Keeney [19], several methods are presented to describe how these events could be defined and automatically triggered in an unanticipated manner.

Findings and Further Adaptations

This case study was fully implemented and functions as expected. This case study demonstrates the use of the Chisel dynamic adaptation framework to adapt an arbitrary application in a context-aware manner for use in a mobile computing environment, without accessing its source code. The telnet application was not prepared in any way to have the particular adaptation applied. Only when the adaptation was deemed necessary did the user have to create a set of adaptation rules, similar to the ones above, embedding any necessary context information. Only when these rules triggered application of the adaptation would the adaptation be necessary, so it could be loaded and applied to the unprepared location deep inside the compiled application, without any requirement to change, interrupt, or restart the application. This case study also demonstrates how the operation of a complex compiled application was changed dynamically according to the environment and user's needs.

Using the Chisel framework, further adaptations are also possible to both the application and the ALICE middleware framework. This mechanism of dynamically redirecting Java socket connections to ALICE MASocket socket connections could also be used to dynamically adapt the Java RMI middleware model, similar to the approach discussed in Biegel et al. [3] and Haahr [15], but in an unanticipated manner. This possible approach could enable the adaptations described by Biegel [3] by intercepting the instantiation of both the java.net.Socket and sun.rmi.server.UnicastRef classes. An alternative approach could intercept the operations of the java.rmi.server.RMISocketFactory interface when it is requested to create the actual sockets used to perform remote object invocations [41].

Although a mobile computing scenario was chosen to demonstrate the Chisel dynamic adaptation framework, this case study equally applies to any environment or operation mode where unanticipated dynamic adaptation is required for satisfactory operation. A mobile computing environment was seen as a perfect example because the state, resources, and requirements of the application, the environment, and the user can all change to extreme values in an unanticipated manner.

Conclusions

This chapter has presented a discussion of dynamic adaptation for mobile middleware. The chapter began with a discussion of how unanticipated dynamic adaptation of applications and middleware is required in a mobile computing environment. A number of reflective and aspect-oriented techniques for dynamic adaptation were discussed, paying particular attention to support for unanticipated dynamic adaptation. The chapter then discussed the use of policy-based management to control unanticipated dynamic adaptation in a manner that was itself dynamically adaptable. The chapter then continued with an introduction to the Chisel dynamic adaptation framework. Chisel was discussed in terms of how the ALICE middleware for mobile computing could be used to adapt an off-the-shelf network application to operate in a mobile computing environment in a completely unanticipated manner.

References

[1] Aiken, R. et al., *Network Policy and Services: A Report of a Workshop on Middleware*, Request for Comments 2768, Internet Engineering Task Force (IETF), 2000 (http://www.ietf.org/rfc/rfc2768.txt).

[2] *Quality Objects (QuO)*, BBN Technologies, http://quo.bbn.com.

[3] Biegel, G., Cahill, V., and Haahr, M., A dynamic proxy-based architecture to support distributed Java objects in mobile environments, in *Proc. of the Int. Symp. on Distributed Objects and Applications (DOA 2002)*, Irvine, CA, October 28–30, 2002.

[4] Bouraqadi-Saâdani, N., Ledoux, T., and Südholt, M., *A Reflective Infrastructure for Coarse-Grained Strong Mobility and Its Tool-Based Implementation*, Technical Report 01-7-INFO, École des Mines de Nantes, Nantes, France, 2001.

[5] Capra, L., Reflective Mobile Middleware for Context-Aware Applications, Ph.D. thesis, Department of Computer Science, University College London, 2003.

[6] Capra, L., Emmerich, W., and Mascolo, C., CARISMA: Context-Aware Reflective mIddleware System for Mobile Applications, *IEEE Trans. Software Eng.*, 29(10), 929–945, 2003.

[7] Chiba, S., Load-time structural reflection in Java, in *Proc. of the 14th European Conf. on Object-Oriented Programming (ECOOP 2000)*, Sophia Antipolis/Cannes, France, June 12–16, 2000.

[8] Damianou, N. et al., The Ponder specification language, in *Proc. of IEEE Int. Workshop on Policies for Distributed Systems and Networks (Policy 2001)*, Bristol, U.K., January 29–31, 2001.

[9] David, P.-C. and Ledoux, T., An infrastructure for adaptable middleware, in *Proc. of the Int. Symp. on Distributed Objects and Applications (DOA 2002)*, Irvine, CA, October 28–30, 2002.

[10] David, P.-C. and Ledoux, T., Towards a framework for self-adaptive component-based applications, in *Proc. of the 4th Ifip Wg6.1 Int. Conf. on Distributed Applications and Interoperable Systems (DAIS 2003)*, Paris, November 17–21, 2003.

[11]   David, P.-C., Ledoux, T., and Bouraqadi-Saâdani, N.M., Two-step weaving
       with reflection using AspectJ, in *Proc. of OOPSLA Workshop on Advanced
       Separation of Concerns*, Tampa Bay, FL, October 14–18, 2001.

[12]   Ferber, J., Computational reflection in class based object-oriented languages,
       in *Proc. of the Conf. on Object-Oriented Programming Systems, Languages,
       and Applications (OOPSLA 1989)*, New Orleans, LA, October 1–6, 1989.

[13]   Filman, R. and Friedman, D., Aspect-oriented programming is quantification
       and obliviousness, in *Proc. of OOPSLA Workshop on Advanced Separation
       of Concerns*, Minneapolis, MN, October 15–19, 2000.

[14]   Forman, G.H. and Zahorjan, J., *The Challenges of Mobile Computing*, Uni-
       versity of Washington, Seattle, 1994.

[15]   Haahr, M., Supporting Mobile Computing in Object-Oriented Middleware
       Architectures, Ph.D. thesis, Department of Computer Science, Trinity Col-
       lege, Dublin, 2003.

[16]   Jørgensen, B.N. et al., Customization of object request brokers by application
       specific policies, in *Proc. of Middleware 2000*, New York, April 4–7, 2000.

[17]   Jugel, M.L. and Meisner, M., *The Java Telnet Application/Applet*, http://jav-
       atelnet.org.

[18]   Kasten, E.P. and McKinley, P.K., *Adaptive Java: Refractive and Transmutative
       Support for Adaptive Software*, Technical Report MSU-CSE-01-30, Department
       of Computer Science and Engineering, Michigan State University, East
       Lansing, 2001.

[19]   Keeney, J. Completely Unanticipated Dynamic Adaptation of Software, Ph.D.
       thesis, Department of Computer Science, Trinity College, Dublin, 2004.

[20]   Kiczales, G., des Rivieres, J., and Bobrow, D., *The Art of the Metaobject
       Protocol*, MIT Press, Cambridge, MA, 1991.

[21]   Kiczales, G. et al., Aspect-oriented programming, in *Proc. of the 11th
       European Conf. on Object-Oriented Programming (ECOOP'97)*, Jyväskylä,
       Finland, June 9–13, 1997, pp. 220–242.

[22]   Maes, P., Computational Reflection, Ph.D. thesis, Artificial Intelligence Lab-
       oratory, Vrije Universiteit, Brussels, Belgium, 1987.

[23]   Mascolo, C. et al., XMIDDLE: a data-sharing middleware for mobile com-
       puting, *Personal Wireless Comm.*, 21, 77–103, 2001.

[24]   COM+, Microsoft Corp., http://www.microsoft.com/com/tech/COMPlus.asp.

[25]   OMG, *Common Object Request Broker Architecture: Core Specification*,
       Object Management Group, Needham, MA, 2002.

[26]   Popovici, A., Alonso, G., and Gross, T., Just-in-time aspects: efficient
       dynamic weaving for Java, in *Proc. of the 2nd Int. Conf. on Aspect-Oriented
       Software Development (AOSD 2003)*, Boston, MA, March 17–21, 2003.

[27]   Popovici, A., Frei, A., and Alonso, G., A proactive middleware platform for
       mobile computing, in *Proc. of the 4th ACM/IFIP/USENIX Int. Middleware
       Conf. (Middleware 2003)*, Rio de Janeiro, Brazil, June 16–20, 2003.

[28]   Popovici, A., Alonso, G., and Gross, T., Spontaneous container services, in
       *Proc. of the 17th European Conf. for Object-Oriented Programming (ECOOP
       2003)*, July 21–25, 2003, Darmstadt, Germany.

[29] Popovici, A., Gross, T., and Alonso, G., Dynamic weaving for aspect oriented programming, in *Proc. of the 1st Int. Conf. on Aspect-Oriented Software Development (AOSD 2002)*, Enschede, The Netherlands, April 22–26, 2002.

[30] Prakash, R., Education: mobile computing, *IEEE Distributed Systems Online*, 2(6), 2001.

[31] Redmond, B., Supporting Unanticipated Dynamic Adaptation of Object-Oriented Software, Ph.D. thesis, Department of Computer Science, Trinity College, Dublin, 2003.

[32] Redmond, B. and Cahill, V., Supporting unanticipated dynamic adaptation of application behaviour, in *Proc. of the 16th European Conf. on Object-Oriented Programming (ECOOP 2002)*, Malaga, Spain, June 10–14, 2002.

[33] Robben, B. et al., *Building a Meta-Level Architecture for Distributed Applications*, Technical Report CW 265, Department of Computer Science, Katholieke Universiteit Leuven, Belgium, 1998, p. 17.

[34] Robben, B. et al., Non-functional policies, in *Proc. of the Second Int. Conf. on Metalevel Architectures and Reflection*, Saint-Malo, France, July 19–21, 1999.

[35] Sadjadi, S.M. and McKinley, P.K., ACT: an adaptive CORBA template to support unanticipated adaptation, in *Proc. of the 24th IEEE Int. Conf. on Distributed Computing Systems (ICDCS'04)*, Tokyo, Japan, March 23–26, 2004.

[36] Sadjadi, S.M. and McKinley, P.K., Transparent self-optimization in existing CORBA applications, in *Proc. of the Int. Conf. on Autonomic Computing (ICAC'04)*, New York, May 17–18, 2004.

[37] Sadjadi, S.M. et al., *TRAP: Transparent Reflective Aspect Programming*, Technical Report MSU-CSE-03-31, Computer Science and Engineering Department, Michigan State University, East Lansing, 2003.

[38] Sato, Y., Chiba, S., and Tatsubori, M., A selective, just-in-time aspect weaver, in *Proc. of the 2nd Int. Conf. on Generative Programming and Component Engineering (GPCE 2003)*, Erfurt, Germany, September 22–25, 2003.

[39] Java 2 Platform, Standard Edition (J2SE), Sun Microsystems, http://java.sun.com/j2se/.

[40] Truyen, E., Vanhaute, B., and Joosen, W., Integrating flexible middleware solutions with applications through non-functional policies, in *Proc. of OOPSLA Workshop on Object-Oriented Reflection and Software Engineering (OORaSE'99)*, Denver, CO, November 1, 1999.

[41] Wall, T., Mobility and Java RMI, M.Sc. thesis, Department of Computing Science, Trinity College, Dublin, 2000.

[42] The AspectJ Project, Xerox PARC, http://aspectj.org.

Section 3

REQUIREMENTS AND GUIDELINES FOR MOBILE MIDDLEWARE

Chapter 16

Naming and Discovery in Mobile Systems

Guanling Chen, Kazuhiro Minami, and David Kotz

CONTENTS

Introduction

Much of the technology necessary to realize Mark Weiser's vision of ubiquitous computing [37] is now available. Small portable devices, and the wireless networks to support them, are pervasive. Sensors capable of location tracking [23] and environmental monitoring [24] will soon be small, inexpensive, and plentiful. The resulting mobile and pervasive-computing environments are crowded, heterogeneous, and always changing. To succeed without distracting the user, middleware systems must provide naming and discovery methods to facilitate anytime, anywhere service access.

Let us first consider a simple scenario to help understand the problem at hand. Suppose Alice is visiting a university to meet several of her colleagues, and she needs directions to various offices in different buildings. The campus is covered by an 802.11 wireless network and Alice can use her dual-mode mobile phone to gain access. A location service can provide rough location information for the entire campus; for example, it can tell which buildings Alice is nearby based on which access point (AP) Alice's phone is currently associated with [26]. Individual buildings may also have separate location services, based on extensive radio fingerprint maps that allow the service to return a precise location by comparing the signal strength perceived by Alice's mobile phone [6]. As Alice moves around, the map application on her phone must find, select, and maintain communication with the appropriate location services depending on her current location. Alice should also be able to redirect the map to a nearby display or print it on a nearby printer for better readability. This scenario clearly illustrates the need for a middleware to bridge the services and clients, but the mobile environment poses several challenges.

In this chapter, we investigate the question of how the middleware can effectively and efficiently enable clients, such as Alice's mobile phone, to discover and interact with desired services, such as the location and display services mentioned here, in a mobile environment. As a framework for our discussion, we first give a general naming and discovery model and then identify the challenges confronting mobile middleware trying to provide naming and discovery functionalities.

A General Model

Here we define a general model for naming and discovery, as shown in Figure 16.1. In this model, the services *register* with the middleware and the clients *query* the middleware to get *results* (the desired services). The service registrations include a *name* describing the service properties and some other information. Although it is possible to provide type-only service discovery, we believe that service names are more expressive and necessary to distinguish services with the same type. A service may have to *update* its registration to reflect changes. A client's query could be persistent; when it is stored, the middleware will notify the client whenever

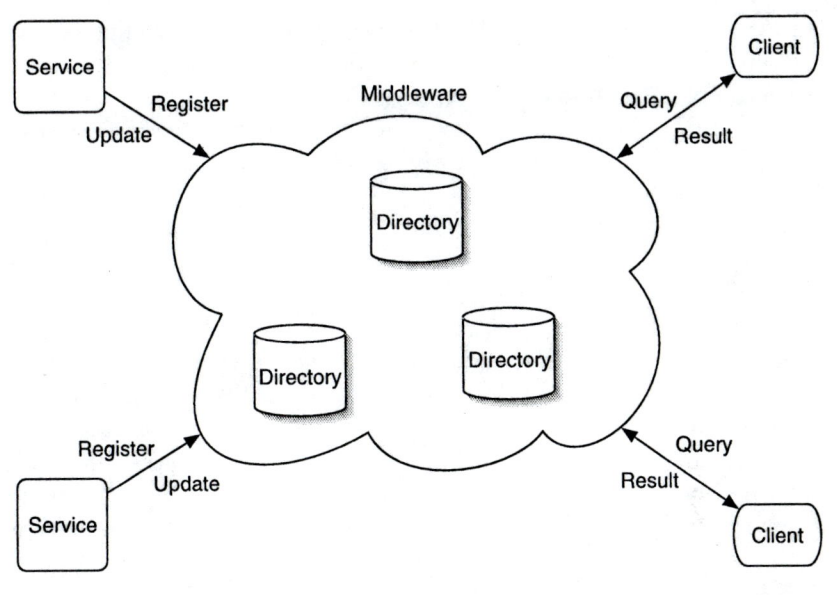

Figure 16.1   A general model for naming and discovery middleware.

the result of that query changes. All service names form a *name space*, and a *directory* is a data structure used to store the service names and other information. A middleware may contain a centralized directory storing the entire name space or a set of distributed directories, each of which stores either the full name space or portion of it.

A name is a description of the service, which could be a simple string, a set of attributes, or a full-blown eXtensible Markup Language (XML) document [10]. A string may or may not have a syntactic structure, such as "printer120" or "/device/printer/120." A string representation typically has limited expressiveness, and the syntax could be awkward with a complex description. Another commonly used representation is attribute-based naming. An *attribute* is a key–value tuple, and a name contains a set of attributes. Note that an attribute could be the child of another attribute, forming a hierarchy; for example, the printer could be named as [service=printer,   color=true,   building=sudikoff [floor=1 [room=120]]]. XML-based names are also expressive, and the standard syntax facilitates interoperability between services and clients. In addition to the properties of the service, XML can also be used to specify the syntax and semantics of the functional interface of the service, allowing a client to directly invoke these methods by inspecting the XML names. A disadvantage of XML-based naming is its verbose syntax, leading to increased processing overhead.

A query is typically specified in a similar syntax similar to that of name encoding, using a simple string, a set of attributes, or an XML description. Given a query, the middleware returns all names that match the query. The matching criteria could be simple equality tests, wild-card comparisons, a subset of matching attributes, or a complex evaluation of XPath and XQuery for XML-based names.

When the middleware returns the names that match a client's query, the client can then directly communicate with the selected services; however, the communication could also be mediated through the middleware, in which case the middleware acts as a broker by passing the messages back and forth between the client and service. We discuss the tradeoffs between these two approaches later in this chapter. Note that *middleware* is a general term, and it could be either a distributed infrastructure, as shown in Figure 16.1, or simple some library code embedded in services and clients so they can speak the same protocols. In this chapter, we are mainly concerned with infrastructure-based middleware that contains one or more directories.

Challenges for a Mobile Environment

One goal of middleware for naming and discovery is to provide interfaces or protocols to facilitate interactions between clients and services with minimum administrative and configuration overhead. Four challenges must be explicitly addressed in a mobile environment. The first problem is the *mobility* of services and clients. Device movement may cause an update of service names, a change of communication location or addresses, changes in the attachment points to the middleware, weak network connections, or temporary disconnections. For example, Alice's phone may roam from AP to AP as Alice moves on campus; the communication between Alice's phone and the location service must be maintained, even when the network address of the phone has changed.

The second challenge is *scalability* — that is, whether the middleware can scale beyond the local network to support a large number of services and clients while avoiding excessive network traffic. In our example, thousands of devices on campus may query the middleware to find an appropriate location service, and the middleware must also support tens of thousands of devices in the environment, such as displays, printers, and sensors.

A third issue, often, is that the service registration and client query are *context dependent*; that is, the middleware must also support context monitoring to dynamically adapt naming and discovery processes by itself rather than leaving this burden to services and clients. In our example, the middleware must be able to return an appropriate location service that best resolves Alice's current location as she moves on the campus.

Finally, *security* and *privacy* must also be addressed by the mobile middleware. In our example, although Alice may allow her colleagues to view her current location, she may not want to reveal her location to the general public. Many trust, authentication, and access control issues are involved.

In the following sections, we first survey the existing service-discovery standards. These standards, however, are not necessarily designed for mobile environments and do not address the previous challenges sufficiently. We then provide a detailed discussion of the approaches, addressing each of those challenging issues. Finally, we conclude and identify future work.

Existing Standards

Jini™ is a Java-based architecture, introduced by Sun Microsystems in 1998, which enables a dynamic federation of users and resources [35]. Network services first use a *discovery* protocol to locate a Jini *lookup* service to which the services register using a *join* protocol. The lookup service then stores both the service attributes (describing the services) and the service objects (implementing appropriate interfaces). A client queries the Jini lookup service and downloads the service object matching the query. The service object acts as the proxy between the client and the service provider, and this interaction does not go through the lookup service. One advantage of Jini, then, is that the proxy service object is downloaded to the client on the fly, and no device driver has to be manually installed for service interaction. Both service registration with lookup and client access to Jini services leverage *leases*. The lease must be renewed before its expiration for continuous registration and access. The lease-based access control gives Jini another advantage: It is robust against abrupt device failures and departures. Jini also supports remote *events* to notify the clients about changes in the system, such as the arrival of new services or state changes of existing services.

Universal Plug and Play (UPnP™), an extension of Plug and Play (PnP) by Microsoft®, aims at automatic discovery and control of devices [34]. UPnP uses the Simple Service Discovery Protocol (SSDP) for service discovery, which is similar to the discovery, join, and lookup protocols used by Jini. The counterpart of Jini's lookup service in UPnP is called a *control point*. Unlike Jini, however, UPnP leverages standard Transmission Control Protocol (TCP)/Internet Protocol (IP) and the Hypertext Transfer Protocol (HTTP) for communication and XML for service description, eliminating the need to rely on a single programming language. The XML-based description is much richer than Jini's simple attributes, and it contains a universal resource locator (URL) to which clients can send their control messages (also encoded as XML). Note that this approach removes the need to install any service-specific code on the client. UPnP further reduces administrative overhead by enabling devices to obtain an IP address automatically through AutoIP, when the Dynamic Host Configuration Protocol (DHCP) is not available, without explicit network configuration.

The Salutation Consortium is developing another service-discovery standard called Salutation [33]. The key component is the Salutation Manager (SLM), which acts as a broker between services and clients, much like the lookup service of Jini. Multiple SLMs can communicate reliably with each other using transportation managers (TMs) to exchange service registrations, and TMs understand different network transports to enable seamless integration. To facilitate interoperation between heterogeneous services and clients, SLM may work in the middle to define the message formats used by the communicating endpoints. Unlike Jini and UPnP, however, Salutation does not provide lease-based service access; instead, it allows a client to ask SLM to periodically check the status of desired services to see if they are still available. Salutation also provides a slim version of its architecture for devices with small footprints.

The Service Location Protocol (SLP) is a standard from the Internet Engineering Task Force (IETF) [20]. The main components include the user agent (UA), service agent (SA), and directory agent (DA). On behalf of the services, SAs advertise their location URLs and descriptive attributes, while UAs discover this information for the clients. A DA acts as the central cache storing the registration information for the services on the network. Note that the service registration with a DA is lease based and must be renewed periodically. SAs and UAs can discover DAs either actively (with a multicast request) or passively (by listening for periodic DA announcements). They may also use DHCP, if appropriately configured to support SLP, to find local DAs.

ZeroConf, also known as Rendezvous, is another IETF standard. Like UPnP, it can automatically configure the IP addresses of devices [3]. It has two extensions to provide service-discovery functionalities: one to leverage SLP and the other to use the Domain Name System (DNS). For the second approach, each service sends its DNS resource record to a known multicast address (224.0.0.251) on port 5253, and so does the query from a client. A small DNS responder runs on each ZeroConf host of the local network and processes these multicast DNS records and queries to allow hosts and services to operate without the presence of a DNS server.

We now summarize these service-discovery standards. Jini is a Java-based architecture, but the other protocols do not enforce a single language implementation. UPnP uses XML-based service description, but others use less-expressive attributes. On the other hand, SLP allows query operators other than equality tests. All protocols except Salutation have a concept of leasing, either for service registration or access, to improve system robustness against abrupt failures. All protocols except SLP support asynchronous event notifications to improve system reliability and scalability. None of these protocols requires manual installation of service-specific driver on clients, and all of the service brokers (such as Jini's lookup service) can be automatically discovered by services and clients. The control points of UPnP and the directory agents of SLP are optional components; that is, the clients may directly discover services using multicast on a local network. Finally, UPnP does not explicitly support security, but Jini has an optional encryption package. The only security mechanism of SLP is to authenticate the source of service registration. For further details of these protocols, both Helal [21] and Richard [31] have written excellent surveys.

Mobility

Both the client and the service-provisioning devices may move in the environment. At least three significant issues are raised by mobility. First, the network addresses of the devices may change, disrupting existing communications between services and clients. Second, the device may have to register with a new directory in the distributed middleware. We call this *handoff* between directories. Finally, device mobility may also cause a change in the name of a service to reflect the movement. In this section, we discuss the first two issues and leave the third to later in the chapter. For simplicity, here we discuss only the situation of service mobility. The handling of client mobility is similar.

When a client has located a desired service through the discovery middleware, it may begin to communicate with that service directly or through the middleware that acts as a message broker. If the client communicates with the service directly, the endpoints must handle the change of network address by themselves. The middleware could send notification events to a client as its service moves and registers with the middleware for a new address. It is up to the client to track its service and send messages to the current address of the server. One solution is to use Mobile IP [29], but the client may still have to resend the messages that were undeliverable during movement of the service.

On the other hand, if the middleware acts as a message broker between the client and service, the mobility of the service can be shielded from the client. A client simply sends its messages to the middleware and specifies the destination service. If the service is on the move, the middleware will buffer the message and deliver it as the service connects back. The Intentional Naming System (INS) takes this approach to provide combined functionalities of service discovery and message delivery [1]. INS consists of a set of Intentional Name Resolvers (INRs), to which a service registers an attribute-based name. The service name propagates through the INR overlay so eventually every INR will have the name of all services. A client sends a message to an INR and specifies its destination as a name query. The message traverses through the overlay with the name query resolved hop-by-hop, eventually reaching the service whose name matches the query. INS supports both unicast and multicast (so all services whose names match the query will receive that message). Similarly, Service-Oriented Network Sockets (SoNS) allow a client to specify a destination by name query, which can be resolved by any service-discovery system [32]. The message is then delivered to the identified service using direct socket communication, and all these details are hidden from the application layer.

For scalability reasons, a mobile middleware may contain a distributed set of directories, each responsible for the services in one region. As a service moves from one region into another, it may need to handoff from its previous registered directory to one in the new region. The handoff process includes moving the registration information from the old directory to the new one and triggering a notification event about this movement to the client. Designed for discovering moving network services, the Mobiscope system dedicates one directory for each rectangular region of geographic space [17]. As the service moves across regions in the environment, its name registration will always be routed to the directory responsible for its current location. The amount of traffic caused by handoff is determined by the region granularity, the number of mobile services, and their mobility patterns. To cope with failures, Mobiscope does not explicitly remove the service registration from the old directory. Instead, it employs a soft-state approach similar to the leasing concept. Because the service has to periodically renew its registration, the old registration will eventually expire and be removed from the old directory. Although it causes more network traffic, this leasing approach is simpler to implement and reduces the chance of inconsistency due to failures during the handoff process.

It is worthwhile to distinguish between physical mobility and network mobility, which are two orthogonal issues. As a service moves physically, it may or may not change its network address. For many of the service-discovery protocols surveyed earlier, a service has not moved as long as it is still in the same subnet. Using our example of Alice's mobile phone, communication between the location service and Alice's mobile phone will not be disrupted as long as the APs with which the phone associates are all in the same subnet. In other words, these discovery protocols define a region as the network subnet. Other protocols, such as Mobiscope, define a region as geometric rectangles that have nothing to do with network boundaries.

In summary, mobility may cause changes in the system. The mobile middleware may choose to only notify the clients about the changes and leave the handling to the clients, or the middleware may choose to handle the mobility itself and hide the changes completely from the clients. The first approach increases the complexity of client development, and the second approach increases the complexity of the middleware system. A designer using the second approach, however, also needs to be careful about the assumptions it makes for the clients, as it takes some control away from clients. For example, if the middleware receives a message from a client while the specified service is on the move, how long should the message be buffered before declaring a failure, or should the message be routed to another service that is newly available?

Scalability

If we consider tracking and monitoring the services that may move in a large area, scalability is an inevitable design issue for naming and discovery middleware. For our simple scenario, thousands of mobile clients may query the middleware to find location services, and the middleware may also have to support registrations from tens of thousands of service-provisioning devices, such as displays, printers, and sensors in the environment. The general approach is to partition the name space into several subspaces, each of which is handled by a directory. Two issues must be addressed here: how to partition the name space and how to route the service registrations and client queries to the corresponding directory.

One natural approach is to divide the entire area into geometric regions and put a dedicated directory in each region for services that have a "location." Each directory is responsible for the name registrations of all the services in its own region. In the Globe system, directories form a hierarchy according to a region-containment relationship [5]; namely, a directory for building A is the child of the directory responding for the whole campus. If a query for printers in building B arrives at the directory of building A, the query is propagated upward to the campus directory and then pushed downward to the directory of building B. Such a hierarchical structure exploits the locality of registration and queries to partition the workload, assuming the queries from one region are mostly to find services in the same region. The queries that do have to cross regions can be further reduced using extensive caching. DataSpace employs a similar approach where each three-dimensional *data cube* has its own directory [25]. The directories for the data cubes form a hierarchy as a big cube encompasses smaller cubes. By also partitioning the name space into directories using physical regions, Mobiscope routes registrations and queries in a peer-to-peer fashion instead of using a hierarchical structure [17]. In Mobiscope, each directory maintains a *spatial routing table* that records all regions of the directories. Given a query, the Mobiscope directory examines its routing table to find all directories intersecting with the region specified in the query.

Instead of grouping services based on their current location, it is also possible to group them using other properties. For example, INS/Twine and CAMEL divide the name space into a set of subspaces, each containing a single directory for the names in the corresponding subspace. Each name is then mapped onto one or more (for redundancy) of these subspaces using some kind of transformation. Both systems consist of a set of directories connected through a distributed hash table (DHT) using peer-to-peer (P2P) protocols [7,18]. A directory has a unique ID, and the DHT layer allows clients to send messages to a specified ID rather than an IP address. The message is then be routed to the directory whose ID is closest to the destination ID of the message in the P2P network. Given a name registration specified as a set of attributes, the system hashes the name into several IDs to which the name will be sent for processing. A similar process occurs for query resolutions. A query is specified as another set of attributes, one of which is selected and hashed into a query ID to which the query is sent. Because a name is replicated in all the directories responsible for the hashed attributes of that name, the query is guaranteed to be sent to a directory containing all the matching names. Here, the matching criterion is that the name contains all the attributes specified in the query.

In summary, naming and discovery middleware must partition the name space into a set of distributed directories for large-scale processing of service registrations and client queries. Most of the existing service-discovery standards surveyed earlier in this chapter partition the name space based on network topology, but they do not provide a scalable mechanism to allow interdirectory queries. The service-discovery systems discussed in this section propose to partition the name space based on location or on a hash of other attribute values. The directories are organized into a hierarchical or a peer-to-peer structure for scalable registration and query routing. The directories cooperatively store the names and resolve the queries, reducing the workload on each individual. By increasing the number of directories, the naming and discovery middleware scales up to handle more services and clients. The tradeoff for the increased scalability generally is the increased request processing latency, as the registrations and queries must be routed to corresponding directories which may take several hops; for example, in a DHT-based directory system, a message may take $\log(N)$ hops to reach a destination directory, where N is the number of directories in the system.

Context Awareness

Mobile clients must discover and use services based on the current context. *Context* is the circumstance in which an application runs and may include physical state (such as noise and light levels), computational state (such as network latency and bandwidth), and user state (such as location and current task). Consider our sample scenario, in which Alice finds nearby services (such as displays and printers) depending on her current location. Note that Alice's query is both a *persistent query*, which is continuously evaluated as the name space changes, and a *context-sensitive query*, which changes itself according to the context. On the other hand, the names for the display and printer may include their current locations, which should be automatically updated when the devices move (although not frequently). This requires a *context-sensitive name* that changes itself according to the context.

These scenarios impose several requirements on the naming and discovery service. It must be flexible, so names can characterize the resource and so queries can express the desired characteristics; it must be scalable, to handle many names and queries; it must be fast, to support frequent name and query updates; and it must be responsive, to quickly notify applications about changes to the set of matches for their persistent query. Some of the existing systems are designed to address these problems, such as the scalability as we discussed earlier. These scenarios, however, also place several requirements on the clients and services. Services must actively track their context so they may update their name. Clients must also track their context so they may update their persistent query. The context tracking and computation sometimes are not trivial and may well exceed the capability of mobile devices attached to a low-bandwidth network.

It is thus desirable to support both context-sensitive names and queries inside the infrastructure. A context-sensitive name registered by a service or a context-sensitive query requested by a client specifies how it should be updated according to the context. The middleware is responsible then to track context data sources and perform context computation to appropriately update the names and queries. Every time the names or the queries are updated, the queries should also be re-evaluated to determine whether the answer to the query is also changed. By offloading these duties from the services and clients, naming and discovery middleware improves performance and facilitates the development of both services and clients.

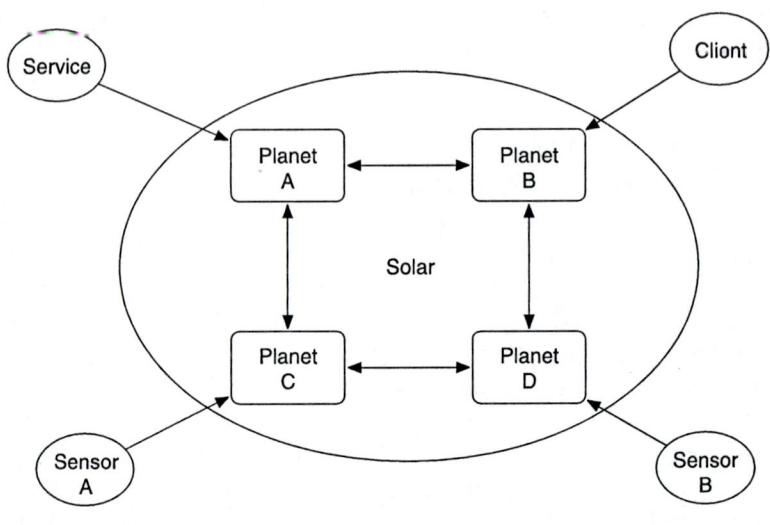

Figure 16.2 Architecture of the Solar system.

We have built a distributed system, called *Solar*, to support context-sensitive names and queries [11]. The core of Solar is a set of functionally equivalent nodes, called *Planets*, that connect with each other using a DHT-based peer-to-peer network [12]. Solar employs an attribute-based naming specification, and each Planet hosts a name directory. We use an attribute-hashing scheme, similar to INS/Twine [7], to partition the name space onto all the Planets. We envision that context is computed by aggregating data from one or more physical or virtual *sensors*. These sensors, like other services, connect to a Planet and register a name.

Figure 16.2 shows the architecture of Solar. A service registers with Planet A, which forwards the service name to the appropriate Planets based on attribute hashing. Now imagine that Alice's phone sends a context-sensitive query to Planet B to find printers in the building:

```
$campus-locator = [ sensor=locator,
    scope=campus] ;
$alice-locator = @filter ("00022dd54817")
    <- $campus-locator;
query [ service=printer, building=$alice-
    locator:building] .
```

The context-sensitive query is defined in such a way that the value of some attribute is some context dynamically computed from other sensors. Here, it first defines a campus-wide locator, $campus-locator, that tracks all devices in the wireless network. Next, it defines a locator that only tracks the location of Alice's phone by filtering through the output of campus locator using the Media Access Control (MAC) address of the phone. Finally, the query asks for all the printers in the same building of Alice's current location, where the value of the "building" attribute equals the one in the output of Alice's locator.

Planet B handles the client's context-sensitive query, and it must compute context from data of sensors that may connect to other Planets, such as on C and D. The planetary overlay thus also serves as an application-level routing layer for the sensor data [12]. The filter, as discussed earlier, is called an *operator*. An operator is a data-processing component that takes one or more sensor datastreams and outputs another one. A context-sensitive name or query can specify a graph of operators to specify the logic for combining data from multiple sensors [11]. Solar is responsible for loading and deploying these operators onto the Planets. Our experimental evaluation of the Solar system demonstrated two advantages of middleware support for context-sensitive names and queries: reduced query latency and improved system scalability [11]. Later, we further evaluated the impact of context updates (changes in the attribute values) on the system using a more realistic naming structure, and we showed that Solar's resource discovery generally performed well in a typical dynamic environment [36].

Other systems also support limited context awareness in their service-discovery approaches. Instead of sending messages to a fixed address, both INS and SoNS allow clients to send messages to a destination controlled by a name query [1,32]. The name query is always resolved with the current set of registered names, so the message may reach a different service at different times. To support context-sensitive names, however, services must monitor their context and update their name registrations. The iQueue system uses a nonprocedural language for an application to specify data composition [13]. The data sources used for composition are determined by name queries. Like Solar, the iQueue system continuously re-evaluates the queries based on current context and rebinds the data sources, if necessary [14].

Security and Privacy

In this section, we discuss security requirements for a resource-discovery system (such as Solar [11]) that supports context-sensitive names. Figure 16.3 shows the information flow among a client, a resource, and a resource-discovery system. Each resource (e.g., the camera) advertises its description as a set of attributes to the resource-discovery system. The

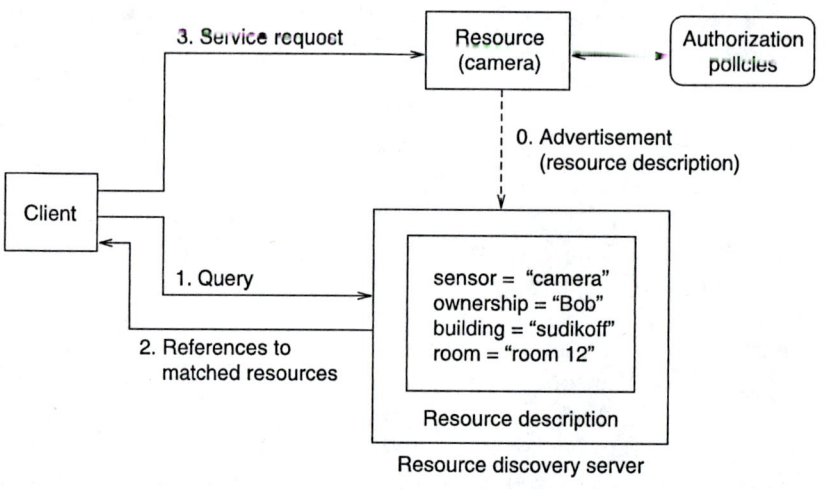

Figure 16.3 Information flow among a client, a resource, and a resource discovery system. The resource (camera) advertises its description with the four attribute–value pairs to the resource discovery system. The numbers in the labels of the solid arrows specify the sequence of the message flows.

building and room attributes of the resource in the example in Figure 16.3 could be dynamic; that is, the values of those attributes will change dynamically as the person carries the camera around. When a client issues a query that specifies conditions on attributes of a resource to the system to locate that resource, the system returns a list of references to the matched resources. For example, a query [building = sudikoff] matches the resources in the Sudikoff building. The client obtains from the server the values of the other attributes of the resource as well as the reference to the resource. The client then chooses a resource (i.e., a camera) from the list and accesses it to receive service. The client's access is authorized by the resource if the client satisfies the resource's authorization policies, which are usually defined by the resource owner.

We make a few assumptions about the resource-discovery system in Figure 16.3. First, we assume that the attribute–value pairs of a resource are flat; that is, no hierarchical structure is involved in the attributes. Second, we assume that the resource-discovery system is administered by a single authority, although the system might consist of multiple servers in a distributed environment. Third, clients, resources, and the resource-discovery system establish secure channels with a session key to communicate with each other while ensuring the secrecy and authenticity of the messages.

Two kinds of information must be protected in the example shown in Figure 16.3. One is the resource's advertisement that contains the description of a resource. The other is the client's query containing the conditions on a resource. The security requirements for the resource-discovery system depend on whether the clients and resources trust the system not to disclose their confidential information. We first discuss the case with a trusted system. After that, we discuss the case with an untrusted system.

In the first case, we only need to consider the protection of resource descriptions from an unauthorized client, because the information in each client's query is shared only with the system. Several existing resource-discovery systems [16,30] require a client issuing a query to have the privilege to access a resource to obtain the reference to that resource; that is, the client will not be aware of existence of the resources to which the client does not have access. Those systems assume that the authorization policies on each resource are static; the policies do not change dynamically. In those systems, the client avoids accessing a list of received resources sequentially until an accessible resource is found.

Pervasive computing, however, has two additional requirements for protecting the resource descriptions. First, the system must protect each attribute of a resource description with a different authorization policy, because an attribute of a resource description might contain confidential or private information. For example, suppose Bob advertises his mobile camera with a wireless network interface as a resource to the resource-discovery system, and he describes the device with the attributes ownership and location. The location attribute contains the current location of the camera. A query that specifies the value of location attributes produces a list of resources, including Bob's camera, if the value in the query matches the current location of the camera. Bob may be willing to allow other people to access the video stream from the camera if his identity is not disclosed to them. However, if the value of the ownership attribute of the camera is also accessible to the client, then the client is able to keep track of Bob's location, and Bob's privacy might be compromised. Because many people are concerned with location privacy [8,19,22,28] in pervasive computing, the location attribute of a device should be protected appropriately to preserve user privacy.

Second, the resource-discovery system must support context-sensitive authorization policies that consider the situation of the users and environments. For example, suppose that a patient in a hospital carries a mobile device that monitors his medical condition and that the description of that device, which contains the attributes ownership and location, is maintained in the resource-discovery system. The administrator of the system could define a context-sensitive policy that states: "A caregiver of the hospital is granted access to the attribute location of the patient's mobile device only if the patient's medical condition is critical." This authorization policy is necessary to protect the patient's privacy in a noncritical situation.

Making a context-sensitive authorization decision requires the resource-discovery system to refer to context information (e.g., a patient's medical data). Several existing authorization systems [2,4,15,28] that support context-sensitive policies take a centralized approach; that is, those systems have a central server that collects all the context information necessary to make authorization decisions. However, the centralized approach does not work if context information is maintained by different servers that do not trust the resource-discovery system to protect their confidential information. Instead of collecting detailed context information, the resource-discovery system could make the authorization decision without obtaining a patient's medical data by collaborating with a medical server that maintains the patient's medical data and that tells the resource-discovery system whether the patient's condition is critical. This collaboration is possible only if the resource-discovery system trusts the medical server to provide correct information. Minami and Kotz [27] generalized this idea and built a secure context-sensitive authorization system that does not require a centralized server trusted by every principal. The system performs an authorization query involving multiple principals in different multiple administrative domains. The proof tree for the query is decomposed into multiple subproof trees produced by different hosts so each participating principal's confidentiality and integrity policies concerning the rules and facts contained in the proof tree are preserved.

We now analyze two existing systems that protect resource descriptions. Berkeley's Secure Service Discovery Service (SSDS) [16] protects the description of each resource by checking whether the client is granted access to that resource. In SSDS, the capability manager converts an access control list (ACL) that is used to protect a resource into a capability for accessing the corresponding resource description in the directory server. If a client is in the ACL, the client receives from the capability manager a capability that can only be used by that client. Because the clients maintain their privileges, the system does not have to maintain any authorization policies to protect the resource descriptions.

Raman et al. [30] extended the Intentional Naming System (INS) [1] to route a client's message only to resources that the client has the privilege to access. The major difference from SSDS is that INS maintains the authorization policies of each resource as the hidden attribute of that resource. In INS, each Intentional Name Resolver (INR) traverses an INS name tree that encodes the hierarchical arrangement of the attribute–value pairs of resources to map a resource name (name specifier) to a destination node to which a client's message is routed. Each node in an INS name tree is associated with an ACL to reduce the search space in the tree. When the matching process in the INR reaches a node whose ACL does not contain the client, it prunes the subtree under that node from the search space. INS associates each resource's original ACL with the corresponding leaf node in the INS name tree and derives the ACLs for intermediate nodes by taking the union of the ACLs in its children nodes.

We now discuss the second case where the resource-discovery system is not trusted by the clients and resources. The clients need to protect their queries and credentials (or capabilities) from the system, because queries might imply the clients' interests, which should be kept confidential, and credentials might imply confidential relationships with the resources. Similarly, the resources need to protect their advertisements from the system. Because it is difficult to hide that information completely from the resource-discovery system, some existing research projects focus on minimizing information to be disclosed to the system.

Splendor [40] is a directory service in pervasive computing, and its focus is to protect the privacy of resource owners from the directory server. In Splendor, each resource advertises its description to the directory service through a proxy server trusted by that service; therefore, the directory service cannot associate attributes of the resource that contains confidential information with the identity of the resource owner. Splendor, however, does not address the issue of traffic analysis by an adversary who is capable of monitoring all the network traffic.

PrudentExposure [41] minimizes disclosure of a client's credentials to a directory service using the technique of hash summarization with a Bloom filter [9]. In PrudentExposure, each credential corresponds to a domain that contains multiple resources, and the credential allows a client to access all the resources in that domain. This connectivity is not appropriate; therefore, a client wants to know which domains the directory service supports while minimizing the disclosure of his credentials. The client sends a bit vector of a fixed length to the service instead of sending the actual credentials. The bit vector represents the client's credentials; that is, if a bit at a certain position in the vector is set, that means that the client possesses the corresponding credential. The position for each credential is decided by computing $mod(h(domain\ identity),\ n)$ where mod is the mod function, h is a hash function, n is the length of the bit vector, and $domain\ identity$ is the identity of the domain specified in the credential. We assume that the bit vector is long enough to experience no collisions between two different credentials.

Similarly, the directory service computes a bit vector that represents a set of domains it supports, using the same hash function. When the directory service receives the bit vector from the client, it intersects it with its own bit vector and returns it to the client. The returned vector represents the list of domains the client can access through the directory service. The service can figure out which credentials the client has only if the service supports the corresponding domain. The client is thus able to minimize the disclosure of his credentials; however, this scheme is not applicable to a resource-discovery system that supports context-sensitive policies, because it is impossible to distribute static credentials to clients in advance.

In summary, resource descriptions maintained by a resource discovery system in pervasive computing contain confidential information to be protected appropriately. The major difference from traditional authorization systems is the necessity to support context-sensitive policies. The protection of a client's query and resource description from an untrusted directory service remains largely left for future research. The literature of trust negotiation [39,38] may provide some insight here, because these protocols often focus on limiting exposure of information to the other party.

Summary and Future Work

Middleware support for naming and discovery is crucial for anytime, anywhere service access in a mobile and pervasive-computing environment. In this chapter, we have identified four challenging issues that existing service-discovery standards do not address well: mobility, scalability, context awareness, and security and privacy. While individual challenges are being addressed by the research community, an integrated system addressing all of these aspects remains to be seen. Our experience tells us that these issues are interwoven with each other, and a new system must have global design principles; for example, scalability must be considered by many components, including overall architecture, name storage, message multicast, and security protocols. The end system should not become so complex that it is difficult to diagnose and maintain, thus defeating the goal of naming and discovery middleware to facilitate discovery and interactions.

Acknowledgment

This research has been supported by NSF Award EIA-9802068, by DARPA contract F30602-98-2-0107, by DoD MURI contract F49620-97-1-03821, by Microsoft Research, by the Cisco Systems University Research Program, and by USENIX Scholars Program. This research program is a part of the Institute for Security Technology Studies, supported under Award No. 2000-DT-CX-K001 from the U.S. Department of Homeland Security, Science and Technology Directorate. Points of view in this document are those of the authors and do not necessarily represent the official position of the U.S. Department of Homeland Security.

References

[1] Adjie-Winoto, W., Schwartz, E., Balakrishnan, H., and Lilley, J., The design and implementation of an intentional naming system, in *Proc. of the 17th ACM Symp. on Operating System Principles*, Charleston, SC, December, 1999, pp. 186–201.

[2] Al-Muhtadi, J., Ranganathan, A., Campbell, R., and Mickunas, D., Cerberus: a context-aware security scheme for smart spaces, in *Proc. of the First IEEE Int. Conf. on Pervasive Computing and Communications (PerCom 2003)*, Dallas, TX, March, 2003, pp. 489–496.

[3] AutoConf, http://www.autoconf.org/.

[4] Bacon, J., Moody, K., and Yao, W., A model of OASIS role-based access control and its support for active security, *Proc. of the Sixth ACM Symp. on Access Control Models and Technologies*, Monterey, CA, June, 2002, pp. 492–540.

[5] Baggio, A., Ballintijn, G., van Steen, M., and Tanenbaum, A.S., Efficient tracking of mobile objects in Globe, *Comput. J.*, 44(5), 340–353, 2001.

[6] Bahl, P. and Padmanabhan, V.N., RADAR: an in-building RF-based user location and tracking system, in *Proc. of the 19th Annual Joint Conf. of the IEEE Computer and Communications Societies,* Tel Aviv, Israel, March, 2000.

[7] Balazinska, M., Balakrishnan, H., and Karger, D., INS/Twine: a scalable peer-to-peer architecture for intentional resource discovery, in *Proc. of the First Int. Conf. on Grid and Pervasive Computing,* Zurich, Switzerland, August, 2002.

[8] Beresford, A.R. and Stajano, F., Location privacy in pervasive computing, *IEEE Pervasive Comput.*, 2(1), 46–55, 2003.

[9] Bloom, B.H., Space/time trade-offs in hash coding with allowable errors, *Commun. ACM*, 13(7), 422–426, 1970.

[10] Chen, G. and Kotz, D., Context aggregation and dissemination in ubiquitous computing systems, in *Proc. of the Fourth IEEE Workshop on Mobile Computing Systems and Applications (WMCSA'02)*, Callicoon, NY, June 2002, pp. 105–114.

[11] Chen, G. and Kotz, D., Context-sensitive resource discovery, in *Proc. of the First IEEE Int. Conf. on Pervasive Computing and Communications (PerCom 2003)*, Dallas, TX, March, 2003, pp. 243–252.

[12] Chen, G., Li, M., and Kotz, D., Design and implementation of a large-scale context fusion network, in *Proc. of the First Annual Int. Conf. on Mobile and Ubiquitous Systems: Networking and Services*, Boston, MA, August, 2004, pp. 246–255.

[13] Cohen, N.H., Lei, H., Castro, P., Davis II, J.S., and Purakayastha, A., Composing pervasive data using iQL, in *Proc. of the Fourth IEEE Workshop on Mobile Computing Systems and Applications (WMCSA'02)*, Callicoon, NY, June, 2002, pp. 94–104.

[14] Cohen, N.H., Purakayastha, A., Wong, L., and Yeh, D.L., iQueue: a pervasive data composition framework, in *Proc. of the Third Int. Conf. on Mobile Data Management (MDM'02)*, Singapore, January, 2002, pp. 146–153.

[15] Covington, M.J., Long, W., Srinivasan, S., Dey, A.K., Ahamad, M., and Abowd, G.D., Securing context-aware applications using environment roles, in *Proc. of the Sixth ACM Symp. on Access Control Models and Technologies*, Chantilly, VA, May, 2001, pp. 10–20.

[16] Czerwinski, S.E., Zhao, B.Y., Hodes, T.D., Joseph, A.D., and Katz, R.H., An architecture for a secure service discovery service, in *Proc. of the 5th ACM/IEEE Int. Conf. on Mobile Computing and Networking (MOBICOM'99)*, Seattle, WA, August, 1999, pp. 24–35.

[17] Denny, M., Franklin, M.J., Castro, P., and Purakayastha, A., Mobiscope: a scalable spatial discovery service for mobile network resources, in *Proc. of the ACM Int. Conf. on Mobile Data Management (MDM'03)*, Melbourne, Australia, January, 2003, pp. 307–324.

[18] Gao, J. and Steenkiste, P., Design and evaluation of a distributed scalable content discovery system, *IEEE J. Selected Areas Commun.*, 22(1), 54–66, 2004.

[19] Gruteser, M. and Grunwald, D., Anonymous usage of location-based services through spatial and temporal cloaking, in *Proc. of the First Int. Conf. on Mobile Systems, Applications, and Services (MobiSys'03)*, San Francisco, CA, May, 2003.

[20] Guttman, E., Service Location Protocol: automatic discovery of IP network services, *IEEE Internet Comput.*, 3(4),71–80, 1999.

[21] Helal, S., Standards for service discovery and delivery, *IEEE Pervasive Comput.*, 1(3), 95–100, 2002.

[22] Hengartner, U. and Steenkiste, P., Access control to information in pervasive computing environments, in *Proc. of the 9th Workshop on Hot Topics in Operating Systems (HotOS)*, Lihue, HI, May, 2003, pp. 157–162.

[23] Hightower, J. and Borriello, G., Location systems for ubiquitous computing, *IEEE Comput.*, 34(8), 57–66, 2001.

[24] Hill, H., Szewczyk, R., Woo, A., Hollar, S., Culler, D., and Pister, K., System architecture directions for network sensors, in *Proc. of the Ninth Int. Conf. on Architectural Support for Programming Languages and Operating Systems (ASPLOS'00)*, Cambridge, MA, November, 2000, pp. 93–104.

[25] Imielinski, T. and Goel, S., Dataspace: querying and monitoring deeply networked collections in physical space, *IEEE Pers. Commun.*, 7, 4–9, 2000.

[26] Kotz, D. and Essien, K., Analysis of a campus-wide wireless network, in *Proc. of the 8th ACM/IEEE Int. Conf. on Mobile Computing and Networking (MOBICOM'02)*, Atlanta, GA, September, 2002, pp. 107–118.

[27] Minami, K. and Kotz, D., Secure context-sensitive authorization, *J. Pervasive Mobile Comput.*, 1(1), 257–268, 2005.

[28] Myles, G., Friday, A., and Davies, N., Preserving privacy in environments with location-based applications, *IEEE Pervasive Comput.*, 2(1), 56–64, 2003.

[29] Perkins, C.E., Mobile networking through Mobile IP, *IEEE Internet Comput.*, 2(1), 32–41, 1998.

[30] Raman, S., Clarke, D., Burnside, M., Devadas, S., and Rivest, R., Access-controlled resource discovery for pervasive networks, in *Proc. of the 18th ACM Symp. on Applied Computing (SAC'03)*, Melbourne, FL, March, 2003.

[31] Richard, G.G., Service advertisement and discovery: enabling universal device cooperation, *IEEE Internet Comput.*, 4(5), 18–26, 2000.

[32] Saif, U. and Paluska, J.M., Service-oriented network sockets, in *Proc. of First Int. Conf. on Mobile Systems, Applications, and Services (MobiSys'03)*, San Francisco, CA, May, 2003.

[33] Solutation Architecture, http://www.salutation.org/.

[34] Universal Plug and Play, http://www.upnp.org/.

[35] Waldo, J., The Jini architecture for network-centric computing, *Commun. ACM*, 42(7), 76–82, 1999.

[36] Wang, J., *Performance Evaluation of a Resource Discovery Service*, Technical Report TR2004-513, Dartmouth College, Hanover, NH, 2004.

[37] Weiser, M., The computer for the 21st century, *Sci. Am.*, 265(3), 66–75, 1991.

[38] Winsborough, W.H. and Li, N., Protecting sensitive attributes in automated trust negotiation, in *Proc. of ACM Workshop on Privacy in the Electronic Society (WPES'02)*, New York, NY, November, 2002, pp. 41–51.

[39] Yu, T., Winslett, M., and Seamons, K.E., Interoperable strategies in automated trust negotiation, in *Proc. of the 8th ACM Conf. on Computer and Communications Security (CCS'01)*, Philadelphia, PA, November, 2001, pp. 146–155.

[40] Zhu, F., Mutka, M., and Ni, L., Splendor: a secure, private, and location-aware service discovery protocol supporting mobile services, in *Proc. of the First IEEE Int. Conf. on Pervasive Computing and Communications (PerCom 2003)*, Dallas, TX, March, 2003, pp. 235–242.

[41] Zhu, F., Mutka, M., and Ni, L., PrudentExposure: a private and user-centric service discovery protocol, in *Proc. of the Second IEEE Int. Conf. on Pervasive Computing and Communications (PerCom 2004)*, Orlando, FL, March, 2004, pp. 329–338.

Chapter 17

Efficient Data Caching and Consistency Maintenance in Wireless Mobile Systems

Sajal K. Das and Mohan Kumar

CONTENTS

Introduction

Rapid advances in wireless mobile communications technologies and successful deployment of a multitude of wireless network infrastructures (e.g., 3G cellular, *ad hoc*, wireless local area networks, or mesh) have made ubiquitous information access a reality. Mobile users desire online access to a variety of data such as news headlines, stock quotes, weather forecasts, traffic or flight information, games, and so on. Efficient caching at mobile devices has the potential to deal with traditional constraints of wireless mobile networks, including limited bandwidth, battery power, user and device mobility, and disconnections. This chapter focuses on a very challenging issue — the design and deployment of efficient and scalable schemes for data caching — as well as cache consistency maintenance in mobile computing systems.

Data broadcasting offers a simple mechanism to disseminate and share information among a large number of users in a wireless communication environment. Data broadcast schemes can be classified into three categories: push-based, pull-based (or on-demand), and hybrid. In a *push-based* scheme, data is periodically broadcast from the server to the clients or mobile users (MUs) via the downlink channel according to a schedule that usually makes use of data access patterns. An MU must tune into the broadcast channel to acquire the desired data. The access latency involved here is mainly due to the waiting period for tuning to the next broadcast cycle. In a *pull-based* scheme, on the other hand, queries made by the MUs on uplink channels are satisfied on-demand by the server via the downlink channel. Due to the asymmetric nature of communications in wireless systems, the downlink channel has higher bandwidth than the uplink channel; hence, the query delay in push-based schemes depends on the data size and system load, among other factors. In a *hybrid* scheme, more popular items are pushed and less popular items are pulled. Determining an optimal cutoff point that partitions the entire dataset dynamically into push and pull subsets is an important problem [2,16,29].

As is well known, *caching* more frequently accessed data at the client's end (i.e., mobile devices) is an efficient way to reduce the average access or query delay, save bandwidth, and improve the overall performance of wireless mobile systems [4,38,48]. An efficient cache management strategy can substantially improve the hit ratio. In most of the mobile devices, such as laptops, palmtops, and cellular phones, wireless communication is a major source of energy consumption that reduces battery life [14]. As an example, the wireless transmission of 1000 bits of information consumes approximately the same amount of energy as the execution of 3 million instructions [28]; thus, caching can potentially minimize communication overhead and conserve battery power. Due to frequent mobility and the possible disconnection of mobile users, the data in device caches is prone to becoming stale quite often; therefore, *cache consistency* is another challenge to be tackled in wireless mobile environments.

Cache and System Architecture

Figure 17.1 illustrates a typical wireless mobile data communication system architecture consisting of a wired backbone network (e.g., the Internet), data servers, mobile support stations (MSSs), and MUs. Each (hexagonal) cell has a base station (BS) or an MSS connected to the servers through the backbone network. Each MSS serves multiple MUs via wireless links (channels) and has a server cache (SC) to store data items. Each MU has a local cache (MUC) that stores the retrieved data items. An MSS is also responsible for cache consistency between the servers and MUCs. Because this chapter focuses on efficient data caching and cache consistency between the SC and MUCs, it is assumed that the consistency between the SC and servers is maintained in the wired network by existing methods [10,22,49].

Scarce resources (e.g., bandwidth) of clients remain as barriers to the full utilization of mobile computing capabilities. Client data caching offers a good solution to cope with the inefficiency of wireless data dissemination by reducing the amount of traffic over the communication channel. The three main issues in cache design for mobile devices are [48]: (1) a *replacement* policy, which chooses to discard a set of victim data currently in the cache in order to accommodate new incoming data; (2) a *prefetching* policy, which prefetches data into the cache in anticipation of possible future accesses; and (3) a cache *invalidation* scheme, which maintains data consistency between the local cache and the original server. The first two issues are discussed in this section, and the third issue is the focus of the next section.

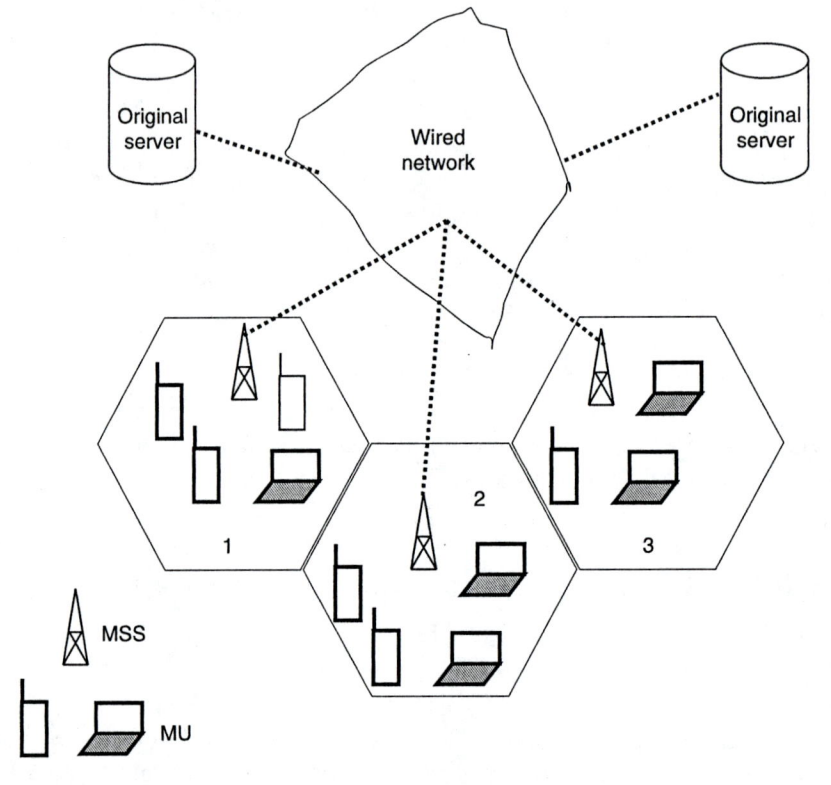

Figure 17.1 Wireless data communication system architecture.

Replacement

Replacement policies in Web proxy caching have been studied widely, and several deterministic schemes have been proposed in the literature [9,31,36,42]; however, such algorithms cannot be adopted directly to manage mobile caches due to the poor connectivity, limited energy, and memory of mobile devices. The cache replacement algorithm for wireless data dissemination was first studied in the context of *broadcast disks* [1]. A replacement policy was proposed in which the data item with the minimum value of p/f was evicted for replacement, where p is the access probability of the item and f is its broadcast frequency; thus, an evicted item either has a low access probability or a short retrieval delay. A cache update policy that attempts to minimize average access latency is presented in Shim et al. [36]. Here, the broadcast channel is divided into time slots of equal size, which are equal to the broadcast time of a single item.

Caching algorithms for broadcast disk systems were also investigated in Khanna and Liberatore [21]. In contrast to previous approaches, this caching algorithm took into consideration the factors of both access history and retrieval delay for cache replacement. Most of the above schemes must work cooperatively with a specific broadcast schedule algorithm to achieve good performance. Xu et al. [48] proposed a cache replacement algorithm that considers access probability, update frequency, data size, retrieval delay, and cache invalidation cost for minimizing the access latency by mobile devices. In particular, a cache replacement policy called *Min-SAUD* accounts for the cost of ensuing cache consistency before each cached item is used. Min-SAUD uses *stretch* as the major performance metric, which accounts for the data service time and thus is fair for data items of variable sizes. Because this approach requires an aging daemon to periodically update the estimated information of each data item, the computational complexity and energy consumption of the algorithm are too high for mobile devices.

Prefetching

Prefetching is a technique that can reduce access latency and improve the cache hit ratio by fetching remote data *a priori*. The idea of prefetching stems from the fact that, after retrieving a page or data item from the remote server, a user usually spends some time viewing the page, and during this period the network link is generally idle. Files that are likely be accessed in the future can be fetched in anticipation, with a view to reduce the transmission delay when the user actually requests these files. In addition, prefetched files can be immediately processed if decryption or decompression is required, allowing further reduction in the delay of loading a page in the mobile device. The difficulty of realizing efficient prefetching lies in accurate anticipation. If some of the prefetched files are never used, prefetching increases the system load. At some point, this increase in load may increase the delay of on-demand requests significantly and may eventually cause the overall system performance to degrade. In the wireless network, prefetch may increase energy consumption due to increased communication, particularly on downlink channels; therefore, the key challenge is to determine what to prefetch for performance improvement.

Prefetching has been investigated to reduce Web access latency in wire-line networks. Existing works [13,15] have investigated prefetching schemes involving a point-to-point session transmission model, which is different from the broadcast communication model in wireless mobile networks. Prefetching and data invalidation have also been studied for performance tradeoff. For example, in Cao [7], mobile devices calculate the prefetch access ratio (PAR), the number of prefetches is divided by the number of accesses for each data item, and a threshold PAR is used to determine whether to prefetch a data item. In Yin et al. [51], a value function is used instead of PAR. These works use prefetching to retrieve the desired data by sending an uplink message to the base station. Due to channel contention and data transmissions at the mobile devices, such proactive prefetching schemes are generally not energy efficient [14].

An adaptive network prefetching scheme proposed in Jiang and Kleinrock [19] utilizes prediction and threshold modules that compute access probabilities and prefetch thresholds, respectively. Whenever a new page is displayed, the prediction module updates the local access history and computes the access probability of each link on that page. At the same time, for each server with at least one link on this page, the threshold module computes its prefetch threshold based on the network and server conditions as well as the costs of time and bandwidth to the user. Finally, all the files with access probability greater than the prefetch threshold of the server are prefetched. The prediction algorithm may also be run at the server, in which case access probabilities will be sent to the user along with the requested page.

Mobile Cache Consistency Schemes

The existing literature for cache consistency maintenance in wireless mobile environments can be broadly categorized into three classes: stateless, stateful, and hybrid. In the *stateless* approach [4,7,17,18,25,39,47,50], the server is unaware of the cache content of the client and hence periodically broadcasts invalidation reports (IRs) to all the MUs. The client must check the validity of cached entries from the server before each query. Even though stateless approaches employ simple database management, they lack in scalability and addressing mobility and hence disconnections. In the *stateful* approach [20], on the other hand, the base station maintains the state of the data item for each mobile cache and only broadcasts invalidation reports for cached items. Although stateful approaches [8,20] are scalable, they incur significant overhead due to server database management; therefore, *hybrid* approaches [43–46] have been proposed that offer scalable and efficient solutions for maintaining data consistency in wireless cellular networks. In the following, we summarize existing schemes followed by our own approach for cache consistency.

Stateless Approach

In the stateless approach, the server assumes no knowledge of the cache contents of the client and hence periodically broadcasts invalidation reports.

At the mobile device, a data item request cannot be serviced until the next IR from the server is received. Barbara and Imielinksi [4] proposed three stateless algorithms: time stamps (TS), amnesic terminals (AT), and signature (SIG), in which the MSS broadcasts IR messages every L seconds. An IR message includes all data item IDs updated during the past kL seconds, where k is a positive integer. The advantage of these algorithms is that an MSS does not maintain any state information about its MUCs, thus allowing simple management of the SC; however, they suffer from the following drawbacks:

- They do not scale well to large databases or fast updating data communication systems due to the large number of IR messages.
- The average access latency is always higher than half of the broadcast period simply because all requests can be answered only after the next IR.
- When the sleep time (when an MU is disconnected from its MSS) is longer than kL, all cache entries are deleted, leading to unnecessary bandwidth consumption, particularly if the data is still valid.

To handle the long sleep–wake-up patterns, several algorithms have been proposed. In the bit-sequence algorithm [18], all cache entries are deleted only when half or more of the data items in the cache have been invalidated; however, the model requires the broadcast of a larger number of IR messages than TS and AT schemes. Although the uplink validation check scheme [47] can deal with long sleep–wake-up patterns, it requires more uplink bandwidth. In order to reduce the IR messages, Hu and Lee [17] developed adaptive methods to broadcast different IRs based on the update frequency, MU access, and sleep–wake-up patterns. Yuen et al. [50] employed an absolute validity interval for each data item; however, it fails to reduce the access delay introduced by periodic broadcast cycles.

In the above approaches, all MUs can benefit from the broadcast only when they retrieve the same data items from the MSS in the same broadcast cycle; however, if they retrieve in separate broadcast cycles, they cannot share the broadcast data. This makes the broadcast inefficient and sensitive to the number of MUs in the cell. Cao [7] modified the TS strategy to keep the invalidated data items in an MUC such that the MU can update data received from the broadcast channel. This approach increases the broadcast channel utilization; however, keeping invalid data in an MUC wastes precious cache memory. A comprehensive performance evaluation of the existing stateless algorithms is studied in Tan et al. [39].

Stateful Approach

Very few stateful cache consistency maintenance algorithms have been proposed in the literature for wireless mobile environments. Kahol et al. [20] presented an asynchronous stateful (AS) algorithm to maintain cache consistency in which an MSS records all retrieved data items for each MU. When an MU first retrieves a datum after it wakes up, based on the MUC content and sleep–wake-up time, the MSS sends an IR to that particular MU. Whenever an MSS receives an update from the original server for each recorded data item, it immediately broadcasts the IR of that item to MUs. The advantage of the AS scheme is that the MSS avoids unnecessary IR broadcast to MUs. Moreover, MUs can deal with any sleep–wake-up pattern without losing valid data items; however, to maintain each MUC, the MSS must record all cached data for each MU, so an MU can only download data requested through the uplink. This makes the broadcast channel utility inefficient and sensitive to the number of MUs. More recently, Cao [8] used a counter-based scheme to identify the frequently accessed data and save unnecessary IR traffic. Whenever the content of an MUC is changed, the MU must piggyback the change to the server, thus consuming battery power and uplink bandwidth.

As mentioned earlier, stateless approaches employ simple database management schemes but lack in scalability in terms of database size and number of MUs; moreover, they cannot handle mobility and disconnections. Although the stateful approaches are scalable, they incur significant overhead due to server database management. Furthermore, existing caching schemes assume reliable communication between the MSS and MUs for IR broadcast; however, any reliable communication mechanism requires acknowledgment from the MUs. After an IR is broadcast, the increased competition for the uplink channel between the MSS and MUs will have an impact on the uplink queries and, hence, on the average access delay and battery consumption of the MU. If an MU is disconnected during the IR broadcast, the server cannot get the acknowledgment back and must retransmit the IR because it does not know whether the IR is lost or the MU is disconnected. Also, the existing schemes do not investigate the possible inconsistency and performance loss due to wireless channel errors; therefore, scalable and efficient algorithms are required for maintaining cache consistency in the error-prone wireless channels.

Scalable Asynchronous Cache Consistency Scheme

Wang et al. [43,44] proposed a hybrid approach called the *Scalable Asynchronous Cache Consistency Scheme* (SACCS) for maintaining cache consistency between the server and mobile devices. SACCS relies on three key features: (1) use of flag bits at the server and mobile device cache, (2) use of an identifier for each entry in mobile device cache after its invalidation, and (3) rendering of all valid entries of a device cache to an *uncertain* state upon wake-up. These features make SACCS a highly efficient (in terms of bandwidth) and scalable algorithm with minimum data management overhead. A dynamic version of this scheme for a multi-cell environment that considers user mobility is presented in Wang et al. [45,46].

In SACCS, each data item in a server is associated with a flag bit. When a device retrieves an item, the corresponding flag bit is set, indicating that a valid copy may be available in the mobile cache. When the data item is updated, the server immediately broadcasts its IR to mobile devices and resets the flag bit. The reset flag bit implies that there is no valid copy in any mobile cache; hence, subsequent updates do not broadcast IRs until the flag bit is set again. A device is either in an *awake* state (i.e., connected with the base station) or in a *sleep* state (i.e., disconnected from the base station) at the time of the IR broadcast. If a device is in the awake state, it deletes the valid copy and sets the entry to an *invalidated* state (i.e., the item is deleted, but an ID is kept) after it receives the IR. If the device is in the sleep state, it misses the IR, and upon wake-up it sets all valid cache entries to an *uncertain* state. An uncertain entry must be refreshed or checked before its usage, thus guaranteeing the cache consistency. All entries with data items in uncertain or invalidated states can be used to identify useful broadcast messages for validating and triggering of data item downloading.

Each data item in the system is considered to be in one of three states: invalid, certain (or valid), or uncertain. The *invalidated* state is defined for items that are not cached at the device. The *uncertain* state is defined for items cached at the mobile device but their validity is not confirmed. The *certain* state is defined for items whose validity is confirmed and can be used to satisfy the data request by an application.

A highly scalable, efficient, and low-complexity algorithm, SACCS provides only weak cache consistency with a small probability of stale cache hits under unreliable IR broadcast environments. Unlike synchronous periodic IR broadcast schemes (for example, see Barbara and Imielinksi [4]), most of the unnecessary IRs can be avoided and, consequently, substantial bandwidth is saved. SACCS maintains only *one extra flag bit* for each data item in the SC. In contrast, the asynchronous stateful algorithm [20] requires $O(MN)$ space in the MSS to maintain the states of MUCs, where M is the number of MUs and N is the number of data items in SC. Moreover, the database management overhead is minimal, requiring only a single bit check and set/reset. Once an item is invalidated, its entry in an MUC is set to the ID-only state; that is, the object is deleted but its ID is kept. All the valid MUC data entries are set to the uncertain state

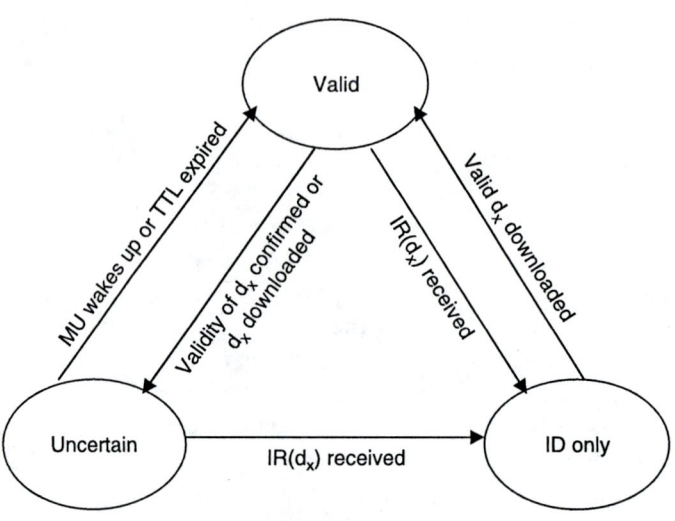

Figure 17.2   State diagram of cache entry *x*.

(i.e., the validity of a cache entry is not clear) after an MU wakes up. This mechanism makes the handling of disconnections and mobility very simple. The MSS sets an estimated time to live (TTL) for each data item based on its update history. This TTL is also cached together with the data in the MUC when acquired from the MSS. The cached data entry in the MUC is automatically set to an uncertain state when its TTL expires. This will protect stale data from being used for an arbitrarily long time due to IR loss, which means that a connected MU cannot correctly receive a broadcast IR. All entries in the uncertain or ID-only state can be used to identify useful broadcast messages for validation and triggering of data object downloading; hence, all the MUs strongly cooperate in sharing the broadcast resource. Figure 17.2 shows how a data entry (d_x) changes from one state to another.

MUC Management

Because the focus here is on cache consistency maintenance, we use a least recently used (LRU)-based replacement algorithm for the management of MUCs in which a newly cached data item or one that receives a hit is moved to the head of the cache list. If an item must be cached when the cache is full, data entries from the tail whose flags do not indicate waiting are deleted to accommodate this new item. Any refreshed items from the uncertain or ID-only state are placed in their original location and again, if necessary, enough data entries from the tail are removed. We limit the number of ID-only entries that can be used at any given instant; for example, if a cache can hold C items, then the number of ID-only entries is limited to C. Because the MU has limited cache size, during a broadcast it caches only those items with IDs in the cache.

MUC Consistency Maintenance

In the following, we discuss data consistency between an SC and MUCs. For simplicity, we assume an error-free channel in which all MUs in the awake state receive IRs (i.e., the MU is connected to the MSS). For details on IR loss due to fading wireless channels, see Wang et al. [44].

For each cached data item d_x, SACCS uses a single flag bit (f_x) in the SC to maintain the consistency between the SC and MUC. When d_x is retrieved by an MU, f_x is set indicating that a valid copy of d_x may be available in an MUC. If and when the MSS receives an updated d_x, it broadcasts an IR(x) and resets f_x. This action implies there are no valid copies of d_x in any MUC. Furthermore, for $f_x = 0$, subsequent updates do not entail broadcast of IR(x). The flag f_x is set again when the MSS services a retrieval (including request and confirmation) for d_x by an MU.

As mentioned before, an MUC can be in one of two states: *awake* or *sleep*. If an MU is awake at the time of an IR(x) broadcast, the copy of d_x is invalidated and an ID-only entry is maintained by the MU. The data items of an MU in the sleep state are unaffected until it wakes up. When an MU wakes up, it sets all cached valid items (including d_x) into the uncertain state; consequently, MUs and their cached entries are unaffected if an IR(x) broadcast occurs during their sleep times. A TTL is associated with each cache entry. When the TTL of a cache entry expires, an MU automatically sets it into an uncertain state. There are some probabilities that an MU can get a stale cache hit in the case of IR loss.

Efficiency and Cooperation

A good cache consistency maintenance algorithm must be scalable and efficient in terms of the database size and the number of MUs. In this regard, SACCS can handle large and fast updating data systems (such as news or stock quotes) because the MSS has some knowledge of the MUC. Only data entries with flag bits set result in the broadcast of IRs when data items are updated; consequently, the IR broadcast frequency is the minimum of the uplink query (i.e., confirmation frequency) and the data update frequency. In this way, the consumption of broadcast channel bandwidth for IRs is minimized. Besides IR traffic, all other traffic in SACCS is also minimized due to the strong cooperation among the MUCs. This is specifically due to the introduction of the uncertain state and the ID-only state for the MUCs. The retrieval of a data item, d_x, from the MSS issued by any given MU brings the entries of x in the uncertain or ID-only state in all of the awake MUCs to a valid state. Moreover, a single uplink confirmation for entry x causes all entries of x in the uncertain state for all the awake MUCs to be in either a valid or an ID-only state. The addition of the uncertain state also allows an MUC to keep all the valid data objects when it wakes up after an arbitrary sleep time. In contrast, in the AS and TS algorithms, all invalidated data is completely deleted from the MUC. This allows little cooperation among the MUs, resulting in a dramatic increase of traffic volume between the MSS and the MUs as the number of MUs increases. Although the scalability of the TS scheme can be improved by retaining the invalid data [7], the cache efficiency is reduced by having to keep in the MUC the invalid data objects rather than IDs, as in the SACCS approach.

The TTL expiration of a valid cache entry is checked only when its data object is accessed or available on the channel. When the data item of a valid entry is accessed or available on the channel, its TTL is first checked. If the TTL has not expired, the entry is treated as valid; otherwise, it is handled as an uncertain entry. The cost of a TTL expiration check is less than that of a query or download of an item.

Mobility

The SACCS algorithm can handle the mobility of MUs as effectively as the AS scheme. When an MU roams, it is in either the awake or sleep state. In the sleep state, no extra work is required in SACCS or AS. Upon wake-up, in SACCS an MU sets all its valid cache entries to the uncertain state. If a roaming MU is awake, SACCS treats it as if it just woke up from the sleep state; thus, consistency is guaranteed and all valid data objects are retained. In AS, when a roaming MU wakes up, its first query will be sent to the new MSS, which in turn retrieves the cache state from the previous MSS. Two alternatives are possible in AS:

- The MU notifies the MSS when it roams. In this case, the MSS manages the handoff and transfers the cache state to the new MSS, requiring no extra action for cache consistency maintenance.
- The MU does not notify the MSS when it roams. A wake-up event is forced in this case. The next query message to the MSS includes a cache state request message that contains the previous MSS and roaming time. The current MSS retrieves its state from the previous MSS and sends the IRs of items (that were updated after the MU roamed) to the requesting MU, thus maintaining cache consistency.

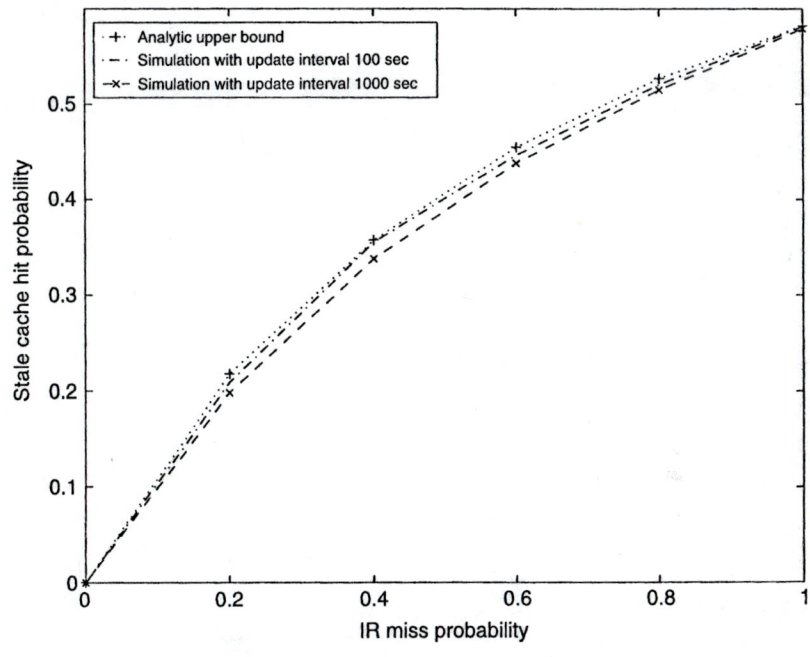

Figure 17.3   Upper bound on stale cache hit probability versus IR miss probability.

Failure Handling

The SACCS algorithm treats an MU failure as a disconnection. When an MU recovers from a failure, it sets all cached valid data entries into the uncertain state. This situation is similar to wake-up from a sleep state. The handling of server failures is also simple. When an MSS server is back after a failure, it simply broadcasts a server-down message to all MUs, which in turn set all valid data entries into the uncertain state. The MUs in the sleep state miss the server-down message, but upon wake-up all valid entries are automatically set to the uncertain state; thus, cache consistency is maintained even if some cached data items are updated during the MSS server failure. This is because the validation of any cached data object must be refreshed or checked before its usage. Finally, all valid data is retained after a server failure due to the fact that it is set to the uncertain state, thus avoiding the download of extra data. Figure 17.3 shows the stale cache hit probability for various IR miss probabilities. The results show that the stale cache hit probability is about 10 percent (21 percent) for an MU with 10 percent (20 percent) IR miss probability, independent of the data object update frequency. This analytical model assumes that an MU is always awake. If we consider the sleep–wake-up

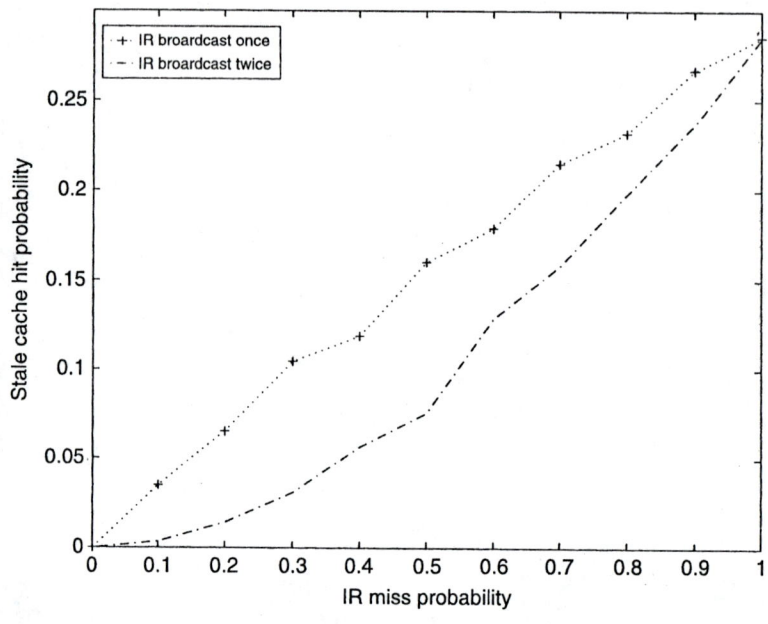

Figure 17.4   Simulation of stale cache hit probability versus IR miss probability.

event for an MU, then the stale hit probability is reduced because, when the MU wakes up from the sleep state, all valid entries will be checked prior to their usage. The stale cache hit probability for a frequently disconnected MU is much smaller than the upper bound.

In a simulation of a system with 100 MUs, each MU has a sleep–wake-up period randomly picked from the set of values (600, 1200, 1800, 2400, 3000) seconds, the sleep ratio picked from (0.2, 0.35, 0.5, 0.65, 0.8), and the request arrival rate picked from (1/20, 1/40, 1/60, 1/80, 1/100). Figure 17.4 shows the simulation results for the stale cache hit probability of the system, which is reduced significantly as compared to that of MUs that are always awake. For example, when the IR miss probability is 10 percent the stale hit probability is only about 4 percent, and for an IR miss probability of 20 percent the stale hit probability is about 7 percent. To reduce the stale hit probability, we can broadcast the IR multiple times when an object is updated earlier than its TTL expiration. Figure 17.5 shows the stale cache hit probability for broadcasting each IR twice if its update is earlier than the TTL expiration. The results indicate that, if the IR miss probability is less than 40 percent, the stale cache hit probability is less than 5 percent. These results demonstrate that SACCS exhibits very small stale cache hit probability by broadcasting IR multiple times when the update is earlier than the TTL expiration.

Performance Study

Wang et al. [44] proposed an analytical model to estimate the upper bound on the stale cache hit probability. A two-state Markov chain model was introduced to characterize the Rayleigh-fading, error-prone wireless channel [12,30]. Simulation experiments were also conducted to evaluate SACCS in terms of stale cache hit probability and average access delay under varying MU speeds. Simulation results demonstrated that SACCS offers superior performance over existing stateful and stateless algorithms in both single- and multi-cell environments. For example, in a single-cell system with five types of MU access and sleep–wake-up patterns and 10 types of data update frequencies, the average query delay for SACCS is less than 50 percent of the asynchronous stateful algorithm and less than 88 percent of the time stamp scheme, a stateless scheme. In a seven-cell system, SACCS achieves 30 percent more gain in terms of average access delay, compared to AS in a wide range of MU roaming frequencies [45,46].

In the following, we describe the performance of SACCS and compare it with the TS and AS schemes. For meaningful comparison, we also use an extended version of TS (called ETS) with some advanced features of SACCS, including: (1) introduction of the uncertain state for an MU keeping its valid data entry after long disconnection, (2) use of the ID-only state in the MUC to trigger data object download, and (3) use of flag bits in the SC to reduce the IR broadcast traffic. For SACCS, an IR of a data object is broadcast twice if its update time is earlier than the TTL expiration to reduce the stale hit probability. The other error recovery costs, such as data retransmission, are ignored in all three algorithms. The TTL (l_x) of a data item is dynamically calculated as $l_x = (l_x + l_{interval}) \times 0.5$, where $l_{interval}$ is the current update interval for the data.

Single-Cell Environment

We consider a single-cell system with one SC and multiple MUs of identical cache size. The parameters are defined in Table 17.1. The request and data update processes of each MU are assumed to follow Poisson distributions. The sleep–wake-up process is modeled as a two-state Markov chain. The state transition probability from awake to sleep state is $\alpha = 1/(1 - s)T_p$, and that from sleep to awake state is $\beta = 1/sT_p$.

In the simulations, we use two channels with bandwidth W_d and W_u for downlink and uplink data transmission, respectively. In the uplink channel, all messages are buffered as a first-in, first-out (FIFO) queue. The downlink has two FIFO queues, one having higher priority than the other. The IR messages are buffered in the higher priority queue. All other messages are buffered in the lower priority queue; this queue can be

Table 17.1   Parameter Definitions

M	Number of MUs in the system
N	Number of data objects in the system
C	MU cache size (bytes)
λ	Average arrival rate of request for an MU
T_u	Average update time interval for a data object (seconds)
T_p	MU sleep–wake-up cycle period (seconds)
s	Ratio of sleep time to sleep–wake-up period of an MU
b_o	Data object size (bytes)
b_u	Uplink message size (bytes)
b_d	Downlink invalidation of or confirmation message size (bytes)
D	Average query delay (seconds)
UPQ	Uplink per query, defined as the total number of queries through uplink channel divided by the total number of queries
L	Invalidation broadcast period for TS scheme (seconds)
wsz	Broadcast window size for TS scheme

scheduled only if the higher priority queue is empty. All requests are ignored when an MU is in the sleep state. When a requested data object is available at an MUC, the query delay is counted as 0. We consider a Zipf-like distribution for the MU access pattern [6,53] such that the access probability (p_x) for data item d_x is proportional to its popularity rank, rank(x). More specifically, $p_x = k/\text{rank}(x)^z$, where k is a normalization constant and z is the Zipf coefficient.

The performance comparison is made in terms of two metrics, D and UPQ, for three different cases. The average waiting time (i.e., half of the IR broadcast period, $L = 2$) is removed from D for ETS to make a better comparison with SACCS and AS in all figures. As shown, for the same sleep–wake-up pattern, the stale cache hit probability is less than 5 percent if the IR miss rate is smaller than 40 percent; hence, the stale hit probability is not presented as a metric in the result. In each case, $b_u = b_d = 20$ bytes for both SACCS and AS, and $b_u = b_d = 10$ bytes for TS and ETS. The bandwidth is set as $W_d = 200$ Kbps and $W_u = 1$ Kbps. The other parameters may be changed in some cases. Some default values are set as $N = 10,000$, $M = 100$, $C = 5$ MBytes, $z = 0.9$, $L = 20$ sec, and $w_{sz} = 5$.

In all cases, we consider a system with 10 types of data objects. The data update rate (T_u), size, and percentage of each type over the total objects are shown in Table 17.2. The chosen parameter values are based on the understanding that a faster updated object usually has a smaller size. The average data object size is about 25 Kbytes, as in Internet measurements [5].

Table 17.2 Ten Types of Data Objects in Database

	Data Type				
Parameter	1	2	3	4	5
Size (bytes)	1 K	5 K	10 K	15 K	20 K
T_u (sec)	50	100	200	400	800
Percentage (%)	5	5	10	10	20
	Data Type				
Parameter	6	7	8	9	10
Size (bytes)	25 K	30 K	35 K	40 K	45 K
T_u (sec)	1600	3200	6400	12,800	25,600
Percentage (%)	20	10	10	5	5

The MUs may differ from one another in terms of the sleep ratio (s), sleep–wake-up period (T_p), and inter-arrival time (T_r) of requests. These parameters for each MU can take values from the corresponding given sets. Each value has an equal probability of being chosen for each MU. The following sets of values are used: arrival rate (1/20, 1/40, 1/60, 1/80, 1/100), sleep ratio (0.2, 0.35, 0.5, 0.65, 0.8), and sleep–wake-up period (600, 1200, 1800, 2400, 3000) seconds.

The query patterns for each MU are assumed to follow a Zipf-like distribution. The access popularity rank of each MU is shifted by a random number between 0 and 99. For example, an MU picks up a shift number 50, which means the MU has the highest access popularity for data object numbered 51. The popularity decreases from 51 to N (the total number of objects), then from 1 to 50. The data object 50 has the lowest access popularity.

Effect of the Number of MUs

In this case, we study the impact of the novel features of SACCS on the system performance as compared with TS, ETS, and AS. We also consider three variants of SACSS as follows: (1) SACCS-nfg, with no flag bit set in the SC; (2) SACCS-nid, with no ID in the MUC; and (3) SACCS-nuc, without the uncertain state in the MUC. (Recall that ETS is an extension of TS with all of the SACCS features.) Table 17.3 and Table 17.4 present the average delay (D) and uplink per query (UPQ) values for a varying number (M) of MUs. For all algorithms, D increases with M. The variants of SACCS have much shorter D as compared with AS, TS, and ETS, especially when

Table 17.3   Average Access Delay (*D*) Versus Number of MUs

Algorithm	Average Access Delay (D) (sec)					
	20 MUs	40 MUs	60 MUs	80 MUs	100 MUs	120 MUs
SACCS	0.907	1.006	1.129	1.329	1.836	2.999
SACCS-nfg	0.912	1.021	1.153	1.372	1.932	3.293
SACCS-nid	0.968	1.129	1.346	1.736	3.128	13.400
SACCS-nuc	1.044	1.149	1.391	1.693	2.674	10.033
AS	0.969	1.139	1.376	1.824	3.619	18.429
TS	13.242	14.444	15.818	17.585	27.244	125.164
ETS	12.774	13.779	14.674	15.488	16.587	18.767

$M > 100$. Moreover, turning off the flag bit in SC has the least impact on D. This is due to the fact that the IR message is very small compared to the data object size. SACCS has about 10 percent less delay than SACCS-nfg when $M = 120$. Turning off the ID-only or uncertain state makes SACCS less scalable and leads to a larger D as M increases. This is because the ID-only and uncertain states allow MUs to share the broadcast data objects, thus saving the downlink bandwidth and, consequently, reducing the access delay. The AS scheme has smaller D than TS, but it does not scale as much as ETS, which allows strong cooperation among MUs because ETS incorporates all three features of SACCS.

For SACCS-based algorithms and ETS, the *UPQ* metric decreases as M increases, but, for AS and TS, it is almost constant. This is due to the cooperation among MUs in SACCS-based algorithms and ETS. Note that SACCS has the least *UPQ*, and turning off the ID produces the largest increase in *UPQ*. Simulation results validate our claims — namely, that the ID-only entry and uncertain state in the MUC are critical for SACCS, and use of the flag bit in the SC reduces IR traffic. Because ETS performs better than TS, in the following cases we will use ETS instead of TS.

Effect of Database Size

Figure 17.5 and Figure 17.6 present the simulation results showing the effects of database size. For ETS, the average query waiting time ($L/2 = 10$ seconds) is not counted. In other words, only the queue delay and transmission time are counted for ETS in all cases. SACCS outperforms AS and ETS in both D and *UPQ* metrics by avoiding unnecessary IR traffic while retaining all the valid data objects in the MUCs. As expected, with an increased number of

Table 17.4   Uplink per Query (UPQ) Versus Number of MUs

Algorithm	Uplink per Query (UPQ)					
	20 MUs	40 MUs	60 MUs	80 MUs	100 MUs	120 MUs
SACCS	0.902	0.894	0.874	0.866	0.854	0.838
SACCS-nfg	0.904	0.895	0.876	0.868	0.856	0.839
SACCS-mid	0.927	0.926	0.927	0.925	0.920	0.917
SACCS-nuc	0.909	0.896	0.884	0.871	0.862	0.850
AS	0.900	0.904	0.906	0.910	0.909	0.901
TS	0.926	0.925	0.929	0.930	0.930	0.922
ETS	0.914	0.892	0.889	0.872	0.861	0.847

data objects, the performance metrics also increase for all algorithms, but SACCS has much smaller D than AS and ETS. Additionally, the average gain (in terms of D) of SACCS over AS and ETS is more than 50 percent throughout the range of database sizes. The UPQ of SACCS is slightly less than that of ETS and about 6 percent less than that of AS.

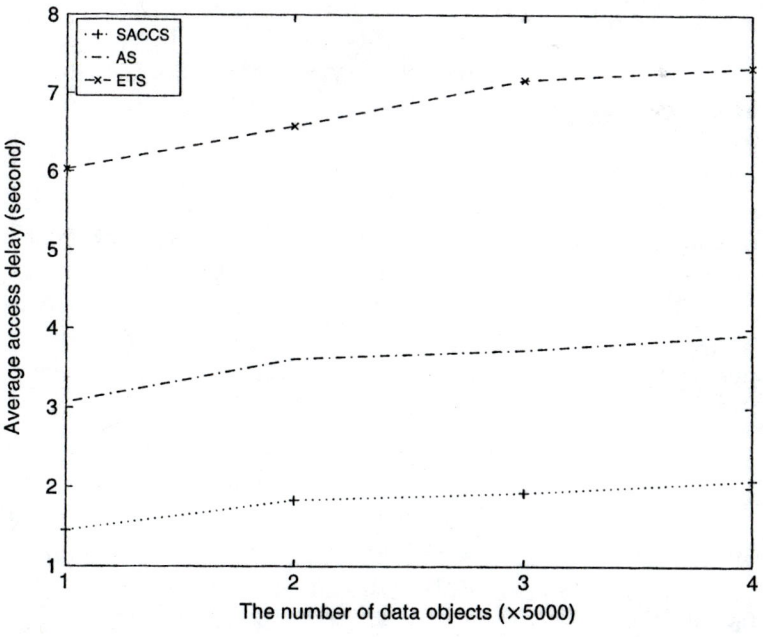

Figure 17.5   Average access delay versus number of data objects.

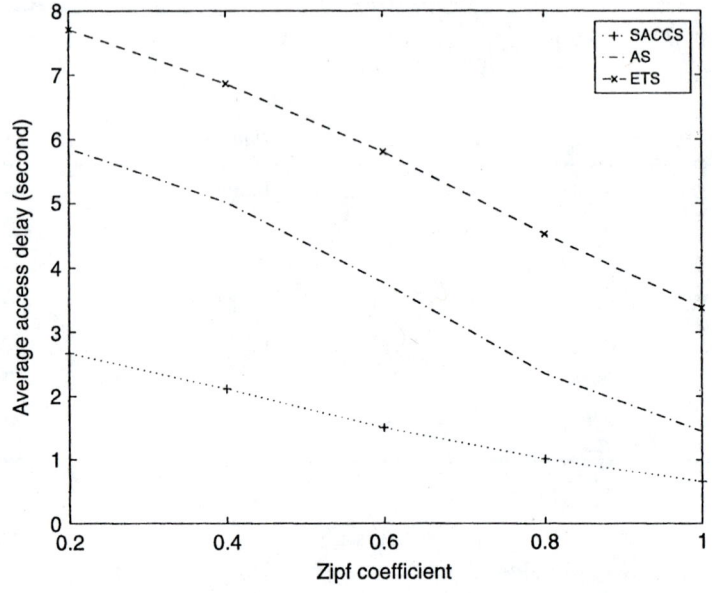

Figure 17.6   Average access delay versus Zipf coefficient.

Effect of Zipf Coefficient

Here, we choose a small database with $N = 1000$ objects. This is because, for a small Zipf coefficient, the access frequencies for different data objects are very close to each other; hence, a large database size results in very few cache hits, which make the comparison meaningless. Figure 17.7 and Figure 17.8 show that SACCS has much smaller D than AS or ETS. The average gain is more than 50 percent over the others. The AS scheme has the largest UPQ, and SACCS has the lowest UPQ when $z > 0.6$ and slightly more than ETS when $z < 0.6$. All three schemes perform better as z increases because the data accesses are more concentrated for larger z, thus increasing the cache hit ratio and then reducing access delay.

Multi-Cell Environment

Figure 17.9 shows a wireless system of seven hexagonal cells. Initially, we assume each cell has an equal number of MUs (i.e., $M = 7$). An MU roaming process is assumed to be Poisson with an average sojourn time of T_r (seconds) in a cell. An MU from cell 1 has equal probability (1/6) of roaming into any of its six neighbors. Similarly, an MU from any other cell has 1/6 probability of roaming into cell 1 and 5/12 probability of

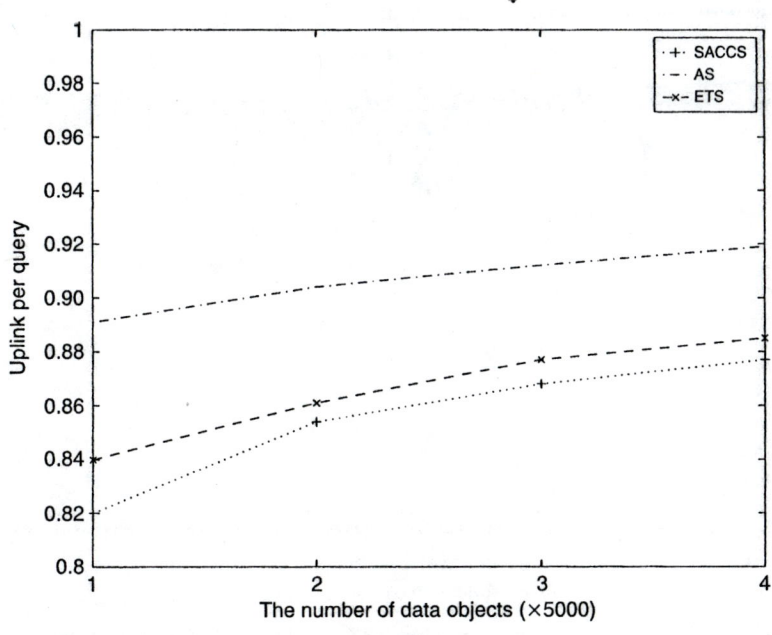

Figure 17.7   Uplink per query versus number of data objects.

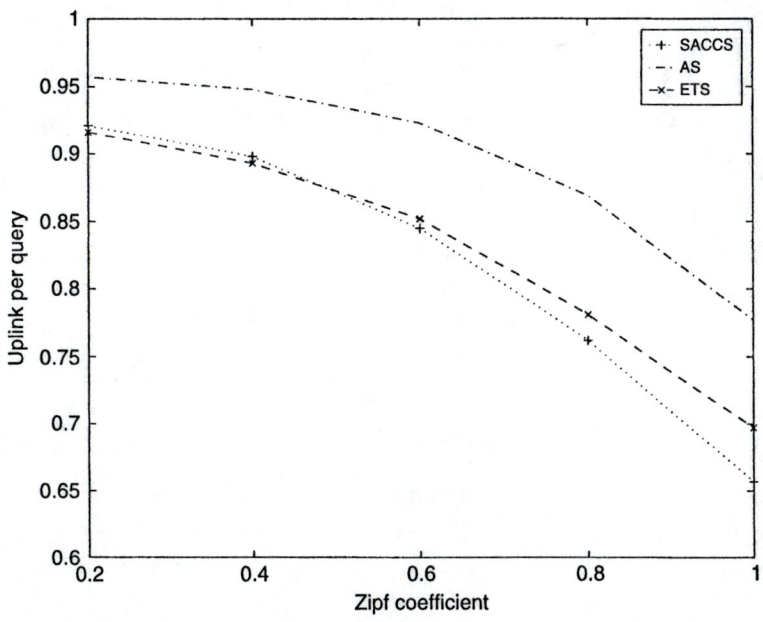

Figure 17.8   Uplink per query versus Zipf coefficient.

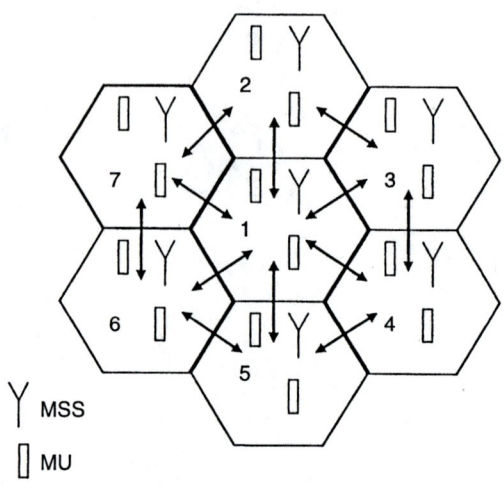

Figure 17.9   Seven-cell configuration; each cell has one MSS.

roaming to another neighboring cell. This balances the number of MUs in each cell having one uplink channel with bandwidth W_u (bps) and one downlink channel with bandwidth W_d (bps).

Simulation results in the previous subsection demonstrated that the performance of TS (or ETS) is much worse than that of SACCS and AS. In addition, no good mobility handling scheme exists for TS. In the following, we study the impact of mobility on SACCS and AS. As stated earlier in this section, a roaming MU is treated as a forced wake-up event. In SACCS, all valid cache entries are set to the uncertain state after the roaming. In AS, the first query and roaming time of the MU are sent to the MSS after the roaming. Here, we focus on the cache consistency scheme in the multi-cell environment in our simulations, rather than the in-session data transactions. In other words, when an MU roams from one cell to another, all requests sent to the former MSS are dropped. The values of the parameters are the same as shown in Table 17.2.

Effect of Average Sojourn Time

Figure 17.10 demonstrates that SACSS has much better performance in terms of D (average query delay) for all ranges of average sojourn time. Moreover, SACCS has a lower UPQ for long average sojourn times (e.g., 800 seconds) but a higher UPQ than that of AS for very short average sojourn times. In SACCS, all cache entries are set to the uncertain state after each roaming. In AS, the first query after each roaming has to be forwarded to the MSS to retrieve the cache state. For a very short average

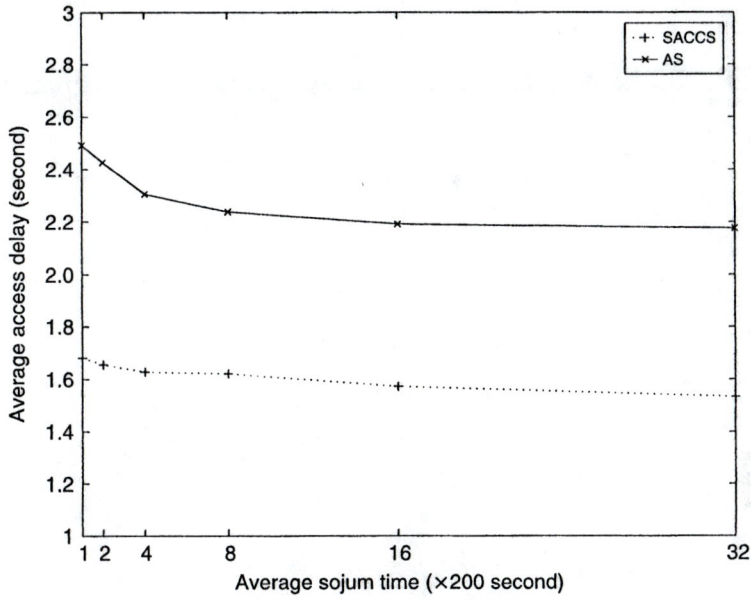

Figure 17.10 Average access delay versus MU average sojourn time.

sojourn time, the uncertain entries in SACCS have very few chances to be refreshed before their usage, thus resulting in a larger *UPQ* for SACCS than for AS. For both SACCS and AS, the performance gets better as the average sojourn time increases (roaming frequency decreases). The value of *D* in SACCS decreases from 1.7 to 1.55 seconds. In AS, it decreases from 2.5 to 2.1 seconds. This is due to fewer forced wake-up events and hence fewer extra uplinks. These results show that the impact of mobility is not significant in either scheme due to the fact that all valid data objects are retained after roaming. In other words, the uncertain state is a powerful method for SACCS to treat the disconnectedness and mobility of MUs.

Utility-Based Data Caching and Prefetching

In mobile computing environments, vital resources such as battery power and wireless channel bandwidth impose significant challenges in ubiquitous information access. Shen et al. [34,35] have developed a novel energy- and bandwidth-efficient data caching mechanism called the *GreedyDual Least Utility* (GD-LU), which enhances dynamic data availability while maintaining consistency. This approach employs the SACCS scheme for consistency between mobile devices and server. The analytical model

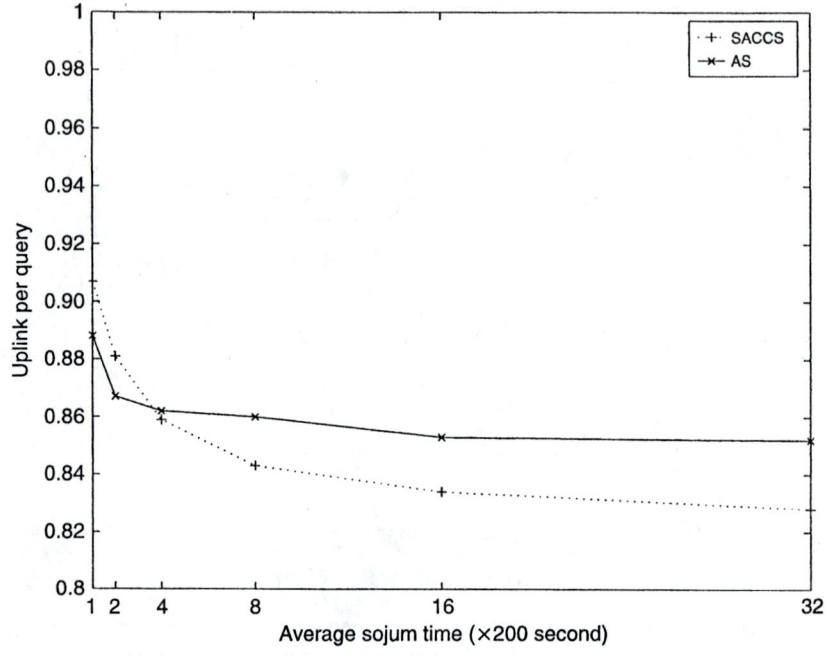

Figure 17.11   Uplink per query versus MU average sojourn time.

considers different events such as data request, data update, connection–disconnection, and mobility handoff that occur at mobile devices and affect energy consumption due to data retrievals. Based on this model, *autility* function is derived to evaluate each data item in terms of energy savings. The GD-LU algorithm has two components: (1) a GD-LU cache replacement algorithm that selects data items with the most utility to cache in local memory, and (2) a GD-LU passive prefetch algorithm in which mobile devices acquire data items from broadcast channels for future requests based on the *relative utility* values of the items. Based on priority queue management, these cache replacement and prefetching algorithms achieve a time complexity of $O(\log N)$, where N is the number of data items in the cache.

As depicted in Figure 17.13, four kinds of events in mobile devices are related to data management: (1) data request, (2) data update (i.e., receiving data invalidation report), (3) disconnection, and (4) mobility hand-off. Data update, disconnection, and hand-off events may occur zero or more times between two consecutive request events. Each mobile device retrieves the data from an original server through a unicast or broadcast channel. Data consistency is maintained by mobile devices as well as the original server.

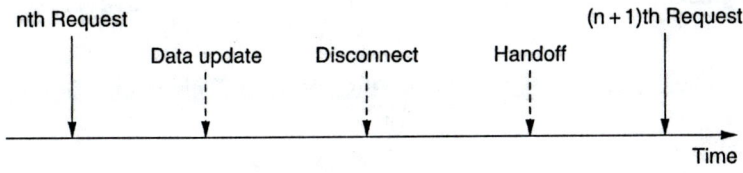

Figure 17.12   Mobile device events.

GreedyDual Least Utility Caching Mechanism

This section briefly describes the cache replacement and passive prefetching algorithms, followed by some performance results.

Cache Replacement Algorithm

The GreedyDual (GD) cache replacement is an efficient, online, optimal algorithm [52] devised to deal with systems that exhibit heterogeneous data retrieval costs. In essence, GD allows a bias to be applied to each item in a cache so as to give higher priority to items that incur higher retrieval costs. Enhanced GD algorithms have been deployed for Web

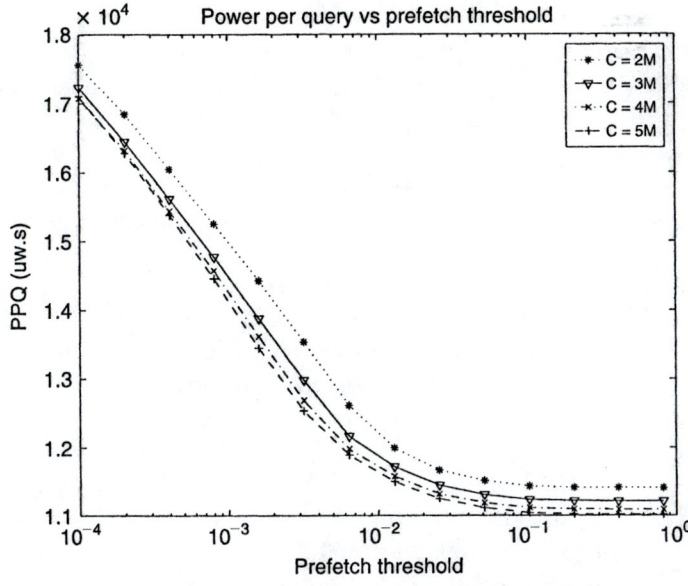

Figure 17.13   Power consumption with passive prefetching.

proxy caching [9] and multimedia stream caching [23]. Based on the GD concept, Shen et al. [34,35] proposed the GreedyDual Least Utility, a cache replacement algorithm for mobile devices in which each data item is associated with a metadata that contains the necessary parameters of that item. Because each metadata contains history information of the corresponding data item, a queue is used to store the metadata of replaced data items.

When an application requests a data item, the GD-LU algorithm first checks the state of the item in the cache. If the data item is valid, it is returned and the corresponding metadata is updated. If the item is in an uncertain state, an uncertain message is sent to the server to check if the data is valid or not since the last retrieval. In the event of a cache miss, the message is sent to the server to retrieve the data. When the mobile device receives the confirmation message, the data item is set to the certain state and returned to the application. If the entire data item is received, the GD-LU replacement algorithm chooses the victim data set to make space to accommodate the incoming data.

Passive Prefetching Algorithm

The base station and the mobile devices can communicate through a broadcast, multicast, or unicast channel. In multicast or broadcast, the devices in one group can cooperatively retrieve data from the server, so the GD-LU replacement can be expected to achieve a better performance than in the case of unicast. Additionally, the data item requested by one mobile device is available to all others in the data dissemination channel. Although a device may not request the data item, prefetching the item into the device cache may reduce future access latency. Acquiring the item from the dissemination channel still costs energy at the mobile device; therefore, blind prefetching of the data that was never requested may result in a waste of energy and also replacement of some recently accessed data, thus degrading the cache performance.

To avoid blind prefetching, the GD-LU algorithm uses a passive prefetching scheme for cache management of mobile devices in multicast and broadcast communications. A threshold (TH) is set at the cache to admit data appearing at the broadcast or multicast channel. According to the GD-LU replacement algorithm, a data item for which the metadata is kept in the queue may have a higher probability to be requested again by the mobile device, so this data item is considered to be a valuable candidate for passive prefetching. To evaluate the future utility of an item at the mobile device, the relative utility (RU) is defined as the ratio of the utility (U_i) of data (d_i) to the utility of cached data. Thus,

$$RU(d_i) = \frac{U_i^c(d_i)}{\mathrm{MAX}_{j \in C}(u_j) - \mathrm{MIN}_{j \in C}(u_j)}$$

where C is the set of data items in the cache, and u_j is the utility value of cached data item j. In Shen et al. [35], a median utility threshold scheme is proposed to achieve a near-optimal performance tradeoff between access latency and power consumption. In passive prefetching, a mobile device does not send an uplink request, thus reducing the burden on the server while improving on cache performance at the device and saving scarce wireless bandwidth.

Performance Evaluation

Extensive experiments show that the GD-LU cache replacement algorithm achieves a greater than 10 percent energy saving than existing approaches. The GD-LU passive prefetching algorithm achieves a near-optimal performance tradeoff between access latency and energy consumption for various cache sizes. In the following, we evaluate the performance of passive prefetching under various system settings and environments. As mentioned, a large threshold means few data items have to be prefetched. When the threshold approaches zero, all data items in the downlink channel are prefetched. Figure 17.13 shows the performance of the power per query (PPQ) metrics against the prefetch threshold for various cache sizes. As shown in Figure 17.13, the energy consumption is high when the threshold is small. For TH $\leq 10^{-2}$, the energy consumption significantly decreases as the threshold increases, but, if TH $> 10^{-2}$, the energy consumption is almost a constant. This is because more data items are prefetched at lower threshold values and some of them are not useful to users. For a large threshold, most of the prefetched items are useful, leading to constant energy consumption.

In Figure 17.14, if we consider energy stretch (EST) as the main performance metric for passive prefetching, we can get an optimal threshold value. Because energy stretch factors into both energy consumption and access latency, the optimal value of threshold can be considered as the best point for the tradeoff between energy consumption and access latency. As shown in Table 17.5, if we choose points with minimum EST, the energy stretch performance of the median utility threshold setting (MU_EST) achieves near-optimal performance under different cache sizes. This demonstrates that the median relative utility threshold setting is a good heuristic method to achieve the optimal performance tradeoff between energy consumption and access latency.

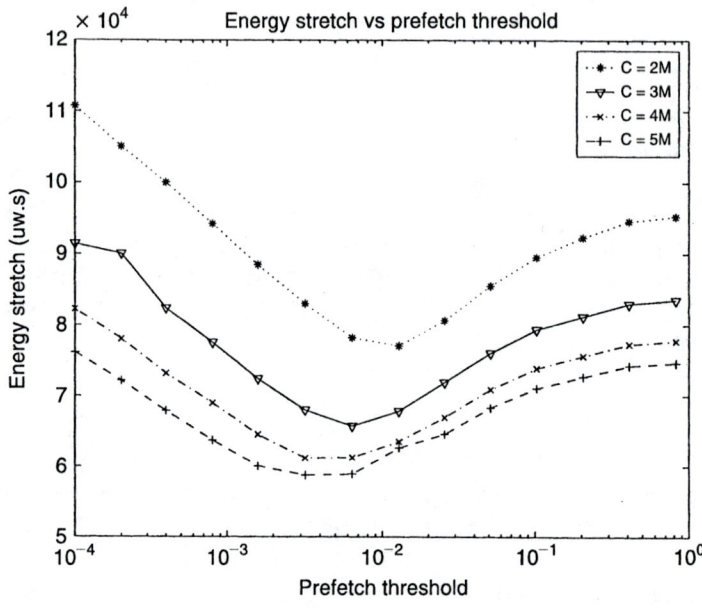

Figure 17.14 Energy stretch with passive prefetching.

Conclusion

In this chapter, we discussed the main issues and candidate solutions for mobile caching, replacement, prefetching, and consistency. We also proposed a scalable and efficient scheme, called Scalable Asynchronous Cache Consistency Scheme (SACCS), for mobile environments and evaluated its performance. Unlike existing methods, SACCS provides weak cache consistency under realistic environments for an MU with IR broadcast miss. The basic idea involves the use of flag bits at the server cache (SC) and mobile user cache (MUC), an identifier (ID) in MUC for each entry after its invalidation, and an estimated time to live (TTL) for each cached entry, as well as the rendering of all valid entries of MUC to the uncertain state when an MU wakes up. Unlike stateful algorithms, SACCS maintains only one flag bit for each data item in the mobile support station to determine when to broadcast the IRs. Furthermore, unlike existing synchronous stateless approaches, SACCS does not require the periodic broadcast of IRs, thus significantly reducing IR messages sent through the downlink broadcast channel. Comprehensive simulation results show that this scheme exhibits significantly better performance than the TS and AS schemes in both single- and multi-cell environments. An LRU-based cache

Table 17.5 Points of Minimal Energy

Cache Size (M)	MU PPQ (10^4 uw/sec)	MU HR	MU ST	MU_EST (10^4 uw/sec)	Min EST (10^4 uw/sec)
2	1.20368	0.256328	6.39952	7.70297	7.7069
3	1.2057	0.271676	5.45998	6.5831	6.57452
4	1.21802	0.281747	4.99865	6.0845	6.12083
5	1.23306	0.289208	4.75577	5.86415	5.87524

replacement algorithm is used in SACCS. It would be interesting to study the impact of other replacement algorithms on its performance. Further study is also needed for the MSS cache management algorithm and effective transfer of cached data among MSSs in response to the roaming of MUs among different MSSs.

This chapter also presented a utility model that investigates energy conservation by deploying caches at mobile devices in a dynamic data environment. Based on the utility function derived, a novel caching mechanism was proposed, called GreedyDual Least Utility (GD-LU), which combines a cache replacement algorithm with a passive prefetching scheme to reduce energy consumption and access latency of various applications such as Web caching at mobile devices. Future work includes the analysis of GD-LU performance under multi-cell data dissemination environments.

Acknowledgment

This work is supported by Texas Advanced Research Program grant 14-771032 and U.S. National Science Foundation grants IIS-0326505 and STI 0129682.

References

[1] Acharya, S. et al., Broadcast disks: data management for asymmetric communications environments, in *Proc. of the ACM SIGMOD Int. Conf. on Management of Data*, San Jose, CA, May, 1995, pp. 199–210.

[2] Acharya, S., Franklin, M., and Zdonik, S., Balancing push and pull for data broadcast, in *Proc. of the ACM SIGMOD Int. Conf. on Management of Data*, Tucson, AZ, May, 1997, pp. 183–194.

[3] Aksoy, D. and Franklin, M., Scheduling for large scale on-demand data broadcast, in *Proc. IEEE INFOCOM'98*, San Francisco, CA, March, 1998.

[4] Barbara, D. and Imielinksi, T., Sleeper and workaholics: caching strategy in mobile environments, in *Proc. of the ACM SIGMOD Int. Conf. on Management of Data*, Minneapolis, MN, May, 1994, pp. 1–12.

[5] Barford, P. and Crovella, M., Generating representative web workloads for network and server performance evaluation, in *Proc. of ACM SIGMETRICS'98*, Madison, WI, June, 1998.

[6] Breslau, L. et al., Web caching and Zipf-like distributions: evidence and implications, in *Proc. IEEE INFOCOM'99*, New York, March, 1999, pp. 126–134.

[7] Cao, G., A scalable low-latency cache invalidation strategy for mobile environments, in *Proc. of ACM Int. Conf. on Mobile Computing and Networking*, Seattle, WA, August, 2001, pp. 200–209.

[8] Cao, G., On improving the performance of cache invalidation in mobile environments, *Mobile Networks Appl.*, 7(4), 291–303, 2002.

[9] Cao, P. and Irani, S., Cost-aware WWW proxy caching algorithms, in *Proc. of the USENIX Symp. on Internet Technologies and Systems (USITS'97)*, Monterey, CA, December, 1997, pp. 193–206.

[10] Cao, P. and Liu, C., Maintaining strong cache consistency in the World-Wide Web, in *Proc. of the 17th Int. Conf. on Distributed Computing Systems (ICDCS'97)*, Baltimore, MD, May, 1997, pp. 12–21.

[11] Che, H., Tung, Y., and Wang, Z., Hierarchical web caching systems: modeling, design, and experimental results, *IEEE J. Selected Areas Comm.*, 20(7), 1305–1315, 2002.

[12] Chockalingam, A. et al., Performance of a wireless access protocol on correlated Rayleigh-fading channels with capture, *IEEE Trans. Comm.*, 46, 644–655, 1998.

[13] Fan, L., Cao, P., Lin, W., and Jacobson, Q., Web prefetching between low-bandwidth clients and proxies: potential and performance, in *Proc. ACM SIGMETRICS'99*, Atlanta, GA, May, 1999, pp. 178–187.

[14] Feeney, L. and Nilsson, M., Investigating the energy consumption of a wireless network interface in an *ad hoc* networking environment, in *Proc. IEEE INFOCOM'01*, Anchorage, AK, April, 2001.

[15] Gitzenis, S. and Bambos, N., Power-controlled data prefetching/caching in wireless packet networks, in *Proc. IEEE INFOCOM'02*, New York, June, 2002.

[16] Guo, Y., Pinotti, M.C., and Das, S.K., A new hybrid broadcast scheduling algorithm for asymmetric communication systems, *ACM Mobile Comput. Comm. Rev.*, 5(3), 39–54, 2001.

[17] Hu, Q. and Lee, D.K., Cache algorithms based on adaptive invalidation reports for mobile environments, *Cluster Comput.*, 1(1), 39–50, 1998.

[18] Jing, J. et al., Bit-sequences: an adaptive cache invalidation method in mobile client/server environments, *Mobile Networks Appl.*, 2(2), 115–127, 1997.

[19] Jiang, Z. and Kleinrock, L., An adaptive network prefetch scheme, *IEEE J. Selected Areas Comm.*, 16(3), 358–369, 1998.

[20] Kahol, A. et al., A strategy to manage cache consistency in a distributed mobile wireless environment, *IEEE Trans. Parallel Distributed Syst.*, 12(7), 686–700, 2001.

[21] Khanna, S. and Liberatore, V., On broadcast disk paging, *SIAM J. Comput.*, 29(5), 1688–1702, 2000.

[22] Li, D. and Cheriton, R., Scalable web caching of frequently updated objects using reliable multicast, in *Proc. of the USENIX Symp. on Internet Technologies and Systems (USITS'99)*, Boulder, CO, October, 1999, pp. 1–12.

[23] Lau, W.H.O., Kumar, M., and Venkatesh, S., A cache-based mobility-aware scheme for real-time continuous media delivery in wireless networks, in *Proc. IEEE Int. Conf. on Multimedia and Expo (ICME'01)*, Tokyo, Japan, August, 2001.

[24] Lin, C.W. and Lee, D.L., Adaptive data delivery in wireless communication environments, in *Proc. IEEE Int. Conf. on Distributed Computing Systems (ICDCS'00)*, Taipei, Taiwan, April, 2000, pp. 444–452.

[25] Liu, G.Y. and McGuire, G.Q., A mobility-aware dynamic database caching scheme for wireless mobile computing and communications, *Distributed Parallel Databases*, 4(5), 271–288, 1996.

[26] Lau, W., Kumar, M., and Venkatesh, S., A cache-based mobility-aware scheme for real-time continuous media delivery in wireless networks, in *Proc. IEEE Int. Conf. on Multimedia and Expo (ICME'01)*, Tokyo, Japan, August, 2001.

[27] Nuggehalli, P. et al., Energy efficient caching strategies in *ad hoc* wireless networks, in *Proc. ACM MobiHoc*, Annapolis, MD, June, 2003, pp. 25–34.

[28] Pottie, G.J. and Kaiser, W.J., Wireless integrated network sensor, *Comm. ACM*, 43(5), 551–558, 2000.

[29] Saxena, N., Pinotti, M.C., Basu, K., and Das, S.K., A new hybrid scheduling framework for asymmetric wireless environments with request repetition, in *Proc. Int. Symp. on Modeling and Optimization in Mobile, Ad-Hoc, and Wireless Networks (WiOpt'05)*, Trento, Italy, April, 2005, pp. 368–376.

[30] Rappaport, T.S., *Wireless Communications: Principles and Practice*, Prentice Hall, Englewood Cliffs, NJ, 1996.

[31] Rizzo, L. and Vicisano, L., Replacement policies for a proxy cache, *IEEE/ACM Trans. Networking*, 8, 158–170, 2000.

[32] Shen, H., Das, S.K., Kumar, M., and Wang, Z., Cooperative caching with optimal radius in hybrid wireless networks, in *Proc. Int. IFIP Networking Conf.*, 3042, 841–853, 2004.

[34] Shen, H., Kumar, M., Das, S.K., and Wang, Z., Energy-efficient caching and prefetching with data consistency in mobile distributed systems, in *Proc. of the IEEE Int. Conf. on Parallel and Distributed Processing Symp. (IPDPS'04)*, Santa Fe, NM, April, 2004.

[35] Shen, H., Kumar, M., Das, S.K., and Wang, A., Energy-efficient data caching and prefetching for mobile devices based on utility, *Mobile Networks Appl.*, 10, 475–486, 2005 (special issue on mobile services).

[36] Shim, J., Scheuermann, P., and Vingralek, R., Proxy cache design: algorithms, implementation and performance, *IEEE Trans. Knowledge Data Eng.*, 11(4), 549–562, 1999.

[37] Stathatos, K., Roussoppulos, N., and Baras, J.S., Adaptive data broadcast in hybrid networks, in *Proc. 23rd Int. Conf. on Very Large Data Bases (VLDB'97)*, Athens, Greece, August, 1997, pp. 326–335.

[38] Su, C. and Tassiulas, L., Joint broadcast scheduling and user's cache management for efficient information delivery, *Wireless Networks*, 6(4), 279–288, 2000.

[39] Tan, K., Cai, J., and Ooi, B., An evaluation of cache invalidation strategies in wireless environments, *IEEE Trans. Parallel Distributed Syst.*, 12(8), 789–807, 2001.

[40] Vaidya, N., and Hameed, S., Scheduling data broadcast in asymmetric communication environments, *Wireless Networks*, 5(3), 171–182, 1999.

[41] Wong, J.W., Broadcast delivery, *Proc. IEEE*, 76(12), 1566–1577, 1988.

[42] Wooster, R. and Abraams, M., Proxy caching that estimates edge load delays, in *Proc. Int. World Wide Web Conf.*, Santa Clara, CA, April, 1997.

[43] Wang, Z., Das, S.K., Che, H., and Kumar, M., SACCS: scalable asynchronous cache consistency scheme for mobile environments, in *Proc. Int. Workshop on Mobile Wireless Networks (MWN'03)*, Quebec, Canada, August, 2003, pp. 797–802.

[44] Wang, Z., Das, S.K., Che, H., and Kumar, M., A scalable asynchronous cache consistency scheme (SACCS) for mobile environments, *IEEE Trans. Parallel Distributed Syst.*, 15(11), 983–995, 2004.

[45] Wang, Z., Das, S.K., Kumar, M., and Shen, H., Dynamic cache consistency schemes for wireless cellular networks, *IEEE Trans. Wireless Comm.* (in press).

[46] Wang, Z., Kumar, M., Das, S.K., and Shen, H., Investigation of cache maintenance strategies for multi-cell environments, in *Proc. of ACM Int. Conf. on Mobile Data Management (MDM'03)*, Melbourne, Australia, January, 2003.

[47] Wu, K.L., Yu, P.S., and Chen, M.S., Energy-efficient caching for wireless mobile computing, in *Proc. of the 12th Int. Conf. on Data Engineering (ICDE'96)*, New Orleans, LA, February, 1996, pp. 336–345.

[48] Xu, J. et al., Performance evaluation of an optimal cache replacement policy for wireless data dissemination, *IEEE Trans. Knowledge Data Eng.*, 16(1), 125–139, 2004.

[49] Yu, H., Breslau, L., and Shenker, S., A scalable web cache consistency architecture, in *Proc. ACM SIGCOMM'99*, Cambridge, MA, August, 1999, pp. 163–174.

[50] Yuen, J.C., Chan, E., Lam, K., and Leung, H.W., Cache invalidation scheme for mobile computing systems with real-time data, *SIGMOD Record*, December, 2000.

[51] Yin, L. et al., Power-aware prefetch in mobile environments, in *Proc. of the IEEE Int. Conf. on Distributed Computing Systems (ICDCS'02)*, Vienna, Austria, July, 2002, pp. 571–578.

[52] Young, N.E., The K-server dual and loose competitiveness for paging, *Algorithmica*, 11(6), 525–541, 1994.

[53] Zhang, J., Izmailov, R., Reininger, D., and Ott, M., Web cache framework: analytical models and beyond, in *Proc. IEEE Workshop on Internet Applications (WIAPP'99)*, San Jose, CA, July, 1999, pp. 132–141.

Chapter 18

Code-on-Demand and Code Adaptation for Mobile Computing

Francis C.M. Lau, Nalini Belaramani, Vivien W.M. Kwan, Pauline P.L. Siu, Wai-Kwong Wing, and Cho-Li Wang

CONTENTS

Introduction

The Ultimate Thin Client

The ultimate mobile device should be thin, lean, and mean. Being thin, it should be physically small enough to fit in a person's pocket. Being lean, it should have only those functionalities that do just what the user needs to do. Being mean, it is so affordable that one could replace the device without hesitation — to the extent that purchasing such mobile devices is as natural and convenient as buying a pack of beer [1]. But, is a thin, lean, mean device sufficient for all the computing needs of a mobile user in the future? The answer depends on what we will do as users with mobile devices in the future. Advancements in technologies continue to give birth to more and more powerful mobile devices. Today, we are witnessing the proliferation of very powerful handhelds blending everything (3G phone, digital camera, PDA, media player, massive storage) in a single device, and the trend seems to be that these devices will evolve to become more lightweight, less expensive, and stronger in communication.

In terms of computational power, however, we do not foresee the closing of the gap that separates the mobile devices from the PC. The latter has always to meet up with the rapidly growing software algorithm complexity, whereas mobile devices would tend more to target applications at the lower end of the complexity spectrum. If mobile devices were to be used in place of the PC, even with the anticipated many-folds increase in power within a short time, software performance would improve only along a plateau. The mobile device, after all, is bound by its form factor and limited power; therefore, the wish that one day the mobile device will replace the PC as the ultimate personal computing device appears to be unrealistic. But, what will a thin, lean, mean device be good for?

Mobile computing has created new usage paradigms. Tasks in the mobile computing world can be much more dynamic than those running on a desktop because of the changing context of the mobile user. Much research on location and context awareness is now taking place which is less relevant in the non-mobile computing world. The need for a mobile device to accommodate many dynamic tasks implies that the device must come preinstalled with all kinds of software, which seems to go against the thin, lean, mean principle.

Ubiquity of Connectivity

Advances in communication are coming fast and strong. In a recent fourth-generation (4G) mobile communications field test, a maximum downstream data rate of 300 Mbps was recorded for a receiver in a car running at 30 km/hr and 800 m to 1 km away from the base station. Before 4G can be deployed, we are already enjoying the abundance of WiFi hotspots as well as the most free form of wireless communication via 3G. An emerging wireless technology is WiMax, which provides metropolitan area network connectivity at speeds of up to 75 Mbps and covering a practical distance of three to five miles. In a few years' time, the wireless network infrastructure will become completely mature, offering an abundance of bandwidth and better coverage to connect any mobile device anywhere in the worldwide network.

Communication is probably the only parameter in the configuration of a mobile device that is not constrained when compared to a PC. In fact, there are many reasons for making a mobile device more powerful in terms of communication capability than a desktop PC. If that is the case one day and if the devices are backed by a mature advanced wireless infrastructure, then wireless devices will become very much an integral part of the global network, and many of the current network-based computing paradigms such as client–server and peer-to-peer computing will become applicable to wireless devices.

The Where, What, and When of Computations

When user tasks are dynamic and diversified, it is not feasible to determine *a priori* which software to install in a mobile device. Ideally, new functionalities should be made available to the device when they are needed or they should be dynamically composed. Four design paradigms are related to this requirement: client–server, mobile agent, code-on-demand [2], and remote evaluation. In the client–server paradigm, the client asks a server implementing the service to access some resources accessible by the server. When someone implements the service, we say that the service holds the know-how. Remote evaluation is similar to the client–server paradigm, but this time it is the client that holds the know-how that will be sent to the server to carry out the service. In the mobile agent paradigm, computing may be carried out by any device in a network. In the code-on-demand paradigm, a client device gets the know-how from a peer or a server and carries out the computation by itself.

In terms of capability, a mobile device may function between two extremes. In one extreme, it acts as a remote display terminal; in the other extreme, it functions as a fully capable computational device. Remote display software, such as virtual network computing (VNC) [3], terminal services client, and g-cluster [4], require a stable network connection between the client and the server. A stable connection, however, is not always available in a mobile environment. When wired or wireless network connections improve, it is likely that the improvement is on the bandwidth but not the latency. For a user not to perceive any delay, real-time applications should have a responsive latency no longer than 50 ms [5], which is not easily achievable in a device–server operation mode where the connection could be multi-hop. When user commands and user screens are sent back and forth between the device and the server, an application would not be responsive enough.

The other extreme is to perform all functions in the device. In this role, the device is preinstalled with all the applications and data required to handle all the user tasks. The problem is that it might not be feasible to determine the kinds of software to install *a priori*. Another problem is that the device might not have enough room to accommodate the entire collection of software.

An alternative exists in between the two extremes, which is that the mobile device would collaborate with some server or peers to carry out a user task. This would be more suitable in the mobile environment where user tasks are dynamic and device functionalities are composed dynamically.

A mobile device should be able to continue working when it is disconnected from the network; therefore, client–server, mobile agent, and remote evaluation are not suitable for carrying out user tasks in the mobile environment. Code-on-demand is better because the program code is downloaded and executed in the mobile device. This paradigm would be more tolerant to various issues such as performance bottlenecks, fault tolerance, availability, service customization, user interface responsiveness, and device mobility; therefore, in the context of mobile computing, code-on-demand paradigms are suitable for carrying out dynamic and diversified user tasks.

The Future

Much of the potential of mobile devices has yet to be exploited. Current mobile applications tend to be simple. Comparing the desktop version and personal digital assistant (PDA) version of a word processor, for example, we can see that many of the functionalities are not available on the PDA. Besides the small form factor, which makes maneuvering difficult and hence some of functionalities not appropriate, there is no reason why feature-rich software has to be out of reach for PDAs in the future. We believe that even the most complex actions could be carried out on mobile devices by improving the input/output modality and redesigning the software engineering approach.

So, what can the thin, lean, mean device do for us in the future? In the not so distant future, it will probably play the role of a full-function computing device, but a lesser device than the PC. Software or software components can be installed on the fly from a nearby server wirelessly upon request, which can be discarded after use. For this to be feasible, the software must itself be suitably lean so downloading and installing it on demand will be efficient and the software will not take up more resources than is commensurate with the user's needs. Also, lean software will naturally be more affordable, or mean. In the longer term, as the wireless infrastructures around us become sufficiently powerful and stable, it can be envisioned that mobile devices will offload more or even all of their computations to the servers, and the devices will be reduced to very thin wireless remote display terminals. Users then will be free of all the trouble of managing a computing system at home or in the pocket. This chapter addresses the near-term solution of code-on-demand.

Small Codes for Small Devices

A New Notion of Application

Traditionally, applications are built as huge monolithic chunks. These applications provide lots of functionalities, yet they are too big to fit into the small devices that are now prevalent in the mobile environment. The current solution is to develop other versions of an application that would fit, and such versions would most likely be downgraded, meaning that some of the functionalities would not be available in these versions. If we compare a Word processor in a PC with the PDA counterpart, we would see that the functionalities of the PC version are overkill, and the PDA version appears to be much deprived. Our solution to the problem is to build an application not as a monolithic chunk but out of components that implement the various functionalities required by the application. The PC version and the PDA version could in fact draw from the same collection of components, and as such the idea of "version" would become blurred.

A prototype to demonstrate our idea, called *Sparkle*, was introduced to support dynamic component composition and dynamic application reconfiguration. Sparkle is a component-based middleware for mobile computing. It allows an application to be dynamically composed at runtime and reconfigured according to changes in the context. In Sparkle, applications are built from small functional units, called *facets*. These functional components can be implemented in different ways to fulfill the same functionality, and one of them would be chosen at runtime based on certain contextual information. When an application runs, suitable functional units are downloaded from the network from a peer or from a server to the device. After the application has finished using the components, they may be cached or discarded. When a facet requires another facet, the latter would also be selected dynamically based on the context. Whenever the context changes because of changes in the environment, new functional components may be brought in to adapt to the environment. When an application is moved from one device to another, components of same functionality but using different implementations may be installed in the new device. Although the implementations are different, the same application will run and resume from the previous execution state. This is achieved by a mechanism that supports state migration of facets [6].

The dynamic composition and configuration of an application according to the context is a kind of code adaptation. An application consists also of a user interface (UI) and data or *contents*. By imposing a clear separation between them, adaptation can be applied to the code, the UI, and the data individually. The focus of the Sparkle project is on the adaptation of code. UI adaptation is temporarily handled via what is called a *container*. Contents can be considered a kind of data that is viewable. We have done some preliminary work in contents adaptation, and some of the techniques should be applicable to UI adaptation [7].

Facets of Functionalities

In Sparkle, applications are built from small functional units called *facets*. The main purpose of the facets is to support dynamic component composition. Separation of functionality from data and the user interface is the fundamental philosophy behind the facet model. Applications allow users to carry out certain tasks, and they provide certain functionalities to carry out these tasks. These functionalities are embodied in facets.

Functionality is a single well-defined task in an application. The task could be as small as a matrix multiplication or as large as detecting meaningful features of an image. It is mainly up to the programmer to decide what the constituent functionalities of an application are or how large they are. Given a set of inputs, the functionality determines what changes are made and the outputs attained. Essentially, functionality can be seen as a contract defining what should be done. The contract includes:

- The set of input parameters (i.e., the number and types of the inputs)
- The set of output parameters (i.e., the number and types of the outputs)
- A description of what is carried out (i.e., valid outputs for a set of inputs)
- Preconditions, if any (e.g., the ranges of input parameters supported)
- Post-conditions, if any (e.g., which values are nullified and error conditions)
- Side-effects, if any (e.g., I/O or changes to state in the container)

The contract defines the functionality to be achieved but not how it should be achieved. Implementations can use different algorithms, each with different performance characteristics or resource requirements. As long as they abide with the contract, they can be considered to be achieving the same functionality. As a consequence, functionality defines the interface for interaction and is independent of the implementation. To make things simpler, every functionality is assigned a globally unique identifier, the functionality ID (funcID); thus, a functionality ID uniquely identifies a contract.

Facets are entities that implement the functionalities. They contain code components that follow the contract of their corresponding functionality. In our prototype, for simplicity, a facet implements only a single functionality; in other words, a facet cannot provide two or more functionalities. In future extensions, a facet may implement multiple functionalities that are related. This would achieve a better scale of economy; also, from the user's perspectiv, related functionalities should be loaded together anyway.

Given this limitation and the nature of functionalities, a facet has only one single programming entry point and is stateless. A facet being stateless means that a facet is independent of any previous invocations. When the execution of a facet is finished, it is either discarded or reset so it does not affect the execution of the next invocation of the facet. These features make facets dispensable — a facet can be discarded from the runtime as soon as it is no longer needed. To keep some application states, an internal data structure called the *container* is used.

In short, the container is used to bring up an interface for user interaction, which in turn will request the appropriate functionalities based on the input of the user. A container provides a place where facets can run. Each container is associated with a particular application and contains a set of functionalities that the application can offer. These functionalities are stored in the container as facet specifications. When a particular functionality is required, the corresponding facet specification is sent as a request to the proxy located in a server or a peer. It has a storage area to store the execution state and application data. By having such a storage area, facets may communicate with each other to obtain the application data they need. It also enables process migration when a user switches to another device.

Our approach differs from conventional distributed systems in which objects are used as a unifying abstraction for both data and functionality. Because functionality is bound with the specific data implementation on which it can act, the object paradigm may not be a good fit for a mobile or pervasive environment [8–10]. Our design separates the data from the functionality so it is possible to use different implementations of the functionality in different devices to operate on the data. It has also been argued that application functionality changes more frequently than data implementation and data layout; therefore, it is preferable to store and communicate passive data rather than active objects. Because a clean separation between data and functionality allows them to be managed and to evolve independently, we have separated data from implementation. We use the container to store the data.

Facets provide pure functionality. They take in some input and carry out their functionality, resulting in the corresponding outputs. The user interface is just a means to access functionality. It is highly dependent on external factors such as display capabilities of the device and user preferences, rather than on the application or task at hand. Different UIs can be used to access the same functionality or task. In fact, the UI changes more often than the essential functionality of an application. Because the UI changes from device to device and version to version, it is desirable to keep it separate from functionality. As they are not bound to each other, this separation allows developers to change the UI without changing the functionality and *vice versa*, thus attaining a more intuitive and flexible software model.

Infinitely and Runtime Composable Software

A special branch of code adaptation is called *functionality adaptation*. Functionality adaptation involves changing the way the task is carried out to respond to changes. It selects different code implementations for execution depending on the context; for example, if a device does not have sufficient computation power, an application can execute another implementation of encryption (such as DES instead of RSA).

Dynamic component composition provides a flexible mechanism for achieving functionality adaptation. Functionality adaptation is made possible by the following: First, the component model allows functionalities to be composed at runtime and discarded after use. Second, the context manager in a client device maintains information about the physical

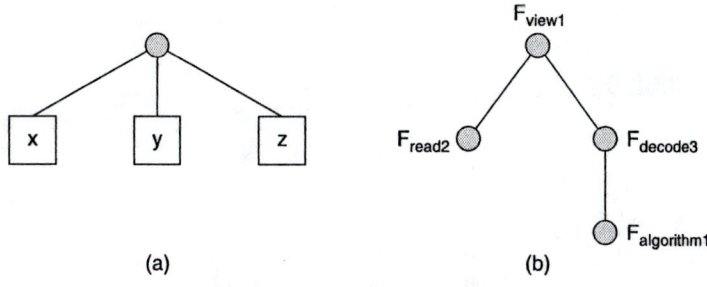

Figure 18.1 (a) Facet dependency tree, (b) facet execution tree.

resource, network connectivity, and context of devices. Third, proxies match requests with suitable facets for clients to execute. A proxy is executed in a peer or a server to help the mobile device choose a suitable facet. They are the main active entities for adaptation. Functionality adaptation, in this approach, is achieved by choosing the appropriate component among different ones that have the same functionality.

Developers need only specify which functionalities they require in an application and provide different versions of them. The adaptation mechanism is transparent to the programmer. Which component gets picked depends on the system and the matching mechanism of the proxy. In addition, because applications are linked by functionalities, rather than specific components, as new technologies or devices emerge developers need only write a newer version of the affected functionalities. The proxies will automatically match these components under the appropriate conditions. Rewriting or reinstallation of the entire program is not required. Because the components are thrown away at runtime after use, even the biggest programs can be used in a small device, depending on the size and runtime behavior of each component.

During the course of application execution, a facet may call upon other facets to help achieve its functionality. To allow dynamic and flexible adaptation, facet providers do not specify the location of a requesting facet; instead, they specify the functionalities that the facet requires for dynamic binding. These functionalities required by the facet are referred to as its *facet dependencies* (or simply *dependencies*). Facet dependencies, therefore, represent a local point of view of the facet. The dependencies on which a facet depends can be represented as a facet dependency tree, as shown in Figure 18.1a. Facet dependency trees are only one level, as a facet knows only the dependencies it requires.

At runtime, when a facet requires another functionality for execution, the client sends a request for a facet with the required functionality. The returned facet, in turn, can have its own dependencies to help achieve its functionality. These dependencies are used as requests for the actual facets only when they are required for execution. If we draw lines between a facet and the actual facets it calls at runtime, we come up with a *facet execution tree*. This facet execution tree cannot be determined statically, but can only be known at runtime. This is because different facets may be selected under different contexts. The facet tree shows the relationship between a facet and all the facets required at runtime to achieve its specified functionality, thus representing a global view of the facet. Figure 18.1b shows an example of a facet execution tree.

Functionality Adaptation

Context Sensitivity

In Sparkle, applications are able to take advantage of the context of the user, including location, the device being used, time, preferences, and nearby services, to provide customized and relevant services to the user; for example, a facet for printing will be sent to a mobile device when there is a printer nearby. Such functionality is not present in the mobile device before the device approaches the printer; furthermore, applications may adapt to the capabilities of the printer. For the same printing functionality, different facets may be implemented, one for monochrome output and one for color. In this way, different facets may be sent to the mobile device, depending on the color of the source image. Our system is adaptive to four types of contextual change. The first type is a change in *device resources*; these include factors internal to the device, such as the working memory available, processing power, and energy. The second type is a change in *network properties*; these are changes in the network characteristics, such as the network bandwidth, network type, and protocol. The third type is a change in the *environment*, which includes factors in the surrounding environment such as location, entities available nearby, and time. The last type is a change in *user preferences*; these are specific choices that the user has made in relation to the execution of a particular application.

Portfolio of Functionalities

Software is commonly distributed to users, at least conceptually, as monolithic applications with a fixed set of functionalities. The major drawback of this model is the rigid boundary placed on the accessibility of functionalities. A functionality can only be accessed in the context of the application with which it is associated and not by another application. We therefore introduce the concept of personalized software. Instead of applications being the focal point of software development and distribution, software is treated as functionalities associated with and used by a particular user. Users have their own portfolios of functionalities. Each user decides which of the functionalities are needed and puts them in the portfolio. When they are in the portfolio, these functionalities are always accessible, in a sense blurring the application boundary.

The actual implementation of these functionalities is distributed at runtime. The functionalities are composed from various components that are brought in at runtime and discarded after they have been used. The system essentially plays the role of a corkboard, pinning up components when they are being used and unpinning them when they are no longer required.

The advantage of such a model is its flexibility. Functionalities are not confined to being invoked only under the realms of certain applications. The model facilitates incorporation of new functionalities, updates to current functionalities, and adaptation to new environments. The revenue model is more accommodating to the specific needs of different users. In addition, this component-based model allows for more code reusability and easier maintenance of the code base for developers.

Sparkle is a demonstration of the feasibility of such a software distribution scheme. The basic foundation of Sparkle is facets, the software components used to build the functionalities. These functionalities are categorized according to the functions or services they provide to users. Users can choose which functionalities to put into their portfolios from among various categories. For example, a user who is viewing an image decides to make changes to it. If a mobile device has not specified any image editing tool, the user should be able to go through a categorization process to add the tool to the portfolio and use it immediately. The image can then be edited regardless of the application context in which the image was opened. Such a scenario blurs application boundaries to a certain extent. Functions and tools are not bounded to a certain application, which enhances the productivity of users. In addition, it is possible for the underlying system to have some intelligence or rules to predict what functions may be required in the near future based on information regarding the functions the user has called recently. These functions can either be displayed to the user or brought in beforehand to enhance the user experience.

Even though the portfolio may contain several functionalities, these functionalities are brought in only as they are required. The functions are brought in from the network, loaded, linked to the system, executed, unloaded, and discarded at runtime. At start-up, different users receive different sets of components according to their portfolios. During computing, functionalities are brought in as they are required. Different versions of the same functionality may be suitable for different resource environments. The version brought in would be the one most suitable for the current execution environment of the user. When the functionality has been used, it is discarded from the system.

Discarding a function from the system is not equivalent to removing that function from the portfolio. The portfolio of the user contains the list of functionalities that the user can access, rather than the actual components by which they are implemented. When a function is discarded from the system, it can be brought in again from the network, when the user again invokes that function. All of this is transparent to the user.

Basically, this takes modular programming a step further. Software not only is made of separate components but is distributed separately, as well. Programmers create components, each of which is small and carries out one thing well. They can create multiple versions to carry out the same functionality within different resource scenarios. Composers can leverage the different components to achieve a certain functionality or to provide a group of functionalities. Users receive the components only when they are needed.

The trend toward mobile and pervasive computing is unstoppable. Users today are employing such devices as PDAs and mobile phones to carry out computing. These devices are heterogeneous, limited in resources, and connected to different networks, such as a wireless local area network (WLAN) or a Bluetooth® *ad hoc* network. Because of its modular design and the fact that components are brought in only when they are needed, this model requires a smaller working memory than larger, monolithic applications. Only components that are suitable for the current computing environment are brought in, in essence achieving runtime compositional adaptation. Being able to discard unnecessary functionalities is important for resource usage efficiency as it frees up resources for currently executing functionalities or to bring in other functionalities. In effect, small devices can run a group of functionalities that they normally would not be able to run if those functionalities had been distributed among applications in a monolithic fashion.

Such a software architecture provides a convenient basis for context-dependent applications. When users move from one place to another (say, into a shopping mall), their devices may have to incorporate functionalities that are required to operate in that particular environment. For example, new functions could be incorporated that allow users to securely book tickets for the cinema as soon as they enter the mall or perhaps to remotely order food in a restaurant to pick up later. To do so may require devices to use proprietary protocols to talk to the shopping mall server. This protocol can be incorporated temporarily as functionalities provided by the shopping mall server and discarded and removed from the portfolio when users leave the mall.

Facet Architecture Based on Ontology

Facet functionalities are specified by ontology, which has been introduced for bridging the knowledge gaps between different domains [11]. Ontology represents the semantics of different concepts. It provides a formal, explicit specification of a shared conceptualization of a domain that can be communicated between people and heterogeneous application systems [12]. Ontologies for their applications are defined in the stationary environment; thus, devices can only communicate using the same ontologies. Sparkle separates data and functionality. To describe the functionality, we use ontology to prescribe the semantics of the user-perceivable task description and provide a formal, explicit specification of shared conceptualization.

A facet consists of two entities: the *code segment* and the *shadow*. The code segment is the part where the executable code lies and has only one publicly callable method to be called by others. This code, when executed, performs a predefined specific functionality. A facet is described by metadata, or a shadow, which is used to identify a facet. A shadow contains the properties of a facet, such as the vendor, version, the functionality it performs, the resources it requires to provide the functionality, and the functionalities that the facet requires for execution. It includes information about the facet, such as the function a facet provides, input and output specifications, vendor and versioning information, resource requirements (e.g., device memory), functionally capability (e.g., rendering monochrome images), functionality dependencies (any other facets this facet would invoke), and the charging scheme for the use of facets. It is represented by an ontology-based task description language extended from the Web ontology language [13] and thus is human and machine readable. Figure 18.2 shows an example of a shadow. In the example, FlipVertical is a functionality of a facet, and all facet vendors implementing this functionality use this name in their shadows.

After facet functionality is described and the facet is located in a proxy, the proxy can use a two-phase adaptation technique to choose which facet should be sent to the client after the client has specified some functionality requirements. Clients do not have to rely on the servers for executing the services. The aim of the two-phase adaptation is to adaptively select a best-suited facet from all the available facets in a proxy. The first phase, the *filtering phase*, filters the facets that satisfy the requirements of the client. These requirements include, at the very least, the functionality required by the client and the amount of resources available in the client device for executing the specified functionality. All the facets filtered by the first phase have satisfied the client's requirements and are eligible for further processing. The second phase, the *selection phase*,

```
<?xml version="1.0"?>
<facet>
        <identifier>FlipVertical</identifier>
        <functionality_id>FlipVertical<functionality_id>
        <vendor>Sparkle</vendor>
        <version>
                <major>1.0</major>
                <minor>a</minor>
        </version>
        <resource>
                <memory>
                <static unit="kbytes">2</static>
                <dynamic unit="kbytes">40</dynamic>
                </memory>
        </resource>
        <dependencies></dependencies>
</facet>
```

Figure 18.2   A shadow example.

selects a facet that best suits the device user. This decision is based on user preferences and other execution contexts of the client. The facet resulting from the two-phase adaptation is considered functionality adapted and is returned to the client.

The three key techniques in the two-phase adaptation are functionality filtering, resource filtering, and context selection. Functionality filtering ensures that facets achieve the functionality requirement of the client, resource filtering ensures that the functionality provided can be completed in the device, and context selection selects a facet that best suits the user and other execution contexts of the client. To allow more flexibility for the proxy system to choose among the facets, requests are specified in terms of queries instead of the exact locations of facets; furthermore, the proxy system maintains personal proxy caches for the users so facets can be better adapted to the device user. With all these supports, a good amount of adaptability can be provided by the proxy system.

Design and Implementation

The Sparkle Project

Sparkle aims to build an infrastructure that is suitable for pervasive computing environments. The infrastructure is based on the existing Internet infrastructure, with adaptability, mobility support, and peer-to-peer cooperation as its main features. The adaptability feature addresses the problem of computing with heterogeneous devices in different execution contexts. Mobility support addresses the problem of continuing the current session in another device or at another location. Peer-to-peer cooperation avoids single points of server failure by allowing facets to be downloaded from nearby peers. To support pervasive computing, service implementation takes the form of a facet, the mobile code component. Facets are downloaded on-demand to the client devices, executed, and then discarded. In fact, when other facets are required in the course of program execution, they will be downloaded incrementally. The constituent code components of a service are not fixed at compile time but are dynamically bound to form a service. Facets are mainly downloaded from proxies running in peers or servers. Facet servers, clients, and the intermediary proxy system are the three main components of Sparkle. Each of them plays a different role in the infrastructure. The following is a brief introduction; for further details, please refer to Belaramani et al. [14].

Facet servers are places for storing facets. They are similar to existing Web servers in that both are used as main storage servers for keeping up-to-date originals. They are used by the proxy servers for retrieving updated information. There is no restriction on the number of copies of a facet that can be placed on these servers. A facet can be placed on more than one facet server, meaning that facets are not unique among the facet servers. Facets can be added to or removed from the facet servers by facet providers. These updates are usually quite frequent in terms of software maintenance. To keep track of the updates made to the facets being stored, each facet server must keep a log of updates of the facets they store locally.

The clients are mainly mobile devices such as PDAs or mobile phones, although the computing model could be applied to non-mobile devices. Each of these computing devices has the capability of on-demand down-loading, executing, and discarding the facets after use. This allows services of any sophistication to be executed on the client device. Whenever a service is required, a request is sent to a nearby proxy for a suitable facet. The request consists of a description of the required service, information about the resources in the client device that can be used for executing the required service, and some user information. A facet satisfying the request is then returned for execution. During execution, other subservices might be required to help provide the service. Requests are then sent to the proxies when these subservices are required at runtime. This enables an unlimited chain of application functionalities.

The proxy system is a main component between the clients and the facet servers. It could be a mobile device peer or a dedicated server. Facets are cached in the proxies for fast retrieval and to reduce the workload of the facet servers. Client requests for facets are therefore sent to the proxy system instead of directly to the facet servers. Apart from being a caching device, the proxy system also acts as a recommender. It makes decisions on behalf of the clients and returns a suitable facet for each request according to the runtime execution contexts of the clients. Because it chooses suitable facets for the client, the proxy system is intelligent; thus, the proxy system is the key enabler in Sparkle.

As noted previously, to allow more flexibility for the proxy system to choose among the facets, requests are specified in terms of queries instead of exact locations of facets; furthermore, the proxy system maintains personal proxy caches for users so facets can be better adapted to individual device users. With all these supports, the proxy system provides greater adaptability. Also, the proxy system supports user mobility by cooperating with lightweight mobile agent systems in the client devices [15]. With this support, the same user in a different location is treated equally, independent of the location and without affecting the computing experience; therefore, it is possible to continue ongoing services in another device with suitable facets being adapted. The proxy system also prepares the personal proxy cache to be used with suitable facets in support of user mobility.

Universal Browser

The Universal Browser is not a traditional Web browser. It is a browser designed for mobile environments. It invokes any function that a user wants on demand. It is a special graphical user interface (GUI) implemented in Sparkle. This special GUI allows users to dynamically retrieve the functionalities they want. As shown in Figure 18.3, a user can use the Universal Browser to browse Web pages, play games, and edit images. These functionalities are retrieved from the network as needed by the user; in fact, at start-up, the device shows an empty GUI. Moreover, these functionalities can be thrown away after use to reclaim resources that may be required by other functionalities. The Universal Browser, because it is supported by Sparkle, can help a user to find the suitable facet (i.e., functionality) that matches the device and discard it when the facet is no longer needed.

Furthermore, the Universal Browser is a context-aware and extensible application. The context awareness of the Universal Browser is totally different from that of state management. The former is at the application level, and the latter is at the system level. "Context awareness" here refers to the downloading of different functionalities under different contexts; for example, facets that have higher memory demands are downloaded to browsers in notebook PCs because they can render an image more quickly than memory-thrifty ones.

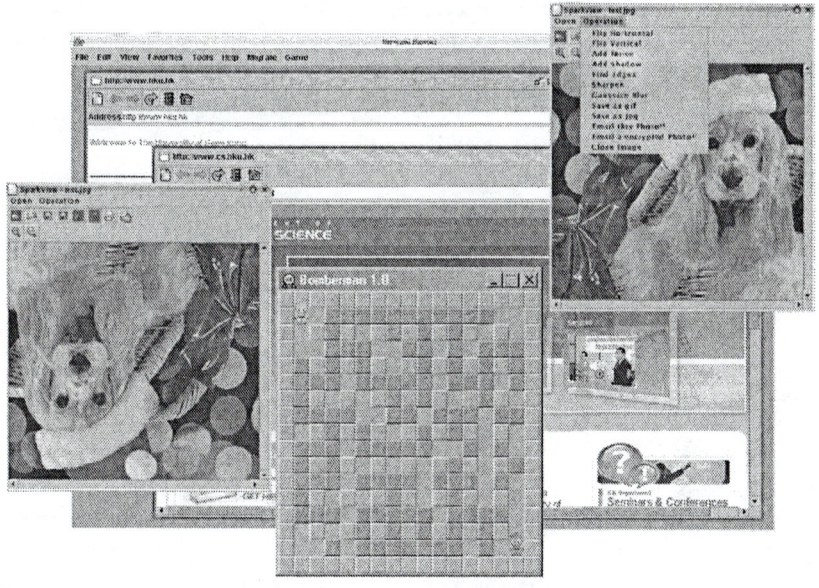

Figure 18.3   Screenshot of the Universal Browser.

Referring to the image viewer application, called SparkleView (showing the puppy in Figure 18.3), if the facet is cached on the device, the user interface will indicate to the user that the functionality is available locally and the background color of the icon changes to dark blue. If the facet is discarded by the underlying system, the UI will indicate it to the user accordingly, and the background color of the icon will change to light blue. SparkleView was run on a Pentium III Mobile CPU, 1133 MHz, with 384 MB RAM and Windows XP operating system, a configuration that is not much more powerful than some of the latest PDAs. The proxy had the following configuration: Pentium 4, 2.26 GHz PC with 512 MB RAM and Fedora Core 2 operating system. The underlying Sparkle system is roughly 650 kB. The SparkleView application is 115 kB, the majority of which is accounted for by the user interface, and the facets are 44 kB in total. Figure 18.4 shows how long it takes to run different functionalities, including the time to bring in the required facets. It can be seen that Gaussian blur takes a very long time. This can be attributed to the fact that Gaussian blur has two levels of dependencies. It calls at least three other facets on every execution; thus, network delay to bring in the facets has an impact. Also, Gaussian blur involves significantly more mathematical calculations than the other functionalities so it takes more time.

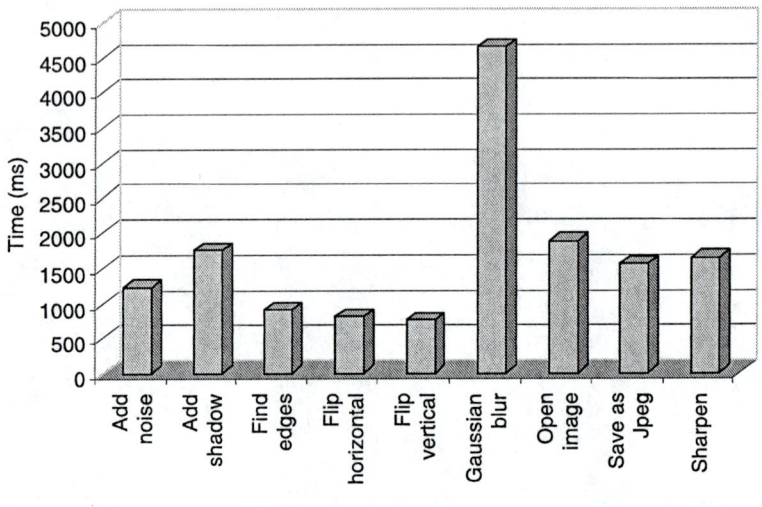

Figure 18.4   Invocation of different functionalities.

Related Work

Several types of component-based middleware exist. Not many of them, however, consider resource constraints in mobile devices. It seems that the code-on-demand design paradigm to handle dynamic functionality composition is rare. In addition, most of them do not consider multiple implementations of the same functionalities [8]. Multiple implementations are necessary for different contexts arising from such factors as device capabilities and user preferences; for example, a user wanting to catch just a quick glimpse of an image would prefer a faster image rendering component that trades image quality. Even if multiple implementations are allowed, some of them require the implementations to be programmed as a single component [16]. This results in other parts of a component being superfluous and occupying memory resources. For some systems, adaptation is application specific [17,18], whereas we employ a systemwide adaptation scheme in our middleware.

The Code Collection Project [19] shares our belief that software components for resource-constrained devices should be easily plugged and unplugged. They proposed an approach to optimize the use of memory via a garbage-collection-like algorithm. The algorithm unloads (i.e., discards) methods that are not likely to be needed in the near future. Sparkle, on the other hand, focuses more on the loading aspect, offering many choices for a requested functionality, and some components could be loaded automatically due to their dependencies.

The Open Services Gateway Initiative (OSGi) Alliance Service Platform [20] is similar to our facet architecture in that it utilizes software components called *bundles*, which are selected and downloaded to Internet appliances on demand at runtime. Bundles may also be removed after use. Although this platform has a bundle specification similar to our facet shadow, the specification is generally not migratable; this leaves the job of selection to the home server of a bundle. In contrast, our model facilitates a proxy-based, code-on-demand paradigm for better runtime performance. This enables a peer-to-peer collaboration for sharing facets; furthermore, our facet selection scheme is more flexible. In the OSGi platform, bundles are selected based on a service ID and a predefined selection preference, whereas our facets are selected based on ontological specifications and the range of functionalities facets are capable of providing.

A distributed OS rather than a middleware, 2K [17] supports reconfiguration of component systems at runtime. Similar to our model, when a component is required in 2K the component and its prerequisite components are brought in from the component repository, which may be located locally or in the network. After that, the resource manager is contacted to allocate the required resources for the components. Developers provide specialized components, called *configurator objects*, to handle dynamic reconfiguration. If any changes occur to the runtime resources, the resource manager will request the component configurator to make an adaptation. The adaptation policy, however, is application specific. Every application implements its own adaptation policy, which places some burden on the application programmers. The facet model instead employs a systemwide adaptation policy that provides a better picture of the resource requirements of all of the running applications. This allows programmers to focus on application logic rather than adaptation details, making it easier to develop mobile applications.

Aspect-oriented programming (AOP) [21] is a programming technique designed to ease the development and maintenance of software. It facilitates common or similar program code pieces for cross-cutting concerns being neatly applied in appropriate places in different programs. For example, to optimize memory used when processing an image via a series of image filters, the filter codes typically are fused together into a compound code segment. In AOP, each filter is specified individually and then fused by a program called the *aspect weaver*, which generates the tangled code. Maintenance and management become easier because individual image filters are addressed, not the tangled code. Sparkle is similar in the sense that it fuses small components into tangled code. It is different from AOP, however, in that the fusion is not limited to components implementing cross-cutting concerns and that the actual choices of components to be fused depend on contextual information. AOP complements other programming paradigms; for example, the component model of AspectCCM [22] utilizes AOP. Sparkle could benefit from AOP in the same manner.

Plug-in, as popularized by Web browser software, is a code-on-demand paradigm where new functionalities are downloaded on the fly and then executed. The fundamental difference between our model and plug-in is that, in plug-in, the entire program binary for the plug-in is downloaded before execution occurs, whereas our facets are downloaded one by one incrementally, and a facet can be extremely fine grained. Although Sparkle may cache or prefetch facets for better performance, it does not need to download the entire functionality at once. Also, facets may be discarded after use to reclaim resources. Many plug-in implementations, unfortunately, require manual uninstallation; furthermore, plug-in does not consider adaptation, whereas Sparkle considers code adaptation based on contextual changes.

No-touch deployment [23] and ClickOnce [24] are other code-on-demand design paradigms where program assemblies are downloaded incrementally on demand. Similar to plug-ins, the exact locations of the assemblies are predetermined beforehand, so the dynamic adaptation of functionalities based on contexts may not be feasible.

Economics of Code-on-Demand and Adaptation

Under Sparkle, users only request what they need. They receive in return an application with just the right amount of functionality, and extra functions may be added later. In fact, the concept of an application is blurred, as there is no limit on the number of functions that could be added to an application. Because only the required functionalities are downloaded, users can pay for only what they actually use. This is fundamentally different from the current off-the-shelf software model, where users pay for a large software package but end up using only a small portion of its functionalities. Runtime-installable components could come from any vendors as long as they are compliant with the corresponding facet contracts. To perform a user task, the user does not need to be concerned with which vendor provides the functionality; rather, facets are retrieved based on the user context. Facet providers, therefore, would compete with each other to create facets offering the greatest value for the money.

If software development adopts the Sparkle model, the software industry may benefit in a number of ways. First, every vendor has an equal opportunity to reach potential customers. This is because the customers would be looking for suitable facets, not necessarily a particular vendor. Facets are based on open standards (the contracts and specifications), thus creating an open market for any vendor capable of producing the required facets. An actual running application would likely be composed of facets by a multitude of different vendors. As applications are dynamically composed from facets, the same collection of facets could probably be used to produce, at a very late stage of the production cycle or even at runtime, different versions of an application suiting different computing platforms or devices. This differs from the current practice where different versions of an application are predominantly standalone and self-contained pieces of software; any sharing of code would have occurred in the very early stages, and later it becomes difficult to add new functionalities to all of the versions. This has a bearing on how vendors roll out new, killer functions to attract additional revenue. These functions have to be bundled in a new version of the software or adopt the plug-in model; either way, it is not the best solution in terms of cost and user convenience. Finally, with a Sparkle-like architecture, Internet services providers may take on a new business model that includes hosting a software repository for facets.

Conclusion

Sparkle is a code-on-demand adaptive mobile middleware that allows feature-rich applications to run on resource-constrained devices. This is achieved by introducing a new programming model where functionalities are made up of small functional units (facets). In the current implementation, each facet performs exactly one functionality and may be downloaded on demand and discarded after use. Because of this easy come-and-go mechanism for facets, an application may have an unlimited number of functionalities over time. This fundamentally changes the concept of conventional mobile applications that have bounded features. In addition, facets are not statically linked to an application. Facets are chosen based on various contexts, such as the environmental context or device context, during the course of application execution. This means that different facets may be downloaded for execution in different contextual situations. This makes Sparkle adaptive.

Sparkle is an experiment for proving a concept and inevitably has not touched upon many related important issues that would have to be addressed in a real design, including security and UI adaptation. Security could be a problem when we mix and match facets from diverse sources or vendors. A misbehaving facet could be most damaging if it has been adopted by a large number of applications. Some applications, such as computer games, have a complex UI; whether or not they can be composed from facets and how to synthesize the resulting dynamic UI are interesting questions for future research.

Acknowledgments

This research is supported in part by two CERG grants (HKU7146/04E and HKU7140/04E) from the Hong Kong government.

References

[1] Weiser, M., *Computer Science Challenges of the Next 10 Years*, http://www.ubiq.com/hypertext/weiser/UbiHome.html.

[2] Fuggetta, A. and Vigna, G., Understanding code mobility, *IEEE Trans. Software Eng.*, 24(5), 342–361, 1998.

[3] RealVNC, http://www.realvnc.com/.

[4] g-cluster, http://www.g-cluster.com.

[5] Hardenberg, C.V. and Bérard, F., Bare-hand human-computer interaction, in *Proc. of the 2001 Workshop on Perceptive User Interfaces*, ACM Press, New York, 2001, pp. 1–8.

[6] Siu, P.P.L., Wang, C.L., and Lau, F.C.M., Context-aware state management for ubiquitous applications, in *Proc. of Int. Conf. on Embedded and Ubiquitous Computing (EUC-04)*, Aizu, Fukushima, August, 2004, pp. 776–785.

[7] Lum, W.Y. and Lau, F.C.M., User-centric content negotiation for effective adaptation service in mobile computing, *IEEE Trans. Software Eng.*, 29(12), 1100–1111, 2003.

[8] Grimm, R., Anderson, T., Bershad, B., and Wetherall, D., A system architecture for pervasive computing, in *Proc. of the 9th ACM SIGOPS European Workshop*, Kolding, Denmark, September, 2000, pp. 177–182.

[9] Grimm, R., Davis, J., Hendrickson, B., Lemar, E., MacBeth, A. et al., Systems directions for pervasive computing, in *Proc. of the 8th Workshop on Hot Topics in Operating Systems (HotOS)*, Schloss Elmau, Germany, May, 2001, pp. 128–132.

[10] Grimm, R., Davis, J., Lemar, E., MacBeth, A., Swanson, S. et al., *Programming for Pervasive Computing Environments*, Technical Report UW-CSE 01-06-01, Department of Computer Science and Engineering, University of Washington, Seattle, 2001.

[11] Heflin, J., *Web Ontology Language (OWL) Use Cases and Requirements*, World Wide Web Consortium (W3C), http://www.w3.org/TR/2004/REC-webont-req-20040210/.

[12] Gruber, T.R., A translation approach to portable ontology specifications, *Knowledge Acquisition*, 5(2), 199–220, 1993.

[13] *Web Ontology Language*, World Wide Web Consortium (W3C), http://www.w3.org/2004/OWL.

[14] Belaramani, N.M., Chow, Y., Kwan, V.W.M., Wang, C.L., and Lau, F.C.M., A component-based software architecture for pervasive computing, in *Intelligent Virtual World: Technologies and Applications in Distributed Virtual Environments*, World Scientific, Singapore, 2004, pp. 191–212.

[15] Chow, Y., Zhu, W.Z., Wang, C.L., and Lau, F.C.M., The state-on-demand execution for adaptive component-based mobile agent systems, in *Proc. of the Tenth Int. Conf. on Parallel and Distributed Systems (ICPADS'04)*, Newport Beach, CA, July, 2004, pp. 46–53.

[16] Yau, S. and Karim, F., Component customization for object-oriented distributed real-time software development, in *Proc. of the 3rd IEEE Int. Symp. on Object-Oriented Real-Time Distributed Computing*, Newport Beach, CA, March, 2000, pp. 156–163.

[17] Kon, F., Campbell, R.H., Mickunas, M.D., Nahrstedt, K., and Ballesteros, F.J., 2K: a distributed operating system for dynamic heterogeneous environments, in *Proc. of the 9th IEEE Int. Symp. on High-Performance Distributed Computing (HPDC'00)*, Pittsburgh, PA, August, 2000, pp. 201–210.

[18] Litiu, R. and Prakash, A., DACIA: a mobile component framework for building adaptive distributed applications, *Operating Syst. Rev.*, 35(2), 31–42, 2001.

[19] Popa, L., Athanasiu, I., Raiciu, C., and Pandey, R., and Teodorescu, R., Using code collection to support large applications on mobile devices, in *Proc. of the 10th Int. Conf. on Mobile Computing and Networking (MOBICOM'04)*, Philadelphia, PA, September 26–October 1, 2004, pp. 16–29.

[20] Open Services Gateway Initiative, http://www.osgi.org.

[21] Kiczales, G., Lamping, J., Mendhekar, A., Maeda, C., Lopes, C.V. et al., Aspect-oriented programming, in *Proc. of the 11th European Conf. for Object-Oriented Programming (ECOOP'97)*, Jyvëskylë, Finland, June, 1997, pp. 220–242.

[22] Clemente, P.J., Hernández, J., Murillo, J.M., Pérez, M.A., and Sánchez, F., Component-based system design and composition: an aspect-oriented approach, in *Component-Based Software Development: Case Studies*, Lau, K.K., Ed., World Scientific, Singapore, 2004, pp. 109–128.

[23] *No-Touch Deployment in the .NET Framework*, http://msdn.microsoft.com/library/default.asp?url=/library/en-us/dv_vstechart/html/vbtchNo-Touch DeploymentInNETFramework.asp.

[24] Noyes, B., ClickOnce: deploy and update your smart client projects using a central server, *MSDN Mag.*, May, 2004.

Chapter 19

Session Maintenance

Oliver Haase

CONTENTS

Introduction

The term *session* is used in many different contexts and on many different layers; however, in general, it stands for the *temporal usage of a service*. Services can range from simple, local services, such as file access, to complex, distributed services, such as video conferencing and the like. In this chapter, we primarily have rather complex, distributed services in mind, even though the following considerations are not restricted to one particular class of services. A typical service session consists of three phases:

- Session establishment
- Session continuity
- Session termination

A mobile environment poses significant challenges for session establishment and continuity. For one, service discovery in a dynamic environment, where both servers and clients can be mobile, is a difficult task. A client that has roamed into a new region requires a means to look up local services (e.g., print and directory services) but also restaurant guides, the local weather forecast, and other location-based services. Roaming services, on the other hand, require a means to advertise their service to the local community. In both directions, the rendezvous between services and clients requires commonly agreed-upon communication channels, protocols, and procedures. As far as session continuity is concerned, changing network attachment points, temporary connection losses, and varying network characteristics, especially over the air interface, constitute the biggest obstacles. A good solution will shield service users from temporary service degradation to the greatest extent possible. Another significant challenge is to provide session continuity when nomadic users switch from one device to another and expect their ongoing sessions to be seamlessly handed off to the new device. Such a switch of devices may be beneficial in terms of CPU power, I/O capabilities, and network bandwidth. Ideally, a transferred session would adapt its resource consumption to the new situation.

It is rather obvious that, under these circumstances, session maintenance (i.e., the combination of session establishment, session continuity, and session termination) is a very complex task in a mobile environment. Without adequate middleware support, both service providers and service users would have to spend an enormous amount of time and effort to solve the same set of problems over and over again. With adequate middleware support, software developers can focus on application-specific problems and in the best case develop a sophisticated, distributed service as if it was a centralized service.

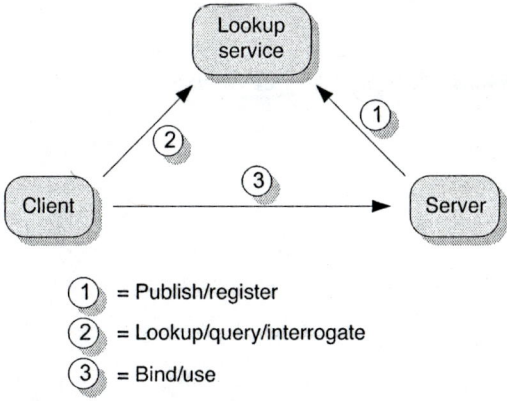

Figure 19.1   General scheme for centralized service lookup.

Service Discovery

The dynamic discovery of a service in an environment where both servers and clients can be mobile requires a *lookup* (or *directory*) service that sits in the middle between the two parties and helps them find each other. The resulting service — that is, the capability of the server and client to move *before* the service session is set up — is referred to as *pre-session mobility*. In general terms, we have two categories of lookup services: the more classical category of centralized lookup services and the more recent category of decentralized, peer-to-peer-based lookup services.

Centralized Service Lookup

As mentioned earlier, the key idea of this class of approaches is to establish a central entity that services register with and that clients query to find services. The name of the central entity varies with the specific middleware approach. Typical names are *lookup service, service registry,* or *(service) directory.* A service is said to *register* with the lookup service, or to *publish* itself. A client *queries* or *interrogates* the lookup service, or it *discovers* a service from the lookup service. When the client has discovered a service, it can *use* it, or *bind* to it. In any instance, the general procedures flow is depicted in Figure 19.1.

The introduction of a centralized lookup service reduces the problem of finding one or several appropriate services to a problem of finding one dedicated service (i.e., the lookup service itself). A service has to find the lookup service to be able to register itself with it, and a client has to find the lookup service to be able to query it for services. This initial boot-strapping process can be approached in several different ways:

- The lookup service can periodically broadcast its existence within a limited network range. These broadcasts are also called *advertisements* or *beacons*. Technically, the advertisement can be done via Ethernet broadcast (within the same IP subnet) or IP multicast with a small time-to-live (TTL) value so as to avoid flooding the entire network.
- A client (as well as a service that seeks to register itself with the lookup service) can send out a range-limited broadcast and ask for a lookup service. The broadcast can again be realized as an Ethernet broadcast or IP multicast.
- Clients and servers can learn the address of a location service through out-of-band means. A hosting network can, for example, publish its various server addresses on a Web page and users can manually configure their devices, or the server addresses in the hosting network can be learned through the Dynamic Host Configuration Protocol (DHCP) [1] when the IP connection has been established.

Peer-to-Peer Lookup

Peer-to-peer networks are a rather novel concept dealing with the problem of finding services in a highly dynamic environment where nodes join and leave frequently. In the pure sense, a peer-to-peer network is a completely decentralized network of identical nodes each of which can act both as a server (providing services to other peer nodes) and as a client (using the services of other peer nodes). In practice, the first peer-to-peer networks chiefly provided file sharing functionality, in particular for music. From a more general point of view, however, the essential services of a peer-to-peer network are the storage of (*key, value*) pairs on its nodes and a decentralized lookup service that can locate such a pair for a given key. Obviously, if *keys* represent service descriptions and *values* represent the actual services, then peer-to-peer technology can be employed to resolve the service–client rendezvous problem discussed in this section.

Early peer-to-peer networks flooded location requests into the entire network or parts of it, until the desired (*key, value*) pair was found; however, because flooding does not scale for large and very large networks, many recent peer-to-peer approaches use *distributed hash tables* (DHTs) for storage and retrieval of (*key, value*) pairs [2,3]. For DHT storage,

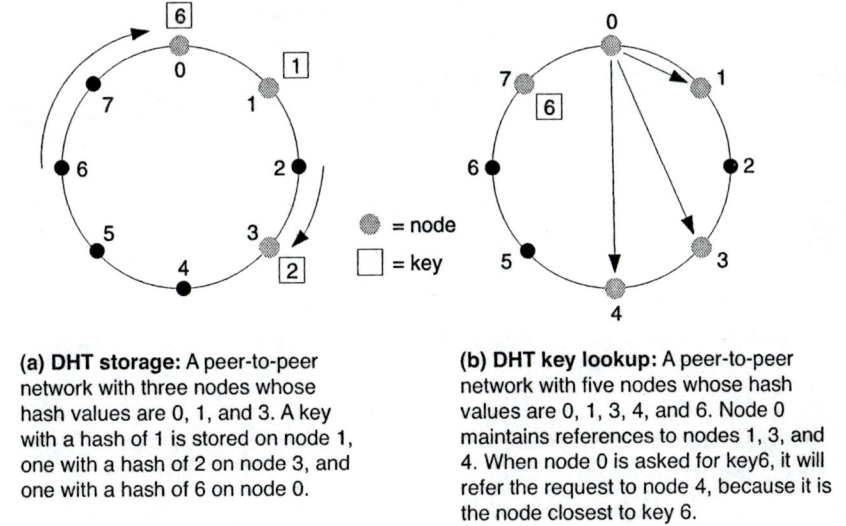

(a) DHT storage: A peer-to-peer network with three nodes whose hash values are 0, 1, and 3. A key with a hash of 1 is stored on node 1, one with a hash of 2 on node 3, and one with a hash of 6 on node 0.

(b) DHT key lookup: A peer-to-peer network with five nodes whose hash values are 0, 1, 3, 4, and 6. Node 0 maintains references to nodes 1, 3, and 4. When node 0 is asked for key6, it will refer the request to node 4, because it is the node closest to key 6.

Figure 19.2   Storage and lookup in a distributed hash table.

the core idea is to hash nodes as well as keys onto the same value range and to store a (*key, value*) pair on the first node whose hash value is greater than the hash value of the key. DHT lookup is based on a distributed version of binary search. For this purpose, each node maintains *log N* references to other nodes, where *N* is the number of nodes in the network. One reference points to a node that is half the value range apart, another one to a node that is a quarter of the value range apart, another one to a node that is an eighth of the value range apart, and so on. When a node searches for a key, it determines the closest node with a hash value less than the hash of the key from its list of references, and delegates the search to that node. This technique guarantees an upper bound of $O(\log N)$ for a key search (measured in delegation steps) and yields an average search effort of $(\log N)/2$ delegation steps. Figure 19.2 illustrates storage and key lookup in a DHT-based peer-to-peer network.

Mid-Session Mobility

As the name suggests, mid-session mobility denotes mobility at the time of an ongoing session. This mobility can occur on two different layers: (1) A user can move with his or her device from one network or one access point to another, or (2) a user can switch from one device to another, potentially more powerful device. The former kind of mobility is *device mobility*, and the latter is *user mobility* (or *personal mobility*). The challenge in situations of mid-session mobility is to seamlessly hand off the ongoing sessions from one network attachment point or from one device to another. This feature is referred to as *session continuity*.

Device Mobility

The Internet Protocol (IP) not only is the predominant communication mechanism between PCs but also, at an ever-increasing rate, is taking over the role of traditional circuit-switched technology for communication between land-line phones as well as cellphones; consequently, we focus our discussion about device mobility entirely on IP. With standard, so-called simple IP, every time a device moves from one access network to another and re-establishes IP connectivity it gets a new IP address assigned. To maintain ongoing sessions, either each application has to be able to react to changes on the IP communication layer, or a middleware layer has to sit in between the communication and application layers to shield the latter one from changes in the former one. Because shielding applications from changing IP addresses is such a common and important task, the Internet community has defined a mechanism for mobility support that is integrated right into the IP stack, so applications that use this Mobile IP stack are kept agnostic of any underlying changes [4,5]. Support of Mobile IP is optional in IP version 4 (IPv4) and mandatory in IPv6. Because Mobile IP is an essential technology in virtually any mobile environment, it is explained in some detail in the next section.

Mobile IP

In a nutshell, Mobile IP works as follows:

- The mobile host gets assigned a static IP address (*home address*), which rarely changes.
- As long as the mobile host stays in its *home network*, regular IP routing takes place and the mobile host receives packets to its address as usual.
- When the mobile host moves to a visited network, it contacts the local foreign agent, which assigns a temporary IP address (*care-of address*) to it.
- The care-of address is sent to the home agent in the home network of the mobile node.

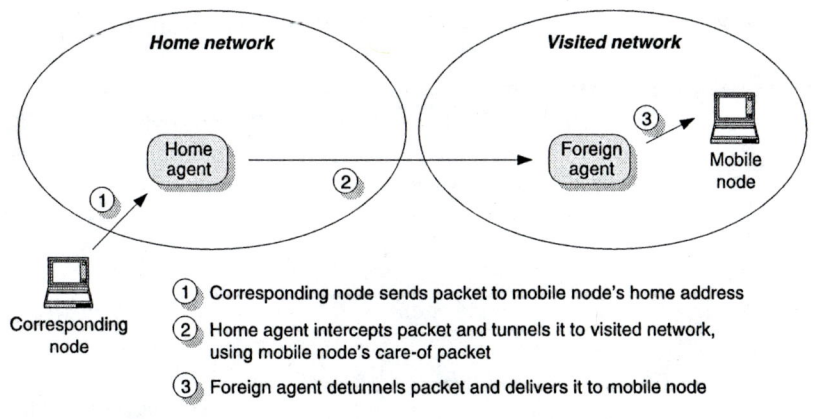

Figure 19.3   Routing of IP packets to a visited domain with Mobile IP.

■ When a *corresponding node* sends an IP packet to the home address of the mobile node, the home agent intercepts it and tunnels it (i.e., packs the entire IP packet into a new, enclosing packet) to the foreign agent in the visited network, using the temporary care-of address of the mobile node.
■ The foreign agent unpacks the tunneled packet and passes on the original IP packet to the mobile node.

The above mechanism for routing an IP packet to a foreign network is illustrated in Figure 19.3. Outgoing traffic (i.e., IP packets sent from the mobile node to any corresponding node) is still routed directly between the two nodes, bypassing both foreign and home agents.

Mechanisms on Top of Simple IP

Obviously, the challenge with changing IP addresses is that other nodes in the network, the services of which the mobile device uses or to which the mobile device is providing service, have to be informed of the new IP address so they do not lose their ability to communicate with the mobile device. For services following a stateless, simple request/response paradigm, a changed IP address is only a problem if the change occurs between the sending of a request and the receipt of a response. Simple Web services fall into that category, because the Hypertext Transfer Protocol (HTTP) is a simple request/response protocol; however, especially for e-commerce applications, even Web services often implement a session concept on top of HTTP, in which case a change of the client's IP address breaks the application layer session. For communication services such as Voice-over-IP, video telephony, and messaging, changing IP addresses always pose an issue because the ability to send (signaling) messages without interruption to the communication peer is a key part of the service.

A middleware component that resolves the above-mentioned issues with device mobility has to sit between the communication (IP) stack and the application on all parties that are involved in the service session. The steps the middleware must complete include the following:

- When the mobile device gets a new IP address assigned, the middleware has to be notified. (Alternatively, the middleware can periodically poll for changes in the device's IP address; polling, of course, is more resource consuming than event-based notification and should only be used if the communication stack has no notification mechanism.)
- As soon as the middleware learns about the new IP address, it sends a notification message to all communication peers.
- The middleware on the peer node receives the notification with the new IP address, and directs all future communication with the mobile node to the new address.

Obviously, the above procedure requires: (1) the application not to communicate directly with the communication peers, and (2) the middleware to maintain a mapping table that maps all communication peers to their current IP address. Figure 19.4 shows the procedure for the case with one mobile device and one communication peer.

User Mobility

As mentioned earlier, the term *user mobility* suggests that users are able to maintain their services as they switch from one device to another. A typical example is when an employee has both an office at work and a home office and wants to have access to the same data in both offices or wants to continue working on the same customer presentation at home after hours. User mobility involves the following aspects:

- All communication peers that participate in a service currently used by the mobile user have to redirect their communication to the new device. This task is very similar to the case of device mobility with simple IP explained earlier, as in both cases the communication peers have to be informed of a new IP address. However, in the case of user mobility, not only can the network parameters such as bandwidth and delay change with the new point of attachment but also the device capabilities. Mobile middleware can

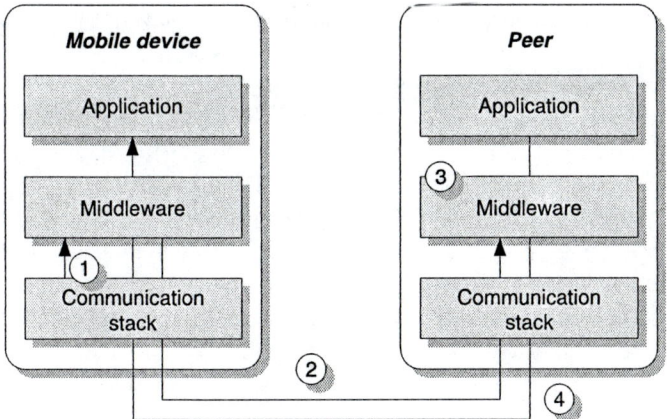

1. Communication stack notifies middleware of IP address change
2. Mobile device notifies peer
3. Peer updates mapping table
4. Future application-level communication is directed to the mobile device's new IP address

Figure 19.4 Middleware hiding IP address changes for simple IP.

take these new parameters into account and adapt the service accordingly.

■ The states of all sessions that are to be handed off to the new device have to be transferred. This issue is even more challenging than the previous one, as the following discussion will illustrate. The state of a service session is determined by the collective session states of all participating peers (that is, the state of the server and the state of the client). The session state can be moved from the old to the new device in several ways:

■ *Directly from the old to the new device* — This is the simplest if for a certain time window the user has control over both devices. This does not necessarily mean that the user has to be in front of both devices, but only that the destination device is allowed to receive state information on his or her behalf. One way of triggering state transfer from a source to a destination device is through explicit user request. That request might or might not specify the destination device. If it does not, the middleware has to take a guess, based on the usual user movement patterns, local proximity of devices (e.g., in a smart house where the user's whereabouts are monitored through smart tags), and such. In any case, there must be a recovery mechanism to manually pull state information to a certain device in case the middleware's guess was wrong.

- *From the old device to the network and then on to the new device* — Evidently, this option solves the above problem at the expense of additional network support.
- *From the old device to a home server and then on to the new device* — Technically, this option is very similar to the network-based option above, except that the home server runs in the user's administrative domain. Also, this option can be considered a fall-back for the first option in case the user has not specified a new device yet at the time of the device switch.

Regardless of how exactly the session state is transferred from one device to another, several issues are common to all of these options:

- They require all applications that are involved in an ongoing session to be installed on both source and target devices; moreover, both machines must run the same or similar versions of the respective applications with the same user configurations. To avoid this kind of problem, not only the states but also the entire applications must be shipped to the new device.
- The session state can contain a large amount of data, and shipping it to the new device may take considerable time. This is even more true if the entire applications are shipped (see earlier discussion).

The shipping of state information from one device to another can be avoided if the state information or part of it is kept in the network rather than on the actual device. One way to implement a service this way is to follow the *model–view–controller* (MVC) design pattern for graphical user interfaces [7]. The MVC model keeps separate the state of the application (model), its graphical representation (view), and the control software that interprets mouse and keyboard actions and translates them into actions that modify the model and the view (controller). To support session handoff in the case of user mobility, the model (i.e., state) should run in the network, and the view and the controller should run on the user device. Using that kind of separation, the software on the user device becomes completely stateless. Then, when the user moves from one device to another, the view and the controller on the new device connect to the model in the network to graphically represent the session state to the end user. In the MVC model, even multiple views and controllers can connect to the same model at the same time from different devices. Mid-session mobility through the use of a network-based MVC implementation is illustrated in Figure 19.5.

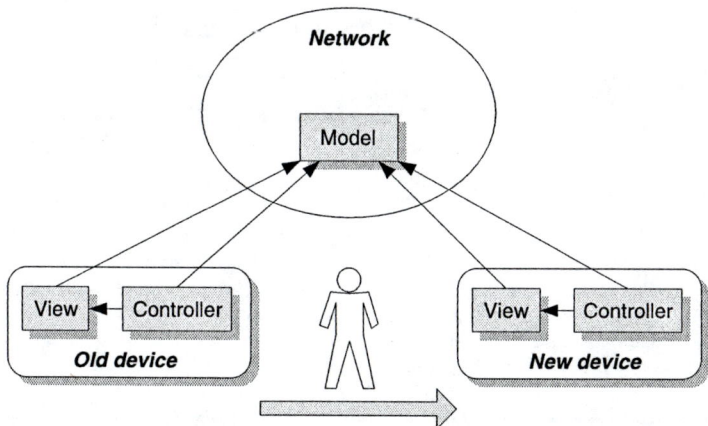

When the user moves from the old to the new device, the view and
controller connect to the network-based model (i.e., state of the application).
In this way, the session is handed off to the new device without the need to
transport any state information.

Figure 19.5   User mobility using a network-based MVC implementation.

Sample Middleware Approaches

In this section, we compare several middleware approaches with respect
to their mobility support. For each approach, we describe the key concepts
only to the extent needed to discuss their mobility-related features.

Common Object Request Broker Architecture

The Common Object Request Broker Architecture (CORBA™) is middleware that allows client objects to remotely invoke methods at server
objects, across machine and language boundaries [7]. For example, a C++
client object can invoke methods on a remote server object without even
knowing that the server object is written in Java. All the client knows
about the server object is the interface it provides; this interface is specified
in a programming-language-independent format, the *Interface Definition
Language* (IDL).

The centerpiece of CORBA is the *Object Request Broker* (ORB) (see Figure
19.6), to which clients connect via client stubs and servers connect via server
skeletons. When a client issues a remote invocation, the client stub marshals
the request into the language-independent *Common Data Representation*
(CDR) format, the ORB transports the request to the remote server, and the
server skeleton demarshals the request into the implementation language of

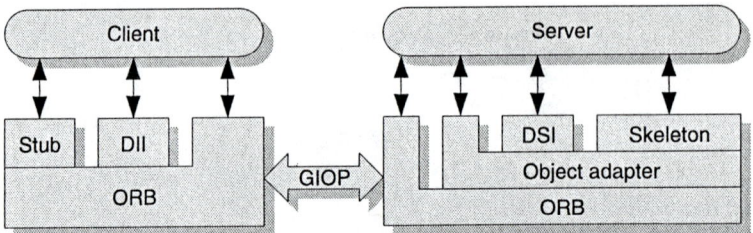

A client *object request broker* (ORB) communicates with a server ORB over the *General Inter-ORB Protocol* (GIOP). A client can connect to a server *stub*, it can invoke remote services through the *dynamic invocation interface* (DII), and it can use ORB services. A server can connect to a server *skeleton*, it can be used through the *dynamic skeleton interface* (DSI), and it also can use ORB services.

Figure 19.6   Architecture of a CORBA Object Request Broker.

the server object. Also, the dynamic invocation interface — together with its counterpart, the dynamic skeleton interface, at the server side — allows clients to dynamically inspect the interface of a server at runtime. Two ORBs can exchange requests and responses over the *General Inter-ORB Protocol* (GIOP).

In addition to the support of remote invocation, an ORB provides a set of common services that both clients and servers can use to implement their service logic. The most important common services in this context are the *naming* and the *trading* services, because they offer support for the service discovery problem. The naming service helps a client find a server by its name, whereas the trading service helps find services by means of the interfaces they implement as well as other properties that are defined by the service providers; however, how a server or a client finds a naming or a trading service is not specified in CORBA. This initial bootstrapping process requires additional mechanisms that are outside the scope of CORBA.

As far as mid-session mobility is concerned, the situation is asymmetric for CORBA clients and for CORBA servers. Because of the strict request/ response communication model of CORBA, a change of the client's IP address is only a problem if it occurs during the time period between sending out a request and receiving the response. In these circumstances, the server cannot deliver the response, and the client has to issue the same remote method invocation again. This client-side functionality has to be implemented inside the client service logic. Whether the change of IP address is due to device or user mobility is irrelevant in this context. For CORBA server objects, however, the situation is more complex. A server object is known to a client through its *interoperable object reference* (IOR). CORBA supports two categories of IORs: *transient* and *persistent*. A transient

IOR is only valid for the lifetime of a server object. It contains, among other things, the host that the server runs on and the port that it listens on. Evidently, when the host or port changes, because of either device or user mobility, the IOR becomes invalid and the server object can no longer be reached. Persistent IORs, on the other hand, are valid longer than the lifetime of a server object. The idea behind this is that a CORBA ORB can shut down a server object when it is not needed and restart it when needed (potentially even on a different host machine). Obviously, this leads to different IP addresses each time the server is restarted, so the persistent IOR cannot contain endpoint information about the IP address of the server. Instead, it contains the address of an *implementation repository*, which basically is a lookup server that stores mappings from persistent IORs to transient IORs. Each time a server object is restarted, it registers its current transient IOR with the implementation repository. A mobility-aware server implementation can use the same mechanism to update the information repository with a new transient IOR after the IP address of the server has changed due to mobility; however, it should be noted that this reregistration is not done automatically by the middleware but has to be implemented in the application logic of the CORBA server object.

Architecture for Location-Independent CORBA Environments

The Architecture for Location-Independent CORBA Environments (ALICE) is a CORBA-based mobile middleware with the main focus on support of device mobility, for both CORBA clients and servers [8]. The key idea is to deploy mobility gateways at the edges of the fixed network that mobile hosts can connect to via wireless communication. From a software architecture point of view, ALICE puts an additional mobility layer between the Transmission Control Protocol (TCP)/IP stack on the mobile host and the Internet Inter-ORB Protocol (IIOP) layer (i.e., the IP-specific GIOP implementation). The mobility layer on a mobile host communicates with the mobility layer on the mobility gateway to hide device mobility from the upper layers, including IIOP. When a mobile host communicates with a remote CORBA object, the current mobility gateway relays the requests and responses between the two CORBA peers. When the mobile host gets a new IP address assigned, it signals this new address to its mobility gateway, which directs any further communication to the new address. For server objects hosted on the mobile device, the mobility layer replaces the IP endpoint information in the (transient) IOR with information that refers to the mobility gateway instead. In this way, the distributed set of all mobility gateways takes on the same task as the centralized CORBA implementation repository explained above. As a mobile device moves from one network access point to another, it can switch mobility gateways, causing a handoff from the old to the new gateway. As a result, the state information from the old gateway is transferred to the new one, and all ongoing connections are tunneled to the new gateway for their remaining lifetime.

Java Remote Method Invocation

Java Remote Method Invocation (RMI) extends the Java language with the ability to invoke a method at a remote object, similar to CORBA, except that both client and server objects have to be implemented in Java [9]. Also like CORBA, Java RMI uses a combination of client stub and server skeleton to marshal and demarshal requests and responses. Before a client can invoke a method at a remote server, it must get a remote object reference of the server. Java RMI provides the RMI registry, a naming service that servers can bind with and that clients can look up to get a remote object reference. A server binds with its local registry, and a client looks up a server by its RMI Uniform Resource Identifier (URI), which has the format `rmi://host/object-name`. By means of the host part of the RMI URI, the client's local naming service finds the remote RMI registry and interrogates it for the object denoted by `object-name`. As long as the client uses a Domain Name System (DNS) name for the host component of an RMI URI, changes to the address of the IP server device prior to the naming lookup can be taken care of by the DNS. Whether the remote object reference contains the IP address or the DNS name of the host depends on the Java version and the way the remote object reference was created. For remote object references that contain the DNS name of the host, changes in the IP address can be hidden by the DNS. User mobility, however, is not supported by Java RMI because moving a server object to another device will break previously created remote object references.

Jini

Jini™ is a middleware technology with the primary goal of solving the service discovery problem, especially in highly dynamic, *ad hoc* networks where both clients and services can join and leave at a high rate [10]. The two key concepts behind Jini are:

- The separation of a service into a service proxy and a stationary part
- A lookup service that services can register with and that clients can query to look up services

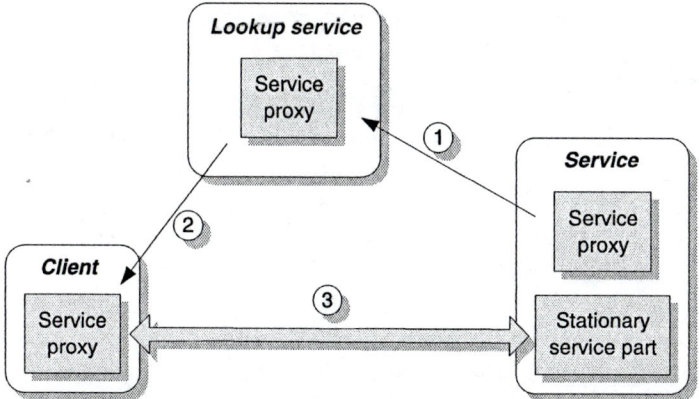

(1) When a service registers with the lookup service, it sends a copy of its Java service object. When a client looks up a service (2), it gets downloaded a copy of the service proxy. When the service is executed, it can communicate with the stationary service part using any arbitrary communication protocol (3).

Figure 19.7   Service lookup and proxy download in Jini.

Jini requires the service proxy to be implemented in Java. This is because when a service registers with a lookup server, the proxy object is shipped to the lookup server; later, when a client retrieves a service from the lookup server, the lookup server sends a copy of the service proxy to the client. The ability to dynamically download code over the network is an integral feature of Java and can readily be exploited by Jini without additional mechanisms. In addition to its proxy object, a service registers a service description with the lookup service. The service description is a combination of the service type and a set of user-defined values. The client specifies the desired service by means of the service description.

The service proxy, when executed on the client machine, can use any standard or proprietary mechanism to communicate with the stationary service part. If the stationary part is written in Java as well, communication could, for example, be done with Java RMI; however, the stationary service part does not have to be written in Java but can be written in any programming language. In these cases, communication could use CORBA or plain IP socket communication. In the extreme case, it is even possible to implement the complete service logic in the service proxy and not implement a stationary service part at all. Figure 19.7 illustrates the above mechanisms.

It is relatively easy to Jini-enable legacy services by writing a thin, portable Java proxy for them that remotely controls the legacy, stationary service, because the stationary service part can be written in any language and communication between the proxy and stationary service part can use any protocol. Jini supports the use of three mechanisms to discover a lookup service:

- A lookup service can periodically advertise its existence through IP broadcast on a well-defined broadcast address.
- Clients and services can send lookup discovery messages on another well-defined IP broadcast address to which the lookup services belong.
- A client or a service can directly address a lookup service whose address it knows through an out-of-band mechanism; this mechanism can be used to remotely connect to a home lookup service while roaming in a visited network.

In contrast to pre-session mobility, however, the support of mid-session mobility — both for device and user mobility — is out of the scope of Jini, as a result of the communication between the service proxy and the stationary service part being left to the service provider. Service providers can and have to implement any support of session mobility themselves; Jini does not offer any kind of support for this task.

Session Initiation Protocol

Strictly speaking, the Session Initiation Protocol (SIP) is a signaling protocol rather than middleware [12,13]; however, a brief discussion of SIP should be included in this chapter for at least the following two reasons:

- SIP is the predominant signaling protocol for all kinds of media sessions, not only in the Internet community but also for next-generation IP-based cellular networks.
- SIP has very interesting built-in mechanisms for the support of pre-session as well as mid-session mobility.

In SIP, users are addressed with their SIP URI, which has the format sip:user@domain. At registration time, the SIP *user agent* of user A registers with the *registrar* in A's home domain. The registrar uses a domain-wide *location service* to store a mapping from the user's SIP URI to his or her current *address of record* (AoR) (i.e., the current IP endpoint information).

When another user, B, wants to contact A, his or her user agent sends a SIP INVITE message to the inbound proxy of A's domain. The inbound proxy is stored in and can be retrieved from the DNS. The inbound proxy now can either redirect the INVITE to the current AoR or forward the INVITE itself. Figure 19.8 shows both registration and session set-up

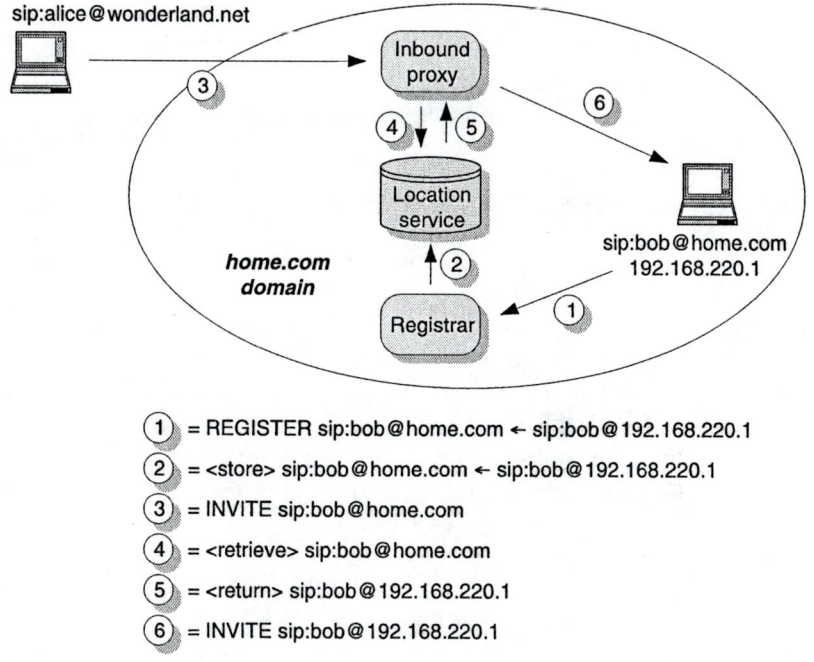

sip:alice@wonderland.net

home.com
domain

sip:bob@home.com
192.168.220.1

(1) = REGISTER sip:bob@home.com ← sip:bob@192.168.220.1

(2) = <store> sip:bob@home.com ← sip:bob@192.168.220.1

(3) = INVITE sip:bob@home.com

(4) = <retrieve> sip:bob@home.com

(5) = <return> sip:bob@192.168.220.1

(6) = INVITE sip:bob@192.168.220.1

Figure 19.8   User agent server lookup with SIP.

message flows for the common case where the inbound proxy directly forwards the incoming INVITE message to the addressed user agent.

The combination of registrar, location service, and inbound proxy transparently hides pre-session mobility because users contact each other using their symbolic SIP URIs. If, however, the IP endpoint changes during the course of an ongoing session, be it due to device or user mobility, another mechanism is necessary. In these circumstances, a user agent whose IP information has changed can reinvite the other parties with which it is currently engaged in a conversation. A SIP reINVITE message is a special INVITE that contains the identifier of the already ongoing session to indicate to the other party that this session is to be modified. The modification can affect the IP address but also redefines the kinds of media to be transmitted (audio, video) as well as the media encoding format.

To summarize, when a SIP user agent changes its IP address:

■ It reregisters the new address of record with its registrar so the user remains reachable for incoming session set-up requests (pre-session mobility).

- It reinvites all parties that it is engaged in a SIP session with by means of a SIP reINVITE message. If need be or if desired, the reINVITE can not only inform the peers of the new IP address but also renegotiate the media formats (mid-session mobility).

Mobile-Agent-Based Ubiquitous Multimedia Middleware

Mobile-Agent-Based Ubiquitous Multimedia Middleware (MUM) is middleware that is specifically designed to support nomadic users [11]. In terms of mobility, the primary focus is on mid-session mobility. MUM does not address pre-session mobility (i.e., the service discovery problem). As far as mid-session mobility is concerned, the scope of MUM goes beyond transparent reconnection of server and client by also providing mechanisms to automatically adapt the service quality to the new situation with respect to network bandwidth and delay, as well as device capabilities. The key idea behind MUM is the use of service proxies that are on the path between the server and the client and that adapt the media stream to fit the device capabilities. When a user moves to a new device, the service path changes, and the proxies on the path adapt their behavior to make the served media fit the new requirements. Because MUM focuses on streaming applications, media adaptation is the greatest challenge, and it is not necessary to transfer state information from the old to the new device.

Remote Desktop

By *remote desktop* we are referring to a whole family of similar systems that allow users to remotely access a server PC from a local client PC. More specifically, the client PC connects to the server PC and represents its desktop as a window on the local PC. Then, every mouse and keyboard input on that window has the same effect as performing the action directly on the remote server. Examples of this category of approaches are Windows XP Remote Desktop, GoToMyPC, and pcAnywhere.

A remote desktop solution aims at supporting user mobility rather than device mobility. In a sense, the idea is a particular instantiation of the MVC approach described before, except that the MVC model is not applied to an individual service but to the entire PC. The remote server PC — which typically runs at the edge of the network, in the end user's home domain — hosts the model, and the local visited PC runs the view and the controller, because any input received at the local PC is translated into commands being sent to the remote PC.

The biggest advantage of a remote desktop solution is that, because the entire home desktop is represented on the local PC, users get access to all their services and their user-specific configuration settings without having to install and configure the same applications on various machines. This is a very beneficial feature, because maintaining versioning control for different installations of the same service is a tough challenge; however, remote desktop solutions have several drawbacks, as well:

■ They require the server PC to always be on and connected to the Internet. These are serious requirements with regard to the stability of the server PC and the reliability of its network connection.
■ Because the model itself is a graphical representation (i.e., the desktop of the server PC), the amount of data that has to be shipped between the server and the client PCs is large. Even for broadband connections, this leads to significant latency and to a very noticeable difference between working directly on a server PC and working remotely via remote access. A per-service separation of the model, view, and controller results in a considerably reduced amount of data being exchanged between the model, view, and controller.

Discussion

As we have seen in this chapter, establishing and maintaining a session in a mobile environment are very challenging tasks. The use of mobile middleware can help liberate application programmers from mobility-related issues so they can focus their attention on the actual application logic.

The first problem to be solved in a mobile environment is *session establishment*, in particular *service discovery*. We have discussed more traditional solutions that involve the use of a centralized lookup service, as well as a more recent, decentralized approach that is based on peer-to-peer techniques. After a service session has been successfully established, the next challenge is *service continuity* (i.e., maintaining a session even in mobile situations). Two kinds of mobility have to be distinguished:

■ *Device mobility*, where the user moves with a mobile device that switches from one network access point to another
■ *User mobility*, where the user switches from one device to another

In the case of device mobility, *Mobile IP* is the standard way to shield communication peers from changing IP addresses. In a nutshell, Mobile IP is based on the concept of a home agent that receives all packages that are addressed to the mobile client and forwards them appropriately. Without Mobile IP (i.e., in *simple IP* mode), a mobile middleware has to sit between the IP stack and application so it can inform its communication peers about an IP address change and the middleware on the peer machine can redirect any future communication to the new address.

In the case of user mobility, two issues have to be addressed. First, all communication peers have to be informed about the changed IP address, just as in the case of device mobility. Second, the session state has to be transferred from the old source to the new destination device. The session state can be transferred in several ways. These include moving the state directly or via a network or home server location, as well as whether only the service state or the entire service is moved. Moving only the session state is more bandwidth efficient, but moving the entire service avoids versioning and personalization inconsistencies. An elegant way to avoid transferring the session state altogether is to implement the service in a distributed model–view–controller architecture, with the model running in the network (or on a home server) and the view and the controller running on the user device.

In this chapter, we also discussed several sample mobile middleware approaches, including CORBA, ALICE (an extension of CORBA for mobility), Java RMI, Jini, SIP, MUM, and the remote desktop. All of these middleware technologies provide a certain degree of mobility support, but they vary significantly with respect to their primary focus. CORBA, one of the most prominent distributed middleware approaches, does not directly provide comprehensive mobility support but does offer services that can be used to build a true mobility-supporting system. The naming and trading services can be used to implement service discovery, and persistent IORs can be used to implement mid-session mobility on the server side. CORBA does not, however, provide any mobility support on the client side.

ALICE, an architecture based on CORBA, uses so-called mobility gateways to implement full mid-session mobility support. Java RMI is primarily a remote method invocation mechanism with some support of service discovery but no mid-session mobility support. SIP is probably the most important mobility-supporting infrastructure in the communication arena. It has built-in mechanisms for service discovery as well as mid-session mobility support; however, SIP is not a fully fledged middleware but rather a communication protocol. The primary focus of MUM is mid-session mobility, and in this respect it goes beyond pure reconnection of a mobile user but also adapts any ongoing media sessions to the new device and network capabilities. Remote desktop solutions, finally, aim at user mobility by allowing access from a client PC to a remote server PC. The remote desktop can be considered a particular instantiation of a distributed model–view–controller solution, with the remote server PC constituting the model and the local client PC the view and the controller.

To summarize, different mobile middleware technologies have different goals and thus different strengths and weaknesses. At least at this point in time, no single, comprehensive mobile middleware covers all aspects of mobility and is widespread and commonly used. Which approach best suits one's needs depends on the specific requirements and has to be decided on a case-by-case basis.

References

[1] Droms, R. and Lemon, T., *The DHCP Handbook*, Macmillan, Indianapolis, IN, 1999.

[2] Stoica, I. et al., Chord: a scalable peer-to-peer lookup service for Internet applications, in *Proc. ACM SIGCOMM'01*, San Diego, CA, August, 2001.

[3] Ratnasamy, S. et al., A scalable content-addressable network, in *Proc. ACM SIGCOMM'01*, San Diego, CA, August, 2001.

[4] Wisely, D., Eardley, P., and Burness, L., *IP for 3G: Networking Technologies for Mobile Communications*, John Wiley & Sons, New York, 2002.

[5] Perkins, C., *IP Mobility Support*, Request for Comments 2002, Internet Engineering Task Force (IETF), 1996 (http://www.ietf.org/rfc/rfc2002.txt).

[6] Gamma, E. et al., *Design Patterns: Elements of Reusable Object-Oriented Software*, Addison-Wesley, Boston, MA, 1995.

[7] Vogel, A. and Duddy, K., *Java Programming with CORBA*, John Wiley & Sons, New York, 1998.

[8] Haahr, M., Cunningham, R., and Cahill, V., Supporting CORBA applications in a mobile environment, in *Proc. of the 5th ACM/IEEE Int. Conf. on Mobile Computing and Networking (MOBICOM'99)*, Seattle, WA, August, 1999.

[9] Pitt, E. and McNiff, K., *Java.rmi: The Remote Method Invocation Guide*, Addison-Wesley, Boston, MA, 2001.

[10] Edwards, W.K., *Core Jini*, Prentice Hall, Upper Saddle River, NJ, 2000.

[11] Bellavista, P., Corradi, A., and Foschini, L., MUM: a middleware for the provisioning of continuous services to mobile users, in *Proc. of the 9th IEEE Int. Symp. on Computers and Communications (ISCC'04)*, Alexandria, Egypt, June, 2004.

[12] Sinnreich, H. and Johnston, A.B., *Internet Communication Using SIP*, John Wiley & Sons, New York, 2001.

[13] Rosenberg, J. et al., *SIP: Session Initiation Protocol*, Request for Comments 3261, Internet Engineering Task Force (IETF), 2002 (http://www.ietf.org/rfc/rfc3261.txt).

Chapter 20

Openness and Interoperability in Mobile Middleware

Eiko Yoneki and Jean Bacon

CONTENTS

Introduction

Today's distributed systems are composed of a diverse range of devices spread across different sites. Communication between devices is increasingly wireless. Providing open distributed processing between geographically and organizationally distributed work groups significantly helps to gain the flexibility to adjust robustly to business reengineering and environmental/technological change. This mission has depended on middleware during the past decade, and with the increase of network complexity, such as the introduction of various wireless networks, the demand on middleware is significantly increasing (see more mobile computing specific requirements in the next section). Such systems must be supported by effective, flexible, simple and comprehensible processes, with requirements being captured through modeling tools that provide readily usable notations for their clear representation.

Middleware is a term that defines the set of services, APIs, and management systems, which support a distributed, networked computing environment. RFC 2768[1] summarizes the functionality of middleware. For application programmers, everything below the API is middleware, and for networking specialists, anything above TCP/IP is middleware. Middleware can be categorized as follows: application-specific middleware for generic purposes residing between applications and operating systems and resource-specific lower middleware. The architecture of mobile middleware will be spread across these different types of middleware to support a lightweight and flexible component structure. Middleware is a

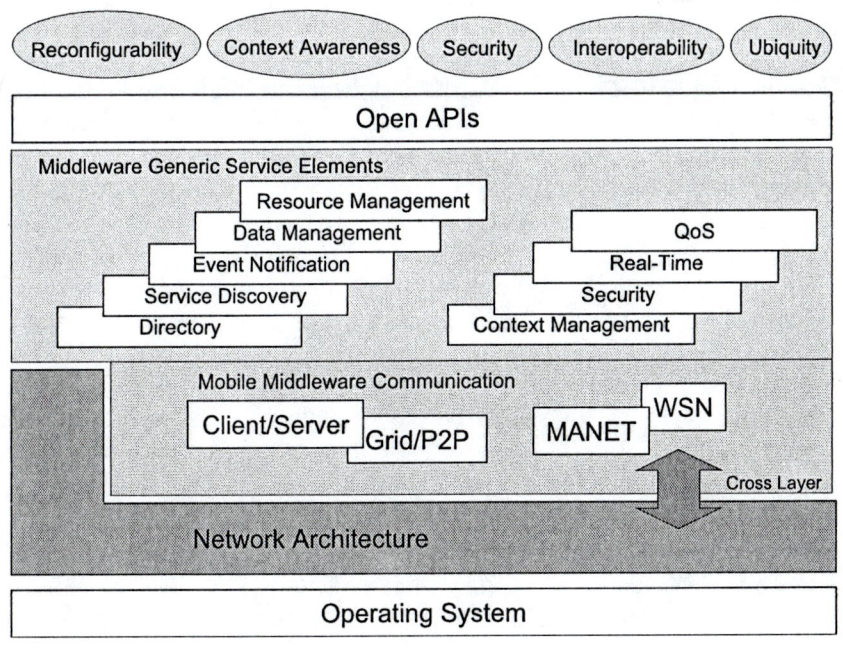

Figure 20.1   Mobile middleware stack.

contextual framework and an umbrella term for a number of concepts and architectures for distributed computer systems. The primary goals of middleware are:

- Simplification of distributed programming,
- Integration of heterogeneous systems,
- Abstraction from the transport facility,
- Masking the heterogeneity of networks, end-systems, operating systems and programming languages.

There is no absolute core set of middleware services that all applications require and a design of middleware can be hierarchical or layered. Considering that characteristics of mobile devices, wireless networks, and operational environments in mobile computing vary, a better design of middleware will be an unstructured collection of components that can be used either individually or in various subsets. One of the important middleware tasks is to support heterogeneous environments over distributed computing systems. Figure 20.1 depicts the architecture of a software stack. Middleware consists of two parts: generic service elements and synchronous and asynchronous communication support in mobile middleware. An application interacts with common Application Programming Interfaces (APIs) for accessing services provided by middleware. Most of the applications in wireless network environments are distributed, and standard protocols for interactions among applications are essential. A set of generic services like service discovery and event notification are desirable to allow the use of components from various middlewares in the design and development of computer systems, and common standards for developing interoperable software will help significantly. Sealing complexity within the middleware helps to simplify the application logic and, to provide openness and interoperability, the semantics of middleware should be unambiguously defined. First, interactions among components need to be modeled. Second, defining interfaces, types, semantics, constraints, evolution and non-functional properties will rule modelled interactions. The issues to consider in relation to such semantics for middleware are listed below.

- *Openness* eases adding new components.
- *Interoperability* links across different platforms and protocols.
- *Reconfigurability* allows the system to be reconfigurable with changing requirements for open-ended, incremental and continually evolving system.
- *Portability* ports programs without change on different computer systems.
- *Mobility* supports both physical and logical mobility of system components.
- *Scalability* ensures efficiency with an increasing number of users.
- *Adaptability* adapts to changing environments maintaining required services.
- *Transparency* hides complexity from users.
- *Brokering* uses agents/brokers for the distribution of services.
- *Distributability* extends the interoperability idea to allow processes and information to be provided and migrated automatically to the most convenient point of an interconnected set of computing systems.
- *Fault tolerance* avoids failure of the whole system resulting from a partial failure.
- *Security* ensures security among collaborating devices.

In the mid 1990's, a significant effort has been made to standardize middleware architecture for distributed object environments. The Reference Model of Open Distributed Processing (RM-ODP[2]) is an effort by the ISO (International Standardization Organization) and ITU-T

(International Telecommunication Union). RM-ODP defines architecture for distributed processing, interoperability and portability of applications. It is considered the most complete specification. The Object Management Group's Common Object Request Broker Architecture (CORBA[52]) supports an environment for open distributed processing by addressing the requirements for ODP. The motivation for developing a standard for Open Distributed Processing was the opportunity provided by a range of computing and communication options, which allowed diverse systems to work in a cooperative fashion. A further factor was the availability of units of computing power at economic prices coupled with the liberalization of telecommunication services. This has been accompanied by the emergence of a paradigm shift in the approach to systems building by using the object-oriented approach.

The Open Distributed Processing standard is a reference model standard that provides a framework in which component standards are placed. The framework provides concepts and languages for describing the functional components of an open distributed processing infrastructure. However, successful development of distributed applications requires more than an enabling infrastructure of computing and communications.

Standardization of middleware for mobile computing requires a large amount of work. Each potential solution attempts to extract common interfaces (see the third section for details). References[95, 96, 97] give good summaries of heterogeneity support in middleware in mobile computing. The increasing diversity of devices, network elements and applications brings a need for different middleware. Thus, various middlewares will be coordinated in the future. This requires interoperability on two levels: between middleware platforms and between applications components running over different middleware.

Characteristics of Mobile Middleware

In a mobile/wireless network environment, latency is high, bandwidth is low, and the connection can be interrupted at any time. Mobile devices have a small footprint and different transport mechanisms to connect to the network, and many devices are not programmable. Most current middleware assumes continuous availability of high-bandwidth network connections. Also mobile middleware needs to compensate for device disparities by rendering standard application screens in device-specific ways, interacting with client business logic in the handheld device.

The biggest challenge for middleware in mobile computing is the management of dynamicity. Existing middleware cannot meet the demands of new application areas such as embedded systems, real-time, and physical and logical movement of the device. The mobility concept in mobile computing can be:

- A device changes its location within the topology.
- A user changes the communication methods of her device.
- An application element dynamically changes service/session.

Supporting mobility is required not only over different network environments but also for separate computations in time and space and explicit control over resources, such as supporting Quality of Service (QoS) and context awareness. Cooperation in time can have synchronous and asynchronous modes, and in space can have either a single data resource or shared data, which requires replication and consistency control.

Moreover, distributed computing is moving from a static architecture (e.g. client server model) towards more dynamic and decentralized setting, such as peer-to-peer (P2P) systems, where applications are required to adapt to changes in topology and available data.

Network Environments

In mobile computing, middleware has a much tighter relationship with the network protocols; they often collaborate to deliver the data via a cross layer approach. The network environments of mobile computing can be classified into the following categories.

- *Nomadic Computing* — Nomadic systems consist of a set of mobile devices, which connect to a fixed network. Mobile devices move around, maintaining an intermittent connection to the fixed network.
- *Ad-hoc Networks* — Mobile Ad-Hoc Network (MANET) is a dynamic collection of nodes with rapidly changing multi-hop topologies that are composed of wireless links. The combination of mobile devices and ad-hoc networks allows the creation of highly dynamic, self-organizing, mobile peer-to-peer systems.
- *Wireless Sensor Networks* — The emergence of sensor networks brought further complex network environments. Wireless Sensor Networks (WSNs), which are composed of wireless sensor nodes distributed in the environment, include various sensors (e.g., cameras as vision sensors, microphones as audio sensors, or temperature sensors). Each node is equipped with a wireless communication transceiver, sensor, power supply unit, machine controllers, and microcontrollers on a MEMS (Micro Electro Mechanical System) chip, which is only a few millimeters square.

These WSNs can cover a large space by integrating data from many sensors, and produce diverse and precise information on the environment. Based on such a technological vision, new types of applications become possible, including environmental control and medial assistance.

■ *Ubiquitous (Pervasive) Computing* — Ubiquitous computing is about networked microprocessors embedded in everyday objects, surrounding us, talking to each other over wireless links. These networked microprocessors with sensing form a network for data collection and distribution[4]. They can organize themselves in an ad-hoc manner and new types of applications will appear where data is transmitted to nearby nodes to establish multi-hop routes in unstructured networks. Data and messages are propagated by dynamic routing mechanisms among out-of-range nodes. Thus, instantaneous data delivery is not guaranteed, and network stability and resource cannot be expected.

Specific Requirements of Mobile Computing

We will look at the specific requirements with respect to the following aspects.

Heterogeneity/Transparency

Supporting heterogeneity/transparencies with middleware frees the application developers from dealing with the complexities of distributed systems. Transparencies vary in different aspects and defining common requirements for transparencies and describing the computational refinements is essential. The heterogeneity and transparencies may be classified as follows:

■ *Access transparency* masks the data representation difference and programming procedure calling mechanisms to enable inter-working between heterogeneous systems.
■ *Location transparency* hides the use of physical addresses, and the distinction between local and remote environments.
■ *Relocation and migration transparency* hides the change of an object location including interfaces bound to it.
■ *Persistence transparency* hides the reactivation and deactivation of an object.
■ *Failure transparency* masks the failure and recovery of objects for supporting fault tolerance.

- *Replication transparency* maintains consistency of replica objects by providing a common interface.
- *Communication transparency* abstracts over the communication protocol and network types (ad-hoc, access-point, sensor networks).
- *Transaction transparency* masks the coordination for the transactional operations.
- *Device transparency* masks the different mobile devices.
- *Programming language transparency* hides the differences between the various programming languages in use.
- *Context transparency* maintains consistent operation through changes in context.

One of the most common goals of middleware for distributed systems is to hide that the system is in fact distributed and present it as if it is a single machine. In such transparent systems, resources (e.g. computation, memory and storage) from different hosts may be shared without awareness by the user. Applying this to mobile computing significantly increases the range of potential application domains. Mobile devices can delegate their tasks to other devices, facilitating load balancing and power consumption distribution. In order to increase transparent operation over distributed systems, a unified interface must be established.

Self-Configuration

In the near future an increase of communication devices will bring large-scale ubiquitous services and applications. The various mobile devices (e.g. handheld computers, laptops, or phones) are increasingly powerful; the standard mobile phone of the future will have a large screen, camera, and various communication units. It will look around to search displays, storages, and access points for communicating to the Internet. It might also search nearby devices for establishing temporal ad-hoc communities and embedded sensors for obtaining context information. Supporting context awareness is a necessary issue to support such mobile devices. Autonomous self-configuration, adapting to changing environments, is an important function of the middleware supporting such environments, requiring monitoring the physical world, event notification, event filtering, system modeling, group communication and decision rules for reconfiguration.

Asynchronous Communication

Asynchronous communication is essential to support mobile computing, where connections are short-lived, disconnected and intermittent. Group communication provides efficient consistency for applications that need

multi-participant interactions. In the wired environments, group communication has been shown to be useful in many application domains. Similar
applications will most likely require the same characteristics in the wireless
environment. However, traditional membership functionality designed for
the wired environment will not perform well with a large number of node
participations, where maintaining membership becomes expensive. Thus,
mobile computing is a hostile environment for strongly consistent interactions among multiple participants.

Service Discovery

In existing middleware, service discovery is provided on the basis of well-
known fixed name services, such as RPC (e.g. Jini) or open standards
(e.g. UPnP[37]). The more dynamic the environment becomes, the more
difficult the discovery of a host becomes. It is challenging to determine
the most appropriate service for the device within the range of device.
The interfaces may be diverse and a device may provide the same content
via several different interfaces. This makes it impossible to provide all the
drivers and communication protocols in every mobile device. Thus, a
desirable mechanism will be that mobile devices can reconfigure themselves at run-time in a dynamic manner and download the appropriate
interfaces.

Context Adaptation

Context awareness is considered as an important functionality in mobile
applications. For example, a context aware phone may switch to a silent
mode when it recognizes being in the lecture room. Context adaptation
requires definition and interpretation of context and design of contextual
information processing including modeling and framework that will result
creating middleware services. Context awareness creates a new type of
user interface to support ubiquitous computing, where a wide range of
information is detected by surrounding embedded devices. According to
obtained contexts, applications dynamically adapt to the situations that
provide more efficient usability. The context information can be speed of
networks, location, identity, temporal information (e.g., time and day of
week), environmental values (e.g., temperature and humidity), social occa-
sion (e.g., with whom), communication resource (e.g., access point and
hotspot), physical condition (e.g., blood pressure and heart rate), activity
(e.g., eating), schedules and so forth. Creating an effective context-modeling
framework is complex task, involving many aspects such as interpretation,
sharing, reasoning about and querying context. Furthermore, different

applications have different aspects on the same context information. Context adaptation encompasses various research topics including sensor fusion, context modeling, human-computer interaction (HCI), and distributed data management. To acquire and utilize information about context requires the following elements in middleware:

- Extraction of context.
- Interpretation of context.
- Reasoning about contextual situation and verifying it.
- Adaptation of application behavior.
- Distributed data management.
- Context modeling.

Security

Mobile computing, especially in ubiquitous computing, expects a large scale of networked environment supporting a wide range of different types of autonomous mobile entities. This introduces new security challenges and novel approaches are required that differ from existing security models and mechanisms. National Institute of Standard and Technology (NIST) addresses the following issues:

- User authentication may be disabled, a common default mode, disclosing the contents of the device to anyone who possesses it.
- Even if user authentication is enabled, the authentication mechanism may be weak or easily circumvented.
- Denial-of-Service attacks using radio noise can easily interrupt wireless communication.
- Wireless transmissions may be intercepted and, if unencrypted or encrypted under a flawed protocol, their contents made known.
- Connection with no authentication of the parties within interconnected device groups may introduce viruses or other types of malicious code and also other forms of attack such as a man-in-the-middle attack.

Another issue to address is mobile agents, which are autonomous software entities, and they can continue their execution and decision making in the foreign machine. The mobile agent based computing paradigm raises several security concerns, which prevent spreading of this new technology. Security issues include: authentication, identification, secure messaging, certification, resource control, non-repudiation, trusted third parties, and denial of service.

The current Internet can easily be attacked by worms, viruses, and spam. Security has to be addressed including privacy, access control, and trust. Security applies on hardware, protocols, operating systems, and middleware. In an ad-hoc communication, there is no centralized authority to judge integrity of information, thus trust measurement among the group members should be used. However, trust requires complex tasks on interpretation of trust value.

Group security aims to establish a common security aspect to all members of a group including confidentiality and integrity. Group communication will form different forms such as broadcasting, multi-casting and any-casting. In general, secure collaboration in ubiquitous computer systems requires security models to accomplish mutual trust and to minimize the need for human interaction. The security policy must encompass many potential collaborators based on the size of the infrastructure. Mobile devices will become disconnected from their networks. Thus, they need to be able to make autonomous security decisions independently from any specific centralized security infrastructure. Furthermore, the dynamic infrastructure makes mobile devices encounter previously unknown services, and the services will receive commands from unknown users. Thus, program validation, protection of the system against unauthorized modification, as well as key and certificate management for the complex implications of ad-hoc communities should be carefully designed. See also[6, 7] for privacy and security in sensor networks.

Trends in Mobile Middleware Solutions

There have been many attempts to support mobile computing through the enhancement of existing middleware approaches. IIOP (Internet Inter ORB Protocol) is an essential part of CORBA allowing communication among devices. IIOP has been ported to dynamic environments and used as a minimal ORB. Also RPC has been extended to add semi-asynchronous paradigms with buffering capabilities to deal with unannounced disconnection (e.g. Mobile DCE). In general, the traditional object-oriented client-server communication needs to be replaced by event-based reflective solutions. Note that existing middleware such as CORBA and .NET mask system and network heterogeneity problems and succeed in hiding the complexity of distributed systems. Thus, openness, the ability to adapt to any new components and services, is achieved with great efficiency.

In industry, the first extension from wired networks required dealing with device heterogeneity. Sun has J2ME[94] (Java Micro Edition) as a basic

Java Virtual Machine running on the various mobile devices. Microsoft® has .NET Compact Framework supporting XML[5] data and Web Service connectivity. These platforms support heterogeneity of devices, but they rely on the application itself to handle non-functional requirements. This requires further different types of services from middleware such as disconnected operation, context aware services, location based services etc. In general, middleware technologies can be categorized into the following models:

- *Remote procedure calls (RPCs)* support heterogeneous distributed platforms by extending traditional procedure. Application developers can use RPCs transparently to access local and remote procedures. The RPC standard has been extended into the Distributed Computing Environment (DCE) standard and more recently Mobile DCE. Most operating systems support basics of DCE's services.
- *Message-oriented middleware (MOM)* provides asynchronous rather than synchronous interactions. MOM encourages loose coupling between message senders and receivers with a high degree of anonymity, thereby removing static dependencies in distributed environments. MOM's characteristics (an intuitive programming model, latency hiding, guaranteed delivery, and store-and-forward messaging) are highly appealing for mobile applications.
- *Object-oriented middleware* enables the remote invocation of object methods. Java remote method invocation (Java RMI[8]), CORBA, the Distributed Component Object Model (DCOM[66]), and .NET are popular OO middleware platforms. Wireless RMI is an enhancement of Java RMI for mobile environments.
- *Component-oriented middleware* evolved from OO middleware allowing the reuse of components. It gives reconfigurable capabilities into distributed applications. Enterprise JavaBeans (EJB[9]) and the CORBA Component Model (CCM[3]) are important examples.
- *Transaction-oriented middleware* supports essentially distributed transactions operations, which interconnect heterogeneous database systems across platforms. It offers better data availability and integrity including high performance.

Any middleware system often needs to interact with another middleware system, where both support similar services and functionalities. Thus common API to interoperate those middleware is desirable. Message-oriented middleware aims to provide communication mechanisms and the Java Message Service (JMS[10]) is a semantic API for the message and object paradigms. CORBA Notification Service is an OO platform with an event-based communication. Component based approaches are attractive for modeling resource management by abstracting and encapsulating the object. Object and component oriented middleware provides a natural way for encapsulating and integrating resources. Both approaches can hide the complexity of resource management operations.

Emergence of Event-Based, Reflective and Service Architectures

Different styles of mobile middleware have emerged to solve the challenges of mobile computing. In wireless network environments, the time-triggered approach is expensive when the expected rate of data communication is low. Event-based middleware that is based on a publish/subscribe communication paradigm became popular because asynchronous and multipoint communication is well suited for constructing reactive distributed computing applications. Especially with the distributed event-based middleware over peer-to-peer (P2P) overlay network environments, the construction of event broker grids will extend the seamless messaging capability over scalable heterogeneous network environments. Event-based middleware is discussed in more detail in Section 3.2.

A significant increase of the event monitoring capability in the real world by wireless devices and sensors led to a further evolution of ubiquitous computing. The middleware in sensor networks can be defined as software that provides data aggregation and management mechanisms, adapting to the target applications' need, where data are collected from sensor networks. This functionality must be well integrated within the scheme of ubiquitous computing. The middleware should offer an open platform for users to seamlessly utilize various resources in physically interacting environments, unlike the traditional closed network setting for specific applications. One of the important issues here is to support an unambiguous event correlation mechanism over time and space in heterogeneous network environments, where middleware should take an active role (see[100] for requirement on event correlation). The trend of system architecture to support such platforms is towards service broker grids based on service management. When designing the middleware for sensor networks, heterogeneity of information over global distributed systems must be considered. The information sensed by the devices is aggregated and combined into higher-level information or knowledge and may be used as context. Adaptation allows the middleware to choose the resources in the given environments such as communication/discovery protocols. Reflection provides more flexibility in middleware on location, context and device characteristics. OpenORB[3] supports adaptation. The context-aware adaptation approach exposes the context information (e.g. Nexus[17], CHARISMA[18], OpenORB[3], and Odyssey[19]). See Chapter 2 for Reflective Middleware.

A key issue here is how to orchestrate all these heterogeneous environments. This gave rise to an architecture based on a service-oriented model; model-driven with metadata objects that support middleware-middleware heterogeneity. See more about SOA in Section 3.3.

IPv6 Advantages

IPv6[91] is becoming the protocol for the next generation of networks. The IPv6 standard is designed to meet future demands on address space, multicast, encryption, QoS and better support for mobile computing. For example, the IPv6 numbering mechanism could simplify dynamic mergers and acquisitions of network service components. Three major advantages are bigger address space, mobility support and security support.

- *Bigger address space* supports 128-bit address space and better address aggregation properties. NATs (Network Address Translators) are no longer needed, and this allows global IP connectivity including mobile devices. All can benefit from full IP access through end-to-end services.
- *Mobility support* is a significant advantage. Mobile IPv6[92] enables to be accessed using general IPv6 APIs, appearing transparent to the application layer.
- *Built-in Security* is supported including IPsec[93], which allows authentication and encryption of IP traffic. This brings a full end-to-end IPsec based secure communication.

Event-Based Communication

Event-based communication, usually based on the publish/subscribe paradigm, is an appropriate mechanism for loosely coupled interactions in mobile computing, where interactions between devices are often only possible for short periods of time, when the devices are nearby in the transmission range. Therefore, it is essential to avoid complex handshake processes, which may consume too much available bandwidth, power and time. In the past decade many event-based middlewares in wired networks have been reported on both centralized and decentralized models, including CEA[67], SCRIBE[68], IBM Gryphon[14], JEDI[13], Hermes[12] and SIENA[69]. SCRIBE is a topic-centric publish/subscribe messaging system using Distributed Hash Tables (DHT) over Pastry[70]. SIENA has an extension to support mobility by an explicit operation to relocate clients. Tuple Spaces are supported in LIME[16], TSpaces and JavaSpaces™. None of these systems support extremely dynamic mobile environments.

Examples of middleware systems developed to support wireless ad-hoc network environments are STEAM[15], ECCO[99] and IBM WebSphere MQ[60]. STEAM provides a proximity based group communication, and ECCO takes a cross layer approach integrating publish/subscribe semantics with a Mobile Ad-hoc Network (MANET) on-demand protocol. However, the majority of these systems construct publish/subscribe above existing transport protocols. In MANET environments, much research currently focuses on datagram routing in both unicast and multicast routing. However, no definite solution to define publish/subscribe semantics using these protocols has been provided. The collaborative style of application in MANET offers loosely coupled components to communicate and work together in a spontaneous manner, and they may require real-time functionality for timeliness and reliability such as inter-vehicle communication.

In general most middleware for wireless networks has focused on nomadic applications. Such applications allow mobile nodes to connect to the wireless network primarily for being a part of a fixed network infrastructure. However node mobility may suffer a disconnection period while moving from a connecting location to a next location.

Future challenges of distributed event-based systems are supporting applications based on a heterogeneous communication system, where characteristics of networks vary, for example, from scheduled real-time buses to wireless ad-hoc networks. Event-based middleware supporting such applications must seamlessly disseminate messages generated by deeply embedded processors to all interested nodes in the global networks. Improving group communication is important to support the semantics of new membership and useful weakened semantics. Current approaches include using a WAN-of-CAN structure[71], where a global network (WAN) comprises substructures subsumed by the abstraction of a Controller Area Network (CAN). The idea is that grouping nodes reflects potentially heterogeneous network architectures.

In industry, as an open-standard messaging technology, open-JMS[10] frameworks and robust J2ME solution for connectivity to the mobile enterprise are thus far the best practice. Softwired's iBus//Mobile[90] supports the development of mobile solutions that make applications accessible on various mobile networks and offers good mobile connectivity, with support for disconnected operation and integration with most common carrier technologies. JMS is a de-facto industry standard; however the interoperability of JMS compliant products is not yet available, and some products support an open JMS framework, which enables enterprises to leverage and integrate existing systems including IBM WebSphereMQ[60] and TIBCO Rendezvous[6] to build robust, flexible Java applications based on JMS, J2EE[28], and Web Services standards.

Middleware has to mask temporary disconnection states from applications. A possible approach is buffering RPC invocations for transmission until connection is re-established. However, mobility and changes of IP address (e.g. IPv4) can also cause problems when fixed hosts perform RPCs to mobile hosts. JMS provides the semantics of disconnected operation. The solutions provided by Mobile IP[92] may not fully satisfy the requirements, because multiple network addresses are used. This condition will be changed by the emergence of IPv6[91].

Service Oriented Architecture

Service Oriented Architecture (SOA) is a well-proven concept for distributed computing environments. It decomposes applications, data, and middleware into reusable services that can be flexibly combined in a loosely coupled manner. SOA maintains agents that act as software services performing well-defined operations. This paradigm enables the users to be concerned only with the operational description of the service. All services have a network-addressable interface and communication via standard protocols and data formats (i.e., messages). SOA can deal with aspects of heterogeneity, mobility and adaptation, and offers seamless integration of wired and wireless environments. It helps component-oriented design on mobile devices and improves the design and development process. SOA for mobile computing requires supporting the heterogeneity of mobile devices and networks and adaptation to mobile platforms. SOA can offer an interoperable service-oriented framework with standard, self-describing interfaces, which hide details of the underlying service.

Generic service elements are context model, trust and privacy, mobile data management, configuration, service discovery, event notification, and the following are the key requirements for SOA.

- *Flexible discovery mechanisms for ad-hoc networks*, which provide the reliable discovery of newly or sporadically available services.
- *Support for adaptive communication modes*, which provides an abstract communication model underlying different transport protocols. Notably, event-based communication is suitable for asynchronous communication.

Peer-to-peer networks and grids offer promising paradigms for developing efficient distributed systems and applications. Grids are essentially P2P systems. The grid community recently initiated a development effort to align grid technologies with Web Services: the Open Grid Services Architecture (OGSA)[72] allows developers integrate services and resources across distributed, dynamic, heterogeneous, environments and communities. The

OGSA model adopts the Web Services Description Language (WSDL[41]) to define the concept of a grid service using principles and technologies from both the grid and Web Services. The architecture defines standard mechanisms for creating, naming, and discovering persistent and transient grid-service instances. The convergence of P2P and Grid computing is a natural outcome of the recent evolution of distributed systems, because many of the challenging standards issues are quite closely related. This creates best practice that enables interoperability between computing and networking systems for the P2P community at-large.

One important element of SOA is service discovery. In typical service discovery architectures, a dedicated directory agent stores service information. A set of protocols enables users to find a directory agent and to register with it, and also a provided naming convention for services. For example, the Service Location Protocol (SLP), Jini, Web Service Description Language, and UPnP proved service discovery functions. On the other hand, the Semantic Web addresses service description and discovery mechanisms in a different way (e.g. DAML). Open Mobile Alliance (OMA[47]) is addressing the issue of global service discovery for wireless networks. In ad-hoc networks, an agent-based approach[98] is attempted.

Model Driven Architecture

The service architecture is the most promising approach for mobile computing, and the model driven architecture (MDA) supports the construction of modular service components. The MDA promises efficient program development through modeling. Wireless World Research Forum (WWRF[73]) has defined an architectural framework. There are local regional activities such as mITF[75] in Japan. Moreover the industry, NTT Docomo[76] and Nokia[77] for example, have linked their proposals to various forums. The framework of OMA highlights the end-to-end view and the OMA is a conceptual architecture that does not indicate a specific topology or location of servers. Thus, no specific hierarchy of protocol stacks between domains is required. The service architecture defined in the WWRF uses an I-centric approach, where a communication system adapts to the demands of each individual (I-centric). Future services will individually adapt requirements raised for a large scale of personalization.

Heterogeneity Support

This section summarizes the heterogeneity support in mobile middleware from the following aspects: reconfigurability, context modeling, security, programming paradigm, and wireless communication, which are discussed in reference[96].

Reconfigurability

In order to construct a self-aware or reflective system, current state and configuration has to be kept so that configuration management can provide capability for reconfiguration. W3C's CC/PP[20] and OMA's UAProf[21], which are based on W3C's RDF[22] address terminal capabilities for wireless devices. In serialization CC/PP uses XML[5] and UAProf WBXML[24]. FIPA's Quality of Service Specification[25] gives an ontology definition for message transport. Device Ontology Specification[26] is also included in FIPA. Furthermore OMA have kept working on Device Management[27] service. XML is used for configuration of CORBA Component Model[3] and J2EE[28]. W3C's Device Independence Activity[29,30] contributes configuration management. Configuration management further requires group membership management in ad-hoc communications[31]. Recent progress of MobileIP allows more flexible membership management. Application level of multicast also has to be integrated as part of middleware, which requires dynamic group management and possibly content-based routing. For event-based communication, it is desirable to support subscriptions with filter expressions, so that only the events that match its filter are delivered to an application. The management of subscription must be done dynamically, which is similar to the group membership management in multicast. Detection mechanisms such as finding new devices, or detecting devices, where their states have changed, are necessary information for reconfiguring the environments.

Many service discovery mechanisms exist including IETF's Service Location Protocol (SLP[32]), Jini[33], OMG's Trading Object Service[34], Universal Description, Salutation[36] and Universal Plug and Play (UPnP[37]). Organization for Bluetooth on the other hand offers its own protocol of Service Discovery[38]. Each solution is self-contained, thus establishing common interface for interoperability is necessary. Another aspect of reconfiguration is software update by downloading and uploading including on-line upgrades and rollbacks.

Context Modeling

UML[23] from OMG is universal tool for system modeling. Model Driven Architecture (MDA[39]) is a design approach introduced by OMG. With MDA, system functionality is defined as a Platform Independent Models (PIMs), using an appropriate Domain Specific Language and translated to platform-specific models for actual implementation. This approach promotes the flexible abstraction level in software and allows developers to implement the defined model on the various platforms in mobile computing. The MDA model architecture are related to multiple standards such as United Modeling Language (UML), the Meta Object Facility

(MOF), the XML Metadata interchange (XMI), and the Common Warehouse Metamodel (CWM).

In W3C, Web Services, Web Services Description Language (WSDL[40]) and Semantic Web (RDF[22] and OWL Web Ontology Language[41]) address modeling. Advancement of Structured Information Standards (OASIS[42]) supports some aspects in Universal Description Discovery and Integration (UDDI[35]). MDA requires model transformation, for example MOF defines a specific standard for model transformation MOF/QVT. MDA-supporting tools are developed by vendors and open source projects such as Sun's Java EE or Microsoft .NET. Modeling needs to be provided in a coherent way so that the conceptual models in future systems can be satisfied. Context modeling has to be light weight in mobile computing, and model transformation will help to design less complex and efficient context modeling.

Security

The most widely used security protocol is SSL/TLS (Transport Layer Security, RFC2246[43]). TLS[44] is used to protect communication in the Internet, and it relies on certificates from an external X509[45] Public Key Infrastructure. TLS is designed to provide privacy and data integrity between applications. TSL is extended to support wireless Internet as KSSL (KiloByte SSL[46]) and more recently SLL for sensor actuators using ECC (Elliptic Curve Cryptography). Alternatively the IPsec (RFC2401, 2407, 2408) protocol can be used. Confidentiality of data is built from Encapsulating Security Payload (ESP, RFC2406) and data integrity is built from Authentication Header (AH, RFC2402). Both WAP[24] and Palm.net[48] use security protocols (e.g., WTLS[47] for WAP) for their mobile clients and insert a proxy/gateway in their architecture to perform protocol conversions.

The Trusted Computing Group (TCG[50]) originally aims to develop Trusted Platform Modules (TPM) and TCG Software Stack (TSS) for hardware and software interface that enables trusted computing. TCG recently released Trusted Network Connect (TNC) protocol specification based on AAA, adding the ability to authorize network clients on the basis of hardware configuration. Future wireless devices need to address the mapping of identities of users in distributed environments. It is necessary to profile attributes in order to establish the identity of an individual. The future Internet will take the form of a web of relationships among different identities that must be accepted and trusted. The Liberty Alliance consortium[51] is specifying architecture to offer federated network identity management including trust modeling, efficient fragmentation of information, and key/certificate management.

The SECURE project[49] aims to provide a self-configuring security system based on the notion of trust in human, providing mutual trust between mobile devices, which share information and work to present an unobtrusive interface to their users.

Programming Paradigm

The client-server programming model still dominates in application development in spite of recent emergence of peer-to-peer networks. However, ad-hoc communication in wireless networks will not fit into a client-server paradigm and the functionality of server needs to be distributed among member nodes for scalability and robustness. Moreover, supporting sensor network programming requires a way to propagate code to the appropriate node in the network and to collect sensed data, leading to a data centric programming approach.

OMG's CORBA[52], Sun's Java 2 Enterprise[28], Standard[53] and Micro[54] Editions, and Microsoft's .NET[55] Object Oriented Middleware (OOM) are typical current middleware for object-oriented platforms. Another type of middleware is message-oriented middleware (MOM), which supports event-based reactive applications. CORBA is OOM and also MOM including Event Service[56], and Notification Service[57]. Java Message Service (JMS[58]) and Sun Microsystems' Java System Message Queue[58] become de facto standard of publish/subscribe messaging in the enterprise computing. IBM's WebSphere MQ[60], Bea Systems' MessageQ[62], Microsoft Message Queuing (MSMQ[61]), and TIBCO's Active Enterprise (including TIBCO Rendezvous[63]) are MOM products providing JMS API. Open Source are ObjectWeb's JORAM[64] and xmlBlaster[65]. Mobile computing will mostly use an event-based and reflective middleware paradigm, where mobile applications will be adaptive, context-aware and personalized. Thus, a novel programming paradigm is desirable.

Wireless Communication

In general, wireless links are unstable and the degree of instability depends on location, time, and other contexts. Devices can be disconnected anytime. Network coverage, node mobility, node density, network dimension and throughput differ depending on the network architectures. Moreover, there exist various wireless networks: Wireless LANs, cellular networks, satellite, and short-range radio and communication protocols (e.g. WiFi, ZigBee, IEEE 802.11, and Bluetooth) are not unified. Furthermore network coordination will vary such as pure ad-hoc or nomadic. Thus, the hybrid systems need to support different types of communication links. Requirements on Quality of Service (QoS) will be diverse, that may not be satisfied through the traditional laying protocol architecture. The system performance of future networks will be enhanced by cross-layer design. Designing cross layer protocol requires careful consideration: when the protocol component improves performance for mobile communications, the application components could introduce another layer that cancels out the enhancements of the protocol components.

Standardization efforts are in progress to integrate various architectures for wireless communication. Among physical, MAC and higher layers, cross layer design issues are addressed, such as 3G standards such as CDMA2000 and QoS support by the Data Link Layer (DLC) and physical layer. Interoperability between wireless networks is another issue for user adoption and management of a wide scale development of services over wireless networks.

Standardization Activities

Several standards and drafts have been developed by the International Standard Organization (ISO), the International Engineering Consortium (IEC), and the International Telecommunication Union (ITU). There are also many specifications on the operation and interaction of devices and software, and they are considered standards such as Request for Comments (RFCs) by the Internet Engineering Task Force (IETF[59]). Usually the standardization contains three aspects: compatibility, interoperability and commonality. A standard can be open or proprietary. Open standards can be implemented by anyone without any restrictions. On the other hand, proprietary standards are only available under restrictive contracts from the organization that owns the specification. Some examples of open standards are:

- GSM (Global System for Mobile Communications - a mobile communications systems developed by (3GPP[85]))
- HTML/XML (Structured hyperlinked document format developed by the W3C)
- SQL (Structured Query Language developed by ANSI and ISO)
- OpenDocument (office document specified by OASIS[42])
- TCP (implementing stremas of data over IP developed by IETF)

The IETF is considered as a backbone organization of the standard Internet operating protocols. The primary purpose of standards by IETF is developing Internet Protocol Suites and the protocols do not cover middleware functionality. Thus, there are many organizations such as OMG (Object Management Group), OMA (Open Mobile Alliance) and W3C (World Wide Web Consortium) that contribute to provide middleware services (see Section 4.1-7), and the list is not exhaustive. Third Generation Partnership Project (3GPP[85]) and Third Generation Partnership Project 2 (3GPP2[86]) develop an open system for mobile communication and the Wireless World Research Forum (WWRF[73]) addresses issues relevant to 4G. These standardization activities include redundancy, and overlapping standards organizations tend to cooperate purposefully by defining boundaries between the scopes of those organizations.

Service description is addressed in the Organization for the Advancement of Structured Information Standards (OASIS[42]), specifically in Universal Description, Discovery and Integration (UDDI[35]). UPnP Forum[37] addresses discovery and auto configuration, Web Services Interoperability Organization (WS-I[86]) promotes Web Services Interoperability, which aims to provide interoperability across platforms and programming languages, Trusted Computing Group (TCG[87]) specifies the security issues in the network layer, and Digital Living Network Alliance (DLNA[88]) addressing interoperability of personal computers and mobile devices at home. Foundation of Intelligent Physical Agents (FIPA[25, 26]) provides a Device Ontology Specification.

In this section, known standard organizations and outline of standardization activities related to mobile middleware is given. References[5,27,72,86,96] provide more details of activities.

IETF (Internet Engineering Task Force)

The IETF is a large international community of network operators, designers, and researchers working on the Internet architecture and the operation. The IETF is supervised by the Internet Society Internet Architecture Board. Standards are expressed in the specification called Requests for Comments (RFCs). Under IETF there are various working groups organized by subject in different areas where standards are discussed and adopted. The middleware needs to coordinate required Internet protocols, where there exist many protocols for different infrastructures. Related protocols for mobile computing in IETF are listed in Table 20.1. IPv6 (stateless auto-configuration and neighbor discovery), DHCPv6, MANET (ad-hoc routing), IP QoS for wireless links and mobile networks, TCP enhancements for wireless links, IP multicast and multi-homing, network mobility, DCCP (Datagram Congestion Control Protocol), SLP (Service Location Protocol), XMPP (Extensible Messaging and Presence Protocol – jabber), SIP[78] (Session Initiation Protocol) and its extension (e.g., SIMPLE working group) are essential. An incorporated solution between the basic Internet mechanism and MANET is required for establishing IP connectivity such as SIP and DHCPv6. Figure 20.2 shows the protocols that are influential to mobile computing by mapping OSI stack. The protocol such as MANET split into the transport and network layer providing routing algorithms and a network interface.

BGP (Border Gateway Protocol) runs over TCP and UDP and is considered part of the application or network layer.

OMG (Object Managing Group)

The Object Management Group (OMG[79]) was formed in 1989 for creating standard architecture for distributed objects (i.e. components). OMG produced Common Object Request Broker Architecture (CORBA), UML and metadata modeling. A central element in CORBA is the Object Request Broker (ORB) and CORBA itself is for wired networks. Wireless CORBA[80], Super Distributed Objects and Smart Transducers appeared recently for wireless network environments. OMG currently includes over 500 member companies. Both International Organization (ISO) and X/Open have sanctioned CORBA as the standard architecture for distributed objects. Microsoft has its own distributed object architecture, the Distributed Component Object Model (DCOM). In wireless network environments, component based design will be important such as The Model Driven Architecture (MDA[89]). MDA helps to construct platform-independent models of application. It can be realized on any platform including Web Services, .NET, CORBA, J2EE, and others. Automatic code generation from the model specification will provide efficiency for development.

OMA (Open Mobile Alliance)

The Open Mobile Alliance (OMA[74]) is a consolidation of six industry groups; Wireless Village, SyncML initiative, WAP forum, Location Interoperability Forum, Mobile Wireless Internet Forum, and Mobile Gaming Interoperability Forum. It addresses specific end-to-end interoperability for mobile computing addressing.

W3C (World Wide Web Consortium)

The World Wide Web Consortium (W3C[40]) is an industry consortium which aims to promote standards for the WWW and interoperability between WWW products. W3C mainly produces specifications and reference software and its products are freely available. Specifications are in form of Recommendation (RFC) culminated from Working Drafts. Recommendation may be updated by separately-published Errata until enough substantial edits accumulate, at which time a new edition of the Recommendation may be produced (e.g., XML is now in its third edition). The W3C was initially formed by CERN, where the Web originated, and by DARPA and the European Commission. W3C Activities are organized into groups such as Document

Table 20.1 Mobile Computing Related Working Group in IETF

Area	Working Group
Application	(apparea) Applications Open Area
	(geopriv) Geographic Location/Privacy
	(imapext) Internet Message Access Protocol Extension
	(opes) Open Pluggable Edge Services
	(simple) SIP for Instant Messaging and Presence Leveraging Extension
	(slrrp) Simple Lightweight RFID Reader Protocol
	(xmpp) Extensible Messaging and Presence Protocol
Internet	(autoconf) Ad-hoc Network Configuration
	(dhc) Dynamic Host Configuration
	(eap) Extensible Authentication Protocol
	(ipv6) IP Version 6 Working Group
	(l2vpn) Layer 2 Virtual Private Networks
	(l3vpn) Layer 3 Virtual Private Networks
	(magma) Multicast & Anycast Group Membership
	(mip4) Mobility for IPv4
	(mip6) Mobility for IPv6
	(mipshop) MIPv6 Signaling and Handoff Optimization
	(nemo) Network Mobility
	(pana) Protocol for carrying Authentication for Network Access
	(send) Securing Neighbor Discovery
	(slp) Service Location Protocol
Operations and Management	(aaa) Authentication, Authorization and Accounting
	(multi6) Site Multihoming in Ipv6
	(netconf) Network Configuration
	(policy) Policy Framework
	(rap) Resource Allocation Protocol
	(ssm) Source Specific Multicast
	(v6ops) IPv6 Operations
Routing	(bgmp) Border Gateway Multicast Protocol
	(forces) Forwarding and Control Element Separation
	(manet) Mobile Ad-hoc Networks
	(mpls) Multiprotocol Label Switching
	(vrrp) Virtual Router Redundancy Protocol
Security	(ipsec) IP Security Protocol
	(ipseckey) IPsec KEYing information resource record
	(ipsp) IP Security Policy
	(kssl) KiloByte SSL
	(pkix) Public-Key Infrastructure (X.509)
	(sacred) Securely Available Credentials

Table 20.1   Mobile Computing Related Working Group in IETF

Area	Working Group
Transport	(avt) Audio/Video Transport
	(beep) Blocks Extensible Exchange Protocol Core
	(dccp) Datagram Congestion Control Protocol
	(diffserv) Differentiated Services
	(enum) Telephone Number Mapping
	(iptel) IP Telephony
	(megaco) Media Gateway Control
	(midcom) Middlebox Communication
	(mmusic) Multiparty Multimedia Session Control
	(nsis) Next Steps in Signaling
	(pilc) Performance Implications of Link Characteristics
	(rsvp) Resource Reservation Setup Protocol
	(rohc) Robust Header Compression
	(sip) Session Initiation Protocol
	(sipping) Session Initiation Proposal Investigation
	(spirits) Service in the PSTN/IN Requesting Internet Service
	(tsvwg) Transport Area Working Group

OSI	TCP/IP	Protocols
Application	Application	HTTP, SIP, SLP, DNS, RSVP, RAP, SNMP, BGP
Presentation		
Session	Transport	TCP, UDP, BEEP, BGP, MANET, IPSEC, SSL, KSSL, PKIX, RTP
Transport		
Network	Network	IPv6, MobileIPv6, DHCPv6, MANET, BGP, 802.11x
Data Link	Interface	
Physical	Physical	Physical media (e.g. radio), Encoding

Figure 20.2   Internet Protocol suite.

Object Model, XML, URI, Web Services, Graphics, HTML, Math, Semantic Web, Device Independence, Mobile Web Initiative, Voice Browser, and WAI activities. Web Services activity consist of Web Services Architecture, Web Services Description, XML Protocol (SOAP[81]), Web Services Choreography and so forth. Semantic Web activity consists of RDF, and the Web Ontology. W3C is one of the most influential organizations for mobile computing.

JCP (Java Community Process)

Java Community Process (JCP[82]) has produced several JSRs to extend Java 2 Micro Edition (J2ME) functionality below.

- J2ME RMI (JSR-66)
- J2ME Location API (JSR-179)
- Mobile Media API (JSR-135)
- Wireless Messaging API (JSR-120)
- Security/Trust Services (JSR-177)
- Graphics and User Interface for J2ME (JSR-184)
- Event Tracking API (JSR-190)
- Java Speech API (JSR-113)
- JDBC (JSR-169)
- J2ME SIP API (JSR-180)
- J2ME Web Service (JSR-172)
- Bluetooth API (JSR-82)

The J2ME architecture consists of configurations, profiles, and optional packages. The Connection Limited Device Configuration (CLDC) and the Connection Device Configuration (CDC) are available. The Mobile Information Device Profile (MIDP) is available for CLDC. JMS (Java Messaging Service) provides the industry standard for messaging, and JXTA[83] and JXTA for J2ME introduce the service-oriented architecture over P2P network environments.

OSGi (Open Service Gateway Initiative)

The Open Services Gateway Initiative (OSGi[84]) was founded in 1999. Its mission was to create open specifications for the delivery of multiple services over wide-area networks to local networks and devices. The Open Services Gateway Initiative focuses on the application layer and is open to almost any protocol, transport or device layers. The three key aspects of the OSGi mission are multiple services, wide area networks, and local networks and devices. Main benefits of the OSGi are that it is platform independent and application independent. In other words, the OSGi specifies an open, independent technology, which can link diverse devices in the local home network. The central component of the OSGi specification effort is the services gateway. A services gateway is a server that is inserted into the network to connect the external Internet to internal clients. The services gateway enables, consolidates, and manages voice, data, Internet, and multimedia communications to and from the home, office and other locations.

Liberty Alliance

The Liberty Alliance Project[51] is a consortium for building open standard specification for federated identity management. It supports security, and privacy-improving trust. It eliminates excess passwords and single sign-on is implemented. Recently identity-based Web Services are implemented so that it authorizes a service provider to access location information.

Future Challenges

The recent evolution of ubiquitous computing, with a dramatic increase of event monitoring capabilities by wireless devices and sensors, requires complex and sophisticated mobile middleware. This new platform enables users to seamlessly deploy various resources in physically interacting environments. The future vision of ubiquitous computing requires progress of devices and technologies, which will be ubiquitous and form a smart environment.

To support such heterogeneous environments in networks, devices, programming models, and data contexts, designing a service model based on a service-oriented architecture will be a key issue. The most difficult challenge will be to determine the appropriate level of detail in the definition of services and components. The aim of the service architecture is to provide a sustainable modular framework, where any module can be replaced without causing any impact. The important role of service architecture can be seen in the global interest in service architectures for future systems. Once the backbone of the service architecture is established, specific aspects of heterogeneity support must be added including the data model, communication model, and programming model. The network environments will be more heterogeneous than ever, and open, peer-to-peer based networking environments will become common.

As the first step to realize interoperability in such heterogeneous environments, it will be important to define mobile middleware architectures that achieve wide acceptance as existing middleware in wired networks. A common set of interfaces and service associated semantics will constitute a solid base for the faster emergence of future mobile applications deployed in a wide range of devices. Security research in mobile computing, especially access control, will be a major issue for the design of mobile middleware.

Acknowledgement

This research is funded by EPSRC (Engineering and Physical Sciences Research Council) under grant GR/557303. We would like to thank members of the System Research and Opera Groups at the University of Cambridge Computer Laboratory for the valuable comments.

References

[1] Aiken, B. et al., Network policy and services: A report of a workshop on middleware. IETF, RFC 2768, February 2000.

[2] ISO\IEC IS 10746-2, Open Distributed Processing – Reference Model: Foundations, 1996.

[3] CORBA Component Model, v 3.0. OMG document formal/2002-06-05, June 2002.

[4] Banavar, G. et al., Challenges: An Application Model for Pervasive Computing, in Proc. MobiCom, 266-274, 2000.

[5] W3C, XML Information Set. W3C Recommendation, 24 October 2001.

[6] Haowen, C. and Perrig, A., Security and Privacy in Sensor Networks, IEEE Computer, Vol. 36, No. 10, 103-105, October 2003.

[7] IEEE Spectrum, Special report, Sensor nation, July 2004.

[8] Java remote method invocation (Java RMI), http://java.sun.com/j2se/1.3/docs/guide/rmi/index.html.

[9] Enterprise JavaBeans (EJB), http://java.sun.com/products/ejb.

[10] Sun Microsystems, Java Message Service. (JMS), http://java.sun.com/products/jms/.

[11] Bacon, J. et al., Using Events to build Distributed Applications, in Proc. IEEE SDNE Services in Distributed and Networked Environments, 148-155, 1995.

[12] Pietzuch, P. and Bacon, J., Hermes: A Distributed Event-Based Middleware Architecture, in Proc. Int. Workshop on Distributed Event-Based Systems (ICDCS/DEBS'02), 611-618, 2002.

[13] Cugola, G. et al., The JEDI Event-Based Infrastructure and its Application to the Development of the OPSS WFMS, IEEE Transactions on Software Engineering (TSE), vol. 27, pp. 827-850, 2001.

[14] IBM, Gryphon: Publish/Subscribe over Public Networks. In http://research-web.watson.ibm.com/gryphon/gryphon.html.

[15] Meier, R. et al. STEAM: Event-based middleware for wireless ad hoc networks, in Proc. Int. Workshop on Distributed Event-Based Systems (ICDCS/DEBS'02), 639-644, 2002.

[16] Murphy, A. et al., Lime: A Middleware for Physical and Logical Mobility, *in Proc. 21st Int. Conf. Distributed Computing Systems*, 524-233, 2001.

[17] Fritsch, D., Klinec, D., and Volz, S., NEXUS Positioning and Data Management Concepts for Location Aware Applications, *in Proc. 2nd Int. Symposium on Telegeoprocessing*, 171-184, 2000.

[18] Capra, L. , Emmerich, W., and Mascolo, C., *CARISMA: Context-Aware Reflective Middleware System for Mobile Applications, IEEE Transactions on Software Engineering*, 29(10):929-945, 2003.

[19] Satyanarayanan, M., Mobile Information Access, *IEEE Personal Communications*, 3(1): 26–33, 1996.

[20] W3C, Composite Capability/Preference Profiles (CC/PP): Structure and Vocabularies, W3C Working Draft, 25 March 2003.

[21] OMA User Agent Profile Version 2.0, http://www.openmobilealliance.org/documents.html.

[22] W3C, Resource Description Framework (RDF). http://www.w3.org/RDF/.

[23] Unified Modeling Language (UML) Specification 1.5, OMG document formal/2003-03-01, March 2003.

[24] WAP Forum, WAP Binary XML Content Format. Document: WAP-192105-WBXML-20011015-a.

[25] FIPA Quality of Service Specification, FIPA document number SC00094A, December 2002.

[26] FIPA Device Ontology Specification, FIPA document number SC00091E, December 2002.

[27] OMA Device Management Version 1.1.2, http://www.openmobilealliance.org/documents.html.

[28] Sun Microsystems, Java 2 Platform, Enterprise Edition (J2EE), http://java.sun.com/j2ee/.

[29] W3C Device Independence Activity, http://www.w3.org/2001/di/.

[30] W3C, Device Independence Principles, W3C Working Group Note, 01 September 2003.

[31] Chockler, G. V., Heidar, I., and Vitenberg, R., Group Communication Specifications: A Comprehensive Study, *ACM Computing Surveys*, 427-469, 2001.

[32] Service Location Protocol, Version 2, IETF RFC 2608.

[33] Sun Microsystems, Jini Network Technology, http://wwws.sun.com/software/jini/.

[34] Trading Object Service Specification, OMG document formal/2000-06-27, 2000.

[35] OASIS UDDI Specification: Universal Description, Discovery and Integration of Business for the Web. (UDDI Version 3), http://www.oasis-open.org/committees/uddi-spec/doc/tcspecs.htm.

[36] Salutation, http://www.salutation.org/.

[37] UPnP Forum, Universal Plug and Play, http://www.upnp.org/.

[38] Bluetooth Service Discovery Protocol, https://www.bluetooth.org/spec/.

[39] OMG, Model Driven Architecture, http://www.omg.org/mda/.

[40] W3C, Web Services Description Language (WSDL) Version 2.0, Two W3C Working Drafts, 10 November 2003.

[41] W3C, OWL Web Ontology Language, A set of W3C Candidate Recommendations, 18 August 2003.

[42] Organization for the Advancement of Structured Information Standards (OASIS), http://www.oasis-open.org/home/index.php.

[43] Frier, A., Karlton, P., and Kocher, P., "The SSL3.0 Protocol Version 3.0", http://home.netscape.com/eng/ssl3/.

[44] Dierks, T. and Allen, C., The TLS Protocol Version 1.0, http://www.ietf.org/rfc/rfc2246.txt

[45] CCITT. Recommendation X.509: "The Directory – Authentication Framework". 1988

[46] Gupta, V. and Gupta, S., KSSL: Experiments in Wireless Internet Security, Sun Microsystems Technical Report: TR-2001-103.

[47] WAP Forum, "Wireless Transport Layer Security Specification", http://www.wapforum.org/tech/documents/WAP-261-WTLS-20010406-2.pdf

[48] Palm, Inc., "The Palm.Net Wireless Communication Service", see http://www.palm.com/products/palmvii/wireless.html

[49] Cahill, V. et al., Using trust for secure collaboration in uncertain environments, *IEEE Pervasive Computing*, 2(3): 52-61, 2003.

[50] Trusted Computing Group, https://www.trustedcomputinggroup.org.

[51] Liberty Alliance Project, http://www.projectliberty.org/.

[52] Common Object Request Broker Architecture (CORBA/IIOP). OMG document formal/2002-12-06, December 2002.

[53] Sun Microsystems, Java 2 Platform, Standard Edition (J2SE), http://java.sun.com/j2se/.

[54] Sun Microsystems, Java 2 Platform, Micro Edition (J2ME), http://java.sun.com/j2me/.

[55] Microsoft, .NET Framework, http://www.microsoft.com/net/.

[56] OMG, Event Service Specification, OMG document formal/2001-03-01, 2001.

[57] OMG, Notification Service Specification, OMG document formal/2002-08-04, August 2002.

[58] Sun Microsystems, Sun Java System Message Queue, http://wwws.sun.com/software/products/message/fiqueue/index.html.

[59] IETF TSVWG home page. http://www.ietf.org/html.charters/tsvwg-charter.html.

[60] IBM, WebSphere MQ, http://www.ibm.com/software/integration/wmq/.

[61] Microsoft Message Queuing (MSMQ), http://www.microsoft.com/windows2000/technologies/communications/msmq/default.asp.

[62] Bea Systems, MessageQ, http://www.bea.com/framework.jsp?CNT=index.htm\&FP=/content/products/more/messageq/.

[63] TIBCO, ActiveEnterpris, http://www.tibco.com/solutions/products/default.jsp.

[64] ObjectWeb Consortium, JORAM, http://joram.objectweb.org/.

[65] xmlBlaster.org, Open Source for MOM, http://www.xmlblaster.org/.

[66] Distributed Component Object Model (DCOM), http://www.microsoft.com/com/dcom.asp.

[67] Bacon, J. et al., Generic Support for Distributed Applications, *IEEE Computer*, vol. 33, pp. 68-76, 2000.

[68] Rowstron, A. et al., SCRIBE: The design of a large-scale event notification infrastructure, *in Proc. Int. Workshop of Networked Group Communication (NGC)*, 30-43, 2001.

[69] Siena, http://www.cs.colorado.edu/users/carzanig/siena/.

[70] Rowstron, A. et al., Pastry: Scalable Decentralized Object Location, and Routing for Large-Scale Peer-to-Peer Systems, *in Proc. IFIP/ACM Middleware*, 329-350, 2001.

[71] Veríssimo, P. et al., CORTEX: Towards Supporting Autonomous and Cooperating Sentient Entities, *in Proc. European Wireless*, 2002.

[72] Open Grid Services Architecture (OGSA) Working Group, http://www.ggf.org/ogsa-wg/".

[73] Wireless World Research Forum, *Book of Visions 2001*, http://www.wireless-world-research.org/.

[74] Open Mobile Alliance, http://www.openmobilealliance.org/.

[75] Mobile IT Forum, http://www.mitf.org/index\fie.html.

[76] Yumiba, H. et al., IP-Based IMT Platform. *IEEE Personal Communications*, October 18-23, 2001.

[77] Nokia, Mobile Internet Technical Architecture, Parts 1-3. ISBN 951-826-671-9, IT Press, 2002.

[78] SIP Web site. http://www.softarmor.com/sipwg/.

[79] Object Management Group, http://www.omg.org/.

[80] Wireless Access and Terminal Mobility in CORBA 1.1 Specification, OMG document dtc/04-04-02, April 2004.

[81] SOAP Version 1.2. W3C Recommendation, June 2003.

[82] Java Community Process (JCP), http://jcp.org/.

[83] JXTA. www.jxta.org.

[84] OSGi, http://www.osgi.org.

[85] 3GPP, http://www.3gpp.org/.

[86] Web Services Interoperability Organization (WS-I), http://www.ws-i.org/.

[87] Trusted Computing Group (TCG), https://www.trustedcomputinggroup.org/home.

[88] Digital Living Network Alliance (DLNA), http://www.dlna.org/.

[89] Model Driven Architecture (MDA), http://www.omg.org/mda/)

[90] Softwired, http://www.softwired-inc.com/

[91] Deering, S., Internet Protocol, Version 6 (Ipv6) Specification, RFC 2460, 1998.

[92] Fristche, W., Mobile Ipv6 – the Mobility Support for Next Generation Internet, Ipv6 forum, http://www.ipv6forum.com, 2000.

[93] Kent, S., Security Architecture for the Internet Protocol, RFC2401, 1998.

[94] Arora, A. et al., JXTA for J2ME – Extending the Reach of Wireless with JXTA Technology, Whitepaper, Sun Microsystems, 2003.

[95] Raatikainen, K., A New Look at Mobile Computing Proceedings of ANWIRE Workshop, 2004.

[96] Raatikainen, K., Columns in OT Land Expert's Corner on Middleware in Mobile World, 2004.

[97] Grace, P. et al., Interoperating with Services in a Mobile Environment, Technical Report (MPG-03-01), Lancaster University. 2003.

[98]   Perich, F. et al., Query Routing and Processing in Mobile Ad-Hoc Environments, Technical Report, UMBC, 2001.

[99]   Yoneki, E. and Bacon, J., Distributed Multicast Grouping for Publish/Subscribe over Mobile Ad Hoc Networks, *in Proc. IEEE Wireless Communications and Networking Conference (WCNC)*, 2005.

[100] Yoneki, E. and Bacon, J., Unified Semantics for Event Correlation over Time and Space in Hybrid Network Environments, *in Proc. IFIP International Conference on Cooperative Information Systems (CoopIS)*, LNCS 3760, 366-384, 2005.

Chapter 21

Trust in Pervasive Computing

Jim Parker, Anand Patwardhan, Filip Perich,
Anupam Joshi, and Tim Finin

CONTENTS

Introduction

The idea of *ad hoc* networking and pervasive environments is now more than a decade old. A significant amount of research on trust and privacy has been accomplished in the area of social sciences; however, because *ad hoc* networks have thus far not been popularly adopted in commercial products, little application research on trust and privacy has been performed in this area. Although recent advances in wireless and storage technology and the consequent proliferation of highly capable portable devices and wireless appliances are expected to lead to widespread use of *ad hoc* networking technologies, practical solutions for achieving security, privacy, and trust are still lacking. The highly invasive nature of some of these technologies poses a threat to the security and privacy of personal data and the area of pervasive computing.

Mobile devices with small form factors, yet with computing power comparable to desktops only years old, are now common. Enhanced multimodal user interfaces such as touch screens, biometric security devices, and accelerometers have significantly improved the usability of these devices. The integration of global position system (GPS) receivers, cameras, and recorders in cellphones and personal digital assistants (PDAs) has ushered in a new generation of *converged* mobile devices. We are now witnessing a continuous proliferation of wireless appliances in everyday life, such as crib monitors, home security alarms, fire alarm annunciators, and surveillance cameras. These technological advances are helping create resource-rich environments in which personal mobile devices can seamlessly integrate to utilize and provide services; moreover, these mobile devices will be capable of sharing their capabilities via wireless means. Peer-to-peer relationships will enable devices to dynamically form collaborative relationships and perform complex tasks leveraging available resources either shared among the peers or present in their surrounding environment.

Thus far, wireless networking has primarily served to extend the reach of the Internet. Most of the prevalent wireless technologies and their applications are infrastructure based. In traditional mobile computing environments, devices primarily adhere to the basic client–server model in which the devices act as clients and access stationary information on trusted servers. In the client–server model, the server is anchored, and a client can verify through several authentication and integrity schemes that the information has originated from the server, forcing accountability. Mobile devices lack the common sense that people often employ to determine the reliability of both the source and information provided by the source; consequently, devices require a mechanism to evaluate the integrity of their peers and the accuracy of information provided by their peers, as otherwise there is no scheme for protecting a device from malicious peers that deliberately provide unreliable information.

A mobile *ad hoc* network (MANET) is a self-organized collection of wireless mobile nodes lacking a fixed network infrastructure and having no central authority. The flexibility and openness of MANETs make them very appealing as an information-gathering and exchange medium; however, these two properties can also lead to security vulnerabilities. Fully realizing the potential of the mobile *ad hoc* paradigm requires an autonomous approach to mitigating risk and placing users in control of risk evaluation and usage. Along with enabling devices to estimate the trust they have in other devices and the accuracy of the information obtained from them, a mechanism must be provided that enables devices to detect and distinguish among *malicious peers*, which purposely provide incorrect information; *ignorant peers*, which are unable to guarantee a reliable level of provided information; and *uncooperative peers*, which have reliable information but refuse to make it available to other devices. This mechanism would also implicitly support an *incentive model*, in which all devices must provide only reliable information and provide this information often; otherwise, they risk losing the ability to communicate with other devices in the environment.

In MANETs, a server-centric mechanism of identification and authentication is not suitable. Even with limited Internet connectivity, total reliance on conventional security mechanisms involving key distribution centers (KDCs), certificate authorities (CAs), or similar forms of remote *trusted* sources imposes serious limitations on the functioning of these devices, in effect limiting them to function only when those remote sources can be contacted. In pervasive environments, the number of devices embedded in the surrounding infrastructure and personal mobile devices will be immense; thus, it will not be possible to predetermine all possible devices that may be encountered nor will it be feasible to centrally register all such devices and then later identify and authenticate them on every encounter.

This chapter presents research work that addresses some of the concerns raised with regard to protecting the privacy and security of mobile devices. The inherent vulnerabilities of pervasive networks have thus far restricted their use. Providing strong assurances of reliability and trustworthiness of information and services with practical implementation considerations for pervasive environments will be the most significant contribution of our research and will be another step toward making the vision of *anytime, anywhere* computing a viable reality.

Social Communities in Pervasive Networks

Pervasive Trust

Using locally available information collected from the surrounding pervasive environment or peers in the vicinity introduces several trust and security issues. Due to the inherent nature of pervasive environments, conventional mechanisms of providing security are not suitable. Devices must be made self-reliant to make trust evaluations and use reputations to guide their behavior; however, because mobile devices are potentially innumerable, it is not possible to cache the identities and reputations of all the encountered devices nor can we expect all devices to be cooperative. The abundant storage capacities of the mobile devices, however, will be sufficient to cache specific device identities; that is, it will be sufficient to remember only those devices that are of future potential value in forming social networks and those that will be most likely to cooperate.

> **Scenario:** Peter is flying a red-eye from LA to NY. His calendar shows a meeting in his NY office at 9:00 a.m. His portable device notices other people present in the airplane and finds his colleagues Clark and Lex, who are also attending the same meeting. Peter can't see them from his seat, but their devices can interface with the personal display screens in front of them, and the built-in cameras in their devices allow the three men to have a live video conversation and exchange notes for the next morning's presentation. They later decide to watch Peter's presentation to provide feedback. Peter grants them the right to access his device and make changes to the presentation. None of the devices belonging to Peter, Clark, or Lex has the capability to edit video content, but Clark discovers that an old friend, Brenda, is also on the airplane and her device has the required capability. On Clark's request and recommendation, Brenda allows Peter to use the video-editing capability. Peter is able to improve and finish his presentation without even leaving his seat.

In this scenario, the participants were initially unaware of the others' presence. Authenticating each other's devices is usually not possible unless prior security associations exist, but predetermining all possible devices that can be encountered is not a feasible option; thus, distributed trust management becomes a necessity for the survival of the network. In the following sections, we propose a distributed trust management scheme that utilizes activity monitoring and reputation management to evaluate trust. We propose employing mobility patterns and distinguishing landmarks or beacons to evolve trust and establish a scalable pervasive reputation management framework. Reputations of known devices in addition to activity monitoring can be used to compute trust in that device.

The networking layers can benefit from knowing who the reliable or trusted peers are within the local neighborhood for preferential consideration in forming routes and for peer discovery. The application layer can benefit from reports of malicious activity detected by the lower layers and then appropriately modify their trust assessments. Further recommendations by trusted devices can then be used to create new trust relationships or modify existing ones.

Connectivity provided by *ad hoc* networking requires that peers in the pervasive environment be cooperative. Due to the security threats posed to individual mobile devices, collaborative efforts in countering intrusive behavior are required. Most of the response mechanisms we have described in Patwardhan et al. [20] are reactionary. Because the scope of intrusion-detection mechanisms deployed on individual devices is limited to their radio range, collaborative mechanisms are required for communicating suspicious activity and intrusions to other devices in the vicinity. We propose using *reputations* to proactively detect and deny resources to devices that have been deemed malicious. Also, in the course of sharing information and services among devices, complex processes of trust evolution can be simplified using recommendations among trusted peers — which are again motivating factors for forming local collaborative groups. We propose the concept of *pack formation*, which uses accounts of prior encounters, evolved trust, and recommendations to form local packs.

Using context information and notions of neighborhoods that can be identified by specific unique landmarks, devices need only store trust information pertaining to the relevant context; for example, a portable device owned by a college student should only remember the most frequently encountered devices in the vicinity of the university campus to deduce that those particular entities are frequent visitors of that neighborhood/community. Furthermore, if malicious activity is attributed to any such known entity, this fact can be reported back to the community where that entity is known to be a frequent visitor.

Without assurances of the reliability and trustworthiness of retrieved data, the utility and effectiveness of the completed tasks are questionable. Metrics to evaluate the reliability of data and trustworthiness for peer-provided information must be available. Further trust evaluation and reputation management mechanisms will allow devices to function autonomously with minimal user intervention. To achieve these goals, it is necessary to take a holistic approach in addressing issues of device security, secure routing, peer discovery, data management, and trust relationships, as these issues are highly interdependent.

We propose giving MANET nodes the ability to independently evaluate *trust* in the nodes with which they interact. This solution involves a reputation management system through which nodes can evaluate, maintain, and distribute information about trust relationships within a MANET. Each node can make autonomous decisions about the trustworthiness of other nodes, providing an alternative to third-party authentication during periods of disconnection. Ding et al. [13] propose using two kinds of trust: *domain trust* and *referral trust*. Nodes can ask other trustworthy nodes to provide information (domain trust) or trust them to provide referrals to other devices that might have that information (referral trust).

Because MANETs rely on cooperation from all nodes, detection and isolation of malicious nodes are necessary for a MANET to function. Malicious and uncooperative nodes can cause disruption in MANETs and potentially disable the network. Each node must be able to identify malicious activity because centralized intrusion-detection (ID) schemes and firewalls cannot be effective in a MANET environment [19,20]. Also, at the application level, devices should be able to make autonomous assessments (i.e., reliable, corrupt, or unknown) about data provided by peers. For our discussion regarding trust management at the application level, we present results from our work in distributed reputation management and accuracy beliefs, followed by a description of several activity monitoring techniques that we use to detect intrusive or malicious behavior at the lower networking levels.

Services To Go

Continuous improvements in compact storage technologies, including semiconductor memory (e.g., CompactFlash, MMC cards) and miniature hard disks and microdrives, have spawned a generation of mobile devices with substantial storage capacities. Abundant onboard storage relieves the burden of requesting services or data from remote servers, thereby freeing devices from the dependency on connectivity to remote servers. Devices guided by their profiles [11,21] can cache large amounts of potentially useful information and keep required information updated by querying other trusted devices in the vicinity and requiring connectivity to the Internet only when absolutely necessary. To guide themselves, the devices will have to be able to sense their contexts (both spatial and temporal). By acquiring local information from reliable sources, the devices could compose locally available services and use their existing knowledge bases to service their needs; in other words, they will be largely self-reliant. Moreover, all such devices will be capable of providing useful services to other (mobile) devices in their vicinity. The collective resources of the individual data storage capacities, the unique sensory and effector capabilities of the devices, and the individual trust relationships will enable complex tasks to be performed and improve the overall performance of collective and individual tasks. Long-range wireless services are often not suitable for high data rates and at times are not cost effective. We propose harnessing the immense storage capacity of mobile devices, optimizing the use of available connectivity to keep the knowledge base updated, and enabling devices to function autonomously.

Pack Formation and Collaborative Queries

As exemplified in our earlier scenario, mobile devices are often bound by commonalities in the physical world. Common goals can be deduced from the profiles of the users and their devices; thus, there exist natural incentives to collaborate. The pack formation mechanism that we have proposed has several advantages: faster response times, increased scope of search, and distributed trust and reputation management. Also, collaborative mechanisms will prove useful when collective action must be taken against colluding adversaries. Here, we present some of the preliminary results from our simulations.

Collaboration in query processing leads to improved response times. We simulated an environment with 50 nodes spread in random locations in a two-dimensional square area using GloMoSim [27]. We present some of the interesting performance results from two separate sets of simulations. In the first case, each device assigned a task set of distinct questions searches individually for answers. Later, the same set of devices with the same task set of questions searches for the answers collaboratively.

For simplicity, we assumed that some initial trust already existed to be able to form collaborative groups. Our results are for pack sizes of five in a total population of 50 devices in a 150-m² area. Each device had a transmission range of 25 m and followed a random waypoint model (speeds varying from 1 to 5 m/sec and pauses of 5 sec). Each device tried to find answers for its assigned task set of 100 questions, and the answers were randomly distributed among the remaining nodes. To simulate the serendipitous nature of the environment, we varied the percentage of the knowledge base present in the neighborhood from 40 to 100 percent in increments of 20 percentage points. We ran the simulation using five different starting positions for the devices, for five runs of the simulation.

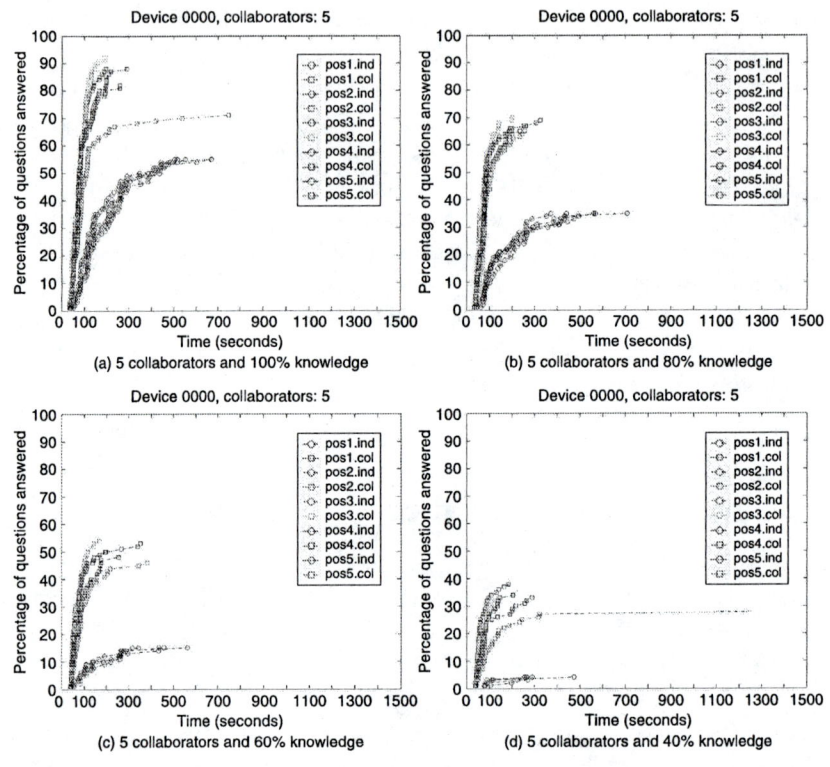

Figure 21.1 Preliminary results from simulations with five collaborators.

Because our focus was on the effectiveness and response time of the search, we assumed that all the sources of information were reliable and would only provide accurate answers. In the collaborative version, pack members helped each other find answers to their questions. When an answer for a collaborator's question was found, the device tried to send it back to that collaborator.

The plots in Figure 21.1 depict two sets of trends, each representing the five different starting positions of the five collaborators, each of which had a task set of 100 questions (not common with other collaborators). Also, the devices themselves did not answer their own questions. Parts a, b, c, and d of Figure 21.1 have decreasing knowledge bases, from 100 to 40 percent, in decrements of 20 percentage points. The collaborative version consistently outperformed the individual searches. In Figure 21.1a, devices in the collaborative scenario were able to find twice as many answers within less than a minute from the start of the querying process. In the non-collaborative version, where the devices independently tried to query other devices in their radio range, they managed to find approximately 50 percent of the answers and required up to 10 minutes to do so. In Figure 21.1d, with only 40 percent knowledge (i.e., answers to only 40 percent of the questions were available), the performance difference is more pronounced. Here, it can be seen that in the non-collaborative setup the devices were able to find no more than 5 percent of the answers, whereas the collaborating devices managed to find as many as 30 percent of the answers in less than 4 minutes.

All the simulations showed promising results in terms of faster responses and search effectiveness with regard to the collaborative models. We observed that, as the pack size was increased from 5 to 10, the control overhead for communication between the pack members increased and introduced minor increases in query response latency, yet the number of successfully answered queries was consistently greater than for the non-collaborative version.

Belief and Reputation in MANETs

This section introduces a *distributed reputation model* that extends the traditional query processing model [22] to allow devices to capture beliefs on the reputations of their peers and accuracy beliefs regarding information obtained from those peers. To mitigate the negative effects of malicious or ill-informed devices, the model categorizes peers as *reliable* and *unreliable*. In the model, the accuracy of an answer is a function of the trustworthiness of the information source and its belief in the accuracy of its answer. Devices assign trust to an information source based on past experience and the recommendations of those devices that it trusts.

Related Work

Trust and belief management models can be divided into two categories: mathematical and logical. Jonker and Treur [15] proposed a mathematical model for capturing trust in multi-agent systems. Their model consists of beliefs, and trust is a function of the values of these beliefs. The trust function is based on initial trust, experiences, and trust dynamics. The types of trust dynamics determine how past experiences affect the newly computed trust value. Richardson et al. [24] presented a mechanism for calculating the trustworthiness of users on the Semantic Web by developing a "web of trust" based on Web algorithms such as Google's PageRank [1]. In this approach, users maintain trust values for a small set of users and use the belief values of these users and their own trust in the users to calculate their own beliefs. Abdul-Rahman and Hailes [2] define a formal trust model based on trust and recommendations. Users store trust values for other users and ask trusted users for recommendations when dealing with unknown users; however, when a trust value is calculated it is not updated.

A significant amount of work has been carried out on developing logical trust models. Blaze et al. [6,7] define trust management as creating policies and assigning credentials. They use a PolicyMaker engine for checking if users' credentials conform to policies before granting them access. KeyNote [5] is designed along the same lines as PolicyMaker, but it has been designed to be simpler, to provide more support for the public key infrastructure (PKI), and to allow policies and credentials to be transported over insecure communication channels. REFEREE is a similar trust management system that is designed to facilitate security decisions for the Web [12]. Kagal et al. [16,17] also described a policy-based infrastructure for security and trust management in multi-agent systems and the Semantic Web. In this system, every entity has a policy that reflects its current binary trust values and exchanges them with other entities via speech acts.

The model described in this section employs a mathematical approach. This is because a mathematical model requires fewer computing resources than logical models, which require reasoning engines and certificate verification. Additionally, unlike logical models that only describe conditions when devices are trusted enough to access a certain information, mathematical models can also be employed to represent answer accuracy and to handle situations when more than one answer is provided for a given query. This model differs from other mathematical models in that the new model proposes several trust learning schemes based on experience and recommendations, allows information sources to specify their trust in the information being provided, and uses both kinds of values to compute beliefs. Most other schemes provide trust learning algorithms based on either experience or recommendations but do not combine the two. They also ignore the believed accuracy of the information source, whereas the proposed model uses it as a factor for rating the trustworthiness of a source.

Reputation Model

A successful model evaluating the integrity and information accuracy of a device must address the inherent limitations of mobile *ad hoc* networks and of mobile devices, including power, memory, and computation constraints as well as network reachability and wired infrastructure support limitations. The reputation model described in this section overcomes these issues because it does not rely on any wired infrastructure nor does it assume connectivity among all devices. The model also does not assume that each device can maintain belief information about every other device or information the other devices can provide.

The model only assumes that every device is able to assign an accuracy degree to any information the device provides to its peers and that every device maintains trust degrees about a subset of devices in the environment representing how much a device trusts the other devices for providing accurate answers to queries. The accuracy degree represents the device's belief about the correctness of the information, which can range in value from *distrust* to *undecided* to *trust*. A device, when asked, can provide a recommendation for some other device in question. Similar to accuracy degree, the recommendation can range in value from *distrust* to *undecided* to *trust*.

The model functions as an extension to a traditional query processing model for mobile *ad hoc* networks. In this reputation-driven model, a querying device collects responses from peers but also computes their trust degrees. It has been suggested that this approach is superior to the alternative where a device first computes the reputation of its peers and then queries those peers for information [23].

Information Source Discovery

When a device needs to obtain an answer for a query, it first attempts to discover which of its peers may have the necessary answer. The device does so by evaluating its cache of advertisements received from its peers and by broadcasting a source discovery request messages to its peers up to n hops away. The discovery message consists of the device's identity (ID_Q), the question (Q), and a nonce for differentiating it from other discovery messages sent by this device. A device may send out the discovery messages more than once based on the responses it receives from its peers.

Information Advertisement

When a cooperating, nonmalicious peer receives a source discovery, it checks its cache to find an answer matching the question. If the peer has a cached answer, it will respond by sending an advertisement message containing the identifier of the device (ID_S) and the question it can answer (Q), where the ID is some globally unique string (e.g., the Media Access Control, or MAC, address) or a cryptographically secure scheme that prevents ID spoofing, such as those presented by Eschenauer et al. [18]. A device may optionally proactively broadcast bulk advertisements at random intervals.

Querying Peers

The querying device evaluates all advertisements in its cache to determine possible sources for its query. If a device is unable to discover a sufficient number of information sources that could provide an answer to its question, the device simply broadcasts the question to all peers in its vicinity, again up to n hops away. If, however, the device is able to collect some information sources, the device sends a query to only those peers.

Collecting Answers

When a cooperating, nonmalicious peer receives a query message and has a matching answer, it will respond with a message containing its ID, the answer (A_i), and the accuracy degree $(A_D(i))$ of the answer from −1 to 1.

Recommendation Request

Each querying device (Q) has a lower limit (n) on the number of trusted peers that must provide an answer to a given query. A trusted peer (R) is any peer for which device Q has a *trust degree* $(T_Q(R))$ above a certain trust threshold (τ). When a device has not received enough answers from at least that many trusted peers, it computes the trust degree of every peer that sends it an answer (ID_R), using its initial trust belief function (α_D) and current trust values. If the device is unable to determine if the answering peer is *trusted* and it has not reached the minimum number of trusted responses, then the device may initiate a recommendation session about the answering peer. In the model, the device can either ask only those devices that it believes are its trusted peers or the device can ask anyone within n-hop distance for recommendations about the answering device. The querying device (ID_Q) does so by sending out a recommendation request message to some remote peer ID_R with the identity of the answering peer.

Recommendation Response

When a cooperating, nonmalicious peer receives a recommendation request, it looks up its trust beliefs to determine if the querying device (ID_Q) is one of its *trusted peers*. If this is the case, then the device responds with its trust degree in ID_S by sending a recommendation response message.

Calculating Final Answer

When a device receives all responses from all peers to whom it sent its query message or when its session time-out period ends, the querying device proceeds by calculating the final answer. For every different answer value it has received, the device calculates the combined accuracy degree of the particular value based on the suggested accuracy degree of the information sources and their trust degrees using its *trust-weighting* and *accuracy-merging* functions ($\otimes$ and $\oplus$, respectively):

$$\forall answer_i : A_{combined}(i) = \oplus_s\left(A_D(i) \otimes T_Q(D)\right) \tag{21.1}$$

The model defines three accuracy-merging functions: *AFV, MIN,* and *MAX.* These functions are similar to computing membership degrees in Boolean combinations of fuzzy variables. The querying device uses the merging function to compute a combined accuracy degree for every distinct value it received as a possible answer:

$$PA = \left\{\left(answer_i, A_{combined}(i)\right)\right\} \tag{21.2}$$

If all trusted devices provide the same answer to the original query (i.e., there is only one tuple in *PA*), the querying device will simply use that answer as the final value if its combined accuracy degree is above a certain threshold (τ). This is similar to the threshold concept for a *trusted peer.*

In many cases, however, the querying device may receive multiple conflicting answers from trusted peers. To address this problem, in this model the querying device can apply two different techniques. The querying device may accept an answer only if precisely one of the suggested answers has a combined accuracy level above τ. This technique is referred to as *only-one* answer (OO). Formally:

$$answer_x \text{ is } OO \Leftrightarrow \tag{21.3}$$

$$\forall\left(answer_i, A_{combined}(i)\right) \in PA : \begin{cases} A_{combined}(i) > \tau & i = x \\ A_{combined}(i) \le \tau & i \ne x \end{cases}$$

This is a pessimistic approach, as the device chooses precisely one answer (e.g., when the device is querying for the current stock price of a certain company), and only one possible answer exists. The querying device will not cache an answer if more than one answer is above the threshold or if no answers are above the threshold. At the same time, this approach will limit the amount of uncertain or distrusted data kept in the cache.

Alternatively, a device may employ a more optimistic technique that considers the possibility of a question having multiple valid answers (e.g., Chinese restaurants in a given location). In this case, the device will choose the answer with the highest combined accuracy degree above τ. If multiple answers have the highest degree of accuracy, then the device will randomly choose one. This is called the *highest-one* answer (HO). Formally:

$$answer_x \text{ is } HO \Leftrightarrow \exists \max:$$

$$(\max > \tau) \wedge \left(\left(answer_x, \max \right) \in PA \right) \wedge \qquad (21.4)$$

$$\left(\forall \left(answer_i, A_{combined}(i) \right) \in PA : A_{combined}(i) \leq \max \right)$$

Updating Trust Belief

If a querying device is able to determine its final answer (Λ), then it uses that fact to evaluate the interaction experiences it had with the answering peers. A querying device has a *positive experience* with an answering peer if the answering peer provided the same answer as the final one and if the answering peer suggested a non-negative accuracy degree for the answer. A querying device has a *negative experience* with an answering peer if the answering peer provides a different answer to the query and suggests a positive degree of accuracy. The querying device also has a negative experience if the querying peer provides the same answer but suggests a negative degree of accuracy. Finally, for all other cases, the querying device has an *undecided experience*. When a querying device has either a positive or negative experience with any answering device, it should update its trust degree for that device for future interactions by using one of the available trust learning functions (Δ) adapted from Jonker and Treur [15].

The first category of trust learning functions employs all history information to predict a future trust degree of a device. In one technique, referred to as *blindly positive*, a device will absolutely trust its peer if it has had at least n positive experiences. One can similarly define a *blindly negative* approach. The other category of trust employs only the current experience of the querying devices with the answering peer and the current trust degree the querying devices has of its peer to calculate the future trust degree. The querying device does so by either increasing or decreasing the trust degree using slow, fast, or exponential steps. The model uses *fast–positive/slow–negative, slow–positive/fast–negative, balanced–slow, balanced–fast,* and *exponential* techniques. The first four techniques use a fixed fast or slow step to increase or decrease a trust degree. The last technique always increases by the half step necessary to move from the current trust degree to an absolute trust degree of 1.0 and similarly to decrease to an absolute distrust degree at −1.0. Updating the querying device will affect to which devices the device sends its queries in the future, as well as how the device recommends the answering devices to its peers.

Answering Peers

For an honest device to return an answer (provided it has an answer), the device must be able to determine whether the querying device is one of its *trusted peers*. The answering device evaluates its beliefs to determine if its local trust degree of the querying device is above the trust threshold (τ); otherwise, the device first initiates a recommendation session by either asking all of its n-hop peers for recommendations about the querying device or by selectively asking devices that it has determined are trusted peers based on its local belief.

Given a set R of recommendation tuples $(Q, D, T_D(Q))$, where Q is the querying source, D is the recommending peer, and $T_D(Q)$ is the recommended trust degree representing how honest Q is according to D, then the combined recommendation trust degree is computed by the answering device A as:

$$R_{combined}(Q) = \oplus_R\left(T_D(Q) \otimes T_A(Q)\right) \tag{21.5}$$

Similar to combining the weighted answer accuracy degree values, the *AVG*, *MIN*, and *MAX* methods can be employed here for combining recommended trust degrees. The device then sends back an answer to the querying device when $T_A(Q) > \tau$ or $R_{combined}(Q) > \tau$. This mechanism implicitly creates an *incentive model*, where it is in the best interests of every device to provide only reliable information and provide this information often because others maintain and share reputation degrees about this device.

The model computes trust and accuracy degrees using a two-level-deep path algebra only; therefore, in the model, a mobile device calculates a trust degree of another device using only its trust degrees of its peers and their proposed trust of the other device. Similarly, a device computes an accuracy degree using only its combined trust degree of answering devices and the suggested accuracy values obtained from these answering devices.

The advantage of using only a limited depth is straightforward. Because each devices combines only up to two trust values, where one value is its own and one is of a remote device, one cannot introduce a possible cycle in the path computation. Second, because each device answers recommendation requests by evaluating only its local trust degrees, this model does not generate additional traffic in the resource-limited wireless *ad hoc* network. Finally, it is a relatively easy exercise for a mobile device to automatically update its beliefs about cached trust degrees and answer accuracy degrees when it changes a trust degree of any of its peers. For these reasons, the model uses only a two-level-deep path algebra.

Malicious Activity Detection and Trust

Mobile wireless networks can be divided into two types of architectures: infrastructure and *ad hoc*. In the infrastructure type, each network has a central node, or access point, through which all traffic must pass. The access points act not only to route traffic between nodes but also to grant or deny access to the network based on policies or access lists. Wireless intrusion-detection devices have been developed to a limited extent for wireless traffic monitoring in infrastructure networks. Because all wireless traffic must transit a central node, the wireless intrusion-detection device can be located at the access point to scan wireless channels and look for malicious activity. Devices trying to gain unauthorized access or disrupt network activity can be easily detected. True MANETs have no central access point. Nodes connect together in an *ad hoc* fashion to form a mesh network for information routing. Single intrusion-detection systems cannot effectively operate in this environment, as the dynamic nature of the network makes a central observation point very unlikely. For the remainder of the discussion, we will assume MANETs are being used for node connections

Current wired security mechanisms require either a third-party authority for authentication or *a priori* distribution of key material. MANETs offer no guarantee of a constant Internet connection, so third-party authentication may not be available. Proposed solutions to this problem for MANETs involve the establishment of a security association (SA) either out of band or with *a priori* knowledge of encounters on which additional secure protocols are enabled [3,8,10,26]. The problem is that SAs cannot be randomly established between two nodes that are previously unknown to one another in a disconnected-Internet scenario. In order for MANETs to become widely accepted and used, some mechanism must give users confidence that security exists within the MANET. We believe the way for security to be established in MANETs is to evaluate and foster *trust* between interacting nodes. Failure to provide this mechanism could have a negative effect on security, quality of service (QoS), and overall willingness to risk information exchange over MANETs.

Malicious Activity Detection

Our research into malicious node detection is based on promiscuous snooping of the network channel. Snooping leverages two properties inherent in most mobile *ad hoc* protocols. The first property is that each node in the network maintains a neighbor list containing the addresses of those nodes with which it is in immediate proximity or on the path from a source to a destination. The second property, as is the case in the IEEE 802.11 [14] and MACAW [4] link-layer protocols, is that a node is able to "hear" the ready-to-send (RTS)/clear-to-send (CTS) negotiation of its neighbors. Accordingly, each node participates in the intrusion-detection process and snoops on its neighbors' transmissions to ensure that the network packets have not been modified or misrouted. The notion of snooping is also employed in Dynamic Source Routing (DSR), which is used for reflecting shorter routes to optimize the route maintenance process.

Our initial research involved an extension that is viable for many *ad hoc* routing protocols (e.g., DSR or *Ad Hoc* On-Demand Distance Vector [AODV]), where the snooping nodes listen to all other nodes in their proximity. In Figure 21.2, promiscuous snooping is shown in an example where node A is sending traffic to node E via nodes B and D. In this

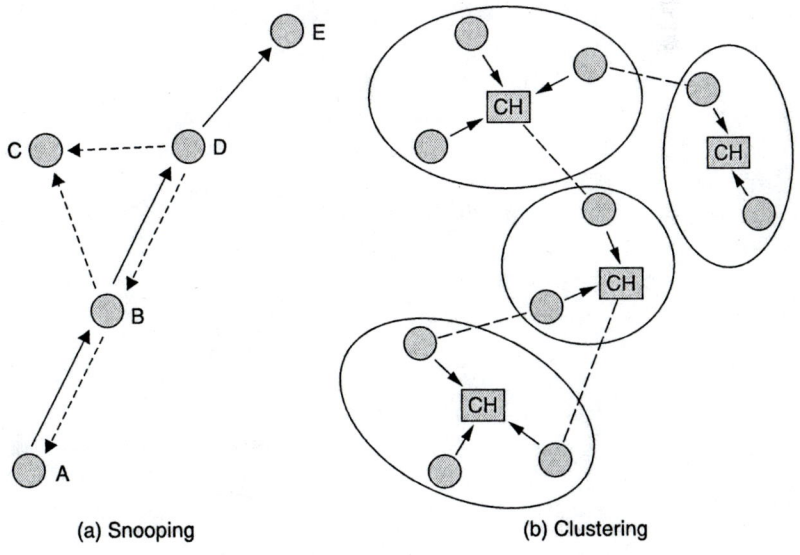

(a) Snooping (b) Clustering

Figure 21.2   Malicious node detection.

example, node C can monitor the traffic as it transits from node B to node D and is then forwarded to node E. Node C is in the position to determine whether node D changes, reroutes, or drops any packets in the datastream.

We drew a distinction between our work and others, such as Watchdog [25] and Neighborhood Watch [9], which work primarily with DSR, as they watch the forward node on the patch from source to destination. We experimented with two response mechanisms for observing node recognition of ongoing malicious activity. In the first, a *passive* response mode, a node, upon determining that another node is aberrant, unilaterally decides to cease interacting with the offending node. Although each node acts independently, eventually the intrusive node will be blocked from using all of the network resources. In the second, an *active* response mode, each node relies on a *cluster-based* hierarchy (Figure 21.2). When a node detects an aberrant neighbor, it informs its *cluster head*, which in turn initiates a voting procedure. If a voting majority determines that the suspect node is in fact intrusive, an alert will be broadcast throughout the network and the intrusive node will be denied network resources. The algorithms were written and simulated using GloMoSim, version 2.03. The simulation environment can be summarized as follows:

- *Number of nodes* — 50; 16 nodes were involved in constant bit rate (CBR) connections, and the number of malicious nodes was varied
- *Grid size* — 2000 by 2000 m
- *Application traffic* — 10 CBR connections generated simultaneously, with four nodes serving as the source for two streams each and two nodes as the source for a single stream each (destination nodes receive only one CBR stream)
- *Mobility* — Random Way-Point Model (maximum speed, 20 m/sec; pause time, 15 sec)
- *Radio* — No-fading radio model, with a range of 376 m
- *MAC layer* — 802.11, peer-to-peer mode
- *Routing protocol* — AODV or DSR
- *Simulation time* — 200 sec
- *Neighbor "hello" period* — 30 sec
- *Dropped packet time out* — 10 sec
- *Dropped packet threshold* — 10 packets
- *Clear delay* (event expiration timer) — 100 sec (the amount of time that a node considers an event without coming to a final determination)
- *Misroute threshold* — 5 events; detectable only in routing protocols using source routes such as DSR
- *Modification threshold* — 5 events

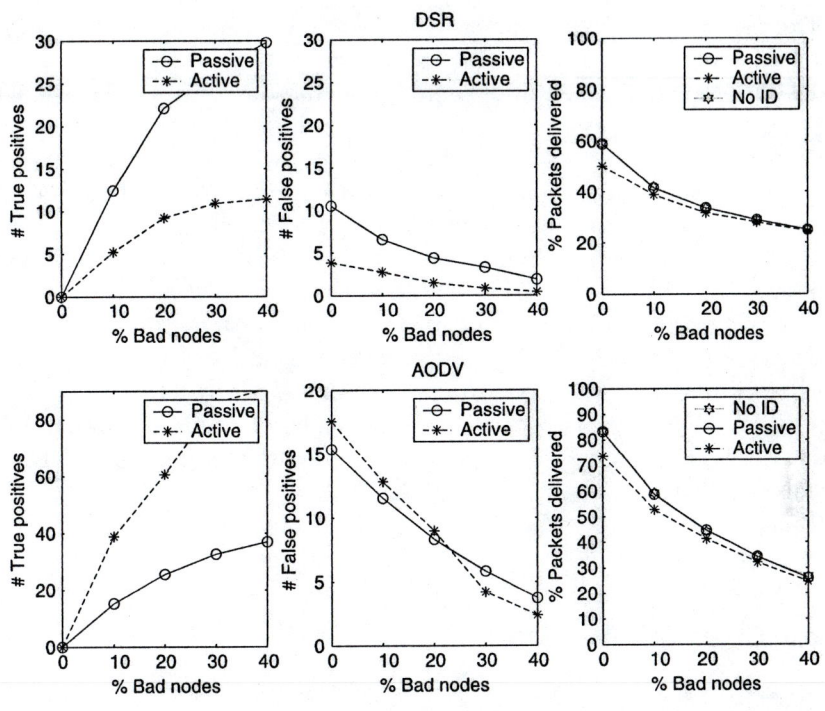

Figure 21.3 Simulation results of the active and passive response protocols for DSR (top) and AODV (bottom). (From Parker, J. et al., On intrusion detection in mobile *ad hoc* networks, in *Proc. of the 23rd IEEE Int. Performance Computing and Communications Conf.*, ©2004 IEEE. With permission.)

Results were obtained by averaging 100 simulation runs for 200 sec each [19,20]. The true positives, false positives, and successfully delivered packets as a percentage of the number of bad nodes in the network for DSR and AODV were measured (Figure 21.3).

The node density of both malicious and normal nodes was a significant factor in the rate of true positives. For a malicious node to be detected, it must act maliciously within the proximity of a good node. As expected, the performance of both the passive and active response protocols improved with respect to both true positives and false positives as the density of the malicious nodes increased. Likewise, and also as expected, the number of successfully delivered packets decreased as the density of malicious nodes increased. This is attributable to the increased bandwidth requirements for the voting mechanism. It became apparent through these experiments that the voting scheme worked better in a denser environment of network nodes as compared to a sparse environment.

During development, a number of issues were raised that require further work. The first issue was security and privacy concerns with regard to nodes snooping network traffic. Even though there is always the possibility of malicious activity, users may tend to dislike the idea of their traffic being monitored by all other nodes in their vicinity. Another issue was performance while snooping and processing network traffic. Nodes such as PDAs that cannot do anything else while performing malicious node detection are of little use to their owners. Certainly there is a concern for malicious nodes raising alarms and causing problems such as denial-of-service attacks against accused nodes. The active response with cluster voting made such attacks more difficult. Finally, reliably detecting and separating malicious users from normal mobile node disconnection in a MANET is quite difficult. Thresholds for the various parameters of malicious detection depend on node density, average speed, etc. The idea of readmittance into the network by falsely accused nodes is something that could be very important if snooping schemes are to be used.

Cross-Layer Information Processing

Malicious node detection is an important step toward developing trust relationships among MANET nodes. Nodes found to be acting maliciously should definitely not be trusted; however, because MANET nodes can also be malicious in other ways outside of the network layer, we decided to start looking at observable events within other layers of the Open Source Initiative (OSI) stack.

Perich et al. [23] developed a simulation for an application-layer query processor in MANETs. Information is assumed to be probabilistically distributed throughout the MANET, and nodes then query other nodes for the information. Results are determined to be correct or incorrect to make trust evaluations about the replying nodes. After a period of time, each node maintains a table of nodes with whom it has interacted, along with trust ratings reflecting *reliable* or *unreliable.*

We believe that, by combining network-layer malicious node detection with the application layer query, a more robust approach to rating individual trust can be achieved. Network-layer events are more numerous but less specific than queried information gathered at the application layer. The ultimate goal is to combine information from the two layers to obtain a more accurate representation of trust. One of the results from Perich et al. [23] showed that evolving trust is heavily based on the initial trust rating. In our work, we are using the initial trust ratings from observation information gathered at various layers of the OSI stack. Additionally, the information value and subsequent risk were the same for all nodes in

Perich et al. [23]. We are studying how trust can be evolved through first using low-risk encounters with low-value exchanges and gradually progressing to higher risk situations; in this case, the overall metric of success or failure would be a function of how much total value is lost over a period of time.

To expand on the original idea for querying we have divided nodes into classes; for example, a tourist on vacation may be more willing to trust information provided by a local citizen than by another tourist. As another example, a potential buyer might be unwilling to trust information provided by someone having a vested interest in a sale, as compared to information provided by an independent, experienced third party.

Throughout our research we have used a combination of simulation and test-bed implementations to measure the time required for the convergence of trust, accuracy in responses, malicious node detection time, and true and false positives for malicious node accusations. Our results are compared with test runs for malicious node detection and trusted query applications separately to determine the relative contribution of each to the combined results. Evaluations are conducted with scenarios for low, medium, and high mobility and with varying degrees of malicious nodes.

The issue of resource management as it pertains to performance is very important. As part of our research, we are trying to determine resource usage at each node as related to scalability. Resource usage depends on the number of nodes in the MANET, the number of node encounters during some time period, and the number of new nodes entered into the MANET during the interaction.

We believe that additional information can be used in the development of trust from other OSI layers. We are exploring MAC layer management and control messages (e.g., beacons, RTS, CTS) for patterns indicating malicious use. Other approaches include exploring the physical layer to provide signature information associated with a particular node's physical address. Depending on the granularity of the attribute measurement, nodes may be able to match indicator attributes and sound an alarm when malicious nodes try to spoof existing trusted node addresses.

There is also room for modification of the query model to reflect certain real-world actions; for example, trusted nodes do make mistakes and may not be as accurate as a particular situation requires. In the case of malicious nodes, they may not always act maliciously but may instead restrict their malicious behavior to specific situations or circumstances. We recognize that the development of trust is a difficult task and would like to examine how to best judge trust within MANETs, given these scenarios.

Discussion

Trustworthy data management in pervasive environments is a challenging task. The trend of continuous improvements in the capabilities of embedded devices and their widespread acceptance are indications of a highly interconnected society where pervasive environments and portable devices form integral parts of our daily lives, creating cultures of their own, and unforeseen applications are emerging. Due to the inherent open, dynamic, and distributed nature of these environments, they cannot be secured by conventional security practices.

Our experience in trust management for pervasive computing has, thus far, only been in the laboratory setting. We have made attempts to anticipate user issues, as well as developer issues, in our simulations and experiments. By implementing algorithms on actual handheld devices, we have shown that such ideas can be viable with current technology, at least in a limited capacity. What we do not know, and can only speculate on, is how the devices will be used in a widespread pervasive network.

From our simulations and experiments, we have determined that nodes must be expected to adhere to some predetermined sets of rules for each layer (physical through application) that govern acceptable behavior, and they must be capable of detecting anomalies and attributing misbehavior to particular nodes. At the lower layers, the monitored data can be quite overwhelming to process directly; however, such data can be filtered using the rules of acceptable behavior and provided to higher layers for further processing. Data acquired from across layers can be processed and subject to reputation management. By aggregating the data from multiple layers, application-layer protocols can use monitored behavior from lower layers to report and react to overtly malicious acts. To protect the network from malicious or faulty devices, observed misbehavior or noncompliance with protocols can be used to penalize the guilty nodes by denying them access to resources or excluding them from the network. To be successful, a majority of the nodes must participate. Nodes failing to fully participate in information routing or message storing and forwarding mechanisms can lead to disruptions in the *ad hoc* network. Nodes failing to participate in monitoring for malicious activity can also leave the network vulnerable to attack.

A difficult problem is how to overcome false alarms. Our experiments have shown the importance of identifying one-hop neighbors to monitor traffic effectively. Incorrect neighbor table information can lead to false accusations or ignoring actual malicious activity. An increase in "hello" packet rates can make neighbor tables more accurate over the short term but also lead to increased traffic congestion. With congestion comes lower data throughput and potential interference with nodes monitoring for malicious activity.

We believe social communities will play an important role in reputation management and trust evolution. Our experiments with pack formation and results from our preliminary work validate our approach. We have experimented with a distributed reputation management system that uses a mathematical model, enabling individual devices to compute the accuracy of peer-provided information. These concepts can be used to reliably determine and manage reputation and trust in a pervasive environment.

We recognize that trust management is not only a technical issue but also one of social acceptance. Our experience has led us to believe that, even though it may be technically possible to overcome the challenges of trust and reputation management in pervasive computing, success or failure will be determined by the willingness of people to purchase and use these devices. To fully realize the potential of the mobile *ad hoc* paradigm there must be an autonomous approach to mitigating risk and placing users in control of risk evaluation and usage. We have presented techniques that can be applied from the application layer to the lower-level networking layers to help mitigate risks stemming from the open, dynamic nature of pervasive environments; however, many challenges remain to be overcome.

References

[1] Google, http://www.google.com/technology/.
[2] Abdul-Rahman, A. and Hailes, S., A distributed trust model, in *Proc. of New Security Paradigms Workshop (NSPW'03)*, Escona, Switzerland, August 18–21, 2003.
[3] Balfanz, D., Smetters, D., Stewart, P., and Wong, H., Talking to strangers: authentication in *ad hoc* wireless networks, in *Proc. of Symp. on Network and Distributed Systems Security (NDSS'02)*, San Diego, CA, February 6–8, 2002.
[4] Bharghavan, V., Demers, A., Shenker, S., and Zhang, L., MACAW: a media access protocol for wireless LANs, in *Proc. ACM SIGCOMM'94*, London, August 31–September 2, 1994, pp. 212–225.
[5] Blaze, M., Feigenbaum, J., Ioannidis, J., and Keromytis, A., *The KeyNote Trust Management System Version*, Request for Comments 2704, Internet Engineering Task Force (IETF), 1999 (http://www.ietf.org/rfc/rfc2704.txt).
[6] Blaze, M., Feigenbaum, J., Ioannidis, J., and Keromytis, A., The role of trust management in distributed systems, in *Security Issues for Mobile and Distributed Objects*, Vitek, J. and Jensen, C., Eds., Springer-Verlag, New York, 1999.
[7] Blaze, M., Feigenbaum, J., and Lacy, J., Decentralized trust management, in *Proc. of IEEE Symp. on Privacy and Security*, Oakland, CA, May 6–8, 1996.
[8] Bobba, R.B., Eschenauer, L., Gligor, V., and Arbaugh, W., Bootstrapping *Security Associations for Routing in Mobile Ad Hoc Networks*, Technical Report 2002-44, Institute for Systems Research, University of Maryland, Baltimore, 2002.

[9] Buchegger, S. and Boudec, J.L., Nodes bearing grudges: towards routing security, fairness, and robustness in mobile *ad hoc* networks, in *Proc. of the Tenth Euromicro Workshop on Parallel, Distributed, and Network-Based Processing (PDP'2002)*, Canary Islands, Spain, January 9–11, 2002, pp. 403–410.

[10] Capkun, S., Buttyan, L., and Hubaux, J., Self-organized public-key management for mobile *ad hoc* networks, in *Proc. of the 8th ACM/IEEE Int. Conf. on Mobile Computing and Networking (MOBICOM'02), ACM Workshop on Wireless Security*, Atlanta, GA, September, 2002.

[11] Cherniak, M., Franklin, M., and Zdonik, S., Expressing user profiles for data recharging. *IEEE Pers. Commun.*, July, 2001.

[12] Chu, Y.-H., Feigenbaum, J., Lamacchia, B., Resnick, P., and Strauss, M., REFEREE: trust management for Web applications, *Comput. Networks ISDN Syst.*, 29(8–13), 953–964, 1997.

[13] Ding, L., Zhou, L., and Finin, T., Trust-based knowledge outsourcing for Semantic Web agents, in *Proc. of the 2003 IEEE/WIC/ACM Int. Conf. on Web Intelligence (WI'03)*, Halifax, Canada, October, 2003.

[14] IEEE, *LAN/MAN Wireless LANs*, Std. 801.11, Institute of Electrical and Electronics Engineers, Piscataway, NJ, 1999 (rev. 2003).

[15] Jonker, C.M. and Treur, J., Formal analysis of models for the dynamics of trust based on experiences, in *Proc. of the 9th European Workshop on Modelling Autonomous Agents in a Multi-Agent World: Multi-Agent System Engineering (MAAMAW'99)*, Garijo, F.J. and Boman, M., Eds., Vol. 1647, Lecture Notes in Artificial Intelligence, Springer-Verlag, Heidelberg, 1999, pp. 221–231.

[16] Kagal, L., Finin, T., and Joshi, A., A policy-based approach to security for the Semantic Web, in *Proc. of the 2nd Int. Semantic Web Conf. (ISWC2003)*, Sanibel Island, FL, October 20–23, 2003.

[17] Kagal, L., Finin, T., and Peng, Y., A delegation based model for distributed trust, in *Proc. of the 17th Int. Joint Conf. on Artificial Intelligence (IJCAI'01), Workshop on Autonomy, Delegation, and Control: Interacting with Autonomous Agents*, Seattle, WA, August 4–10, 2001.

[18] Eschenauer, L., Gligor, V.D., and Baras, J.S., On trust establishment in mobile *ad hoc* networks, in *10th Int. Security Protocols Workshop, Cambridge, U.K., April 2002*, Christianson, B., Crispo, B., Malcolm, J.A., and Roe, M., Eds., Vol. 2845, Lecture Notes in Computer Science, Springer-Verlag, Heidelberg, 2004, pp. 47–66.

[19] Parker, J., Undercoffer, J.L., Pinkston, J., and Joshi, A., On intrusion detection in mobile *ad hoc* networks, in *Proc. of the 23rd IEEE Int. Performance Computing and Communications Conf. (IPCCC'04), Workshop on Information Assurance*, Phoenix, AZ, April 14–17, 2004.

[20] Patwardhan, A., Parker, J., Joshi, A., Iorga, M., and Karygiannis, T., Secure routing and intrusion detection in *ad hoc* networks, in *Proc. of the 3rd Int. Conf. on Pervasive Computing and Communications (PerCOM'05)*, Kauai Island, Hawaii, March 8–12, 2005, pp. 191–199.

[21] Perich, F., *MoGATU: Data Management in Pervasive Computing Environments*, UMBC eBiquity Research Group, Department of Computer Science and Electrical Engineering, University of Maryland, Baltimore County (http://mogatu.umbc.edu/), 2001.

[22] Perich, F., Joshi, A., Finin, T., and Yesha, Y., On data management in pervasive computing environments, *IEEE Trans. Knowledge Data Eng.*, 16(5), 621–634, 2004.

[23] Perich, F., Undercoffer, J.L., Kagal, L., Joshi, A., Finin, T., and Yesha, Y., In reputation we believe: query processing in mobile *ad hoc* networks, in *Proc. of the First Int. Conf. on Mobile and Ubiquitous Systems: Networking and Services (MobiQuitous'04)*, Boston, MA, August 22–26, 2004.

[24] Richardson, M., Agrawal, R., and Domingos, P., Trust management for the Semantic Web, in *Proc. of the 2nd Int. Semantic Web Conf. (ISWC'03)*, Sanibel Island, FL, October 20–23, 2003.

[25] Marti, T.J., Giuli, K.L., and Baker, M., Mitigating routing misbehavior in mobile *ad hoc* networks, in *Proc. of the 6th ACM/IEEE Int. Conf. on Mobile Computing and Networking (MOBICOM'00)*, Boston, MA, August, 2000.

[26] Zapata, M. and Asokan, N., Securing *ad hoc* routing protocols, in *Proc. of the 8th ACM/IEEE Int. Conf. on Mobile Computing and Networking (MOBICOM'02), ACM Workshop on Wireless Security*, Atlanta, GA, September, 2002.

[27] Zeng, X., Bagrodia, R., and Gerla, M., GloMoSim: a library for parallel simulation of large-scale wireless networks, in *Proc. of the 12th Workshop on Parallel and Distributed Simulation (PADS'98)*, Banff, Alberta, Canada, May 26–29, 1998.

Section 4

MOBILE MIDDLEWARE FOR SEAMLESS CONNECTIVITY

Chapter 22

Seamless Connectivity in Infrastructure-Based Networks

Michael E. Kounavis and Andrew T. Campbell

CONTENTS

Introduction

As the wireless Internet rolls out over the next several years, there will be an increasing demand for new mobile devices (e.g., new cellular phones, PDAs, laptop and palmtop computers), services (e.g., m-commerce, wireless Web, high-quality wireless multimedia), and technologies (e.g., W-CDMA, Wi-Fi, WiMax) that can meet the needs of mobile users. Recent trends indicate that a wide variety of mobile devices will be available, each requiring specialized services and protocols. Connectivity between mobile devices and emerging diverse wireless service providers is likely to hinge on the capability of each provider's infrastructure to support a variety of mobile devices as well as the capability of mobile devices to connect to heterogeneous access networks. Today, the incompatibility of signaling systems and physical-layer radio technologies prevents mobile devices from roaming between heterogeneous wireless networks. Such factors limit the vision of seamless mobility. For example, a CDMA-based cellular phone cannot be easily connected to an IEEE 802.11-based wireless local area network (WLAN). Although a variety of handoff algorithms have been proposed and investigated in the past, these algorithms are mostly tailored toward the needs of some specific type of mobile device or access network. The diversity in signaling systems that characterize wireless access network architectures poses a challenge in realizing intersystem handoff. Furthermore, access network protocols make specific assumptions about the capability of mobile devices; for example, many Mobile IP-based approaches assume that handoff control is located at the mobile device. Such mobile-controlled handoff schemes may not be suitable for many low-power devices that are incapable of continuously monitoring channel quality measurements.

In this chapter, we present solutions to the problem of seamless connectivity where the implementation details of mobility management algorithms are hidden from handoff control systems, allowing the handoff detection state (i.e., the best candidate access point for a mobile device) to be managed separately from the handoff execution state (i.e., mobile registration information). The same detection algorithms operating in mobile devices or access networks can interface with multiple types of mobility management architectures, operating in heterogeneous access networks. Handoff control systems issue a number of generic service requests, which mobility management systems execute according to their own programmable implementation. In one case, when the location of the handoff control system is at the mobile device, different mobility management protocols can be dynamically loaded into mobile devices, allowing them to roam between heterogeneous access networks in a seamless manner.

Our approach to seamless connectivity requires mobile devices and access networks to be programmable, thus supporting dynamic service creation. Existing mobile and wireless networks have limited service creation environments, however. Typically, service creation is a manual, *ad hoc*, and costly process in wireless networks. Mobile network services and protocols cannot be easily extended or modified because they are generally implemented using dedicated firmware or embedded software, or they constitute part of the low-level operating system support. For example, it is difficult to dynamically modify the handoff prioritization strategy in PCS networks or to introduce new handoff control algorithms (e.g., mobile-assisted handoff) in wireless LANs. We believe that these observations call for new communication methodologies and software technologies for mobile networks.

Software engineering has progressed to the point where systems and standards can be used for implementing platform-independent, component-based, distributed software. Such advances have enabled the development of programmable [4] and active [5] network toolkits for the deployment of new services. Programmable networks [4] separate the communications hardware from the control software which allows the modeling of hardware resources using open programmable interfaces. In this manner, third-party software providers can enter the market for telecommunications software.

Work on software radios [6,7] has shown that wireless physical layers can be created dynamically by introducing code into programmable base stations with wideband, tunable front ends. In this work, we focus on control-plane compatibility between mobile devices and access networks and discuss how advances in modern software engineering, programmable networking, and mobile middleware can help with achieving intersystem handoff.

The main results of our research are presented here. Our research was conducted at the COMET networking laboratory at Columbia University from 1996 to 2000 [9]. Our results include the following:

- We designed, implemented, and evaluated a *reflective handoff* service that allows access networks to dynamically inject signaling systems into mobile devices before handoff; thus, mobile devices can seamlessly roam between wireless access networks that support radically different mobility management systems.
- We showed how a *multi-handoff* access network service can simultaneously support different styles of handoff control over the same wireless access network. This programmable approach can benefit service providers who must satisfy the mobility management needs of a wide range of mobile devices, from cellular phones to more

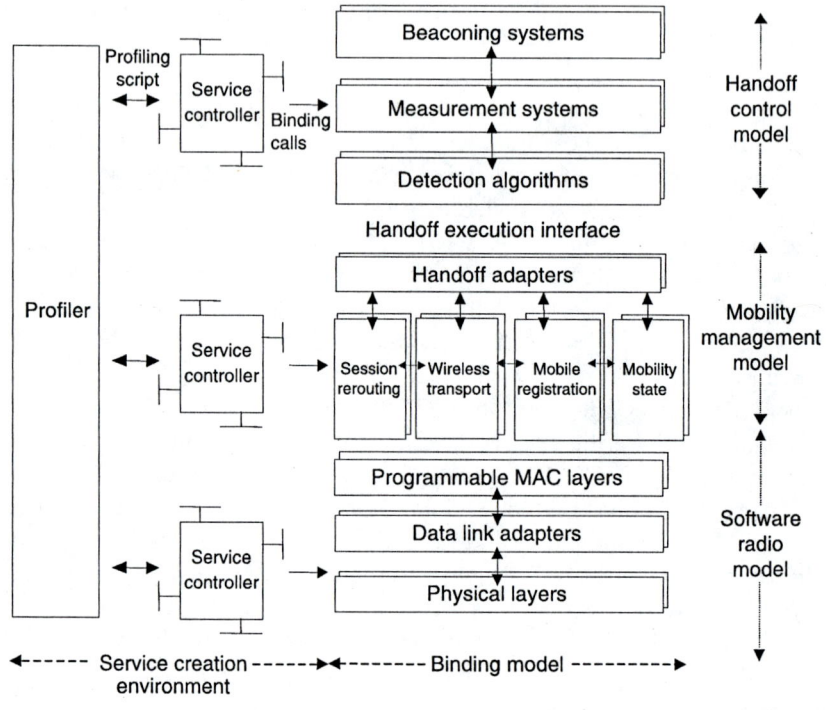

Figure 22.1 Binding model and service creation environment for seamless connectivity.

sophisticated palmtop and laptop computers. To allow a range of mobile devices to connect to programmable mobile networks, we further decomposed the handoff control process into programmable objects, separating the transmission of beacons from the collection of wireless channel quality.

Principles for Seamless Connectivity

Principles for seamless connectivity in infrastructure-based networks include the separation of handoff control from mobility management, decomposition of the handoff control process, decomposition of the mobility management process, and programmability of the physical and data link layers. Such principles have been embodied in the design of a binding model for programmable wireless networks, as shown in Figure 22.1. In the following text, we discuss each principle in detail.

Separation between Handoff Control and Mobility Management

The most important principle for seamless connectivity is the separation between the handoff control process (i.e., the mobility detection process) and the mobility management process (i.e., the mobility execution process). Handoff control is the process of deciding the most suitable points of attachment for a mobile device. Handoff control is typically supported on a per-mobile-device basis by collecting channel quality information and by comparing the quality associated with alternative points of attachment. Mobility management is the process of redirecting the flow of information to and from mobile devices through new points of attachment each time the handoff control process determines that the points of attachment of mobile devices should change. By hiding the implementation details of mobility management algorithms from handoff control systems, the handoff detection state (i.e., the best candidate access points for a mobile device) can be managed separately from the handoff execution state (i.e., mobile registration information). This software approach can be used to enable intersystem handoff between different types of wireless access networks.

The basic idea behind realizing intersystem handoff is that the same detection mechanisms operating in mobile devices and access networks can interface with multiple types of mobility management architectures that operate in heterogeneous access networks (such as the Mobile IP [2], Cellular IP [13], Mobiware [12], and HAWAII [23] experimental wireless access networks). Handoff control systems can issue a number of generic service requests through a standard *handoff execution interface* (shown in Figure 22.1), which mobility management systems execute according to their own programmable implementation. For example, a generic "pre-bind" method call to a candidate access point would be executed by establishing a signaling channel in an experimental Mobiware architecture [12] or by joining a multicast group specific to a mobile device and buffering packets in an experimental Daedalus/BARWAN [24] architecture. In one extreme case, where the location of the handoff control system is at the mobile device, different mobility management protocols can be dynamically loaded into mobile devices, allowing them to roam between heterogeneous access networks in a seamless manner.

A handoff execution interface, illustrated in Figure 22.1, separates handoff control from mobility management. Handoff control and mobility management systems are implemented as separate programmable architectures. Examples of handoff execution methods include:

- *Handoff* methods, which map down to mobility management services that execute handoff (e.g., register the care-of address of a mobile device with its home agent)
- *Pre-bind* methods, which initiate priming actions at candidate access points associated with a mobile device (e.g., load an active filter for transport adaptation [12], start buffering packets)
- *Configure* methods, which bind new signaling systems or delete existing ones (e.g., replace the mobile Mobiware control plane of the device with a Cellular IP one during reflective handoff)

It is difficult to consider a single handoff execution interface that is capable of encompassing all existing and future wireless systems; rather, it is more likely that standard handoff execution interfaces will support particular families of wireless technologies. In this case, network architects can select execution methods that satisfy sets of handoff control and mobility management systems that they wish to program. Another essential component of architectures that separate handoff control from mobility management is a layer of *handoff adapters*, shown in Figure 22.1. Handoff adapters represent a set of distributed objects that serve as the glue between handoff control systems and mobility management services. Handoff adapters and mobility management services collectively implement handoff execution algorithms.

Handoff adapters are an important part of any programmable handoff architecture. First, handoff adapters control the handoff execution process. Handoff adapters invoke mobility management services in an order that is specific to the handoff style being programmed. Many different mobility management services (e.g., session rerouting, wireless transport, mobile registration, mobility state management services) can be invoked as part of the handoff execution process. For example, in a forward, mobile-controlled handoff, an adapter would invoke a radio link transfer service before session rerouting. In the backward, mobile-assisted handoff, the order of this execution would be reversed. Second, adapters distribute method invocations to the network nodes or hosts where mobility management services are offered. Handoff is usually detected at a single host or network node (e.g., a mobile device, access point, mobile capable router/switch). In contrast, mobility management services can be offered at multiple hosts or network nodes inside a wireless access network (e.g., at wireless access points or by mobility agents in the network).

Decomposition of the Handoff Control Process

To further facilitate the customization of the handoff control process across heterogeneous wireless access networks, we suggest that the handoff control process should be decomposed into its basic building blocks using open and standard interfaces between these components. A handoff control model that separates the algorithms that support beaconing, channel quality measurement, and handoff detection is illustrated in Figure 22.1. Typically, these functions are supported as a single, monolithic software structure in existing mobile systems. By separating the handoff detection from wireless channel quality measurements, we allow for new detection algorithms to be dynamically introduced in access networks and mobile devices. For example, detection algorithms specific to overlay networks can be introduced into mobile devices to allow them to perform vertical handoffs, or detection algorithms specific to microcellular networks can be selected to compensate against the street-corner effect [20]. By separating the collection of wireless channel quality measurements from the beaconing system, mobile networks can support different styles of handoff control over the same wireless infrastructure. For example, a wireless service provider may want to offer a network-controlled handoff service (e.g., supported in the Advanced Mobile Phone System, or AMPS [21]) for simple mobile devices, a mobile-assisted handoff service (e.g., as in GSM [21]) for more sophisticated mobile devices involved in the process of measuring channel quality, or a mobile-controlled handoff service (e.g., the handoff scheme considered by the Mobile IP Working Group) for more sophisticated laptop or palmtops mobile computers. The following services comprise the handoff control model:

- *Detection algorithms*, which determine the most suitable access points to which a mobile device should be attached. Wireless access points can be selected based on such factors as channel quality measurements, resource availability, and user-specific policies [22]. A mobile device can be attached to one or more access points at any moment in time.
- *Measurement systems*, which create and update handoff detection state. By handoff detection state, we mean the data used by detection algorithms to make decisions about handoff. Detection algorithms and measurement systems use the same representation for handoff detection state.
- *Beaconing systems*, which assist in the process of measuring wireless channel quality. Programmable beacons can be customized to support service-specific protocols such as quality of service (QoS)-aware beaconing [12] or reflective handoff.

Decomposition of the Mobility Management Process

We believe that, like the handoff control process, the mobility management process should be decomposed into basic software building blocks to allow rapid customization where and when needed. The mobility management model shown in Figure 22.1 reflects the decomposition of mobility management into basic services that execute handoff. We have adopted a generalized architectural model that is capable of supporting the design space of different mobile networking technologies. To program mobility management systems, one needs to be able to introduce new forwarding functions at mobile-capable routers/switches (e.g., Cellular IP [13] or HAWAII [23] forwarding engines) as well as distributed controllers that manage mobility (e.g., Mobile IP foreign agents). The mobility management model discussed in this chapter is limited to supporting handoff services only. Other mobility management functionality (e.g., location, fault, and account management) typically found in mobile networks can be considered as future work. We identify the following services as part of the handoff execution process:

- *Session rerouting mechanisms* control the datapath in access networks to forward data to or from mobile devices through new points of attachment. Rerouting services may include admission control and QoS adaptation for the management of wireless bandwidth resources.
- *Wireless transport objects* interact with the physical and data link layers in mobile devices and access points to transfer active sessions between different wireless channels. A channel change may be realized through a new time slot, frequency band, code word, or logical identifier. Transport objects can provide value-added QoS support, such as Transmission Control Protocol (TCP) or snooping [24].
- *Mobile registration* is associated with the state information a mobile device exchanges with an access network when changing points of attachment.
- *Mobility state* can be expressed in terms of a connectivity, addressing and routing information, bandwidth, and name-space allocations of the mobile device, as well as user preferences.

Programmability of the Physical and Data Link Layers

The last principle for seamless connectivity is to implement the physical and data link layers in software so as to allow their dynamic customization. The software radio model shown in Figure 22.1 defines the composition of physical and data link layer services. The software radio model supports functions such as the dynamic assignment of channel locations and widths and the selection of modulation and coding techniques used on each channel. Software radios allow mobile devices to dynamically tune to the appropriate air interface of the serving access network while roaming between heterogeneous wireless environments. Media Access Control (MAC) layer protocols can be made programmable [8], allowing for services that support different QoS requirements. Physical and data link layer modules can be implemented in various ways [6,7,14,15]. Data link adapters separate data link layer modules from the lower physical layer components; for example, data link adapters allow programmable MAC protocols to operate on top of any type of channel coding or modulation scheme, as discussed in Kounavis et al. [9].

Building a Prototype

By applying the design principles presented above, we were able to build a prototype system demonstrating seamless connectivity in two novel wireless services: (1) a multi-handoff access network service allowing diverse types of mobile devices to be connected to the same physical network infrastructure; and (2) a reflective handoff service allowing mobile devices to roam across heterogeneous wireless access networks. In what follows, we describe the elements of our prototype system (i.e., our service creation environment and services deployed), and we discuss our experiences with building this prototype.

Service Creation Environment

To support the dynamic customization of handoff control and mobility management services, we have implemented a service creation environment that explicitly supports transportable code by dynamically selecting, deploying, and binding distributed objects. The service creation environment shown in Figure 22.1 is implemented using Common Object Request Broker Architecture (CORBA™) [26] middleware. The service creation environment offers interfaces for the dynamic introduction and modification of network services through transportable code. Programmable handoff services are implemented as collections of distributed objects. In our framework, middleware technologies enable networkwide system programmability and interoperability, separating the definition of programmable handoff objects from their implementation. Programmable handoff objects expose control interfaces, thus allowing the creation of bindings at runtime. Binding is the process through which an object obtains a reference to another object to request its services. An object reference can be described in many different ways; for example, an object reference can be represented by a hostname where the object is activated, and by a TCP/IP port number where the object listens for service requests. Many software technologies support binding, including distributed systems platforms such as CORBA, Distributed Component Object Model (DCOM), and Java Remote Method Invocation (RMI), as well as localized mechanisms (e.g., dynamic link libraries).

The service creation environment is composed of a profiler and a set of service controllers. The profiler drives the service creation process, creating representations of programmable handoff services over a defined access network topology. As illustrated in the figure, the profiler interacts with a set of service controllers using a well-defined scripting language to create or modify handoff services. Profiling scripts are blueprints of network services. Service controllers compile profiling scripts, resolve object bindings, and create handoff control and mobility management systems. Modification of programmable handoff services takes place after bindings are removed or objects are deleted. The service creation process allows the network architect to create new objects using inheritance of abstract classes (e.g., an abstract handoff detection algorithm class). Service controllers activate objects, invoking binding calls on object control interfaces for the deployment of services.

The scripting language we used in our prototype is simple and supports command, assignment, and exception handling statements. Command statements can be used for declaring programmable handoff services; adding, deleting, and customizing programmable handoff objects; and creating and removing object bindings. Network architects can customize objects during the profiling process. In this case, parameters characterizing the operation of a service (e.g., user, service-specific, or environmental parameters) can be passed in objects at runtime through the profiler and service controllers.

Multi-Handoff Access Network Service

Using this service creation environment, we designed and implemented a multi-handoff access network service that simultaneously supports three styles of handoff control over the same physical wireless access network: *network-controlled handoff* (NCHO), *mobile-assisted handoff* (MAHO), and *mobile-controlled handoff* (MCHO). Figure 22.2 shows the implementation of the handoff control model for the multi-handoff access network service. Objects shown in Figure 22.2 are grouped into beaconing systems, measurement systems, and detection algorithms, which are the components of the handoff control model. Figure 22.2 also shows object interactions and their invocation order (e.g., NCHO-1, NCHO-2). The handoff control objects include:

- *Beacon producer* (BeaconProducer) and *measurement producer* (MeasurementProducer) objects, which invoke low-level wireless LAN utility functions. Beacon producer objects transmit beacons

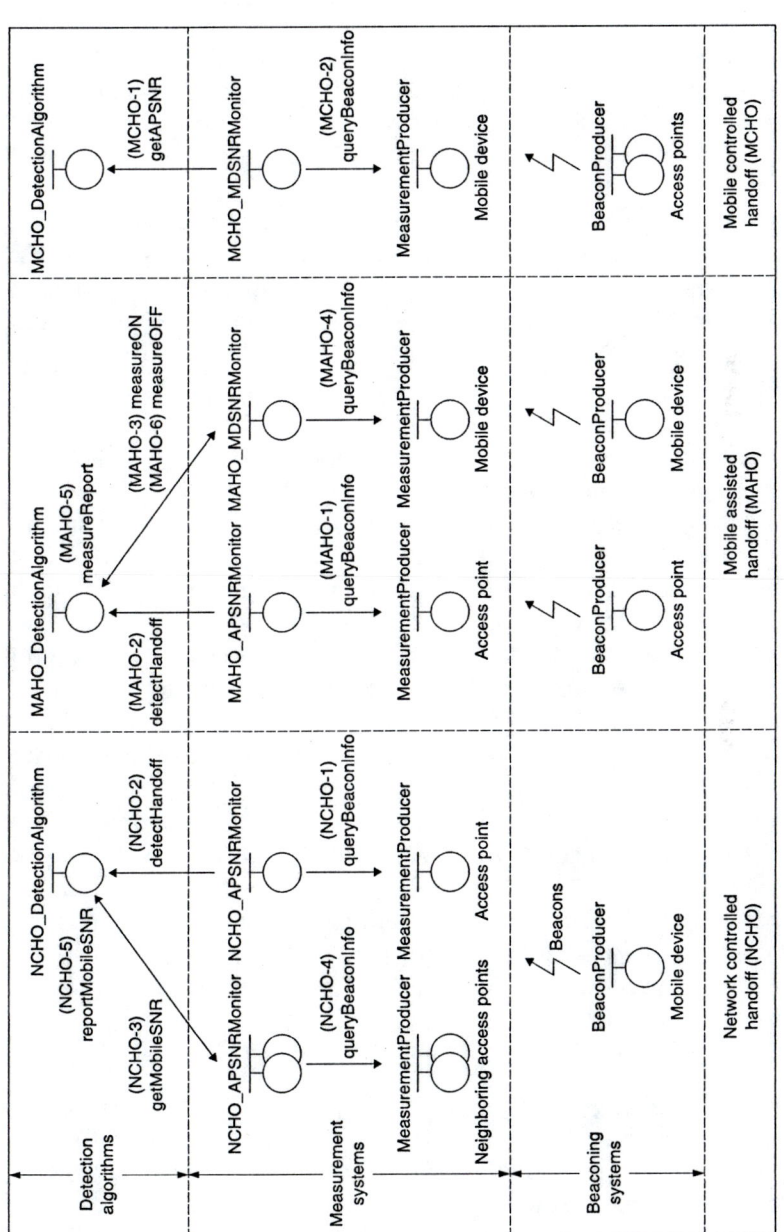

Figure 22.2 Handoff control in a multi-handoff access network.

at specified frequencies. Measurement producer objects generate raw channel quality measurements. Measurement and beacon producer objects can simultaneously participate in multiple styles on handoff control.

■ *Signal strength monitor* (*_APSNRMonitor, *_MDSNRMonitor) objects, which collect and average wireless signal strength measurements. The signal-to-noise ratio (SNR) represents only one of the many measurements that can be used for handoff decision making.

■ *Detection algorithm* (*_DetectionAlgorithm) objects, which make decisions for handoff based on signal strength measurements. Each handoff style employs its own set of signal strength monitors and detection algorithms in order to determine the best access points that mobile devices should be attached to.

Our implementation of the mobility management model is based on an extended Mobiware [12] architecture. Mobiware is programmable, promoting the separation between signaling, transport, and state management. Mobility management services are separated as discussed in the previous section. All sessions that operate between a mobile device and an associated Internet gateway are abstracted as a single state entity called a *flow bundle* in a Mobiware wireless access network. Flow bundles are used during handoff to switch IP flows in Mobiware access networks and provide general-purpose encapsulating and routing services similar to Asynchronous Transfer Mode (ATM) virtual paths or IP tunnels. Open programmable switches allow the establishment, removal, rerouting, and adaptation of flow bundles; thus, the access network behaves as a pool of resources allowing different handoff styles to operate in parallel. The proof-of-concept implementation of the multi-handoff access network service showed that wireless networks can be built that support seamless connectivity not only for one type of mobile device but also for many different types of devices. Further information about the handoff algorithms developed as part of our multi-handoff access network service can be found in Kounavis et al. [9].

Reflective Handoff Service

We have implemented reflective handoff as a mobile, controlled-handoff scheme. Access points transmit beacons that additionally carry globally unique identifiers designating specific access networks. A reflective detection algorithm uses access network identifiers to determine whether a mobile device is likely to move to the coverage area of a new access network. Each mobile device maintains a local cache of signaling system

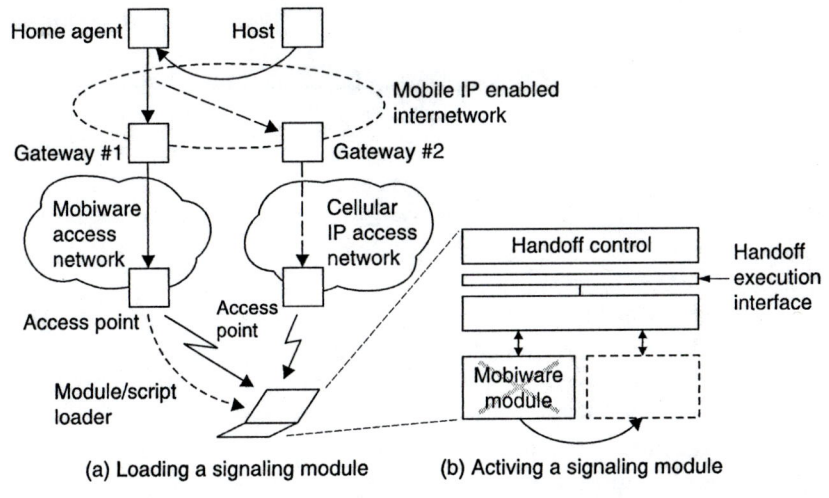

Figure 22.3 Reflective handoff service.

modules. Signaling system modules are collections of objects supporting mobility management services in mobile devices. Before a mobile device performs a handoff to a new access network, it checks whether a signaling module associated with the new candidate access network is cached. If a signaling module is not cached, it is dynamically loaded. Access points support module loaders deployed during the service creation process. A signaling system is loaded from the old access network. A two-way handshake mechanism is used for loading signaling modules, which are loaded before reflective handoff, is executed. Access networks schedule the transmission of signaling modules over the air interface, to avoid flooding the wireless network.

Reflective handoff is managed by handoff adapters, which activate or deactivate signaling modules on demand. The handoff execution interface for the reflective handoff service includes a *configure* method, which is used for binding new signaling systems. Parameters associated with access network state (e.g., the address of the gateway to the Mobile IP Internet) are passed into signaling modules upon activation. Module loaders transmit the access network state when loading signaling system support into mobile devices. Reflective handoff may involve the loading of entire mobility management protocol stacks or configuration scripts, which customize objects already cached at mobile devices.

Two distinct types of access networks support reflective handoff in our testbed, as shown in Figure 22.3: an experimental Mobiware and an experimental Cellular IP access network. Cellular IP [13] delivers fast local handoff control in datagram-oriented access networks. In addition, Cellular

IP supports per-mobile host state, paging, routing, and handoff control in a set of access networks interconnected to the Internet through gateways. In Cellular IP, packets sent from mobile hosts create routing caches pointing to the downlink path so packets destined to a mobile device can be routed using the route cache. Mobiware and Cellular IP signaling modules use the Internet Protocol to communicate with access networks. Mobiware and Cellular IP access networks support the same wireless data link and physical layers (WaveLAN) in our testbed but use different mobility management systems.

In our testbed, a Mobile IP-enabled Internetwork connects the Mobiware and Cellular IP wireless access networks via gateways. Mobile IP is used for managing macro-level mobility between gateways, whereas the Cellular IP and Mobiware wireless access supports fast local handoff control. Hierarchical mobility management in IP-based mobile networks has been widely reported in the literature [13,16,23,29]. A mobile device attached to an access network uses the IP address of the gateway as its care-of address. Access networks provide mechanisms for initiating Mobile IP-based inter-gateway handoffs and for establishing datapaths between gateways and access points where mobile devices are attached.

Signaling modules implement generic handoff execution functions as dynamic link libraries. Three types of handoff are supported: (1) *internal handoffs*, which take place between access points of the same access network; (2) *entry handoffs*, which take place when a mobile device is attached to a new access network; and (3) *exit handoffs*, which take place when a mobile device leaves an access network. The handoff execution interface method calls for internal, entry, and exit handoffs.

Reflective handoff requires that all the signaling modules associated with the handoff process are loaded and activated. Reflective handoff has been implemented as the process for invoking an exit handoff on the signaling system of the old access network and an entry handoff on the signaling system of the new wireless access network. Access networks realize execution calls in different ways and support mechanisms for registering the care-of address of mobile devices (i.e., gateway IP address) with their corresponding home agents. Care-of address registration has been realized as part of entry handoff or exit handoff. When an entry handoff takes place, access networks check whether the care-of address of a mobile device has been registered with its home agent. If the care-of address is not registered, then the access network initiates registration. Registration support during exit handoff is optional.

We have programmed our handoff control system to load signaling modules as soon as the mobile device detects that it is inside the coverage area of an access network where the signaling system is not cached. This loading algorithm minimizes the probability of handoff failure due to the absence of a signaling module at the mobile device. A soft state mechanism used for managing stored signaling modules is used to avoid having large caches. A timer associated with a module is refreshed while a mobile device remains inside the coverage area of an access network associated with a particular module or set of modules. Mobile devices can permanently cache signaling modules associated with access networks, however.

Seamless Connectivity in the Future

Future Wireless Access Networks

There is a growing consensus [34] that mobile networks will eventually be capable of partitioning the wireless spectrum dynamically to support different wireless applications in the future. Today, the amount of wireless spectrum, licensed for use by wireless applications, is fixed and cannot be easily modified. Spectrum is allocated on a wireless application basis (e.g., broadcast television, cellular telephony) by regulatory bodies that enforce regional and global harmonization. While this practice has proved to work well, it has made spectrum a scarce resource, forcing engineers to build communication systems on top of fixed and, in most cases, limited-frequency bands. Furthermore, the need for backward compatibility in the transition from one cellular technology to another has resulted in further waste of radio resources because different wireless systems must be supported simultaneously in different frequency bands (e.g., 2G and 3G cellular systems).

We suggest that with recent advances in distributed software, mobile middleware, and software radios, spectrum allocations can be made on a wireless service provider basis instead of a wireless application basis. Each service provider can be the owner of a programmable mobile network. We speculate that programmable mobile networks will be able to modify their physical layers at runtime by controlling their radio devices (i.e., wideband analog-to-digital converters [ADCs]). In this way, programmable mobile networks will enable wireless service providers to negotiate the wireless spectrum between each other, adjust the bands that support their wireless and mobile services, or reprogram their access networks from the physical to the application layer. When some amount of the spectrum is allocated to a service provider, it will be negotiated according to technical factors, user demands, or market strategies. Negotiation can take place between spectrum brokers, as illustrated in Figure 22.4. We believe that programmability and deregulation in the use of wireless spectrum will impact wireless service providers, improving spectrum utilization and accelerating the migration to new wireless technologies. Deregulation does not necessarily exclude national bodies from contributing to spectrum harmonization. Instead of

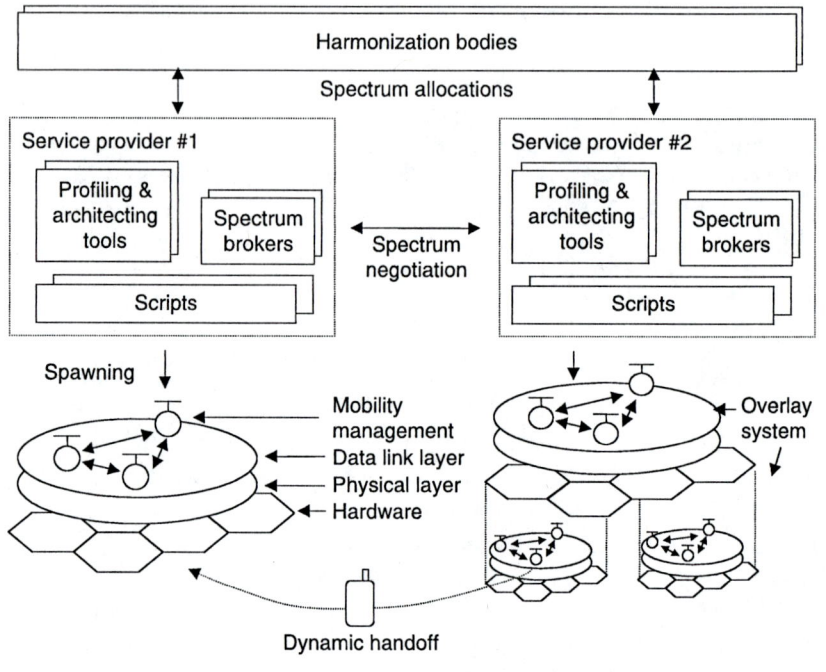

Figure 22.4   Future wireless access networks.

specifying the frequency bands used by wireless applications, harmonization bodies can set the rules of a distributed game for buying and selling the spectrum. The players of the game would be wireless service providers. For example, in Figure 22.4, harmonization bodies are shown as being responsible for initial spectrum allocations made to cellular wireless service providers. When some amount of the spectrum has been bought by a service provider, it can be programmed, sold to another provider, or returned to its source.

Another possible feature of future wireless access networks is *nested radio etiquettes*. Radio etiquettes are sets of radiofrequency bands, protocols, and high-level interaction rules for sharing the wireless spectrum among competing users. Radio etiquettes can also determine procedures for buying and selling spectrum dynamically. We believe that radio etiquettes can be used for specifying spectrum-sharing protocols among wireless service providers offering different types of applications (e.g., a cellular telephony service provider, a wireless Web service provider). Future mobile networks can use a hierarchy of nested etiquettes to utilize the wireless spectrum in a stable manner. Nested etiquettes are shown in Figure 22.5. A *parent* radio etiquette, shown in the figure, incorporates protocols and high level rules for selling spectrum to *child* wireless service

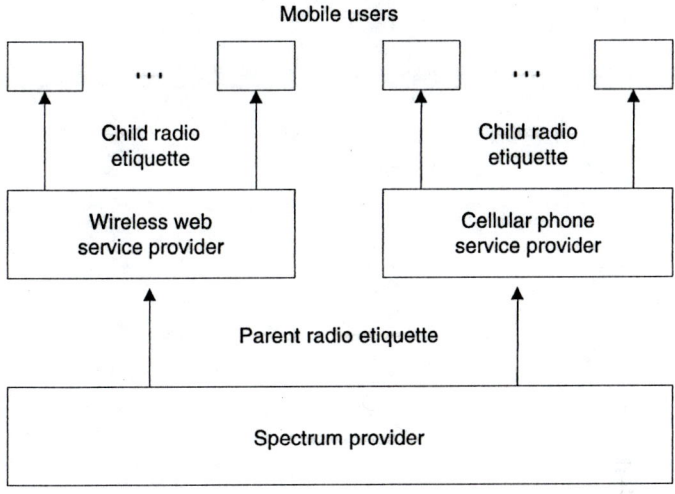

Figure 22.5 Nested radio etiquettes.

providers. A child wireless service provider obtains a piece of spectrum over some specified geographical area and a set of software handlers to program this band. Using software handlers, the service provider can program its own mobile network architecture over the piece of spectrum the service provider has bought. A child mobile network architecture can be programmed in accordance with a child radio etiquette, which can be different from the etiquette of the parent. Radio etiquettes may differ among child service providers that have the same parent as well. Child service providers will use the etiquette of the parent to negotiate spectrum between each other.

Last, we speculate that future wireless access networks will be capable of automating the creation, deployment, and management of other mobile networks. In the future, mobile networks may be capable of spawning other networks using dynamic code that implements physical layer, wireless transport, mobile signaling, control, and management algorithms. Spawned mobile networks may be composed of multiple physical layers, MAC protocols, or signaling systems. In one case, illustrated in Figure 22.4, a spawned mobile network can be an overlay system consisting of multiple physical layers that support their own mobility management architectures. When programmable mobile networks are spawned, their structure can be observed, analyzed, or modified. During this architecting phase, physical layer resource management may take place among wireless service providers to adjust the amount of spectrum offered to mobile users. Physical layer resource management may be carried out in accordance with the hierarchy of nested radio etiquettes.

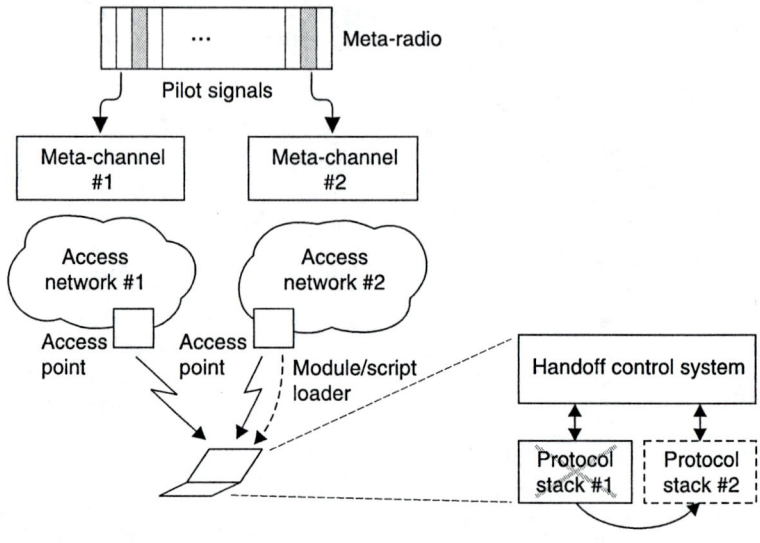

Figure 22.6 Dynamic handoff.

Seamless Connectivity via Metaradios and Metachannels

The future wireless networking environment presented above is highly heterogeneous. In such an environment, the incompatibility of signaling systems and radios is likely to prevent mobile devices from roaming across heterogeneous wireless environments. One approach to deal with this heterogeneity is to allow mobile devices to reprogram their communication protocol stacks to interact with different mobile network architectures. In this section, we discuss how the reflective handoff service described earlier can be extended to support seamless connectivity in future mobile networks that program the spectrum dynamically.

Our *dynamic handoff* approach is illustrated in Figure 22.6. A signaling channel, or *metachannel*, is associated with each wireless service provider and can be used for loading software components into mobile devices. Each service provider has a different metachannel. A metachannel may operate on a separate frequency band, time slot, code word, or logical identifier; for example, a separate frequency band can be allocated for dynamic handoff and programmed using software radios. Each mobile device maintains a local cache of signaling systems and physical and data link layer modules. Signaling modules are used for interacting with the different mobility management architectures. Data link and physical layer modules are used for composing different types of air interfaces dynamically.

Dynamic handoff may involve the loading of all software components required to compose new protocol stacks or just the loading of configuration scripts, which bind modules together. Module and script loaders are deployed into access networks as a part of the spawning process.

Mobile devices do not need to be aware of the metachannels of wireless service providers in advance. Metachannels can be discovered dynamically using a *metaradio* (shown in Figure 22.6). A metaradio is a separate channel where programmable mobile networks transmit pilot signals, thus allowing mobile devices to discover metachannels and tune their radios accordingly. The information that pilot signals carry depends on the way the metachannels are implemented. Different pilot signals can be associated with different wireless service providers and metachannels; for example, a pilot signal could contain a globally unique identifier associated with a wireless service provider and a code word identifying the provider's metachannel.

Challenges

Seamless connectivity via metaradios and metachannels seems to be an exceedingly difficult task. Software radio base stations suffer today from poor ADC and digital signaling processing (DSP) performance. In addition, programmable mobile networks require large transmission and reception bandwidth. Typically, the wider the band a software radio base station supports, the higher the ADC resolution and DSP processing power the base station requires. To accomplish dynamic spectrum partitioning, wireless service providers may have to control software radio base stations with multiple tunable front ends and to program these base stations with dynamic code; however, the cost of software radio base stations increases with the number of front ends that the base stations support. Moreover, the division of a large reception bandwidth into sub-bands limits the number of air interfaces that can be programmed. We believe that new hardware techniques must be investigated to allow programmable mobile networks to manage the wireless spectrum dynamically.

Another middleware challenge is associated with the design of the dynamic code that implements programmable mobile networking support. Programmable mobile networks require a comprehensive architectural model for binding distributed components to compose mobile network services. The binding model should ideally specify interfaces for composing any mobile network algorithm. One can consider many different combinations of application programming interfaces (APIs) constituting a binding model for programmable mobile networks. Finding the optimal binding model that reduces the programming labor associated with different mobile network architectures is an open problem.

A final challenge is related to physical layer resource management between heterogeneous wireless access networks. Physical layer resource management must take place over slow timescales; otherwise, it is unlikely that programmable mobile networks will operate in a stable manner. Physical layer resource management requires the design of a distributed game that would maximize the revenue of a parent wireless service provider, taking into account the incentives of its wireless service subscribers. An interesting problem arises when wireless service subscribers are child mobile networks. In this case, it is important that the parent wireless service provider is aware of the architectural characteristics of its child mobile networks. Quantifying the behavior of mobile network architectures during the spectrum negotiation process represents a major resource management challenge.

Conclusions

In this chapter, we have presented a solution to the problem of seamless connectivity in infrastructure-based networks where the implementation details of mobility management algorithms are hidden from handoff control systems, allowing the handoff detection state (i.e., the best candidate access point for a mobile device) to be managed separately from the handoff execution state (i.e., mobile registration information). The same detection algorithms operating in mobile devices or access networks can interface with multiple types of mobility management architectures operating in heterogeneous access networks. Handoff control systems issue a number of generic service requests, which mobility management systems execute according to their own programmable implementation. A framework for the programmability of wireless access networks was described, and proof-of-concept design, implementation, and evaluation of programmable handoff architecture presented. Two new services were designed and programmed using the programmable handoff architecture: multi-handoff access network and reflective handoff services. Clearly, many new services that support seamless wireless connectivity in wireless networks can be considered, and we have presented only a small but interesting set in this chapter. The multi-handoff access network service and reflective handoff service use the same profiling, service creation, and programming environments. This indicates that our approach is more generally applicable to the introduction of new in mobile networks.

References

[1] Buchanan, K., Fudge, R., McFarlane, D., Phillips, T., Sasaki, A., and Xia, H., IMT-2000: service provider's perspective, *IEEE Pers. Commun. Mag.*, 4, 8–13, 1997.
[2] Bagwat, P., Perkins, C., and Tripathi, S., network layer mobility: an architecture, and survey, *IEEE Pers. Commun. Mag.*, June, 1996.

[3] Raychaudhuri, D., Wireless ATM networks: architecture, system, design and prototyping, *IEEE Pers. Commun. Mag.*, 3, 42–49, 1996.

[4] Lazar, A.A., Programming telecommunications networks, *IEEE Network*, 11(5), 8–18, 1997.

[5] Tennehouse, A. and Wetherall, D., Towards an active network architecture, in *Proc. of Multimedia Computing and Networking*, San Jose, CA, January, 1996.

[6] Mitola, J., Technical challenges in the globalization of software radio, *IEEE Commun. Mag.*, 37(2), 84–89, 1999.

[7] Bose, V., Ismert, M., Wellborn, W., and Guttag, J., Virtual radios, *IEEE J. Selected Areas Commun.*, 17(4), 591–601, 1999.

[8] Bianchi, G. and Campbell, A.T., A programmable medium access controller for adaptive quality of service control, *IEEE J. Selected Areas Commun.*, March, 2000 (special issue on intelligent techniques in high-speed networks).

[9] Kounavis, M.E., Campbell, A.T., Ito, G., and Bianchi, G., Design, implementation and evaluation of programmable handoff in mobile networks, in *Proc. of the Third Int. Conf. on Open Architectures and Network Programming (OPENARCH'00)*, Tel-Aviv, Israel, March, 2000.

[10] Kounavis, M.E., Campbell, A.T., Ito, G., and Bianchi, G., Supporting programmable handoff in mobile networks, in *Proc. of the Sixth Int. Workshop on Mobile Multimedia Communications (MoMuC'99)*, San Diego, CA, November, 1999.

[11] Chan, M.C. and Lazar, A.A., Designing a CORBA-based high performance open programmable signaling system for ATM switching platforms, *IEEE J. Selected Areas Commun.*, 17(9), 1537–1548, 1999.

[12] Campbell, A.T., Kounavis, M.E., and Liao, R.R.-F., Programmable mobile networks, *Comput. Networks*, 31(7), 741–765, 1999.

[13] Valko, A.G., Gomez, J., Kim, S., and Campbell, A.T., On the analysis of cellular IP access networks, in *Proc. of the Sixth IFIP Int. Workshop on Protocols for High-Speed Networks*, Salem, MA, August 25–27, 1999.

[14] Bose, V., Wetherall, D., and Guttag, J., Next century challenges: radioactive networks, in *Proc. of the Fifth ACM/IEEE Int. Conf. on Mobile Computing and Networking (MOBICOM'99)*, Seattle, WA, August, 1999.

[15] Mitola, J., Cognitive radio for flexible mobile multimedia communications, in *Proc. of the Sixth Int. Workshop on Mobile Multimedia Communications (MoMuC'99)*, San Diego, CA, November, 1999.

[16] Liao, R.R.-F. and Campbell, A.T., On programmable universal mobile channels in a cellular Internet, in *Proc. of the Fourth ACM/IEEE Int. Conf. on Mobile Computing and Networking (MOBICOM'98)*, Dallas, TX, October, 1998.

[17] Liao, R.R.-F., Bocheck, P., Campbell, A.T., and Chang, S.-F., Utility-based network adaptation in MPEG-4 systems, in *Proc. of the Ninth Int. Workshop on Network and Operating System Support for Digital Audio and Video (NOSSDAV'99)*, Basking Ridge, NJ, June, 1999.

[18] Kulkarni, A.B. and Minden, G.J., Active networking services for wired/wireless networks, in *Proc. IEEE INFOCOM'99*, New York, March, 1999.

[19] ARRCANE Project, http://www.docs.uu.se/arrcane/.

[20] Tripathi, N.D., Reed, J.H., and Van Landingham, H.F., Handoff in cellular systems, *IEEE Pers. Commun. Mag.*, 5(6), 26–37, 1998.

[21] Goodman, D.J., *Wireless Personal Communications Systems*, Addison-Wesley, Boston, MA, 1997.

[22] Wang, H.J., Katz, R.H., and Giese, J., Policy-enabled handoffs across heterogeneous wireless networks, in *Proc. of the Second IEEE Workshop on Mobile Computing Systems and Applications (WMCSA'99)*, New Orleans, LA, February, 1999.

[23] Ramjee, R., La Porta, T., Thuel, S., and Varadhan, K., HAWAII: a domain-based approach for supporting mobility in wide-area wireless networks, in *Proc. of the Seventh Int. Conf. on Network Protocols*, Toronto, Canada, October 31–November 3, 1999.

[24] Seshan, S., Balakrishnan, H., and Katz, R.H., Handoffs in cellular networks: the Daedalus implementation and experience, *Wireless Pers. Commun.*, 4(2), 141–162, 1997.

[25] Gamma, E., Helm, R., Johnson, R., and Vlissides, J., *Design Patterns: Elements of Reusable Object-Oriented Software*, Addison-Wesley, Boston, MA, 1995.

[26] Object Management Group, www.omg.org.

[27] Kounavis, M.E. and Campbell, A.T., *The Metabus: Breaking the Monolith of the Software Bus*, Technical Report, Center for Telecommunications Research, Columbia University, New York, 1999.

[28] Smith, B.C., Procedural Reflection in Programming Languages, Ph.D. thesis, Massachusetts Institute of Technology, Cambridge, MA, 1982.

[29] Caceres, R. and Padmanabhan, V., Fast and scalable wireless handoffs in support of mobile Internet audio, *ACM J. Mobile Networks Appl.*, 3(4), 351–363, 1998.

[30] Der Merwe, J., Rooney, S., Leslie, I., and Crosby, S., The Tempest: a practical framework for network programmability, *IEEE Network*, 12(3), 2–10, 1998.

[31] Cellular IP source code distribution, comet.columbia.edu/cellularip.

[32] OMG, *Minimum CORBA*, joint revised submission, OMG Document orbos/98-08-04 ed., Object Management Group, Needham, MA, 1998.

[33] Campbell, A.T., Kounavis, M.E., Villela, D., Vicente, J., De Meer, H. et al., Spawning networks, *IEEE Network*, 13(4), 16–29, 1999.

[34] Staple, G. and Werbach, K., The end of spectrum scarcity, *IEEE Spectrum*, 41(3), 41–44, 2004.

Chapter 23

Peer-to-Peer Computing in Mobile *Ad Hoc* Networks

Marco Conti, Franca Delmastro, and Giovanni Turi

CONTENTS

Introduction

Mobile *ad hoc* networks (MANETs) represent complex distributed systems composed of wireless mobile nodes that can freely and dynamically self-organize into arbitrary and temporary *ad hoc* network topologies. This spontaneous form of networking allows people and devices to seamlessly exchange information in areas with no preexisting communication infrastructure (e.g., disaster recovery environments). Although the early MANET applications and deployments have been military oriented, civilian applications have grown substantially. Especially in the past few years, due to rapid advances in mobile *ad hoc* networking research, mobile *ad hoc* networks have attracted considerable attention and interest from commercial business industries, as well as the standards community. The introduction of new technologies, such as Bluetooth® and IEEE 802.11, greatly facilitated the deployment of *ad hoc* technology outside of the military domain, generating a renewed and growing interest in the research and development of MANETs.

While *ad hoc* networking applications have appeared primarily in specialized fields such as emergency services, disaster recovery, and environment monitoring, the flexibility of MANETs makes this technology attractive for several other scenarios, such as, for example, in personal area and home networking, law enforcement, search-and-rescue operations, commercial and educational applications, and sensor networks. Currently developed mobile *ad hoc* systems adopt the approach of not having middleware but instead relying on each application to handle all the services it needs. This represents a major complexity and inefficiency in the development of MANET applications. Indeed, most MANET research has concentrated on the enabling technologies and on networking protocols (mainly routing [5]), and research on middleware platforms for mobile *ad hoc* networks is still in its infancy.

Recently, in research circles, some middleware proposals for mobile *ad hoc* environments have appeared [4,16,19,21,22]. Their emphasis is on supporting transient data sharing [21] between nodes within communication range or data replication for disconnected operations [4,20], or both [15]. To achieve this, classical middleware technologies have been adopted. These include tuple spaces, mobile agents, and reactive programming through the usage of event publishing and subscribing [1,20,22]. Although these technologies provide service abstractions that highly simplify application development, their efficiency in *ad hoc* environments is still an open issue.

Ad hoc networking shares many concepts, such as distribution and cooperation, with the peer-to-peer (P2P) computing model [24]. A defining characteristic of P2P systems is their ability to provide efficient, reliable, and resilient message routing between their constituent nodes by forming virtual *ad hoc* topologies on top of a real network infrastructure. The difference with traditional distributed computing systems is the lack of a central authority that controls the various components; instead, nodes form a dynamic and self-organizing system. The applications best suited for P2P implementation are those where centralization is not possible, relations are transient, and resources are highly distributed [22]. In particular, the range of applications covered by the P2P model includes file sharing, distributed search and indexing, resource storage, and collaborative work. The key aspect of P2P systems is the ability to provide inexpensive but time-scalable, fault tolerant, and robust platforms; for example, file sharing systems, such as Gnutella [17], are distributed systems where the contribution of many participants with small amounts of disk space results in a very large database distributed among all participant nodes.

The distributed nature of *ad hoc* networks fits well the P2P model of computation. Systems based on the P2P model are those where centralization is not possible, relations are transient, and resources are highly distributed. MANET environments have similar requirements and characteristics, and this duality suggests that exploiting the P2P paradigm for designing a middleware platform for MANETs would be a very promising direction to take.

In this chapter, we investigate the efficiency of P2P middleware platforms when implemented in mobile *ad hoc* networks. Specifically, we focus on two well-known platforms: Gnutella and Pastry. Through simulations, we show the limitations and inefficiencies that these P2P systems exhibit in MANET environments. Finally, the chapter ends by discussing the potential of an innovative protocol stack architecture for *ad hoc* networks, where the emphasis is on cross-layering. The key idea here is that the information collected by each protocol could be used inside the stack to optimize the tasks of other protocols. The focus in this work is on the cross-layer interaction between a proactive routing protocol and a P2P platform that offers the same functionalities and semantics of Pastry [13]. We provide perspectives on how costs and the complexity of building and maintaining a Pastry-like overlay network can be reduced through cross-layer interactions.

Performance of Peer-to-Peer Platforms in *Ad Hoc* Environments

A key challenge to the usability of a data-sharing, peer-to-peer system is implementing efficient techniques for the search and retrieval of shared data. The best search techniques for a system depend on the requirements of the distributed application; for example, applications such as group multicasting, Web caches, or archival systems focus on availability and require guarantees on content location (if such exists). These requirements are usually met at the expense of flexibility — for example, organizing search indexes by data *identifiers*, which allow quick lookup procedures by limiting the subject space, by imposing strict rules on their format, and by exactly controlling how the search index should be organized in the distributed system.

In contrast, other kinds of applications, such as file sharing or publish–subscribe systems, require the ability to issue rich queries, such as *regular expressions*, meant for a wide range of users from autonomous organizations. Moreover, this second class of applications requires a greater respect for the autonomy of individual peers and not requiring them to host parts of the distributed search index. These requirements clearly relax assumptions and expectations on the performance of the P2P system. The aforementioned differentiation regarding the requirements of distributed data-sharing applications led to two P2P computational models: *structured* and *unstructured* platforms.

In *structured* platforms, peers organize themselves in a distributed search index (also called a *structured overlay network*) that usually contains information on the exact location of each shared piece of data. In these systems, each peer maintains only a partial knowledge of the index and establishes key network relationships with other peers to allow almost complete coverage of the search structure. The main idea is to map both peers and data identifiers on the same logical space, assuming that a peer with a logical identifier P gets relevant information about data logically close to P. This approach allows for *subject-based* lookup procedures, where a peer with identifier D of the wanted data item (e.g., a file name or a multicast group identifier) initiates a distributed search algorithm that hop after hop in the structured overlay ends up on the peer logically closest to D. Among the various proposals in the area of structured data sharing, Pastry [13] and Chord [25] organize the overlay as a distributed ring of identifiers, while a content-addressable network (CAN) [23] uses the concept of a distributed quad-tree on an n-dimensional space. All of these platforms achieve optimal lookup performance, guaranteeing the retrieval of shared content information in a logarithmic number of hops in the overlay and requiring each node to establish a small number of relationships in the overlay.

In *unstructured* platforms, peers do not self-organize in a distributed search index and are not required to maintain relevant information about shared content owned by other entities (e.g., a distributed search index). Peers establish network relationships in a pseudo-random fashion starting from a given entry point (i.e., a boot peer) and look for shared data, initiating flood search procedures. This approach does not match availability guarantees, as in the case of structured platforms, but it does allow for *content-based* lookup procedures based on regular expressions, to retrieve shared data. Content-based lookups are directly applied on the published content and assume that a large number of peers get hit by search requests (e.g., through query propagation schemes based on flooding). Platforms such as Gnutella [17] or Kazaa [16] demonstrate the flexibility offered by the unstructured approach in supporting very large-scale file-sharing applications on the Internet. Moreover, the characteristics of open platforms, with their discussion and development forums, have brought existing systems to maturity; available protocol specifications make it easy to adopt and deploy these platforms with new implementations, thus introducing innovative optimizations directly in real testbeds.

Within the context of P2P computing for mobile *ad hoc* networks, it is advisable to consider both unstructured and structured approaches, as they could better support distributed data sharing for particular applications. Furthermore, an initial evaluation of the capacity and performance of existing data-sharing platforms in *ad hoc* environments would provide an important starting point for future discussions and new proposals. To this end, in the following discussion we provide algorithmic details regarding Gnutella and Pastry, representatives of unstructured and structured approaches, respectively. Through simulation and experimental results, we demonstrate their capacity and performance when employed in *ad hoc* scenarios.

The Gnutella Protocol

In this section, we describe the Gnutella protocol for overlay maintenance and data lookup. For more details, please refer to the latest specification [17]. Some of the information in this section is not part of the original protocol (e.g., the behavior of a peer according to its connectivity in the overlay) but represents details added for clarity. Note that the Gnutella specification makes a distinction between *ordinary* and *super* peers. Super peers make up the overlay and provide the search infrastructure, and they are usually represented by nodes with a permanent Internet connection. In contrast, ordinary (or *leaf*) peers have intermittent connectivity, so they are not part of the overlay formation but simply attach themselves to an arbitrary number of super peers, proxying queries through them. In the context of this chapter, we are interested in the general properties of the overlay formation protocol, so we do not consider ordinary peers in the following discussion.

State Maintenance

Gnutella operations rely on the existence of an unstructured overlay network. Peers open and maintain application-layer connections among them, forming logical links in the overlay. Messages dedicated to peer discovery, link control, and data lookup are then sent exclusively along the overlay. As each peer is allowed to open only a limited number of connections and establish direct relationships with only a few other peers, message forwarding is a cooperative task necessary to achieve broad coverage of the overlay. The message lifespan is controlled by assigning a bounded time to live (TTL) at the application layer which decrements at each logical hop.

To establish a connection, as shown in Figure 23.1, a peer (P_1) initiates a handshaking procedure with another peer (P_2) by sending a request message (C-request). Peer P_2 could then reply sending either a connection accept (C-accept), to signal its willingness to link with P_1, or a connection reject (C-reject), to signal that it has no more available connection slots. The handshaking is a successful one for P_1 when it receives a C-accept message and for P_2 when it receives a confirm message (C-confirm) from P_1. These events trigger an update of the connection tables, and both peers will then behave according to their internal status. This mechanism assumes that each peer is given a *boot server* to establish its entry point (or first connection) in the overlay. In real Gnutella networks, this information comes directly from the user or through a lookup against a Gnutella Web cache (see www.gnucleus.com/gwebcache/ for an example), where peers that are already part of the overlay publish their addresses.

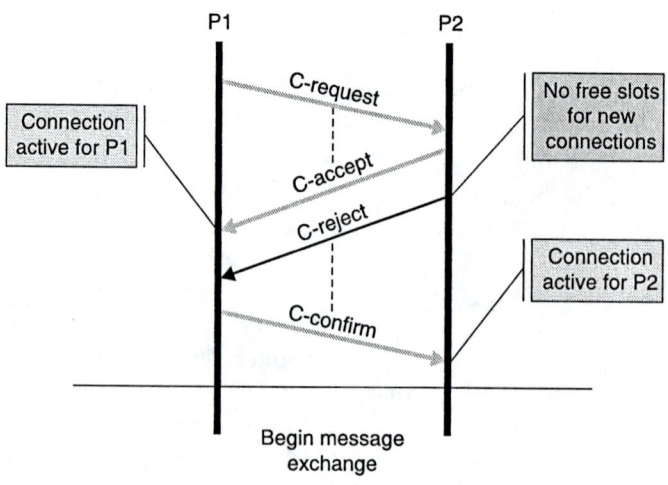

Figure 23.1   Gnutella handshaking procedure.

Gnutella peers usually store information on the state of active connections in a dedicated table, which we refer to as *connection table*. The size of the connection table is expected to range inside a lower and an upper bound (LB and UB). These bounds are not directly reported in the Gnutella specification [17] but provide a systematic way of modeling the behavior of Gnutella peers [7]. In particular, they guarantee that each peer opens a minimal amount of links (at least the LB), without overdoing the overall connectivity (exceeding the UB) and consequently abusing network resources. These two limits influence peer behavior by leading to three operational states. When the table size is smaller than the LB, the peer remains in a *connecting* state, where it (1) performs peer discovery, (2) initiates connections toward discovered peers, and (3) accepts incoming connection requests from foreign peers. As the table size reaches the LB, the peer enters a *connected* state, where it stops doing peer discovery and does not initiate connection requests anymore. In this state, it still accepts incoming connection requests from remote peers, as long as its table has free slots available. Clearly, if the table size falls back down under the LB, the peer transits back to the connecting state. Finally, when the table size reaches the UB, the peer enters a *full* state, where it also stops accepting incoming connection requests. Again, as the table size drops down under the UB, the peer returns to a *connected* state.

In each of the above states, the system performs data lookup on demand (i.e., driven by the user) and periodically probes its active connections using one-hop probe Ping messages. A connection becomes inactive when peers are not able to probe-Ping each other for more than a specified amount of time. Finally, a peer can intentionally drop active connections by issuing "bye" messages.

Peer Discovery, Pong Caching, and Queries

We now briefly describe how peers discover each other (see Table 23.1 for a detailed pseudo-code description). After having established their first connection, peers discover other agents by sending over the connection multi-hop discovery Ping messages with a TTL equal to seven [17]. At each hop in the overlay, the TTL fields of the discovery Pings are decremented before being forwarded over each active connection listed by the current peer (except the one where the Ping came from). This bounds the horizon of the application-layer broadcast to seven hops; at the edge, discovery Pings are discarded due to TTL expiration. Peers receiving a valid discovery Ping reply back with a Pong message containing credentials for future connections (i.e., network address and port number). Note that this last step is executed only if the peer is not in the full state. In fact, in this state the peer could not even accept incoming connection requests. Pong

Table 23.1 Pseudo-Code for Gnutella Peer Discovery and Pong Caching

Symbols and Constants

PT (Pong threshold)	The minimum number of valid entries in the Pong cache that a peer should have to directly answer a query; the Gnutella specification suggests 10 as a fair value.
FULL	This indicates the full state of the peer behavior.
TTL_{max}	This is the maximum TTL value that can be assigned to a message; the Gnutella specification suggests 7 as a value.

Objects

Name	Field or Method
PingMsg or PongMsg	ttl = message time to live
PingMsg or PongMsg	hops = number of hops in the overlay performed by the message
PingMsg or PongMsg	msgID = unique message identifier
PingMsg	IsProbePing() = checks whether this Ping is a Probe Ping (i.e., ttl = 1 and hops = 0)
PingMsg	IsDiscoveryPing() = checks whether this Ping is a valid discovery Ping (ttl + hops <= TTL_{MAX})
PeerTable	Represents the connection table
PeerTable	state = current peer behavior

```
AddPong(pongmsg) ⇒ add Pong to cache
PurgePongCache() ⇒ delete stale cache entries

HandlePing(pingmsg, connection)
if (pingmsg.isProbePing()) then
    pong = newPongMsg(local credentials)
    pong.ttl = 1
    SendPong(connection, pong)
else if (pingmsg.isDiscoveryPing()) then
    PurgePongCache()
    if (PongCache.size() ≥ PT) then
        pongs = PongCache.SelectPongs(PT)
        for all pong in pongs do
            pong.ttl = pingmsg.hops + 1
            SendPong(connection, pong)
        end for
    else
        if (PeerTable.state ≠ FULL) then
            pong = newPongMsg(local credentials)
            pong.ttl = pingmsg.hops + 1
            SendPong(connection, pong)
        end if
        if (pingmsg.ttl > 1) then
            pingmsg.ttl --
            pingmsg.hops ++
            for all c in PeerTable such that c ≠ connection
            do
                ForwardPing(c, pingmsg)
            end for
        end if
    end if
end if
...
HandlePong(pongmsg)
if (pongmsg.ttl + pongmsg.hops ≤ TTL_MAX) then
    AddPong(pongmsg)
    if (pongmsg.ttl + pongmsg.hops > 1) then
        connection = GetMsgOriginator(pongmsg.msgID)
        ForwardPong(connection, pongmsg)
    end if
end if
```

replies are given enough TTL so they are able to reach the Ping originator, which can then use the embedded credentials to open new connections. Note that each Pong reply is back propagated along the overlay path of the related Ping. In fact, the originators of discovery Pings associate unique identifiers with the messages, making it possible for intermediate peers to remember where Pong replies should be forwarded.

The standard discovery procedure can be enhanced with Pong caching. Upon receiving Pong messages, a peer stores the embedded credentials in a local cache. Incoming Ping messages could then be directly answered if the local cache contains enough items (i.e., up to a predefined threshold), without further forwarding the discovery Ping. In this case, a certain number of items are selected from the Pong cache and returned all together in a series of replies to the originator; otherwise, if the cache does not contain enough items, the peer performs the standard Ping forwarding procedure. This caching scheme significantly reduces the discovery overhead.

Queries are handled similarly to discovery Pings. Upon receiving a query message, a peer looks up the locally shared content using the constraints contained in the query. If one or more matches are found, the peer replies back with a query hit, providing pointers to local results. In any case, the peer decrements the TTL field and forwards the query message to its neighbors. Subsequent data downloads are carried out outside the overlay through direct file transfers.

Performance of Gnutella in Mobile Ad Hoc Environments

To better understand the capacity and limitations of Gnutella when employed in mobile *ad hoc* environments, we performed a set of simulations to subject the platform to typical *ad hoc* scenarios using the Network Simulator (*ns2,* version 2.27). In this chapter, we report only results related to (1) scenarios with mobile nodes moving with different patterns, and (2) scenarios where we recreated partitioning of the physical network (refer to Conti et al. [9] for a complete analysis and a detailed discussion). In these scenarios, we analyzed the overhead of the protocol as the amount of network traffic generated in a particular time unit and its capacity for building the overlay, measured as the average number of per-peer connections (i.e., *average peer degree*).

The plots in Figure 23.2 show the average degree achieved by Gnutella peers under three different mobility patterns. This study revealed that node mobility severely impacts the overlay formation capacity of Gnutella. Only in *static* scenarios were the peers able to reach (on average) the minimum amount of connectivity (LB), fixed at four connections per peer [8,18]. When we introduced mobility patterns based on the random way-point mobility model, the average peer degree fell to 3.7 connections per

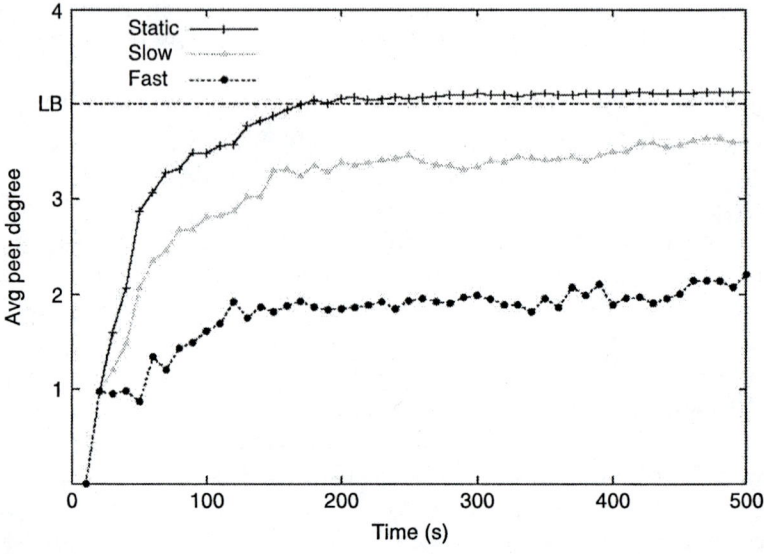

Figure 23.2 Average Gnutella peer degree under increasing node mobility with the OLSR routing protocol.

peer in a *slow* configuration, where nodes moved up to 5 m/sec, and down to 2 connections per peer in *fast* configuration, where nodes moved up to 15 m/sec. These conditions resulted in high rates of overlay partitioning and, consequently, low rates of information discovery.

Figure 23.3 highlights the reaction of Gnutella peers under network partitioning situations. These experiments were carried out for a simulation time of 600 sec; an overlay of 30 peers stabilized network activity in the first 200 sec, network partitioning was introduced around time 270, and, finally, the original network was resumed around time 430. This was achieved by placing the 30 mutually visible nodes on a static grid and letting those in the center move in opposite directions so as to divide the network into two halves. The plots report the reactions (in term of the amount of network traffic generated [kB/sec]) of Gnutella peers with regard to the network partitioning and rejoining events (see vertical lines). We could observe bursts of networking activity corresponding with the beginning and termination of network partitioning — up to 300 percent increase in traffic at the beginning of the network partitioning, and up to 200 percent at the termination. The straightforward explanation for this behavior is that peers transit back to the connecting state when the network is divided into two halves and perform even broader discovery procedures when the connectivity is restored at the physical layer. The plots also show that Gnutella behavior is independent of which routing

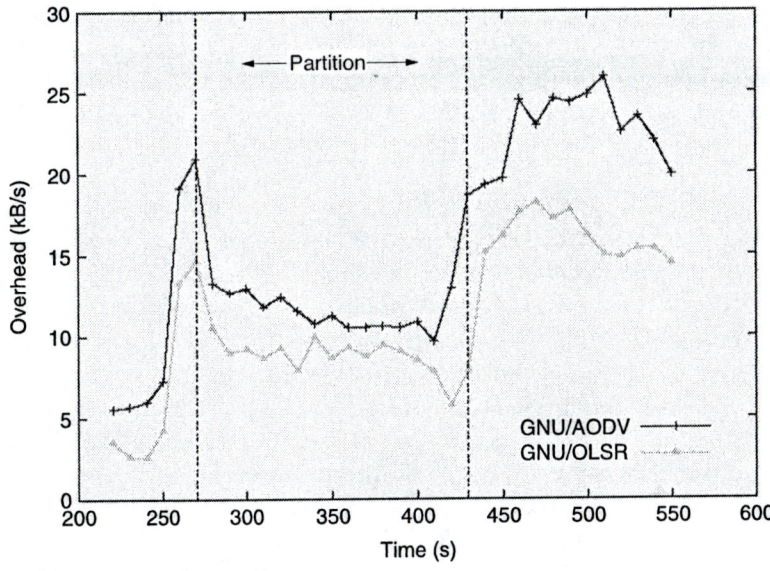

Figure 23.3 **Effects of network partitioning on the overhead generated by Gnutella peers, with both OLSR and AODV.**

algorithm — Optimized Link State Routing (OLSR) or *Ad Hoc* On-Demand Distance Vector (AODV) — is in use at the network layer; however, implementation of OLSR in these simulations provided routes of better quality. The overhead generated by Gnutella on top of AODV results in a bigger factor, ranging between 10 and 20 percent.

From these simulations, we could verify that, although Gnutella meets important requirements for the management of data-sharing overlay networks, it was not designed for *ad hoc* networks and suffers from node mobility, causing peers not to achieve minimum connectivity requirements. Moreover, upon network partitioning, the protocol generates traffic bursts in response to the topological reconfigurations. This is clearly not desirable in mobile *ad hoc* situations, where the network partitions frequently or where groups of nodes enter and leave the network.

The Pastry Protocol

The overlay network defined by Pastry [13] is represented by a large circular space of 2128-1 logical identifiers, also called a *ring overlay*. Each Pastry node chooses a 128-bit identifier (nodeId), which represents a logical position in the ring. The nodeId is calculated at join time, when the node hashes one of its physical identifiers (e.g., IP address, hostname, public key) through a strong hash function (H), which uniformly distributes inputs in the ring space and reduces the chances of two different physical identifiers being mapped on the same nodeId and increases the chances of scattering nodes with closer nodeIds far apart in the ring.

Subject-Based Message Routing

The fundamental service offered by a Pastry ring is that peers exchange messages through a *subject-based* mechanism. The idea is to associate a logical *subject* (or a key) with an application-layer message and route it hop-by-hop in the ring until it arrives at a peer with a nodeId that is closest to the subject of the message. This final peer is considered the *root* for the message and is responsible for handling the message content at the application layer. The association between messages and subjects occurs through the same hash function used for mapping nodes to logical addresses, which guarantees the same aforementioned distribution properties. To give an example, consider a file-sharing application where each file is represented by its name. Typically, with Pastry, two types of messages would be created, one to *publish* in the ring the sharing of a file associated with a given name and another to *lookup* the node (or the nodes) sharing a file with a given name. In this scenario, publish and lookup messages get routed through the same logical identifiers, and the corresponding root peers associate them at the application layer.

The subject-based routing policy used by Pastry is based on a numerical proximity metric between message subjects and nodeIds. From the algorithmic standpoint, consider logical identifiers to be represented as a sequence of digits with base 2^b, where parameter b is defined *a priori*. At each step of a routing procedure, a Pastry peer (P) forwards the message with subject K to another peer (Q), whose nodeId shares with K a prefix that is at least one digit (b bits) longer than the prefix shared with P. If no such node is known, P tries to forward the message to peer L, which has the same common prefix with K but is numerically closer to K with respect to P (this can be easily identified by looking at the digit after the common prefix). The expected maximum number of hops in the overlay between a source and a destination peers is equal to $\lceil \log_2^b N \rceil$ in an overlay of N nodes [13].

State Representation

To support this routing procedure, each peer maintains information related to other peers in the overlay, using the following data structures:

- *Routing table.* This structure is organized into $\log_2^b N$ rows with $2^b - 1$ entries each. Each entry at row n of the routing table refers to a node whose nodeId shares with the local nodeId the first n digits, but whose $(n + 1)$th digit differs (it has the same value of the column index). If there are no nodeIds with this characteristic, the entry is left empty. In practice, a destination node is chosen among those known by the local node based on the proximity of its logical identifier to the value of the key. This choice provides good locality properties but only in the logical space. In fact, nodes that are logically neighbors have a high probability of being physically distant. In addition, the choice of parameter b represents a tradeoff between the size of this data structure and the maximum number of hops in subject-based routing procedures that is expected to be equal to $\lceil \log_2^b N \rceil$, and simulations results confirmed this [13].

■ *Neighborhood set.* This structure represents the set of nodes that are physically close to the local node. The neighborhood set is not normally used in routing messages but could be useful for maintaining physical locality properties among nodes.

■ *Leaf set.* This structure represents the set of nodes with the closest logical identifiers to the current node. The leaf set is centered on local node P, with half of the identifiers larger than P and the other half smaller than P. The leaf set represents the perfect knowledge that each peer has of its logical contour.

In routing a given message, the node first checks to see if the related subject falls within the range of nodeIds covered by its leaf set. If so, the message is directly forwarded to the destination node — namely, the leaf set entry whose nodeId is logically closest to the message subject. If the subject is not covered by the leaf set, then the routing table is used, and the message is forwarded to a node that shares a common prefix at least one digit longer than the local nodeId. Sometimes, it is possible that the appropriate entry in the routing table is empty or that the associated node is currently disconnected from the network, but the overlay is still not updated; in this case, the message is forwarded to a node (if any exists) that shares the same prefix as the local node but is numerically closer to the subject. Each Pastry data structure entry maintains a correspondence between the logical identifier of each node and its credentials (IP address and port number) to allow the establishment of direct peer connections driven by application needs.

State Management

The main procedures used by Pastry to establish and maintain the overlay network (i.e., the previously presented data structures) consist of join and disjoin operations. First, when a new node (say, X) decides to join the overlay, it must initialize its internal data structures and then inform other nodes of its presence. It is assumed that the new node knows at least one of its physical neighbors (say, A) which already is part of the overlay. Typically, this *bootstrap* node can be located automatically (by sending "expanding ring" queries) or can be obtained by the system administrator through outside channels. Node X then asks A to route a special "join" message with the key equal to X. Like any message, Pastry routes the join message to the existing node Z whose ID is numerically closest to X, passing through some intermediate nodes. In response to the "join" request, nodes A, Z, and all the intermediate peers send the contents of their tables to X. At this point, the new node X processes the received information and initializes its own structures in the following way:

- The neighborhood set is initialized with the contents of node A, as it is a physical neighbor of X.
- The leaf set is initialized with node Z, which has the closest existing nodeId to X.
- The ith row of the routing table is initialized with the corresponding row of the routing table of the ith node (B_i) encountered in the routing path from A to Z (as it shares a prefix of length i with X).

At the end, X informs all the newly known nodes about its arrival, transmitting a copy of its resulting state. This procedure ensures that X initializes its state with appropriate values and that the states in all other involved nodes are updated.

The management of departure nodes is another important feature of Pastry. In Druschel and Rowston [13], it is assumed that Pastry nodes may fail or depart without warning. In particular, a node is considered failed when its logical neighbors can no longer communicate with it. To this aim, nodes in the leaf set are periodically probed with User Datagram Protocol (UDP) Ping messages. Leaf entries that do not reply to probe Pings are considered failed and get replaced by entries of the leaf set relative to the live node with the largest index on the side of the failed node. In this way, each node can easily repair its leaf set, and the delay with which it becomes aware of logical neighbor failure depends on the probing frequency. A similar probing mechanism is used to maintain a consistent neighbor set. In this case, a node realizes that an entry in its routing table has failed only when it attempts to connect to it to forward an application message. This event does not normally delay message routing, as another destination node could be selected. In any event, the failed routing table entry has to be replaced. To this end, the peer contacts the entries belonging to the same row of the failed one, asking for a nodeId that can replace it. If none of them has a pointer to a live node with the appropriate prefix, the local node has to contact nodes belonging to successive rows of the routing table. In this way, many remote connections could be required to manage single entries of the routing table.

The maintenance procedures explained above highlight the complexity associated with the management of a structured overlay network. The many remote connections required to check the validity of table entries considerably increase the overhead introduced on the underlying network. In addition, peers that are considered failed have no way to get back into the ring apart from performing once again a join procedure; for example, if a node temporarily loses its network connection, it has to reboot the system and join the overlay again. This limitation represents a major problem in mobile *ad hoc* networks, where frequent topology reconfigurations could cause situations of intermittent connectivity that are not compatible with the low tolerance offered by Pastry maintenance procedures.

To better understand the overhead introduced by ring-maintenance operations in Pastry, we analyzed its behavior in a small real testbed. Specifically, we evaluated an open-source implementation of Pastry called FreePastry [14]. During our study, we focused on both reactive and proactive solutions for the routing protocol at the network layer, using AODV in the former case and OLSR in the latter. A complete analysis of the results obtained from our real testbed has been presented in Borgia et al. [2].

For our study, we set up a real *ad hoc* network of eight nodes inside the CNR campus in Pisa, where the structural characteristics of the building and the nearby presence of access points and measurement instrumentations limit the transmission capabilities of nodes, thus forcing the establishment of a multi-hop *ad hoc* network. The network topology configuration used for the experiments related to FreePastry is shown in Figure 23.4, where all nodes ran one of the routing protocols and only six ran a distributed application on top of FreePastry. In particular, nodes B and G worked just as routers, allowing packet forwarding from the source to the destination through the optimal path.

From our experimental results we noticed that, due to the limited number of peers participating in the FreePastry ring, the overlay data structures maintained the logical identifiers of almost all peers. For this reason, there were rare generations of multi-hop subject-based routing procedures; however, the operations required to create and maintain the overlay imposed high overheads on the underlying *ad hoc* networks, causing errors depending on the routing protocol used. In fact, FreePastry implements maintenance operations using both UDP and Transmission Control Protocol (TCP) connections to remote peers, introducing further overhead on the network as overlay relationships do not take into account neighborhood information.

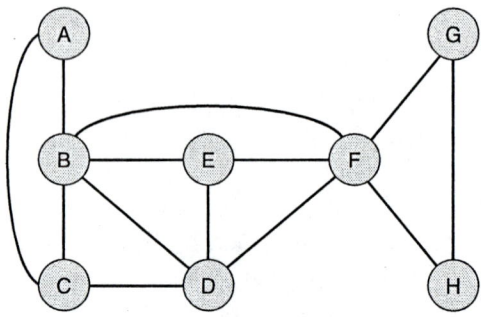

Figure 23.4 Experimental network topology.

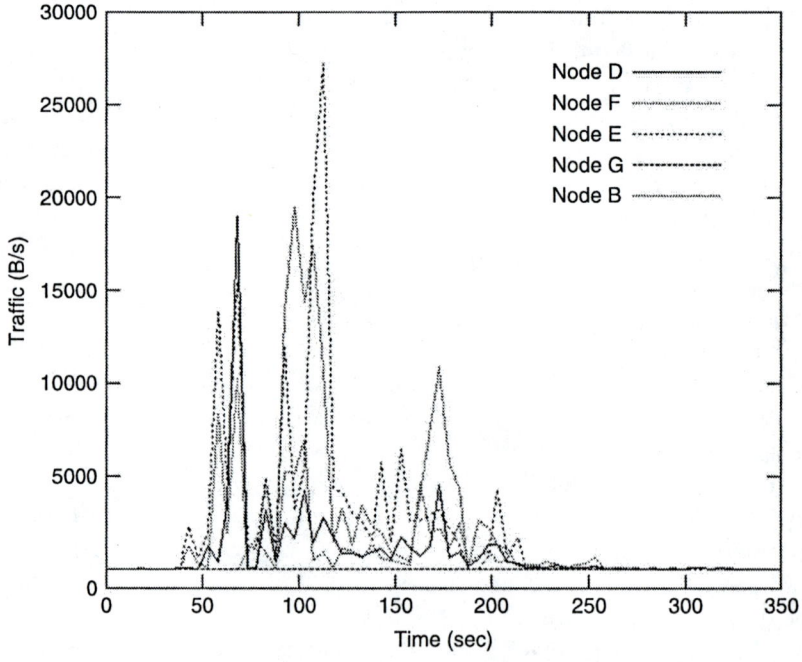

Figure 23.5 Pastry on OLSR: traffic related to main nodes.

To obtain a performance evaluation of Pastry, we calculated the traffic generated to maintain the overlay, in addition to the traffic produced by the routing protocol at the network layer. Figure 23.5 and Figure 23.6 show the amount of traffic observed by some nodes of the network running OLSR and AODV, respectively. In both cases, the amount of traffic generated by node B and G, which just worked as routers, was negligible. In fact, the proactive protocol does not introduce a high overhead on the

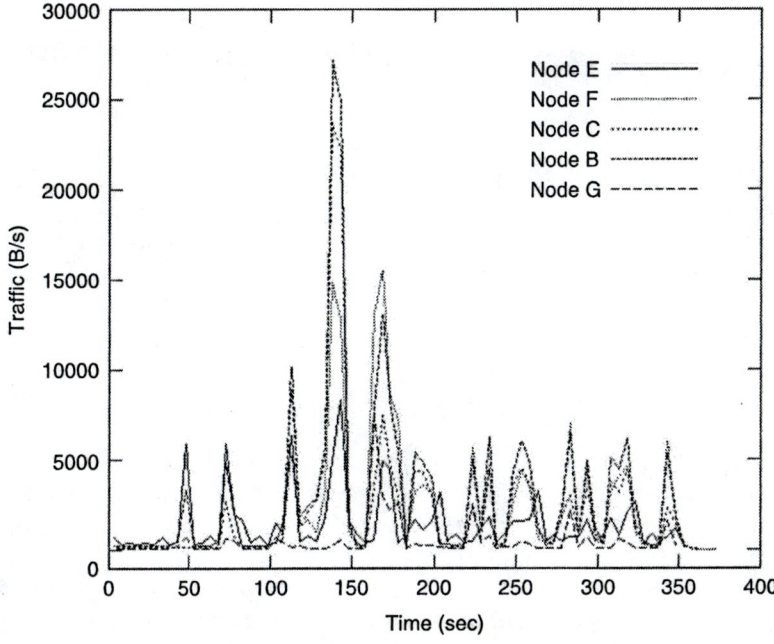

Figure 23.6   Pastry on AODV: traffic related to main nodes.

network, as it encapsulates different routing messages (Hello and Topology Control) into single routing packets. By configuring an incremental ring formation, UDP and TCP connections used by FreePastry peers to collect initial overlay information generate evident traffic peaks. Even if the total amount of traffic is not so high, relative to the WiFi bandwidth, those peaks can negatively influence the additional transmission of application data. Furthermore, we could observe that the successful creation of the overlay depends on the reliability of the path discovery process implemented by the routing protocol. In the case of AODV, a high number of TCP retransmissions and connection failures occurred, primarily due to the delay introduced by discovering routes toward a destination and the use of unidirectional links as valid routes.

To summarize, experimentations carried out with FreePastry showed the generation of heavy overheads related to overlay management procedures, due to a high number of remote connections opened between peers. The lack of attention to the usage of network resources greatly reduces the overall system performance in *ad hoc* networks. In order to improve the system, Pastry nodes should become aware of the underlying network topology, maintaining a tight correspondence between physical and logical address space. The work in Castro et al. [8] presents a reorganization of the overlay in Pastry that exploits network proximity information to improve application performance and network usage. This solution, based on an additional location discovery protocol that estimates physical distances between nodes, shows the potential improvement introduced by proximity information at the expense of running another protocol in conjunction with the P2P platform. The last section of this chapter discusses an alternative solution based on a full cross-layer protocol stack architecture, which allowed us to greatly optimize ring management procedures by exploiting interactions between the P2P platform and routing protocols at the network layer.

Performance of Pastry in Mobile Ad Hoc Environments

As in the case of Gnutella, we performed *ns2* simulations to understand the capacity of the platform in typical *ad hoc* scenarios. By putting Pastry through the same mobility scenarios used for Gnutella, we were able to study the capacity for building the overlay, measured as the average number of entries in Pastry routing tables, and the rate of *unsuccessful* subject-based message routing procedure. For the latter, we created a simple subject distribution model in which each Pastry peer was configured to be the root for one subject, and we performed a route procedure toward the subject maintained by other peers. We considered a route procedure for a subject k to be successful if it terminated at the peer responsible for k.

In our simulation models, we configured a network of 30 mobile nodes and then varied the density of Pastry peers to be exactly 100 and 50 percent of the network size (30 and 15 Pastry peers, respectively). The plots reported in Figure 23.7 show the average number of routing table entries collected by Pastry peers under patterns of increasing node mobility. The Pastry protocol [13] defines a logarithmic lower bound (i.e., $\log_2^b(n)$, where n is the size of the overlay) on the number of entries that each peer should collect in internal tables to guarantee high rates of successful subject-based routing procedures (the original Pastry paper [13] reports a maximum of about 5 percent failed routing procedures as acceptable values). The horizontal lines in Figure 23.7 indicate the lower bounds relative to the two overlay sizes ($\log_2^2(30)$ and $\log_2^2(15)$, respectively). As already observed in the case of Gnutella, node mobility has a big negative impact on the protocol capacity of building the overlay. The scenarios with mobile nodes (in both *slow* and *fast* configurations) exhibited a sharp reduction in the amount of ring overlay knowledge that each Pastry peer was able to collect (1.5 for the slow scenario and 1.2 for the fast scenario) in the case of a one-to-one node–peer correspondence.

The above results were confirmed by a study on the failure rate of subject-based routing procedures. As applications based on structured

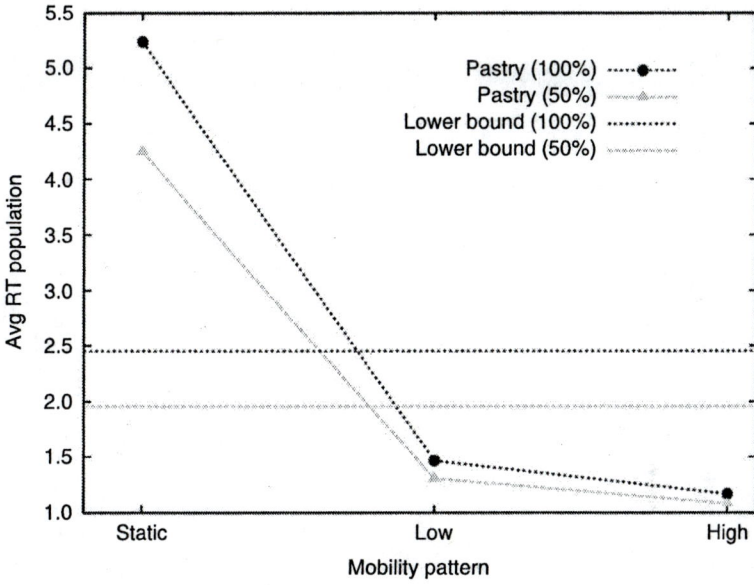

Figure 23.7   Average number of Pastry routing table entries collected by peers under pattern of increasing node mobility.

platforms focus on availability, subject-based routing procedures should maintain very low failure rates — for example, below 20 percent in the case of mobile *ad hoc* environments. Medium or higher failure rates suggest that the platform is not usable. The results reported in Figure 23.8 show that the performance of a straightforward implementation of the Pastry protocol quickly degrades under patterns of increasing node mobility, from around 5 to 10 percent in the case of static network to 70 to 80 percent and more in the case of slow and fast mobility patterns.

Cross-Layering

From the results illustrated in the previous sections, we can conclude that the two analyzed P2P platforms result in significant performance degradation when operating over an *ad hoc* network. Specifically, the simulation results indicate that only in static configurations are the protocols we analyzed able to construct overlays that meet necessary quality criteria and guarantee good performance. Furthermore, experiments with real testbeds indicate that, in static configurations, when the FreePastry implementation has correctly operated on top of our multi-hop *ad hoc* network severe problems can be identified from the performance standpoint. These

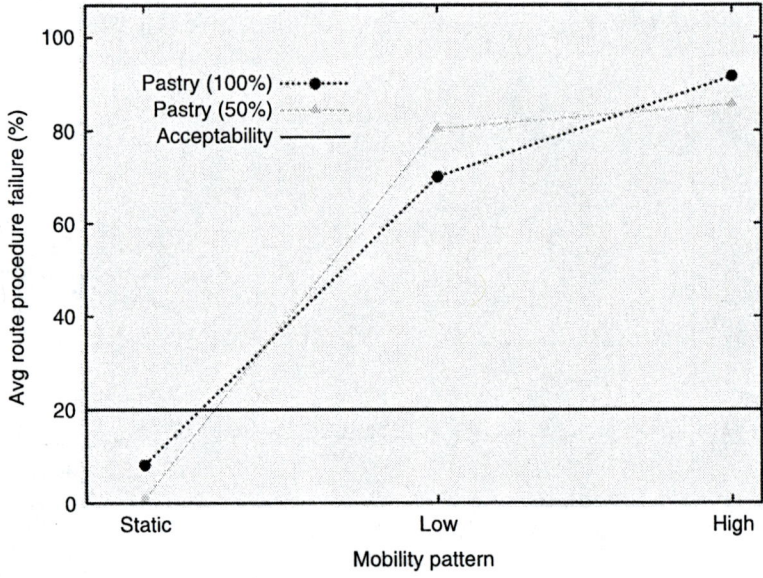

Figure 23.8 Effects of patterns of increasing node mobility on the rate of unsuccessful subject-based routing procedures.

are mainly due to several factors that affect MANET behavior in a real environment. IEEE 802.11 operates in the industrial, scientific, and medical (ISM) spectrum and hence experiences a lot of noise from external sources. The quality of the wireless links is therefore highly variable, and this affects the higher layer protocols (e.g., routing, forwarding, transport) behavior. In the end, this affects the overlay construction and management; for example, because FreePastry maintenance operations are also based on TCP services, the platform performance is negatively affected by poor TCP performances. In addition, because Pastry operates its own subject-based routing ring independently from the underlying *ad hoc* network, it introduces a significant and bursty overhead on the *ad hoc* network.

Both simulation results and experiences with a real implementation provided us with some indications of how to solve the performance problems in MANET environments. Specifically, results related to FreePastry indicate that significant performance benefits can be expected by exploiting routing information (extended with services information) at the middleware layer. This allows realizing ring overlay maintenance operations and avoiding the big overheads connected with implementations via middleware operations. Similar indications have been obtained for Gnutella [9].

These results provide strong arguments for exploiting, in a MANET, cross-layer interactions according to the reference architecture proposed

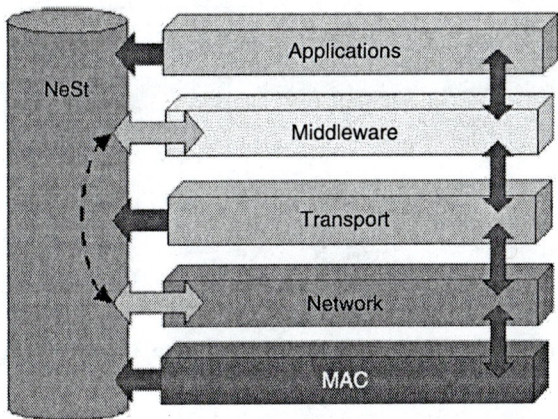

Figure 23.9   Cross-layer architecture.

in Conti et al. [6,10]. In this architecture, as shown in Figure 23.9, a vertical component, called the *Network Status* (NeSt), is introduced to enable indirect interactions between protocols belonging to different layers of the stack. This approach combines the flexibility of a layered architecture (protocols at different layers are designed and maintained independently) with the performance optimizations of a full cross-layer design.

The legacy TCP/IP architecture allows interactions only between adjacent layers, and each protocol knows only its own data structures. In the cross-layer architecture, protocols use the state information flowing throughout the stack to adapt their behavior accordingly. In this way, the overall network performance is optimized. Data sharing among the protocols is mediated by the NeSt, which maintains an abstraction of selected information required to be shared among some protocols belonging to different layers. Additionally, each protocol can subscribe its interests to the NeSt, such that it will be notified every time an associated event occurs. Producer protocols must notify the NeSt of their agreement to share their contents and provide timely updates.

One of the main advantages of the cross-layer architecture is making protocols aware of the current state of the network from the point of view of the local node. This facilitates higher level tasks, such as middleware and applications, where knowledge of the network topology can potentially improve system performances. Both reactive and proactive routing protocols can be used, but a proactive routing protocol guarantees better support to higher layer protocols by periodically flooding topology updates. In this way, each node has a complete and updated knowledge of the current state of the network; however, in the case of overlay platforms, network routing table contents are not sufficient to simplify management procedures, as each overlay has to be associated with a specific service, and only nodes providing that service are a part of the overlay. For this reason, a service discovery (SD) protocol is necessary to identify all nodes providing the same service. To avoid the introduction of an additional protocol sending service information over the network (causing an increase of the total overhead), we defined an embedded solution that exploits the proactive approach of sending periodic topology updates. To this aim, service information is added as optional fields to routing packets and automatically sent as link-state update (LSU) messages. In addition, the service information is associated with each IP address in the NeSt abstraction of the network routing table such that the middleware layer can directly access this information and autonomously build its overlay. A detailed definition of a possible cross-layer service discovery protocol is presented in Reference 12.

To verify the effectiveness of this cross-layer approach, we have designed and implemented CrossROAD (Cross-Layer Ring Overlay for *Ad Hoc* Networks) [11], a ring overlay platform with Pastry-like semantics that is optimized through cross-layer interactions for operating in *ad hoc* networks. CrossROAD exploits the cross-layer architecture shown in Figure 23.9 to directly interact with a proactive routing protocol that guarantees a complete knowledge of the network topology. In this way, it exploits routing information to optimize the construction and management of the overlay. Following Pastry basic principles, CrossROAD assigns to each node a logical address. Then, by exploiting the service discovery protocol, the local node knows the IP addresses of the others taking part to the same service, directly accessing the related NeSt data structures. For this reason, CrossROAD defines the logical identifier of each node as the result of the hash function applied to its IP address. In this way, at the startup, CrossROAD can autonomously build the complete overlay, simply hashing the IP addresses of all participants.

This is only the first step toward realizing an optimized solution for *ad hoc* networks — that is, avoiding initial remote connections to establish the ring and collect routing information. To do this, when a node running CrossROAD wants to join the ring, the system has to subscribe the associated service to the NeSt and consequently recover the IP addresses of all the other participants. Naturally, to maintain a consistent view of the network topology, the routing protocol has to update the NeSt abstraction with the same frequency as it manages a link-state update packet. At this point, CrossROAD locally computes the logical identifiers of the other nodes and stores this information in its local data structures. The overlay management is highly simplified, as CrossROAD does not require any remote connection. In addition, the routing protocol, periodically sending its LSU packets, collects all the topology changes and directly updates its own routing table and the related abstraction in the NeSt.

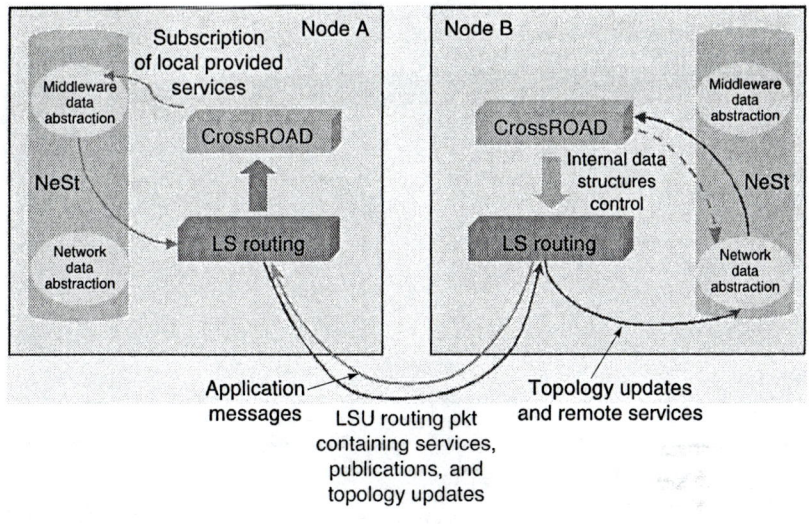

Figure 23.10 Middleware routing interactions in the cross-layer architecture.

Figure 23.10 shows an example of two nodes, A and B, connected to the same network. Let assume that node B is already connected to the overlay and that node A tries to participate by starting the same service. The CrossROAD agent on node A recognizes the provided service and sends this information to the NeSt, which will notify the routing protocol of that event. At this point, the proactive protocol encapsulates the service information in the next LSU packet to be sent through the network. Then, when node B, or any other, receives the LSU, it updates the network topology and the related NeSt abstraction, taking into consideration the additional service information. At this point, when the application running on node B has to send a message, CrossROAD first verifies if the content of its routing table is consistent with the current network topology, updating it before sending the message if necessary. In this way, CrossROAD updates its view of the overlay every time it has to send a message, thus reducing the possibility of considering wrong destinations. With this solution, the overlay management is enormously simplified and no additional overhead is imposed on the *ad hoc* network functionalities. Finally, CrossROAD further optimizes the subject-based routing protocol defined by Pastry. Generally, Pastry implements this message forwarding, following the proximity logic, as a multi-hop routing, mainly due to the high number of nodes taking part to the service and to the limited dimensions of its data structures. CrossROAD turns this multi-hop routing into a direct peer-to-peer connection, maintaining the same algorithm. In fact, because each node knows

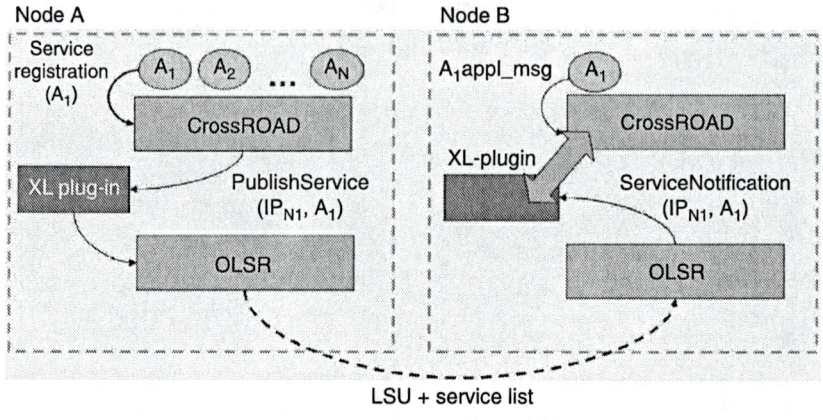

Figure 23.11 Cross-layer architecture implementation.

the others, the sender of a specified message can directly identify the closest destination for the selected key and send the message through a simple peer-to-peer connection. Subsequently, the forwarding protocol at the network layer delivers the message through the shortest path.

To analyze CrossROAD performances on top of a real multi-hop *ad hoc* network, we exploited OLSR, version 0.4.8 [26]. This implementation allows the development of an internal plug-in for the definition of additional information to be sent on the network through the routing protocol. This represents a subset of the NeSt aimed at implementing interactions between the middleware and routing layers. The plug-in we developed (*XL-plugin*) consists of a dynamic library loaded by the routing daemon at startup. To implement the cross-layer interaction, an exchange protocol between the middleware and the routing of a message has been defined.

Figure 23.11 represents a simple example of how cross-layer interactions can be exploited in the creation of the overlay between two nodes (A and B). The interaction between CrossROAD and the plug-in begins when application A_1, running on node A, registers the related service identifier, creating a new instance of CrossROAD. In this way, the local node joins the overlay in sending to the plug-in a message of *Publish-Service* containing its IP address and the service identifier associated with the specific application. When the plug-in receives this message, it encapsulates that information in the first routing packet that will be sent on the network, and it stores this content in the *local services table*, which maintains a list of services provided by the local node. In contrast, when this message is received by another node (e.g., B), the plug-in processes the additional information and stores it in the *global services table*, selecting all services currently provided by every node of the network. At this point, when the application decides to send a message on the overlay, first CrossROAD checks the consistency of its internal data structures with the plug-in. It then selects the optimal destination for that message among nodes currently running the overlay and contacts that node through a simple P2P connection. In this way, all remote connections required by Pastry to build and maintain the overlay data structures are eliminated, and every node of the overlay knows all the other participants, thus avoiding the multi-hop middleware routing introduced by the subject-based policy of Pastry. To compare and contrast the Pastry model with its cross-layer enhancement, in our prototype we integrated on top of CrossROAD a simple application of distributed messaging (DM). Nodes running DM set up and maintain an overlay network related to this service. When a node has created or joined the overlay, the application provides the possibility of creating or deleting one or more mailboxes distributed on the other nodes and to send messages to them. The physical location of a mailbox is randomly selected by applying the hash function to the associated identifier, which is used as the key value of the related messages. CrossROAD evaluation has been performed under the same conditions with which we evaluated FreePastry. Specifically, we considered the same eight-node network shown in Figure 23.4. In the first set of experiments, we measured the overhead introduced by CrossROAD to maintain the overlay (Figure 23.12) to compare it with that of FreePastry (see Figure 23.5 and Figure 23.6).

The figure illustrates that the overhead traffic for all nodes is less than 100 B/sec, which is much lower than that observed with FreePastry. In addition, traffic peaks introduced by FreePastry, corresponding to TCP and UDP connections used to initialize and maintain the overlay data structures, completely disappear in CrossROAD. Another important feature of CrossROAD is the timeliness with which every node becomes aware of the other participants [3]. This is an important property to guarantee appropriate behavior of the overlay when network partitioning and rejoining occur.

To highlight this, we analyzed a possible network partitioning and the consequent reaction of CrossROAD in the overlay management. To do this, a new network topology, shown in Figure 23.13, has been set up. The network consists of five nodes, and only nodes in adjacent positions are in the transmission range of each other. All nodes run OLSR enhanced with XL-plugin and CrossROAD. When all nodes are correctly connected to the overlay, node C begins periodically sending an application message with a specified key value (period equal to 1 sec). Initially, the key value results are logically closest to the node identifier of B, hence node C

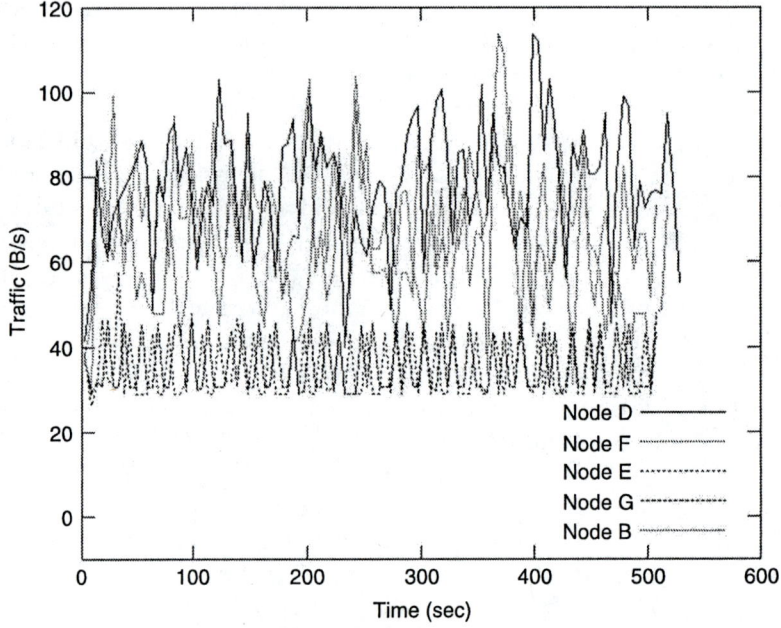

Figure 23.12   CrossROAD throughput related to main nodes.

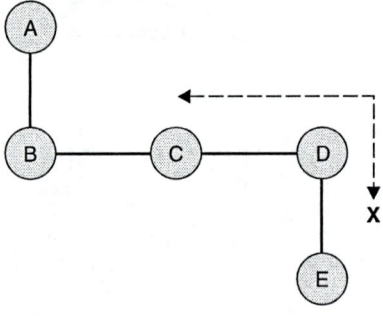

Figure 23.13   Network partitioning topology.

sends those messages directly to B. Then, after about 30 sec, node C begins moving toward position X at a speed of about 1 m/sec, generating network partitioning, and nodes A and B create an independent *ad hoc* network as well as nodes C, D, and E. Because the direct link from B to C is lost, the cross-layer interaction between CrossROAD and the routing protocol allows node C to become aware of the network partitioning and the consequent removal of nodes A and B from the overlay. Hence, the

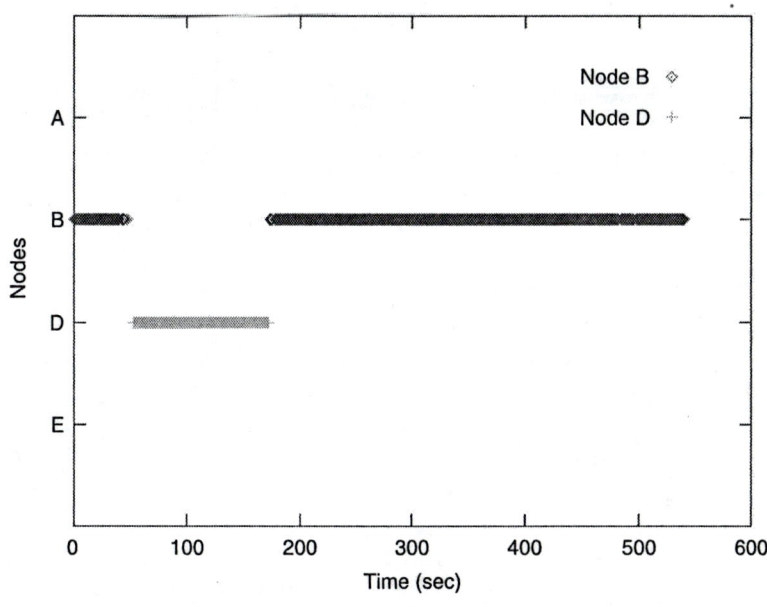

Figure 23.14   CrossROAD data distribution during network partitioning.

successive messages sent by node C on the overlay with the same key value are directly sent to the new best destination: node D. After 2 minutes, node C begins coming back to the initial position, reestablishing a single *ad hoc* network. At this point, subsequent messages are sent again to node B. As shown in Figure 23.14, CrossROAD correctly manages data distribution in the case of overlay and network partitioning. Specifically, the figure shows that (1) during the first phase (a single *ad hoc* network), node C data is stored on node B; (2) when the partition occurs, after a transient, node C data is delivered to node D; and (3) when the network is again connected, C data is again stored on node B.

Summary and Conclusions

Ad hoc networks are distributed systems composed of self-organized wireless nodes. As these systems cannot benefit from any centralized infrastructure, networking functionalities, such as packet forwarding, routing, and network management, as well as application services, should be distributed among user devices. The distributed nature of *ad hoc* networking finds in the peer-to-peer (P2P) interaction its natural model of computation. Recently, several self-organizing overlay platforms have been proposed for building decentralized and distributed applications for the Internet. The variety of applications and services realizable on top of these overlays also suits *ad hoc* scenarios; thus, having them work in *ad hoc* environments would be an advantage for the MANET technology. However, it is not clear how these overlays should be ported and how they will perform on *ad hoc* networks.

In this chapter, we focused both on structured and unstructured P2P platforms by investigating their effectiveness when operating on top of a mobile *ad hoc* network. Specifically, we investigated via simulation the performance of Pastry and Gnutella, as representatives of the structured and the unstructured classes. Our results indicated that only in static scenarios are these platforms able to construct effective overlays. Furthermore, evaluations of the performance of FreePastry (an open-source implementation of Pastry) on a real testbed suggested that overlay management procedures may introduce a significant overhead on *ad hoc* networks.

To summarize, our results indicated that in MANET environments, implementing an overlay that is completely independent from the physical topology of the network results in poor system performance. To avoid this, we explored an innovative protocol stack architecture for *ad hoc* networks that exploits cross-layering. The idea behind cross-layering is that the information collected at each layer could be used by the rest of the stack to optimize the overall functioning of single nodes and consequently the entire network. To maintain the flexibility of legacy layered architectures, cross-layer interactions are implemented via a shared memory (the Network Status, or NeSt), through which protocols share their internal information. In this chapter, we showed how a Pastry-like platform could be efficiently ported on *ad hoc* environments by exploiting the cross-layer architecture. Specifically, to verify the effectiveness of this cross-layer approach, we designed and implemented CrossROAD, a ring overlay platform with Pastry-like semantics that is optimized through cross-layer interactions for operating in *ad hoc* networks. CrossROAD exploits the cross-layer architecture to directly interact with a proactive routing protocol that guarantees complete knowledge of the network topology. In this way, it exploits routing information to optimize construction and management of the overlay.

A comparison between the performance of Pastry and CrossROAD in a real testbed indicated that, by exploiting cross-layer interactions with routing agents, peers become completely autonomous, and applications experience and optimize data distribution and lookup services without the introduction of additional network overhead. In addition, by exploiting routing updates, the resulting platform is able to quickly react to network partitioning and consolidation.

Acknowledgments

This work was partially funded by the Information Society Technologies program of the European Commission, Future and Emerging Technologies, under the IST-2001-38113 MOBILEMAN project, and by the Italian Ministry for Education and Scientific Research in the framework of the FIRB-VICOM project.

References

[1] Anceaume, E., Datta, A.K., Gradinariu, M., and Simon, G., Publish/subscribe scheme for mobile networks, in *Proc. of the First ACM Int. Workshop on Principles of Mobile Computing (POMC 2002)*, Toulouse, France, October 30–31, 2002, pp. 74–81.

[2] Borgia, E., Conti, M., Delmastro, F., and Pelusi, L., Lessons from an *ad hoc* network test-bed: middleware and routing issues, *Ad Hoc Sensor Wireless Networks*, 1(1–2), 125–157, 2005.

[3] Borgia, E., Conti, M., Delmastro, F., and Gregori, E., Experimental comparison of routing and middleware solutions for mobile *ad hoc* networks: legacy vs. cross-layer approach, in *Proc. of SIGCOMM Workshop on Experimental Approaches to Wireless Network Design and Analysis*, Philadelphia, PA, August 22, 2005.

[4] Bellavista, P., Corradi, A., and Magistretti, E., Lightweight replication middleware for data and service components in dense MANETs, in *Proc. of the 6th IEEE Symp. on a World of Wireless Mobile and Multimedia Networks (WoWMoM 2005)*, Taormina, Italy, June 13–16, 2005.

[5] Chlamtac, I., Conti, M., and Liu, J., Mobile *ad hoc* networking: imperatives and challenges, *Ad Hoc Networks*, 1(1), 13–64, 2003.

[6] Conti, M., Crowcroft, J., Maselli, G., and Turi, G., A modular cross-layer architecture for *ad hoc* networks, in *Handbook on Theoretical and Algorithmic Aspects of Sensor, Ad Hoc Wireless, and Peer-to-Peer Networks*, Wu, J., Ed., CRC Press, Boca Raton, FL, 2005.

[7] Castro, M., Costa, M., and Rowstron, A., *Peer-to-Peer Overlays: Structured, Unstructured, or Both?*, Technical Report MSR-TR-2004-73, Microsoft Research, Cambridge, U.K., 2004.

[8] Castro, M., Druschel, P., Hu, Y.C., and Rowstron, A., *Exploiting Network Proximity in Peer-to-Peer Overlay Networks*, Technical Report, http://free-pastry.rice.edu/PAST.

[9] Conti, M., Gregori, E., and Turi, G., A cross-layer optimization of Gnutella for mobile *ad hoc* networks, in *Proc. of the Sixth ACM Symp. on Mobile Ad Hoc Networking and Computing (MobiHoc 2005)*, Urbana-Champaign, IL, May, 2005.

[10] Conti, M., Maselli, G., Turi, G., and Giordano, S., Cross-layering in mobile *ad hoc* network design, *IEEE Computer*, 37(2), 48–51, 2004 (special issue on *ad hoc* networks).

[11] Delmastro, F., From Pastry to CrossROAD: cross-layer ring overlay for *ad hoc* networks, in *Proc. of IEEE PerCom 2005 Workshop on Mobile Peer-to-Peer*, Kauai Island, HI, March, 2005.

[12] MobileMAN Deliverable 13, http://cnd.iit.cnr.it/mobileMAN.

[13] Druschel, P. and Rowston, A., Pastry: scalable, distributed object location and routing for large-scale peer-to-peer systems, in *Proc. of IFIP/ACM Int. Conf. on Distributed Systems Platforms (Middleware)*, Heidelberg, Germany, November, 2001.

[14] FreePastry, www.cs.rice.edu/CS/Systems/Pastry/FreePastry.

[15] Hermann, K., MESHMdl: a middleware for self-organization in *ad hoc* networks, in *Proc. IEEE Workshop on Mobile and Distributed Computing (MDC 2003)/ICDCS'03*, Providence, RI, May, 2003.

[16] Kazaa, http://www.kazaa.com.

[17] Klinberg, T. and Manfredi, R., *Gnutella Protocol Specification v0.6*, Network Working Group, 2002 (http://rfc-gnutella.sourceforge.net/src/rfc-0_6-draft.html).

[18] Meier, R. and Cahill, V., STEAM: event-based middleware for wireless *ad hoc* networks, in *Proc. IEEE Int. Conf. on Distributed Computing Systems (ICDCS'02)*, Vienna, Austria, July, 2002.

[19] Mascolo, C., Capra, L., and Emmerich, W., Middleware for mobile computing (a survey), in *Advanced Lectures in Networking*, Gregori, E., Anastasi, G., and Basagni, S., Eds., Vol. 2497, Lecture Notes in Computer Science, Springer-Verlag, Heidelberg, 2002.

[20] Mascolo, C., Capra, L., Zachariadis, S., and Emmerich, W., XMIDDLE: a data-sharing middleware for mobile computing, *Wireless Pers. Commun.*, 21, 77–103, 2002.

[21] Murphy, A.L., Picco, G.P., and Roman, G.-C., Lime: a middleware for physical and logical mobility, in *Proc. IEEE Int. Conf. on Distributed Computing Systems (ICDCS'01)*, Phoenix, AZ, April, 2001, pp. 524–233.

[22] Pratt, I. and Crowcroft, G., Peer-to-peer systems: architectures and performance (tutorial session), in *Proc. of Networking 2002*, Pisa, Italy, May, 2002.

[23] Ratsanami, S., Francis, P., Handley, M., Karp, R., and Schenker, S., A scalable content-addressable network, in *Proc. ACM SIGCOMM'01*, San Diego, CA, August, 2001, pp. 161–172.

[24] Schollmeier, R., Gruber, I., and Finkenzeller, M., Routing in mobile *ad hoc* and peer-to-peer networks: a comparison, in *Proc. of Networking 2002*, Pisa, Italy, May, 2002.

[25] Stoica, I., Morris, R., Karger, D., Kaashoek, M.F., and Balakrishnan, H., Chord: a scalable peer-to-peer lookup service for internet applications, in *Proc. ACM SIGCOMM'01*, San Diego, CA, August, 2001, pp. 149–160.

[26] Tonnesen, A., *OLSR: Optimized Link State Routing Protocol*, Institute for Informatics at the University of Oslo, Norway (http://www.olsr.org).

Chapter 24

Supporting Continuous Services to Roaming Clients

Ashutosh Dutta, Henning Schulzrinne, and K. Daniel Wong

CONTENTS

Introduction

Lately, streaming real-time multimedia content over the Internet has been gaining momentum in the communications, entertainment, music, and interactive game industries. Real-time applications include interactive services such as IP telephony, multiplayer games, and streaming services such as broadcasting multimedia content, multiparty conferences, and collaborations. Multimedia streaming applications are far more demanding in terms of bandwidth, latency, and reliability than traditional TCP/IP-based applications and are thus ideal drivers for the next-generation Internet. In addition, they may require multicast support to provide flexibility and take care of bandwidth bottlenecks. As personal communication and ubiquitous access become more important, it is necessary to come up with flexible network technologies that can support multiple applications such as Mobile IP telephony, multimedia, and other streaming applications over a wireless IP network. To support multimedia applications for roaming users over the wireless Internet one has to consider several factors such as signaling, registration, configuration, quality of service, bandwidth management, mobility management, and authentication, among others. Thus, it is desirable to design a mobility management framework that can take care of location management, quality of service, and end-to-end security while providing personal, session, service, and terminal mobility features to the end users. In this chapter, we highlight some of the mobility management mechanisms and describe an application-layer framework that helps provide the desired roaming features over a heterogeneous access network. We also provide some experimental results obtained while prototyping this application-layer framework in a mobile multimedia testbed.

Wireless Internet Roaming Scenario

Wireless Internet roaming involves movement between different types of networks while the mobile device is subjected to cell, subnet, and domain mobility. Figure 24.1 shows an example of how a roaming user moves out of a personal area network (PAN) and then makes a transition to a local area network (LAN) and wide area network (WAN). During the roaming process, the mobile device transitions between different cells, subnets, and domains and traverses different types of access networks (e.g., Bluetooth®, 802.11, CDMA, WiMAX). It is important to identify the issues and requirements with regard to building an application-layer framework to support roaming users. This application-layer framework can use a set of standard Internet Engineering Task Force (IETF) protocols that can help support real-time and non-real-time multimedia applications on the mobile terminals of next-generation (3G/4G) wireless networks. We summarize below the essential requirements and issues that must be addressed with regard to supporting roaming users on the wireless Internet in a ubiquitous way. While designing any architecture, wireless service providers should keep these factors in mind. In general, a mobility management scheme for wireless IP networks should fulfill the following requirements:

- It must support personal, service session, and terminal mobility.
- It must support global roaming, independent of the underlying wireless technology (e.g., W-CDMA, CDMA2000, 802.11b, TDMA).
- It must support both real-time and non-real-time services such as mobile telephony and mobile Web access. To achieve this, the mobility management scheme should interact effectively with the quality of service (QoS) management and authentication, authorization, and accounting (AAA) schemes to verify the user's identities and rights, as well as to ensure that the QoS requirements and applications are satisfied and maintained as users roam.
- It must transparently support both Transmission Control Protocol (TCP)- and Real-Time Transport Protocol (RTP)/User Datagram Protocol (UDP)-based application. It should support the TCP as is, without requiring any changes to the TCP or TCP-based application.
- It must support multicast services efficiently as mobile stations and users move around.
- To be able to support wireless Internet telephony, it must address many important issues, such as registration, configuration, dynamic binding, and location management on a need basis.

Handoff is the most important factor for supporting wireless Internet telephony. Handoff, often referred to as *handover*, is a process that allows an established call/session to continue when a mobile station (MS) moves from one cell to another without interruptions in the call/session. This handoff process can be either hard or soft. In the hard handoff case, the mobile device receives and accepts only one radio signal from a radio channel or base station within a single cell; as the mobile device moves into a new cell, its signal is abruptly and rapidly handed over from its current cell (or base station) to the new one, within a few seconds. With soft handoff [1], the MS continues to receive and accept radio signals from base stations within its previous as well as its new cell for a limited period of time. The MS signal is also received at multiple base stations. To ensure the layer-two independence requirement of a mobility management scheme, a maximum acceptable handoff time (MAHT) of 2 to 3 seconds is required. In the end-to-end wireless IP paradigm, three logical levels of handoff procedure can be defined:

- *Cell handoff,* which allows an MS to move from one cell to another in a subnet within an administrative domain; one subnet may consist of multiple cells, in which case the IP address of the mobile host remains the same
- *Subnet handoff,* which allows an MS to move from a cell within a subnet to an adjacent cell within another subnet that belongs to the same administrative domain
- *Domain handoff,* which allows an MS to move from one subnet within an administrative domain to another in a different administrative domain

The handoff process is built upon the registration, configuration, dynamic address binding, and location management functions. The handoff process is transparent to users and should satisfy the following requirements so it can ensure the integrity, privacy, and confidentiality of the user's location and perform the necessary AAA process to verify users' identities. It should ensure the service mobility as the MS roams around by making sure that it maintains the QoS of the ongoing sessions through minimizing the loss of transient data during the handoff, as well as satisfying the delay requirements of real-time applications. Registration and configuration involves registering with the network and configuring the endpoint itself. The IETF developed the Dynamic Host Configuration Protocol (DHCP) [2] and Point-to-Point Protocol (PPP) [2], and Mobile IP [2] for both IPv4 and IPv6 networks provide several standard ways of registration for the end clients; however, several variations of DHCP (such as Dynamic Rapid Configuration Protocol, or DRCP [3]) and other extensions of DHCP [4–6] can take care of configuring IP addresses much more quickly while making efficient use of scarce wireless bandwidth.

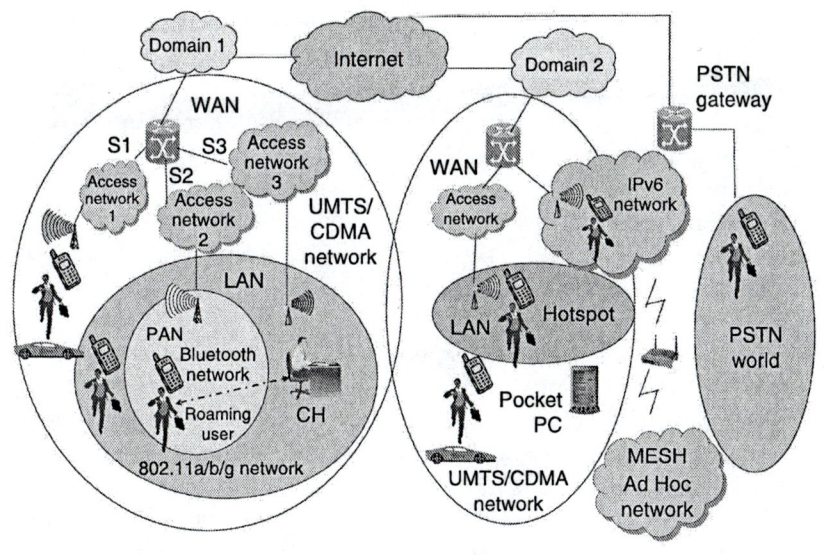

Figure 24.1   Wireless Internet roaming scenario.

Mobility Management Taxonomy

Supporting mobility on the Internet is primarily intended to allow a mobile device to move between different cells, subnets, and domains while maintaining an ongoing multimedia session independent of the point of attachment of the device. Several protocols and mechanisms have been developed to support intra-domain and inter-domain mobility on the Internet. Dutta et al. [7] provide a short survey of mobility management techniques currently available. Here, we provide a brief description of these mobility protocols. Current mobility management techniques can be implemented at several layers of the protocol stack, such as the networking layer, transport layer, and application layer. Depending on the type of movement, it can be considered either micro- and macromobility. Mobile IP (MIP) is a mechanism developed for the network layer to support mobility [6]; however Mobile IPv4 introduces network elements such as home agents and foreign agents and suffers from triangular routing and extra IP–IP encapsulation of 8 or 20 bytes, thus causing performance degradation. Mobile IPv4 usually works in two different modes, foreign agent mode and colocated mode. In colocated mode, a new address in the foreign network is obtained via services such as DHCP or its faster variants, such as DHCP with a rapid commit option [4], DRCP [3] in a LAN, or PPP in a wide area scenario.

Several proposals could help take care of the triangular routing problem by means of direct binding update, regional registration, and other smooth handoff techniques [8,9], but many of these solutions require kernel modification, making it difficult to deploy them. The Cellular IP approach [10] and the Handoff Aware Wireless Access Internet Infrastructure (HAWAII) [11] are network layer micromobility management protocols. These take care of the inefficiency of Mobile IP by supporting intra-domain mobility and host-based routing. Both of these approaches separate local and wide-area mobility (i.e., adopt a domain-based approach) and use Mobile IP for inter-domain (wide area) mobility.

Mobile IP with Location Registers (MIP-LR) is another network layer scheme developed to avoid encapsulation of packets [12] and to provide survivability in an *ad hoc* network such as military networks. It does so by replicating multiple location registers (LRs). Address management is carried out by DHCP servers, and home location registers (HLRs) provide the location updates to each corresponding host wishing to communicate with any mobile user in the beginning. MIP-LR can be implemented at both the network layer and application layer, thus avoiding kernel independence.

Telecommunications-Enhanced Mobile IP (TeleMIP) is an intra-domain mobility framework that uses two layers of scoping within a domain and is based on the Intra-Domain Mobility Management Protocol (IDMP) [13]. By specifying an intra-domain termination point — a mobility agent (MA) — this protocol helps to reduce the signaling updates due to movement within a domain and thus reduces the loss of transient traffic due to frequent handoffs within a domain. Mobile IPv6 [2] provides a network-layer mobility framework for IPv6. Because address autoconfiguration is a standard part of MIPv6, the mobile host (MH) will always obtain a care-of address (COA) that is routable to the foreign network; thus, it is not necessary to have a foreign agent (FA) in the MIPv6 framework. When the mobile node moves to a new foreign network, it acquires a temporary COA using stateless autoconfiguration [2] or DHCPv6 [2].

Among the few transport-layer mobility solutions, the TCP-Migrate approach [14] proposes a new set of migrating options for TCP that provide a pure end-system alternative to network-layer solutions. With this extension, established TCP connections can be suspended by a TCP peer and reactivated from another IP address without a third party (except for the involvement of dynamic Domain Name System updates). This approach, however, requires modifying the transport protocol at the end terminals. MSOCKS [15] is another transport-layer solution that introduces proxy in the middle of a network and is built on the top of the SOCKS protocol for firewall traversal. Upon movement of the mobile device and its address change, the intermediary proxy helps splice the TCP connection. The

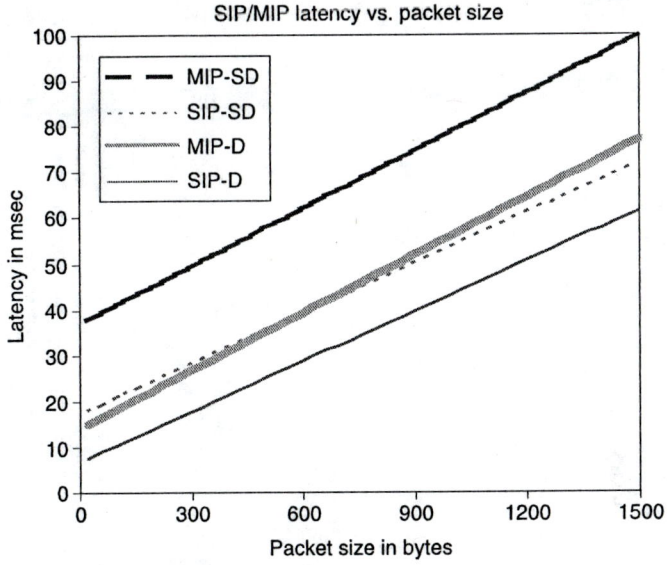

Figure 24.2   Quantitative comparison of SIP and MIP.

recently developed Stream Control Transmission Protocol (SCTP) [16] has a built-in add-IP feature that helps support continuity when the IP address of the mobile device changes.

Application-layer mobility uses the Session Initiation Protocol (SIP) as the signaling mechanism [17]. This mechanism does not depend on the home agent or foreign agent in the middle of the network nor does it require schanges in the end hosts; thus, it will provide easier deployment of mobility management solutions for the wireless Internet. Table 24.1 provides a qualitative comparison of some of the available mobility management protocols described above. Figure 24.2 provides a latency comparison between SIP-based and MIP-based terminal mobility during the subnet handoff. The curves illustrate the relative performance improvement of SIP over MIP for different packet sizes, as analyzed from both Network Simulator 2 (ns2)-based simulation and laboratory experiments. By using SIP for mobility management, one can expect to achieve 50 percent latency improvement in real-time (RTP/UDP) traffic, thus providing a reduction in latency from a baseline of ~27 msec to ~16 msec for large packets and a 35 percent utilization increase of 60 bytes/packet size compared with a baseline of 80 bytes/packet size with IP-in-IP encapsulation in Mobile IP.

Table 24.1 Qualitative Comparison of Mobility Management Protocols

Mobility Protocol	Intra-Domain Encapsulation	Inter-Domain Encapsulation	Changes to End Systems	Triangular Routing	Infrastructure Change	Fast Handoff	Layer	Mobility Type
Mobile IPv4	Yes	Yes	Yes	Yes	No	No	Network	Macro
MIPv6	No	No	Yes	No	No	Yes	Network	Macro
MIP-RO	Yes	Yes	Yes	No	No	No	Network	Macro
MIP-RR	Yes	Yes	No	Yes	No	Yes	Network	Macro
MIP, FA assisted	Yes	Yes	No	Yes	Yes	Yes	Network	Macro
TeleMIP	Yes	Yes	No	Yes	Yes	Yes	Network	Macro
MIP with LRs	No	No	Yes	No	No	Yes	Network	Macro
Cellular IP	No	Yes	No	Yes	No	Yes	Network	Micro
HAWAII	No	Yes	Yes	Yes	Yes	Yes	Network	Micro
MSOCKS	No	No	Yes	No	Yes	No	Transport	Macro
TCP Migrate	No	No	Yes	No	Yes	No	Transport	Macro
SIP	No	No	No	No	No	Yes	Application	Macro

Note: MIPv6, Mobile IP version 6; MIP-RO, Mobile IP route optimization; MIP-RR, Mobile IP with Regional Registration; FA, foreign agent; TeleMIP, Telecommunications-Enhanced Mobile IP; LRs, location registers; HAWAII, Handoff-Aware Wireless Access Internet Infrastructure; TCP, Transmission Control Protocol; SIP, Session Initiation Protocol.

Application-Layer Mobility Management Framework

Some of the functional components that are needed to support wireless Voice-over-IP (VoIP) in a SIP-based environment are presented below. These logical components are used to provide different atomic functions such as network detection, IP address configuration, registration, authentication, and accounting. We describe briefly some of these functional components.

Signaling

The application-layer mobility management framework is based on a vision of a fourth-generation (4G) network where the end systems are assumed to be IP endpoints, although a possible transition from a second-generation (2G) to a third-generation (3G) network can also be devised with interaction between IP and non-IP endpoints by implementing a soft switch. Because of the distributed nature of these networks, SIP [18] has been used to perform session management, including initiation and termination of a multimedia call, between clients. The SIP server and SIP user agent are part of the signaling architecture, although the SIP server functionality can easily be integrated into the softswitch for demonstrating IP-Public Switched Telephone Network (IP-PSTN) call features. SIP handles the signaling of multimedia calls among multiple parties. Numerous proposals have been made to extend SIP so it can take care of mobility for mid-session calls. SIP-based mobility techniques as defined in Dutta et al. [19,20] provide an alternative approach to Mobile IP for maintaining mid-session mobility and can also take care of pre-session mobility by means of unique Uniform Resource Identifier (URI) registration. The Host Mobility Management Protocol (HMMP) is a framework based on an extension of SIP that provides mechanisms to support a TCP-based application using SIP signaling [19].

Registration

Registration is a process by which a network is made aware of the existence and location of an MS and its associated user. When an MS becomes active or roams into a network for the first time, it registers with the network. This process involves such steps as the MS sending a registration request to the network and the network performing an AAA process and sending appropriate responses to the MS, as well as location management entities to ensure that the network is aware of the current location of the MS. Depending on the extent of registration, it can be categorized as complete, expedited, or partial registration. Complete registration usually takes more time than the expedited registration. Variants of AAA protocols [2] help take care of the security association between the mobile station and home AAA server when a client moves between the subnets within a domain. The home AAA server or an intermediate broker agent such as an SIP central point of contact is contacted when the user moves into a new domain for the first time to establish the credentials. It is important to complete the registration process in a timely manner during the handoff process.

Mobility Binding

Wireless Internet roaming must take care of several kinds of mobility, such as service mobility, personal mobility, session mobility, and terminal mobility. *Service mobility* refers to the end user's ability to maintain ongoing sessions and obtain services in a transparent manner regardless of the end user's point of attachment. *Personal mobility* refers to the ability of end users to originate and receive calls and access the subscribed network services on any terminal in any location in a transparent manner and the ability of the network to identify end users as they move across administrative domains. This is achieved by personal mobility features inherent in SIP; its URI scheme, registration mechanism, and dynamic DNS [2] are some of the main components of SIP that help provide personal mobility. *Mid-session mobility* is the same as *terminal mobility* and requires that smooth handoff is achieved as the mobile device moves among several heterogeneous networks. It requires maintaining proper binding between the mobile device and new points of attachment without affecting the end-to-end communication.

Binding allows continuous connectivity of TCP and UDP streams when the communicating end nodes are moving around. Binding between the mobile host and corresponding host when the mobile host is moving is typically taken care of by Mobile IP [2], although Mobile IP in its original form suffers from some drawbacks such as triangular routing and encapsulation. Variants of Mobile IP, such as IDMP, MIP-RO, and other micro-mobility management protocols (e.g., Cellular IP and HAWAII), have addressed the shortcomings associated with Mobile IP. MIPv6 also takes care of drawbacks associated with MIPv4. An application-layer technique using SIP-based mobility management provides another approach to handle personal and terminal mobility. Schulzrinne and Wedlund [17] and Dutta et al. [19,20] have proposed some extension of SIP whereby the mobility of multimedia calls (RTP/UDP-based stream) and the TCP application can be addressed without using underlying Mobile IP and any network components in the middle of the network.

Authentication, Authorization, and Accounting

An application-layer mobility management scheme can use Diameter [21] or RADIUS [22] as an AAA protocol running on Network Access Servers (NASs) and AAA servers to provide AAA-related services such as profile verification and charging. A new authentication protocol called PANA (acronym for Protocol for Carrying Authentication for Network Access) [23] is a lightweight protocol used between mobile hosts and NASs as a user front end of Diameter or RADIUS. PANA is implemented as an application-layer protocol to enable a flexible access control that works independent of any layer-two technology (e.g., 802.11, ADSL, cable), on both IPv4 and IPv6, and with any configuration protocol such as DHCP or PPP.

Security

Although it is important to maintain the mobility binding and optimize the handoff during movement of the mobile device between different subnets and networks, it is also important to encrypt the signaling and media and authenticate the user as it moves around. A multilayered security scheme helps protect both the signaling and data at various parts of the network, including the last-hop wireless access. SIP clients may use several authentication schemes (e.g., Digest) when registering with SIP servers. PANA provides a user-level authentication procedure between the mobile client and the first-hop access router and works in conjunction with the AAA server in the back end while providing access control. Packet-based encryption can be applied to provide over-the-air wireless security between the mobile client and edge router or end-to-end security between the communicating nodes. Internet Protocol Security (IPsec) has been considered a candidate for providing packet-based encryption over the last hop at the IP layer. Secured RTP (SRTP) [24] can be deployed to provide end-to-end encryption for real-time traffic (media) in a mobile environment. SRTP also helps avoid the problem of the IPsec tunnel setup associated with an IP address change. SIP signaling between the end points can be protected by using Secure/Multipurpose Internet Mail Extensions (S/MIME) [25].

SIP-Based Terminal Mobility

Primarily, terminal mobility can be categorized as pre-session and mid-session. Pre-session mobility generally does not contribute to the delay for media delivery associated with the ongoing session but may add delay to any new session. Mid-session terminal mobility provides a means of cell, subnet, and domain handoff while the session is in progress. Traditionally, terminal mobility is taken care of by network-layer mechanisms such as Mobile IP and its variants. The application-layer mobility management framework enhances SIP-based terminal mobility for RTP/UDP traffic to support roaming users on the wireless Internet.

According to the International Telecommunication Union (ITU)-T and ITU-E standards, real-time multimedia applications such as RTP/UDP-based traffic typically have a significant delay and loss budget (e.g., 500-msec round-trip delay; up to 3 percent packet loss) to support reliable communication. Thus, it is advisable to avoid the triangular routing and any kind of encapsulation mechanism that may contribute to performance degradation. As part of our proof-of-concept effort, we have implemented SIP-based terminal mobility in a comprehensive wireless multimedia testbed [26] and have added support for various types of mobility (e.g., cell, subnet, domain), heterogeneous access, fast handoff, QoS, and security. SIP signaling can support subnet and domain handoff, and cell handoff is completely taken care of by the link-layer mechanism. SIP-based terminal mobility will, however, benefit from a layer-two triggering mechanism using cross-layer optimization techniques. We augment SIP-based terminal mobility with a complete handoff process that is supported by a combination of the network detection, registration, configuration, dynamic address binding, security, and location management functions described earlier. Figure 24.3 shows several logical components associated with the multimedia testbed based on the application-layer mobility management framework. The figure shows the interaction of the logical components that provide such functionalities as configuration, registration, security, mobility binding, QoS, and AAA services during a mobile device's trajectory, covering cell, subnet, and domain mobility. Schulzrinne and Wedlund [17] describe how SIP can be used to provide terminal-mobility solutions during subnet handoff for real-time traffic such as RTP/UDP traffic. Dutta et al. [19,20] have proposed several mechanisms to address mobility for TCP traffic using SIP signaling.

Subnet handoff delay includes a layer-two association delay and a delay due to IP address acquisition and binding update. In the case of subnet movement, the typical time for acquiring an IP address will depend on the protocol being used. DHCPv4 in a LAN environment takes about 5 to 15 sec [27]. On the other hand, PPP takes about 15 sec to complete the negotiation. Address acquisition by means of MIP, DHCPv6, or stateless autoconfiguration takes less time. Table 24.2 provides a survey of IP address discovery methods under the Linux™ operating system; however, the time required for IP address discovery will depend on the operating system and processing power of the end hosts. Domain handoff involves movement between administrative domains and requires additional steps

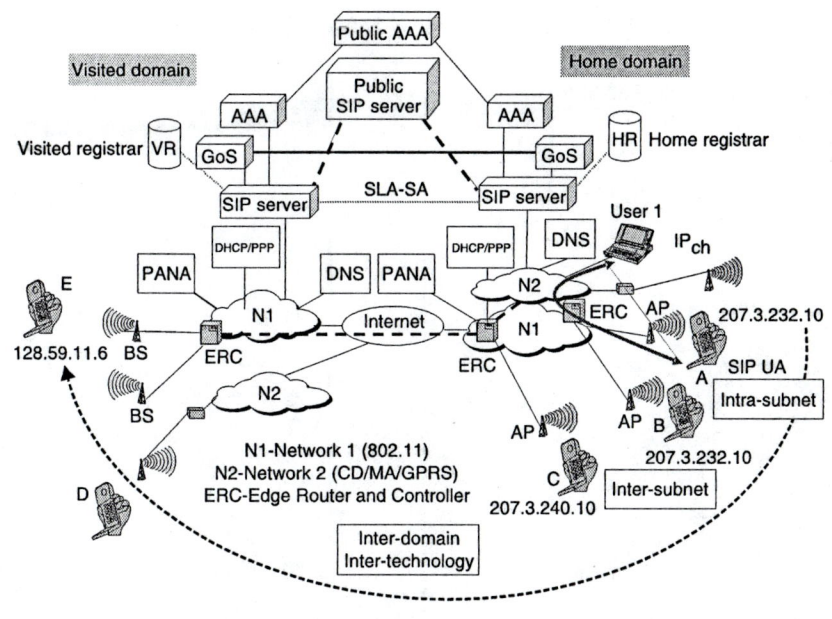

Figure 24.3 SIP-based wireless Internet roaming.

such as local authentication and profile verification, thus it will contribute to the additional delay. In the case of domain handoff, a complete registration takes place where there is an interaction between the AAA servers in each domain.

The SIP URI scheme, registration mechanism, and dynamic DNS have been implemented in a testbed to determine the pre-session mobility and location management features. The registration process includes sending a registration request from the MS to the SIP server with the new IP address after the mobile device has moved. But, before a registration is successful, it will be subjected to local access authentication based on PANA and will be subjected to AAA verification by the network. A successful registration updates the current location of the mobile device for any new upcoming session.

Inter-Domain Secured Terminal Mobility

In a wireless Internet roaming scenario, it is important to preserve the continuity of the session while maintaining the security association between the communicating entities. Very often when a mobile device moves among different heterogeneous networks, it is subjected to different layer-two security mechanisms; thus, it is desirable to have a multilayer

Table 24.2 Survey of IP Address Acquisition Time

				Acquisition Method				
	DHCP/ ARP	DHCP (w/o)	DHCPv6 (Stateless)	DHCP v6 (Stateful)	PPP	FA COA	Auto IP	Static	Proactive IP
Time (Δ2)	4–5 sec	300–400 msec	160 msec	500 msec	7–8 sec	1–2 sec	4–5 sec	100 msec	L2 delay

Note: DHCP/ARP, Dynamic Host Configuration Protocol (DHCP)/Address Resolution Protocol (ARP); PPP, Point-to-Point Protocol; FA COA, foreign agent care-of address; IP, Internet Protocol.

security mechanism to make sure that both signaling and media are protected throughout the process and interruptions due to mobility and new security associations are minimized. A multilayer security framework based on an SIP-centric architecture has been designed and implemented that provides an end-to-end secured mobile multimedia communication. It provides access control to the network using PANA, profile verification using Diameter, and last-hop over the air protection using packet-based encryption such as IPsec. End-to-end security for multimedia traffic (e.g., audio, video) is provided by SRTP. SRTP avoids the periodic setting up of end-to-end IPsec tunnels between the endpoints during movement of the mobile device. The SRTP key is distributed securely using an INVITE exchange, and S/MIME takes care of securing the MIME bodies of SIP signaling.

Mobile multimedia communication supporting inter-domain mobility has been emulated by creating two different AAA domains. According to the SIP–AAA model defined in the testbed, when the SIP server receives an SIP register message from the mobile host, it consults with the home AAA server for authentication and authorization by using Diameter as the back-end protocol. Basilier et al. [28] describe the entire scenario of SIP–AAA interaction in detail. We also use PANA for distributing Internet Key Exchange (IKE) credentials to an authorized host. The credentials are then used for establishing an IPsec tunnel between a host and the first-hop access router that provides a secure communication channel in the access network. The dynamic distribution of the IKE credentials enables hosts to roam among different administrative domains as it is not necessary for a host to preconfigure the credentials.

Dutta et al. [26] have provided their experimental results involving inter-domain secure mobility. Domain handoff delay is composed of several components, such as delay due to 802.11b channel change, subnet and domain discovery, IP address acquisition, local authentication by means of PANA, profile verification using AAA Diameter, and delay due to binding updates such as SIP reINVITE or MIP registration. It is note-worthy to mention that the delay parameters strongly depend on the media used, number of routers in the path, background traffic, number of hops, authentication methods, and processing speed of the correspon-dent and mobile hosts. Timing associated with SIP signaling depends on the processing power of the end hosts and specific SIP stack implemen-tation (e.g., JAVA, C, Tcl/Tk). Table 24.3 shows the timing associated with the execution of different protocols during subnet and domain handoff. As expected, inter-domain handoff requires more time than inter-subnet handoff because of the additional steps associated with profile verification in the visited domain.

Table 24.3   Timing for Inter-Domain, Inter-Subnet Handoff

	Operation			
	DRCP	PANA	SIP	Media RTP
Subnet handoff (msec)	79	2	228	1490
Domain handoff (msec)	81	45	289	1656

Note: DRCP, Dynamic Registration and Configuration Protocol; PANA, Protocol for Carrying Authentication for Network Access; SIP, Session Initiation Protocol; RTP, Real-Time Transport Protocol.

Figure 24.4 shows the sequence of protocols including RTP packets received on the mobile device during inter-domain handoff using an SIP-based mobility scheme. These protocol sequences represent the execution of each operation as the mobile device makes subnet and domain hand-offs. A comparison of inter-domain secure mobility while using MIP as the binding protocol shows that the mobile device loses fewer packets when SIP is used as the binding protocol. In the case of MIP with IPsec, additional time is spent due to the registration of the mobile device with the home agent during the subnet change.

SIP-Based Mobility Across Heterogeneous Networks

Secure and seamless universal roaming requires mobility support that involves movement between heterogeneous access networks. We describe experiments involving both Mobile IP and SIP-based mobility schemes to provide secure and seamless universal roaming involving heterogeneous access, such as 802.11 and CDMA1XRTT. Dutta et al. [29] provide an overview of how SIP-based mobility management can work over hetero-geneous networks (e.g., 802.11b, W-CDMA, GPRS). They have also looked into various key issues such as network detection, active interface iden-tification, registration, retransmission of signaling in the event of rapid handoff, SIP support with network address translations (NATs), session continuity, fast handoff, and asymmetry of data delivery. In an experiment involving mobility across heterogeneous access networks, 3G connectivity was provided by Verizon Wireless's CDMA1XRTT access network. During the experiment, it was observed that the time taken to obtain an IP address over PPP using CDMA1XRTT access over a WAN environment takes more time (about 15 sec) than obtaining the address using DHCP without the ARP option (less than 1 sec) in a LAN environment. It was found that the average throughput is about 8 kbps over a cellular digital packet data

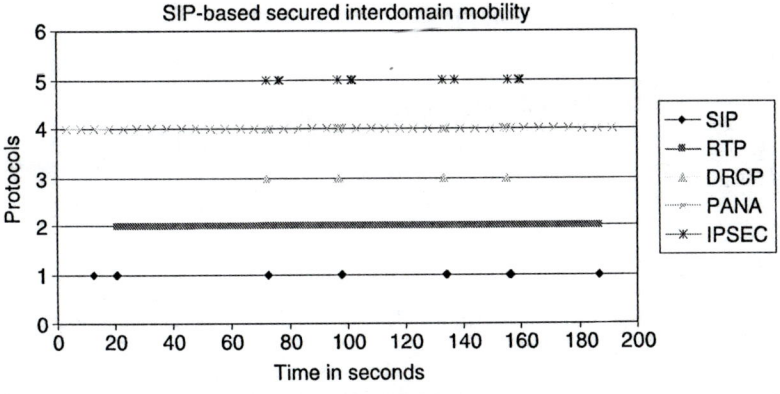

Figure 24.4 Secured inter-domain mobility protocol sequence.

(CDPD) network and about 60 kbps over the CDMA1XRTT network. But, newer technologies such as CDMA1XEVDO can provide data rates up to 384 kbps.

To reduce the packet loss due to a delay in IP address acquisition, we have implemented a make-before-break algorithm that sets up a PPP connection while the mobile device is still communicating with 802.11 in LAN. A policy-based approach is used to define the trigger that determines the active interface that the client would use to communicate with the corresponding host. This handover policy can be based on the link condition, QoS of the received traffic, or other server-based advertisement. In addition, we have also designed and tested a secured multiple-interface mobility management scheme where a mobile device with a dual interface moves among an enterprise network (e.g., 802.11), cellular network (e.g., CDMA, GPRS), and hotspot (e.g., 802.11). This scheme provides secured seamless roaming support without the need to tear down IPsec tunnels during each subnet move. Dutta et al. [30] described the details of the implementation using both MIP- and SIP-based approaches. Both constant bit rate (CBR) traffic (audio) and variable bit rate (VBR) traffic (video) have been tested, and we have analyzed the packet loss, delay, and inter-packet gap during the handoff. Figure 24.5 shows the results of an experiment involving secured mobility across heterogeneous networks. The low gradient in the graph indicates the low speed within a cellular network. The mobile device received few out-of-order packets during its movement from the cellular network to the 802.11 network because of the transient packets in the path that arrived in the WAN interface at a later point. Dutta et al. [30] also have described a mechanism where SIP-based mobility management can be used with the recently proposed Mobile Internet Key Exchange Protocol (MOBIKE) [31] for real-time traffic.

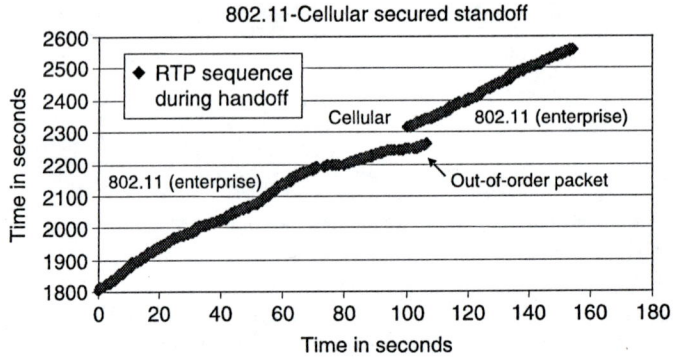

Figure 24.5 RTP sequence during heterogeneous handoff.

SIP Mobility over IPv6

Next-generation networks based on 3GPP and 3GPP2 standards have adopted IPv6; thus, it is quite important to consider mobility management for IPv6 networks. SIP-based terminal mobility has a lot of similarities with Mobile IPv6, as both of these mechanisms provide direct binding updates to the communicating hosts without depending much on elements within the network, such as home agents or foreign agents. We have tested mobility binding for wireless telephony over an IPv6 network using both Mobile IPv6 and SIP-based terminal mobility. We used Linux kernel 2.4.9 with a patch from the USAGI projects and MIPL Mobile IPv6 for the experimental testbed. Several experiments were carried out for real-time voice traffic to analyze the effect of Duplicate Address Detection (DAD) in the disruption of SIP-based multimedia calls. An experimental handoff analysis of SIP mobility with IPv6 and MIPv6 involving signaling and media redirection is shown in Table 24.4. The SIP mobility timing was measured for both DAD and no-DAD cases, whereas only the no-DAD

Table 24.4 Effect of Duplicate Address Detection (DAD) on SIP and MIPv6

	Signaling (msec)			Media (msec)		
Handoff Case	SIP (DAD)	SIP (No-DAD)	MIPv6 (No-DAD)	SIP (DAD)	SIP (NDAD)	MIPv6 (No-DAD)
H12	3829	171.4	1.5	3854	420.8	21.1
H23	3932	161.6	2.0	4187.7	418.6	30.3

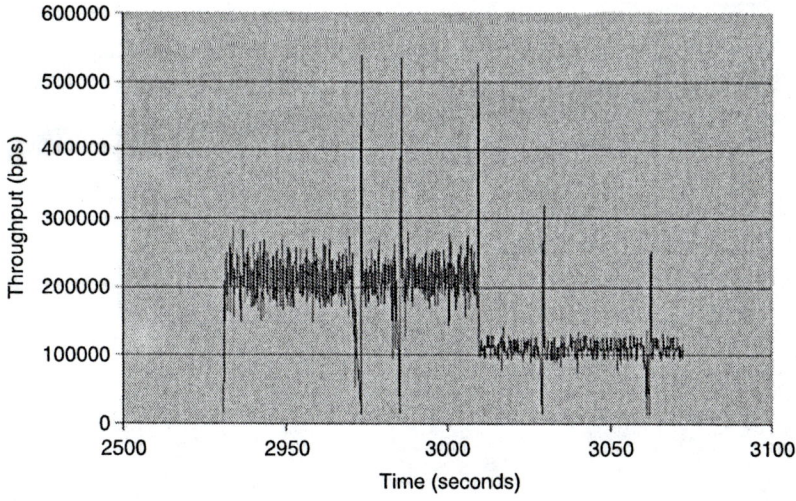

Figure 24.6 SIP mobility with QoS.

case is shown for Mobile IPv6. The aggressive router selection (ARS) method was also added to the no-DAD case. The ARS procedure forces a mobile device to bind to the new router quickly enough to avoid a neighbor-unreachability detection [2].

Details of the experimental comparison between MIPv6 and SIP-based mobility and IPv6 can be found in Nakajima et al. [32]. To summarize, however, we found that by eliminating DAD and adopting an aggressive router selection process we were able to minimize the signaling delays to 200 msec and media delays to less than 500 msec.

SIP-Based Mobility with Quality of Service

To make sure that the SIP-based media sessions, such as audio and video streaming traffic, maintain the same level of quality of service during the movement of a mobile device between two subnets, we integrated SIP-based terminal mobility with Dynamic SLS Negotiation Protocols (DSNP) [33]. DSNP uses a combination of Integrated Service and Diffserv techniques to make reservations in the target access router before movement of the mobile device into the target network. Dutta et al. [26] have described the details of how the quality of service is achieved as the mobile device moves between subnets. Figure 24.6 illustrates how the QoS for multimedia traffic is handled during the subnet handoff. Results show the throughput at the mobile during its movement between subnets. Slight fluctuation in bandwidth is observed as soon as the mobile device switches to a new subnet.

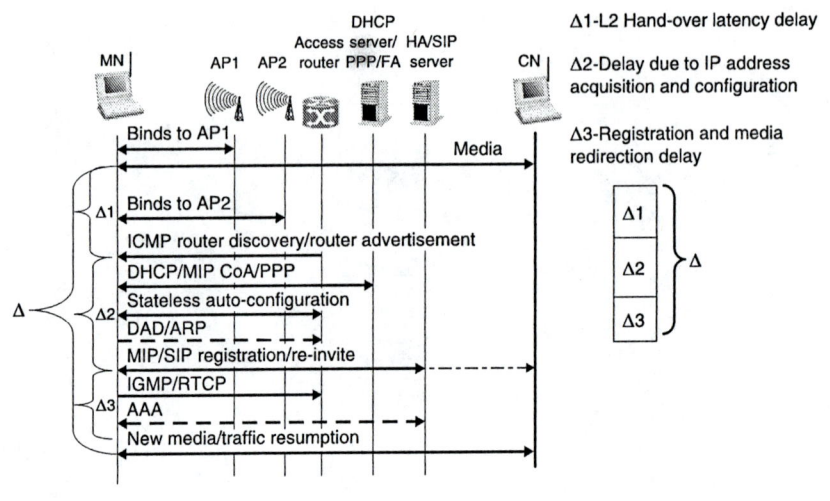

Figure 24.7   Handoff latency factors.

SIP-Based Fast-Handoff Mechanism

Several factors at different layers contribute to handoff delays resulting in transient data loss during mid-session mobility. Figure 24.7 shows the latency factors associated with different layers during a handoff. These factors include layer-two access point handoff, as well as layer-three triggering time, time required to obtain an IP address (using methods such as DHCP, PPP, or MIP COA), and time required for media redirection. The times are denoted by $\Delta 1$, $\Delta 2$, and $\Delta 3$, respectively. Fast-handoff techniques can be deployed at different layers to help reduce the transient data loss due to the delay-associated latency with the macro handover. The IETF is currently considering several alternative approaches for supporting fast handoff within the Mobile IPv4 and MIPv6 context. Layer-three-based, intra-domain mobility management solutions, such as the Hierarchical Mobile Internet Protocol (HMIP) [8,33,34], help reduce the transient data loss when a mobile host moves between subnets within a domain. Similar fast-handoff mechanisms have also been proposed for Mobile IPv6. Soliman et al. [36] introduced an agent called Mobility Anchor Point (MAP) to localize the intra-domain mobility management. Cellular IP [10] and HAWAII [11] provide mechanisms to make handoffs faster in an intra-domain scenario. Park et al. [5] and Han et al. [6] described some of the techniques required to carry out DAD optimization for IPv6 clients.

Vakil et al. [37] developed a virtual soft-handoff approach for code-division multiple access (CDMA)-based wireless IP networks. It takes into account the fact that both the access points receive the stream during

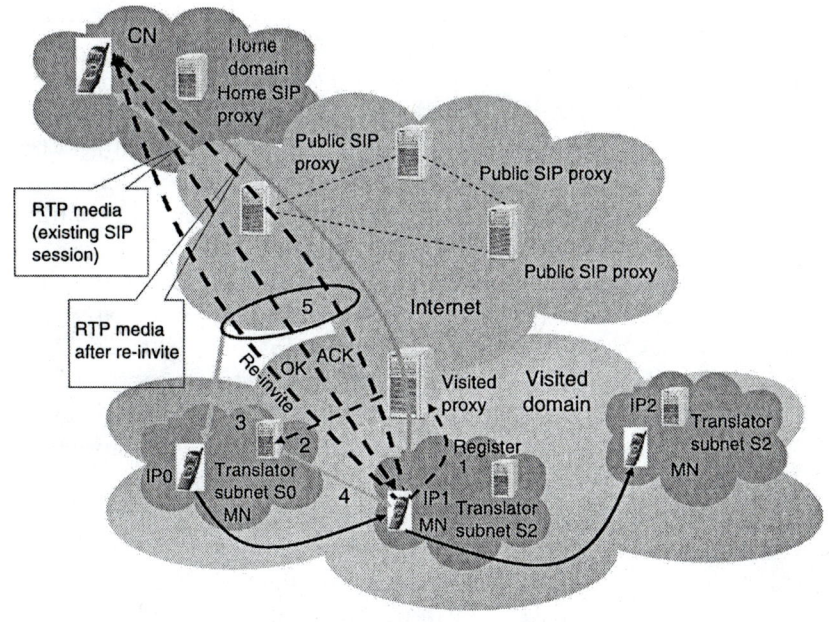

Figure 24.8   SIP-based fast-handoff scenario.

movement of the mobile device; however, this scheme does not provide a generalized solution suitable for other type of access network such as 802.11. It is important, then, to design a generic framework that can provide fast handoff for both real-time and non-real-time sessions. We describe below different SIP-based fast-handoff mechanisms supporting secured inter-domain mobility (see Figure 24.8). These mechanisms help reduce the delay associated with media redirection when the mobile device is far away from the corresponding host and is subjected to frequent handoff. We also describe the benefits of having a proactive configuration and preauthentication mechanism to help speed up the handover by acquiring an IP address ahead of time.

When the corresponding host is very far from the mobile device, transient packets would be lost because of the large delays associated with the SIP reINVITE; thus, it is necessary to retrieve the transient packets. Each visited domain may consist of several subnets. Every move to a new subnet causes the mobile host to send a reINVITE to the corresponding host containing its new COA. If the reINVITE request gets delayed due to path length or congestion, transient media packets will continue to be directed to the old address. We assume that the visited network has an outbound proxy. We enhance this proxy with the ability to temporarily register visitors. The visitor obtains a temporary, random identity from the visited network and uses it as its address-of-record to register with the registrar in the visited network. The mobile host informs the home registrar of this temporary address. It then only updates that registration with its current local IP address. This speeds up registration but does not address the delayed binding update issue raised by the reINVITE feature of SIP. We have considered several ways to achieve fast handoff using SIP — namely, using a SIP registrar and RTP translator or NATs or using the outbound proxy and a back-to-back user agent (B2BUA) as a mobility agent. In-transit packets can be redirected to a unicast or multicast address based on the movement pattern of the mobiles and usage scenario. These proposed methods help alleviate transient data loss related to continuous handoffs within a domain, thus minimizing the delay contributed by $\Delta 3$. Dutta et al. [38] provide more details about the fast-handoff mechanism.

In our first approach, each subnet within a domain is equipped with an RTP translator [2] that provides application-layer forwarding of RTP packets for a given address and UDP port to a given network destination. The visited-network registrar described earlier receives the registration updates from the mobile host that has just moved and immediately sends a request to the RTP translator in the network that the mobile host just left. The request causes the RTP translator to bind to the old IP address used by the mobile host and forward any incoming packets to the new mobile host address. After a set interval or after no media packets have been received by the RTP translator, the RTP translator relinquishes this old address and removes the forwarding table entry, assuming that the reINVITE has reached the corresponding host.

The second approach uses an SIP outbound proxy. SIP requests typically traverse a SIP proxy in the visited network, the outbound proxy. This outbound proxy can also support fast handoff by using the data in the mobile host-to-corresponding host reINVITE to configure the RTP translator or NAT. The advantage of this approach is that the outbound proxy usually has access to the Session Description Protocol (SDP) information containing the mobile host media address and port, thus simplifying the configuration of the translator or NATs. On the other hand, this outbound proxy has to remember the INVITE information for an unbounded amount of time and become call stateful, as it requires the old information when a new reINVITE is issued by the mobile host.

Another way of providing fast-handoff is by using a back-to-back SIP user agent. A B2BUA consists of two SIP user agents; one user agent receives a SIP request, possibly transforms it, and then has the other part of the B2BUA reissue the request. A B2BUA in each domain has to be addressed by the mobile host in the visited domain. The B2BUA issues a new request to the corresponding host containing its own address as the media destination and then forwards the packets, via RTP translation

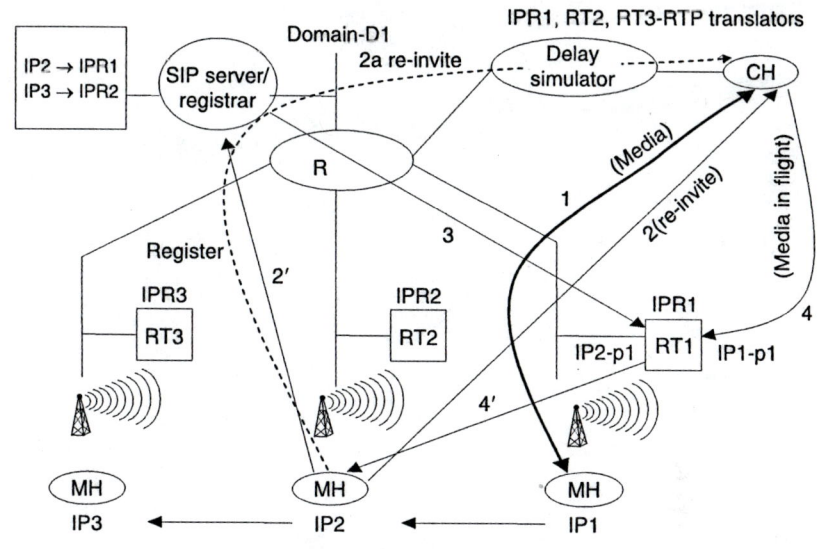

Figure 24.9 SIP-based fast-handoff method.

or NAT, to the mobile host. Locally scoped multicasts may also help avoid packet losses if the mobile host can predict that it is about to move to a new subnet shortly. In that case, it informs the visited registrar or B2BUA of a temporary multicast address as its contact or media address. When the mobile host has arrived in its new subnet, it updates the registrar or B2BUA with its new unicast address, while continuing to listen to multicast addresses. The use of scoped multicast is only effective if the mobile host can quickly acquire a multicast address and there is an inherent multicast infrastructure.

In Figure 24.9, RT1, RT2, and RT3 are RTP translators in the respective subnets. These RTP translators forward the traffic associated with one IP address/port number to another IP address/port number. RTP translators in each of these subnets intercept the traffic meant for the mobile host and send it to the new address of the mobile host after capturing it. This message can be sent via SIP Common Gateway Interface (CGI). In our tests, the reINVITE signal was delayed to simulate network congestion or distance between the corresponding host and mobile host. The Videoconferencing Tool (VIC) and Robust Audio Tools (RATs) were used to measure the performance of the audio and video streaming traffic, respectively.

Two methods such as rtptrans and NAT-based iptables were used to direct the transient traffic from the previous subnet to the new one. In some test runs, we delayed the reINVITE signals by 100 msec, 200 msec, 500 msec, 1 sec, 2 sec, and 3 sec to study how RTP translators improve

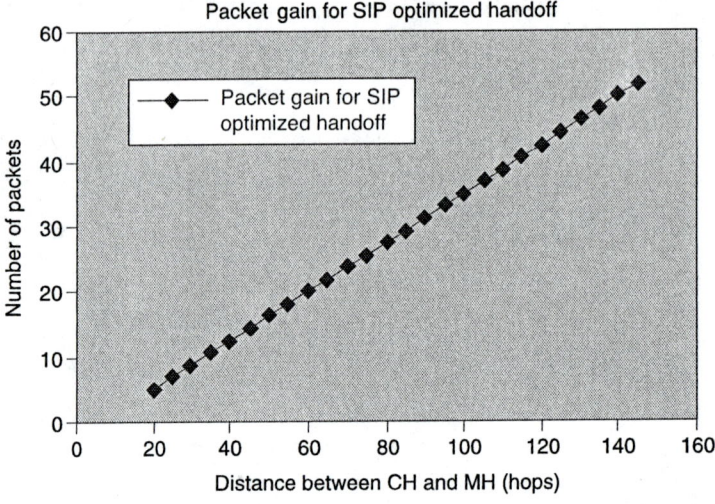

Figure 24.10   Packet gain for optimized SIP handoff.

the delivery of RTP packets and enhance the smooth handoff mechanism during movement of a mobile device. We found that the packet-forwarding delay due to redirection at the registrar was less than 1 msec when the iptables-based NAT approach was used, whereas the rtptrans approach added 4 msec of delay. Figure 24.10 shows the effect of a SIP-optimized handoff compared to a regular SIP-based handoff. SIP-based fast handoff looks promising when the distance between the corresponding host and mobile host becomes greater.

Proactive Handoff Using Preconfiguration and Preauthentication

The fast-handoff approaches described earlier help reduce the effect of delay associated with media redirection after the mobile device has been configured with the new IP address. A mobile device can also benefit from faster handoff if the IP address configuration and authentication mechanisms are carried out in an efficient manner. We have looked into expediting the handoff process by considering proactive IP address acquisition methodologies and preauthenticating a mobile device while it is still in the previous network. This specific approach takes advantage of network discovery and selection methods, where the client discovers the neighboring elements (e.g., routers, DHCP servers, SIP servers) and communicates with these entities before it actually moves into these networks. This will help expedite the authentication and IP address acquisition part of the handoff process that usually takes place after the mobile device has moved into the new network. This specific method actually reduces the delay due to IP address acquisition and binding update. Details of proactive handoff schemes and associated results are described in Dutta et al. [39].

SIP-Based Simultaneous Mobility

In a roaming scenario, it is highly likely that both the communicating hosts may be moving at the same time. In a purely *ad hoc* environment, mobility is addressed by the proposed IETF *ad hoc* routing protocols, such as Dynamic Source Routing (DSR) and *Ad Hoc* On-Demand Distance Vector (AODV) routing; thus, the *ad hoc* routing protocols have the inherent ability to take care of the simultaneous mobility problem. But, in an infrastructure environment, *ad hoc* routing protocols are not used and a mobile user may be affected if the corresponding user is also moving at the same time. Both signaling and media transport will be affected because of simultaneous movement of both of the communicating parties. MIP does not have a problem with simultaneous mobility, as corresponding hosts are unaware of the mobility of mobile hosts. The mobile host's home agent functions as an anchor point for the mobile host. No matter where the mobile host moves, packets intended for the mobile host always go first to its home network for interception and tunneling by its home agent; however, mobility protocols that use direct binding updates between the mobile and corresponding hosts are prone to simultaneous mobility problems. MIP-RO-, MIPv6-, and SIP-based mobility are mobility protocols that fall into this category.

Wong et al. discuss the problems associated with simultaneous mobility for SIP and MIP-LR-based mobility [40] and for Mobile IPv6 in [41]. We highlight here certain basic problems associated with simultaneous mobility of the end users and provide some of the solutions associated with an SIP-based mobility scheme. Communicating hosts may be subjected to simultaneous mobility problems both during pre-session and mid-session mobility. During pre-session mobility, the communicating hosts may have trouble establishing a session because the signal may get lost, and during mid-session mobility ongoing data may be lost due to the loss of binding updates. If both the clients A and B are subjected to handoff pre-session or mid-session, the vulnerability interval for the simultaneous mobility will be determined in the following manner: A mobile host loses contact with the previous point of attachment at a specific time (e) and cannot be reached with the old IP address. It also takes a certain amount of time (say, γ) before it configures its interface and gets connected to the new network. Depending on the path it traverses, the binding update may take

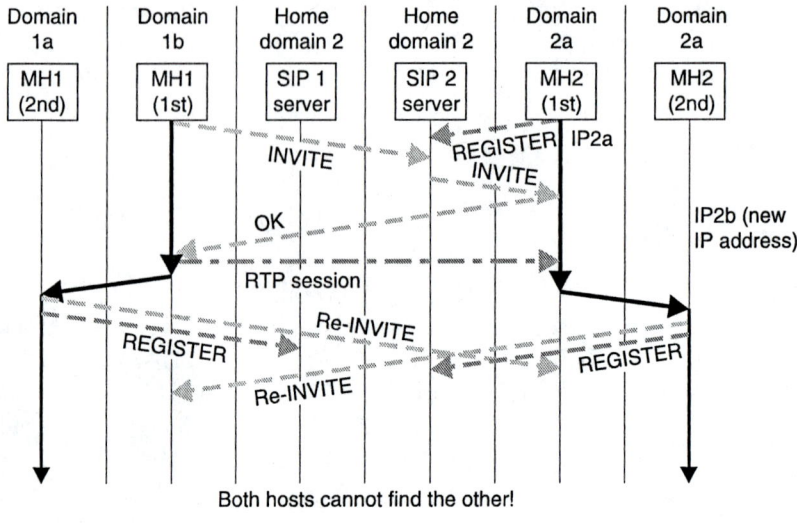

Figure 24.11   SIP-based call flow scenario.

a different amount of time in either direction. If it takes α amount of time for the binding update to reach from A to B, and β amount of time to reach from B to A, then the vulnerability interval is $(\alpha + \beta)$. If the inter-handoff interval has a Poisson distribution with a mean value of λ, then the probability that one of the mobiles is subjected to simultaneous mobility problem can be denoted as:

$$ P_0 \approx \frac{E[\alpha + \beta]}{\lambda} $$

and the probability that one out of N handoffs is subjected to simultaneous mobility problem is denoted as:

$$ P_N = 1 - \left(1 - P_0\right)^N $$

From the equations above it is evident that the probability of a simultaneous mobility problem is greater when the average handoff rate is greater (i.e., the inter-handoff arrival time is less) and the latency associated with the binding update is also greater.

Figure 24.11 shows a failed-call scenario for SIP-based mid-session simultaneous mobility. When the client is subjected to a simultaneous

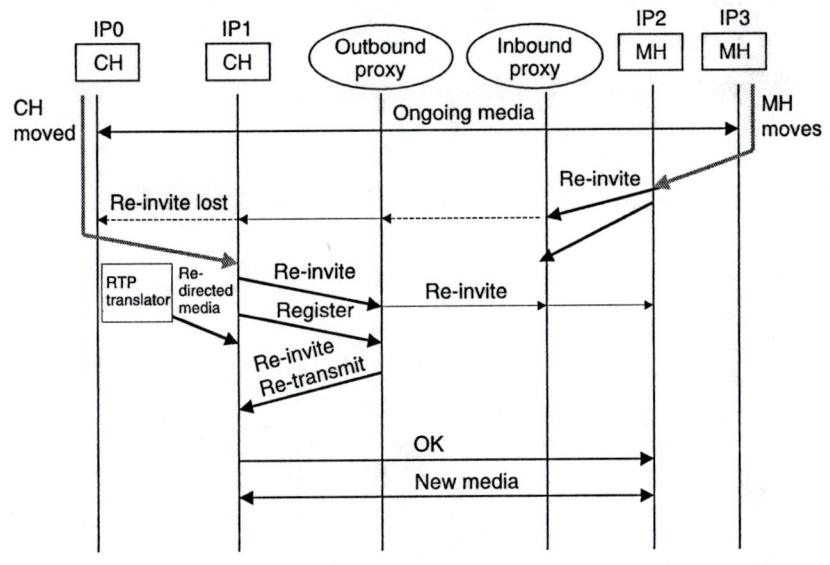

Figure 24.12 Successful mid-session mobility flow with server-assisted retransmission.

mobility problem during SIP-based mid-session mobility, the existing RTP media is also affected. As a result, the endpoints do not get to communicate properly.

Similarly, SIP-based pre-session mobility can also be affected such that the initial session may never get set up because the initial signaling messages never reach the endpoints. Mobile-originated retransmission and the use of a forwarding mechanism from the previous networks are two proposed ways for handling simultaneous mobility associated with SIP (see Wong et al. [40] for alternative solutions). SIP has an inbuilt retransmission mechanism where signaling messages are retransmitted after a timeout if the acknowledgment is not received. These retransmissions can be originated from the inbound servers closer to the mobile device. One problem with the timer-based retransmissions is that significant latency could be added to the handoff when messages are lost, such as when simultaneous mobility occurs. The forwarding agent-based solution proposes forwarding agents, similar to rtptrans as described earlier. These agents capture the signals and forward them to the new location via the proxy in the current network. Figure 24.12 and Figure 24.13 illustrate these two mechanisms for handling simultaneous mobility associated with SIP mobility.

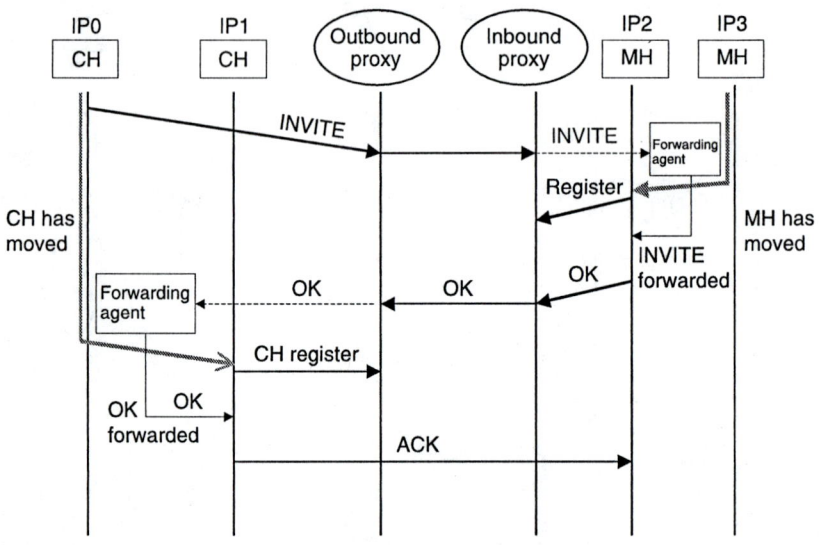

Figure 24.13 Successful completion of session initiation signaling.

Integrated Mobility Management

Each of the mobility management schemes described earlier has certain shortcomings associated with it. Some mobility management schemes are suitable for real-time communication, some are suitable for inter-domain mobility, and some are more suited for a micromobility environment; thus, it may be desirable to design an integrated mobility management scheme that is policy driven. We have designed a multilayer integrated mobility management architecture and have prototyped several components in the testbed. Our architecture consists of three mobility protocols — SIP, MIP-LR, and Micromobility Protocol (MMP; a variation of Cellular IP) — that work in conjunction with a policy manager and provide the desired functionality based on the type of application and mode of movement. SIP-based mobility management is used for real-time communication, and application-layer MIP-LR is used for non-real-time traffic during movement of a node between two different domains, while MMP takes care of the movement within a domain. Figure 24.14 shows a testbed illustrating how these three mobility protocols interact with each other in an integrated fashion.

We evaluated the performance of our integrated mobility management approach under a simulated *ad hoc* environment while the mobility protocols interact with the dynamic DNS, DRCP/DCDP protocol suite. Refer to Dutta et al. [42] for further details about how the integrated mobility management scheme was realized in our testbed. In this approach, when

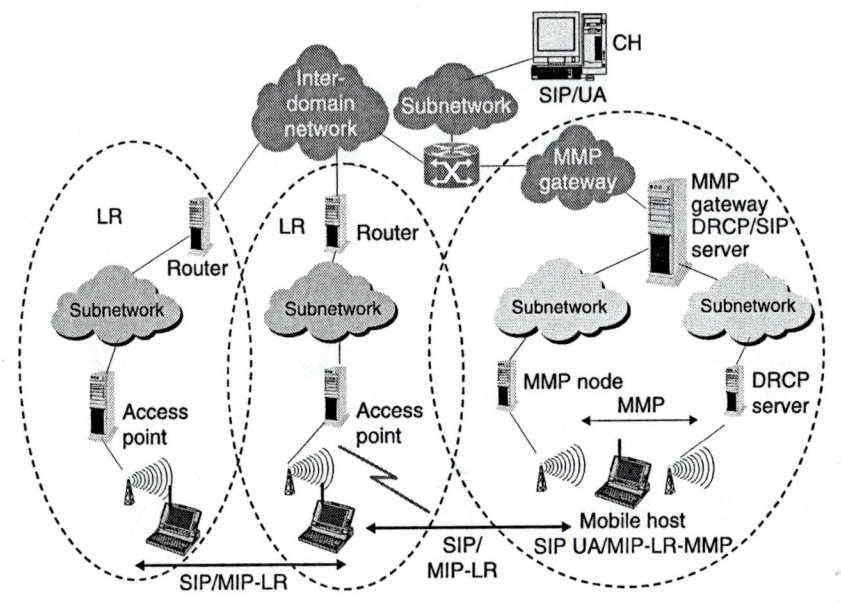

Figure 24.14   Integrated mobility management.

the mobile node moves to a new domain for the first time, it obtains a new IP address and registers with the SIP server or Visiting Location Register (VLR). This registration gets propagated to other SIP servers or home location registers (HLRs) spread across the network. The corresponding host obtains the new IP address of the mobile host from the SIP redirect server or HLRs. To support real-time communication (RTP/UDP) traffic during movement of the mobile device between the domains, a reINVITE is sent to the corresponding host to keep the session active; similarly, an MIP-LR update message is sent to the corresponding host for the TCP/IP traffic. But, for any subsequent move within the new domain, reINVITE or update messages are not sent, because MMP takes care of routing the packets properly within that domain. Figure 24.15 illustrates the integrated mobility management scenario and interaction of all three protocols.

Mobile Content Distribution over Multicast

Content distribution to end users may include both mobile and non-mobile clients over wired and wireless media. Because multimedia streaming applications are bandwidth intensive, it is desirable to add multicast support to the framework as well. Proposed IETF protocols that provide native IP multicast routing over a wide area network include Protocol

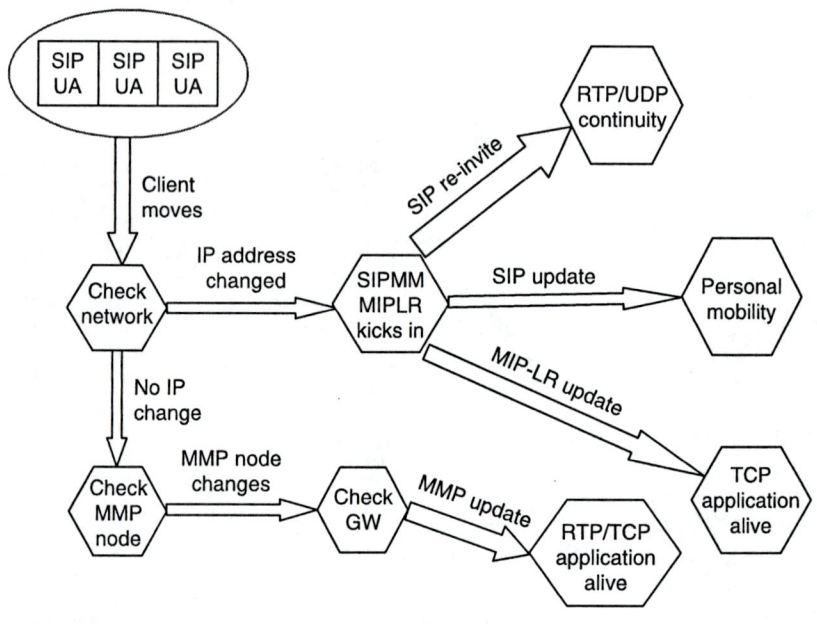

Figure 24.15 Interaction of several components within integrated mobility management.

Independent Multicast (PIM), Multicast Open Shortest Path First (MOSPF), Distance Vector Multicast Routing Protocol (DVMRP), Core Base Tree (CBT), and Border Gateway Multicast Protocol (BGMP). Explicit Multicast and Source-Specific Multicast (SSM) are some of the recently developed protocols ideal for broadcast application. Explicit multicast is useful for maintaining many multicast groups when the membership in each group is small, unlike traditional multicast, which supports a limited number of large multicast sessions. SSM is ideal for Internet broadcast applications because a specific content can be identified as the source address (S) and the multicast group address (G). Recently developed techniques such as the UDP Multicast Tunneling Protocol (UMTP) [43] and Agent Message Transport (AMT) [44] provide mechanisms to support multicast in non-multicast-enabled networks.

Mobility Support for Multicast

Multicasting to mobile users has primarily been divided into two categories: home-subscription-based solutions and remote-subscription-based solutions. Romdhani et al. [45] discussed some of the challenges associated with multicasting to mobile users and have provided a comparison of available

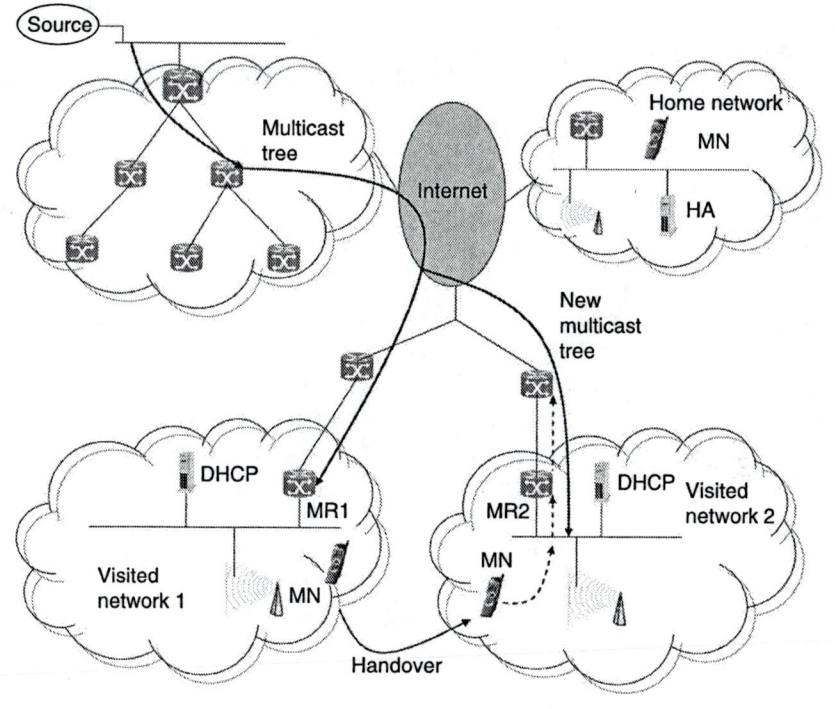

Figure 24.16   General scenario for multicast mobility.

protocols that can support mobile multicast using both the approaches. Xylomenos and Polyzos [46], Varshney and Chatterjee [47], and Acharya et al. [48] describe many of the architectural issues associated with mobile hosts in a multicast environment. To provide fast handoff of the multicast streams to a mobile client moving between subnets, Wu [49] proposed handover with preregistration by deploying mobility support agents. McAuley et al. [50] proposed a multicast proxy approach to reduce the transient data loss of multicast communication during handoff; in this situation, the proxy's clients do not themselves directly participate in the multicast tree. Figure 24.16 shows a very basic scenario for multicast mobility.

The MIP-based bidirectional tunneling solution puts the multicasting burden on the home agent (HA) by creating tunnels between the HA and the mobile device using the Internet Group Management Protocol (IGMP); however, tunneling multiple multicast packets to the foreign network is inefficient. Mobile Multicast (MoM) [53] proposes to reduce the explosion problem in bidirectional tunneling by electing one designated HA. Range-based MoM [54] takes the MoM approach one step further and elects a multicast agent close to the foreign agent to tunnel multicast packets to the foreign network. The remote subscription approach takes the burden off the home agent and eliminates tunneling by avoiding the duplication of multicast packets being tunneled to foreign networks; however, this requires that after each handoff the user must rejoin a multicast group. In addition, the multicast trees used to route multicast packets will be updated after every handoff to track the multicast group members. To limit the tree updates or limit duplication of multicast packets, proxy- or agent-based solutions have been proposed [51,52].

Commercial content distribution networks are already using multicasting as the core technology. Most recently, Packet Video in conjunction with DoCoMo has begun providing wireless multicast streaming services to end users, but it has not taken into account the subnet mobility factor. Also, iBEAM's product, ActiveCast, has been used to distribute streaming application over the Internet; however, it lacks in being able to take care of user mobility, uses geosynchronous orbit (GEO) satellites for content distribution, and does not provide flexible methods for advertisement insertion or a variety of local and global content. As an alternative to a wired core network, companies such as Inktomi and Coolcast are already providing multicast services through satellite to reach a wide range of users.

Similar to the unicast scheme, latency associated with receiving a continuous multicast stream from a single source while the client moves to the next cell consists of several components, such as detection of a new cell/subnet/domain ($\Delta 1$), address acquisition and network configuration ($\Delta 2$), triggering of the multimedia stream to be delivered in the new subnet ($\Delta 3$), and actual delivery of the multimedia stream ($\Delta 4$). Because the multicast communication is receiver initiated, triggering techniques play an important role for multimedia stream delivery. To minimize loss and latency during the client's movement it is desirable to minimize the handoff time and to provide almost instantaneous flow of the multicast stream by adopting some proactive triggering mechanism. Similarly, it may be desirable to avoid the waste of bandwidth in a wireless environment due to continuity of the multicast traffic associated with the leave latency during a mobile's movement.

Fast Handoff in MarconiNet

In this section, we describe several methods associated with faster stream delivery under a MarconiNet [55] environment. MarconiNet proposes an integrated streaming architecture to support multimedia applications such as broadcasting streaming content over the Internet using both wired and wireless access. It makes use of extensions of IETF-based protocols such as the Session Announcement Protocol (SAP), Real-Time Streaming Protocol (RTSP), and Session Description Protocol (SDP). Mobility and stream delivery in MarconiNet take advantage of localized IP multicast and uses

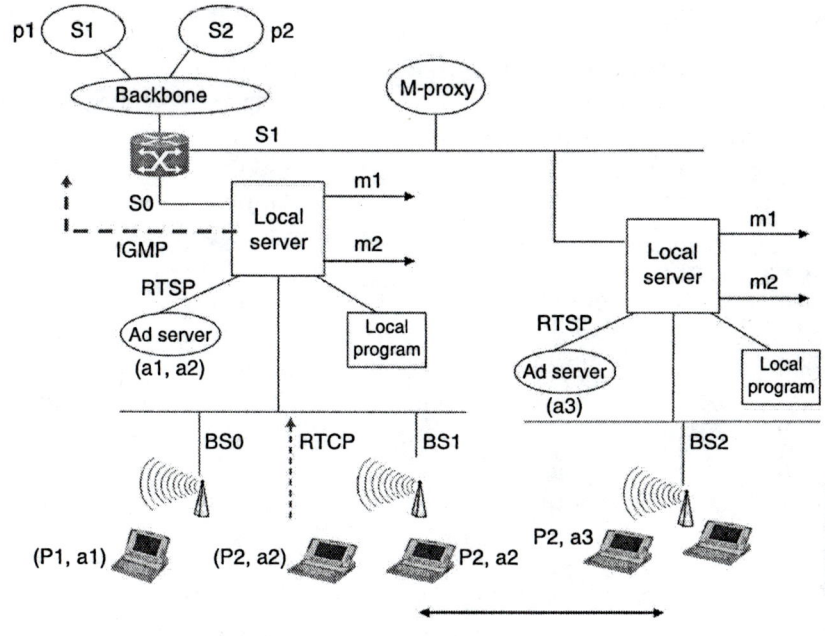

Figure 24.17   Multicast mobility in MarconiNet.

application-layer techniques to provide flexible services such as localized advertisements, news broadcasts, location-specific information, QoS guarantees, and optimized intra-domain handoff for mobile users. Figure 24.17 shows a laboratory implementation of MarconiNet that uses hierarchical scope-based multicast and multicast proxy to provide streaming services to the mobile users. Following are methods for providing fast handoff for multicast streaming in a MarconiNet environment.

Postregistration

In a postregistration scenario, when a client moves to a new subnet it obtains the new IP address and then sends the join query via the IGMP method. In some cases, IGMP could be modified to provide an aggregated group report to minimize the join latency, but in the MarconiNet environment we propose a new mechanism where the client uses an application-layer triggering mechanism based on the Real-Time Transport Control Protocol (RTCP) to facilitate the join and leave. Triggering at the lower hierarchy is accomplished by an RTCP feedback mechanism, but the local server may trigger the multicast flow from the upstream router using the IGMP method. Using an RTCP-based triggering technique offers a solution in the user space and thus is best suitable when the kernel of the end client is not multicast enabled.

Preregistration

The preregistration method has the advantage over postregistration of reducing the join latency for an impending client at the expense of extra flows of the multicast stream in the adjacent cell for a certain time. Two kinds of preregistration schemes have been described. In the first approach, for each of neighboring stations sharing an overlapping area with another station, we provision an associated multicast announcement (address). Each local station can find out what program (group address) the impending mobile host is subscribed to by looking at the common multicast address. Just before a mobile node leaves the cell, a local policy decision (e.g., signal-to-noise ratio) will trigger an RTCP message to the local announcement address. The server in turn announces that message to the admin-scoped shared multicast addresses to which the neighboring stations are listening. The neighboring stations (servers) look up the multicast address and check it against their own databases to determine the group association of this multicast stream. In the absence of this specific subscription, the server sends an IGMP "join" message to the upstream router and passes the stream to the local cells using a locally scoped multicast address even before the client has moved to the new cell, thus minimizing the interruption. Similarly, the client sends an RTCP "bye" message to the server as it moves away from the previous server.

Another approach is to deploy proxy agents in each subnet. These proxy agents join the upstream multicast tree on behalf of the servers, even before the clients move into the cell. The neighboring proxy servers then listen to a common multicast address to figure out the impending host's subscribed multicast address. In this case, the multicast proxy sends the IGMP query messages beforehand on behalf of the local servers. Similarly, a multicast proxy agent within each upstream router can help forward the global stream to the respective global multicast addresses (e.g., for areas where these clients are intending to move) in each subnet for a specific period of time that is determined by the client's entry to the cell. Thus, each neighboring server can receive the stream regardless of whether the mobile node is moving into that cell or not. In either case, as soon as the mobile node moves into the new cell it notifies the proxy agent to leave the tree and the multicast proxy agent stops forwarding the traffic. The preregistration method helps reduce the join latency and the cost of bandwidth in the previous cell or subnet.

During Registration

Group membership information can also be passed along during the client's registration to the new network. During the movement of a node between subnets, it can send a request for a particular multicast address

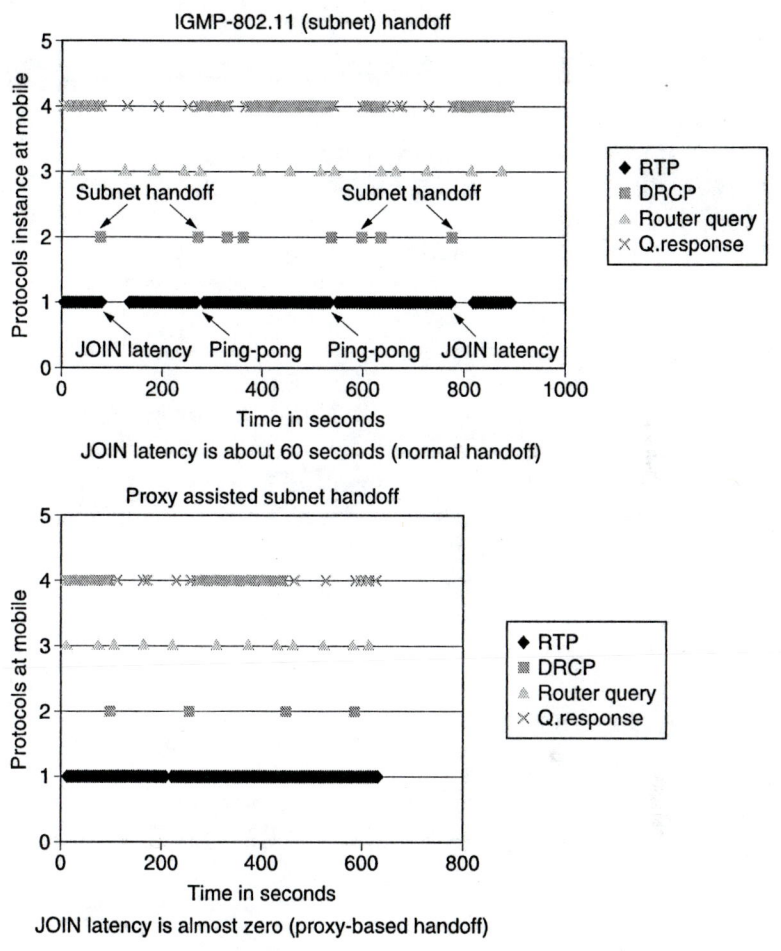

Figure 24.18 Effect of proxy-based handoff on multicast join latency.

in its DHCP discover option or PPP server option message about the local multicast address it has been listening to in the previous subnet. During the process of obtaining the IP address from the DHCP server, the client can send the unsolicited "join" request for the desired locally scoped multicast address to the server; thus, the server can join the desired group during the time the client is in the process of being configured. Figure 24.18 shows how the join latency during the movement of a mobile device between subnets can be minimized by using a proxy-assisted handoff approach. A normal join latency of 60 sec is reduced to almost 0. Similarly, a proxy-based approach can also reduce the leave latency associated with mobility.

Conclusions

We have discussed several issues associated with wireless Internet roaming. We identified the functional elements that can help make such roaming possible, described various mobility management protocols at different layers, and provided an application-layer mobility framework that uses SIP as the base signaling protocol. Experimental results from a mobile multimedia testbed were presented that demonstrated the proof-of-concept of several functional components, such as seamless mobility, security, configuration, and quality of service, including IPv6 features. Also presented were several fast-handoff approaches for SIP-based terminal mobility, including a simultaneous mobility scenario. An integrated mobility management framework was introduced that uses a policy-based mobility management approach and invokes a different mobility protocol based on the type of traffic and movement involved. We provided an overview of the available mobility protocols for multicast streaming that help support mobile content distribution on the wireless Internet. Finally, we described a hierarchical scope-based streaming architecture and its fast-handoff techniques.

Acknowledgment

The authors would like to acknowledge other colleagues, namely Prathima Agrawal, Subir Das, Moncef Elaoud, Dave Famolari, Tony McAuley, Sunil Madhani, Yoshihiro Ohba, Kenichi Taniuchi, Faramak Vakil, Ken Young, and Tao Zhang, for their useful feedback during numerous mobility-related discussions over the years.

References

[1] Wong, D. and Lim, T.J., Soft handoffs in CDMA mobile systems, *IEEE Pers. Commun. Mag.*, 4(6), 6–17, 1997.
[2] Internet Engineering Task Force, www.ietf.org.
[3] McAuley, A., Das, S., Madhani, S., Baba, S., and Shobatake, Y., *Dynamic Registration and Configuration Protocol (DRCP)*, Internet Draft, Internet Engineering Task Force (IETF), 2000.
[4] Park, S.K., *Rapid commit option for DHCPv4*, Internet Draft, Internet Engineering Task Force (IETF), 2003.
[5] Park, S.K. and Han, Y.W., *IPv6 DAD Optimization Goals and Requirements*, Internet Draft, Internet Engineering Task Force (IETF), 2003.
[6] Han, Y. et al., *Advance Duplicate Address Detection*, Internet Draft, Internet Engineering Task Force (IETF), 2003.

[7] Dutta, A., Altintas, O., Chen, W., and Schulzrinne, H., Mobility approaches for all IP wireless networks, in *Proc. of the 6th World Multi-Conference on Systemics, Cybernetics, and Informatics (SCI'02)*, Orlando, FL, July, 2002.

[8] Calhoun, P., Montenegro, G., and Perkins, C.E., *Mobile IPv4 Regional Registration*, Internet Draft, Internet Engineering Task Force (IETF), 2003.

[9] Perkins, C. and Wang, K.-Y., Optimized smooth handoffs in Mobile IP, in *Proc. of the Fourth IEEE Symp. on Computers and Communications (ISCC'99)*, Red Sea, Egypt, July, 1999, pp. 340–346.

[10] Campbell, A., Gomez, J., Kim, S., Valk, A.G., Wan, C.-Y., and Turnyi, Z.R., Design, implementation, and evaluation of cellular IP, *IEEE Pers. Commun. Mag.*, 7, 42–49, 2000.

[11] Ramjee, R., LaPorta, T.F., Salgarelli, L., Thuel, S., Varadhan, K., and Li, L., IP-based access network infrastructure for next-generation wireless networks, *IEEE Pers. Commun. Mag.*, 7, 34–41, 2000.

[12] Jain, R., Raleigh, T., Yang, D., Chang, L.F., Graff, C.J., Bereschinsky, M., and Patel, M., Enhancing survivability of mobile Internet access using Mobile IP with location registers, in *Proc. IEEE INFOCOM'99*, New York, March, 1999.

[13] Das, S., Dutta, A., McAuley, A., Misra, A., and Das, S.K., IDMP: an intra-domain mobility management protocol for next generation, wireless networks, *IEEE Wireless Commun.*, 9(3), 38–45, 2002.

[14] Snoeren, A.C. and Balakrishnan, H., An end-to-end approach to host mobility, in *Proc. of the 6th ACM/IEEE Int. Conf. on Mobile Computing and Networking (MOBICOM'00)*, Boston, MA, August, 2000, pp. 155–166.

[15] Maltz, D.A. and Bhagwat, P., MSOCKS: an architecture for transport layer mobility, in *Proc. IEEE INFOCOM'98*, San Francisco, CA, March, 1998, p. 1037.

[16] Koh, S.N. et al., *Use of SCTP for Seamless Handover*, Internet Draft, Internet Engineering Task Force (IETF), 2003.

[17] Schulzrinne, H. and Wedlund, E., Application-layer mobility using SIP, *Mobile Comput. Commun. Rev.*, 4, 47–57, 2000.

[18] Rosenberg, J., Schulzrinne, H., Camarillo, G., Johnston, A.R., Peterson, J. et al., *SIP: Session Initiation Protocol*, Request for Comments 3261, Internet Engineering Task Force (IETF), 2002 (http://www.ietf.org/rfc/rfc3261.txt).

[19] Dutta, A., Vakil, F., Chen, J., Tauil, M., Baba, S., and Schulzrinne, H., Application layer mobility management scheme for wireless Internet, in *Proc. of IEEE 3G Wireless*, San Francisco, CA, May, 2001, p. 7.

[20] Hsieh, P.-Y., Dutta, A., and Schulzrinne, H., Application layer mobility proxy for real-time application, in *Proc. of IEEE 3G Wireless*, San Francisco, CA, May, 2003

[21] Calhoun, P., Loughney, J., Guttman, E., Zorn, G., and Arkko, J., *Diameter Base Protocol*, Request for Comments 3588, Internet Engineering Task Force (IETF), 2003 (http://www.ietf.org/rfc/rfc3588.txt).

[22] Rigney, C. et al., *Remote Authentication Dial-In User Service (RADIUS)*, Request for Comments 2138, Internet Engineering Task Force (IETF), 1997 (http://www.ietf.org/rfc/rfc2138.txt)

[23] Forsberg, D., *Protocol for Carrying Authentication for Network Access (PANA)*, Internet Draft, Internet Engineering Task Force (IETF), 2004.

[24] Baugher, M. et al., *The Secure Real-Time Transport Protocol*, Internet Draft, Internet Engineering Task Force (IETF), 2003.

[25] Dusse, S., Hoffman, P., Ramsdell, B., Lundblade, L., and Repka, L., *S/MIME Version 2 Message Specification*, Request for Comments 2311, Internet Engineering Task Force (IETF), 1998 (http://www.ietf.org/rfc/rfc2311.txt).

[26] Dutta, A., Agrawal, P., Chen, J.-C., Das, S., Famolari, D. et al., Realizing wireless Internet telephony and streaming, *Comput. Commun.*, 27, 725–738, 2005.

[27] Vatn, J.-O. and Maguire, G.C., The effect of using co-located care-of addresses on macro handover latency, in *Proc. of the 14th Nordic Teletraffic Seminar*, Technical University of Denmark, Lyngby, Denmark, August, 1998, p. 32.

[28] Basilier, H., Calhoun, P., Holdrege, M., Johansson, T., Kempf, J., and Rajaniemi, J., *AAA Requirements for IP Telephony/Multimedia*, Internet Draft, Internet Engineering Task Force (IETF), 2002.

[29] Dutta, A., Altintas, O., Schulzrinne, H., and Chen, W., Multimedia SIP sessions in a heterogeneous access environment, in *Proc. of IEEE 3G Wireless*, San Francisco, CA, May, 2002.

[30] Dutta, A., Zhang, T., Madhani, S., Taniuchi, K., Fujimoto, K. et al., Secure universal mobility for wireless Internet, in *Proc. of the Second ACM Int. Workshop on Wireless Mobile Applications and Services on WLAN Hotspots (WMASH 2004)*, Philadelphia, PA, October, 2004.

[31] Kivinen, T. et al., *Design of the MOBIKE Protocol*, Internet Draft, Internet Engineering Task Force (IETF), 2005.

[32] Nakajima, N., Dutta, A., Das, S., and Schulzrinne, H., Handoff delay analysis and measurement for SIP based mobility in IPv6, in *Proc. of the 8th IEEE Int. Symp. on Computers and Communications (ISCC'03)*, Kemer-Antalya, Turkey, June, 2003.

[33] Chen, J. et al., *Dynamic Service Negotiation Protocol (DSNP)*, Internet Draft, Internet Engineering Task Force (IETF), 2002.

[34] Malki, K., *Low Latency Handoffs in Mobile IPv4*, Internet Draft, Internet Engineering Task Force (IETF), 2004.

[35] Calhoun, P., Hiller, T., Kempf, J., McCann, P., Pairla, C., Singh, A., and Thalanany, S., *Foreign Agent Assisted Handoff*, Internet Draft, Internet Engineering Task Force (IETF), 2000.

[36] Soliman, H., Castelluccia, C., Malki, K., and Bellier, L., *Hierarchical Mobile IPv6 Mobility Management (HMIPv6)*, Internet Draft, Internet Engineering Task Force (IETF), 2003.

[37] Vakil, F., Famolari, D., Baba, S., and Famolari, D., Virtual soft handoff in IP-Centric wireless CDMA networks, in *Proc. of IEEE 3G Wireless*, San Francisco, CA, May, 2001.

[38] Dutta, A., Madhani, S., Chen, W., Altintas, O., and Schulzrinne, H., Optimized fast-handoff schemes for application layer mobility management, *Mobile Comput. Commun. Rev.*, 7(1), 17–19, 2003.

[39] Dutta, A., Zhang, T., Taniuchi, K., Ohba, Y., and Schulzrinne, H., MPA assisted optimized proactive handoff scheme, in *Proc. of MobiQuitous*, San Diego, CA, July, 2005.

[40] Wong, K.D., Dutta, A., Young, K., and Schulzrinne, H., Managing simultaneous mobility of IP hosts, in *Proc. of IEEE Military Communications Conf. (MILCOM'03)*, Boston, MA, October, 2003.

[41] Wong, D. and Dutta, A., Simultaneous mobility in MIPv6, in *Proc. of the IEEE Electro/Information Technology Conf. (EIT 2005)*, Lincoln, NE, May, 2005.

[42] Dutta, A., Wong, D., Burns, J., Jain, R., Young, K. et al., Realization of integrated mobility management for *ad hoc* networks, in *Proc. of IEEE Military Communications Conf. (MILCOM'02)*, Anaheim, CA, October, 2002.

[43] Finlayson, R., *The UDP Multicast Tunneling Protocol*, Internet Draft, Internet Engineering Task Force (IETF), 2003.

[44] Thaler, D. et al., *IPv4 Automatic Multicast without Explicit Tunnels (AMT)*, Internet Draft, Internet Engineering Task Force (IETF), 2002.

[45] Romdhani, I., Kellil, M., Lach, H.-Y., Bouabdallah, A., and Bettaher, H., *IP Mobile Multicast: Challenges, Solutions and Open Issues*, Technical Report, Motorola Labs, Paris, France, 2002.

[46] Xylomenos, G. and Polyzos, G.C., IP multicast for mobile hosts, *IEEE Commun. Mag.*, 35, 54–58, 1997.

[47] Varshney, U. and Chatterjee, S., Architectural issues in IP multicast over wireless networks, in *Proc. of the IEEE Int. Wireless Communications Networking Conf. (WCNC'99)*, New Orleans, LA, September, 1999.

[48] Acharya, A., Bakre, A., and Badrinath, B.R., *IP Multicast Extensions for Mobile Internetworking*, Technical Report LCSR-TR-243, Department of Computer Science, Rutgers University, New Brunswick, NJ, 1995.

[49] Wu, J.-L.C., An IP mobility support architecture for 4GW wireless infrastructure, in *Proc. of the 1999 Personal Computing and Communication Workshop (PCC'99)*, Lund, Sweden, November, 1999.

[50] McAuley, A., Bommaiah, E., Misra, A., Talpade, R.R., Thomson, S., and Young, K.C., Mobile multicast proxy, in *Proc. of IEEE Military Communications Conf. (MILCOM'99)*, Atlantic City, NJ, November, 1999.

[51] Tan, C.L. and Pink, S., Mobicast: a multicast scheme for wireless networks, *Mobile Networks Appl.*, 5, 259–271, 2000.

[52] Mysore, J. and Bharghavan, V., A new multicasting-based architecture for Internet host mobility, in *Proc. of the 3rd ACM/IEEE Int. Conf. on Mobile Computing and Networking (MOBICOM'97)*, Budapest, Hungary, September, 1997.

[53] Williamson, C., Harrison, T., Mackrell, W.L., and Bunt, R.B., Performance evaluation of the MoM mobile multicast protocol, *ACM Mobile Networks Appl. J.*, 3, 189–201, 1998.

[54] Lin, C. and Wang, K.-M., Mobile multicast support in IP networks, in *Proc. IEEE INFOCOM'00*, Tel Aviv, Israel, March, 2002.

[55] Dutta, A. and Schulzrinne, H., MarconiNet: overlay mobile content distribution network, *IEEE Commun. Mag.*, 42(2), 64–75, 2004.

Chapter 25

Impact of Mobility on Resource Management in Wireless Networks

Majid Ghaderi and Raouf Boutaba

CONTENTS

Introduction

Wireless Communications

The coverage area in an infrastructure-based wireless network is divided into small regions called *cells*. Each cell in a cellular network is equipped with a base station (BS), and the number of radio channels is assigned according to the transmission power constraints and availability of spectrum. A channel can be a frequency, a time slot, or a code sequence. Any mobile terminal (MT) residing in a cell can communicate through a radio link with the base station located in the cell, which in turn is connected to the core of the network through a base station controller, as shown in Figure 25.1. A group of base stations is controlled by a base station controller. The core network consists of circuit-switched (CS) and packet-switched (PS) domains. The former basically provides voice service over a circuit-based infrastructure that has evolved from analog technologies to more advanced digital technologies. The latter has recently been added to the infrastructure to provide packet-based data services. Both voice calls and data packets follow the same path until reaching the base station controller. In the core, voice calls are handled by the CS domain, which is also connected to the Public-Switched Telephone Network (PSTN), and data packets are transmitted through the PS domain, which is connected to external data networks such as the Internet.

The rapid expansion of wireless networks and proliferation of wireless devices are the deriving forces behind research and development activities in the field of wireless communications. User mobility is a distinguishing characteristic of wireless communications and, certainly, the primary factor contributing to their success. In a wireless network, mobile users are free to roam the coverage area of the network while receiving the same service regardless of their location in the network. This seamless mobility is the result of careful network planning and design that applies *radio resource management* (RRM) mechanisms capable of anticipating and dealing with user mobility.

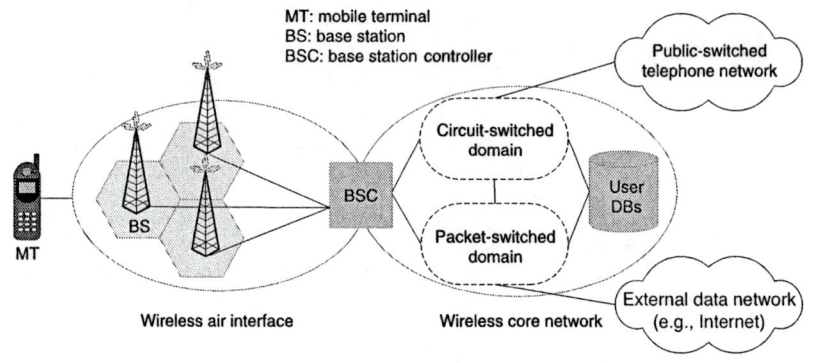

Figure 25.1   A wireless cellular system.

When an MT requests service, it may either be granted or denied service. The latter is known as *call blocking*, and its probability is known as the *call blocking probability*. An active MT in a network may move from one cell to another. The continuity of service to the MT in the new cell requires a successful *handoff* from the previous cell to the new cell. A handoff is successful if the required resources are available and allocated to the MT. The probability of a handoff failure is the *handoff failure probability*. During the life of a call,* an MT may cross several cell boundaries and hence may require several successful handoffs. Failure to get a successful handoff at any cell in the path forces the network to discontinue service to the MT. This is known as *call dropping* or *forced termination* of the call, and the probability of such an event is known as *call dropping probability*.

In general, dropping a call in progress is considered to have a more negative impact from the user's perspective than blocking a newly requested call; therefore, handoff calls are typically given higher priority than new calls with regard to access to the wireless resources. This preferential treatment of handoffs increases the blocking of new calls which has an adverse effect on bandwidth utilization [1]. The most popular approach to prioritize handoff calls over new calls is by reserving a portion of available bandwidth in each cell to be used exclusively for handoffs. Fundamental issues are how many resources must be reserved and when the reservation must be made to support seamless mobility and avoid forced termination of handoff calls while maximizing network resource utilization. Solving this resource management problem is extremely challenging and has been subject to significant research effort. The goals of this chapter are to (1) present the challenges in radio resource management, (2) review and discuss proposed solutions, and (3) give possible research directions.

---

* In this chapter, the terms "call", "connection," and "flow" are used interchangeably.

Resource Management Challenges

In general, resource management (RM) is a challenging problem due to the stochastic nature of resource demands in a network. In wireless networks, RM is even more challenging due to additional sources of uncertainty — namely, randomness in user movement and wireless channel behavior. In addition, radio resources are limited and expensive; hence, efficient RRM is vital for service providers as well. In particular, the most important factors contributing to the complexity of radio resource management are as follows:

- Radio resources (i.e., radio channels) are limited and subject to time-varying and location-dependent errors. Due to the variable signal-to-noise ratio (SNR), wireless channels have variable bit-error rates (BERs). To provide quality of service (QoS), sophisticated resource management mechanisms that take into consideration the nature of wireless channels are needed.
- Users in a wireless network are free to roam in the network coverage area. A cell may become congested due to excessive handoffs from neighboring cells and fail to accommodate some handoffs, which results in premature call termination. In particular, the current trend in cellular communications is to reduce the cell size to achieve higher capacity by reusing radio frequencies more effectively. The smaller the cell size, the higher the volume of handoffs, which consequently increases the probability of premature call termination. To avoid network performance degradation due to excessive call termination, resource management schemes must account for user mobility by advance resource reservation or other appropriate means such as queuing handoff calls temporarily.
- Mobile devices have limited battery life, memory, and processing capabilities. Although memory and processing capabilities have witnessed exponential advances during the past few years, as predicted by Moore's law, battery technology has remained roughly the same over a decade. Power-aware resource management mechanisms are generally more complicated.

Among the above factors, in this chapter our primary focus is on the mobility of users and its impact on resource management. In particular, we will emphasize *call admission control* (CAC) as a core component of resource management in wireless networks.

Call Admission Control

Call admission control restricts the access to the network based on resource availability to prevent network congestion and service degradation for already supported users. A new call request is accepted if there are enough idle resources to meet the QoS requirements of the new call without violating the QoS for already accepted calls. Although reserving resources for handoffs can prevent handoff dropping, over-reservation will degrade the radio resource utilization. Typically, the goal of a CAC scheme is to maintain a prespecified target call dropping probability while minimizing the call blocking probability. Let us assume that the available bandwidth in each cell is channelized. Let us also focus on call-level QoS measures. Therefore, the call blocking probability (p_b) and the call dropping probability (p_d) are the QoS parameters highlighted in this chapter. Three CAC-related problems can be identified based on these two QoS parameters [2]:

- *MINO* — Minimizing a linear objective function of the two probabilities (p_b and p_d)
- *MINB* — For a given number of channels, minimizing the new call blocking probability subject to a hard constraint on the handoff dropping probability
- *MINC* — Minimizing the number of channels subject to hard constraints on the new and handoff call blocking/dropping probabilities

As mentioned before, channels could be frequencies, time slots, or codes, depending on the radio technology used. Each base station is assigned a set of channels, and this assignment can be static or dynamic [3].

Because MINO aims at minimizing penalties associated with blocking new and handoff calls, it appeals to the network provider, as minimizing penalties results in maximizing net revenue. MINB places a hard constraint on handoff call dropping, thereby guaranteeing a particular level of service to already admitted users while trying to maximize the net revenue. MINC is more of a network design problem, where resources must be allocated *a priori*, based on, for example, traffic and mobility characteristics [2].

In general, the two categories of CAC schemes in cellular networks are:

- *Deterministic CAC* — QoS parameters are guaranteed with 100 percent confidence [4,5]. Typically, these schemes require extensive knowledge of the system parameters such as user mobility, which is not practical, or they sacrifice the scarce radio resources to satisfy the deterministic QoS bounds.

- *Stochastic CAC* — QoS parameters are guaranteed with some probabilistic confidence [2,6,7]. By relaxing QoS guarantees, these schemes can achieve a higher utilization than deterministic approaches can.

Most of the CAC schemes investigated in this chapter fall into the stochastic category.

Prioritization Schemes

In this section, we discuss various handoff prioritization schemes, focusing on reservation schemes. Channel borrowing, call queuing, and reservation are studied as the most common techniques.

Channel Borrowing Schemes

In a channel borrowing scheme, a cell (an acceptor) that has used all its assigned channels can borrow free channels from its neighboring cells (donors) to accommodate handoffs [3]. A channel can be borrowed by a cell if the borrowed channel does not interfere with existing calls. When a channel is borrowed, several other cells are prohibited from using it. This is called *channel locking* and has a great impact on the performance of channel borrowing schemes [8]. The number of such cells depends on the cell layout and the initial channel allocation. The proposed channel borrowing schemes differ in the way a free channel is selected from a donor cell to be borrowed by an acceptor cell. A survey on channel borrowing schemes is provided by Katzela and Naghshineh [3].

Call Queuing Schemes

Queuing of handoff requests when no channel is available can reduce the dropping probability at the expense of higher new call blocking. If the handoff attempt finds all the channels in the target cell occupied it can be queued. If any channel is released, it is assigned to the next handoff call waiting in the queue. Queuing can be done for any combination of new and handoff calls. The queue itself can be finite or infinite. Although finite queue systems are more realistic, systems with infinite queue are more convenient for analysis. Figure 25.2 depicts a classification of call queuing schemes.

Hong and Rappaport [6] analyzed the performance of the simple *guard channel* (GC) scheme with queuing of handoffs, where handoff attempts

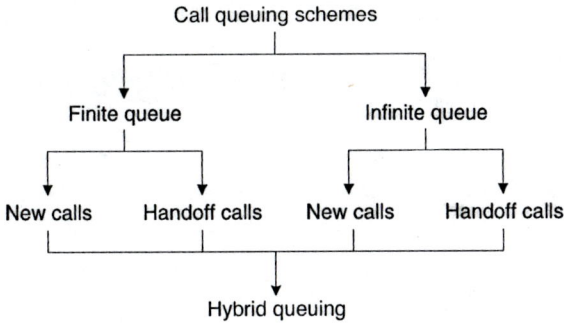

Figure 25.2 Call queuing schemes.

can be queued for the time duration in which a mobile dwells in the handoff area between cells. They used a first-in/first-out (FIFO) queuing strategy and showed that queuing improves the performance of the pure guard channel scheme; that is, p_d is lower for this scheme, but there is essentially no difference for p_b.

The tolerable waiting time in queues is an important parameter. The performance of queuing schemes is affected by the reneging of queued new calls due to caller impatience and the dropping of queued handoff calls as they move out of the handoff area before the handoff is accomplished successfully. Chang et al. [9] analyzed a priority-based queuing scheme in which handoff calls waiting in queue have priority over new calls waiting in queue to gain access to available channels. They simply assumed that those calls waiting in the queue cannot handoff to another cell. Recently, Li and Chao [10] investigated a general modeling framework that can capture call queuing as well. They proved that the steady-state distribution of the equivalent queuing model has a product form solution. Queuing schemes have been proposed primarily for circuit-switched voice traffic. Their generalization to multiple classes of traffic is a challenging problem. Lin and Lin [11] analyzed several channel allocation schemes, including queuing of new and handoff calls. They concluded that the scheme with new and handoff call queuing has the best performance.

Reservation Schemes

The notion of guard channels was introduced in the mid-1980s as a call admission control mechanism to give priority to handoff calls over new calls. In this policy, a set of channels, called the *guard channels*, is permanently reserved for handoff calls. Hong and Rappaport [6] showed that this scheme reduces handoff dropping probability significantly

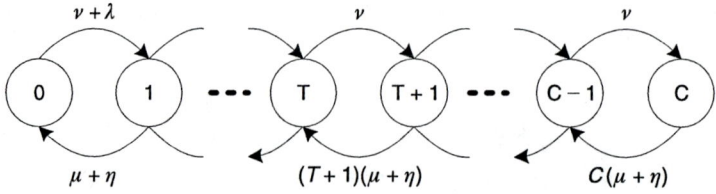

Figure 25.3 State transition diagram of the guard channel scheme.

compared to the nonprioritized case. They found that p_d decreases by a significantly larger order of magnitude compared to the increase of p_b when more priority is given to handoff calls by increasing the number of handoff channels.

Consider a cellular network with C channels in a given cell. The guard channel (GC) scheme reserves a subset of these channels (say, $C - T$) for handoff calls. Whenever the channel occupancy exceeds a certain threshold T, the GC rejects new calls until the channel occupancy falls below the threshold. Assume that the arrival process of new and handoff calls is Poisson with rates λ and v, respectively. The call holding time and cell residency for both types of call are exponentially distributed with means $1/\mu$ and $1/\eta$, respectively. Let $p = (\lambda + v)/(\mu + \eta)$ denote the traffic intensity. Further assume that the cellular network is homogeneous, so a single cell in isolation is representative of the network.

Let us define the state of a cell as the number of occupied channels in the cell; therefore, the cell channel occupancy can be modeled by a continuous-time Markov chain with C states. The state transition diagram of a cell with C channels and $C - T$ guard channels is shown in Figure 25.3. Given this, it is straightforward to derive the steady-state probability (P_n) that n channels are busy:

$$P_n = \begin{cases} \left(\dfrac{\rho^n}{n!} \right) P_0 & 0 \le n \le T \\[3ex] \rho^T \left(\dfrac{v^{n-T}}{n!} \right) P_0 & T \le n \le C \end{cases} \tag{25.1}$$

where:

$$P_0 = \left[\sum_{n=0}^{T} \frac{\rho^n}{n!} + \rho^T \sum_{n=T+1}^{C} \frac{v^{n-T}}{n!} \right]^{-1} \tag{25.2}$$

and then:

$$p_b = \sum_{n=T+1}^{C} P_n \text{ and } p_f = P_C$$

However, Fang and Zhang [12] showed that when the mean cell residency times for new calls and handoff calls are significantly different (as is the case for nonexponential channel holding times), the traditional one-dimensional Markov chain model may not be suitable and a two-dimensional Markov model, which is more complicated, must be applied.

A critical parameter in this basic scheme is the optimal number of guard channels. There is a tradeoff between minimizing p_d and minimizing p_b. If the number of guard channels is conservatively chosen, then admission control fails to satisfy the specified p_d. A static reservation typically results in poor resource utilization. To deal with this problem, several dynamic reservation schemes [7,13–16] were proposed in which the optimal number of guard channels is adjusted dynamically based on the observed traffic load and dropping rate in a control time window. If the observed dropping rate is above the guaranteed p_d, then the number of reserved channels is increased. On the other hand, if the current dropping rate is far below the target p_d, then the number of reserved channels is decreased. The next section investigates dynamic reservation schemes.

A different variation of the basic GC scheme is known as the *fractional guard channel* (FGC) [2]. Whenever the channel occupancy exceeds threshold T, the GC policy is to reject new calls until the channel occupancy falls below the threshold. In the fractional GC policy, new calls are accepted with a certain probability that depends on the current channel occupancy; thus, we have a randomization parameter that determines the probability of acceptance of a new call. Note that both GC and FGC policies accept handoff calls as long as some channels are free. One advantage of FGC over GC is that it distributes the newly accepted calls evenly over time, which leads to a more stable control [17].

Dynamic Reservation Schemes

The two approaches in dynamic reservation schemes are *local* and *distributed* (collaborative), depending on whether the scheme uses local information or gathers information from neighbors to adjust the reservation threshold. In local schemes, each cell estimates the state of the network using local information only, and in distributed schemes each cell gathers network state information from its neighboring cells.

Local Schemes

We categorize local admission control schemes into *reactive* and *predictive* schemes. By reactive approaches we refer to those admission policies that adjust their decision parameters (i.e., threshold and reservation level) as a result of an event such as call arrival, completion, or rejection. Predictive approaches refer to those policies that predict future events and adjust their parameters in advance to prevent undesirable QoS degradations.

Reactive Approaches

The well-known guard channel (cell threshold, cut-off priority, or trunk reservation) scheme is the first one in this category. GC has a reservation threshold, and when the number of occupied channels reaches this threshold no new call requests are accepted. One natural extension of this basic scheme is to use more than one threshold (*e.g.*, two thresholds [13]) to have more control over the number of accepted calls. It has been shown that the simple guard channel scheme performs remarkably well [18], often better than more complex schemes during periods in which the load does not differ from the expected level. For a discussion on various reservation strategies, refer to Epstein and Schwartz [19].

Predictive Approaches

Local admission control schemes are very simple, but they suffer from the lack of global information about changes in network traffic. In turn, distributed admission control schemes have access to global traffic information at the expense of increased computational complexity and signaling overhead induced by information exchange between cells. To overcome the complexity and overhead associated with distributed schemes and benefit from the simplicity of local admission schemes, predictive admission control schemes were proposed. These schemes try to estimate the global state of the network by using some modeling or prediction techniques based on information available locally; for example, the prediction-based scheme proposed in Ghaderi et al. [20] uses online measurements to forecast future bandwidth demands using a *minimum mean square error* predictor.

Distributed Schemes

The main idea behind all distributed schemes [7,14–17,21] is that every mobile terminal with an active wireless connection exerts an influence

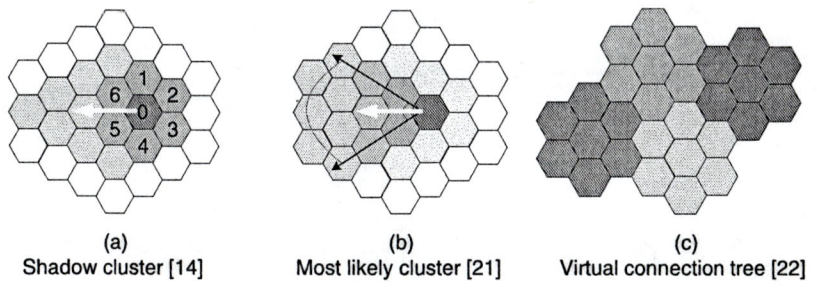

(a)	(b)	(c)
Shadow cluster [14]	Most likely cluster [21]	Virtual connection tree [22]

Figure 25.4 Three examples of cluster definition.

upon the cells in the vicinity of its current location and along its direction of travel [14]. Cells that are geographically or logically close together form a *cluster*, as shown in Figure 25.4. Either each mobile terminal has its own cluster independent of other terminals or all the terminals in a cell share the same cluster. Typically, the admission decision for a connection request is made in cooperation with other cells of the cluster associated with the mobile terminal asking for admission. In Figure 25.4a, a cluster is defined assuming that a terminal affects all the cells in the vicinity of its current location and along its trajectory, and in Figure 25.4b it is assumed that those cells that form a sector in the direction of the trajectory of the mobile terminal are most likely to be affected (visited) by the terminal. Figure 25.4c shows a static cluster that is fixed regardless of the terminal mobility.

Each user currently in the system may either remain in the cell it is in or move to a neighboring cell; hence, the system can be modeled using a binomial random variable. We approximate the joint behavior of binomial distributions with a normal distribution, and the number of active calls in a cell at any time follows a Gaussian distribution. Also, we neglect the possibility of users having moved a distance of two or more cells and of a user arriving or completing a call during a time interval of length T.

Now, consider a hexagonal cellular system similar to those depicted in Figure 25.4. Assume that at time $t = t_0$ a new call has arrived. New calls are admitted into the system provided that the predicted handoff failure probability of any user in the home and neighboring cells at time $t = t_0 + T$ is below the target threshold P_{QoS}. Let $n_i(t)$ denote the number of active calls in cell i at time t. Assuming that handoff failure in each cell can be approximated by the overload probability, we obtain:

$$p_f = \mathbb{P}\{n(t_0 + T) > c\} \qquad (25.3)$$

Therefore, the handoff failure in cell i is given by:

$$P_f(i) = \frac{1}{2} \text{erfc}\left(\frac{c_i - E[n_i(t_0 + T)]}{\sqrt{2\text{Var}[n_i(t_0 + T)]}}\right) \tag{25.4}$$

where c_i is the capacity of cell I, and $\text{erfc}(x)$ is the complementary error function defined as:

$$\text{erfc}(x) = \frac{2}{\sqrt{\pi}} \int_x^\infty e^{-t^2} dt \tag{25.5}$$

and the expected and variance in the number of calls at time $t_0 + T$ in cell i are given by:

$$E[n_i(t_0 + T)] = n_i(t_0)p_s + p_b \sum_{j=1}^{6} n_j(t_0)$$

$$\text{Var}[n_i(t_0 + T)] = n_i(t_0)v_s + v_b \sum_{j=1}^{6} n_j(t_0) \tag{25.6}$$

where p_s is the probability of staying in the current cell, and p_b is the probability of handing off to another cell during time period T, given by:

$$p_s = e^{-(\mu+b)T}, \; p_b = \frac{1}{6}\left(1 - e^{-bT}\right) \tag{25.7}$$

Similarly, v_s and v_b are, respectively, the variances of the binomial processes of stay and handoff with parameters p_s and p_b, which are expressed as:

$$v_s = (1 - p_s)p_s, \; v_b = (1 - p_b)p_b \tag{25.8}$$

The idea of distributed admission control was originally proposed by Naghshineh and Schwartz [7]. They proposed a collaborative admission control known as *distributed call admission control* (DCAC). DCAC periodically gathers some information (namely, the number of active calls) from the adjacent cells of the local cell to make the admission decision in combination with the local information. The analysis we presented earlier is slightly different from the original DCAC and is based on the work by Epstein and Schwartz [16]. DCAC is very restrictive in the sense that it takes into consideration information from direct neighbors only and assumes at most one handoff during the control period.

It has been shown that DCAC is not stable and violates the required dropping probability as the load increases [17]. Levin et al. [14] proposed a more complicated version of the original DCAC based on the *shadow cluster* concept, which uses dynamic clusters for each user based on its mobility pattern instead of restricting itself (as does DCAC) to direct neighbors only. A practical limitation of the shadow cluster scheme in addition to its complexity and overhead is that it requires a precise knowledge of the mobile trajectory. The so-called *active mobile probabilities* and their characterization are very crucial to the CAC algorithm. Active mobile probabilities for each user give the projected probability of being active in a particular cell at a particular instance of time.

Wu et al. [17] proposed a dynamic, distributed, and stable CAC scheme referred to as SDCA (stable dynamic call admission), which extends the basic DCAC [7] in several ways, such as by using a diffusion equation to describe the evolution of the time-dependent occupancy distribution in a cell instead of the widely used Gaussian approximation. SDCA is a distributed version of the fractional guard channel in that it computes an acceptance ratio a_i for each cell i to be used for the current control period.

Classification of Distributed Schemes

Distributed CACs can be classified according to two criteria:

- Cluster definition
- Information exchange and processing

A cluster can be either static or dynamic. In the static approach, the size and shape of the cluster are the same regardless of the network situation. In the dynamic approach, however, the shape or size of the cluster changes according to the congestion level and traffic characteristics. The virtual connection tree in Acampora and Naghshineh [22] is an example of a static cluster, and the shadow cluster introduced in Levine et al. [14] is a dynamic cluster. A shadow cluster is defined for each individual mobile terminal based on its mobility information (e.g., trajectory) and changes as the terminal moves. Table 25.1 shows a tradeoff between the cluster type and the corresponding CAC performance. Typically, dynamic clusters have a better performance at the expense of increased complexity.

In general, distributed CACs can be categorized as *partially distributed* or *completely distributed* based on the decision-making process:

- *Partially distributed* — In this approach, the necessary information is gathered from the neighboring cells, but the processing is centralized. The virtual connection tree concept introduced in Acampora and Naghshineh [22] is an example of a partially distributed scheme. In this scheme, each connection tree consists of a specific set of base stations where each tree has a network controller. The network controller is responsible for keeping track of the users and resources. Despite the fact that information is gathered from a set of neighboring cells, the final decision is made locally in the network controller.
- *Completely distributed* — In this approach, not only is information gathered from the neighboring cells but the neighboring cells also collaborate in the decision-making process. The shadow cluster concept introduced in Levine et al. [14] is an example of a completely distributed scheme. In this scheme, a cluster of cells, the shadow cluster, is associated with each mobile terminal in a cell. Upon admitting a new call, all the cells in the corresponding cluster calculate a preliminary response which after processing by the original cell will form the final decision.

Although it is theoretically possible to involve all the network cells in the admission control process, it is expensive and sometimes useless in practice. To consider the effect of all the cells, analytical approaches involve large matrix exponentiations. In Wu et al. [17] and Mitchell and Sohraby [23], two different approximation techniques have been proposed to compute these effects with a lower computational complexity. Table 25.2 provides a comparison of the different classes of dynamic CAC schemes. In general, there is a tradeoff between the efficiency and the complexity of local and distributed schemes.

Recall that the fundamental questions a CAC algorithm faces are how many resources must be reserved and when the reservation must be made to support seamless mobility. Clearly, if the CAC algorithm has complete knowledge about the mobility of the users, it can answer these questions

Table 25.1 Cluster Type Versus Call Admission Control Performance

Cluster Type	CAC Efficiency	CAC Complexity
Static	Moderate	Moderate
Dynamic	High	High

Table 25.2   Comparison of Dynamic Call Admission Control Schemes

Scheme		Efficiency	Overhead	Complexity	Adaptivity
Local	Reactive	Low	Low	Low	Moderate
	Predictive	Moderate	Low	Moderate	Moderate
Distributed	Partially distributed	High	High	High	High
	Completely distributed	High	Very high	Very high	High

in an optimal manner; however, in practice, it is difficult or even impossible to accurately predict the future location of a mobile user. In other words, such global knowledge is not available. Nevertheless, it is still possible to use some partial knowledge in probabilistic form about the mobility of users which can be helpful in making the correct admission decision. This can be achieved using mobility modeling and prediction techniques. In the following sections, we discuss the issues involved in mobility modeling and prediction while reviewing some related literature.

Mobility Modeling

A mobility model is a representation of a certain real or abstract world that contains moving entities (i.e., mobile users). The world is said to exist for some finite amount of time during which each moving entity has one unique but changing location of presence as defined in the location granularity of the world [24]. In essence, the location of some mobile entities is a function of time and some heuristics are inherent to the mobility model. These inherent heuristics differentiate mobility models from one another. A number of various mobility models have been proposed in the literature [25–27].

A growing concern in the research community is that the simplistic random walk and random way point models are no longer suitable for obtaining meaningful results from simulations for several reasons [28], one of which is the fact that certain features of random mobility are particularly difficult to justify. These would include the lack of geographic dependencies in the simulated environment and the purposelessness of the user motion.

It may be possible to exploit these properties and improve existing wireless systems, but any such potential improvements will not be visible when testing with a simulator that uses a random motion model. The properties spoken of are mainly expected to be some regularities in user movement; thus, various mobility models have been presented, each with a specific goal and suitable for a specific scenario. The following discussion attempts a brief overview of the various scenarios for which mobility models have been developed and presents a few examples of existing models.

Geography

Mobility models fall into one of two classes when geography is concerned. What is meant by geography is the underlying geographic constraints that any mobile user in the real world has to deal with, such as roads, rivers, and walls. The two classes are those that include geography and those that do not. While this seems obvious, it is important to keep the distinction in mind. A number of complex mobility models exist that have no notion of any geography [29] and may be quite incomplete for the validation of certain studies.

Mobility Scale

One of the main classification mechanisms for mobility models is the scale for which they are designed to function. The smallest scale mobility models are designed for the indoor environment. Although the indoor environment is often labeled as the most appropriate environment to be modeled by a random model, other models such as the picocellular model presented in Voigt and Fettweis [30] are specifically designed to handle such small-scale mobility. The next scale up is the street unit model, such as that presented in Markoulidakis et al. [24]. This model allows for modeling streets with such a level of detail that stop signs and traffic lights are included. The next scale up includes models that deal with area zones; these models are primarily concerned with the idea that certain mobility, such as vehicular mobility, only happens on streets and certain areas are more densely populated than others [24,31]. The last scale is the city area type of model. This model is concerned with the different parts of a city area and the mobility of users between these parts. Much variation exists among models of this scale, due to the many simplifications or complications a particular model may involve. Most mobility models fall into this category [24,26,32]. The scale of the model is important to note because different problem studies require different granularities of mobility; for example, studies that deal with location management often deal with sections of the network as opposed to individual base stations and thus require no more detail than that provided by a city area scale model.

User Individuality

Another main classifier when making distinctions between mobility models is the treatment of users, or mobile terminals. The simplest user treatment consists of aggregating the users into a density function that defines how many users are in a given location at a given time. The next step involves the definition of multiple user classes. Each class differs from the others in some way, such as the speed at which the users in that class travel. This approach can be seen in the work involving poles of gravity [26,28], where the users are modeled as a flow from one location to another. Users are still represented by a density function; however, a number of different density functions each represent a user class. The next step is to make each user an individual entity. This increases the complexity of the model considerably, as each user has an individual state. An example of this can be seen in Nousiainen et al. [32], where individual users move around various world maps. When individual users exist, each user can differ from the others in a number of ways. Most models still classify users such that each one belongs to one of a small number of classes (e.g., highly mobile users, business users, residential users) [25,27]. Each of these users, however, can have a different behavior that is unique to that user by some randomization of that user's properties, such as speed of motion or the location the user considers home. These are the most complex user models.

Mobility Prediction

Intuitively, mobility prediction would be the determining of the future location of a mobile terminal. Although this is the general idea, many details must be specified to truly understand what is being described. The first item that requires further definition is *location*. The location of the user carrying the mobile terminal is primarily considered to be its geographic coordinates. It has been noted, however, that problems arise when associating the location of the MT directly with the user's geographic location [33,34]. It is better to consider the motion of the MT through the network as the successive list of connections that the MT experiences. What this means exactly is that the MT location is always one of a finite set of locations representing one of the possible base stations in the network. This is illustrated in Figure 25.5, where the dashed line represents the user mobility, and the location of the MT changes as the connection to the network changes from one base station to another. In this particular example, the MT starts at BS 3, moves through BSs 7, 8, and 9, and ends up at BS 6. It is evident that the idea of the location of a mobile terminal is greatly simplified by adopting this abstraction from the user's location. The main issue to note here is that a wireless network is concerned not

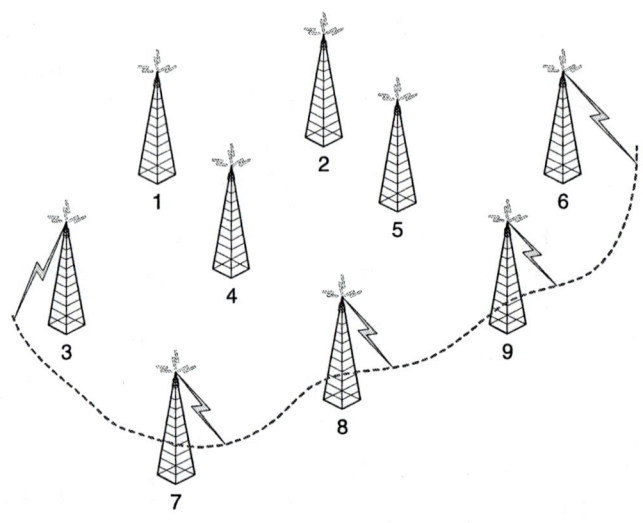

Figure 25.5 Mobile location.

with the location of a user of this network but with the base stations that this user is and will be using to connect to this network.

The second item worth mentioning is the notion that a prediction is usually based on some previous knowledge. The exact specification of what knowledge is used to make a prediction is very crucial in determining the appropriateness of that prediction scheme. If a prediction is based on data that is simply not available in a given situation, then that prediction scheme is useless in that scenario regardless of how well it performs in other scenarios. An example of this is a requirement of privacy. The data it requires may be present in the system but simply not accessible.

The last item requiring discussion is the scale of the prediction. If the prediction mechanism is predicting the time of an event, to what accuracy is the time predicted (seconds, minutes, etc.)? If the prediction mechanism is, on the other hand, predicting the event at a time, this event is most likely defined at least in part by a location and thus the granularity of the location requires discussion. This means that a location can be specified in geographical coordinates, as a single base station, as a group of base stations, etc.

Mobility Prediction Mechanisms

The many approaches to solving the problem of mobility prediction have resulted in many prediction mechanisms. Each is unique and developed to solve a specific problem or specific type of problem but they are all related and many can be used in scenarios other than those for which they are proposed. Some of them are heuristic based [15,21], some are based on the geometrical modeling of user movements [29] and street layouts [35], and still others are based on artificial intelligence techniques [36]. The following provides a general overview of attempts to solve the problem of mobility prediction based on fundamental features found in most prediction mechanisms.

Presence of Infrastructure

Mobility is important in two main types of wireless networks: (1) a system supported by infrastructure, such as a cellular system supported by base stations, and (2) a system that has no supportive infrastructure, such as *ad hoc* networks. The main difference is that an infrastructure-supported system can refer to a fixed base station for location but an infrastructure-less system requires an abstract location reference.

Extent of Prediction

Mobility prediction research has mainly focused on supporting the next expected handoff. In reality, MTs will be able to move throughout the network and experience multiple handoffs during the lifetime of a call. It may therefore be necessary to predict more than just the next location the MT will visit or the next event the MT will experience.

User–Mobile Terminal Relationship

A large portion of recent research still assumes that user mobility and the connection trace for a mobile terminal are strongly dependent. A large number of prediction systems have been proposed that attempt to measure or capture some regularity of the user's mobility to extrapolate from this knowledge about the future behavior of the user's MT. Real-life mobility traces have shown that this assumption of user mobility and the connection trace of the MT is not as valid as most researchers believe [33,34]. This raises the issue that it may not be possible to accurately predict the behavior of an MT by studying the behavior of the user. It will most likely be necessary to study the behavior of the MT and its interaction with the network directly.

Per-User Versus Aggregate Knowledge

Another main distinction between prediction systems is whether data is stored on a per-user basis or aggregated into some structure. One of the most common per-user types of prediction mechanisms is based on the idea of path recognition [21]. The general argument when using a per-user prediction system is that, although the mobility patterns seen in the network as a whole are complex, these patterns become much simpler and more regular when viewed on a per-user basis and can thus be exploited more easily. Prediction schemes that use aggregation [35] argue that user mobility is subject to geographical constraints at the place of each BS so all users will exhibit similar behavior at a given BS; therefore, it is possible to predict the future location of such a user knowing the aggregate behavior of all or similar users at that BS.

Measurement Versus Pattern Matching

Another difference in the approaches to solving the problem of mobility prediction seen in current work is whether the prediction produced is based on measurement of user or mobile terminal behavior or on matching the pattern of this behavior with previous behavior. A measurement-based approach will typically compute a probability of events occurring, depending on the value of some parameters. In Aljadhai and Znati [21], the direction of an MT is measured and used in combination with previous direction measurements to forecast the future motion of the MT. Pattern-matching techniques, on the other hand, attempt to match the observed user behavior with some previously observed behavior and forecast the future based on the observed patterns. Erbas et al. [37] and Kim [38] provide an example of pattern-matching prediction mechanisms — specifically, a path-matching mechanism. This distinction is most evident in the per-user type of prediction mechanisms, as most of the schemes that use aggregation will attempt to capture patterns in an overall sense but perform each individual prediction using a measurement of some kind.

Conclusion

Due to the unique characteristics of wireless networks, mainly mobility and limited resources, resource management has received tremendous attention. As a result, a large body of work has been done extending earlier work in fixed networks as well as introducing new techniques tailored for wireless networks. A large portion of this research has been in the area of call admission control. In this chapter, we have discussed several issues related to mobility and its impact on admission control schemes. In the first part, we presented several CAC schemes proposed to support seamless mobility in a cellular network. We concluded that mobility information is essential to assist CAC in making the best admission decision; therefore, in the second part of the chapter, we focused on mobility modeling and prediction techniques applicable in cellular networks.

One of the interesting observations stemming from such research, as illustrated in Gao and Acampora [39], is how comparable the performance of simple reservation-based CAC schemes (e.g., GC) are to more complex ones. This is particularly true when the traffic conditions are known *a priori* [18]. Yet, a large body of research in this area has focused on designing more and more sophisticated schemes in the hope of improving the CAC performance. Many assumptions about mobility and traffic characteristics made in CAC-related research are often not practical, and most of the schemes proposed in the literature are difficult to deploy in current and future cellular systems.

Some of the lessons learned from surveying and analyzing the literature and from which recommendations can be drawn are as follows:

- Use more realistic (nonexponential) mobility [40] and traffic (packet-based) models in designing and analyzing CAC schemes. New mobility models may not necessarily preserve the Markovian property. Meanwhile, new traffic modeling and engineering techniques are aiming at a more accurate description of traffic dynamics not only at the call level but also at the packet level. From this perspective, recent findings in traffic analysis such as self-similarity must be taken into consideration. To avoid complex schemes and eliminate impractical assumptions about traffic and mobility, measurement-based CAC schemes must be further studied for wireless networks.

- Apply cross-layer design to improve the performance of CAC schemes and achieve bit-level, packet-level, and call-level QoS; in particular, scheduling mechanisms at the packet level and control mechanisms at the call level can benefit from information about the state of the wireless channel to achieve superior performance.

- Wireless links are unpredictable and behave randomly due to physical-layer factors such as interferences, shadowing, multipath fading, etc.; therefore, it is extremely challenging for CAC algorithms to provide fine-grained, per-connection QoS guarantees such as those offered in wireline networks without sacrificing the radio resource utilization. A practical CAC scheme in wireless environments should provide only a coarse-grained, aggregate QoS guarantee and shift the burden of per-connection QoS all the way up to the application layer. For such a design approach to be successful, *adaptive* applications that can adjust their resource requirements according to network condition should be considered.

We believe that the most challenging problems to be solved are mobility and wireless channel effects, particularly when considering multiservice networks. Existing mobility models cannot reflect real user mobility patterns. They commonly make assumptions that are not realistic and have a significant influence on the performance of mobility prediction mechanisms. One common assumption is that wireless users move randomly, which is a scenario that presents no mobility patterns and would thus make the pattern-based mobility predictors and CAC mechanisms based on such predictors questionable.

References

[1] Valko, A.G. and Campbell, A.T., An efficiency limit of cellular mobile systems, *Comput. Commun. J.*, 23(5-6), 441–451, 2000.

[2] Ramjee, R., Towsley, D., and Nagarajan, R., On optimal call admission control in cellular networks, *ACM/Kluwer Wireless Networks*, 3(1), 29–41, 1997.

[3] Katzela, I. and Naghshineh, M., Channel assignment schemes for cellular mobile telecommunication systems: a comprehensive survey, *IEEE Pers. Commun. Mag.*, 3(3), 10–31, 1996.

[4] Talukdar, A.K., Badrinath, B., and Acharya, A., Integrated services packet networks with mobile hosts: architecture and performance, *ACM/Kluwer Wireless Networks*, 5(2), 111–124, 1999.

[5] Lu, S. and Bharghavan, V., Adaptive resource management algorithms for indoor mobile computing environments, in *Proc. ACM SIGCOMM'96*, Palo Alto, CA, August, 1996, pp. 231–242.

[6] Hong, D. and Rappaport, S.S., Traffic model and performance analysis for cellular mobile radio telephone systems with prioritized and nonprioritized handoff procedures, *IEEE Trans. Veh. Technol.*, 35(3), 77–92, 1986. (See also CEAS Technical Report No. 773, College of Engineering and Applied Sciences, State University of New York, 1999.)

[7] Naghshineh, M. and Schwartz, M., Distributed call admission control in mobile/wireless networks, *IEEE J. Select. Areas Commun.*, 14(4), 711–717, 1996.

[8] Chu, T.-P. and Rappaport, S.S., Generalized fixed channel assignment in microcellular communication systems, *IEEE Trans. Veh. Technol.*, 43(3), 713–721, 1994.

[9] Chang, C.-J., Su, T.-T., and Chiang, Y.-Y., Analysis of a cutoff priority cellular radio system with finite queuing and reneging/dropping, *IEEE/ACM Trans. Networking*, 2(2), 166–175, 1994.

[10] Li, W. and Chao, X., Modeling and performance evaluation of a cellular mobile network, *IEEE/ACM Trans. Networking*, 12(1), 131–145, 2004.

[11] Lin, P. and Lin, Y.-B., Channel allocation for GPRS, *IEEE Trans. Veh. Technol.*, 50(2), 375–384, 2001.

[12] Fang, Y. and Zhang, Y., Call admission control schemes and performance analysis in wireless mobile networks, *IEEE Trans. Veh. Technol.*, 51(2), 371–382, 2002.

[13] Moorman, J.R. and Lockwood, J.W., Wireless call admission control using threshold access sharing, in *Proc. IEEE GLOBECOM'01*, San Antonio, TX, November, 2001, pp. 3698–3703.

[14] Levine, D., Akyildiz, I., and Naghshineh, M., A resource estimation and call admission algorithm for wireless multimedia networks using the shadow cluster concept, *IEEE/ACM Trans. Networking*, 5(1), 1–12, 1997.

[15] Choi, S. and Shin, K.G., Predictive and adaptive bandwidth reservation for handoffs in QoS-sensitive cellular networks, in *Proc. ACM SIGCOMM'98*, Vancouver, Canada, October, 1998, pp. 155–166.

[16] Epstein, B.M. and Schwartz, M., Predictive QoS-based admission control for multiclass traffic in cellular wireless networks, *IEEE J. Select. Areas Commun.*, 18(3), 523–534, 2000.

[17] Wu, S., Wong, K.Y.M., and Li, B., A dynamic call admission policy with precision QoS guarantee using stochastic control for mobile wireless networks, *IEEE/ACM Trans. Networking*, 10(2), 257–271, 2002.

[18] Peha, J.M. and Sutivong, A., Admission control algorithms for cellular systems, *ACM/Kluwer Wireless Networks*, 7(2), 117–125, 2001.

[19] Epstein, B. and Schwartz, M., Reservation strategies for multi-media traffic in a wireless environment, in *Proc. IEEE Vehicular Technology Conf. (VTC'95)*, Chicago, IL, July, 1995, pp. 165–169.

[20] Ghaderi, M., Capka, J., and Boutaba, R., Prediction-based admission control for DiffServ wireless Internet, in *Proc. IEEE Vehicular Technology Conf. (VTC'03)*, Orlando, FL, October, 2003, pp. 1974–1978.

[21] Aljadhai, A. and Znati, T.F., Predictive mobility support for QoS provisioning in mobile wireless networks, *IEEE J. Select. Areas Commun.*, 19(10), 1915–1930, 2001.

[22] Acampora, A. and Naghshineh, M., An architecture and methodology for mobile-executed handoff in cellular ATM networks, *IEEE J. Select. Areas Commun.*, 12(8), 1365–1375, 1994.

[23] Mitchell, K. and Sohraby, K., An analysis of the effects of mobility on bandwidth allocation strategies in multi-class cellular wireless networks, in *Proc. IEEE INFOCOM'01*, 2, Anchorage, AK, April, 2001, pp. 1005–1011.

[24] Markoulidakis, J.G., Lyberopoulos, G.L., Tsirkas, D.F., and Sykas, E.D., Mobility modeling in third-generation mobile telecommunications systems, *IEEE Pers. Commun. Mag.*, 4(4), 41–56, 1997.

[25] Cavalcanti, D.A., Kelner, J., Cunha, P.R., and Sadok, D.H., A simulation environment for analyses of quality of service in mobile cellular networks, in *Proc. IEEE Vehicular Technology Conf. (VTC'01)*, Atlantic City, NJ, October, 2001, pp. 2183–2187.

[26] Basgeet, D.R., Irvine, J., Munro, A., Dugenie, P., and Kaleshi, D., SMMT: scalable mobility modeling tool, in *Proc. IEEE Vehicular Technology Conf. (VTC'02)*, Vancouver, Canada, September, 2002, pp. 102–106.

[27] Scourias, J. and Kunz, T., An activity-based mobility model and location management simulation framework, in *Proc. of ACM Int. Workshop on Modeling, Analysis, and Simulation of Wireless and Mobile Systems (MSWiM)*, Seattle, WA, August, 1999, pp. 61–68.

[28] Basgeet, D.R., Irvine, J., Munro, A., and Barton, M.H., Importance of accurate mobility modeling in teletraffic analysis of the mobile environment, in *Proc. IEEE Vehicular Technology Conf. (VTC'03)*, Seoul, Korea, April, 2003, pp. 1836–1840.

[29] Zonoozi, M.M. and Dassanayake, P., User mobility modeling and characterization of mobility patterns, *IEEE J. Select. Areas Commun.*, 15(7), 1239–1252, 1997.

[30] Voigt, J. and Fettweis, G.P., Influence of user mobility and simulcast-handoff on the system capacity on picocellular environments, in *Proc. of IEEE Wireless Communications and Networking Conf. (WCNC 1999)*, New Orleans, LA, September, 1999, pp. 712–716.

[31] Tugcu, T. and Ersoy, C., Application of a realistic mobility model to call admissions in DS-CDMA cellular systems, in *Proc. IEEE Vehicular Technology Conf. (VTC'01)*, Rhodes, Greece, May, 2001, pp. 1047–1051.

[32] Nousiainen, S., Kordybach, K., and Kempi, P., User distribution and mobility model for cellular network simulations, in *Proc. of IST Mobile & Wireless Telecommunications Summit*, June, 2002, pp. 518–522.

[33] Chan, J. and Seneviratne, A., A practical user mobility algorithm for supporting adaptive QoS in wireless networks, in *Proc. of IEEE Int. Conf. on Networks (ICON'99)*, Brisbane, Australia, September, 1999, pp. 104–111.

[34] Chan, J., Landfeldt, B., Seneviratne, A., and Sookavatana, P., Integrating mobility prediction pre-allocation into a home-proxy based wireless internet framework, in *Proc. of IEEE Int. Conf. on Networks (ICON'00)*, Singapore, September, 2000, pp. 18–23.

[35] Soh, W.-S. and Kim, H.S., QoS provisioning in cellular networks based on mobility prediction techniques, *IEEE Commun. Mag.*, 41(1), 86–92, 2003.

[36] Capka, J. and Boutaba, R., Mobility prediction in wireless networks using neural networks, in *Proc. of IFIP/IEEE 7th Int. Conf. on Management of Multimedia Networks and Services (MMNS'04)*, San Diego, CA, October, 2004.

[37] Erbas, F., Steuer, J., Eggesieker, D., Kyamakya, K., and Jobmann, K., A regular path recognition method and prediction of user movements in wireless networks, in *Proc. IEEE Vehicular Technology Conf. (VTC'01)*, Atlantic City, NJ, October, 2001, pp. 2183–2187.

[38] Kim, H. and Jung, J., A mobility prediction handover algorithm for effective channel assignment in wireless ATM, in *Proc. IEEE GLOBECOM'01*, San Antonio, TX, November, 2001, pp. 3673–3680.

[39] Gao, Q. and Acampora, A., Performance comparisons of admission control strategies for future wireless networks, in *Proc. of IEEE Wireless Communications and Networking Conference (WCNC 2002)*, Orlando, FL, March, 2002, pp. 317–321.

[40] Capka, J. and Boutaba, R., A mobility management tool: the realistic mobility model, in *Proc. of IEEE Int. Conf. on Wireless and Mobile Computing, Networking, and Communications (WiMob'05)*, Montreal, Canada, August, 2005.

Chapter 26

Seamless Consistency

Evaggelia Pitoura, George Samaras,
and Can Türker

CONTENTS

Introduction

Mobile devices such as laptops, palmtops, and smart phones have now-adays become more widespread than desktop computers. The possibilities that they offer for accessing and processing information on the move, everywhere and at any time, account for their broad acceptance. The need for information access and processing in mobile computing environments presents great challenges to database researchers. Apart from advances in wireless communications and device technology, special data management techniques are necessary to provide a transparent interaction between fixed and mobile devices in global information systems.

Data management must hide the constraints of mobile wireless computing to make access to data appear seamless. Such constraints include limitations imposed by wireless network connectivity such as disconnections, variable connectivity, and limited bandwidth, as well as such resource constraints as battery and computational power. Furthermore, handling mobility introduces additional challenges. The focus of this chapter is on the central data management aspect of providing a flexible and powerful consistency management infrastructure that takes into account the peculiarities of mobile computing.

We present an overview of the main issues related to consistency in terms of three different settings. First, we present research problems and solutions related to supporting disconnections. Often, mobile hosts operate autonomously due to disconnections from the network. In this case, any data updates performed either locally at the mobile host or at other sites must be reconciled so consistency is achieved. Second, we consider consistency management in the case of mobile hosts operating with less than full connectivity. In this case, consistency management must take into consideration such connectivity limitations. Finally, we consider consistency in the case where servers are provided with a broadcast facility and use it to broadcast (push) data to their clients. This is the case for a number of wireless technologies.

We also describe the model and protocols for consistency provided by four state-of-the-art commercial mobile database systems: IBM DB2® Everyplace®, Oracle® Lite™, Microsoft® SQL Server CE, and Sybase® Anywhere™.

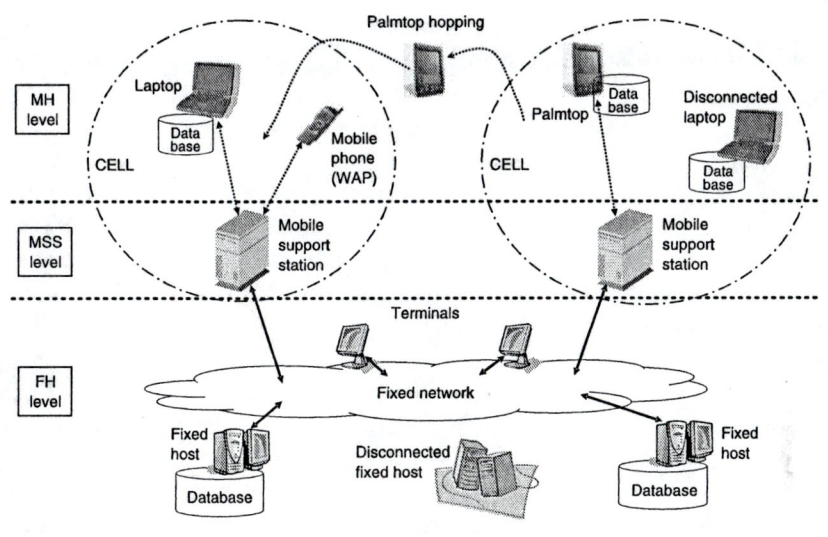

Figure 26.1   Reference model of a mobile computing architecture.

Although, many research efforts have focused on weak connectivity and hybrid delivery, such research has not yet been adapted by the major commercial products. Thus, we present how each of these products handles consistency in the case of disconnections, focusing on synchronization and conflict resolution.

Consistency in Mobile Computing

In this section, we present our reference architecture for mobile computing and introduce the issue of consistency maintenance. We also highlight those characteristics of mobile computing that affect the enforcement of seamless consistency.

System Architecture and Limitations

Figure 26.1 depicts our reference model for mobile computing environments. Desktop computers, workstations, and servers are the fixed hosts (FHs) that are interconnected by means of a fixed network. Large databases run on such FHs that guarantee efficient processing and reliable storage of data. Mobile hosts (MHs) such as palmtops, laptops, notebooks, or cellular phones are not always connected to the fixed network. They may be disconnected for different reasons. Furthermore, mobile hosts may differ in their capabilities with respect to computing power and storage space. Although notebook computers can run databases as desktop computers, other mobile devices have more limited computational resources; for example, palmtops have much less memory than a notebook computer. Still, it is possible to use small-footprint databases that allow data processing even on mobile devices with very restricted resources. After having uploaded data from an FH to their MHs, mobile users can disconnect from the fixed network. While moving (physically) they might want to run some queries and update their local data copies. The connection between the mobile and fixed hosts is performed by means of mobile support stations (MSSs), which are dedicated fixed hosts that provide the link between an MH and any FH via wireless local area network (LAN) connections (using protocols such as Bluetooth®), cells (as in the Global System for Mobile Communications [GSM] cellular phone system), or connections to the network using standard modems.

Wireless communication technology offers users the possibility of reconnecting to the fixed network anytime and anywhere to share the results of their work; however, some difficulties relevant to this task include:

- Bandwidth variability occurs as the MH changes location (e.g., due to different traffic loads or changing networks that interface with the MH).
- Unreliability can be introduced by local interruptions in signal strength or by MH hardware failures or switch-offs that cause the MH to be disconnected from the network.
- Tariff policies can make it uneconomical to be connected at certain times or in certain locations.
- MHs tend to have fewer computational and storage resources than FHs.
- Battery life is limited, causing devices to restrict their connections.
- MHs are inherently less secure and reliable due to the possibility of being damaged or even stolen.

Such limitations often result in MHs operating while disconnected from the network or connected with less than full connectivity. In addition to disconnections and weak connectivity due to power limitations, it is less expensive to receive than transmit; also, the transmission bandwidth of the mobile device tends to be lower than the transmission bandwidth of the MSSs. This situation is often referred to as *communication asymmetry*. Protocols addressed later in this chapter exploit these characteristics with regard to maintaining consistency.

Maintaining Consistency

In the simplest mode, the MH simply sends requests for data to the server; the MH has no local data, and there is no local processing. This model induces large response times and makes the operation of the MH fully dependent on the existence of a stable network connection. In this chapter, we assume that local data exists at the MH and that transaction execution is possible at both the MH and the FH. This introduces consistency issues, as updates at the various sites must be integrated to produce a single consistent database state.

A *transaction* is a unit of operations that changes a database from a consistent state into another consistent state. A database is in a *consistent state* if it satisfies all defined semantic integrity constraints and all (previous and current) database states are achieved by performing correct database transitions. A transaction model defines the framework for the definition and execution of transactions. A traditional flat transaction satisfies the well-known atomicity, consistency, isolation, durability (ACID) properties, meaning that a transaction is an atomic, a consistent, and a recoverable unit that does not interfere with other transactions that are executed concurrently. Database consistency is ensured through *serializability*, which ensures that the concurrent execution of a number of transactions is equivalent to some serial execution of these transactions; however, for performance reasons, in some application scenarios the ACID properties must be relaxed.

To be able to process information on the MH in the disconnected mode, data must be replicated on the MH. *Data replication* is the process of maintaining a defined set of data in more than one location. It involves copying designated changes from one location (source or master) to another (target or remote) and synchronizing the data in both locations. An MH has its own complete or partial copy of the master database (that resides on the FH) so it can process data locally. In general, data replication provides increased data availability and faster response times.

The increased flexibility permitted by replication introduces several challenges. Whenever an update is performed on any of the databases, the update must be propagated to the other databases to maintain a consistent state in all databases. A strategy to propagate the update can be:

- *Eager* — All replicas are updated synchronously as part of the originating transaction. This approach is often also called *synchronous replication*.
- *Lazy* — A replica is updated by the originating transaction, and propagation to other replicas is asynchronous, typically as a separate transaction for each replica. This approach is often also referred to as *asynchronous replication*.

Update propagation can also be controlled by means of data ownership:

- *Master* — Each data object has a master site; only the master can update the primary copy of the data object on its local and remote databases.
- *Group* — Any site with a copy of a data item can update it.

In this chapter, we focus on consistency at both the transaction level (i.e., maintaining database integrity constraints) and the single-item level (i.e., replica coherency, or ensuring that all copies of an item converge to a single value). We do not, however, delve into issues related to advanced transaction models and their treatment; recent surveys on this topic include Bernard et al. [6] and Serrano-Alvarado et al. [30].

Consistency and Disconnected Operation

Disconnected operation refers to the autonomous operation of an MH when network connectivity becomes unavailable or undesirable. Prior to a disconnection, data items are preloaded to the MH to allow operation during disconnection. Data preloading or prefetching to sustain a forthcoming disconnection is termed *hoarding*. During the time the MH is disconnected from the network, data may be created, modified, or deleted at either the MH or the FH. After reconnection, any operations performed by the MH should be integrated with relevant operations at other sites to achieve seamless consistency. In general, disconnected operation can be described as a transition between three states: (1) hoarding, (2) disconnected operation, and (3) reintegration or synchronization [15] (see Figure 26.2a). In terms of database systems, synchronization can be performed in basically two ways:

- *Session-based* — An MH database directly connects to the FH database. During connection, synchronization takes place. Only when synchronization is complete can the MH and the FH disconnect. When disconnecting, both MH and FH have a consistent database state. This approach is also referred to as *connection-based* synchronization.
- *Message-based* — A messaging system (such as e-mail) transfers the information required to perform the reconciliation. Synchronization is performed when the MH and the FH are disconnected. This strategy is also known as *store-and-forward* or *file-based* synchronization.

In general, synchronization may be either *one way*, in which case the MH communicates its operations to the FH (or *vice versa*), or *two way*,

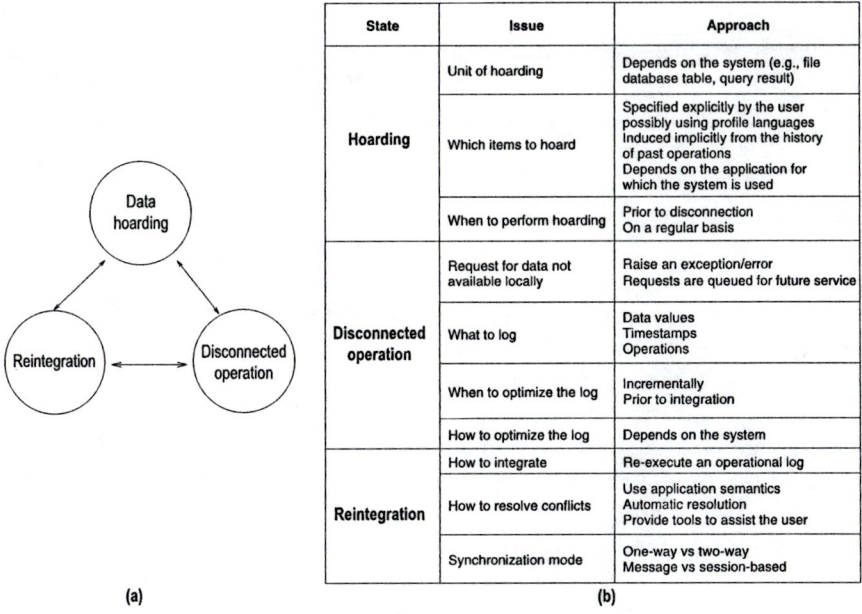

State	Issue	Approach
Hoarding	Unit of hoarding	Depends on the system (e.g., file database table, query result)
	Which items to hoard	Specified explicitly by the user possibly using profile languages Induced implicitly from the history of past operations Depends on the application for which the system is used
	When to perform hoarding	Prior to disconnection On a regular basis
Disconnected operation	Request for data not available locally	Raise an exception/error Requests are queued for future service
	What to log	Data values Timestamps Operations
	When to optimize the log	Incrementally Prior to integration
	How to optimize the log	Depends on the system
Reintegration	How to integrate	Re-execute an operational log
	How to resolve conflicts	Use application semantics Automatic resolution Provide tools to assist the user
	Synchronization mode	One-way vs two-way Message vs session-based

(a)

(b)

Figure 26.2   Disconnected operation: (a) states and (b) related issues.

in which case both the MH and the FH exchange information about their operations.

Data Hoarding

In the *data hoarding state*, in anticipation of a disconnection, data items required for operation are either cached (replicated) or moved (partitioned) at the MH to allow the MH to operate while disconnected. For foreseeable disconnections (for example, when a user enters a region with expensive communication prices and decides to operate disconnected), data hoarding may be performed just before the disconnection. To sustain less predictable disconnections, hoarding may have to be deployed on a regular basis (e.g., periodically); for example, the Coda file system [15] runs a process called *hoard walk* periodically to ensure that critical files are in the cache of the MH. The granularity of hoarding varies. As an example, in the case of file systems the data items prefetched may be files, directories, or volumes, but in the case of database management systems they may be relations (tables) or views. Additional information such as a set of allowable operations or a characterization of data fidelity may also be cached along with the data. Figure 26.2b depicts some of the issues related to each of the three states of disconnected operation, which we will now explore further.

An important issue is determining which items to hoard. This decision may be either (1) assisted by instructions explicitly given by the user, or (2) made automatically by the system by utilizing implicit information, which is most often based on the past history of data references. Maintaining consistency in the case of shared resources adds to the complexity of hoarding. Taking into consideration the probability of conflicting operations when deciding which items to hoard may improve the effectiveness of disconnected operation. In the case of databases, a logical approach would be to hoard by issuing queries (i.e., by prefetching the data items that constitute the answer to a given query, as in semantic caching) [28]. This, in a sense, corresponds to loading on the MH materialized views. Operation during disconnection is then supported by posing queries against these views.

Badrinath and Phatak [3] proposed an extended database organization to efficiently handle hoarding queries from MHs. Under the proposed organization, the database designer can specify a set of *hoard keys* along with the primary and secondary keys for each relation. Hoard keys are supposed to capture typical access patterns of MHs. Each hoard key partitions the relation into a set of disjoint logical horizontal fragments. Hoard fragments constitute the hoard granularity; that is, MHs can hoard and reintegrate within the scope of these fragments. Finally, in *profile-driven data prefetching* [8], a simple profile language is introduced to allow users to specify the items of interest to be prefetched along with their relative importance through weighting.

Disconnection

While *disconnected*, an MH uses only local data. Operations performed at the MH during disconnection are reported in a system log. A related issue is what information to keep in the log. The type of information affects the effectiveness of reintegration of updates upon reconnection as well as the effectiveness of log optimizations. Optimizing the log by keeping its size small is important for at least two reasons: (1) to save local memory at the MH, and (2) reduce the time for update propagation and reintegration during reconnection.

Consistent operation during disconnected operation has been extensively addressed in the context of *network partitioning*. In this context, a network failure partitions the sites of a distributed database system into disconnected clusters. Various approaches have been proposed and are excellently reviewed in Davidson et al. [9]. In general, such approaches can be classified along two orthogonal dimensions. The first concerns the tradeoff between consistency and availability, for which two general approaches have been devised. In the *pessimistic approach*, updates are performed only at one site using locking or some form of check-in/check-out. In the *optimistic approach,* updates are allowed at more than one site with the possible danger of conflicting operations. The second dimension concerns the level of semantic knowledge used in determining correctness. Network partitioning is usually concerned with peer-to-peer models where transactions executed in any partition are of equal importance; however, in mobile computing, transactions at the mobile host are most often considered second class. Furthermore, network partitions correspond to failure behavior, whereas disconnections in mobile computing are common. Finally, most disconnections in mobile computing can be considered foreseeable.

Reintegration

Upon reconnection, the MH enters the *reintegration state.* In this state, operations performed at the mobile host are reintegrated with updates performed at other sites. Update reintegration is usually performed by re-executing the log at the FH. Whether or not the operations performed at the disconnected sites are accepted depends on the consistency semantics adopted by the particular system. Such consistency semantics vary from enforcing transaction serializability to resolving only concurrent updates of the same object.

In the case of files systems, the only conflicts detected are write/write conflicts because they produce divergent copies; read/write conflicts are not considered. Such conflicts occur, for example, when the value of a file read by a disconnected MH is not the most recent one, because the file has been updated at the FH after disconnection of the MH. In Coda, the system log is executed as a single transaction [16]. All objects referenced in the log are locked. File resolution is based on *application-specific resolvers* (ASRs) per file. An ASR is a program that encapsulates the knowledge necessary for file resolution and is invoked at the MH when divergence among copies is detected.

In transaction-oriented systems, a common trend is to tentatively commit transactions executed at the disconnected MH and make their results visible to subsequent transactions in the same MH. Upon reconnection, a certification process takes place, during which the execution of any tentatively committed transaction is validated against an application- or system-defined correctness criterion. If the criterion is met, the transaction is committed; otherwise, the execution of the transaction must be aborted, reconciled, or compensated. Such actions may have cascading effects on other tentatively committed transactions that have seen the results of the transaction.

Isolation-only transactions (IOTs) [18] provide support for transactions in file systems. An IOT is a sequence of file access operations. Transaction execution is performed entirely on the MH, and no partial result is visible on the FHs. A transaction (*T*) is considered a *first-class transaction* if it does not have any partitioned file access (i.e., the MH maintains a connection for every file it has accessed); otherwise, T is considered a *second-class transaction*. Whereas the result of a first-class transaction is immediately committed to the FHs, a second-class transaction remains in the pending state until connectivity is restored. The result of a second-class transaction is held within the local cache of the MH and visible only to subsequent accesses on the same MH. Second-class transactions are guaranteed to be locally serializable among themselves. A first-class transaction is guaranteed to be serializable with all transactions that were previously resolved or committed at the FH. Upon reconnection, a second-class transaction T is validated against one of two proposed serialization constraints. The first is global serializability, which means that, if the local result of a pending transaction was written to the FH as is, it would be serializable with all previously committed or resolved transactions. The second is a stronger consistency criterion referred to as *global certifiability*, which requires that a pending transaction be globally serializable not only with but also after all previously committed or resolved transactions.

In the *two-tier replication* [11], replicated data has two versions at mobile nodes: master or tentative. A master version records the most recent value received while the site was connected. A tentative version records local updates. There are two types of transactions analogous to second- and first- class IOTs: tentative transactions and base transactions. A *tentative transaction* works on local tentative data and produces tentative data. A *base transaction* works only on master data and produce master data. Base transactions involve only connected sites. Upon reconnection, tentative transactions are reprocessed as base transactions. If they fail to meet the application-specific acceptance criteria, they are aborted.

In *promotion* [36,37], the unit of caching and replication is a *compact*. When an MH requires data, it sends a request to the FH database. The FH sends a compact as a reply. A compact is an object that encapsulates the cached data, operations for accessing the cached data, state information (such as the number of accesses to the object), consistency rules that must be followed to guarantee consistency, and obligations (such as deadlines). Compacts provide flexibility in choosing consistency methods from simple check-in/check-out pessimistic schemes to complex optimistic criteria.

Consistency and Weak Connectivity

Mobile hosts often operate with less than full connectivity; in particular, wireless network connectivity may be slow or expensive. In addition, connectivity is often lost for short periods of time or varies in the bandwidth provided or reliability level offered. Disconnections correspond to the extreme case of total lack of connectivity. Connectivity constraints affect the protocols for enforcing consistency. The majority of such protocols are also asynchronous, as in the case of disconnections; however, they are more general, because in handling disconnections the focus is mainly on synchronizing data upon reconnection. Here, update propagation may be lazy or based on epidemic propagation, and the replication protocols used are not necessarily based on primary copy schemes. We discuss weak connectivity in the context of both cache and transaction-oriented consistency.

Cache-Related Consistency

In file systems, weak connectivity is handled by appropriately revising those operations whose deployment involves the network. In particular, for caching, approaches to weak connectivity are centered around the following three topics that affect bandwidth consumption: (1) handling cache misses, (2) the frequency of propagation to the server of updates performed at the MH cache, and (3) the validity of the value of cached items. Several design choices must be made with regard to these issues. We discuss them in the context of caching, but the discussion is directly applicable to replication as well.

Servicing a cache miss may incur very long delays in slow networks or excessive costs in expensive ones; thus, cache misses may be serviced selectively based on how critical the required item is and on the quality of the current network connection. Determining when to propagate cache updates and integrate them at the server requires an interplay among various factors. Aggressive reintegration reduces the effectiveness of log optimizations, because records are propagated to the server early; thus, they have less opportunity to be eliminated at the MH. Short-lived temporary files, for example, are usually eliminated if they stay in the log long enough. Early reintegration can also affect the response times of other traffic, especially in slow networks. On the other hand, it achieves consistent cache management and timely propagation of updates and reduces the probability of conflicting operations. Furthermore, early reintegration keeps the log in the memory of the MH short, thus saving space. In addition, lazy reintegration may overflow the MH cache, because cached data that has been updated cannot be discarded before being committed at the server. Regarding the validity of cached items, notifying the MH each time an item is changed at the server may be too expensive in terms of bandwidth. Postponing the notification results in cache items having obsolete values and affects the value returned by read operations. Another possibility is to update cache items on demand (i.e., each time an MH issues a read operation on an item). Alternatively, a read operation may explicitly contact the server to attain the most recent value.

In Coda, cache misses are serviced selectively [20]; in particular, a file is fetched only if the service time for the cache miss (which depends, among other things, on bandwidth) is below the user's patience threshold for this file (e.g., the time the user is willing to wait to get the file). Reintegration of updates to the servers is done through trickle reintegration, an ongoing background process that propagates updates to servers asynchronously. A record becomes eligible for reintegration only after spending a minimal amount of time (called the *aging window*) in the log. Because transferring the log in one chunk may saturate a slow network for an extended period, the reintegration chunk size is made adaptive, thus bounding the duration of communication degradation.

The variable-consistency approach [34] deploys a client–server architecture with replicated servers that follow a primary/secondary schema mainly to avoid global communication but also to handle weak connectivity. The client communicates with the primary server only. The primary server makes periodic pickups from its clients and propagates updates back to the secondaries asynchronously. When some number N of secondaries have acknowledged receipt of an update, the primary informs the client that the associated cached update has been successfully propagated and can be discarded. The traditional read interface is split into *strict* and *loose* reads. Loose read returns the value of the cache copy, if such a copy exists; otherwise, loose read returns the value of the copy at the primary or any secondary, whichever it finds. In contrast, the strict read call returns the most consistent value by contacting the necessary number of servers and clients to guarantee retrieving the most up-to-date copy.

Ficus and its descendant Rumor [23] are examples of file systems following a peer-to-peer architecture. No distinction is made between copies at the mobile and copies at the fixed host; all sites store peer copies of the files they replicate. Updates are applied to any single copy. The file system is organized as a directed acyclic graph of volumes. A volume is a logical collection of files that are managed collectively. A pair-wise reconciliation algorithm is executed periodically and concurrently with respect to normal file activity. The state of the local replicated volume is compared to that of a single remote replica of the volume to determine which files must have updates propagated. The procedure continues until updates are propagated to all sites storing replicas of the volume.

Transaction-Oriented Consistency

Similar to cache-based approaches, the approaches to ensure transaction-oriented consistency under weak connectivity aim at minimizing the communication cost and surviving short disconnections; however, due to the complicated dependencies among database items, the problem is more complex. Again as in the case of disconnected operation, the focus is on asynchronous replication protocols that allow data updates to be propagated outside transaction boundaries.

The two-layered transaction model [38] distinguishes between transactions running at the MH and the FH. Transactions that run solely at the mobile host are *weak*, and the rest are *strict*. A distinction is drawn between weak copies and strict copies. In contrast to strict copies, weak copies are only tentatively committed and hold possibly obsolete values. Weak transactions update weak copies, and strict transactions access strict copies located at any site. Weak copies are integrated with strict copies either when connectivity improves or when an application-defined limit to the allowable deviation among weak and strict copies is passed. Before reconciliation, the results of weak transactions are visible only to weak transactions at the same site. Strict transactions are slower than weak transactions because they involve the wireless link but guarantee permanence of updates and currency of reads. During disconnection, applications can use only weak transactions. In this case, weak transactions have similar semantics with second-class IOTs [18] and tentative transactions [11]. Adaptability is achieved by restricting the number of strict transactions depending on the available connectivity and by adjusting the application-defined degree of divergence among copies.

The Bayou system [10,35] does not support full-fledged transactions. Bayou is built on a peer-to-peer architecture with a number of replicated servers weakly connected to each other. In this schema, a user application can read any and write any available copy. Writes are propagated to other servers during pair-wise contracts called *anti-entropy* sessions. When a write is accepted by a Bayou server, it is initially deemed tentative. As in two-tier replication [11], each server maintains two views of the database: a copy that reflects only committed data and another full copy that also reflects the tentative writes currently known to the server. Eventually, each write is committed using a primary-commit schema; that is, one server designated as the primary takes responsibility for committing updates. Because servers may receive writes from clients and other servers in different orders, servers may have to undo the effects of some previous tentative execution of a write operation and reapply it. The Bayou system provides dependency checks for automatic conflict detection and merge procedures for resolution. Instead of transactions, Bayou supports *sessions*. A session is an abstraction for a sequence of read and write operations performed during the execution of an application. *Session guarantees* are enforced to avoid inconsistencies when accessing copies at different servers; for example, a session guarantee may be that read operations reflect previous writes or that writes are propagated after writes that logically precede them. Different degrees of connectivity are supported by individually selectable session guarantees, by choices of committed or tentative data, and by placing an age parameter on reads. Arbitrary disconnections among Bayou's servers are also supported because Bayou relies only on pair-wise communication; thus, groups of servers may be disconnected from the rest of the system yet remain connected to each other.

Deno [7] extends Bayou in that no primary server owns an item and serializes the updates on that item. This is achieved through a combination of weighted voting and epidemic propagation. The use of voting allows the system to have better availability than primary copy schemes. Weighted voting improves performance by adapting currency distribution to site availabilities, update activities, or other parameters. Deno supports both weak and strong consistency in the following sense. A read-only transaction sees *weak consistency* if it is serialized with respect to all update transactions, but possibly not with other read-only transactions. A read-only transaction sees *strong consistency* if it is serialized with respect to both read-only and update transactions.

Consistency in Hybrid Environments

In traditional client–server systems, data is delivered on a demand basis. A client explicitly requests data items from the server. When a data request is received at a server, the server locates the information of interest and returns it to the client. This form of data delivery is referred to as *pull-based*. In wireless computing, the stationary server machines are provided with a relative high-bandwidth channel that supports broadcast delivery to MHs in their cell. This facility provides the infrastructure for *push-based* data delivery. In push-based data delivery, the server sends data to a client population without a specific request. Clients monitor the broadcast and retrieve the data items they require as they arrive. Data of interest may also be cached locally at the client. Push-based delivery is important for a wide range of applications that involve dissemination of information to a large number of clients. We consider a hybrid (push and pull) environment in which each MH reads data from (1) the broadcast channel, (2) its local cache or database, or (3) directly from the FH (database) server through pull requests (Figure 26.3). Data broadcast is often periodic.

When the value of the broadcast data is allowed to be updated, the need for consistency control protocols arises. Such protocols vary depending on various parameters. First, protocols depend on the assumptions made about data delivery — for example, the existence of a backchannel for on-demand data delivery or the cache capability of clients. We focus on maintaining both cache- and transaction-oriented consistency. We distinguish two cases based on whether clients are allowed to read data

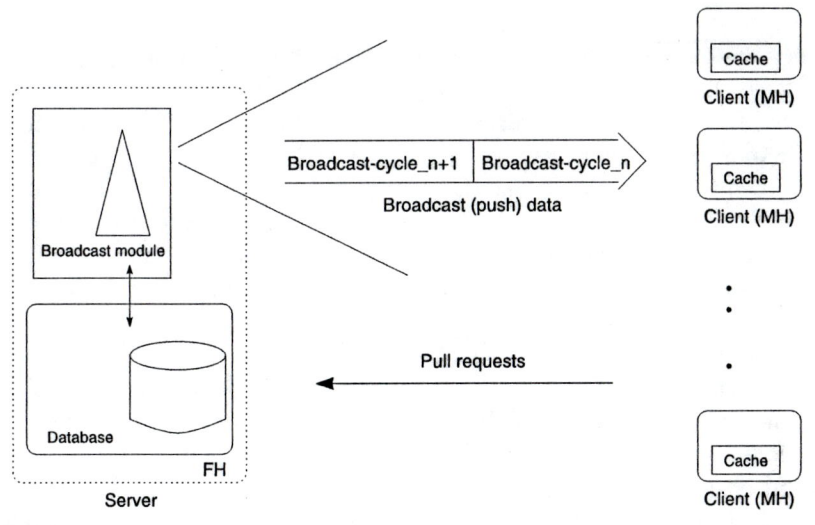

Figure 26.3   Hybrid data delivery.

directly from the broadcast or the broadcast is used only to communicate consistency-related information to clients.

Cache-Related Consistency

We next present protocols that aim at ensuring that the data values read by the MHs are kept up to date with respect to the values of the items at the server. We first discuss the case in which the server at the FHs uses the broadcast channel to push control information to its clients, then we discuss the case where the MHs also read data items from the broadcast channel.

Pushing Control Information

The server can use the broadcast to inform its clients of updates of items in their cache. Invalidation of cache entries may be performed either asynchronously or synchronously [5]. In *asynchronous* methods, the server broadcasts an invalidation report for a given item as soon as this item changes its value. In *synchronous* methods, the server broadcasts an invalidation report periodically. A client has to listen for the report first to decide whether its cache is valid or not; thus, each client is confident of the validity of its cache only as of the last invalidation report. This adds some latency to query processing, because to answer a query a client has to wait for the next invalidation report.

Invalidation reports vary in the type of information they convey to the clients; for example, the reports may contain the values of the items that have been updated or just their identity and the time stamps of their last updates. Including in the report the updated values can be wasteful of bandwidth, especially when the corresponding items are cached at only a few clients. On the other hand, if the values are not included, the client must either discard the item from its cache or communicate with the server to receive the updated value. The reports can provide information for individual items or aggregate information for sets of items.

Cache invalidation protocols make different assumptions about whether or not the server maintains any information about which clients it is serving, what the contents of their cache are, and when their cache was last validated. Servers that hold such information are *stateful*, and servers that do not are *stateless*.

Three synchronous strategies for stateless servers are proposed in Barbará and Imielinski [5]. In the broadcasting *time stamps* (TS) strategy, the invalidation report contains the time stamps of the latest change for items that have had updates in the last w seconds. In the *amnestic terminals* (AT) strategy, the server only broadcasts the identifiers of the items that changed since the last invalidation report. In the *signatures* strategy, signatures are broadcast. A signature is a checksum computed over the value of a number of items by applying data compression techniques similar to those used for file comparison. Each of these strategies is shown to be effective for different types of clients. Clients that are often connected are *workaholics*, and clients that are often disconnected are *sleepers*. Signatures are best for long sleepers — that is, when the period of disconnection is long and difficult to predict. The AT method is best for workaholics. Finally, TS is shown to be advantageous when the rate of queries is greater than the rate of updates, provided that the clients are not workaholic.

An asynchronous method based on bit sequences is proposed in Jing et al. [14]. In this method, the invalidation report is organized as a set of bit sequences with an associated set of time stamps. Each bit in the sequence represents a data item in the database. A bit "1" indicates that the corresponding item has been updated since the time specified by the associated time stamp, and a bit "0" indicates that the item has not changed. The set of bit sequences is organized in a hierarchical structure. It is shown that the algorithm performs consistently well under conditions of variable update rate and client disconnection times. A scalable version of the algorithm is also presented. In this version, rather than increasing the number of bits in the report for large databases, the granularity of each bit is increased so it represents groups of rarely changed items, instead of just one item.

Care must be taken so disconnected clients that miss invalidation reports can reuse part of their cache. A client may miss cache invalidation reports because it is disconnected. Synchronous methods surpass asynchronous ones in that clients must only periodically tune in to read the invalidation report instead of continuously listening to the channel; however, if the client remains inactive longer than the period of the broadcast, the entire cache must be discarded, unless special checking is deployed.

Reading from the Broadcast

When, besides invalidation reports, clients also read data items from the broadcast, several different data consistency models are reasonable [1]. For example, if clients do not cache data, if the server always broadcasts the most recent values, and if there is no backchannel for on-demand data delivery, then the *latest value* model [1] is a model that arise naturally. In this model, clients read the most recent value of a data item. This model is weaker than serializability because there is no notion of transactions; that is, operations are not grouped into atomic units. When clients cache data but are not allowed to perform any updates, an appropriate consistency model is *quasi-caching* [2]. In this model, although the value of the cached data may not be the most recent one, this value is guaranteed to be within an allowable deviation as specified through per-client coherency conditions. Quasi-caching is a reasonable choice in the case of long disconnections or weak connectivity.

Transaction-Oriented Consistency

Whereas cache consistency protocols focus on ensuring the currency (or temporal coherency) of the data read by the clients, the protocols presented next seek to ensure in addition some form of serializability among transactions executed at the clients and transactions executed at the server.

Pushing Control Information

The broadcast facility can be exploited in various algorithms for concurrency control. Using the broadcast facility in optimistic concurrency control protocols to invalidate some of the client transactions is suggested in Barbará [4]. In optimistic concurrency control, the transaction scheduler at the server checks at commit time whether or not the execution that includes the client's transaction to be committed is serializable. If it is, it accepts the transaction; otherwise, it aborts it. In the proposed enhancement of the protocol, the server periodically broadcasts to its clients a *certification report* (CR) that includes the readset and writeset of active transactions that have declared their intentions to commit to the server during the previous period and have successfully been certified. The MH uses this information to abort from its transactions those transactions whose readsets and writesets intersect with the current CR; thus, part of the verification is performed at the MH. Only when the MH cannot detect any conflict is the server involved in completing the verification. If the transaction can commit, the server will install the values in the central database and notify the MHs via broadcast.

Reading from the Broadcast

In this case, clients perform transactions that involve data items at their local cache, the server, and the broadcast channel. Protocols in this category differ on whether clients are allowed to perform updates. In this case, updates at the client are performed at local copies. Client transactions are subsequently validated at the server. Read-only client transactions are often allowed to commit locally.

A characterization of currency (temporal coherency) and consistency (semantic coherency) of broadcast data as well as general techniques for enforcing them are given in Pitoura et al. [27]. Currency characterizes the freshness of the values seen by the clients with regard to the values at the server as well as the temporal discrepancy among the values read by the same transaction. Five different forms of consistency are presented based on relaxing serializability. Similar isolation levels for read-only transactions are proposed in Seifert and Scholl [29].

Different methods for enforcing the currency and consistency of client read-only transactions are evaluated in Pitoura et al. [25]. With the *invalidation* method, the server broadcasts an invalidation report with the data items that have been updated since the broadcast of the previous report. Transactions that read obsolete items are aborted. With the *serialization graph testing* (SGT) method, the server broadcasts control information related to conflicting operations. Clients use this information to ensure that their read-only transactions are serializable with the server transactions. With *multiversion broadcast* [24,26], multiple versions of each item are broadcast, so client transactions always read a consistent database snapshot. Multiple versions of data items are also explored in Seifert and Scholl [29] for supporting read-only transactions with varying degrees of isolation.

The approach reported by Shanmugasundaram et al. [31] allows updates at the client. It enforces a weaker form of serializability called *update consistency*. Update consistency requires that (1) all update transactions are serializable, and (2) each read-only transaction is serializable with respect to the subset of update transactions it (directly or indirectly) reads from. The values read are current as of the beginning of the broadcast cycle. To ensure update consistency, besides serializing all transactions submitted to it (including those possibly originated at the clients), the server transmits a *control matrix* to its clients. The control matrix includes information about conflict operations and is used by the client to determine whether a read operation can proceed. To minimize the size of the control matrix, two approximations are proposed: the F-matrix and the R-matrix. When compared to the SGT method, besides supporting client updates, the control matrix differs in that conflict information is broadcast per pair of data items, whereas with SGT conflict information is represented in the form of a serialization graph.

To support partial validation of client transactions locally at the clients, an approach based on *time-stamp ordering* is proposed in Lee et al. [17]. The server broadcasts along with the data items their read and write time stamps, so clients can partially validate their transactions locally before sending them to the server for final validation.

Consistency Support in Commercial Mobile Database Systems

Commercial mobile database systems focus mainly on handling disconnected operation. All commercially available mobile databases employ a similar data replication approach: one or more *remote* databases on the mobile hosts have replicas of the *master* database stored in the fixed host. Changes are captured locally to each database. When reconnected to the fixed network, the MH may initiate a synchronization that reconciles the copies of the databases by creating a unique consistent database state on remote and master databases. This section provides an overview of the following current commercial mobile database solutions: SyncML® and HotSync®, IBM DB2® Everyplace®, Oracle Lite™, Microsoft SQL Server CE, and Sybase Anywhere™. We briefly describe these systems with respect to data synchronization, conflict resolution, and transaction execution. Table 6.1 summarizes briefly the approach taken by each system.

SyncML and HotSync

Due to the great importance of synchronization in mobile computing, and especially on palm-like devices, an industry consortium of several major software firms has defined a standard synchronization protocol called *SyncML* [33]. This protocol specifies cross-format and cross-platform synchronization capabilities. SyncML works with a standard set of messages represented as eXtensible Markup Language (XML) documents and defines seven different synchronization modes:

Table 26.1 Connectivity and Commercial Mobile Database Systems

Commercial Mobile DB	Hoarding Unit	Updates at the Client	Synchronization; Protocol	Conflict Resolution
IBM DB2 Everyplace	Synchronization objects; single tables, or unions and joins of tables	Only on updatable synchronization objects defined on a single table with neither aggregate functions nor duplicate elimination	Two-way; message-based	Transaction level
Oracle Lite	Snapshots (materialized views)	Only on updatable snapshots that correspond to simple snapshots that do not contain aggregate functions, grouping elements, duplicate elimination, subqueries, joins, and set operations	Two-way; complete and incremental	Update, uniqueness, and delete conflicts (either the MH or the FH wins)
Microsoft SQL Server CE	Tables and subsets of rows and columns	Permitted	One- and two-way; message-based	Priority-based
Sybase Anywhere	Any (SQL queries)	Permitted	Two-way; message-based and session-based	Business rules

- *Two-way sync* — The MH and FH exchange information about modified data in these devices. The MH sends the modifications first.
- *Slow sync* — All items are compared with each other on a field-by-field basis. In practice, this means that the MH sends all its data to the FH and the FH does the sync analysis (field by field).
- *One-way sync from MH only* — The MH sends its modifications to the FH, but the FH does not send its modifications back to the MH.
- *Refresh sync from MH only* — The sends all its data from a database to the FH (i.e., exports). The FH is expected to replace all data in the target database with the data sent by the MH.
- *One-way sync from FH only* — The MH gets all modifications from the FH, but the MH does not send its modifications to the FH.
- *Refresh sync from FH only* — The FH sends all its data from a database to the MH. The MH is expected to replace all data in the target database with the data sent by the FH.
- *FH-alerted sync* — The FH triggers the MH to start a synchronization.

Palm's HotSync is an example of a widely used synchronization protocol. HotSync supports a two-way synchronization that enables updates to be performed on the MH and the FH. Every data record on the MH has a set of status bits that indicate whether the record has been created, modified, or deleted after the last synchronization. The FH maintains a copy of each MH that connected to that FH to perform synchronization. The synchronization process is initiated by the MH. Two synchronization modes are distinguished:

- *Fast*, if the MH was last synchronized with the FH. The MH sends to the FH only the records that have changed from the last synchronization depending on the value of the status bits. The FH updates its copy and sends updates to the MH.
- *Slow*, if the last synchronization times of the MH and FH do not match. In this case, the synchronization software performs a field-by-field comparison to capture the changes.

Depending on the status bit values, HotSync determines which actions to perform (see Table 26.2).

When updates are in conflict, *conflict resolution* is performed. For example, a row in a database table, present in the master and some remote databases, could be modified by two different users, one working on its MH copy and another working on the master copy residing on the FH. The two users would request a modification to the master database, thus causing a dilemma: Which of the two updates is the one that should be saved to the FH and propagated to the other database replicas? Several

Table 26.2 HotSync Logic

Status on MH	Status on FH	Action
Created	Not present	Send to FH
Not present	Created	Send to MH
Deleted	No change	Delete from FH
Deleted	Updated	Send to MH
No change	Updated	Send to MH
Updated	Updated	Conflict resolution

conflict-resolution techniques can be devised; for example, Intellisync® for PalmPilot™ by Pumatech implements five options:

- All conflicting items are added; this option creates entirely new records on both the MH and the FH for the conflicting fields.
- All conflicting items are ignored; this option ignores any differences between fields in the MH and the FH.
- A user notification is sent.
- MH wins and overrides data on the FH.
- FH wins and overrides data on the MH.

Note that the choice of the conflict-resolution strategy is application dependent.

IBM DB2 Everyplace

IBM's solution [12,13] for mobile computing relies on a three-tier architecture, as depicted in Figure 26.4. The following components of this architecture are relevant for mobile usage:

- *DB2 Everyplace* is a small-footprint relational database for mobile and embedded devices. It provides a Structure Query Language (SQL) interface to create, manipulate, and query tables; however, stored procedures, user-defined functions, triggers, views, and subqueries are not supported. An additional graphical user interface (GUI) supports querying by example (QBE).
- *Sync Client* is an application residing on the MH that triggers synchronization between the DB2 Everyplace and the master database; for that, it communicates with the Sync Server. The synchronization follows the specification of SyncML.

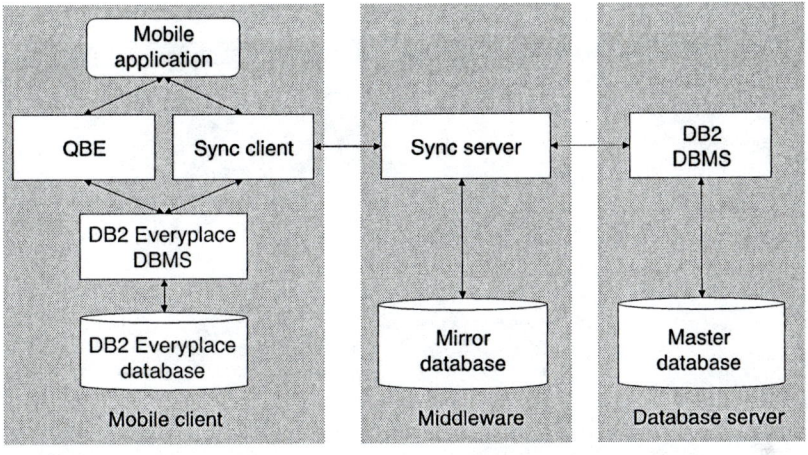

Figure 26.4   IBM DB2 Everyplace architecture.

- *Sync Server* is a middle-tier application that coordinates synchronization between the mobile user and the master database. It authenticates the mobile user and propagates all changes of the mobile database to the master database and *vice versa*.

DB2 Everyplace allows synchronization objects to be defined that represent the part of the master database to be replicated. Synchronization objects can correspond to single tables or unions and joins of tables. Filter rows and columns can be selected. Two types of synchronization objects are distinguished:

- *Updatable* synchronization objects rely on a single table and neither contains aggregate functions nor duplicates elimination. Also, integrity constraints defined on the master tables cannot be replicated on the MH.
- *Read-only* synchronization objects allow arbitrary (materialized) views on the master database.

The updates on updatable synchronization objects are performed within the frame of transactions. DB2 Everyplace supports savepoints but no transaction nesting; however, transactions can be executed on the MH during disconnections. The consistency between the mobile and the master database is ensured by a synchronization process. DB2 Everyplace allows a two-way synchronization using a message-based protocol. Replication is asynchronous. It is possible to control how frequently the changes are applied to the target by specifying time intervals, events, or both. But, for mobile environments that have occasionally connected clients, data is replicated *on demand*.

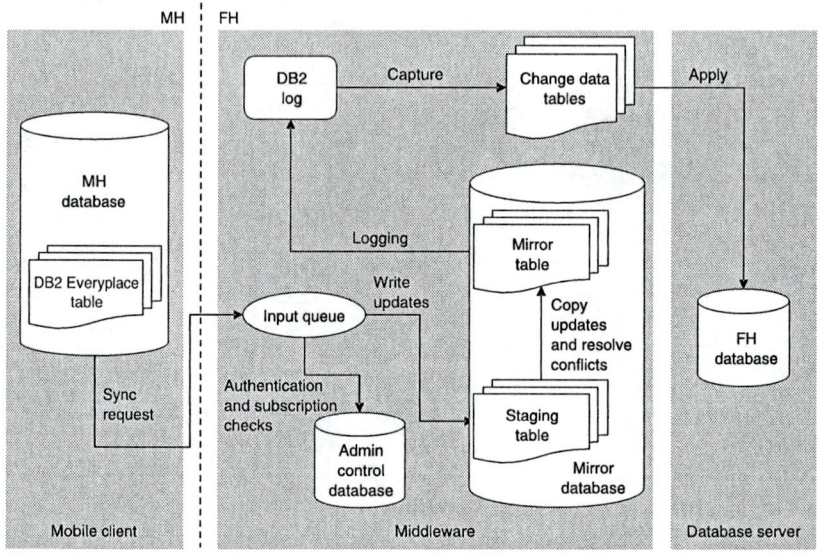

Figure 26.5 Sync from MH to FH.

The propagation of the changes of the mobile database to the master database can be initiated either by the mobile user or the mobile application using the Sync Client. Figure 26.5 depicts the data synchronization flow when the MH database submits changes to the FH database. After a change has been made to the MH database, the user selects synchronization from the device display. The synchronization request (holding the data to be synchronized) is saved in an *input queue* on the FH, waiting for a reply from the FH database. After the request is acknowledged, the request data is written in a *staging table*, where the data waits for other updates to be completed. The data is then copied in the *mirror table*, where potential conflicts are resolved. Changes to the mirror table are recorded in the *DB2 log* files. The *DataPropagator* capture program records all the changes in the mirror table and sends them to a *change data table*, where the FH database collects its updates. The *administration control* database checks the authentication and subscriptions.

Figure 26.6 illustrates the reverse process — the synchronization flow when the FH database submits its changes to the MH databases. We do not describe again here all the steps involved. It is worth pointing out, though, that the synchronization server sends to an MH database only the data to which the MH is subscribed.

Conflicts are handled by the Sync Server running on the FH. This component logs and checks the version of each record in each table in the replication subscription for the FH and MH databases. The Sync Server

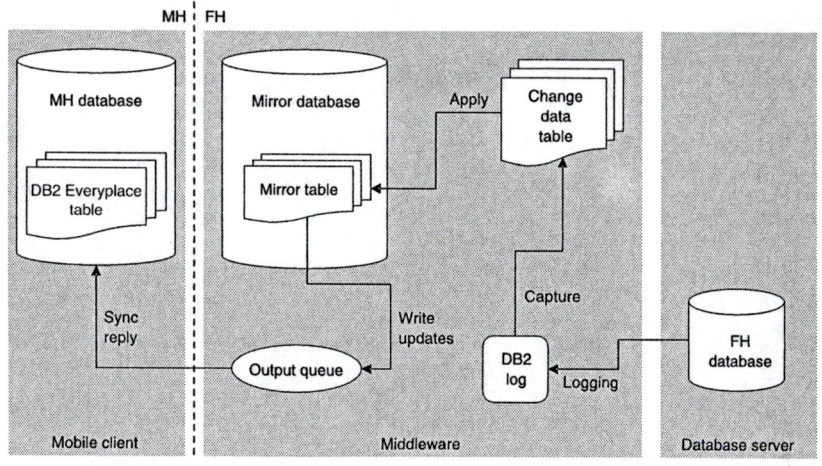

Figure 26.6 Sync from FH to MH.

can therefore determine whether or not an MH is trying to update a row based on an obsolete version of the data for that row. In such cases, the update is rejected. The conflict resolution is performed when data is staged to the mirror table.

Oracle Lite

Oracle Lite is a small-footprint, Java-enabled database that can run on mobile clients. Currently, it is available as Oracle Database Lite 10g [21,22]. As illustrated in Figure 26.7, Oracle Lite relies on a three-tier architecture. In the following, we present Oracle Lite in cooperation with an Oracle 10g master database on the server side. The following components of the Oracle Lite architecture are relevant with respect to mobility and consistency issues:

- *Mobile SQL* provides an interactive SQL interface to an Oracle Lite database. This interface allows the execution of SQL commands on the mobile database (e.g., tables and views can be created, queried, and manipulated).
- *Mobile Sync* is a small-footprint application that runs on the mobile client. It enables a synchronization between the Oracle Lite and the master database based on *materialized views* (also called *snapshots*).
- *Mobile Server* is a middleware application that maintains the materialized views. These materialized views are created for each user. When a synchronization is triggered, it coordinates the synchroni-

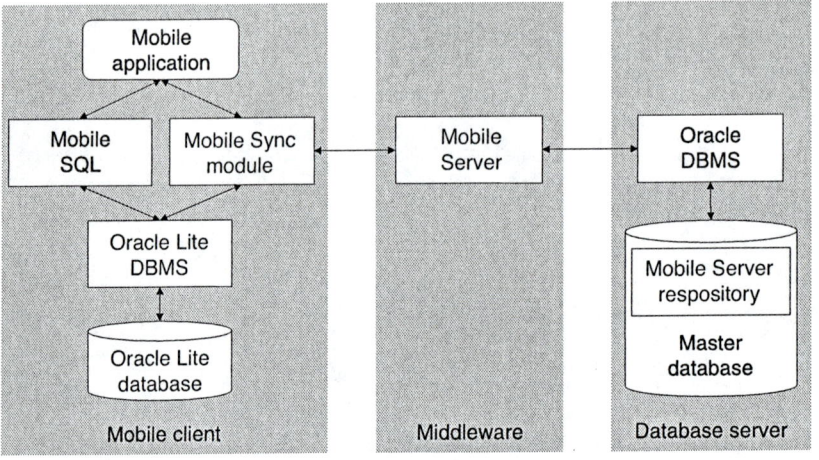

Figure 26.7 Oracle Lite architecture.

zation process. In particular, it authenticates the user and propagates all relevant changes from Oracle Lite to the master database and *vice versa*.

■ *Mobile Server Repository* contains all the information required to run the Mobile Server (e.g., the user profiles and definitions of the materialized views). The information is usually stored in the master database.

Data on a mobile database is replicated from the master database. The replication takes place in the form of snapshots that are created by the Mobile Server as part of the application installation. Snapshots allow the specification of parameterized SQL queries. In this way, it is possible to replicate entire tables, subsets of columns, and subsets of rows, as demanded. A snapshot is *simple* if it does not contain aggregate functions, grouping elements, duplicate elimination, subqueries, joins, and set operations; otherwise, it is *complex*. Oracle Lite distinguishes two types of snapshots:

■ *Read-only* snapshots are used when only local queries will run on the mobile database. In this case, no changes can be done by local processing. Only changes on the master database can be written to the snapshot. This approach implements a *unidirectional* replication.

■ *Updatable* snapshots are used when both queries and updates will be supported on the mobile database. Changes to these snapshots are propagated to the master database and *vice versa*. This approach hence represents a *two-way* or *bidirectional* replication. An updatable snapshot is restricted to a simple one.

Users can make changes in a mobile database while the MH is disconnected and can synchronize them with the master database. In other words, transactions can be executed on the MH while the MH is disconnected. Oracle Lite has built-in Java components that support savepoints but no nested transactions. Consistency between client and server is ensured by a *snapshot refresh* process which propagates and applies the relevant changes to the corresponding databases. This process involves three steps:

- Changes to updatable snapshots are sent to the master database.
- The valid transactions of the mobile client are applied on the master database (in the same sequence as they were performed at the snapshot).
- Snapshots are synchronized with the data in the master database.

Three modes of refreshes are supported:

- *Fast* refresh transmits only the changes (deltas); this option is viable only for simple snapshots.
- *Complete* refresh completely replaces the last snapshot by a new one.
- *Optimum* refresh attempts to perform a fast refresh; if this is not possible (e.g., due to a snapshot being complex), a complete refresh is executed.

During synchronization, conflicts are detected by the master database server. The following types of conflicts are handled:

- *Update*, when the snapshot old row values do not match the current FH row values
- *Uniqueness*, detected if a unique constraint is violated during an insert or update of a replicated row
- *Delete*, when a row is deleted from an updatable snapshot, and the old values of the deleted row do not match the current values at the master database

The Mobile Server uses winning rules to automatically resolve conflicts; either the client or the server wins. When the client wins, the Mobile Server automatically applies client changes to the server. When the server wins, the Mobile Server automatically composes changes for the client. The conflict-resolution mechanism can be changed by the user by setting the winning rule to "client wins" and attaching before triggers to the corresponding database tables. The triggers compare old and new row values and resolve client changes as specified.

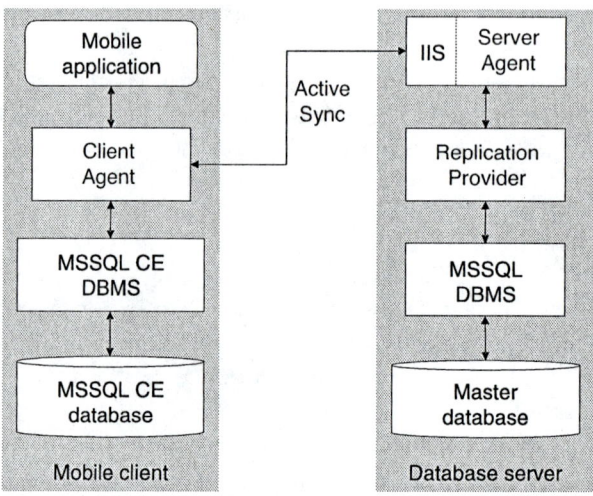

Figure 26.8 Microsoft SQL Server CE architecture.

Microsoft SQL Server CE

Microsoft relies on a two-tier approach. As depicted in Figure 26.8, a Microsoft SQL Server CE [19] database is running on the mobile client. On the server side, a usual Microsoft SQL Server maintains the master database. The architecture of Microsoft SQL Server CE consists of the following components that are relevant for supporting consistent replication:

- *SQL Server CE* is a small-footprint SQL database for mobile and embedded devices that supports the creation and manipulation of tables. Stored procedures, user-defined functions, triggers, views, and subqueries are not supported.
- *Client Agent* is an application residing on the mobile client. Using ActiveSync®, it triggers synchronization between the mobile and the master database. For that, it connects through the Hypertext Transfer Protocol (HTTP) to the Internet Information Server (IIS) residing on the FH.
- *ActiveSync* is an application that is used to perform communication between the mobile client and database server. ActiveSync is responsible for the synchronization between the mobile and the master databases.
- *Server Agent* is a module of IIS on the FH. It receives the database statements from the mobile clients and unpacks and propagates them to the corresponding master database. For database queries, it sends back the results to the corresponding mobile clients.

■ *Replication Provider* defines the publications that specify the tables of the master database to be replicated on the mobile clients. SQL Server CE allows the replication of entire tables and subset of rows and columns, as well; however, views cannot be replicated.

The SQL Server CE allows two types of replication: (1) message-based replication and (2) merge replication. The former relies on a one-way synchronization from the FH to the MH, and the latter allows two-way synchronization. In both cases, the synchronization between the mobile and the master databases is initiated and controlled by the Windows CE application. For merge replication, changes on the mobile database occur under transactional control. These transactions can be executed on the mobile client even when the MH is disconnected. SQL Server CE supports savepoints and nested transactions. Transactions can be nested up to a depth of five levels. Updates made within the nested transaction are not visible to the top-level transaction. Results become visible to the parent subtransaction after the nested subtransaction has committed. Changes are not visible outside the top-level transaction until that transaction commits.

Data synchronization is performed through a message-based communication protocol. An application on an MH subscribes to the publication through the *replication object*. Updates from the subscribers are sent to the publisher that merges these updates with the updates from all the other subscribers. Eventually, the publisher sends the updates back to the subscribers to achieve consistency in the MH replicated databases. Publications are tailored to specific groups of users, so an MH will receive only the updates to which it is subscribed.

When synchronization occurs, the client agent extracts all modifications to data records and sends them to the server agent through HTTP. Conversely, when the server agent propagates data changes at the publishing FH database back to the subscribing MH, it is the client agent that applies these changes to the subscribing database. When synchronization occurs, the server agent creates a new input message file on the IIS server and writes the data modification requests sent by the client agent. When the input message file has been written, the server agent initiates a reconciler process. The reconciler detects and resolves conflicts. Conflict resolution can be devised to behave with simple strategies. The default conflict resolver is priority based. A subscriber can be set to have priorities. The reconciler then checks update priorities and chooses the update with the highest priority. Other strategies consider that the first change to propagate to the publisher software always wins the conflict. More complex conflict-resolution rules can be customized by the application programmer. In addition, an interactive resolver is also supplied.

Sybase Anywhere

Adaptive Server Anywhere (ASA) [32] is a small-footprint relational database. The architecture of Sybase Anywhere relies on a two-tier client-server model. On the mobile client side, an instance of Adaptive Server Anywhere is running. On the server side, the master database can be a Sybase Enterprise Edition database. Sybase Anywhere supports three types of replication between server and client:

- *SQL Remote* is intended for a two-way, message-based replication involving a consolidated database server and large number of remote databases.
- *Mobilink*® is intended for a two-way, session-based replication. It also supports non-Sybase databases. It is designed for replication between a central consolidated database and a large number of remote databases. At the end of each synchronization session, the databases are consistent.
- *Replication Server* is intended for a two-way, connection-based replication. It is suited to replicating data between a small number of enterprise databases connected by a high-speed network.

The Replication Server is not suited for occasionally connected devices. In contrast, Mobilink and SQL Remote provide more flexible solutions in environments where the remote machines are mobile or are only occasionally connected. For these reasons, we will discuss only SQL Remote and Mobilink because they are more suited to mobile computing needs. In both replication technologies, the replicated data can be tables, subsets of rows and subsets of columns, joins, and SQL queries.

SQL Remote uses a publish and subscribe process. Synchronization data is saved in publications at the FH database, where the administrator subscribes the MH databases to the publications they require. The publications are sent to the relevant MHs. The MHs in turn send their synchronization data to the FH to be incorporated in the FH database. The synchronization is message based, as shown in Figure 26.9. The synchronization process in SQL Remote consists of the following steps:

- At the MH and FH database side, a *message agent* coordinates the synchronization using a transaction log. The transaction log tracks all committed changes to the corresponding database.
- Periodically, the message agent on the FH database scans the transaction log and sends the relevant changes to MH databases that are subscribed to those publications.
- The message agent at the MH accepts the messages sent from the FH and applies the transactions to its own database.

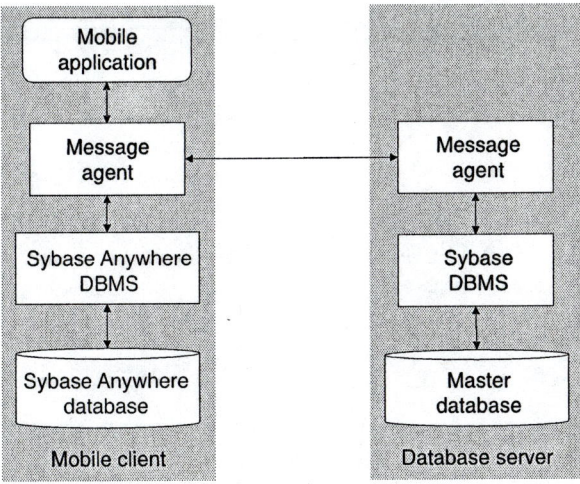

Figure 26.9 Sybase SQL remote architecture.

- At any time, the MH can run the message agent to collect and send to the FH database the transactions made locally.
- The message agent at the FH processes the messages received from the MH and applies the transactions to its own database.

Conflict resolution is performed at the FH through the use of conflict triggers.

Mobilink supports a session-based synchronization. The logic required to keep track of changes to the database, to prepare data to send to the server, and to incorporate data that it receives in reply can be injected either into the mobile application or into the ASA database management system (DBMS). As indicated in Figure 26.10, the synchronization is then coordinated either by the mobile application or the mobile database. Mobilink's session-based synchronization performs the following steps:

- The MH application prepares and transmits to the FH a list of all rows that have been modified since the last synchronization.
- The synchronization server, running on the FH, prepares and transmits data changes from the FH database to the MH application.
- The MH application then sends a confirmation message to the synchronization server confirming that the changes have been made to the data. The transaction is recorded in the FH database.
- The synchronization server sends the MH application a message confirming that synchronization is complete. The connection between the MH application and the FH is shut down.

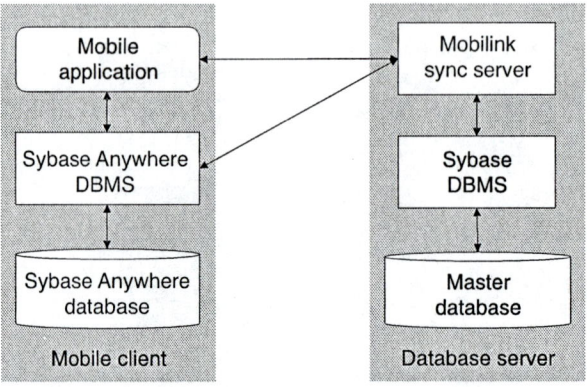

Figure 26.10   Sybase Mobilink architecture.

In situations where the data rarely changes, Mobilink uses a time-stamp technique to synchronize only data that has changed since the last synchronization. It uses a snapshot technique when many changes are frequently carried out on the data. In that case, the software synchronizes all the data in the database.

To resolve conflicts, business rules must be implemented to determine which data is correct. Conflict resolution is performed on the FH database, and then conflict-resolution actions are replicated to the remote databases. Transactions can use savepoints. All savepoints are released when a transaction ends. Transactions can be nested. Changes to the database are permanent only after the uppermost transaction is committed. The transactions can run on the mobile clients even when the MH is disconnected. In this case, savepoints are not considered.

Conclusions

In this chapter, we discussed consistency management in mobile computing. Our focus was on both file-based and transaction-oriented consistency. An important issue that requires further investigation is building applications and experimenting with the various consistency models because performance results are lacking. In particular, while research in consistency maintenance related to disconnected operation has already been integrated in a number of commercial products, research in weak connectivity and hybrid delivery are still not part of current practice. An interesting direction for future work also includes seamless consistency in the case of *ad hoc* network architectures with no fixed hosts. In this case, most of the consistency models and protocols presented in this chapter must be revised. Finally, another important new direction is the study of seamless consistency in the case of very small devices, such as in sensor networks.

References

[1] Acharya, S., Franklin, M.J., and Zdonik, S., Disseminating updates on broadcast disks, in *Proc. of the 22nd Int. Conf. on Very Large Data Bases (VLDB'96)*, Bombay, India, September 1996.

[2] Alonso, R., Barbará, D., and Garcia-Molina, H., Data caching issues in an information retrieval system, *ACM Trans. Database Syst.*, 15(3), 359–384, 1990.

[3] Badrinath, B.R. and Phatak, S., Data partitioning for disconnected client server databases, in *Proc. of the 1st Int. Workshop on Data Engineering for Wireless and Mobile Access (MobiDE'99)*, Seattle, WA, August, 1999.

[4] Barbará, D., Certification reports: supporting transactions in wireless systems, in *Proc. of the IEEE 17th Int. Conf. on Distributed Computing Systems (ICDCS'97)*, Baltimore, MD, May, 1997.

[5] Barbará, D. and Imielinski, T., Sleepers and workaholics: caching strategies in mobile environments, *VLDB J.*, 4(4), 567–602, 1995.

[6] Bernard, G. et al., Mobile databases: a selection of open issues and research directions, *SIGMOD Rec.*, 33(2), 78–83, 2004.

[7] Cetintemel, U., Keleher, P.J., Bhattacharjee, B., and Franklin, M.J., Deno: a decentralized, peer-to-peer object-replication system for weakly connected environments, *IEEE Trans. Comput.*, 52(7), 943–959, 2003.

[8] Cherniack, M., Galvez, E.F., Franklin, M.J., and Zdonik, S.B., Profile-driven cache management, in *Proc. of the 19th Int. Conf. on Data Engineering (ICDE'03)*, Bangalore, India, March, 2003, pp. 645–656.

[9] Davidson, S.B., Garcia-Molina, H., and Skeen, D., Consistency in partitioned networks, *ACM Comput. Surv.*, 17(3), 341–370, 1985.

[10] Demers, A., Petersen, K., Spreitzer, M., Terry, D., Theimer, M., and Welch, B., The Bayou architecture: support for data sharing among mobile users, in *Proc. of IEEE Workshop on Mobile Computing Systems and Applications (WMCSA'94)*, Santa Cruz, CA, December, 1994, pp. 2–7.

[11] Gray, J., Helland, P., O' Neil, P., and Shasha, D., The dangers of replication and a solution, in *Proc. of the ACM SIGMOD Int. Conf. on Management of Data*, Montreal, Canada, June, 1996, pp. 173–182.

[12] IBM, *IBM DB2 Everyplace Application and Development Guide Version 8.2*, IBM Corporation, Armonk, NY, 2004.

[13] IBM, *IBM DB2 Everyplace Sync Server Administration Guide Version 8.2*, IBM Corporation, Armonk, NY, 2004.

[14] Jing, J., Elmagarmid, A.K., Helal, A., and Alonso, R., Bit-sequences: an adaptive cache invalidation method in mobile client/server environments, *MONET*, 2(2), 115–127, 1997.

[15] Kistler, J.J. and Satyanarayanan, M., Disconnected operation in the Coda file system, *ACM Trans. Comput. Syst.*, 10(1), 213–225, 1992.

[16] Kumar, P. and Satyanarayanan, M., Flexible and safe resolution of file conflicts, in *Proc. of the USENIX Winter Conf. on Unix and Advanced Computing Systems*, New Orleans, LA, January, 1995.

[17] Lee, V.C.S., Lam, K.-W., Son, S.H., and Chan, E.Y.M., On transaction processing with partial validation and timestamp ordering in mobile broadcast environments, *IEEE Trans. Comput.*, 51(10), 1196–1211, 2002.

[18] Lu, Q. and Satyanarayanan, M., Improving data consistency in mobile computing using isolation-only transactions, in *Proc. of the 5th Workshop on Hot Topics in Operating Systems (HotOS)*, Orcas Island, WA, May, 1995.

[19] Microsoft Corporation, http://msdn.microsoft.com/library/.

[20] Mummert, L.B., Ebling, M.R., and Satyanarayanan, M., Exploiting weak connectivity for mobile file access, in *Proc. of the 15th ACM Symp. on Operating System Principles*, Copper Mountain, CO, December, 1995.

[21] Oracle, *Oracle Database Lite, Administration and Deployment Guide 10g (10.0.0)*, Oracle Corporation, Redwood Shores, CA, 2004.

[22] Oracle, *Oracle Database Lite, Developer's Guide 10g (10.0.0)*, Oracle Corporation, Redwood Shores, CA, 2004.

[23] Page, T.W., Guy, R.G., Heidemann, J.S., Ratner, D.H., Reiher, P.L. et al., Perspectives on optimistically replicated peer-to-peer filing, *Software Pract. Exper.*, 28(2), 155–180, 1998.

[24] Pitoura, E. and Crysanthis, P.K., Exploiting versions for handling updates in broadcast disks, in *Proc. 25th Int. Conf. on Very Large Data Bases (VLDB'99)*, Edinburgh, Scotland, September, 1999.

[25] Pitoura, E. and Crysanthis, P.K., Scalable processing of read-only transactions in broadcast push, in *Proc. of the IEEE Int. Conf. on Distributed Computing Systems (ICDCS'99)*, Austin, TX, May, 1999.

[26] Pitoura, E. and Crysanthis, P.K., Multiversion data broadcast, *IEEE Trans. Comput.*, 51(10), 1224–1230, 2002.

[27] Pitoura, E., Crysanthis, P.K., and Ramamritham, K., Characterizing the temporal and semantic coherency of broadcast-based data dissemination, in *Proc. of the 9th Int. Conf. on Database Theory (ICDT'03)*, Siena, Italy, January, 2003.

[28] Dunham, M.H., Ren, Q., and Kumar, V., Semantic caching and querying, *IEEE Trans. Knowl. Data Eng.*, 15(1), 192–210, 2003.

[29] Seifert, A. and Scholl, M.H., Processing read-only transactions in hybrid data delivery environments with consistency and currency guarantees, *Mobile Networks Appl.*, 8(4), 327–342, 2003.

[30] Serrano-Alvarado, P., Roncancio, C., and Adiba, M.E., A survey of mobile transactions, *Distributed Parallel Databases*, 16(2), 193–230, 2004.

[31] Shanmugasundaram, J., Nithrakashyap, A., Sivasankaran, R.M., and Ramamritham, K., Efficient concurrency control for broadcast environments, in *Proc. of the ACM SIGMOD Int. Conf. on Management of Data*, Philadelphia, PA, June, 1999.

[32] Sybase, *Sybase Adaptive Server Anywhere Reference, Version 6.0.3*, Sybase, Inc., Boulder, CO, 2000.

[33] SyncML Consortium, *SyncML Sync Protocol*, 1999–2000, http://www.open-mobilealliance.org/tech/affiliates/syncml/syncmlindex.html.

[34] Tait, C.D. and Duchamp, D., An efficient variable-consistency replicated file service, in *Proc. of USENIX File Systems Workshop*, Ann Arbor, MI, May, 1992, pp. 111–126.

[35] Terry, D.B., Theimer, M.M., Petersen, K., Demers, A.J., Spreitzer, M.J., and Hauser, C.H., Managing update conflicts in Bayou, a weakly connected replicated storage system, in *Proc. of the 15th ACM Symp. on Operating System Principles*, Copper Mountain, CO, December, 1995.

[36] Walborn, G. and Chrysanthis, P.K., PRO-MOTION: support for mobile database access, *Pers. Ubiquitous Comput.*, 1(3), 171–181, 1997.

[37] Walborn, G. and Chrysanthis, P.K., Transaction processing in PRO-MOTION, in *Proc. of the 14th ACM Symp. on Applied Computing (SAC'99)*, San Antonio, TX, March, 1999.

[38] Pitoura, E. and Bhargava, B.K., Data consistency in intermittently connected distributed systems, *IEEE Trans. Knowl. Data Eng.*, 11(6), 896–915, 1999.

Chapter 27

Seamless Service Access via Resource Replication

Paulo Ferreira and Luís Veiga

CONTENTS

Introduction

Replication is a well-known technique for improving data availability and application performance as it allows the collocation of data and code. Data availability is ensured because, even in the presence of network failures, data remains locally available; in addition, application performance is potentially better (when compared to a remote invocation approach), as all accesses to data are local. Several significant issues must be addressed to take full advantage of replication. In this chapter, we address the following: (1) replica management, (2) memory management, and (3) adaptability. Note that many other issues are equally important [1], such as how to merge or reconcile replicas that have diverged due to updates being performed independently, but such issues are considered elsewhere in this book.

Replica management is related to the fundamental issues of knowing which data and how the data should be replicated. Memory management addresses the need to ensure that the memory of mobile devices (e.g., PDAs, laptops) is occupied with useful data, which involves: (1) freeing the memory occupied by useless replicas, which can be achieved by garbage collecting such replicas, and (2) swapping out useful data to disk or to remote computers. In addition, for object-based applications, memory management is responsible for ensuring the referential integrity of the objects graph. Adaptability is the capability applications have to control and adapt to the resources that they use (e.g., memory, network) to better deal with the variability of mobile environments; such variability affects network bandwidth, network connection or disconnection, amount of available memory, etc.

The above-mentioned issues are becoming more and more relevant as we move from a traditional wired network of desktop computers to an environment formed by mobile devices able to wirelessly connect to the fixed network or take part in *ad hoc* networks. As a matter of fact, mobile devices, when compared to desktop computers, are much more resource constrained in terms of memory, network availability and bandwidth, battery, etc. Also, applications running on such devices face a much more dynamic environment given the natural movement of users and devices.

The fact that mobile devices impose severe constraints in terms of such resources emphasizes the importance of the following: (1) The data to be replicated should be data that is really needed, so memory is not wasted; (2) replicated data that is no longer needed must be automatically detected and garbage collected, thus releasing the memory occupied; and (3) the underlying middleware must support flexible mechanisms so applications can react and adapt to the dynamics of mobile environments (e.g., variable network availability, amount of free memory on the device).

Finally, portability and programmability are also relevant aspects that must be taken into account by the mobile middleware. Mobile environments are characterized by the heterogeneity of devices, operating systems, virtual machines, etc., so, the mobile middleware should be, as much as possible, independent from such differences to be portable to a wide range of platforms. Programmability means that the middleware should release applications programmers from having to deal with system-level issues and should provide a familiar application programming interface (API); thus, the mobile middleware should not imply modification of either the operating systems or the virtual machines and should not impose radically new APIs.

In the remainder of this chapter, we first clarify the programming model being considered, then we present the archetypical architecture for mobile middleware that is used in this chapter, the mechanisms supporting how and which data is replicated (for both object and file models), the algorithms for the garbage collection (GC) of replicas, and the policies allowing applications to control objects replication.

Programming Models

In general, applications can be developed according to several different programming paradigms. These depend on the different abstractions supported by the underlying mobile middleware that, accordingly, provides the corresponding API. For example, the middleware may simply provide a file system API, or it may support more complex abstractions such as tuples, objects, or relational entities. With regard to the issues addressed in this chapter, the relevant characteristic of such paradigms is their ability, or lack of it, to support arbitrary graphs of data. In fact, as explained later, the existence of data graphs has a strong impact on deciding which and how data must be replicated.

The object-oriented paradigm naturally supports the notion of data graphs; the same applies to structured files whose contents include references to other files (e.g., graphs of HTML files connected by URLs). In this chapter, we consider two cases: (1) applications using arbitrary graphs of data, and (2) applications that simply use plain unstructured data (i.e., without containing references that allow data graphs to be built). In the first case, we use the term *object* to designate a datum that can be an instance of a class, a Hypertext Markup Language (HTML) file, etc.; thus, an object is simply a set of bytes that may contain references to other objects. In the second case, we use the term *file* to designate a datum that is a set of unstructured bytes (thus, without the notion of reference).

Given that the programming model in which applications handle arbitrary data graphs is the most widely used and is highly flexible, in this chapter we focus our attention on this latter case, which we refer to as the *object model*; however, we do also consider the file model, because file system support is widespread and is well known both by users and applications programmers. In this model, the mobile middleware offers a file-based API (possibly extended with replication-specific functions) in which there are no references between files.

The object model is arguably the most widely adopted programming model. Object replication has been addressed by several projects, such as Thor [2,3], OceanStore [4], Deno [5], OBIWAN [6,7], M-OBIWAN [8,9], DERMI [10], Javanaise [11,12], Gold-Rush [13], Alice [14], and replicated CORBA™ [15], among others. Not all of these, however, have addressed with the same level of concern the challenges raised by mobility environments.

Relevant projects supporting file replication include Coda [16,17], which was the first to address the issue of disconnected work in distributed file systems; also, Ficus [18], Rumor [19], and Roam [20,21] represent a line of very interesting work regarding distributed file systems with growing concerns of mobility, leading to the concept of *selective replication*, which is a mechanism by which only the files that applications really need to access are replicated, rather than a whole volume.

We do not consider the programming paradigm based on relational databases given that applications developed according to this paradigm are primarily query oriented, instead of navigation based (on a graph). Operations on data are declaratively defined by SQL queries for insertion, update, and removal of records. Data on different tables can be joined by matching field values. Several queries can be composed into transactions guaranteeing the properties of atomicity, consistency, isolation, and durability (ACID) [22]. In mobile computing, such ACID properties are relaxed to provide acceptable consistency requirements while allowing concurrent update on replicas placed in different nodes. One influential work regarding database replication is the Bayou project [23–25], in which the merging of replicas is specially considered. Replication of relational databases in mobile computing has also been addressed in Mobisnap [26].

Architecture

Replication can be used in either client–server (CS) [27] or peer-to-peer (P2P) [28] distributed systems. In CS systems, data is replicated from a server into the mobile client device, where it is accessed, then, if needed, data is sent back to the server. The servers have the fundamental purpose of storing data persistently. In a CS architecture, data is shared among mobile clients always by intermediation of a server. With a P2P architecture, any computer may behave either as a client or as a server at any moment. In particular, for concerning replication, this means that a process (P) running on a mobile device can either request the local creation of replicas of remote data (P acting as a client) or be asked by another process to provide data to be replicated (P acting as a server). Hybrid architectures consider the coexistence of both approaches.

The P2P approach is generally more flexible than a CS architecture, given that mobile devices are free to replicate data among them (e.g., epidemically [29]); this raises further problems in terms of consistency that are considered elsewhere in this book. For CS and P2P architectures, the issues of replica management, memory management, and adaptability are equally relevant. Figure 27.1, presents an archetypical architecture illustrating the most important data structures with regard to the replication of an objects graph. Objects X, Y, A, B, and C are created by the application; their replicas (A', B', etc.) are created either upon the programmer's request or automatically by the middleware, without having been explicitly required by the application code but resulting from its execution. Proxies-in and proxies-out, as well as references pointing to them, are part of the middleware and are transparent to the programmer.

Without loss of generality, we assume that processes P1 and P2 run in two different computers, and the initial situation is the following: (1) P2 holds a graph of objects Y, A, B, and C; (2) object A has been replicated from P2 to P1, thus we have A' in P1; (3) A' holds a reference to AproxyIn (for reasons that will be made clear later); (4) given that B has not been replicated yet, A' points to BproxyOut instead (note that object A does not distinguish B' from B as they offer the same interface); and (5) object X (created in P1) points to Y.

The most relevant data structures are the following:

- *Proxy-out/proxy-in pairs* — A proxy-out stands in for an object that is not yet locally replicated (e.g., BproxyOut stands for B' in P1). For each proxy-out there is a corresponding proxy-in [30].
- *GC-stubs and GC-scions* — A GC-stub describes an outgoing interprocess reference, from a source process to a target process (e.g., from object X in P1 to object Y in P2). A GC-scion describes an incoming interprocess reference from a source process to a target process (e.g., to object Y in P2 from object X in P1).

GC-stubs and GC-scions do not impose any indirection on the native reference mechanism. In other words, they do not interfere with either the underlying structure of references or the invocation mechanism. They are simply GC-specific auxiliary data structures. Thus, GC-stubs and GC-scions

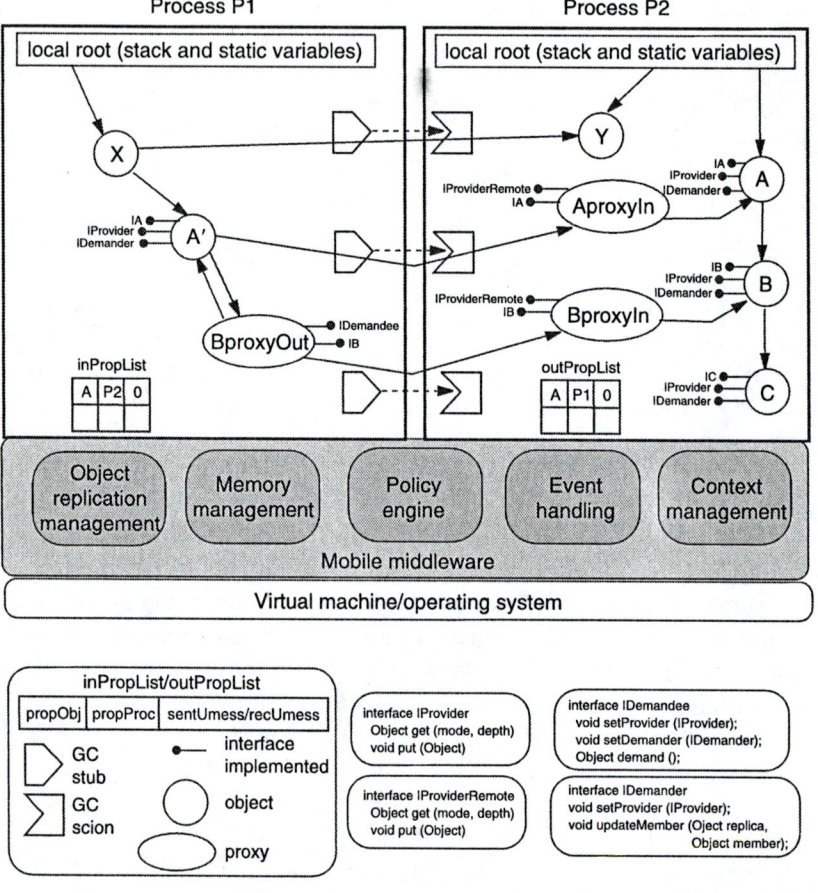

Figure 27.1 Archetypical architecture for mobile middleware supporting object replication.

should not be confused with (virtual machine) native stubs and scions (or skeletons) used for Remote Method Invocation (RMI).

■ *InPropList and OutPropList* — These lists indicate the process from which each object has been replicated, and the processes to which each object has been replicated, respectively. Thus, each entry of the InPropList/OutPropList contains the following information: prop-Obj is the reference of the object that has been replicated into/to a process; propProc is the process from/to which the object propObj has been replicated; sentUmess/recUmess is a bit indicating if an "unreachable" message (for distributed GC purposes) has been sent or received (more details later).

Another important aspect of this archetypical architecture is the functionality supported by the above-mentioned data structures as well as the interfaces they implement. In particular, proxies-out, being one of the main entities of the mobile middleware responsible for the replication mechanism, have to implement the same interface of the object they are replacing. In addition, for reasons that will be made clearer later, application objects also have to implement a few methods that are, in fact, middleware code; however, these methods are not written by the application programmer but are automatically generated either at compilation or at runtime.

The interfaces implemented by each object and proxy-out/proxy-in pairs are the following:

- *IA*, *IB*, and *IC* are the remote interfaces of objects A, B, and C, respectively, designed by the programmer; they define the methods that can be invoked on these objects. (The same reasoning applies to objects X and Y; however, given that they are not involved in the replication scenario, we do not consider them.)
- *IProvider* is an interface with methods get and put that supports the creation and update of replicas; method get results in the creation of a replica and method put is invoked when a replica is sent back to another process (possibly to the process where it came from to update its master replica).
- *IDemander* and *IDemandee* are interfaces that support the incremental replication of an objects graph.
- *IProviderRemote* is a remote interface that inherits from IProvider so its methods can be invoked remotely.

In addition to the data structures already presented are five other modules in the archetypical architecture (addressed in the following sections): (1) *object replication management*, which provides the mechanisms supporting data replication; (2) *memory management*, which is responsible for the distributed garbage collection of replicas; (3) the *policy engine*, which triggers or mediates responses to events occurring in the system; (4) *event handling*, which registers the relevant events occurring in the system; and (5) *context management*, which abstracts resources and manages the corresponding properties whose values vary during applications execution.

When considering a file model (i.e., mobile middleware that provides a possibly extended, well-known, traditional, file-based API), the archetypical architecture does not contain some of the data structures mentioned, for obvious reasons. In particular, the GC-stubs/GC-scions are not required. Figure 27.1 is still valid, though; for example, the lists InPropList/

OutPropList are still needed so the system keeps track of which files were replicated into or from which process, and the policy engine still allows the application to deal with mobile environment dynamics.

Replica Management

A fundamental issue concerning data replication is the impact of this mechanism on the API. In other words, it is crucial that applications programmers are not forced to deal with system details concerning which and how data is effectively replicated. Such system issues (e.g., object faulting and resolving) must be handled transparently by the underlying middleware when data gets replicated from one computer to another. For example, when considering the object model, in which distributed applications access a data graph, the referential integrity of the graph must be ensured to avoid dangling references. In this section, we address the following issues: (1) the mechanisms supporting objects and files replication, and (2) how objects and files are chosen to be replicated. These issues are analyzed taking into account the fact that mobile devices impose strong constraints in terms of available memory and network bandwidth.

How To Replicate

Object replication differs significantly from file replication. The difference results from the fact that, with the object model, applications access data by navigating on objects graphs. Such navigation does not occur when applications access plain unstructured files. This difference has an important impact on deciding how and which data should be replicated.

When an application running on a mobile device navigates on an objects graph, its execution proceeds normally as long as all objects are local. When the target object is not yet locally replicated, however, this generates an object fault that must be resolved by the middleware; therefore, the above-mentioned aspects regarding the replication of objects (how and which) are strongly dependent on the navigation performed by the application.

Regarding the file model, in which files are plain unstructured streams of bytes, applications do not navigate on a graph while executing (in contrast to the case where the object model is used); thus, applications access local files by means of traditional system calls, such as "open," "read," or "write," which are offered by most operating systems and possibly extended by the mobile middleware with replication-specific capabilities. Among other functionalities, these extensions improve the memory usage in mobile devices by compressing file contents or replicating just the blocks that are really necessary, for example. An interesting situation in which files are replicated, possibly resulting from an explicit request from the user, is the moving of files from a desktop computer into a laptop or PDA.

Object Model

Given the memory restrictions imposed by mobile devices (when compared to desktop computers), the replication of objects cannot be done simply by serializing all the objects graphs in the originating computer and sending them to another one. Such an approach (which is available when using Java [31] or .NET™ [32] platforms) is clearly inappropriate due to the high level of computing, communication, and memory resources required; thus, the mobile middleware must handle incremental replication (i.e., the partial replication of an objects graph). The incremental replication of an objects graph has two clear advantages when compared to the replication of the whole reachability graph in one step: (1) the latency imposed on the application is lower because it can immediately invoke the new replica without waiting for the whole graph to be available, and (2) only those objects that are really needed are replicated, thus reducing the memory and network bandwidth required.

Situations when an application does not have to invoke every object of a graph or when the computer on which the application is running has limited memory or network bandwidth available are those in which incremental replication is most useful. In some situations, however, it may be better to replicate the whole graph; for example, if all objects are really required for the application to work, if there is enough memory, and the network connection will not be available in the future, then it is better to replicate the transitive closure of the graph. The mobile middleware must allow the application to easily make this decision at runtime (i.e., choosing between incremental or transitive closure replication mode).

Taking into account the archetypical architecture presented in Figure 27.1, we now describe how objects can be incrementally replicated from process P2 into another process (P1). (Note that, in the initial situation, A' was replicated the same way that B will be, as explained later.) Starting with the initial situation, the code in A' may invoke any method that is part of interface IB, exported by B, on BProxyOut (which A' sees as being B'). For transparency, this requires the system to detect and resolve the corresponding object fault. All IB methods in BProxyOut simply invoke on itself a method (which is part of the mobile middleware) called BProxyOut.demand (belonging to interface IDemandee); this method runs as follows (see Figure 27.2, in which the numbers refer to the items below):

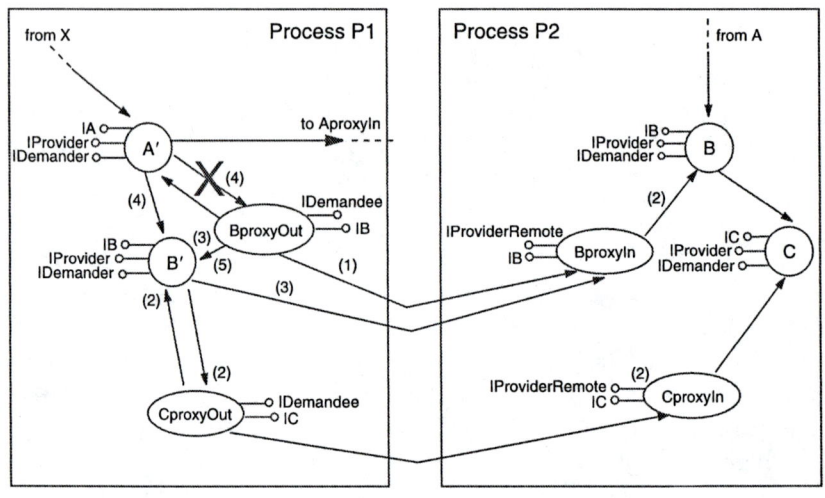

Figure 27.2 Incremental object replication.

1. It invokes (remotely) method BProxyIn.get in P2.
2. BProxyIn.get invokes method B.get, thus creating B', CProxyOut, and CProxyIn and setting the references between them. Once this method terminates, B', BProxyOut, and CProxyOut are serialized and sent to P1 (CProxyIn remains in P2). Note that A' still points to BProxyOut (and *vice versa*).
3. BProxyOut invokes B'.setProvider(this.provider) so B' also points to BProxyIn; this is necessary because the application can later decide to update replica B (by invoking method B'.put, which in turn will invoke BProxyIn.put) or to refresh replica B' (method BProxyIn.get).
4. BProxyOut invokes A'.updateMember (B',this) so A' replaces its reference to BProxyOut with a reference to B'.
5. Finally, BProxyOut invokes the same method on B' that was invoked initially by A' (which triggered this whole process) and returns accordingly to the application code.
6. From this moment on, BProxyOut is no longer reachable in P1 and will be reclaimed by the garbage collector of the underlying virtual machine.

It is worth noting that, when B gets replicated in P1, as described above, further invocations from A' on B' will be normal direct invocations with no indirection at all. Later, if and when B' invokes a method on CProxyOut (standing in for C', which is not yet replicated in P1), an object fault occurs; this fault will be resolved with a set of steps similar to those previously described. In addition, note that this mechanism does not imply modification of the underlying virtual machine. This fact is key to the portability of the mobile middleware supporting incremental replication.

The replication mechanism just described is very flexible in the sense that it allows each object to be individually replicated; however, this process has a cost, resulting from the creation and transfer of the associated data structures (i.e., proxies). To minimize this cost, the mobile middleware must allow an application to replicate a set of objects as a whole (i.e., a cluster of objects having only a proxy-in/proxy-out pair).

A cluster is a set of reachable objects that are part of a reachability graph; for example, if an application holds a list of 1000 objects, it is possible to replicate part of the list so only 100 objects are replicated and a single proxy-in/proxy-out pair is effectively created and transferred between processes. Thus, the middleware must allow the amount of objects in the cluster to be determined at runtime by the application. The application specifies the depth of the partial reachability graph that it wants to replicate as a whole. This is an intermediate solution between incrementally replicating each individual object and replicating the whole graph.

File Model

Concerning file replication, the previously mentioned memory and network bandwidth restrictions of mobile devices are obviously equally valid; thus, the mobile middleware, possibly with cooperation from the operating system, must minimize both the space occupied by replicated files and network usage. This has been done in previous distributed file systems for fixed networks, either by compressing file contents or by replicating just the needed blocks (instead of whole files) [33].

A very interesting and promising approach aimed at reducing the amount of memory consumed by replicated files explores replica content similarities [34–36]. The idea consists of applying the SHA-1 hash function [37] to portions of the contents of each replica; each portion is called a *chunk*. The probability of two distinct inputs to SHA-1 producing the same hash value is far lower than the probability of hardware bit errors. Relying on this fact, the obtained hash values can be used to univocally identify their corresponding chunk contents. From this assumption, if two chunks produce the same output upon application of the SHA-1 hash function, then they are considered to have the same contents. If both chunks are to be stored locally at the same computer, then only the contents of one of them must be effectively stored. Note that, in a similar way, if one of the chunks is to be sent to a remote machine that is holding the other chunk, then the actual transference of the contents over the network can be avoided.

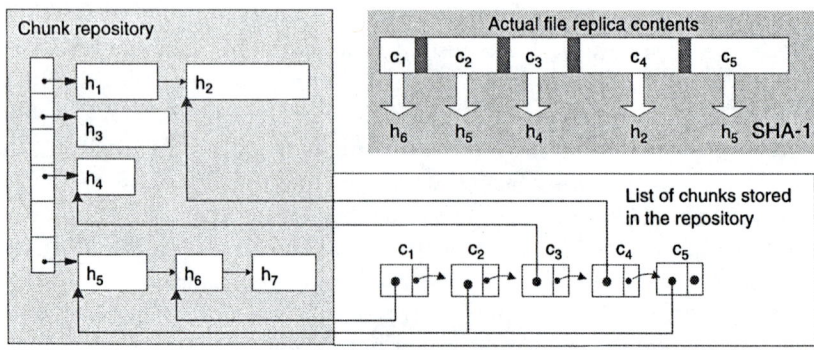

Figure 27.3 A file replica contents, the corresponding list of chunks, and the repository including them.

A content-based approach is employed to divide replica contents into a set of nonoverlapping chunks, based on Rabin's fingerprints [38]. As a result, chunks may have variable sizes, depending on their contents. An important property of such a chunk division approach is that it minimizes the consequences of insert-and-shift operations in the global chunk structure of a replica. The expected average chunk size may, however, be parameterized by controlling the number of low-order bits from the output of the fingerprint that are considered by the chunk division algorithm. Moreover, to prevent cases where abnormally sized chunks might be identified, a minimum and maximum chunk dimension is imposed.

On each file system peer is a common chunk repository that stores all data chunks, indexed by their hash value, that comprise the contents of the files that are locally replicated (see Figure 27.3). The data structures associated with the content of locally replicated files simply store pointers to chunks in the chunk repository. Hence, the contents of an update to a replica, or the whole replica, consists of a sequence of data chunks stored in the chunk repository.

When a file replica is written, a data chunk is either created or modified, then its hash value is calculated and the chunk repository is examined to determine if an equally hashed chunk is already stored. If not, a new entry corresponding to the new chunk is inserted in the repository. If a similar chunk already exists, a new pointer to that chunk is used. So, if different file replicas or versions of the same file replica contain data chunks with similar contents, they share pointers to the same entry in the chunk repository, thus reducing memory usage by the file system.

Replicating a file from one computer to another also makes use of the chunk repositories at each peer. When a chunk has to be sent across the network to another peer, only its hash value is sent initially. The receiving peer then looks up its chunk repository to see if that chunk is already stored locally. If so, it avoids the transference of the contents of that chunk and simply stores a pointer to the already existing chunk; otherwise, the chunk contents are sent and a new chunk is added to the repository. To deal with the deletion of unused chunks from a repository, each chunk maintains a counter that is incremented each time a new pointer is set to that chunk. Conversely, that counter is decremented when a pointer to it is removed from the structures of the system. This can occur when a previously replicated file is removed from the set of replicated files.

What To Replicate

Previous research done on replication has primarily focused on reducing the latency perceived by applications or utilizing replication as a means to increase the fault tolerance of servers; however, in a mobile environment, the main goal of replication is to allow applications to keep on doing useful work in disconnected mode or when the network bandwidth is rather small. At the same time, the mobile middleware must optimize memory usage; thus, deciding which data should be replicated is an important issue, as replicating data that will not be needed by applications means that memory and network are being wasted.

Object Model

For the object model, in which applications access objects by navigating on the graph, replication arises naturally when resolving an object fault. In other words, when the application running in a mobile device invokes an object that is not yet locally mapped, it generates an object fault that is automatically resolved by the underlying middleware. The resolution of the object fault involves requesting the faulted object from another computer, as described earlier. This computer answers with the object faulted and possibly a few more that are anticipated to be needed shortly. This anticipation (i.e., replication in advance of some objects) is based on the fact that objects are accessed by means of graph navigation. The objects that must be replicated, then, are those that can be accessed from the ones already replicated. An important aspect is the level of control and flexibility that the middleware supports with regard to the number of objects that are replicated in advance — for example, allowing an application to specify at runtime which branch of a graph and how many objects should be replicated (discussed further later in this chapter).

The particular values of such options (i.e., branch and depth of the graph to be replicated) depend on many factors. Such a decision can be made manually by the programmer or automatically by the middleware (or both combined). In the first case, the programmer may annotate the code with hints that the middleware will use accordingly; in the second case, the middleware bases its decisions on the past behavior of applications. On the one hand, programmer annotations can be difficult to construct and are error prone; on the other hand, past behavior may not be available and it is not necessarily a good indicator of future application behavior. This issue is out of the scope of this chapter; however, an interesting possibility is to base the above-mentioned values on context information [39].

File Model

With regard to the file model, the issue of deciding which files to replicate is equally important; however, given that there is no graph in which applications navigate, deciding which files to replicate must be done differently. In particular, the lack of a data graph means that no path exists on which the middleware can rely to predict which files will be accessed in the future by an application.

Solutions to this replication problem (also known as hoarding) can be grouped as follows: (1) solutions based on actions explicitly performed by the user stating which files should be replicated, or (2) solutions provided by the middleware, which may take advantage of user-provided hints. The first case is concerned with a scenario in which, before disconnecting a PDA or a laptop from the network, a user explicitly replicates a set of files (from a desktop computer or a server), so he can still do his work while disconnected. Note that this explicit replication can also be done using specific information provided by the user describing the files to be hoarded; based on this information, the system automatically replicates such files. Systems such as Coda [17] and SPY [40], for example, require users to explicitly specify their hoarding set. So, user actions are decisive with respect to both aspects of file replication: grouping files together given that they are strongly related and deciding which files (or groups of files) should be replicated given that they will certainly be used in the future. We will not consider this case further, as it relies mostly on the explicit actions of the user. In the second case, the middleware takes on a much more active role, and the following approaches can be used:

- While working, the user provides the middleware with specific intervals of time during which a specific task is being performed; the middleware, during each interval, detects which files are accessed and assumes that they are all strongly related and needed for the task under consideration.

■ The middleware performs file replication based on the notion of semantic distance [41], with no help from the user. This notion relies on the temporal data usage patterns of file accesses; basically, those files that are accessed during a certain time interval are assumed to be strongly semantically related. (Note that the access patterns are, in fact, more sophisticated than this; for example, spatial storage locality can be used, as well.) For this purpose, the middleware continuously monitors file accesses and clusters files accordingly.

Using file accesses to group related files is difficult because often it is not clear whether or not a sequence of file accesses is related. Furthermore, files that are related but are not accessed, at least for some time, may lose their privileged relation with other files.

Note that these two approaches mainly address the problem of finding which files are strongly related so they should be replicated together; however, the problem remains of predicting which files (or groups of files) will be used in the future (in particular, while the mobile device is completely disconnected or has severe network bandwidth limitations). Such forecasting can be done assuming that recently used files will certainly be accessed in the near future. This is, in fact, a least recently used (LRU) approach that has been used extensively in the past for virtual memory support and has been adopted by a hoarding system [42] with some refinements (e.g., by allowing the user to bound the time interval under consideration by the LRU). Such enhancement of the LRU approach requires user intervention; it basically consists of a user-provided hoarding profile that requires intensive user intervention. Users, however, should not have to worry about issues other than their work, so a replication system should minimize user intervention and especially the amount of user attention required.

Memory Management

In this section, we address the issue of memory management, particularly: (1) how the mobile middleware detects and deletes useless replicas, and (2) how to move useful data to disk or to remote computers. This is a very relevant issue because memory is a scarce resource in mobile devices. In addition, when considering the object model in which applications navigate on a graph of objects, it is fundamental that the mobile middleware ensures the referential integrity of the objects graph. As described in this section, this is ensured by means of automatic memory management, also known as garbage collection (GC), which also detects and deletes useless replicas.

It is widely recognized that manual memory management (explicit allocation and freeing of memory by the programmer) is extremely error prone, leading to memory leaks and dangling references. Memory leaks consist of data that is unreachable to applications but still occupies memory because its memory was not properly released. Dangling references are references to data whose memory has already been (erroneously) freed; later, if an application tries to access such data by following the reference to it, it fails. Such failure occurs because the data no longer exists or, even worse, the application unknowingly accesses other data (that has replaced the data erroneously deleted).

Memory leaks in servers and desktop computers are known to cause serious performance degradation. In addition, memory exhaustion arises if applications run for a reasonable amount of time. In mobile devices, such memory leaks are even more serious given the limited amount of memory available when compared to desktop computers.

Dangling references are well known to occur in centralized applications when manual memory management is used. Such errors are even more common in a classical distributed environment (i.e., in a fixed network of computers with no data replication). In a mobile environment supporting distributed applications accessing replicated objects, correct manual memory management is certainly more difficult. In conclusion, manual memory management leads not only to application performance degradation and fatal errors but also to reduced programmer productivity; thus, distributed garbage collection must be provided by mobile middleware.

Current middleware (e.g., Java and .NET) does not support distributed garbage collection. In fact, the approach taken is a simplified one, based on leases that favor liveness at the expense of safety. Objects still reachable remotely from other objects (possibly replicated) in other processes may be discarded (reclaimed, in GC terms) if they are not invoked for a certain period of time. This is clearly incorrect, as leases may expire too soon and cause dangling references; thus, applications will fail later when trying to access such objects. In addition, defining the leasing time is left to the application programmer, possibly leading to errors that could compromise the referential integrity of the objects graph.

When considering mobile middleware supporting data replication, the challenging requirements that mobile computing pose on distributed garbage collection (DGC) include the following: (1) safety and completeness of the DGC algorithms used (i.e., real distributed GC algorithms, not just lease mechanisms); (2) support for correct handling of replicated objects, for both local and distributed GC, notwithstanding data inconsistency; and (3) adaptation of local garbage collection (LGC) algorithms to resource-constrained devices used in mobile computing. We now address the distributed garbage collection of replicated objects and files.

Distributed Garbage Collection of Replicated Objects

Several of the classical DGC algorithms, designed for function-shipping-based distributed systems (i.e., with no support for replication), build upon some common elements found in algorithms such as indirect reference counting (IRC) [43] or stub–scion pair (SSP) chains [44] (in particular, GC-stubs and GC-scions). Most of these solutions [45,46] are hybrids, as each process has two components: a local tracing collector and a distributed collector. Each process does its local tracing independently from any other process. The local tracing can be done by any mark-and-sweep-based collector. The distributed collectors, based on reference listing, work together by changing asynchronous messages.

The local and distributed collectors depend on each other to perform their job in the following way. A local collector running inside a process traces the local objects graph starting from that process's local root (stack and static variables) and set of GC-scions. A local tracing generates a new set of GC-stubs; that is, for each outgoing interprocess reference it creates a GC-stub in the new set of GC-stubs. From time to time, possibly after a local collection, the distributed collector sends a message called *New-SetStubs*; this message contains the new set of GC-stubs that resulted from the local collection and is sent to the processes holding the GC-scions corresponding to the GC-stubs in the previous GC-stub set. In each of the receiving processes, the distributed collector matches the just-received set of GC-stubs with its set of GC-scions; those GC-scions that no longer have the corresponding GC-stub are deleted.

As described earlier, GC-scions and GC-stubs represent incoming and outgoing remote references, respectively, among objects residing in different processes (i.e., interprocess references). GC-scions and GC-stubs are created as a result of the export and import of references. A reference to an object in process P1 is said to be exported by P1 when it is sent on a message to process P2. A reference is said to be imported by a process P2 when it is received as the contents of a message delivered at P2. Note that the message just mentioned may carry one or more objects to be replicated.

Every time an object reference is exported by process P1 to process P2, the corresponding GC-scion and GC-stub must be created in P1 and P2, respectively. As long as an object is targeted by (at least) a GC-scion in a process, it must be preserved even when it is locally unreachable (from the local process stack and static variables). This is due to the fact that such an object may still be invoked from other processes through an interprocess reference.

Due to the activity of the mutator (i.e., the application in GC terms) in the referring process, some remote references may disappear (and their

Table 27.1   GC-Related Messages

Message	Sent/Received by	Sent When
NewSetStubs	DGC/DGC	A new set of GC-stubs is available

corresponding GC-stubs) because the objects enclosing them are no longer reachable (either locally in that process or via other remote references). Therefore, the processes holding the objects targeted by those remote references (and, correspondingly, their GC-scions) are informed that the GC-stubs no longer exist so they can delete their counterpart GC-scions. This ensures liveness, in the sense that objects that are no longer referenced remotely cease to be protected by the distributed GC component running in the process. Such objects are, from this moment on, at the mercy of the local GC. If they are also unreachable locally, then they will be reclaimed when the next LGC occurs.

Note that, to avoid competition regarding GC-scion creation and deletion between processes, there should be no explicit messages to delete GC-scions; instead, a particular type of message, the NewSetStubs message, is sent periodically by processes. The receiving processes may then detect the GC-scions for which the corresponding GC-stubs are no longer included in the message received and delete them accordingly. The algorithm operation is summarized in Table 27.1 and Table 27.2.

Table 27.2   GC-Related Events

Event	Occurs When	Action Taken
Reference exported	Replicate an object from a process	Create GC-scion.
Reference imported	Replicate an object into a process	Create GC-stub.
New set of GC-stubs available	Local GC runs	Send NewSetStubs message to the processes holding the GC-scions corresponding to the previous set of GC-stubs.
NewSetStubs message received	NewSetStubs message sent	Compare GC-stubs with set of GC-scions; delete GC-scions with no corresponding GC-stubs.

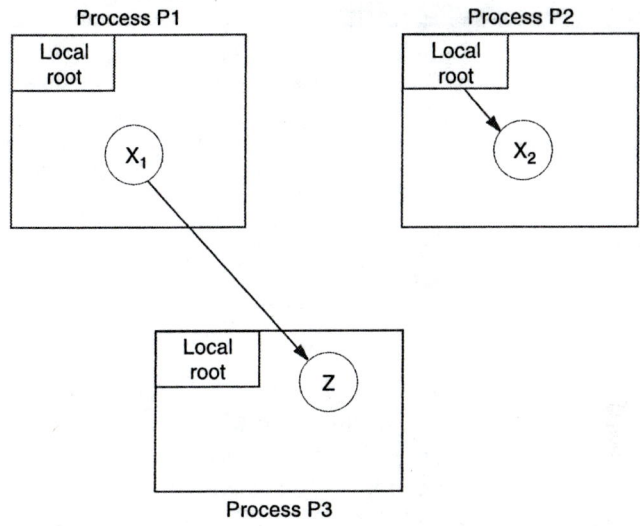

Figure 27.4   Safety problem of current DGC algorithms that do not handle replicated data: Z is erroneously considered unreachable.

DGC Correctness with Replication

In the previous section, we described how DGC algorithms work in distributed systems with no replication support; however, algorithms designed for function-shipping-based systems are not correct when applied to replicated objects. This affects all the classical DGC algorithms that were designed for function-shipping-based systems, such as IRC or SSP chains. These algorithms are not safe in the presence of replicated data, as explained next.

Consider Figure 27.4 in which object X is replicated in processes P1 and P2; each replica of X is denoted as X_1 and X_2, respectively. Now, suppose that X_1 contains a reference to object Z in another process P3, X_2 points to no other object, X_1 is locally unreachable, and X_2 is locally reachable. Then, the question is should Z be considered garbage? Classical DGC algorithms (designed for function-shipping systems) consider that Z is effectively garbage; however, this is wrong because, in a middleware object-replication system, it is possible for an application in P2 to acquire a replica of X from some other process, in particular, X_1. The fact that X_1 is *locally* unreachable in process P1 does not mean that X is *globally* unreachable; as a matter of fact, the contents of X_1 can be accessed by an application in process P2 by invoking method get on the corresponding proxy-in.

Table 27.3   GC Messages Related to Replication

Message	Sent/Received by	Sent When
Unreachable	LGC/DGC	Object replica is reachable only from the InPropList.
Reclaim	LGC/DGC	Object replica is reachable only from the OutPropList (with all recUmess bits set).

In a middleware object replication system, target object Z is considered unreachable only if the union of all the replicas of the source object (X in this example) do not refer to it. This is the Union Rule, introduced in Larchant [47,48] and implemented by resorting to a union message. This message is exchanged among processes and ensures the correct order of GC-scion creation and deletion, when objects are replicated through a number of processes. The causal delivery of such messages is required to ensure the safety of the algorithm with respect to races between the creation and deletion of GC-scions.

The Larchant algorithm is the first to correctly take into account GC-stubs, GC-scions, and the lists InPropList and OutPropList (with information about objects replicated from and to different processes). As described later, this prevents replicas from being prematurely discarded (as would happen with a non-replication-aware DGC algorithm).

Scalability Issues

Unfortunately, the previous solution imposes rather severe constraints on scalability. The Larchant algorithm is not scalable because it requires the underlying communication layer to support causal delivery [49]; therefore, DGC algorithms specific to object replication systems, such as Larchant, successfully address replication but lack scalability. To address this limitation on scalability, the requirement of causal message delivery must be dropped while still enforcing the Union Rule presented earlier. This is the approach introduced in DGC for wide area replicated memory (WARM) [50].

In this algorithm, the Union Rule is preserved but is enforced differently by making use of the special bits sentUmess and recUmess present in InPropList and OutPropList, as shown in Table 27.3. Whenever a replica located in a process becomes unreachable, except by an entry in the InPropList of the process, an "unreachable" message is sent to the process holding the corresponding OutPropList entry. The sending process registers this fact by setting the sentUmess bit present in the InPropList. This

Table 27.4 GC Events Related to Replication

Event	Occurs When	Action Taken
Object replica reachable only from the InPropList	LGC runs	Send "unreachable" message to the process with the corresponding OutPropList entry; set the sentUmess bit accordingly.
"Unreachable" message received	"Unreachable" message sent	Set the recUmess bit accordingly; if all recUmess bits for a particular object are set, then send the corresponding "reclaim" messages and delete the OutPropList entry.
"Reclaim" message received	"Reclaim" message sent	Delete corresponding InPropList entry.

prevents this message from being sent after every LGC in the process. Conversely, the receiving process registers this fact by setting the recUmess bit present in the corresponding OutPropList.

Whenever an object is reachable solely from entries in the OutPropList of the process and all the corresponding "unreachable" messages have been received, it is determined that none of the replicas is still reachable, and it is now safe to delete all of them (as a union). This is performed by sending "reclaim" messages to every process holding the replicas, followed by deletion of the OutPropList entries. Upon reception of the "reclaim" messages, the processes delete the corresponding InPropList entries. Finally, this enables the reclamation of all the replicas, when the next LGC takes place in each process. Table 27.3 and Table 27.4 present all the events related to GC replication and the corresponding actions taken. These four tables, then, summarize the way GC is performed by middleware supporting object replication.

A particularly interesting case in which the DGC just presented is useful is Web-content replication [51] with referential integrity ensured by the underlying middleware. The middleware supports HTML files (residing at Web servers) that can be incrementally replicated to other computers during browsing sessions. Replicated files can be subject to editing (e.g., translation, composition with other content) and further replicated. The DGC algorithm ensures referential integrity of the Web content still of interest to users, such as files that are reachable from a root (may include bookmarks, subscription lists, etc.). Therefore, the system correctly deals with possible inconsistencies among replicas and enforces the Union Rule already presented.

Other Memory Management Techniques

Mobile devices are so memory constrained that, in some circumstances, even the memory occupied by useful reachable objects must be freed. This may occur because, at a particular instant, there are other more relevant replicas for which no memory is available. Freeing the memory occupied by useful objects is a delicate process. Given that such objects can be accessed by applications through navigation of the objects graph, the middleware must still ensure the referential integrity while freeing such memory. This can be achieved as follows: Some object replicas (namely, all of those belonging to the same class) are migrated to a nearby machine (called a *surrogate*) [52]; later, if needed, such objects are remotely invoked. To provide transparency to applications, the underlying virtual machine transforms object accesses into remote invocations. This off-loads processing demands as well as memory occupation at the extra cost of frequent remote invocations.

An alternative approach proposed in Chen et al. [53] and Veiga and Ferreira [54] consists of swapping-out replicas to other computers with more resources available, particularly free memory; such replicas can be refetched later, if needed, and are always invoked locally. In Chen et al. [53], a modified virtual machine manages object location and records information to assess spatial and temporal locality for each object. Taking into account the management of replicas described earlier, object clusters are natural candidates to be swapped-out (and later swapped-in, if needed), as they have previously been incrementally replicated to the mobile device as a whole. This approach is adopted in Veiga and Ferreira [54] with small overhead, because information is kept only for clusters (which contain several objects) instead of individual objects. Unlike the previous approaches, it does not require the use of a specially modified virtual machine, making it rather portable.

Finally, Chen et al. [55] have proposed a mechanism to perform compression of the Java Virtual Machine heap. To minimize application disruption, large objects (greater than 1.5 kB) are compressed and decompressed; in addition, large array objects are broken down into smaller subobjects, each being "lazily" allocated upon its first write access.

Garbage Collection of Replicated Files

Distributed algorithms for garbage collection have also been applied in the context of replicated file systems (e.g., Ficus [18,56], Rumor [19]); however, when compared to the case in which the middleware supports the object model, DGC is not as relevant, the reason being that no referential integrity is maintained in the file model. In Ficus and Rumor, a

GC algorithm is used to reclaim disk space occupied by replicas of files found to have been deleted. When a process deletes a file replica, it initiates a two-phase, distributed-consensus algorithm to determine the global inaccessibility of all of the file replicas. Special care must be taken to resolve ambiguities between the creation and deletion of different replicas of the same file (due to race conditions regarding the order of these operations).

Adaptability

Due to their intrinsic nature, execution environments in mobile computing suffer from great and diverse variations during application execution. These variations can be qualitative (e.g., network connection or disconnection, specific devices such as printers present in the device neighborhood, consistency and security constraints) or can be related to quantitative aspects (e.g., amount of usable bandwidth or memory available). Applications should be able to deal with this variability; however, applications programmers should not be forced to account for every possible scenario in their coding, as this approach is inefficient, error prone, and limited to situations accounted for *a priori*.

Dealing with such variability can be achieved through automatic adaptation of applications; thus, the middleware must provide flexibility for application development and runtime adaptability. Applications can then cope with the multiple requirements and usage diversity encountered in mobile settings.

Adaptability is provided through the enforcement of declaratively defined policies supported by the mobile middleware [54]. In this way, policies do not have to be hard-coded in applications and can be deployed, enforced, and updated at any time. Mobile middleware relies strongly on the following features: (1) the extensible capability to support the specification and enforcement of runtime management policies, (2) a pluggable set of basic mechanisms supporting object replication, and (3) a set of predefined policies to control the mechanisms previously mentioned.

The policy engine (see Figure 27.1) is the inference component that triggers or mediates responses to events occurring in the system, and it holds a variable set of policies to be enforced in the system. The policy engine receives events generated by middleware modules and applications, evaluates policy rules, and triggers events, which are handled by actions taken based on results of these evaluations; in particular, object replication is performed according to specified policies.

The context management module (see Figure 27.1) performs resource abstraction and manages properties whose values vary during execution. Resource abstraction enables the representation of physical computer resources as sets of primitive context properties. Examples include memory, connectivity, and bandwidth available. The actual mappings between basic or primitive resources and resource designations is performed by the context manager. Each of these resources implies an architecture-dependent type of measuring. This heterogeneity is masked, to the rest of the system, by a low-level component in the context manager.

Situations such as appearing devices, discovering remote resources, or application counterparts are also handled by the context manager. Detecting these situations allows the middleware and the applications to decide whether to replicate data from different sources, swap out some data, etc. In general terms, any change to the properties considered (resources, middleware state, or user-defined properties) managed by the context manager can potentially trigger associated events defined by the policies loaded.

Replication Policies

A set of predefined policies and policy-driven modules is provided to manage specific execution mechanisms. In particular, the mobile middleware must support the specification and enforcement of policies concerning the replication of objects. Object replication is incremental and adaptive. Unless otherwise specified, it is performed transparently to applications but can also be flexibly configured. In particular, the mobile middleware must allow specification of the following: (1) the best moment to create a replica, (2) when to merge two or more replicas of the same object, (3) the number of objects to replicate at a given time (a cluster of objects), (4) which branch of a graph should be further replicated, and (5) which objects should be swapped-out (i.e., dynamically replaced by a proxy-out, transferring the remaining objects to a neighboring device).

In addition, the middleware must also support the definition and enforcement of consistency-related policies. Although addressed elsewhere in this book, maintaining the consistency of replicas must take into account the variability of mobile environments. This allows applications to deal, for example, with situations in which it is impossible to access the most up-to-date replica of an object but it is possible to obtain a slightly out-of-date replica that is still adequate enough for the application to proceed. Thus, regarding object replication and consistency, the adaptability supported by the mobile middleware support enables specification of the following: (1) alternative sources from which to replicate objects; (2) whether specific objects, clusters, or graphs, should be replicated from their authoritative home nodes or from other peers with outdated replicas, or should not be required at all for the application to proceed; (3) whether to cache changes made locally to the data; and (4) how failures are handled (e.g., automatically weaken some of the requirements to be able to replicate back some of the work performed).

Conclusion

In this chapter, we addressed several fundamental challenges concerning the support for replica management, memory management, and adaptability that must be considered by mobile middleware aiming at providing seamless service access via resource replication. Among the several programming paradigms available, we focused on two: object model and file model. The first is widely used and is highly flexible; the second is well known by both users and applications programmers and is supported by most operating systems. Although the object model allows applications to navigate on a graph of objects, the file model does not support such a concept. As explained in this chapter, this difference has important consequences on replica management and memory management.

With regard to mobile middleware supporting the object model, we presented an archetypical architecture and described how object replication can be achieved taking into account the limited amount of memory available in mobile devices (when compared to desktop computers). The mechanism of incremental replication provides the necessary flexibility for applications to deal with the variability of network availability and bandwidth and the amount of free memory.

With regard to the file model, we addressed how files can be replicated while minimizing the space occupied by taking advantage of similarities in the contents of the replicas. This mechanism also contributes to reducing the amount of network communication required for replica creation and updating. Deciding which and when files must be replicated is a difficult problem that still raises interesting research issues. We mentioned some relevant work in the area of file hoarding, but too much intervention is required from the user. Ideally, mobile middleware should automatically discover which files should be replicated and when, so users are never prevented from doing their work.

Concerning distributed garbage collection, this chapter has presented the most relevant algorithms that are able to deal correctly with replicated objects; however, they all lack the capability to reclaim distributed cycles of garbage replicas (i.e., the algorithms are not complete). This is a significant problem on which more research is needed. Lessons may be learned from recent complete DGC algorithms [57–60] developed for function-shipping systems (without replication support). The garbage collection in middleware supporting the file model aims at reclaiming disk space occupied by deleted files; thus, the relevant problem that must be solved is that of disk space management. The object model does not require the middleware to ensure referential integrity, as it happens with the object model.

Finally, we addressed the issue of adaptability (i.e., how the mobile middleware allows applications to control the resources they use). In particular, we focus on replication policies whose relevance results from the high variability of mobile environments in terms of network quality and memory available. A basic aspect is the need to clearly separate policies from mechanisms and allowing such policies to be dynamically instantiated. Although some solutions have been proposed by several projects, much research remains to be done so applications programmers can focus on the application logic without having to bother with system-level issues.

A successful mobile middleware platform has to deal with many other aspects besides those addressed in this chapter; however, we believe that replica management, memory management, and adaptability are among those that are most crucial. This chapter has provided potential solutions to such problems and has highlighted some research topics deserving additional attention.

Acknowledgments

We thank the institutions that have supported the authors: FCT (Fundação para a Ciência e a Tecnologia, Portugal) and Microsoft Research.

References

[1] Satyanarayanan, M., Fundamental challenges in mobile computing, in *Proc. of the 16th Int. Conf. on Distributed Computing Systems (ICDCS'96)*, Hong Kong, May, 1996, pp. 1–7.

[2] Liskov, B., Day, M., and Shrira, L., Distributed object management in Thor, in *Proc. of Int. Workshop on Distributed Object Management (IWDOM'92)*, Edmonton, Canada, August, 1992, pp. 1–15.

[3] Gruber, R., Kaashoek, F., Liskov, B., and Shrira, L., Disconnected operation in Thor object-oriented database system, in *Proc. of IEEE Workshop on Mobile Computing Systems and Applications (WMCSA'94)*, Santa Cruz, CA, December, 1994.

[4] Kubiatowicz, J., Bindel, D., Chen, Y., Czerwinski, S., Eaton, P. et al., Ocean-Store: an architecture for global-scale persistent storage, in *Proc. of the Ninth Int. Conf. on Architectural Support for Programming Languages and Operating Systems (ASPLOS'00)*, Cambridge, MA, November, 2000, pp. 190–201.

[5] Çetintemel, U., Keleher, P.J., Bhattacharjee, B., and Franklin, M.J., Deno: a decentralized, peer-to-peer object-replication system for weakly connected environments, *IEEE Trans. Comput.*, 52(7), 943–959, 2003.

[6] Veiga, L. and Ferreira, P., Incremental replication for mobility support in OBIWAN, in *Proc. of IEEE Int. Conf. on Distributed Computing Systems (ICDCS'02)*, Vienna, Austria, July, 2002, pp. 249–256.

[7] Ferreira, P., Veiga, L., and Ribeiro, C., OBIWAN: design and implementation of a middleware platform, *IEEE Trans. Parallel Distributed Syst.*, 14(11), 1086–1099, 2003.

[8] Veiga, L., Santos, N., Lebre, R., and Ferreira, P., Loosely coupled, mobile replication of objects with transactions, in *Proc. of Workshop on QoS and Dynamic Systems, Tenth Int. Conf. on Parallel and Distributed Systems (ICPADS'04)*, Newport Beach, CA, July, 2004.

[9] Santos, N., Veiga, L., and Ferreira, P., Transaction policies for mobile networks, in *Proc. of the 5th IEEE Workshop on Policies for Distributed Systems and Networks (Policy 2004)*, Yorktown Heights, NY, June, 2004.

[10] Pairot, C., García, P., and Skarmeta, A.F.G., Dermi: a decentralized peer-to-peer event-based object middleware, in *Proc. of IEEE Int. Conf. on Distributed Computing Systems (ICDCS'04)*, Tokyo, Japan, March, 2003, pp. 236–243.

[11] Caughey, S.J., Hagimont, D., and Ingham, D.B., Deploying distributed objects on the Internet, in *Recent Advances in Distributed Systems*, Krakowiak, S. and Shrivastava, S.K., Eds., Lecture Notes in Computer Science, Springer-Verlag, Heidelberg, 2000.

[12] Hagimont, D. and Boyer, F., A configurable RMI mechanism for sharing distributed Java objects, *IEEE Internet Comput.*, 5(1), 36–44, 2001.

[13] Butrico, M., Chang, H., Cocchi, A, Cohen, N., Shea, D., and Smith, S., Gold Rush: mobile transaction middleware with Java-object replication, in *Proc. of the Third USENIX Conf. on Object-Oriented Technologies (COOTS)*, Portland, OR, June, 1997.

[14] Haahr, M., Cunningham, R., and Cahill, V., Towards a generic architecture for mobile object-oriented applications, in *Proc. of IEEE Workshop on Service Portability and Virtual Customer Environments*, San Francisco, CA, December, 2000, pp. 91–96.

[15] Siegel, J., *CORBA Fundamentals and Programming*, John Wiley & Sons, New York, 1996.

[16] Kistler, J.J. and M. Satyanarayanan, M., Disconnected operation in the Coda file system, *ACM Trans. Comput. Syst.*, 10(1), 3–25, 1992.

[17] Satyanarayanan, M., The evolution of Coda, *ACM Trans. Comput. Syst.*, 20(2), 85–124, 2002.

[18] Popek, G.J., Guy, R.G., Page, Jr., T.W., and Heidemann, J.S., Replication in Ficus distributed file systems, in *Proc. of the First IEEE Workshop on Management of Replicated Data*, Los Angeles, CA, November, 1990, pp. 20–25.

[19] Guy, R.G., Reiher, P.L., Ratner, D., Gunter, M., Ma, W., and Popek, G.J., Rumor: mobile data access through optimistic peer-to-peer replication, in *Proc. of Int. Workshops on Data Warehousing and Data Mining: Advances in Database Technologies, 17th Int. Conf. on Conceptual Modeling (ER'98)*, Singapore, November, 1999, pp. 254–265.

[20] Ratner, D., Reiher, P., Popek, G.J., and Kuenning, G.H., Replication requirements in mobile environments, *Mobile Netw. Appl.*, 6(6), 525–533, 2001.

[21] Ratner, D., Reiher, P., and Popek, G.J., Roam: a scalable replication system for mobility, *Mobile Netw. Appl.*, 9(5), 537–544, 2004.

[22] Gray, J.N. and A. Reuter, A., *Transaction Processing: Concepts and Techniques*, Morgan Kaufmann, San Francisco, CA, 1993.

[23] Demers, A.J., Petersen, K., Spreitzer, M.J., Terry, D.B., Theimer, M.M., and Welch, B.B., The Bayou architecture: support for data sharing among mobile users, in *Proc. of IEEE Workshop on Mobile Computing Systems and Applications (WMCSA'94)*, Santa Cruz, CA, December, 1994, pp. 2–9.

[24] Terry, D.B., Theimer, M.M., Petersen, K., Demers, A.J., Spreitzer, M.J., and Hauser, C.H., Managing update conflicts in Bayou, a weakly connected replicated storage system, in *Proc. of the 15th ACM Symp. on Operating System Principles*, Copper Mountain, CO, December, 1995, pp. 172–182.

[25] Terry, D.B., Petersen, K., Spreitzer, M., and Theimer, M., The case for non-transparent replication: examples from Bayou, *IEEE Data Eng. Bull.*, 21(4), 12–20, 1998.

[26] Preguiça, N., Martins, J.L., Cunha, M., and Domingos, H., Reservations for conflict avoidance in a mobile database system, in *Proc. of First Int. Conf. on Mobile Systems, Applications, and Services (MobiSys'03)*, San Francisco, CA, May, 2003.

[27] Sinha, A., Client-server computing, *Commun. ACM*, 35(7), 77–98, 1992.

[28] Androutsellis-Theotokis, S. and Spinellis, D., A survey of peer-to-peer content distribution technologies, *ACM Comput. Surv.*, 36(4), 335–371, 2004.

[29] Petersen, K., Spreitzer, M.J., Terry, D.B., Theimer, M.M., and Demers, A.J., Flexible update propagation for weakly consistent replication, in *Proc. of the 16th ACM Symp. on Operating System Principles*, Saint-Malo, France, December, 1997.

[30] Shapiro, M., Structure and encapsulation in distributed systems: the proxy principle, in *Proc. of the 6th Int. Conf. on Distributed Systems*, Boston, MA, May, 1986, pp. 198–204.

[31] Campione, M. and Walrath, K., *The Java Tutorial: Object Oriented Programming for the Internet*, 2nd ed., Sun Java Series, Addison-Wesley, Boston, MA, 1996.

[32] Platt, D.S., *Introducing Microsoft .NET*, Microsoft Press, Redmond, WA, 2001.

[33] Nelson, M.N., Welch, B.B., and Ousterhout, J.K., Caching in the Sprite network file system, *ACM Trans. Comput. Syst.*, 6(1), 134–154, 1988.

[34] Muthitacharoen, A., Chen, B., and Mazieres, D., A low-bandwidth network file system, in *Proc. of the 18th ACM Symp. on Operating System Principles*, Lake Louis, Alta, Canada, October, 2001, pp. 174–187.

[35] Barreto, J. and Ferreira, P., A replicated file system for resource constrained mobile devices, in *Proc. of IADIS Applied Computing (IADIS'04)*, Madrid, Spain, October, 2004.

[36] Barreto, J. and Ferreira, P., A highly available replicated file system for resource-constrained Windows CE .NET devices, in *Proc. of the 3rd Int. Conf. on .NET Technologies*, Pilsen, Czech Republic, May 30–June 1, 2005.

[37] NIST, *FIPS PUB 180-1: Secure Hash Standard*, Technical Report, National Institute of Standards and Technology, Gaithersburg, MD, 1995.

[38] Rabin, M., *Fingerprinting by Random Polynomials*, Technical Report TR-15-81, Center for Research in Computing Technology, Harvard University, Boston, MA, 1981.

[39] Chen, G. and Kotz, D., *A Survey of Context-Aware Mobile Computing Research*, Technical Report TR2000-381, Department of Computer Science, Dartmouth College, Hanover, NH, 2000.

[40] Tait, C., Lei, H., Acharya, S., and Chang, H., Intelligent file hoarding for mobile computers, in *Proc. of the 1st ACM/IEEE Int. Conf. on Mobile Computing and Networking (MOBICOM'95)*, Berkeley, CA, November, 1995.

[41] Kuenning, G.H and Popek, G.J., Automated hoarding for mobile computers, in *Proc. of the 16th ACM Symp. on Operating System Principles*, Saint-Malo, France, December, 1997, pp. 264–275.

[42] Kuenning, G.H., Ma, W., Reiher, P., and Popek, G.J., Simplifying automated hoarding methods, in *Proc. of the 5th ACM Int. Workshop on Modeling, Analysis, and Simulation of Wireless and Mobile Systems (MSWiM)*, Atlanta, GA, September, 2002, pp. 15–21.

[43] Piquer, J.M., Indirect reference-counting, a distributed garbage collection algorithm, in *Proc. of Int. Parallel Architectures and Languages Europe (PARLE'91)*, Eindhoven, The Netherlands, June, 1991, pp. 150–165.

[44] Shapiro, M., Dickman, P., and Plainfossé, D., *SSP Chains: Robust, Distributed References Supporting Acyclic Garbage Collection*, Rapport de Recherche 1799, Institut National de Recherche en Informatique et Automatique, Rocquencourt, France, (http://www-sor.inria.fr/SOR/docs/SSPC_rr1799.html).

[45] Abdullahi, S.E. and Ringwood, G.A., Garbage collecting the Internet: a survey of distributed garbage collection, *ACM Comput. Surv.*, 30(3), 330–373, 1998.

[46] Plainfossé, D. and Shapiro, M., A survey of distributed garbage collection techniques, in *Proc. of Int. Workshop on Memory Management (IWMM'92)*, Kinross, Scotland, September, 1995.

[47] Ferreira, P. and Shapiro, M., Larchant: persistence by reachability in distributed shared memory through garbage collection, in *Proc. of the 16th Int. Conf. on Distributed Computing Systems (ICDCS'96)*, Hong Kong, May, 1996.

[48] Ferreira, P. and Shapiro, M., Modelling a distributed cached store for garbage collection: the algorithm and its correctness proof, in *Proc. of the 12th European Conf. on Object-Oriented Programming (ECOOP'98)*, Brussels, Belgium, July, 1998.

[49] Cheriton, D.R. and Skeen, D., Understanding the limitations of causally and totally ordered communication, in *Proc. of the 14th ACM Symp. on Operating Systems Principles*, Austin, TX, November, 1993, pp. 44–57.

[50] Sanchez, A., Veiga, L., and Ferreira, P., Distributed garbage collection for wide area replicated memory, in *Proc. of the Sixth USENIX Conf. on Object-Oriented Technologies (COOTS)*, San Antonio, TX, January, 2001.

[51] Veiga, L. and Ferreira, P., Repweb: replicated web with referential integrity, in *Proc. of the 18th ACM Symp. on Applied Computing (SAC'03)*, Melbourne, FL, March, 2003.

[52] Messer, A., Greenberg, I., Bernadat, P., Milojicic, D., Chen, D. et al., Towards a distributed platform for resource-constrained devices, in *Proc. of IEEE Int. Conf. on Distributed Computing Systems (ICDCS'02)*, Vienna, Austria, July, 2002, p. 43–51.

[53] Chen, D., Messer, A., Milojicic, D., and Dwarkadas, S., Garbage collector assisted memory offloading for memory-constrained devices, in *Proc. of the Fifth IEEE Workshop on Mobile Computing Systems and Applications (WMCSA'03)*, Monterey, CA, October, 2003.

[54] Veiga, L. and Ferreira, P., Poliper: policies for mobile and pervasive environments, in *Proc. of the 3rd Int. Workshop on Reflective and Adaptive Middleware, 5th ACM/IFIP/USENIX Int. Middleware Conf.*, Toronto, Canada, October, 2004.

[55] Chen, G., Kandemir, M., Vijaykrishnan, N., Irwin, M.J., Mathiske, B., and Wolczko, M., Heap compression for memory-constrained Java environments, in *Proc. of ACM Conf. on Object-Oriented Programming Systems, Languages, and Applications (OOPSLA 2003)*, Anaheim, CA, October, 2003, pp. 282–301.

[56] Ratner, D., Reiher, P.L., Popek, G.J., and Guy, R.G., Peer replication with selective control, in *Proc. of the First Int. Conf. on Mobile Data Access (MDA'99)*, Hong Kong, December, 1999, pp. 169–181.

[57] Rodrigues, H.C.C.D. and Jones, R.E., Cyclic distributed garbage collection with group merger, in *Proc. of the 12th European Conf. on Object-Oriented Programming (ECOOP'98)*, Brussels, Belgium, July, 1998, pp. 249–273.

[58] Fessant, F.L., Detecting distributed cycles of garbage in large-scale systems, in *Proc. of the 20th ACM Symp. on Principles of Distributed Computing (PODC 2001)*, Newport, RI, August, 2001.

[59] Veiga, L. and Ferreira, P., Complete distributed garbage collection: an experience with Rotor, *IEE Res. J. Software*, 150(5), 283–290, 2003.

[60] Veiga, L. and Ferreira, P., Asynchronous complete distributed garbage collection, in *Proc. of IEEE Int. Conf. on Parallel and Distributed Processing Symp. (IPDPS'05)*, Denver, CO, April, 2005.

Section 5

MOBILE MIDDLEWARE FOR LOCATION-DEPENDENT SERVICES

An Overview of the Location Management Problem for Mobile Computing Environments

Javid Taheri and Albert Y. Zomaya

CONTENTS

Introduction

Numerous companies and service providers are pursuing a fully integrated service solution for wireless mobile networks. Current voice, fax, and paging services will be combined with data transfer, video conferencing, and other mobile multimedia services to build the next generation of wireless mobile networks. Basically, such networks will be designed to support a true combination of both real-time and non-real-time services and then form a global personal communication network. To support such a wide range of data transfer and user applications, *mobility management* has to be considered when designing infrastructures for wireless mobile networks.

Mobility management involves two processes: *location management* and *handoff management*. Location management allows the wireless network to discover the current point of attachment of a mobile terminal and deliver calls to it, and handoff management allows the mobile network to locate roaming mobile terminals for call delivery and to maintain a connection as the mobile terminal moves around. During the first stage of location management, known as *location registering* or *location update*, the mobile terminal periodically informs the network of its new access point and helps the network to authenticate the user and revise user location profiles. The second stage is *call delivery*, in which the wireless mobile network is queried for a mobile terminal location and the current position of that terminal is found [1,2].

On the other hand, handoff primarily represents a process of changing some of the parameters of a channel (frequency, time slot, spreading code, or a combination of these) while the current connection is in use [3,4]. The handoff process usually consists of two phases: *handoff initialization* and *handoff enabling phase*. During the handoff initialization phase, the quality of the available communication channel is considered to decide when the handoff process should be triggered, and in the handoff enabling phase the allocation of additional resources by a new base station is

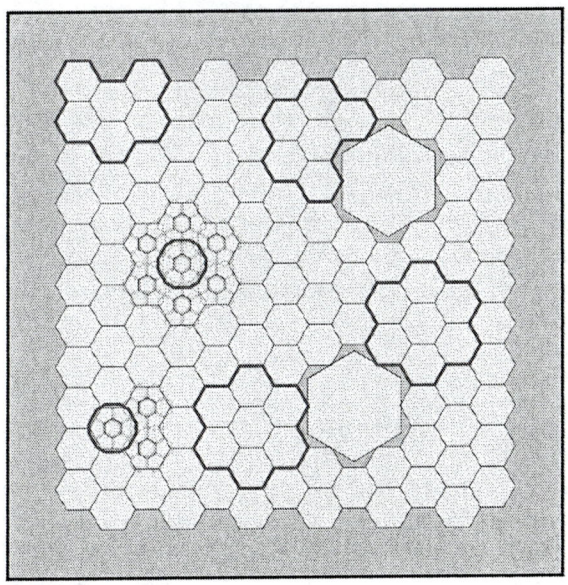

Figure 28.1 An example of a GSM network.

initiated and processed. Poorly designed handoff schemes tend to generate huge signaling traffic, and thereby a dramatic decrease in quality of service (QoS) of integrated services of the wireless network.

Mobility management requests are often initiated either by a mobile terminal movement (crossing a cell boundary) or by deterioration in the quality of a signal received on a currently allocated channel. Due to the anticipated increase in the use of wireless services in the future, the next generation of mobile networks should be able to support a huge number of users and their bandwidth requirements. Furthermore, more frequent handoffs will result when the size of the cells becomes smaller or a drastic change occurs in the propagation condition of a signal; therefore, mobility management becomes more crucial in the next generation of mobile networks.

Figure 28.1 shows an example of a Global System for Mobile Communications (GSM) network [1–6]. In this network, cells are grouped together into regions. Each region contains the entire allotted frequency spectrum, and each cell of the group uses a part of the allocated frequency. The same frequency can be used in other regions by carefully considering the minimum distance between cells to avoid cross-talking [5,6]. On the other hand, as the demand for wireless services increases, the size of the cells becomes smaller and the reusability of the allocated frequencies becomes more problematic. As a result, network management and, consequently, location management become significant issues, and efficient techniques will be required to ensure delivery of all incoming calls, even in the tiniest of cells.

As mentioned earlier, location management consists of location update and location inquiry. In location update, every mobile terminal updates its location in the network and notifies the network of its current location. Location inquiry is performed by the network itself; in this process, the network tries to locate the user based on the last known location. Location update is usually performed when a user changes places in the network, and location inquiry is usually performed when the network tries to direct an incoming call to a customer.

Location Update Strategies

Location update strategies can be categorized into two main groups: *dynamic* and *static*. In dynamic schemes, different network topologies are considered for different users [7–18]. These topologies are highly related to the movement patterns and calling behavior of each user. On the other hand, in static schemes, the network has a unique behavior for all users, like the current GSM networks. It is obvious that dynamic schemes are much more complex than static ones and require more computation capabilities in the network; thus, implementing static schemes is the more popular approach [19–40]. A location update strategy, however, must use minimum network resources to manage user tracking and should not require massive computations. The following techniques are the most common ones for static location updates:

- *Always update strategy.* The user updates its location whenever it crosses a cell boundary. In this case, the network is able to locate the user in minimum time for each incoming call; that is, the network will page the user only in the last updated cell. This scheme generates a massive number of unnecessary location update signals.
- *Never update strategy.* The user never updates its location; consequently, for each incoming call, the network must page the user in all cells of the network. It is obvious that this strategy is highly inefficient, especially when the number of cells in the network is large. Moreover, as the number of users in the system increases, the number of paging signals would also increase.
- *Time-based strategy.* The user updates its location after a predefined time span [19]; therefore, in case of an incoming call, the network will page the user in the last updated cell and the possible cells that the user might be in after the last location update. Although this strategy seems better than the other two, some disadvantages

still remain [19]; for example, if the user changes its location too rapidly after the last location update, the network will not be able to locate the user.

■ *Movement-based strategy.* The user makes an update after passing a predefined number of cells (namely, *M*). In this case, the network should page the user in a radius of *M* cells from the last update cell [20,21]. The main disadvantage of this strategy applies to the case of cyclic users who periodically move between adjacent cells of a network (e.g., move in a zigzag path between two cells).

■ *Distance-based strategy.* The user updates its location after moving out of the last update cell for a predefined distance (namely, *R*). In this case, the network can page the user in all cells that are physically within the *R* radius of the last updated location [22,23]. The main drawback of this method is the need for having geographical information about the network for each mobile terminal. Clearly, each mobile terminal should have the complete information about the network topology and must be able to determine its geographical location as well, which makes the design of mobile terminals so complicated and probably more expensive than the current ones.

■ *Location area scheme.* This methodology is used in current GSM networks. Adjacent cells are grouped together to form a *location area* (LA) [25,29,37–40]. In this case, the user performs an update when it leaves its current LA and moves into a new one; therefore, in the case of an incoming call, the network must page the user in all cells of the last updated LA. Figure 28.2 shows a typical GSM network layout. In this case, if the user updates its location in cell *X*, then for each incoming call all cells of the current LA (shown in gray) will page the user.

■ *Paging cell scheme.* The user makes an update whenever it passes through some predefined cells known as paging cells [13,26,28–35, 37]. In this case, a paging neighborhood is defined for each paging cell. Basically, the paging neighborhood of each paging cell contains all non-paging cells of the network that must page the user in case of an incoming call. For example, consider the paging cells shown in gray in Figure 28.3. The paging neighbors of paging cell *X* are all cells marked as *N*. Note that the user can be in any cell marked as *N* without passing through any other paging cell marked as *X*; therefore, all of the cells marked as *N* must be included in the user paging process for any incoming call. In this strategy, paging cells are assigned so they split the entire network into some smaller networks. The never update strategy is then used in each of these subnetworks.

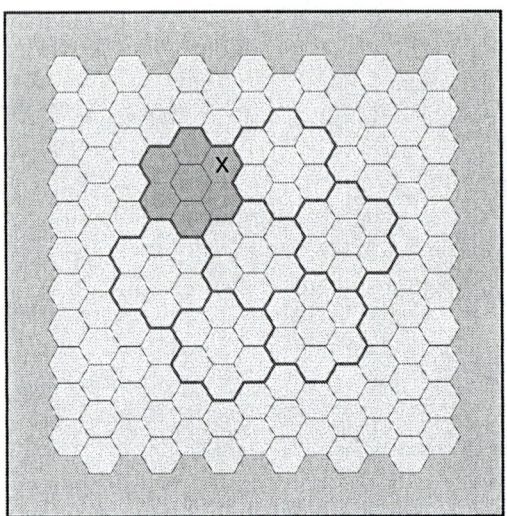

Figure 28.2   Location area configuration.

The question that still arises in all of the above approaches is what configuration of cells provides the best performance? For example, what is the optimal predefined distance to be used in the distance-based strategy? Or, what is the best predefined number of cells to travel for the movement base strategy? Similarly, what is the best topology for location areas in the location area scheme to have the best network performance with minimum cost? And, finally, which cells should be defined as paging cells in the paging cell scheme to split the network in the best possible way? In this work, the aim is to explain the basics of the location management problem and some related solution strategies. Finally, several algorithms to solve this problem using the location area and paging cell schemes are briefly presented.

Location Management Cost

Many algorithms have been proposed to solve the location management problem, although it is still necessary to develop a framework that can be used to compare these techniques [34–40]. A location management cost should be defined to evaluate every approach. The location management cost usually consists of two main parts: *updating cost* and *paging cost*. The updating cost is the portion of the total cost due to location updates performed by mobile terminals in the network, and the paging cost is incurred by the network during a location inquiry when the network

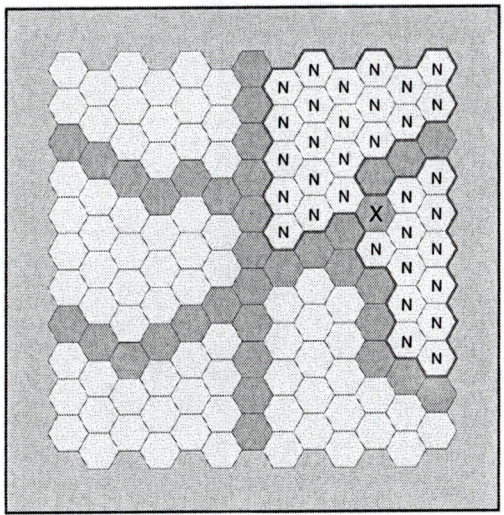

Figure 28.3 Paging cell configuration.

tries to locate a user. The total cost of location management involves other parameters, such as the cost of database management to register users' locations, the cost of the wired network (backbone) that connects the base stations to each other, and the cost of switching between base stations in the case of handoff and call diverting, among others. Nevertheless, in general, these costs are assumed to be the same for all location management strategies. As a result, the combination of location update and paging costs is considered to be sufficient to compare the different approaches; therefore, the total cost of a location management scheme can be given as follows:

$$Cost = \beta \times N_{LU} + N_P$$

where N_{LU} is the total number of location updates, N_P represents the total number of paging transactions, and β is a constant representing the cost ratio of a location update to a paging transaction in the network. The number of location updates is usually caused by the movement of the user in the network, and the number of the paging transactions is greatly related to the number of incoming calls.

Evidence shows that the cost of each location update is much higher than the cost of a paging transaction because of the complex procedures that have to be executed every time a location update is performed. On the other hand, most of the calls of a mobile user are incoming calls (two thirds); therefore, if the user moves in the network without making any

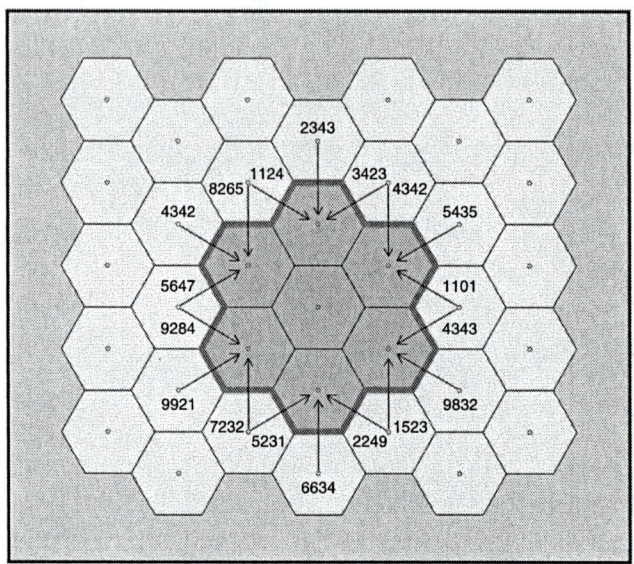

Figure 28.4   Incoming user flow for a sample location area.

call, the network will undergo a huge number of useless transactions. Thus, in most cases the cost of a location update is considered to be 10 times than that of a paging transaction (i.e., β = 10) [34–40].

Location Management Cost for the Location Area Strategy

To calculate the total cost of network management in the LA strategy, a few issues must be considered. Based on the LA scheme, a location update transaction takes place when a user changes its current LA; therefore, all mobile terminals (users) entering the LAs of a network should be added to compute this cost. On the other hand, the paging cost must be considered when a user has an incoming call. As a result, all cells are involved in calculating this cost. To clarify this point, consider the LA strategy shown as gray in Figure 28.4. In this configuration, the flow of the total number of users who enter the LA via the boundary cells can be calculated as:

Number of location updates = 2343 + 4323 + 4342 + 5435 + 1101 + 4343 + 9832 + 1523 + 2249 + 6634 + 5231 + 7232 + 9921 + 9284 + 5647 + 4342 + 8265 + 1124 = 92,271

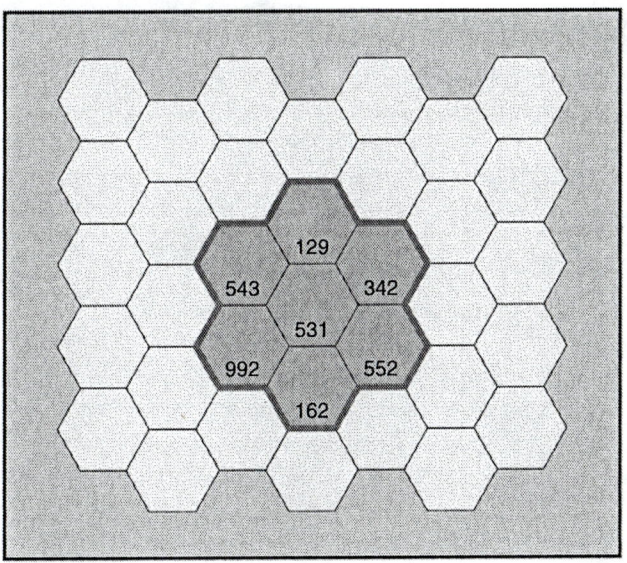

Figure 28.5 User call arrivals for a sample location area.

Calculating the paging cost is much easier. This can be simply done by counting the number of incoming calls in the LA and multiplying it by the number of cells (of the LA). The total number of incoming calls for the LA in Figure 28.4 is given in Figure 28.5, and the total number of paging signals for this LA would be:

The number of paging transactions = (129 + 543 + 531 + 342 + 992 + 162 + 552) × 7 = 22,757

Finally, if this procedure is repeated for all the LAs of the network, the total number of location updates and paging signals can be calculated. For the marked LA in Figure 28.4 and Figure 28.5 these costs would be as follows:

Total number of location updates = 92,271

Total number of paging signals = 22,757

Total cost = (92,271 × 10) + 22,757 = 945,467

In the above calculation, the cost of a location updates is assumed to be 10 times more expensive than that of a paging transaction (i.e., $\beta = 10$).

Location Management Cost for the Paging Cell Strategy

Calculating the total network management cost for the PC scheme is a little different. As mentioned earlier, the paging cell scheme has two types of cells: paging cells and non-paging cells. As a first step, a paging neighborhood is assigned for each paging cell; this neighborhood includes all non-paging cells that must page the user in the case of an incoming call. Also, another factor, the *vicinity factor*, is defined for each cell of the network, for both paging and non-paging cells. This factor is basically the maximum number of paging neighbors for each cell that must page the user in case of an incoming call. Obviously, the vicinity factor of each paging cell is the number of its paging neighbors; however, for non-paging cells, the way the vicinity factor is calculated is different. Based on the fact that each non-paging cell might be in the paging neighborhood of more than one paging cell, the maximum number of paging neighbors that the cell is a part of is considered as its vicinity factor.

To clarify this definition, consider the paging cell configuration of the network given in Figure 28.6, where the paging cells are marked in gray. The cell marked N is in the paging neighborhood of at least three paging cells (shown in Figure 28.6a,b,c). The number of paging neighbors for cell X is 25, 17, and 22 for Figure 28.6a,b,c, respectively; that is, cell N is part of at least three paging neighborhoods with 25, 17, and 22 cells. To consider the worst case, the vicinity factor of non-paging cell N is considered to be 25, which is the maximum of 25, 17, and 22. If a non-paging cell is a part of more than two paging cell neighborhoods, such as the one in Figure 28.6, then this calculation must be performed for all of them, and their maximum number is then considered as the vicinity factor for that non-paging cell. In Figure 28.6d, for example, the non-paging cell marked N is a part of the paging neighborhood of all cells marked as X.

Finally, to calculate the total cost of the network location management, the general cost function is modified as:

$$Cost = \beta \times \sum_{i \in S} N_{LU}(i) + \sum_{i=0}^{N} N_P(i) \times V(i)$$

where $N_{LU}(i)$ is the number of location updates for paging cell number i, $N_P(i)$ is the number of arrived calls for cell i, $V(i)$ is the vicinity factor for cell i, S is the set of cells defined as paging cells, and N is the total number of cells in the network.

Consider the paging cell configuration given in Figure 28.7. The number of location updates for each paging cell appears at the center of the cell. The total number of location updates for this configuration would be:

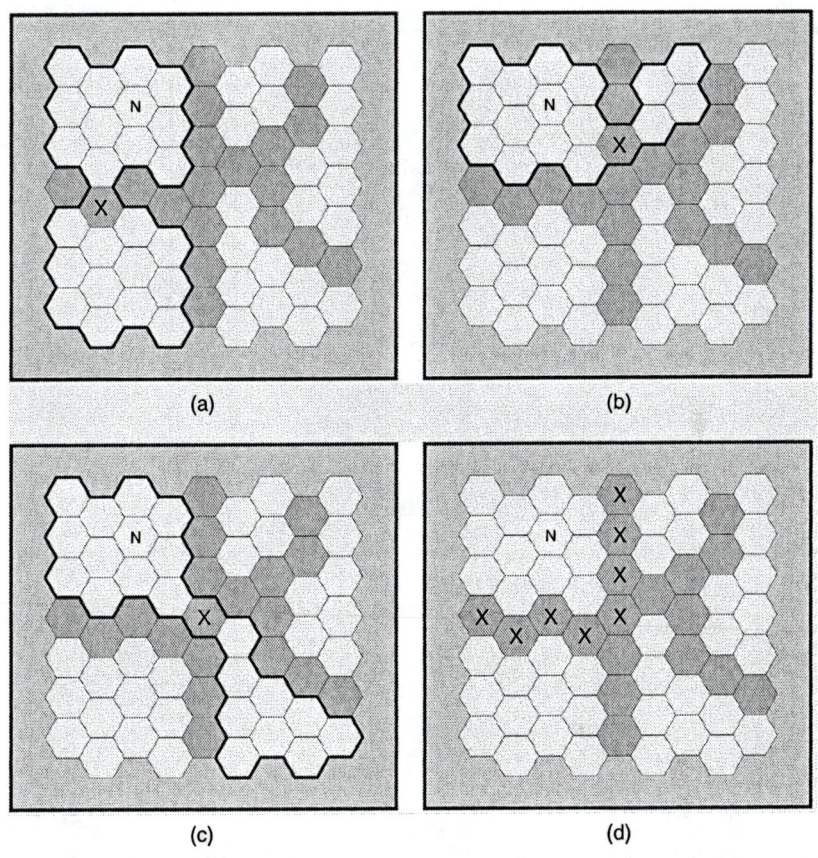

(a)   (b)

(c)   (d)

Figure 28.6 An instance of a non-paging cell belonging to more than one paging cell neighbor.

Total number of location updates = 453 + 395 + 541 + 492 + 432 + 361 + 409 + 123 = 3206

Calculating the paging cost of the network in this scheme is different. Initially, the vicinity factor of each cell, including paging cells and non-paging cells, must be computed and then the total paging cost can be determined. In Figure 28.8, two numbers appear in each cell. The number at the center of the cell is the number of incoming calls and the number that appears near the edge of the cell is the vicinity factor for that cell. Finally, the total number of paging transactions for this network would be:

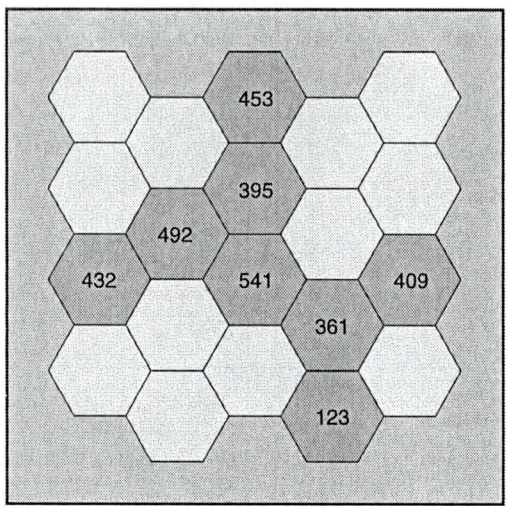

Figure 28.7 Number of location updates.

Total number of paging transactions = $(51 \times 8) + (43 \times 8) + (48 \times 8) + (92 \times 10) + (62 \times 8) + (83 \times 8) + (71 \times 10) + (45 \times 10) + (35 \times 8) + (12 \times 8) + (38 \times 9) + (73 \times 10) + (83 \times 10) + (53 \times 10) + (82 \times 10) + (58 \times 6) + (45 \times 10) + (76 \times 10) + (57 \times 6) + (19 \times 10) = 10,076$

If each paging transaction costs 1 unit and each location updates costs 10 units (i.e., $\beta = 10$), then the total cost of the network would be:

Total cost = $(3206 \times 10) + 10,076 = 42,136$ units

Network Simulator

Simulating the network is one of the most important tasks when dealing with this class of problems. In almost all of the available literature on solving this problem, the cell attributes of the network are generated randomly. In general, two independent attributes for each cell are considered: the number of call arrivals and the number of location updates, and both are randomly selected. In other words, the number of call arrivals and location updates of adjacent cells are not correlated; however, these numbers are highly correlated in a real network. Therefore, to generate these numbers for each cell in this work, a set of sophisticated routines is used to generate user profiles and, consequently, the network attributes of each cell [36]. As a

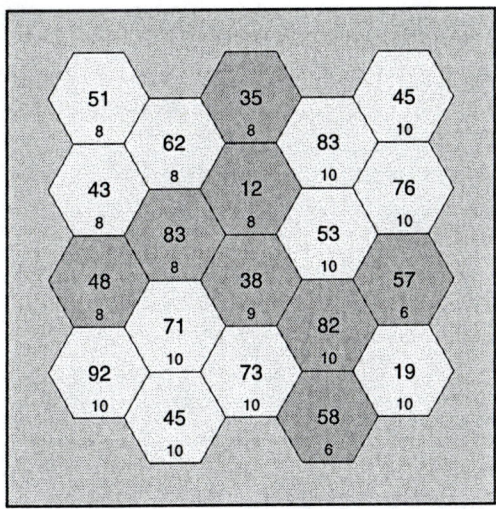

Figure 28.8   Number of call arrivals.

result, the generated network configurations better match real-world traffic. Moreover, two relatively independent aspects — network generation and population generation — are considered in this work when simulating an artificial environment to evaluate designed algorithms.

In the network generation stage, the aim is to model the backbone of the network; Figure 28.9 shows the wizard dialog developed in the software to generate a simple network. The aim of the population generation stage is to simulate the behavior of mobile network customers; Figure 28.10 shows the wizard dialog developed in the software to generate a sample population, and Figure 28.11 presents a sample histogram of the population. This histogram represents the total number of location updates for each cell.

Network Modifier

After generating a sample network with predefined cell attributes and mobile terminal (user) characteristics, it is necessary to modify the network backbone to support the users based on the maximum capacity of each cell. In this case, two restrictions are considered for each cell: the maximum number of location updates and the maximum number of call arrivals. Now, if these numbers exceed the predefined values of each cell, then the cell is split into smaller cells to support the traffic. Figure 28.12 and Figure 28.13 show a sample network that is modified based on the above argument.

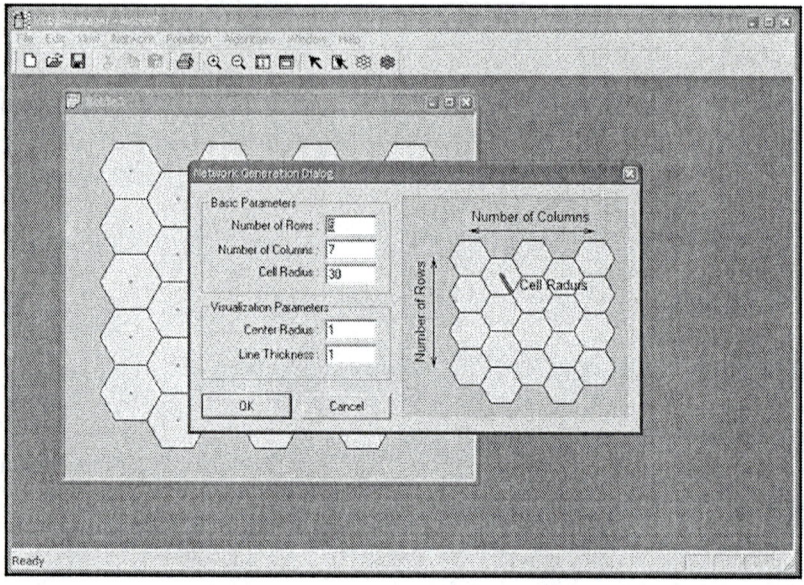

Figure 28.9   Network generation wizard dialog.

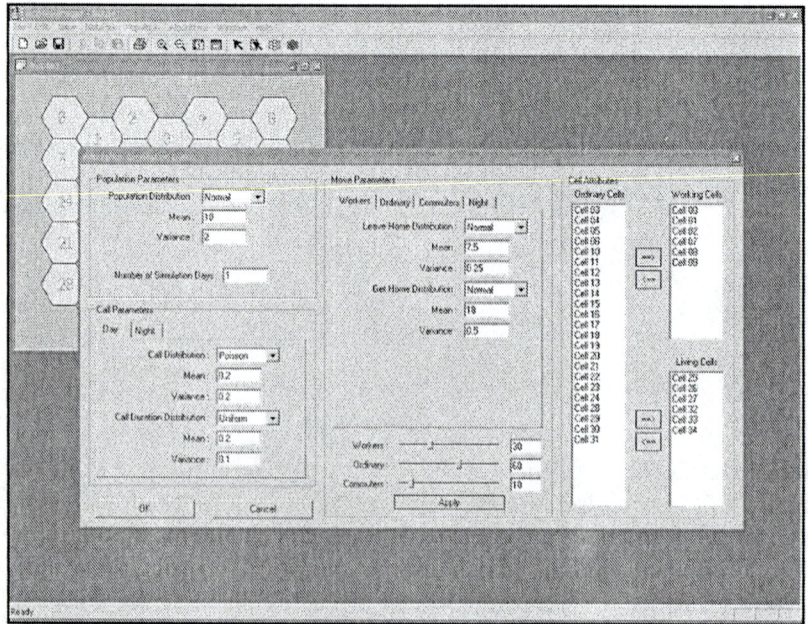

Figure 28.10   Population generation wizard dialog.

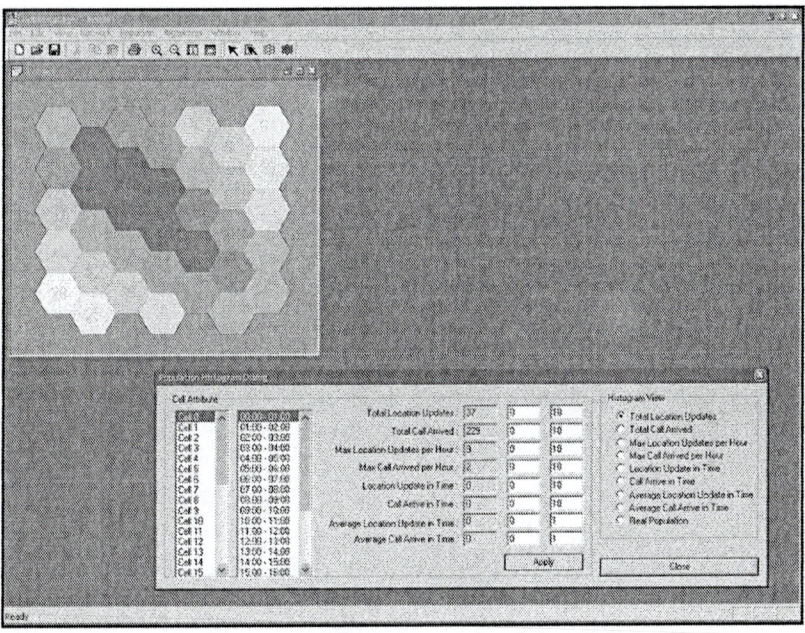

Figure 28.11   Histogram for a sample population.

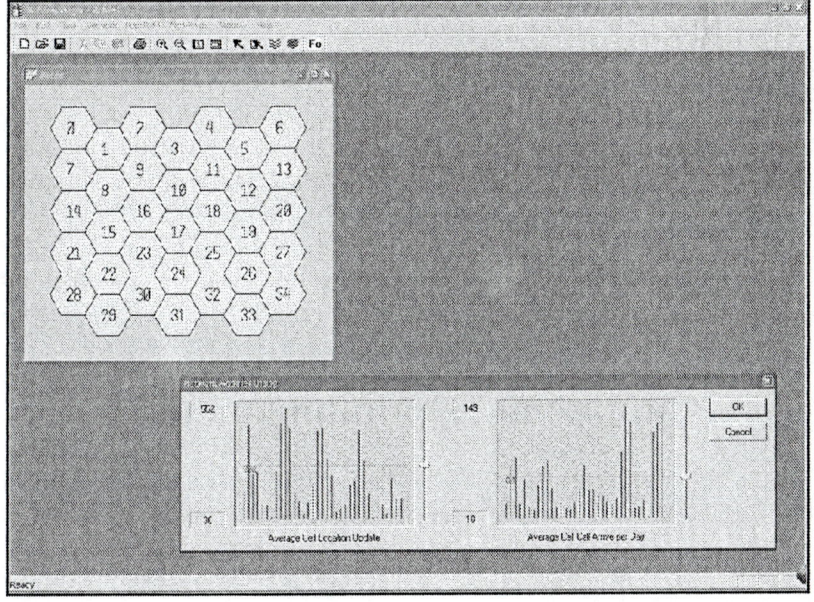

Figure 28.12   A sample network before modification.

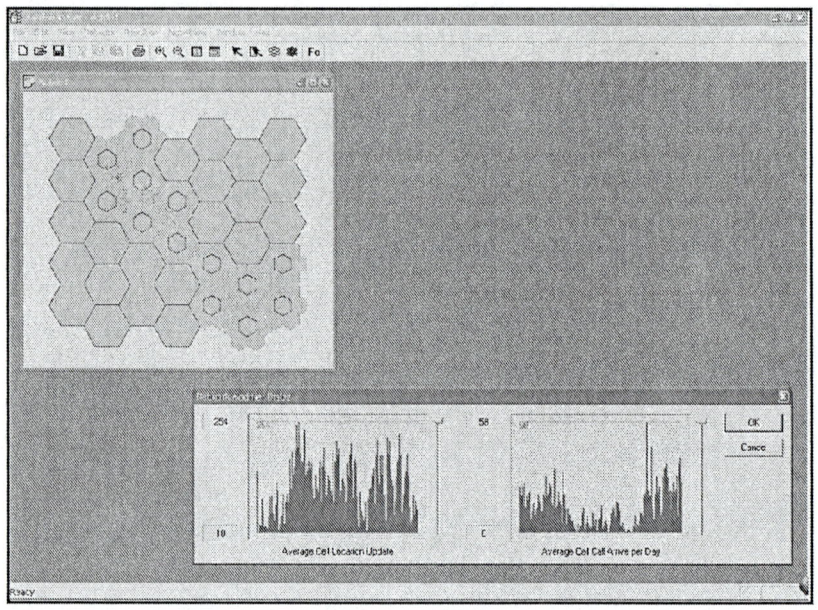

Figure 28.13 A sample network after modification (split cell).

Solving the Location Management Problem

The approaches used to solve this problem are either *static* or *dynamic*. In the static case, the assumption is made that the user profiles are all known before trying to find the best configuration for the network. In contrast, in the dynamic case, the network has an initial configuration that is not necessarily optimal, and the algorithm is supposed to improve the network configuration when it is in use. To solve this problem in the static case, a sophisticated profile is gathered from the users, and then that data is used to determine the optimal configuration of the network. The network configuration is set up once and used afterward; however, in the dynamic mode, the network configuration undergoes continuous modification even after deployment of the network. As a result, the configuration of the network might change many times during network operation. It is worth mentioning that the current GSM network uses a static network. Because of the higher computational load associated with handling dynamic networks, these networks are not used now but might be deployed in the future; therefore, most of the research today focuses on examining static networks, such as GSM networks, and improving their performance. In this regard, two different approaches are used: *analytical* and *heuristic*.

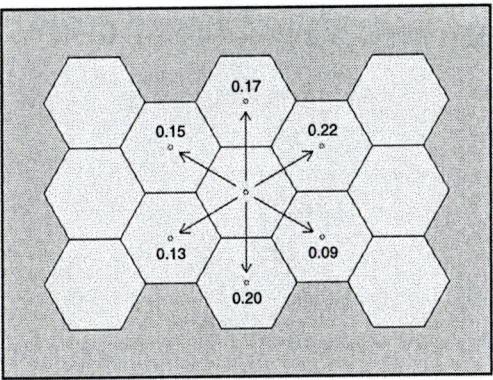

Figure 28.14   User movement flow for a sample cell.

Analytical Approaches

For an analytical approach, the problem is modeled as a stochastic one. A probabilistic model is used to model the movement patterns of users in the network. Figure 28.14 depicts the probable flow of users for an instance cell; for example, it is assumed that 20 percent of the users in this cell move to a cell below and 4 percent (1−0.15−0.17−0.22−0.09−0.20−0.13 = 0.04) of them never leave this cell. The aim of such a modeling process is to have a mathematical model that attempts to capture reality as much as possible. The use of stochastic models seems to be a suitable approach for trying to reach optimal solutions; however, the use and development of such models are never that straightforward. To clarify, for a 50-cell network, the total number of parameters that must be considered is 7 × 50 = 350; there are six probabilities (one probability per neighboring cell) and also a probability that the user stays in the cell.

Heuristic Approaches

Just like analytical approaches, a cost function is chosen for a problem (network) and heuristic algorithms are designed to find the optimal solution for the network under consideration. In heuristic approaches, unlike analytical ones, a single or several seed solutions are considered as initial points to start the algorithm, and then the heuristic approach is launched to improve the quality of these solutions gradually. The mechanisms that heuristics use are normally simple but difficult and cannot ensure that an optimal solution can be found. Toward this end, in ill-defined problems a process based on probabilistic selection or modification is incorporated in

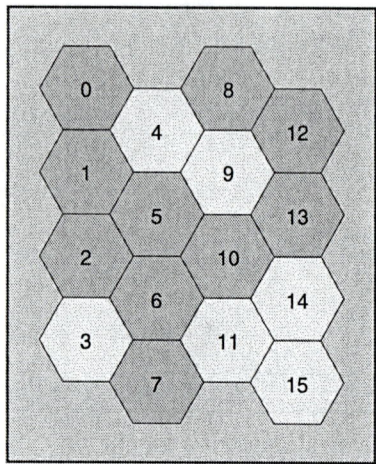

Figure 28.15 Solution for Test Network 1.

the heuristic. Also, heuristic-based methods are usually executed several times before the best answer is found, especially in the case of complex problems. In the following text, several such approaches are implemented to solve the location management problem.

Hopfield Neural Network in the Paging Cell Scheme

In this approach, the wireless network is modeled as a plate with troughs and crests. The general cost function of the problem is related to the energy level of the Hopfield neural network (HNN) optimizer, so lower energy values represent better configuration of the network and, consequently, lower costs associated with mobility management. In this case, a HNN optimizer starts from an initial condition and tries to improve the network configuration step by step. To clarify the way this algorithm actually works, several test networks are described below [37]:

■ *Test Network 1* — This 4 × 4 network is shown in Figure 28.15, and the cell attributes are given in Table 28.1. Note that two numbers are given for each cell in the table: the number of location updates (N_{LU}) and the number of arrived calls (N_P). The final answer (optimal paging cell configuration) is shown as gray cells in Figure 28.15, and Figure 28.16 shows how the total energy of the network is reduced during the algorithm optimization process.

Table 28.1 Cell Attributes of the Network in Figure 28.15

Cell	N_w	N_r
0	518	517
1	774	573
2	153	155
3	1696	307
4	1617	642
5	472	951
6	650	526
7	269	509
8	445	251
9	2149	224
10	1658	841
11	952	600
12	307	25
13	385	540
14	1346	695
15	572	225

- *Test Network 2* — This 7×9 network has the cell attributes given in Table 28.2. Figure 28.17 shows the network and the optimal answer, and Figure 28.18 provides the convergence results.
- *Test Network 3* — This 13×15 network has the cell attributes given in Table 28.3. Figure 28.19 shows the network and the optimal answer, and Figure 28.20 provides the convergence results.

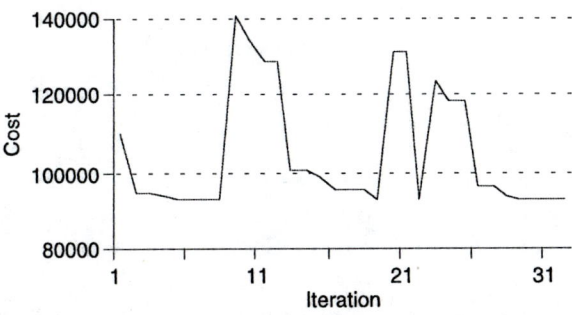

Figure 28.16 Energy level for Test Network 1.

Table 28.2   Cell Attributes of the Network in Figure 28.17

Cell	N_{LU}	N_P	Cell	N_{LU}	N_P	Cell	N_{LU}	N_P
0	120	67	21	577	66	42	395	39
1	345	68	22	433	51	43	340	55
2	173	58	23	527	89	44	134	60
3	307	67	24	377	38	45	234	83
4	111	10	25	207	38	46	445	80
5	289	42	26	130	30	47	562	64
6	184	39	27	143	43	48	378	46
7	323	78	28	332	49	49	345	48
8	121	35	29	381	58	50	366	33
9	202	52	30	589	106	51	460	34
10	462	64	31	745	69	52	379	78
11	517	75	32	602	109	53	182	57
12	426	30	33	331	64	54	153	59
13	287	51	34	248	43	55	167	60
14	370	45	35	110	29	56	350	58
15	401	44	36	172	48	57	125	69
16	325	67	37	389	45	58	244	58
17	199	64	38	440	49	59	126	55
18	148	61	39	505	48	60	381	63
19	335	51	40	642	82	61	173	65
20	541	65	41	478	51	62	121	73

Hopfield Neural Network in Location Area Scheme

The HNN is used in this case to solve our problem by adopting the location area scheme [38]. As in the previous approach, an initial configuration (initial answer) is considered for a network, then the HNN optimizer tries to improve the quality of this configuration gradually. The total cost of the network management of the system is related to the energy level of the HNN optimizer. Now, the general form of the HNN is not efficient enough for solving this problem, and several modifications were necessary to adjust the original HNN. The following test networks illustrate how this approach works [38]:

- *Test Network 4* — The 5 × 5 network has cells with the attributes given in Table 28.4. In this table, each row (cell) contains three sections: number of incoming users (UpP), number of arrived calls

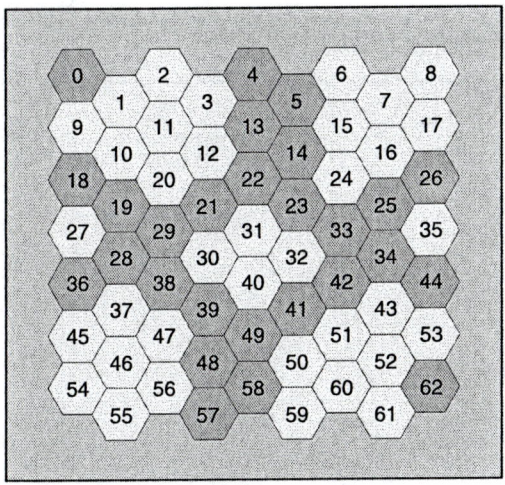

Figure 28.17   Solution for Test Network 2.

(CAr), and number of users entering the cell from a specific neighbor. The legend used to show the neighbor attributes, $(x:a,b)$, is as follows: x is a sequential ID used to count the number of neighbors for this cell, a is the cell number of the xth neighbor, and b is the number of users enters this cell from the xth neighbor. The optimal solution found by the algorithm is marked with solid lines in Figure 28.21. Figure 28.22 shows the energy value (cost) and the probability factor for each algorithm cycle.

■ *Test Network 5* — The 5×7 network (Figure 28.23) has cells with the attributes given in Table 28.5, and Figure 28.24 demonstrates the solution process.

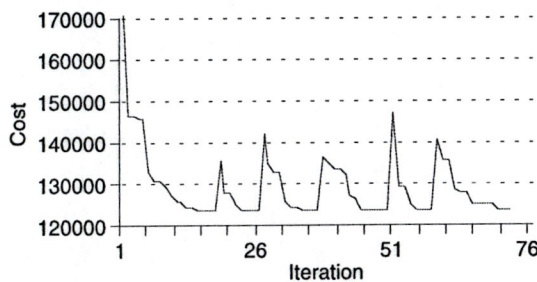

Figure 28.18   Energy level for Test Network 2.

Table 28.3 Cell Attributes of the Network in Figure 28.19

Cell	N_{LU}	N_P	Cell	N_{LU}	N_P	Cell	N_{LU}	N_P
0	41	36	65	807	44	130	729	68
1	109	55	66	824	34	131	631	74
2	64	50	67	974	52	132	656	89
3	143	61	68	938	54	133	279	48
4	82	59	69	908	67	134	159	41
5	167	57	70	767	55	135	234	47
6	70	34	71	702	64	136	554	60
7	155	51	72	710	59	137	751	62
8	75	44	73	446	45	138	759	43
9	168	35	74	178	72	139	732	87
10	79	34	75	119	35	140	627	46
11	273	66	76	306	67	141	735	93
12	166	66	77	457	81	142	690	43
13	275	39	78	535	59	143	791	67
14	131	63	79	702	36	144	727	50
15	143	60	80	911	88	145	677	79
16	331	57	81	919	69	146	568	31
17	316	95	82	991	40	147	448	52
18	490	67	83	1016	63	148	187	48
19	268	36	84	951	56	149	80	32
20	398	72	85	828	63	150	263	51
21	275	34	86	688	41	151	499	54
22	463	77	87	590	76	152	741	89
23	267	49	88	434	66	153	596	58
24	486	83	89	151	67	154	596	57
25	317	50	90	122	41	155	489	69
26	533	79	91	342	49	156	537	57
27	461	50	92	418	48	157	537	38
28	456	50	93	505	56	158	716	34
29	187	50	94	813	69	159	766	95
30	177	46	95	1051	62	160	684	44
31	436	76	96	1109	76	161	628	70
32	575	89	97	1136	92	162	356	42
33	740	87	98	1081	81	163	142	55
34	532	39	99	851	48	164	71	38
35	574	21	100	795	50	165	211	74
36	599	72	101	873	71	166	358	50
37	752	70	102	687	53	167	572	84
38	653	79	103	526	48	168	388	65

Table 28.3 Cell Attributes of the Network in Figure 28.19 (cont.)

Cell	N_{LU}	N_P	Cell	N_{LU}	N_P	Cell	N_{LU}	N_P
39	749	58	104	220	46	169	389	63
40	559	35	105	139	38	170	290	49
41	713	70	106	367	52	171	409	52
42	693	74	107	450	47	172	340	33
43	559	84	108	617	56	173	609	77
44	215	54	109	845	73	174	522	71
45	202	63	110	959	60	175	614	53
46	424	86	111	1124	67	176	449	62
47	648	68	112	1119	81	177	368	78
48	656	50	113	985	66	178	122	42
49	680	38	114	827	48	179	55	14
50	651	55	115	836	78	180	114	43
51	766	54	116	834	80	181	131	32
52	951	84	117	798	77	182	304	40
53	896	103	118	456	54	183	82	26
54	838	60	119	200	51	184	207	62
55	736	81	120	154	24	185	92	63
56	722	62	121	503	64	186	213	62
57	775	96	122	580	48	187	87	34
58	581	52	123	759	57	188	304	39
59	232	59	124	800	49	189	161	33
60	145	41	125	849	61	190	306	39
61	323	43	126	1031	69	191	142	55
62	514	46	127	896	72	192	207	58
63	601	47	128	897	42	193	52	59
64	590	40	129	756	59	194	46	57

Simulated Annealing in the Location Area Scheme

The simulated annealing (SA) technique can be combined with the location area scheme [39]. The network begins with an initial configuration, and the total cost of the network management is related to the temperature level of the SA optimizer. The SA optimizer tries to find the optimal setup of the network by heating and reheating the network temperature. The following test networks, which were already solved by the HNN optimizer earlier, show how this optimizer manages to solve the problem stated in the previous section [39].

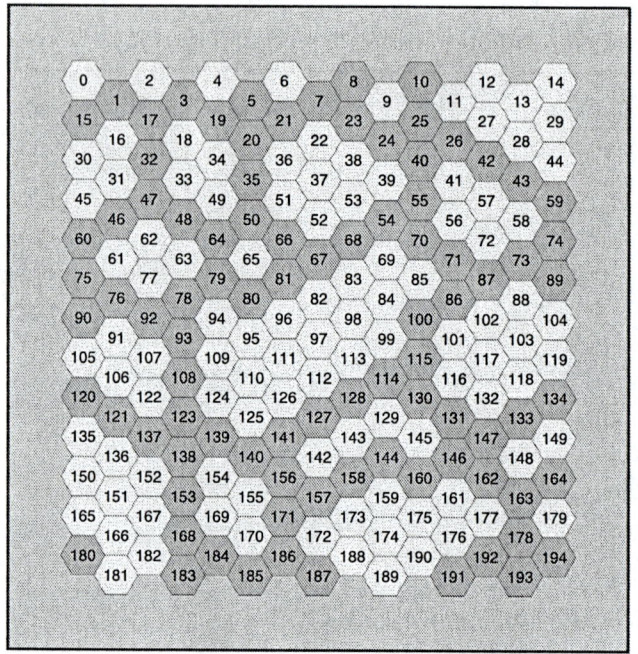

Figure 28.19 Solution for Test Network 3.

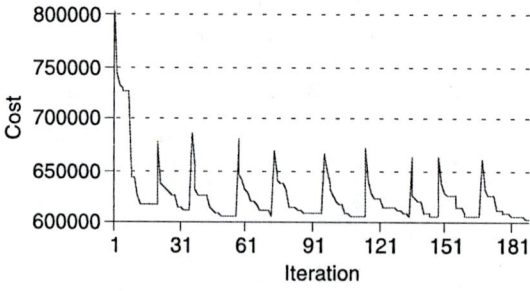

Figure 28.20 Energy level for Test Network 3.

- *Test Network 4* — This network is shown in Figure 28.25 and has the cell attributes shown in Table 28.4. Figure 28.26 shows the general profile for the energy (cost) function during the optimization process, and Figure 28.27 characterizes the reheating pattern.
- *Test Network 5* — The details of this network are given in Table 28.5 and Figure 28.28, Figure 28.29, and Figure 28.30.

Table 28.4 Cell Attributes for Test Network 4

No.	UpP	CAr	Neighbors
0	129	50	(0:1,70) (1:5,46)
1	279	73	(0:0,76) (1:2,41) (2:5,31) (3:6,69) (4:7,55)
2	100	44	(0:1,29) (1:3,35) (2:7,22)
3	265	52	(0:2,31) (1:4,61) (2:7,63) (3:8,73) (4:9,27)
4	120	73	(0:3,63) (1:9,50)
5	202	52	(0:0,42) (1:1,29) (2:6,66) (3:10,59)
6	341	44	(0:1,77) (1:5,60) (2:7,32) (3:10,22) (4:11,63) (5:12,74)
7	284	34	(0:1,66) (1:2,19) (2:3,52) (3:6,38) (4:8,33) (5:12,65)
8	347	46	(0:3,70) (1:7,42) (2:9,60) (3:12,79) (4:13,61) (5:14,25)
9	199	52	(0:3,34) (1:4,44) (2:8,72) (3:14,45)
10	167	69	(0:5,51) (1:6,27) (2:11,29) (3:15,46)
11	327	41	(0:6,54) (1:10,37) (2:12,66) (3:15,26) (4:16,85) (5:17,47)
12	454	84	(0:6,83) (1:7,61) (2:8,71) (3:11,77) (4:13,51) (5:17,101)
13	336	55	(0:8,68) (1:12,65) (2:14,40) (3:17,44) (4:18,76) (5:19,29)
14	151	69	(0:8,20) (1:9,45) (2:13,33) (3:19,34)
15	158	52	(0:10,39) (1:11,32) (2:16,29) (3:20,42)
16	365	92	(0:11,83) (1:15,42) (2:17,83) (3:20,47) (4:21,61) (5:22,43)
17	401	56	(0:11,37) (1:12,96) (2:13,49) (3:16,79) (4:18,76) (5:22,49)
18	364	80	(0:13,98) (1:17,71) (2:19,25) (3:22,46) (4:23,59) (5:24,53)
19	135	51	(0:13,34) (1:14,30) (2:18,21) (3:24,36)
20	124	63	(0:15,34) (1:16,60) (2:21,24)
21	150	82	(0:16,61) (1:20,25) (2:22,57)
22	253	59	(0:16,41) (1:17,46) (2:18,34) (3:21,50) (4:23,68)
23	159	52	(0:18,71) (1:22,49) (2:24,33)
24	138	59	(0:18,72) (1:19,40) (2:23,20)

A Genetic Algorithm in the Location Area Scheme

A genetic algorithm (GA) is combined with the location area scheme to solve the location management problem [40]. In this case, each LA configuration of a network is modeled as a chromosome. By using the basic operators of a GA optimizer, the general condition of the chromosomes improves generation by generation until the final solution is found. The following test networks have been solved by the HNN and SA approaches and are presented here to indicate the performance of this approach [40]:

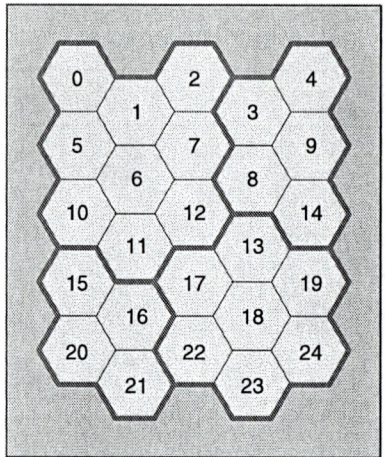

Figure 28.21 Solution for Test Network 4 found using the Hopfield neural network.

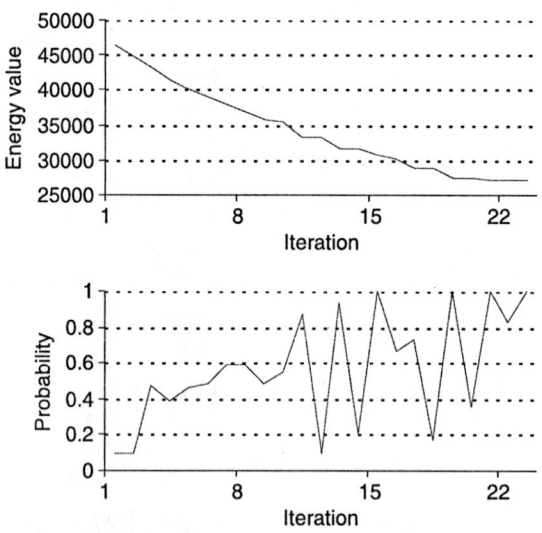

Figure 28.22 HNN solving process for Test Network 4.

■ *Test Network 4* —The network shown in Figure 28.31 has the cell attributes given in Table 28.4. Figure 28.32, in this case, shows the energy levels (cost) of the best chromosome and worst chromosome and the average energy value of all chromosomes for each population.

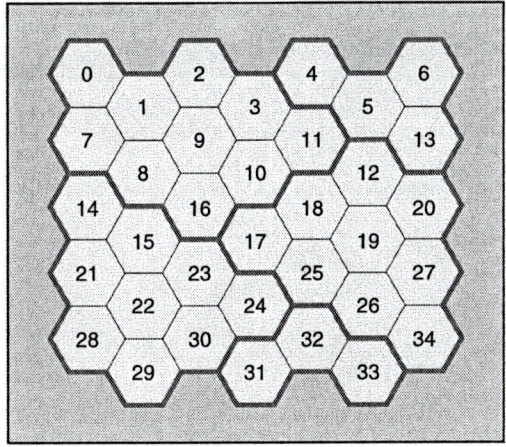

Figure 28.23   Solution for Test Network 5 found using the Hopfield neural network.

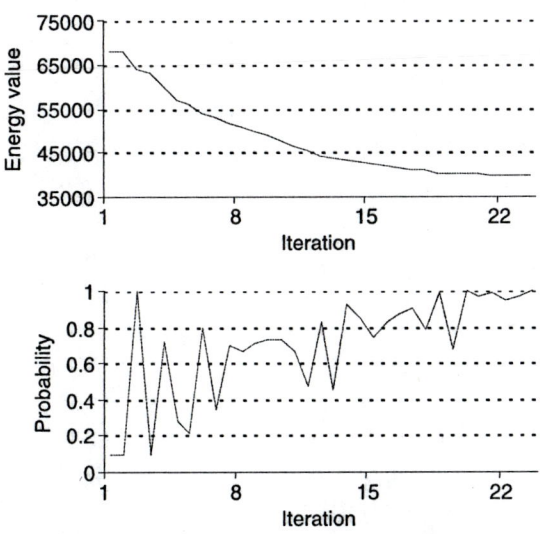

Figure 28.24   HNN solving process for Test Network 5.

■ *Test Network 5* — The details of this network are given in Figure 28.33 and Table 28.5. Figure 28.34 shows the performance of the algorithm for this test network.

Table 28.5 Cell Attributes for Test Network 5

No.	UpP	CAr	Neighbors
0	115	43	(0:1,75) (1:7,36)
1	315	46	(0:0,61) (1:2,43) (2:7,33) (3:8,61) (4:9,112)
2	161	59	(0:1,31) (1:3,69) (2:9,50)
3	229	43	(0:2,47) (1:4,29) (2:9,43) (3:10,41) (4:11,57)
4	69	34	(0:3,23) (1:5,15) (2:11,24)
5	115	46	(0:4,15) (1:6,18) (2:11,32) (3:12,20) (4:13,16)
6	35	33	(0:5,22) (1:13,5)
7	213	71	(0:0,41) (1:1,33) (2:8,86) (3:14,44)
8	368	51	(0:1,63) (1:7,73) (2:9,54) (3:14,21) (4:15,71) (5:16,73)
9	475	95	(0:1,96) (1:2,56) (2:3,52) (3:8,52) (4:10,122) (5:16,84)
10	420	54	(0:3,23) (1:9,115) (2:11,37) (3:16,63) (4:17,58) (5:18,109)
11	248	41	(0:3,54) (1:4,18) (2:5,29) (3:10,44) (4:12,45) (5:18,42)
12	218	46	(0:5,24) (1:11,49) (2:13,17) (3:18,37) (4:19,25) (5:20,54)
13	54	35	(0:5,17) (1:6,8) (2:12,9) (3:20,7)
14	142	59	(0:7,47) (1:8,18) (2:15,26) (3:21,34)
15	311	45	(0:8,72) (1:14,29) (2:16,42) (3:21,32) (4:22,81) (5:23,41)
16	403	40	(0:8,76) (1:9,80) (2:10,39) (3:15,49) (4:17,76) (5:23,69)
17	431	53	(0:10,55) (1:16,74) (2:18,71) (3:23,70) (4:24,42) (5:25,105)
18	450	49	(0:10,113) (1:11,38) (2:12,49) (3:17,66) (4:19,118) (5:25,53)
19	461	92	(0:12,31) (1:18,122) (2:20,70) (3:25,109) (4:26,52) (5:27,69)
20	182	69	(0:12,53) (1:13,9) (2:19,73) (3:27,42)
21	133	57	(0:14,38) (1:15,25) (2:22,25) (3:28,34)
22	420	99	(0:15,97) (1:21,23) (2:23,108) (3:28,67) (4:29,57) (5:30,59)
23	410	58	(0:15,34) (1:16,60) (2:17,78) (3:22,111) (4:24,54) (5:30,57)
24	408	90	(0:17,36) (1:23,66) (2:25,94) (3:30,110) (4:31,15) (5:32,58)
25	526	63	(0:17,107) (1:18,58) (2:19,106) (3:24,116) (4:26,98) (5:32,23)
26	374	70	(0:19,52) (1:25,116) (2:27,46) (3:32,55) (4:33,15) (5:34,84)
27	200	46	(0:19,77) (1:20,32) (2:26,57) (3:34,29)
28	136	62	(0:21,37) (1:22,67) (2:29,26)
29	173	87	(0:22,56) (1:28,26) (2:30,82)
30	346	58	(0:22,60) (1:23,46) (2:24,112) (3:29,78) (4:31,41)
31	99	48	(0:24,20) (1:30,27) (2:32,36)
32	214	41	(0:24,48) (1:25,36) (2:26,51) (3:31,33) (4:33,35)
33	84	63	(0:26,12) (1:32,37) (2:34,14)
34	143	83	(0:26,85) (1:27,27) (2:33,24)

Figure 28.25 Solution for Test Network 4 found by using the simulated annealing optimizer.

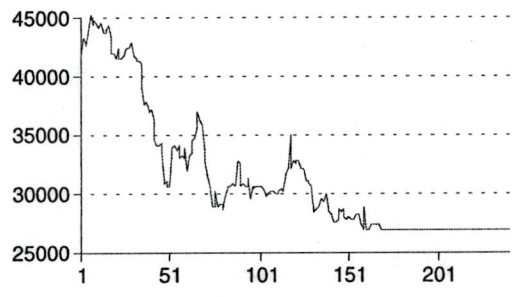

Figure 28.26 Energy levels for Test Network 4.

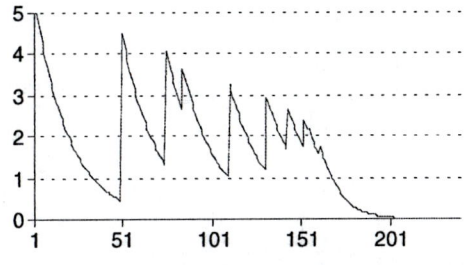

Figure 28.27 Reheating curve for Test Network 4.

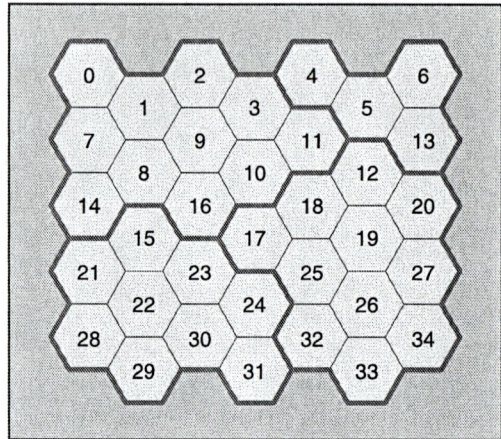

Figure 28.28 Solution for Test Network 5 found by using the simulated annealing optimizer.

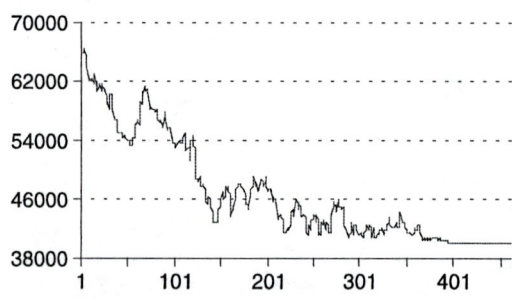

Figure 28.29 Energy levels for Test Network 5.

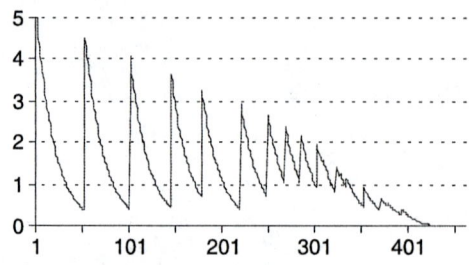

Figure 28.30 Reheating curve for Test Network 5.

Figure 28.31 Solution for Test Network 4 found by using the genetic algorithm optimizer.

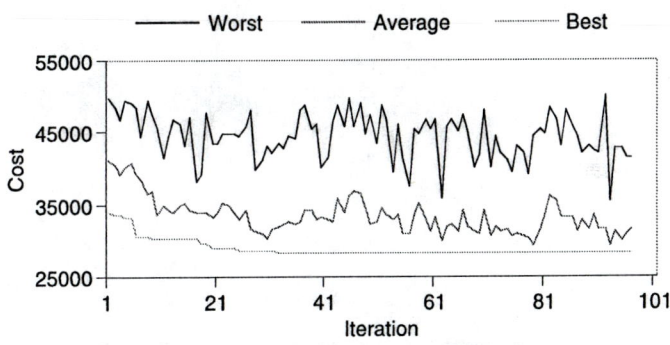

Figure 28.32 GA chromosome quality for Test Network 4.

Results Explanation

A number of observations can be made in relation to the use of heuristics in solving the location management problem.

Effect of Selecting the Initial Population

The most outstanding result is the correlation of the final answer and the initial population or cell attributes of the network in both the PC and LA cases. In almost all of the available literature, the cell attributes of the

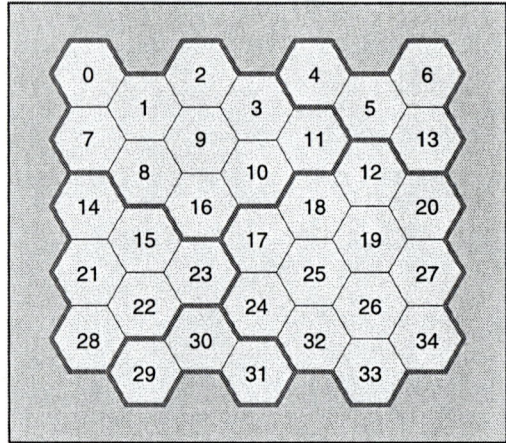

Figure 28.33   Solution for Test Network 5 found by using the genetic algorithm optimizer.

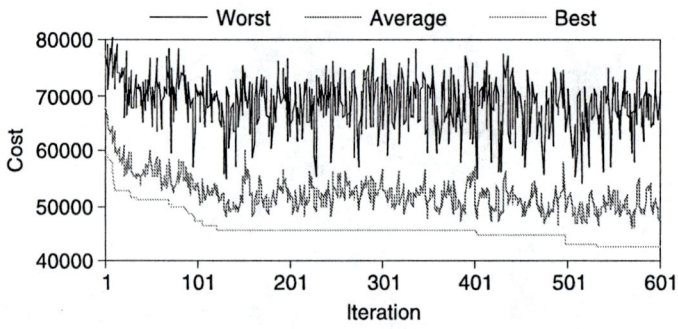

Figure 28.34   GA chromosome quality for Test Network 5.

network are generated randomly. This means that two independent attributes for each cell are considered: the number of call arrivals and the number of location updates. These numbers, however, are highly correlated in real-life scenarios. A more robust and logical approach has been used to generate the population and, consequently, the network attributes of each cell. In summary, the users are classified into three main categories: ordinary users, workers, and commuters, and each group has its own characteristics and follows a certain pattern of moving and call making. Also, the cells of network are classified into the three main categories of ordinary cells, working cells, and living cells. As a result, the final cell configurations (i.e., solutions) appear to be more realistic [36–40].

Algorithm Robustness

One of the most important criteria by which to judge an algorithm is its ability to generate the same solution regardless of how many times it is launched. This factor is very important for metaheuristics (e.g., SA, HNN, GAs). To examine the robustness of the proposed approaches, these algorithms were executed many times (on average ten or more) for each test network and for all PC- and LA-based algorithms. In some cases, the algorithm managed to find identical answers for different numbers of iterations; in some other cases, similar (but not identical) answers were found. It is worth mentioning, however, that the total energy costs of the network in all the cases did not deviate from each other by more than 2 percent [36–40].

PC Scheme: Separation of the Network

As was expected, the optimal paging cell configurations for all test networks led to the splitting of the network into smaller subnetworks (Figure 28.15, Figure 28.17, and Figure 28.19).

PC Scheme: Thin and Thick Boundary Cells

It should be noted that in some cases the network was split by a thin boundary of paging cells and in other cases by thick boundaries (Figure 28.17) [37].

LA Scheme: Shape of the Location Areas

Despite the fact that the current GSM networks employ circular LA, current work shows that more efficient LAs could have a variety of configurations, such as rectangular or triangular, as can be seen in Figure 28.21, Figure 28.23, Figure 28.25, Figure 28.28, Figure 28.31, and Figure 28.33 [38–40].

LA Scheme: Number of Cells in a Location Area

Another interesting result is the number of cells in each LA. In GSM networks, cells are grouped in 7×7 configurations to make a location area, although different LAs should have different numbers of cells. It is worth noting that the number of cells in a location area is highly correlated with the number of users passing through the cells and the number of calls they receive in each cell (see Figure 28.21, Figure 28.23, Figure 28.25, Figure 28.28, Figure 28.31, and Figure 28.33) [38–40].

LA Scheme: Number of Neighbors for Each Location Area

In almost all the cases, each LA is adjacent to two to four other LAs, although this number is six for the currently implemented GSM networks. The main reason for this phenomenon is the reduction in the number of location updates and consequent reduction in the cost of network management. The smaller the number of neighbors of a LA, the fewer updates are made by mobile users (see Figure 28.21, Figure 28.23, Figure 28.25, Figure 28.28, Figure 28.31, and Figure 28.33) [38–40].

LA Scheme: Boundary Cells

Boundary cells of a LA tend to be less busy and have lower traffic than other cells in the same location area; for example, cells 19, 20, 26, 27, and 34 in Figure 28.23 are busy cells in comparison with cells 12, 18, 17, 25, 32, and 33, which have fewer location updates and call arrivals, as seen in Table 28.4 [38–40].

Conclusions

This chapter introduced the basic concepts of the location management problem. Several location management strategies were introduced followed by sophisticated approaches to solve such a problem. These approaches were based on metaheuristics used in combination with location area and paging cell schemes. The Hopfield neural network, simulated annealing, and genetic algorithms were used as representatives of the different classes of metaheuristics. Finally, several general observations were made regarding the different experiments.

References

[1] Lin, Y.-B. and Chlamatac, I., *Wireless and Mobile Network Architecture*, John Wiley & Sons, New York, 2001.

[2] Agrawal, D.P. and Zeng, Q.-A., *Introduction to Wireless and Mobile Systems*, Thomson Brooks/Cole, Pacific Grove, CA, 2003.

[3] Austin, M.D. and Stuber, G.L., Direction based handoff algorithms for urban microcells, in *Proc. IEEE Vehicular Technology Conf. (VTC'94)*, Stockholm, Sweden, June, 1995, pp. 101–105.

[4] Zeng, Q.A. and Agrawal, D.P., Performance analysis of a handoff scheme in integrated voice/data wireless networks, in *Proc. IEEE Vehicular Technology Conf. (VTC'00)*, Boston, MA, September, 2000, pp. 845–851.

[5] Al-Tawil, K., Akrami, A., and Youssef, H., A new authentication protocol for GSM networks, in *Proc. of the IEEE 23rd Annual Conf. on Local Computer Networks (LCN'98)*, Lowell, MA, October, 11–14, 1998, pp. 21–30.

[6] Spirito, M.A. and Mattioli, A.G., On the hyperbolic positioning of GSM mobile stations, in *Proc. of Int. Symp. on Signals, Systems, and Electronics (ISSE'98)*, September, 1998, pp. 173–177.

[7] Yuen, W.H.A. and Wong, W.S.W., A dynamic location area assignment algorithm for mobile cellular systems, in *Proc. of IEEE Int. Conf. on Communications (ICC'98)*, Atlanta, GA, June, 1998, pp. 1385–1389.

[8] Ho, J.S.M. and Xu, J., History-based location tracking for personal communications networks, in *Proc. IEEE Vehicular Technology Conf. (VTC'98)*, Ottawa, Canada, May, 1998, pp. 244–248.

[9] Lei, Z. and Rose, C., Wireless subscriber mobility management using adaptive individual location areas for PCS systems, *Proc. of IEEE Int. Conf. on Communications (ICC'98)*, Atlanta, GA, June, 1998, pp. 1390–1394.

[10] Gu, D. and Rappaport, S.S., A dynamic location tracking strategy for mobile communication systems, in *Proc. IEEE Vehicular Technology Conf. (VTC'98)*, Ottawa, Canada, May, 1998, pp. 259–263.

[11] Naor, Z. and Levy, H., Minimizing the wireless cost of tracking mobile users: an adaptive threshold scheme, in *Proc. IEEE INFOCOM'98*, San Francisco, CA, March, 1998, pp. 720–727.

[12] Jie, L., Kameda, H., and Keqin, L., Optimal dynamic location update for PCS networks, in *Proc. of IEEE Int. Conf. on Distributed Computing Systems (ICDCS'98)*, Amsterdam, The Netherlands, May, 1998.

[13] Lee, B.-K. and Chong-Sun Hwang, C.S., A predictive paging scheme based on the movement direction of a mobile host, in *Proc. IEEE Vehicular Technology Conf. (VTC'99)*, Houston, TX, May, 1999, pp. 2158–2162.

[14] Liu, H.-I. and Liu, C.-P., A geography based location management scheme for wireless personal communication systems, in *Proc. IEEE Vehicular Technology Conf. (VTC'00)*, Boston, MA, September, 2000, pp. 1358–1361.

[15] Bera, A. and Das, N., Performance analysis of dynamic location update strategies for mobile users, in *Proc. of IEEE Int. Conf. on Distributed Computing Systems (ICDCS'00)*, Taipei, Taiwan, April, 2000, pp. 428–435.

[16] Wu, H.-K., Jin, M.-H., Horng, J.-T., and Ke, C.-Y., Personal paging area design based on mobile's moving behaviors, *Proc. IEEE INFOCOM'01*, Anchorage, AK, April, 2001, pp. 21–30.

[17] Lin, Y.-B., Lee, P.-C., and Chlamtac, I., Dynamic periodic location area update in mobile networks, *IEEE Trans. Veh. Technol.*, 51(6), 1494–1501, 2002.

[18] Akyildiz, I.F. and Wang, W., A dynamic location management scheme for next-generation multitier PCS systems, *IEEE Trans. Wireless Commun.*, 1(1), 178–189, 2002.

[19] Rose, C., Minimizing the average cost of paging and registration: a timer based method, *Wireless Networks*, 2, 109–116, 1996.

[20] Akyildiz, I.F. and Ho, J.S.M., Movement based location update and selective paging for PCS networks, *IEEE/ACM Trans. Networking*, 4, 629–638, 1996.

[21] Casares, G.V. and Mataix, O.J., On movement based mobility tracking: an enhanced version, *IEEE Commun. Lett.*, 2, 45–47, 1998.

[22] Wong, V.W.S. and Leung, V.C.M., An adaptive distance based location update algorithm for next generation PCS networks, *IEEE J. Selected Areas Commun.*, 19, 1942–1952, 2001.

[23] Wong, V.W.-S. and Leung, V.C.M., An adaptive distance-based location update algorithm for PCS networks, in *Proc. of IEEE Int. Conf. on Communications (ICC'00)*, New Orleans, LA, June, 2000, pp. 2001–2005.

[24] Suh, B., Choi, J.-S., and Kim, J.-K., Mobile location management strategy with implicit location registration, in *Proc. IEEE Vehicular Technology Conf. (VTC'99)*, Houston, TX, May, 1999, pp. 2129–2133.

[25] Gu, D. and Rappaport, S.S., Mobile user registration in cellular systems with overlapping location areas, *Proc. IEEE Vehicular Technology Conf. (VTC'99)*, Houston, TX, May, 1999, pp. 802–806.

[26] Kim, T.K. and Leung, C., Generalized paging schemes for cellular communication systems, in *Proc. of 7th IEEE Pacific Rim Conf. on Communications, Computers, and Signal Processing (PACRIM'99)*, Bali, Indonesia, October, 1999, pp. 217–220.

[27] Tonguz, O.K., Mishra, S., and Josyula, R., Intelligent paging in wireless networks: random mobility models and grouping algorithms for locating subscribers, in *Proc. IEEE Vehicular Technology Conf. (VTC'99)*, Houston, TX, May, 1999, pp. 1177–1181.

[28] Subrata, R. and Zomaya, A.Y., Location management in mobile computing, in *Proc. of the 3rd ACS/IEEE Int. Conf. on Computer Systems and Applications*, Cairo, Egypt, January, 2001, pp. 287–289.

[29] Demirkol, I., Ersoy, C., Caglayan, M.U., and Delic, H., Location area planning in cellular networks using simulated annealing, in *Proc. IEEE INFOCOM'01*, Anchorage, AK, April, 2001, pp. 13–20.

[30] Shirota, M., Yoshida, Y., and Kubota, F., Statistical paging area selection scheme (SPAS) for cellular mobile communication systems, in *Proc. IEEE Vehicular Technology Conf. (VTC'94)*, Stockholm, Sweden, June, 1994, pp. 367–370.

[31] Curle, P. and Colombo, G., Sub-optimal solutions for location and paging areas dimensioning in cellular networks, in *Proc. of the Fourth IEEE Int. Conf. on Universal Personal Communications (ICUPC'95)*, Tokyo, Japan, November, 1995, pp. 672–676.

[32] Lei, Z., Saraydar, C.U., and Mandayam, N.B., Mobility parameter estimation for the optimization of personal paging areas in PCS/cellular mobile networks, in *Proc. of Signal Processing Advances in Wireless Communications (SPAWC'99)*, Annapolis, MD, May, 1999, pp. 308–312.

[33] Wu, H.-K., Jin, M.-H., Horng, J.-T., and Ke, C.-Y., Personal paging area design based on mobile's moving behaviors, in *Proc. IEEE INFOCOM'01*, Anchorage, AK, April, 2001, pp. 21–30.

[34] Subrata, R. and Zomaya, A.Y., A comparison of three artificial life technique for reporting cell planning in mobile computing, *IEEE Trans. Parallel Distributed Syst.*, 14(2), 142–153, 2003.

[35] Subrata, R. and Zomaya, A.Y., Evolving cellular automata for location management in mobile computing networks, *IEEE Trans. Parallel Distributed Syst.*, 14(1), 13–26, 2003.

[36] Taheri, J. and Zomaya, A.Y., Realistic simulations for studying mobility management problems, *Int. J. Wireless Mobile Comput.*, 1(8), 2005.

[37] Zomaya, A.Y. and Taheri, J., A modified Hopfield network for mobility management, 2006 (in press).

[38] Taheri, J. and Zomaya, A.Y., The use of a Hopfield neural network in solving the mobility management problem, in *Proc. of the IEEE/ACS Int. Conf. on Pervasive Services (ICPS'04)*, Beirut, Lebanon, July, 2004, pp. 141–150.

[39] Taheri, J. and Zomaya, A.Y., A simulated annealing approach for mobile location management, in *Proc. of the 8th Workshop on Nature Inspired Distributed Computing (NIDISC'05), Int. Parallel and Distributed Processing Symp.*, Denver, CO, April, 2005.

[40] Taheri, J. and Zomaya, A.Y., A genetic algorithm for finding optimal location area configurations for mobility management, in *Proc. of IEEE Conf. on Local Computer Networks (LCN 2005)*, Sydney, Australia, November 15–17, 2005, pp. 568–577.

Chapter 29

Location Privacy Protection in Mobile Wireless Networks

Jieyan Fan, Dapeng Wu, Qi He, and Pradeep Khosla

CONTENTS

Introduction

The convergence of a wireless communication infrastructure, mobile computing devices, and embedded systems is causing a profound shift in the way we live and work, offering the promise of bringing us closer to the Holy Grail of information technology: ubiquitous computing (computing at any place and at any time) [3]. To fulfill the promise of ubiquitous computing, information about mobile users' locations is a critical and valuable resource that has to be utilized. Many efforts have been directed toward making such information available as a key service in the ubiquitous computing environment. The location information service or functionality can act as a double-edged sword, however, as it can make our life more convenient while providing criminals with powerful weapons to compromise the privacy of mobile users. Computer scientists have realized that, unless the use of this information is strictly controlled, it can be applied to a variety of unsavory situations [3,4].

To address the location privacy issue, an architecture for location privacy control [4] was designed and tested on the wireless Andrew network, an IEEE 802.11 wireless local area network (WLAN) that covers the entire campus of Carnegie Mellon University. The architecture implemented in He et al. [4] is illustrated in Figure 29.1. This architecture has a centralized location server where a mobile user can register and submit location information along with a *permission rule set* regarding the user's privacy preferences. Others can send queries to the server for location information about mobile users whose location information is stored in the server. The server processes the query according to the queried user's preferences specified within the set of rules, and then the server may return the queried information, deny the query, or return a fake location [4].

This simple architecture is essentially identical to the one described in Weiser [3], who suggested that a distributed architecture could be of benefit to the control of mobile location privacy, because a centralized architecture has the following drawbacks:

- The location privacy of mobile users is not completely under their control because the system administration maintains a central server where the location information of mobile users is stored.
- The central server is a single-failure point; that is, the location privacy of mobile users would be compromised if an attacker successfully hacked into it.
- The centralized architecture is not scalable.

To achieve complete personal control over location privacy by replacing a centralized architecture with a distributed one is not trivial. For example, the system administration, for the sake of system maintenance and management,

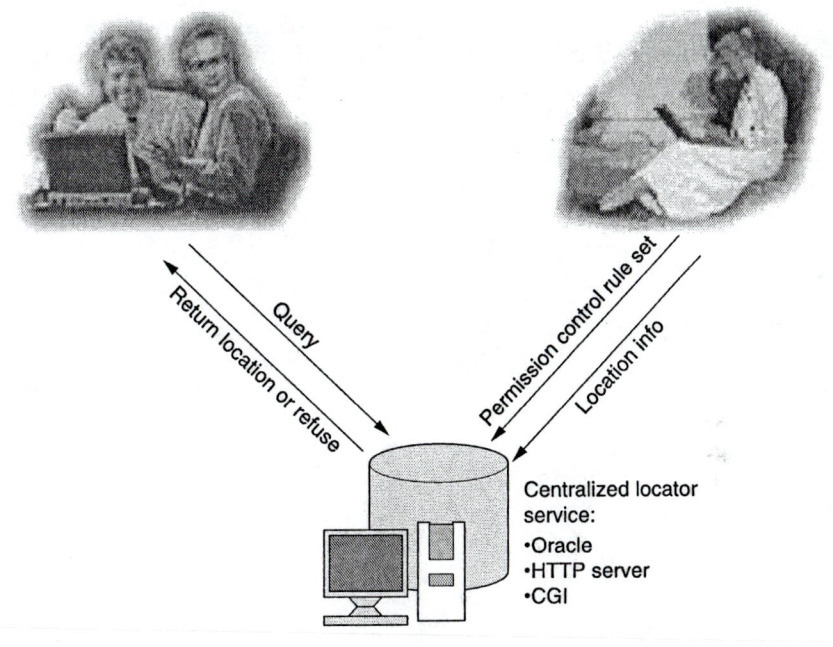

Figure 29.1 WaveGuard architecture.

has the privilege of checking any access point and obtaining a list of Internet Protocol (IP) addresses and corresponding Media Access Control (MAC) addresses of the mobile devices that are connecting to the checked access point. The administration also has the data that can indicate a bijection relationship between MAC addresses (or IP addresses) of authorized mobile devices and registered legitimate mobile users. The location information about a mobile user can be easily figured out by the administration. So, we face a dilemma: On the one hand, the administration would like to require all legitimate users to provide information for authentication to grant them permission to use their wireless service; on the other hand, mobile users would prefer not to expose any of their information (e.g., IDs and MAC addresses) that would enable anyone, including the administration, to get clues regarding their whereabouts.

To resolve the above dilemma, He et al. [1] proposed an authorized-anonymous-ID-based scheme. In this scheme, an authorized-anonymous-ID generated by a cryptographic technique called *blind signature* [5] is used to replace the real ID (e.g., a MAC address) of an authorized mobile device (e.g., a WLAN card). An anonymous ID can tell nothing more than whether the provider of the ID is an authorized user. This authorized-anonymous-ID is then used as the key for packet authentication, and the message authentication code (generated by the key) [6] is used for access control. In this way, the administration can grant authorized mobile users access to the wireless communication infrastructure, while mobile users need not divulge their real IDs during authorization, which could otherwise lead to compromising their location privacy. Built on an agent-based architecture, the authorized-anonymous-ID-based scheme gives mobile users complete control over their location privacy. Here, the agent is a computer program or code that is autonomous.

System Architecture

To address the location privacy issue in a ubiquitous computing environment, we need to understand the key components for ubiquitous computing. Based on this understanding, we can then design a system to protect location privacy. Below, we first discuss key components in ubiquitous computing from a security perspective and then present our agent-based system architecture for location privacy protection.

Ubiquitous Computing

For quite some time, a ubiquitous computing environment has been depicted as dynamically changing self-organized networks formed by resource-constrained mobile devices that occasionally join and leave the networks. It is difficult to believe, however, that this *ad hoc* infrastructure-less system could be the basis of what we call ubiquitous computing. We believe that identifying reasonable formations of ubiquitous computing and exploring security implication of ubiquitous computing based on the discovered formations are fundamentally significant research approaches. In Brewer et al. [7], we learn that a ubiquitous computing environment should be formed by a powerful infrastructure that is highly available, cost effective, and sufficiently scalable to support millions of users and their low-power mobile devices, which are small and lightweight; it does not matter very much what a device can do, but what does matter is the possibility that the device can harness terabytes of data and the power of supercomputers even while mobile — as long as it has access to a ubiquitous network [7]. This understanding of the formation of ubiquitous computing guides the design of the system architecture for location privacy protection.

Security, already a thorny problem on the Internet, is greatly complicated by ubiquitous computing, not only because of its vulnerability due to the sharing nature of the wireless medium and computational limitations resulting from requirements for low weight, compact size, and good ergonomics of mobile device and embedded systems, but more importantly also because of the following challenges:

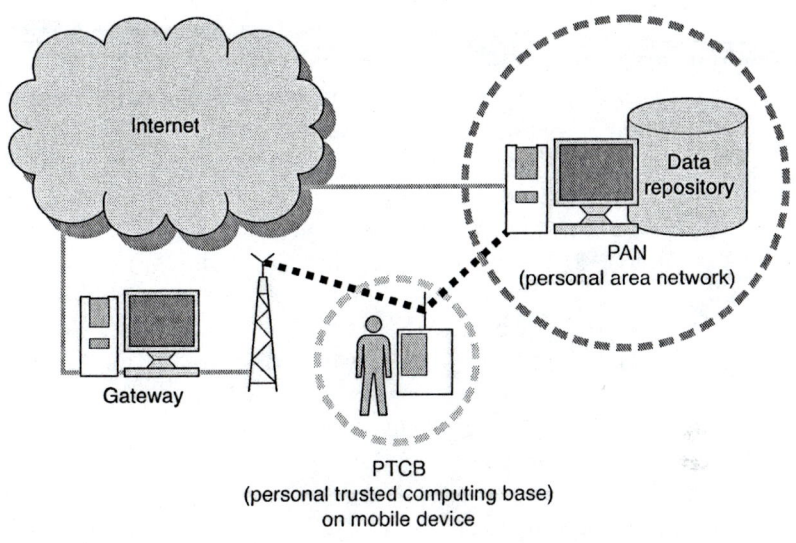

Figure 29.2 Ubiquitous computing.

- Geographically distributed systems are connected to form heterogeneous networks of unlimited scale which lack a central authority, a homogeneous security policy, and a ubiquitous security infrastructure for security enforcement or guarantee.
- Ubiquitous computing creates an environment full of computing and communication devices that must gracefully integrate with human users [3], so the electronic security mechanisms must be user centered and cannot rely on or be controlled by network infrastructure operators.
- The design of security mechanisms for ubiquitous computing must follow the end-to-end principle [8].

The agent technology [9] can effectively address the aforementioned challenges for three reasons. First, agents are autonomous and distributed, so no central authority is required for security enforcement. Second, agents can act on the users' behalf; hence, agent-based security mechanisms can be designed to be user centered. Third, agents are application oriented, which naturally satisfies the end-to-end principle (i.e., agents communicate on the application layer).

Based on the agent concept, we perceive, from a security perspective, that a ubiquitous computing environment should consist of the following three key components (see Figure 29.2):

- A *personal trust computing base* (PTCB) is a personal-held computing device, such as personal digital assistant (PDA) and laptop; a PTCB is under the full control of the owner, and only the owner with proper authentication information such as personal identification number (PIN) or biometrics information can activate a PTCB to work on behalf of the owner.
- A *personal area network* (PAN) is an architecture that consists of a main home PC, which has a connection to an Internet gateway, and a wide range of appliances (i.e., PTCBs) connected to the main home PC in many ways. Each PTCB is associated with some kind of autonomous software, called *agent* (or *proxy*). The agent runs on the PTCB if the PTCB is computationally capable of running the agent; otherwise, the agent runs on the main home PC. To secure the communication within a PAN, we need to consider two cases: (1) an agent runs on the PTCB, and (2) an agent runs on the main home PC. For the first case, a protocol can be designed to initialize a symmetric key shared by the PTCB and the main home PC; the messages between the PTCB and the main home PC can then be encrypted and authenticated with the shared secret key. For the second case, a method known as *resurrecting duckling* [10] can be employed whereby a PTCB and its agent share a symmetric key to secure their communication; the PTCBs can then securely communicate with each other through their agents, which can negotiate keys for encryption or authentication. Hence, the communication within a PAN can be secured by symmetric cryptosystems. An engineering practice that addresses the second case is described in Burnside et al. [11], who designed and implemented a device-to-proxy protocol and a proxy-to-proxy protocol.
- The *Internet* provides a communication channel between a PTCB and a PAN; however, the channel cannot be trusted.

Agent-Based System Architecture

Because the properties of agents meet the security requirements imposed by ubiquitous computing, He et al. [1] designed an agent-based system architecture for location privacy protection. The following agents act on behalf of the players (devices or users) under this architecture:

- *Administrator* (*A*) is an agent that acts on behalf of an administration to authenticate legitimate users and grant them access to the wireless infrastructure.

- *Rover* (*R*) is an agent running on a PTCB and acts on behalf of the owner of a mobile device. It is responsible for determining the location of the mobile device, automatically updating the location information stored in the home PC (managed by another agent called manager, described below), and interacting with users for privacy permission settings [4].

- *Manager* (*M*) is an agent running on a home PC that can be delegated to act on behalf of a mobile user. It manages the location information submitted by the rover and executes the user's control policy [4] for location privacy when it processes location information queries from other users.

- *Connector* (*C*) is an agent running at an access point and is delegated by the administrator agent to authenticate mobile devices and control wireless connections between mobile devices and the access point.

- *Lookup* (*L*) is an optional agent that provides Internet users with public lookup services. Lookup agents, acting as well-known public service providers, will listen for location information queries from users and forward the queries to the queried user's manager agent running at home.

With the above agents, He et al. [1] proposed a multi-agent system architecture, as illustrated in Figure 29.3. Under the architecture, agents communicate with each other through three types of protocols: *registration*, *controlled connection*, and *location query/response* (numbered 1, 2, and 3, respectively, in Figure 29.3). In the next section, we present the registration protocol and the controlled connection protocol. Readers can find more information about the location query/response protocol in He et al. [4]. Before we present the authorized-anonymous-ID-based scheme for mobile location privacy protection, in the next section, we first describe the blind signature technique that is essential in the design of the authorized-anonymous-ID-based scheme.

Blind Signature

The blind signature scheme, first introduced by Chaum [5], allows a person to receive a message signed by another party (a signer) without revealing any information about who the signer is. This scheme is critical for the authorized-anonymous-ID-based scheme. To illustrate the concept of blind signature, we present the following example, which was originally given in Chaum [5].

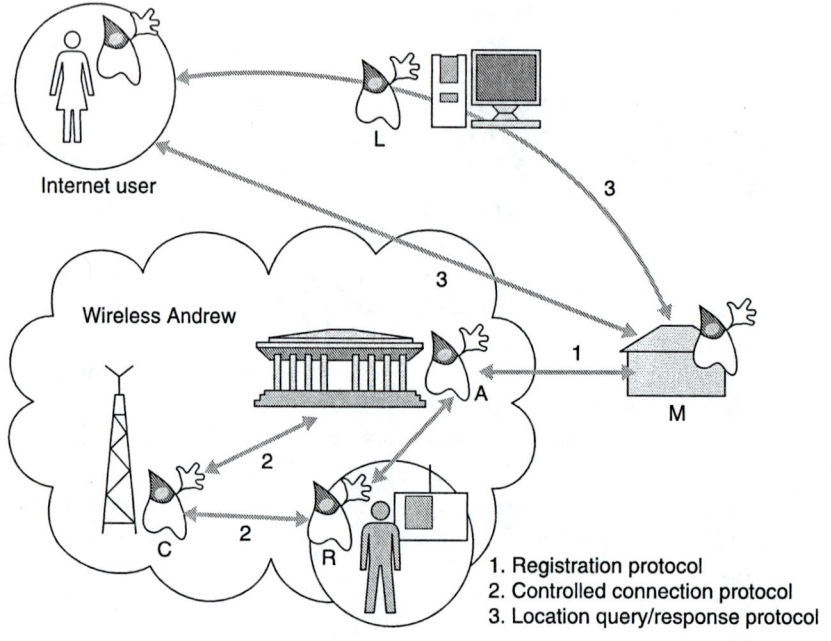

Internet user

L

Wireless Andrew

3

3

1

A

M

2

C

2

R

1. Registration protocol
2. Controlled connection protocol
3. Location query/response protocol

Figure 29.3   Agent-based system architecture.

Voting Example

Consider a committee that needs to elect a new chairman. An administrator is in charge of the election, and all members of the committee will vote; however, for some reason, the members cannot gather together to drop their ballots into a single hat. Each member wants to keep his vote secret. Each member also wants to verify that his vote is counted. One solution is to use a special carbon-paper-lined envelope. The envelope is designed in such a way that writing a signature on the outside of the envelope leaves a carbon copy of the signature on a piece of paper inside the envelope. Thus, voting would proceed as follows:

■ Each member casts his vote on a ballot and places the ballot into a carbon-paper-lined envelope. He then inserts the carbon-paper-lined envelope into a regular outer envelope bearing the member's return address and mails the nested envelope to the administrator.
■ The administrator verifies the member by his name and address. If the member is qualified to vote, the administrator removes the carbon-paper-lined envelope and signs on it. Due to the carbon

paper, the signature is copied to the ballot inside the carbon-paper-lined envelope. The administrator then inserts the carbon-paper-lined envelope into another regular outer envelope and mails it to the member.

■ When the member receives the envelope, he removes the ballot from the envelope. The member can then mail the ballot to the administrator as his vote. This time, however, the member does not write his return address on the envelope.

■ In the end, the administrator displays all the ballots.

Because the administrator does not know the contents of the ballot when signing on the carbon-paper-lined envelope, no correspondence takes place between the administrator and the member who votes. However, because only the authorized member will receive the signed carbon-paper-lined envelope and the signed ballot, the validity of the ballot is guaranteed. Here, we assume that the signature is well designed for this vote so the signature cannot be imitated. Because all the ballots are displayed, a member can verify that his vote was counted by identifying some particular features of the ballot, such as the fiber pattern.

Protocol

As in the previous example, suppose a provider wants a message to be signed by a signer but does not want the signer to learn anything about the message. The provider can use blind signature to solve this problem. Blind signature requires three set of functions:

■ Function S, which is known only to the signer, and function S^{-1}, which is known to the public — It is impractical to infer S from S^{-1}. The pair of functions S and S^{-1} can be implemented by public key cryptographic algorithms. Specifically, S corresponds to encryption with a private key and S^{-1} corresponds to decryption with a public key; hence, $S^{-1}(S(x)) = x$, where x is the message.

■ Function C and its inverse C^{-1} — Both are known only to the provider. C and C^{-1} satisfy the property that $C^{-1}(S(C(x))) = S(x)$. In addition, it is impractical to infer x from $C(x)$ and $S(x)$.

■ Redundancy checking function r, which checks whether redundancy is enough to make search for valid signatures impractical — Function r takes a signed matter as input, and the value of r is Boolean; that is, $r(x) = true$ indicates sufficient redundancy of x to make the search for valid signatures impractical, whereas $r(x) = false$ indicates insufficient redundancy of x.

Using the above functions, the protocol for the provider and the signer is analogous to the previous voting example:

- The provider chooses a random number x, which satisfies $r(x)$ = *true*. It then forms $C(x)$ and sends $C(x)$ to the signer. This step is analogous to a member inserting his ballot into the carbon-paper-lined envelope, placing the carbon-paper-lined envelope in an outer envelope with his own return address, and mailing it to the administrator.
- The signer signs $C(x)$ to generate $S(C(x))$ and returns the signed matter, $S(C(x))$, to the provider. This step is analogous to the administrator validating the return address, signing on the outside of the carbon-paper-lined envelope, wrapping the carbon-paper-lined envelope in a new outer envelope, and sending it back to the return address.
- The provider strips the signed matter $(S(C(x)))$ by applying C^{-1}, yielding $C^{-1}(S(C(x))) = S(x)$. This step is analogous to the member opening the carbon-paper-lined envelope and removing the signed ballot.
- Anyone can check that the stripped matter $S(x)$ was formed by the signer by applying the publicly known function S^{-1} and checking that $r(S^{-1}(S(x))) = $ *true*. This step is analogous to the member sending the signed ballot to the administrator and everyone being able to check the ballot without knowing who voted.

From the above protocol, we see that blind signature has following features:

- Anyone can validate the stripped signature $S(x)$ by performing $r(S^{-1}(S(x)))$.
- The signer knows nothing about the correspondence between the stripped signed matter $S(x)$ and the unstripped signed matter $S(C(x))$; thus, the provider's message is blind to the signer.
- The provider can create at most one stripped signature for each thing signed by the signer; that is, even with $S(C(x_1))$, $S(C(x_2))$, ..., $S(C(x_n))$ and a choice of C, C^{-1}, and x_1, x_2, ..., x_n, it is impractical to produce $S(y)$ such that $r(y) = $ *true* and $y \neq x_i$ for any i.

Next, we present the authorized-anonymous-ID based scheme, which uses the blind signature technique.

Authorized-Anonymous-ID-Based Scheme

This section presents the authorized-anonymous-ID-based scheme — specifically, the registration protocol and the controlled connection protocol. To authenticate users when they request access to the wireless infrastructure, we must first assign a valid ID to a legitimate user (i.e., authorizing a user), and then only authorized users are allowed to access the network infrastructure (i.e., access control). Hence, our scheme has two phases: (1) the registration phase, specified by a registration protocol, and (2) the controlled connection phase, specified by a controlled connection protocol.

The registration phase authorizes users and the controlled connection phase controls access. In the first phase, the manager (or rover) of a mobile user applies for an authorized-anonymous-ID from the administrator of the wireless infrastructure. After the first phase, the obtained authorized-anonymous-ID is carried by the rover of the mobile user and will be presented when the mobile device is requesting connection through an access point. In the second phase, the rover presents the ID to request connection and the ID is also used by the access point to authenticate the packets from the mobile device thereafter for the purpose of access control.

Table 29.1 lists the notations used in the description of the protocols. Note that because both R_u and M_u have the private key of U, both of them can represent U. If R_u and M_u are interchangeable in the protocol, we can use U to represent them. In other words, U in the following protocols can be replaced by either R_u or M_u.

Registration Protocol

In the registration protocol, the initial authentication of users is required. We assume that an infrastructure supports the initial authentication of users. This infrastructure could be either a public key infrastructure (PKI) or a Kerberos-based system [12]. When a PKI is in place, to obtain authentication a user must sign a request with the user's digital signature and send the request to the administrator.

The registration protocol is based on authorized-anonymous-ID. With an authorized-anonymous-ID as a digital token, a legitimate mobile device can be granted permission to access the wireless infrastructure after a successful authentication; yet, the association between the token and the real ID of a legitimate user is eliminated. The registration protocol is as follows (see Figure 29.4):

Table 29.1   Notations

U	A mobile user, identified by her public key. The corresponding private key is held by her rover running in her PTCB and Manager in home-PC of PAN.
R_u	Rover of mobile user U
M_u	Manager of mobile user U
E_x	Public key of X
D_x	Private key of X
$K_{xy}(m)$	Encrypt m by using symmetric cryptosystem with a key shared by x and y
$K_{xy}^{-1}(c)$	Decrypt c by using symmetric cryptosystem with a key shared by x and y
$H(x)$	One-way hash function with input x
$E_x(m)$	Encrypt m by using asymmetric cryptosystem with the public key of x
$D_x(c)$	Decrypt a cipher c with the public key of x
r_0, r_1	Random numbers
ack	Acknowledgment for the last received message

- The U generates two random numbers, r_0 and r_1.
- The U encrypts r_0 by applying the A's public key (e.g., $E_A(r_0)$).
- The U generates $c_0 = E_A(r_0) \times H(r_1)$ and sends c_0 to the A.
- When the A receives c_0 from the U, it authenticates the U.
- If the authentication passes, the A generates $c_1 = DA(c_0) = r_0 \times D_A(H(r_1))$ and sends c_1 back to the U.
- The U receives c_1.
- The U removes the blind factor by dividing c_1 by r_0 ($id = c_1/r_0$).
- The U verifies id by checking $E_A(id) = H(r_1)$. If $id = D_A(H(r_1))$, then the verification will pass. Notice that we have assumed that it is impractical to find an id such that $id \neq D_A(H(r_1))$ and $E_A(id) = H(r_1)$.
- The U keeps $id = D_A(H(r_1))$ as its authorized-anonymous-ID.

As we mentioned previously, the role of a U in the registration protocol could be played by either R_u or M_u, depending on the environment where the U is currently staying. Usually, when at home with a mobile device, the user can have the M_u initiate the protocol to get the authorized-anonymous-ID ($r_1, D_A(H(r_1))$) and then convey the authorized-anonymous-ID to the R_u via a secure channel between the R_u and the M_u, which is protected by a symmetric cryptosystem. When the mobile device already has a connection to the administrator, the rover can also initiate the

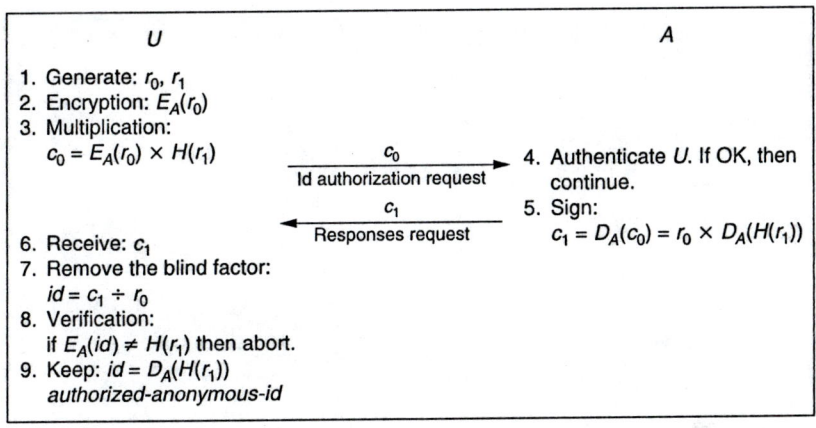

Figure 29.4 Registration protocol.

registration procedure to get an ID to make the mobile user disappear with the new ID (the reconfusion protocol is discussed later). No matter who initiates the protocol, the ID must be passed to the rover for the mobile device to be authenticated at access points.

Controlled Connection Protocol

Once an R obtains an ID, the mobile device can use the controlled connection protocol to gain access to the wireless infrastructure via an access point. First, the R sends an access request by presenting its authorized-anonymous-ID (encrypted with the administrator's public key) to the C at the access point. The C then forwards the message to its A for verification. The A decrypts the message, verifies the authenticity of the embedded authorized-anonymous-ID, signs the ID (if it is a valid one), encrypts the ID and the signature with the key shared by the A, and sends the encrypted message back to the C. When the C receives the encrypted message from the A, it decrypts it and checks the signature signed by the A and sends an "ack" message to the R if the signature is valid. Thereafter, the R and the C share the ID as a secret for packet authentication, and only successfully authenticated packets can get through the access point to the Internet. This protocol is outlined in Figure 29.5.

Improvements

The basic protocols presented earlier can be improved by the following methods.

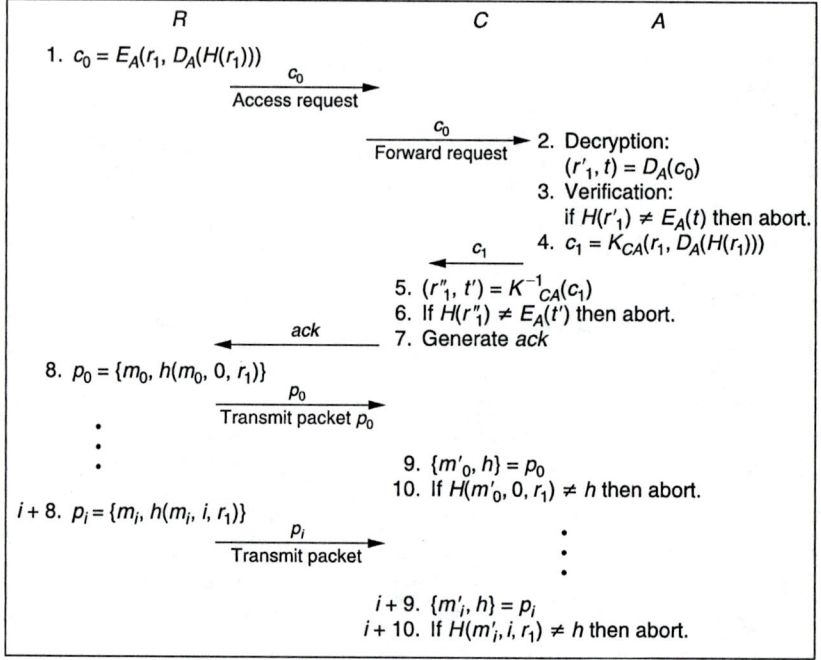

Figure 29.5 Controlled connection protocol.

Reconfusion

It is known that the longer an ID exists, the greater the chances of the association between the ID and the corresponding mobile user being exposed. To mitigate this problem, we can use a method called *reconfusion*, the objective of which is to generate a new authorized-anonymous-ID to replace the old authorized-anonymous-ID. Figure 29.6 outlines the protocol for the reconfusion method. First, an R sends the administrator a request (encrypted with the public key of the administrator) for a new authorized-anonymous-ID. Different from the registration protocol, in which a real identification (e.g., a public key certificate) is required to be presented, a request for a new authorized-anonymous-ID in reconfusion contains

- One of the mobile user's current or previous authorized-anonymous-IDs
- A random number multiplied by a factor that is blind or unknown to the administrator
- A symmetric encryption key suggested for this communication session between the R and the A

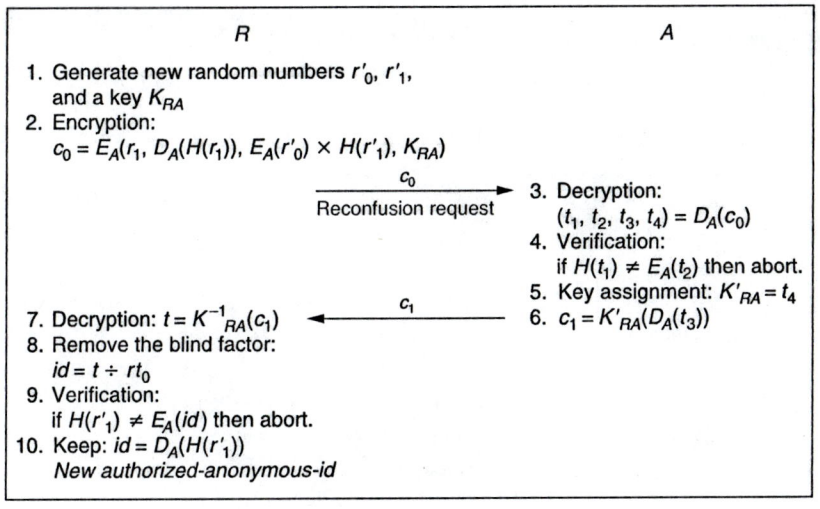

Figure 29.6   Reconfusion protocol.

After successful verification of the presented ID, the A signs a blind signature on the random number, encrypts it with the suggested key, and sends it back to the requesting rover. The R decrypts the message from the A, removes the blind factor, and retrieves the new authorized-anonymous-ID. The nice feature of the reconfusion protocol is that any disclosure of the previous ID would not compromise the anonymity of the new ID.

Access Authorization Revocation

It is not desirable, from an administration perspective, for an authorized-anonymous-ID to give a mobile device eternal rights to access the infrastructure; hence, the administration may want to have a function that can revoke or invalidate an issued authorized-anonymous-ID. One way to add this revocation function is for the administrator to periodically expire and change its own keys for access authorization. The anonymous IDs signed by the revoked keys will no longer be valid for authentication. This solution has a drawback, however, in that the mobile users need to periodically update their anonymous IDs, which introduces much communication overhead if the keys of the administrator expire too quickly. Another solution is to attach an expiration time stamp to the ID. The expiration time stamp should not be unique to the mobile user; otherwise, the unique association between the expiration time stamp and the ID can reveal the identity of a mobile user.

Untraceable Routing Infrastructure

Frequent communication between a home computer and a mobile device could also contribute to exposing an association between a mobile device and its stationary home. An untraceable routing infrastructure [13] can be used to erase the track but at a certain communication cost. A restriction on the protocols is that the standards of current wireless technologies, such as IEEE 802.11 and Bluetooth®, require manufacturers to assign an identification number (i.e., MAC address) to every device. The MAC address is essentially an annoying tag attached at all times to a mobile device. The custom of assigning a number to each wireless communication device is adopted from numbering every network interface card (e.g., Ethernet card) for stationary computers, where location privacy does not matter. In a ubiquitous computing environment, however, such a practice exposes the ID of a mobile device at the MAC layer. An ideal way to remedy this is to replace the MAC address with the authorized-anonymous-ID. ID collision should not be a serious problem in this case and can be prevented in many ways — for example, by adding a time stamp.

Related Work

Mobile IP [14] resembles the structure of our system. To support both mobility and privacy, these two systems must interact, but they are essentially different in two senses. First, they serve different purposes — Mobile IP is aimed at packet routing and forwarding, whereas the location information service/control system is targeted at providing location service under personal control. Second, they are implemented at different layers — Mobile IP is used at the network layer, whereas the location information service system is implemented at the application layer. As suggested before, the authorized-anonymous-ID can replace the hardware MAC address, but it is not necessary to change other layers (except for the application layer).

The basic idea behind the *privacy extension* for Mobile IPv6 [15,16] is to replace the MAC address of a mobile device with a random one, called a *temporal mobile identifier* (TMI) [15] or *pseudo-random interface identifier* (PII) [16]. In these schemes, personal mobile location privacy control relies on either home administration or foreign administration, or both. Moreover, it is necessary for home administration to share some secrets with foreign administration to prevent eavesdroppers from having any knowledge about binding users' temporal identifiers and real identifiers. These efforts do not allow mobile location privacy to be completely controlled by a mobile user, as the administration can associate any identifier (PII or TMI) with the corresponding real ID of the mobile user (or device).

Under the authorized-anonymous-ID-based scheme, the dilemma arising from two seemingly conflicting expectations — security (connection access control) and privacy (location information confidentiality) — is resolved by using an authorized-anonymous-ID created with a cryptographic technique: blind signature. The authorized-anonymous-IDs are used by mobile users as permission tokens for connection access controlled by the administration, but the authorized-anonymous-IDs, embedded in the packets transmitted to access points, would not reveal any information about the mobile users because the IDs being used are completely disassociated from the real IDs of the users.

Because efforts, such as those by Castelluccia and Dupont [15] and Escudero [16], have been directed toward addressing the location privacy issue at a lower layer (e.g., IP layer) rather than at the application layer, it may be worth mentioning that, according to our rationale study, a machine equipped with a lower layer technique may not be able to effectively achieve personal control over location privacy, because the lower layer technique will depend on the operators of the infrastructures to hide the identity of a mobile user. Also, in contrast to our solution at the application layer, the solutions at the IP layer are even more difficult to deploy. A detailed justification can be found in the Personal Ubiquitous Multi-Agent (PUMA) project report [17].

Summary

In this chapter, we investigated the problem of protecting location privacy of mobile users in the setting of ubiquitous computing. We pointed out that location privacy protection is particularly challenging due to the different requirements imposed by the administration and mobile users. To address this issue, we introduced an authorized-anonymous-ID-based scheme. In this scheme, an authorized-anonymous-ID is created by the blind signature technique and is used to replace the real ID of an authorized mobile device. With an authorized-anonymous-ID architecture, mobile users maintain complete control over their location privacy but the administration can still authenticate legitimate mobile users. The system introduced in this chapter has been built on the wireless Andrew network, a WLAN covering the campus of Carnegie Mellon University.

References

[1] He, Q., Wu, D., and Khosla, P., Quest for personal control over mobile location privacy, *IEEE Commun. Mag.*, 42(5), 130, 2004.

[2] Hills A., Wireless Andrew, *IEEE Spectr.*, 36(6), 49, 1999.

[3] Weiser, M., Some computer science issues in ubiquitous computing, *Commun. ACM*, 36(7), 75, 1993.

[4] He, Q. et al., WaveGuard: secure location service for wireless Andrew, in *Proc. Int. Conf. on Wireless Communications*, 2001.

[5] Chaum, D., Blind signatures for untraceable payments, in *Proc. of Crypto82*, Plenum Press, New York, 1982.

[6] Krawczyk, H., Bellare, M., and Canetti, R., *HMAC: Keyed-Hashing for Message Authentication*, Request for Comments 2104, Internet Engineering Task Force (IETF), 1997 (http://www.ietf.org/rfc/rfc2104.txt).

[7] Brewer, E. et al., A network architecture for heterogeneous mobile computing, *IEEE Pers. Commun.*, 5(5), 8–24, 1998.

[8] Saltzer, J.H., Reed D.P., and Clark, D., End-to-end arguments in system design, *ACM Trans. Comput. Syst.*, 2(4), 277–288, 1984.

[9] Chorafas, D.N., *Agent Technology Handbook*, McGraw-Hill, New York, 1997, chap. 13.

[10] Stajano, F. and Anderson, R., The resurrecting duckling: security issues for *ad hoc* wireless networks, in *Proc. of 7th Int. Workshop of Security Protocols*, Cambridge, U.K., April, 1999.

[11] Burnside, M. et al., Proxy-based security protocols in networked mobile devices, in *Proc. of the 17th ACM Symp. on Applied Computing (SAC'02)*, Madrid, Spain, March, 2002.

[12] Kaufman, C., Perlman, R., and Speciner, M., *Network Security: Private Communication in a Public World*, 2nd ed., Prentice Hall, Upper Saddle River, NJ, 2002.

[13] Reed, M., Syverson, P., and Goldschlag, D., Anonymous connections and onion routing, *IEEE J. Selected Areas in Commun.*, 16(4), 482, 1998.

[14] Perkins, C. and Johnson, D., Mobility support in IPv6, in *Proc. of the 2nd ACM/IEEE Int. Conf. on Mobile Computing and Networking (MOBICOM'96)*, White Plains, NY, November, 1996.

[15] Castelluccia, C. and Dupont, F., *A Simple Privacy Extension for Mobile IPv6*, Internet Draft, Network Working Group, 2001 (http://www.ietf.org/internet-drafts/draft-dupont-mip6-privacyext-03.txt).

[16] Escudero, A., Location privacy in IPv6: tracking binding updates, in *Proc. of the 8th Int. Workshop on Interactive Distributed Multimedia Systems (IDMS 2001)*, Lancaster, U.K., September, 2001.

[17] He, Q., Khosla, P., and Su, Z., A practical study on security of agent-based ubiquitous computing, in *Proc. of Deception, Fraud, and Trust in Agent Societies Workshop, First Int. Conf. on Autonomous Agents and Multi-Agent Systems (AAMAS'02)*, Bologna, Italy, July, 2002.

Chapter 30

Location-Based Service Differentiation

Spyros Panagiotakis and Nancy Alonistioti

CONTENTS

Introduction

Personalization remains a challenging issue in the upcoming mobile communications era. The provision of applications and services that are aware of individual preferences and characteristics of subscribers of mobile networks contributes to mobile communication with a personalized touch. In that context, user profiles and localization play a critical role. The former contributes to collecting, linking, and publishing various user-specific attributes, the latter to adapting service offerings to the user's location. Location is important with regard to how people organize and relate to their world. Knowing where a person or object is at any time adds a powerful new dimension to the range of information services that can be offered.

In this chapter, we introduce a more flexible and innovative model for user profiling. This innovation is based on the enrichment of common user profiling architectures to include the location and other contextual attributes so enhanced adaptability and personalization can be achieved. In the proposed architecture for context-sensitive user profiling, the user preferences are primarily dependent on the location (e.g., home, office) of the user. Other parameters that can also be taken into account are the type of the terminal device, the radio access network, the user presence, the time of day, and the mood of the user. These attributes constitute the user context. For each context, an associated user profile is created, and service provision is adapted to the user profile instance that best applies to the current user context. In particular, this chapter focuses on location and context awareness in service provisioning and proposes a framework that allows efficient management of location and context information. Furthermore, the concept of context-sensitive user profiling is introduced. Also discussed are the generic model, structure, and content of context-sensitive user profiles, along with some related implementation issues and a general architecture proposed for managing location and user profiles.

Context Awareness and Profiling

Within the context of future mobile communications, users will be able to access an abundance of services that typically will be developed by many cooperating entities. Moreover, the diversity of service access contexts (inevitable in an era of pervasive, anywhere computing) and the coexistence of different technologies resulting from the evolutionary character of the transition to next-generation systems will lead to heterogeneity of the networks and systems that support the provision of end-user applications. As a result, applications will have to be optimally delivered and executed over a large diversity of infrastructures and configurations, and services must be able to adapt dynamically to changing conditions and contexts. Because it would not be feasible to develop distinct versions for the various possible execution contexts, applications should be to a large extent independent of the environment they run on. Intelligent mechanisms should exist for identifying the context and particular high-level requirements of an application and mapping them to appropriate reconfiguration operations on the underlying hardware and software infrastructure [1].

Consider a mobile user who is receiving real-time news video from an application provider on a large-screen terminal at home. When the mobile user enters a vehicle to go to work and changes from the large-screen terminal to a portable phone, a location information server in the service-support layer notifies the application of the location update. Based on the user's profile or by interacting with the user, the application successfully recognizes the user's wish to change contents (media) from video to text, and a media converter executes content conversion. This is only one example of requirements for flexible, customized, context-aware, and ubiquitous multimedia service provision to mobile users that the evolution of mobile communication systems to 3G and beyond introduces. In a system that aims to provide flexible and context-aware service provision and adaptation, knowledge of the system status as well as the various entities' states and events is a significant factor. It is necessary to know at any given time the network status, the user location, the profiles of the various entities (users, terminals, network equipment, services) involved, and the policies that are employed within the system. In other words, the system must be able to cope with a large amount of context information.

We define *context* as the combination of information relevant to the nearest environment of a user, such as the user location, the serving network, and the user's terminal device. The contextual information is encoded in various related profiles, such as the user preferences profile and the terminal, ambient, network, and service profiles. The combination of all of these profiles constitutes the *user profile* [2]. Because profiling information is exchanged among different administrative boundaries, to ensure interoperability the eXtensible Markup Language (XML) [3] and the other XML-based languages (e.g., WSDL [4], SMIL [5], OWL [6]) are the best candidates for describing profiles. Moreover, profiles should be concise so they are transmitted efficiently. In situation-aware architectures, user profiles must be dynamically composed because their constituting segments may be distributed.

Contextual profiles greatly influence deployment and execution of a service, as context-aware services should be able to adapt to context and related updates. In the above scenario, for example, the service logic must be able not only to detect the user moving from one place to another but also to adapt its content to the new location the user has entered. This can be accomplished only if the user preferences of each subscriber are location sensitive, so a different set applies to each location or geographical zone. In such a situation, as the user moves from one geographical area to another, the service must be able to follow the user preference mobility, along with the user's physical mobility, and adapt the service offering accordingly. This advanced feature of service provisioning introduces the requirement for location (and contextual, in general) sensitivity in profiling architectures; hence, it is not enough for user profiles to simply encode location and other contextual information in static profiles. It is more important for the user-specific information included in the profiles to be organized on the basis of location and context attributes. In the sections that follow, a detailed analysis of that concept is presented.

The profiling information included in the user profile can be classified as summarized below:

- *Terminal profile* — According to the *user agent profile* (UAProf) specification [7], the terminal profile includes device attributes specific to (1) the hardware platform (e.g., CPU, screen size), (2) the software platform (e.g., Java platform and virtual machine version; operating system name, vendor, and version), (3) the network characteristics (e.g., current bearer service, supported bearers), (4) the browser user agent (e.g., support for Javascript, tables, frames), (5) the Wireless Application Protocol (WAP) characteristics (e.g., WML version, WML deck size), and (6) the push characteristics (e.g., maximum push message size supported).
- *Network profile* — The network profile can include: (1) identification and general description information for the surrounding networks, such as type (e.g., GSM/GPRS, UMTS, WLAN), physical location, and network operator data; (2) variations in the quality of service (QoS) provided by the network infrastructure (e.g., the available bandwidth), monitored and detected in real time; (3) technical characteristics, such as the type of bearer, supported bandwidths, topological and coverage information, and QoS levels supported; and (4) characteristics of a hybrid business/technical nature, such as support of open interfaces (e.g., Parlay/OSA [8,9], JAIN [10]) that enable access to selected network functionality.
- *Ambient profile* — The ambient profile includes information such as location, temperature, presence of other people, and whether the user is outdoors or indoors and in a suburban, urban, or rural

area. It also includes date and time information, which has particular significance for certain types of applications and may affect the service functionality as well as management tasks (e.g., billing).

■ *Service profile* — The service profile is a Web document that describes the different multimedia content elements and objects that each service consists of. When a multimedia element within a service document is offered in multiple content alternatives, the full collection of those alternatives is specified inside the document. For example, if an image inside a Web document is offered in three different versions according to the screen size of the targeted device, a listing with those three alternatives is given. The profile can also include all the required information for publishing the service and supporting its provisioning to the end users: the necessary service identification information (service name, service provider identifier), as well as data needed for discovering (category, keywords, language, valid location), accessing (available service versions, service client location, minimum terminal capabilities, and QoS required), and managing (application server IP address and ports, tariffs, pricing model) the service. Although XML and Web Services Description Language (WSDL) have been widely used for publishing such documents, the attributes and facilities offered by SMIL (e.g., the switch element) make it an ideal candidate for describing service profiles.

■ *Charging profile* — The charging profile collects charging-, pricing-, and billing-specific user information; for example, it records whether the user is willing to pay additional charges to ensure a better quality of the provided services. It can include: (1) user subscription information (e.g., user identities, subscription status), (2) subscribed charging characteristics (i.e., whether the subscriber is a normal, prepaid, flat rate, or hot billing subscriber), (3) subscribed charging services (e.g., the location-based charging service), and (4) the user charging, pricing, and billing preferences (i.e., the applicable charging rules, pricing, and billing models).

■ *User preferences profile* — The user preferences profile encodes the user preferences, which specify desirable service provision features that are particular to an individual user. User preferences can be categorized into service-independent, which apply to all services that are accessed by the user, and service-specific, which pertain to a particular application. The user preferences may include a broad range of attributes related to (1) the perceived QoS requirements, such as desired voice quality in phone calls, audio/video quality for streaming applications, or the degree of resolution for images; (2) the languages that the user prefers; (3) the content and media presentation characteristics (e.g., text versus audio versus video); (4) font sizes; (5) fees and billing; (6) privacy and security; (7) favorite geographical zones; (8) user identity, including name, gender, and profession; (9) user presence information; and (10) user history and calendar.

Service-to-Context Adaptation

The challenge with mobile, distributed computing is enhancing the user's dynamic environment with a new category of applications that are aware of the context in which they run. A context-aware service takes into account the current context of the user and, based on this information, it adapts its behavior to the respective user's needs including personal preferences and environment's capabilities [11]. Context-aware applications present information and services to users, in addition to automatically executing services and commands according to the context and its changes. Changes in the contextual environment are modeled as events and are communicated to the application for real-time service adaptation.

To enable third parties to develop context-aware services, various efforts have been undertaken by standardization work groups and forums toward the introduction of open, network-independent interfaces allowing context retrieval [8–10]. These interfaces provide applications with transparent access to network functionality (e.g., call control, location information, messaging, profiles retrieval), thus offering third-party application developers the opportunity to create advanced, network-independent, and context-full services with standard software engineering tools and general-purpose programming languages.

Service-to-Terminal/Network Adaptation

A context-aware application would be responsible for editing the service presentation to consumers, taking into account the user's terminal in conjunction with the network capabilities available. For example, a user participating in a teleconference using a PC can receive high-quality sound and video input, but from a mobile phone it is possible to continue participating in the discussion without the video input. Furthermore, if the user specifies a maximum load time for a page, the transcoding system should sense the end-to-end bandwidth to derive the ideal data size for the content. The effectiveness of this service-to-terminal/network adaptation depends on the dynamic creation of a generic service presentation description; therefore, a service should provide multiple views to users, modifying and transcoding the user interface multimedia components for appropriate presentation based on the user's terminal and network capabilities. Moreover, a context-aware application should aim at reducing communication needs but should also provide roaming capabilities so the user can transparently switch between networks and adapt to the changing QoS of the network [11].

Service-to-Location Adaptation

Location has a lot to do with how people organize their lives. In that context, the importance of localization as a primary concept for service customization, network reconfiguration, and operator differentiation has influenced many telecommunication companies and institutions to develop or integrate positioning systems into their networks. The exploitation of location information introduces a powerful new dimension to the range of information services that can be offered. By combining positional mechanisms with user-specific location and mobility information, it is possible to offer truly customized personal communication services through a mobile phone or other mobile devices [12]. For example, a user can find services according to geographical criteria utilizing different information sources at the same time. A service, on the other hand, may send location-, situation-, or event-related information on the basis of a user profile, so someone driving a car can receive information about the location of the closest gas station and a traveler has access to well-targeted information about suitable overnight accommodations. Location-sensitive applications can retrieve location and presence information, as well as receive notifications for related events through the associated open application programming interface (API) that the underlying network provides and which enables authorized applications to access the location and presence server of the network operator [8,9].

Service-to-User Preferences Adaptation

Adapting service provision and providing personalized services based on user preferences were introduced by the 3rd Generation Partnership Project (3GPP) to the Virtual Home Environment (VHE) [13]. VHE is a concept for personal service environment (PSE) portability across network boundaries and between terminals. The primary aim of VHE is to consistently present the same personalized features, user interface customization, and services in whatever network and terminal (within the capabilities of the terminal and the network) wherever the user may be located. VHE is enabled by user profiles as they encode parameters that are essential to the user, such as the user's communication preferences and service presentation on the terminal.

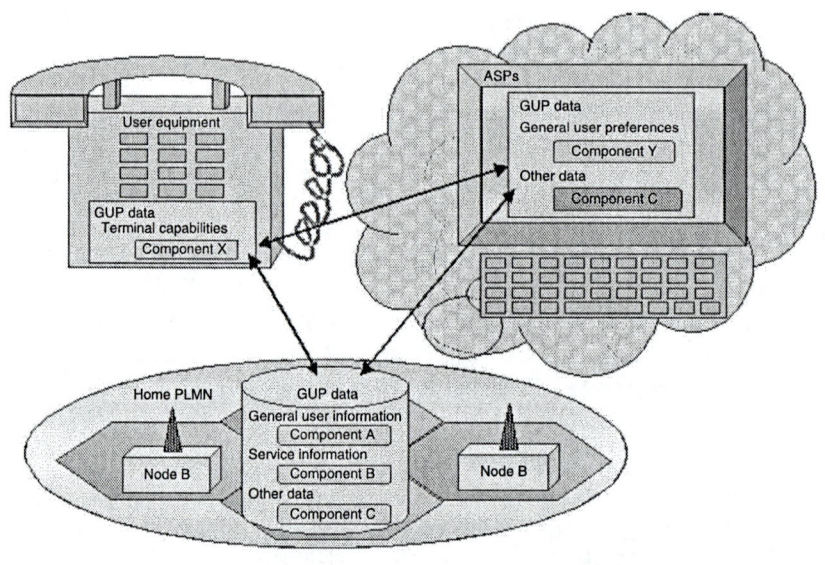

Figure 30.1 Conceptual view of the GUP distribution.

The 3GPP Generic User Profile

The 3GPP specifies the service requirements for the generic user profile (GUP) [14] and GUP architecture [2]. The objective is to provide a conceptual description that promotes harmonized usage of shared user-related information distributed among different entities, such as the user equipment (UE), home network (home public land mobile network [PLMN]), or third-party application/service providers (ASPs). 3GPP introduced the GUP as a solution to the increasing amount of user-related data. The conceptual view of GUP distribution can be seen in Figure 30.1. GUP allows intra-network and inter-network usage, which results in its access by different applications in a standardized way. Intra-network usage involves the exchange of data between applications within the mobile operator's network, and inter-network usage deals with the communication of profiling information between the mobile operator's network and ASPs. Each entity can hold a copy of a component that can be originally located in another entity (e.g., in Figure 30.1, component C in the ASP domain).

Figure 30.2 shows the GUP reference architecture. GUP data repositories are located in various nodes across networks and are distributed from the UE (e.g., Universal Subscriber Identity Module [USIM]) to the home PLMN (e.g., home subscriber service [HSS]/home location register [HLR]) or the third-party ASPs. Each GUP data repository stores the primary

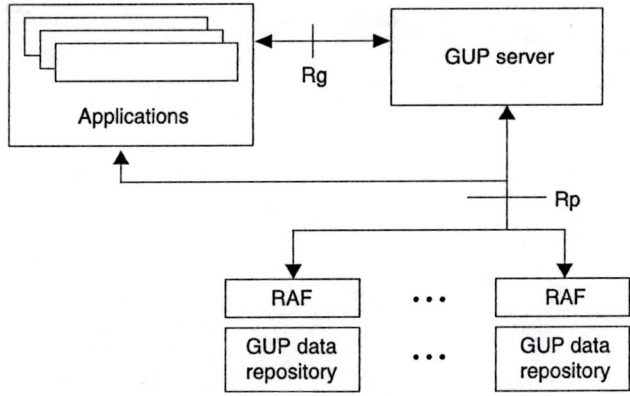

Figure 30.2   GUP reference architecture.

master copy of one or several profile components. The repository access function (RAF) provides standardized access to the GUP data repository. It hides the implementation details of the data repositories from the GUP infrastructure; neither the storage format nor the interface between the RAF and GUP data repository is specified. It is presumed that a GUP data repository and its RAF are usually colocated. The Rg and Rp are standardized interfaces providing harmonized access to the GUP server and GUP data repositories, respectively.

The GUP server is a functional entity providing a single point of access to the GUP data of a particular subscriber. It authenticates and authorizes profile requests from applications, identifies the profile components relevant to the request, and localizes them at the various GUP data repositories. The GUP reference architecture does not specify or limit the physical location of the GUP server.

As it is depicted in Figure 30.3, the GUP server supports two operation modes: acting either as a proxy server or as a redirect server. Although 3GPP tends to restrict the use of the redirect mode to applications in the home PLMN, other criteria to select either mode are still under discussion. Depending on the operation mode, the GUP server is involved directly or not in conveying profile data between repositories and applications, including optional synchronization of the slave copy of an application with the master copy.

Definitely, the 3GPP GUP architecture provides an integrated framework for user profile management. The solution it provides is ideal for managing the storage, distribution, and propagation of the profiling documents; however, it does not include the required flexibility to deal with location- and context-sensitive profiles because it monolithically and indiscriminately

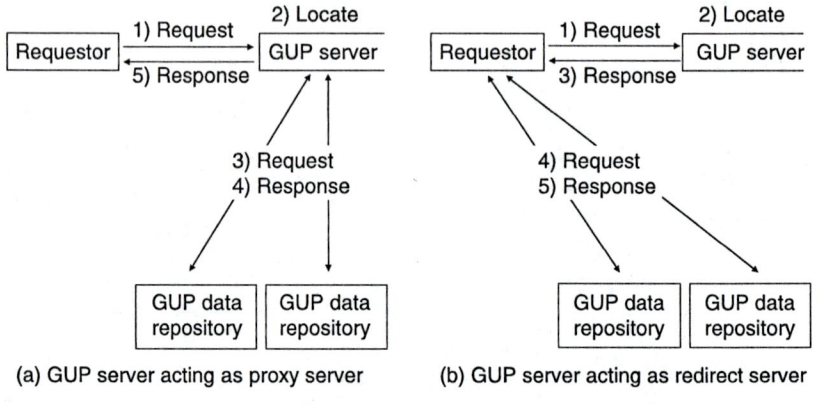

(a) GUP server acting as proxy server   (b) GUP server acting as redirect server

Figure 30.3 GUP server operation modes.

addresses all profiling documents, context sensitive or not. Hence, additional intelligence that takes into account the user context and the content of those documents is required to efficiently manage location- and context-sensitive profiles. The add-on intelligence has to deal with the location- and context-based parameterization, storage, and management of those profiles, so upon request only the profile associated with the current location and context of the targeted user is retrieved. The following sections elaborate in detail how the GUP framework can be complemented to include the required flexibility.

Proposed Framework for Context-Aware Service Provision

To support the demand of future communications networks for efficient and personalized service provisioning, the standard Universal Mobile Telecommunications System (UMTS) infrastructure should be supplemented with intelligent components. Figure 30.4 illustrates the general architecture and an example physical placement of the proposed service provisioning platform to meet requirements for 3G and the era beyond 3G. It is assumed that the independent ASPs will be offering their applications and services using the transport service provided by the underlying UMTS network.

The platform constitutes an integrated distributed software framework for context-aware management of applications and services offered to mobile users [1,12,15]. The platform can be viewed as an intelligent service middleware that mediates between ASPs and the network operator to

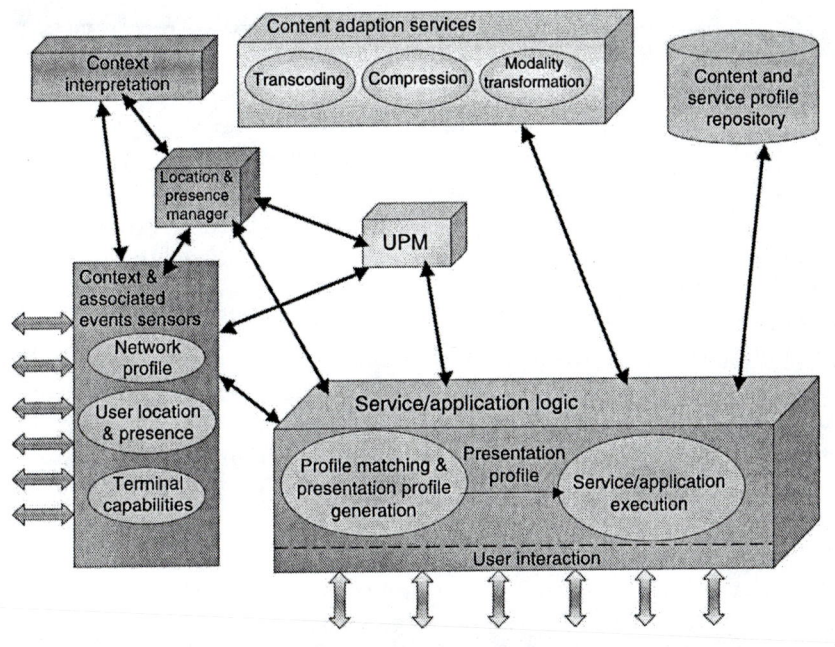

Figure 30.4   Generic framework for context-aware service provision.

provide ASPs and end users with advanced context-aware services and enable the dynamic composition of mobile services and applications. To accomplish this, the platform takes into account context information such as location and mobility information for the subscribers, preferences from their user profiles, and the current terminal and network capabilities to adapt service provisioning accordingly and perform reconfiguration actions on the network nodes and end-user terminal equipment.

The framework comprises the service/application logic component that orchestrates the adaptability of services to context. Whenever an authorized user requests access to a specific service, this component takes as input the updated contextual information for the user from the user profile manager (UPM), the location manager, and other context sensors in the form of profiles (e.g., user profile, user location, terminal capabilities, network profile), along with the presentation profile of the requested service (the service profile). In parallel, the service logic registers itself with the UPM, location manager, and context sensors to receive event notifications concerning updates to the current context of the requesting user. Then, based on the encoded information included in the contextual profiles, it filters the service profile document to generate a customized service presentation profile adapted to the current context.

The generated presentation profile is propagated to the service execution engine that is responsible for compiling and executing the commands in the incoming profile. This task can include the retrieval of appropriate multimedia content from the content repositories in the exact format indicated in the presentation profile, interaction with the content adaptation services for transcoding or translating some media files, composition of media files according to the indicated presentation format, and packaging and forwarding the produced result to the requesting user. If a context parameter changes while the user executes the service, an associated notification reaches the service logic and the operation described above is repeated. Based on the updated context information, a new service presentation profile is generated for the user and the service execution is adapted accordingly. For example, if the incoming event notification indicates that the user has exited his home to drive to work, the service logic is advised by the UPM to be aware of the new user preferences for service consumption in the new situation or geographical area the user enters, modifies the service presentation profile accordingly, and instructs the service execution component to perform the required adaptations (e.g., a video-to-text conversion).

The UPM is responsible for managing the user profiling data distributed among several data repositories across the network and providing user-specific information to the requesting applications/services. The sections that follow provide a detailed analysis of the UPM. The location and presence manager [12] is an independent module of the framework responsible for retrieving, managing, and exploiting information related to the location, presence, and mobility of the subscribers. Hence, it interacts with the sources of location and presence information of the underlying network infrastructure (e.g., 3GPP GMLC [8,9,17], 3GPP Presence Server [16], a private location and presence sensing network) to track the location and presence of subscribers. Detailed analysis of the location manager is also provided in the sections that follow.

The context-sensing components are considered to be the service/application interface to the context sources. The context information (e.g., network performance metrics, mobile equipment [ME]- or radio access technology [RAT]-specific, location- and presence-specific) can be equally retrieved through either a private sensor network or the OSA/Parlay APIs that might be provided by the underlying network infrastructure to the authorized applications (applicable only to OSA/Parlay-aware applications).

The context interpretation components translate the raw contextual information retrieved from the context sources to the high-level and user-specific information required for personalized adaptation. For example, location interpretation can translate the geographical coordinates taken from a global position system (GPS) device to a street address or applicable favorite zone (e.g., home zone [12]) of the positioned user. A geographical information system (GIS) could be used to this end.

The content adaptation services are responsible for adapting media content to the current content. They include various media processing technologies used to increase content accessibility and improve the user's experience within heterogeneous networking environments. They can transcode, compress, or convert media content according to the characteristics of the client display (e.g., screen size and color depth), current network parameters (e.g., available bandwidth), and specific user preferences (e.g., low-resolution images and video or audio tracks instead of video.)

The Location Manager

The location manager, as shown in Figure 30.4, is an independent module of the platform responsible for retrieving, managing, and exploiting information related to the location and mobility of subscribers. It interacts with the sources of location and presence information of the underlying network infrastructure (e.g., 3GPP GMLC [8,9,17] and 3GPP Presence Server [16], a private location and presence sensing network) to track the location, presence, and mobility of the subscribers. To translate the retrieved location information into a recognizable and usable format instead of the geographical coordinates or network areas that the underlying location sensor or server (e.g., the Gateway Mobile Location Center [GMLC]) provides, the location manager interacts with the appropriate interpretation component of the platform. Then, location, presence, and mobility data and events along with the preferences of the corresponding subscriber (taken from the user profile [2]) are processed for the user in the recently entered new location.

Figure 30.5 illustrates the environment of the location manager. The user's location can be used to determine, based on the user's preferences, the reconfiguration policies that are propagated from the location manager to the underlying network infrastructure. By combining location information with user preferences, the location manager is able to provide end users and any authorized entity internal (e.g., application/service logic component) and external (e.g., application/service providers) to the platform with new advanced location-aware services. Furthermore, it enriches the service provisioning approach of the platform with location information features, enabling better customization and personalization of the service offering. Figure 30.6 depicts the service logic behind the functionality of the location manager.

The location manager retrieves the required location information by accessing the location sensors of the underlying network infrastructure through the associated open API; for example, the 3GPP location server

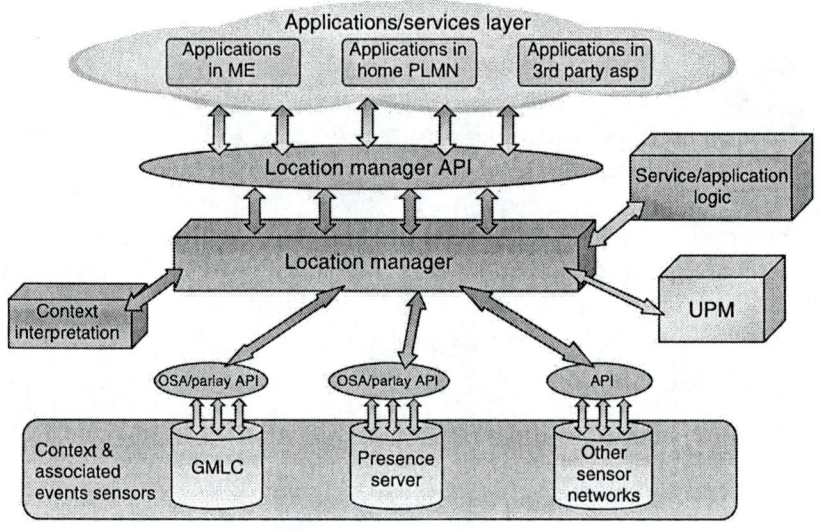

Figure 30.5 Location manager environment.

[17] supports most of the positioning technologies specified for location measurement and computation in mobile networks (UMTS and GPRS), such as cell location, the observed time difference of arrival (OTDOA), enhanced observed time difference (E-OTD), and network-assisted GPS. The positioning method selected and the accuracy with which the location measurements are performed depend on the quality of the location information requested by the location service client for each request.

For transparency and independence, the functionality provided by the location manager is accessed by internal modules of the framework as well as authorized application/service providers through an open API provided by the framework to the authorized entities. That open interface includes methods that provide:

■ Retrieval of the location of the specified user — Location retrieval can be immediate (when the current location of the user is requested) or deferred (when the location of the user is requested after a specific event takes place) [17]. Making use of appropriate spatial databases or GIS systems, the location manager is able to map the current user location as determined by the underlying location sensor (expressed in geographical coordinates or network areas) to the requested format (e.g., street address or predefined geographical zones). Hence, the retrieved location information will be in a recognizable and usable format. The service/application logic component of the platform uses these methods to achieve

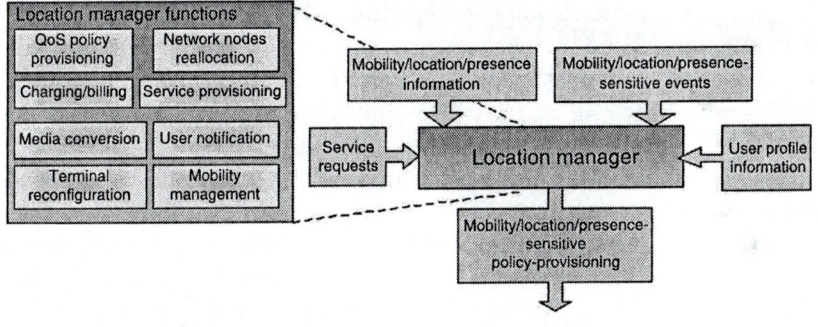

Figure 30.6   Location manager service logic.

location-based service provisioning in terms of location-sensitive service deployment, discovery, and execution.

- Activation or cancellation of the location-based charging and billing mechanism [12] for the provision of flexible and personalized billing schemes
- Activation or cancellation of the location- and mobility-aware QoS management mechanism directed by the location manager — With this service, the QoS provided to the subscribers follows the location updates that are induced due to their mobility. Hence, whenever a mobile user enters or leaves a predefined geographical zone, the QoS that the user receives is automatically adapted to the user's subscribed QoS preferences associated with the new zone as they are included in his user profile.
- Creation, modification, and deletion of location-sensitive policy classes
- Activation and deactivation of location-sensitive policies
- Creation, modification, and deletion of policy events
- Registration and deregistration for receiving location-sensitive event notifications
- Handling of event notification from the network
- Notifications to registered end users and applications for the available policies, restrictions, updates, tariffs, reconfigurations, and other events associated with the current location of the user or induced due to the location updates that occur

The primary goal of the location manager is to enable easier development of location-based applications and services. To accomplish this objective, we have designed the methods and built the services that we expect a location-based application or service to require frequently. Furthermore, the location manager collects raw location information from the location sensors of the underlying network infrastructure and maps it into the required higher level format by using the appropriate spatial databases (represented in Figure 30.4 and Figure 30.5 by the context interpretation modules, such as a GIS). In this way, we can provide applications with an execution environment and some reusable and customizable location-sensitive building blocks for structuring their functionality, while we hide from application developers the complexity and physical interactions required for retrieving and adapting the location information. The design of our platform does not focus on the development of a specific application or service (location-based or not). Instead, our primary goal is to build a generic framework for service development and deployment that can accommodate any service or application, gathering the necessary informational resources and building blocks for facilitating development and deployment.

Location tracking takes place with respect to the user privacy settings located in the user profile. The privacy policies are an implementation of the settings recommended in References 17 and 18. Each user, upon registration with the platform, defines the policies that will govern the retrieval of his location by third parties. These include: (1) a list of the services and applications that are authorized to retrieve the location of the user, (2) a list of the persons or groups (known as *requestors*) that are authorized to request the retrieval of his location, (3) the password given by each user to authorized requestors (a unique password for each requestor) for authenticating themselves with the location platform, (4) an indication of when the location manager should prompt the user for explicit authorization prior to any location-retrieving action (i.e., never, all actions, or only when previously specified criteria are not met). Whenever an application (or the requestor/consumer of that application) requests the location of a user, the location manager examines whether it is authorized by the user to do so. If the application is not authorized to make the request, then the end user's preferences determine whether the request is simply denied or the end user is asked for explicit authorization.

By further deploying the operations provided by the location manager, the functionality of our platform could be enhanced to include novel, advanced, location-aware services such as advanced mobility management that provides for optimized network planning, location registration, paging, and handover management. This could be accomplished by monitoring and storing the user mobility and location data for a long period and then processing the mobility and location historical data by applying specific mobility and location prediction algorithms. Other services would include terminal device reconfiguration with software and protocol updates and upgrades, in addition to performance-enhancing mediation and, finally, network node selection and reallocation based on user mobility for network performance optimization and better service provision to the end user.

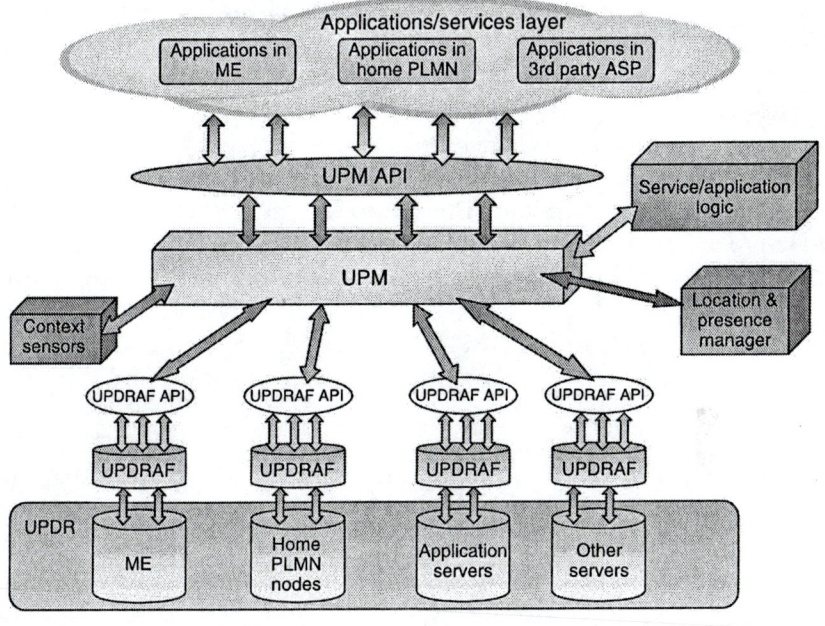

Figure 30.7   The UPM environment.

The User Profile Manager

The architecture we propose for user profile management enhances the 3GPP GUP architecture with the concepts of *home zones* and *context zones*. It adopts the distribution and information model of the GUP, incorporating in its logic a provision for enhanced context sensitivity. The enhanced GUP server proposed here is referred to as the *user profile manager*, and it is a structural part of the innovative architecture (see Figure 30.4), providing context-aware service provision [15]. The UPM is responsible for managing the user profiling information distributed among several data repositories across the network and disseminating the user-specific information to the requesting applications and services. It mediates between applications and services and the *user profile data repositories* (UPDRs), hiding from applications the underlying infrastructure and facilitating interaction with the profiling sources. The applications that may request access to the user profile data can vary from applications in the ME to applications in the home PLMN or third-party application/service providers. The UPM allows end users and authorized applications to insert, delete, or modify user profiling data in the UPDRs; retrieve user profiling data upon request; and receive profiling-dependent event notifications each time a registered event occurs. Figure 30.7 illustrates the environment of the UPM.

The user profiling information is distributed and stored in various UPDRs. Each UPDR stores the primary master copy of one or several profile components. Possible candidates for the UPDR include the ME, HSS/HLR, and various application and management servers in a home PLMN or third-party ASPs. Synchronization among profiling data in the UPDRs and UPM is required.

Access to UPDRs is accomplished through the associated *user profile data repositories access functions* (UPDRAFs). Each UPDRAF can be viewed as the front end of the underlying repository that realizes the harmonized access interface. It hides the implementation details of the UPDR from the rest of the UPM infrastructure. UPDRAFs perform protocol and data transformation where needed. The protocol between UPDRAFs and the UPDR is implementation dependent and not standardized. The UPDRAFs can also take part in the authorization of access to the UPDR. Through UPDRAFs, the UPM can insert, delete, or modify the underlying profiling data; read the data; and receive synchronization notifications whenever a change in any profiling data occurs. The interfaces between the UPM and UPDRAFs and between the UPM and applications are discussed in following sections.

The UPM interacts also with the location and presence manager and the context sensors of the proposed framework to retrieve the location, presence, and other contextual information required to compute context zones and provide context-sensitive user profiling. The concepts of home zones, context zones, and context-sensitive user profiling are discussed in the following text.

Context-Sensitive User Profiling

The key to offering truly context-aware and customized services to subscribers is the user profile. In contrast to common architectures for user profiling that consider profiles as static collections of user preferences, we propose here that the user profile can also be context aware. Context awareness of the user profile can be based on the concept of home zones. Each home zone is composed of a geographical area in which a user wishes to experience personalized and customized service provisioning (e.g., the home, office, car) [12]. Ideally, a home zone should be as large as the user wishes, so true customization can be achieved. If a user wants to experience different service provision in each room of his home or office, then each room of the home or office would be considered as a distinct home zone for that user. Limitations in the accuracy of location measurement induced by current position estimation technologies do not allow location-based services to distinguish among very narrow home zones. For this reason, minimum distances between the defined home zones of a user, depending on the accuracy of the location measurements, should exist so the position-detection system can follow the moves of subscribers from one home zone to another. In the near future, when location estimation technologies mature further, location-based services will be able to provide users with more accurate positioning.

Taking into account that within a single home zone a subscriber can switch from one type of mobile equipment to another (e.g., from a UMTS mobile handset to a PDA or laptop) and can access different radio environments (e.g., from GPRS or UMTS networks to WLAN/WiFi or Bluetooth®), further classification of user profiles within a home zone can be achieved. The current home zone, current terminal equipment, and current serving radio access technology (RAT) of a subscriber are considered to be the three key context attributes that uniquely identify the current context of a user. Each triplet of type {current home zone, current terminal device, current radio access technology} defines a user-specific *context zone* that can be used for identifying the user status and customizing service provisioning accordingly. As the user moves to a different location (home zone) or switches between mobile devices or access networks, a different context zone is assigned.

Because within the geographical area that defines a home zone a subscriber can move from one terminal device to another or change the RAT being utilized, obviously a home zone is considered to be broader than a context zone. Thus, within a home zone, multiple context zones can coexist which differ according to the terminal device used or the serving RAT. If the user changes terminal devices while remaining attached to the same RAT, a different context zone is assigned. The same occurs when the user transits to a different RAT while using the same mobile equipment. Figure 30.8 illustrates the relationships among home zones and context zones. Taking into consideration that within the illustrated home zone (home zone A) a user can access two different RATs (e.g., UMTS and WLAN RATs) with each of his terminal devices (ME1, ME2), then within home zone A four different context zones can be identified. In particular, context zone 1 (home zone A, ME1, WLAN), context zone 2 (home zone A, ME2, WLAN), context zone 3 (home zone A, ME1, UMTS), and context zone 4 (home zone A, ME2, UMTS) can be defined for the targeted subscriber.

A user may define multiple preference profiles for a single home zone. Better personalization and customization can be achieved during service provisioning by differentiating the user preferences in the three basic ways — location of the user (e.g., home, office, car), terminal device used, and serving radio access network — and by maintaining in parallel different user profiles instances for each context zone instance, so they are {location, terminal, network}-dependent profiles or context-zone-dependent profiles [12]. Profiles can be also further classified by and associated with specific

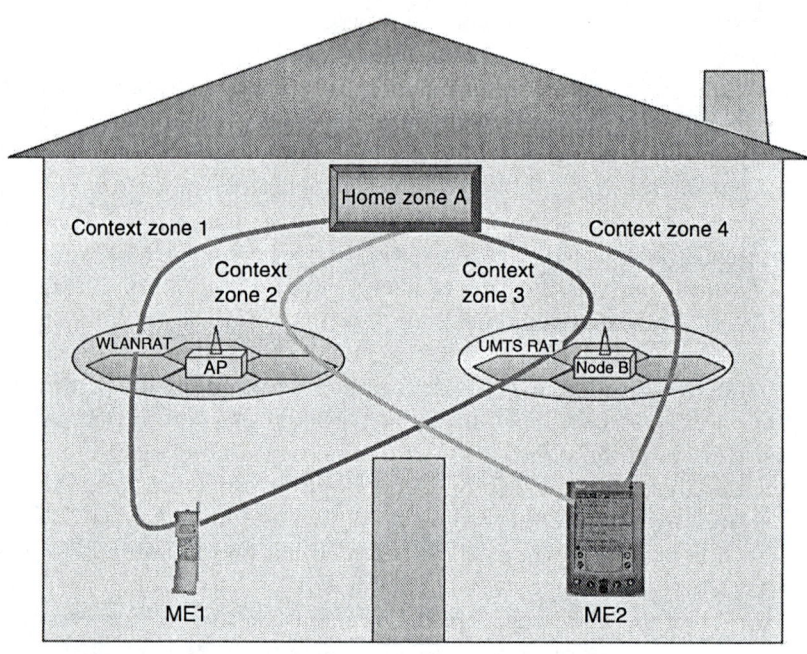

Figure 30.8   Relationships among home zones and context zones.

presence attributes of subscribers, such as the time of day (e.g., lunch time) or a specific ambient attribute (e.g., temperature, velocity, height) or even a certain mood of the subscriber (e.g., happy or sad), thus broadening the concepts of context zones and context-zone-dependent profiles accordingly. The presence information required in such presence-sensitive user profiling can be retrieved by either specific presence sensors or the associated presence server [16] of the underlying network infra-structure. A presence-aware context zone includes the presence attribute in addition to the three aforementioned ones, so each quaternary of type: {current home zone, current terminal device, current radio access technology, current presence status} defines a context zone.

Whenever a user enters a context zone where multiple profiles have been defined, the system prompts the user to indicate which profile to activate. Each user profile instance can be considered as a user preferences customization set that includes the user interface preferences, the browser appearance, the preferred memory usage, etc., as well as the application/ service subscription profiles with the preferred settings for the subscribed applications/services. The services provisioned to a user depend on the current context zone and, hence, the associated active profile.

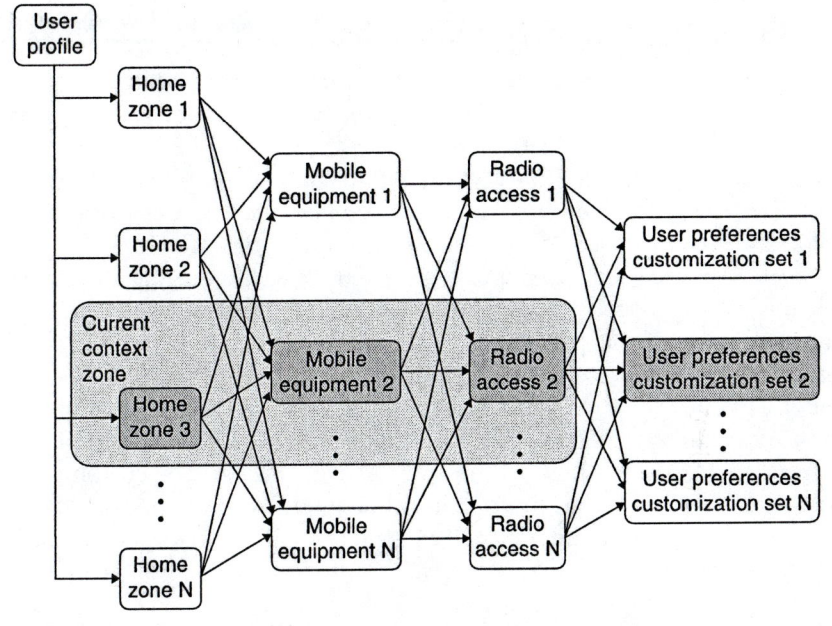

Figure 30.9   Context-sensitive user profile.

Figure 30.9 illustrates the concept of context-sensitive user profiling. The rounded rectangles in blue represent the triplet of the current context zone of the subscriber (home zone 3, ME2, RAT2) and the associated active user preferences customization set (user preferences customization set 2). From that point of view, each context-sensitive user profile can be considered as a tree for which the subscriber's identity (e.g., IMSI, e-mail, or SIP identity) is the root, the three context zone attributes are nodes, and the user preferences customization sets are leaves. By storing locally such a tree-like user profile for each subscriber, finding the user's current context zone, and traversing the tree from the top down, the serving manager (e.g., UPM, ASPs) can retrieve the most up-to-date user profiling data each time it is required. Although in Figure 30.9 only the three basic context zone attributes are taken into account for user preferences selection and service differentiation, further contextual parameters could be used (e.g., presence, time, ambience, mood attributes) for better elaboration and specification.

Each subscriber, upon registration with a service provisioning platform, specifies his applicable home zones by describing each one with real-world addresses or street names. It is then up to the location manager of the platform [12] to translate the home zones specified by the user to the appropriate geographical coordinates or network areas (e.g., cell IDs, location or routing areas), making use of databases with the appropriate spatial and geographical information. Although the particular ME and RAT types are dynamically retrieved as the user accesses the network, optionally the user can declare to the platform the various types of ME he possesses along with the different radio access networks he usually utilizes. The specified home zones, along with the ME and RAT types declared, are used as the initial tree nodes of the user profile. Hence, a tree-like and context-sensitive user profile is created and stored for each user (see Figure 30.9). Then, each time the subscriber specifies a user preferences customization set, it is associated with the current context zone of the subscriber (communication with the context sensors of the platform is assumed here for retrieving the current context zone) or the context zone indicated by the subscriber.

For each new context value (i.e., new home zone, new ME, new RAT), a new node in the tree is inserted. Equally, for each new customization profile a new leaf is added to the tree. A user preferences customization set can be associated with more than one context zone, if the user wishes. For each active subscriber, the UPM stores locally a data structure that represents the tree of that subscriber's context-sensitive user profile. The data itself is not included; instead, a reference to the UPDR that stores each profile is kept, along with a unique data reference generated upon data creation and storage. The three context values (home zone, terminal type, RAT type) are used by the UPM to identify the path to the active profile data. To this end, a pointer that crosses the user profile tree and points to the active customization profiles is created for each user. The pointer is updated each time a change occurs.

Whenever a user enters the platform, the UPM retrieves the user's current location to identify the home zone in which the user is currently situated, along with the applicable terminal and network profiles, to identify the context zone of the user and retrieve the user profile instance that applies to the specific context. It is implicit that the user is always prompted to confirm the user profile instance selection or alternatively to choose the one desired. The UPM stores locally the current context zone of each user for later use and faster searching in the user profile tree. With context-aware user profiling, only the profile that best applies to the current home zone and status of the user is taken into account, thus customizing the service offering accordingly. Figure 30.10 illustrates the user profile instance transition induced by the context zone change. The user moving from home zone 1 to home zone 3 initiates a change of the current context zone from context zone 1 (home zone 1, ME1, RAT2) to context zone 2 (home zone 3, ME1, RAT2) and hence to a different user preferences customization set (from user preferences customization set N to user preferences customization set 2). Such a change of the active user profile instance triggers the UPM to generate and propagate an associated alert event, so the components registered for receiving such alerts are properly informed.

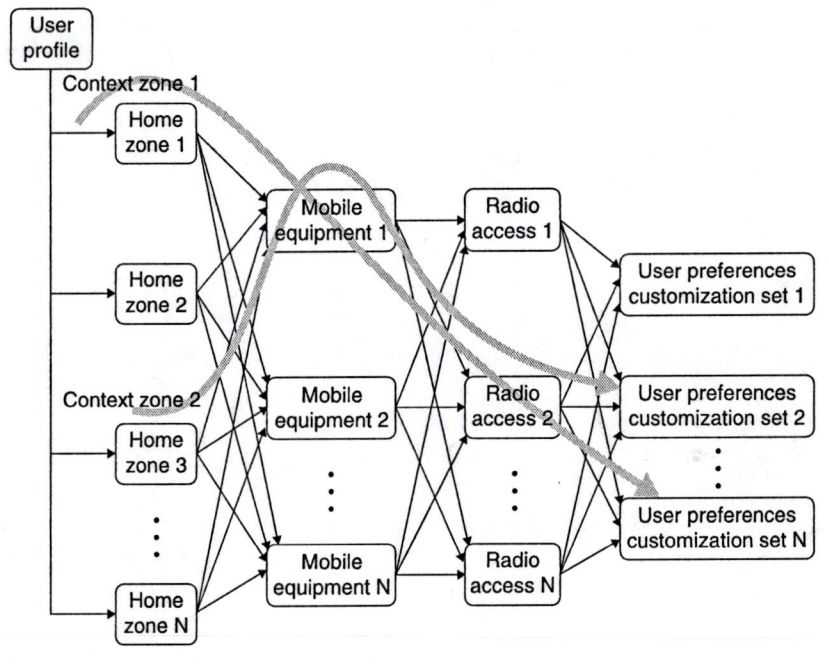

Figure 30.10 User profile instance transition induced by the context zone change.

Interfaces of the UPM

The UPM–Applications Interface

To hide the implementation details of the profiling architecture from applications and services and ensure service transparency, interaction between internal modules or services of the profiling architecture, as well as authorized third-party applications, and the UPM is implemented through an open API provided by the UPM to authorized entities. The API can be used, for example, to create the user profile or some components of it, to read any piece of data in the profile, or to modify those using the harmonized access interface. Furthermore, it is possible to authorize all requests and protect the user's privacy in all operations. Protocol and data transformations are performed when necessary. The UPM locates the data repositories responsible for the storage of the requested profile components and, in proxy mode, carries out the requested operation on the data; in redirect mode, the UPM returns the locations of the corresponding UPDRs and the application can then send the requested operations directly to the corresponding UPDRAFs.

Profiling data stored in the various user profiling components is identified based on the identity of the associated subscriber or user, the corresponding context zone of the user, and the component type. In these interactions, the UPM acts as the server and the applications as the client. The proposed open interface for accessing the UPM services includes methods that provide:

■ Creation, deletion, or updating of user profile data — These procedures are always related to a single subscriber and context, which are identified in the request (through the subscriber identity and context zone parameters). If the context zone attribute is missing, the current context zone of the identified subscriber is assumed. New or updated data is also provided.

■ Retrieval of the user profile data or some specific components — The queried data is identified based on the subscriber identity, context zone, and data reference. If the context zone is not passed in the request, then the UPM retrieves the current context zone of the targeted user and returns to the requesting application the requested profile data that corresponds to the current context of the user.

■ Listing of the existing profile items in the various UPDRs that are associated with the specified context zone of the targeting user

■ Creation, modification, or deletion of profile-dependent policy events

■ Registration or deregistration for receiving profile-dependent event notifications

■ Submission of event-driven notifications to the registered applications whenever some of the registered events occur — Synchronization of the profile data kept by an application can be performed by the last three procedures (i.e., creation of a synchronization event, registration for receiving synchronization-dependent events, and receipt of associated notifications when synchronization is required).

■ Notifications to the users for the profile-dependent policies, restrictions, and updates associated with the current context of the user

The UPDRAF–UPM Interface

For the UPDRAF–UPM interface, in addition to the UPM some authorized applications (e.g., the home PLMN applications) can also be assigned direct access on UPDRAFs (redirect mode of the UPM operation). Furthermore, the 3G trend to distribute and decentralize operations, thus opening mobile service provisioning to new players, may allow the UPM to fall within a different business domain (e.g., an authorized third-party service provider) than the UPDRs. To provide the UPM and other applications with a harmonized access interface to the underlying UPDRs that also hide the implementation details of the underlying infrastructure, we propose providing access to UPDRs through open APIs implemented by the associated UPDRAFs (see Figure 30.7). In this case, the UPM (and other authorized applications) acts as the active requestor toward the UPDRAFs (e.g., to read or modify the data). It is assumed that both ends initially share the same data structure definitions. Through this API, the UPDRAFs would be responsible for authorizing incoming requests to protect the user's privacy in all operations while also performing protocol and data transformations where necessary.

Similar to the UPM–applications interface, the profiling data stored in the various UPDRs is identified based on the identity of the associated subscriber or user, the corresponding context zone of the user, and the component type. In those interactions, the UPM acts as the client and the UPDRAFs as the server. The proposed open interface for accessing UPDRAFs includes methods that provide:

■ Creation, deletion, or updating of user profile data — These procedures are always related to a single subscriber and context, which are identified in the request (through the subscriber identity and context zone parameters). New or updated data is also provided. Updated data is identified by a data reference generated upon first data creation and storage.
■ Retrieval of the user profile data or some specific components — The queried data is identified by the subscriber identity, the context zone, and the data reference.
■ Listing of the existing profile items in the underlying UPDRs associated with the specified context zone of the user
■ Creation, modification, or deletion of profile-dependent policy events
■ Registration or deregistration for receiving profile-dependent event notifications
■ Submission of event-driven notifications to the UPM and registered applications whenever some of the registered events occur — The synchronization of profile data between the UPM and UPDRs can be performed by the last three procedures (i.e., creation of a synchronization event by the UPM, registration for receiving synchronization-dependent events, and receipt by the UPM of associated notifications when synchronization is required).

Example Scenarios

In this section, some example scenarios and interactions that demonstrate the enhanced context-sensitive nature of our platform are presented. First, we will examine how a third-party ASP can deploy its applications with the platform and how location-aware service discovery and execution are achieved. Then, we will present how an application can retrieve the current user preferences of a user and how that application can remain informed about changes in the active user preferences of the subscriber when some change in the current context zone attributes of the user takes place.

Location-Aware Service Deployment

With the proposed architecture, third-party ASPs of the platform have the ability to offer their applications in the form of downloadable applications. An application cannot be made available to mobile users before the owning ASP registers it to the service provisioning platform [1,12]. This can be accomplished dynamically through a Web-like API provided by the platform to the authorized ASPs. The API facilitates and automates the service deployment process, making it transparent to the ASPs. Through this API, the authorized third-party ASPs are able to register new applications with the platform, as well as delete and update existing ones. The platform performs all reconfiguration actions required to accommodate the new or updated application. These actions can include storage of the service profile document in the service profile repository, as well as the configuration or reconfiguration of the metering devices that monitor the IP traffic flows to and from that application (for charging purposes).

During the application registration process, the requesting ASP provides to the platform the XML-encoded service profile document that includes all the required information for the support and provisioning of that application to the end users. That service profile can include the necessary service identification information (service name, service provider identifier), as well as the data required for discovering (category, keywords, language, valid location), accessing (available service versions, service client location, minimum terminal capabilities and QoS required), and managing (application server IP address and ports, tariffs, pricing model, available formats of multimedia content) the application. A simplified example of such a location-aware XML document containing service information is provided in Figure 30.11.

The "valid location" field in this XML document specifies the geographical area or extent of coverage for which an application is offered. By including location information in these documents, the platform becomes

```
<?xml version="1.0" encoding="UTF-8"?>
<!DOCTYPE SERVICE_DEFINITION "service_definition.dtd">
<SERVICE_DEFINITION>
      <Service>
            <SAPPublicKey>365636</SAPPublicKey>
            <ServiceID>1</ServiceID>
            <ServiceName>Soccer results</ServiceName>
            <ServiceVersion>2.2</ServiceVersion>
            <Description>Results of national soccer leagues</Description>
            <Language>English</Language>
            <ValidLocation>Greece</ValidLocation>
            <Category>Sports/Soccer</Category>
            <Keywords>Soccer</Keywords>
            <Availability>Yes</Availability>
            <LocationBasedChargingAvailability>Yes</LocationBasedChargingAvailability>
            <ServiceVersion>
                        <ServiceVersionID>2</ServiceVersionID>
                        <ServiceVersionName>Light Edition</ServiceVersionName>
                        <VersionDescription>Plain text version</VersionDescription>
                        <MexeClassmark>2</MexeClassmark>
                        <SoftwareReq>
                              <JVM>
                                    <Edition>Personal Java</Edition>
                                    <Version>1.2</Version>
                              </JVM>
                        </SoftwareReq>
                        <TransportProtocol>TCP</TransportProtocol>
                        <QoSIndicator>23</QoSIndicator>
                        <URL>http://www.soccer.gr/results</URL>
                        <IPAddr>195.138.67.173</IPAddr>
                        <IPPort>8080</IPPort>
                        <DefaultPricingModelNumber>1</DefaultPricingModelNumber>
                        <DefaultTariffClassNumber>1</DefaultTariffClassNumber>
                        <CostDescription>8 cents per 15 seconds</CostDescription>
                        <LocationBasedCharging>
                              <InHomeZonePricingModelNumber>2</InHomeZonePricingModelNumber>
                              <InHomeZoneTariffClassNumber>3</InHomeZoneTariffClassNumber>
                              <OutHomeZonePricingModelNumber>3</OutHomeZonePricingModelNumber>
                              <OutHomeZoneTariffClassNumber>4</OutHomeZoneTariffClassNumber>
                              <CostDescription>In Home Zones:5 cents per 15 seconds,
                                          Out of Home Zones: 10 cents per 15
                                          seconds</CostDescription>
                        </LocationBasedCharging>
                        <DSERTimeout>12</DSERTimeout>
            </ServiceVersion>
      </Service>
</SERVICE_DEFINITION>
```

Figure 30.11   Example location-aware XML document for application deployment.

aware of the location characteristics of registered applications. Hence, whenever a user requests the platform for discovering services and applications, for example, the platform is able to match the location of the requesting user with the valid location area of the available applications, thus providing the user with a listing of only those applications offered at the user's current location. Furthermore, implicit location information can be included in the XML document as part of the location-based charging service offering label. More specifically, this information indicates to the platform whether location-based charging applies for that application along with the applicable pricing policies.

Location-Aware Service Discovery and Execution

Taking into consideration the huge range of expected applications available to mobile users of the forthcoming 3G networks and the competitive nature of the new era, the need to provide mobile subscribers with an efficient and simple mechanism for personalized service and application discovery and provisioning is growing. The proposed platform provides users with the ability to discover and choose the applications they wish to access by requesting application listings through a personalized, Web-portal-like user interface offered by the service discovery client. The application listings would be presented to the users after the service profile repository records are filtered according to the applicable terminal device and network capabilities, the current user location, and the applicable user preferences [1,12,15]. Each application in the service discovery listing is associated with a short description and fee information so the user may choose an application based on the criterion of lowest cost, for example. Following the service discovery phase, the user is able to select the desired application and begin downloading the associated client to his device.

During service access, the platform has the ability to track the location of the user to perform the reconfiguration actions required by the mobility of the user. When a user enters or exits a home zone, for example, the user receives notification of the new home zone or radio environment he has entered, the new local applications offered, and the applicable fee schedule therein. At the same time, the applicable user preferences in the new home zone are retrieved through the UPM so the service/application logic component tailors the service offerings accordingly.

Retrieving User Preferences

Figure 30.12 illustrates the interactions among various components of the profiling architecture when an application accesses the UPM to retrieve the current user preferences of a subscriber. Only the direct mode of the UPM operation is illustrated. Such interactions include the following:

- An authorized application requests to retrieve some user profile components of a specific subscriber. The application does not include the context zone parameter in the request.
- The UPM authenticates the application and checks its authorization to receive the requested data.
- Because the context zone is not provided, the UPM presumes that the profile data requested is associated with the current context zone of the targeting user. The UPM then contacts the location manager of the architecture to retrieve the current home zone of the user.
- The UPM retrieves the type of mobile equipment from the appropriate context sensor.
- The UPM retrieves the type of radio access technology serving the user from the appropriate context sensor.

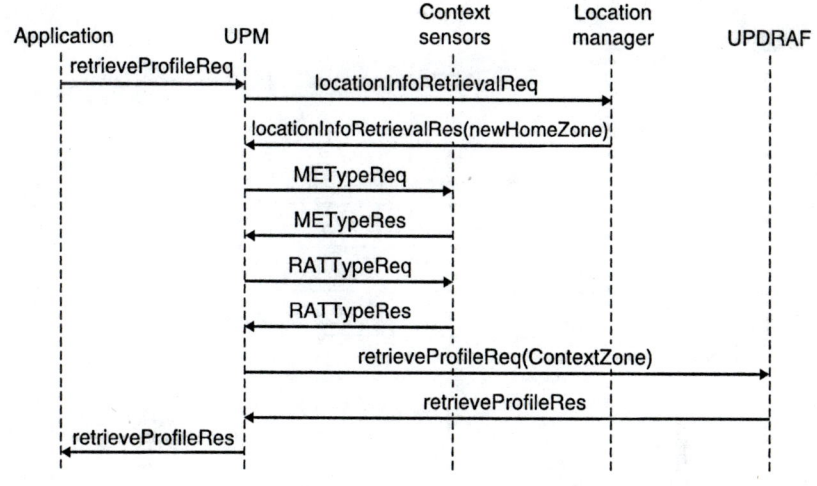

Figure 30.12   Interactions for retrieving user preferences.

- The UPM identifies the current context zone of the user and, based on that information, it crosses the context-zone-dependent structure of the user profile to identify the location of the requested components.
- The UPM accesses the API of the identified UPDR to request the specified data. The UPM includes the context zone and data reference in the request.
- The UPDRAF searches the context-zone-dependent structure of the stored profiling information to retrieve the requested data and responds to the UPM.
- The UPM responds to the application with the requested profiling data. Because the data requested by the application may have been stored in several UPDRs, it is likely that the UPM will have to interact with all of the involved repositories to retrieve the data. In that case, the UPM should properly consolidate the returned data before responding to the application.

Profiling-Dependent Event Notifications

The example below presents how an authorized application could remain informed about changes in the active user preferences of a subscriber that are induced by changes in the current context zone of the user. Figure 30.13 illustrates the interactions required among the various components of the profiling architecture. Such interactions include the following:

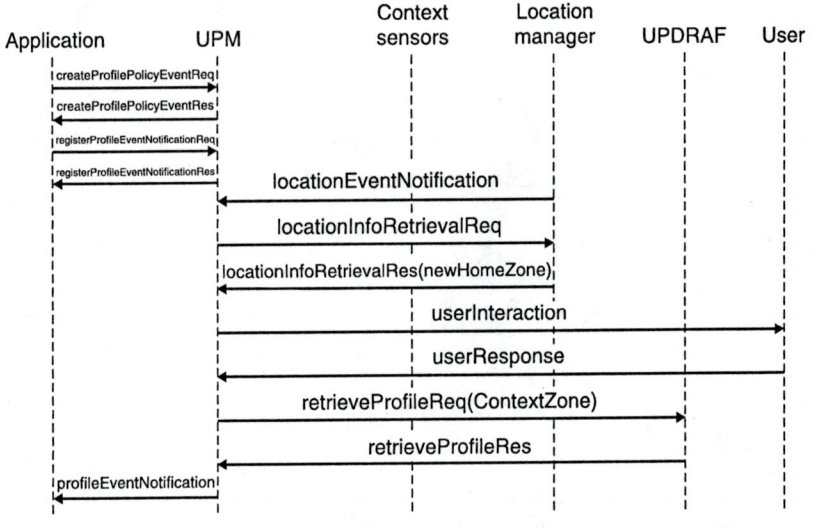

Figure 30.13 Interactions for profiling-dependent event notifications.

- The application accesses the API provided by the UPM to create a profiling-dependent policy event related to a specific subscriber. The specific event reflects the interest of that application in receiving notifications whenever some change in the current user preferences of the subscriber occurs.
- The UPM authenticates the application and checks its authorization for the requested operation.
- The application registers itself to receive notifications related to the aforementioned event.
- The UPM receives an event from the location manager of the architecture indicating that the specified subscriber has entered a new home zone. It is assumed that registration of the UPM for receiving such events from the location manager has occurred.
- The UPM contacts the location manager of the architecture to retrieve the current home zone of the user. This step can be skipped if the location event received includes the new home zone of the subscriber.
- Because the UPM has not received notifications from the associated context sensors for changes in the ME or the serving RAT of the subscriber, it assumes that only the geographical location of the user has changed. The UPM identifies the new context zone of the user and, based on that information, crosses the context-zone-dependent structure of the user profile to identify the available profiles of the user in the new context zone.

- The UPM interacts with the user to request him to identify among the service customization profiles available in the new context zone the one he wishes to activate.
- The user selects and activates the desired profile.
- The UPM updates the user profiling pointer to point to the newly activated profile of the user and locates the UPDR that stores the new profile components. The UPM then accesses the API of the identified UPDR to request the specified data. The UPM includes the context zone and data reference in the request.
- The UPDRAF searches the context-zone-dependent structure of the stored profiling information to retrieve the requested data and responds to the UPM.
- The UPM notifies the requested application that a change has occurred. The notification to the application can be a simple announcement of a change occurring in the user preferences, unless the application has requested to also receive the new profiling data along with the notification (illustrated in Figure 30.13). In the first case, the application should again contact the UPM to receive the new data (see first example above). In the latter case, the UPM should contact the UPDRAF of the appropriate UPDR to retrieve the profiling data associated with the new context zone of the user before responding to the application.

Conclusion

This chapter has focused on location and context awareness in service provisioning and has proposed a framework that allows for efficient management of location and context information. Furthermore, a more flexible and innovative model for user profiling was introduced. This innovation is based on the enrichment of common user profiling architectures to include location and other contextual attributes so enhanced adaptability and personalization can be achieved. For each context instance, an associated user profile instance is created and service provisioning is adapted to the user profile instance that better applies to the current context. The generic model, the structure, and the content of that context-sensitive user profile, along with some implementation issues, were also discussed.

References

[1] Alonistioti, N., Houssos, N., and Panagiotakis, S., A framework for reconfigurable provisioning of services in mobile networks, presented at the Int. Symp. on Communications Theory and Applications (ISCTA'01), Ambleside, Cumbria, U.K., July, 2001.

[2] 3rd Generation Partnership Project (3GPP), TS 23.240: 3GPP Generic User Profile (GUP); Architecture (Stage 2), version 6.2.0, 2003-12.

[3] Extensible Markup Language (XML), http://www.w3.org/XML.

[4] Web Services Description Language (WSDL) 1.1, http://www.w3.org/TR/wsdl.

[5] Synchronized Multimedia Integration Language (SMIL) 2.0 specification, http://www.w3.org/TR/smil20/.

[6] Web Ontology Language (OWL), http://www.w3.org/2004/OWL/.

[7] User Agent Profile (UAProf) specification, http://www.wapforum.org/what/technical.html.

[8] Parlay Group, Parlay API specification, http://www.parlay.org/specs/index.asp.

[9] 3rd Generation Partnership Project (3GPP), TS 29.198: Open Service Access (OSA); Application Programming Interface (API), Part 1-12, version 5.0.0, 2002-06.

[10] Keijzer, J., Tait, D., and Goedman, R., JAIN: a new approach to services in communication networks, *IEEE Commun. Mag.*, 38(1), 94–99, 2000.

[11] Kovacs, E., Rohrle, K., and Schiemann, B., Adaptive mobile access to context-aware services, in *Proc. of the 1st Int. Symp. on Agent Systems and Applications and 3rd Int. Symp. on Mobile Agents*, Palm Springs, CA, October 3–6, 1999, pp. 190–201.

[12] Panagiotakis, S. and Alonistioti, A., Intelligent service mediation for supporting advanced location and mobility aware service provisioning in reconfigurable mobile networks, *IEEE Wireless Commun. Mag.*, 9(5), 28–38, 2002.

[13] 3rd Generation Partnership Project (3GPP), TS 23.127: Service Aspects; The Virtual Home Environment, version 5.2.0, 2002-06.

[14] 3rd Generation Partnership Project (3GPP), TS 22.240: Service Requirement for the 3GPP Generic User Profile (GUP); Stage 1, version 6.4.0, 2004-09.

[15] Alonistioti, N., Panagiotakis, S., and Kaloxylos, A., A framework for dynamic and context-aware composition of adaptable mobile services, in *Proc. of 3rd Int. Conf. on Computer Science, Software Engineering, Information Technology, e-Business, and Applications (CSITeA 2004)*, Cairo, Egypt, December, 2004.

[16] 3rd Generation Partnership Project (3GPP), TS 23.141: Presence Service; Architecture and Functional Description, version 6.7.0, 2004-09.

[17] 3rd Generation Partnership Project (3GPP), TS 23.271: Functional Stage 2 Description of LCS, version 6.10.0, 2004-12.

[18] 3rd Generation Partnership Project (3GPP), TR 23.871: Enhanced Support for User Privacy in Location Services, version 5.0.0, 2002-07.

Chapter 31

Location-Dependent Database Access

Faïza Najjar, Sean Kelley,
and Margaret H. Dunham

CONTENTS

Introduction

In a mobile and wireless environment, efficient and consistent data access is a challenging research area because of the weak connectivity and resource constraints. The mobile data access strategies can be essentially distinguished by delivery modes. The modes for server delivery can be described as client pull, server push, or hybrid:

- *Client pull* (sometimes called *on-demand access mode*) — A mobile client first submits a query via the uplink channel and then "pulls" data, from the server through a wireless network (the downlink channel), in the same manner as in a traditional client–server system.
- *Server push* (also called *broadcast* [15]) — The mobile client receives information as a result of his or her whereabouts without having to actively submit a query. The information sent to the mobile client may be on a public wireless channel (e.g., a welcome message when entering a new town) or may be subscription based (e.g., alert system).
- *Hybrid delivery* integrates both server push and client pull delivery.

Location-dependent data access assumes that a mobile user queries data where the value is dependent on the user's location. A typical query of this type is: "Where is the closest restaurant?" This query may be stated explicitly (in the client pull environment) or implicitly (in the server push environment). In a broadcast setting, a content provider can broadcast restaurant information for a local area near the broadcast station. A *location-dependent query* (LDQ) is a mobile query where the result of the query depends on the location of the user making the query [4,23,26]. This definition can be expanded to include a broadcast environment. Here, the content of the broadcast is based on the location of the broadcast station; thus, as a mobile user roams, the result of the user's implicit query (i.e., the data that he receives) changes.

We conclude this section by reviewing some terms commonly used within location-dependent data and queries in wireless environments. We then briefly overview temporal and spatial database concepts, as these are important to understanding location-dependent database access. Subsequent sections of this chapter examine architectural issues related to

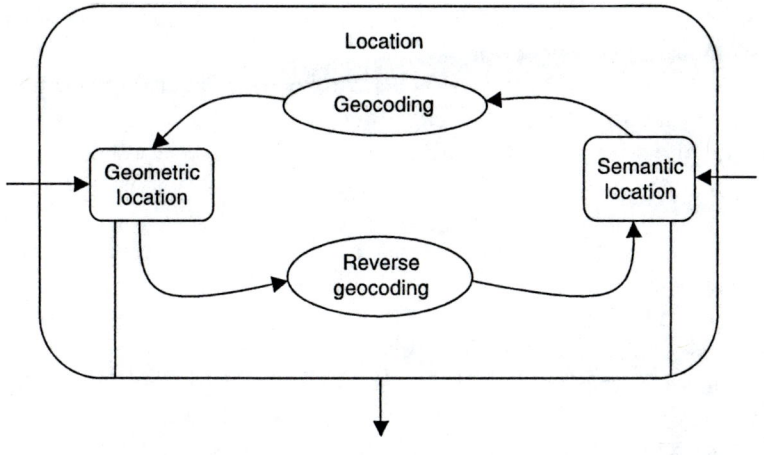

Figure 31.1   Location model architecture.

location-dependent data access, an overview of location-dependent queries, an overview of moving object databases and queries, location modeling and translation, and nearest-neighbor queries and indexing.

Terminology

As long as people move across the Earth's surface, the need to know their
current location anywhere, anytime has become a very important constraint
in mobile databases. The term *location* refers to the position of a point
relative to a geometric subdivision or a given set of disjoint geometric
objects [1,16]. Before studying queries and data access approaches in mobile
databases, it is important to emphasize some fundamental properties of
location data.

With regard to location models, we distinguish two kinds of location:
geometric (or geographic) and *semantic* (or symbolic) (see Figure 31.1).
The available mechanisms for identifying the *geometric* location can be
divided into two basic classes:

- In the World Geodic System 1984 (WGS84), a location is three
  dimensional and unique and has three coordinates: latitude, longitude, and altitude. These coordinates can be easily provided by
  a satellite-based positioning system, such as the widely known
  global position system (GPS).
- A location can also be specified as a set of coordinates defining
  the bounding geometric shape (such as polygon) of an area.

Geometric location can be considered in heterogeneous systems and is commonly used in the outdoor domain [14]; however, mobile users are often interested in the "meaning" of a location than its geometric coordinates.

Semantic location is the logical representation of the real-world entities describing the location space. Entities can be cities, street address, Zip Codes, or system-defined elements such as cell IDs in cellular phone networks, infrared beacons, or wireless local area network (WLAN) access point IDs in the indoor domain. These last entities are uniquely identifiable by a hierarchical naming system such as location trees.

Finally, both geometric and symbolic locations are present and have to be considered for LDQs. The process of converting a given symbolic location to (x,y) coordinates (latitude and longitude) is called *geocoding* (see Figure 31.1) [22]. The opposite function of geocoding is *reverse geocoding*, which is the process of deriving the semantic location of a specified longitude/latitude coordinate.

Precision depends on the measurement errors in geometric coordinates and on the completeness and accuracy of address names (normalized). *Scope* is the geometric area of potential coordinates (it often takes the shape of polygon). A valid scope (also called a *data region*) of a spatial data was introduced by Zheng et al. [15]; it defines the region within which the result is valid with respect to the query. In addition to the location, more spatial information sometimes is required, such as orientation or speed.

The properties of location can affect the processing of queries significantly when users change their position in a mobile and wireless environment.

Location-Dependent Data Versus Spatial Data Versus Temporal Data

Space and time are two powerful forces that have mesmerized scientists, theologians, astrologists, and philosophers for 2500 years. Einstein's theory of general relativity described the "effect of gravitation on the shape of space and the flow of time" [11]. Einstein's work led to the development of what is referred to as *spacetime*: "a universe in which space and time are woven together into a single fabric" [11]. The relationship between space and time will be a key theme throughout this chapter because this relationship serves as a foundation for queries stated in a mobile computing environment.

The adjective *spatial* is defined as "relating to, involving, or having the nature of space." A *spatial database* (SDB) offers spatial data types in its data model and query language and supports spatial data types

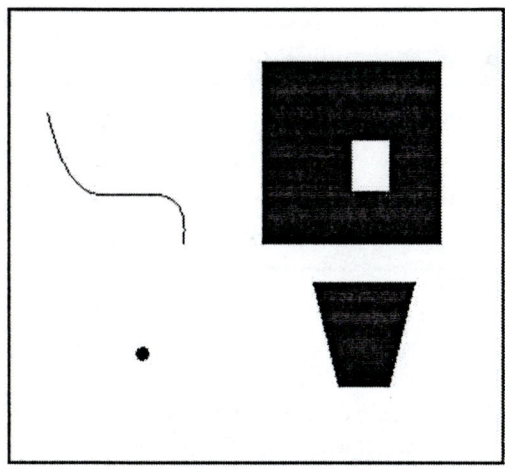

Figure 31.2 Point, line, region, and region with gaps.

in its implementation, providing at least spatial indexing and spatial join methods. Spatial data is oftentimes geographic or geometric in nature, and the space of interest is dependent on the problem being solved [10]. The problem space may be, for example, the surface of Mars (geographic space), the layout for a very-large-scale integration (VLSI) design (manmade space), or the structure of a DNA sample taken from an extinct species (subatomic space) [10]. The complexity of the space being studied can often become easier to understand if it is broken down into large collections of relatively simple geometric objects [10]. For example, points, lines, and regions (see Figure 31.2) are often found in spatial databases and are used to model real-world entities. A line can be used to track movement through space or create a connection between two or more points (e.g., roads, boundaries). A region can be two or three dimensional, may have holes, and can consist of many disjoint pieces [10].

The position of these objects may be viewed at different levels of granularity, and the geometric representation of an object may change as the granularity changes; for example, a city may be viewed as a line or region, depending on the level of detail required by the model [13]. In addition, spatial objects often have descriptive properties that may or may not change, depending on the position of an object within space. Furthermore, these objects may be associated with one another using spatial relationships. These relationships can be subdivided into three categories: topological, directional, and metric (see Table 31.1) [10].

Table 31.1   Spatial Relationship Types

Relationship Type	Examples
Topological	Inside, disjoint, overlaps
Directional	Above, below, north of, south of
Metric	Distance < 30, distance = 10

A spatial query language would then contain special operators that would facilitate the inspection and manipulation of these objects. Specifically, one might use the ADJACENT operator to identify two regions that touch each other; likewise, it may be necessary to use an ABOVE or BELOW operator to determine the direction in which one point lies with respect to another point. Various index techniques are used to support the querying of spatial objects in an effort to decrease the access time. Spatial indexes are composed of keys that allow for a less complex representation of the underlying objects. This, in turn, decreases the amount of I/O or processing needs for spatial queries.

Temporal databases (TDBs) allow for the management of time-varying data [12]. Most applications are temporal in nature with different degrees of support for formal temporal semantics. These applications include inventory management, scientific analysis, asset management, and budgeting/forecasting [12]. The two most important concepts associated with TDBs are *valid time* and *transaction time*. Valid time is defined as a set of "collected times — possibly spanning the past, present and future — when some fact is true in the modeled reality" [12]. Frequently, valid time is not recorded in a database because it may not be known or it may not be relevant to the application [12]. *Transaction time* can be defined as the period of time that a fact is current within a database. Transaction time is associated with a window of time, beginning with the time that a record is inserted into the database and ending with the time that a record is deleted from the database. It is important to note that deletion might be logical, and a time stamp is sufficient to invalidate a fact. Invalidation implies that the fact will no longer be associated with the current state of the database. As a result, transaction time allows us to navigate between the various states through which a database passes. Finally, it is important to note that transaction time and valid time might be the same, and much of the research around temporal databases involves understanding and modeling their differences.

One additional temporal concept that is important to mention is the concept of *current time* or *now*. Some of the peculiarities of *now* include

Table 31.2   Bitemporal Conceptual Data Model

Customer Name	DVD ID	Transaction Time, Valid Time
John Smith	1	{(10/2,10/2), (10/2,10/3), (10/3,10/2), (10/3,10/3), (10/3, 10/4)}
Jim Kelly	2	{(10/5,10/5), (10/6,10/5), (10/6,10/6), (10/7,10/5), (10/7,10/6), (10/7,10/7)}

Source: Jensen, C.S., An Introduction to Temporal Database Research, Ph.D. dissertation, University of Arizona, Tucson, 2000. With permission.

the fact that it is always moving forward and that it serves as a moving boundary dividing the past and the future [12]. The idea of current time makes it challenging to integrate temporal techniques with techniques from other research areas. In addition, tracking current time often requires unique data management and access strategies.

The complexity and inefficiencies of *ad hoc* temporal data management have led to the development of a variety of formal data models [12]. For the purposes of this chapter, only two of these models will be discussed. The first model is the *Bitemporal Conceptual Data Model* (BCDM). This model involves associating each tuple with a series of transaction time/valid time pairs. This series of pairs is stored in a single attribute, allowing the full history of a fact to be stored in one tuple [12]. Table 31.2 illustrates the BCDM model using a relation that contains data pertaining to DVD rentals. The downsides to this approach include the varying lengths and large size of each tuple. In addition, storing a string of time stamps in a single attribute makes querying and displaying this information unintuitive to the end user.

The second strategy involves storing each time-stamp pair in separate tuples; as a result, the lifetime of a fact is spread across multiple records (see Table 31.3). Although this fixed-length strategy may be easier to store and manipulate, queries spanning the entire lifetime of a fact may require additional logic. This strategy also leverages the idea of "now," which decreases the amount of updates necessary to maintain a complete picture [12].

Many existing query languages (e.g., SQL) can be used to manipulate temporal data, but the logic required can be overly complex. As a result, extensions for existing data manipulation languages (e.g., TSQL2) and hundreds of temporal languages have been developed to allow for the natural manipulation of complex temporal ideas [12].

Table 31.3 Fixed-Length Strategy

Customer Name	DVD ID	Transaction Time	Valid Time
John Smith	1	10/2	10/2
John Smith	1	10/2	10/3
John Smith	1	10/3	10/2
John Smith	1	10/3	10/3
John Smith	1	10/3	10/4
Jim Kelly	2	10/5	10/5
Jim Kelly	2	10/6	10/5
Jim Kelly	2	10/6	10/6
Jim Kelly	2	10/7	10/5
Jim Kelly	2	10/7	10/6
Jim Kelly	2	10/7	10/7

Source: Jensen, C.S., An Introduction to Temporal Database Research, Ph.D. dissertation, University of Arizona, Tucson, 2000. With permission.

Architecture

Global Position System

The global position system (GPS) is the most prominent example of satellite navigation systems [22]. Many location-based services use GPS to determine the current location. The main advantages of GPS are its accuracy and high level of coverage, but it fails in indoor environments. Today, 24 satellites are moving on six different orbits with four satellites per orbit [22]. They are all one-way communication and they are sending signals continuously. The architecture of GPS can be divided into three segments as follows:

- *User segment* (GPS receivers, which can be plug-in cards or separate devices with a serial interface connection)
- *Space segment* (defined by the satellites)
- *Control segment* (administration of satellites as well as the corrections)

Mobile users who want to know their current location can use GPS signals which are now free of charge, and the receivers are not expensive. GPS services are classified into two types:

- *Precise positioning service* (PPS), which is not accessible by civilian users but only by the military, allows positioning with a precision of about 3 meters.
- *Standard positioning service* (SPS) has been available for civilian users since 2000 and provides the current location with a precision of 25 meters.

LDS Middleware Architecture

The middleware architecture discussed in this section comes primarily from Seydim et al. [24], whose work is typical of that done in this area. We use the term *location-dependent services* (LDSs) to refer to the software that is responsible for processing location-dependent data access. Any LDS design must at least support the following basic functions:

- Bind location to query.
- Determine content provider(s) for processing the query.
- Translate query location to location used by data.

We assume that the LDS software uses a *location service* (LCS) to determine the location for the query. As we assume that the LDS is independent of both the content provider and the wireless provider, the LDS must also determine where to process the query. This could be at multiple content provider locations. The LDS must then translate the query into a query format understood by each content provide. We make no assumptions about the type of data and system used by any content provider.

Three architectural approaches can be used to support LDS applications:

- *Content side* — Here, all LDS support functions are provided by the content provider with the support of an LCS module. The content provider is responsible for binding the mobile user's query to a location and then for processing the query itself. Location is assumed to be the current position of the mobile user. The LCS estimates the mobile user's most accurate position by using the information stored in the network or the device itself. If the granularity provided by the LCS is not compatible with the granularity of the stored data, then the content provider may have to customize its database accordingly; otherwise, the content provider can ask for the location of the mobile user at a certain granularity. This approach is simple and suitable for thin clients with scarce resources, such as mobile phones and personal digital assistants (PDAs). Ericsson's Mobile Positioning System (MPS) is an example of this type of architecture [5].

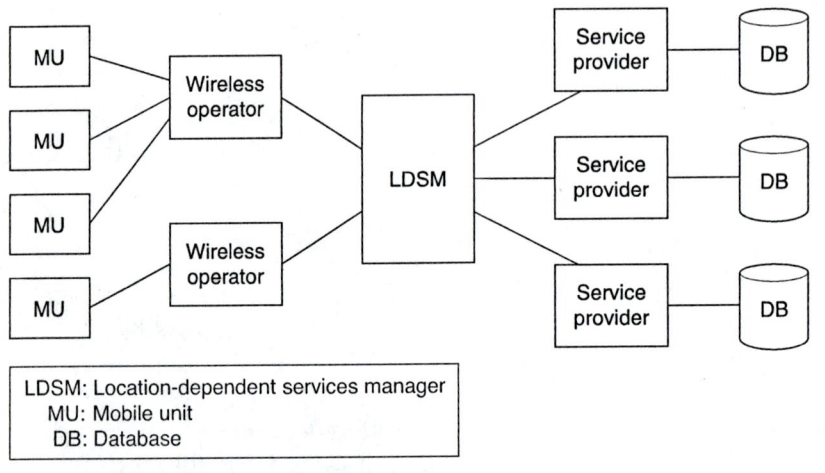

Figure 31.3   LDSM Middleware.

- *Wireless side* — Here, support for LDS applications is moved to the wireless operator side. The wireless operator provides all the functionality to the mobile user in a well-defined and limited way. The mobile user does not know anything about the content provider; all he sees is the menu provided by his wireless operator.
- *Middleware*— This approach assumes that a special software agent, the *location-dependent services manager* (LDSM), sits between the wireless operator and the content (service) provider. This approach is shown in Figure 31.3. This software agent performs all LDS functions, with the help of LCS software, independent of both the wireless operator and the content provider. Currently, SignalSoft Corp. [28], and Mobilaris [18] support this type of architecture; however, these implementations are limited in the type of location binding and translation supported.

Although the first two approaches are relatively simple and straightforward, many problems are associated with them. First of all, they are not flexible enough to support complicated LDQ requests. Each request must be well defined and sent to a specific content provider directly. Although this may be true for current LDS applications, we envision future applications that are composed of queries to be sent to multiple content providers. In addition, it may not be known *a priori* which content provider is to receive and process the query. Indeed, the content provider choice may itself be location based; for example, one provider in the United States may perform yellow pages applications while another does the same in Europe. Thus, the same query could be sent to two different providers. The matching of the query to a provider should be dynamic, not prewired. Users should be able to use any LDS from any service provider, not simply the ones presented by default from the wireless network.

Besides being independent of the underlying cellular technology, the middleware approach provides a more flexible and transparent framework for LDS applications. Different location identification software or LCSs can be used in the architecture to locate the mobile user. Unlimited content options can be provided by this approach, which allows access to many different localized information services. We envision LDS application software providers competing with each other for users. These LDS providers will use different functionalities and approaches to implement location-dependent applications. Users from different wireless operators may subscribe to the same LDS services.

The middleware design facilitates the implementation of a flexible LDS support service that could work with multiple wireless operators and content providers. In addition, very complex location binding can be supported. LDQs of the future will have to have the query bound to locations other than the current position of the *mobile unit* (MU). Some types of queries must be repeatedly requested; for example, a user who wishes to spend the night only at a Brand A hotel could issue a query that is requested periodically to find a Brand A hotel. This sophisticated type of LDQ can be supported by middleware software but is not easily supported by any of the other two approaches. Other types of queries may be fragmented and sent to different content providers. These subquery results could then be merged together and returned to the user. The architecture could also use intelligence for caching results for frequent disconnection cases or use access patterns for prefetching data and for efficient use of resources.

Overview of Location-Dependent Queries

In this section we provide the definition of location-dependent data/query and then classify the queries. The term *location-dependent data* (LDD) can be defined as the data whose value is dependent on some reference location, which in most cases is the current position of the query's issuer. A *data item* refers to one type of LDD (e.g., restaurants) and usually has different instances; thus, a *data instance* is an answer to the query. Before a mobile user can access information, it is important to consider the location model, in which location information specified in the user's query is either explicit or implied in the query.

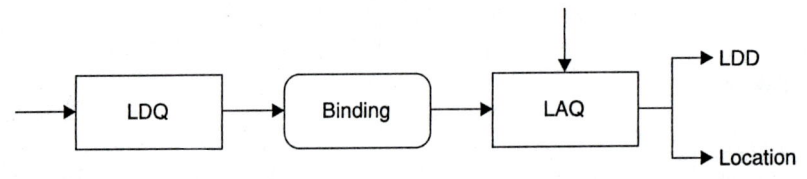

Figure 31.4 Location-dependent query (LDQ) versus location-aware query (LAQ).

The query-retrieving LDD is referred to as a *location-dependent query* (LDQ), and the result set of the query changes depending on the location of the query issuer. The location information is always hidden and implied by "current location or here" (e.g., "find the nearest gas station").

When the user's location is known (e.g., by GPS), the LDQ can be converted into a *location-aware query* (LAQ) [23] with an explicit indication of this special location attribute in the query. The process of obtaining the location information of the query issuer is called *location binding* (see Figure 31.4). Examples of location binding include: "Find the nearest hotel of position $x°$ latitude and $y°$ longitude" or "Find all Chinese restaurants in Dallas." So, LAQ retrieves the same query results independent of the query issuer. Sometimes, LAQs also require location binding if the query results depend on the location of a mobile client — for example, "Find friend application" [22] or "Find a mall close to my friend."

LDQ Versus Point Location Queries

A formal definition of LDQ [32] is as follows:

> **Definition 1.** Given a set of data instances $P = \{P_1, P_2, ..., P_N\}$ and the corresponding set of valid scopes $R = \{R_1, R_2, R_N\}$, an LDQ issued at location q returns the data instance P_j from P if and only if q is located in the region R_j where $1 \le j \le N$.

One of the most fundamental query problems is the *point location problem*, in which a subdivision of space into disjoint regions is given, and the problem consists of identifying which region contains a given query point. A more technical definition is given as follows:

> **Definition 2.** Given a partition of data regions $R = \{R_1, R_2, ..., R_N\}$ and a point query q, a (planar) point location query (PLQ) returns the data region R_i from R that covers q.

For example, "Where am I?" is a PLQ; thus, a LDQ is reduced to a PLQ when the data region covering the point query maintains a pointer pointing to its associated data instance.

Classification of Queries

There are various query classifications in the literature [9,15,26]. In Seydim et al. [24], the authors provide a formal query classification for mobile database queries. In this subsection, we will summarize these previous works (see Table 31.4). At the highest level, queries may contain location information (but no movement) or may be mobile queries, or neither of these. Queries with no movement (but location) can request data from local or non-local locations. Traditional queries involve no locations. Mobile queries involving location can be divided into two types: *stationary-location-related queries* (SLRQs), and *mobile objects database queries* (MODQs). We can also identify the *continuous queries* (CQs) inside an MODQ; for example, a client in a moving car who would like to receive continuously updated information to find the least expensive hotel can submit the following query: "Find the room rates of all the hotels within 1 mile from me."

Overview of Moving Object Queries and Databases

A *moving object database* (MOD) is defined as a type of "spatiotemporal database supporting spatial objects with continuously changing position" [7]. This definition permits us to think of MODs as a kind of *spatiotemporal database* (STDB); however, spatiotemporal databases are not typically associated with continuous movement. Adding continuous movement to an application usually requires additional modeling, access, and update strategies.

A database must have a couple of important characteristics to be classified as a MOD. These databases must "store spatial information whose position and extent changes over time" [7]. This is a broad definition because it includes two types of spatial objects: (1) those that are stationary or move slowly with respect to time, and (2) those that move continuously with respect to time. The first category includes objects such as states, counties, lakes, and roads. The second category encompasses objects such as automobiles, planes, and people [7]. The first category is generally associated with spatiotemporal applications and databases. MODs are responsible for storing and manipulating both types of objects described above, and it is the second class of objects that makes this technology so unique.

Table 31.4 Classification of Queries in Mobile Databases

Criteria	Query Type	Query Subtype	Location Model	Query Examples
Space: access to location information, no movement	Local	NNQ	Explicit/implicit	"Find the nearest train station."
		Simple versus general	Explicit/implicit	"Where is the Chinese restaurant with respect to my hotel?"
	Non-local	Geographically clustered	Explicit/implicit	"What's the weather in Tunis?"
		Geographically dispersed	Explicit/implicit	"List all hotels with a room rate below $100."
Space and time: mobility of the issuer and/or databases	SLRQ	LDQ/NNQ	Implicit	"Where is the closest airport?"
		LAQ	Explicit	"Identify all ambulances that are within 2 miles of my current location."
		PLQ	Implicit/explicit	"Where am I located?"
	MODQ/CQ	SLRQ + (time, speed, direction)	—	"Find the hotels that I will reach within 5 minutes"
No space or time	Traditional	—	Non-location	"List all the actors' names in the movie 'National Treasure'."

Note: LAQ, location-aware query; LDQ, location-dependent query; NNQ, nearest-neighbor query; PLQ, point location query; SLRQ, stationary location-related query.

Moving object databases store the position of objects in space. Some of those objects move through space and others are stationary; for example, an application used to manage a fleet of delivery trucks would focus on tracking the delivery trucks (moving objects) as they travel through the streets (stationary objects) of a particular city (stationary object). As time moves forward, an object may change its position within space, affecting the spatial relationship between that object and other objects found within the database [6]. A moving object has a starting position, which is a point in space denoting the beginning of a route. That starting position can be coupled with a start time. In addition, a line through space is used to represent the route of an object. The end position of a route may or may not be known beforehand. Other characteristics such as trajectory, velocity, and acceleration may also be associated with a moving object [27].

Many operators are commonly used to manipulate and understand moving objects. The operator IN can be used to determine whether or not an object is within a region at a given point in time. It may also be necessary to determine if an object is *entering* or *exiting* a region. Furthermore, it might be important to identify whether two objects are moving *away* or *toward* one another. Similarly, two objects moving toward each other may *collide*. *Catching up* may be used to describe two objects that are moving in the same direction with the distance between those objects decreasing. *Opposite direction* would imply that two objects are moving away from each other. *Meet* requires that the faster of two objects must have a negative acceleration and the slower object must have a positive acceleration, or both [25].

Because the entities being tracked in MODs are continuously changing with regard to space and time, the location of an object is intrinsically uncertain, and this idea is generally referred to as *location uncertainty*. Moreover, regardless of how frequently the current position of a moving object is updated within the database, the database location will never be the same as the actual location. As a result, it may be necessary to qualify or quantify that uncertainty using "must" or "may" semantics. Specifically, it might be necessary to differentiate between all objects that *must* satisfy a query predicate and all objects that *may* satisfy a query predicate. Furthermore, a probability could be used to identify the likelihood that a result *may* satisfy a query [28].

Queries against MODs can fall into a number of categories. First, *range queries* can be described as requesting objects that fall within a given spatio-temporal range; for example, one might request the identification of all vehicles within 5 miles (spatial predicate) of a bank robbery between 2 and 3 p.m. (temporal predicate) [18]. Next, *nearest-neighbor* (kNN) queries are interested in objects that are within a defined proximity to some point in space. The nearest-neighbor predicate is typically applied to the spatial dimension or the temporal dimension, but not to both. The kNN predicate is then coupled with a range predicate to be applied to the other dimension. To clarify, a temporal kNN query might be something like: "Retrieve the first ten vehicles that were within 100 yards [spatial range] from the scene of the accident ordered based on the difference between the time the accident occurred and the time the vehicle crossed the spatial region [temporal kNN predicate]" [18]. Generally, temporal kNN queries are interested in the most recent past or some point in the future but rarely in both, allowing the query to concentrate on one portion of the time dimension. A spatial kNN query might ask: "Retrieve the five ambulances that were nearest to the location of the accident [spatial kNN predicate] between 4 and 5 p.m. [temporal range]" [18]. *Join* queries are traditionally used to detect or predict the collision or the overlap of either two moving objects or a moving object and a stationary object. For example, join queries can be used by flight controllers to estimate the effects of a flight pattern reconfiguration during an emergency in an effort to avoid any in-flight collisions.

Several nuances that are typically associated with querying MODs are worth mentioning. First, one may choose to limit the perspective of a query with regard to the time dimension, and it may be necessary to focus only on the historical positions of an object. In contrast, it is also common to focus on the current and future positions of an object which requires associating that object with a function of time. That function could take velocity and trajectory into consideration when estimating the future position of an object. It may also be feasible to obtain a "reservation" from each object that specifies the spatial destination. This greatly reduces the complexity of predicting the future positions of objects. Splitting queries along these lines is often necessary because the techniques regarding indexing and optimization can be different. In addition, because the locations of moving objects are continuously changing, these locations are inherently uncertain [28]; for example, one may wish to "retrieve the friendly helicopters that are expected (according to the current speed and direction information) to enter a stated region in the next hour." It might be necessary to apply "must" or "maybe" semantics to this query. The latter would be coupled with some uncertainty bound or probability dictating the acceptable threshold for incorrect predictions. The reply to a query using an uncertainty bound might be "M is on Route 968 at location (x,y) with an error or deviation of at most 2 miles." Furthermore, the database engine might commit to sending a location update when some deviation is reached [28].

Moving object databases and applications are being used extensively in the field of *mobile workforce management*. Location tracking is combined with predetermined route information to manage mobile devices such as planes, trains, delivery trucks, and taxis. Call centers can use these applications to optimize resource utilization, determine delivery times, and react to extenuating circumstances such as bad weather or traffic congestion [27]. As an example, a database used to track the location of taxis might have to know all the free cabs within a one-mile radius of a given customer. Likewise, a delivery service might have to alter a route for a special delivery and would do so by first determining all routes that fall within a certain distance of that address.

Another application for moving object databases is called *location-aware content delivery*. The origins of this technology lie in a field called *context-aware computing*. These applications use a mobile device such as a cellphone to provide location information. That information can then be combined with contextual information such as the time of day, current weather conditions, nearby locations, and the mobile users' current activities [2]. Ultimately, content is tailored based on that contextual information and sent to that mobile device. This content might include coupons for local stores, information on surrounding tourist attractions, and restaurant reviews [2].

Finally, moving object technology is often used in the digital battlefield. Military leaders have realized that the key to winning a war is to create a fast and accurate information delivery system [19]. The aim of battlefield analysis is to collect and analyze information about such things as enemy location and movement, characteristics of the terrain, and weather conditions [19]. These applications leverage distributed computing solutions that seek to track mobile objects in an environment where network bandwidth is scarce [3]. Data may come from a wide variety of sources including satellites, reconnaissance drones, and mobile devices worn by ground troops. The application should integrate that data into a unified view and allow key decision makers to adjust their strategies accordingly; for example, it may be necessary to determine all friendly aircraft that can be expected to enter a region in the near future [27]. In the event of an emergency, this type of information would allow decision makers to determine if it is best to pull out due to a lack of adequate air support or stay in place because assistance will arrive in a timely manner.

Location Modeling and Translation

As indicated earlier, any LDQ must involve either an explicit or implicit location; however, a location can be viewed in many ways: latitude and longitude, street address, Zip Code, etc. An even bigger problem is that different system entities involved in processing the LDQ may have different views. Different content providers could store data at different location granularities. The LCS could bind the location to yet a third. This difference

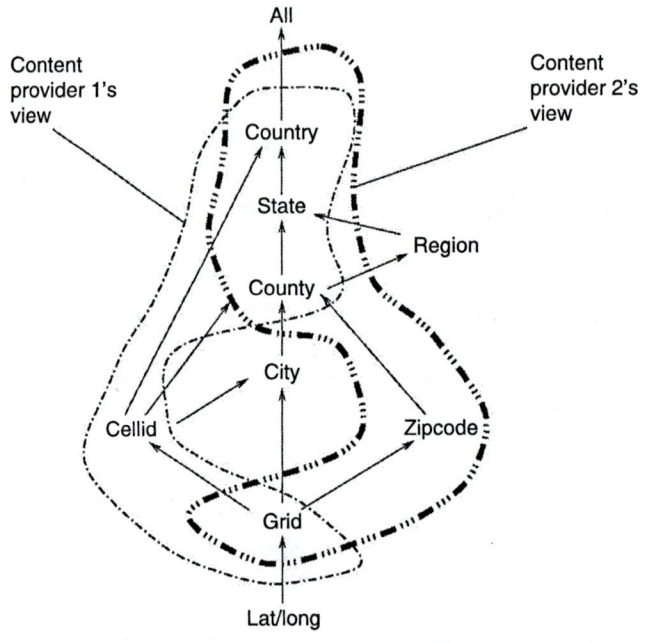

Figure 31.5 Concept hierarchy to support leveling. (From Seydim, A.Y. et al., in Proc. of the 2nd ACM Int. Workshop on Data Engineering for Wireless and Mobile Access (MobiDe'01), Santa Barbara, CA, May, 2001, pp. 47–54. With permission.)

between location granularities is referred to as the *location granularity mismatch problem*. A layered translation approach to solve this problem, *location leveling* (LL) has been proposed [23]. LL is an extension of the geocoding/reverse geocoding concept introduced earlier. It is a software technique used to convert any location to which the mobile unit is bound to a location provided by the content provider. The LL technique assumes that concept hierarchies are provided by all content providers which indicate the relationships between their data and other data. Some of the translation approaches may be algorithm based, and others are table based. Figure 31.5 illustrates one such concept hierarchy.

Here, two different content provides have two different views of the data. It is assumed that the content providers provide translation algorithms between any two locations for which an edge is present. Thus, if a user submits a query that is bound to a Zip Code to content provider 2, where the data in the corresponding database is at the county granularity, then a translation from Zip Code to county would be required to determine which counties satisfy the proposed query.

Nearest-Neighbor Queries Through Point Location and Indexing

Nearest-neighbor (NN) searching is an important problem in a variety of applications, including multimedia databases, document retrieval, knowledge and data mining, pattern recognition and classification, machine learning, and statistics. The NN search (e.g., finding the nearest gas station) can be seen as an LDQ problem when the solution space is precomputed (e.g., by Voronoi diagram). So, to determine the closest site, it suffices to first compute the subdivision induced by the Voronoi diagram and then generate a point location data structure for the Voronoi diagram. In this way, the *nearest-neighbor queries* (NNQs) are reduced to PLQs. We begin this section by providing some preliminaries about Voronoi diagrams and then explore the use of indexes to facilitate efficient evaluation of LDQs in mobile and wireless environments.

Voronoi Diagram

The concept of the Voronoi diagram is used extensively in a variety of applications, including robotics, knowledge discovery and data mining, classification, multimedia databases, document retrieval, and statistics, among many other fields. The Voronoi diagram is a versatile geometric structure. Given a set of *n* points (referred to as *sites*) in the plane, a fairly intuitive definition of a Voronoi diagram is a partitioning of the space into *n* regions (closed and convex, or unbounded) according to the *nearest-neighbor rule*: Each site is assigned to the region to which it is closest (see Figure 31.6). More formally, the Voronoi diagram [17] can be defined as follows:

> Given a set P of n points ($n \geq 3$) in a plane that are in general positions (that is, no three of them are colinear and no four points are cocircular), we associate all locations in the space with the closest members of the point set with respect to the Euclidean distance.

The set of points that are closest to a particular site p_i forms the so-called *Voronoi cell* (see Figure 31.7.a) and is denoted by $\upsilon(p_i)$. Thus, mathematically we have:

$$\upsilon(p_i) = \{p \in I\!R^2 \mid \text{dist}(p,p_i) < \text{dist}(p,p_j) \text{ for all } j \neq i\}$$

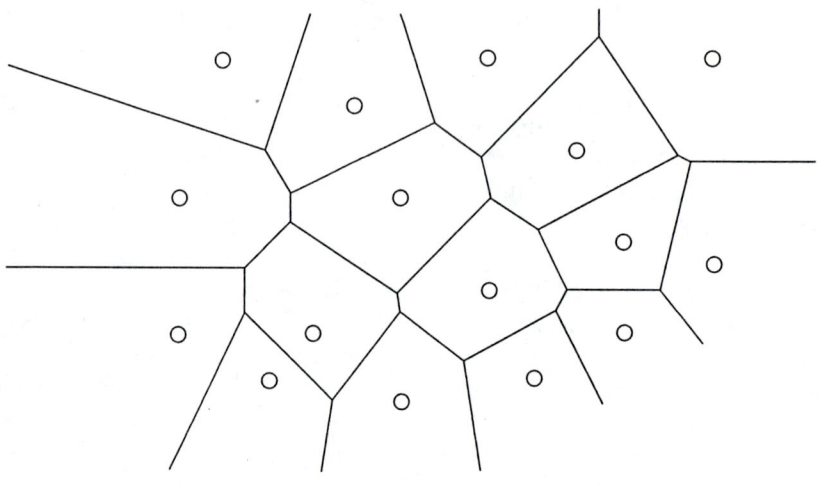

Figure 31.6   The Voronoi diagram of a set of points.

The boundary of a Voronoi diagram consists of Voronoi edges (i.e., line segments, half lines), which are equidistant from two point sites, and Voronoi vertices, which are equidistant from at least three sites (see Figure 31.7.b). Because the Voronoi diagram, $Vor(P)$, is viewed as a planar graph, every vertex in a $Vor(P)$ has degree 3 (i.e., exactly three Voronoi edges are incident to it). The union of the boundaries of the Voronoi cells is the *Voronoi diagram*, which we refer to as $Vor(P)$. Note that adjacent Voronoi cells overlap only on their boundaries.

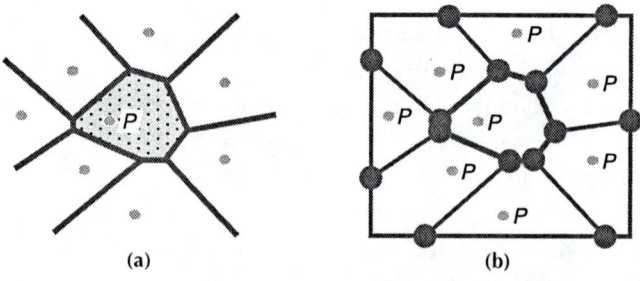

(a)   (b)

Figure 31.7   (a) A Voronoi cell of P_i (filled region); (b) a Voronoi diagram in a bounded box.

Now that we understand the structure of the Voronoi diagram, we next discuss its complexity in terms of the total number of vertices and edges [1,17]:

Property: For $n \geq 3$, the number of vertices in the Voronoi diagram of a set of n point sites is at most $2n - 5$, and the number of Voronoi edges is at most $3n - 6$.

The Voronoi diagram, then, has a linear complexity. The Voronoi diagram serves as the basis for nearest-neighbor queries. Each query is a point in the space containing N sites, and we are required to report the closest site to the query point. In such cases, it suffices to compute first the Voronoi diagram and then to generate a point location data structure for the Voronoi diagram. Consequently, the problem of LDQ can be reduced to the problem of PLQ.

Related Work in Indexing Techniques

We now explore the use of indexing on simple shapes which in general performs efficiently in mobile and wireless environments. The indexing problem of LDQ can be defined as: Given the valid scopes (Voronoi cells) of all data instances of a certain query type, how can we index them to allow efficient processing of LDQs through PLQs? The goal of PLQs is to preprocess a subdivision into a data structure that provides an optimal $O(n)$ space and $O(\log n)$ query time for answering PLQs in the plane. Several indexing techniques for PLQs exist in the literature. The two most well-known types of index techniques are object decomposition and object approximation. The former is based on the precomputed solution space and is also referred to as the *solution-based index*. Especially, for NN search, the solution space can be represented by valid scopes (e.g., Voronoi diagrams). Whereas the latter represents a simple approximation of each data region, it is also referred to as the *object-based index*, because it is built on object locations. The most commonly used index in geographical information systems (GISs) is the *minimal bounding box* (MBB) or *minimal bounding rectangle* (MBR).

In the following subsections, we briefly review the existing indexing approaches and then introduce our new index structure for LDQ. Most of the decomposition techniques are based on the principle of recursive hierarchical partition. We assume for the following subsections that a polygonal subdivision (that is, a Voronoi diagram) contains n vertices and m edges.

Kirkpatrick's Technique: Triangulation

Kirkpatrick's algorithm is based on the triangulation of the current subdivision. The construction of the index begins by building a finite sequence of triangulations $(T_0, \ldots, T_p)$, where T_0 is the initial triangulation of the original subdivision. The main idea is to recursively remove an independent set of vertices (that is, a set of mutually nonadjacent vertices) along with all the incident edges. The resulting subdivision is then retriangulated until T_p, which consists of the single triangle forming the external face of the original triangulation. If T_p is not a triangle, we compute the convex hull and triangulate the pockets between the subdivision and the convex hull. For all T_{i+1} ($1 \leq I \leq p-1$), each triangle intersects a constant number of triangles in T_i, and the number of vertices in T_{i+1} is smaller by a constant fraction than the number of vertices in T_i. A rooted directed acyclic graph is the index structure, where the root corresponds to the last triangulation T_p and the leaves represent the triangles of T_0. Figure 31.8 illustrates both the process of triangulation and the corresponding index data structure [30]. The search of the location of the point query proceeds level by level through the hierarchical *directed acyclic graph* (DAG), visiting the nodes representing the triangles that contain the query point q. Given a query point q, the searching of the point location proceeds by first locating the point inside the outer triangle (that is, the root). At any step in the search approach, one has located the query point in a triangle t on some level in the DAG. One then follows the pointers in the DAG to search all the children triangles, which were eliminated to form t. Figure 31.8 shows the unique triangle (that is, 3) containing q. The visiting nodes are in the path of 20, 18, and finally triangle 3.

Trapezoidal Map

The *trapezoidal map* or *decomposition* (sometimes also known as the *vertical decomposition*) of space S is viewed as a collection of line segments. The trapezoidal map is obtained by passing a vertical line through each endpoint p_i of each segment in S, going upward and downward until it hits another segment of S. Some of these lines will continue to infinity, because they do not hit any other line segments. We assume henceforth that:

- The entire domain is enclosed in a large bounding box to avoid infinite lines.
- The x-coordinates of the segments are all distinct to avoid degeneracy.

Thus, the process of randomized incremental construction of the trapezoidal map is described by the following steps:

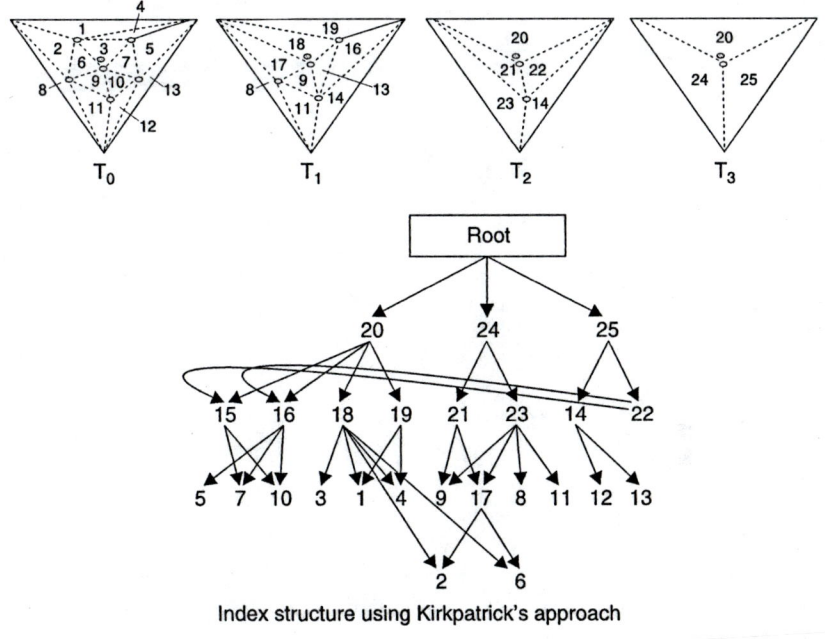

Figure 31.8 The sequence of triangulation generated by Kirkpatrick's approach and its corresponding DAG for searching.

- *Input* a set S of m planar line segments; that is, $S = (s_1, s_2, \ldots, s_m)$ in a random order.
- *Output* the trapezoidal map $T(S)$ and a search data structure D (rooted DAG):
  - For i from 1 to m, find the set $\Delta_0, \ldots, \Delta_k$, of trapezoid T properly intersected by s_i.
  - Remove the set from T and replace it with the new trapezoids that appeared after the insertion of s_i.
  - Remove the leaves from D, and create new leaves from the currently found trapezoids; that is, link internal nodes to new leaves with respect to the history of the randomized construction (explained below).

We now describe the point location data structure, which is based on a rooted DAG. Each internal node consists of two types of nodes: x-nodes and y-nodes. Each x-node, represented by circles, contains the x-coordinate x_0 of an endpoint of one of the segments, and its two children correspond to the neighbor points lying to the left and right of the x-coordinate (that is, $x = x_0$). Each y-node, represented by a hexagon,

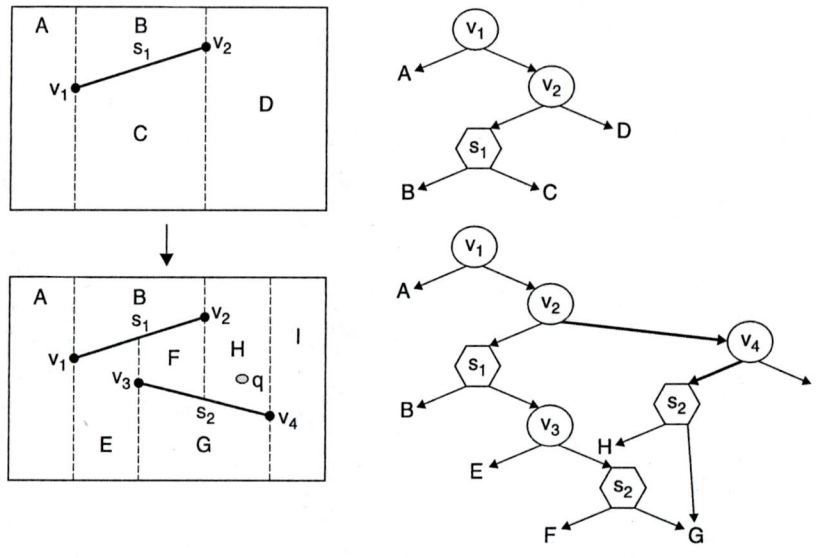

Figure 31.9 Construction of trapezoidal map and its associated DAG.

contains a pointer to a line segment of the subdivision. The left and right children of a y-node correspond to the spaces above and below the line segment represented by the y-node (see Figure 31.9). The leaves represent the trapezoids.

Given a query point q, the search process begins at the root and terminates when a leaf node is reached. At the x-node, we evaluate whether q lies to the left or to the right of the vertical line defined by the stored x-coordinate. At a y-node, we evaluate whether q lies above or below the segment. This is illustrated in Figure 31.9. For further details see Berg et al. [1] and Lee et al. [15].

kd-Trees

The k-dimensional tree (kd-tree) is one of the most prominent d-dimensional data structures [8,15]. Perhaps the most popular class of indexing technique for the NN search structure involves some sort of hierarchical space decomposition. It is a binary search tree that represents a recursive subdivision of the universe into subspaces by means of $(d-1)$-dimensional hyperplanes. The hyperplanes are iso-oriented, and their direction alternates between the d possibilities. For $d = 3$, for example, splitting hyperplanes are alternately perpendicular to the x-, then y-, then z-axis and then back to x- and so on. The choice of the splitting rule is an important issue

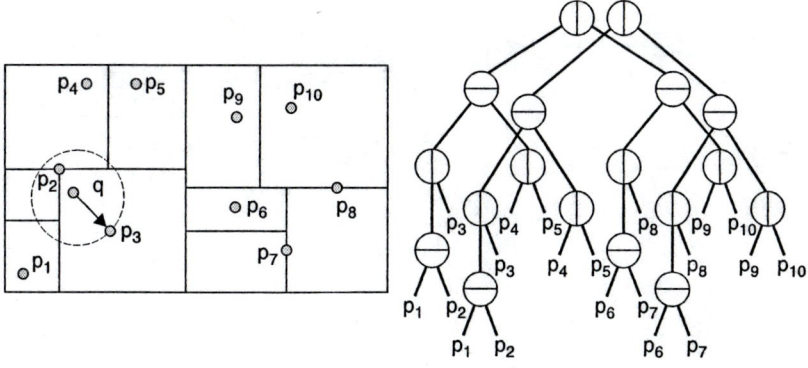

Figure 31.10   An example of a *kd*-tree (right) and its corresponding spatial subdivision (left).

in the implementation of the *kd*-tree. A good split is one that divides the points into subsets of roughly equal sizes and which produces cells containing at least one data point. Internal nodes contain the axis-orthogonal splitting hyperplane and have one or two children representing the rectangular subcells (see Figure 31.10). The leaves store all the data points. Searching and insertion of new points are straightforward operations. Deletion is somewhat more complicated and may cause a reorganization of the subtree below the data point to be deleted. Note that it is difficult to keep the tree balanced in the presence of frequent insertions and deletions. The structure works best if all the data is known *a priori* and if updates are not frequent. One disadvantage of a *kd*-tree is that for certain distributions no hyperplane can be found that splits the data points evenly.

The construction of a *kd*-tree can be briefly described as follows: We first compute the median of the point set in one of the dimensions (say, the *x*-axis) and partition the point set into two subsets based on the median point; that is, all the points having coordinates less than the median point along the *x*-axis are placed in one subset (left) and the remaining points are placed in the other subset (right). This process is then recursively continued along the *y*-axis in the resulting cells. When the partitioning is completed all along that axis, it is repeated back to *x*-axis and so on until only one data point remains. The general advantage of the *kd*-tree is that the decision of which subtree to use is always unambiguous.

The *kd*-tree search is not the most efficient, as the search algorithm visits all the nodes containing query point *q* and maintains the closest point to *q*. Because the root represents the entire space region, it has to go first to a leaf (say, p_i) that is the initial closest point to *q*. It then would visit the parent and all the nodes intersecting the circle centered at *q* of

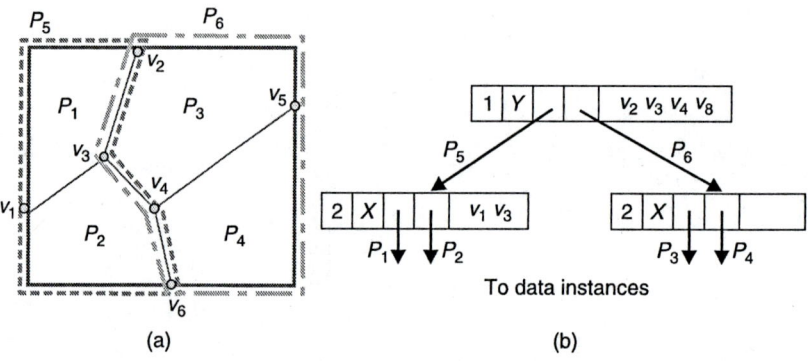

(a)   (b)

Figure 31.11 Index construction using the D-tree: (a) divisions in the index, and (b) D-tree structure. (From Wolfson, O. et al., in Proc. of the 10th Int. Conf. on Scientific and Statistical Database Management (SSDBM'98), Capri, Italy, July, 1998. With permission.)

radius distance (q,p_i). If the distance is greater than the distance of the closest point encountered so far, the search returns immediately. In Figure 31.10, with query point q, the search algorithm first finds the closest point p_3 (with a deep search). The visited points in the kd-tree are italicized. The final answer is p_2.

D-Tree

The D-tree is reported to have a better performance for indexing solution-space than traditional indexes [30]. The D-tree is a binary, height-balanced tree. It is similar to the kd-tree; however, the kd-tree is built based on hyperplanes, and the D-tree is constructed based on the divisions that form the boundaries of the valid scopes. For a space containing a set of valid scopes that are disjoint and complementary, it recursively partitions the space according to similar numbers of scopes until a space has one scope only. The partition between two subspaces is represented by one or more polylines. Figure 31.11a shows a valid scope for four objects. Polyline $pl(v_2, v_3, v_4, v_6)$ partitions the original space into P_5 and P_6, and $pl(v_1, v_3)$ and $pl(v_4, v_5)$ further partition P_5 into P_1 and P_2 and P_6 into P_3 and P_4, respectively.

Figure 31.11b shows the corresponding D-tree. Each node of the D-tree contains a header attribute, the partition that divides the space into two complementary subspaces, and two pointers (left and right). Searching on the D-tree begins from the root and recursively follows either the left pointer or the right pointer according to the partition and the position of the query point until a leaf is reached that contains a pointer to the data instance corresponding to the region.

Dim: Partition dimension
LSF: Left sites frontier
RSF: Right sites frontier
Left_ptr: Pointer to the left child
Right_ptr: Pointer to the right child

Figure 31.12   Data structure of the N-tree node.

The N-Tree: A New Index Structure for LDQ

One of the original motivations for the Voronoi diagram was nearest-neighbor searching. We propose a new index structure, called *N-tree* (neighbors tree), which indexes data regions based on neighbors in the solution space represented by a Voronoi diagram. N-tree, however, can also be applied to any planar subdivision or valid scopes. N-tree is a balanced binary space partitioning (BSP) tree, similar to D-tree [29,30]; however, the partition in D-tree is based on polylines, but N-tree is based on the frontier represented by a set of neighbors that cover the other sites. Because each point in two-dimension space has two values (*x*- and *y*-coordinates), we have to sort and split on the *x*-coordinate or *y*-coordinate according to the minimum cardinality of the frontier. We recursively partition a space consisting of a sorted set of data instances (sites) into two complementary subspaces (left subspace and right subspace) containing about the same number of sites until each subspace has one site only. The data structure of the N-tree is given by Figure 31.12.

The N-tree index is based on both object and space solution; that is, instead of storing the boundaries of valid scopes, it stores the object locations of adjacent sites. Our index structure can be considered as a hybrid index, because it combines both object- and solution-based indexes. We are inspired by the Delaunay triangulation [1,17] and the straight-line duality of the Voronoi diagram, where the objects are represented by vertices and two objects are connected (edge) if and only if their valid scopes are adjacent. Thus, we define the frontier set of neighbors to be the set of edges $<p_i, p_j>$ with one endpoint in each of the left and right site frontiers (LSF and RSF, respectively). We have to verify that LSF (or RSF) covers all the sites in the left subspace (or right subspace); that is, LSF and RSF are sufficient to guide a query point to the appropriate subspace: LS (left sites) or RS (right sites). Before describing the partition algorithm, we first illustrate it by providing an example (see Figure 31.13).

The idea of the construction of the hierarchical partition and its associated index N-tree is as follows:

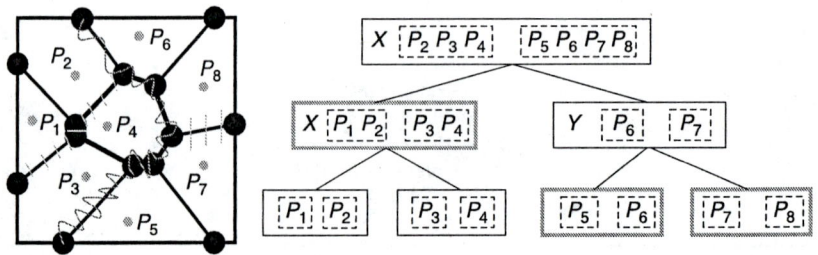

Figure 31.13 Construction of N-tree based on the example given in Figure 31.7.

- Input —
    S, a set of sites sorted by increasing order of x-coordinates (Sx) and simultaneously sorted by decreasing order of y-coordinates (Sy); the cardinality of S is N.
    E, a set of neighbors $<p_i, p_j>$ sorted, respectively, on x-coordinates (Ex) and y-coordinates (Ey); the cardinality of E is m (the number of edges in the Voronoi diagram generated by N points).
- *Output* — N-tree

For each partition dimension (x,y):

- Partition S into two complementary subsets, LS and RS.
- Find the frontier (F) and retrieve from E the edges with one endpoint in each of the two sets LS and RS.
- Choose the partition with the minimal cardinality of the F.
- Find LSF, the left site frontier (or RSF, the right site frontier).
- Update LE and RE, the edges, respectively, of the left subset and the right subset.
- Verify that LSF (or RSF) covers the remaining sites of LS (or RF).
- Partition on LS and LE (recursive partition on the left subset).
- Partition on RS and RE (recursive partition on the right subset).

The search algorithm starts from the root and computes the Euclidean distance from query point q to the associate points of this node then recursively follows either the left pointer or the right pointer according to the minimal distance found (in LSF or RSF). If the number of sites in an internal node is equal to the sum of all points in this node, the search returns immediately the closest point instead of reaching the leaf. In the example of Figure 31.13, for any query point, the search stops at level one (for the best case) and level two (for the worst case) of the tree because the number of sites in the subtree at level one is equal to four and the number of points in that node is equal to four. So, in general, the worst case to get the closest site is (logN − 1).

Summary of Indexing Techniques

Indexing techniques are briefly summarized in Table 31.5, where N is the number of regions (data instances or sites), n is the number of vertices, and m is the number of segments in the polygonal subdivision.

Conclusions

In this chapter, we have provided a brief overview of location-dependent data access, including LDQs, MODs, NN queries, and point queries. We have also examined indexing techniques to be used for them. A major emphasis of this chapter was on nearest-neighbor queries.

References

[1] Berg, M., Kreveld, M.V., Overmars, M., and Schwarzkopf, O., *Computational Geometry: Algorithms and Applications*, 2nd ed., Springer-Verlag, Heidelberg, 2000.

[2] Chen, G. and Kotz, D., *A Survey of Context-Aware Mobile Computing Research*, Technical Report TR2000-381, Department of Computer Science, Dartmouth College, Hanover, NH, 2002.

[3] Chamberlain, S., Model-based battle command: a paradigm whose time has come, in *Proc. of the First Int. Symp. on Command and Control Research and Technology*, National Defense University, Washington, D.C., June, 1995, pp. 31–38.

[4] Dunham, M.H. and Kumar, V., Location-dependent data and its management in mobile databases, in *Proc. of the Ninth Int. Workshop on Database and Expert Systems Applications (DEXA'98)*, Vienna, Austria, August, 1998, pp. 414–419.

[5] Ericsson, *Ericsson Mobile Positioning System*, 2000, http://www.ericsson.se/wireless/products.

[6] Erwig, M. and Schneider, M., Developments in spatio-temporal query languages, *Proc. of the Tenth Int. Workshop on Database and Expert Systems Applications (DEXA'99)*, Florence, Italy, September, 1999, pp. 441-449.

[7] Forlizzi, L., Guting, R.H., Nardelli, E., and Schneider, M., *A Data Model and Data Structures for Moving Objects Databases*, Technical Report Informatik 260, Department of Communication Systems, FernUniversität, Hagen, Germany, 1999.

[8] Gaede, V. and Gunther, O., Multidimensional access methods, *ACM Comput. Surv.*, 30(2), 170–231, 1998.

[9] Gupta, M., Tang, N., and Pasos, A., *Query Processing Issues in Mobile Databases*, term paper on distributed database systems, University of California, Davis, 2003 (http://sirius.cs.ucdavis.edu/teaching/).

[10] Guting, R.H., An introduction to spatial database systems, *VLDB J.*, 3(4), 357–399, 1994.

Table 31.5 Summary of Different Indexing Techniques for LDQ

Characteristic	Kirkpatrick	Trapezoidal	kD-Tree	D-Tree	N-Tree
Index data structure	Rooted DAG	Rooted DAG	Binary tree	Balanced BSP tree	Balanced BSP tree
Construction approach (recursive partitioning)	Triangulation	Randomized incremental	Hierarchical space decomposition	Hierarchical space decomposition	Hierarchical space decomposition
Construction complexity	$O(n\log n)$	$O(m\log m)$	$O(N\log N)$	$O(N^2\log N + N^2 + mN)$	$O(N^2\log N + Nm\log m + Nm)$
Search time complexity (number of nodes visited)	$O(\log n)$	$O(\log m)$	$>O(\log N)$ $O(\sqrt{N}+k)$	$O(\log N)$	$O(\log N)$
Remarks	Easy to understand; not practical	Simple; practical	General purposes; not efficient	Good performance	Efficient storage and retrieval
Applications (point location problems)	Binary search; nearest-neighbor search	Nearest-neighbor search	Geometric retrieval problems	Nearest-neighbor search	Nearest-neighbor search

[11] Board of Trustees, *General Relativity*, Science for the Millennium Expo, University of Illinois, Urbana-Champaign, 1995 (http://archive.ncsa.uiuc. edu/Cyberia/NumRel/GenRelativity.html)

[12] Jensen, C.S., An Introduction to Temporal Database Research, Ph.D. dissertation, University of Arizona, Tucson, 2000.

[13] Koubarakis, M. et al., Eds., *Spatio-Temporal Databases: The Chorochronos Approach*, Springer, New York, 2003.

[14] Kubach, U., Becker, C., Stepanoo, I., and Tian, J., Simulation models and tools for mobile location-dependent information access, in *Mobile Computing Handbook*, Ilyas, M. and Mahgoub, I., Eds., Auerbach, Boca Raton, FL, 2005.

[15] Lee, D., Lee, W.-C., Xu, J., and Zheng, B., Data management in location-dependent information services: challenges and issues, *IEEE Perv. Comput.*, 1(3), 65–72, 2002.

[16] Mehta, D.P. and Sahni, S., *Handbook of Data Structures and Applications*, Chapman & Hall, Boca Raton, FL, 2005.

[17] Narr, J.S. and LaRocque, G., *Approaching the Digital Battlefield* [transcript], Center for Defense Information, Washington, D.C., 1996.

[18] Location services, Mobilaris AB, Stockholm, Sweden, 2000 (http://www. mobilaris.se).

[19] Okabe, A., Boots, B., Sugihara, K., and Chiu, S.N., *Spatial Tessellations: Concepts and Applications of Voronoi Diagrams*, 2nd ed., John Wiley & Sons, New York, 2000.

[20] Porkaew, K., Lazaridis, I., and Mehrotra, S., *Querying Mobile Objects in Spatio-Temporal Databases*, Springer-Verlag, Heidelberg, 2003.

[21] Ryu, K.H. and Ae Ahn, Y., *Application of Moving Objects and Spatiotemporal Reasoning*, Technical Report TR-58, TimeCenter, Aalborg Øst, Denmark, 2001 (http://www.cs.aau.dk/TimeCenter/pub.htm).

[22] Schiller, J. and Voisard, A., *Location-Based Services*, Morgan Kaufmann, San Francisco, CA, 2004.

[23] Seydim, A.Y., Dunham, M.H., and Kumar, V., Location-dependent query processing, in *Proc. of the 2nd ACM Int. Workshop on Data Engineering for Wireless and Mobile Access (MobiDe'01)*, Santa Barbara, CA, May, 2001, pp. 47–54.

[24] Seydim, A.Y., Dunham, M.H., and Kumar, V., An architecture for location-dependent query processing, in *Proc. of the Twelfth Int. Workshop on Database and Expert Systems Applications (DEXA'01)*, Munich, Germany, September, 2001, pp. 549–555.

[25] Seydim, A.Y. and Dunham, M.H., Location leveling, *IEEE Trans. Knowledge Data Eng.* (submitted).

[26] Seydim, A.Y. and Dunham, M.H., Location-dependent query processing in mobile computing, in *Mobile Computing Handbook*, Ilyas, M. and Mahgoub, I., Eds., Auerbach, Boca Raton, FL, 2005, pp. 255–274.

[27] Su, J., Xu, H., and Ibarra, O., Moving objects: logical relationships and queries, in *Proc. of the 7th Int. Symp. on Spatial and Temporal Databases (SSTD'01)*, Los Angeles, CA, July, 2001, pp. 3–19.

[28] Wireless location services, SignalSoft Corp., Boulder, CO, 2000 (http://www. signalsoftcorp.com).

[29] Wolfson, O., Chamberlain, S., Dao, S., and Jiang, L., Location management in moving object databases, in *Proc. of the Second Workshop on Satellite-Based Information Services (WOSBIS'97)*, Budapest, Hungary, October, 1997.

[30] Wolfson, O., Chamberlain, S., Xu, B., and Jiang, L., Moving object databases: issues and solutions, in *Proc. of the 10th Int. Conf. on Scientific and Statistical Database Management (SSDBM'98)*, Capri, Italy, July, 1998.

[31] Xu, J., Zheng, B., Lee, W.-C., and Lee, D.L., Energy efficient index for querying location-dependent data in mobile broadcast environments, in *Proc. 19th Int. Conf. on Data Engineering (ICDE'03)*, Bangalore, India, March, 2003, pp. 239–250.

[32] Xu, J., Zheng, B., Lee, W.-C., and Lee, D.L., The D-tree: an index structure for planar point queries in location-based wireless services, *IEEE Trans. Knowledge Data Eng.*, 16(12), 1526–1542, 2004.

Chapter 32

Location-Dependent Service Accounting

Michael Georgiades, Christos Politis,
Nadeem Akhtar, and Rahim Tafazolli

CONTENTS

Introduction

Accounting constitutes one of the most important functions in any tele-communication network. The provisioning of accounting requires very complicated functionality due to the large number of users involved as well as differences in the tariff schemes applied to different user groups. The accounting functionality consists of four subfunctions: metering, charging, accounting, and billing. The metering function is responsible for identifying and recording information relevant to the usage of a resource in a meaningful way; for example, the information could be about access to a resource or the time and duration of usage. The records created by the metering function are referred to as *usage metering records* (UMRs). The charging function collects the metering records pertaining to specific service usage events and combines them into service transaction records, which include pricing information. Charges are calculated for the resources used from the metered information by applying appropriate tariff schemes. The accounting function assigns charges in terms of service transaction records to customers' accounts. The charges for service usage may be distributed to service providers that cooperate together to provide the services to the users. Finally, the billing function collects service transaction records and selects from the account that pertains to a particular service subscriber, accumulates the charges, and finally sends an invoice to the subscriber.

The accounting and billing models used by traditional wired Internet Service Providers (ISPs) have been based on a flat-pricing model, with the prices being determined by the user subscription and the time they are connected. For subscribers in mobile communication networks, the charging scheme also takes into account additional parameters; for example, voice and video calls have different rates, with the latter being typically more expensive. Similarly, the user location also affects charging as a roaming user has to be billed on a different tariff compared to a user connected to the home network. Location-dependent accounting will play an increasingly important role in next-generation networks that aim to provide a rich collection of services as well as support for seamless mobility. Mobile users will want to access anytime, anywhere services and the service providers will have to implement appropriate accounting methods for charging in such scenarios.

A new dimension to the charging problem is added by the emergence of user-operated networks in the form of personal area networks (PANs), body area networks (BANs), etc. These low-complexity networks consist of a number of communication devices under a common administrative control. The owner of such a network can use these multiple devices to connect diverse networks. Furthermore, the internal structure of such networks may be dynamic in the sense that PAN members may join and leave the network when required. It is also possible that a PAN may be distributed over space with different devices connected via interconnection networks such as the Universal Mobile Telecommunications System (UMTS), wireless local area networks (WLANs), etc. These new developments have to be taken into account when designing accounting solutions for location-dependent services and applications. The accounting solution has to be scalable to perform efficiently under various scenarios.

In the next section, we present the main trends in the area of location-dependent accounting and discuss the main issues involved, including security aspects. This is followed by a discussion of how middleware can support mobility-aware accounting, accounting, billing, and charging models for location-dependent services as well as new security challenges that arise.

Accounting, Billing, and Charging for the Mobile Internet: Trends in Service Accounting

Recent years have seen several new advances in service accounting and some of them are highly relevant for location-dependent accounting. A number of standardization bodies are working actively in this area, in the mobile telecommunications world as well as in the Internet Engineering Task Force (IETF) community. Examples include, the SA5 Group of the 3GPP, the UMTS Forum, the Mobile IP and Authentication, Authorization, and Accounting (AAA) Working Groups of the IETF, and the Authentication Authorization Accounting ARCHitecture Research Group (AAARCH) Working Group of the Internet Research Task Force (IRTF).

The Mobile IP and AAA Working Groups have worked together in defining mobility-related requirements for authentication, authorization, and accounting. Glass et al. [2] describe the requirements that have to be supported by the AAA service to aid in providing Mobile IP services. Within these working groups, a differentiation is made between a home domain and a foreign domain, which is an administrative domain visited by a Mobile IP client and containing the AAA infrastructure required to carry out the necessary operations. The solutions and models proposed by these working groups have mainly been constrained by the protocol stack designed for the traditional fixed architectures. The Terminal Access Controller Access Control System (TACACS) also offers an accounting model with initiation, termination, and interim update messages [3]. TACACS has been designed to run over the Transmission Control Protocol (TCP) and, although offering a number of advantages such as scalability, these approaches are not suitable for handling accounting events in environments that support high mobility.

Other traditional solutions include the Common Management Information Protocol (CMIP) and Simple Network Management Protocol (SNMP). SNMP [4] has been widely deployed for a variety of intra-domain accounting applications, usually using the polling data collection model, which allows data to be collected for multiple accounting events simultaneously. Remote Authentication Dial-In User Service (RADIUS) accounting [6] is useful where low processing delay is required, such as credit risk management or fraud detection; nevertheless, RADIUS accounting implementations are vulnerable to packet loss as well as application-layer failures, network failures, and device reboots. These deficiencies are magnified in inter-domain accounting, something which is essential for roaming. The Diameter Base Protocol [7] aims to provide an AAA framework for applications such as network access or IP mobility. Diameter is also intended to work in both local AAA and roaming situations and has been the most suitable solution so far.

The majority of these protocols and systems have been mainly designed to offer accounting solutions on the traditional fixed architectures providing general-purpose management solutions and in some cases have been extended to also handle mobility issues. As a result, they lack specific management functions for location-aware accounting and do not provide mobility-enabled metering of resource usage. In the following, we highlight some of these trends in more detail and discuss their impact on accounting for location-depended services and applications.

RADIUS

Remote Authentication Dial-In User Service (RADIUS) [5] is an AAA client–server protocol between a Network Access Server (NAS) and a centralized authentication server. The latter is referred to as the RADIUS server. RADIUS was initially designed for intra-domain use, where the NAS and RADIUS server exist within a single administrative domain. The protocol messages are carried over the User Datagram Protocol (UDP) on top of IPv4 as well as IPv6 and use the attribute–value pairs (AVPs) format for data delivery. RADIUS provides hop-by-hop security and supports a variety of authentication methods, including the Challenge Handshake Authentication Protocol (CHAP), Password Authentication Protocol (PAP), Extensible Authentication

Protocol (EAP), and UNIX log-in. A shared secret key is used between the different servers or the end users and the server. This means that transactions between the client and RADIUS server are authenticated through the use of a shared secret, which is never sent over the network. In addition, any user passwords are sent encrypted between the client and RADIUS server to eliminate the possibility that someone snooping on an unsecured network could determine a user's password. RADIUS, however, does not provide end-to-end security on the data path between two corresponding hosts.

TACACS

The Terminal Access Controller Access Control System (TACACS) is a remote authentication protocol developed by Cisco Systems for use with its own routers and NAS systems. This protocol has been reengineered over the years by Cisco and is supported on many terminal servers, routers, and NAS devices found in enterprise networks today. The current version is called TACACS+, which includes a number of enhancements made to the original TACACS protocol. Like RADIUS, TACACS+ is also a client–server protocol and supports a similar set of services; however, it uses TCP as the transport protocol and encrypts not only the password in the access-request packet but also the remainder, such as username, authorized services, and accounting. Contrary to RADIUS, which combines authentication and authorization (the access-accept packets sent by the RADIUS server to the client contain authorization information), TACACS+ separates authentication, authorization, and accounting in a AAA activity. This allows separate authentication solutions that can still use TACACS+ for authorization and accounting.

Diameter Protocol

The Diameter Protocol is being developed by the Authentication, Authorization, and Accounting (AAA) Working Group of the IETF [2]. This protocol aims to provide an AAA framework for a diverse set of services such as network access or IP mobility. Diameter supports both local AAA functionality and also works in roaming scenarios. The Diameter Protocol is eventually expected to replace its precursor, RADIUS. As the Internet grew and new business and networking models were being deployed, a number of shortcomings of RADIUS were exposed [6]. Some examples are listed below:

- Failover support
- Transmission-level security
- Reliable transport
- Agent support
- Server-initiated messages

The Diameter Protocol design aims to eliminate these shortcomings as well as support additional features such as capability negotiation, peer discovery, and roaming support. In contrast to RADIUS, Diameter is a peer-to-peer AAA protocol, and it also supports inter-domain accounting, which was lacking in RADIUS. Diameter follows a modular design approach and consists of a base protocol and a set of extensions. The base protocol is not intended to be used alone but must be used with a service-specific extension. Examples of extensions include the Network Access Server Requirements (NASREQ) for network access and Mobile IP. The protocol provides for reliable message transport by using TCP or Stream Control Transmission Protocol (SCTP) as transport protocols. It provides hop-by-hop security, end-to-end security, and a mechanism for congestion control. AVPs are used to deliver all the protocol data. The AVP values may be associated with the Diameter Base Protocol or they may also be specific to the Diameter applications. The AVPs that are used by Diameter may carry different types of information, such as:

- User authentication information
- Service specific authorization information, between client and servers
- Resource usage information

The AVPs are also used to support relaying, proxying, and redirecting of Diameter messages through a server hierarchy.

IPDR

The IP Detail Record (IPDR) is a joint initiative promoted by leading vendors of charging and billing solutions, equipment vendors, system integrators, and Internet Service Providers (ISPs). The purpose of the IPDR initiative is to create a single industry standard. It also aims to define accounting interfaces for new equipment and services as they emerge and to consolidate existing nonstandardized IP services. The IPDR Streaming Protocol for accounting has been proposed as part of this initiative.

Traffic flowing from an exporter to a collector is composed primarily of data records. A data record is a collection of information gathered by the service element for various purposes (e.g., accounting). An exporter is an implementation on the data-producing side of the Streaming Protocol, and a collector is an implementation on the data-receiving side of the protocol. In addition to data records, a small portion of the traffic includes bidirectional control message exchanges.

The protocol can be used for the delivery of different types of accounting-related data, mainly data records from service elements to different systems, such as mediation systems and the business support system (BSS)/

operations support system (OSS). The protocol is being developed to address the needs for exporting a high volume of data records from the service element. In the IPDR Reference Model, a service element is the logical entity that senses usage of services by the service consumer. Typical examples of service elements may include various components traditionally considered infrastructure network elements, such as:

- VoIP gateways
- Web servers
- Application servers
- Streaming media servers
- Game servers
- Location-based wireless services

The IPDR Streaming Protocol runs over a transport layer protocol that must be connection oriented and reliable. Like Diameter, TCP or SCTP maybe used. Another possible candidate transport protocol is the Blocks Extensible Exchange Protocol (BEEP; RFC3080), which sits between low-level transport protocols such as TCP or SCTP and higher level application protocols such as the Hypertext Transfer Protocol (HTTP). The reliability of data delivery is provided at both the transport-layer level and the IPDR Streaming Protocol level.

The IPDR Streaming Protocol specifications also include IPDR service definitions, based on eXtensible Markup Language (XML) schema, by reference. These definitions describe the properties of the different accounting records and their fields. Furthermore, the IPDR streaming templates identify the Uniform Resource Identifiers (URIs) of the schema, which describe a given accounting record structure and identify each field using its qualified name, as defined in the referenced XML schema or in subordinate imported schemas.

Middleware Accounting Solutions for Location-Dependent Services

The evolution of emerging and heterogeneous wireless access technologies and the increase in demand for multimedia and mobility services has forced the wireless industry to evolve toward a pervasive integrated system. The idea is to develop a global core network that will accommodate Internetworking among the various access networks and facilitate the creation and deployment of innovative applications and services, which will also include location-based services. Different forms of mobility must be supported, and users should be able to connect to the Internet from any location offering access availability. Networks should be able to offer anytime, anywhere access services to a variety of mobile devices (e.g., mobile phones, PDAs, wireless pagers). For this, mobile users need to maintain connectivity any time during their roaming.

Regarding service provisioning, it is important to differentiate between two classes of possible services that must be supported for mobile clients. On the one hand, a mobile client should be provided with traditional Internet service designed for the fixed network infrastructure, but, on the other hand, that mobile client should be provided with mobility-related services depending on the current position of the mobile user (also referred to as location-dependent services). The main challenges that must be dealt with for roaming in such a heterogeneous environment and the provisioning of location-dependent services are the design and implementation of accounting systems.

Mobility-aware accounting solutions require tracking the location of mobile users while they roam in this global environment and coordinating resources of interest possibly present in a number of heterogeneous networks. Like other services, mobility-aware accounting systems must also continue operation regardless of possible temporary disconnections of mobile users from the Internet. Also, a number of different ISPs and network operators contribute input to the pricing and charging strategies, and mutual agreements formed between these players must be monitored and organized by an accounting system.

Intra-Domain and Inter-Domain Accounting

The majority of the research and implementations on accounting management have focused mainly on intra-domain accounting applications; however, with the increasing demand of services such as Voice-over-IP (VoIP), global roaming, and location-based services, applications requiring inter-domain accounting are becoming more and more common. Inter-domain and intra-domain accounting differ in a number of ways. Intra-domain accounting aims only to collect resource information within an administrative domain. Accounting protocol packets and session records do not have to cross administrative boundaries, resulting in low packet loss and data to be transferred between trusted entities [1]. On the other hand, inter-domain accounting collects resource information within an administrative domain to be used within another administrative domain. In this case, accounting protocol packets and session records will cross administrative boundaries. This could result in higher packet loss and less trust between the involved entities requiring additional security measures.

Mobile-Middleware-Based Accounting Solutions

Both service provisioning and systems management will face many challenges and therefore changes due to the merging of mobile communications, wireless communications, and the Internet. Traditional solutions for fixed networks do not suit the new scenario where users, devices, and even service components can change their location during service provisioning. The IETF AAA working group has worked on many issues related to middleware, including defining processes for access and admission control, identification, authentication for validating that identity, authorization for determining an eligibility for resource requests, resource utilization, and accounting, at least to the degree that resource utilization is recorded [8]. Accounting solutions for pervasive services require the fixed Internet infrastructure to be extended with mobility-enabled monitoring, processing, pricing, and charging functions. Several middleware solutions have been proposed that can support local operations while allowing mobile devices to maintain continuous connectivity with remote centralized home managers [9,10]. In the following sections, we describe some middleware technologies that, besides other services, can be used to support pervasive accounting solution by deploying an accounting infrastructure where required.

Mobile Agents for Usage-Based Accounting

One efficient technology that can resolve the accounting issues introduced by mobility described earlier is the introduction of mobile agents (MAs) [11]. MAs can track the movements of a mobile device to maintain their colocation with the clients they are responsible for, or they can dynamically move close to required resources and service components, hence preserving locality for the operations. MAs can be a very suitable candidate technology for supporting session-dependent usage-based accounting. They can be used to install new monitoring and charging behavior dynamically and can maximize locality for access to monitoring data. They can be used to support accounting even in the case of temporary network disconnections.

Currently, service provisioning and accounting do not deal with the location of the user's device at provisioning time. Location awareness is very important in adapting services during roaming, depending on the available local resources and allowing for location-conscious accounting strategies. Location awareness is central to the MA-based approach, which makes allocation of resources visible up to the application layer. In general, MAs offer the following advantages for pervasive accounting:

- *Mobility awareness* — MAs can maintain colocality with the user by following the movements of the portable device or at least preserve locality by moving close to needed resources and service components.
- *Dynamicity* — MAs offer the possibility of providing an extended flexible support infrastructure by adapting to requirement changes at runtime and influencing pricing and charging accordingly.
- *Location awareness* — MAs can adapt services to the currently available local resources and enable accounting strategies depending on the resources available.
- *Personalization* — MAs have the ability to adapt services depending on the users preferences and device characteristics.
- *Security* — MAs also support authentication of mobile users, monitor and register usage of system resources, and support secrecy and integrity in communications.
- *Interoperability* — Users may roam across several heterogeneous networks, and MAs can support pervasive accounting by monitoring and controlling the consumption of resources.

Mobile-agent-based Internet applications rely on the Secure and Open Mobile Agent (SOMA) distributed programming framework, which provides a layered service infrastructure. The SOMA architecture consists of four layers: mobility middleware layer, core services layer, Java Virtual Machine layer, and a heterogeneous distributed system. The mobility middleware layer consists of a user virtual environment, a mobility virtual terminal, and a virtual resource management services [10]. The core services layer consists of services such as communication, migration, naming, security, interoperability, persistency, and quality of service (QoS) adaptation. SOMA groups several places into domain abstractions that correspond to network localities. Each node provides at least one place for agent execution, and places are grouped into domains. Each domain has a default place in charge of inter-domain routing and Common Object Request Broker Architecture (CORBA™)-based interoperability. The accounting layer is responsible for all issues related to metering, storing, and processing information regarding resource usage. Metering collects and carries out measurements regarding resource consumption. Metering data is stored either locally or in a remote administration site among with other service information. This information is processed depending on the accounting strategy for access control, auditing, capacity planning, and billing.

Portable-device usage-based pervasive accounting (PUPA) has been proposed in Bellavista et al. [12] and is a usage-based accounting management platform designed to provide a layered infrastructure for network operators, system administrators, and service providers for accounting support in mobile and dynamic environments. It operates on top of the SOMA platform to provide accounting services based on the MA technology. It consists of two layers: a metering layer and a pervasive accounting layer. The metering layer collects information about resource consumption at the system, network, and application level. For the system level, this includes information on the processes working on local resources and on their usage of the communication infrastructure such as process name, process identifier, CPU usage time and percentage, and memory allocation. Network-level information includes metering data regarding the number of UDP/TCP packets sent and received, the number of TCP connections, etc. For the application level, the metering layer can collect information about all service components accessed from within the Java execution environment. The pervasive accounting layer exploits the metering layer and determines how, where, and when to perform accounting management.

Application Domain Accounting for Roaming Services

Application domain roaming accounting proposed in Bellavista et al. [13] requires dynamic and flexible middleware solutions to support a variety of different billing schemes at service provisioning time. Application domain accounting solutions can provide us with a solution where users can also be charged for the:

- Accessed service contents
- QoS levels received
- Dynamic adaptations of content formats

The MA metering service, therefore, must monitor attributes such as the bandwidth used in the access network locality and how much memory is consumed on the content adaptation nodes. The metering and charging functions proposed are mobile middleware components capable of migrating during service provisioning to follow the movement of the roaming users. Figure 32.1 shows how the roaming accounting middleware utilizes the SOMA platform described earlier. The architecture of our roaming accounting middleware consists of three main services: the location service, the metering service, and the charging service.

The metering service holds the metering logic that is to be applied on the user's resource usage. The charging service holds the charging logic used to produce the billing reports for the users according to their service-level agreements. The location service is used to achieve real-time location visibility of the mobile device and passes this information to other middleware services.

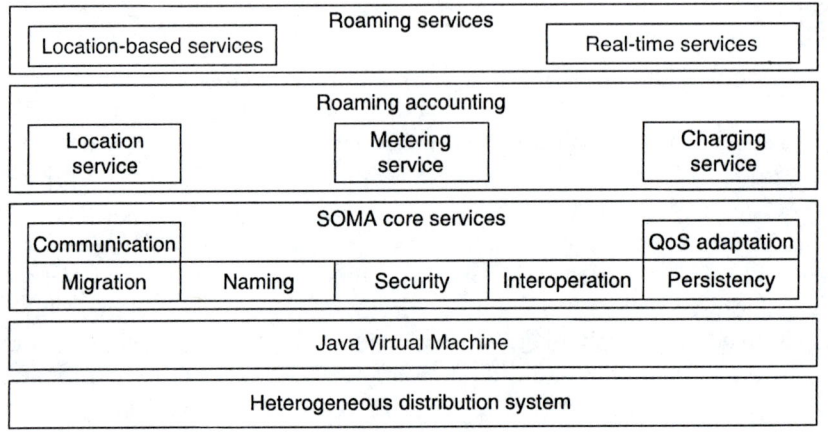

Figure 32.1 Architecture for roaming accounting middleware.

QoS-Aware Accounting for Roaming Services

The location of a mobile device during roaming will have an impact on the level of the QoS offered during service provisioning, and this needs to be considered by accounting solutions. The QoS offered to a user may depend not only on the best-effort model of the Internet communication but also on the dynamic change of the user's location, which results in reshuffling of the resources available. The initial QoS-level agreement is no longer sufficient to fairly account for roaming users but should be dealt with at runtime during service provisioning.

The active middleware for quality-aware accounting of mobile services (AQuAM) proposed in Bellavista et al. [14] is a layered support infrastructure capable of accounting final users for the total QoS level offered during service provisioning for roaming users. It works at the middleware level to provide information related to the service such as customer preferences, security profiles, and low-level accounting information to the application level for modifying the QoS level offered accordingly.

Accounting, Billing, and Charging Models for Location-Dependent Services

Functional Model for the Internet

This section presents a functional model for collecting and processing information related to resource usage and billing and charging of users [15]. In this model, metering is the function of capturing all data related to consumption of the network resources (e.g., volume of exchanged data) and is performed by network devices. To achieve this functionality it is necessary for the network devices to export flow information in a standardized way. An IP flow information export system includes a data model, which represents the flow information, and a transport protocol. The IETF Real-Time Traffic Flow Measurement (RTFM) Working Group has as its main objective the review of existing work and the development of an improved traffic flow model [21]. The RTFM Working Group defined the Traffic Flow Measurement Architecture [22] and introduced FlowMeter MIB as a "standards track" document within the IETF [21].

Another working group formed by the IETF is the IP Flow Information eXport (IPFIX) Working Group, whose goal is to define a "standard IP flow" and choose a protocol to which IP flow information will be transferred [23]. Regarding RTFM, the IPFIX framework specifies a complete and similar architecture for gathering flow information. Both architectures consider the same definition of flow (i.e., a set of packets sharing a common set of endpoint address-attribute values), their basic entities are similar, and both leave parts of the system unspecified. The resource usage data, after being collected by the network, is sent to an accounting server for further processing. Transferring the accounting data is done with the use of an appropriate accounting protocol. The collection of metering data can be initiated either by the network device itself (push model) or by the accounting server, which plays the role of the collector entity (pull model).

The accounting function, which is performed by the accounting server, is responsible for the collection and storage of the accounting data. Accounting may also include summarization of interim information, elimination of duplicated data, and generation and processing of session records. Moreover, session records and their related IDs are also produced and handled. Accounting is also responsible for forwarding these records to other peer entities in the case of roaming terminals. The accounting attributes depend on the applied charging model (e.g., flat-rate, session-oriented, time-based, volume-based). According to the standards [21], these attributes contain the identity of the user, the service type, the volume of the transmitted data, the start and stop time for a session, the reserved network resources during a session, etc. When different organizations are involved in an accounting process, accounting data and session records will typically cross administrative boundaries. The accounting process that involves the collection of information on resource usage of an entity within an administrative domain for use within another administrative domain is called *inter-domain accounting*. Different accounting systems are often employed in a multi-domain environment; therefore, different ways of representing accounting information are likely to exist in each domain. Thus, the standardization of accounting and session records as well as the acceptance of a general AAA protocol is fundamental [24].

The billing function deals with bill preparation and presentation to the party that is responsible for payment. This function receives session records or processed accounting data from the accounting function via a transfer protocol such as the Simple Mail Transfer Protocol (SMTP), File Transfer Protocol (FTP), or HTTP. Finally, it prepares an invoice according to the appropriate billing policy (e.g., computing a special discount for a user, addition of a monthly fee).

Further to this functional model, the AAAARCH group proposes the introduction of a more flexible model that enables a policy-based accounting model [16,19]. Based on this proposal, the network devices can be configured to monitor all flows equally, or they can be configured to collect data only for specific flows. The implementation of NeTraMet sets up policy-based metering [25]. NeTraMet is an open-source implementation of the RTFM architecture for network traffic flow measurement. More specifically, a NeTraMet meter is an SNMP agent that implements the RTFM Meter MIB. In such a flexible model, the accounting server can be dynamically configured to handle resource usage data according to the communication type (e.g., voice sessions, HTTP requests). Furthermore, the pricing policy that determines the price to be charged for a used service could also be static or dynamic. In the case of static pricing, predefined values stored in pricing tables are used for computing the charges [19]. These prices are the same under congested and normal traffic conditions. For dynamic pricing, several parameters such as network and service utilization are also taken into account for the computation of corresponding prices.

Finally, on top of the aforementioned dynamic model is current interest in investigating the possibility of supporting content accounting. Although most content providers offer their contents free of charge, it is expected that in the near future some of them will eventually market their content. Content accounting consists of the following functions: metering content consumption, collecting and storing the metered data, defining pricing schemes to perform the charging policy, and billing the customers. The main difficulties in the case of content accounting are the definition of metering strategies, the specification of a standard way to store metered data, and the support of several payment methods such as micropayments or the use of a credit or debit card. In the IETF, the Internet Open Trading Protocol Working Group, which is an interoperable framework for Internet commerce, investigates content accounting issues [26].

Functional Model for Wireless Communications

This section presents a functional model for charging, billing, and accounting used in wireless networks [18]. The charging function collects information related to a chargeable event from several network nodes. The charging information generated by network nodes is structured in the form of a charging data record (CDR) and is transferred via standard charging protocols. This is a well-established procedure in use since the establishment of the Global System for Mobile Communications (GSM), and until recently the acronym CDR stood for "call detail records." The charging function is responsible for further processing and storing temporarily the generated CDRs to correlate any partial records and transfer them securely to the billing function. The CDRs are transferred to the billing function via a transfer protocol such as File Transfer, Access, and Management (FTAM), FTP, or the Trivial File Transfer Protocol (TFTP). The charging functionality is accomplished by two entities, the charging gateway functionality (CGF) and the charging collection functionality (CCF), which are described in more detail later.

The billing function processes the records arriving from the charging functional entity according to the respective tariffs stored in the home location register (HLR), the main register that stores detailed information related to every subscriber, or in the billing system and calculates the charge for which the user will be billed.

In the case of roaming users, the accounting function is responsible for apportioning charges among the home environment, the serving network, and the user and then calculating the portion that is due to each operator. The billing record concerning a roaming user is forwarded to its home network operator using the transferred account procedure (TAP) and a specific TAP format. The transfer of TAP records between the visited and the home mobile networks may be performed directly or via a clearinghouse. Clearinghouses are independent business players responsible for TAP records creation, tariffing, and re-tariffing. The calculation of apportioning revenue between the operators normally happens once per month. In an open marketplace, where many independent entities (mobile operator, network and application/service providers) are involved in the service provision process, a well-specified accounting function will be responsible for apportioning their revenue automatically.

Network Entities Involved in Charging

To facilitate an understanding of the charging, accounting, and billing processes in wireless/mobile networks, a brief description of the overall architecture is provided here, together with a presentation of the network components involved in these functions. Based on the 3GPP's release 5, the 3G mobile network infrastructure is logically divided into core network (CN) and access network (AN) infrastructures. The CN is composed of a circuit-switched domain (CS domain), a packet-switched domain (PS domain), an IP multimedia subsystem (IMS), and a service domain. Two different types of access networks are used by the CN: the base station system (BSS) for GSM and the radio network system (RNS) for UMTS:

- The *CS domain* refers to the set of all the core network entities offering a CS-type of connection for user traffic as well as all the entities supporting the related signaling. A CS-type of connection is a connection for which dedicated network resources are allocated at the connection establishment and released at the connection release. Involved entities are the mobile-service switching center (MSC), the MSC server, the media gateway (MGW) function, the gateway MSC (GMSC), the GMSC server, and the interworking function (IWF).

- The *PS domain* refers to the set of all the core network entities offering a PS-type of connection for user traffic as well as all the entities supporting the related signaling. A PS-type of connection transports the user information using autonomous concatenation of bits called *packets*; each packet can be routed independently from the previous one. Involved entities are the serving General Packet Radio Service (GPRS) support node (SGSN), the gateway GPRS support node (GGSN), and the border gateway (BG).

- The *IMS* has been recently introduced in to enable person-to-person real-time services (e.g., voice calls) over the PS domain. The introduction of a multimedia call model based on the Session Initiation Protocol (SIP) creates the ability to deliver IP-based real-time multimedia services including VoIP. The IMS includes all the core network elements providing IP-based real-time and non-real-time person-to-person and person-to-machine multimedia services, such as audio, video, text, and chat, and a combination of them that is delivered over the PS domain. Involved entities are the Call Session Control Function (CSCF), the Media Gateway Control Function (MGCF), the Media Gateway (MGW), the Multimedia Resource Function Controller (MRFC), the Border Gateway Control Function (BGCF), and the Application Server (AS).

- The *service domain* incorporates a wide range of services deployed by the network provider, such as multimedia messaging service (MMS) or location service (LCS), or by independent application/service providers making use of the Open Service Access (OSA) architecture.

The *OSA architecture* specified by the 3GPP and adopted by the European Telecommunications Standards Institute (ETSI) opens up 3G networks through open standardized application programming interfaces (APIs), enabling independent service application developers to make use of the underlying network functionality without exposing the underlying communication infrastructure to unauthorized business entities. The two nodes that handle this information are the CCF and CGF. The CGF provides a mechanism to transfer charging information from the GPRS support nodes (GSNs) to the billing function. In addition, it acts as a storage buffer for real-time CDR collection and is able to perform consolidation of CDRs and preprocessing of their fields. The CDRs are transferred from the GSNs to the CGF via the standardized Ga interface using the GTP's charging protocol.

The CGF can be implemented in a separate network element or the charging gateway (CG) or can be integrated in the GSNs. The CCF is responsible for the IMS domain and is a logical function equivalent to the CGF. Additionally, it is able to validate, combine, aggregate, and consolidate the received charging information, generate partial records, remove duplicate charging data, and support load sharing, redundancy, high availability, and efficient management of the generated CDRs. The IMS network entities send the CDRs to the CCF through the standardized Rf interface (except for the AS that uses the Ra interface). Regarding the charging protocol, used over the Rf interface, the Diameter Protocol was initially proposed but a decision has not yet been finalized.

The network elements that generate charging records are:

- The MSC for the CS domain — Information related to the call duration, the source (calling number), the destination (called number), the cause for termination, etc. [27]
- The GSNs for the PS domain — Information related to the use of a radio interface and the general GPRS resources; these CDRs include attributes such as the usage duration (PDP context duration), the source (served IP address), the destination (access point name [APN]), the volume of transferred data, QoS-related information, etc. [28]
- The proxy-CSCF, the serving-CSCF, the interrogating-CSCF, the MGCF, the MRFC, the BGCF, and the application server for the IMS that establish and control IP multimedia sessions between the users [29]

Concerning the service domain, application services such as the multimedia messaging service (MMS) and the location service (LCS) are able to generate charging information in the form of CDRs [20]. The CDRs generated by the MMS relay/server contain information regarding the usage of the MMS resources, the source and the destination, the usage of PS domain resources, and the usage of external data networks [30]. Note also that the generation of the CDRs and their respective information for LCS usage is currently under investigation by the SA5. In the service domain, we can also include the OSA framework and the independent application providers. The OSA architecture enables application providers to add application and content charges via the OSA charging service capability feature (SCF) [31]. Finally, two components responsible for content-charging issues [17] have been recently introduced. In particular, the subscriber content charging function (SCCF) handles content-charging requests produced when a subscriber accesses a specific content. Content-charging requests are typically sent to the SCCF from the content provider charging function (CPCF). The CPCF receives content-charging requests from the content server, processes them, and relays them to the SCCF. Additionally, the CPCF maintains and manages the account for the content provider.

Charging Protocols

The CGF and the CCF are connected with the GSNs and the IMS network entities. The transfer of CDRs over these standardized interfaces is performed with the use of standardized protocols. The charging protocol used over the Ga interface is GTP. This protocol is based on the GPRS Tunneling Protocol (GTP), which is used for packet data tunneling in the backbone network. GTP contains functionality for transferring CDRs, redirecting them in another CGF, detecting communication failures, and preventing the transfer of duplicate CDRs during redundancy operations. The SA5 of the 3GPP proposes to use this protocol over the Rf interface as well. However, after taking into account the work produced by the IETF, another option is still under consideration for transferring charging information between the IMS and the CCF — namely, the use of the Diameter Protocol. This protocol is suitable for handling authentication, authorization, and charging. Moreover, the Diameter Protocol is also capable of supporting prepaid users.

Security Framework Challenges

One of the most fundamental rights in a healthy society is the right of every citizen to be left alone. Article 12 of the U.N. Universal Declaration of Human Rights states that "no one shall be subjected to arbitrary interference with his privacy, family, home or correspondence." In reality, though, this right is increasingly being trod upon. Along with undreamed-of comforts and conveniences for the population in general, the digital revolution has made it possible to gather and store information about human behavior on a massive scale. We leave electronic footprints everywhere we go, footprints that are being watched, analyzed, and sold without our knowledge or even control.

Within this context, security and privacy are mandatory aspects when developing new pervasive (ubiquitous) technologies. The security services for a next-generation network (B3G) should be bound together by a security framework, and the architecture for the integration of different security services (e.g., authentication, cryptographic key management) must be properly defined. This section describes the actual and low-profile security services required for defining a security framework for the next-generation wireless networks. The following security aspects are worth discussing, due to the challenges they impose on B3G systems:

- *Trust* — Current wireless and mobile communications systems take for granted certain and preestablished trust relationships, which is unlikely to be generally true for future systems because of their dynamic nature.
- *Privacy* — The requirements concerning privacy, imposed by legislation and user expectations, should be fulfilled within any wireless communications system. To ensure that the system is trusted, implementing this in a way that the end user can relate to and which safeguards the end-user's privacy is a very important task.
- *Authentication and authorization* — The most important research issues regarding authentication and authorization are the following:
  - Systems must be able to detect maliciously injected or spoofed packets; this requires the design and development of source authentication mechanisms to verify the packet originator.
  - Mechanisms must be in place for preventing compromising networks or systems from using the secret keys of legitimate networks or systems to be authenticated.
  - Various types of attacks can compromise the availability of any system or network; addressing this issue requires the design and development of mechanisms to accomplish graceful degradation in the presence of malicious activity.
  - Mechanisms necessary to detect adversaries that may disrupt operation, leading to denial of service (DoS), must be designed and developed.
- *Dependability* — The ability to deliver service that can justifiably be trusted; dependability is an integrative concept that encompasses the following attributes:
  - *Availability* — readiness for correct service
  - *Reliability* — continuity of correct service

- *Safety* — absence of catastrophic consequences for users and the environment
- *Confidentiality* — absence of unauthorized disclosure of information
- *Integrity* — absence of improper system state alterations
- *Maintainability* — ability to undergo repairs and modifications
- *Confidentiality* and *integrity*
  - *Secrecy* — Confidentiality is about controlling the ability of any given subject to extract information from an object. Under certain circumstances, it is possible that attacks against confidentiality affect user anonymity.
  - *Service integrity* — It will ensure that the content of a message or services was not altered during transit by an adversary. Integrity can be provided by either classical or keyed hash functions. Some integrity check must be done, but attention has to be paid to the overhead that it implies and to the time required to compute the check. Because packets are encrypted before transmission, the most likely forgery that can be imagined is a random change of bits in a packet.
- *Secure group management* — Each network/system exhibits limited computational and communication capabilities; however, in-network data aggregation and analysis can be performed by groups of systems. Secure protocols for group management are therefore required, securely admitting new group members and supporting secure group communication. The outcome of the group computation is usually transmitted to a base station. The output must be authenticated to ensure it comes from a valid group. Any solution must also be efficient in terms of computational and power resources, preventing many classical group-management approaches.
- *Intrusion detection* — Wireless networks and systems are susceptible to many forms of intrusion. In wired networks, traffic and computation are typically monitored and analyzed for anomalies at various concentration points. This is often expensive in terms of network memory and energy consumption, as well as its inherently limited bandwidth. It is important to understand how cooperating adversaries might attack the system. The use of secure groups may be a promising approach for decentralized intrusion detection.
- *Data aggregation* — One of the major benefits of wireless networks is the provisioning of fine-grained data.
- *Self-organization* — The wireless networks topology must be adapted in case of node or system compromise and failure. If a

malicious node discloses the network topology, routing establishment paths may be affected as well. One critical factor is that many wireless systems are mobile, and this mobility affects self-organization.

- *Anonymity/pseudonymity* — In wireless systems, the disclosure of personal information will be inevitable, even with the use of policies and well-defined user preferences. This is due to the enormous size of the information space that must be constructed for all possible scenarios of everyday interactions that must be taken into consideration; therefore, it is necessary to have interactive mechanisms and interfaces for the control of personal information disclosure with the simultaneous necessary transparency.

- *Security management* — Due to potentially harsh, uncertain, and dynamic environments, as well as the energy, power, and bandwidth constraints of the participating systems, wireless networks pose additional challenges in the context of network discovery, network control and routing, collaborative information processing, querying, and tasking. Security management is a very difficult and multi-dimensional problem.

Conclusions

The integration of the Internet with telecommunication networks promises a distributed computing infrastructure that provides globally available services. Due to the continuous increase in the number of portable devices connected to the Internet, the demand for users to have the ability of accessing any kind of information services regardless of their location will increase. The main challenges that must be addressed for roaming in such a heterogeneous environment and the provisioning of location-dependent services are the design and implementation of accounting systems. Currently, particular charging and billing models are used for the Internet and different ones for wireless communications, but these models will have to be revised for dealing with such a global integrated system. Accounting is the subject of various research projects and is being standardized by the IETF. Recently, the Authentication, Authorization, and Accounting Working Group of the IETF selected Diameter as the preferred protocol for transport accounting, although other protocols such as SNMP are also considered. Mobile middleware (specifically, mobile agents) can also be a very suitable candidate technology for supporting location-based accounting. Mobile agents provide mobility awareness, dynamicity, location awareness, personalization, interoperability, and security, which can be real benefits when it comes to location-based accounting. Furthermore, changing location also raises significant security issues for the authentication of mobile users and terminals, authorization to access system resources, communication secrecy, and integrity assurance, as well as many other security issues.

References

[1] Aboba, B., Arkko, J., and Harrington, D., *Introduction to Accounting Management*, Request for Comments 2975, Internet Engineering Task Force (IETF), 2000 (http://www.ietf.org/rfc/rfc2975.txt).

[2] Glass, S., Hiller, T., Jacobs, S., and Perkins, C., *Mobile IP Authentication, Authorization, and Accounting Requirements*, Request for Comments 2977, Internet Engineering Task Force (IETF), 2000 (http://www.ietf.org/rfc/rfc2977.txt).

[3] Finseth, C., *An Access Control Protocol, Sometimes Called TACACS*, Request for Comments 1492, Internet Engineering Task Force (IETF), 1993 (http://www.ietf.org/rfc/rfc1492.txt).

[4] Levi, D., Meyer, P., and Stewart, B., *Simple Network Management Protocol (SNMP) Applications*, Request for Comments 3413, Internet Engineering Task Force (IETF), 2002 (http://www.ietf.org/rfc/rfc3413.txt).

[5] Rigney, C., Willens, S., Rubens, A., and Simpson, W., *Remote Authentication Dial In User Service (RADIUS)*, Request for Comments 2865, Internet Engineering Task Force (IETF), 2000 (http://www.ietf.org/rfc/rfc2685.txt).

[6] Rigney, C., *Radius Accounting*, Request for Comments 2866, Internet Engineering Task Force (IETF), 2000 (http://www.ietf.org/rfc/rfc2866.txt).

[7] Calhoun, P., Loughney, J., Guttman, E., Zorn, G., and Arkko, J., *Diameter Base Protocol*, Request for Comments 3588, 2003 (http://www.ietf.org/rfc/rfc2866.txt).

[8] Aiken, B. et al., *Network Policy and Services: A Report of a Workshop on Middleware*, Request for Comments 2768, Internet Engineering Task Force (IETF), 2000 (http://www.ietf.org/rfc/rfc2768.txt).

[9] Bellavista, P., Corradi, A., and Stefanelli, C., An integrated management environment for network resources and services, *IEEE J. Selected Areas Commun.*, 18(5), 676–685, 2000.

[10] Bellavista, P., Corradi, A., and Stefanelli, C., Mobile agent middleware to support mobile computing, *IEEE Comput.*, 34(3), 73–81, 2001.

[11] Bellavista, P., Corradi, A., and Vecchi, S., Mobile agent solutions for accounting management in mobile computing, in *Proc. of the 7th IEEE Int. Symp. on Computers and Communications (ISCC'02)*, Taormina, Italy, July, 2002, pp. 753–760.

[12] Bellavista, P., Corradi, A., and Vecchi, S., Mobile agents for usage-based accounting in wireless ubiquitous environments, in *Proc. of WOA 2002*, Bologna, Italy, November, 2002.

[13] Bellavista, P., Corradi, A., and Vecchi, S., Application domain accounting for roaming services, in *Proc. of the 9th IEEE Workshop on Future Trends of Distributed Computing Systems (FTDCS'03)*, San Juan, Puerto Rico, May, 2003, pp. 359–366.

[14] Bellavista, P., Corradi, A., and Vecchi, S., QoS-aware accounting in mobile computing scenarios, in *Proc. of the 11th Euromicro Workshop on Parallel, Distributed, and Network-Based Processing (PDP'03)*, Genoa, Italy, February, 2003, pp. 537–543.

[15] Aboba, B., Arkko, J., and Harrington, D., *Introduction to Accounting Management*, Request for Comments 2975, Internet Engineering Task Force (IETF), 2000 (http://www.ietf.org/rfc/rfc2975.txt).

[16] Jonkers, H. and Hille, S., *Accounting Context: Application and Issues*, Authorization, Authentication, and Accounting Architecture Research Group, 2000, www.aaaarch.org/doc06/file-11249.pdf.

[17] 3G TR 23.815 Version 5.0.0 (2002-03), 3rd Generation Partnership Project; Technical Specification Group Services and System Aspects; Service Aspects; Charging Implications of IMS Architecture (Release 5).

[18] 3G TS 22.115 Version 5.2.0 (2002-03), 3rd Generation Partnership Project; Technical Specification Group Services and System Aspects; Service Aspects; Charging and Billing (Release 5).

[19] Carle, G., Zander, S., and Zseby, T., *Policy-Based Accounting*, Request for Comments 3334, Internet Engineering Task Force (IETF), 2002 (http://www.ietf.org/rfc/rfc3334.txt).

[20] 3G TS 32.200 Version 5.2.0 (2002-12), 3rd Generation Partnership Project; Technical Specification Group Services and System Aspects; Telecommunication Management; Charging Management; Charging Principles (Release 5).

[21] Brownlee, N., *RTFM: New Attributes for Traffic Flow Measurement*, Request for Comments 2724, Internet Engineering Task Force (IETF), 1999 (http://www.ietf.org/rfc/rfc2724.txt).

[22] Brownlee, N., *Traffic Flow Measurement: Architecture*, Request for Comments 2722, Internet Engineering Task Force (IETF), 1999 (http://www.ietf.org/rfc/rfc2722.txt).

[23] IP Flow Information Export (IPFIX), www.ietf.org/html.charters/ipfix-charter.html.

[24] Bhushan, B. et al., Federated accounting: service charging and billing in a business-to-business environment, in *Proc. of IEEE/IFIP Int. Symp. on Integrated Network Management (IM 2001)*, Seattle, WA, May, 2001, pp. 107–121.

[25] Brownlee, N., *Traffic Flow Measurement: Experiences with NeTraMet*, Request for Comments 2123, Internet Engineering Task Force (IETF), 1997 (http://www.ietf.org/rfc/rfc2123.txt).

[26] Internet Open Trading Protocol (TRADE), www.ietf.org/html.charters/trade-charter.html.

[27] 3GPP TS 32.205 Version 5.2.0 (2002-12), 3rd Generation Partnership Project; Technical Specification Group Services and System Aspects; Telecommunication Management; Charging Management; 3G Charging Data Description for the Circuit-Switched (CS) Domain (Release 5).

[28] 3GPP TS 32.215 Version 5.2.0 (2002-12), 3rd Generation Partnership Project; Technical Specification Group Services and System Aspects; Telecommunication Management; Charging Management; Charging Data Description for the Packet-Switched (PS) Domain (Release 5).

[29] 3GPP TS 32.225 Version 5.1.0 (2002-12), 3rd Generation Partnership Project; Technical Specification Group Services and System Aspects; Telecommunication Management; Charging Management; Charging Data Description for the IP Multimedia Subsystem (Release 5).

[30] 3GPP TS 32.235 Version 5.1.0 (2001-12), 3rd Generation Partnership Project; Technical Specification Group Services and System Aspects; Telecommunication Management; Charging Management; Charging Data Description for Application Services (Release 4).

[31] 3GPP TS 29.198-12 Version 5.1.0 (2002-09) 3rd Generation Partnership Project; Technical Specification Group Core Network; Open Service Access (OSA); Application Programming Interface (API); Part 12: Charging (Release5).

Section 6

MOBILE MIDDLEWARE FOR CONTEXT-DEPENDENT SERVICES

Chapter 33

Mobile Middleware: Processing Context-Related Data in Mobile Environments

Yih-Farn (Robin) Chen and Rittwik Jana

CONTENTS

Introduction

Real-world scenarios in which human–computer interactions are guided by the surrounding context have created a new realm of computing — the era of *context-aware computing*. As the man–machine interface becomes more pervasive and intimate, it will have to rely on cognitive science as a basis for understanding what humans are capable of doing under different situations and modify the application behavior accordingly. Context information gathered from sensors, networks, user profiles, and other sources can be used to adapt mobile applications to enhance the user experience. Much emphasis has been placed lately on research to enrich the user experience with more natural inputs such as voice, gestures, emotions, and facial expressions via video and situation awareness of the environment captured using a myriad of sensor-enabling technologies [1–3]. Unfortunately, obtaining contextual knowledge is a nontrivial task. Sensors can be distributed and may communicate their data using a variety of protocols and formats. Raw sensor data may have to undergo complex postprocessing before it can become useful in any applications [7]. The overarching goals in researchers' minds are, first, to facilitate the understanding of these complex input processes by computers and, second, to enhance human–computer interaction.

Context can be defined in a number of different ways. In Abowd and Dey [4], context is defined as any information that can be used in an entity's situation. An entity can include a person, place, or object relevant to the conversation between the end user and the application. Many researchers have also tried to define context by enumerating examples of context, namely computing context, physical context, and user context [5]. We take a further look at the various definitions of context in the next section from device, user, and environment perspectives. One goal of context-aware computing is to acquire and utilize this information to provide services that are appropriate to the particular place, people, space, or time; for example, if the system knows that a user is in a meeting, the service may want to notify that user's cellphone using the vibration mode instead of ringing. By taking into account the relevant context (i.e., the different exposed situations that are relevant to a requested service) applications can be streamlined and customized to provide a more satisfying overall user experience. It is, however, the responsibility of the application designer to present and use context intelligently to provide such an experience.

Examining and reacting to context are becoming critically important with the increasing popularity of mobile computing. Today's information era requires people to be constantly on the move; yet, they must remain productive and informed. Mobile computing puts many constraints on the clients' ability to utilize services that they are accustomed to in a regular desktop environment; in particular, client resource constraints imposed by display width and size and slow or fluctuating bandwidth connections are typical. Specifying required input parameters to request a service can be a tedious and frustrating experience; for example, when a cellphone user makes a request for the current weather, the service should figure out where the mobile user is and what device limitations exist — short message service (SMS) or multimedia messaging service (MMS) — and then deliver the information appropriately.

Acquiring context is not a trivial task. This is one of the main reasons why it is difficult to use context. Quite often this means dealing with sensor-based technologies that detect various phenomena about the environment. Acquiring context can be either explicit (i.e., provided by an end user by means of textual, voice, or video input) or implicit through automatic inference. Apart from the lack of proliferation of this kind of nonstandard technology (e.g., sensors, global positioning receivers), there is also some hesitation to ubiquitously deploy such technologies due to privacy and other considerations. A consensus must be reached among public utilities to standardize the rollout of such an infrastructure. Of course, a viable economics model must be developed to allow third-party vendors to participate competitively. The recent mandatory E911 initiative by the Federal Communications Commission of the U.S. government aims at deploying location sensing technologies in coordination with cellular carriers to report the telephone number of a wireless 911 carrier and the precise location of the antenna to within a 50- to 100-meter resolution [9].

In this chapter, we examine a particular methodology of modeling context information with reference to a framework of a service platform that uses context information to provide value-added services. Next, we describe the various forms of context — namely, device, user, and environment — followed by a reference model or framework that describes context-aware service platforms. We also examine the important interactions between the middleware platform and the context processing platform. In particular, we highlight the service-level adaptations that occur while connecting these entities. Two examples of context-aware services are also explained later, followed by some current research activities.

What Is Context?

Many mobile services today are delivered without consideration for the context in which a mobile user currently resides. A mobile user in an important meeting may prefer to have important notifications communicated through text messages rather than actual phone calls. The experience of mobile users can be greatly enhanced if the surrounding context information is somehow automatically collected, processed, and used intelligently to modify the service that is being delivered to them.

Typically, a context is defined as a situation within which something exists or happens [4]. To facilitate our discussions on context processing, we use a model that separates the concerns specific to devices, users, and environments. This allows us later on to employ an entity-relationship model [51] on contextual data for ease of processing and reasoning. Figure 33.1 gives a typical scenario in which mobile users might reside and demonstrates the three different kinds of context. We discuss how we can build a richer context-processing environment by reasoning from the separate pieces of context information collected.

Device Context

We define device context as the state in which a device exists, not including context information that can be derived (perhaps through the context reasoning engine, to be described later) by its surrounding environment or from information about the mobile user that owns the device. A device may have a built-in mechanism to collect its own context; for example, the location of the device may be determined by an embedded global position system (GPS) receiver. The location-tracking mechanism of the cellphone owned by the girl walking the dog in Figure 33.1 may help us determine that she is in Central Park, New York, if we can determine that she has the cellphone with her. The separation of the device context and

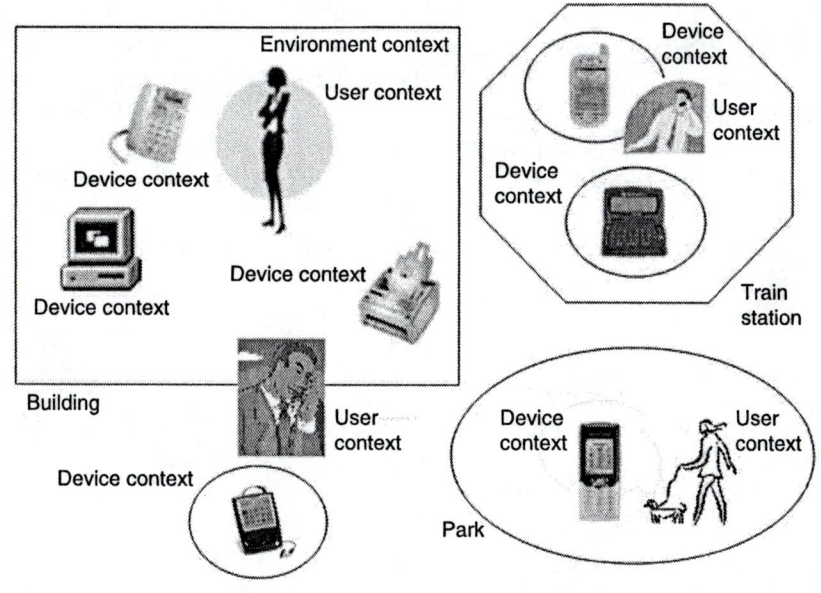

Figure 33.1   User context, device context, and environment context.

user context is important, as a user may have multiple devices and may not always have all of them in proximity.

Some internal and changeable state of the device may also be considered context; for example, a device may currently be in the vibration mode only, with the audio turned off completely. In addition, the device may have its own sensors to monitor and display its own temperature, battery power, etc. Although we can try to infer the temperature of a cellphone from the general temperature in Central Park, it may not always be accurate because the temperature of the cellphone itself may be quite different, depending on whether it is kept in a pocket or exposed to the sunlight.

User Context

We define user context as the state in which a mobile user exists at a given time. A user may be driving, jogging, in a meeting, or talking on the phone or may even be emotionally disturbed. Some state may be derived from the user's calendar (with some probability of being correct) and some may be obtained by attaching a radiofrequency identification (RFID) tag on the person of the user; yet, some may be specified explicitly by the user. Note again that the location of users is not necessarily the same as the locations of their devices. It also depends on the accuracy of the location determination technology; for example, drivers may choose to carry their cellphones with them but leave their wireless local area network (WLAN)-enabled personal digital assistants (PDAs) in a briefcase in the car trunk. Depending on how the location-determination technology is used in an application, the two devices may well be considered to be in two different locations: one that can communicate with other dashboard devices through the Bluetooth® technology and one, in the vehicle trunk, that cannot.

Environment Context

We define environment context as the state in which a particular *space* exists. A space may be a room, a building, a park, a train station, an area covered by a Zip Code, or a *virtual* space (such as a chat room) in which a user or device resides. Some typical attributes of a space may be the temperature, the number of people in it, or the noise level. The way in which we communicate with a user in a noisy train station may be quite different from how we would commutate with a coworker sitting in an office [39].

Device–User–Environment Interactions

By assembling the device-specific, user-specific, and environment-specific information and reasoning from it, we can derive very rich contextual information. We may infer, for example, that a particular mobile user is having a meeting in a very noisy place (the man at the train station in Figure 33.1 or someone at a party) and is holding a multimedia-enabled device in his hand by assembling all relevant context information from the environment, the user, and the device itself. To alert this mobile user of an emergency situation occurring in his proximity, the best way to communicate with him is probably to alert him through device vibration followed by a video message with text, but not by relying on audio. On the other hand, if we can infer that the mobile user is driving a car, then we can probably deliver the same message with a ring tone followed by audio automatically without requiring him to push any buttons to pick up the message.

It is also important to determine that a user has left a space and entered another one; for example, suppose the user with a BlackBerry™ device in Figure 33.1 is leaving the building and is on his way to Central Park. For the purpose of seamless mobility [43], the mobile device must be able to sense the environment change and may have to move from a WLAN (WiFi) connection to a cellular General Packet Radio Service

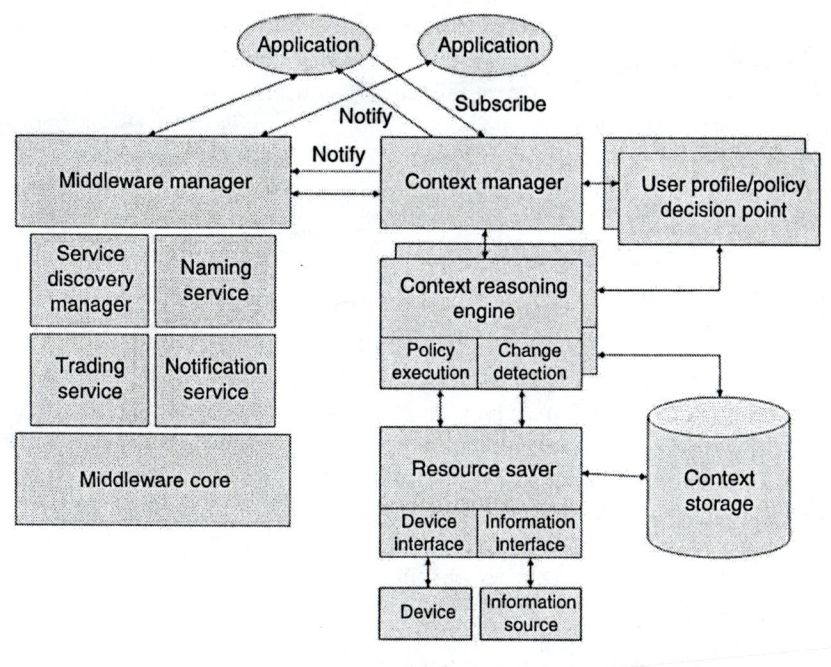

Figure 33.2 A reference model for context-aware service platforms.

(GPRS)/Global System for Mobile Communications (GSM) connection or another form of communication, depending on what is available outside the building. Later, we give more details on how we can infer from different contexts.

A Reference Model for Context-Aware Service Platforms

To facilitate the discussion of context-aware service platforms, we propose a reference model for such platforms. The model provides a framework on how context information can be acquired and delivered in a reliable and secure way to middleware platforms that utilize it to make decisions on service adaptations. We also examine current standards and technologies that facilitate the use of context information during application delivery. An illustration of the context reference framework is shown in Figure 33.2. The framework consists of two major clusters of components interacting with each other and the applications: *contextware* [1] on the right-hand side and middleware on the left-hand side. Our model is similar to the context management architecture proposed by Conan et al. [47], with the following major differences: (1) notions of policy decision points and policy enforcement points, (2) a resource server that has plug-in modules to interface with both devices and information sources, and (3) a context storage database to hold both dynamic and static context states. The following sections describe the constituent components of the framework.

Context Manager

The *context manager* must perform two functions to enable a pervasive environment. First, the context manager must describe or model the context information. Second, the manager must support queries and subscriptions from applications and notify the corresponding entities when an event of interest occurs. The manager provides a plug-in context registration service which lets applications share the abstracted context information. For example, a conference call application may be interested in transferring a cellular call originating at a car in the parking lot to the driver's WiFi-enabled cellphone as he is walking down the office corridors and eventually to a desktop phone when he enters his office. It can subscribe to an event on "Smith entering room B259" with the context manager.

Context Acquisition Through the Resource Server

A wide selection of sensors and sensing technologies can be applied to collect contextual information. Some of the sensing technologies can be further categorized into:

- Temperature and humidity
- Vision and light
- Location, orientation, and presence
- Magnetic and electric fields
- Touch, pressure, and shock
- Audio
- Weight
- Smell (gas sensors)
- Acceleration (motion detection)

Often it is not enough to collect all the contextual information from a single sensor. In mobile computing, sensors can be in one place or distributed spatially. In an array of sensors, communication protocols have to be engineered to facilitate the fusion of such data. This fusion can involve rather complex postprocessing; for example, in wearable computing, sensors are placed all over the human body [22], and placement of these sensors is crucial to the contextual observations. In any application, the designer has to ultimately choose the correct sensor type and its relevant positioning. In our model, we assume that a resource server is equipped with the appropriate device interfaces that can facilitate the collection of sensor information from various sources.

A large number of research papers, however, concentrates on location-based services using either indoor or outdoor location information as context [13–16,49]. The location-determination technology may provide an interface that allows our resource server to collect location information of interested parties. Location-aware computing systems or applications are often designed to respond to a user's location, either spontaneously or when explicitly activated by a user request.

Context Storage

Context data storage is important for the use of historical information to establish trends or predict future context values. Ideally, a context-based storage system consists of a logical context data model and a physical data storage space. By using the entity-relationship model [51], the context information can be easily represented using a variety of *entities* — people, place, things, attributes — that describe some property of an entity and *relationships*, which involve two or more entities. It is necessary to have an access policy mechanism that enforces who is allowed to access what data; for example, the location data of a truck driver may be accessible by the trucking company from 8 a.m. to 5 p.m. but not outside of the driver's work hours. We discuss how policies can be deployed and enforced in context-aware services later in this section.

Several other parameters must be considered regarding context storage for ease of data management:

- *Persistence* — Potentially, we can associate each data item with a temporal attribute. We can designate its time-to-live (TTL) value or associate the context data with a particular communication session (such as a conference call). The data should be removed as soon as the session is terminated.
- *History* — Logging the history of sensor readings has more implications on what can be done with the data than having only the last instance available. For example, the history of location information and orientation can help us calculate the speed, direction, and even acceleration of a vehicle. Data points may be associated with precise time stamps or may just be ordered.
- *Granularity* — This parameter determines the frequency of context snapshots or freshness of the latest reading.

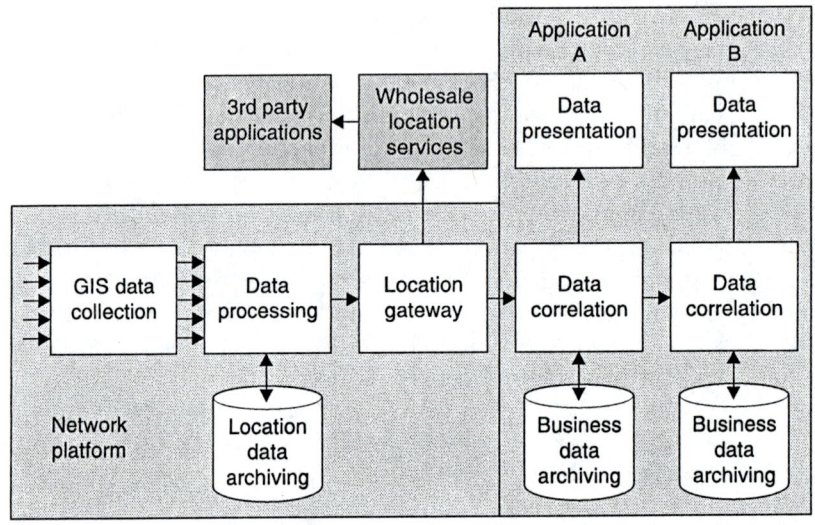

Figure 33.3 Typical data management architecture of context-aware service supporting location-based services.

Management of Contextual Data

Figure 33.3 shows a typical data management architecture for a context-aware service that supports location-based services (LBSs). In essence, the location data is ingested and processed at the network platform. This postprocessed information is then made available to application providers via an open interface, thereby enabling applications to adhere to a unified and consistent mechanism to offer location information. The location gateway acts as the intermediary between wholesale services and the data-processing engine. Data correlation involves combining or correlating with other data points (e.g., call detail records, user work schedule) and presenting an appropriate view of the combined data to an application. An example data presentation would be a map that shows the current locations of all fleet vehicles of a rental car company.

Comparing this architecture to the reference model, the data collection is achieved by the resource servers, the data processing and correlation engine will sit within the context reasoning block, and the location gateway acts as the context manager for vending location information to interested consumers. The example also demonstrates that context reasoning (see below) may be performed inside contextware or in external applications that handle their own business logic.

Context Reasoning

We consider context reasoning as a means to deduce new and relevant information to an application or user from the various sources of context data, following the definition by Przybilski et al. [44]. The authors considered several approaches (among others) to context reasoning:

- *Low-level context processing* — Context reasoning frequently involves *context inference*, which maps low-level data (such as GPS signals and sensor data) to high-level data (such as location and the level of danger). Logic reasoning and ontology are frequently employed here. Another common example is *sensor data fusion*. The primary motivation here is to reduce communication costs by the integration of similar data sources or to recognize falsely calibrated sensors through simple reasoning so outliers can be rejected in further analyses.
- *Application-centric view* — This approach focuses on how we use context data in applications. One of the key considerations here is how an application and the underlying contextware [1] agree on how and what context data can be used in the applications and what inferences can be made to enhance the applications; for example, which application and which other users have the rights to know the location information of a truck driver, and what period during the day? We will discuss the issue of policies further in the next section.
- *Context monitoring* — This approach can be critical for proactive applications that want to know or predict what the context will likely change to and that want to use the prediction to change the way a service is delivered; for example, a conference call service may proactively monitor a mobile user's location and switch the conference call to an office phone or desktop as soon as the mobile user enters the office.

Context reasoning can provide additional information that does not exist in the context storage, but care must be taken to maintain data consistency before adding such information to the context storage, as the underlying raw context information may change continually.

Policy Management

The acquisition and storage of context information may trigger privacy concerns. Any context-processing mechanisms must adhere strictly to the security and privacy policies. The Internet Engineering Task Force (IETF)

Open Pluggable Edge Services (OPES) group [21] has developed an architecture framework to authorize, invoke, and trace application-level services on the network between the endpoints. In particular, RFC 3838 describes how policies are authored, deployed, and enforced in the network. We will briefly describe these concepts and how they might apply to context-aware services that are frequently deployed on the network between target applications and clients.

Following the OPES model, the policy functions of the reference architecture are decomposed into three components: rule author (RA), policy decision point (PDP), and policy enforcement point (PEP). Multiple RAs may provide rules to the PDP. These rules control the invocation of context-aware services on behalf of the rule author. Each PDP has the authority to decide what policies should be used that concern a particular party or a group of parties in an application. PEPs execute policies deployed from PDPs. The PDPs and the PEPs (as shown in Figure 33.2) interpret the collected rules and appropriately enforce them.

Enabling Context-Aware Services

The use of context information by applications in a mobile environment poses a number of challenges arising from the distributed and dynamic nature of sensors, the accuracy and resolution of sensors, and the fusion of output of multiple sensors to determine context. In addition, the mobile environment poses further challenges with regard to the dependability, predictability, and timeliness of communication. Middleware is required that provides abstractions for the fusion of sensor information to determine context, representation of context, and intelligence inference. Essential services that provide support for operation in a mobile environment, such as supporting the reliability of communication, are also required. In this section, we examine how a middleware service platform can aid in the delivery of context-aware services.

Application to Context Manager

The context manager can make the context information accessible either directly to external applications or through the middleware service platform. The context manager provides a query interface for other agents to obtain the current context. Depending on the type of logic or learning mechanism used, the context provider agents may have different ways of evaluating queries; however, the context consumers query the manager in a uniform way. This greatly aids development of context-aware applications. The context manager is the administrator of the system, responsible for monitoring ongoing context-based sessions and managing the environment resources.

Applications can subscribe to context change notifications, where the clients essentially register with the manager, asking that "when something about this happens, let me know" and to obtain the authorization needed to use and negotiate context information. The manager stores relevant information in the context storage for inference, consistency, and knowledge sharing. This context storage is used to retrieve information about entities in the environment. The manager accepts subscription requests from user, service, and context provider agents to be added to the agent directory. This subscription phase allows the manager to search and select the best available context that will satisfy the agreements required by applications and users. In addition, the manager binds the provider agents to applications according to their context agreement. It maintains this information in the context storage for future use so resources can be reconfigured as the context changes.

Context Manager to Middleware Manager Interactions

Having external applications to process changes in context and to reconstruct a model of the mobile user environment may prove to be daunting to application programmers. A middleware platform may interact with the context manager and help simplify the task of application adaptation. The context manager can notify the middleware manager of changes in contexts that are detected. This is usually implemented in a rule-based notification server that decides when the desired situation has been fulfilled in order to activate user rules, producing notifications. The algorithmic processing in the context manager undertakes most of the intensive task of event aggregation and correlation. A sentient notification service [31] achieves the following goals by connecting the context manager to the middleware manager:

- It provides a model of producing logical abstractions over physical entities and their associated context.
- It senses the environment and produces time-varying inferences.
- It gives the middleware manager a powerful pattern-matching ability for large-scale events. The Sentient Computing project [32] uses sensors and resource status data to maintain a model of the world that is shared between users and applications. Based on the Bat location system, Sentient Computing can help store and retrieve

context-sensitive data about mobile users. Whenever information is created, the system knows who created it, where they were, and who they were with. This contextual metadata can support the retrieval of multimedia information.

Context-Aware Mobile Service Platform

Now we discuss one example (see Figure 33.4) of a reference implementation from a system perspective of how context is acquired and provided to a mobile service platform. This information is then supplied to applications running on the middleware platform. Using location as an example piece of context information, different positioning technologies are used to locate a subscriber. These interfaces are facilitated via a location-based services platform with the standardized Mobile Location Protocol (MLP) interface. Varying degrees of accuracy are achieved due to the complexity of the positioning technologies used. The LBS platform also has other responsibilities (e.g., interfacing with geographical information system [GIS] engines for mapping, geocoding, and routing).

Central to the connection between the middleware platform and the LBS platform is a network resource gateway standardized by Open Service Access (OSA) in the Parlay framework [20]. The Parlay Group aims to intimately link IT applications with the capabilities of the telecommunications world by specifying and promoting application programming interfaces (APIs) that are secure, easy to use, rich in functionality, and based on open standards. Parlay integrates telecommunications network capabilities with IT applications via a secure, measured, and billable interface. Providing this abstraction offers several advantages, namely:

- Being open and technology independent
- Eliminating the need for programmers to learn difficult telecommunications protocols, thus lowering costs, and raising the programming abstraction level to the point where telecommunications capabilities become just normal IT APIs
- Making it possible for external application servers to interact with telecom network capabilities
- Reducing business risk for all parties involved via the open API model
- Allowing the creation of innovative and rapid applications that function across multiple networks
- Supporting 2G, 2.5G, and 3G wireless networks with the same APIs, providing a future-proof evolution path for network services

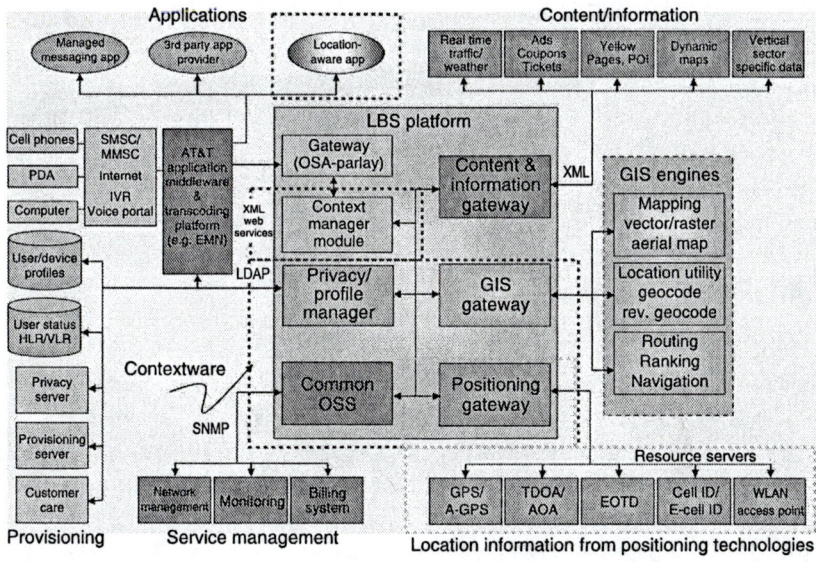

Figure 33.4 Context-aware mobile service platform.

The middleware service platform Enterprise Messaging Network (EMN®) [38] has been developed at AT&T Labs Research. This middleware platform allows a set of disparate mobile devices to communicate with each other and to securely access corporate and Internet content and services. If we were to look inside EMN, it consists of a set of distributed and cooperative component containers. The role of a container is defined by the set of components it manages. We distinguish two main roles, namely *gateways* and *servers*. Gateways handle protocol-specific interfaces to mobile devices and perform authentication, device profiling, and session management functions. Gateways host particular types of components, called *devlets,* which have the main role of providing the device access to the protocol adaptation layer. EMN provides devlets for a multitude of protocols: e-mail, HTTP, pager, voice, fax, SMS, and instant messaging.

The servers are responsible for accessing and processing (such as content repurposing and alert scheduling) the information. The components that make up the behavior of an EMN server are called *infolets.* Infolets implement the associated application logic and usually provide access to one or more sources of information. The context information can be accessed from the EMN gateways and servers using ParlayX Web service calls exposed via the terminal location API. The application logic can consist of tying the different information gathered from several infolets and repurposing this information to a variety of user interfaces. In the latter example, the OSA gateway acts as the context manager and implements a policy management and enforcement engine. The mobile service platform queries the OSA gateway for context information and utilizes this in its realm of deployed applications. The middleware platform also presents the end user with a consistent look and feel for the context-aware application regardless of the destination device.

Middleware Manager to Applications Interactions: Context Markup Language

Scandon and Sadeh [41] introduced a Semantic Web architecture aimed at supporting the automated discovery and access of personal resources; support for these exist in a variety of context-aware applications. Within this architecture, each source of contextual information (e.g., a calendar, location tracking functionality, collections of relevant user preferences, organizational databases) is represented as a Semantic Web service. A central element of the architecture is its *semantic e-Wallet*, which acts as a directory of contextual resources for users while enforcing their *privacy preferences*. Privacy preferences enable users to specify what information can be provided to whom in different contexts. The security community has developed powerful languages to capture access control privileges such as the Security Assertion Markup Language (SAML), the XML Access Control Markup Language (XACML), and the Enterprise Privacy Authorization Language (EPAL); however, Scandon and Sadeh [41] extend this to incorporate semantic technologies.

Privacy Requirements

Privacy protection remains a serious impediment to the widespread adoption and deployment of context-aware and ubiquitous computing environments. It is important that privacy protection mechanisms are in place from the start so a relationship of trust can be formed between technology users and service providers. "Privacy is the claim of individuals, groups, or institutions to determine for themselves when, how, and to what extent, information about them is communicated to others" [45]. The ownership of information is with the subject, who alone should control the release of information. Owners can also delegate this function to a service provider, via a service-level agreement, who can then release this information on the owner's behalf. Several privacy-enhancing mechanisms have been proposed, all of which have a role-based, access-control mechanism as their foundation.

Traditionally, with respect to location-based services, operators have kept subscriber location information contained within their own networks. With roaming, the subscriber's location will be transferred over network boundaries and outside networks. The basic requirements include the following:

- The identity of the target (e.g., the Mobile Station International ISDN Number [MSISDN]) should be made anonymous to the service provider.
- The user will have control over the level of granularity of positioning to protect privacy.
- The user will have control over who is allowed or which application is able to locate the user:
  Control statically by user for each application
  Control dynamically based on application
  Control dynamically per location per request

Context-Aware Mobile Services

In this section, we will try to utilize all the concepts we have explained in the chapter by giving examples of context-aware mobile services that can be built on top of the reference model and the architecture that we proposed earlier in this chapter. We will focus on two examples: an audio/video conference call example and an emergency alert example.

Seamless Transition of Audio/Video Conference Call

Assume that our friend Smith (the one with a BlackBerry in Figure 33.1) is engaged in an audio/video conference call with his colleagues as he walks from the park back to the office building. Suppose Smith was tagged with an RFID, and the context acquisition software in the building immediately senses the presence of Smith and begins tracking his location and streaming the information to the context resource server, as shown in Figure 33.2. As soon as the context resource server/reasoning engine senses that Smith is in his office, the conference call is transferred to his desktop speaker phone or the video/audio conference call software on his PC. In addition, because the contextware also senses that an RFID-tagged fax machine is nearby (through the RFID resource server), it notifies Smith's colleagues that they can now send faxes of offline documents to that machine if necessary.

Emergency Alerts

Suppose our other friend, Adam, while chatting on the phone at the train station (Figure 33.1), is not aware that a bomb threat exists at a federal office building that he plans to visit after he gets off the train. While monitoring Adam's calendar, our context reasoning engine collects news sources through the Really Simple Syndication (RSS) protocol and figures out that the entire building has been evacuated. Because our alert application (Figure 33.2) subscribes to any events that might affect Adam's calendar, it is notified that the appointment should be canceled. It requests the presence information of the user and his devices from the middleware manager, which then determines that an alert should be sent to Adam's pager because Adam is still chatting on his phone.

Research Challenges and Emerging Technologies

This section highlights some of the more recent projects in several institutions that are related to context-aware computing. This is by no means an exhaustive list and does not indicate the relative importance of one project over another. We found these projects interesting in their own right and found it necessary to include them to provide a sense of the different flavors of context-aware computing.

Location Tracking Systems

The *Cricket* location support system from MIT provides an indoor analog of GPS to provide information about location, orientation, and geographic spaces [15]. Cricket beacons, mounted on walls or ceilings, transmit ultrasound and RF signals; compact listeners, attached to mobile or static devices, use the difference in signal arrival times to determine where they are. Cricket V2 has a new software stack that runs on TinyOs [50], an open-source operating system designed for wireless embedded sensor networks, and provides better support for continuous object tracking. Many context-aware applications in pervasive and sensor computing environments, such as human/robot navigation, physical/virtual computer games, and patient tracking and monitoring, can benefit from an indoor location determination system such as Cricket.

The *Bat ultrasonic location system* was originally developed in AT&T Labs. The first experiments with context-aware systems used room-scale information generated by infrared active badges [31]. This small device, worn by personnel, transmits a unique infrared signal every 10 seconds. Each office within a building is equipped with one or more networked sensors that detect these transmissions. The location of the badge (and hence its wearer) can thus be determined on the basis of information provided by these sensors, but many applications require fine-grained three-dimensional (3D) location and orientation information, which active badges cannot supply. AT&T Labs have since developed a 3D ultrasonic location system (known as the *Bat system* [13]), which is low power, wireless, and inexpensive. Recently, the authors reported their experience in deploying the Bat system over a large area and running it for a period exceeding two years [49]. The survey revealed that the two most important incentives for users to increase their Bat system usage and to wear their Bat devices were the implementation of more Bat applications (35 percent of personnel wear their Bats daily) and adoption of the system by other users.

Context Acquisition and Discovery

In the *Intentional Naming System* (INS) [30] from MIT, names are *intentional;* they describe application *intent* in the form of properties and attributes of resources and data, rather than simply network addresses (e.g., URLs) of objects. Potentially, the use of an INS would allow a context acquisition system to address resources easily in a dynamic environment lacking preconfigured support for describing, locating, and gaining access to available resources.

The *Smart Dust* [33] project from the University of California–Berkeley is probing the limitations of microfabrication technology to determine whether an autonomous sensing, computing, and communication system can be packed into a tiny device called a *mote*; this mote element would form the basis of an integrated, massively distributed sensor network. Only a few cubic centimeters in size, the motes collect light, temperature, humidity, and other data about their physical environment. The data is then relayed from one mote to a neighboring mote (mesh network) until it reaches its desired destination for processing.

Context Reasoning and Model Building

The *Sentient Computing project* [32] uses sensors and resource status data to maintain a model of the world that is shared between users and applications. Based on the Bat location system, Sentient Computing can help store and retrieve context-sensitive data about mobile users. Whenever information is created, the system knows who created it, where they were, and who they were with. This contextual metadata can support the retrieval of multimedia information.

Other Developments

Mobile computing devices such as palmtop computers, mobile phones, and personal digital assistants have gained widespread popularity. Even though devices and networking capabilities are becoming increasingly powerful, the design of mobile applications will continue to be constrained by the physical limitations. Mobile devices will likely continue to be battery driven, and wide-area wireless networks will have fluctuations in bandwidth depending on the physical location. Traditional middleware for fixed distributed systems cannot be used in the mobile environment, particularly for the reasons outlined earlier. Many researchers have investigated systems that collect context information and help applications adapt to changes. Research must continue, and middleware must be developed so a better quality of services can be delivered to mobile applications. Capra et al. [35] studied reflective middleware that maintains an updated representation of the context. Acquisition of context is also very important and has largely focused to date on obtaining location information. Progress has been made in the acquisition of other forms of context such gesture, voice, ambience, etc. Researchers have recognized the need to create context toolkits [39,40] or contextware that provides important abstractions and support for context-aware computing. We expect the seamless integration of contextware and middleware service platforms to emerge in the next decade to greatly enrich mobile users' experience.

The representation of context information in a universal way is important for systems to interoperate. The W3C community is also actively involved in defining standards such as Composite Capability/Preference Profiles (CC/PP), which has the role of representing delivery context in assisting device-independent presentation for the Web [36]. Recently, W3C annouced the Mobile Web Initiative; its goal is to make mobile devices first-class citizens of the World Wide Web. Research on context-aware services for mobile devices has the potential to make unique contributions to this initiative [52].

Conclusions

This chapter investigated the issues associated with context-related data processing in a mobile environment. The importance of context was discussed, followed by a definition of the central context entities: user, device, and environment. A reference model for context-aware service platforms was introduced which is responsible for the management, acquisition, storage, and reasoning of the collected data. As described in the chapter, providing a context-aware service is often facilitated by interactions between a middleware manager and applications using a context markup language or a vendor-independent API. Two such examples of context-aware services were highlighted toward the end of the chapter, followed by emerging research challenges and projects that continue to illustrate the importance of this field.

References

[1] Ferscha, A., Contextware: bridging physical and virtual worlds, in *Proc. of the 7th Int. Conf. on Reliable Software Technologies, ADA-EUROPE 2002*, Vienna, Austria, June 17–21, 2002.

[2] Hightower, J. and Boriello, G., Location systems for ubiquitous computing, in *IEEE Comput.*, 34(8), 57–66, 2001.

[3] Schmidt, A. and Beigl, M., There is more to context than location: environment sensing technologies for adaptive mobile user interfaces, in *Proc. of Interactive Applications of Mobile Computing (IMC'98)*, Rostock, Germany, November 24,1998.

[4] Abowd, G.D. and Dey, A.K., Towards a better understanding of context and context-awareness, in *Proc. of the 1st Int. Symp. on Handheld and Ubiquitous Computing*, Vol. 1707, Lecture Notes in Computer Science, Springer-Verlag, Heidelberg, 1999, pp. 304–307.

[5] Schilit, B., Adams, N., and Want, R., Context-aware computing applications, in *Proc. of IEEE Workshop on Mobile Computing Systems and Applications (WMCSA'94)*, Santa Cruz, CA, December, 1994, pp. 85–90.

[6] Cox, R.V., Kamm, C.A., Rabiner, L.R., Schroeter, J., and Wilpon, G.J., Speech and language processing for next-millennium communications services, *Proc. IEEE*, 88(8), 1314–1337, 2000.

[7] Urnes, T., Hatlen, A.S., Malm, P.S., and Myhre, O., Building distributed context-aware applications, *Pers. Ubiquitous Comput.*, 5, 38–41, 2001.

[8] Schilit, B., System Architecture for Context-Aware Mobile Computing, Ph.D. dissertation, Columbia University, New York, 1995.

[9] Wireless Enhanced 911, http://www.fcc.gov/911/enhanced/.

[10] Weiser, M., The computer for the 21st century, *Sci. Am.*, 265(3), 66–75, 1991.

[11] Pascoe, J., Ryan, N., and Morse, D., Issues in developing context-aware computing, in *Proc. of the 1st Int. Symp. on Handheld and Ubiquitous Computing*, Karlsruhe, Germany, September, 1999, pp. 208–221.

[12] Dey, A.K., Providing Architectural Support for Building Context-Aware Applications, Ph.D. dissertation, Georgia Institute of Technology, Atlanta, 2000.

[13] Ward, A., Jones, A., and Hopper, A., A new location technique for the active office, *IEEE Pers. Commun.*, 4(5), 42–47, 1997.

[14] Bahl, P. and Padmanabhan, V., RADAR: an in-building RF-based user location and tracking system, in *Proc. IEEE INFOCOM'00*, Tel Aviv, Israel, March, 2000, pp. 775–784.

[15] Priyantha, N.B., Chakraborty, A., and Balakrishnan, H., The cricket location-support system, in *Proc. of the 6th ACM/IEEE Int. Conf. on Mobile Computing and Networking (MOBICOM'00)*, Boston, MA, August, 2000, pp. 32–43.

[16] Want, R. and Russell, D.M., Ubiquitous electronic tagging, *IEEE Distributed Syst. Online,* 1(2), 2000.

[17] Harter, A., Hopper, A., Steggles, P., Ward, A., and Webster, P., The anatomy of a context-aware application, in *Proc. of the 5th ACM/IEEE Int. Conf. on Mobile Computing and Networking (MOBICOM'99),* Seattle, WA, August, 1999, pp. 59–68.

[18] *PulsON Technology Time Modulated UWB Overview,* Time Domain Corp., Huntsville, AL, 2001.

[19] Hightower, J. and Borriello, G., *Location Sensing Techniques,* Technical Report UW-CSE-01-07-01, Computer Science and Engineering Dept., University of Washington, Seattle, WA, 2001.

[20] The PARLAY Group, www.parlay.org.

[21] OPES (Open Pluggable Edge Services), http://www.ietf-opes.org.

[22] *MIThril: The Next Generation Research Platform for Context-Aware Wearable Computing,* MIT University, Cambridge, MA (http://www.media.mit.edu/wearables/mithril/).

[23] Schilit, B.N. and Theimer, M.M., Disseminating active map information to mobile hosts, *IEEE Network,* 8(5), 22–32, 1994.

[24] Turner, R.M., Context-sensitive reasoning for autonomous agents and cooperative distributed problem solving, in *Proc. of IJCAI Workshop on Using Knowledge in Its Context,* Chambéry, France, August, 1993.

[25] Pascoe, J., Adding generic contextual capabilities to wearable computers, in *Proc. of the 2nd IEEE Int. Symp. on Wearable Computers (ISWC'98),* Pittsburgh, PA, October 19–20, 1998, pp. 92–99 (http://cs.ukc.ac.uk/pubs/1998/676/content.zip).

[26] Rao, H., Chen, Y., Chang, D., and Chen, M., iMobile: A proxy-based platform for mobile services, in *Proc. of the 1st ACM Workshop on Wireless Mobile Internet (WMI 2001),* Rome, July, 2001.

[27] W3C, *The Platform for Privacy Preferences 1.0 (P3P1.0) Specification,* World Wide Web Consortium, 2001, http://www.w3.org/TR/2001/WD-P3P-20010928.

[28] W3C, *A P3P Preference Exchange Language 1.0 (Appel 1.0)* [working draft], World Wide Web Consortium, 2002, http://www.w3.org/TR/P3P-preferences.

[29] Myles, G., Friday, A. and Davies, N., Preserving privacy in environments with location-based applications, *IEEE Pervasive Comput.,* 2(1), 56–64, 2003.

[30] Adjie-Winoto, W., Schwartz, E., Balakrishnan, H., and Lilley, J., The design and implementation of an intentional naming system, in *Proc. of the 17th ACM Symp. on Operating Systems Principles,* Charleston, SC, December, 1999, pp. 186–201.

[31] The Active Badge System, http://www.uk.research.att.com/ab.html.

[32] Sentient Computing, http://www.cl.cam.ac.uk/Research/DTG/research/sentient/.

[33] Warneke, B., Last, M., Liebowitz, B., and Pister, K.S.J., Smart dust: communicating with a cubic-millimeter computer, *Computer,* 34(1), 44–51, 2001.

[34] Tiny OS, http://webs.cs.berkeley.edu/tos/.

[35] Capra, L., Emmerich, W., and Mascolo, C., Reflective middleware solutions for context-aware applications, in *Proc. of the Third Int. Conf. on Metalevel Architectures and Separation of Crosscutting Concerns (REFLECTION 2001)*, Kyoto, Japan, September, 2001, pp. 126–133.

[36] W3C, *Composite Capability/Preference Profiles*, World Wide Web Consortium, 2003, http://www.w3.org/Mobile/CCPP/.

[37] The WASP Project, http://www.w3.org/2003/p3p-ws/pp/utwente.pdf.

[38] Chen, Y.F., Huang, H., Jana, R., Jim, T., Hiltunen, M. et al., iMobile EE: an enterprise mobile service platform, *ACM J. Wireless Networks*, 9(4), 283–297, 2003.

[39] Salber, D., Dey, A.K., and Abowd, G.D., The context toolkit: aiding the development of context-enabled applications, in *Proc. of ACM SIGCHI Conf. on Human Factors in Computing Systems (CHI'99)*, Pittsburgh, PA, May 15–20, 1999, pp. 434–441.

[40] Schmidt, A., Ubiquitous Computing: Computing in Context, Ph.D. dissertation, Lancaster University, Lancaster, U.K., 2002.

[41] Scandon, F. and Sadeh, N., *Semantic Web Technologies To Reconcile Privacy and Context Awareness*, Computer Science Technical Report CMU-ISRI-03-107, School of Computer Science, Carnegie Mellon University, Pittsburgh, PA, 2003.

[42] Nurmi, P. and Floreen, P., *Reasoning in Context-Aware Systems*, position paper, University of Helsinki, Finland, 2004, http://www.cs.helsinki.fi/u/ptnurmi/papers/positionpaper.pdf.

[43] IETF Seamless Mobility Working Group, http://www.ietf.org.

[44] Przybilski, M., Nurmi, P., and Floreen, P., A framework for context reasoning systems, in *Proc. of the IASTED Int. Conf. on Software Engineering (SE2005)*, Innsbruck, Austria, February, 2005.

[45] Westin, A.F., *Privacy and Freedom*, Bodley Head, London, 1970.

[46] Wagealla, W., Terzis, S., English, C., and Nixon, P., On trust and privacy in context-aware systems, in *Proc. of the Second iTrust Workshop on Trust Management in Dynamic Open Systems*, London, September, 2003.

[47] Conan, D., Taconet, C., Ayed, D., Chateigner, L., Kouici, N., and Bernard, G., A pro-active middleware platform for mobile environments, in *Proc. of the IASTED Int. Conf. on Software Engineering (SE2004)*, Innsbruck, Austria, February, 2004.

[48] Barbir, A. et al., *Policy, Authorization, and Enforcement Requirements of the Open Pluggable Edge Services (OPES)*, Request for Comments 3838, Internet Engineering Task Force (IETF), 2004 (http://www.ietf.org/rfc/rfc3838.txt).

[49] Harle, R.K. and Hopper, A., Deploying and evaluating a location-aware system, in *Proc. of the Third Int. Conf. on Mobile Systems, Applications, and Services (MobiSys'05)*, Seattle, WA, June, 2005.

[50] Levis, P., Madden, S., Gay, D., Polastre, J., Szewczyk, R. et al., The Emergence of Networking Abstractions and Techniques in Tiny OS, in *Proc. of the First USENIX/ACM Symp. on Networked Systems Design and Implementation (NSDI 2004)*, San Francisco, CA, March 29–31, 2004.

[51] Chen, P.P.-S., The entity-relationship model: toward a unified view of data, *ACM Trans. Database Syst.*, 1(1), 9–36, 1976.

[52] W3C Mobile Web Initiative, World Wide Web Consortium, http://www.w3.org/2005/MWI/.

Chapter 34

Integrated Profiling of Users, Terminals, and Provisioning Environments

Alessandra Agostini, Claudio Bettini, and Daniele Riboni

CONTENTS

Introduction

Context awareness is emerging as an essential feature for the next generation of Internet mobile services. Context awareness is a desirable feature for many application areas, including natural language understanding, electronic commerce, telemedicine, and e-learning, just to cite a few. It is, however, particularly relevant for mobile and pervasive computing, as mobile devices naturally enable a much wider set of contexts characterized, among other things, by a spatiotemporal dimension, a wide range of device features and networking capabilities, and very different environment situations. In this chapter, we discuss how the many parameters defining the *context* of a specific mobile service request can be acquired from different sources, formally represented, managed by software agents, and integrated in a consistent uniform description to be used for service adaptation.

We consider the context defining a mobile service request as being described by a set of parameter values possibly belonging to different *profiles*. A profile is intended as a structured set of parameters describing an entity. Most common examples of profiles are *user profiles* and *device profiles*. The first type usually contains data about user preferences, interests, and demographics, as well as behavior models. The second type usually contains technical data describing device capabilities such as installed memory, screen resolution, computing power, available user interfaces, and installed software, as well as device status parameters such as the battery level or the available memory.

Modeling the context of a mobile service request also includes considering profile parameters of the *provisioning environment*. These parameters include the availability, type, and status of the network connection between the user and the service provider; the spatiotemporal condition of the user (e.g., time, location, speed, direction); the user's environment (e.g., close-by resources, temperature, weather conditions); and the policies the service provider may enforce for the service request. The different types of profile data are discussed in detail, and we provide for each one a survey of the existing approaches to represent, manage and use these data.

The real challenge we are considering in this chapter is the acquisition from different sources of all the profile parameters defining the context of a service request and their aggregation into a consistent uniform description. The distribution of profile sources imposes two main requirements: (1) a common formalism and a shared vocabulary to be used by the different profile sources to represent the data, and (2) a mechanism to deal with possibly conflicting parameter values provided by different sources. It is indeed possible that different sources have different values for the same profile parameters; for example, the context provider may maintain its own user profiles, storing among other values the user's interests as deduced from previous service requests. On the other side, the user may provide a personal profile that includes the user's own interests as explicitly defined by the user or automatically derived by a software agent. Conflicts may exist even when considering more technical parameters such as, for example, positioning data; different data may be provided by the network operator using triangulation and cell ID and by the user utilizing GPS or other client-side methods.

Other relevant aspects that must be taken into account are related to the relationships among profile parameters and to the dynamics of profile data. The value of a parameter may well depend on the value of other parameters. For example, the preference of a user for receiving high-quality multimedia content on his device may depend on the cost of the connectivity he is using at the time of the request or the status of the device (e.g., battery level). Other user preferences may depend on the user location or on the current user activity. The conditional setting of parameter values can be modeled by the introduction of simple user policies that should be evaluated at the time of the service request. Analogously, the service provider and possibly other profile sources may have their own set of policies. A comprehensive solution for distributed profiling should also take care of possible conflicts both within a set of local policies and among policies from different sources.

Profile data can also be dynamic in the sense that parameter values may change quickly and possibly during service provisioning. Typical examples are tracking applications whose service is actually based on the update of the positioning data. An example of a more advanced service is adaptive multimedia streaming. Streaming may be initiated with a very high bit rate based on profile data acquired at the time of request, but it should progressively degrade if the profile parameters change during the streaming session, suggesting the use of a lower bit rate. Taking into account this aspect requires a mechanism to detect changes in relevant profile parameters at remote different sources as well as defining how much a value should change to require recomputation of the aggregated global profile.

Current Profiling Approaches

This section considers profile data describing device features, profile data describing users, and profile data describing the provisioning environments. For each of these profile categories, we illustrate the main approaches to profiling considering both existing commercial systems and research work. The last subsection briefly illustrates existing delivery platforms based on profiling.

Profile Representation of Devices

A precise definition of the characteristics and capabilities of the device used for accessing an Internet service is essential for performing adaptation. In particular, profile data includes information regarding both software (e.g., browser name and version, Java support) and hardware (e.g., CPU and network interfaces). Although software capabilities remain constant throughout the service provision, data regarding certain hardware parameters can change (e.g., remaining battery lifetime, available memory).

HTTP Headers

The diffusion of mobile devices with low capabilities has spurred the definition of markup languages (e.g., WML, cHTML) targeted at different classes of terminals. The simplest technique (and, actually, still the most adopted by service providers) for choosing the most appropriate markup language and for adapting Web contents to the device that makes the request involves identifying the device by means of the Hypertext Transfer Protocol (HTTP) request headers. It is worth noting that this technique is applicable only to HTTP-based services. The HTTP/1.1 specification defines the syntax and semantics of all standard HTTP/1.1

Table 34.1 HTTP Request Headers Provided by the Internet Explorer Browser for Various Devices

	Windows XP	PocketPC	SmartPhone
User-Agent	Mozilla/4.0 (compatible; MSIE 6.0)	Mozilla/4.0 (compatible; MSIE4.01)	Mozilla/4.0 (compatible; MSIE4.01)
Accept	application/ msword, ...	image/gif, \ldots	image/gif, ...
UA-CPU	—	i486	ARMOMAP710
UA-OS	—	WinCE (PocketPC)	WinCE (SmartPhone)
UA-pixels	—	240 × 320	176 × 220

header fields. Unfortunately, the information conveyed by HTTP/1.1 headers that can be useful for representing device capabilities is quite limited and includes only the user agent (i.e., browser) and media types (MIME types) accepted by the user agent (e.g., the supported markup languages), charsets, and encodings. Hence, this information only allows the service provider to determine how to mark up the content. The *User-Agent Display Attributes Headers* Internet Draft [51] has been widely adopted by browser developers for extending the set of information provided by HTTP headers with data regarding display characteristics, such as screen size and resolution, and color capabilities. Moreover, some browsers include in the HTTP request other undocumented header fields for representing information, such as the operating system, CPU, and voice capabilities. As an example, Table 34.1 shows some of the HTTP request headers provided by the Internet Explorer® browsers of Windows® XP, Pocket PC, and SmartPhone devices.

Obviously, the header approach has a number of shortcomings. First, the provided information is limited to static characteristics of the device, but a mobile computing scenario requires knowledge of the current status of devices, such as available memory and battery charge status. Moreover, because some of the headers are not well documented, it is necessary to perform a sort of reverse engineering to understand their meaning. In fact, different browsers provide different HTTP headers for the same device.

Composite Capability/Preference Profiles and UAProf

To overcome the limitations of the HTTP headers approach, the World Wide Web Consortium (W3C) defined the structure and vocabularies of *Composite Capability/Preference Profiles* (CC/PP) [34]. CC/PP uses the eXtensible Markup Language (XML) serialization of Resource Description Framework (RDF) graphs to create profiles that describe the capabilities of the device and, possibly, the preferences of the user. CC/PP profiles are structured as sets of *components* that contain various *attributes* with associated values. Components and attributes are defined in *CC/PP vocabularies* (i.e., RDF schemas that formally define their semantics and allowed values). Data-type support in CC/PP is quite limited; in fact, attribute values can be either simple (string, integer, or rational number) or complex (set or sequence of values, represented as `rdf:Bag` and `rdf:Seq`, respectively).

Currently, CC/PP is used primarily for representing device capabilities and network characteristics. UAProf [42] is the most renowned CC/PP-compliant vocabulary. It has been proposed by the Open Mobile Alliance for representing the hardware, software, and network capabilities of mobile devices. Some components defined within UAProf have been extended with new attributes by Intel [13]. In particular, UAProf defines seven components:

- *HardwarePlatform* provides a detailed description of the hardware capabilities of the terminal, including input/output capabilities, CPU, memory, battery status, and available expansion slots.
- *SoftwarePlatform* describes the device operating system, its Java support, supported video and audio encoders, and the user's preferred language.
- *BrowserUA* describes in detail the browser features, providing not only the browser name and version but also information regarding its support for applets, JavaScript, voiceXML, and text-to-speech and speech-recognition capabilities, as well as the user's preference regarding frames.
- *NetworkCharacteristics* provides information about the network capabilities and environment (such as the supported Bluetooth version), support of security protocols, and the current bearer signal strength and bit rate.
- *WAPCharacteristics* contains a set of attributes regarding the device Wireless Application Protocol (WAP) capabilities, including the supported WAP, Wireless Markup Language (WML), and WMLScript versions, as well as the WML deck size.
- *PushCharacteristics* and *MMSCharacteristics* provide information regarding the device WAP push capabilities and multimedia messaging service (MMS) support, respectively.

```
<?xml version="1.0"?>
  <rdf:RDF xmlns:rdf="http://www.w3.org/1999/02/22-rdf-syntax-ns#"
    ...

    <rdf:Description ID="HardwarePlatform">
      <rdf:type resource="http://www.w3.org/2000/01/rdf-schema#Class"/>
      <rdfs:subClassOf rdf:resource="#Component"/>
      ...

    </rdf:Description>
    ...

    <rdf:Description ID="NumberOfSoftKeys">
      <rdf:type rdf:resource="http://www.w3.org/2000/01/rdf-schema#Property"/>
      <rdfs:domain rdf:resource="#HardwarePlatform"/>
      <rdfs:comment>
        Description:  Number of soft keys available on the device.
        Type: Number       Resolution: Locked       Examples:       "3", "2"
      </rdfs:comment>
    </rdf:Description>


    <rdf:Description ID="ScreenSize">
      <rdf:type rdf:resource="http://www.w3.org/2000/01/rdf-schema#Property"/>
      <rdfs:domain rdf:resource="#HardwarePlatform"/>
      <rdfs:comment>
        Description:  The size of the device's screen in units of pixels,
        composed of the screen width and the screen height.
        Type: Dimension       Resolution: Locked       Examples: "640x480"
      </rdfs:comment>
    </rdf:Description>
    ...
```

Figure 34.1   Excerpt of the UAProf definition of the *HardwarePlatform* component.

A small excerpt of the UAProf definition of the *HardwarePlatform* component can be seen in Figure 34.1.

Currently, many hardware vendors make publicly available on their Web sites the UAProf profiles of their devices (several examples can be found at http://w3development.de/rdf/uaprof_repository/). At the time of this writing, the list of UAProf descriptions provided by important vendors such as Nokia, Sony Ericsson, and BlackBerry™ are kept up to date with the new models, suggesting that this technology is considered interesting by hardware vendors. Figure 34.2 shows an excerpt of the UAProf profile of a mobile phone, as published on the supplier Web site.

```
<rdf:Description rdf:ID="Profile">
  <prf:component>
    <rdf:Description rdf:ID="HardwarePlatform">
      <rdf:type rdf:resource="http://www.openmobilealliance.org/..." />
      <prf:ScreenSize>128x128</prf:ScreenSize>
      <prf:Model>6820</prf:Model>
      <prf:InputCharSet>
        <rdf:Bag>
          <rdf:li>ISO-8859-1</rdf:li>
          <rdf:li>US-ASCII</rdf:li>
          <rdf:li>UTF-8</rdf:li>
          <rdf:li>ISO-10646-UCS-2</rdf:li>
        </rdf:Bag>
      </prf:InputCharSet>
      <prf:ScreenSizeChar>18x5</prf:ScreenSizeChar>
      <prf:BitsPerPixel>12</prf:BitsPerPixel>
      <prf:ColorCapable>Yes</prf:ColorCapable>
      <prf:TextInputCapable>Yes</prf:TextInputCapable>
      <prf:ImageCapable>Yes</prf:ImageCapable>
      <prf:Keyboard>Qwerty</prf:Keyboard>
      <prf:NumberOfSoftKeys>3</prf:NumberOfSoftKeys>
      ...
```

Figure 34.2 Excerpt of the UAProf profile of a mobile phone.

The CC/PP and UAProf specifications also define the communication protocol of profiles to the service provider. These protocols are based on *profile defaults* and *profile diffs* [41,42]. The client should send HTTP requests containing a reference to the device default profile and attribute–value pairs that describe the variations from the default profile (e.g., the insertion of a new memory card or volatile information such as the current available memory). Possibly, the CC/PP profile can be updated with new information by firewalls and proxies encountered by the HTTP request.

User Profiling

Within the user modeling literature, a *user model* is intended as the system representation of the *characteristics* of a user, including, for example, knowledge and beliefs, skills and expertise, and interests and preferences. Our definition of a *user profile*, as a structured set of parameters, can be considered in all respects a user model, as it represents many relevant user characteristics; however, it is only the whole context profile (composed by different integrated profiles coming from different sources) that contains, among other things, the complete user model. Research prototypes and commercial systems exploiting user profiles can be categorized taking into account different relevant aspects. With respect to our needs, we consider as *primary dimensions*: (1) the adopted method for modeling users and (2) the richness and generality of user data modeled. We also consider, as *secondary dimensions*: (3) the kind of user data acquisition (e.g., explicit or derived data collection) and (4) the type of user adaptation (e.g., content or presentational adaptation). In the following, we report on some academic research and commercial systems that provide seminal solutions or are considered to be well-established providers with regard to one or more of the above dimensions.

Early research adopted a simple user model expressed in the form of records of command usage or data access; the user adaptation was directly connected to the frequency of such usage. No attempt was made to infer or represent any other information about the user. The user model was embedded in the application, and it was not possible to distinguish specific user modeling components from the other application modules.

The forerunner of future user modeling shell systems is the *General User Modeling System* (GUMS) [20] (i.e., a system providing user modeling services at runtime that can be configured during development time). In general, user modeling shell systems [35] feature quite sophisticated approaches to modeling users and include rich categories of user data (i.e., the primary dimensions provided above). They support both explicit user data acquisition and derivation of implicit user characteristics, as well as the handling of contradictions (also referred to as *truth maintenance*). For example, GUMS allows the definition of simple stereotype hierarchies, and for each stereotype a set of Prolog facts describes stereotype members and a set of rules defines the system's reasoning. The final application can also communicate new facts about the user at runtime. In contrast, the *Belief, Goal, and Plan Maintenance System* (BGP-MS) [36] provides two integrated formalisms for representing users' beliefs and goals. Assumptions about the user and stereotypical assumptions about user groups can be represented in a first-order predicate logic. A subset of these assumptions is stored in a terminological logic. Inferences are defined in a first-order modal logic.

The *SeTA* prototype [4], a toolkit for the construction of adaptive Web stores, includes state-of-the-art solutions in many dimensions. With regard to dimension 4, SeTA integrates the personalized suggestion of items (content adaptation) with the adaptation of the layout based on user preferences and expertise. This is especially made possible due to the richness of the user model (dimension 2) and to a specific representation technique (dimension 1). The SeTA user model contains four main types of data:

1. Explicitly provided personal data such as age, gender, job, education level
2. Domain-independent user features, such as the user's receptivity, expertise, and interests
3. Domain-dependent preferences regarding product properties that are used by SeTA to select the items most suited to the user
4. Information relative to the classification of the user in the stereotypical customer classes

User data acquisition (dimension 3) is both explicit and dynamically computed taking into account the user's behavior during the current session. With respect to dimensions 1 and 3, SeTA integrates knowledge representation (KR)-based user modeling techniques with machine learning mechanisms. In particular, although personal data is composed of simple attribute–value pairs, the other profile attributes have a more sophisticated representation. For example, for each user features attribute, a probability distribution is associated to its possible values (e.g., expertise about phones: low = 0.1; medium = 0.2; high = 0.7).

Recommender systems, such as *GroupLens* [45], should also be mentioned, as they are heavily based on user profiling. In these systems, the affinity between users is evaluated considering explicit ratings of items provided by users, implicit ratings derived from navigational behaviors, and transaction history data. A comprehensive comparison of products for e-commerce and Customer Relationship Management (CRM) can be found in Fink and Kobsa [21].

Profiling Provisioning Environments

As outlined earlier, the set of context data useful for performing a better adaptation goes beyond device capabilities and user information. In this section, we present some profiling methods for gathering information regarding the network status, the position of the user and of people and objects in the user's surroundings, and the user's environment.

Bandwidth Estimation Techniques

An estimate of the data rate that can be transmitted by the network link that connects the service provider to the user is important for determining the adaptation parameters of a wide spectrum of Internet services. As a consequence, a number of techniques have been proposed in the last years to estimate available bandwidth. A survey regarding metrics, techniques, and tools can be found in Prasad et al. [43].

Application-level approaches try to estimate quantities of interest (especially available bandwidth) at the communication endpoint observing packet dispersion [37] either in probing traffic or existing transmissions. As an example, some techniques estimate the end-to-end available bandwidth by means of streams of *probing packets* that the source (server) sends to the receiver (client). Similar application-level approaches require a strict cooperation between the sender and the receiver, as the receiver has to give explicit feedbacks. Other techniques require the receiver to perform the estimation itself to improve the system scalability. Application-level approaches have a number of known weaknesses; in particular, even in wired networks, the main weakness resides in the estimation accuracy itself and in convergence times. The application of these techniques in a mobile computing scenario poses new issues, mainly due to the required cooperation of clients having low power and network capabilities. In fact, client-side cooperation determines power consumption and loss of bandwidth (which in many mobile network technologies is a very valuable and costly resource). Moreover, due to their particular characteristics, being able to obtain a reliable end-to-end measurement in some mobile networks (e.g., in WiFi networks) is questionable.

To overcome these weaknesses, various network-level approaches have been devised. Network-level techniques exploit explicit network feedback to monitor available resources, as described in Kazantzidis et al. [33]. The main disadvantage of these techniques resides in the fact that for nodes to operate in the network they must provide specific support for each given architecture and technology; clearly, this limits scalability and ease of deployment of such techniques. Moreover, these techniques are unsuitable in end-to-end networks such as Universal Mobile Telecommunications System (UMTS) and General Packet Radio Service (GPRS). For a more in-depth discussion of bandwidth monitoring issues in the context of mobile service adaptation, we refer the interested reader to Maggiorini and Riboni [38].

Location

Currently, a number of mobile computing applications provide services targeted to the user's location. Navigation systems, emergency services, mobile tourist guides, and proximity marketing are only few examples of location-aware applications. To support such applications, many different location systems and technologies have been developed to provide users and devices with information about their physical location and other people and items located in their surroundings [28].

It is worth noting that the various techniques may specify the locations of objects and people using different representation schemes. Generally, outdoor systems provide a *physical position* (e.g., coordinates), and indoor systems provide a *symbolic position* (e.g., in room R32, in the living room). The physical position of an object can be naturally expressed through a two- or three-dimensional coordinate system (latitude, longitude, and optionally altitude) in a given spatial reference model. For example, the National Marine Electronics Association (NMEA) 0183 is a standard for communicating physical location information based on asynchronous *sentences* that provide the latitude and longitude (expressed by degree, minute, and second triplets), and other data (e.g., velocity). Unfortunately, representing symbolic positions is much more difficult. One possible solution consists in defining an ontology of symbolic locations as done for example in Millard et al. [40]. This ontology defines classes, such as country, city, street, building, floor, and room, using the Ontology Web Language (OWL) [39]. An ontology-based representation of symbolic locations also allows some simple forms of reasoning. Mapping between physical and symbolic locations is generally executed by an external spatial-aware application.

Localization techniques differ in many aspects, such as the accuracy of the provided position, the physical medium exploited for determining location, and power and infrastructure requirements. Probably, the most renowned outdoor positioning technology is the global position system (GPS). GPS is a worldwide positioning infrastructure formed by 24 satellites, together with ground stations in charge of maintaining the precise position of satellites. Satellites transmit signals encoded with timing information obtained from an atomic clock. Signals are used by GPS receivers to calculate their position by means of *trilateration*. Basically, to determine its position, a GPS receiver uses an estimate of the distance from four or more satellites, obtained analyzing the travel time of radio signals. Given the particular nature of these signals, the GPS technology is generally unavailable indoors. From the user's perspective, GPS receivers are small, relatively economical widgets integrated into vehicles and mobile devices or easily connectable to mobile devices through a wireless link (usually a Bluetooth connection). The communication between GPS receivers and mobile devices is based on the NMEA standard. GPS accuracy can vary depending on a number of factors, including the particular receiver, electronic interferences, atmospheric effects, and the presence of tall buildings or other surfaces that reflect signals before they reach the receiver. Currently, low-cost GPS receivers have an accuracy of 10 meters or less. Even if GPS accuracy is not a problem for a number of location-based services, various modifications to the basic GPS technology have been devised for improving accuracy (e.g., DGPS and AGPS [7]).

A method that offers lower accuracy with respect to GPS but is available both outdoors and indoors is *cell ID*. Cell ID exploits the GSM base station to which the user is connected for approximating the user's position. As a consequence, accuracy depends on the cell size and varies from hundreds of meters in densely populated urban areas to tens of kilometers in rural areas. The main advantage of this technique, currently used by many operators, is that it is employable with no modifications to the network infrastructure, and it does not require new functionalities to be added to mobile devices; however, the localization accuracy provided by this technique is inadequate for a number of location-based services, and various improvements have been proposed for increasing its accuracy [1].

One of the first indoor positioning infrastructures, *Active Badge* [49], was developed between 1989 and 1992 at Olivetti Research Labs. The Active Badge proximity system is based on infrared transmitters carried by people and receivers located in buildings that are in charge of determining symbolic locations (typically, the room that people are in). More recently, similar techniques have been proposed that adopt ultrasound instead of infrared beacons and determine the location of users and objects by means of triangulation, thus obtaining greater accuracy. In the *Active Bat* system [27], the user's location is calculated by a centralized module that collects and analyzes data retrieved from sensors. In contrast, the *Cricket* system [44] utilizes emitters that are spread in the environment, and user-side widgets are in charge of receiving beacons and performing triangulation, thus protecting the user's privacy.

Radiofrequency identification (RFID) systems utilize a set of readers that can read data through electromagnetic transmission from RFID tags. RFID tags can be either active or passive. Active tags have radio capabilities and ranges of up to hundreds of meters. Passive tags only reflect signals received from readers; thus, their communication range is smaller, but passive tags are considerably less expensive than active ones. The advantage of RFID systems is that they are easily deployable and tags are relatively inexpensive.

Other approaches have tried to exploit general-purpose wireless networks for implementing location systems. Various techniques (e.g., RADAR [6]) propose a solution based on WiFi networking for tracking users inside buildings. The user's position is determined by analyzing signal strengths at multiple overlapping base stations that cover a certain area. Even though such techniques have the advantage of being implementable on top of a widespread wireless network infrastructure, the accuracy they provide is not entirely satisfactory. As an example, the RADAR system is able to determine the location of users to within 3 meters of their true position with 0.5 probability. Similar considerations hold for positioning systems deployed on top of Bluetooth network infrastructures.

Various commercial systems (generally called *location servers*, such as Microsoft's MapPoint® location server, Geodan's Movida location server, and SiRF's SiRFLoc® server) offer the opportunity to nicely integrate location information collected by different means (e.g., GPS, cell ID). These systems allow application providers to access location information in a uniform way, independent of the specific technique used to derive it.

Environment Conditions

Various interesting reports exist that are focused on gathering information about the user's surrounding environment on the basis of sensors. The *AmbieSense* (http://www.ambiesense.com/) project is based on the use of *context tags*. Context tags are small electronic widgets that can be spread all over the environment (for example, within shops, hotels, furniture, and even clothes). They automatically send contextual information about the surrounding environment to mobile users. Interestingly, context tags can also be attached to *users*. In this case, they provide context information about the users to which they are attached. User context information includes sociocultural data such as a user's interests and status, as well as other spatiotemporal aspects. This information is of paramount importance, because each user belongs to the sociocultural environment of other users who interact with him.

Other projects (e.g., TEA [23]) are focused on the integration of simple and inexpensive sensors that measure raw data, such as presence, temperature, sound, and light level, to derive more complex, implicit context conditions (e.g., the action performed by the user). An application of similar techniques is the analysis of human eye-blinking and other factors to determine the fatigue states of car drivers [12].

Profile-Based Delivery Platforms

Several delivery platforms that take into account users' profile data have been developed, by both academic and industrial groups. In this section, we briefly present two research proposals adopting the CC/PP approach, and we discuss the customization mechanisms of some well-known commercial application servers.

CC/PP-Based Architectures

Even if CC/PP and UAProf provide a satisfactory solution to the issue of representing both static and dynamic properties of mobile devices, currently the adoption of these technologies is not yet widespread. The key requirement for implementing the CC/PP approach is to enable browsers to recognize the current device and communicate (through HTTP headers) its UAProf profile to the service provider (either building the UAProf profile from scratch or pointing to the profile stored on the vendor Web site). Moreover, to keep parameters up to date regarding the current status of the device, the profile should be updated on the client side by a proper monitor application. The Dynamic Execution Layer Interface (DELI) and the CC/PP Software Development Kit (SDK) provide two experimental platforms supporting the CC/PP technology.

DELI [14] is an open-source Java library developed by HP Labs that allows the resolution of HTTP requests containing references to the CC/PP profile of the client device. DELI adopts the profile integration approach of UAProf in that it associates a *resolution rule* with every attribute. Whenever a conflict arises (i.e., when the *default profile* and *profile diffs* provide different values for the same attribute), the resolution rule determines the value to be assigned to the attribute by considering the order of evaluation of partial profiles. DELI is fully integrated with Cocoon, the well-known XML-based application server.

Intel's CC/PP SDK [13] proposes an architecture utilizing client- and server-side modules for the management of UAProf profiles. Client-side modules execute on Microsoft Pocket PC 2002 devices. The CC/PP profile of the device is kept up to date by a monitoring module that is in charge of retrieving static as well as dynamic information about the device status and capabilities. Communication of the CC/PP profile with server applications is achieved by means of the *CC/PP client proxy*, which intercepts HTTP requests (e.g., originated by the microbrowser of the device) and inserts profile information into the HTTP headers. Server-side, the main component of this architecture is the *CC/PP Content Customization Module*, a module of the Apache Web server that is in charge of retrieving partial profiles by analyzing the HTTP request headers and of combining them to obtain the merged profile. This profile is used by the application logic for adapting the content and its presentation. The CC/PP SDK framework provides three different mechanisms for personalization. *Content selection* involves building different representations of the same content and choosing the most appropriate representation on the basis of profile data. *Stylesheet conversion* is used for adapting XML-based content using a different eXtensible Stylesheet Language (XSL) stylesheet for various classes of profiles. Finally, *script processing* uses a script language to dynamically build an interface suited to the profile.

Commercial Application Servers

Today, most commercial application servers provide personalization and content adaptation solutions that take into account at least the characteristics of the user's device. As an example, the personalization scheme of IBM's *WebSphere®* portal is based on the creation of Web pages and services using XML Device-Independent Markup Extensions (XDIME). Depending on the specific device, XDIME contents are transformed by properly predefined eXtensible Stylesheet Language Transformations (XSLT) into the most appropriate format (e.g., WML, XHTML Basic), evaluating policies that take into account the capabilities of the particular device that issued the request. The framework also includes a repository of mobile device profiles describing the capabilities of a broad range of terminals. Similar solutions are provided by other well-known application servers such as BEA's WebLogic and OracleAS Wireless.

CARE Mobile Middleware for the Integration of Distributed Profile Data

Taking into account the issues outlined in the introduction, we have proposed the *Context Aggregation and REasoning* (CARE) middleware for context awareness in mobile environments [2]. CARE has been defined based on a specific list of requirements obtained by the analysis of a wide spectrum of Internet services that would benefit from adaptation. In particular, the requirements were identified considering the data required for implementing highly adaptive services, the infrastructure available now and that will be available in the near future, as well as the issues of data privacy and accessibility. The main requirements we identified are:

- *Interoperable context representation* — A representation formalism is necessary for the specification of a very broad set of profile data that integrates device capabilities with spatiotemporal context, device, and network status, as well as user preferences and semantically rich context data. Because such data must be exchanged among various entities, it is highly advisable to use a standard language, a shared vocabulary, and unambiguous semantics.
- *Support for context dynamics* — It must be possible for multiple entities (e.g., users, providers, agents) to define how some changes in context reflect in other context data; for this reason, a representation formalism is required for the specification of policies that can dynamically determine the value of some profile data based on other values. Moreover, changes in context must be asynchronously communicated to the interested entities; therefore, the architecture should provide a configurable mechanism for intra-session adaptation based on real-time updates of certain profile data (e.g., location).

- *Support for distributed context data* — Context data is naturally provided by different sources, in some cases delivering conflicting information. The architecture should support the distributed storage and management of profiles and policies, with information stored and managed close to its source.
- *Conflict resolution* — The architecture should provide a mechanism for aggregating profile data and policies from different sources that supports a flexible and finely grained conflict-resolution mechanism.
- *Privacy* — The architecture should rely on an advanced system for privacy protection that allows users to precisely control the partial sharing of their profile data.
- *Efficiency* — The time required for adaptation should not significantly affect the final user.

Architecture Overview

Clearly, the specification and implementation of a full-fledged architecture satisfying all of these requirements are long-term goals. The approach illustrated here is intended to be a first step in this direction. In CARE, three main entities are involved in the task of building an aggregated profile: the *user* and the user's devices, the network *operator* with its infrastructure, and the *service provider* with its own infrastructure. The architecture has been designed to handle an arbitrary number of entities. A profile manager devoted to managing profile data is associated with each entity, and they are referred to as the UPM, OPM, and SPPM, respectively. We assume that the user's location is kept up to date by an external location server that communicates that location to the UPM. The UPM and SPPM are also in charge of interacting with ontology services for managing and reasoning with sociocultural contextual data. Adaptation and personalization parameters are determined by policy rules defined by both the user and the service provider and are managed by their corresponding profile managers.

In Figure 34.3 we illustrate the system behavior by describing the main steps involved in a service request. In step 1, a user issues a request to a service provider through his device and the connectivity offered by a network operator. The HTTP header of the request includes the Uniform Resource Identifiers (URIs), which are used to contact the UPM and the OPM. Then, in step 2, the service provider queries the Context Provider module to retrieve the profile information necessary to perform adaptation. In step 3, the same module queries the profile managers to retrieve distributed profile data and user policies. The profile data is aggregated by the Merge module in a single profile which is given, together with

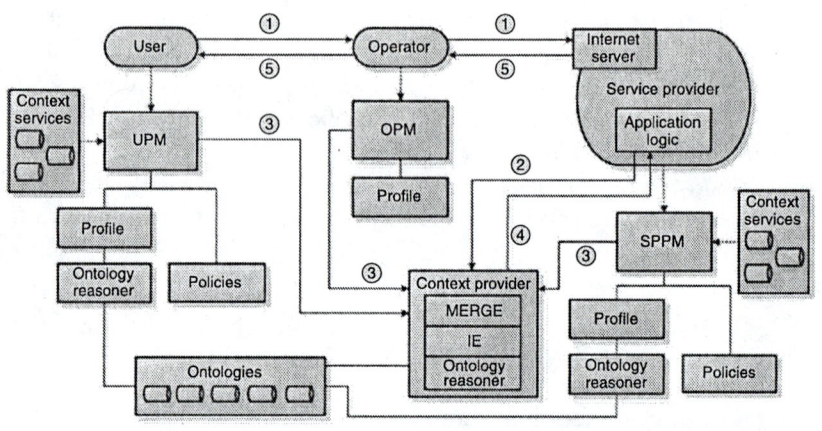

Figure 34.3 Architecture overview and data flow upon a user request.

policies, to the Inference Engine (IE) for policy evaluation. Ontological reasoning is performed on demand (i.e., at the time of the service request) only if the integrated profile lacks values for the ontology-based profile data necessary for providing the service. In this case, the Context Provider populates the ontology with the integrated profile, performs ontological reasoning, and adds the new context information to the integrated profile. In step 4, the integrated profile is returned to the service provider. Finally, profile data is used by the application logic to properly adapt the service before it is provided (step 5).

Profile Management and Aggregation

In the following, we explain the mechanism of profile management and address the issue of how to aggregate possibly conflicting data in a single profile.

Profile Managers

As outlined above, each profile manager is responsible for managing profile attributes provided by the entity it pertains to. In addition, the UPM and SPPM manage user and service provider policies, respectively, and can perform ontological reasoning. In particular:

- The UPM stores information related to the user and the user's devices. This data includes, among other things, location, personal data, interests, sociocultural context information, and device capabilities. The UPM periodically performs ontological reasoning for deriving new semantically rich context information; for example, from the user's calendar and contacts it may be possible to derive the user's current activity. The UPM also manages policies defined by the user that describe the content and presentations he wants to receive under particular conditions.

- The OPM is responsible for managing attributes describing the current network context (e.g., available bandwidth, connection profile, network status, user's location).
- The SPPM is responsible for managing service provider proprietary data, including information about users derived from previous service experiences, as well as adaptation policies. Similarly to the UPM, the SPPM can also perform ontological reasoning.

Profile Representation

In our framework, we divide profile data into two classes: *shallow profile data* and *ontology-based profile data*. We consider as shallow profile data those attributes that can be modeled in a natural and simple way by using attribute–value pairs, provided their semantics are clear to all entities. This class contains data about environmental conditions and the technological infrastructure; however, only a few attributes regarding the user and related sociocultural information can be modeled in this way. We represent shallow profile data by CC/PP profiles. Because existing CC/PP vocabularies primarily cover only hardware, software, and network capabilities of mobile devices, we have extended them to include a much richer set of context information.

Unfortunately, CC/PP has many shortcomings when it comes to modeling non-shallow profile data, such as user activities. Indeed, CC/PP vocabularies define both the semantics of each attribute and the list of its possible values by using the < rdfs : comment> resource and a description in natural language, leading to possibly different interpretations. Moreover, the two-level structure (components and attributes) imposed by CC/PP greatly affects its expressive power. For representing non-shallow profile data, a natural choice is ontologies; in fact, they have a higher expressive power than CC/PP, and, in most cases, they offer reasoning services. The introduction of ontologies in our framework has two main purposes: First, public/shared ontologies support knowledge sharing among the various involved entities. Second, ontologies are used for consistency checking of contextual data instances and for other reasoning tasks; for example, they are used to automatically derive, based on other context data, that the user is busy in a specific type of activity such as "InternalMeeting." In this second case, ontologies can be private to a specific profile manager. We currently use OWL-DL [39] as the ontology language, because we want to take advantage of the reasoning services it supports and it is becoming a *de facto* standard in various application domains. However, primarily for interoperability purposes, different from other approaches [16,48], we decided to continue storing all of the profile data in CC/PP profiles but to link those attributes modeling non-shallow context data to ontology concepts that formally define their semantics. To adhere to the CC/PP specification, the mapping between a CC/PP attribute and an ontology concept is defined in the vocabulary that defines the attribute, using the `<rdfs:comment>` resource.

Profile Aggregation and Conflict Resolution

When the Context Provider has obtained profile data from the other profile managers, this information is passed on to the Merge module, which is in charge of profile aggregation. Conflicts can arise when different values are given by different profile managers for the same attribute; for example, the UPM could assign to the *Coordinates* attribute a certain value x (obtained through the GPS of the user's device), and the OPM could provide for the same attribute a different value y, obtained through triangulation. To resolve this type of conflict, the service provider has to specify resolution rules at the attribute level in the form of priorities among entities. Priorities are defined by *profile-resolution directives*, which associate with every attribute an ordered list of profile managers.

Example 1

Consider the following profile-resolution directives:

> setPriority */* = (SPPM, UPM, OPM)
> setPriority NetSpecs/* = (OPM, UPM,SPPM)
> setPriority UserLocation/Coordinates = (UPM, OPM)

In the first one, a service provider gives highest priority to its own profile data and lower priority to data given by the other entities; clearly, if no value is present in the service provider profile, then the value is taken from other profiles following the priority directive. The second and third directives give the highest priority to the operator for network-related data and to the user for the single *Coordinates* attribute, respectively. The absence of SPPM in the third directive indicates that values for that attribute provided by the SPPM should never be used.

The semantics of priorities actually depend on the type of the attribute. When the attribute is *simple*, the value to be assigned to the attribute is the one retrieved from the first entity in the list that supplies it. When the attribute is of type `rdf:Bag`, the values to be assigned are the ones retrieved from all entities present in the list. If some duplication occurs, only the first occurrence of the value is taken into account (i.e., we apply union). Finally, if the type of the attribute is `rdf:Seq`, then the values assigned to the attribute are the ones provided by the entities present in the list, ordered according to the occurrence of the entity in the list. All duplicates are removed, keeping only the first occurrence.

Policies for Supporting Adaptation

As noted earlier, policies can be declared by both the service provider and the user. The evaluation of policies against aggregated profile data determines the adaptation parameters applied by the service provider; for example, the provider of a streaming service can determine the resolution of a video considering explicit profile data such as available bandwidth and screen resolution. Similarly, users can dynamically change their preferences regarding content and presentation on the basis of some profile data. For example, a user may prefer a visual medium (e.g., text or video) when working on a PDA but would prefer to switch to an audio medium when using a WAP phone.

Policy Representation

The choice of a representation language is a compromise between simplicity, expressiveness, and efficiency. The policy language must also support the definition of a mechanism for handling conflicts that could arise when user and service provider policies determine different values for the same attribute. Each policy rule can be interpreted as a set of conditions on profile data that determine a new value for a profile attribute when satisfied. A policy in our language is composed of a set of rules of the form:

$$\text{If } C_1 \text{ and } \ldots \text{ and } C_n, \text{ then set } A_k = V_k$$

where A_k is an attribute, V_k is either a value or a variable, and C_1 is either a condition such as $A_i = V_i$ or *not* A_i, with the meaning that no explicit nor derived value for A_i exists; for example, the informal user policy *"When I am in the main conference room using my palm device, any communication should occur in textual form"* can be rendered by the following policy rule:

> If *location* = "MConfRoom" and *device* = "Pda" then set *PreferredMedia* = "Text."

Rules can also be labeled, and expressions of the form $R_1 > R_2$ can be specified to state that rule R_1 has higher priority than rule R_2. Because priorities are introduced in the language only for managing conflicts between rules, we restrict priorities to being assigned to rules setting a value to the same attribute.

Conflicts and Resolution Strategies

Because policies can dynamically change the value of an attribute that may have an explicit value in a profile or may be changed by some other policies, they introduce nontrivial conflicts. They can be determined by policies or by explicit attribute values given by the same entity or by different entities. A categorization of possible conflicts is useful for determining the system behavior. We summarize the desired behavior of the system, in the presence of possible conflicts, considering each case as follows:

1. *Conflict between explicit values provided by two different entities when no policy is given for the same attribute.* In this case, the priority over entities for that attribute determines which value prevails. This kind of conflict is totally handled by the Merge submodule of the Inference Engine.
2. *Conflict between an explicit attribute value and a policy given by the same entity that could derive a different value.* A simple example of a conflict of this type is the use of policies to override default attribute values when specific events occur or specific conditions are verified. In this case, a policy given by an entity, deriving a value for an attribute, intuitively has higher priority over an explicit value for that attribute given by the same entity; thus, the value derived from the policy must prevail.
3. *Conflict between an explicit attribute value and a policy given by a different entity that could derive a different value.* Conflicts of this type can occur, for instance, when a provider is not able or does not want to agree with a user explicit preference and sets up a policy rule to override the values explicitly given by the user. This kind of conflict is solved by considering the priority rules adopted earlier for explicit attribute values. Based on the priority over entities for that attribute, if the entity giving the explicit value has priority over the other, then the policy can be ignored; otherwise, the policy should be evaluated and, if a value is derived, this value prevails over the explicit one.

4. *Conflict between two policies given by two different entities on a specific attribute value.* Similarly to conflict 3, the priority over entities for that attribute states the priority in firing the corresponding rule. If a rule fires, no other conflicting rule from different entities should fire.

5. *Conflict between two policies given by the same entity on a specific attribute value.* When no intuitive way exists to solve such a conflict, we assume that the entity gives a priority over these rules, using the syntax provided by the policy language. If this is not given, a default ordering will be used. The priority over rules for that attribute is used to decide which one to evaluate first. If a rule fires, no other conflicting rule from the same entity should fire.

Although the actual implementation of conflict resolution is beyond the scope of this chapter (see Bettini and Riboni [11] for a complete description), we outline here the rationale of the mechanism. The intuitive evaluation strategy is to proceed, for each attribute A, from the rule having $A(\ )$ in its head with the highest priority and to continue considering rules on $A(\ )$ with decreasing priorities until one of them fires. If none of them fires, the value of A is the one obtained by the Merge module on A or none if such a value does not exist. In our case, a direct evaluation algorithm has been devised and implemented that is linear in the number of rules.

Intra-Session Profile Updates

The dynamic nature of some profile attribute values requires a mechanism for keeping up to date the profile information used by the service provider during a session. Consider the case of user profile data; although some attributes do not change during a session (e.g., the user personal data), other information may change depending on device status (e.g., available memory), user interaction with the device or application (e.g., turning a feature on or off), and user behavior (e.g., change of location). Data owned by the network operator is possibly even more unstable. Different mechanisms can be adopted to address this requirement, with the usual approaches being based either on polling techniques or on asynchronous notifications fired by triggers. Polling, especially when involving properties of the user device, poses problems of cost and bandwidth consumption.

Our choice is to include in the CARE middleware a trigger mechanism to obtain asynchronous feedback on specific events (e.g., available bandwidth dropping below a certain threshold, user location changed by more than 100 meters). When a trigger fires, the corresponding profile manager sends the new values of the modified attributes to the Context Provider

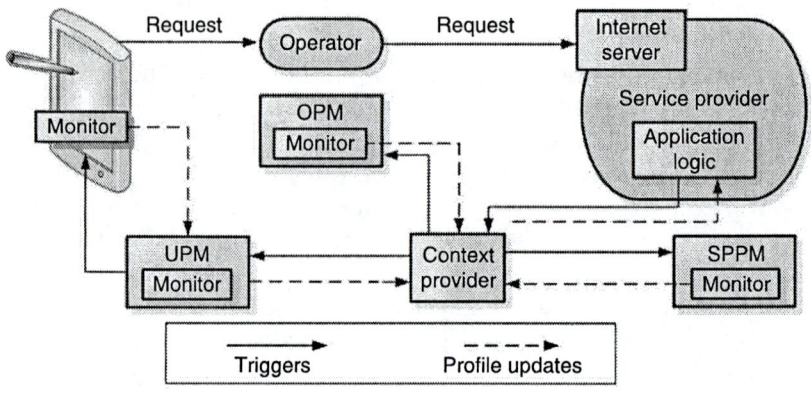

Figure 34.4   Trigger mechanism.

module, which should then reevaluate the profile attributes. Referring to the steps of a typical service request described earlier, this is equivalent to restarting the process from step 3. Of course, various optimizations are possible to avoid recomputing those values that are completely independent from the changed attributes. Figure 34.4 illustrates the trigger mechanism. To ensure that only useful update information is sent to the service provider, a deep knowledge of the service characteristics and requirements is needed. As a consequence, the conditions for notifications are set by the service provider's application logic and are communicated (in their entirety at the beginning of the session and as updates during the session) to the Context Provider module, which forwards them to the appropriate profile managers. Because most of the events monitored by trigger settings sent to the UPM are generated by the user device, our choice is to have the UPM properly communicate the settings to a server module resident on the user's device. To keep the information owned by the UPM up to date, a specific software module must be running on the user's device, monitoring its state accordingly to the triggers and communicating values to the UPM when necessary.

Case Study

To test the CARE middleware, we have developed various prototype services that take advantage of it; as an example, an adaptive architecture for the management of context-aware points of interest has been demonstrated in Agostini et al. [3]. In this section, we illustrate a Web-based adaptive proximity marketing service [2]. The prototype service we developed is an adaptive proximity marketing Web application. The main goal of the service is to provide personalized, location-aware advertisements for sales on items contained in a user's personal shopping list. Items in the shopping list are ranked according to the distance of shops that have them on sale. When the user clicks on an item, the service shows a Web page providing multimedia information about the target product. Multimedia contents are further adapted considering the user's device and network conditions. We want to point out that we were not the first to consider such a service (see, for example, ELBA [19]); as a matter of fact, our emphasis is on service adaptation, based on both user and service provider policies. The service discriminates between users as being either paying or non-paying service subscribers. Non-paying subscribers also receive *unsolicited advertisements* regarding items that are not on their shopping list but are chosen by standard Customer Relationship Management (CRM) software that considers the explicit preferences of the user as retrieved from the aggregated profile. Currently, the service is browser based and provided on a per-request basis. The list of ads is periodically refreshed, and the refresh interval is dynamically chosen by the server application on the basis of context.

In the following, we report one of the test cases we have considered to illustrate some of the profile-resolution directives and policies that determine the adaptation parameters. We consider the case of John, a hypothetical user strolling around a hypothetical town with a PDA in his hands. The town is divided into bidimensional cells identified by pairs of coordinates. We assume that some of the cells are covered by a GPRS connectivity service (providing low bandwidth), and others by a broadband WiFi HotSpot service. Movements of our user and context changes are simulated. The service continuously adapts to John's changes of context (e.g., location, network conditions, time of the day), showing different ads and properly adapting the presentation.

Profile-Resolution Directives

Directive 4 in Figure 34.5 is an example of a profile-resolution directive. In this case, the directive is intended to solve conflicts due to different estimations of the user's current speed given by different entities. The service provider gives higher confidence to the value provided by the UPM, because speed can be estimated very precisely by user-side sensors (e.g., supplied by car appliances or GPS-enabled devices). If no value for speed is given by the user, the value provided by the operator (if present) is taken into account; otherwise, the value inferred by the service provider analyzing the history of the user's location is chosen.

Profile Resolution Directive

(1) *setPriority AllowRecommendations* = (*SPPM, UPM*)
(2) *setPriority Coordinates* = (*UPM, OPM*)
(3) *setPriority MediaQuality* = (*SPPM, UPM*)
(4) *setPriority UserSpeed* = (*UPM, OPM, SPPM*)
(5) *setPriority RefreshTime* = (*UPM, SPPM*)

Policy	Owner
(6) **If** *DeviceType* = 'Pda' **Then Set** *MediaQuality* = 'High'	User
(7) **If** *AvailableBandwidth* < 56kbps **Then Set** *MediaQuality* = 'Low'	Service provider
(8) **If** *UserCurrentTime* > 11:30 **And**	
UserCurrentTime < 13:30 **Then Set** *Phase* = 'Noon'	Service provider
(9) **If** *UserClass* = 'Non-Paying' **Then Set**	
AllowRecommendations = 'Yes'	Service provider
(10) **If** *UserSpeed* = 'Slow' **Then Set** *RefreshTime* = '15min'	Service provider
(11) **If** *UserSpeed* = 'Fast' **Then Set** *RefreshTime* = '3min'	Service provider
(12) **If** *LocationArea* = 'HomeNeighbourhood' **Then Set**	
RefreshTime = 'Infinite'	User

Figure 34.5 Excerpt of profile-resolution directives and policies.

User and Service Provider Policies

Suppose John declares policy 6 in Figure 34.5 to request high-quality multimedia content when using his PDA; similarly, service providers can declare policies for determining content and presentation directives. Possibly conflicting policy 7 is declared by the service provider, who wants to deliver low-quality multimedia contents when the available bandwidth drops below a certain threshold. Policies can also be declared to enrich the profile; in our scenario, policy 8 induces the phase of the day based on the current time. The refresh rate of the service is determined by policies 10, 11, and 12. Policy 10 determines a long refresh interval when the user is moving slowly; policy 11 shortens the refresh interval when the user is moving fast; and policy 12 sets the refresh interval to infinite (thus disabling the service) when the user is close to home and presumably already knows the shops in the area.

A Specific Scenario

In this section, we illustrate via a simple scenario the resulting behavior of the system determined by the evaluation of profile-resolution directives and policies. Suppose John decides to go shopping downtown in the morning. His personal shopping list is stored on his UPM together with the rest of his profile data. At first, he activates the local proxy that will attach profile manager references to each service request (see Figure

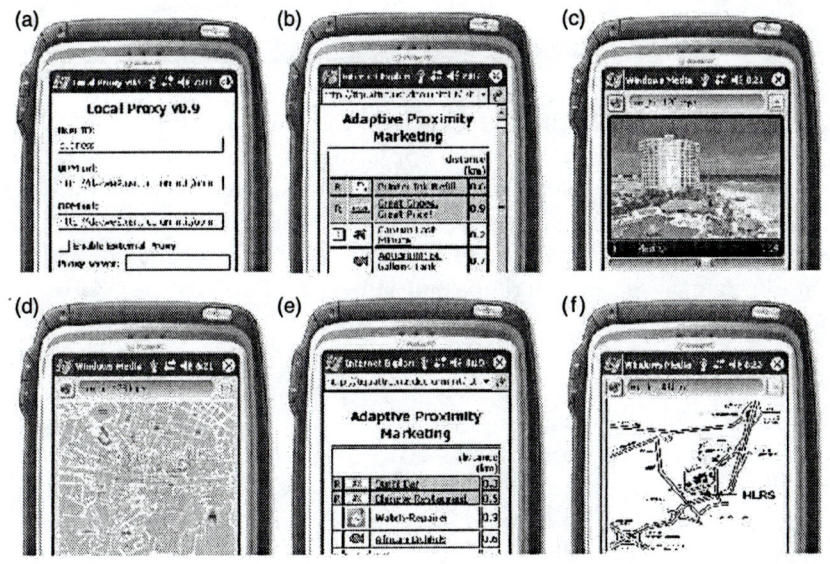

Figure 34.6   Screenshots of the Web application prototype.

34.6A). Suppose also that he starts the service by accessing the proximity marketing Web page. After profile data is gathered from the network by the SPPM, the Merge module obtains two different values for the *Coordinates* attribute: One is provided by John's GPS device, and the other is provided by the network operator on the basis of the user's current cell. Applying resolution directive 2, the Merge module keeps only the value provided by the user's GPS.

Suppose John hates unsolicited advertisements and therefore sets the value to "No" for the *AllowRecommendations* attribute on his UPM; however, because he is a non-paying service subscriber, policy 9 fires, overriding his explicit preference. Note that this override is due to profile-resolution directive 1, which gives higher priority to the service provider for this attribute. As a consequence, on top of the list of ads, John receives a couple of recommendation messages (Figure 34.6B). While looking at the list, John notices an interesting last-minute offer by a nearby travel agency regarding Cancun, one of his preferred destinations. When John selects the corresponding link, evaluation of policy 6 sets the attribute *MediaQuality* to "High," thus adapting the service to deliver a high-quality multimedia presentation of the resort (Figure 34.6C). John also receives a map to the travel agency (Figure 34.6D). Note that policy 7 cannot fire, because John is currently in a WiFi hot-spot. John is walking, so his speed (estimated by his GPS-enabled PDA) is slow and policy 10 fires; thus, the list is updated every 15 minutes.

At 12 p.m., while John is still downtown, the list of advertisements he receives is quite different. Restaurant recommendations appear at the top of the list (see Figure 34.6E). This change in the list is due to the service provider's policy 8, which based on current time sets the value of the *Phase* attribute to "noon" and the application logic adds the restaurant category on top of the user's inferred interests. Although restaurants were not included in his explicit interests, John is particularly attracted by an advertisement regarding a sushi bar. Note that, when not explicitly present in profile data, preference criteria among restaurants may be derived in various ways; for example, a guess regarding John's culinary preference for sushi bars may be made by finding in his profile an interest in oriental culture. When John selects the "Sushi Bar" link, this time he receives a low-resolution map for reaching the restaurant (see Figure 34.6F). This is due to the fact that WiFi connectivity is not available in the current location, and policy 7 can fire; note that policy 6 could also fire, creating a conflict regarding the value of the *MediaQuality* attribute. This conflict is solved by the IE module by considering profile-resolution directive 3, which gives higher priority to the service provider for that attribute; for this reason, policy 7 fires and the attribute *MediaQuality* is set to "Low."

After lunch, John decides to go back home; while in the taxi, his rate of movement makes policy 11 fire, determining a refresh interval of 3 minutes. When the taxi enters John's neighborhood, policy 12 can also fire, thus deriving the value of "Infinite" for the *RefreshTime* attribute and conflicting with policy 11. Once again, the conflict is resolved by the IE considering the profile-resolution directive for the attribute. In this case, directive 5 gives higher priority to the user, so the value determined by his policy 12 overrides the value of the service provider's policy 11, disabling the service.

Alternative Middleware Proposals

Many research groups and companies have been working to define and implement middleware for supporting service adaptation and personalization in a multiple-device, mobile environment. The Houdini middleware [30], developed at Bell Labs, has the main goal of efficiently enabling context-aware mobile applications while preserving the user's privacy. In Houdini, the sharing of context information is controlled by policies declared by the user. The key component of the architecture is a module that evaluates the requests of profile data issued by service providers against the privacy policies declared by the user. The Houdini policy language and the conflict-resolution strategy of its inference engine are similar to ours; however, because it is primarily focused on adaptation, our policy mechanism is different, as policies are declared by both the user and the service provider to determine customization parameters, and the privacy of data is ensured by the appropriate UPM module. From an architectural point of view, in Houdini the profile information is divided into two categories: *static* data (e.g., personal data, buddies list, address book, calendar) and *dynamic* data (e.g., preferences, location, device status). These classes of data, provided by various sources, are handled by two distinct modules that are responsible for their integration on a per-request basis. In our architecture, a single Merge module is in charge of integrating distributed context data and solving possible conflicts.

CARMEN [8] is a middleware for supporting context-aware mobile computing. In CARMEN, service access is mediated by context-aware mobile proxies. These intermediate proxies execute directives obtained from Ponder [17] policies, which manage migration, binding, and access control. The Ponder language turns out to be a good choice for the class of policies used in this middleware; our policy language, although less expressive, is well suited for adaptation rules, as it is extremely efficient and has mechanisms for solving conflicts. CARMEN adopts a profile management mechanism that is different from ours, as profile data is stored by Lightweight Directory Access Protocol (LDAP)-compliant directory services, but in our framework we make use of CC/PP repositories.

A further interesting architecture for supporting context-aware systems in mobile environments is the Context Broker Architecture (CoBrA) [15]. Context awareness in CoBrA is based on a formal model of context — represented by an OWL ontology [16] — that is shared by all the system components. The Context Broker module is in charge of gathering context information from sensors spread throughout the environment. The Context Broker Inference Engine performs ontology-based reasoning to derive new context information from raw data and to detect and resolve inconsistencies in profile data. The privacy-enforcing mechanism of CoBrA is based on ontologies, as well; in particular, privacy policies are represented through an extension of the Rei [32] policy language. It is worth noting that, because the main goal of this architecture is to support knowledge sharing and interoperability in ambient intelligence scenarios, reasoning efficiency is not the main focus of this work. In contrast, because our middleware is intended to support the provision of Internet services possibly accessed by a huge number of users at a time, we perform ontological reasoning in advance of the service provision to preserve system scalability.

Service-Oriented Context-Aware Middleware (SOCAM) [25] supports context-aware services in intelligent environments similar to CoBrA, because its context model is also based on ontologies. In particular, the adopted ontology [48] is composed of a general-purpose upper ontology and application-specific lower ontologies. Context reasoning is performed to check the consistency of context information and to derive higher level context information, and it is based on both description logic and user-defined logic rules. Experiments have shown that this reasoning approach is computationally expensive and unfeasible for time-critical applications that make use of a huge amount of context information. Rules in SOCAM are declared by a single entity (which can be the final user or a service provider), and rule inconsistencies are not addressed; thus, a scenario in which service constraints and user preferences interact (the major focus of our work) is not considered in this architecture.

An architecture for the user-side adaptation of applications in a mobile environment is described in Efstratiou et al. [18]. This architecture contains a single profile manager, which is in charge of discovering *context services* (i.e., services that provide profile data). Context data is kept up to date on the profile manager by means of asynchronous notifications of changes sent by context services. An Adaptation Control module on the user device is in charge of evaluating adaptation policies against profile data, modifying the behavior of local applications as necessary. Policies are declared by users who define priorities among applications as well as among the resources of their devices; consequently, the behavior of applications is adapted to obtain the optimal level of service with respect to the user's requirements. It is worth noting that this architecture does not support server-side adaptation.

Riché and Brebner [46] proposed a different approach to profile management that implements a distributed and replicated storage system on user devices. This approach is useful for preserving the privacy of data; however, the intermittent connectivity of mobile devices, along with their limited CPU, storage, and power resources, makes it difficult to guarantee the availability of profiles, even if sophisticated techniques are adopted.

Open Issues and Challenges

Several issues, both practical and theoretical, still have to be investigated further to make advanced distributed profiling an effective and viable solution for service adaptation. We identify the following as the main open issues:

- Automatic recognition and representation of complex profile attributes
- Preservation of privacy for personal user data including location data

- Optimization and caching techniques to provide real-time integrated profile data
- Scalability with respect to the number of users and profile managers

In this section, we briefly discuss the first two issues, which are probably the most challenging ones and involve more theoretical aspects. The third issue requires: (1) optimization of the algorithms for conflict resolution and policy evaluation, and (2) addressing the crucial issue of devising caching techniques and algorithm refinements to reduce the time delay due to network latency when retrieving profile data from remote sources. The last issue is also relevant because the middleware should be able to scale to a large number of users and profile managers. This issue also requires extensive realistic experiments, possibly using synthetic data from profile generators and movement trace generators, as well as simple real-user testbeds.

Ontologies and Ontological Reasoning

As mentioned earlier in this chapter, the development of profile representations that are far more expressive than CC/PP is necessary. Our favorite example is the representation of user activities, but many other complex profile attributes call for more expressive definitions. Ontologies are a natural candidate, as they are emerging for a variety of services, such as knowledge sharing and semantics disambiguation, which are very relevant even in the context we are considering.

We have two main concerns with the use of current ontology languages for the representation of complex profile attributes and their automatic recognition: (1) expressiveness and (2) efficiency of reasoning. Regarding expressiveness, we encountered several difficulties when trying to use the well-known ontology language OWL [39] (a W3C-recommended language) to represent user activities. In particular, OWL is very weak in reasoning with properties, as it lacks a constructor for property composition. A typical example is that in OWL it is not possible to define "uncle" as a composite of "parent" and "brother." It also lacks the ability to express feature agreement; for example, it is not possible to force the value of the property "has-employer" to be the same in a class of persons (without specifying which one) to represent "colleagues." This is essentially due to the fact that the underlying description logic does not support *role-value maps*, not even in limited forms that preserve decidability [5]. Another relevant issue regarding expressiveness is the lack of support for the representation of rules and the integration of rule-based reasoning with ontological (subsumption-based) reasoning.

The introduction of first-order rules into OWL would greatly augment the expressiveness of the language, and a number of projects are currently addressing this issue [22,29]; for example, the Semantic Web Rule Language (SWRL) is based on a combination of OWL-DL with the Unary/ Binary Datalog RuleML sublanguages of the Rule Markup Language (RuleML). In SWRL, the set of OWL axioms is augmented with Horn-like rules, where unary and binary predicates correspond to OWL classes and properties, respectively. A prototype implementation has been developed. Unfortunately, the introduction of rules into OWL or even OWL-DL can easily lead to the undecidability of the basic reasoning tasks, making languages such as SWRL unsuited for real-time services that make use of large and complex ontologies.

Efficient reasoning with ontologies is indeed the second general concern mentioned above. Very good progress has been made in recent years in the development of modal logic theorem provers that can be used to compute subsumption and related reasoning tasks in ontologies. However, the underlying reasoning problems are inherently difficult, and the classification task becomes unfeasible in real time even for small- to medium-size ontologies and not very expressive languages such as OWL-Lite [26].

Privacy Issues

Another major issue involved in advanced distributed profiling is the privacy of user data. The two main approaches to privacy preservation are enforcement of privacy policies and the use of anonymization techniques. The first approach essentially considers each request for access to personal information and decides whether or not to grant or deny access based on a specific policy. Policies usually consider the entity making the request, the information requested, and the modality of the access; in the case of mobility, spatiotemporal conditions may also be considered. In our architecture, the distribution of user profile data may be restricted by such a set of privacy policy rules enforced at each profile manager. For example, a set of UPM policies could allow the user client interface and user trusted agents to update personal data and policies, allow the user GPS module to update location data, and allow a set of service provider profile managers to read the profile attribute values in certain CC/PP components. Note that profile managers are considered trusted agents by their corresponding entities. Several formalisms that have been proposed for database access control can be easily adapted to the context we are considering [31]. Extensions to these basic models that have been proposed by Bertino et al. [9] include temporal constraints that specify periodic time windows where access is denied or accepted, as well as qualitative temporal relationships among accesses (using operators such as *aslongas*, *whenevernot*, and *until*).

The presence of spatiotemporal constraints in policies to preserve mobile user privacy has been recently identified as a requirement by Youssef et al. [50]. They propose an access control system for moving objects and customer profiles where each access rule is composed of a triple <s, o, +/–> and a spatiotemporal constraint <stc>. The triple specifies the subject (s), which may be a specific service provider; the object (o), which may be a specific user profile; and a flag (+/–) specifying if it is a positive or negative access rule. The implicit access mode is *read*. The constraint <stc> defines the spatiotemporal context of access rule application, and it is composed of a location and a time interval. For simplicity in the definition of rules, the rule components can be defined at different levels of granularity. Although this is still a preliminary study, it is an example of adapting the general idea of database access control to the release of mobile users' profile data. This approach can probably be applied without much effort to the middleware described earlier.

A general concern about the access control approach is the Boolean result returned by the evaluation of rules at each access request. The access is either granted or denied; this means the entire profile is either released or not released, and, in some cases, the denial may lead to loss of service. Bettini et al. [10] proposed an extension to the classical approach by introducing *conditional* granting. In such a model, the access can be granted provided the requester satisfies some conditions at the time of access and accepts fulfilling certain *obligations* in the future — for example, notifying the owner when the information is used for certain purposes.

A totally different approach is based on anonymization [24,47]. In this case, instead of denying access to the information, the information is properly manipulated so it preserves a form of anonymity. The main idea here is to make it impossible for any sensitive information released to an untrusted entity to be connected by this entity to the specific individual to which it refers; that is, it should be impossible for this entity to distinguish among *k* individuals to whom the released information potentially refers. This can be achieved by appropriate middleware using pseudonyms instead of real user names, hiding real network addresses, and appropriately obfuscating information that may reveal the identity of the user. Obfuscation techniques are usually based on generalization of values using, for example, granularity hierarchies, or they can be based on truncation of values. As trivial examples, consider the truncation of Zip Codes to three or four digits, the truncation of Social Security numbers or phone numbers, as well as the obfuscation of location data by, for example, releasing the name of the closest city instead of the precise GPS coordinates.

We believe that it would be very interesting and challenging to further investigate, in the context of advanced profiling for mobile users, both the access control and the anonymization approaches and their possible integration in a single privacy protection solution.

Acknowledgments

The authors wish to thank Dario Maggiorini and Nicolò Cesa-Bianchi for their contributions to the general design of the CARE middleware and all the students and professional programmers who worked at its implementation. This work has been supported by Italian MIUR (FIRB "Web-Minds" project N.RBNE01WEJT_005).

References

[1] Adusei, I.K., Kyamakya, K., and Jobmann, K., Mobile positioning technologies in cellular networks: an evaluation of their performance metrics, in *Proc. of Military Communications Conf. (MILCOM 2002)*, Orlando, Fl, October 7–10, 2002, pp. 1239–1244.

[2] Agostini, A., Bettini, C., Cesa-Bianchi, N., Maggiorini, D., Riboni, D. et al., Towards highly adaptive services for mobile computing, in *Proc. of IFIP TC8 Working Conf. on Mobile Information Systems (MOBIS 2004)*, Oslo, Norway, September 15–17, 2004, pp. 121–134.

[3] Agostini, A., Bettini, C., and Riboni, D., Demo: ontology-based context-aware delivery of extended points of interest, in *Proc. of the 6th Int. Conf. on Mobile Data Management (MDM'05)*, Aiya Napa, Cyprus, May 9–13, 2005, pp. 322–323.

[4] Ardissono, L. and Goy, A., Tailoring the interaction with users in Web stores, *User Modeling User-Adapted Interaction (UMUAI)*, 10, 251–303, 2000.

[5] Baader, F., Calvanese, D., McGuinness, D.L., Nardi, D., and Patel-Schneider, P.F., Eds., *The Description Logic Handbook: Theory, Implementation, and Applications*, Cambridge University Press, Cambridge, U.K., 2003.

[6] Bahl, P. and Padmanabhan, V.N., RADAR: an in-building RF-based user location and tracking system, in *Proc. IEEE INFOCOM'00*, Tel Aviv, Israel, March, 2000, pp. 775–784, 2000.

[7] Bajaj, R., Ranaweera, S., and Agrawal, D.P., GPS: location-tracking technology, *IEEE Comput.*, 35(4), 92–94, 2002.

[8] Bellavista, P., Corradi, A., Montanari, R., and Stefanelli, C., Context-aware middleware for resource management in the wireless Internet, *IEEE Trans. Software Eng.*, 29(12), 1086–1099, 2003 (special issue on wireless Internet).

[9] Bertino, E., Bettini, C., Ferrari, E., and Samarati, P., An access control model supporting periodicity constraints and temporal reasoning, *ACM Trans. Database Syst.*, 23(3), 231–285, 1998.

[10] Bettini, C., Jajodia, S., Wang, X., and Wijesekera, D., Provisions and obligations in policy rule management, *J. Network Syst. Manage.*, 11(3), 351–372, 2003.

[11] Bettini, C. and Riboni, D., Profile aggregation and policy evaluation for adaptive Internet services, in *Proc. of the First Int. Conf. on Mobile and Ubiquitous Systems: Networking and Services (MobiQuitous'04)*, Boston, MA, August 22–26, 2004, pp. 290–298.

[12] Bittner, R., Smrcka, P., Vysok'y, P., Hána, K., Pousek, L., and Schreib, P., Detecting fatigue states of a car driver, in *Proc. of Int. Symp. on Medical Data Analysis (ISMDA 2000)*, Frankfurt, Germany, September 29–30, 2000, pp. 260–273.

[13] Bowman, M., Chandler, R.D., and Keskar, D.V., *Delivering Customized Content to Mobile Device Using CC/PP and the Intel CC/PP SDK*, Technical Report, Intel Corp., Santa Clara, CA, 2002.

[14] Butler, M., Giannetti, F., Gimson, R., and Wiley, T., Device independence and the Web, *IEEE Internet Comput.*, 6(5), 81–86, 2002.

[15] Chen, H., Finin, T., and Joshi, A., Semantic Web in the context broker architecture, in *Proc. of the Second IEEE Int. Conf. on Pervasive Computing and Communications (PerCom 2004)*, Orlando, FL, March, 2004, pp. 277–286.

[16] Chen, H., Perich, F., Finin, T.W., and Joshi, A., SOUPA: standard ontology for ubiquitous and pervasive applications, in *Proc. of the First Int. Conf. on Mobile and Ubiquitous Systems: Networking and Services (MobiQuitous'04)*, Boston, MA, August 22–26, 2004, pp. 258–267.

[17] Damianou, N., Dulay, N., Lupu, E., and Sloman, M., The Ponder policy specification language, in *Proc. of the 2nd IEEE Workshop on Policies for Distributed Systems and Networks (Policy 2001)*, Bristol, U.K., January, 2001, pp. 18–38.

[18] Efstratiou, C., Cheverst, K., Davies, N., and Friday, A., An architecture for the effective support of adaptive context-aware applications, in *Proc. of the Second Int. Conf. on Mobile Data Management (MDM'01)*, Hong Kong, January, 2001, pp. 15–26.

[19] European Location Based Advertising (ELBA), European Project No. IST-2001-36530, http://www.e-lba.com/.

[20] Finin, T.W. and Drager, D., A general user modeling system, in *Proc. of the 6th Canadian Conf. on Artificial Intelligence*, Montreal, Canada, pp. 24–29, 1986.

[21] Fink, J. and Kobsa, A., A review and analysis of commercial user modeling servers for personalization on the World Wide Web, *User Modeling User-Adapted Interaction (UMUAI)*, 10, 209–249, 2000.

[22] Gandon, F.L., Sheshagiri, M., and Sadeh, N.M., *ROWL: Rule Language in OWL and Translation Engine for JESS*, Technical Report, Carnegie Mellon University, Pittsburgh, PA, 2004.

[23] Gellersen, H.-W., Schmidt, A., and Beigl, M., Multi-sensor context awareness in mobile devices and smart artifacts, *MONET*, 7(5), 341–351, 2002.

[24] Gruteser, M. and Grunwald, D., Anonymous usage of location-based services through spatial and temporal cloaking, in *Proc. of the First Int. Conf. on Mobile Systems, Applications, and Services (MobiSys'03)*, San Francisco, CA, May, 2003, pp. 42–47.

[25] Gu, T., Wang, X.H., Pung, H.K., and Zhang, D.Q., An ontology-based context model in intelligent environments, in *Proc. of Communication Networks and Distributed Systems Modeling and Simulation Conf. (CNDS'04)*, San Diego, CA, January 2004.

[26] Guo, Y., Pan, Z., and Heflin, J., An evaluation of knowledge base systems for large OWL datasets, in *Proc. of the Third Int. Semantic Web Conf. (ISWC'04)*, Hiroshima, Japan, November 7–11, 2004, pp. 274–288.

[27] Harter, A., Hopper, A., Steggles, P., Ward, A., and Webster, P., The anatomy of a context-aware application, in *Proc. of the 5th ACM/IEEE Int. Conf. on Mobile Computing and Networking (MOBICOM'99)*, Seattle, WA, August, 1999, pp. 59–68.

[28] Hightower, J. and Borriello, G., Location systems for ubiquitous computing, *IEEE Comput.*, 34(8), 57–66, 2001.

[29] Horrocks, I., Patel-Schneider, P.F., Boley, H., Tabet, S., Grosof, B., and Dean, M., *SWRL: A Semantic Web Rule Language Combining OWL and RuleML*, W3C Member Submission, World Wide Web Consortium (W3C), May 2004 (http://www.w3.org/Submission/2004/SUBM-SWRL-20040521/).

[30] Hull, R., Kumar, B., Lieuwen, D., Patel-Schneider, P., Sahuguet, A. et al., Enabling context-aware and privacy-conscious user data sharing, in *Proc. of the 5th Int. Conf. on Mobile Data Management (MDM'04)*, Berkeley, CA, January 19–22, 2004, pp. 187–198.

[31] Jajodia, S., Samarati, P., Sapino, M.L., and Subrahmanian, V.S., Flexible support for multiple access control policies, *ACM Trans. Database Syst.*, 26(2), 214–260, 2001.

[32] Kagal, L., Finin, T.W., and Joshi, A., A policy language for a pervasive computing environment, in *Proc. of the 4th IEEE Workshop on Policies for Distributed Systems and Networks (Policy 2003)*, Como, Italy, June, 2003, pp. 63–75.

[33] Kazantzidis, M., Slain, I., Chen, T., Romanenko, Y., and Gerla, M., End-to-end versus explicit feedback measurement in 802.11 networks, in *Proc. of the 7th IEEE Int. Symp. on Computers and Communications (ISCC'02)*, Taormina, Italy, July, 2002, pp. 429–434.

[34] Klyne, G., Reynolds, F., Woodrow, C., Ohto, H., Hjelm, J., Butler, M.H., and Tran, L., *Composite Capability/Preference Profiles (CC/PP): Structure and Vocabularies 1.0*, W3C Recommendation, World Wide Web Consortium (W3C), 2004, http://www.w3.org/TR/2004/REC-CCPP-struct-vocab-20040115/.

[35] Kobsa, A., Generic user modeling systems, *User Modeling User-Adapted Interaction (UMUAI)*, 11, 49–63, 2001.

[36] Kobsa, A. and Pohl, W., The BGP-MS user modeling system, *User Modeling and User-Adapted Interaction (UMUAI)*, 4(2), 59–106, 1995.

[37] Lai, K. and Baker, M., Measuring bandwidth, in *Proc. IEEE INFOCOM'99*, New York, March, 1999, pp. 235–245.

[38] Maggiorini, D. and Riboni, D., Continuous media adaptation for mobile computing using coarse-grained asynchronous notifications, in *Proc. of Int. Symp. on Applications and the Internet (SAINT2005)*, Trento, Italy, January 31–February 4, 2005, pp. 162–165.

[39] McGuinness, D.L. and van Harmelen, F., *OWL Web Ontology Language*, W3C Recommendation, World Wide Web Consortium (W3C), 2004, http://www.w3.org/TR/owl-features/.

[40] Millard, I., Roure, D.D., and Shadbolt, N., The use of ontologies in contextually aware environments, in *Proc. of UbiComp'04 First Int. Workshop on Advanced Context Modelling, Reasoning and Management*, Nottingham, U.K., September, 2004, pp. 42–47.

[41] Ohto, H. and Hjelm, J., *CC/PP Exchange Protocol*, W3C Note, World Wide Web Consortium (W3C), 1999, http://www.w3.org/1999/06/NOTE-CCPPexchange-19990624.

[42] OpenMobileAlliance, *User Agent Profile Specification*, Technical Report WAP-248-UAProf20011020-a, Wireless Application Protocol Forum, 2001, http://www.openmobilealliance.org/.

[43] Prasad, R.S., Murray, M., Dovrolis, C., and Claffy, K., Bandwidth estimation: metrics, measurement techniques, and tools, *IEEE Network*, 17(6), 27–35, 2003.

[44] Priyantha, N.B., Chakraborty, A., and Balakrishnan, H., The Cricket location-support system, in *Proc. of the 6th ACM/IEEE Int. Conf. on Mobile Computing and Networking (MOBICOM'00)*, Boston, MA, August, 2000, pp. 32–43.

[45] Resnick, P., Iacovou, N., Sushak, M., Bergstrom, P., and Riedl, J., GroupLens: an open architecture for collaborative filtering of netnews, in *Proc. of the Fifth Conf. on Computer Supported Cooperative Work (CSCW'94)*, Chapel Hill, NC, October 22–26, 1994, pp. 175–186.

[46] Riché, S. and Brebner, G., Storing and accessing user context, in *Proc. of the Fourth Int. Conf. on Mobile Data Management (MDM'03)*, Melbourne, Australia, January, 2003, pp. 1–12.

[47] Samarati, P., Protecting respondents' identities in microdata release, *IEEE Trans. Knowledge Data Eng.*, 13(6), 1010–1027, 2001.

[48] Wang, X.-H., Gu, T., Zhang, D.Q., and Pung, H.K., Ontology-based context modeling and reasoning using OWL, in *Proc. of the Second IEEE Int. Conf. on Pervasive Computing and Communications (PerCom 2004)*, Orlando, FL, March, 2004, pp. 18–22.

[49] Want, R., Hopper, A., Falcao, V., and Gibbons, J., The active badge location system, *ACM Trans. Inform. Syst.*, 10(1), 91–102, 1992.

[50] Youssef, M., Atluri, V., and Adam, N.R., Preserving mobile customer privacy: an access control system for moving objects and customer profiles, in *Proc. of the 6th Int. Conf. on Mobile Data Management (MDM'05)*, Aiya Napa, Cyprus, May 9–13, 2005.

[51] Mutz, A., Montulli, L., and Masinter, L., *User-Agent Display Attributes Headers*, HTTP Working Group, 1996, http://mirrors.isc.org/pub/www.watersprings.org/pub/id/draft-mutz-http-attributes-00.txt.

Chapter 35

QoS-Aware Resource Discovery in Mobile Environments

Yun Huang, Shivajit Mohapatra, Qi Han,
and Nalini Venkatasubramanian

CONTENTS

Introduction

Recent advances in high-quality digital wireless network technologies coupled with the unprecedented growth of mobile computing devices, such as personal digital assistants, laptop computers, mobile phones, etc., are enabling new classes of mobile applications with diverse quality-of-service (QoS) requirements. Today, mobile applications span a variety of domains — from business and entertainment to education, command and control, and crisis response [1–3]. Mobile gaming, audio/video streaming, and collaborative multimedia applications are becoming ubiquitous and are projected to be the dominant applications in next-generation mobile systems. These applications have distinctive performance and processing requirements that tend to make them extremely resource hungry. They also have diverse QoS requirements that determine the utility of the (perceived) information to the end user. QoS needs can be expressed as user-perceived quality needs (e.g., video quality) that translate into lower level application/system parameters. In addition, QoS statements may specify constraints on timing, availability, security, and resource utilization at various levels of abstraction. For example, timing-based QoS requirements can be specified using abstract properties such as correct and timely data delivery and uninterrupted service. These properties can be translated to concrete application parameters, such as jitter, end-to-end delay, and synchronization skew, or concrete resource requirements, such as network and disk bandwidth and buffer requirements [4]. The notion of QoS can include bandwidth management, throughput control, timeliness, reliability (e.g., mean time to failure, mean time to repair), perceived quality and cost (e.g., communication cost, service cost), and even battery energy management [5]. Resources required to support these multidimensional notions of QoS in mobile applications can be in the form of computation (CPU), storage, bandwidth, memory, or services that must continue to be available as the user moves in the mobile infrastructure.

One approach is to overallocate and reserve resources to meet peak demands at all times. Overallocation is impractical in mobile environments because (1) it results in low resource utilization, and (2) it is difficult to predetermine where and when resources are needed. If resource availabilities are known in advance, static admission control techniques combined with resource reservation protocols can be used to admit requests if the QoS demands of the services can be met adequately; however, in mobile environments resource availabilities can change over time in a very erratic manner rendering static reservations invalid.

Resource discovery to ensure sustained QoS for mobile applications presents several interesting research challenges. These challenges arise at different levels (e.g., network, server, device) and can be summarized as follows:

- *Bandwidth-limited wireless networks* — Wireless networks are often bandwidth constrained (e.g., 10 Kbps for cellular, 10–100 Mbps for WLAN); they are also characterized by irregular connectivity, transmission errors, and frequent disconnections. Supporting high-quality, data-intensive flows (e.g., multimedia) requires predictable mobile network behavior in terms of bandwidth availability, network losses, and transmission delays.
- *Uncertainty due to user mobility* — User mobility introduces uncertainty in user locations and consequently in bandwidth usage at different points. This implies that resource discovery must be able to cope with the uncertainties and ensure consistent resource availabilities that satisfy application QoS.
- *Insufficient resources in mobile devices* — Portable devices have limited processing capabilities, memory, and energy, and mobile applications have significant resource needs.
- *Cost/quality tradeoffs* — An inherent tradeoff exists between application QoS and resource consumption; for example, dedicated network resources could provide higher quality multimedia applications, but this may result in low network utilization and hence higher cost.
- *Lack of accurate context* — Discovering optimal resources requires knowledge of the underlying context. Collecting and maintaining accurate context information in mobile environments is challenging due to user mobility and tradeoffs between context accuracy and maintenance overhead.

Extending existing approaches for resource discovery that have been developed in the context of wired networks directly for QoS-aware mobile applications is problematic. For example, Common Object Request Broker Architecture (CORBA™)-based approaches are too heavyweight for mobile applications [6]; furthermore, they are designed for relatively stable environments where disconnections are not the norm. Java-based middleware solutions such as JXTA™ [7] incorporate service-discovery techniques that are based on object class matching. These approaches are primarily focused at the application level, and QoS guarantees are difficult to meet because the class files of mobile objects may be distributed over the network. In peer-to-peer (P2P) systems [8], resources are discovered through the collaboration among peer nodes; however, applying the P2P approach to mobile applications suffers from high connection costs, high network traffic for overlay maintenance, low network efficiency, and high latency [9]. In addition, the unpredictability of peer-based networks makes it difficult to ensure application QoS. In general, none of these approaches is designed with a focus on ensuring QoS for mobile applications; hence, it is desirable to investigate effective resource discovery strategies appropriate for mobile environments.

In this chapter, we elaborate on why intelligent resource discovery and provisioning are necessary to guarantee QoS requirements for mobile applications. We argue that context (application, network, resource, device) plays a crucial role in effective resource discovery for mobile applications. We discuss static and dynamic aspects of the context collection and resource allocation problem and discuss how dynamic adaptation can sustain QoS guarantees under changing network, device, and system conditions in a cost-effective manner. We first present approaches for performing static resource allocation (network and server resources); subsequently, we describe how dynamic resource reprovisioning can be effectively used to handle dynamic changes in resource availability and system state. Using mobile multimedia as the driving application for a case study, we illustrate how to integrate the dynamic changes nonintrusively into a wide-area mobile infrastructure.

Mobile applications are typically run in two types of environments: mobile *ad hoc* networks and infrastructure-based networks. Mobile *ad hoc* networks (MANETs) are wireless networks consisting entirely of mobile nodes that communicate on the move without any infrastructure support such as base stations or access points. Nodes in these networks will both generate user and application traffic and carry out network control and routing protocols. Rapidly changing connectivity, network partitions, higher error rates, collision interference, and bandwidth and power constraints together make routing [10–13] and topology management [14,15] interesting and difficult problems in MANETs. To provide focus, in this chapter, we mainly address the issues involved in resource discovery in infrastructure-based wireless networks (e.g., cellular or wireless local area networks). Typically, these networks are composed of mobile devices with wireless interfaces and a core infrastructure with fixed and wired hosts. Mobile users move and connect to the fixed network via wireless access points or base stations. They can access services and possible resources provided by the end systems from the fixed network.

A Mediation-Based Architecture for Resource Discovery

We can address the challenges that arise in cost-effective resource discovery for QoS-aware mobile applications from the following three perspectives:

- *Adaptive context collection* — The success of resource allocation relies on timely and accurate knowledge of underlying context. Efficient context collection and monitoring techniques are therefore needed to keep track of the current global and local states and possibly even predict future changes.

- *Static resource discovery* — QoS-based static resource allocation mainly solves the problem of scheduling and admission control at the initiation of an end-to-end interaction. Given an application request from a mobile device with corresponding QoS needs, the resource discovery and provisioning process allocates the resources to establish end-to-end interactions by evaluating resource availabilities and estimating the system performance. For example, this process may include determining a network path with minimum delay, choosing a video server with compressed data that matches quality needs, or allocating a Web server that is in close proximity to a mobile user.
- *Dynamic resource reprovisioning* — Dynamic service adaptation aims to maintain optimal QoS for mobile users during the service period. When users are mobile, their wireless network connectivity might change dynamically, causing some resources or data to be inaccessible. This implies that resource reprovisioning is crucial to ensure continuous services for mobile applications.

To provide coordinated resource discovery for multiple mobile applications, we introduce the notion of a mediation-based middleware architecture (Figure 35.1). In this architecture, a mediator maintains the necessary system

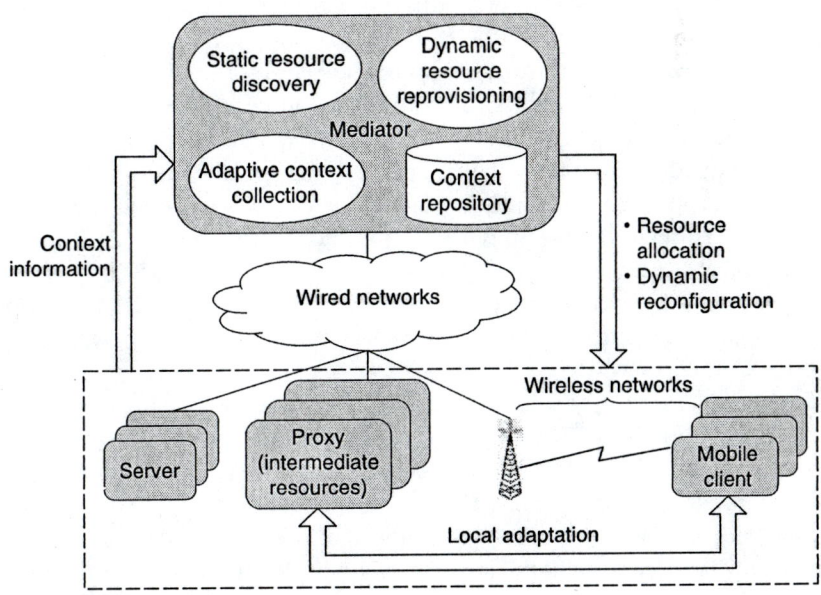

Figure 35.1   A mediator-based architecture for resource discovery in mobile environments.

context in a *context repository* using whichever admission control, resource allocation, and reconfiguration decisions are made to ensure QoS for mobile applications. Traditionally, much of the system context information is gathered and maintained independently; for example, network topology can be maintained by routing information exchange, a replica map can be obtained from a distributed domain name service, network management software keeps track of topology and link parameters, load-balancing services monitor server load patterns, and content management and replication services manage data placement and distribution.

Integrating the above information into a common context repository (database or directory service) has several advantages. First, effective resource discovery and provisioning algorithms can exploit information from various levels for better system utilization. Second, keeping track of dynamic changes in servers, networks, and content availabilities can be decoupled from policies for resource discovery. This provides for a clean separation of concerns in system design. Third, using well-defined and uniform representations for the information allows easier manageability of data. Fourth, knowledge of cross-layer information (from the network, application, and devices) allows for flexible and efficient context collection. Such knowledge allows us to tailor the accuracy of the data in the context repository based on application needs, collection overhead, and connectivity conditions; for example, we may relax collection parameters when user QoS needs are not stringent, which will reduce collection overhead in an already congested network.

We also advocate the use of proxy nodes as intermediate resources in the mediation-based architecture. Due to resource constraints in mobile devices and dynamic conditions in wireless networks, achieving sustained QoS between the service provider (server) and mobile device is difficult. The proxy approach attempts to use available resources in the wired networks within close proximity to the mobile device to support strategies such as proxy caching [16,17], proxy-based transcoding [18,19], or task-offloading [20] that can alleviate stringent resource needs of mobile applications; however, these intermediate resources must be discovered and deployed effectively.

A generalized architecture supports QoS-aware mobile applications as follows: A request containing QoS parameters is initiated at a mobile device. The mediator utilizes the resource availability information stored in the context repository (maintained by the adaptive context collection module) to decide an optimal allocation of network, server, or proxy resources. Significant changes in the availability of the allocated resources will trigger the resource reprovisioning process to adapt the resource allocation accordingly. When the request terminates, the resources are reclaimed along the connection and the context collection module updates the resource availability status in the context repository. Using mobile multimedia as a driving application, we next describe techniques for addressing the key issues in resource discovery for QoS-based mobile applications — adaptive context collection, static resource discovery, and dynamic resource reprovisioning.

Adaptive Context Collection for Effective Resource Discovery

Resource discovery and reprovisioning algorithms utilize information about the current system context to ensure that applications meet their QoS requirements. For QoS-aware resource discovery, the relevant context can be classified into the following three categories:

- *Network core parameters* include link bandwidth, link delay, loss rate, wireless channel conditions, mobility-related parameters, etc.
- *Network edge characteristics* include parameters for servers (e.g., current server load, server CPU usage), clients (e.g., device resource constraints, power levels, client locations), and content-specific attributes (e.g., number of replicas of video files and their locations).
- *Intermediate resources* include information on the capabilities of intermediate proxies, knowledge about when proxies are available, and the relative stability of the proxy.

Accurate knowledge of context such as that described above enables optimal discovery and allocation of resources; furthermore, knowledge of changing system context can be used by resource reprovisioning techniques for better QoS and performance. Therefore, the dynamic changes in system and network context must be captured rapidly with low overhead without interfering with the resource discovery and reprovisioning process.

The accuracy of context information can play a significant role in the efficacy of resource discovery techniques; for example, resource provisioning algorithms may select a network path for a flow or connection based on current resource availability, reserve the chosen path, and subsequently admit the request. In this process, imprecise system-state information can lead to two types of failures. A *routing* failure may occur when a feasible path cannot be found for the new connection, and a *setup* failure may occur when a seemingly feasible path is selected that ultimately does not have enough resources for the new connection. Neither failure is desirable; in particular, setup failures incur extra overhead to reserve resources that may never be used along the path. Maintaining accurate system and network status information can therefore help in making the correct decisions, thus ensuring the desired application QoS and consequently better user experience for mobile applications. However, maintaining accurate system context implies more frequent and tight monitoring, which in turn introduces significant network traffic, resulting in poor utilization of underlying computation, communication, and storage resources. The challenge then is to obtain *sufficiently* accurate state information to reduce the cost of collection while meeting user QoS needs.

There are various strategies for context representation and collection that address the cost/accuracy tradeoff. Information representation and collection strategies are often intertwined. Any parameter in the context repository can be represented by either a single instantaneous value (the last measured value) [21] or a range-based representation that approximates the value of a parameter by using an interval with an upper and a lower bound. The range size may remain static [22] or change dynamically [23,24]. Corresponding collection policies [23] determine when and how often to sample the network components for current status information and whether to update the database with collected samples. Sampling periods may be fixed or may vary over time.

The need for flexible information collection is further aggravated in mobile environments for the following reasons: (1) Mobile devices roam across access points that connect them to wired networks, and the constant user mobility causes significant variations in the resource availability on various network links; (2) handheld devices typically have highly limited storage and computing resources; and (3) the resource availability can be substantially affected by computation and communication profiles of the applications executing on the device which implies that capturing device limitations and the changes in device status as a part of the system image is necessary.

Collection and Maintenance of Location Information for Mobile Hosts

Of particular interest in mobile applications is user location information. With accurate user location information, continuous service despite user mobility becomes possible, and nearby available resources can be discovered more effectively. Location information can be collected in two ways. Fine-grained approaches maintain the current location of each individual mobile client [25,26], and coarse-grained collection captures information at an aggregated level for multiple clients; for example, *client aggregation* (i.e., population of mobile clients in each cell at a particular time instant) can be used as a coarse measurement for location information.

Fine-grained location information management has gained a lot of attention from researchers over the last few years. It typically involves three issues [25,26]: location update strategies, which decide when mobile users should inform the network about their current locations; paging strategies, which decide when the base station should send out queries to search for the mobile user; and location information maintenance architectures, which decide how to store and disseminate the location information. In addition, user mobility patterns (such as that in Haas [27]) are studied concurrently to better capture current user location; however, gathering individual user location does not benefit the overall goal of efficient system status collection. Perturbation of residual resources caused by a single mobile user is almost negligible. Furthermore, keeping track of individual user mobility may entail significant overhead, as each mobile host must be probed separately and constantly.

A key observation that can be exploited for cost-effective collection of location information is that the movement of a large number of users may lead to nonuniform distribution of mobile users across cells. Previous work [28] has explored the use of coarse-grained mobility information that captures the distribution or aggregation of users in the mobile network to support cost-effective resource provisioning. In addition to lower overhead, collection using coarse-grained mobility information is independent of individual mobility models, so it avoids inaccuracies introduced in modeling or predicting individual user's mobility. One measure of macro-level changes in mobile settings is the client aggregation status (i.e., number of users in a cell), which has the potential to significantly affect resource availability in the network or system. Client aggregation status can be obtained from cellular access points, such as base stations, that manage the communication of the mobile hosts residing within each cell. Base stations can apply either a simplistic strategy (e.g., an update from the mobile host is triggered when handoff occurs) to maintain the total population of mobile hosts or more complex prediction-based approaches. In the absence of this information from base stations (which requires tight coordination with the service provider), it may be possible to predict the aggregate mobility status from an individual mobility model. Prediction of aggregation status requires some knowledge about the distribution of mobile hosts and their mobility patterns in a region.

When the coarse location information is obtained (either from base stations or via model-based predictions), it can be used to enhance system status collection. A family of collection strategies [28] have been proposed that use client aggregation status to drive the adjustment of sampling frequency and range size. The basic idea is as follows. The underlying topology is first partitioned into nonoverlapping regions. Each region is equipped with a collection point that accumulates all the state information

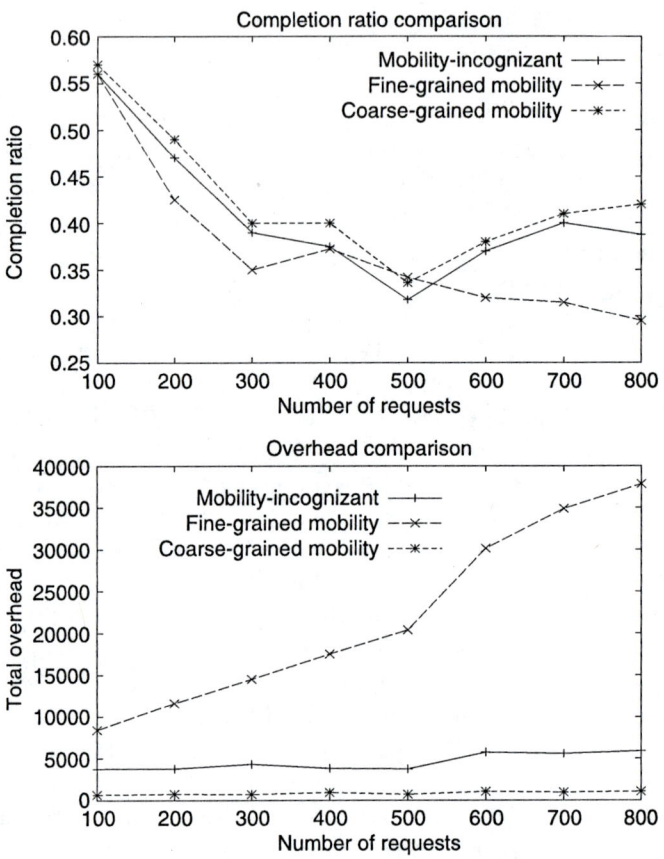

Figure 35.2 Performance comparison of various system context collection approaches.

of the mobile hosts, servers, links for that region. A range with an upper bound and a lower bound is used to represent the mobile host aggregation status (e.g., number of hosts in a cell) in the directory service. The collection algorithm itself consists of two phases: Phase 1 derives the aggregate mobility patterns from individual user mobility patterns and utilizes the aggregation status and current resource utilization status to adjust the collection parameters such as sampling frequency and range size; Phase 2 utilizes feedback from mobile devices and the resource provisioning process for further customization of the collection process.

Figure 35.2 demonstrates the performance of resource discovery and cost involved in maintaining resource availability information by using three different approaches to system status collection: mobility-incognizant collection, collection using fine-grained mobility information, and collection using coarse-grained mobility information. The *request completion ratio* (the percentage of requests that successfully complete) is used as a metric to measure the application performance. The request completion ratio is different from the *request admission ratio*. Admitted requests may not complete for several reasons: No route has sufficient resources (a path failure), the locating mobile host fails (a location failure), or the alternate rescheduling server may not have sufficient resources if the path to the original server is not available. We observe that the request completion ratios of resource discovery under the three approaches are close to each other; however, the use of fine-grained mobility information introduces significantly higher overhead, and the use of coarse-grained mobility information incurs the lowest overhead. This demonstrates the effectiveness of utilizing coarse-grained mobility information. With context collection strategies in place, we now describe how context information can be used in static resource discovery and dynamic resource reprovisioning to provide improved QoS for mobile applications.

Static Resource Discovery

In this section, we discuss the static resource discovery problem for mobile QoS-based services. Static resource discovery addresses the issue of discovering available resources (network, server, proxy) to provide acceptable services for a mobile user when the request is initiated. In this model, an incoming request from a mobile device expresses the service desired along with the QoS requirements (e.g., bandwidth needs, startup latencies, end-to-end delays) for the service. The incoming request goes through an admission control process that determines if end-to-end QoS can be satisfied under current (or predicted) conditions. If resources are unavailable to satisfy the QoS needs of the request, the incoming request is rejected.

Resource discovery mechanisms must ensure end-to-end application QoS while achieving an optimized resource allocation at the system level. To begin with, server and network resources must be discovered and provisioned. Selecting both network and server resources for multiple concurrent requests with varying QoS needs from a limited set of underlying resources is a challenging problem. In addition, resource-constrained mobile devices can benefit significantly by using fixed resources within the wired networks (e.g., proxies, idle machines, peer nodes) that are accessible through the wired/wireless infrastructure. Hence, the resource discovery mechanism must also address the discovery of proxy resources along the path of the service, preferably close to the mobile device. We advocate this approach because accessing data from nearby resources reduces network and server traffic; the proxy-based solution makes possible personalized services with high QoS satisfaction while improving overall system resource utilization.

A fundamental performance/quality tradeoff exists that must be addressed to provide effective resource discovery for mobile applications. For example, a solution that lowers the overall network traffic (for better performance) in the system might initiate the selection of the nearest resources; however, this can introduce frequent switches in the multimedia stream if the user moves rapidly, leading to increased jitter (i.e., lower QoS). Also, knowledge of the future (of application needs and resource availabilities) plays a useful role in supporting continued service with sustained QoS. Static resource discovery mechanisms that exploit knowledge of user and system needs (possibly through prediction techniques) in the long term can help better provisioning in two ways. First, it allows the system to choose a suitable QoS service level that can be sustained for the entire duration of service. This is especially relevant when application QoS is constrained by available system resources and device energy limitations. This will minimize frequent changes in the QoS level, leading to better user satisfaction. Furthermore the selection of network, proxy, and server resources can be globally optimized for larger service durations which will reduce frequent switching, thereby reducing service jitter and improving QoS.

There has been significant effort directed toward discovering network and server resources for QoS-based applications. Server selection algorithms [29–32] are often used to direct user requests to the optimal server based on chosen metrics (such as proximity or load) when data is replicated across multiple servers. These mechanisms often treat the network path leading from the client to the server as static. Although this is useful for computation-intensive applications, interactive applications such as mobile multimedia must guarantee the availability of network resources as well. QoS-based routing techniques [33–37] typically aim to select the optimal path between a source–server pair and ignore the situation where multiple servers might be able to serve the same request. Combined path and server selection (CPSS) [38] is an integrated approach that allows load balancing not only between replicated servers but also among network links. This has the potential to achieve higher systemwide utilization and allow more concurrent users.

In the remainder of this section, we present techniques for the static discovery of proxy resources through a case study in infrastructure-based wireless networks (e.g., cellular or wireless local area networks). This case study illustrates how one might use idle grid resources as intermediate nodes or proxies for mobile applications. We also discuss how knowledge or predictions of user mobility patterns, device capabilities, and system resource availabilities can play important roles in proxy resource discovery for mobile applications.

Case Study: Using Grid Resources as Proxies for Mobile Multimedia Applications

Grid computing [39] is a distributed, high-performance computing and data-handling infrastructure that incorporates geographically and organizationally dispersed, heterogeneous resources. Traditional grid-based research has focused on facilitating computation-intensive and data-intensive applications, such as AppLes [40], GrADS [41], and Nimrod [42]. Leveraging grid resources to facilitate mobile applications is motivated by the fact that the proliferation of freely available idle grid resources can be exploited efficiently to compensate a mobile computing environment that is short of resources. A grid computing environment provides an ideal setting where grid resources can act as proxies to improve the power or performance of low-power mobile devices [43]; however, we need to address the challenge of identifying available grid resources that can be used as proxies.

Grid resource discovery for efficient mobile services poses distinct challenges due to the intermittent availability of heterogeneous grid resources. The system must apply its knowledge of resource availability (e.g., using time maps to quantify grid resource availability) when selecting a grid resource to build an end-to-end service channel for a mobile user. Note that the availability of grid resources is unpredictable and the amount of available grid resources may fluctuate. Their stability features may also have to be investigated when resource discovery is performed. We illustrate how to apply knowledge of user mobility patterns to discover the nearby intermittently available grid resources and how the system exploits knowledge of device information (e.g., energy sufficiency) to perform better resource discovery. The proposed approaches have been implemented and evaluated in the context of the MAPGrid system [44]. The prototype system is illustrated in Figure 35.3.

Designing a Grid Resource Discovery Algorithm

In the MAPGrid system, we define a grid volunteer server (VS) as a machine that participates in the grid by supplying idle resources; that is, VSs are intermittently available and geographically distributed. A VS (used interchangeably with proxy) in our case can be a wired workstation, server, cluster, etc. that provides high-capacity storage for storing multimedia data and a CPU for multimedia transcoding, decompression, or buffer memory. VSs are fixed machines and connect to the network using wired connections,

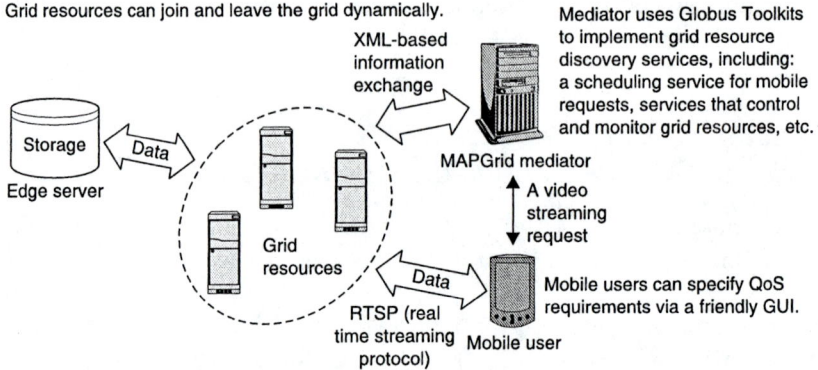

Grid resources can join and leave the grid dynamically.

XML-based information exchange

Mediator uses Globus Toolkits to implement grid resource discovery services, including: a scheduling service for mobile requests, services that control and monitor grid resources, etc.

Storage — Data

Edge server

MAPGrid mediator

A video streaming request

Grid resources

Data

RTSP (real time streaming protocol)

Mobile user

Mobile users can specify QoS requirements via a friendly GUI.

1. Mobile client sends a MAPGrid video request to the mediator;
2. The mediator performs grid resource discovery for this request;
3. The mobile client receives video data from selected grid resources;
4. Java Media Framework is used to show the video object.

Figure 35.3 The MAPGrid system prototype.

whereas mobile hosts connect to the infrastructure using a locally available wireless network. A mobile user initiates a multimedia request, $R<VID, T, itinerary(opt)>$, where VID identifies the requested video object, T represents the entire service period, and *itinerary* contains the user's mobility information (null, if no mobility information is available). Given the mobile requests and information regarding grid resource availabilities, static resource discovery tries to increase the overall acceptance of requests in the system by selecting the optimal grid resources for each mobile request while satisfying users' QoS requirements. The QoS requirements (e.g., required network transmission bandwidth) will be determined by streaming a particular video streaming object.

The approach proposed in Huang and Venkatasubramanian [45] is to divide the entire service period (T) into nonoverlapping chunks (possibly of different sizes), each of which is mapped to an appropriate VS (e.g., the one that is geographically close and lightly loaded). Video objects are also divided into equal-sized segments. Corresponding video segments are downloaded onto selected VSs. The selected VS processes the request by transcoding the video segment and transmits the video stream via wireless links to bandwidth-limited and performance-limited mobile clients, such as a PDA. Below we discuss a phased approach that exploits knowledge of user mobility patterns and grid resource availabilities.

In the first step, a time-based approach or a distance-based approach is applied to partition the service period (T) into chunks [45]. The time-based policy attempts to minimize the number of VS switches, and the distance-based policy uses knowledge of user mobility patterns and applies a well-known unsupervised neural learning technique called a *self-organizing map* (SOM). If *itinerary* information is given, after service partitioning the *Focus* of each chunk is calculated to identify the ideal resource location for each service period. Specifically, if (a_i, b_i) represent the coordinates of the center for region i, and D_i represents the time duration spent in region i, we convert the problem of locating the chunk *Focus* into a minisum planar Euclidean location problem [46]. The objective is to minimize the overall distance cost $f(x,y)$, where (x,y) represent the coordinates of the *Focus* position of this chunk. The minimum value of the objective function $f(x,y)$, specified in Equation 35.1, determines the *Focus* position of this chunk, which is composed of N regions:

$$f(x, y) = \sum_{i=1}^{N} Di \sqrt{(x - a_i)^2 + (y - b_i)^2}$$

(35.1)

In the second step, a graph theoretic technique is applied for selecting an optimal set of grid resources to service each chunk. Decisions should be made by taking into consideration all of the following factors: (1) intermittent availability of grid resources, (2) currently allocated workloads and predicted future workloads on grid resources, and (3) user's distance to grid resources. Basically, the goal is to select a lightly loaded and nearby grid resource to service each chunk. To deal with heterogeneous grid resources in a unified way and to represent how much a request affects a server during each service period (chunk), a *VSFactor* is defined to measure the desirability of a VS as a grid resource for one service chunk of the mobile user. According to this definition, shown in Equation 35.2, a VS with a larger *VSFactor* value is a better choice for servicing a particular service period [45]:

$$VSFactor(VS, chunk) = \frac{Availability\ of\ VS}{VS\ workload \times distance(VS,\ focus\ of\ this\ service\ chunk)}$$

(35.2)

The problem of discovering intermittently available grid resources is further cast as a maximum flow problem, illustrated in Figure 35.4. Nodes O and F are artificial nodes and represent the source vertex and sink node, respectively. Node C represents the mobile client, and the VS nodes represent volunteer servers. A set of time nodes (TNs) is introduced, and each TN represents a period of service time. Weights for directed edges are also assigned as illustrated in Figure 35.4. A feasible maximum flow solution that meets resource constraints corresponds to a possible scheduling solution; the basic solution has been adapted to develop a family of policies catering to various application QoS needs [47].

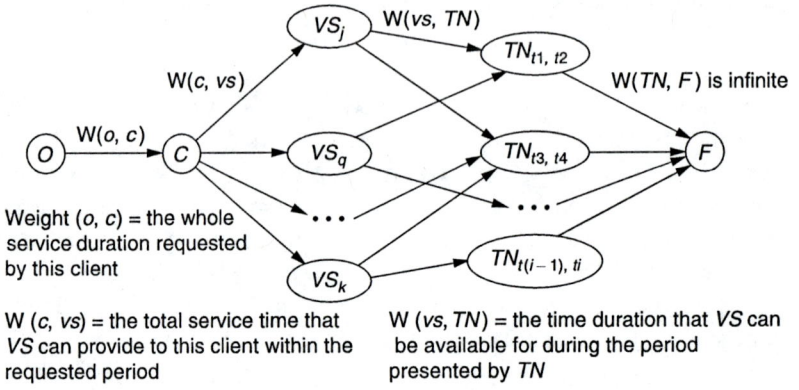

Weight (o, c) = the whole service duration requested by this client

W (c, vs) = the total service time that VS can provide to this client within the requested period

W (vs, TN) = the time duration that VS can be available for during the period presented by TN

Figure 35.4   Modeling the resource discovery problem.

The Role of Device Constraints in Resource Discovery

One of the major constraints of executing multimedia applications on thin mobile devices is energy insufficiency. A tradeoff exists between QoS and energy that can be exploited to overcome energy limitations on mobile devices. This tradeoff is based on the fact that streaming lower quality video to power-deficient mobile clients results in lighter traffic over the network and less computation for decoding video frames and therefore less energy consumption on the mobile device. One possibility is to use proxy resources to perform dynamic transcoding, which can help balance application QoS based on the residual energy of the device. Because degrading video quality directly affects user perception (QoS), it is important to understand the notion of video quality for a handheld device and its implications on power consumption. Figure 35.5 illustrates an energy/quality (E/Q) matrix for handheld computers (Compaq iPaq 3650) [19] to identify video quality parameters (a combination of bit rate, frame rate and video resolution) that produce user perceptible changes in video quality and noticeable shifts in power consumption for handheld computers.

Using the E/Q matrix, we can map each video quality level to a network transmission bandwidth and a power cost value (or vice versa). When a mobile request $R<VID, T, QMIN, QMAX, ER, itinerary(opt)>$ specifies the lowest QoS level (Q_{MIN}) and the highest QoS level (Q_{MAX}) and gives information about current residual energy E_R, the E/Q matrix can be used to determine the best QoS level for this service. A straightforward extension of the static grid resource discovery algorithm to an energy-aware admission control algorithm using the E/Q matrix is shown in Figure 35.6. Detailed explanations of the algorithm are presented in Huang et al. [48].

Quality	Video transformation parameters	Average power Windows CE (w)	Average power Linux (w)
Q8 (original)	SIF, 30fps, 650Kbps	4.42	6.07
Q7 (excellent)	SIF, 25fps, 450Kbps	4.37	5.99
Q6 (very good)	SIF, 25fps, 350Kbps	4.31	5.86
Q5 (good)	HSIF, 24fps, 350Kbps	4.24	5.81
Q4 (fair)	HSIF, 24fps, 200Kbps	4.15	5.73
Q3 (poor)	HSIF, 24fps, 150Kbps	4.06	5.63
Q2 (bad)	QSIF, 20fps, 150Kbps	3.95	5.50
Q1 (terrible)	QSIF, 20fps, 100Kbps	3.88	5.38

Figure 35.5   Energy/quality (E/Q) matrix for handheld computers (Compaq iPaq 3650).

Figure 35.7 shows one experimental result that illustrates the performance of the energy-aware resource discovery algorithm [48]. Three different approaches are compared: (1) no QoS adaptation during resource discovery, (2) proxy-based transcoding only, and (3) the energy-aware admission control (EAC) algorithm described in Figure 35.6. With the first approach, the system streams the video with the

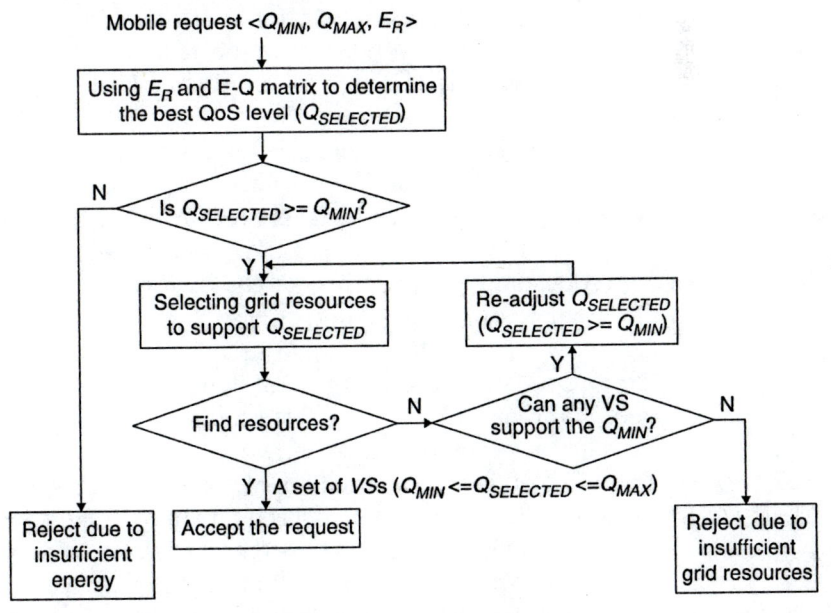

Figure 35.6   Energy-aware admission control (EAC) algorithm.

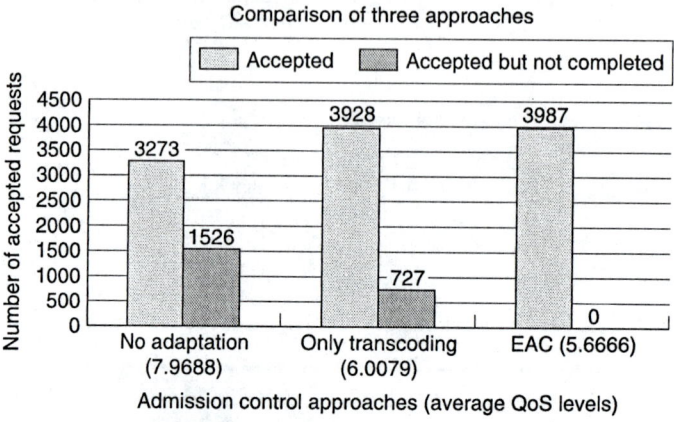

Figure 35.7 Experimental results for static resource discovery; the energy-aware admission control (EAC) techniques increases request acceptance and completion rates.

highest QoS level (original quality) to the devices (i.e., multiple quality levels are not supported at the servers). With the second approach, QoS degrades when the proxy (VS) resources in the system are not sufficient for the highest QoS level. The result shows that the first approach leads to the lowest request acceptance rate but the highest QoS provisioning for each service; however, as the residual energy of the device is not considered, it also results in the highest level of incomplete services due to insufficient device energy. The second approach reduces the average QoS levels for requests and therefore accepts more requests while reducing the number of incomplete services. The third approach takes the residual energy of the device into account while also performing quality transcoding; thus, it accepts the largest number of requests and completes all the accepted requests, assuming no other dynamic changes thereafter. Other experimental results [45,48] also show that intelligent static resource discovery not only increases users' QoS satisfaction but also significantly improves the acceptance rate, completion ratio, and system throughput.

Approaches for static resource discovery, however, can often lead to the overprovisioning of resources; that is, they may choose to deliberately overestimate the number of resources a service is likely to require and thereby sacrifice resource utilization. To a limited extent, static resource discovery methods can be further optimized by using profiled or historical information, such as demand for a particular service, bandwidth and latency requirements for a service, or mobility patterns. In practice, though, it is still difficult to predict resource utilization and user mobility patterns in a mobile and wireless infrastructure. In the next section, we describe approaches using dynamic resource reprovisioning, which allows a system to track dynamic changes and adapt to these changes on the fly.

Dynamic Resource Reprovisioning

For mobile environments, it is difficult to accurately discover and provision resources using static methods for the entire duration of a service. This can be due to several reasons: (1) In wireless networks, disconnected operation and bandwidth fluctuations are common, making it impossible to discover/ provision network bandwidth for the entire service duration. (2) Device mobility makes it particularly difficult to provision resources, especially when there is no prior knowledge of how the user is expected to move. (3) Intermediate nodes might suddenly become unavailable (user unplugs the system) or system resources such as CPU, memory of servers, and intermediate proxies might change unpredictably (starting or stopping applications can affect these resource availabilities). (4) Finally, mobile hosts may have unexpected changes in resource availability (e.g., new applications are started) which makes it difficult to predict how resources are being consumed. These issues can be effectively addressed using dynamic resource provisioning, where allocations of resources to services are automatically and continuously adjusted in response to either changing demands for a service or dynamic changes in resource availabilities in the system. This leads to a more accurate provisioning of resources and greater resource utilization.

Dynamic reprovisioning in mobile environments complicates the problem of context collection and resource management, and several difficult challenges must be addressed. What happens when an intermediate node (proxy) suddenly becomes unavailable or severely resource constrained? In this case, services might be degraded, improved, or terminated to free resources or improve QoS. Strategies have to be designed to determine which services should be affected and how. Techniques to support on-the-fly assignments and revocations of resources must be developed. Satyanarayanan [49] outlines several issues that must be addressed during resource revocation, such as characterizing the impact of revocation on services, handling deadlocks, and designing revocation strategies. The above decisions are impossible to make without accurate knowledge of the system context. As the global state of the system changes dynamically, it is difficult to maintain accurate context information. Who should maintain context information and how often should the context be updated? What represents accurate context? Should context collection be distributed or centralized? Can we make global state estimation from local states? If so, what is the accuracy (or error bound) on these estimates? Does a general rule or systematic way exist for quantifying adaptations? All of these open research issues and challenges are very pertinent to modern mobile environments and good solutions, and insights to these problems will strongly impact mobile computing systems of the future.

To illustrate potential approaches to dynamic provisioning for mobile services, we build upon our earlier case study. Specifically, we address the dynamic discovery and adaptation of resources for mobile multimedia applications that use grid resources as proxies. We focus on three aspects of the framework that are points of dynamic changes: (1) the proxy, (2) the network, and (3) the mobile device.

Dynamic Changes in Proxy Resources

Proxies are participating machines on which applications can be randomly started or stopped, causing fluctuations in resource availability. Allocation mechanism must be capable of dealing with proxy failures and changes in proxy resources. The worst case occurs when a proxy is disconnected from the grid or switched off (e.g., unplugged), resulting in the unavailability of the proxy itself. To deal with this problem, the broker needs to reallocate other available proxies to the interrupted requests to complete the interrupted services. When a specific proxy becomes unavailable, the broker retrieves information from the directory service about requests that are scheduled on the failed proxy and triggers the rescheduling process for each invalidated service. To reduce service failures and minimize service recovery time, the solution determines the order in which to migrate the disrupted services onto available proxies. To minimize service recovery time, invalid services are classified into two categories by the broker: (1) services that have been started, and (2) services that are not yet started. Services in the first category receive a higher rescheduling priority than those of the second type, for which service rescheduling can be postponed with an acceptable delay. Furthermore, within each category, requests with shorter remaining times of service and lower resource requirements receive higher rescheduling priority. If requests cannot be rescheduled, the broker downgrades a number of the disrupted services to accommodate them in the available proxies; if they still cannot be rescheduled, even after downgrading the service, the broker notifies the clients that the service has failed and releases preallocated resources for the services on the other proxies. The rescheduling process is then triggered for each invalidated service in order of decreasing priority. If requests cannot be rescheduled or postponed (for category two requests), the broker reports a request failure. Note that, in the case of request failure, any resources reserved for this request on other selected proxies for the remaining service time should be released.

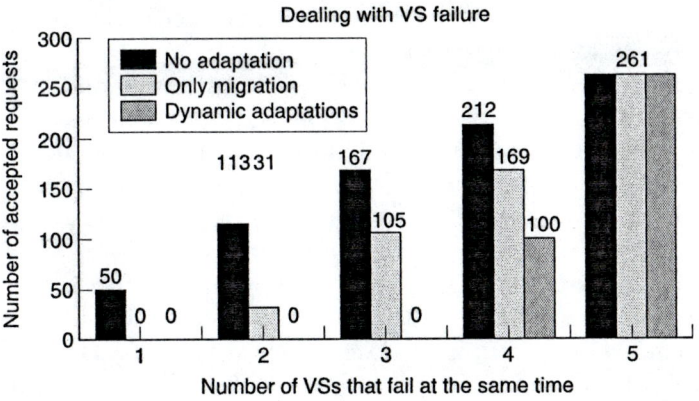

Figure 35.8 Dealing with proxy failures.

The proxy also must perform dynamic adaptations when its own resources reduce unpredictably (e.g., applications are started dynamically). If the resource changes are small, the proxy performs local adjustments to satisfy the QoS requirements of the current set of services. This might require downgrading the QoS (e.g., video quality) of an existing subset of services or allocating fewer resources temporarily to local applications; however, if a significant change occurs in the resource availability at the proxy that affects completion of certain services, then a subset of services must be dynamically migrated away to another less loaded proxy. In this case, the proxy signals the broker to initiate a service migration algorithm that migrates a set of services to another less loaded proxy. Given system resource limitations, however, services that cannot be scheduled on other proxies result in service failures. After successful migration of a service, the proxy readjusts the released resources and distributes them among the remaining services to minimize the number of migrations. The maximum number of migrations over the lifetime of a service can be achieved by placing an upper bound on the total number of migrations possible for a service. Figure 35.8 shows that, after proxies and the broker perform the above adaptations, a significant decrease can be seen in the number of requests that fail to complete due to dynamic changes in proxy availability [48].

Disconnections/Fluctuations in Wireless Networks

Wireless network behavior depends closely on several factors such as wireless signal strength, congestion, and noise. Each of these factors contributes to the connectivity and bandwidth availability of the network and can vary erratically over time, thereby making network provisioning a difficult problem. To effectively provision resources and adapt services for wireless networks, we need to (1) get accurate context information about changing network congestion and noise levels, and (2) predict device mobility patterns. The congestion and noise information can be gathered from the feedback from the device and by querying the wireless access points. With this information, the proxy can perform two different adaptations to improve QoS for multimedia applications: (1) proactive resource allocation and service adaptation based on device mobility and network congestion, and (2) adaptive network traffic management.

We explain the proactive adaptation approach by first differentiating it from the traditional reactive approaches that are representative of current best-effort systems. In a more traditional *reactive* adaptation approach, a change in resource availability is first detected (possibly due to dropped packets, increased noise or congestion levels, or low power at the device) at a potential loss of QoS (video jitter). The proxy then reacts to this dynamic change by adapting the video stream (by either lowering or improving stream quality) to improve performance. However, the video/data packets already communicated might get dropped if a mobile device suddenly enters a cell that is highly congested.

In dynamic environments, a proactive scheme can perform significantly better than a reactive scheme. In such a scheme, the proxy proactively predicts future system conditions and can determine how services can be adapted in advance. Specifically, the scheme exploits knowledge of system context and device mobility model to predict the number of users in a future target cell. With the knowledge of average traffic generated by each user, it can predict the dynamic congestion and noise levels within each target cell. This knowledge can be used in conjunction with feedback from the device to proactively adapt either the video stream or the buffering (burst sizes) to maximize the application QoS; for example, the proxy predicts the noise or congestion level of a cell just before a user moves into the cell and determines how to adapt the stream as the user enters the cell. In such a scenario, two factors significantly influence the performance of the schemes: predicted dynamic noise levels within each cell and the mobility (velocity) of the device. Mohapatra and Venkata-subramanian [50] made a comparison between the proactive and reactive schemes for multimedia applications. Their study also concluded that the nature of distribution of noise induced by each mobile device has very little effect on the overall adaptations.

Figure 35.9 shows that the proactive adaptation results in a much smoother video when compared to reactive adaptation. We see in Figure 35.10 that a significant improvement in the overall system utilization is achieved using a proactive adaptation scheme. An additional benefit is that proactive proxy-based adaptations can facilitate dynamic power

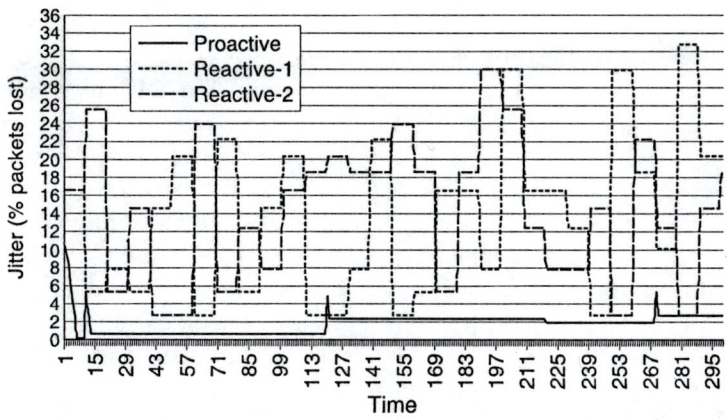

Figure 35.9 Proactive versus reactive adaptation. Proactive adaptation results in much smoother video with fewer quality fluctuations as opposed to the reactive scheme.

management of the network interface card at the mobile device. Multimedia applications due to their periodic and predictable behavior present opportunities to use proxy-based reprovisioning for improving the periods of inactivity for a wireless radio without affecting the application QoS. Mohapatra et al. [19] described an approach where a proxy can buffer video data (as opposed to sending data on a per-frame basis) and send

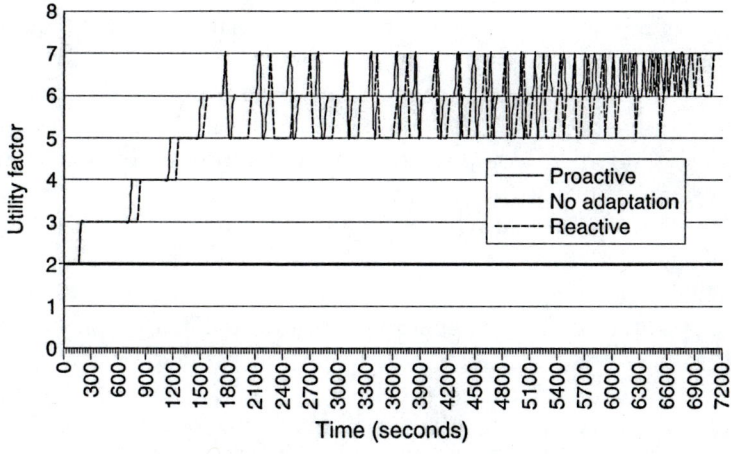

Figure 35.10 Utility factor versus adaptation scheme. A proactive scheme is able to provide higher utility to the system in terms of overall improvement in video quality as well as battery energy savings.

data to the mobile device in bursts along with control information containing the size of the bursts. This simple adaptation facilitates network card optimization, as the device can now transition the radio to a low-duty cycle for longer periods, thereby saving energy. The radio can then be switched back to active mode before the arrival of the next burst from the proxy by using the control information. The size of the bursts is dependent on the network congestion level and the buffering capabilities of the wireless access point, as well as those of the mobile device and the quality of the video stream. The mobile host saves energy during periods of network inactivity as the radio consumes significantly less energy when operating in low-duty-cycle mode. Energy thus saved can result in improved QoS for the mobile service.

Fluctuations at the Mobile Device

On the mobile device, resource availabilities depend on both the number of applications and their demands on certain resources, either of which can change dynamically. We discuss two possible adaptations employed by the mobile client. One approach is to dynamically migrate computationally expensive tasks from a mobile device to the proxy. Although migrating tasks can reduce the computational load on a device, saving both battery energy and reducing the load on the CPU, they often add extra communication overhead. Tasks such as media composing, encryption, and decryption can be potential candidates for task migration. Such an approach can be profitable if the benefits of migration outweigh the overheads of migration and communication. Mohapatra and Venkatasubramanian [20] present a graph theoretic analysis of how tasks can be partitioned and migrated from a local device to a proxy. When and how often such migrations have to happen are dictated by the number of applications and their computation and communication characteristics. Battery energy can also be incorporated into the adaptation by further modifying the algorithm to favor migrations when the residual battery power is low.

In addition to adapting to local variations in resources, a proxy can perform dynamic service adaptations based on feedback from the device as well as network state; for example, in the MAPGrid framework, we assume that a communication protocol is available that allows a proxy (grid volunteer server) to continually monitor the resource availability on a mobile device to which it is connected. This allows the proxy to dynamically adapt services to handle changes in the resource availability at a mobile device. We present a specific example of how a proxy can adapt the video streaming service in response to changes in device battery energy, but the concept can be extended to other resources.

If the residual battery energy of the device changes (e.g., either due to starting or stopping of applications on the device or as a result of energy optimizations on the device), the proxy reacts by changing the quality of the video stream to accommodate the changes. If the proxy determines that the device does not have sufficient battery energy to support the entire duration of the current service, it performs dynamic reprovisioning by streaming video at a lower quality to adjust to the lower residual energy at the device. If the proxy determines that the device cannot support even the lowest acceptable quality, it notifies the device of its depleted battery energy state. This implies that the device is consuming too much power (maybe due to other applications) and local proxy-based adaptations cannot complete the service. Conversely, if an increase occurs in the residual battery life of the mobile device due either to energy optimization strategies or a reduction in the number of executing applications, then proxy-based reprovisioning scheme can respond by increasing the video stream quality. The device can also employ dynamic adaptations at the operating system and hardware levels to optimize resource usage; for example, techniques such as dynamic voltage scaling (DVS) [19] can be used to improve CPU utilization dynamically while saving energy. Note that these optimizations ultimately manifest themselves as higher QoS for applications executing on the mobile device.

Dynamic Changes in Device/User Mobility

As discussed earlier, the static resource discovery solution optimizes the assignment of proxies and their corresponding chunks for a particular service on the assumption that the overall mobility pattern of the mobile user is known. Specifically, a proxy that is geographically closest to the mobile device is preferred over more distant proxies; however, in real life, a user might dynamically choose to follow a different path in the middle of a service, thereby making the assignment of proxies suboptimal (assuming the proximity criteria for optimality).

Initial studies have tried to study this problem by developing certain policies regarding how proxy assignments can be made when user mobility cannot be predicted in advance. One simple policy is to use a single proxy (single VS) for the entire duration of the service. A variation of this is the FastStartup policy, where a proxy is assigned some initial video segments to begin the service immediately while the algorithm searches for an ideal proxy or a set of proxies. Figure 35.11 shows that policies such as FastStartup that systematically use multiple proxies perform better than the single-proxy strategy under varying device mobility patterns [45]. A seamless way of incorporating dynamicity is to initiate the FastStartup policy when the mobile device has significantly departed from the assumed

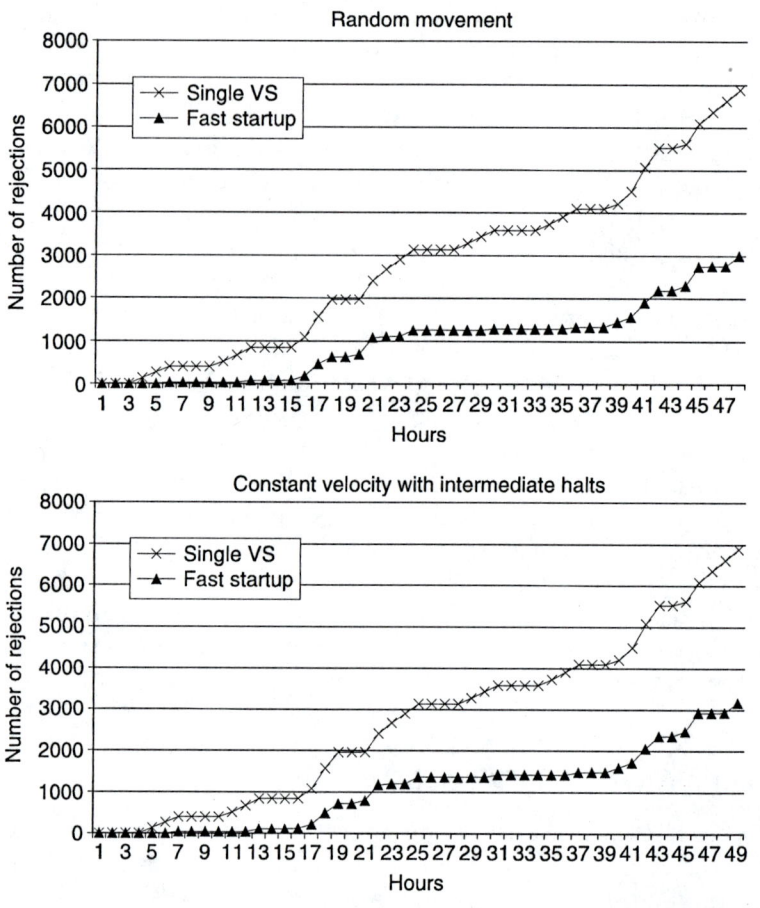

Figure 35.11 Number of rejections over time with different device mobility patterns.

(predicted) path. Intelligent mobility prediction techniques can also be applied to further improve dynamic resource reprovisioning; for example, knowledge of (predicted) mobility patterns can be used to design space- and time-based partitioning techniques for proxy allocation.

Summary

In this chapter, we have described the problem of QoS-aware resource discovery for mobile applications and illustrated various elements of a solution to this issue using a generalized mediation-based architecture. The chapter discussed in more detail three key issues that must be addressed for resource discovery: (1) mechanisms to cost effectively capture and maintain context information to enable resource discovery for mobile services; (2) algorithms for static resource discovery when a request is first initiated, using the available context information; and (3) techniques for dynamic resource reprovisioning to deal with unanticipated changes in the applications, devices, and distributed infrastructure. Through a case study focusing on mobile multimedia services, we showed how to take advantage of *a priori* knowledge of resource availabilities and mobility patterns to tailor the discovery and provisioning policies for better overall performance and enhanced user QoS. Research has also shown how context-aware intelligent policies for static and dynamic resource provisioning can support better application QoS, higher request acceptance rates, and longer device lifetimes in mobile environments. One of the key observations in this chapter is the role of in-network proxy resources in enabling effective solutions to address the tradeoffs among quality, energy, and performance for QoS-based mobile applications. Although service providers may choose to install dedicated proxy resources in-network, recent efforts have shown the possibility of leveraging additional heterogeneous machines within the proximity of mobile devices.

Although the above steps are enabling technologies for the seamless execution of mobile applications under highly dynamic conditions, several challenges still remain to be addressed. Enabling QoS-based services in MANETs where a core wireless network infrastructure is unavailable poses new challenges. Issues of power-aware routing, QoS in the presence of topology changes, and the ability to switch transparently between *ad hoc* and infrastructure modes are still challenges further exacerbated in always-best-connected (ABC) networks where multiple access technologies (WiFi, cellular, Bluetooth®, wired) may all be available simultaneously to varying degrees. Future work will also have to address the degree of location awareness required by the middleware for efficient allocation of mobile and fixed resources in a scalable fashion. Security (or lack thereof) of applications executing on wireless infrastructures is a big hurdle to the pervasive deployment of mobile services. Tradeoffs arise when timeliness requirements interfere with other application requirements, such as security and reliability. Many of these challenges must be addressed to truly realize the eventual goal of widespread mobile services.

References

[1] Chen, G. and Kotz, D., *A Survey of Context-Aware Mobile Computing Research*, Technical Report TR2000-381, Department of Computer Science, Dartmouth College, Hanover, NH, 2000.

[2] Chalmers, D. and Sloman, M., QoS and context awareness for mobile computing, in *Proc. of Int. Symp. on Handheld and Ubiquitous Computing (HUC'99)*, Karlsrhue, Germany, September 27–29, 1999, pp. 380–382.

[3] *ERCIM News*, 54, 2003 (special issue on applications and service platforms for the mobile user).

[4] Venkatasubramanian, N., Talcott, C., and Agha, G.A., A formal model for reasoning about adaptive QoS-enabled middleware, *ACM Trans. Software Eng. Methodol.*, 13(1), 86–147, 2004.

[5] Chalmers, D. and Sloman, M., A survey of quality of service in mobile computing environments, in *IEEE Online Commun. Surv.*, 2(2), 1–10, 1999.

[6] Adwankar, S., Mobile CORBA, in *Proc. of the Third Int. Symp. on Distributed Objects and Applications (DOA'01)*, Rome, Italy, September 17–20, 2001.

[7] Yuan, M.J., *Enterprise J2ME: Developing Mobile Java Applications*, Prentice Hall, Upper Saddle River, NJ, 2003.

[8] Hefeeda, M., Xu, D., Habib, A., Bhargava, B., and Botev, B., CollectCast: a peer-to-peer service for media streaming, *ACM Multimedia Syst. J.*, 11(1), 68–81, 2005.

[9] Bakos, B., Farkas, L., Jukka, N., and Csucs, G., Peer-to-peer protocol evaluation in topologies resembling wireless networks. an experiment with Gnutella query engine, in *Proc. of IEEE Int. Conf. on Networks (ICON'03)*, Sydney, Australia, September 28–October 1, 2003.

[10] Perkins, C. and Bhagwat, P., Highly dynamic destination-sequenced distance-vector routing (DSDV) for mobile computers, in *Proc. of ACM SIGCOMM'94*, London, August 31–September 2, 1994.

[11] Park, V.D. and Corson, M.S., A highly adaptive distributed routing algorithm for mobile wireless networks, in *Proc. of IEEE INFOCOM'97*, Kobe, Japan, April 7–11, 1997.

[12] Johnson, D.B. and Maltz, D.A., Dynamic source routing in *ad hoc* wireless networks, in *Mobile Computing*, Vol. 353, Imielinski, T. and Korth, H., Eds, Kluwer Academic, Norwell, MA, 1996.

[13] Perkins, C., *Ad hoc* on-demand distance vector routing, in *Proc. of Military Communications Conf. (MILCOM 1997)*, Monterey, CA, November 2–5, 1997.

[14] Bao, L. and Garcia-Luna-Aceves, J.J., Topology management in *ad hoc* networks, in *Proc. of ACM MobiHoc*, Annapolis, MD, June, 2003.

[15] Godfrey, P.B. and Ratajczak, D., Naps: scalable, robust topology management in wireless *ad hoc* networks, in *Proc. of IEEE Information Processing in Sensor Networks (IPSN'04)*, Berkeley, CA, April 26–27, 2004.

[16] Singh, A., Trivedi, A., Ramamritham, K., and Shenoy, P., Ptc: proxies that transcode and cache in heterogeneous Web client environments, *World Wide Web*, 7(1), 7–28, 2004.

[17] Raunak, M.S., Shenoy, P., Goyal, P., and Ramamritham, K., Implications of proxy caching for provisioning networks and servers, *SIGMETRICS Perform. Eval. Rev.*, 28(1), 66–77, 2000.

[18] Chandra, S. and Vahdat, A., Application-specific network management for energy-aware streaming of popular multimedia formats, in *Proc. of USENIX Annual Technical Conf.*, Monterey, CA, June 10–15, 2002.

[19] Mohapatra, S., Cornea, R., Dutt, N., Nicolau, A., and Venkatasubramanian, N., Integrated power management for video streaming to mobile handheld devices, in *Proc. of the Eleventh ACM Int. Conf. on Multimedia (ACMMM 2003)*, Monterey, CA, November 2–8, 2003.

[20] Mohapatra, S. and Venkatasubramanian, N., PARM: power-aware reconfigurable middleware, in *Proc. of IEEE Int. Conf. on Distributed Computing Systems (ICDCS'03)*, Providence, RI, May, 2003.

[21] Moy, J., *OSPF Version 2*, Request for Comments 1247, Internet Engineering Task Force (IETF), 1991 (http://www.ietf.org/rfc/rfc1247.txt).

[22] Apostolopoulos, G., Guerin, R., Kamat, S., and Tripathi, S., Quality of service based routing: a performance perspective, in *Proc. of ACM SIGCOMM'98*, Vancouver, Canada, October, 1998.

[23] Han, Q. and Venkatasubramanian, N., Autosec: an integrated middleware framework for dynamic service brokering, *IEEE Distributed Systems Online*, 2(7), 2001.

[24] Fu, Z. and Venkatasubramanian, N., Adaptive parameter collection in dynamic distributed environments, in *Proc. of IEEE Int. Conf. on Distributed Computing Systems (ICDCS'01)*, Phoenix, AZ, April, 2001.

[25] Akyildiz, I., McNair, J., Ho, J., Uzunalioglu, H., and Wang, W., Mobility management in next-generation wireless systems, *Proc. IEEE*, 87(8), 1347–1384, 1999.

[26] Wong, V.W.-S. and Leung, V.C., Location management for next-generation personal communications networks, in *IEEE Network Mag.*, 14(5), 18–24, 2000.

[27] Haas, Z., A new routing protocol for the reconfigurable wireless networks, in *Proc. of the Sixth IEEE Int. Conf. on Universal Personal Communications (ICUPC'97)*, San Diego, CA, October 12–16, 1997.

[28] Han, Q. and Venkatasubramanian, N., Information collection services for QoS-aware mobile applications, *IEEE Trans. Mobile Comput.*, 5(5), 518–535, 2006.

[29] Guyton, J. and Shwartz, M.F., Locating nearby copies of replicated Internet services, in *Proc. of ACM SIGCOMM'95*, Cambridge, MA, August 28–September 1, 1995.

[30] Fei, A., Pei, G., Liu, R., and Zhang, L., Measurements on delay and hop-count of the Internet, in *Proc. of Global Telecommunications Conf. (GLOBE-COMM'98)*, Sydney, Australia, November, 1998.

[31] Francis, P., Jamin, S., Pasxon, V., Zhang, L., Gryniewica, D., and Jin, Y., An architecture for a global Internet host distance estimation service, in *Proc. of IEEE INFOCOM'99*, New York, March, 1999.

[32] Myers, A., Dinda, P., and Zhang, H., Performance characteristics of mirror servers on the internet, in *Proc. of Global Telecommunications Conf. (GLOBECOM'99)*, Rio de Janeiro, Brazil, December, 1999.

[33] Chen, S. and Nahrstedt, K., Distributed quality of service routing in *ad hoc* networks, *IEEE J. Special Areas Commun.*, 17(8), 1488–1505, 1999 (special issue on *ad hoc* networks).

[34] Zhao, W. and Tripathi, S.K., Routing guaranteed quality of service connections in integrated service packet network, in *Proc. of Int. Conf. on Network Protocols (ICNP'97)*, Atlanta, GA, October 28–31, 1997.

[35] Cidon, I., Rom, R., and Shavitt, Y., Multipath routing combined with resource reservation, in *Proc. of IEEE INFOCOM'97*, Kobe, Japan, April 7–11, 1997.

[36] Breslau, L. and Shenker, S., Best-effort versus reservations: a simple comparative analysis, in *Proc. ACM SIGCOMM'98*, Vancouver, Canada, October, 1998.

[37] Ma, Q., Steenkiste, P., and Zhang, H., Routing high-bandwidth traffic in max-min fair share networks, in *Proc. of ACM SIGCOMM'96*, Palo Alto, CA, August, 1996.

[38] Fu, Z. and Venkatasubramanian, N., Directory based composite routing and scheduling policies for dynamic multimedia environments, in *Proc. of IEEE Int. Conf. on Parallel and Distributed Processing Symp. (IPDPS'01)*, San Francisco, CA, April, 2001.

[39] Foster, I. and Kesselman, C., Eds., *The Grid: Blueprint for a New Computing Infrastructure*, 2nd ed., Morgan Kauffman, Boston, MA, 1998.

[40] Berman, F. and Wolski, R., The AppLeS project: a status report, in *Proc. of the 8th NEC Research Symp.*, Berlin, Germany, May, 1997.

[41] Berman, F., Chien, A., Cooper, K., Dongarra, J., Foster, I. et al., The GrADS project: software support for high-level grid application development, *Int. J. High-Performance Comput. Appl.*, 15(4), 327–344, 2001.

[42] Abramson, D., Giddy, J., and Kotler, L., High performance parametric modeling with Nimrod/G: killer application for the global grid?, in *Proc. of IEEE Int. Conf. on Parallel and Distributed Processing Symp. (IPDPS'00)*, Cancun, Mexico, May 1–5, 2000.

[43] McKnight, L., Howison, J., and Bradner, S., Wireless grids: distributed resource sharing by mobile, nomadic, and fixed devices, *IEEE Internet Comput.*, 8(4), 24–31, 2004.

[44] MAPGrid, http://mapgrid.ics.uci.edu/.

[45] Huang, Y. and Venkatasubramanian, N., Supporting mobile multimedia services with intermittently available grid resources, in *Proc. of Int. Conf. on High-Performance Computing (HiPC'03)*, Hyderabad, India, December 17–20, 2003.

[46] Kaminsky, P.M., *IEOR 251: Logistics Modeling* [lecture], 2002.

[47] Huang, Y. and Venkatasubramanian, N., QoS-based resource discovery in intermittently available environments, in *Proc. of the 11th IEEE Int. Symp. on High-Performance Distributed Computing (HPDC'02)*, Edinburgh, Scotland, July 23–26, 2002, pp. 246–254.

[48] Huang, Y., Mohapatra, S., and Venkatasubramanian, N., An energy-efficient middleware for supporting multimedia services in mobile grid environments, in *Proc. of the 6th IEEE Int. Conf. on Information Technology (ITCC2005)*, Las Vegas, NV, April 11–13, 2005.

[49] Satyanarayanan, M., Fundamental challenges in mobile computing, in *Proc. of the 15th ACM Symp. on Principles of Distributed Computing (PODC 1996)*, Philadelphia, PA, May 23–26, 1996.

[50] Mohapatra, S. and Venkatasubramanian, N., Proactive energy-aware video streaming to mobile handheld devices, in *Proc. of the Fifth IFIP TC6 Int. Conf. on Mobile and Wireless Communications Networks (MWCN 2003)*, Singapore, October 27–29, 2003.

Chapter 36

QoS Control and Management

Xia Gao

CONTENTS

Introduction

In recent years, the growth in mobile computing technology has been explosive, and new wireless technologies have rapidly emerged. The desire to be connected anytime, anywhere, and in any way has led to an increasing array of heterogeneous systems, applications, devices, and service providers. It is envisioned that this heterogeneity is unlikely to disappear in the foreseeable future for two reasons. One is that the variety of application requirements makes it difficult to find a single optimal and universal solution. The other is that, in their eagerness to capture the market, competing organizations are releasing proprietary systems. As a result, the key to the success of next-generation mobile communication systems is the ability to provide seamless services in such a heterogeneous environment.

Internet Protocol (IP) is a universal network-layer protocol for the Internet and is becoming a promising universal network-layer protocol over all wireless systems, as well. IP provides unique addressing and packet routing and forwarding services and acts as a common platform for services and applications. It appears that an all-IP network layer can eventually integrate wireless communication networks and the Internet into the so-called "mobile Internet." To provide users with satisfactory services, however, ubiquitous connectivity and corresponding best-effort services are not enough. In a heterogeneous wireless network environment, application performance could easily deteriorate for a variety of reasons, and this performance fluctuation could be widespread. In response to this issue, quality of service (QoS) is designed to hide low-level application variation and to provide necessary service guarantees.

To provide QoS in the mobile Internet, many unique issues related to heterogeneity and mobility must be addressed. Consider a user moving from one network to another. The user may interact with a variety of service providers with different terms of service-level agreements (SLAs), network capacity, topology, and policies. The user may have a choice of wireless access technologies with different channel characteristics (e.g., bandwidth, loss, delay) and QoS-supporting capabilities. The user may switch to a new terminal with different computing power, display size, and data rate, or the user may adapt applications to meet new service requirements or network conditions. These factors can complicate the end-to-end service provision and limit the ability of service adaptation.

A considerable amount of research has targeted QoS-related issues. Most of the early work in this field focused on developing QoS frameworks, such as integrated services (Inte-Serv) and differentiated services (Diff-Serv), for the legacy best-effort Internet. Some progress has recently been made in addressing the wireless-related QoS issues in wireless access, mobility management, and portable devices. The primary research, however, is still in the context of individual architectural components, and much less progress has been made in addressing the issue of an overall QoS architecture. To address the need for an overall QoS architecture, we are investigating existing QoS research and are working to identify the main design challenges and principles and propose a generalized QoS architecture for the future mobile Internet.

This chapter first summarizes state-of-the-art QoS techniques and standardization activities, then examines in detail important challenges in building a ubiquitous QoS framework over the heterogeneous environment, and finally proposes a QoS framework integrating a three-plane network infrastructure and a unified terminal cross-layer adaptation platform to provide seamless support for future applications.

Current Status of QoS Research

To provide reliable and sustained QoS in the mobile Internet, it is necessary to efficiently manage wireless resources, adaptively cope with both temporal and spatial resource dynamics, and effectively address practical implementation issues. The ultimate solution for QoS support requires an integrated design effort that spans every layer in the network protocol stack. Research on the *user layer* and *application layer* focuses on the specification and mapping of application and user QoS preferences on evolving network service profiles. Different methods have been developed to elicit a user's cognitive and perceptual processes for network QoS. At the *middleware* level, new architectures (such as the agent-based model) are proposed to create systems that are robust, adaptive, and reconfigurable. New middleware lies between applications and the operating system (OS) and should provide applications with better support of multimedia processing, seamless mobility, and QoS adaptation.

At the *transport* layer, numerous modifications of the Transmission Control Protocol (TCP) are proposed to improve TCP performance over a wireless link. Some well-recognized characteristics of a wireless link are random channel error, large and varying delay, low bandwidth, path asymmetry, and temporary disconnection. At the *network* layer, Inte-Serv and Diff-Serv are two resource allocation architectures that allow for resource assurances and service differentiation for traffic flows and users. Multiprotocol label switching (MPLS) and related traffic engineering (TE) techniques such as constraint-based routing and multipath load sharing give Internet operators a set of management tools for bandwidth provisioning and performance optimization. At the *link* layer, work is underway to add QoS support in Ethernet-type local area networks (LANs) such as

802.11. Two active research areas are link-layer error recovery and wireless scheduling. Finally, at the *physical* layer, many channeling coding, modulation, and power control schemes are proposed to increase the communication success ratio and decrease interference and power consumption. Software-defined radio gives the physical layer the flexibility to access different wireless systems with one single interface. Interested readers may refer to Huston [1], Chalmers and Sloman [2], and Aurrecoechea et al. [3] for more details on the related works mentioned here. The rest of this section focuses primarily on the state-of-the-art research that forms the basis for our framework.

The *policy management framework* [6] is one of the efforts intended to simplify the definition and deployment of network behavior, including the automatic provisioning of QoS mechanisms. This framework includes four main elements. The *policy management tool* (PMT) is an interface assisting the administrator in creating network policies. These policies are stored using standard schema in the *policy repository* (PR). The *policy decision point* (PDP) is responsible for retrieving policy rules from the policy repository and generating policy decisions to be executed by the controlled *policy enforcement point* (PEP).

The Common Open Policy Service (COPS) protocol [7,8] is defined to support policy control in an IP QoS environment. The COPS protocol is a simply query-and-response protocol allowing the PDP to communicate policy information with the PEP. COPS has two main models: outsourcing (COPS-RSVP [7]), and provisioning (COPS-PR [8]). The policy management framework provides the opportunity to combine policy control, QoS signaling, and resource control in a unified framework. Some research is addressing the use of different types of the COPS protocol to combine RSVP and Diff-Serv networks, to allow dynamic SLA negotiation and deployment, and to integrate QoS signaling with Session Initiation Protocol (SIP) application signaling.

The *application network* refers to an application-specific overlay network over the Internet. It extends the capabilities of network intermediaries to provide additional services such as content adaptation, personalization, and location-aware data insertion. The content distribution network (CDN) and the content services network (CSN) [4] are two examples. Unlike a CDN, the main functions of which are storage and caching, the main focus of a CSN is to provide process ability to users, Internet Service Providers (ISPs), and content providers. CSN interacts collaboratively with user agents, content servers, and other network intermediaries, including ISP caching proxies and CDN surrogates, in the content delivery process to provide value-added services. CSN is composed of *application proxy servers, redirection servers,* and *service distribution and management servers.* An application proxy server hosts the software of value-added services and provides computational abilities. A redirection server directs a service request to an application proxy server according to a number of attributes and measurements. A service distribution and management server measures the demand of services and communicates with redirection servers to route the requests.

The *next-step-in-signaling* (NSIS) signaling framework is being developed to investigate the requirements, architecture, and protocols for QoS signaling across different network environments [5]. QoS signaling is defined as a way to communicate QoS parameters and management information among hosts, end systems, and network devices. It may include request and response messages to facilitate negotiation or renegotiation, asynchronous feedback, and QoS querying. QoS signaling supports per-flow and per-class QoS granularities. Different QoS signaling requirements may apply to different parts of the network, such as end-to-end, end-to-edge, edge-to-edge, or network-to-network, depending on where the QoS initiator and QoS controller are located. When the signaling runs across several QoS domains, NSIS allows the use of different signaling protocols but requires the universal QoS control information.

QoS Issues in Heterogeneous Networks

As indicated earlier, a user in the converging mobile Internet may utilize an application with different kinds of terminals across heterogeneous wireless access technologies and among different administrative domains. Thus, to provide seamless services for these users, in addition to solving common problems such as time-varying and location-dependent wireless link loss, limited bandwidth, and mobility, QoS management in the next-generation network has to face new challenges caused by the diversity of technologies.

Different Hyper Handovers

Handover is defined as a capability for managing the mobility for a mobile terminal or a moving network in active state. Handover in a heterogeneous network environment is different from that in a homogeneous wireless access system where it occurs only when a user moves from one base station to another. Handover within a homogeneous system is defined as *horizontal handover*, but handover between different administrative domains, access technologies, user terminals, or applications is defined as *hyper handover*. (The handover between different wireless access technologies is usually called *vertical handover*; our definition of hyper handover extends the dimensions of vertical handover.)

Table 36.1 Four Categories of Hyper Handovers

Categories	Differences in Handover Peers
Administration domain	ISP, ASP, AAA, SLA, policy, network topology, application context, network traffic, available services
Access technology	Bandwidth, loss, delay, coverage area, mobility support, QoS support, suitable application, cost, security
Terminal	CPU, memory size, display, input/output, battery, network interface, built-in applications, software platform
Application	Traffic specification, QoS requirements, user preference, user sensitivity, adaptation ability, network connection

Table 36.1 shows the main differences users might experience when encountering the different kinds of hyper handover. Note that this table separates the effect of each type, but in actual usage a user may experience a combination of hyper handovers; for example, when moving from one administrative domain to another, a user may also switch to a different access technology required in the destination domain. Assuming that the access technology in the new domain has lower bandwidth, the user may decide to switch to another terminal to be able to use applications with lower bandwidth requirements. In this scenario, the user experiences handover that includes all four categories of hyper handovers.

QoS Issues in Hyper Handovers

Many works have focused on QoS provision in a wireless access network with a single wireless access technology in the same administrative domain and on the same terminal. These traditional QoS management methods try to hide the transient QoS variation and violation from applications. Based on the time scale concerned, these functions can be classified as static or dynamic [2,3]. *Static QoS functions* usually activate in the application initiation period and remain constant for a long time. Specification, translation, negotiation, admission control, and resource reservation are static functions:

■ *Specification* is the definition of the QoS requirements or capabilities of applications, which encompass user preference, flow performance metrics, allowed variation, adaptation policy, expected service level, and interaction format with lower layer.

■ *Translation* performs the mapping between representations of QoS at different system levels (e.g., user level, application level, middleware, transport layer) and thus automate the derivation of a low layer's more detailed specification from an upper layer's more abstract specifications.

■ *Negotiation* is the process of reaching an agreed-upon specification among all parties on the end-to-end path. It could involve modification of the QoS specification in case of failure during the admission control procedure or the resource reservation when an agreement is reached.

■ *Admission control* is the procedure of comparing the resource requirement arising from the QoS specification against the available resources in the system. The decision regarding whether or not a new request can be accommodated generally depends on a systemwide resource management algorithm and current resource availability.

■ *Resource reservation* arranges the allocation of prenegotiated end-system and network resources to corresponding application flows. It is achieved by appropriately configuring related packet processing components in the QoS transport plane.

Contract specifications are often inexact because resource usage and flow characteristics are not generally completely defined in advance. *Dynamic QoS functions* allow the contract to be fulfilled on an ongoing basis. The most important dynamic functions include monitoring, policing, maintenance, renegotiation, adaptation, and feedback:

■ *Monitoring* measures the QoS actually provided. It can operate in different layers and over different time scales and plays an integral part in the QoS control feedback loop that tracks the QoS achieved by resource modules.

■ *Policing* ensures that all parties adhere to the QoS contract and satisfy their part. It can occur at edge routers where the inflow traffic is checked against its traffic specifications. It can also occur between two administrative domains where bilateral contracts must be satisfied.

■ *Maintenance* includes the modification of the parameters of the system to shield a lower layer QoS variation from applications.

■ *Renegotiation* is the process to renegotiate service contract when maintenance functions cannot satisfy the SLA. This usually happens as a result of major changes such as handover or failures in the system.

■ *Adaptation* refers to the situation that applications resort to their own specific adaptation techniques to adjust to the changes of QoS in the system when QoS maintenance functions fail to sustain specified QoS contract.

■ *Feedback* is selectively provided by a system to a user and waits for the user to intervene. The frequency and trigger of the feedback are predefined by the user according to the severity of the event or other policy issues such as cost and security.

Table 36.2 shows how these QoS-related components are influenced by the hyper handovers.

Unlike horizontal handovers, hyper handovers introduce large-grained changes in QoS. This results in more complicated QoS management in hyper handovers than in horizontal handovers. Here are some of the main challenges:

■ *Mobility support* — Currently, no QoS support is available during the handover period. When a terminal moves from one base station to another, packets that arrived at the previous station are either dropped or forwarded to the new station without QoS support. A number of mobility support protocols such as Fast Handover for Mobile IP (FMIP), Hierarchy Mobile IP, and Cellular IP [9] attempt to shorten the handoff delay and to reduce the packet loss rate; however, these protocols do not currently support the QoS parameters required by specific applications. Applications with different QoS parameters for bandwidth, delay, and loss belong to different QoS classes. They should receive differentiated service according to their classes. Current mobility-supporting schemes treat applications the same way. This sometimes violates the philosophy of differentiation and results in unnecessary system overuse.

■ *Application network* — An application proxy can help applications to shorten response latency and to adapt to network variation; however, an application proxy may change the end-to-end path and complicate the resource reservation procedure. It can also change the configuration of applications sessions by adding or switching services which influences the QoS negotiation and adaptation procedure.

■ *Dynamic QoS functionalities* — Resource reservation and admission control algorithms that were originally developed in the wired networks are modified for horizontal handover. The changes occurring during hyper handover challenge the dynamic QoS functionalities. For example, consider two adjacent domains with different cost or policy requirements for application QoS.

When a user moves from one domain to another, the QoS specification may have to be changed and some QoS adaptation functionalities may not be feasible.

Necessity of a Unified QoS Infrastructure

Previous sections have pointed out numerous QoS issues regarding providing users with a seamless mobile Internet experience. A large number of schemes have already been proposed to solve one issue or another. Due to a lack of coordination when deployed in the same network or device, these schemes may conflict with each other and make the system performance unpredictable. Furthermore, interest is increasing in protocol designs that rely on interactions between different layers to improve the performance of wireless networks. Generally termed *cross-layer design*, many of these proposals are aimed at achieving performance improvements, though often at the cost of good architecture design. Note that when the layering is broken, the luxury of designing a protocol in isolation is lost, and the effect of any single design choice on the entire system must be considered. Moreover, in many cases such undesired interactions are not easily foreseen. Hence, unbridled cross-layer design can lead to a spaghetti design that stifles future technology innovation and proliferation because every protocol update may require a complete system overhaul [16,18].

A unified QoS infrastructure is required to coordinate various QoS schemes deployed in the network or on the devices and optimize user perceived performance. A number of QoS principles guide our design of a generalized QoS framework for the heterogeneous network:

- *Adaptation* states that adaptation support should be included in the design of every component because of the limited and dynamically varying available resources, stringent application requirements, heterogeneous executing platforms, and user mobility. The QoS specification could include adaptive parameters such as loss percentage and service disruption probability. Seamless modification of already reserved QoS should be allowed.
- *Reconfiguration* states that the configuration of the whole system or part of the system should be allowed to dynamically change to adapt to the changes of application QoS requirements, user preferences, network maintenance, or resource variation. Both hardware and software organizations should be modular.
- *Robustness* states that the system should be resistant to and able to easily recover from temporary network disconnections and failures. The QoS should decrease gracefully in case of sustained

Table 36.2 QoS Issues for Each Kind of Hyper Handover

	Administrative Domain	Access Technology	Terminal	Application
QoS specification	Users have different SLAs with each domain; users define different adaptation policies.	Change of access link can make users adapt; different QoS framework requires different formats.	Processing power and display size change so traffic specifications might change.	Specifications change due to changes in applications.
QoS negotiation	Problems include network traffic load, end-to-end path, or SLA changes.	Problems include network traffic load, application adaptation, end-to-end path.	Problems include network traffic load, application adaptation, end-to-end path.	Problems include new QoS specification, end-to-end path change, network traffic load.
QoS adaptation	Different policies and costs lead to different adaptation techniques.	Varying bandwidth, delay, loss, jitter, coverage area, and physical layer cause problems.	QoS adaptation in the OS and middleware could change.	Changes are based on the choice of new applications.
Resource reservation	Changes in QoS specifications, contexts, and application network adaptations make it expensive and inaccurate.	Proactive reservation is based on Mobile ReSource reserVation Protocol (MRSVP); suffers from path changes.	Suffers from changes in the end-to-end path and traffic specifications.	This is not a significant issue because a long reservation delay is tolerable.

Admission control	Problems are the same as above.	When changing from high bandwidth to low bandwidth, system could fail.	This is not a significant issue.	System could fail if new application's QoS cannot be satisfied.
Monitoring, policing, and maintenance	It is necessary to transfer the current application profile between domains to continue tracking.	Profiles may not be compatible and policies could change.	Policies about how to combine profiles must be defined.	Policies about how to combine profiles must be defined.
Load balancing	Load sharing among servers, applications, and the network might change.	Load is redistributed because of the changes of wireless links.	Load is redistributed because of the changes of terminal.	Load balancing must be done from scratch because of application changes.
Content-distribution network	Cache can change the end-to-end path for the same flow.	This is not a significant issue.	This is not a significant issue.	Different caches could be used.
Context-aware computing	New location-based services can replace old location-based services; the same flow might have a different traffic load.	Application nodes such as transcoding proxy are used in the access network to adapt to link change.	New kind of context could be required; application proxy could also change.	Different kinds of context information could be used; application proxy could also change.
Service discovery and service composition	Applications could add new services; the same service can change application/service provider (ASP).	Applications change to match the link properties.	Applications change to match the terminal properties.	Complementary service changes to match application's properties.

(continued)

Table 36.2 QoS Issues for Each Kind of Hyper Handover (cont.)

	Administrative Domain	Access Technology	Terminal	Application
Flow reconfiguration and flow synchronization	Properties of flows of the applications can change; relative priorities can change.	Properties of flows of the applications can change; relative priority can change.	Properties of flows of the applications can change; relative priority can change.	Properties of flows of the applications can change; relative priority can change.
Others	Micromobility-based handover schemes such as cellular IP and Hawaii do not work without a common foreign agent.	Transport protocols that are good for one kind of link may not be good for the other, but transport layer has no knowledge of link-layer changes.	Mobile agents migrating from one terminal to another must adjust to the new OS and choose new QoS adaptation techniques.	Change of application could lead to reconfiguration of the underlying protocol layers to optimize QoS adaptation ability.

system failure. Automatic release of resources after failure or automatic setup of resources after recovery should be possible.

■ *Separation* states that data transfer, QoS control signaling, and QoS management signaling are functionally distinct architectural activities that should be separated in the architectural QoS framework.

■ *Integration* states that each resource module on the end-to-end path must provide a QoS guarantee and maintenance of ongoing flows. It requires the QoS support of OSs on both the client and server sides and includes every single communication segment on the end-to-end communication path.

■ *Transparency* states that the upper layer should be shielded from the complexity of lower layer QoS adaptation and management techniques.

■ *Inter-layer interaction* states that each protocol layer should cooperate with others in a systematic way to achieve globally optimal QoS adaptation. Still under the constraints of transparency principle, the interactions between the different layers should be limited, well defined, and standardized.

■ *User orientation* states that users should be considered in the end-to-end path. The user's perception of QoS is the optimizing objective of the overall QoS framework and is captured through user profiling. Users should be able to set up when and how they should be prompted when a QoS violation, adaptation, or exception occurs.

■ *Maintenance* states that the QoS framework should be easy to maintain, and dynamic QoS deployment should be done in a time- and energy-efficient manner. The error-prone and labor-intensive manual maintenance of network equipment should be replaced by a policy-based management system.

■ *Scalability* states that the QoS framework should behave normally under extreme heavy loads. It should be scalable in the number of active flows, signaling messages, handovers, and states per entity, as well as in CPU usage.

In the next section, we present our QoS framework for mobile Internet designed according to these principles. As is pointed out later, our key observation is that the QoS frameworks of the network and terminal should be based on different principles because of their uniqueness. As a result, our architecture has two interacting network and terminal subsystems that achieve a good balance between performance and architecture integrity.

New QoS Network Infrastructure for Heterogeneous Networks

This section presents our frameworks for coping with the main QoS challenges for hyper handovers and enabling seamless services. The discussion focuses on the new functional blocks and new functionalities added to the conventional framework. These new functional blocks are proposed to alleviate the influences of hyper handovers. When examining the QoS design principles [3], note that some principles contradict each other, such as the transparency and user-orientation principles. Also note that not all principles are feasible for both terminals and networks, such as the inter-layer interaction and separation principles. For the terminal, scalability may not be a significant issue and the performance of each layer can be easily measured. Measurements can be used as feedback to help achieve optimal cross-layer adaptation, so it is desirable to have optimal inter-layer interaction and user-oriented feedback. For the network, the main concerns are scalability and robustness; thus, the separation and transparency principles are important for simplifying the design, control, and maintenance of large-scale networks. Based on this basic observation, we propose two QoS frameworks for the network and terminal that support end-to-end QoS provision in the heterogeneous mobile Internet. These two frameworks can be designed using somewhat differing principles but should have a well-designed interface between them to allow seamless end-to-end services [16].

The main design principles for the QoS network framework are scalability, transparency, separation, robustness, maintenance, and adaptation. User-orientated feedback and cross-layer interaction principles are secondary. As shown in Figure 36.1, the network framework is designed to have three separate planes: *management and functional signaling* (or, simply, *management*), *QoS control and signaling* (or, simply, *control*), and *data*. This structure conforms to the separation principle and allows scalable QoS framework and adaptation techniques to be designed transparently in each layer. The policy framework is then used to provide limited and systematic inter-layer interactions. This unifies the three functional planes into a universal architecture.

The main hyper-handover-related QoS function blocks are listed in Figure 36.1. Some are completely new, but others simply have newly added functionalities in addition to the common functions. This section highlights some of the key blocks, either new or differing from those in existing QoS frameworks. Note that this is an extensive list and includes the functions that may reside in different components of the network. Any special component, such as an access router or core router, may only implement a subset of all the blocks or a fraction of all the functions available in a block.

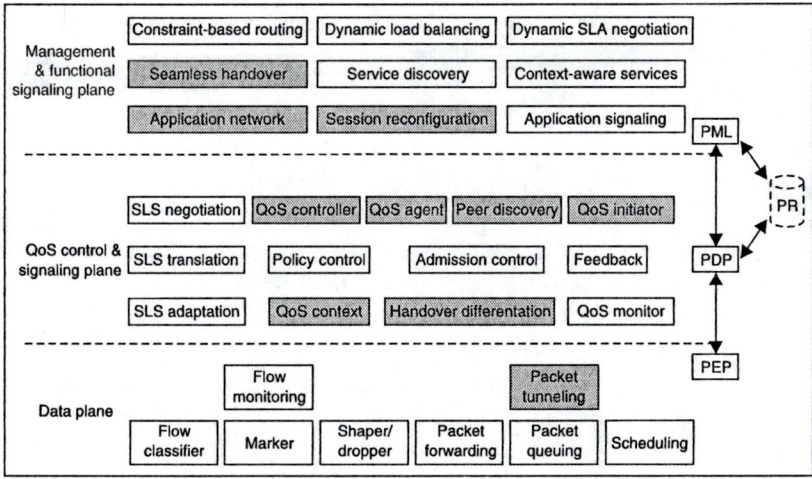

Figure 36.1 QoS metwork infrastructure.

Data Plane

Most of the blocks in the data plane exist in other QoS frameworks, such as Inte-Serv and Diff-Serv; the only new block is packet tunneling. In the data plane, however, these blocks all have different functionalities controlled by the control plane to enable seamless handovers. Here are some of the most important differences:

- *Flow classifier* — The common flow classifier uses 5-tuples (source/destination IP address, source/destination port, and protocol type) to identify a flow and to differentiate packets; however, this format is no longer suitable in the mobile Internet because in many cases either the source or destination of a flow may change due to handover. For example, when a mobile host (MH) subscribing to a location-based service moves to a new domain, it must switch its connection to a new server providing a similar service. In this case, the same flow uses different destination IP addresses during the transmission. Other similar cases include changes in the intermediate content caching proxy and transcoding proxy in a CDN.
- *Flow monitoring* — Common tasks of the block are to guarantee that the coming traffic obeys its service-level specification (SLS) and to specify how the marking block should mark a packet if it violates the SLS. This process usually happens at the edge router, which could change also during the hyper handover as an MH

migrates from one domain to another. Without the necessary history information, such monitoring cannot be done properly so the transfer of traffic history from the old edge router to the new one is needed.

■ *Packet tunneling* — This is a new block. It not only deals with IP tunneling in the mobile IP but also tries to provide QoS support during a hyper handover. Each application can ask for different QoS support during the handover. The QoS requirement during handover may be different from the QoS requirement for the normal data transmission. If this is the case, when handover is detected the packets already in the forwarding queue are encapsulated first, then they are put into a new queue that could belong to a different out-interface and receive different scheduling treatment.

Control Plane

The data plane as a PEP has only limited QoS adaptation ability, realized primarily through the scheduling block. Because it does not have global knowledge, to adapt to the heterogeneous environment it requires the control plane to set up corresponding operation policies and parameters. The components in the data plane then process the packets in the way specified by the control plane. The control plane uses end-to-end signaling to integrate every host on the end-to-end path to achieve QoS negotiation, adaptation, and feedback. It allows a mobile host to communicate with any intermediate application server to set up or tear down temporary connections. It also allows any pair of interacting nodes to negotiate a QoS contract between each other. Using a policy-based framework speeds up systemwide implementation of the QoS framework by automating the equipment setup time; however, because the control plane focuses primarily on the nonfunctional part (improving performance), it further relies on the management plane to decide how existing QoS support could be used to achieve application-specific tasks.

QoS Initiator, QoS Agent, and QoS Controller

The QoS initiator (QI) and QoS controller (QC) blocks are the initiator of and responder to the QoS signaling message, respectively. As in a more detailed example to be given later, different signaling protocols could be used and different QoS parameters could be transferred depending on the network topology and traffic type (per-flow or flow aggregation). The QoS agent block acts as the PDP. It gets input from the policy control and admission control blocks, cooperates with other QoS agents through

QoS signaling protocol, and configures the PEPs in its domain to provide large-time-scale QoS support to normal data transmissions. It further works together with QoS context and handover differentiation blocks to provide short-time-scale QoS during the handover period.

Peer Discovery

This block gets information from the management plane regarding what the destination host is and which kind of negotiation should go on. QoS signaling assumes that the QC and QI have already existed and know each other before the signaling happens. In the common IP network, this assumption is valid, and, based on the destination IP address, routing and relations between a QoS agent and other nodes can be predefined. However, in hyper handover cases where flows keep changing their source or destination points, a QoS agent must dynamically find out which node is its signaling peer. This is decided by the management plane.

QoS Context

Significant QoS signaling exchanges between interacting nodes in the network may be required in order to establish the initial QoS treatment for the packets of an MH [10]. Preliminary studies have indicated that to reestablish QoS in the new path from scratch can be very time consuming. So the QoS context block is helpful in transferring the QoS context of an MH from one node to another. Such a QoS transfer can happen not only between two access points but also between any nodes with similar functionalities such as edge routers. However, for end-to-end QoS support, transferring context between parts of the nodes may be insufficient to completely reinitiate the QoS treatment of the MH.

Handover Differentiation

Current handover schemes treat applications in a way that violates the philosophy of differentiation and sometimes leads to unnecessary system overuse. Although application QoS requirements will influence the handover QoS requirements, they are not necessarily the same. To enable handover differentiation, a number of handover classes have been defined in the literature:

- *Fast handover* is a handover that can satisfy strict delay bounds (e.g., for real-time services).
- *Smooth handover* is a handover that can minimize a loss of packets.
- *Seamless handover* is a handover with minimum perceptible interruption of the services.

This block provides the applications with differentiated QoS support during the handover period. It sets up the necessary handover tunneling path and controls the packet tunneling block in the data plane to separate normal QoS requirements of data transmission from handover QoS requirements.

Management Plane

The management plane behaves as the policy management tool (PMT). The PMT produces policies based on user SLA, application network server properties, and network load condition. Policies are stored in the policy repository, which can be physically located anywhere in the network. Policies can be either per-user based or per-class based. They can specify the QoS rules during the handover and adaptation period. When a hyper handover occurs, before the QoS signaling for the adaptation is initiated, different entities in the network can be involved to provide seamless service or value-added services; for example, constraint-based routing could be used to reroute the flow bypassing the congested nodes. Servers of location-based services, context-aware services, or content caching could be used to improve user perception of the services. Applications could use an application-level signaling protocol such as SIP to add or remove servers from the session. Instead of static SLA, a different protocol such as COPS could be used to dynamically set up user SLAs. Based on these services, QoS signaling is then used to further negotiate the service contract and guarantee the seamless services. Three key blocks are:

- *Session reconfiguration* — In many cases mentioned earlier, the configuration of an application session is changed. These changes happen in many ways. The number of flows of one session could be changed during the handover if new services in the new domain are added to the current session. The traffic of one flow could have different QoS characteristic if the flow attaches to a different server after handover. The addresses of the source and destination nodes can be different if intermediate proxies change. In all the cases, the application must reconfigure the relative priorities among flows of its session to optimize overall session performance.
- *Application network* — An application network can improve user perception by moving the server closer to the client or adapting the traffic to the current context. It works together with the IP-layer QoS framework to improve application performance. Many issues need to be considered when evaluating an application network; for example, because an application network may change the traffic characteristics of flows going through it, QoS translation is necessary and must take into account such long-term traffic

changes. Another issue is possibly increased service variation during the handover if one domain has an application network server that another domain does not have.

- *Seamless handover* — In addition to mobility support that is based on Mobile IP and Cellular IP are many other types of mobility support. One mechanism is an SIP-based application mobility scheme that provides application-specific handover support [10]. Another mechanism is the predefined Handover Support Overlay Network, which provides QoS support during the handover. All of these different mechanisms must cooperate with the QoS signaling protocols to perform admission control and resource reservation.

Unified Cross-Layer Adaptation Platform

Cross-layer adaptation algorithms, which are discussed here in different contexts, are considered to be promising techniques for hiding the complexity of the underlying heterogeneity from mobile applications. The common themes of these algorithms are an understanding of the user, application, or system performance requirements, as well as adjustment of the behavior of configurable components to adapt to various heterogeneities. As discussed earlier, optimal inter-layer interaction and user feedback are desirable for a terminal QoS framework. Furthermore, terminal-based adaptation is feasible in hyper handovers because a handover usually happens at the last wireless hop; hence, terminals can have good knowledge about the context changes and QoS variations. It is natural for a terminal to initiate the QoS adaptation or cooperate with the network to adjust to the environment heterogeneity

Most of the cross-layer adaptation algorithms improve some performance index to some extent; however, they usually only focus on the design of the algorithm itself. Also, they assume that underlying assumptions are reasonable and that the overhead incurred is small compared with the performance improvement. Unfortunately, as the number of cross-layer adaptation algorithms on one terminal increases, the chances that outputs from the different algorithms will conflict also increase. At the same time, as the time-varying mobile environment changes, some adaptation algorithms may become inappropriate.

After carefully evaluating existing solutions, this section briefly lists and discusses the main problems encountered in previous works and which also serve as the motivation behind our terminal QoS framework:

- *No systemwide coordination* — When multiple cross-layer adaptation algorithms coexist at the same terminal, how they interact with each other is not well studied. More specifically, possible conflicts between different schemes, validation of each scheme's assumptions, and the feasibility of each scheme under the current running environment are not considered.

■ *No systematic way to achieve cross-layer communications* — Nearly all of the cross-layer adaptation schemes rely on sharing important information among different layers to achieve the performance goal. Most of them focus on the design of the algorithm itself and use some *ad hoc* ways to exchange information, such as specialized APIs or header extensions.

■ *Difficult to modify, extend, and interconnect* — Because of the *ad hoc* approach to designing cross-layer communications, the modification, extension, and interconnection with other components become time consuming and error prone. Unnecessary details on each layer have to be exposed to allow few variations.

To address the problems of unifying cross-layer adaptation and communication but keeping the architecture expandable, manageable, and powerful, we propose the cross-layer adaptation platform (CLAP) shown in Table 36.3.

Cross-Layer Adaptation Algorithm Abstraction and Policy Validation

Before describing the CLAP architecture, this section first defines and gives an expression for cross-layer adaptation algorithms. Only after the expression of the algorithms is well understood will the design choices of the architecture become obvious. Cross-layer adaptation algorithms can be defined in a hierarchical way as shown in Figure 36.2. *Service* abstraction is used to define the behavior or functionality provided by a component. To fully specify a service, one must define: (1) the functions, (2) the information (parameters) required to perform these functions, and (3) the information made available by this component to other components of the system. To support dynamic configuration, a *component* also must define: (1) the service choices inside the component, and (2) the information needed to select the service. *Cross-layer adaptation algorithms* can then be abstracted as: (1) components involved in each layer; (2) policies used to configure each component, including policy conditions using the output from some components and policy actions using configuration parameters as the output to control some other components; (3) priority of the algorithm in case of policy conflict; and (4) assumptions of the algorithm (i.e., under which conditions the algorithm should be invoked), which are expressed as another set of policies used for coordination among algorithms.

Table 36.3 Cross-Layer Adaptation Platform (CLAP)

Layer	Input/Trigger	Cross-Layer Adaptation	Layer policy repository (PR)/policy decision point (PDP)	Policy Enforcement Point (PEP)
User/point-to-point management layer (PML)	Management; authentication, authorization, and accounting (AAA); user preference		Layer policy repository (PR)/policy decision point (PDP)	Graphical user interface (GUI) adaptation, interaction/feedback adaptation
Application	Relative importance, specific requirements	System PDP Cross-layer adaptation algorithms Policy checks: consistency, feasibility, dominance Policy translation Policy distribution	Layer PR/PDP	Application adaptation, application redirection
Middleware	Adaptation ability	Common policy repository (CPR)	Layer PR/PDP	Session management, flow synchronization
Transport	Supporting protocols		Layer PR/PDP	Link-aware protocols, application-aware loss recovery protocols
Internet Protocol	Network topology, traffic load		Layer PR/PDP	Mobility support, Inte-Serv, Diff-Serv, multiprotocol label switching (MPLS), traffic engineering (TE)
Link	Link availability, quality, capability		Layer PR/PDP	Link selection, scheduling, layer-two QoS mapping, layer-two mobility
Physical layer	Signal-to-noise ratio (SNR), spectrum allocation		Layer PR/PDP	Software-defined radio, channel coding, power control

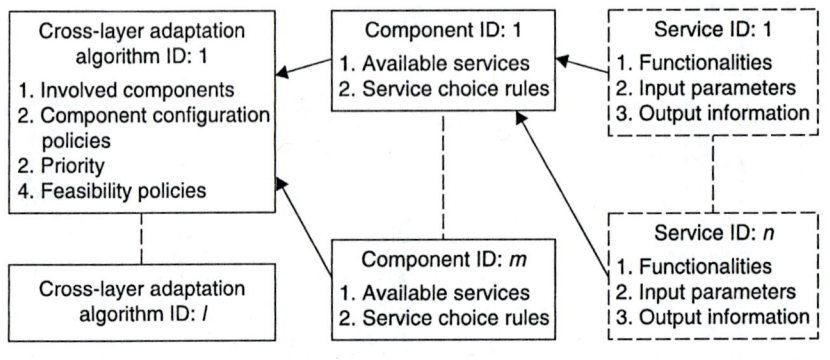

Figure 36.2 Abstraction of cross-layer adaptation algorithms.

Next we provide a description of how to express these three entities as policies. This discussion is in line with the Policy Common Information Model (PCIM) [6] proposed by the Distributed Management Task Force (DMTF). A policy rule has "If [condition] ... then [action]" semantics. A policy rule condition, in the most general cases, is represented as either an ORed set of ANDed conditions (disjunctive normal form, or DNF), or an ANDed set of ORed conditions (conjunctive normal form, or CNF). Two steps are required to make a policy decision. The first step is evaluation of a policy rule condition. The second step deals with the actions for enforcement when the conditions of a policy rule are TRUE. For each cross-layer adaptation algorithm, a number of policies could be produced, and these policies are aggregated into a *policy group*. Each policy group has a unique group ID. This systemwide ID will be used later for applications as the index to refer to corresponding cross-layer adaptation algorithms and interpret the attached parameters.

One of the main functionalities of the system PDP is to validate that the policy outputs of different algorithms are consistent with each other and feasible in the current environment and to coordinate the behavior of each algorithm, if necessary. The policy validation algorithms carried by the system PDP may include the following checks [3]: (1) *bounds checks*, which validate that the values taken by an attribute in the policy specification are within specific limits determined by the system; (2) *relation checks*, which validate that the value taken by any two parameters in the policy specification satisfy a relationship determined by the specific algorithm; (3) *consistency checks*, which validate that any two policies defined by different algorithms do not conflict each other; (4) *feasibility checks*, which ensure that the policies of each algorithm are feasible under current conditions; and (5) *dominance checks*, which find the "dominant policies" when inconsistencies occur between policies.

Functionalities of the System PDP

The system PDP takes the cross-layer adaptation algorithms as input and then transfers them as policies stored at the common policy repository (CPR) or sends them to the layer PDPs to execute. In addition to these, it also takes inputs from other layers as well as the QoS network infrastructure and adds management polices to the CPR. Such management policies may modify or limit the behavior of existing algorithms if needed.

As shown in the "Input/Trigger" column of Table 36.3, each layer may send different information to the system PDP. The information may include the capability of the system, as discussed before, or the information may include predefined events or triggers that require the system to intervene. At the user level, one may use the point-to-point management layer (PML) to input high-level policies such as user preferences, service-level agreements, adaptation preferences, handover preferences, business goals, security levels, or environmental parameters. The policies may be expressed in language closer to natural communication rather than in terms of the specific technology implementing it. Such high-level policies at first should be checked to ensure consistency, correctness, and feasibility. They should then be translated to technology-oriented policies also stored in the CPR. More generally, our schemes can easily allow remote configuration and adaptation by taking remote control policies into account.

Such newly added policies should also be compared with existing cross-layer adaptation policies to determine whether conflicts will occur. If so, new policies should be added to the CPR to guide the system as to how to operate in such cases. This may lead to new policies being also installed in the layer PDP and PR. The comparison between high-level polices and cross-layer adaptation policies is expedited by the use of "feasibility polices" in the algorithm abstraction shown in Figure 36.2. In Figure 36.2, each cross-layer adaptation algorithm expresses its assumptions regarding the surrounding environment and limitations. For elements not covered by these policies, some defaulted or observed values can be used.

Extra event or error triggers could also be implemented. Keep in mind that statistics and parameters specific for the cross-layer adaptation algorithm are not exposed to the system PDP because the policy framework solves only system configuration problems. The system PDP could work together with the layer PDP to define the globally important information that should be reported by the layer PDP. The information could include the change of location, network, or current battery capability. Such parameters are important in terms of system performance and will lead to system reconfiguration if triggered.

The update of already installed policies in the layer PDP and PEP can be speeded up by using "enable" attributes included in each policy rule. Basically, the policies of each algorithm need not be uninstalled or

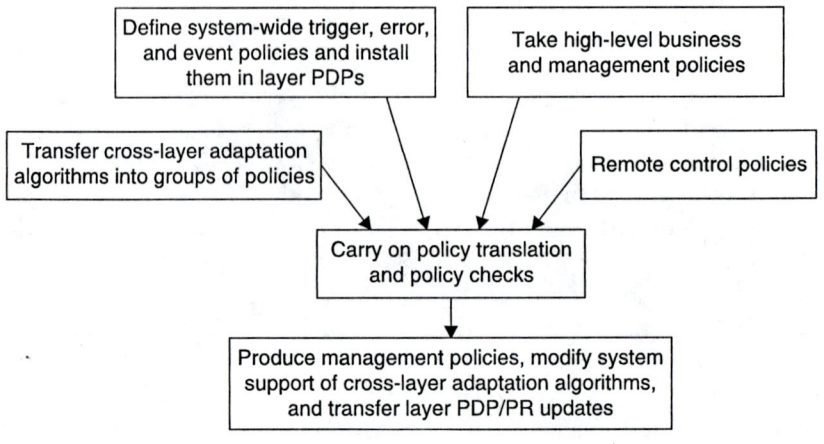

Figure 36.3   System PDP functionalities.

substituted by the policies of another one. By simply resetting the "enable" attributes, the system PDP can easily and flexibly change support for a specific algorithm; furthermore, the application does not have to be changed. It still uses the same policy group number, although the support is no longer the same. To summarize the important functionalities of the system PDP, this section includes the general running sequence of the system PDP shown in Figure 36.3.

Configuration of Layer PDP, PR, and PEP

Our architecture has a hierarchical PDP setup to maintain the transparency of each layer and make the system scalable, flexible, and easy to manage. As partly shown in Table 36.2, the PEPs of each layer are the components involved in the cross-layer adaptation algorithm and have the abstraction as shown in Figure 36.2. Each PEP is controlled by the layer PDP on its own layer and can install policies locally. For PEPs supporting multiple functions at the same time, the PEPs could evaluate the parameter values specified by the policies and invoke corresponding procedure. Each PEP could collect and output policy-specified statistics and parameters for cross-layer coordination. Local PR maintains only local policies that are either produced by layer PDP or transferred from the system PDP. Layer PDP is the key element of the structure that allows proper operation of the system. It has following main functionalities:

■ *Maintain local adaptation abilities* — Not all of the adaptation abilities have to be cross-layer, so layer PDP is used to coordinate adaptation on its own layer. Furthermore, because local adaptation may also influence some of the cross-layer components, the layer PDP can implement local policy checks to guarantee policy consistency.

■ *Keep simple and well-defined interfaces* — As terminals become more and more complicated and cross-layer adaptation techniques keep improving, more components of a layer will participate in cross-layer coordination. To make the system scalable and extendable, layer PDP could choose to expose only limited components to the system PDP by encapsulating these components.

■ *Collect layered triggers of systemwide importance* — As discussed before, the system PDP could work together with layer PDP to define important triggers on each layer. These triggers are not for performance adaptation but for systemwide reconfiguration or modification.

Cross-Layer Information Exchange

Cross-layer adaptation algorithms require two types of support from the system. The first type of support is dynamically choosing services from the same component and guaranteeing systemwide feasibility. The proposed CLAP architecture provides this type of support. The second type is supporting cross-layer information exchange. This support is application specific and flow based and is not provided by the CLAP architecture itself. Here, we propose a data structure referred to as a *cross-layer tag* (CLT), for the information exchange. CLT is similar to the IPv6 [11] extension header and its format is <Next Header, Header Length, Policy Fields>.

Although the CLT is mainly for intra-terminal usage, by using the same fields of IPv6 such as Next Header and Header Length, it can be integrated into the IPv6 header for external communications. In this case, the CLT header type could be zero, the same as the hop-by-hop header in IPv6, or 60, the same as the Destination Option header in IPv6. This serves as a mechanism to carry the end-to-end adaptation parameters to the communicating nodes. Depending on the specific algorithm, the CLT itself or some modification can be used in the end-to-end communication.

Each CLT can have multiple policy fields with the same format as <Policy Group ID, Data Length, Data Fields>. The unique Policy Group ID is assigned to a corresponding cross-layer adaptation algorithm. This ID is used as the index by the PEPs on each layer to understand the usage and format of data in the data fields. The Data Length field is designed to allow parameters of variable lengths. Based on the application requirements and system capability, one or more cross-layer adaptation algorithm could be used by one application. Because the system PDP has guaranteed the consistency of each policy and modified the policy support when necessary, applications are released from such responsibilities.

Data fields contain both normal data types, such as string, integer, Boolean, or float numbers, and specific location pointers for cross-layer data uploading. For information exchanges from the upper layer to the lower layer, CLT could be appended to the normal packets and processed by related components. On the other hand, for information uploaded from the lower layer to the upper layer, such a channel may not exist, so the system allocates a shared memory area in the CPR to allow information exchange. During the application initialization period, when the adaptation policies are chosen, the related parameter exchange can be decided. Thus, if an uploading channel is necessary, a piece of shared memory is assigned and the pointer is returned. The shared memory is indexed by the unique Flow ID or Process ID. The pointer is then included in the CLT data fields and received by the PEPs. The PEPs can then use the pointer to access the memory to exchange information with upper layer components.

One of the functionalities of user-/application-layer PEPs is to maintain and assign an appropriate Policy Group ID to each application. Because the policy granularity could be an application, a flow, or a group of packets, we propose usage of the Flow ID field in IPv6 that supports flexible granularity adjustment. The application-layer PEP collects information about the application, checks the cross-layer adaptation algorithm availability (which is modified by the system PDP), and matches the policies with application requirements.

An Instance of QoS Network Infrastructure

The first part of this chapter introduced the main functionalities of the general QoS network infrastructure and terminal CLAP architecture. Here, we provide an example of where these functionalities could reside and how they interact with each other. Figure 36.4 shows an example of an all-IP mobile Internet. Two administrative domains are attached to the core network through dedicated gateways. All kinds of hyper handovers occur when a user moves from domain 1 to domain 2, and we assume that all networks have some IP-layer QoS support and domain 2 uses application networks to further improve its performance.

The data plane consists of the user terminal, access router, core network routers, gateways, special application network servers, and corresponding nodes. Some of the application network servers shown in Figure 36.2 can be designed for AAA, service discovery, service composition, and application adaptation proxy. QoS support for handover is provided. Extra

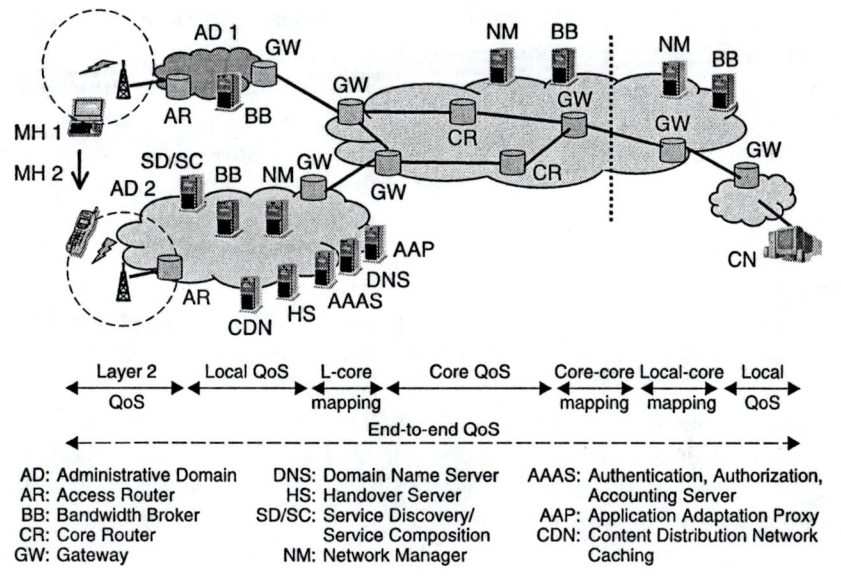

Layer 2   Local QoS   L-core   Core QoS   Core-core Local-core Local
QoS   mapping   mapping   mapping   QoS

End-to-end QoS

AD: Administrative Domain   DNS: Domain Name Server   AAAS: Authentication, Authorization,
AR: Access Router   HS: Handover Server   Accounting Server
BB: Bandwidth Broker   SD/SC: Service Discovery/   AAP: Application Adaptation Proxy
CR: Core Router   Service Composition   CDN: Content Distribution Network
GW: Gateway   NM: Network Manager   Caching

Figure 36.4   An instance of QoS framework.

forwarding, tunneling, and reservation abilities are implemented in specialized handover servers that could have predefined QoS paths with handover servers in other domains.

The *bandwidth broker* (BB) in each domain is the most important component in the *control plane*. The BB has a global knowledge of the network traffic so it can perform admission control for the network. If the end-to-end path traverses through several administrative domains, the BBs of different domains can contact each other to arrange the end-to-end path and to negotiate bilateral service contracts between two domains.

The *network manager* (NM) is the central controller of the *management plane* that manages the network and has knowledge of each specific application server in the network. The NM could be physically located distributively throughout the network and carry on different functionalities. Similar to the manager server in the CSN, the NM has a well-known address and provides entry points for the BB or mobile host seeking to acquire QoS-related information.

The *mobile host* is one end of the end-to-end communications. An MH must support any necessary protocols to communicate with other components in the three planes of QoS network infrastructure, including the BB and NM. Furthermore, the MH has the flexibility to implement any cross-layer adaptation algorithms to adapt to network variations. For the purpose of systemwide coordination, the MH abstracts all protocols and cross-layer adaptation algorithms as a set of functional PEPs and coordination policies using the methodology discussed earlier. Functional PEPs reside in different layers and are managed by a layer PDP. Some local policies are installed in layer PR and other global policies are in the CPR. CLAP then carries on all types of policy checks to ensure coordination among algorithms and reconfigures the behavior of the MH through the hierarchical PDP if a predefined event or trigger occurs.

When an MH moves to a new domain, it can send inquiries to the NM about specific application servers. Based on the current load and other performance and policy issues, the NM could send back the response with the destined server address and related communication context. The access point or access router on behalf of the MH could also send such inquiries. The NM could work together with the BB to set up the desired path. QoS signaling is used to facilitate messages of negotiation or renegotiation, asynchronous feedback, and QoS querying. As shown in Figure 36.2, the end-to-end path could pass many network segments. QoS mapping and signaling should be performed whenever necessary — for example, MH-to-AR signaling, local edge-to-edge signaling in each administrative domain, and signaling between gateways in the core network. Notice that our framework allows all types of signaling schemes to be used.

Conclusions

This chapter has summarized the main QoS challenges regarding the seamless support of the various categories of hyper handover. Our QoS framework includes a three-plane network infrastructure and a terminal-based hierarchical policy management system. The three-plane network framework is based on a QoS signaling architecture and policy framework and integrates the QoS functionalities of the IP layer with the abilities of other layers, including application networks. A system implementation based on network managers and bandwidth brokers was used to describe how the framework supports end-to-end QoS provision adaptively and seamlessly. The terminal-based hierarchical policy management system is designed to coordinate the behavior of different cross-layer adaptation algorithms on one terminal to achieve optimal systemwide performance. Compared with other related work [12–15] on seamless handover support, this framework is more generic and has some important advantages, such as support of all kinds of hyper handover and the differentiation capability of handover QoS for different traffic classes.

Due to the heterogeneity of the mobile Internet and the need to maintain QoS, numerous issues remain to be addressed. Here are some of the most relevant:

- *Interactions between application-layer signaling, mobility signaling, and QoS signaling* — These different kinds of signaling are used for different purposes but share some common path or properties. How to integrate these signaling schemes is a major challenge.
- *Security* — It is one of the hottest topics in today's IP world. Our discussion does not touch on security issues but, as mentioned in Moore et al. [6], security is mandatory in the design of the QoS signaling protocol to avoid attacks, such as denial of service.
- *Flow identification* — When hyper handover occurs, the IP address is no longer a reliable way to identify flows and related QoS reservation information along the path. How networks can identify flows during the handover is an open question.

References

[1] Huston, G., *Internet Performance Survival Guide*, John Wiley & Sons, New York, 2000.
[2] Chalmers, D. and Sloman, M., A survey of quality of service in mobile computing environments, in *IEEE Online Commun. Surv.*, 2(2), 1–10, 1999.
[3] Aurrecoechea, C. et al., A survey of QoS architectures, *ACM Multimedia Syst. J.*, 6(3), 138–151, 1998.
[4] Ma, W. et al., Content services network: the architecture and protocols, in *Proc. of the 6th Int. Web Caching and Content Delivery Workshop*, Boston, MA, June 20–22, 2001, pp. 83–101.
[5] Brunner, M., http://www.ietf.org/internet-drafts/draft-ietf-nsis-req-02.txt, 2002.
[6] Moore, B. et al., *Policy Core Information Model, Version 1 Specification*, Request for Comments 3060, Internet Engineering Task Force (IETF), 2001 (http://www.ietf.org/rfc/rfc3060.txt).
[7] Boyle, J. et al., *COPS Usage for RSVP*, Request for Comments 2749, Internet Engineering Task Force (IETF), 2000 (http://www.ietf.org/rfc/rfc2749.txt).
[8] Chan, K. et al., *COPS Usage for Policy Provisioning*, Request for Comments 3084, Internet Engineering Task Force (IETF), 2001 (http://www.ietf.org/rfc/rfc3084.txt).
[9] Campbell, A. et al., Design, implementation, and evaluation of cellular IP, *IEEE Pers. Commun.*, 7(4), 42–49, 2000.
[10] Kempf, J., http://www.ietf.org/internet-drafts/draft-ietf-seamoby-context-transfer-problem-stat04.txt, 2002.

[11] Deering, S. and Hinden, R., *Internet Protocol, Version 6 (IPv6) Specification*, Request for Comments 1883, Internet Engineering Task Force (IETF), 1995 (http://www.ietf.org/rfc/rfc1883.txt).

[12] Roos, A. et al., Critical issues for roaming in 3G, *IEEE Wireless Commun.*, 10(1), 29–35, 2003.

[13] Aguier, R.L. et al., An IP-based QoS architecture for 4G operator scenarios, *IEEE Wireless Commun.*, 10(3), 54–62, 2003.

[14] Floroiu, J.W. et al., Seamless handover in terrestrial radio access networks: a case study, *IEEE Commun. Mag.*, 41(11), 110–116, 2003.

[15] Zhuang, W. et al., Policy-based QoS management architecture in an integrated UMTS and WLAN environment, *IEEE Commun. Mag.*, 41(11), 118–125, 2003.

[16] Gao, X. et al., End-to-end QoS provisioning in mobile heterogeneous networks, *IEEE Wireless Commun.*, 11(3), 24–34, 2004.

[17] Kawadia, V. and Kumar, P.R., A cautionary perspective on cross-layer design, *IEEE Wireless Commun.*, 12(1), 3–11, 2005.

Chapter 37

IT-Based Open Service Delivery Platforms for Mobile Networks: From CAMEL to the IP Multimedia System

Thomas Magedanz and Muhammad Sher

CONTENTS

Introduction

Service delivery platforms (SDPs) have always stood at the forefront of telecommunications, as they are the foundation for the creation, deployment, provision, control, charging, and management of telecommunication services provided to end users, thus enabling the generation of revenues. The SDPs represent the programming interface that allows programming of the underlying network capabilities and therefore are primarily based on usage of information technologies (ITs). SDPs have continually changed over the last decades, as innovation has been taking place at a rapid pace in this domain.

Historically, different types of networks (fixed networks, mobile networks, data networks, and the Internet) have been operated by different operators with different business models providing quite different services, and these SDPs have been specifically designed for a given network environment. In such an environment, referred to as the *stovepipe architectural model* (Figure 37.1), the services are designed, deployed, provisioned, controlled, and managed on top of heterogeneous SDPs with no need for service integration. This environment has a clear separation of fixed voice telephony, cellular voice telephony, and fixed or even mobile Internet access for Web browsing and e-mailing.

The emergence of mobile multimedia services (such as unified messaging, click to dial, cross-network multiparty conferencing, and seamless multimedia streaming services) has led to the convergence of networks (e.g., fixed/mobile convergence and voice/data integration) and an overall Internet/telecommunications convergence. This idea is

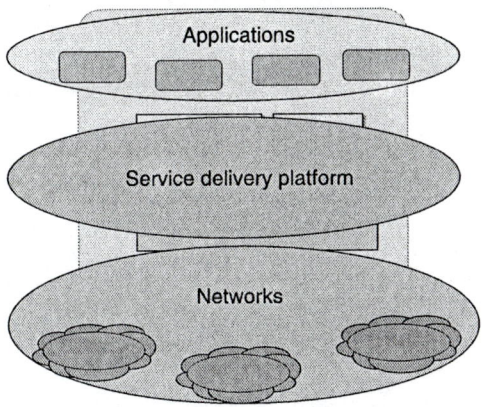

Figure 37.1   SDPs enable applications on networks.

illustrated in Figure 37.2. In the face of such convergence, the need for universal SDPs supporting integrated services has emerged. This means that SDPs should in principle enable the rapid and uniform programming and provision of seamless multimedia services on top of any network environment. Without a doubt, two of today's important trends are pivotal for SDP design — namely, the support of mobile users and the support of (mobile) multimedia data services.

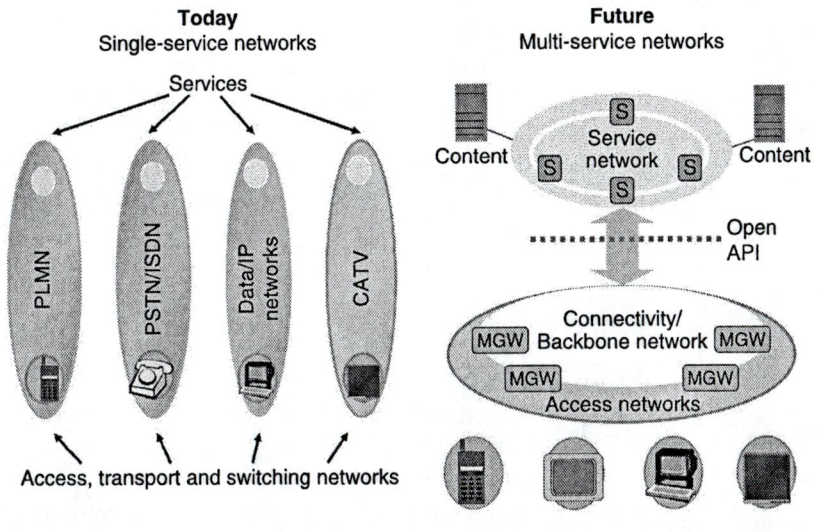

Figure 37.2   Stovepipe architecture versus converged network SDP.

In this chapter, we want to discuss the evolution of SDP concepts and technologies over the last decades. In the following section, we look briefly at the evolution of IT and middleware technologies, followed by an overview of SDP evolution from intelligent networks (INs) toward the Internet Protocol (IP) Multimedia Subsystem (IMS) defined for emerging all-IP networks. The next section then looks at the application of the remote procedure call-based intelligent network concept in the mobile domain, referred to as the Customized Logic for Mobile Enhanced Logic (CAMEL). This is followed by an introduction to the notion of telecommunications application programming interfaces based on Common Object Request Broker Architecture (CORBA™) and Java 2 Enterprise Edition (J2EE) middleware, as well as a discussion of Open Service Access (OSA)/Parlay and Java APIs for Integrated Networks (JAIN). We also examine the impact of Web service technologies on the definition of Parlay X and the Open Mobile Alliance (OMA) Open Services Environment (OSE). We then introduce the IP Multimedia Subsystem, which is today crucial to fixed/mobile convergence and emerging all-IP networks and is regarded as the ultimate SDP, spanning both fixed and Mobile IP networks. Finally, the conclusion provides an outlook on emerging policy-based networks.

Mobile Service Delivery Platforms and Impact of the IT Evolution

Service delivery platforms have had a consistent impact on the evolution of IT and have changed the face of telecommunications and the Internet due to the convergence of fixed and mobile telecommunications and related changes in business models for service provision, as well as the increasing functional complexity of services. Figure 37.3 illustrates the SDP evolution that has been driven by the development of IT, the Internet, telecommunications, and mobile communications toward enabling seamless multimedia services.

IT Evolution in a Nutshell

Information technology encompasses the computer communications, networks, and information systems that allow the exchanges of digital objects. We can also say that IT includes all forms of technology used to create, store, exchange, and use information in its various forms, such as business data, voice conversations, still images, motion pictures, and multimedia presentations. A convenient term that represents both telephony and computer technology, *information technology* is the technology that is

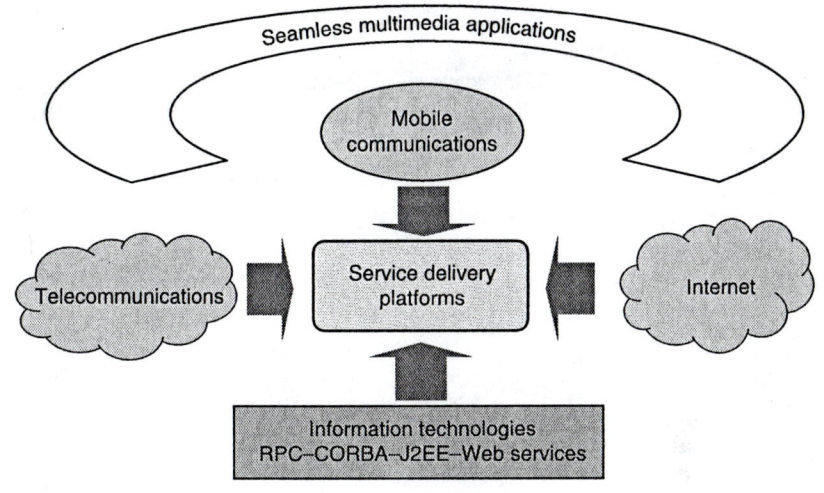

Figure 37.3 Impact of service delivery platforms.

driving what has often been called the information revolution. IT addresses matters concerned with furthering computer science and technology and the design, development, installation, and implementation of information systems and applications.

Historically, the telecommunications world was quite different from the information technology and data communications world. In the telecommunications world, whose history began in the early days of telegraphy toward the end of the 19th century, the prime focus was on providing a highly reliable telephone system for real-time voice transport. This system has evolved subsequently from manually switched phone calls (recall the switchboard at the beginning of the 20th century) to automated switching systems that handled the so-called basic call process. Value-added services were implemented inside these switches by means of dedicated, switch-specific programs (stored program-controlled switches) that interacted via a switch-specific interface with the basic call process. The heterogeneity of switches inside a network has made this a time- and cost-intensive procedure.

In the data communications world, the focus was on the interconnection of remote computers without severe real-time constraints, as mainframe computers began to be replaced by distributed computing systems in the 1960s and 1970s. No doubt the future will evolve toward a mobile-network environment that unites the traditionally separated telecommunications and Internet worlds and will be an all-IP-based one. The reasons for this are that IP-based technologies are much less expensive to deploy and maintain and most of the future information and communication services will be data driven, with content originating from the Internet. This also means that the corresponding signaling and control protocols from the Internet world will have a strong impact on the telecommunications network evolution.

Based on the invention of the remote procedure call (RPC) in the 1960s, it became possible to create programs that can talk to other remote programs without any knowledge of their (possible) distribution. In the face of emerging object-oriented programming languages such as C++ and Java and the related middleware platforms, such as CORBA and J2EE, the notion of open network application programming interfaces (APIs) emerged at the end of the last century as a natural evolution of the intelligent network concept. The principle idea, which has received global attention and acceptance, is to program services in the programming language of choice against abstract service interfaces on application servers connected via the object-oriented middleware to a network gateway, which has to map the API operations onto a specific network protocol.

With the emergence of Web service technologies defined by the World Wide Web Consortium (W3C), these APIs also have been adapted to benefit from this major IT trend. Parlay X represents a Web-services-based, simpler but functionally limited interface compared to the classic OSA/Parlay APIs. Also, OMA's Open Services Environment (OSE) and the Microsoft® Connected Service Framework (MCSF) represent similar but different initiatives in this context.

SDP Evolution at a Glance

The SDPs typically provide so-called value-added services, which extend the basic capabilities of the underlying networks, such as by providing flexible calling options (e.g., call forwarding) or special charging services (e.g., free phone) on top of the plain old telephony service (POTS). Such service capabilities include advanced (multiparty and multimedia) call control, different kinds of messaging, data session control, flexible charging, user location, and presence status, among others. Previously, telecommunication services such as call forwarding and call screening had been provided by the switching nodes through an approach known as *stored program control*. However, the heterogeneity of switching nodes, signaling protocol diversity, and lack of common standardized interfaces for value-added service provision made this an expensive approach with a very slow time to market.

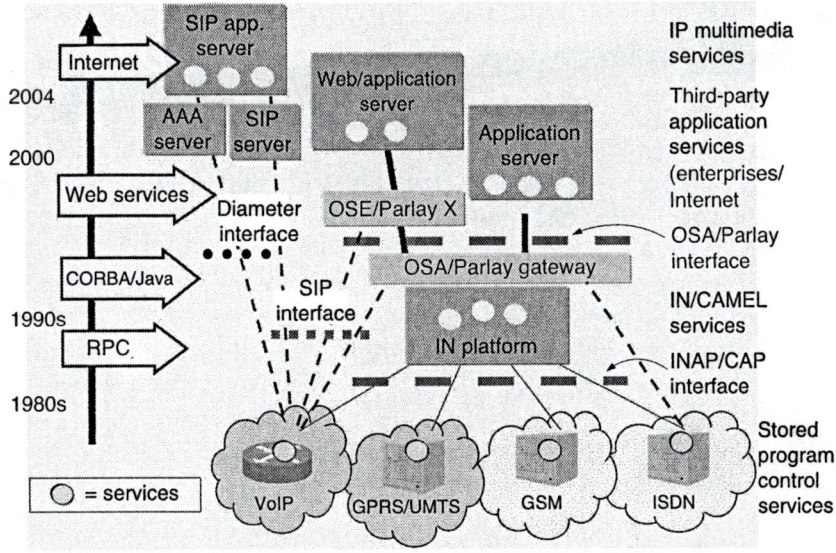

Figure 37.4   Service delivery platform evolution at a glance.

The invention of the common channel signaling systems, particularly signaling system number 7 (SS7), enabled the design of real-time remote service control architectures. Based on the RPC paradigm AT&T and later Bellcore (Telcordia) invented intelligent networks in the 1970s. Based on a standardized call model for switches and the corresponding SS7-based Intelligent Network Application Protocol (INAP), centralized, highly reliable computer systems (so-called service control points, or SCPs) have been able to control switches remotely via the SS7 network. Various global intelligent network standards defined by ITU-T, ANSI T1S1, ETSI, and Telcordia that offered increasing service capabilities were developed and implemented in the 1980s and 1990s (Figure 37.4).

The intelligent networks have changed telecommunications service design and implementation by defining reusable service components for rapidly implementing such new services as free phones, premium rates, prepayment, and virtual private networks (VPNs) in a network-independent way. Service provision times have increased significantly. In addition to exploiting INs in the mobile world, some efforts have been undertaken to make use of INs for controlling Voice-over-IP (VoIP) environments [1]. Despite the global success of INs and CAMEL, it became obvious over time that the IN programming model is limited because of its use of IT and the inherent complexity of the INAP protocol.

The Intelligent Network in the Mobile Domain: CAMEL

Based on the success of the intelligent networks in the fixed network world, the international standardization bodies adapted the intelligent network concept for the mobile world; this new concept is known as the Customized Application for Mobile Enhanced Logic (CAMEL) in Europe and the Wireless Intelligent Network (WIN) in the United States [1]. The CAMEL Application Protocol (CAP) has been defined for the implementation of IN and CAMEL environments. Four versions of CAMEL that were defined in the 1990s extended the scope of the application so all major intelligent network services could be provided within mobile networks.

Some Words on the IN Concept

In an attempt to provide value-added services more rapidly in response to ever-changing user demands in the 1960s, the remote procedure call was developed to enable remote communication and thus more economic programming of switching systems. AT&T and Telcordia originally developed the intelligent network concept. A centralized computer-based system (the service control point) remotely controls in real time the switching nodes (service switching points, or SSPs) to provide value-added services by means of a dedicated signaling protocol (INAP), on top of the basic channel signaling network (SS7). Besides INAP, the call model (CM) was a key component for modeling the behavior of a switch during call processing. The centralization of service logic and data while providing distributed service access made possible much more efficient creation, provision, and management of services. Moreover, the idea of the intelligent network contributed to the development of generic service-independent building blocks (SIBs), as shown in Figure 37.5, which have allowed the creation of a multitude of value-added services; the concept is similar to a distributed operating system that allows many applications to be executed. The combination of IT middleware and telecommunications systems greatly expanded the programmability of the telecommunications network.

The Telco Lego brick system for value-added services has also been standardized by the International Telecommunications Union (ITU-T) in the Q.1200 Recommendation Series for Intelligent Networks Capability Sets and by Telcordia and the American National Standards Institute (ANSI) within the Advanced Intelligent Network (AIN) releases [1]. It should be noted that the IN concept decoupled the service provision from the underlying network, thus allowing, in principle, the provision of IN-based services on top of different bearer networks. The IN concept was originally applied to the Public-Switched Telephone Network (PSTN), subsequently to the Integrated

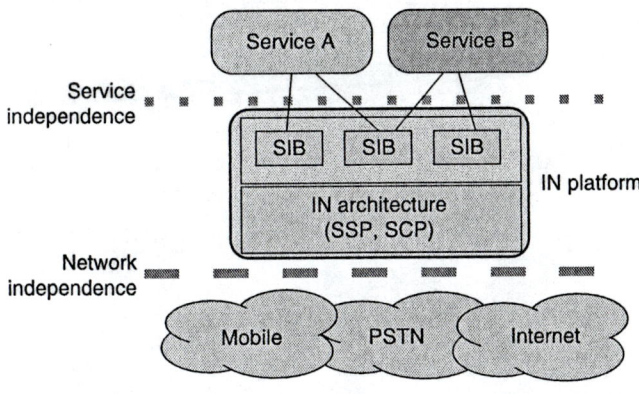

Figure 37.5   The intelligent network concept.

Services Digital Network (ISDN), and then to the Global System for Mobile Communications (GSM) under the name of WIN or CAMEL.

Today, intelligent networks are used all over the world for value-added service provision, such as universal access numbers, virtual private networks, free phones, premium call rates and messaging, and, most particularly, prepaid cards. Most IN sales have been in the form of CAMEL platforms deployed on top of GSM/GPRS networks that replace the traditional but limited service architecture based on value-added service nodes [3]. Furthermore, studies and prototypes have also proved the applicability of IN concepts in all-IP networks; therefore, intelligent networks represent the first open universal value-added service platform for several bearer networks.

CAMEL Principles and Architecture

One major reason for the success of the GSM system is the strict standardization of all service aspects; however, this has also limited innovation and competition among operators. As a result, specific service nodes have been introduced within operator networks that provide IN-like service capabilities, such as prepaid services. Because a user's service-node-based service was limited to the particular operator's network due to the proprietary interfaces, roaming users lost access to these value-added services. Thus, ETSI and later the 3rd Generation Partnership Project (3GPP) have looked for a standardized solution for value-added service provisioning for roaming users. Because of the success of intelligent networks in the fixed network world, they have been adopted for usage in the mobile domain. One important design challenge for CAMEL was to cope with a multi-vendor mobile environment; therefore, a stepwise standardization has been developed that defines so-called CAMEL phases.

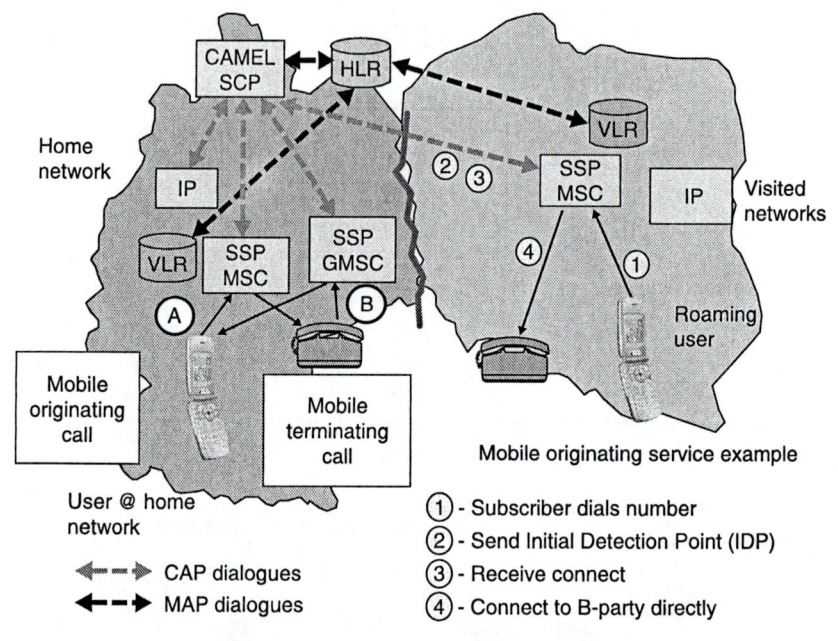

Figure 37.6   CAMEL intelligent network within mobile networks.

The general CAMEL architecture is provided in Figure 37.6. A specialized set call procedure known as the CAMEL Service Environment (CSE) in the home network provides operator-specific value-added services for mobile originating and mobile terminating services, such as prepaid cards and virtual private networks. This CSE can be accessed via a simplified IN protocol — the CAMEL Application Protocol (CAP) — from SSP-enhanced mobile-service switching centers (MSCs), called GSM service switching functions (SSFs); from the home network; and, most importantly, from partnering visited networks. In addition to CAP, CAMEL is dependent on the Mobile Application Protocol (MAP) for the dynamic provision of CAMEL subscription information within GSM subscriber profiles and location-based CAMEL services.

CAMEL Standards and Applications

The driving force for CAMEL standardization and subsequent deployment was the support of prepaid roaming and virtual private networks for postpaid roaming users. CAMEL standards mainly concentrate on the definition of the CAP based on INAP specialization and extensions to the MAP for dynamic CAMEL service provisioning. Figure 37.7 displays the major evolution steps.

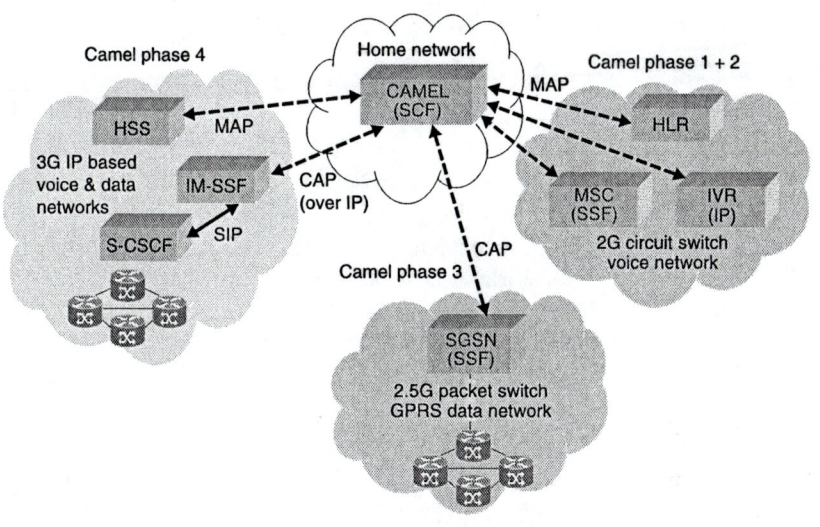

Figure 37.7 Evolution of CAMEL standards.

CAMEL Phase 1 was developed in 1996 as the first but limited standard; it defined mainly call control, location services, and some charging aspects to support a simplified prepaid service and virtual private networks. It defined a simplified IN call model and just six CAP operations. CAMEL Phase 2 was completed in 1998 and added full charging support and thus full prepaid support and user interaction capabilities. The complexity of the call model and protocol were enhanced. CAMEL Phase 3, defined in two versions in 1999 and 2000, extended the CAMEL capabilities to more sophisticated call control, data session control, extended location services, and short message service (SMS), among others; however, it required major upgrades of the infrastructure and thus substantial investments. CAMEL Phase 4, defined in 2001, provided the full IN Capability Set 2, call control, and a modular CAP protocol structure. In addition, it defined the option to use CAMEL on top of the emerging IMS.

Today, CAMEL Phase 1 and Phase 2 are primarily deployed in networks, as these solve major operator needs. CAMEL Phase 3 and Phase 4 are considered too expensive for most operators. Experience has shown that CAMEL deployment is quite costly due to high interoperability testing. Moreover, it has become obvious that both a home network and all visiting networks have to run the same CAMEL version to take advantage of CAMEL; otherwise, the deployment of a higher CAMEL phase in a home network is of very limited use.

Some Words on Wireless Intelligent Networks

Wireless intelligent networks (WINs) differ from CAMEL as they focus only on adopting intelligent networks to provide specific value-added services on top of wireless networks. Service provision of roaming users in other networks is not a goal. WIN is defined by ANSI, TIA, and T1 standards bodies in the United States; ANSI represents U.S. interests abroad in terms of technical and policy positions, and TIA and T1 create standards for wireline and wireless networks. TR45 is a wireless division of TIA, and the TR45.2 subcommittee focuses on standards for mobile and personal communications. ANSI-41 and WIN T1 develop network interconnection and interoperability standards for wireline and wireless networks, and T1S1/T1P1 subgroups develop standards and technical reports related to wireless networks and services [1]. The WIN standards follow a development process different from CAMEL. When WIN standards are conceived, they are assigned a project number and name (e.g., PN-4287 Prepaid Charging). Once adopted by TIA, the PN becomes an interim standard, such as IS-771 WIN Phase 1. After an interim standard has been published by TIA, a three-year period of revision and acceptance follows. When industry adopts the interim standard, it becomes part of ANSI 41 (e.g., IS-771) and is targeted to become part of ANSI 41-E.

Open Network Application Programming Interfaces: Parlay, OSA, JAIN

The notion of distributed broadband IN systems emerged in the 1990s; however, a substantial change of the IN system architecture has not been adopted, because, despite its advantages, the IN system has some inherent limitations. In particular, the IN platform has not provided the desired level of flexibility in service provisioning, as the service platform is still coupled with the underlying network protocols and switching equipment. As a result, full decoupling of the service level and the switching level is not possible; the programming of IN services can be quite complex and achieved only by a limited number of special telecom experts. Additionally, the business models of IN-based telecommunication networks were quite closed, which was considered a major limitation in the face of a changing value chain of multimedia services.

Because of these limitations and the ongoing convergence of telecommunications, IT, and the Internet, a new programming paradigm for telecommunications has emerged: open application programming interfaces (APIs). Driven by the need for a common multimedia service platform for converging networks and the proven commercial usability of distributed object-oriented platforms, new standards for open service platforms are emerging. The main reason is the ability to map the API to different network types (e.g., a call control API to a fixed telephony network and to a VoIP network) and thus to run the services seamlessly across different networks [4]. One way to do so is to implement an OSA/Parlay gateway on top of an IN/CAMEL platform (i.e., to map the APIs onto INAP/CAP); however, direct mappings to the ISDN User Part (ISUP) and Session Initiation Protocol (SIP) are also feasible. Another reason is the ability to provide secure connection of third-party providers and enterprise application servers to the network operator's gateway, thus allowing the operators to flexibly implement different business models for applications.

API Motivation

A primary aspect of the IN concept was to exploit the capabilities of state-of-the-art information technologies to enlarge the developer community and allow more economical implementation of services. In the 1990s, the notion of object-oriented programming, software languages such as C++ and Java, and coincidently distributed object-oriented systems such as OMG's CORBA and Sun's J2EE appeared, and the IN architecture was the subject of many R&D activities centered around the distribution of IN components and ease of service programming. In addition, due to the emerging content-based services, the business value chain grew increasingly complex, and operators had to support more complex business models. Making a network available to third-party service providers and enterprises was considered as an option to generate more market-oriented services and thus revenues. Utilizing software distribution technologies (i.e., middleware) and developing an API offering much easier programming of telecommunications services represent the main design criteria of these merging API-based SDPs.

API Principles and Architecture

Based on the pioneering work of the Telecommunications Information Networking Architecture (TINA) Consortium in the early 1990s, the Parlay Group (consisting of operators, vendors, and IT companies) was organized in 1998 to define an open-network Parlay API. This API is inherently based on object-oriented technology, and the idea is to allow (if desired by the business model) third-party application providers to make use of the network or, better yet, the value-added service interfaces. Today, however, the best way to view Parlay is to consider it as some kind of telecom-specific enterprise application integration (EAI)

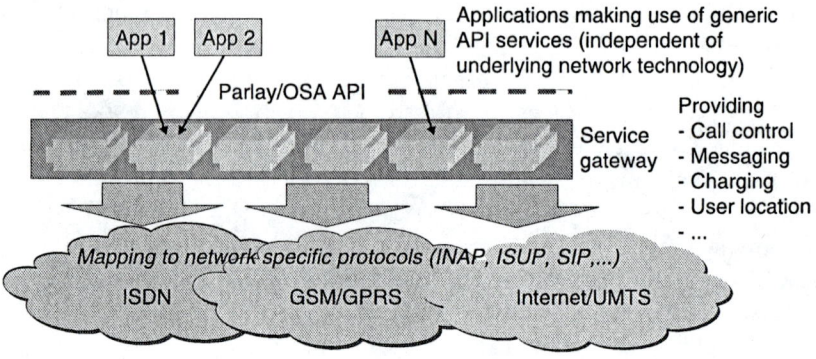

Figure 37.8 Open network APIs (OSA/Parlay).

platform technology, allowing service providers to develop value-added applications on top of a different or changing network environment, as shown in Figure 37.8. This allows a smooth network technology evolution below the developed applications.

Originally designed for use on top of IN systems (i.e., to open up IN systems for third-party developers in fixed networks), the API evolved quickly into a general API to be used on any underlying fixed, mobile, voice, or packet network. In 2001, 3GPP aligned their work on the Open Service Access (OSA) API with Parlay to further their service vision of the Virtual Home Environment (VHE), as did ETSI for their Service Provider Access to Networks (SPAN) APIs in the same year.

Today all three standards are completely aligned and thus represent probably the most accepted standard in this domain. Additionally, in the late 1990s, with the increasing commercial acceptance of the Java language, Sun Microsystems initiated development of the Java APIs for Integrated Networks (JAIN) as a set of specifications for implementing Java-based next-generation IN platforms on top of different bearer networks [2]. Recognizing the similarities in their targets, JAIN joined the Parlay, 3GPP, and ETSI groups in 2002 and developed the JAIN Service Provider Access (SPA) API. Figure 37.9 explains the evolution of open network API standards.

Examining these APIs in more detail, it is important to recognize the open/extensible nature of the API architecture. The main idea is to provide in a dedicated network node, known as the OSA/Parlay gateway and operated by the network operator, an open set of service interfaces that exhibit specific value-added service capabilities, such as call control, messaging, data session control, location, presence, and charging. Applications should be able to access these capabilities, thanks to object-

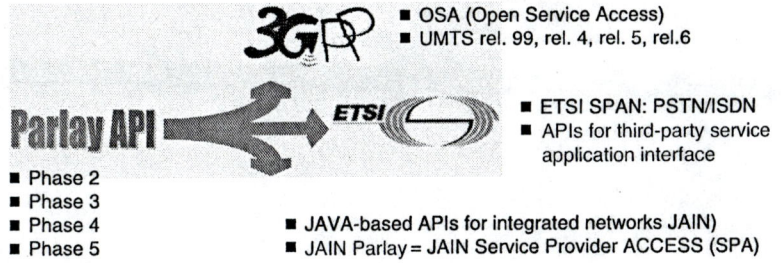

Figure 37.9   Evolution of open network API standards.

oriented technology (i.e., CORBA, C++, Java) and easy-to-use interfaces (optionally via a network), to implement value-added services. A dedicated interface, known as the framework, is in charge to register and discover new service interfaces, perform application and network authentication, monitor service-level agreements, etc.

It is important to note that the API is network independent; that is, in principle each network (the type of the network does not matter) will provide its own OSA/Parlay gateway, and one application can make use of several gateways. This means that an application can run with the same logic simultaneously on top of a fixed telephone network and a VoIP network. Today, OSA/Parlay technology is being deployed slowly all over the world. Typical applications include premium content-delivery services, location-based services, and enterprise mobile office applications. In these cases, the network operators provide the capability to send messages and charge for the service, mobile user location information, and third-party call control, respectively.

API Standards and Applications

The API can be extended functionally over time by the inclusion of new service capabilities, thus enabling some kind of telecommunications enterprise application integration (EAI). The Parlay group, ETSI, and the 3GPP OSA APIs represent the current aligned state of the art in this context. These APIs support both OMG's CORBA and Sun's J2EE [5]. In addition, as noted earlier, Sun has created a similar Java-only telecommunications architecture known as JAIN. The functional capabilities of these network APIs include multimedia call control, messaging, user interaction, charging, user status and location, presence, etc., thus enabling services such as the implementation of click to dial, mobile commerce, and content delivery.

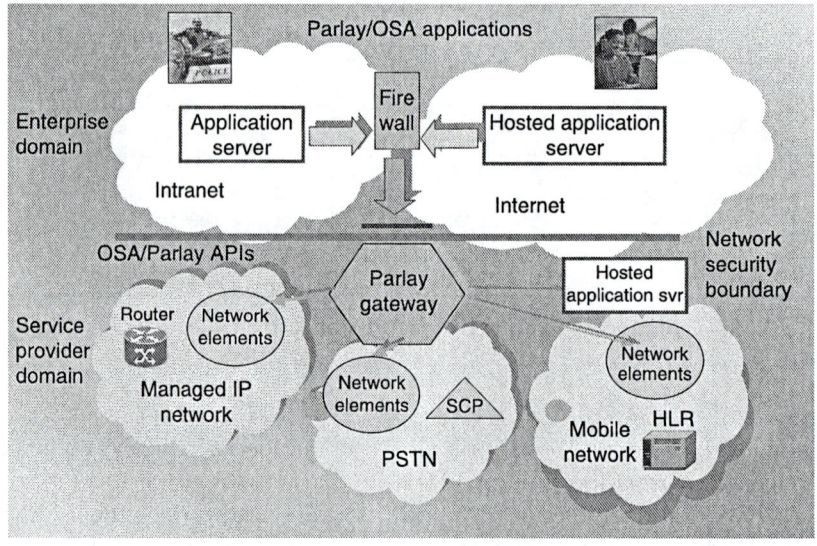

Figure 37.10 The OSA/Parlay API concept.

Parlay

The Parlay Group began in 1998 by defining the Parlay API (Figure 37.10). Originally designed as an extension of the IN within fixed networks, the API has been extended over the years as a generic API for any underlying network (i.e., fixed, mobile, IP). The Parlay Group has studied the impact of new IT on the API design and has examined the use of CORBA, Java, and Web service technologies. The history of the Parlay Group can be summarized as follows:

- *Phase 1* (only for PSTN; finished end of 1998) — The Parlay Consortium consisted of the following five companies: BT, Microsoft, Nortel Networks, Siemens, and Ulticom (DGM&S Telecom). The APIs developed include Framework, Call Control, Messaging, and User Interaction. Version 1.2 of these APIs was released in 1999.
- *Phase 2* (extended scope to wireless and IP; finished end of 1999) — Six new consortium members were added: AT&T, Cegetel, Cisco, Ericsson, IBM, and Lucent. Group completely opened in 2000.
- *Phase 3* (extended toward M-business; finished June 2001) — Alignment with 3GPP Open Service Architecture and Java APIs for Integrated Networks (JAIN).

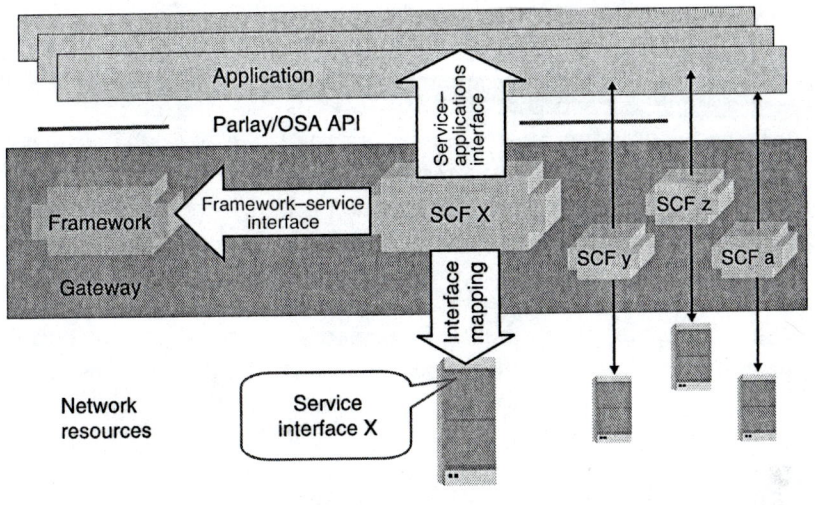

Figure 37.11 Framework versus service interfaces.

- *Phase 4* (presence, policy management; finished in September 2002) — Incorporation of Web services. Simplified interface development: Parlay X.
- *Phase 5* — new Messaging API, enhanced Parlay WS, and Parlay X2.

Today, the Parlay APIs represent state-of-the-art technology for implementing open service delivery platforms on top of various bearer networks, including all-IP networks. The Parlay API specification itself is divided into two main components: the *framework* and the *service capability features* (SCFs), as shown in Figure 37.11. The SCFs are responsible for providing the real mapping to the underlying network resources, whereas the framework logically ties everything together and makes a Parlay installation manageable. The SCFs that are bundled in one server comprise a service capability server (SCS). A minimum Parlay gateway must have a framework and at least one SCF. Because the interfaces between the SCF and the framework are also defined using the chosen middleware (CORBA) technology, it is not necessary that they reside on the same host or even that they are implemented with the same technology or programming language [2]. This means that a Parlay gateway can be a compact box but can also be designed as a heterogeneous distributed architecture. This fact is further enhanced by fault tolerance and load balancing for systems that require consistent availability.

Parlay Framework

The Parlay framework, as a core component in the interface architecture, serves as a single entry point for all applications. Besides the initial access, it also provides authentication (framework to application and *vice versa*), authorization, service discovery, and service agreement. For the SCFs, the framework provides interfaces for registration and lifecycle management. The framework interfaces include:

- *Framework Access Session API* — Contains the trust and security management components that manage the initial entry point, authentication of both framework and client applications, and granting access (authorization) to specific SCFs.
- *Framework to Application API* — Handles general events concerning the relation and provides integrity management (load management, fault management, heartbeat, OAM), service discovery, and service agreement functionality for the application.
- *Framework to Enterprise Operator API* — Provides general events but also service subscription capabilities, which include all aspects of client management, service contracts, service profiling, and even operator account management.
- *Framework to Service API* — Besides general events, this API enables services to register, to discover other SCFs, to manage their lifecycle, and to perform integrity management.

Parlay SCFs

The Parlay Group defines in its version 5.0 the following SCFs that provide a more generic and abstract interface to network resources, as well as network features and the hiding of network specific protocols and entities:

- *Call control* — Call control consists of generic call control (up to two parties), multiparty call control, multimedia call control (e.g., video calls), and conferencing call control (moderated multiparty).
- *User interaction* — One aspect is call related and allows interactive voice response (IVR)-driven applications; another part is not call related and its purpose is mainly for messaging, such as sending and receiving SMS.
- *Mobility* — This SCF allows requesting, triggering, and notification regarding user location and user status information.
- *Terminal capabilities* — This SCF includes querying terminal capabilities and features.
- *Data session control* — Data session control allows for third-party control of packet-switched connections among peers.

- *Generic messaging* — Generic messaging offers a mailbox function or message-box-like access to stored messages; it has a directory-like structure and can store and handle e-mails, SMS, multimedia messaging service (MMS), voicemails, and video mails.
- *Connectivity management* — Connectivity management handles the quality of service (QoS) aspects of connections and services.
- *Account management* — Account management allows querying, creating, and deleting balances and vouchers for accounts.
- *Charging* — Charging makes it possible for applications to charge for their services (online or offline, prepaid or postpaid).
- *Policy management* — Policy management covers the policy-driven parts of a network and allows the management of many aspects related to policy-driven computing, including domains, repositories, groups, rules, and conditions.
- *Presence and availability management* — This SCF provides management of presence and availability information of a user. For example, user A is currently online and can be reached via a certain number of addresses but the same is not true for user B. It allows an application to modify as well as watch this information.

Parlay X

A major goal of OSA/Parlay is to make networks programmable by means of state-of-the-art middleware technologies. To make the network programmable by the application providers, the API has been enhanced by the use of a new Web services paradigm that combines the eXtensible Markup Language (XML), Web Services Description Language (WSDL), and Simple Object Access Protocol (SOAP). The concept of Web services is based on the idea that, generally, the starting point for information services originates on the Internet (the Web). Furthermore, additional services could be constructed from packaging other remote services available in the Web.

Consequently, Web service technology, such as .NET™ from Microsoft, is the basic means for describing, registering, finding, and using Web services. In the telecommunications world, Parlay Web services, and the more simplified version of it called Parlay X (shown in Figure 37.12), represent the state of the art in Web services [5]. These are also considered as starting points for what the OMA is hoping to develop: mobile Web services enablers. OMA is considered today as a super standards forum, bringing together various others such as Wireless Village, the Wireless Application Protocol (WAP) forum, etc. The main target is to approve useful wireless standards for enabling rapid service delivery. Because many Parlay gateways and applications are based on CORBA middleware, the

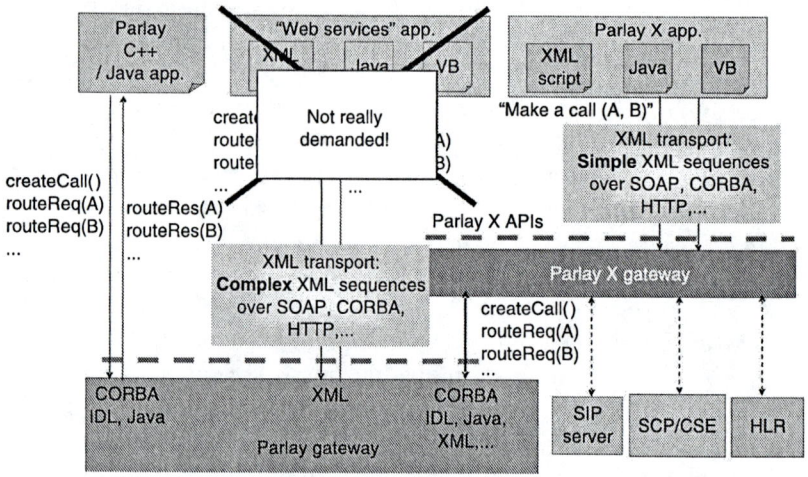

Figure 37.12 Parlay Classic API versus Parlay X APIs.

Web services specification for the classic Parlay interfaces is not really necessary now, whereas the Parlay X Web services is becoming more and more important. The Parlay X specification tries to simplify the Parlay APIs, but this simplification comes at the expense of the functionality that can be provided. It has been decided that the Parlay X specification should follow a scheme of 20 percent functionality and 80 percent simplicity.

The latest specification, Parlay X 2.0, offers the following features:

- Third-party calling
- Call notification
- Short messaging
- Multimedia messaging
- Payment
- Account management
- Terminal status
- Terminal location
- Call handling
- Audio call
- Multimedia conference
- Address list management
- Presence

It is obvious that most of these interfaces can be directly mapped to one or more Parlay interfaces, but more and more implementations today are mapped directly to related components and their corresponding interfaces or protocols (e.g., to SIP and Diameter in case of an IMS). Parlay X Web services have also been adopted by 3GPP CN TSG (2004) for inclusion in the OSA Release 6 in TS 29.199-xx-600.

Open Mobile Alliance

The OMA was originally created in 2002 by consolidating the WAP Forum and the Open Mobile Architecture initiative. Today, the OMA has also incorporated the following organizations:

- Location Interoperability Forum (LIF)
- SyncML initiative
- MMS Interoperability Group (MMS-IOP)
- Wireless Village
- Mobile Gaming Interoperability Forum (MGIF)
- Mobile Wireless Internet Forum (MWIF)

Some of the most prominent members in the OMA are Cisco, Hewlett-Packard, Sun Microsystems, and Sony-Ericsson. The OMA states in their principles: "Open Mobile Alliance aims to enable mobile subscribers to use interoperable mobile services across markets, operators, and mobile terminals by defining an open-standards-based framework to permit application and service to be built, deployed, and managed efficiently and reliably in a multi-vendor environment." To achieve these principles, the OMA has worked with various other organizations from the mobile area, such as 3GPP and 3GPP2, ETSI, ITU-T, and Parlay, among others, to leverage existing and approved standards in their architecture. The overall OMA system architecture is described and defined by the architecture working group in the OMA Service Environment (OSE) specification. It ensures that the OMA service enablers utilize IMS capabilities whenever applicable; therefore, it also describes the consistent usage of capabilities from the IP Multimedia Subsystem. Ultimately, the OSE specification will describe how the architectures from different OMA working groups and external organizations can be reworked and combined to minimize "silos" in the OMA enablers and how all the pieces of the OMA architecture will fit together.

Microsoft Connected Services Framework

This relatively new approach was officially launched by Microsoft in February 2005 for general availability and is built totally on Microsoft technologies (Figure 37.13). Although the access for services is mainly built on standardized technologies and protocols, such as Web services, XML,

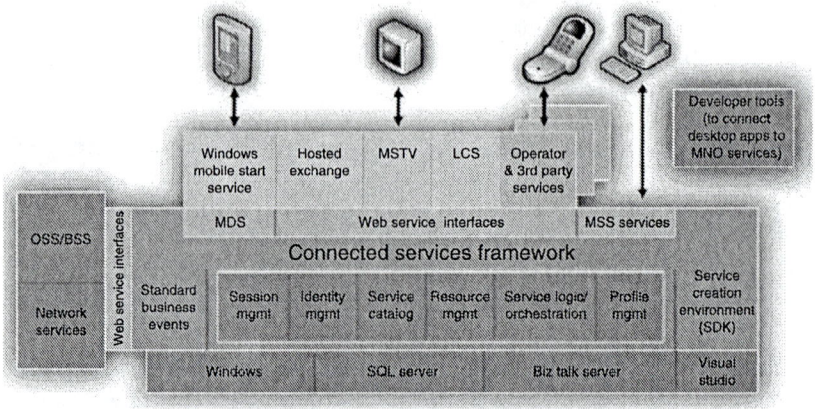

Figure 37.13   Microsoft's Connected Services Framework.

and SOAP, the framework itself is a suite of products and services from Microsoft. What kind of role the Connected Services Framework will play in the future of the SDP market is not foreseeable now, but to ignore such a large company with such a significant market share would be a mistake.

3GPP Open Service Access

The Open Service Access (OSA) was designed by the 3rd Generation Partnership Project (3GPP) to provide intelligent services in the GSM communications systems (a 2G cellular system originating in Europe) and the Universal Mobile Telecommunications System (UMTS), the European 3G cellular standard that followed on the success of GSM. Figure 37.14 depicts the OSA API and how network capabilities become programmable through the OSA/Parlay gateway. Additionally, it shows how these APIs can benefit from advances in in-house as well as externally hosted Internet service creation and hosting technology. Like JAIN, the OSA/Parlay APIs combine two service creation models: the Internet service creation model and the telecommunications service creation model [10]. Parlay and OSA are two closely related APIs, and since 2001 they have been formally merged so they are often referred to as OSA/Parlay.

Java APIs for Integrated Networks

Java APIs for Integrated Networks (JAIN) is an initiative led by Sun Microsoft' Java Community Process (JCP) to create abstractions and

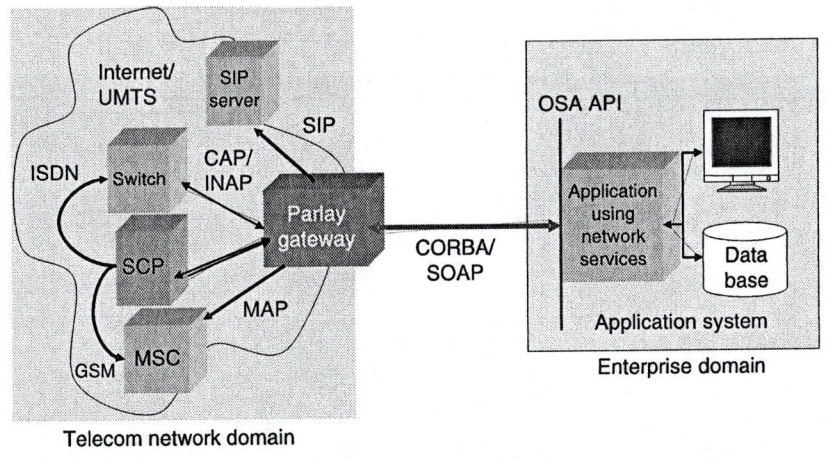

Figure 37.14   Mapping of OSA/Parlay API to real protocols.

associated Java interfaces for service creation across PSTNs, packet switching, and wireless networks. The goal of JCP is to allow the broader Java community to participate in the proposal, selection, and specification process for the Java API. The JAIN standardization effort is organized into two broad areas:

- Proposal specifications that standardize interfaces to PSTN and IP signaling protocols
- Application specifications that deal broadly with the APIs required for service creation within a Java framework

JAIN defines a service creation environment (SCE), a Service Logic Execution Environment (SLEE), a software component library, and a set of development tools, as shown in the Figure 37.15. The JAIN SCE allows the development of new service building blocks and the assembly of services from these building blocks. Services are than deployed into the SLEE. The SLEE is a set of software interfaces that support and simplify the construction of portable communications services. The primary goal of the SLEE is to ensure service portability. To achieve this, SLEE provides a specification for APIs that are required by services and that must be supported by JAIN-compliant SLEE vendors [15]. The second goal of the SLEE is to simplify the services. It does this by specifying a common set of functions or components that must be made available to application developers.

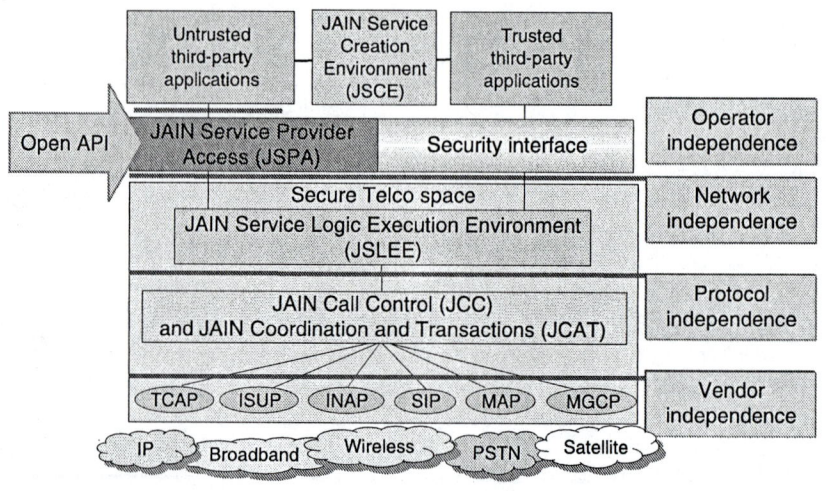

Figure 37.15 JAIN and SLEE environments.

IP Multimedia System for Emerging All-IP Networks

In light of the emerging all-IP network and development of generic Internet protocols defined by the Internet Engineering Task Force (IETF) for multimedia session control (i.e., SIP) and authentication, authorization, and accounting (AAA) (i.e., Diameter), a new overlay service architecture has been defined for fixed and Mobile IP networks: the IP Multimedia Subsystem (IMS), which was standardized by 3GPP and 3GPP2 in 2000 and is being considered for global deployment in 2006. The main idea of the IMS is to allow the flexible connection of the so-called SIP application server (AS) to the IMS core infrastructure, which, when connected via SIP and Diameter, can implement any kind of multimedia control or content services. Besides VoIP and multimedia multiparty services, Push To Talk (PTT) is considered to be an IMS killer application.

IMS Motivation

The all-IP network vision includes the use of fixed and Mobile IP networks for both data and voice/multimedia information and requires a target service control architecture. The IMS is an approach to providing an SDP architecture for IP networks that is built entirely on Internet protocols defined by the IETF and extended by request of 3GPP to support telecommunications requirements, such as security, accountability, quality of service, etc. Mobile operators today face the problem that mobile users can access the Internet and make use of Internet services, such as instant messaging, chat rooms, and content download. It is necessary, then, for the operators to define a minimum SDP architecture for providing QoS, security, and charging for IP-based services while providing maximum flexibility for the realization of value-added and content services.

The IMS provides easy and efficient ways to integrate different services, even from third parties. It enables the seamless integration of legacy services and is designed for consistent interactions with circuit-switched domains. The IMS manages event-oriented QoS policies, such as the use of VoIP and HTTP in a single session (VoIP has QoS, HTTP is best effort). The IMS also has event-oriented charging mechanisms for charging specific events at the appropriate level [6]. These characteristics put the IMS into the position of being the future technology for a comprehensive service- and application-oriented network.

IMS Principles and Architecture

The IMS is based on Internet protocols, basically those for session control (SIP), AAA (Diameter), and media transport (Real-Time Transport Protocol, or RTP), and a clear separation of data transport, session control, and application logic. Figure 37.16 displays a generic IP-based SDP architecture (note that QoS control and charging are not addressed in the figure). In the IMS architecture, SIP is used as the standard signaling protocol that establishes controls and modifies and terminates voice, video, and messaging sessions between two or more participants (Figure 37.17). The related signaling servers in the architecture are referred to as Call State Control Functions (CSCFs) and are distinguished by their specific functionalities [7].

It is important to note that an IMS-compliant end-user system has to provide the necessary IMS protocol support (namely, SIP) and the service-related media codecs for the multimedia applications in addition to the basic connectivity support (e.g., GPRS, WLAN). In general, SIP is a signaling protocol, similar to Digital Subscriber Signaling #1 (DSS1) and ISDN User Part (ISUP) used in circuit-switched networks and the Intelligent Network Application Protocol (INAP) used in the telecommunications world. However, telecom experts may argue that DSS1, ISUP, and INAP are quite different. Indeed, DSS1 and ISUP are just signaling protocols for ISDN telephony in between the end system and the switch and between switches, respectively. By contrast, INAP is a service control protocol and is used to remotely control switches for value-added service provision. SIP is used commonly for both domains, but just in the VoIP world. Most

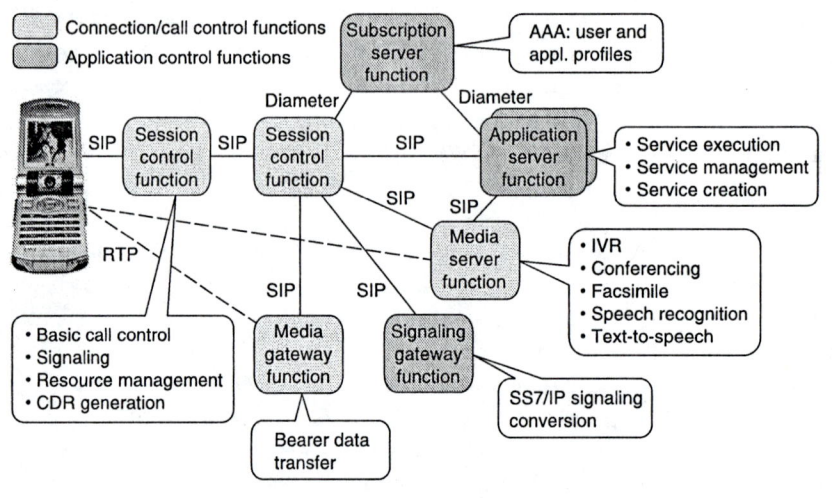

Figure 37.16   IMS high-level functionality diagram.

of all, SIP is a signaling protocol in between VoIP elements, but it can also be used to talk from a SIP server to an SIP application server (SIP AS), which *de facto* is a dedicated SIP element. This also means that SIP has some inbuilt service capabilities, allowing SIP elements to implement some IN-like services (e.g., call forwarding, call screening).

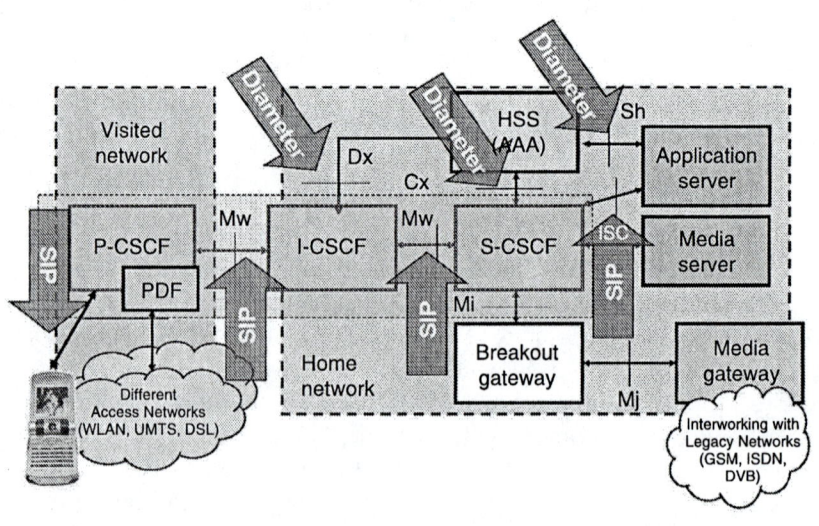

Figure 37.17   SIP and Diameter, key protocols in the IMS architecture.

The functionality related to authentication, authorization, and accounting (AAA) within the IMS is based on the IETF Diameter protocol and is implemented in the home subscriber service (HSS), the CSCFs, and various other IMS components to allow charging functionality within the IMS. To avoid developing the protocol from scratch, Diameter was based on the Remote Authentication Dial-In User Service (RADIUS), which has previously been used to provide AAA services for dial-up and terminal servers across environments. The Diameter protocol has two parts: the Diameter Base Protocol and the Diameter application. The Diameter Base Protocol is necessary for delivering Diameter data units, negotiating capabilities, and handling errors. The Diameter application defines application-specific functions and data units.

The other protocol that is important for multimedia contents is the Real-Time Transport Protocol, which provides end-to-end delivery for real-time data. It also contains end-to-end delivery services such as payload-type (codec) identification, sequence numbering, time stamping, and delivery monitoring for real-time data. RTP provides QoS monitoring using the RTP Control Protocol (RTCP), which also conveys information about media session participants [6].

IMS Components

Figure 37.18 shows the important entities of the IMS. Briefly, they can be described as follows: The first contact point within the IP multimedia core network subsystem is the Proxy Call State Control Function (P-CSCF). The P-CSCF behaves like a proxy, accepting internal requests and services and forwarding them. The next component is the Interrogating Call State Control Function (I-CSCF), which is the contact point within an operator's network for all connections for a subscriber of that network operator or a roaming subscriber currently located within that network operator's service area. The functionalities performed by the I-CSCF assign a Serving Call State Control Function (S-CSCF) to a user performing SIP registration/charging or utilizing resources [7]. The S-CSCF performs the session control services for the endpoint, and it maintains the session state as required by the network operator for support of the services. Its functionality includes user registration and interaction with services platforms for the support of services.

After the core components we have the home subscriber service (HSS), which is the equivalent of the home location register (HLR) in 2G systems; however, it is extended with two Diameter-based reference points. It is the master database of an IMS that stores IMS user profiles, including individual filtering information, user status information, and application server profiles. The other component in the IMS service

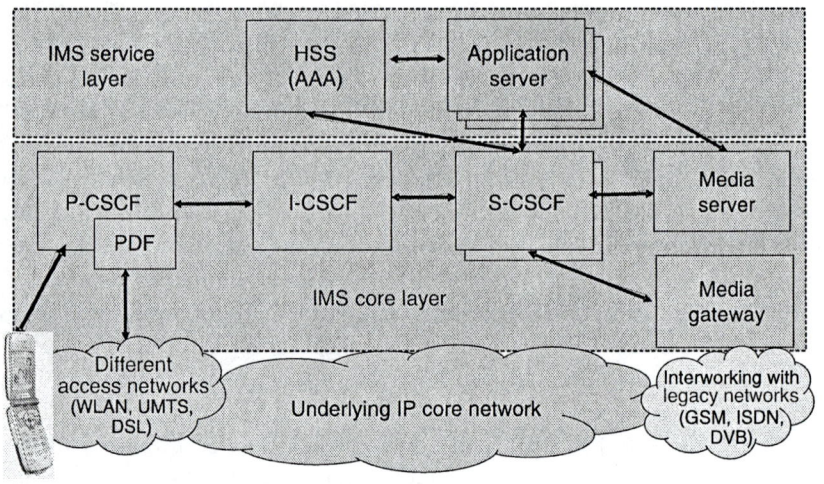

Figure 37.18 The main IMS architecture.

layer is the SIP application server, which is the service-relevant part of the IMS. Only well-defined signaling and administration interfaces (ISC and Sh) and thus SIP and Diameter protocols need to be supported. This enables developers to use almost any programming paradigm within an SIP AS, such as legacy IN servers (e.g., CAMEL support environments), OSA/Parlay servers or gateways, or any proven VoIP SIP programming paradigm, such as SIP servlets, Call Programming Language (CPL), or Common Gateway Interface (CGI) scripts. The SIP AS is triggered by the S-CSCF, which redirects certain sessions to the SIP AS based on the downloaded filter criteria or by requesting filter information from the HSS in a user-based paradigm. The SIP AS itself includes filter rules to decide which of the applications deployed on the server should be selected for handling the session. During the execution of service logic it is also possible for the SIP AS to communicate with the HSS to get additional information about a subscriber or to be notified about changes in the profile of the subscriber.

The Media Resource Function (MRF) in the IMS can be split up into the Media Resource Function Controller (MRFC) and the Media Resource Function Processor (MRFP). It provides media stream processing resources for media mixing, media announcements, media analysis, and media transcoding, as well as speech. The Border Gateway Control Function (BGCF), Media Gateway Control Function (MGCF), and Media Gateway (MG) perform bearer coordination between the RTP and IP and the bearers used in the legacy networks [8].

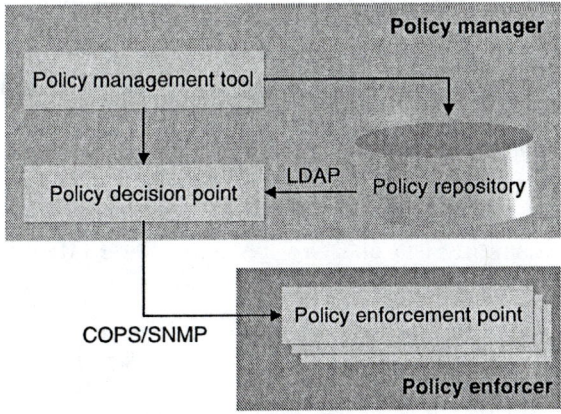

Figure 37.19   Policy-based administration framework.

IMS Standards and Applications

The IMS has been standardized since the beginning of this century within Release 5 and extended in Release 6 within 3GPP and 3GPP2 as an extension to the GPRS/packet domain network. The Release 5 standards were driven by the vision to define the IMS for providing multimedia services, such as VoIP, with the long-term goal of making obsolete the circuit-switched GSM part of 3G networks. Realistically, Release 6 has optimized the IMS to provide the envisaged IMS killer application Push To Talk (over cellular). In addition, ETSI TISPAN is looking at service infrastructures for fixed mobile convergence, and next-generation networks are extending the IMS to make it applicable on top of various access networks (i.e., WLANs) and the fixed Internet (i.e., DSL).

The policy-based QoS control architecture in the IMS is key to providing IP-based multimedia applications and services with an end-to-end QoS guarantee. The reference model of a policy-based network consists of two main elements: the policy decision point (PDP) and the policy enforcement point (PEP) [13]. The PDP weighs the policy request sent by the PEP as a result of a policy event conflicting with a corresponding set of policy rules. As a response to a policy request, the PDP either evaluates the policy rules for the request or retrieves the set of policy rules relevant for the request. The policy decision or the set of policy rules is then transported to a PEP using the policy transaction protocol, which is called the Common Open Policy Service (COPS), as shown in Figure 37.19. The PDP is the final authority that a PEP must refer to for action to be taken. This allows operators to control QoS in a user plane and exchange charging correlation information between the IMS and the GPRS network using the COPS protocol [13]. COPS is a simple query-and-response protocol that allows policy servers (PDPs) to communicate policy decisions to network devices (PEPs) and uses TCP to provide reliable exchange of messages. It supplies the means to establish and maintain a dialogue between the client and the server and to identify the requests.

The IMS architecture supports both online and offline charging capabilities. Online charging is a charging process where IMS entities, such as an application server, interact with the online charging system. The online charging system in turn interacts in real time with the user's account and controls or monitors the charges related to service usage; for example, the AS queries the online charging system prior to allowing session establishment, or it receives information about how long a user can participate in the conference. Offline charging is a charging process where charging information is mainly collected after the session and the charging system does not affect in real time the service being used [6]. In this model, a user typically receives a bill on a monthly basis that shows the chargeable items during a particular period. Due to the varying nature of the charging models, different architecture solutions are possible.

Value-Added Services in IMS

Value-added services can be provided in all IP environments on all involved SIP systems that are interacting via SIP — end systems, such as user agents (UAs); SIP servers, such as proxies or back-to-back user agents (B2BUAs); or the SIP AS. Unfortunately, today no common programming paradigm exists for SIP value-added services. Most frequently encountered is the notion of service scripts — namely, SIP servlets, CPL, and CGI scripts. Compared to IN/CAMEL and OSA/Parlay platforms, these service scripts have severe limitations with regard to functionality and developer support. However, as SIP has been selected as the universal signaling protocol in the 3GPP IMS domain, the notion of SIP application servers, which often offer a combination of CGI and servlet approaches, is emerging. OSA/Parlay, however, can also be used on top of SIP as well as IN/CAMEL, as displayed in Figure 37.20:

- CAMEL services via the CAMEL Service Environment (CSE) are intended for the support of existing IN services (provides service continuation).
- OSA services via an OSA service capability server is intended for the support of third-party application providers. The OSA SCS provides access and resource control.

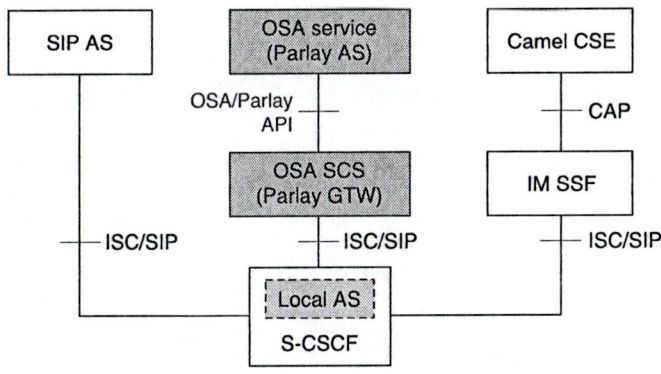

Figure 37.20 3GPP IMS application server options.

- IMS services on an SIP application server are intended for new services; a multitude of widely known APIs (CGI, CPL, SIP Servlets) is available (see Figure 37.21).
- IMS services directly on the CSCF are similar to an SIP AS. The SIP AS is colocated on the CSCF, which seems to be useful for simple services; it may be beneficial for service availability and service performance.

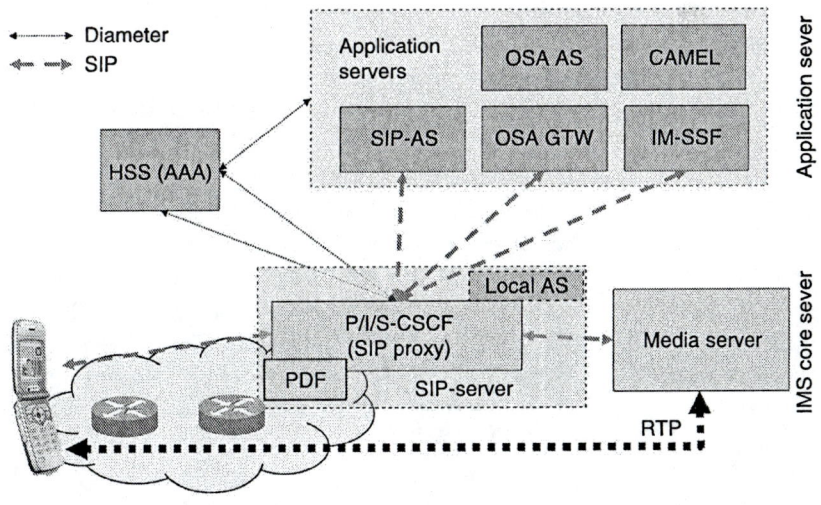

Figure 37.21 IMS application server options.

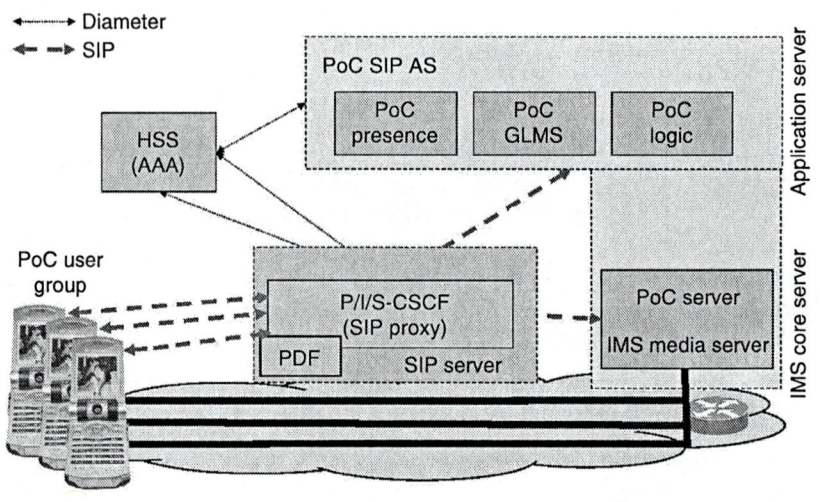

Figure 37.22 IMS-based PTT over cellular realization.

It has to be noted that the IMS itself is designed as a platform providing service enablers. This means IMS-based services are not standardized; therefore, today it is difficult to identify concrete IMS services. The reason is that the standards bodies want to provide as much freedom as possible for service ideas and service implementations. Only absolutely necessary minimum functionality for QoS, security, and charging capabilities has been standardized to enable better value-added IP-based services, compared with classic Internet services; however, IMS services are important for the introduction of IMS as an SDP. The IMS services are assumed to be addressed by the OMA. An excellent example is PTT or PTT over cellular (PoC), as shown in Figure 37.22. The key PTT functions include presence, group list management, PTT media processing, and the PTT application logic (including floor control handling) [9]. These are glued tightly together in the first vendor-specific PoC deployments, but from 2006 onward IMS-based PTT implementation will be deployed. The idea is to enable the reuse of the PTT core ingredients for other service offers, such as presence-based services.

As seen in the PTT example, new services are beneficial combinations of service capability features. Most likely upcoming services will also rely on features such as presence, group-list management, additional logic, and other features on the operator network, such as location, SMS, and MMS. It is obvious that service capability features must be reused for scalability and capital expense reasons. The remaining problems include how to manage and orchestrate services, how to create efficient services that bundle service capability features, and how to open up the network for services in a secure way. As postulated in the 3GPP specifications, the adoption of OSA/Parlay concepts and technologies can contribute a lot. OSA/Parlay already provides an industry standard that provides unified access (with a gateway character) to service capability features of the operators' networks. Even secure access by third parties can be handled by the OSA/Parlay framework. This framework may control resources but could also allow malfunctions or defective services. Assuming secure access for third parties, not only is the network opened up for a variety of service developers but also a milestone in service personalization is reached. By using rapid application development tools optimized for OSA/Parlay, every subscriber can design and specify his or her own set of services.

Summary and Outlook

In this chapter we have examined the evolution of SDPs over the last several decades. It should have become clear by now that, in the face of converging networks and emerging interconnected multiple-domain all-IP networks, the provision of SDPs is a challenging task, requiring much integration work. The IMS is today regarded as the ultimate SDP approach, as it takes advantage of the inherent use of IP-centric protocols; however, the deployment of IMS is driven by applications and the need to coordinate with legacy access networks as well as legacy SDPs, such as IN/CAMEL and OSA/Parlay. In this regard, Figure 37.22 illustrates that the adoption of OSA/Parlay as a unifying service framework on top of both CAMEL and IMS can unify service provision across different and converging networks.

Figure 37.23 and Figure 37.24 extend the above in regard to the ability of OSA/Parlay to flexibly support Fixed Mobile Convergence (FMC) network operators for enterprise application integration (EAI) purposes as well as the optional and flexible support of services originating from third parties and enterprises. The latter is an important aspect, as it is clear that there will be no future killer applications for FMC and next-generation networks (NGNs); rather, a myriad of customized applications will be provided to specific user communities. In general, policies represent a technique to describe by means of scripts the requirements and desired behaviors of a network or network elements in a protocol-independent way. This requires abstracting (similar to an API approach) the specific details of the underlying network components, which may also include service platforms.

The Internet is based on an open best-effort business model, and the overall Internet comprises an open set of subnetworks; that is, various Internet Service Providers (ISPs) have to cooperate with each other. In this multiple-domain environment, end-to-end service provision, particularly augmented with QoS across the different domains, is a challenging

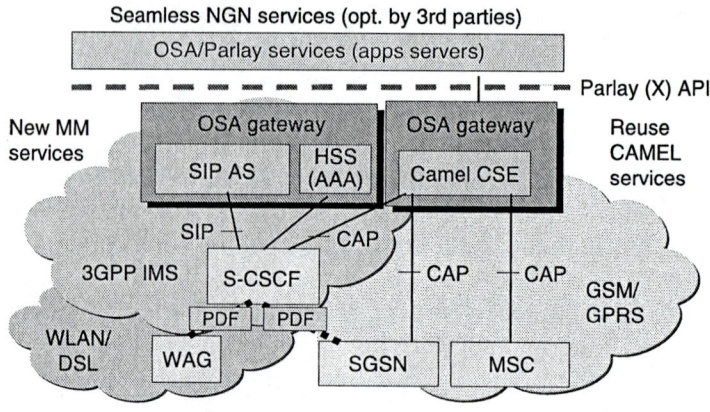

Figure 37.23 Summary of SDP evolution linked to network evolution.

task. Because various networks rely on different technologies and protocols, the provision of a service (e.g., a VPN) must be accomplished in a
network-protocol-independent way. This is what policy-based management and control have been designed for.

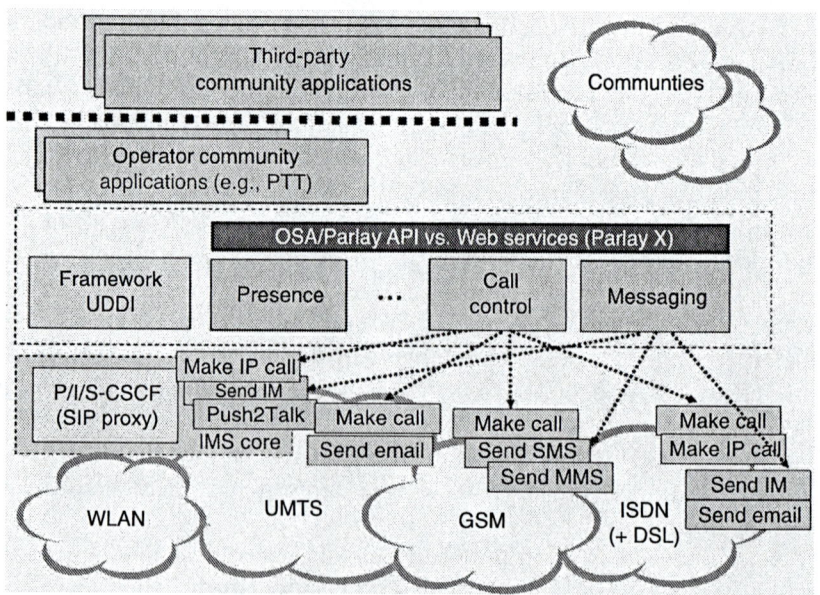

**Figure 37.24 OSA/Parlay on top of IMS and CAMEL enabling flexible service
implementations.**

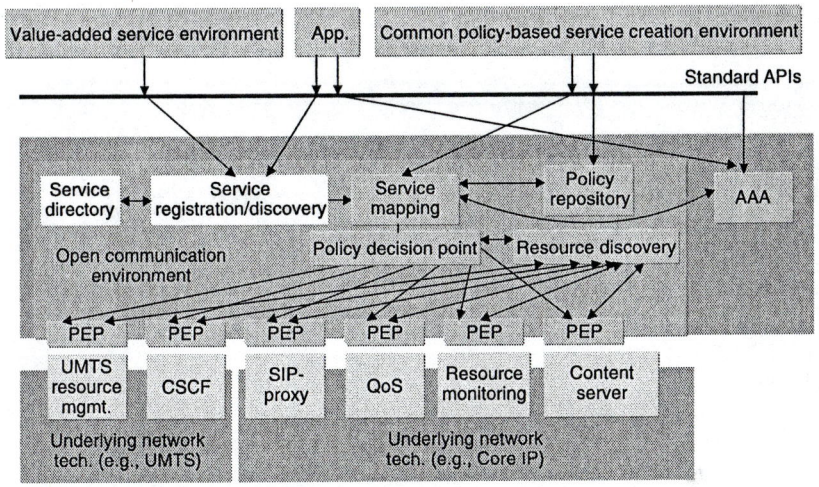

Figure 37.25 Policy-based network management for all-IP network environments.

Today, on the Internet, policy-based service provision is considered the major enabling technology for QoS-enabled multimedia service provision in all IP networks, as shown in Figure 37.25. Policies are a means of programming for programmable networks that provide the necessary abstraction from the underlying network technologies. Note that this is similar to the early intelligent network or the recent open API approach in the telecommunications world. Future networks will be all-IP networks, so policies represent a strong starting point for the development of future value-added services in all IP networks; that is, it can be assumed that policies can also be used to program value-added services on top of SIP servers and other network elements [11]. However, a lot of research still has to be performed, as this young technology is still in its infancy but is very promising. It is important to recognize that the IMS and OSA/Parlay, as well as associated initiatives, such as OMA OSE and Microsoft's CSF, are far from complete, and much R&D work must be performed in the coming years. Corresponding open SDP infrastructures are required to support the prototyping and validation of new SDP concepts and multimedia applications. The FOKUS NGN Testbed, featuring the Open OSA/Parlay playground and the Open IMS playground, represents such a testbed infrastructure.

Acknowledgments

We are very thankful to all members of NGNI Competence Centre of 3Gb Testbed at Fokus Fraunhofer Berlin for their valuable suggestions to improve the draft of this text. We are especially thankful to Simon, Karsten, and Fabricio for providing information about Parlay, open issues, and QoS in IMS.

References

[1] Magedanz, T. and Popescu-Zeletin, R., *Intelligent Networks: Basic Technology, Standards and Evolution*, International Thomson Computer Press, London, 1996.

[2] Venieris, I., Zizza, F., and Magedanz, T., *Object-Oriented Software Technologies in Telecommunications: From Theory to Practice*, Wiley, London, 2000.

[3] Glitho, R. and Magedanz, T., Intelligent networks in the new millennium, *IEEE Commun. Mag.*, 38(6), 82–84, 2000.

[4] Magedanz, T. and Smirnov, M., Eds., Voice/data integration: a snapshot of intelligent networks and Internet convergence, *Computer Commun. J.*, 35(5), 503–505, 2001.

[5] Jain, R., *Programming Converged Networks: Call Control in Java, XML, and Parlay/OSA*, John Wiley & Sons, New York, 2004.

[6] Poikselka, M. et al., *The IMS: IP Multimedia Concepts and Services in the Mobile Domain*, John Wiley & Sons, New York, 2004.

[7] Knüttel, K. and Magedanz, T., IP multimedia subsystem: a system description for a comprehensive service and application oriented network architecture, in *Proc. of the 2nd IASTED Int. Conf. on Communication and Computer Networks (CCN 2004)*, Cambridge, MA, November 8–10, 2004, pp. 67–72.

[8] Knüttel, K., Magedanz, T., and Witszek, D., The IMS Playground@Fokus: an open testbed for next generation network multimedia services, in *Proc. of the First Int. IFIP Conf. on Testbeds and Research Infrastructures for the Development of Networks and Communities (TridentCom 2005)*, Trento, Italy, February 23–25, 2005, pp. 2–11.

[9] Blum, N. and Magedanz, T., Push-to-video as a platform for NGN services, in *Proc. of the 11th European Wireless 2005: Next Generation Wireless and Mobile Communications and Services*, Nicosia, Cyprus, April 10–13, 2005.

[10] Klaus, J.M. and Magedanz, T., Parlay PAM in 3GPP's IP-multimedia subsystem, in *Kommunikation in Verteilten Systemen (KiVS)*, Müller, P., Gotzheim, R., and Schmitt, J.B., Eds., Technische Universität Kaiserslautern, Germany, 2005, pp. 73–80 (http://www.icsy.de/~kiv05/termine/index.html).

[11] Magedanz, T. et al., Self-adaptive service provisioning framework for 3G+/4G mobile applications, *IEEE Wireless Commun. Mag.*, 11(5), 48–57, 2004 (special issue on applications and services for the B3G/4G era).

[12] Cortese, G. et al., CADENUS: creation and deployment of end-user services in premium IP networks, *IEEE Commun. Mag.*, 41(1), 54–69, 2003.

[13] Salsano, S. et al., QoS control by means of COPS to support SIP-based applications, *IEEE Network*, 16(2), 27–33, 2002.

[14] Camarillo, G. and Garcia-Martin, M.A., *The 3G IP Multimedia Subsystem: Merging the Internet and the Cellular Worlds*, John Wiley & Sons, New York, 2004.

[15] Jain, R., Bakker, J.-L., and Anjum, F., *Programming Converged Networks*, John Wiley & Sons, New York, 2005.

Related Web Links

http://www.fokus.fraunhofer.de/ims (FOKUS Open IMS playground home page)
http://www.3gpp.org (3rd Generation Partnership Project)
http://www.3gpp.org/TSG/CN.htm
ftp://ftp.3gpp.org/TSG_CN/WG2_camel
http://www.3gpp.org/ftp/tsg_cn/WG5_osa/
http://www.packetcom.org
http://www.ietf.org
http://portal.etsi.org/tispan/TISPAN_ToR.asp
http://www.protocols.com/pbook/ss7.htm
http://www.parlay.org
http://www.parlay.org/docs/2003_06_01_Parlay_for_IEC_Wireless.pdf
http://www.parlay.org/specs/index.asp
http://www.parlay.org/specs/library/ParlayX-WhitePaper-1.0.pdf
http://www.parlay.org/specs/Parlay_X_Web_Services_Specification_V1_0_1r2.zip
http://www.openmobilealliance.org
http://www.openmobilealliance.org/tech/wg_committees/arc.html
http://www.openmobilealliance.org/release_program/docs/CopyrightClick.
    asp?pck=RD&file=OMA-Service-Environment-V1_0-20040907-A.pdf
http://www.microsoft.com/serviceproviders/solutions/csf.mspx
http://www.microsoft.com/serviceproviders/solutions/csf_resources.mspx
http://download.microsoft.com/download/a/8/e/a8e610d8-fdd9-4529-b5b6-1a37
    b342bf96/CSF_Diagram.ppt

Chapter 38

Mobile Middleware and Context for Service Composition

Soraya Kouadri Mostéfaoui, Zakaria Maamar, and Nanjangud C. Narendra

CONTENTS

Introduction

The growth of Internet technologies is having a tremendous impact on the way businesses interact with their peers, customers, and sometimes competitors. To remain competitive, businesses need to take advantage of the information revolution that the Internet and the Web have both brought about. Most businesses are currently adopting Web-based solutions for their applications, aiming for more automation, efficient business processes, and worldwide visibility. Web services are among the technologies that help businesses be more Web oriented [20]. Web services can be defined as accessible software components that other applications and humans can discover and trigger. A Web service is associated with three properties [2]: (1) independent as much as possible from specific platforms and computing paradigms, (2) developed particularly for interorganizational situations, and (3) easily composable so that the development of complex adapters is not required.

Parallel to the new role of the Internet as a vehicle for delivering Web services, major progress in wireless and mobile technologies has been witnessed. Users are adopting new practices such as surfing the Web from mobile devices. Because users are extensively relying on such devices to conduct their day-to-day operations, both enacting Web services from mobile devices and downloading these Web services from their hosting sites to mobile devices for execution constitute worthwhile research avenues to pursue [18,41]. M-services (M for "mobile") denote the Web services that are intended for deployment in a wireless configuration [21].

Composing services (whether Web services or M-services) rather than accessing a single service offers greater benefits to users. Composition addresses the situation of a client request that cannot be satisfied by any available service, and a composite service obtained by combining a set of available services might be used for fulfilling the request [9]. Discovering and selecting the component services according to user requirements, inserting the selected component services into a composite service, triggering the composite service for execution, and finally monitoring the execution of the composite service are among the operations that users will be responsible for. In addition, the unique characteristics and challenges of mobile computing result in a pressing need of revisiting the fundamental design and development concepts of applications [25].

Because user expectations and requirements constantly change, it is deemed appropriate to include their preferences in the composition of services. Indeed, some users while on the move would like to receive information depending on their current locations. This simple example sheds light on the importance of making applications adjustable. This adjustability depends on features of the environment in which service provisioning is expected to happen. Samples of these features would include those that are about the user (e.g., stationary user, mobile user), computing resources (e.g., fixed device, handheld device), time of day (e.g., afternoon, evening), and physical location (e.g., meeting room, shopping center). Sensing, collecting, assessing, and refining the features of a situation permit definition of its context. Prior to integrating context into service composition and provisioning, various issues must be addressed [32]: How is context structured? How does a service bind to context? Where is context stored? How frequently does a service consult context? How are changes detected and assessed for context update purposes? What is the added load for a service to take context into account?

Applications plunged into a mobile context pose multiple challenges, including how to locate and deliver up-to-date information to users while they are on the move, how to guarantee a reliable delivery despite risks of network disconnections, how to recover from disconnections with less impact on the business process under execution, and how to secure delivered information broadcast over the air. These challenges place an additional burden on application developers. Developers are put on the front line of satisfying businesses' and service providers' promises to deliver Internet content to mobile users anywhere and anytime. To reduce the complexity of some of these technical challenges, mobile middleware technologies must be developed. Abstracting communication details, supporting communications in a transparent way, and recovering from potential crashes are some of the objectives that such middleware must meet. As a result, application developers can now focus on the underlying logic of the functionalities to offer to end users, instead of low-level technical details.

The field of mobile middleware-based development encompasses the convergence of high-speed wireless networks and personal mobile devices. The aim of this development is to provide the ability to compute, communicate, interact, and collaborate anywhere and anytime. Wireless technologies for communication are the link between mobile clients and other system components. It is expected that mobile middleware will trigger the development of a new generation of applications that can be used in different domains and offered to different categories of people. These applications are multiple, including job dispatch based on the location of employees and sending inventory requests to a supply-chain partner using mobile devices.

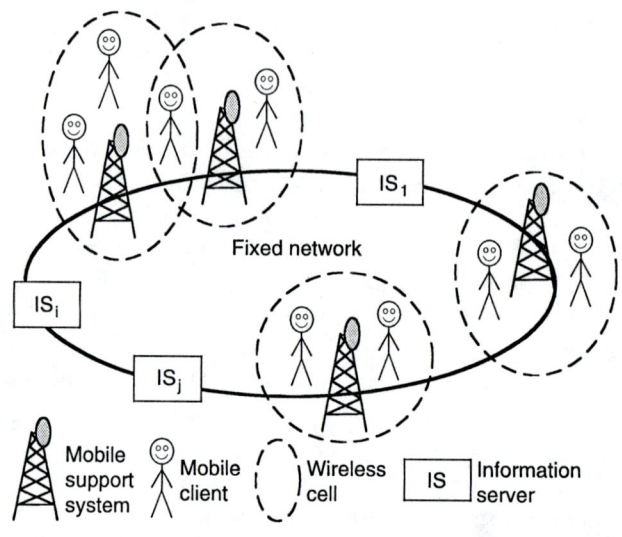

Figure 38.1   Mobile computing model.

After discussing the relevancy of middleware technologies and context for service composition, we then present applications that are based on mobile middleware. We also discuss the principles behind service composition and the value of adding context to service composition. Before drawing our conclusions and highlighting future directions, we report on some experiences of integrating mobile middleware into service-oriented applications.

Mobile Middleware-Based Applications

Mobile Computing Model

The general mobile computing model in a wireless configuration includes two distinct sets of entities (Figure 38.1): mobile clients and fixed hosts. Some of the fixed hosts, called *mobile support stations* (MSSs), are enhanced with wireless interfaces. An MSS can communicate with mobile clients within its radio coverage area, called a *wireless cell*. A mobile client can communicate with a fixed host/server via an MSS over a wireless channel. The wireless channel is logically separated into two subchannels: uplink and downlink. The uplink channel is used by mobile clients to submit queries to the server via an MSS, and the downlink channel is used by MSSs to disseminate information or forward the responses from servers to specific mobile clients. Each cell has an identifier for identification purposes that is periodically broadcast to all mobile clients residing in a corresponding cell.

From an operation perspective, users expect to be provided with information through wireless communication networks. Information is to be made available while considering the following aspects: terrain topography, propagation techniques, and security. A security problem that is inherent to all wireless communication networks is third parties being able to easily capture radio signals in the air; thus, appropriate data protection and privacy safeguards must be ensured. With regard to the network, instances of failure and means of recovery must also be taken into consideration. Some of the requirements that the mobile computing model has to satisfy are listed below:

■ The *information availability requirement* addresses the need for a user to have uninterrupted and secure access to information on the network. Aspects to consider are survivability and fault tolerance, ability to recover from security breaches and failures, network design for fault tolerance, and design of protocols for automatic reconfiguration of information flow after failure or security breach.
■ The *network survivability requirement* addresses the need to maintain the "aliveness" of the communications network in spite of potential failures. Aspects to consider are understanding system functionality in case of failures, minimizing the impact of failures on users, and providing the means to overcome failures.
■ The *information security requirement* addresses the importance of providing reliable and unaltered information. Aspects to consider are confidentiality to protect information from unauthorized disclosure and integrity to protect information from unauthorized modification, ensuring that information is accurate, complete, and reliable.
■ The *network security requirement* addresses information security through network security. Aspects to consider are confidentiality, sender authentication, access control, and identification.

Additional requirements exist. The increasing reliance and growth in information-based wireless services impose three requirements — availability, scalability, and cost efficiency — on the offered services. Availability means that users can count on accessing any wireless service from anywhere, anytime, regardless of the site, network load, or device type. Availability also means that the site provides services that meet some measures of quality such as a short, acceptable, and predictable response time. Scalability means that service providers should be able to serve a fast-growing number of customers with minimal performance degradation. Finally, cost effectiveness means that the quality of wireless services (e.g., availability, response time) should be achieved with adequate expenditures in IT infrastructure and personnel.

Technical Challenges and Role of Context

Schmidt and Beigl [30] suggest that context is more than the current location of a user. Context should consist of the people around the user, the situation (e.g., in a meeting, making a phone call), the environment (e.g., location, temperature, time), and the user's physical condition (e.g., pulse, body temperature). Additional information about the context can help a computing device to act and react more promptly and efficiently.

Occasionally, users will be stymied by a lack of appropriate applications on their mobile devices (e.g., an application that converts a drawing file into a format that the mobile device can display). To avoid such inconveniences, users should be able to search for additional functions when necessary and either invoke them remotely or download them to their mobile devices for local performance. From a user's perspective, it is important to make sure that all of these operations happen in a transparent way. From a developer's perspective, it is important that the necessary tools and techniques exist and can be integrated in an efficient way. A developer should primarily focus on the business logic that implements the offered services to users, rather than on low-level technical details.

Despite the multiple opportunities that mobile computing offers, especially to those who are on the move, various obstacles still hinder the expansion of this model. For example, mobile devices are still limited by their battery power, and current technologies are meant to be used for situations with a permanent and reliable communications infrastructure. To optimize service provisioning in mobile environments, several important issues must be addressed, such as [22]:

- *Handling disconnections during service execution* — In a mobile scenario, disconnections may be frequent due to the lack of coverage areas or devices changing location. As a result, minimizing device disconnection is critical to service execution success.
- *Context-sensitive service deployment* — In addition to such criteria as monetary cost and time of response, service deployment should consider the locations of users and capabilities of computing resources on which the services will operate. Locations and capabilities must be assessed before service deployment.

The traditional usage of mobile computing devices has revolved around merely using these devices to make and answer phone calls and short text messages. The increasing power and versatility of these devices are beginning to enable implementing complex workflow-like scenarios comprising service composition [25]. In the following, a simple yet realistic usage scenario related to vacation planning is presented, aimed at explaining the role of context in service composition.

Melissa is a tourist who is visiting Dubai. After checking-in at the hotel, Melissa browses some of the Web sites that Dubai tourism authorities recommend in their brochures. The top-ranked Web site is built on Web services technology and offers various Web services that can be composed in different ways. Melissa visits this Web site and chooses sightseeing and shopping services. Melissa's plans are to visit outdoor places in the morning and go shopping in the afternoon. The first part of the plan is subject to weather forecasts, as she will not consider any outdoor activities if the weather is too hot. Initially, Melissa is prompted to select some outdoor locations to visit and indicate pick-up/drop-off times for her sightseeing and shopping.

For the first activity in Melissa's plan, the sightseeing service checks with weather service forecasts for the next five days. Because unusually hot weather is not predicted, scheduling the places she wants to visit begins by ensuring that these places are open to the public on these days, and transportation and guides are arranged. The logistics of Melissa's transportation is assigned to a transportation Web service. If a hot-weather warning does arise, then the sightseeing Web service would suggest other places (e.g., museums) offering indoor activities. Similar considerations apply to Melissa's shopping activity and consist of checking out current promotions in the malls that Melissa has selected. It should be noted that the transportation Web service coordinates the beginning time for shopping with the ending time for sightseeing.

The day after her arrival, Melissa is driven to an historical site. When she is caught in an unexpected traffic jam, her PDA compares her current location to the location where she is supposed to be (i.e., the historical site) and notices that she is not where she is supposed to be at that time. Melissa's PDA jumps into action by sending a note to the sightseeing and transportation services so corrective measures can be taken, such as informing the guide about the delay. Figure 38.2 presents a rough representation of the services participating in this scenario. This chronology yields insight into the multiple challenges that a contextual service composition in a mobile configuration faces, including: How is the context related to the services delivered? What information applies to context? How can middleware ease the development of context-aware applications? How can we assess the context for adaptability needs? Is the location of a person sufficient for tracking the execution of the services the person has selected? How much does context of Web services contribute to this tracking?

Managing contextual information becomes crucial in mobile computing scenarios. Context is the information that characterizes the interactions between an entity and its external environment [13]. In a mobile middleware-based configuration, contextual information resides at three levels [5,19]:

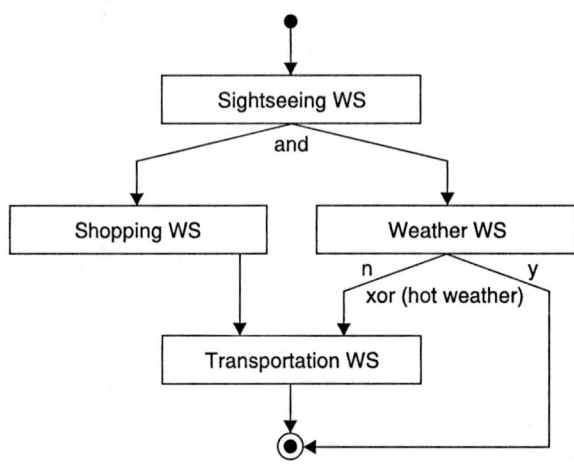

Figure 38.2   Chronology of Web services in Melissa's scenario.

- The *environmental level* enables defining the overall environmental context. Some examples from Melissa's scenario would be device types (mobile, PDA), locations (museum, shopping mall), and weather conditions (raining, sunny day).
- The *service level* models and manages the context surrounding individual services offered over mobile devices. Some examples from Melissa's scenario are performance requirements (maximum execution time per Web service) and architectural requirements (point-to-point messaging, publish–subscribe messaging).
- The *resource level* presents the context of the resources on which the services are to operate. Some examples from Melissa's scenario are screen size of a mobile device, memory capacity, and processing speed.

Principles of Service Composition

Definitions

Yang and Papazoglou [40] suggest that Web-services-based applications developed in terms of Web Services Description Language (WSDL) alone might be isolated and opaque and, moreover, cannot be easily interlinked to express the business semantics of Web services. The authors argued that, to break this isolation and opacity, Web services should be interconnected. Thus, a composition might be defined as an approach that connects Web services together in order to construct composite services. The connection of Web services implements a business logic that depends on the application domain and control flow of the business case for which the composite service is being devised. Examples of business cases are various and include travel planning and journal paper review. Skogan et al. [35] reported that the objective of Web services composition is to provide a method for creating executable compositions. The basic idea is that the composition can be expressed in a language in the form of a model that can be run by an execution engine.

A fine-grained decomposition of the process of composing services reported in Shaparau [31] has five stages:

- *Definition* — Corresponds to specification of the activities to perform, requirements to satisfy, and behaviors to expect; each service is modeled regardless of other peers
- *Scheduling* — Determines how and when services should be run
- *Construction* — Relies on previous steps to develop an executable scheme for service composition
- *Execution* — Converts the aforementioned executable scheme into structures of specific programming languages, in preparation for running the services
- *Monitoring* — Oversees the execution of the services for possible adjustment

Taxonomy

The following text provides an overview of approaches for developing composite services. We mainly rely on the works of Chakraborty and Joshi [11] and Yang and Papazoglou [40].

Proactive Composition Versus Reactive Composition

Proactive composition is an offline process that gathers in advance available component services to include in a composite service; thus, the composite services are precompiled and ready to be executed upon the users' requests. In proactive composition, the component services are usually stable and may possibly be running on resource-rich platforms. Two types of proactive composition can be identified [40]. In the semi-fixed composition, the entire service composition is specified statically, but the actual service bindings happen only at run time. When a composite service is invoked, the actual composition specification is generated on the basis of a matching between the constituent services that are specified in the composition and other available services. In the fixed composition, all the constituents of the composite service are synchronized in a fixed manner. The composition structure and the component services are statically bound. Reactive composition, also called *explorative composition* [40], is the process of creating composite services on the fly. A composite service is devised on a per-request basis from users. Because of the on-the-fly property, a component manager is required to guarantee identification and collaboration of the component services.

Mandatory Composite Service Versus Optional Composite Service

A mandatory composite service corresponds to the compulsory participation of all the component services in the execution process. Because it is expected that the component services will be spread across the network, the reliability of the execution process of each component service affects the reliability of the entire composite service. An optional composite service does not necessarily involve all the component services. Certain component services can be skipped during execution due to the possibility of substitution or for nonavailability reasons.

Requirements

Milanovic and Malek [26] proposed that a composition approach must satisfy several requirements, such as connectivity, nonfunctional quality of service (QoS), correctness, and scalability. Reliable connectivity is useful in determining the services to compose and the input and output messages to exchange. Nonfunctional QoS, such as timeliness, security, and dependability, ensure that the composition produces the expected results. Finally, scalability allows complicated business transactions to scale with the number of composed services.

More recently, another set of service composition requirements was identified [35]. A pattern-oriented design approach was adopted for the identification of the minimal requirements associated with service composition. The result of the approach is a set of patterns, many of which were already described in van der Aalst [38]: sequence, parallel, split, synchronization, exclusive choice, and simple merge. The two new patterns identified are discriminator and selector. Yet another set of requirements for service composition has been proposed by Esfandiari and Tosic [14]. The proposed requirements are split into the following groups: service discovery, service selection, composition, verification, and hot-swapping support.

Composition Approaches in a Non-Mobile Configuration

Many modeling languages and techniques have been proposed to model service composition. The first attempts toward a composition language were the IBM Web Services Flow Language (WSFL) and BEA System's Web Service Choreography Interface (WSCI). These were followed by the Business Process Execution Language (BPEL) in an attempt to combine these languages using Microsoft's XLANG. In the following, we survey the most significant proposals of service composition languages and models that are used in a non-mobile configurations.

- *Workflows* — Workflow methods are used primarily when the process model underlying the composition is already defined. eFlow is a system that adopts workflows [10]. In eFlow, composite services are represented as process schemas consisting of basic services and are modeled by an execution graph.
- *Model-driven approaches* — The main feature of model-driven approaches is the separation of composition logic from composition specifications [31]. Unified Modeling Language (UML) is among the languages that permit modeling composition logic. Skogan et al. [35] suggested a UML-based composition approach that aims at forming UML models to be automatically converted into composition specifications.
- *Artificial intelligence (AI) planning* — Given a user's objective and a set of available services, a planner would find a collection of services that enables reaching the objective. In Bouguettaya et al. [6], an AI planning approach was used to generate composite services from high-level declarative descriptions. When the composability of services has been verified, the composition specification uses the Composite Service Specification Language (CSSL). Another work that has adopted AI planning and particularly the hierarchical task network planning is reported in Sirin et al. [33], but this approach will not be further detailed here.
- *Web components* — Yang and Papazoglou [39] proposed a Web-component-based approach to service composition that includes composition planning, specification, implementation, and execution. The use of Web components is backed by the concepts of reuse, specialization, and extension. A Web component packages together elementary or complex services and presents their interfaces and operations in a consistent and uniform manner in the form of class definitions.

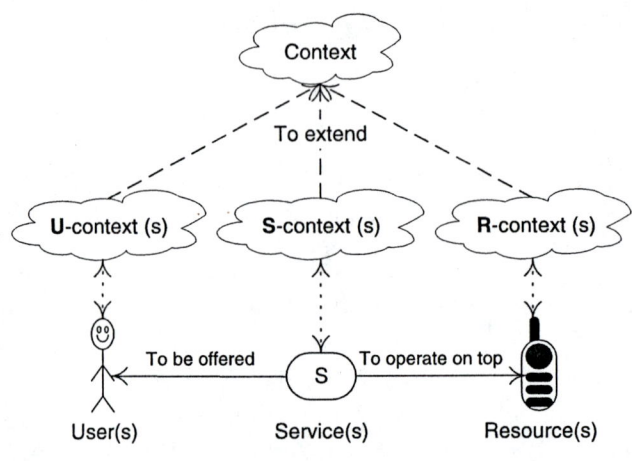

Figure 38.3 Use of context in a configuration of users, services, and resources.

Using Context in Service Composition

In Maamar et al. [23], the authors argue that, despite the tremendous interest in service composition and provisioning, very little has been accomplished to date regarding their context-based provisioning. Several obstacles still hinder this provisioning, including the fact that current services act as passive rather than active components that can be embedded with context-awareness mechanisms; existing approaches for service composition typically facilitate choreography only, while neglecting contextual information on users, services, and resources; and support techniques for modeling and specifying the provisioning of services are lacking in a mobile configuration. Maamar et al. [23] proposed associating users, services, and resources with contextual structures. Users want services to satisfy their needs, and services require resources on which to operate. To track the progress of interactions between users and services, as well as between services and resources, contextual structures are necessary (Figure 38.3). The authors identified three types of context: user (U-context), service (S-context), and resource (R-context).

Muldoon et al. [27] defined the user context of a user as an aggregation of the user's location, previous activities, and preferences. Sun [36] adopted the same definition of user context but added physiological information to this context. Maamar et al. [22] defined the service context of a service as an aggregation of its simultaneous participations in composite services, locations and times of execution, and constraints during execution. The resource context of a resource was defined as an aggregation of its current status, periods of unavailability, and capacities of meeting the execution requirements of services. According to Bellavista et al. [3], provisioning a service in a specific context highlights the logical set of resources that are accessible to a user during a service session. This accessibility depends on various factors, such as user location, access device capabilities, subscribed services, user preferences, and level of trust.

A service is context aware if it can transparently adapt its behavior according to the requirements of the entities for which this service works [15]. Users or peers of this type of service for composition are examples of such entities. The execution of a service is adjusted (e.g., postponed) when details on resources are known. These details concern the status of a resource (e.g., idle, operational, busy) and are stored in its R-context, which acts as a context source for the service. When a service is adjusted to the availability of a resource, its respective S-context is updated. Similar consideration applies to supplying services to users. Before a service is invoked, the user's preferences are handled. These preferences regard the user's current location, order of activities, and current time and are stored in the user's U-context, which acts as a context source for the service. When a service is adjusted to the preferences of a user, its respective S-context is again updated.

Scooby is an example of a middleware language that aims at composing services taking into account user preferences [29]. This approach explores the issue of enabling users to employ natural language descriptions to describe how they want their environment to be tailored according to the available services (i.e., it enables users to express their policies). Ontology is used to describe services and their functionalities, and user policies are expressed in terms of the concepts known in the described ontology. Users are able to describe, edit, and browse their policies using a multimodal interface combining speech, text, and diagrammatic representations. A policy compiler is responsible for translating policy specifications in terms of composed services in the Scooby middleware language. Finally, a policy manager uses the policy precedence ordering to route the events and requests to the composed services.

Although the work of Baresi et al. [1] on context-aware service provisioning was not conducted within a mobile configuration, its presentation is relevant in this chapter. The proposed approach has put forward (1) two levels, referred to as *application* and *technology*, and (2) perspectives, referred to as *request* and *provisioning*. The request perspective provides an abstract model of the context in which the service is triggered, and the provisioning perspective represents the available services in terms of functional description and composition. Similarly to BPEL4WS (www. ebpml.org/bpel4ws.htm), services are composed in a recursive manner following workflow logic. Additionally, adaptation rules have been proposed for mapping user-level requirements onto technological constraints. This helps make the service composition specifications context sensitive.

In addition to request and provisioning perspectives, Baresi et al. looked into adaptation rules that handle critical situations during service composition. Service failures or lack of QoS can jeopardize the success of a service execution scenario.

Because of the changing and complex nature of user needs, requirements, and expectations, it is unlikely that a certain provider would develop all types of services. Services will be different at various levels with regard to the functionality they offer, the reliability they guarantee, and the context information they manage, just to cite a few. In their work on context heterogeneity, Maamar et al. [24] suggested developing Ontology Web Language-Based Context (OWL-C) as a language dedicated to context specification. Ignoring the problem of context heterogeneity of services affects the normal progress of their composition. These side effects are multiplied by, for example, adopting the wrong strategy for selecting a component service (e.g., favoring an execution-cost criterion over a reliability criterion, instead of the reverse), delaying the triggering of some urgent component services, or poorly assessing the exact execution status of a service (e.g., service being blocked while under execution). Emphasizing the importance of seriously addressing the issues of context heterogeneity, common attributes that should ensure the quality of contexts are reported in Huebscher and McCann [16]. Some of these attributes are precision, correctness probability, resolution, and refresh rate.

Integration of Mobile Middleware into Service-Oriented Applications

Some Relevant Projects

In a mobile configuration, middleware should permit a highly configurable and adaptive execution environment that reacts to changes when needed and upon request [12]. This reaction depends on the quality of information that context requires for different parties, whether users, services, or resources. Configurability and adaptability requirements are typical of the considerations when developing a middleware implementation in a mobile configuration. This implementation should be modular for adjustability purposes and robust for continuity purposes. By continuity, we mean the capacity of an application to maintain operation and recover to normal levels of operation after a change. Chan and Chuang [12] discussed the concept of computational reflection, where mobile middleware is given mechanisms that allow the middleware to carry out self-observation, model its capabilities, oversee its computation consumption, and possibly change the way in which a business process takes shape.

Some of the issues that Chan and Chuang [12] raised when developing such middleware include determining how the middleware and applications know what context information to react to and how they know how to react to the content changes. The authors considered three levels of adaptation: interservice, intraservice, and application. The MobiPADS framework constitutes the authors' response to these issues and is an implementation of these adaptation levels. The framework consists of two agents: a server residing in the wired part of the network and a client residing in the mobile device. It is expected that the server will always be in the vicinity of the client, according to the wireless access point that is in charge of managing this client. With regard to context awareness, MobiPADS incorporates an event notification model that monitors the status of a particular context and reports changes in terms of events to the appropriate parties.

Challenges that face the technical evolution of wireless networks include the lack of means for developing intelligent middleware and a lack of design approaches for applications called to operate on top of these networks. To address these challenges, Sørensen et al. [37] initiated the CORTEX project. CORTEX is geared toward the following key characteristics: sentience, autonomy, time criticality, safety criticality, geographical dispersion, mobility, and evolution. Sentience is the ability to perceive the state of the surrounding environment. Mobility reflects the fact that applications running on top of mobile devices must have the capacity to physically move and discover new neighbors for interaction and information-sharing purposes. Autonomy means that applications should be capable of acting in a decentralized and independent way. CORTEX includes a sentient object programming model that allows developers to design distributed applications in terms of sentient objects, instead of decomposing them into component parts.

Although Sørensen et al. [37] recognized the difficulty of designing an application that has all of the aforementioned characteristics, they emphasized in their investigation the importance of considering applications that combine the autonomy of application participants (sentience) with the need to maintain a consistent view of the application environment while cooperating with other participants. The application domain of CORTEX is a set of intelligent vehicles. Vehicles today are being equipped with various types of embedded sensors, such as position broadcasts and raindrop detection on windshields. In an environment populated with similarly equipped vehicles, communication could occur for various purposes: dissemination of traffic information derived from the embedded sensors, cooperation among vehicles to assist drivers in critical situations, and interactions among remote cars.

Using Reflective Middleware

The CORTEX project used vehicles, which can easily offer the necessary computing resources that bypass the limitations of mobile devices, but work on the Context-Aware Reflective Middleware System for Mobile Applications (CARISMA) project has helped to reduce these limitations [8]. Mobile devices will continue to be battery dependent and users will continue to resist carrying heavy devices. Capra et al. [8] promote improving user acceptance of these limitations by deploying context-aware applications that adapt to changes in the environment. Overseeing changes in the environment is an intense exercise of querying sensors, filtering the collected data, and detecting changes of interest to an application. To reduce this complexity, a layer of middleware between the network operating system and an application is necessary [8]. Acting on behalf of the application, the role of the middleware would consist of maintaining updated context information and detecting changes of interest to the application that call for a prompt reaction.

Current middleware technologies for traditional wired distributed systems have proven to be successful by adopting the principle of transparency; however, Capra et al. argue that transparency cannot be the guiding principle for developing the new abstractions and mechanisms required by mobile computing middleware to foster a new generation of context-aware applications. Capra et al. adopted the reflection principle of Chan and Chuang [12] as it offers advantages for building mobile computing middleware. A reflective system can modify its own behavior through inspection and adaptation. Behavior modification has to be consistent to prevent multiple types of conflicts. In CARISMA, reflection helps achieve dynamic adaptation to context changes [8].

Applications are offered the possibility of abstracting the middleware as a set of services that can be dynamically customizable. Customization takes place by means of metadata, which encodes the middleware behavior to answer application service requests in various contexts. Through reflection, the metadata can be changed and, thus, the middleware behavior tuned. The risk of conflicts is addressed through a resolution strategy that relies on a particular type of sealed-bid auction [8]. The idea is to permit applications to compete to have the middleware deliver the quality of service they desire. Addressing the requirements of mobility, context awareness, and adaptation for mobile computing has been the driving force for CARISMA.

Customizable middleware for mobile applications is also the theme of the Fluid Computing project [7] at IBM's Zurich Research Labs. The emphasis here, however, has been on developing a middleware architecture that implements a data and application replication mechanism across multiple mobile devices. This mechanism enables application state to seamlessly flow from one device to another, much like a fluid.

The Fluid Computing middleware replicates data on multiple devices and achieves coordination of these devices through synchronization. Each device has a replica of the application state, which allows it to operate autonomously. The role of the synchronization protocol is to keep the replicas consistent, depending on the quality of the network connectivity available (weak consistency). The Fluid Computing synchronization protocol operates in two modes. *Batch mode* replication occurs when connectivity is regained after disconnection, exchanging all updates that have accumulated during the disconnection. *Trickle mode* replication occurs as long as there is some connectivity, propagating updates in real time as soon as they are generated at a replica. Fluid Computing allows disconnected operation; the user can continue working with a device and making updates even when the device has no connectivity. Replicas are therefore allowed to become inconsistent when disconnected.

Using Software Agents

Another approach to mobile middleware development promotes the use of software agents for their various appealing features, such as autonomy and mobility. Bellavista et al. [4] have investigated the value added by mobile agents in the development of middleware for context-aware applications. The investigation of Bellavista et al. was motivated by the fact that service providers and network operators (wired or wireless) face new challenges and state-of-the-art technical issues when attempting to achieve seamless integration of mobile users with traditional fixed Internet-based scenarios. This seamless integration can be hindered by the changing locations of users and the limited capacities of mobile devices.

Bellavista et al. [4] argued that location visibility is not the only crucial concern of applications in a mobile configuration, and they pointed out the need for novel development methods for context-aware applications. This awareness is reflected in the logical set of resources that a client can access, subject to location, permissions, access device capabilities, etc. In addition, Bellavista et al. addressed the combination of what mobile middleware requires and what mobile agents offer. The success of this combination is dependent on addressing the following issues [4]: how to extract context and achieve its visibility, how to refine and aggregate information about the context of mobile clients during their service sessions, how to exploit context-awareness flexibility to support decision making about service adaptation, and how to use full context visibility to manage QoS aspects.

Software agents as necessary components in mobile middleware scenarios were discussed by Sheng et al. [34]. Services are poised to become accessible from mobile devices, and the proliferation of such devices and deployment of more sophisticated wireless communication infrastructures are providing the Web with the ability to deliver data and functionality to mobile users. However, several obstacles still hinder the seamless provisioning of services in wireless environments. Indeed, current service provisioning techniques are inappropriate because of the distinguishing features and inherent limitations of wireless environments such as low throughput and poor connectivity of wireless networks, limited computing resources, and frequent disconnections of mobile devices. In addition, the variability in computing resources, display terminals, and communication channels requires intelligent support for personalized and timely delivery of relevant data to users. This support can be achieved using software agents, which may, for example, act on behalf of users and services.

Conclusions and Future Directions

Service-oriented computing is an emerging paradigm for distributed computing that is changing the way software applications are designed, built, delivered, and consumed. Services are platform-independent computational elements that can be described, published, discovered, and orchestrated using standard protocols. In general, combining multiple services rather than relying on a single service is essential and provides more benefits to users. Composition primarily addresses the situation when a user's request cannot be satisfied by any available service, but a composite service obtained by combining available services might fill the need.

In this chapter, we presented some of the state-of-the-art work on service composition and its implementation in a mobile environment. We also discussed the role and importance of context as a crucial element in facilitating service composition in this environment. Techniques currently under development for utilizing contextual information in service composition were also highlighted. We described several ongoing research projects that are investigating how mobile middleware can be enhanced with service-oriented computing concepts, and, finally, we briefly touched upon the possible usage of software-agent technology for enhancing the autonomy and proactivity of mobile middleware.

Future work in this area would involve investigating the extent to which software agents can be used to enhance context-oriented service composition in mobile middleware. This would be made possible by leveraging some of the advantages of software agents, such as autonomy and collaboration. Other exciting areas for future work would consist of tailoring existing work on context-aware adaptation and exception handling for Web services [28] to mobile middleware and investigating the use of autonomic computing middleware [17] for adaptive service composition. Indeed, autonomic computing aims at making computer systems more adaptive through self-configuration, self-optimization, and self-protection. Service composition might easily benefit from that and be made more adaptable and more manageable if it is based on autonomic middleware.

Acknowledgments

N.C.N. wishes to thank his manager, K. Muralidharan, for his support.

References

[1] Baresi, L., Bianchini, D., De Antonellis, V., Fugini, M.G., Pernici, B., and Plebani, P., Context-aware composition of e-services, in *Proc. of the 4th VLDB Workshop on Technologies for E-Services (TES'2003), held in conjunction with the 29th Int. Conf. on Very Large Data Bases (VLDB'03)*, Berlin, Germany, September 9–12, 2003.

[2] Benatallah, B., Sheng, Q.Z., and Dumas, M., The self-serv environment for Web services composition, *IEEE Internet Comput.*, 7(1), 40–48, 2003.

[3] Bellavista, P., Corradi, A., Montanari, R., and Stefanelli, C., Context-aware middleware for resource management in the wireless Internet, *IEEE Trans. Software Eng.*, 29(12), 1086–1089, 2003 (special issue on software engineering for the wireless Internet).

[4] Bellavista, P., Bottazi, D., Corradi, A., Montanari, R., and Vecchi, S., Mobile agent middlewares for context-aware applications, in *Handbook of Mobile Computing*, Mahgoub, I. and Ilyas, M., Eds., CRC Press, Boca Raton, FL, 2004.

[5] Bellur, U. and Narendra, N.C., Towards service orientation in pervasive computing systems, in *Proc. of the 6th IEEE Int. Conf. on Information Technology (ITCC2005)*, Las Vegas, NV, April 11–13, 2005.

[6] Bouguettaya, A., Medjahed, B., and Elmagarmid, A., Composing Web services on the Semantic Web, *VLDB J.*, 4(12), 333–351, 2003 (special issue on the Semantic Web).

[7] Bourges-Waldegg, D., Duponchel, Y., Graf, M., and Moser, M., The fluid computing middleware: bringing application fluidity to the mobile Internet, in *Proc. of Int. Symp. on Applications and the Internet (SAINT2005)*, Trento, Italy, January 31–February 4, 2005.

[8] Capra, L., Emmerich, W., and Mascolo, C., CARISMA: context-aware reflective middleware system for mobile applications, *IEEE Trans. Software Eng.*, 29(10), 929–945, 2003.

[9] Budak Arpinar, I., Aleman-Meza, B., Zhang, R., and Maduko, A., Ontology-driven Web services composition platform, in *Proc. of the IEEE Int. Conf. on E-Commerce Technology (CEC'04)*, San Diego, CA, July 6–9, 2004.

[10] Casati, F., Ilnicki, S., Jin, L., Krishnamoorthy, V., and Shan, M., Adaptive and dynamic service composition in eFlow, in *Proc. of the 12th Conf. on Advanced Information Systems Engineering (CAISE'2000)*, Stockholm, Sweden, June 5–9, 2000.

[11] Chakraborty, D. and Joshi, A., *Dynamic Service Composition: State of the Art and Research Directions*, Technical Report TR-CS-01-19, University of Maryland, Baltimore County, Baltimore, MD, 2001.

[12] Chan, A. and Chuang, S.-N., MobiPADS: a reflective middleware for context-aware mobile computing, *IEEE Trans. Software Eng.*, 29(12), 1072–1085, 2003.

[13] Dey, A. K., Abowd, G. D., and Salber, D., A conceptual framework and a toolkit for supporting the rapid prototyping of context-aware applications, *Human–Computer Interaction (HCI) J.*, 16(2–4), 97–166, 2001 (special issue on context-aware computing).

[14] Esfandiari, B. and Tosic, V., Requirements for Web service composition management, in *Proc. of the 11th Workshop of the HP OpenView University Association (HPOVUA'2004)*, Paris, France, June 20–23, 2004.

[15] Hegering, H.G., Kupper, A., Linnhoff-Popien, C., and Reiser, H., Management challenges of context-aware services in ubiquitous environments, in *Proc. of the 14th IFIP/IEEE Int. Workshop on Distributed Systems: Operations and Management (DSOM'2003)*, Heidelberg, Germany, October 20–22, 2003.

[16] Huebscher, M.C. and McCann, J.A., Adaptive middleware for context-aware applications in smart-homes, in *Proc. of the 2nd Int. Workshop on Middleware for Pervasive and Ad Hoc Computing (MPAC'2004), held in conjunction with the ACM/IFIP/USENIX 5th Int. Middleware Conf. (Middleware'2004)*, Toronto, Canada, October 18–22, 2004.

[17] Kephart, J.O. and Chess, D., The vision of autonomic computing, *IEEE Comput.*, 36(1), 41–50, 2003.

[18] Kortuem, G. and Segall, Z., Wearable communities: augmenting social networks with wearable computers, *IEEE Perv. Comput. Mag.*, 2(1), 71–78, 2003.

[19] Kouadri Mostéfaoui, S. and Hirsbrunner, B., Context-aware service provisioning, in *Proc. of the IEEE/ACS Int. Conf. on Pervasive Services (ICPS'04)*, Beirut, Lebanon, July 19–23, 2004.

[20] Ma, K.J., Web services: what's real and what's not, *IEEE IT Prof.*, 7(2), 2005.

[21] Maamar, Z. and Mansoor, W., Design and development of a software agent-based and mobile service-oriented environment, *e-Service J.*, 2(3), 2003.

[22] Maamar, Z., Sheng, Q.Z., and Benatallah, B., On composite Web services provisioning in an environment of fixed and mobile computing resources, *Inform. Technol. Manage. J.*, 5(3/4), 2004 (special issue on workflow and e-business).

[23] Maamar, Z., Kouadri Mostéfaoui. S., and Mahmoud, Q.H., On personalizing Web services using context, *Int. J. E-Business Res.*, 1(3), 2005 (special issue on e-services).

[24] Maamar, Z., Narendra, N.C., and Sattanathan, S., Towards an ontology-based approach for specifying and securing Web services, *J. Inform. Software Technol.*, 48(7), 540–548, 2006.

[25] Mahmoud, Q.H. and Maamar, Z., Challenges and possible solutions in wireless application design, *Cutter IT J.*, June, 2005 (http://cutter.com).

[26] Milanovic, N. and Malek, M., Current solutions for Web service composition, *IEEE Internet Comput.*, 8(1), 51–59, 2004.

[27] Muldoon, C., O'Hare, G., Phelan, D., Strahan, R., and Collier, R., ACCESS: an agent architecture for ubiquitous service delivery, in *Proc. of the 7th Int. Workshop on Cooperative Information Agents (CIA'2003)*, Helsinki, Finland, August 27–29, 2003.

[28] Narendra, N.C., Modeling adaptation in Web services execution using context ontologies, in *Proc. of the San Diego Information Systems Conf. (SISC'2005)*, San Diego, CA, July 8–10, 2005.

[29] Robinson, J., Wakeman, I., and Owen, T., Scooby: middleware for service composition in pervasive computing, in *Proc. of the 2nd Int. Workshop on Middleware for Pervasive and Ad Hoc Computing (MPAC'2004), held in conjunction with the 5th ACM/IFIP/USENIX Int. Middleware Conf. (Middleware'2004)*, Toronto, Canada, October 18–22, 2004.

[30] Schmidt, A. and Beigl, M., New challenges of ubiquitous computing and augmented reality, in *Proc. of the 5th CaberNet Radicals Workshop*, Porto, Portugal, July 5–8, 1998.

[31] Shaparau, D., *Approaches to Web Service Composition*, Technical Report, University of Trento, Italy, 2004.

[32] Satyanarayanan, M., Pervasive computing: vision and challenges, *IEEE Pers. Commun.*, 8(4), 10–17, 2001.

[33] Sirin, E., Parsia, B., Wu, D., Hendler, J., and Nau, D., HTN planning for Web service composition using SHOP2, *J. Web Semantics*, 1(4), 377–396, 2004.

[34] Sheng, Q.Z., Benatallah, B., Maamar, Z., Dumas, M., and Ngu, A., Enabling personalized composition and adaptive provisioning of Web services, in *Proc. of the 16th Int. Conf. on Advanced Information Systems Engineering (CAiSE'2004)*, Riga, Latvia, June 7–11, 2004.

[35] Skogan, D., Gronom, R., and Solheim, I., Web service composition in UML, in *Proc. of the 8th IEEE Int. Conf. on Enterprise Distributed Object Computing (EDOC'2004)*, Monterey, CA, September 20–24, 2004.

[36] Sun, J., Information requirement elicitation in mobile commerce, *Commun. ACM*, 46(12), 45–47, 2003.

[37] Sørensen, C.F., Wu. M., Sivaharan T., Blair, G.S., Okanda, P., Friday, A., and Duran-Limon, H., A context-aware middleware for applications in mobile *ad hoc* environments, in *Proc. of the 2nd Int. Workshop on Middleware for Pervasive and Ad Hoc Computing (MPAC'2004), held in conjunction with the 5th ACM/IFIP/USENIX Int. Middleware Conf. (Middleware'2004)*, Toronto, Canada, October 18–22, 2004.

[38] van der Aalst, W., Don't go with the flow: Web services composition standards exposed, *IEEE Intelligent Syst.*, 18(1), 72–76, 2003.

[39] Yang, J. and Papazoglou, M.P., Web components: a substrate for Web service reuse and composition, in *Proc. of the 14th Int. Conf. on Advanced Information Systems Engineering (CAiSE'2002)*, Toronto, Canada, May 27–31, 2002.

[40] Yang, J. and Papazoglou, M.P., Service components for managing the lifecycle of service compositions, *Inform. Syst.*, 29(2), 97–125, 2004.

[41] Yunos, H.M., Gao, J.Z., and Shim, S., Wireless advertising's challenges and opportunities, *IEEE Comput.*, 36(5), 30–37, 2003.

Chapter 39

Mobile Middleware for Situation-Aware Service Discovery and Coordination

Stephen S. Yau and Dazhi Huang

CONTENTS

Introduction

Recent advances in embedded systems, microelectronics, and wireless communication technologies have increased the flexibility of using mobile devices for various practical applications that improve the personal productivity of users. Mobile devices, however, are still resource poor in comparison with computing resources in network infrastructures (NIs), such as the Internet, grid, and enterprise computing environments. Such NIs usually consist of non-mobile computing resources to provide high-performance computing and communication capabilities. Although these NIs have become more flexible and interoperable by adopting a service-oriented architecture (SOA) [1], which can enable rapid composition of distributed applications regardless of the programming languages and platforms used in developing and running different components of the applications, it is still very difficult for these NIs to provide the desired flexibility to individual users, especially to mobile users, due to their immobility and large size. Hence, dynamic integration of mobile devices with NIs [2] is a subject in ubiquitous computing that has attracted much attention. Dynamic integration is the process by which a mobile device can detect, communicate with, and use the required services in nearby NIs in an application-transparent way. The benefit of dynamic integration is that the applications in both a mobile devices and an NI can interoperate with each other as if a mobile device itself is an integral part of the NI or *vice versa* [2]. The dynamic integration of mobile devices and NIs has raised a number of research issues, such as wireless *ad hoc* communication, service discovery and coordination, and distributed trust management. In this chapter, our discussion is focused on techniques for service discovery and coordination for the dynamic integration of mobile devices with service-based NIs.

In service-based NIs, various capabilities, such as storage, computation, and communication, are provided by different organizations as services and are interconnected by various types of networks. We consider a service to be a software/hardware entity with well-defined interfaces to provide certain capability over wired or wireless networks using standard protocols, such as Transmission Control Protocol (TCP)/Internet Protocol (IP), Hypertext Transfer Protocol (HTTP), and Simple Object Access Protocol (SOAP). By dynamically and seamlessly integrating mobile devices with such service-based NIs, users can utilize various capabilities in the NIs anytime and anywhere through their mobile devices. Moreover, the services in the NIs can be integrated following specific workflows, which are series of cooperating and coordinated activities designed to achieve complicated mission goals for users. Service discovery, which is the process of locating services that can satisfy the needs of users, is a prerequisite of accessing the capabilities provided by the NIs. Service coordination, which is required to ensure the correctness of workflow execution, is a process of monitoring the status of participant services, invoking proper participant services, and propagating necessary information to participant services to ensure the correct results obtained from the coordinated participant services.

To enable the effective integration of mobile devices with service-based NIs, service discovery and coordination must be situation aware (SA) for the following reasons: (1) Services may become unavailable or may not be able to provide desirable quality of service (QoS) due to distributed denial-of-service attacks, system failures, or system overload; (2) workflows may have to be adapted when the situation changes to achieve the users' mission goals; and (3) new workflows may be generated in runtime to fulfill users' new mission goals. We consider *situation awareness* (SAW) to be the capability of being aware of situations and adapting the system's behavior based on situation changes [3,4]. SAW is necessary for checking whether a service can meet the users' requirements and should be invoked. A *situation* is a set of contexts in a mobile application over a period of time that affects future system behavior. A *context* is any instantaneous, detectable, and relevant property of the environment, the system, or users, such as time, location, light intensity, wind velocity, temperature, noise level, available bandwidth, and a user's schedule.

Achieving SA service discovery and coordination in mobile computing environments is challenging due to the heterogeneous environments, resource limitations of mobile devices, and user mobility. Furthermore, developing mobile applications that make use of SA service discovery and coordination is difficult without appropriate system support. These challenges can be effectively addressed by developing mobile middleware that provides a set of components that can perform context management, situation analysis, service discovery, and coordination efficiently. In this chapter, we first review the background of mobile middleware for situation-aware service discovery and coordination. We then discuss the requirements, design issues, and enabling techniques for mobile middleware for SA service discovery and coordination.

Background

In this section, we review the background of our topic with respect to situation awareness, service discovery, service coordination, and mobile middleware.

Situation Awareness

In this section, we provide an introduction to the literature on SAW and its related areas. Early work on situation awareness, which was done primarily in the artificial intelligence community, focused on formalizing and reasoning on situations. Situation calculus and its extensions [5–8] were developed for describing and reasoning how actions and other events affect the world, assuming that all actions and events changing the world are known or predictable. Situation calculus considers a situation to be a complete state of the world, which leads to the well-known frame problem and ramification problem [6].

In Barwise [9], an opposite view of situation was considered; a situation was formally defined as a part of "the way the world M happens to be," and a situation supports the truth of a sentence F in M. Barwise [9] defined a *scene* as a "visually perceived situation" and observed that a scene that people perceive consists of not only objects and individual properties associated with the objects but also relationships between any two objects. Based on this concept of a scene, Barwise [10] introduced *situation semantics*, in which basic properties, relations, and situations are defined as objects. Barwise's definition of situation is more practical compared to the definition of situation in situation calculus, as Barwise's definition of situation allows the precise description of situations and can be easily supported by the prevailing object-oriented modeling techniques.

Currently, many researchers have adopted Barwise's definition of situation and have developed their own formalisms of situations for various purposes, such as supporting data fusion [11,12] and SA software development [3,13]. For example, Matheus et al. [11,12] introduced a core SAW ontology based on a similar view of situations as Barwise's which defines a situation as a collection of situation objects, including objects and relations, as well as other situations.

Since the early 1990s, much research on context-aware computing has been conducted by the mobile computing community. *Context* here usually refers to the information that can be used to characterize the situations on users, applications, and environments, although various researchers have used slightly different definitions [3,4,14–18]; hence, context awareness is considered to be a part of SAW. Although a number of issues related to context-aware (or context-sensitive) computing have been discussed in the literature [19–22], the term "context-aware computing" was first introduced by Schilit and Theimer [14]. Since then, several frameworks, toolkits, and infrastructures have been developed to provide support for context-aware application development. Notable results include CALAIS [23], Context Toolkit [17], CoolTown [24], Mobile Platform for Actively Deployable Service (MobiPADS) [25], Gaia [26, 27], TSpaces™ [28], and Reconfigurable Context-Sensitive Middleware (RCSM) [3,29].

CALAIS [23] focuses on applications accessible from mobile devices and supports acquisition of contexts of users and devices, but it is difficult to evolve existing applications when requirements for context acquisition and the capabilities and availabilities of sensors change. Context Toolkit [17] provides architectural support for context-aware applications, but it does not provide analysis of complex situations. CoolTown [24] supports applications that display contexts and services to end users. MobiPADS [25] is a reflective middleware designed to support dynamic adaptation of context-aware services based on which application's runtime reconfiguration is achieved. Gaia [26,27] provides context service, space repository, security service, and other QoS for managing and interacting with active spaces. TSpaces [28] utilizes tuple spaces to store contexts and allows tuple space sharing for application software to read and write, but it ignores the status of the device where the application software executes, network conditions, and the surrounding environment as part of the overall context. RCSM [3,29] provides development and runtime support for SA application software, including a declarative Situation-Aware Interface Definition Language (SA-IDL) and its compiler for automated code generation, a reconfigurable SA processor supporting runtime situation analysis and triggering proper actions of SA applications, and a RCSM Object Request Broker (R-ORB) supporting context discovery, acquisition, and SA inter-object communications. Readers interested in context-aware computing are referred to Pokraev et al. [30] and Mostefaoui et al. [31].

Service Discovery and Coordination

Service discovery, also known as *service location* or *matchmaking*, is the process of locating suitable services that can meet the users' requirements. During the past decade, substantial research has been done on service discovery that has generated various service discovery protocols, such as Jini™ [32], Salutation [33], Service Location Protocol (SLP) [34], Universal Plug and Play (UPnP™) [35], and Universal Description, Discovery, and Integration (UDDI) [36]. Currently, major concerns regarding service discovery include: (1) how to identify the most suitable services that meet the users' needs, and (2) how to utilize resources (e.g., network bandwidth, battery power) efficiently in service discovery, especially in mobile computing environments.

For the first question, most existing service discovery approaches are based on syntactical matching (i.e., keyword or table-based matching [37]), such as UDDI; however, matching keywords or service interfaces are not good enough in many real-world scenarios to find suitable services that meet the users' needs. The semantics of services, the goals of users, the situations of users, systems and environments, the QoS that can be provided by the services, and the security requirements for using the services have to be considered when identifying the most suitable services. To understand the semantics of services, various models and languages [38–40] have been proposed to capture the service semantics. The goals of users are usually described using the corresponding query languages of the service description. Based on these models and languages, techniques for semantic-based service discovery were introduced [41–46]. Recently, security and other aspects of QoS have been incorporated in service discovery [47–49].

It has been shown that incorporating context and situation awareness in service discovery can greatly improve the precision and recall of the discovery results, where *recall* is defined as the number of relevant services retrieved in service discovery divided by the total number of relevant services available, and *precision* is defined as the number of relevant services retrieved in service discovery divided by the total number of services discovered [45,46,50,51]. Hence, it improve the efficiency of mobile applications and reduce distractions to users. For service discovery, *recall* is defined as the number of relevant services retrieved in service discovery divided by the total number of relevant services available, and *precision* is defined as the number of relevant services retrieved in service discovery divided by the total number of services discovered [45,46].

Doulkeridis et al. [50] used contextual information, such as time, location, user name, and device type, to differentiate services belonging to the same service category to increase the precision of service discovery. In addition, they analyzed context history to find useful patterns for predicting future service availability, which further increases the precision of service discovery. Yau et al. [51] used situation information to guide the generation of user profiles and select appropriate user profiles for personalized information retrieval. Users' information retrieval requests are expanded based on their profiles and situations, which capture some implicit but important characteristics of the information that the users want to retrieve and hence improve the precision and recall of information retrieval.

Although the work of Yau et al. [51] was mainly directed toward information retrieval, the idea can easily be applied to service discovery. In Broens [45] and Broens et al. [46], contextual information is used in two ways: (1) It is used to expand service requests with more information to retrieve more relevant services and increase the precision of service discovery, and (2) it is used to complete the contextual inputs required by relevant services when some of these inputs are not provided by service requestors to allow the retrieval of relevant services with missing inputs and to increase the recall of service discovery. So far, however, no unified service discovery approach considers service semantics, security policies, SAW, and other QoS properties in the discovery process. Other noteworthy work on context- and situation-aware service discovery includes the discovery mechanisms in several context- and situation-aware platforms and middleware, such as CoolTown [24,52], Context Toolkit [17], and RCSM [2,53]. Although much progress has been made on SA service discovery, it is still a relatively new area, and much work remains to be done. Later in this chapter, we discuss the design issues and enabling techniques for mobile middleware for SA service discovery.

With regard to contextual inputs, because centralized service directories and registries are not always available in mobile computing environments lacking infrastructure support, service discovery in such environments often has characteristics similar to peer-to-peer (P2P) service discovery [54–57]. Given the very limited communication bandwidth and battery power of mobile devices, the efficiency of service discovery protocols in mobile computing environments is of special interest to researchers in this area. Much research effort has been directed toward efficient service discovery in mobile computing environments [58–63].

Industrial standards on service coordination include Web Services Coordination (WS-Coordination) [64] and the Web Services Coordination Framework (WS-CF) [65], as well as some notable approaches to context-aware service coordination [18,66–68]. Some industrial standards [64,65] aim at providing standard and extensible coordination frameworks to support coordinated workflows and transactions on Web services but do not address situation awareness in service coordinations. Braione and Picco [18] developed a formal specification framework for modeling dynamically changing contexts and rules in contextual reactive systems for coordinating distributed systems. Mobile Agent Reactive Spaces (MARS) [67] aims at promoting context-dependent coordination by incorporating the concept of programmable coordination media in distributed systems. Using coordination contracts to support the construction and evolution of complex service coordination was proposed by Andrade et al. [66]. In EgoSpaces [68], a coordination model focuses on the context of a particular component in a mobile *ad hoc* network and provides a middleware for context specification and runtime reconfiguration. Elsewhere [18,66-68], service coordination is based only on the current context information; however, changes in contexts over a period of time can be useful information and should not be neglected. Later in this chapter, we discuss the design issues and enabling techniques for mobile middleware for SA service coordination.

Mobile Middleware

Existing mobile middleware can be divided into two major categories depending on how it supports coordination among mobile devices: (1) tuple-space based, and (2) message based. Notable work in the first category includes LIME (Linda in a Mobile Environment) [69], TSpaces [28], and Limone [70]. The tuple-space-based coordination model supports location transparency and disconnected operations, and mobility is viewed as transparent changes in the content of the tuple space. Hence, this approach can easily support interactions among mobile devices. For tuple-space-based mobile middleware, the major concerns are the scalability and performance of this type of middleware. Notable work from the second category includes ALICE (Architecture for Location-Independent Computing Environments) [71], Mobiware [72], Gaia [26,27], Reconfigurable Context-Sensitive Middleware (RCSM) [3,4], and MobiPADS [25]. ALICE [71] adds a mobility layer between the transport and the Common Object Request Broker Architecture (CORBA™) Internet Inter-ORB Protocol (IIOP) layers of the CORBA architecture to support both mobile clients and servers. Mobiware [72] provides the facilities for managing an open, active, and adaptive mobile network by utilizing a CORBA-based architecture and using different adaptive algorithms as Java objects, which can be injected dynamically into mobile devices, access points, and mobile-capable network switches or routers. In the following subsections, we provide a brief overview of RCSM and MobiPADS, which provide context and situation awareness.

RCSM

Reconfigurable Context-Sensitive Middleware (RCSM) is a lightweight situation-aware middleware that provides development and runtime support for SAW, dynamic service discovery, and group communication for ubiquitous computing applications [2–4,29]. A conceptual architecture of RCSM is shown in Figure 39.1. RCSM consists of the following major components:

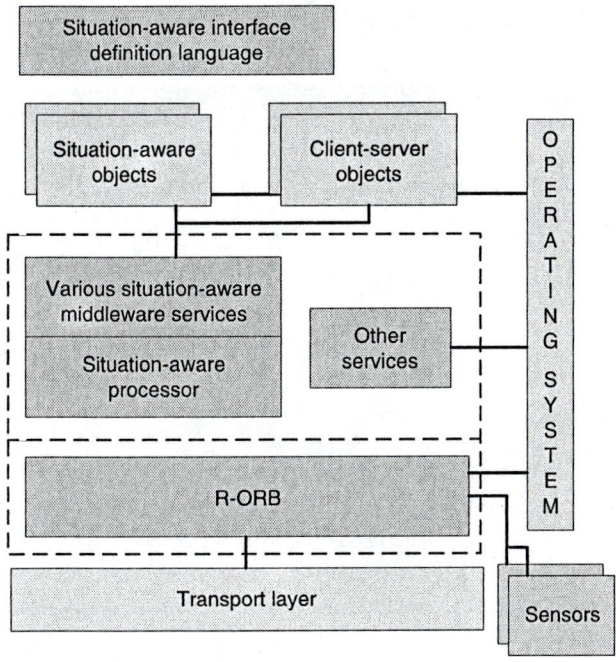

Figure 39.1   RCSM architecture.

- *SA Processor* provides the runtime services for situation analysis and manages the SAW requirements of SA objects. The SAW requirements of situation-aware objects are defined using the Situation-Aware Interface Definition Language (SA-IDL) [3,29]. An SA-IDL compiler has been developed to generate the situation-aware object skeleton codes and corresponding configuration files, which will be used by the SA Processor to perform situation analysis accordingly. The SA object skeleton codes provide the standard interfaces for SA objects to interact with the SA Processor.
- *RCSM Object Request Broker (R-ORB)* provides the runtime services for context discovery and acquisition, as well as SA communication management. The context manager in R-ORB implements an efficient context discovery protocol [73] to support adaptive context discovery and acquisition in ubiquitous computing environments based on the requirements on contexts extracted from the configuration files of SA applications by the SA Processor. SA object discovery protocols have also been developed to enable efficient and spontaneous communication among distributed SA objects [29,53].

A unique feature provided by RCSM is support for SAW. Using SA-IDL, contexts can be precisely described as context objects, and situations can be defined not only by the current values of multiple contexts but also by the historical values of multiple contexts over a period of time. The SA Processor is designed to cache and analyze the context history to determine the situation. In addition, the SAW requirements, such as definitions of situations, can be modified in runtime through the SA Processor. When the requirements have been changed, the R-ORB and SA Processor will reconfigure themselves to collect necessary contexts and perform situation analysis based on the new requirements.

MobiPADS

The Mobile Platform for Actively Deployable Service (MobiPADS) [25] is a reflective middleware that serves as an execution platform for context-aware mobile computing. MobiPADS enables active service deployment and reconfiguration in response to context changes and can optimize the operation of mobile applications when the operating context changes. MobiPADS consists of two types of agents: (1) Mobi-PADS server agents and (2) MobiPADS client agents. MobiPADS server agents reside in the network infrastructure and are responsible for most of the optimization computations. MobiPADS client agents reside in the mobile devices and provide various services for mobile applications. MobiPADS adopts the idea of mobile codes and stores the codes of service objects in MobiPADS agents. Service objects can be deployed on either the client or server agents and can migrate between the client and server agents when needed (e.g., when the device where the client agent resides moves), thus enabling the flexible reconfiguration of mobile applications. Each MobiPADS agent also has a set of system components for managing system configurations (MobiPADS client and server and service objects), migrating service objects between the Mobi-PADS server and client, recording known services, sending contextual event notification, and establishing virtual communication channels between service objects. Each MobiPADS service is a pair of mobilets, which consists of a slave mobilet at the server agent for providing actual processing capabilities and a master mobilet at the client agent for instructing the slave mobilets and presenting results to the client. In MobiPADS, the mobilets can be chained together to support necessary service composition for mobile applications, which is similar to work-flows in workflow systems. An eXtensible Markup Language (XML)-based language has been developed in MobiPADS to describe how service objects interact with each other and how they are configured.

MobiPADS utilizes an event subscription-notification model to provide context awareness. The idea is similar to the event–condition–action (ECA) model in active databases. All contexts are modeled as event sources, which will generate contextual events when certain conditions are satisfied. In MobiPADS, all the entities (system components, mobilets, and mobile applications) can subscribe to contextual events of interests and will be notified when certain events occur, hence achieving context awareness. MobiPADS also supports event composition, which allows combining multiple events from different context sources together to express complex semantics; however, MobiPADS only focuses on the current events and does not consider historical events, which are important for achieving SAW.

Requirements for Mobile Middleware for SA Service Discovery and Coordination

To present mobile middleware for SA service discovery and coordination to enable the dynamic integration of mobile devices with service-based network infrastructures, it is necessary to first identify the requirements for mobile middleware for SA service discovery and coordination. The following is a list of requirements for such a middleware:

■ The capability to achieve SAW. To achieve SA service discovery and coordination, the mobile middleware must be situation aware. Ranganathan and Campbell [27] and Yau et al. [29] identified several requirements for middleware for context and situation awareness. Although the term "context awareness" is used by Ranganathan and Campbell [27], their concept of context is broader than the definitions of context awareness used by other researchers and is closer to our concept of situation awareness [3,4]; hence, we consider the requirements identified by these authors also to be requirements for mobile middleware for situation awareness. The following is a summary of the requirements for mobile middleware for situation awareness [27,29]:

(R1.1) Support for specifying SAW requirements for mobile applications/agents, including the contexts and situations of interests and the behaviors of applications and agents in different situations

(R1.2) Support for discovering contexts from the ambient environments based on the needs of various mobile applications and agents, acquiring contexts from various sources, and delivering acquired context data to mobile applications and agents

(R1.3) Support for analyzing the acquired context data to determine the situation and delivering the results of situation analysis to mobile applications and agents in a timely manner to trigger proper actions or adaptations

(R1.4) Support for sharing situation and context information among different mobile applications and agents

(R1.5) Support for runtime reconfiguration due to changes in SAW requirements of mobile applications and agents

(R1.6) Support for incorporating various reasoning and learning mechanisms for situation analysis

(R1.7) Syntactic and semantic interoperability among various mobile applications and agents

Among these seven requirements for achieving situation awareness, R1.1 through R1.3 are the three basic requirements that a mobile middleware must satisfy for achieving situation awareness; whereas, R1.4 through R1.7 are desirable for better performance, flexibility, interoperability, and extensibility.

■ The capability to accurately and efficiently locate necessary services to satisfy the users' needs. The purpose of service discovery is to locate services that can satisfy the needs of users. The needs of users are determined by the users' goals. Whether a service is necessary for achieving a specific goal of users depends on the service semantics, such as what the service can do and how the service works. In a dynamic environment, such as a mobile computing environment, both the users' goals and the service semantics may depend on the situations; that is, the users' goals may vary and the services may have different behaviors when the situation changes. Also, the users may have certain preferences for different choices of services in different situations; therefore, the mobile middleware must provide the following support for SA service discovery:

(R2.1) Support for developers or system administrators to specify service semantics precisely, especially with regard to how a service behaves when the situation changes

(R2.2) Support for users to specify their goals and preferences in different situations

(R2.3) Support for efficiently and accurately matching the service semantics and users' goals to locate the necessary services based on the situation

■ The capability to adaptively coordinate the execution of workflows consisting of multiple services to achieve the users' goals. Sometimes a user's goal cannot be achieved by invoking any one service alone but can be achieved by invoking multiple services in a coordinated manner. This is usually referred to as *workflow planning* or *service composition*, neither of which is within the scope of this chapter. Here, we simply assume that a workflow planner [74,75] is available and can communicate with the mobile middleware. The workflows generated by existing workflow planners are usually static; that is, it is assumed that the workflow planners have complete knowledge of the planning domain and can precisely (usually by reasoning) determine the status of the entire system in each step of the workflow execution. In real-world applications, however, this assumption tends to be invalid, so SA service coordination is necessary to ensure the correct execution of workflows. Because workflows usually consist of distributed services and the execution of each service may have certain situational constraints (e.g., a service can only be invoked in certain situations or a service may have different behaviors when the situation changes), the following support must be provided by the mobile middleware for SA service coordination:

(R3.1) Support for analyzing a workflow to identify all of the participant services in service coordination

(R3.2) Support for invoking appropriate participant services based on the workflow and the situation, monitoring the status of participant services, and reporting the status of participant services to users or mobile applications and agents

(R3.3) Support for detecting failures (e.g., violations of situational constraints on the execution of participant services, unavailability of participant services) in workflow execution, identifying alternative services that can be used, and reporting the failures and current situation to the workflow planner when no alternative services can be found

Design Issues and Enabling Techniques for Mobile Middleware To Achieve SA Service Discovery and Coordination

In this section, we discuss the design issues and enabling techniques for mobile middleware to achieve SA service discovery and coordination from the following two perspectives:

■ Context management and situation analysis for achieving situation awareness
■ Incorporating situation awareness in service discovery and coordination in mobile middleware

Context Management and Situation Analysis for Achieving Situation Awareness

As discussed in the previous section, mobile middleware for SA service discovery and coordination must be capable of achieving situation awareness. Although existing mobile middleware, such as RCSM [2–4,29,53,73], MobiPADS [25], and Gaia [26,27], utilizes various approaches to achieve context and situation awareness, three important aspects for designing such middleware can be identified as follows.

Modeling and Specifying SAW Requirements of Mobile Applications and Agents

To satisfy R1.1, R1.6, and R1.7, suitable models and languages for SAW requirements of mobile applications and agents must be developed. Based on a model proposed by Yau et al. [3], an interface definition language (SA-IDL [3,29]) was developed to allow developers to specify interfaces of SA objects and support the automated code generation of SA object skeletons. Each SA application is considered to be a set of SA objects [3,4] that will take various actions in various situations. These actions are abstracted as various functions of the SA objects. SA-IDL supports the specification and inheritance of context classes and allows specifying the frequency of context acquisition [29]. For specifying situations, SA-IDL provides operators for context preprocessing and logical connectives for composing complex situations with various contexts within a certain time range. Compilation of SA-IDL specifications will generate SA object skeletons, which can be extended by application developers by adding the implementations of functions to be triggered, and SA files for configuring RCSM. In Chan and Chuan [25], context-awareness requirements are expressed by adaptive policies describing service composition and reconfiguration upon detection of contextual events; XML is used to maintain the system profile, which contains adaptive policies for services and applications.

Ranganathan and Campbell [27] presented a predicate model of contexts that can support the use of various reasoning mechanisms, such as first-order logic and temporal logic by agents, to reason about contexts and determine their behaviors in different contexts. Ontologies are used to describe context predicates for semantic interoperability. Other models and languages for SAW exist; however, several important issues have not been addressed in modeling and specifying SAW requirements: (1) incorporation of model and language primitives for representing spatial properties in SAW to support mobility, (2) analysis of the expressiveness of various SAW models and specification languages, and (3) verification of

SAW requirement specifications. Further investigation for these issues may utilize the results from existing research on ambient logic [76] and model checking [77].

Context Management

To satisfy R1.2 and R1.5, it is necessary to develop components or services in mobile middleware to manage various sensing units (context sources) available on the mobile device, dynamically discover remote sensing units connected to other mobile devices in the neighborhood, and acquire and propagate context data from local or remote sensing units to mobile applications. MobiPADS [25] features an *event register* that enables application objects to subscribe to contextual event sources, and objects are notified when the subscribed events occur. Gaia [27] includes a *context provider lookup service* that allows context providers to advertise the contexts they provide and supports software agents that search for context providers having the necessary contextual information. In RCSM [29,73], a component (context manager in R-ORB) has been developed to manage local sensing units, acquire necessary contexts for applications, and perform context discovery, if necessary.

Because context discovery and context acquisition from remote sensing units require communication among mobile devices, which is likely to cause long delays and consume much energy, a key issue for developing such components or services in mobile middleware is the development of an adaptive, lightweight, and energy-efficient protocol to perform context discovery and remote context acquisition in a timely manner. MobiPADS and Gaia do not address this issue. Yau et al. [73] proposed a context discovery protocol (R-CDP) to address this issue. R-CDP combines pull and push communication paradigms for dynamic context discovery and efficient context retrieval. In R-CDP, context requesters dynamically discover contexts from context providers using the pull communication paradigm, and the context providers proactively push updated context values to the requesters whenever the contexts undergo measurable variations. To improve energy efficiency, R-CDP has been designed to be aware of network conditions, and it adaptively changes the interval of context request advertisements to alleviate network congestion, thus reducing retransmissions of context requests. R-CDP utilizes transmission probabilities to reduce redundant context result transmissions when multiple context providers are in the network, and it adaptively changes transmission probabilities as the number of context providers in the network changes. R-CDP also adopts and extends the refresh priority (RP) function [78] to determine when context providers should send updates of contexts by calculating the RP based on the divergence of contexts (the difference between the previous and current context values) and energy consumption of context providers. However, R-CDP matches contexts using keywords in context advertisements which cannot provide semantic interoperability (i.e., when different providers have different interpretations for the same keyword or use different keywords for the same context). Ontology-based approaches [11,12,27,87–89] could be combined with context discovery protocols such as R-CDP to solve this problem.

Situation Analysis

To satisfy R1.3, R1.4, and R1.5, it is necessary to develop components or services in mobile middleware to manage the SAW requirements of mobile applications and agents, maintain a history of contexts and situations of interests to mobile applications and agents, analyze relevant contexts and situation histories to determine the current situation based on SAW requirements, and notify applications and agents when a situation of interest changes. A common approach to situation analysis is to define situations in the form of logical rules and to use a rule engine to infer situations in runtime. For example, in RCSM, situations the actions to be triggered in different situations are defined by SA-IDL rules, and the SA Processor in RCSM is a rule engine that can process SA-IDL rules to analyze situations and determine what actions should be triggered [29]. The advantage of using a rule engine for situation analysis is that whenever a rule is changed the rule engine will automatically perform reasoning with the new rule, thus allowing SAW requirements to be changed in runtime; however, such an approach also has the disadvantage that it is difficult for normal users to define complicated rules without making mistakes. Furthermore, in a rule-based approach, situations are inferred based on predefined rules that cannot be changed without human interference. Ranganathan and Campbell [27] developed context synthesizers that can deduce higher level contexts (situations) using machine learning techniques such as Bayesian networks. The context synthesizers with learning capability are used with the context synthesizers based on rules. Combining learning with a rule-based approach may overcome the aforementioned disadvantage but also poses a serious performance problem, as machine learning techniques are computationally intensive and are not suitable for resource-poor mobile devices. Further investigation is required with regard to efficient online learning techniques or proper architecture that can offload the computationally intensive learning task from the mobile devices.

Incorporating Situation Awareness in Service Discovery and Coordination in Mobile Middleware

Service discovery and coordination are generally provided as middleware services, such as the naming service and object transaction service in CORBA, to ease the burden of service and application developers. As discussed earlier, the major advantages of incorporating SAW in service discovery and coordination, compared with traditional non-SAW service discovery and coordination techniques, include the following:

- It can greatly improve the *precision* and *recall* of discovery results [45,46,50,51].
- It enables adaptable service coordination [13,25], which allows efficient and reliable workflow execution in unpredictable and dynamic environments.

Because of these advantages, SA service discovery and coordination can greatly improve the capability of mobile applications and agents to utilize various services in service-based network infrastructure effectively, thus improving the performance and robustness of mobile applications and agents; therefore, mobile middleware should provide services for SA service discovery and coordination. In this section, we discuss the issues and enabling techniques for incorporating SAW in service discovery and coordination in mobile middleware.

Incorporating SAW in Service Discovery in Mobile Middleware

Although various middleware may utilize different service discovery mechanisms, service discovery generally occurs in three phases [46]: (1) service request advertising, (2) matchmaking (i.e., matching service semantics with service requests), and (3) discovery result delivery. Situation information can be utilized in many useful ways in service discovery:

- Situation information can be used to expand service requests to provide more relevant information that is not explicitly specified by users [45,46,51].
- Services may behave differently when situations change; that is, services are situation aware. Situation information is required in the matchmaking phase in service discovery for inferring the service semantics based on service descriptions.

- Situation information can be used to further categorize services for retrieving better results [50]; for example, services can be grouped by their locations, and the services close to the user are returned as the results.
- Situation information can be used in describing users' preferences for different services; for example, people usually make appointments with their regular doctors for routine care, but in emergencies they need to find the nearest hospital emergency room as quickly as possible.
- Situation information can be used by service providers to control their willingness to provide services; that is, the service providers can define policies to determine whether their services are allowed to be discovered. This is a very useful feature, especially for collaborative applications in which users may provide some services through their mobile devices. Although many security mechanisms exist to control access to services, a user may not even want other users to know of the existence of a service running on the user's device in some situations due to performance and privacy concerns. For example, when a user's device is being used for other critical tasks, the user may want to make the services on the user's device temporarily invisible to other users to save resources. No existing discovery techniques, however, can provide this kind of feature.

The following issues must be considered when designing mobile middleware for SA service discovery:

- The service semantics, users' goals and preferences, and service providers' policies under different situations should be described by appropriate languages (R2.1 and R2.2).
- Mechanisms for matching service semantics and users' goals and preferences based on situation information must be developed and incorporated in the matchmaking process (R2.3).

Among existing service description languages, the Web Services Description Language (WSDL) [79] and Ontology Web Language for Web Services (OWL-S) [38] are the most popular ones. WSDL is an XML-formatted language for describing the capabilities of a Web service as collections of communication endpoints capable of exchanging messages. It provides a basic and simple abstraction of Web services. It is a contract or complete description that describes the components being exposed and provides names, data types, methods, and parameters required to call them. The overall structure of OWL-S includes three main parts: (1) the service *profile* for advertising and discovering services; (2) the *process model*, which gives a detailed description of a service's operation; and (3) the *grounding*, which provides details on how to interoperate with a service via messages. OWL-S provides primitives for service descriptions in Semantic Web; however, no existing service description language provides the necessary constructs for describing situations.

Some researchers have developed languages for service descriptions to support context-aware service discovery [45,46,80], but their languages provide only limited capability for expressing current context and do not have the capability to define complex situations. As discussed earlier, various languages have been developed for describing the SAW requirements of mobile applications. These languages provide the necessary constructs for describing situations and the behavior of applications in various situations. Although these languages have not been developed for describing service semantics, the way they model SAW can be adopted and combined with existing service description languages [38,79] to support SA service discovery. Different service providers may use different languages to describe the services they provide or may have different interpretations of the terms used in service descriptions, even with the same language; therefore, among the existing models and languages for SAW, the ontology-based approaches [11,12,27,87–89] appear to be good candidates for incorporating with service descriptions.

The mechanisms for matching service semantics and users' goals and preferences based on situation information depend on how the services are described. For services described using OWL, various inference engines, such as F-OWL [81] and Pellet [82], are available to provide formal reasoning support for inferring service semantics and matching service semantics with users' goals. Broens et al. [45,46] use OWL for describing service ontology, and contextual information is expressed as *contextual attributes* of services. In the matching algorithm [45,46], concept lattices [83] are used to rate the resulting services based on their contextual attributes. Doulkeridis et al. [50] used a multidimensional object exchange model (OEM) graph [84] to represent a service directory in which each service is a leaf node on the graph and contextual information is used to determine which service should be used in different contexts. Based on such a representation of the service directory, a breadth-first search algorithm is used to locate necessary services based on users' requests (goals).

Incorporating SAW in Service Coordination

As discussed before, service coordination is a process of monitoring the status of participant services, invoking proper participant services, managing dependencies among participant services, and propagating necessary information to participant services to ensure the correct results obtained from the coordinated participant services. In mobile computing environments, service coordination is difficult due to the mobility of devices, which makes the set of participant services very dynamic. Recent work on context-aware service coordination [18,66–68] has shown the advantages of coordinating services based on contextual information. Incorporating SAW in service coordination will be more beneficial because not only the current context but also the context history are considered.

Existing mobile middleware provides various coordination mechanisms. Middleware based on tuple space, such as LIME [69], TSpaces [28], and Limone [70], adopts the Linda communication model [85], in which distributed processes communicate implicitly through a shared tuple space. Because of such a communication model, tuple-space-based middleware decouples application behavior and communications between application components and uses the shared tuple space to store application data available to mobile units, which represents part of the context for mobile applications [86]. Murphy and Picco [86] provided a case study based on LIME that further exploits usage of this type of middleware to manage physical context, such as the location and battery power of mobile devices, for context-aware computing. A very nice feature of this type of mobile middleware that makes it suitable for SA service coordination is that it is not necessary for applications and agents to know the participant services in advance. Applications and agents simply specify their requests and advertise the requests in the shared tuple space, and the services that can handle these requests automatically react to the requests. Such a process seamlessly combines service discovery and coordination, and it is very easy to adapt to context/situation changes; however, a problem that must be addressed is that applications and agents do not have much control over the participating services. Although applications and agents can specify the desirable characteristics of participant services, services can ignore such specifications (maliciously or carelessly) and fail the entire service coordination process.

Another coordination approach used by many other types of mobile middleware, such as Gaia, MobiPADS, and RCSM, manages coordination based on predefined profiles or rules and explicitly defines how contexts and situations will affect the coordination. In MobiPADS [25], service chains that explicitly define service coordination are defined in system profiles, associated with various contextual conditions. In runtime, service chains can be reconfigured based on current context and system profiles. In RCSM [2–4,29,53], service coordination is supported by inter-object communications. Rules that determine when an object can interact with which object through what interface are specified during the development of applications. In these rules, situations are used as the condition for

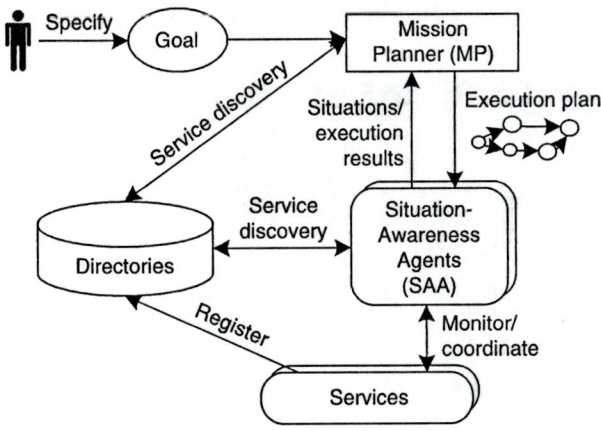

Figure 39.2   SA service coordination.

triggering the communications between objects in runtime to fulfill the application requirements. Object discovery protocols [2,53] are provided in RCSM to address the dynamicity of mobile computing environments. This type of middleware has more control over the participating services, but it is not as flexible as tuple-space-based middleware due to the need to predefine profiles or rules for coordination; also, it requires additional overhead for service discovery.

Yau et al. [13] introduced an agent-based approach combined with artificial intelligence planning techniques to enable more flexible service coordination. Figure 39.2 illustrates the agent-based approach for SA service coordination [13]. It is assumed that the service-based system has a mission planner (MP), which accepts mission goals specified by users and generates execution plans based on available services and current situation. The generated execution plan is a series of service compositions to be executed to fulfill the overall mission goal. A step (service invocation) in the execution plan may have certain dependencies on situations; that is, a step can be executed only when a certain situation is detected. As discussed earlier, it is unlikely that the MP will have complete knowledge of the planning domain in real-world applications, especially in service-based mobile applications where the users are moving, central control over adding or removing services is not available, and services may become unavailable without notifying users. A consequence of planning without complete domain information is that the generated workflows may not be executable due to dependency violations in a service invocation caused by situation changes due to uncontrollable external agents. For this reason, SAW agents must be developed to coordinate the execution of services in the execution plan based on situations. SAW agents are distributed autonomous software entities that have the necessary capabilities to support SA service coordination, including participant service management, agent discovery, context acquisition, and situation analysis.

With these capabilities, SAW agents can adaptively coordinate services in execution plans as follows:

- In each step of the workflow execution, SAW agents check whether all the dependencies on situations are satisfied.
- If the dependency on a situation is not satisfied in a certain step, the SAW agents will check whether this step can be undone.
- If the step is undoable, the SAW agents will first undo the step and then search for an alternative service.
- If an alternative service is found, the SAW agents will resume the execution using the new service; otherwise, the SAW agents will notify the MP of the current situation, and the MP will perform replanning to find another workflow that can fulfill the mission goal.

These approaches to service coordination have their advantages and disadvantages. Further investigation is required to evaluate and compare their complexity and performance to determine which approaches are more effective for SA service coordination.

Summary

In this chapter, we discussed the motivation for developing mobile middleware for situation-aware service discovery and coordination for effective integration of mobile devices with service-based network infrastructures. We presented the current state of the art on situation awareness, service discovery and coordination, and mobile middleware. In particular, we discussed the use of RCSM and MobiPADS mobile middleware for providing context and situation awareness to the application layer. The requirements, design issues, and enabling techniques for mobile middleware for situation-aware service discovery and coordination were also presented. Although context- and situation-aware mobile middleware and various techniques for context- and situation-aware service discovery and coordination have been developed, no mobile middleware exists today that can fully support situation-aware service discovery and coordination. Further investigations on incorporating situation awareness in service description languages, situation-aware matchmaking mechanisms, new coordination models, and an adaptive middleware architecture are necessary for the design and development of mobile middleware supporting situation-aware service discovery and coordination.

Acknowledgments

This work was supported in part by the National Science Foundation under grant number ANI 0123980 and the Department of Defense, Office of Naval Research (DoD/ONR), under the Multidisciplinary Research Program of the University Research Initiative, Contract No. N00014-04-1-0723.

References

[1] Web services architecture, http://www.w3.org/TR/2004/NOTE-ws-arch-2004 0211/.

[2] Yau, S.S. and Karim, F., A context-sensitive middleware-based approach to dynamically integrating mobile devices into computational infrastructures, *J. Parallel Distributed Comput.*, 64(2), 301, 2004.

[3] Yau, S.S., Wang, Y., and Karim, F., Development of situation-aware application software for ubiquitous computing environments, in *Proc. of the 26th IEEE Int. Computer Software and Application Conf. (COMPSAC 2002)*, Oxford, England, August 26–29, 2002, p. 233.

[4] Yau, S.S. et al., Reconfigurable context-sensitive middleware for pervasive computing, *IEEE Pervasive Comp.*, 1(3), 33, 2002.

[5] McCarthy, J. and Hayes, P.J., Some philosophical problems from the standpoint of artificial intelligence, *Machine Intelligence*, 4, 463, 1969.

[6] Pinto, J.A., Temporal Reasoning in the Situation Calculus, Ph.D. thesis, University of Toronto, 1994.

[7] McCarthy, J., *Situation Calculus with Concurrent Events and Narrative*, Stanford University, Stanford, CA, 2000 (http://www.formal.stanford.edu/jmc/narrative/narrative.html).

[8] Plaisted, D., A hierarchical situation calculus, *J. Computing Res. Repository (CoRR)*, cs.AI/0309053, 2003.

[9] Barwise, J., Scenes and other situations, *J. Philosophy*, 77, 369, 1981.

[10] Barwise, J., *The Situation in Logic*, CSLI Lecture Notes 17, Stanford University, Stanford, CA, 1989.

[11] Matheus, C.J., Kokar, M.M., and Baclawski, K., A core ontology for situation awareness, in *Proc. of the 6th Int. Conf. on Information Fusion (FUSION 2003)*, Cairns, Queensland, Australia, July 8–11, 2003, p. 545.

[12] Matheus, C.J. et al., Constructing RuleML-based domain theories on top of OWL ontologies, in *Proc. of the 2nd Int. Workshop on Rules and Rule Markup Languages for the Semantic Web (RuleML 2003)*, Sanibel Island, FL, October 20, 2003, p. 81.

[13] Yau, S.S. et al., Situation awareness for adaptable service coordination in service-based systems, in *Proc. of the 29th IEEE Int. Computer Software and Application Conf. (COMPSAC 2005)*, Edinburgh, Scotland, July 26–28, 2005.

[14] Schilit, B. and Theimer, M., Disseminating active map information to mobile hosts, *IEEE Network*, 8(5), 22, 1994.

[15] Brown, P.G., Bovey, J.D., and Chen, X., Context-aware applications: from the laboratory to the marketplace, *IEEE Personal Commun.*, 4(5), 58, 1997.

[16] Brézillon, P. and Pomerol, J.C., Contextual knowledge sharing and cooperation in intelligent assistant systems, *Le Travail Humain*, 62(3), 223, 1999.

[17] Dey, A.K. and Abowd, G.D., A conceptual framework and a toolkit for supporting the rapid prototyping of context-aware applications, *Human–Computer Interaction (HCI)*, 16(2–4), 97, 2001.

[18] Braione, P. and Picco, G.P., On calculi for context-aware coordination, in *Proc. of the 6th Int. Conf. on Coordination Models and Languages (COORDINATION 2004)*, Pisa, Italy, February 24–27, 2004, p. 38.

[19] Want, R. et al., The active badge location system, *ACM Trans. Inform. Syst.*, 10(1), 91, 1992.

[20] Schilit, B.N., Theimer, M., and Welch, B.B., Customizing mobile application, in *Proc. of USENIX Symp. on Mobile and Location-Independent Computing*, Cambridge, MA, August 1993, p. 129.

[21] Spreitzer, M. and Theimer, M., Providing location information in a ubiquitous computing environment, in *Proc. of the 14th ACM Symp. on Operating System Principles*, Austin, TX, November, 1993, p. 270.

[22] Harter, A. and Hopper, A., A distributed location system for the active office, *IEEE Network*, 8(1), 62, 1994.

[23] Nelson, B.J., Context-Aware and Location Systems, Ph.D. thesis, University of Cambridge, 1998 (http://www.sigmobile.org/phd/1998/theses/nelson.pdf).

[24] Caswell, D. and Debaty, P., Creating Web representations for places, in *Proc. of the 2nd Int. Symp. on Handheld and Ubiquitous Computing (HUC 2000)*, Bristol, U.K., September 24–27, 2000, p. 114.

[25] Chan, A.T.S. and Chuang, S.N., MobiPADS: a reflective middleware for context-aware computing, *IEEE Trans. Software Eng.*, 29(12), 1072, 2003.

[26] Roman, M. et al., A middleware infrastructure for active spaces, *IEEE Pervasive Comput.*, 1(4), 74, 2002.

[27] Ranganathan, A. and Campbell, R.H., A middleware for context-aware agents in ubiquitous computing environments, in *Proc. of the 4th ACM/IFIP/USENIX Int. Middleware Conf.*, Rio de Janeiro, Brazil, June, 2003, pp. 143–161.

[28] Lehman, T.J. et al., Hitting the distributed computing sweet spot with TSpaces, *Comput. Networks*, 35(4), 457, 2001.

[29] Yau, S.S. et al., Development and runtime support for situation-aware application software in ubiquitous computing environments, in *Proc. of the 28th Annual Int. Computer Software and Application Conf. (COMPSAC 2004)*, Hong Kong, September 27–30, 2004, p. 452.

[30] Pokraev, S. et al., *Context-Aware Services: State of the Art*, TI/RS/2003/137, Xerox Corp., Stamford, CT, 2003 (https://doc.telin.nl/dscgi/ds.py/Get/File-27859/Context-aware_services-sota,_v3.0,_final.pdf).

[31] Mostefaoui, G.K., Pasquier-Rocha, J., and Brezillon, P., Context-aware computing: a guide for the pervasive computing community, in *Proc. of IEEE/ACS Int. Conf. on Pervasive Services (ICPS'04)*, Beirut, Lebanon, July 19–23, 2004, p. 39.

[32] Jini architecture specification, Sun Microsystems, Santa Clara, CA, 2001 (http://www.sun.com/software/jini/specs/).

[33] Salutation architecture specification, The Salutation Consortium, 1999 (http://www.salutation.org/spec/Sa20e1a21.pdf).

[34] Guttman, E. et al., *Service Location Protocol, Version 2*, Request for Comments 2608, Internet Engineering Task Force (IETF), 1999 (http://www.ietf.org/rfc/rfc2608.txt).

[35] *Understanding Universal Plug and Play*, White Paper, Microsoft Corp., Redmond, WA, 2000 (http://www.upnp.org/download/UPNP_ UnderstandingUPNP.doc).

[36] UDDI technical white paper, Universal Description, Discovery, and Integration (UDDI), 2000 (http://www.uddi.org/pubs/Iru_UDDI_Technical_White_Paper.pdf).

[37] Klein, M. and Bernstein, A., Toward high-precision service retrieval, *IEEE Internet Comput.*, 8(1), 30, 2004.

[38] OWL-S 1.0, http://www.daml.org/services/owl-s/1.0/.

[39] *W3C Resource Description Framework (RDF)*, World Wide Web Consortium (W3C) (http://www.w3.org/RDF/).

[40] Web service modeling ontology, http://www.wsmo.org/.

[41] Trastour, D., Bartolini, C., and Gonzalez-Castillo, J., A Semantic Web approach to service description for matchmaking of services, in *Proc. of the 1st Int. Semantic Web Working Symp. (SWWS'01)*, Stanford, CA, July, 2001.

[42] Paolucci, M. et al., Semantic matching of Web services capabilities, in *Proc. of the 1st Int. Semantic Web Conf. (ISWC2002)*, Sardinia, Italy, June 9–12, 2002, p. 333.

[43] Paolucci, M. et al., Using DAML-S for P2P discovery, in *Proc. of the First Int. Conf. on Web Services (ICWS'03)*, Las Vegas, NE, June, 2003, p. 203.

[44] Paolucci, M. et al., A broker for OWL-S Web services, in *Proc. of AAAI Spring Symp. Series on Semantic Web Services*, 2004 (http://www.daml.ecs.soton.ac.uk/SSS-SWS04/40.pdf).

[45] Broens, T., Context-Aware, Ontology-Based, Semantic Service Discovery, Master thesis, University of Twente, The Netherlands, 2004.

[46] Broens, T. et al., Context-aware, ontology-based service discovery, in *Proc. of the 2nd European Symp. on Ambient Intelligence (EUSAI'04)*, Eindhoven, The Netherlands, November 8–10, 2004, p. 72.

[47] Czerwinski, S. et al., An architecture for a secure service discovery service, in *Proc. of the 5th ACM/IEEE Int. Conf. on Mobile Computing and Networking (MOBICOM'99)*, Seattle, WA, August, 1999, p. 24.

[48] Yolum, P. and Singh, M., An agent-based approach to trustworthy service location, in *Proc. of the 1st Int. Workshop on Agents and Peer-to-Peer Computing (AP2PC'02)*, Melbourne, Australia, July 14–15, 2002, p. 45.

[49] Liu, J. and Issarny, V., QoS-aware service location in mobile *ad hoc* networks, in *Proc. of the 5th Int. Conf. on Mobile Data Management (MDM'04)*, Berkeley, CA, January 19–22, 2004, p. 224.

[50] Doulkeridis, C., Valavanis, E., and Vazirgiannis, M., Towards a context-aware service directory, in *Proc. of the 4th VLDB Workshop on Technologies for E-Services (TES'03)*, Berlin, Germany, September 8, 2003.

[51] Yau, S.S. et al., Situation-aware personalized information retrieval for mobile Internet, in *Proc. of the 27th IEEE Int. Computer Software and Application Conf. (COMPSAC 2003)*, Dallas, TX, November 3–6, 2003, p. 638.

[52] Debaty, P., Goddi, P., and Vorbau, A., *Integrating the Physical World with the Web To Enable Context-Enhanced Services*, Technical Report HPL-2003-192, HP Labs, Palo Alto, CA, 2003 (http://www.hpl.hp.com/techreports/2003/HPL-2003-192.pdf).

[53] Yau, S.S. and Karim, F., An energy-efficient object discovery protocol for context-sensitive middleware for ubiquitous computing, *IEEE Trans. Parallel Distributed Syst.*, 14(11), 1074, 2003.

[54] Stoica, I. et al., Chord: a scalable peer-to-peer lookup service for Internet applications, in *Proc. of ACM SIGCOMM'01*, San Diego, CA, August, 2001, p. 149.

[55] Balazinska, M., Balakrishnan, H., and Karger, D., INS/Twine: a scalable peer-to-peer architecture for intentional resource discovery, in *Proc. of the 1st Int. Conf. on Pervasive Computing*, Zurich, Switzerland, June, 2002, p. 195.

[56] Crespo, A. and Molina, H.G., Routing indices for peer-to-peer systems, in *Proc. of IEEE Int. Conf. on Distributed Computing Systems (ICDCS'02)*, Vienna, Austria, July, 2002, p. 23.

[57] Kashani, F.B., Chen, C., and Shahabi, C., WSPDS: Web services peer-to-peer discovery service, in *Proc. of Int. Symp. on Web Services and Applications (ISWS'04)*, Las Vegas, NE, June 23, 2004, p. 733.

[58] Nidd, M., Service discovery in DEAPspace, *IEEE Pers. Commun.*, 8(4), 39, 2001.

[59] Chakraborty, D. et al., GSD: a novel group-based service discovery protocol for MANETs, in *Proc. of the Fourth IFIP TC6 Int. Conf. on Mobile and Wireless Communications Networks (MWCN 2002)*, Stockholm, Sweden, September 9–11, 2002, p. 140.

[60] Denny, M. et al., Mobiscope: a scalable spatial discovery service for mobile network resources, in *Proc. of the 4th Int. Conf. on Mobile Data Management (MDM'03)*, Melbourne, Australia, January 21–24, 2003, p. 307.

[61] Helal, S. et al., Konark: a service discovery and delivery protocol for *ad hoc* networks, in *Proc. of IEEE Wireless Communications and Networking Conf. (WCNC 2003)*, New Orleans, LA, March 16–20, 2003, p. 2107.

[62] Berger, S. et al., Towards pluggable discovery framework for mobile and pervasive applications, in *Proc. of the 5th Int. Conf. on Mobile Data Management (MDM'04)*, Berkeley, CA, January 19–22, 2004, p. 308.

[63] Tchakarov, J. and Vaidya, N., Efficient content location in mobile *ad hoc* networks, in *Proc. of the 5th Int. Conf. on Mobile Data Management (MDM'04)*, Berkeley, CA, January 19–22, 2004, p. 74.

[64] Web Services Coordination (WS-Coordination), http://www-106.ibm.com/developerworks/library/ws-coor/.

[65] Web Services Coordination Framework (WS-CF), http://www.oracle.com/technology/tech/webservices/htdocs/spec/WS-CF.pdf.

[66] Andrade, L.F. et al., Coordination for orchestration, in *Proc. of the 5th Int. Conf. on Coordination Models and Languages (COORDINATION 2002)*, York, U.K., April 8–11, 2002, p. 5

[67] Cabri, G., Leonardi, L., and Zambonelli, F., Engineering mobile agent applications via context-dependent coordination, *IEEE Trans. Software Eng.*, 28(11), 1039, 2002.

[68] Julien, C. and Roman, G., Egocentric context-aware programming in *ad hoc* mobile environments, in *Proc. of the 10th Symp. on the Foundations of Software Engineering*, Charleston, SC, November 18–22, 2002, p. 21.

[69] Murphy, A., Picco, G., and Roman, G., LIME: a middleware for physical and logical mobility, in *Proc. of IEEE Int. Conf. on Distributed Computing Systems (ICDCS'01)*, Phoenix, AZ, April, 2001, p. 524.

[70] Fok, C.L. et al., A lightweight coordination middleware for mobile computing, in *Proc. of the 6th Int. Conf. on Coordination Models and Languages (COORDINATION 2004)*, Pisa, Italy, February 24–27, 2004, p. 135.

[71] Haahr, M., Cunningham, R., and Cahill, V., Supporting CORBA applications in a mobile environment, in *Proc. of the 5th ACM/IEEE Int. Conf. on Mobile Computing and Networking (MOBICOM'99)*, Seattle, WA, August, 1999, p. 36.

[72] Campbell, A.T. et al., The Mobiware toolkit: programmable support for adaptive mobile networking, *IEEE Pers. Commun.*, 5(4), 32, 1998.

[73] Yau, S.S., Chandrasekar, D., and Huang, D., An adaptive, lightweight and energy-efficient context discovery protocol for ubiquitous computing environments, in *Proc. of the 10th IEEE Workshop on Future Trends of Distributed Computing Systems (FTDCS'04)*, Suzhou, China, May 26–28, 2004, p. 261.

[74] Doherty, P. and Kvarnstrom, J., TALplanner: a temporal logic-based planner, *AI Mag.*, 22(3), 95, 2001.

[75] Davulcu, H., Kifer, M., and Ramakrishnan, I.V., CTR-S: a logic for specifying contracts in Semantic Web services, in *Proc. of the 13th Int. World Wide Web Conf.*, New York, May 17–22, 2004, p. 144.

[76] Gordon, A.D. and Cardelli, L., Equational properties of mobile ambients, *Math. Structures Comput. Sci.*, 13(3), 371, 2003.

[77] Clarke, E.M., Grumberg, O., and Peled, D.A., *Model Checking*, MIT Press, Cambridge, MA, 2000.

[78] Olston, C. and Widom, J., Best-effort cache synchronization with source co-operation, in *Proc. of ACM SIGMOD Int. Conf. on Management of Data*, Madison, WI, June 3–6, 2002, p. 73.

[79] Web Services Description Language (WSDL) 1.1, World Wide Web Consortium (W3C), http://www.w3.org/TR/wsdl.

[80] Mostefaoui, S.K. and Hirsbrunner, B., Context aware service provisioning, in *Proc. of IEEE/ACS Int. Conf. on Pervasive Services (ICPS'04)*, Beirut, Lebanon, July 19–23, 2004, p. 71.

[81] F-OWL, http://fowl.sourceforge.net/index.html.

[82] Pellet OWL Reasoner, http://www.mindswap.org/2003/pellet/index.shtml.

[83] Ganter, B. and Stumme, G., *Formal Concept Analysis: Methods and Applications in Computer Science*, Technische Universitat Dresden, Germany (http://www.aifb.uni-karlsruhe.de/WBS/gst/FBA03.shtml).

[84] Stavrakas, Y. and Gergatsoulis, M., Multidimensional semistructured data: representing context-dependent information on the Web, in *Proc. of the 14th Conf. on Advanced Information Systems Engineering (CAiSE'2002)*, Toronto, CA, May 27–31, 2002, p. 183.

[85] Gelernter, D., Generative communication in Linda, *ACM Comput. Surv.*, 7(1), 80, 1985.

[86] Murphy, A. and Picco, G., Using coordination middleware for location-aware computing: a LIME case study, in *Proc. of the 6th Int. Conf. on Coordination Models and Languages (COORDINATION 2004)*, Pisa, Italy, February 24–27, 2004, p. 263.

[87] Yau, S.S. and Liu, J., Incorporating situation awareness in service specifications, in *Proc. of the 9th Int. Symp. on Object- and Component-Based Real-Time Distributed Computing (ISORC 2006)*, Gyeongju, Korea, April 24–26, 2006, p. 287.

[88] Yau, S.S. and Liu, J., Hierarchical situation modeling and reasoning for pervasive computing, in *Proc. of the 3rd Workshop on Software Technologies for Future Embedded and Ubiquitous Systems (SEUS 2006)*, Gyeongju, Korea, April 27–28, 2006, p. 5.

[89] Yau, S.S. et al., Support for situation awareness in trustworthy ubiquitous computing application software, *J. Software Pract. Experience*, 36(9), 893, 2006.

Section 7

CURRENT EXPERIENCES
AND ENVISIONED
APPLICATION
DOMAINS FOR
SERVICES BASED ON
MOBILE MIDDLEWARE

Chapter 40

Mobile Middleware for Integration with Enterprise Applications: WebSphere® Everyplace® Access

---

David Reich

CONTENTS

Introduction

This chapter outlines a way to extend traditional enterprise applications and their associated programming models such as Java 2 Enterprise Edition (J2EE) from the desktop client down to mobile devices. These mobile devices are not just laptop or table computers, but rather handheld devices such as PDAs, personal communicators, or even cellular telephones. This technology allows us to extend the applications to these devices, not rewrite them. Herein lies the power of these programming models — the ability to extend applications and the associated middleware to new modes of interaction.

The goal of mobile middleware is to provide information to users anywhere, anytime, and on (almost) any device. Although various segments of the mobile marketplace focus on end users (e.g., location-aware services such as mapping) and consumers (e.g., games and perhaps somewhat integrated PIM and e-mail functions), mobile middleware focuses on knowledge workers. *Mobile knowledge workers*, in this context, are those who conduct business transactions on their mobile devices. They can be sales representatives who need to place orders and query inventory and delivery schedules or insurance adjusters working in disaster areas. They can even be rental car bus drivers taking passengers to a parking garage to pick up their cars; these bus drivers have to be able to check passengers in on the bus and show them where their cars are. All of these applications are different from the consumer or traditional end-user of a mobile device. These are enterprise transactions that have to be conducted on the go, with intermittent and, as far as the application is concerned, indeterminate states of connectivity.

The requirements for mobile enterprise transactions include local data, secure transaction processing, and reliable data synchronization, as well as acceptibility to the user community, seamless and transparent application provisioning and upgrades, transparent synchronization and transaction transmission, and, ideally, the same application user interfaces as those on their desktop machines. Also desirable are alternative access methods depending on the device — for example, voice access for a very restricted screen size (or no screen for that matter) or in-car operation. This chapter focuses on the technologies to bring the enterprise or knowledge worker transactions to mobile devices. First, it presents the big picture and then proceeds to dissect it into its component parts to see how they all interoperate.

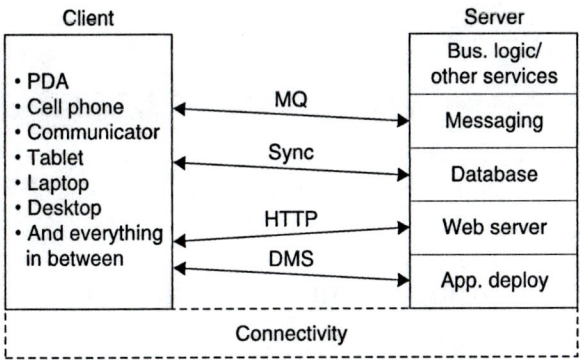

Figure 40.1 Enterprise middleware high-level interaction architecture.

Mobile Enterprise Middleware

Figure 40.1 provides a high-level view of how all of the elements of mobile enterprise middleware come together. The client (the left side) along with the application server (the right side) and the components that enable the mobile enterprise transaction capability are the main elements. A lot is going on inside this seemingly simple diagram, but it is very well componentized and is all based on industry standards, making understanding the interactions (which we will go into shortly) and the programming model very straightforward. Also note that at the bottom is a connectivity layer.

In the mobile space, one must assume that connectivity is intermittent, or occasional. While it is certainly feasible for users to (re)connect wherever they happen to be at the moment, that can be cumbersome. One example of a network "roamer" is the WebSphere® Everyplace® Connection Manager (WECM). WECM is a product from IBM that enables network roaming across intermittent connections. Not only does it provide secure virtual private network (VPN) tunneling, but it also supports a hierarchy of network connectivity, whereby it will suspend a virtual connection when, for example, a user roams between wireless access points or even from a wired to a wireless to a dial-up physical connection. This form of network roaming allows applications or application services to maintain a virtual connection regardless of the physical network, suspending transactions in the middle to be completed when the connection is reestablished, regardless of whether the new connection is through the same physical medium. For the sake of the discussions here, connectivity is assumed, and we will focus on the enterprise transactions for mobile devices.

Let's move on now to the client stack and the server elements. First, understand that, while Figure 40.1 shows the server elements in one box, it is certainly possible and in most cases desirable to separate many of these elements across multiple servers in the infrastructure. They were placed this way in this first figure to ease the understanding of all of the elements and to simplify the graphic.

Elements of a Mobile Enterprise Client

In a mobile enterprise client, many things have to be taken into account. The first thing that comes to mind is the application programming and presentation model. Because of the heterogeneous nature of the mobile market, this seemingly simple decision can be very complex. After all, not only do you have to contend with the differences in the different types of devices, but you also have to deal with the differences between manufacturers of similar devices (e.g., all of the cellphone makers and service providers).

Java provides the Java Micro Edition (J2ME™) specifications to help address this quandary and profiles and configurations to address the granularity in capabilities of the different classes of devices; however, a number of challenges remain:

- How does the user start and stop applications?
- How do the applications get delivered to the device? How are they updated and maintained?
- What kind of overhead is required to run an application? After all, one Java Virtual Machine (JVM), one application.
- What can be done to provide enterprise services to these limited capacity devices?

Figure 40.2 shows a block diagram of the client stack. The native operating system (OS) of the platform along with the JVM form the foundation of the stack. On top of the native OS runs applications the reader is used to. These include the device configuration, games, perhaps calendar or messaging functions, and so on. Where the enterprise part of this comes in is on the Java side of the client. The JVM is generally used to run only one application at a time; however, through the power of the OSGi (http://www.osgi.org) framework and, in this case, the IBM reference platform called the Service Management Framework (SMF), multiple applications and services can be provisioned and dynamically started and stopped on this single JVM instance. This is vitally important in a limited-footprint device.

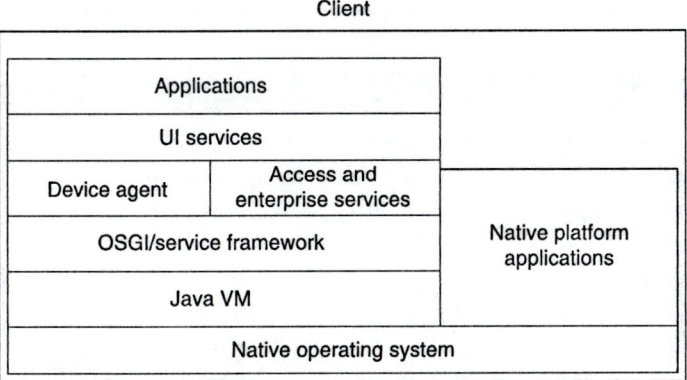

Figure 40.2 Client stack.

OSGi provides a framework for componentizing applications into bundles and also a specification for delivering these bundles to a client through a device management server (DMS). In Figure 40.2, the client part of the device management is shown as the *device agent*. It is through this DMS mechanism that applications or, more specifically, bundles are delivered to the client.

OSGi is designed to facilitate the structure, lifecycle, execution environment, and provisioning of applications (components) to clients. The OSGi specification offers a framework for bundle structure, starting, stopping, allocation, and release of resources and system, as well as application services to be added together to provide and consume services to construct an overall software stack. This stack can provide services to other applications as well as consume services. One of the biggest advances of OSGi is that it enables the sharing of a JVM instance among multiple applications. Traditionally, Java applications are allocated one per JVM; when the application terminates, the JVM is torn down and a new one instantiated for another application. The OSGi framework allows and enables applications to share the JVM, thus creating a new lifecycle model for applications, especially where the teardown and restart of a JVM can be very expensive. The OSGi platform (SMF) lives as an application on the JVM (refer back to Figure 40.2) and hosts the other bundles and applications to provide these services to the system and application code.

Things That Must Be Served

The next part of the diagram as we move up the stack is "Enterprise and Access Services." These are platform extensions that enable applications to access functions such as database, message queuing, and even the presentation container. These are detailed later in Figure 40.4, but for now just understand that this set of extension services is what allows coding to Java Database Connectivity (JDBC) and message queuing (MQ) services and provides services for presentation. These extensions are themselves bundles and can be referred to as *system bundles*, because they extend the platform, as opposed to *application bundles*, which are, as the reader might guess, applications. In addition to the links to enterprise functionality, this set of access services also provides the container for the execution of the J2EE functionality on the device. In many cases, server processing is required to complete a task. In the J2EE model, this processing is done with Servlets, beans, Java Server Pages (JSPs), and the like. This services layer provides the containers that make these services available to the applications.

So how does this all get onto the device? This question can require a complex answer, based not on technological issues but on business ones. You can safely assume a device has a native operating system. You might also be able to assume that it comes with a JVM (this is not always the case, but as time goes on more and more device makers are preinstalling Java). The big question, then, is how does the SMF core platform and extension services get to the device?

We begin by looking at the business issues. SMF does not have a zero footprint. It takes up space. It takes a piece of work to install it on a device. It is not something that the end user of the device should do, or should even want to do. End users should not have to know about anything called SMF; they should just know that they can do their work on their devices. So, it might be up to the operating platform (software) vendor to include it in their image. Perhaps the device makers (who usually customize the native operating system) will add it to the image and preload it on the devices. It could also be the service providers or carriers, such as cellular phone companies, who load it on the devices after they get them from the manufacturers. Still another way the platform could get to the device is in an information technology shop; for example, suppose a large insurance company wishes to equip its 10,000 agents with a device and use SMF as the enterprise application platform. This insurance company might preload the devices with the platform and then send the agents on their way.

When the platform is on the device, the rest is technologically straightforward. OSGi specifies a framework for delivery of bundles to a system. These bundles can be application bundles, further system bundles, or even updates to existing bundles. This is the beauty of OSGi. If the user wants to start a program, the platform looks to see if all of the required bundles are present, checks the server to ensure that the bundles that are already there are the correct version, dynamically downloads the new or

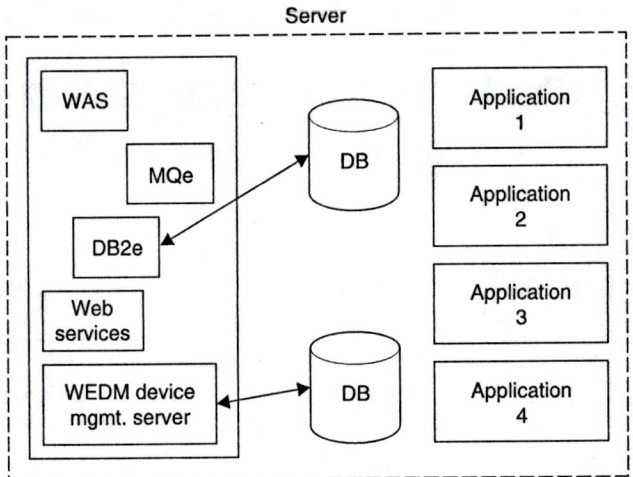

Figure 40.3 Server components.

updated bundles, starts them up, and runs the application, with all of this other work being done out of sight of users. All the users know is that they have started an application. The rest of the work is set up by system administrators and performed dynamically, over the air, with no user interaction required.

The Server Side

Now that we have examined the coarse-level architecture of the client and the basics of how these elements get to the client, let's now turn our attention to the server. Figure 40.3 shows the same coarse-grained architecture for the server as for the client. Recall that, for the sake of explanation, all of the servable elements are on one server. In practicality, this could be implemented across multiple servers in the infrastructure.

The core server platform in this case could be constructed from different products. We could create a DMS server from Tivoli software on, say, a Linux™ platform, with database services from a different place, Web serving from still another place, and so on. Or, IBM has an offering called WebSphere® Everyplace® Access (WEA), in which the server is integrated with all of these components in either a loosely or a tightly coupled environment. The WEA server provides the infrastructure and integrates all of the components seen in Figure 40.3. Although a WEA server can do it all, it is apparent that the server has several distinct duties to perform:

- Serve up the application and system bundles to the clients.
- Provide database access and synchronization gateways to the database engines.
- Run the database engines.
- Provide a messaging gateway for secure transactions to an application server.
- Run the application server engine (Web application server).
- Publish and provide Web services.
- Run server-side application logic.

So, as we can see, the WEA server can do all of this, but we will want to separate these functions into distinct servers in the overall IT infrastructure. For discussion purposes and, in fact, for architectural purposes, we can consider utilizing this one server, especially if we use the WEA product, rather than trying to build all of the parts ourselves.

This leads to an interesting side note: Because the elements of OSGi, Java, and WEA are built on widely accepted industry standards, it is possible to put one's own choice of components into just about any part of this infrastructure. WEA offers the advantages that everything has all been put together and tested as such and it does not lock customers into utilizing the entire product. If other server elements are already in place or some things do not specifically suit particular needs, because these standards are adhered to it is even possible to create one's own infrastructure elements.

From Top to Bottom, Through the Mobile Enterprise Middleware Stack

Let's discuss the server side for a moment, then move on to taking this discussion a little deeper technically and walking through a real-life example to bring the picture back together. At the beginning of the stack, where applications come from, is the device management server. This brings us to Figure 40.4, where the action begins to happen. The figure illustrates a further detailed breakout of both the client and server that enables enterprise applications to perform on mobile devices. In this particular scenario, the client uses a Web container (browser) presentation for the application. The stacks are shown in their functional layers, but the transactions as the application loads and executes are shown from the top down in between the stacks.

In this example, the application is an order entry for, say, a mobile sales representative. Via the browser, the sales rep (user) instantiates an application. This application may or may not be present on the device,

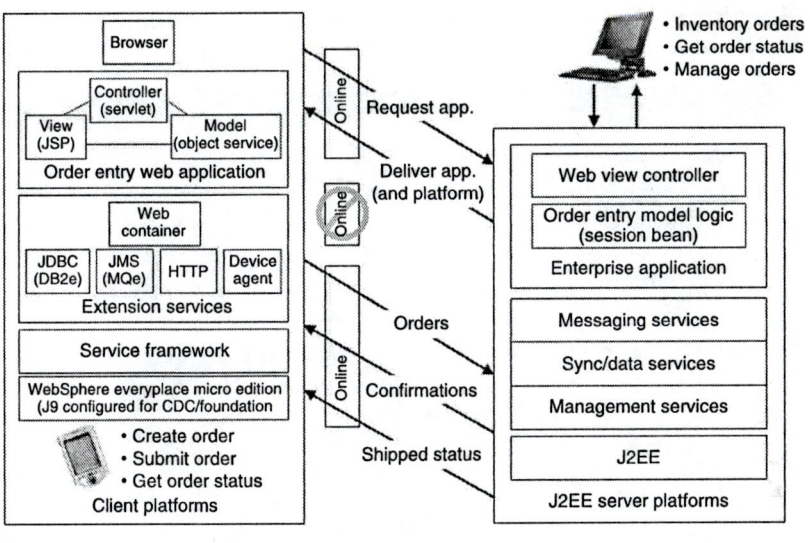

Figure 40.4   Detailed flow across mobile clients and server.

but at least the SMF platform and a link for the application are already there. Recall that the core platform (J9 JVM and SMF) is on the device courtesy of the manufacturer, service provider, or IT shop, so the device is ready to receive and run application bundles. The bundle behind the activation of the application comprises at least part of the application. At this point, the SMF platform queries the server to determine if any system bundles have to be updated and also checks the application bundle manifests to determine dependencies and communicates with the device management services to develop the list of needed application bundles.

The system and application bundles are delivered to the client, and the application is instantiated. Now, the obvious question here is what happens if no live connection to the server is available. If the platform is fully functional, the latest-version dependency checks can be bypassed (dependent on the configuration by the administrators when the platform was loaded and configured). If the application bundle being instantiated has all of its (cascading) dependencies met, then the application can run without a live connection. If any of these tests fail, the platform will return an error telling the user that not enough of the platform or application is present to run the application, and the user will have to wait until he is connected to perform the requested work.

Assume that the application is there or a live connection is available. Now the application is on the device and displays forms to the user. In this scenario, the application is order entry. An important piece of an order entry application is ensuring that inventory exists to satisfy the order (or seeing that inventory is on order) so the sales representative is not placing an order that cannot be filled. This order entry process can take one of several paths.

To enable offline operation, some local data must be present to perform transactions. This, along with all other options, is totally under the control of the application designer and programmer. This data is stored locally in a system such as DB2® Everyplace® (DB2e), designed for small-footprint devices. Using JDBC, applications can access the data stored in DB2e databases. Of course, it is possible to code the application to attempt to contact a live server database, but, depending on the device, connectivity options, or reliance on such a connection, it may be desirable to have a several-step approach whereby the application queries the active connection state and, if the connection does not exist, goes against the local data store (DB2e), with optionally a check to see when the last synchronization was to determine if the data is too stale to be reliable. It may be preferable to not even bother to try to go to a live database server because of the overhead required or the nature of the device, where a connection could go away in the middle of the transaction. These are all tradeoffs based on the nature of the application and the level of offline interaction required.

Looking back at the order entry application, the database can be queried to determine available inventory and show the results to the user. Another function a sales representative must perform is entering new orders. This can be for an existing customer or for a new customer. Again, based on the design decisions of the application programmer, the customer information can be queried from the local database (or a synch will have to be triggered if the data does not reside on the device), or, in the case of a new customer, a new customer order transaction would be submitted

Assuming that the customers for this sales rep's region have been synchronized to the device, the data for an existing customer is readily accessible to the application and the form can be completed with the customer information. (Depending on the nature of the business, the device, and so on, it may have been decided to synch the entire customer database to every client device based on the parameters of the business and client platforms.) If this is a new customer, however, no data exists, and the sales rep will have to decide how to get this new customer record into the company's data systems. This is the same decision process that must be gone through for other types of transactions as well, such as entering new orders.

Several options are available for processing the new data. One is to have the user enter the data and store it in the local database. At some time in the future, this data will makes its way to the company's central servers via a synchronization operation. When a new customer is entered into the system, some other business processes might take place — for example, offering a welcome gift. This requires some processing outside the client (data entry) device and business logic back on the server. (It may be desirable to initiate a welcome gift from the client, but it seems to make more sense to do this from the server.) In a traditional application against a live server, the welcome gift can easily be triggered when the data is entered. Due to the mobility of the client device, however, it is necessary to look at new ways of triggering server-side actions with the application from a device that may or may not be connected at the time of the transaction. This can be accomplished in a number of ways.

A server-side trigger could be set up such that when new data is synchronized to the central database, business logic is activated (such as sending the welcome gift for a new customer). Or, a system such as MQ (or MQe, for MQ Everyplace®), also part of the big WEA picture, could be utilized for secure, reliable delivery messaging; this message could take the form of a "new customer entered and here is the customer ID number" message. Rather than rely on a server-side trigger to notice when new data is there, however, it might be preferable to send the new customer transaction to the server in the MQ message. The data could be taken from the message and added to the server database, and an acknowledgment could be sent back to the client to be added to its database. This begs the question of what happens if the customer places another order before the data is loaded into the central database server and synchronized back to the client devices, but such design decisions are based on the nature of the business and must be made to ensure that the business processes are reliable and no data is lost.

It is important to note that this is not a radical departure from application design seen before. This is just intelligent partitioning of the application assuming intermittent states of connectivity and using an application architecture other than EXE or large JAR files (using bundles instead). The beauty of all of this is that if you use this architecture for your applications, knowing that a device is a device is a device, the application runs on handhelds to desktops and everything in between, with no (or at least minimal) changes to the application. This is really scaling your applications.

So, back to the application. We are now at the point where the inventory information is available, the customer information has been input (or obtained from the customer database), and the order is ready to be input. Using the same theory as for the customer information, a list of available products is displayed (pulled from the local copy of the data), and the order information is entered into the form. Local application logic and form validation can be used to ensure, for example, that:

- Sales reps are not trying to sell more than they have.
- Customers are not trying to order more than their credit limit.
- The requested delivery dates are reasonable based on current inventory information.

Herein again lies the power of the OSGi/SMF platform and application bundles. We can front-end much of the processing into the device to avoid round-trips to the server on intermittent (and usually low-bandwidth) connections and allow users to work anywhere, anytime, and on almost any device, with the same user interfaces as if they had gone to the home office to process orders.

When the order is ready to be placed (the form is completed and validated), the user will press the "submit" button. At this point, the decisions to be made are similar to those made when entering a new customer record. Fundamentally, we have only three types of transactions:

- New record entry (new customers, new orders)
- Query (inventory, customer information)
- Update or delete existing records or data

For all of these, it will be necessary to look at the attributes of the transactions and the business processes behind them to determine how they will get into the enterprise infrastructure. Realize that this is a different type of transaction than a new customer record. It may not be of timely importance to add a new customer record to a central database, but orders must be processed quickly so the inventory can be updated and customer service is prompt. Based on the same parameters analyzed to determine how to process a new customer record, the best solution may be simply to queue up the data for transmission to the central server (see Figure 40.4). It may not be necessary to store the transaction locally, but it would be good to receive notification when the order is received at the central system, because the inventory will have to be adjusted, accounting processes will have to be triggered, and other business processed will have to be initiated. This is not to say that just storing the data locally for later synch and server-side triggering is wrong. There is almost no right and wrong way to do this; it is up to a particular business's needs. Again, coming back around to the WEA server, all of this is standards-based work, so it is possible to plug different elements into the infrastructure and leverage the code across platforms. Users not being locked into a technology and using the standards allows reference implementations on multiple platforms (servers and devices).

To close out this transaction, assume we have a live connection and the message is transmitted quickly. An acknowledgment comes back to the device, and the order application is updated with a confirmation number. At this time, a message may also get pushed to the device, where an application bundle removes the items that were just ordered from inventory, or the server sends a message telling the user that the inventory database now must be synchronized. All of these tools are enterprise functions enabled on the client device through SMF and bundle management and the J9 JVM.

User Interface Considerations

Thus far, we have assumed a Web container presentation for this application, but this class of applications has a number of user access mechanisms. Although a Web container is assumed on every device for presentation, it is possible to have a rich graphical user interface (GUI) experience on the device, as well as even voice or multimodal applications.

Rich Graphical User Interface

The J2ME programming model provides for several classes of GUIs based on the device characteristics. Mobile Information Device Profile (MIDP), CDC/Foundation, and Personal Profile provide for native Java UI widgets and rich user interfaces, but they raise the issue of coding different UIs for different classes of devices. In addition, the Java widgets were designed for desktop systems, and on embedded devices they can be very limiting.

The Standard Widget Toolkit (SWT), introduced with the Eclipse™ platform (more on Eclipse later in this chapter) was designed to be a rich widget toolkit that maps easily to native widgets for a platform. ("Easily" meaning that for application programmers it is easy, but quite a bit of work for the Eclipse development team to port to different platforms. That is really the goal of all of this, though — to make it easy for the application developer.) The SWT runs across multiple platforms and has an embedded version under development at the time of this writing. This means that it is possible to create a user experience much richer than just browser-presented forms and have it consistent across the range of client devices supported. The use of standards-based enterprise elements results in very low maintenance costs across the range of devices in the infrastructure. Looking back at Figure 40.2, this is the UI services layer of the client stack.

Other User Interfaces

Although we can create rich GUI interfaces with the core J2ME API, the IBM product portfolio includes other user interface technologies that can be used with the client stack. One such example is voice-enabled applications. The UI access services layer has a number of extension points that allow plugging in different user access methods. In the case of voice, the WebSphere® Voice Server and Voice Application Access Server provide voice browser and voice portal functions, respectively. They can be put into the infrastructure and, again, because of the adherence to standards, can be used as a user access device in place of, say, the visual browser. Applications could be written to emit VoiceXML (another industry standard) rather than Hypertext Markup Language (HTML), and, with the voice servers connecting to the application via Hypertext Transfer Protocol (HTTP), just like a visual browser, a voice-enabled device could access these enterprise services, as well.

Voice access adds another dimension to the mobile experience because, let's face it, devices are getting smaller but our fingers are not. By putting a voice server in front of this entire mobile infrastructure or, alternatively, putting a voice engine and the browser inside the device, we can extend access to devices that may otherwise be out of reach for enterprise applications. Examples would include devices that traditionally have no keyboard, mouse, or screen or perhaps situations where hand- and eye-free operation is essential (for example, a system built into an automobile).

Because of the componentized nature and standards used by WEA and the enterprise mobility stack, again you can see how you can easily plug in components to enhance the user experience.

Eclipse, the RCP, and SWT

Another facet of the rich user experience is the Workplace Client Technologies, Micro Edition (WCTME). WCTME is the marketing term for the client described in this chapter, with yet another user access alternative. The Eclipse project (http://eclipse.org), an open source project, was originally conceived as a plug-in framework for writing software development tools. It has morphed into a more general-purpose platform — specifically, the core Eclipse platform of a plug-in architecture, tool bar, help subsystem, and update manager framework. On top of that is a graphical editing framework, a source editing framework, project navigators, builders, wizards, and other tooling functions. It was discovered that Eclipse can be a very nice general-purpose, plug-in framework when the software-tooling-specific functions are dropped.

In version 3, Eclipse adopted the OSGi framework for the foundation of the plug-in architecture because of the dynamic load/unload nature of bundles, as well as the JVM sharing and bundle provisioning features mentioned earlier in this chapter. So, now the Eclipse platform turns into a general application platform, a rich user interface alternative to Web container application presentation. This general-purpose platform is called the Rich Client Platform (RCP). Along with SWT, RCP has an embedded version (eRCP) under development to run on small footprint devices.

Eclipse and SWT are ported to different platforms, and when that happens it is possible to write applications with a rich UI to Eclipse on that platform. The Eclipse RCP and SWT allow us to write applications to the Web container user interface, replace the browser with voice access, use the core J2ME UI frameworks, or completely forget about what OS is running on the device and write a rich UI to the RCP and SWT. Doing so further insulates users from the underlying operating system platform, and they do not have to worry whether the application will be deployed to Windows, Linux, or any other platform under Eclipse. Best of all, this is all, once again, standards based (SWT and open source Eclipse, OSGi, and Java), thus providing the flexibility to mix and match as desired.

Application Development Tools

The remaining piece to any application development puzzle is the tools that developers use to write the applications. IBM offers a toolkit based on Eclipse, called the WebSphere® Studio Device Developer (WSDD). This toolkit targets J2ME devices and enables a set of extensions (plug-ins) to allow development of OSGi bundles and linkages to other enterprise middleware functions such as Web services. WSDD provides linkages to the Rational® tools, such as Rational Application Developer (RAD) and Rational Web Developer (RWD). Specifically, if a copy of RAD or RWD is detected when WSDD is installed, WSDD can be installed either separately and distinctly from the Rational tools or as a linkage. If the latter option is chosen, end-to-end application development is enabled in one Integrated Drive Electronics (IDE) interface, from the J2EE, database, and Messaging and Web services artifacts to the RCP/SWT and client/bundle functionality. The full details of the toolset are beyond the scope of this book, but information can be obtained from the IBM Web site or the link mentioned in the chapter summary.

Final Thoughts on Enterprise Integrated Mobile Middleware

This chapter has illustrated how enterprise applications can be made mobile through the OSGi stack; current presentation standards including Web container, Eclipse RCP, and SWT; and advanced UI alternatives such as voice access. This picture has actually been available in varying flavors for a while now, and recently IBM delivered a fully functional server that can serve up any or all parts of this infrastructure in WEA. WEA is the mobile-middleware-enabling server that provides the services for WCTME clients. How the server components are configured in terms of what runs on what boxes in the infrastructure depends on the business processes that will be supported. The net result is that a standards-based, flexible, full-functioned server and client infrastructure exists today for mobile access to enterprise applications. Readers can learn more about this infrastructure and obtain a working case study, complete with code, by looking for the IBM Redbook number SG24-6496.

Chapter 41

Context Middleware for Adaptive Mobile Services

Theo Kanter, Carl-Gustav Jansson, Martin Jonsson,
Fredrik Kilander, Wei Li, Peter Lönnqvist,
and Gerald Q. Maguire, Jr.

CONTENTS

Introduction

Motivations

Increasingly, wireless access becomes heterogeneous, consisting of cellular systems interspersed with IEEE 802.11 wireless local area network (WLAN) access and other access technologies (3G and beyond). This poses new challenges for delivering services to users who move about in such a heterogeneous infrastructure. Users of mobile communication can access services via multiple access networks with varying properties in terms of price, data rates, latency, error rates, etc. Multiple devices may be used to interact with services, which further adds to the complexity faced by a user. Some of these devices will allow truly mobile access due to their form factor, function, etc. (e.g., mobile phones or mp3 players with wireless interfaces); therefore, to facilitate mobile communication, we must enable automatic rearrangement of communication, services, and interaction with services. New usage patterns are expected to emerge as a result of minimizing the effort required by users to arrange and manage their communication, services, and interaction. *Adaptive* services can make *opportunistic* use of available communication. In addition, adaptive services must be centered around the user, as we can no longer rely on a single network or single operator to arrange and manage these services. This has caused us to investigate new architectural choices that move service control and coordination closer to the user.

Challenges

A number of challenges must be addressed as a result of this approach. Relocation of service control and coordination mandates investigating the properties of a service architecture that centers services on users, devices, and objects that people may use. These services as well as the arrangement and the delivery of services must take into account the goal the user is trying to achieve. Information from sensors provides data related to the user's communication or to the user's context; this defines a service *context*. Thus, services can adapt and hence can be context aware. How do we model such context data? How do we create or refine context data from raw sensory data? Given that context data is available, how do we make it available to different entities that participate in arrangement of the service? How can context data help minimize the user's exposure to an increasingly diverse infrastructure? Finally, what are the benefits of adaptive services for mobile users?

Adaptive and Context-Aware Services

The Adaptive and Context-Aware Services (ACAS; see Figure 41.1) research project [1] at the Center for Wireless Systems at KTH is part of the Adaptive Wireless Service and Infrastructure (AWSI) research program [2]. ACAS has investigated the specific challenges involved in creating support for groups of users who move within a heterogeneous wireless infrastructure to utilize adaptive user-centric Internet services. The ACAS project identified three main research areas: (1) adaptive user interaction with services, (2) seamless adaptive services, and (3) a smart adaptive infrastructure.

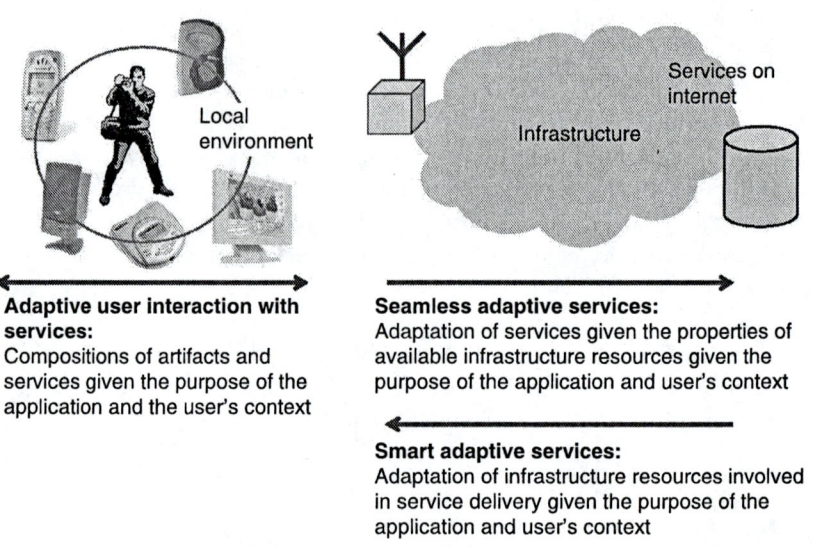

Adaptive user interaction with services:
Compositions of artifacts and services given the purpose of the application and the user's context

Seamless adaptive services:
Adaptation of services given the properties of available infrastructure resources given the purpose of the application and user's context

Smart adaptive services:
Adaptation of infrastructure resources involved in service delivery given the purpose of the application and user's context

Figure 41.1   Overview of the ACAS project.

Related Work

Numerous attempts have been made to create context-aware systems supporting personal mobility and context awareness. Here, we address only those that are most relevant to the challenges listed earlier. The Aura project [3] and related work on SenSay, a context-aware mobile phone [4], are investigating how a small set of sensors may relieve the user from being constantly aware of and having to manage the state of the telephone. The Solar system [5] is a prototype implementation of a graph-based abstraction for context collecting, aggregation, and dissemination of (sensor) generated events passing one or more operators and finally being delivered to a subscribing application. The Context Toolkit [6] supports the development of context-aware applications using context widgets that provide context information to applications. The lowest level interfaces to a physical sensor, the middle layer is concerned with abstracting and combining data, and the highest level coordinates the underlying components and provides a callback interface to applications.

The Web Architectures for Service Platforms (WASP) [7] was designed to support context-aware applications specifically in a 3G environment. Use of Web services technologies and the WASP Subscription Language (WSL) to communicate with the platform connects context-aware applications with context providers (sensors) and third-party service providers.

The Context Recognition by User Situation Data Analysis (Context) Project [8] studied the characterization and analysis of information about users' context and how to use it in adaptation. The project developed a context-aware application on a Nokia smart phone that predicts its owner's behavior as it learns about the user's preferences by logging calls and user location and noting when applications such as cameras are used.

Service Architecture Framework

Introduction

In response to the challenges outlined earlier, the ACAS project endeavored to design mobile middleware that would enable plug-and-play services in both the local area and the wide area with a minimal set of *a priori* knowledge, going beyond the design goals of Universal Plug and Play (UPnP™) or Jini™ [9]. This mobile middleware is intended to enable automatic configuration of services between entities without having to know the services or their properties beforehand.

The Context Information Network

The proposed infrastructure for supporting adaptive and context-aware services and applications is a context information network, which acquires context data from sensors, processes context information, and distributes context information to context managers (CMs) within the context information network to make it available to services and applications via context APIs. The context information network provides context management, context manager discovery, and policy-based management. The proposed service architecture of the context information network consists of application-layer entities (Figure 41.2) with three different abstraction levels:

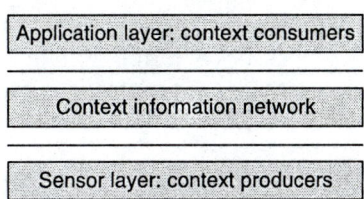

Figure 41.2 Service architecture layers.

- *Application level* — The application level has end consumers of context information closest to the applications. Applications themselves are typically not interested in low-level sensor data; instead, they prefer to react to particular abstractions such as: "A friend is close right now."
- *General level* — The general level is where context is distributed and abstractions of a general nature are created.
- *Sensor level* — This level is closest to physical or virtual sensors. This level transcribes or interprets sensor readings into context information languages that can be subjected to generalization by reasoning.

Context Networks Using the SIP Presence Framework

The previous subsection proposed a context information network as an intermediate middleware layer for the acquisition, processing, and distribution of context information. The purpose of this middle layer is to connect producers and consumers of context information and enable application-level entities (e.g., people, meeting places, robots) to find services and facilitate automatic decision making regarding the allocation and invocation of services, using available devices and resources.

The Internet Engineering Task Force (IETF) SIMPLE working group extended the Session Initiation Protocol (SIP) by creating SIP for Instant Messaging and Presence Leveraging Extensions (SIMPLE) [9]. These extensions, together with presence and watcher information event packages, the publish mechanism for presence, and the event notification extension for collections fulfill important requirements derived from the challenges outlined above; therefore, we chose SIMPLE as the base for our context information delivery infrastructure and extended it to create a context information network. This choice of framework was also important as far as providing input regarding future steps in standardization efforts in the 3rd Generation Partnership Project (3GPP) in liaison with the Open Mobile Alliance (OMA). Both efforts focus on creating an application and service infrastructure based on SIP/SIMPLE. Furthermore, our choice of SIP as the underlying protocol was not only because of its popularity and availability, but also because it provides user addressing, session initiation, and control; however, our middleware goes beyond simply that of a SIP/SIMPLE infrastructure, as will be seen as we examine the details of the middleware.

Context Information Networks

A context information network has context sources and context-aware applications, distributed over a set of mobile and fixed devices. As the core of our context network infrastructure, we propose *context management-*

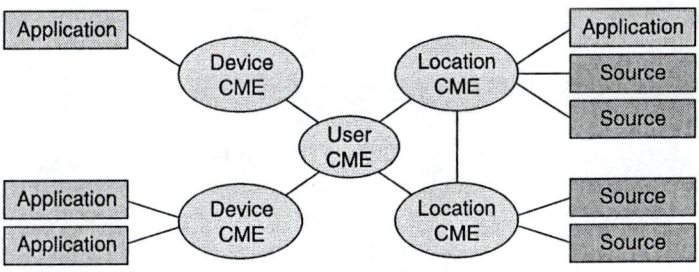

Figure 41.3 Context management entities.

enabled entities (CMEs; see Figure 41.3). Applications connect these CMEs to access relevant context information. By interconnecting these CMEs, requests for context information can be forwarded between entities, allowing for context information to be shared between applications and between devices. Interconnected CMEs form a context information network in which subscribers and publishers may deliver their context information (Figure 41.3).

We argue that general context information distribution can be efficiently achieved by establishing a network of CMEs. These CMEs have a uniform core that is surrounded by the appropriate functionality to make it deployable in applications, have a user and organizational infrastructure, and are on top of sensors (Figure 41.4). Context information in the network

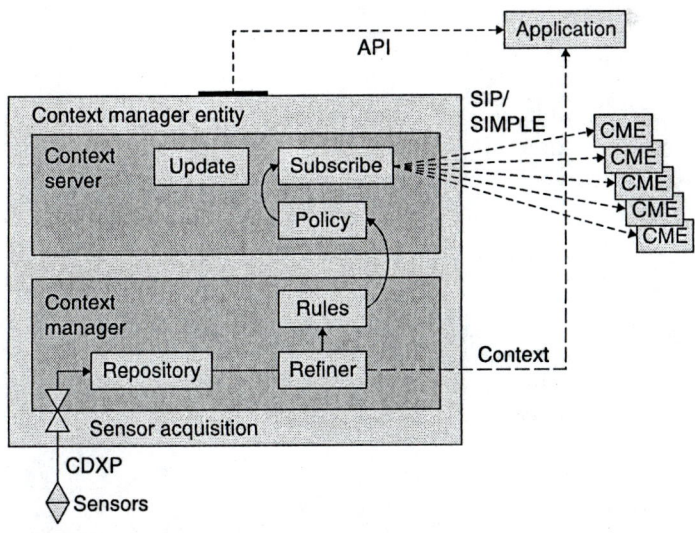

Figure 41.4 Generalized context management entity.

is likewise expressed in a common form to facilitate rule-guided selection, translation, and inference. When access to the context manager is mediated and policed, we say that it is a *context service*, a publicly recognizable, addressable, and tangible entity. Clients wishing to consult a context manager for updated context information use the service interface (API) where they may or may not be served for reasons of security, personal integrity, or workload [11].

The CMEs consist of a context manager and a context server (the client and server part of the context management network). Context-aware applications access managed context information through the context server API. The context server uses service policies to make decisions regarding whether to serve a request or not and passes them to the context manager for processing and management. CMEs use the SIP for addressing and accessing other local or remote CMEs and can use the Context Data eXchange Protocol (CDXP) [12] to transport context information directly between hosts, which is more efficient than SIMPLE; this is important for wireless links.

Context Sensing

Context information originates in sensors. Sensors measure features of the physical world (such as temperature or acceleration) or logical properties of a computer system (such as disk allocation, service availability, or an error condition).

Physical Sensors

The information available from a physical sensor is usually in the form of a number or string label. Very often the raw sensor data must be interpreted according to some model to be meaningful. The conversion to a meaningful temperature that includes the relevant units is assumed to be done by appropriately written, installed, and calibrated driver software. The driver can reuse the same physical reading to present a number of interpretations (e.g., Fahrenheit, Kelvin, or Celsius). Depending on the connection topology, the same driver instance may utilize several physical sensors in different locations or a group of sensors that cluster around a shared power and communications resource.

Software Sensors

Software sensors are typically built into software, and the properties they measure are based on the behavior and particulars of other software. One

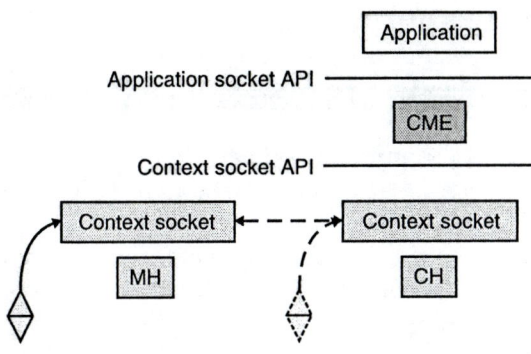

Figure 41.5   Acquisition of sensor data.

particular software sensor of interest is the service sensor. The sensing part of the service sensor has components that are able to detect the presence of services such as file servers and printers. Service sensing and detection techniques are typically employed in client–server systems where clients discover services dynamically.

Distributed and Shared Context Sensing

In addition to SIP, the context manager is also able to communicate using CDXP [12]. This allows for more efficient transportation of context information, which is advantageous when local applications wish to communicate with the context manager, as they may use the loopback interface to achieve inter-process communication. When the IP address of an entity is known, CDXP may also be used to minimize overhead and delay.

The Sensor Sampling Control Protocol (S2CP) [12] and CDXP are two core components of a context-aware infrastructure. CDXP is used for subscribing to and sharing context information between CDXP clients and servers — for example, as shown in Figure 41.5 located on a corresponding host (CH) and a mobile host (MH), respectively. S2CP is used to control sampling of sensors. A CDXP server will trigger a S2CP component as needed to obtain local sensor information to serve requests.

Context Description Language

The context description language used in the ACAS project is based on the eXtensible Markup Language (XML). The central item of the language is the context element, an object composed of the following attributes (all names are in the ACAS namespace):

id — A unique identifier

value — The value of a property of some entity

datatype — The datatype of the value, such as integer, real, string, or XML

unit — The property to which the value refers

entity-reference — The Uniform Resource Identifier (URI) to the entity that the context element describes

time — Time and date when the value was captured or composed

source uri — The URI to the entity that captured or composed the context element

source content — Optionally, a description that explains the context element

Here is an example of a context element formulated in XML:

```
<acas:contextelement id="123c">
    <acas:value datatype="integer"
         unit="temperature/kelvin">292</acas:value>
    <acas:entity-reference rel="acas:dsv.su.se/k2/r7741/t"/>
    <acas:time>Sat Apr 24 00:05:21 CEST 2004</acas:time>
    <acas:source uri="uri:acas:dsv.su.se/k2/csf/apax"/>
</acas:contextelement>
```

Context Managers

Context information originates in sensors; it is made available by publishing and reaches clients through subscriptions. The gap between fundamental sensor information and the abstract relationships convenient to end-user applications is bridged by context refinement. This refinement occurs in the CME's refiner. Adaptation and context-based responses can extend the standard behavior of an application. It is desirable that the effort of extending an application with context awareness is small, thus increasing the probability that application writers will use context information. Another consideration is that many different applications are likely to benefit from the *same* context information. Location, for example, is a powerful concept that can be exploited in many ways; thus, it is likely that several applications concurrently executing for the benefit of a particular user can utilize the same location information and hence should share subscriptions. We hypothesize that endpoint applications in a context information network are best served by subscribing to a minimum of information. In practice, this means that they will receive only context data elements with abstractions that are *directly* and *immediately relevant* to the application (thus using the *correct* minimum of information). This simplifies the matching process in the application, as it reduces the size of the state description that must be maintained, and (for battery-powered and wireless units) it reduces the amount of information that must be exchanged across the network.

Our approach is to avoid duplication of effort and to move context refinement out of the applications and into the context data network. This is achieved by allowing the clients to program the context manager. In the simplest scenario, applications pass a set of context rules (originally provided by the application developer) over the context API and are subsequently notified whenever the state description generated by the output of the rules is updated.

Context-sensitive applications on a laptop or personal digital assistant (PDA) would benefit from an onboard context service. If the context managers embedded in the applications can pass on their subscription requests to a device-specific context service, they will each require less processing and memory to serve their applications. The device context service in turn will be able to collapse subscriptions for common context information (e.g., location) and thus reduce network traffic. The details are given later in this chapter.

Context Servers

The context server provides a gateway to the context manager and thus polices access to the information within. This is necessary because there may be situations when the owner of the context server is willing and ready to share parts of the context repository with certain other clients. Policing subscription requests means that the request must be evaluated, accepted, or modified before being submitted to the context manager; for example, the user may be willing to share fine-grained location information with personal friends and coarse-grained location information with an employer or customers, but may prefer to reveal no location to everyone else.

The context server aggregates, refines, or creates context information for one or more subscribers. In general, the CME has a service interface, a refinement and subscription module, and a client role. Remote subscribers announce themselves and their needs on the service interface. The CME splits the subscription into two parts, one that it believes it can answer locally and one that must be obtained by resubscribing at other CMEs in its client role. When updated context information becomes available, the CME notifies each subscriber according to the local part of the subscription. We believe CMEs have three distinct and vertically ordered classes (see levels provided earlier).

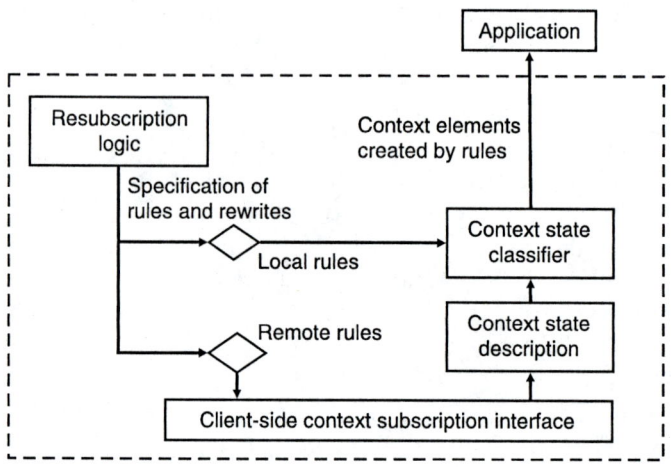

Figure 41.6   Application-level CME.

The application-level CME (Figure 41.6) is loaded or consulted by an application to serve that application with context information. In this case, the CME's service interface is likely to appear as a separate thread of execution, controlled by a set of functions in the application's address space rather than as a remote service, although the latter is not prohibited. Likewise, notifications from the CME are implemented as callbacks to functions specified by the application, again within the address space of the application.

The general CME (Figure 41.7) appears as a stand-alone service, running either in a single instance on a small device or on behalf of a user or organizational element on a persistent and connected server computer. A general CME implements the full remote service interface and client role, as well as the capacity to host multiple subscribers. A sensor CME (Figure 41.7) transforms sensor data to the context description languages in use. This CME has the full service interface and subscription machinery, and the client role is replaced by access mechanisms to whatever sensors this CME supports.

Typically, each context-sensitive user application would be equipped with an instance of an application. Each of the user's computers (PDA, laptop, PC) would have a general CME local to the device which consults a persistently installed general CME dedicated to the user. The user's personal general CME is distinguished, not because of its technology but by its role. Its peers are the personal CMEs of other users and those CMEs that represent parts of the environment (rooms, stores) or an organization.

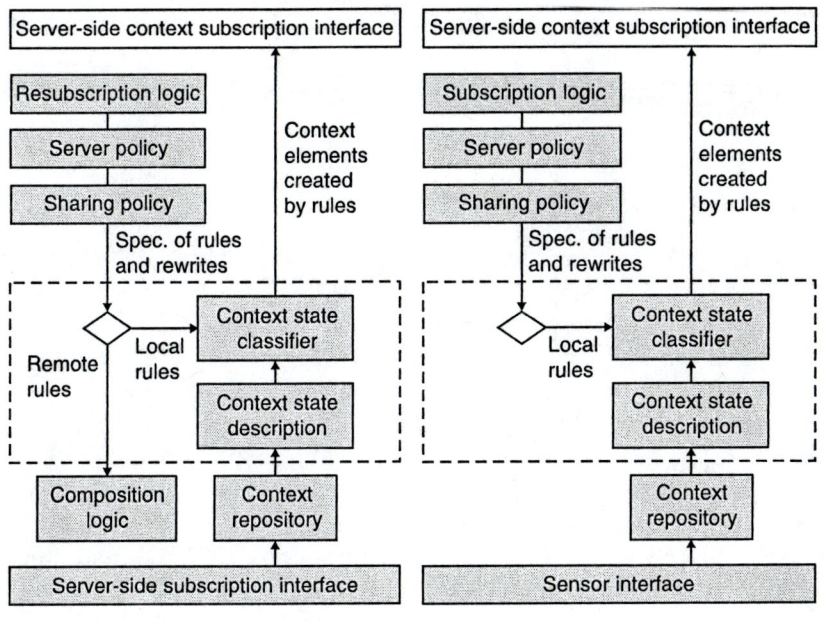

Figure 41.7   General (left) and sensor (right) CMEs.

Distributed throughout the network, running on whatever hardware is appropriate, are sensor CMEs. These provide underlying data that is transformed into information. Access to context information is controlled by a set of policies in the context server of each CME. We do not specify how these are to be defined, expressed, or implemented, nor do we dictate how clients identify themselves or the mechanism used to authenticate them. All this is future work.

Context Repository

To maintain state descriptions for client applications, even though the clients themselves are not online, a user's personal general CME generally executes on a robust, stationary, and main powered processor, with access to nonvolatile storage. This context repository of the CME is the primary store of context information that has been requested by a CME's owner. This repository could be layered on top of a database management system so the CME can be restarted without adversely impacting applications. This context repository is updated based on sensor data or context data arriving from other CMEs. Due to the ephemeral and time-varying nature of the data, the CME is not obligated to retain the information in the repository longer than is needed to serve the current set of subscriptions.

If resources are limited, then the CME should reject new subscriptions, rather than attempting to serve the current set of clients with insufficient resources.

Processing Context Information

Processing of context information ideally occurs on three distinct levels, as presented earlier. At the lowest level, sensor readings are converted into observations expressed in a suitable format, thus becoming context information. On the top level, applications consume these observations or abstractions drawn from them and adjust their behavior to comply with application-specific logic. Between the top and bottom levels, middleware in the form of a network of CMEs ensures that context data is routed and abstracted according to the needs of applications. Our central concept builds upon a context management entity. A CME must have the ability to cater to several subscriptions simultaneously; therefore, at the application library level, multiple threads are used. Below, we provide some of the details about modules present in a context management entity.

Context Refinement

Context refinement and subscriptions are unified in the Tryton rule language, an experimental production language developed by the ACAS project for this purpose. Essentially, when a client wishes to subscribe to context, it composes a Tryton program in which production rules imply the necessary context information. The left-hand side of each rule requires the corresponding context information to be available in the context manager, and the right-hand side specifies the context element that is to be sent to the client. The context manager must examine the left-hand side of each rule and assign it the following properties:

- The context relation is known and can be provided locally (as in the context manager for a sensor).
- The context relation is local to the rule set (an abstraction produced by the right-hand side of other rules in the subscription).
- The context relation is unknown and must be subscribed for elsewhere.

For the last case, the local context manager turns to other context managers that have knowledge and requests subscriptions as needed.

Context Subscriptions

Publishers and subscribers are the writers and readers of context information that have a means to communicate and therefore are able to establish a subscription relationship. A subscription consists of two parts: (1) a header with essential properties (such as the address back to the client) and (2) a Tryton program that specifies the context to retrieve. When the context repository has accumulated a sufficient data set, the Tryton program is executed against it and any productions it yields are sent back to the subscriber. A subscription should include at least the following properties:

- The subscription ID
- Address of the subscriber
- A matching expression provided by the subscriber
- A matching expression used by the publisher
- Expiration time of the subscription
- Confirmation interval of the publisher
- Minimum update interval of the subscriber
- Credentials provided by the subscriber

As can be seen from the example of the context rule provided below, the subscriber needs to subscribe to context elements that contain the location of user fk. The current value of this element is then compared against the temperature readings currently available in the state description. To find a match and thus deduce the temperature in fk's current location, a subscription for the corresponding elements must exist. In our example, we know that there exists a dependency between the result of the user location pattern (the inner select in the Structure Query Language [SQL] code example) and the subscription for temperatures. Whenever the location of user fk changes, the temperature subscriptions must be reevaluated and possibly modified.

```
SELECT value FROM state
    WHERE unit = 'temperature/celsius'
    AND eref = (SELECT value FROM state
            WHERE unit = 'location/room'
            AND eref = 'staff/fk')
```

Reasoning Issues

Each Tryton rule in a subscription can examine a copy of a simple element found in the context repository, or it can also execute a computed element (based on several elements, including the output of other rules from the same subscription). This enables the Tryton language to create abstractions based on the state of the context repository; however, the unbounded, forward-chaining nature of the language makes it a potential efficiency hazard, not to mention the considerable problems with respect to ontologies and how to choose and express them. This is not particular to the Tryton language but is a general concern for any effort to perform automated abstraction. A typical example of a Tryton rule is one that equips an entity that has a location with the Centigrade temperature of that location if a Kelvin reading is available. In the rule, variables correspond to context elements in the context repository; strings such as `location`, `eref`, and `unit` are properties of an element. Running the rule involves making all possible matches of variables to elements and evaluating the condition. With more than two variables and a large context repository, the effort is considerable, and as a result a potential efficiency issue arises.

Context-Aware Service Discovery

Approach

We envision context-aware services that are built upon peer-to-peer networks. Service discovery is an important and required mechanism for CMEs to find each other and to discover which one provides a given service or where context information is located. Context managers should be able to automatically formulate subscription requests for context information by analyzing the rules given to them by applications. To subscribe with useful sources (other context services), CMs must have some notion of where to send them. Proximity-based detection combined with URI exchange is one way to find this information [15]. Another way is to use static configurations and long-term relationships with trusted parties who always subscribe to the same information. The policies of a context service to restrict uninhibited data exchange must be built upon longer-term relationships or the existence of searchable directories over public context information and context information sources. An example of "friend-of-a-friend" information propagation is described in Delgado [16].

For service discovery in a local area network (LAN) or metropolitan area network (MAN), we initially used SIP/SIMPLE for the acquisition and exchange of context information between peers; therefore, we have investigated the conditions and possible design choices involved in using SIP/SIMPLE in peer discovery and handovers for building reconfigurable application networks in a heterogeneous infrastructure. As a result of this work, we propose a Service Peer Discovery Protocol (SPDP) that uses SIP/SIMPLE but is better targeted at supporting service discovery in LANs and WANs than previous protocols (e.g., SLP). Details are provided below.

We also anticipate certain interesting scenarios in which the user moves in and out of different local environments where support for context information exists. Later in this chapter we cover specific requirements and methods for both local and remote service discovery in Bluetooth® personal area networks (PANs) and methods for proximity-based discovery, in which Bluetooth is used as the carrier.

Peer Discovery Protocol in LANs and MANs

The Service Peer Discovery Protocol is designed to discover services and to offer a model for negotiating services between endpoints without requiring third-party negotiation [16,17]. SPDP is defined as an extension of the SIP event framework [9]; thus, its messages are carried in SIP packets and it inherits all the request routing and security features of SIP, as well as using SIP for naming and localization of users. SIP event notification defines two methods, SUBSCRIBE and NOTIFY, to handle signaling of events. All the messages are expressed in XML following the knowledge representation described earlier. In our prototypes, we used a simplified representation of the entities and services organized in taxonomy trees for straightforward and unique identification of the resources. These simplifications did not affect measurements regarding the performance of the discovery process but did avoid addressing issues of ontology.

The protocol identifies two types of entities: (1) *user agents*, which are uniquely identified by their SIP URIs as they move in the network and which are served by zero or more (2) *context registrar servers* (CRSs), which store peer presence and capabilities. A peer can locate a CRS using SIP localization for servers, based on the Domain Name System (DNS). The mapping of CMEs onto SDPD entities is straightforward, as CMEs can be designed to assume the role of a SDPD user agent that is colocated with the SIP user agent, or a CME can assume the role of a CRS that is colocated with a SIP server.

The Service Peer Discovery Protocol has some characteristics that distinguish it from other service protocols. It interacts very closely with SIP, which it uses as a carrier for the protocol messages. SDPD uses XML to represent the protocol information and has a more flexible representation of services as compared to the Service Location Protocol (SLP) [18]. SDPD relies on unicast to perform discoveries, which makes it better suited for peer-to-peer communication in wireless scenarios. Delgado [16] has shown that SDPD, in comparison with other approaches (e.g., SLP), is more traffic and time efficient and much better suited in terms of flexibility and scalability to support searches in peer-to-peer scenarios in wireless networks, leveraging the expressiveness of XML.

Peer Discovery in Personal Area Networks

Service discovery is a basic function in a PAN. The properties of a PAN differ from those of a LAN in two areas: (1) A PAN is a highly dynamic environment, so service discovery must be carried out frequently to maintain an up-to-date picture of the available services of the local network; and (2) the service discovery protocol should minimize the use of communication. The native Bluetooth Service Discovery Protocol (SDP) is inefficient for several reasons [19] and should be enhanced to work with services on top of the IP. In our solution, we considered two different scenarios: (1) a PAN composed only of Bluetooth-capable devices, and (2) a PAN composed of Bluetooth- and wireless local area network (WLAN)-capable devices. In the latter case, context information about Bluetooth devices and additional service providers, attached to an IP infrastructure, are known in the WLAN and distributed via the WLAN to the PAN gateway; thus, this information can be known in both networks.

Bluetooth-Capable Devices

In the case of Bluetooth devices, we enhanced the native Bluetooth SDP with a coordinator-based service discovery [19]. Coordinator-based service discovery relies on a central coordinator node that holds the complete service information for all Bluetooth nodes in range. The coordinator, however, must be a device with sufficient computational, power, and memory resources. The result is that a device entering a PAN must establish only one connection and perform only one service discovery process.

Bluetooth- and WLAN-Capable Devices

When the PAN is composed of Bluetooth- and WLAN-capable devices, context information about Bluetooth devices and additional service providers attached to an IP infrastructure are known in the WLAN. In the first scenario, we assume that one or more devices supporting both Bluetooth and WLAN technology are present to access the services provided by the WLAN devices, and a Bluetooth device sends its request to the coordinator. If the coordinator is not able to answer this request, it forwards it to a node with access to other networks. In the same way, a WLAN device that wants to request a service in the Bluetooth network forwards that request to the node supporting both Bluetooth and WLAN technology [19]. The implemented service discovery allows faster service discovery than is otherwise possible with the native SDP for more than one device and a coordinator in range; for example, if a user entering the PAN discovers the coordinator before the end of the inquiry process, the application can be much faster.

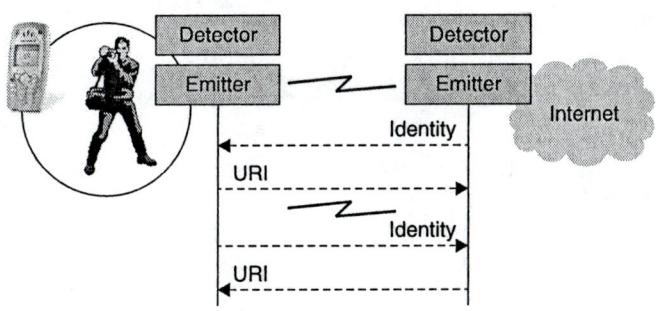

Figure 41.8 Proximity-based discovery.

Proximity-Based Discovery

For a mobile user, we anticipate certain interesting scenarios in which the user moves in and out of different environments where support for context information exists. Detectors and emitters are mounted in fixed locations or worn by mobile users (Figure 41.8). When two emitters and detectors make contact, each notes the identity of the other (Media Access Control [MAC] address or similar) and responds with a URI that is the address of its own context service. The two parties, user and room, are now aware of each other and can receive or exchange whatever context information each sees fit to expose to the other.

Proximity Detection

We have developed a set of detection services for use in a ubiquitous computing environment. The first implementation was based on the MICA mote [20]. We equipped mobile users with motes to locate their positions (based on stationary motes in known locations where users were likely to be or to pass through). This did *not* turn out to be an efficient means to discover the user's location due to the time delay and uncertainty regarding the ability of a mobile mote to establish communication with another (stationary) mote in a new location; however, it is very cost efficient to use motes to acquire context data about a *static* environment.

Another detection service we implemented is based on Bluetooth. It runs on a desktop PC with a 3COM Bluetooth USB dongle with BlueZ (a Linux Bluetooth Stack) installed on Redhat Linux 9.0. A nearby user carrying a Bluetooth mobile phone or PDA will be discovered by this service. This discovery process is reasonably fast (less than 2 seconds for five devices simultaneously), as we only inquire for the Bluetooth device address (BD_ADDR). The detection service then tries to obtain a vCard [22] from the discovered devices using OBEX [22], a standard protocol for information exchange over Bluetooth supported by most Bluetooth mobile phones and PDAs. If that attempt is successful, we can then extract a temporary token from the "home address" field of the vCard file. Privacy issues this approach raises are discussed further later in this chapter.

Please note that we have used the low-level Bluetooth device address only for bootstrapping the association of the local infrastructure with a user's home personal CME (PCME) systems, not for identifying users. The user and the user's PCME are addressed by the temporary token set by the user (or by autonomous applications) which is intentionally changeable according to the user's configuration (such as once an hour); however, we are aware of the possibility of permanently identifying users via Bluetooth device addresses. The reason why we selected Bluetooth for our prototype was the wide support for Bluetooth by many commodity devices, such as mobile phones, and its low power consumption.

The token generated from proximity detection in a Bluetooth cell could be formatted as 12345@anonymizer, which gives the proxy's domain address ("anonymizer") and a pseudonym-enabled neutral reference number ("12345"), which is the PCME's address in that proxy. Only the proxy knows how to map this reference to the user's PCME address.

Service Allocation

One way to make portable devices such as cellular phones, PDAs, and laptops even more useful is to interconnect them in a PAN; via this PAN, a single dual-interface device could enable all local devices to access services in other networks. To be able to reach the devices in the PAN from outside in a simple manner, some location-independent form of addressing may be used. One attractive approach is to use SIP. For a user anywhere in the Internet, SIP enables that user to reach another user within the PAN by an address that does not depend on where the user is located. To make life more convenient for the callee, the selection of device for an incoming multimedia invitation can be made automatically, by utilizing context information when selecting the device; consequently, a method is necessary for allocating services to the devices that are most suitable for the service and for the moment. The current context in the PAN will affect this decision. This context information includes the current status and capabilities of devices, the user, and the user's surroundings. The environment of a PAN is dynamic so this context information may change. This knowledge has to be reflected in the service allocation. Jansson [23] has investigated and proposed a context-driven, policy-based method for how this service allocation is best performed, building on the

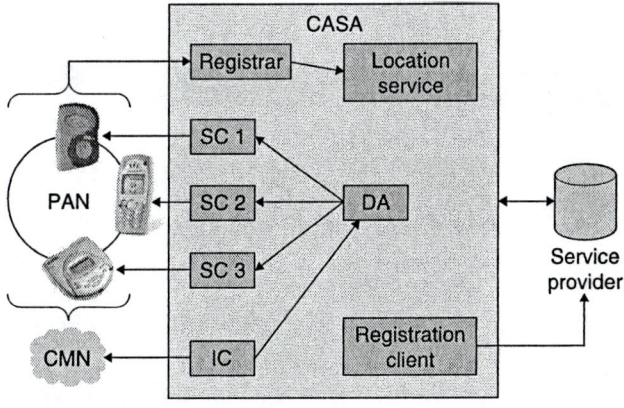

Figure 41.9   Context-aware service allocation.

results of Avgeropoulos [24]. The Context-Aware Service Allocator (CASA) system eases communication between users in two ways.

The allocation will be based on context information from the PAN, user settings, and information contained in the incoming request (e.g., the media descriptor located in the SDP body); consequently, high demands are placed on the accuracy of the context information. The decisions made by the CASA must always be based on up-to-date information. A goal of the system design is to provide the allocating algorithm with information that is as current as possible.

The CASA system acts as an SIP proxy server to devices in the PAN that register with it, hence it provides a registrar server. This local registrar will work closely with the location service, which keeps mappings between the network address and the SIP address stored in a hash table. Observe that the SIP addresses registered at the CASA will only be used *internally*. The location service will later be used when a service allocation occurs and the network address is required to locally route the SIP message to the appropriate device; thus, CASA acts as a proxy. Each device will be managed by a *service controller* (SC) (see Figure 41.9). This SC is responsible for communication with the device, and consequently an incoming message will be passed on to one or more SCs by CASA. The service allocation process determines which SCs will be selected if the message is not part of an existing transaction. If it is, the message will be sent to the SC that is already participating in the transaction. All communication between the PAN and outside nodes will go via CASA, which works as a switch for these SIP messages. When a register request is received from a device in the PAN, it is switched to the registrar.

Discussion

Performance measurements related to CASA considered three different methods for fetching context information:

- *On-demand information fetching* — When an incoming request arrives, the IC will be consulted to get context information. The IC connects to the context management network and queries the CME for the information.
- *Prefetching* — The IC will periodically contact the context information network and fetch context information that will be stored locally.
- *Call-backs* — The CASA will be fed continuously with fresh information in the background, such that when the information is needed by the discovery agent it will already be there, fully updated.

We have seen that the context information fetching method affects the accuracy of the allocation, utilization of the context management network, and delays associated with allocating a service request in different ways.

Jansson's thesis [23] discussed the different tradeoffs and performances. Depending on the algorithm, the allocation time can be as low as 10 msec for 100 devices; for example, prefetching (at a 1-msec interval) can make a difference but should not be done uniformly. CASA should use prefetching for a period of time and then fetch on demand before switching back to prefetching when the delays add up above a certain threshold. CASA would benefit from a service discovery method that could identify available services and include them in the set of services provided in the PAN. Such protocols exist [19] but are not yet integrated.

Service Description Format

The proposed service description format consists of several parts. First, each service is identified with service name and a globally unique service ID:

```
<service name="ProjectorControl7514"
        id="hostname.dsv.su.se:foobar"
        targetNamespace="urn:acas:servicedescription"
        xmlns:acas="urn:acas:servicedescription"/>
```

The next part consists of static metadata concerning this instance of the service, relating mostly to the function of the service. The dominant part is a keyword list that can contain any number of keywords to be used when querying for services. The readable-description tag becomes very important in semiautomatic service discovery tasks, where in the end a user has to make a selection among a selection of services, and this tag is what is presented to the user:

```
<serviceDescription>
        <readable-description>
                "Projector control service for the projector
                        in lab room 7514"
        </readable-description>
<serviceType>
        "projectorControl"
</serviceType>
<keywordList>
        <keyword>projector</keyword>
        <keyword>projectorControl</keyword>
        ...
</keywordList>
...
</serviceDescription>
```

Because the description format is intended to be a metalevel format, it contains only information that is related to the service discovery process and little information regarding how the service is to be invoked, which methods it uses, etc.; however, the document must contain pointers to an instance of the service for the description to be useful. Additional descriptions provide information on how and where the service is implemented and the protocols the service implements for communication.

One of the main reasons for inventing a new description format is to make it possible to provide information about the context in which the service is embedded. This information is encoded in the same context information protocol as the context information for the sensor data used within the rest of the system — namely, in the form of context element records. By using the same context representation, the same rule machinery can be used to process the service description documents to enable context-aware service discovery. The context information embedded with the document could contain information about the location of the service, if it is owned by a specific person, etc.

Managing and Protecting Context Information

In the case of managed networks, the operator can access and collect context information in *context registrar servers* (CRSs) and set access policies to ensure privacy. In an unmanaged network consisting of peers, the users themselves must set policies to protect their privacy. A CRS must, as is true for any service, protect its resources against abuse, mischief, and other kinds of unauthorized intrusions. Because we are dealing with data captured by real-world sensors and the possibility that the data can be used for monitoring and tracking a user, the information must be used appropriately on behalf of the user. To this end, we assume that access to a CRS requires proper authentication by the client and a means is in place for the CRS to verify this authorization. Obviously, this does not prevent a public CRS from accepting anyone as a client (i.e., providing public context information to all who ask). Once the client's identity has been established, three kinds of filters may be applied before the client's subscription is incorporated in the context server's set of requests.

The first of these filters is the server policy, which specifies the general conditions under which to accept a request or not, without any particular focus on the identity of the client; instead, the server policy is concerned with computational resources and possibly priorities between categories of clients. The second filter is the sharing policy, where access to sensitive information is regulated on a per-client basis, very similar to user rights in a database management system. Note that this policy is local to the context service instance; no central authority exists to fall back on, and if a distribution system is in place it must be able to cache policy definitions for use in fragmented networks. Finally, the third filter is the abstraction policy. It is an elaboration of the second one which makes a binary decision. With an abstraction policy, it becomes possible to serve part of a request, without giving complete information. The location of a person, for example, can be given with high accuracy to family, friends, and rescue services, while other people and organizations may have to be content with less detail.

Protecting Privacy in a Public Service Infrastructure

To simplify context acquisition about a physical or virtual environment (e.g., a room, a vehicle, an organization, or some real or abstract boundary), we define a CME that has context generators attached to hardware sensors that measure physical properties or software sensors that measure computation and communication properties of the CME. To protect the user's real identity in their communication with context-aware systems (Figure 41.10), we introduce the Anonymizer proxy, which is designed to hide the real communication address from others even if they have been granted some (context) information access. In other words, through the proxy, the communicating parties can exchange information (such as context) without revealing their communication addresses and real-world identities. This behavior is also referred to as *pseudonymous access*. It is

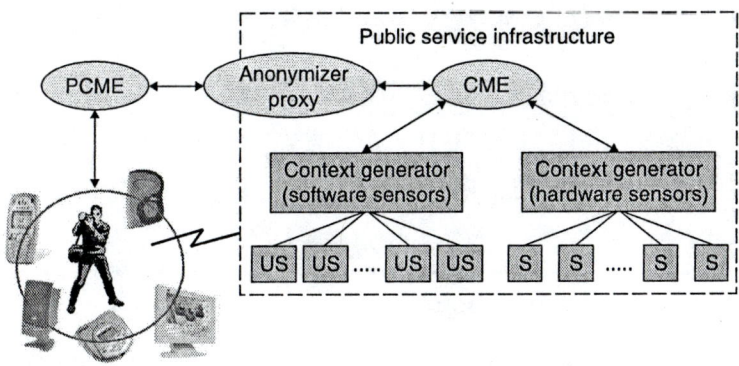

Figure 41.10 Protecting privacy in public service infrastructure.

analogous to the scenario where people in an online chat room establish discussions without knowing each other.

The Anonymizer proxy is situated between the communicating parties (e.g., the context consumer and provider). It translates between some temporary address and the real communication address, and then it delivers the message (reconstructed) to the destination party. To achieve better privacy, more than one proxy can be placed between the communicating parties. This is comparable to mixers in anonymous networks [25]. We expect that an Anonymizer proxy could be provided as a commercial service as with a mixer, with the larger the volume of use, the better.

Mobile Applications and Services

This section discusses results from previous prototypes, works in progress, and a field study in the ACAS project.

Context-Aware Call Delivery

In the case of call delivery in an SIP telephony infrastructure, the user may have several options regarding how to receive the call, depending on available devices in a certain location. Available devices could be shared or personal. Requiring the user to make manual decisions would burden the user. Ayrault [19] demonstrated automatic discovery of devices with mixed Bluetooth and WLAN capabilities that are near the user. As noted earlier, service allocation can be as fast as 10 msec for 100 devices. By colocating a PCME that receives context from sensors with the SIP UA that gets the call invite, the decision about how to take the call can take into account which device is available near the user and automatically

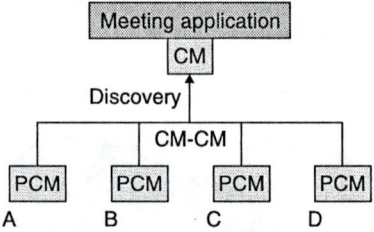

Figure 41.11   Meeting detection with CMEs.

allocate the call to the appropriate device that is near the user according to policies in place.

We can conclude [20,23] that calls (or other invitations, such as content delivery) can be allocated automatically with low delays to an appropriate available device. As multiple devices per user become the common model, this will relieve people from having to manually configure or even considering which device to use for which service.

Automatic Call Diversion

Consider the following scenario in which user A enters the office building and user B enters a conference room that has a detector that determines that a meeting has begun (Figure 41.11). A call for user B comes to B's SIP proxy, which diverts it to B's voice mail server. This could be manually configured by Users A and B pushing a button on their phone, PDA, etc., indicating that they are in a meeting (this is currently the method for most cellphone users). This could be done automatically if:

- The infrastructure can determine that these two users are in a meeting room.
- Users update their proxies regarding their location.
- Local information is exploited to affect the remote state of the user's SIP proxy (this could be centralized versus distributed).

The conclusion is that adding context awareness can free users from having to pay attention to certain communication events (when the user or company has opted for such a policy).

Opportunistic Communication

Escudero's thesis [25] showed how we can use policy to make, at a low cost (computation, power), per-packet decisions for any of the IP protocols on how to use wireless interfaces and thus avoid unnecessary vertical handoffs. Because of the low costs, we can make better use of multiple interfaces. Wennlund's thesis [12] provided context sockets that software (e.g., applications) running in end devices can use to transport context between devices. The theses of Jansson [23] and Avgeropoulos [24] provide the mechanisms for policy-driven allocation of services; therefore, the policy-driven, per-packet decisions on how to use wireless interfaces can be much more informed about the purpose of the communications and what access conditions may be expected over time. The conclusion is that mobile devices can become much smarter in communicating if the policy-based decisions are aware of nearby devices, access networks, and the urgency and timeliness of information to be sent across wireless links.

Context-Aware Mobile Audio

In ongoing work in our context-aware mobile audio prototype [30], we provide a user with a single device allowing the user to use the audio for different purposes, such as streaming audio, Internet radio, or voice calls (VoIP). Vacas's thesis [27] added a speech interface and leveraged information from different context sources (e.g., storage space, battery capacity, access conditions, location, speaker identification) for optimizing the user utility of mobile audio by clever (context-aware) use of available resources. The application does so by utilizing context information when making decisions about fetching content; thus, the device would be able to sustain audio playout much longer, make better use of bandwidth, and pre-cache content whenever applicable and possible. Vacas' thesis [27] presents sample programs and measurements showing the improvements that can be made by adding context awareness to mobile audio.

Field Trial

The ACAS user study was conducted in the spring of 2005 and included 27 students recruited from an international masters program at KTH. Each user was loaned an HP iPAQ 5550 handheld running a standard Microsoft® Pocket PC operating system. The study consisted of the following phases:

- *Phase 1* – Phase 1 took several weeks and allowed the students to familiarize themselves with the iPAQs before answering a questionnaire about the use of the iPAQs and to identify popular applications suitable for context adaptation in the later stages of the study. Not surprisingly, uses of the devices differed among individual users; entertainment (music, games) and instant communication (VoIP and messaging) were among the most popular applications. Several participants complained that the device was heavy, bulky, and difficult to use and that it had poor battery performance. The result indicated that communication was important to a majority of the users and was selected as the application area in which to pursue practical context sensitivity. This was to be based on the Spot system, a prototype localization system similar to other WiFi-based localization systems, except that localization is user relative (meaning that the position of a user or device is not given in absolute coordinates but rather expressed in terms of spots). Locations were created and named by the users.

- *Phase 2* — In Phase 2, we deployed the Spot system and enforced its use (via the iPAQs) by including it as an essential element in a course, informing students that unidentified locations would be registered. Although the Spot system performed as expected, students updated their positions as circumstances dictated, which was often very infrequently.
- *Phase 3* — In Phase 3, we enhanced the prototype with automatic meeting detection of groups of students based on proximity-based discovery. Based on the positive results of this experiment, we had planned on conducting further user studies and extensions of the prototype, capitalizing on the ability to increase users' awareness of location information, but we were unable to do so because of a lack of resources.

Conclusions

Accomplishments

The ACAS service architecture moves decisions closer to the users or objects; therefore, service decisions can be better informed as to what the user or service is doing or intends to do through the acquisition, processing, and management of context data. The ACAS project has made contributions in the following respects:

- Defining a service architecture (the context information network) to sense, manage, store, and exchange context information
- Designing middleware components to prototype this context information network
- Developing context-aware mechanisms for service discovery and service allocation and protecting user privacy
- Evaluating location and proximity sensing
- Conducting early demonstrators of opportunistic wireless and mobile communication and prototypes, including a field trial

Lessons Learned

Our approach differs from related work in that a user's personal CME receives context information from a local or non-local infrastructure instead of simply the user's mobile device; also, we have introduced a context refiner concept to infer high-level context information. Nodes utilizing ACAS middleware can participate in service delivery of mobile applications that are aware of the purpose of the sessions and can thus make more informed decisions about their combined operation, thus improving the usability of the application.

The applications, the service architecture, and its components have developed over several years. They have evolved as a result of the cross-fertilization of concepts, application ideas, and results from measurements and field trials. We continue to recognize that our understanding of what services are useful and how services will be used is not always what we expect. Many of the properties in user-centric mobile applications are emergent in the sense that they stem from the combined operation of several components in the service layer or as a result of interaction between components in different layers. We expect that context information will result in even more combinations of components.

Future Work

For future work, we suggest that more emphasis should be placed on actually building prototypes of context-aware mobile applications and evaluating the contribution of different components or their combined operation for achieving application transparency for users, given the available resources (devices, services, and communication technologies), as well as the selection of appropriate resources in users' work situations and maximizing user utility (QoS or uptime) when situations change.

Third-Generation Mobile Networks

The 3GPP and OMA service architectures build on SIP/SIMPLE for presence-enabled services as does ACAS context middleware; thus, 3GPP and OMA could leverage ACAS middleware components and extend the presence layer with context. Coordination can be achieved via SIP/SIMPLE and XDM [28], allowing optional extensions of PIDF+LO profiles [29]. This would allow 3G mobile applications to include context in service decisions but would require further work to incorporate policy-based context sensing and context dissemination in 3G networks.

Service Aggregator

To better serve service requests in public spaces, we propose a *service aggregator* to:

- Maintain a collection of service descriptions joined by a common property (e.g., a physical location, an organizational entity, the consensus of a user group).
- Answer queries for services and aggregates of services.
- Produce context information; the aggregator should act as a software sensor, thus presenting a context service interface to middleware and applications.
- Consume context information; the aggregator should be able to detect temporary services (e.g., in a mobile user's laptop) so they can be aggregated into the correct context, such as the user's current location.

Security, Privacy, and Trust

Our work brought to light a number of threats to the privacy and integrity of users, and in this chapter we presented a solution for protecting the user's privacy in public places. CASA revealed additional problems related to connections via wireless interfaces on personal devices which pose additional security threats, although these were not describe here. These security threats can lead to passive or active attacks; for example, passive attacks (e.g., eavesdropping, traffic analysis) or active attacks (e.g., masquerading, replay, message modification, denial-of-service attacks) can occur. Moreover, security features are sometimes not enabled, even when installed; cryptographic keys may be too short, shared, or not updated automatically and frequently. All of these problems make networks more vulnerable to attacks; for example, PAN service discovery can face a denial-of-service attack when a malicious user entering the PAN pretends to be a coordinator, thus getting information from other devices within range but hiding it from these devices. The above mandates further apply to developing a policy framework for dissemination, sharing, and accessing context information in wireless networks.

Acknowledgments

This research was conducted with partial support from the Swedish Foundation for Strategic Research (SSF), Ericsson, Hewlett Packard, Telia Sonera, Netlight, and R2M.

References

[1] Adaptive and Context-Aware Services (ACAS) project, http://psi.verkstad. net/acas.

[2] Adaptive Wireless Service and Infrastructure (AWSI) research program, http://www.wireless.kth.se/AWSI/.

[3] Project Aura, distraction-free ubiquitous computing, http://www-2.cs. cmu.edu/~aura/.

[4] Siewiorek, D. et al., *Sensay: A Context-Aware Mobile Phone*, Technical Report, Human–Computer Interaction Institute and Institute for Complex Engineered Systems, Carnegie Mellon University, Pittsburgh, PA, 2004.

[5] Chen, G. and Kotz, D., *Solar: A Pervasive-Computing Infrastructure for Context-Aware Mobile Applications*, Technical Report, Department of Computer Science, Dartmouth College, Hanover, NH, February, 2002.

[6] Dey, A.K., Salber, D., and Abowd, G.D., A conceptual framework and a toolkit for supporting the rapid prototyping of context-aware applications, *Human–Computer Interaction (HCI) J.*, 16(2–4), 97–166, 2001.

[7] *Web Services: The Cement for Mobile, Context-Aware Services*, May 13, 2004, http://www.freeband.nl/projecten/wasp/ENindex.html.

[8] The Context Project, http://www.cs.helsinki.fi/group/context/.

[9] Hedlund, H., A Gateway Between JINI and UPnP, M.Sc. thesis, Royal Institute of Technology (KTH), Stockholm, Sweden, 2000.

[10] IETF Secretariat, *IETF Charter SIP for Instant Messaging and Presence Leveraging Extensions (Simple)*, Internet Engineering Task Force (IETF), 2006 (http://www.ietf.org/html.charters/simple-charter.html).

[11] Li, W., Jonsson, M., Kilander, F., and Jansson, C.G., *Building Infrastructure Support for Ubiquitous Context-Aware Systems*, Lecture Notes in Computer Science, Springer-Verlag, Heidelberg, 2004, pp. 509–518.

[12] Wennlund, A., Context-Aware Wearable Device for Reconfigurable Application Networks, M.Sc. thesis, Microelectronics and Information Technology, Royal Institute of Technology (KTH), Stockholm, Sweden, 2003.

[13] Sedov, I., Preuss, S., Cap, C., Haase, M., and Timmermann, D., Time and energy efficient service discovery in Bluetooth, in *Proc. IEEE Vehicular Technology Conf. (VTC'03)*, Seoul, Korea, April, 2003.

[14] Freeman-Hargis, J., *Rule-Based Systems and Identification Trees*, AI Depot, http://ai-depot.com/Tutorial/RuleBased-Methods.html.

[15] Pradhan, S., Brignone, C., Cui, J.-H., McReynolds, A., and Smith, M.T., Websigns: hyperlinking physical locations to the Web, *IEEE Comput.*, 34(8), 42–48, 2001.

[16] Delgado, D.U., Implementation and Evaluation of the Service Peer Discovery Protocol, M.Sc. thesis, Microelectronics and Information Technology, Royal Institute of Technology (KTH), Stockholm, Sweden, 2004.

[17] Cascella, R., Reconfigurable Application Networks through Peer Service Discovery and Handovers, M.Sc. thesis, Microelectronics and Information Technology, Royal Institute of Technology (KTH), Stockholm, Sweden, 2003.

[18] Guttmanm, E., Perkins, C., Veizades, J., and Day, M., *Service Location Protocol, Version 2*, Request for Comments 2608, Internet Engineering Task Force (IETF), 1999 (http://www.ietf.org/rfc/rfc2608.txt).

[19] Ayrault, C., Service Discovery in Personal Area Networks, M.Sc. thesis, Microelectronics and Information Technology, Royal Institute of Technology (KTH), Stockholm, Sweden, 2004.

[20] Crossbow, http://www.xbow.com/.

[21] Internet Mail Consortium, *vCard Specification*, http://www.imc.org/pdi/.

[22] http://www.irda.org/standards/pubs/OBEX13.zip.

[23] Jansson, J., Context-Aware Service Allocation in Personal Area Networks, M.Sc. thesis, Microelectronics and Information Technology, Royal Institute of Technology (KTH), Stockholm, Sweden, 2004.

[24] Avgeropoulos, K., Service Policy Management for User-Centric Services in Heterogeneous Mobile Networks, M.Sc. thesis, Microelectronics and Information Technology, Royal Institute of Technology (KTH), Stockholm, Sweden, 2004.

[25] Escudero-Pascual, A., Anonymous and Untraceable Communications: Location Privacy in Mobile Internetworking, Licentiate of Technology thesis, Telecommunication Systems Laboratory, Microelectronics and Information Technology, Royal Institute of Technology (KTH), Stockholm, Sweden, 2001.

[26] Mola, G., Interactions of Vertical Handoffs with 802.11b wireless LANs: Handoff Policy, M.Sc. thesis, Microelectronics and Information Technology, Royal Institute of Technology (KTH), Stockholm, Sweden, 2004.

[27] Vacas, I.R., Context Aware and Adaptive Mobile Audio, M.Sc. thesis, Microelectronics and Information Technology, Royal Institute of Technology (KTH), Stockholm, Sweden, 2005.

[28] http://www.openmobilealliance.org/release_program/XDM_v10.html.

[29] IETF Secretariat, *Geographic Location/Privacy (geopriv)*, Internet Engineering Task Force (IETF), 2006, http://www.ietf.org/html.charters/geopriv-charter.html.

[30] Domínguez, M.J.P., Audio for Nomadic Users, M.Sc. thesis, Microelectronics and Information Technology, Royal Institute of Technology (KTH), Stockholm, Sweden, 2003

[31] Forgy, C.L., RETE: a fast algorithm for the many pattern/many object pattern matching problem, *Artificial Intelligence*, 19(1), 17–37, 1982.

Chapter 42

Middleware Support for Autonomous Cellphones

Nayeem Islam, Manuel Roman, and Dong Zhou

CONTENTS

Introduction

Cellphone functionality has evolved tremendously over the last ten years. First, we had just voice transmission, then short message service (SMS) and Web browsing (WAP and iMode) were added. Later, interactions with vending machines (cMode [1]) and multimedia messaging service (MMS) became available. More recently, video conferencing, Internet access, and interaction with the surrounding physical environment (iArea [1]) has become possible. Phones are poised to replace our keys [2], identification cards, and money with digital counterparts. Furthermore, increasing data transmission rates (2 Mbps with UMTS or 14.4 Mbps with Japan's FOMA networks [3]) enable the development of applications that allow cellphones to interact with distributed services (e.g., Web services) and access and share rich multimedia contents.

The increasing number and sophistication of cellphone applications demand a cellphone middleware infrastructure to assist in the development and execution of applications. Examples of middleware services include communication (e.g., RPC), discovery, security, quality of service (QoS), group management, event distribution, and publish–subscribe systems. Furthermore, due to the ubiquity of cellphones, these services must be reliable to guarantee uninterrupted execution. According to existing studies, 10 percent of cellphones are returned due to software problems. With over 1200 million subscribers worldwide, that means that over 120 million phones are returned every year. Asking cellphone users to take their devices to customer support centers to correct software errors is frustrating for cellphone users and too costly for carriers.

Unlike the PC world, the large majority of cellphone users have little understanding of the technical issues involved and simply expect their phones to work flawlessly all the time — which presents an interesting challenge to middleware researchers who have to provide mechanisms guaranteeing that the middleware infrastructure running on cellphones is reliable and capable of reacting to errors. The goal is to build autonomous middleware services that are able to monitor their own status, detect anomalies, and correct problems transparently — that is, without requiring user intervention and minimizing the disruption of normal cellphone functionality.

Advanced Scenarios

Mobile handsets have evolved from voice transmission devices to sophisticated digital assistants that are permanently connected to the Internet and play a central role in users' everyday activities. New mobile handsets provide advanced digital services and include hardware attachments that allow capturing images and video, triggering payments, and tracking the position of the handset accurately. Japan has one of the most advanced cellular phone infrastructures in the world, and it offers services that are described as future ubiquitous computing services in most research papers. We present here a list of these services (http://www.nttdocomo.com):

- *Pay by credit card infrared* — Launched in 2003, this service allows customers to enter their VISA credit card data in their phones and send the information via infrared to authorized establishments.
- *Mobile wallet service* — This service allows users to utilize their phone as an electronic wallet or a digital key to identify users at places such as airports. This service is currently implemented by embedding Sony's IC card (FELICA) into the cellphones. Users charge the digital wallet using the phone's Internet connection and use the IC card to pay at authorized locations. Service trials started in 2004 at designated train and bus stations for purchasing tickets.
- *Mall personalized info distribution (radiofrequency identification, or RFID)* — The trial service began in 2003. This service allows RFID-enabled phones to deliver customized information to users at designated commercial centers based on location and time.
- *Amusement park reservation and payment* — This service was launched in 2003. It allows users to make advanced reservations for various attractions at the LaQua amusement park; furthermore, the service allows users to check real-time congestion at rides to help them decide where to go next.
- *E-tickets and e-coupons* — Launched in 2003, this service offers online shopping for tickets and covers everything from search and purchase to payment and admission; furthermore, the service provides discount coupons.
- *Cmode* — This service started in 2002 and allows users to interact with enhanced Coke® vending machines from their phones. The interactions include buying drinks, downloading content to phones, and printing maps and tickets.
- *Bus location notify service* — This service allows users to check the location of city buses in real-time from their phones. It was launched in 2003.

- *Area information delivery*— Launched in 2003, this service allows users to receive location-specific information from approximately 500 different segments of Japan. Information includes detailed weather forecasts, maps, traffic information, restaurants, and sales at local stores.
- *FOMA remote control* — This service allows videophone-compatible terminal users to check video images of their homes and control their home appliances. The system requires a device at home that transmits the images and uses infrared to control the various appliances. The service was launched in 2004.
- *Barcode shopping*— This service utilizes two-dimensional barcodes (QR codes), which can encode up to 652 full-size characters. Users' phones can scan these QR codes to obtain such data as contact information (business card QR codes) or shopping information. With regard to shopping information, users can obtain details about products and even purchase them using their phones. The service requires phones enabled with cameras and specialized software to process the QR code images.
- *Personal navigation system* — The service allows users with global position system (GPS)-enabled phones to check their positions on a map in real time and receive turn-by-turn directions to reach their destinations; furthermore, the service allows users to locate other users and see their location in real time. This service was launched in Japan in 2003.

All of these advanced scenarios have one point in common: They require middleware services to enable their functionality, but we cannot rely on users to configure the functionality or fix problems; instead, these middleware services must be able to monitor themselves and ensure proper behavior at any time.

Next-Generation Cellphone Challenges

Advanced scenarios described in the previous section present a number of research and development challenges for middleware targeting next-generation cellphones. For example, in most of these scenarios the devices must communicate with services residing outside of the traditional cellular communication world, and in some cases software components need to be downloaded, installed, and executed on the device for the mobile client to best consume the services. Services illustrated by the scenarios typically require one or more means of communication chosen from a collection of communication apparatuses supported by the device, and such choices may vary with different services. Connectivity provided by each means of communication may change with the location of the device and the number of peers located nearby the device. Additionally, a service must be able to support a wide spectrum of mobile devices with very different communication and computation capabilities. Finally, in contrast to functionalities provided by software on today's cellphones (such as preinstalled games or PIM applications), the services offered in these advanced scenarios are shared by a group of users. These advanced scenarios present a number of challenges to next-generation mobile middleware — namely, *platform openness, variations in resource availability,* and *shared access to services.* We next further explain each of these challenges.

Platform Openness

As cellphones evolve from simple voice communication devices into sophisticated personal digital assistants permanently connected to the Internet, they are morphing from closed instruments into open platforms. The openness of such platforms imposes further requirements on the functionalities of next-generation cellphone middleware:

- *Open computing* — Future cellphones will be open computing platforms for downloadable software. Next-generation cellphones are expected to download more and more software components from the Internet or proprietary portals and then install and run them locally. Such downloadable software components include both those written in intermediate language (such as Java) and executed in confined environment as well as those written in native language with fewer constraints in accessing device resources. Such an open-computing platform requires sophisticated device management along with a secure runtime, as downloaded applications generally have lower quality than preinstalled applications and are likely to cause problems for either the entire cellphone or some other application on the cellphone.
- *Open communication* — Next-generation cellphones will gradually employ open protocols such as TCP/IP, XHTML, and SOAP to replace wireless proprietary protocols such as WAP and WML. This openness in communication platform is driven by the need for interoperability with the rest of the Internet and requires the customization of some open specifications because of the relative resource limitations of cellphones (e.g., kUDDI [4] and kXML [5]).
- *Open data and service access* — Cellphones will generate more data. Such data includes pictures, audio/video clips, personal information management (PIM) data, and productivity software data, as well as presence and location information. Such data must be able

to be accessed by other people or by the owner of the cellphone from other devices. In addition, cellphones are beginning to make services such as Web services available to the outside world. Such open access requires proper access control, as well as techniques such as service delegation and network storage for data replication.

■ *Open device provisioning* — A next-generation cellphone should support provisioning the device for different wireless operators, so cellphone owners can freely choose among different carriers without changing their cellphones. *Device provisioning* refers to the processes of bootstrapping the device, restoring the device configuration after failure, and configuring key services provided by operators. Some future cellphones will also support multiple air interfaces (e.g., WCDMA and WLAN) and require devices simultaneously provisioned for multiple access networks. Such open-device provisioning requires flexibility in device management.

Variation in Resource Availability

Compared with stationary hosts, such as desktop PCs, cellphones have considerable variability in the resources available to them; for example, the processing capability of a cellphone may change when energy management scales down CPU frequency to conserve energy. As another example, the connectivity of a cellphone changes when it roams from areas with good coverage into areas with weak coverage or no coverage or when the cumulative wireless traffic within the cell dynamically changes. Such dynamism in resource availability, along with the heterogeneity in the communication and computation capabilities of the spectrum of devices that future mobile middleware must support, requires an advance application model that is more flexible and adaptive than either the straightforward preinstall-based or client–server-based model; that is, the advanced application model should intelligently exploit both the communication efficiency of the preinstall-based model and the computation efficiency of the client–server-based model.

Shared Access to Services

As we mentioned earlier, many services for future cellphones will be accessed by a group of users or by the same users from different devices. Many of these services will maintain states that change as the result of serving client requests or interacting with other server-side software components. It is important to present a consistent view of such states to users of each of these services; on the other hand, the availability of such services to mobile clients is also important. Providing access to such services even when the clients are experiencing weak connections or disconnection helps improve the user experiences of such services. It is a challenge to maintain, as much as possible, both state consistency and service availability under unfavorable environmental conditions.

Functional Requirements

To meet the challenges presented by next-generation cellphones, future mobile middleware should provide the following functionalities:

- *Advanced device management*, which helps cellphones to deal with the potential results of platform openness
- *Adaptive application model*, which intelligently adjusts to deal with changes in resource availability and provides synchronization and consistency support for access-shared services during weak connections or disconnection
- *Autonomous service infrastructure*, which is capable of monitoring itself, detecting anomalies, and correcting or requesting assistance to avoid such anomalies

Other types of functionality that are also required but not further discussed in this chapter include customizability and adaptability in mobile communication, as well as security and access control during device runtime, as required by support for platform openness.

Middleware Support

In this section, we briefly describe three projects that address the major requirements of next-generation cellphone middleware. The first project, called MBB, uses externalization concepts and a state machine approach to enable fine-grained software inspection, diagnosis, and reconfiguration and serves as the foundation for advanced device management. The second and third projects (Mervlet and Replet) dynamically select between a standalone version and a client–server version of an application to adaptively respond to variations in resource availability, thus providing support for state synchronization and consistency of access to shared services.

Supporting Autonomous Middleware Services

According to Dashofy et al. [6], the four requirements for building a repairable autonomous system are:

- Ability to describe the current architecture of the system
- Ability to express arbitrary changes to the architecture
- Ability to analyze the result of the repair
- Ability to execute a repair plan without stopping the system

We add an additional requirement:

- Ability to detect errors in the execution of the system

We refer to the second requirement as *configurability* — that is, the functionality to modify any aspect of the software (state, logic, structure) at runtime; furthermore, we use two terms to refer to structural changes: *updates* and *upgrades*. Updates correspond to the functionality to replace existing pieces of software (e.g., to correct errors). Upgrades refer to the ability to add additional functionality to the existing software. Reflective middleware services [7,8] provide functionality for configurability. They support the replacement and assembly of certain components to adapt to changes and create certain device dependent configurations; however, most reflective systems assume a basic skeleton where only certain pre-defined changes and configurations are allowed. We seek a mechanism that allows modifying every aspect of the system (including the static skeleton) and enables fine-grained customizations.

In this section, we present a new middleware construction approach that assists in the development of configurable autonomic middleware services. These services are assembled dynamically from small execution units (micro building blocks) and can be reconfigured at runtime. Our approach externalizes three key middleware execution elements: *state*, *structure*, and *logic*. As a result, we obtain fine-grained control over running middleware services in terms of runtime modifications. We have used this construction technique to build a communication middleware service that we can manipulate at runtime.

Middleware Externalization

Middleware externalization relies on three key aspects: *state externalization*, *structure externalization*, and *logic externalization*. State externalization exports the internal middleware state attributes. Structure externalization exports the list of components that compose the middleware service, and logic externalization exports the interaction rules among the structural components (logic of the middleware service). Externalization allows inspecting and modifying internal execution parameters of the middleware services, thus giving users the ability to understand the execution state and to introduce arbitrary changes at runtime without stopping the system.

The main benefit of architecture externalization is maintaining the ability to learn, reason, and modify every aspect of a middleware service. The notion of architecture externalization is similar to computational reflection [9], which is a technique that allows a system to maintain information about itself (meta-information) and use this information to change its behavior (adapt); however, the key differences between computational reflection and architecture externalization are the scope of information maintained by the software and the scope of the changes allowed. Existing computational reflection middleware services [10,11] explicitly define the internal aspects they export and the changes they accept; however, middleware services based on architecture externalization export every detail in terms of structure, logic, and state and accept arbitrary changes in any of the three categories.

Building an externalized middleware service requires identifying the functional units of the service (we call them *micro building blocks*) and defining their state, input parameters, output parameters, and interactions (logic) explicitly. Externalized middleware services are assembled at runtime using an architecture descriptor that contains information about the components that compose the system (structure), the interaction rules for these components (logic), and a descriptor with detailed information about each structural component (input parameters, output parameters, and state attributes). The collection of all state attributes corresponds to the global middleware service state. These descriptors are the service blueprints and provide the information required to assemble the service at runtime. We use the descriptors to configure middleware services to different devices. Furthermore, these blueprints constitute a valuable formalism that aids in understanding the composition and behavior of existing middleware services. Developers can access these descriptors (or extract them directly from a running system), determine the internal details of the system, and introduce changes to customize the service without reading a single line of source code.

Another benefit of architecture externalization is that it exports the execution state of the system which includes information about the currently executed internal component. This information becomes essential in determining safe reconfiguration points, which correspond to execution states where it is safe to replace components, modify the logic, and modify state attributes. The system can determine these safe points without requiring any support from the software developers. Finally, another benefit of architecture externalization is the ability to virtualize the software infrastructure and create snapshots of the running system. This functionality is particularly useful to suspend, resume, and migrate software automatically; furthermore, heterogeneous systems can exchange architecture definitions and reconfigure themselves to enable interoperability.

We have built a software construction mechanism that relies on architecture externalization. We refer to this type of software as micro building block (MBB)-based software.

Micro Building Blocks

In this subsection, we provide a detailed description of the abstractions we have defined for MBBs. We also describe the descriptors that specify the details of the system. There are four abstractions and four system descriptors. The four abstractions are *micro building blocks, actions, collections,* and *domains.* The four system descriptors are *domain descriptor, structure descriptor, logic descriptor,* and *MBB descriptor.*

Micro Building Blocks

An MBB is the smallest addressable functional unit in the system. An MBB receives a collection of input parameters, executes an action that might affect its state, and generates a collection of output parameters. An example of an MBB is `registerObject`, which receives two input parameters (a name and an object reference), updates a list of registered objects (its state) with the new entry, and returns the number of registered objects. MBBs store their state attributes as name and value tuples in a system-provided storage area. This mechanism avoids implementing state transfer protocols to replace MBBs. Replacing an MBB requires registering the new MBB instance and providing it with a pointer to the existing state storage area. Storing the state in a designated storage area allows external attribute manipulation. External services operate on the existing state, and the MBBs obtain the new values when they resolve them by name during execution of their algorithm. Furthermore, storing the state as a collection of name and value tuples simplifies state suspension, resumption, and migration. We provide services that implement this functionality transparent to MBBs. Figure 42.1 illustrates the structure of an MBB. The execution model invokes a collection of MBBs according to the action definition (see next section); however, MBBs do not store references to the next MBB in the chain. This mechanism implies that no MBB in the system stores references to any other MBB. This approach allows replacing MBBs easily. It is not necessary to notify any MBB about the replacement because no MBB knows about any other MBB.

Actions

Actions specify the MBB execution order and therefore define the logic of the system. An action is a deterministic directed graph where nodes are MBBs that denote execution states, and edges define the transition order.

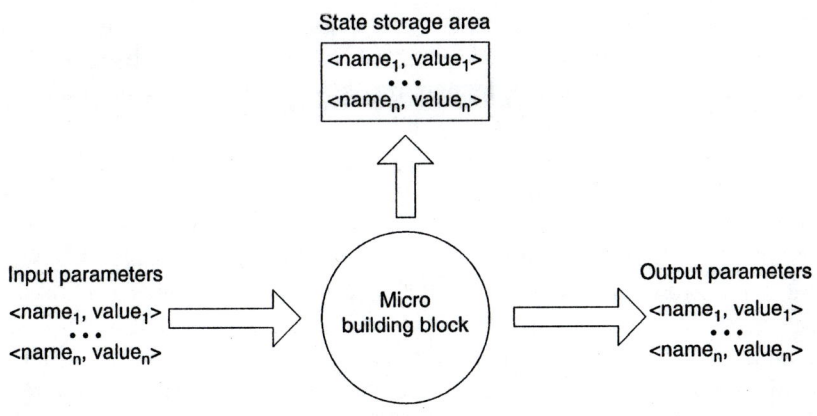

Figure 42.1 Micro building block structure.

The MBB infrastructure provides a runtime component called *MBB scheduler* that parses actions and invokes the appropriate MBB. Executing an action corresponds to traversing the graph; furthermore, every action invocation has an associated tuple container that we refer to as an *action state*, which stores the input and output parameters that the different MBBs generate. Figure 42.2 depicts an action example where MBB1 is the start node. The action begins with invocation of MBB1, continues with invocation of MBB2, then (depending on the value of X) invokes MBB3 or MBB4, and, finally, invokes MBB5. The value of variable X either is provided by the client invoking the action or is an output parameter generated by MBB1 or MBB2. This value is stored in the action execution state.

Action graphs include additional nodes and edges that specify the transitions in case of errors. If no errors are detected, the system uses the

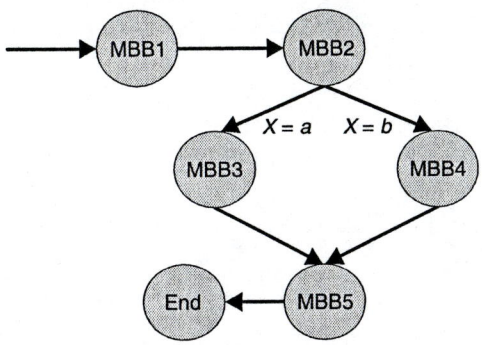

Figure 42.2 Action example.

default action graph (for example, the one depicted in Figure 42.2); however, if execution errors are detected, then the system uses the error nodes and edges. For example, Figure 42.2 has an additional edge that goes from each node to the end state (not included in the figure); thus, if an error is detected then the action simply terminates. Note, however, that it is possible to define more sophisticated behaviors. Action graphs allow cycles to support loop statements, such as "while," "for," and "repeat."

Actions provide reflection at the execution level by exporting information about the current execution state and by providing support to modify the action graph at runtime; furthermore, the explicit representation simplifies reasoning about the logic of the system, supports static analysis, and allows third parties to modify the behavior of the system by adding or removing states and configuring the graph. Actions contribute to MBB replacement. One of the key requirements to automate runtime MBB replacement is detecting when the system has reached a safe execution state. With actions, these safe states can be determined automatically. The safe reconfiguration states correspond to MBB invocations. Finally, actions contribute to the ability of systems to be updated and upgraded. Updating an action corresponds to replacing an existing action or modifying the execution graph. Upgrading the system implies adding new actions or, in the case of interpreted actions, modifying the action graph to incorporate or modify states.

Collections

A collection is an abstraction that allows defining groups of MBBs that can be treated as a single unit. Every collection stores a list of internal MBBs, and the system assigns them a state memory area where its MBBs store their state; furthermore, the system provides the functionality to load collections, unload collections, inspect and modify collections, and migrate collections across machines. By default, every collection has its own state area that it is not shared with other collections; however, the system provides the functionality to access the state memory of other collections. When assembling a system, it is common to have different functional categories; for example, most of our systems require communication middleware and a runtime infrastructure with the functionality to execute actions (MBB scheduler) and to manipulate MBBs, such as loading, unloading, and inspecting their states. Each of these functional categories consists of several MBBs, and we group them together into collections.

Domains

A domain provides the execution environment for MBBs. This execution environment includes a storage area to store the structure of the domain (list of MBBs), the logic of the domain (list of actions), and the state

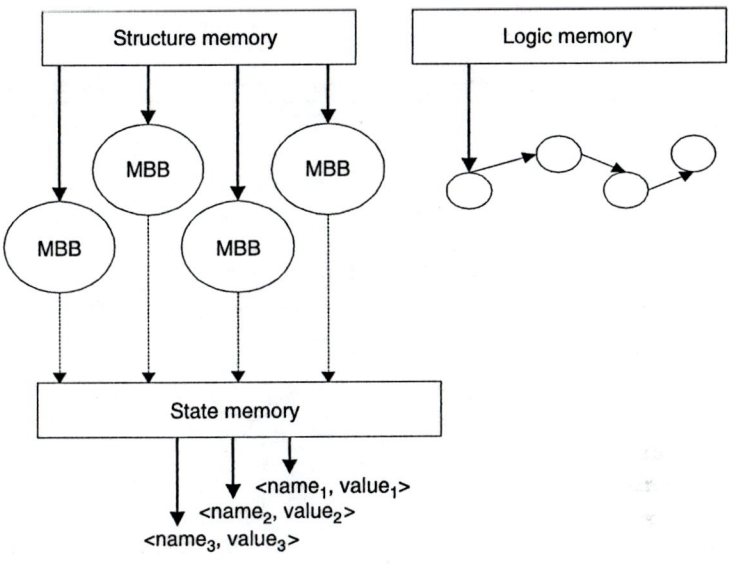

Figure 42.3 Domain components

of the domain (MBB state attributes and execution state values). Figure 42.3 illustrates the components and structure of a domain. All three memories (structure, logic, and state) store name and value tuples. The structure memory maintains a collection of tuples that correspond to MBBs registered in the domain. The tuple name refers to the name of the MBB (every MBB is assigned a name at registration time), and the value stores the reference to the MBB. Note that the reference can be a local pointer or a pointer to a remote MBB. The MBB execution model makes local or remote invocation transparent to developers. The logic memory stores a list of actions exported by the domain. Similar to the structure memory, the logic memory refers to actions by name. Finally, the state memory stores the state attributes for the MBBs registered in the domain. During the MBB registration, the system assigns a pointer to the state memory to the MBB. MBBs belonging to the same collection share the same state memory. We refer to the three memories as the *domain memory*.

Architecture Descriptors

Micro building block systems are assembled at runtime using structure, logic, and MBB description files. We refer to these files as *architecture descriptors*, and they are the blueprints of the system. A domain contains

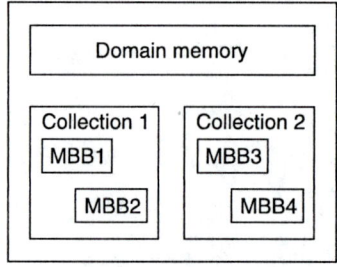

Figure 42.4   Example domain containing two collections.

collections, and each collection has structure and logic. We call the file that describes the overall list of collections the *domain descriptor*, the file that describes the structure of a collection the *structure descriptor*, and the file that describes the logic of a collection the *logic descriptor*. Finally, for each MBB is a file that describes the state, input, and output parameters, as well as the component that implements the MBB (e.g., Java class, .NET™ object, or a native DLL). We refer to this file as an *MBB descriptor*. For the following subsections, we will refer to the example domain illustrated in Figure 42.4.

Domain Descriptor

The domain descriptor provides a list of collections that compose a system. The descriptor is an XML file that contains a number of collection entries. Each entry has the name of a structure file (structure descriptor) and the name of a logic file (logic descriptor). Figure 42.5 depicts the domain descriptor corresponding to the domain presented in Figure 42.4.

Structure Descriptor

The structure descriptor provides a list of MBBs that compose a collection. The descriptor is an XML file that includes a list of MBB names and MBB descriptors. The system uses the MBB descriptor to instantiate the appropriate MBB and to retrieve information about the MBB (state variables, input and output parameters) and uses the MBB name as the key to access the MBB in the domain structure memory. All MBBs are registered in the domain structure memory as a `<name, MBB reference>` tuple. Figure 42.6 illustrates an example of a structure descriptor that specifies the structure of the Collection1 defined in Figure 42.5.

```
<DOMAIN>
    <COLLECTION name= Collection1 >
            <STRUCTURE>Collection1Structure.xml</STRUCTURE>
            <LOGIC>Collection1Logic.xml</LOGIC>
    </COLLECTION>
    <COLLECTION name= Collection2 >
            <STRUCTURE>Collection2Structure.xml</STRUCTURE>
            <LOGIC>Collection2Logic.xml</LOGIC>
    </COLLECTION>
</DOMAIN>
```

Figure 42.5 Domain descriptor example.

```
<STRUCTURE>
    <ELEMENT>
            <NAME>MBB1</NAME>
            <MBBDESCRIPTOR>MicroBuildingBlock1.xml</MBBDESCRIPTOR>
    </ELEMENT>
    <ELEMENT>
            <NAME>MBB2</NAME>
            <MBBDESCRIPTOR>MicroBuildingBlock2.xml</MBBDESCRIPTOR>
    </ELEMENT>
</STRUCTURE>
```

Figure 42.6 Structure descriptor for Collection1.

MBB Descriptor

Every MBB has an associated description that includes a list of input parameters, a list of output parameters, a list of state attributes, and a platform-dependent field that specifies the component that implements the MBB (e.g., a Java class file, a .NET object, or a DLL). Figure 42.7 presents the descriptor file for MBB1 of Collection1. The class tags denote the implementation for each platform. The state tags include a name and type tuple for each state attribute. The input tag describes the input parameters in terms of name and type tuples. Finally, the output tag denotes the parameters the MBB generates in terms of name and type tuples.

```
<MBB>
    <CLASS>
            <OS name="EPOC">MBB1.class</OS>
            <OS name="WIN32">MBB1.class</OS>
            <OS name="WINCE">MBB1.class</OS>
    </CLASS>
    <STATE>
            <NAME>StateAttribute1</NAME>
            <TYPE>int</TYPE>
    </STATE>
    <INPUT>
            <NAME>inputParam1</NAME>
            <TYPE>string</TYPE>
            <NAME>inputParam2</NAME>
            <TYPE>vector</TYPE>
    </INPUT>
    <OUTPUT>
            <NAME>registeredobjects</NAME>
            <TYPE>integer</TYPE>
    </OUTPUT>
</MBB>
```

Figure 42.7   MBB descriptor for MBB1.

Logic Descriptor

The logic descriptor is an XML file with a description of each action. This description includes the name of the action and a list of the action's states, including the name of the state, the name of the MBB to invoke, the next state in normal conditions, and the next state when an exception is detected during the execution of the current state. The logic descriptor allows defining conditional transitions. Figure 42.8 illustrates an example of an action that uses MBB1 and MBB2. State 1 invokes MBB1 and relies on the value of nextState to transition to the next state. If the value of nextState is 1, then the action transitions to state 2; otherwise, if the value is 2, the action ends. The nextState variable is stored in the inputParameters tuple container that the scheduler provides to each MBB invoked during an action. In the example, MBB1 has to provide the

```
<ACTION name="exampleAction">
    <STATE>
            <NAME>1</NAME>
            <MBB>MBB1</MBB>
            <NEXTSTATE variable="nextState">
                    <VALUE>1</VALUE>
                    <NAME>2</NAME>
                    <VALUE>2</VALUE>
                    <NAME>end</NAME>
            </NEXTSTATE>
            <ERRORSTATE>end</ERRORSTATE>
    </STATE>
    <STATE>
            <NAME>2</NAME>
            <MBB>MBB2</MBB>
            <NEXTSTATE>end</NEXTSTATE>
            <ERRORSTATE>end</ERRORSTATE>
    </STATE>
</ACTION>
```

Figure 42.8   Action specification example.

value of `nextState`. According to Figure 42.8, both states 1 and 2 define an error state transition to end. This means that, in case of an exception during the execution of the MBB, the scheduler will end the execution of the action. Note that we can transition to any state and we can also define conditional transition statements.

MBBs and Autonomic Middleware Services

Micro building blocks support the construction of autonomic middleware services. MBBs meet the five requirements we introduced earlier; therefore, the resulting middleware services can be manipulated at runtime. The first requirement concerns describing the current architecture of the system. MBBs use descriptors with information about the architecture, structure, and logic of the system to assemble the system at runtime; furthermore,

MBBs maintain runtime information about the internal details of the system and provide an application programming interface (API) to access this information. The second requirement addresses being able to express arbitrary changes to the architecture. MBBs allow modification of every aspect of the system, including logic, structure, and state. MBBs also meet the third requirement, analyzing the results of a repair. After introducing a change, we can analyze the resulting structure, state, and logic of the system to confirm that the changes meet the expected results. The fourth requirement concerns the ability to execute a repair plan without stopping the system. As we explained earlier, the execution model of MBBs can detect safe reconfiguration points automatically and therefore does not require stopping the system. Finally, the fifth requirement addresses the importance of detecting errors in the execution of the system. The MBB runtime infrastructure relies on a component called the *MBB scheduler* to drive execution of the system. This MBB scheduler uses the actions to determine the MBB invocation order and invokes the MBBs. If an MBB crashes, the scheduler detects it and uses the error state transitions to determine the next MBB to invoke. This behavior allows us to detect errors in the execution of the system and take the required steps to correct them or abort execution of the system. MBBs meet all five requirements and therefore provide the basis for constructing mechanisms to monitor and correct the behavior of middleware services automatically.

MBBs and Flow Languages

Micro building blocks explicitly externalize their logic using graphs where the nodes are MBBs and the links determine the invocation order of these MBBs. This concept is similar to existing flow languages such as the Business Process Execution Language for Web Services (BPEL4WS) [12,13]. BPEL4WS allows defining business processes that make use of Web services. These business processes rely on four key abstractions:

- Partners
- Flows
- Containers
- Properties

Partners are the components that implement the functionality in the business process. They implement a Web service and specify the Web services on which they rely. *Flows* are directed graphs that determine the order in which partners are invoked. A *container* is responsible for storing messages exchanged among partners (WSDL messages); these messages store information about the context of the business process and are key to preserving the business process session. Finally, a *property* is data stored in the container that helps to correlate a message with the appropriate business process.

BPEL4WS provides a language to define the interaction among Web services. On the other hand, MBBs provides an infrastructure to define the interaction of the components that compose each individual Web service. The goal of both solutions is similar, but their scales are different; BPEL4WS operates at a macro scale, and MBBs operate at a micro scale. Both, however, rely on a similar technique (although BPEL4WS does not externalize the state). In MBB terms, the *partner* is an MBB, the *flow* is an action, the *container* is the tuple container with the invocation parameters that traverses the different MBBs that belong to an action, and a *property* is a tuple contained in this tuple container.

Mervlets

The Mervlet project targets dynamic Web applications — that is, applications that use constructs such as Servlets to dynamically generate responses for requests received from Web clients [13]. The primary target of the Mervlet project is to support an adaptive application model for dynamic Web applications under resource variation. The project also addresses auxiliary services (namely, preference management and application and environment profiling) that can be used for middleware autonomy.

Traditional wireless data service systems, such as i-mode [14] and WAP [15], assume a relatively simple client that is browser based and not programmable. Such systems suffer from a variety of problems, including poor interaction performance, inability to allow the user to work under conditions of network disconnection or weak connections, and a poor user interface, thus leading to overall poor user experience. The advent of smarter phones with faster processors, more memory, and persistent storage makes it possible to provide a better user experience and new classes of applications. Taking full advantage of such rich client-side resources, however, is difficult, as no application model closely based on current Web protocols allows full exploitation of client-side resources and is able to deal with the huge divergences in and dynamics of handset capabilities, such as CPU speed, memory and storage sizes, battery power, connection bandwidth and stability, and user interface resources.

The Mervlet project is part of our Agile Operating Environment (AOE), which provides generalized yet agile support for the improvement of the user Web experience and construction of new types of Web applications by judiciously exploiting client device resources for system functions such as partial application caching, optimized logging for fault tolerance, and application user interface presentation generation. It is agile as these

Figure 42.9   Symmetric AOE model.

facilities can be customized for devices of different capabilities or reconfigured by users or applications to adjust to runtime resource availability.
Also, they are self-adaptive to changes in resource availability, thus further
enhancing the user experience in mobile environments.

AOE Architecture

The AOE supports Web-based applications where users request services
through a browser from their user client devices. It specifically targets
applications that involve dynamic, personalized content. A typical AOE
application consists of a cluster of Mervlets that can dynamically generate
Web pages, where a Mervlet is similar to a Servlet except that it can be
replicated and executed on client devices, its user interface can be dynamically attached, and it may recover from faults in the client device, network,
or server. Each Mervlet within an AOE application implements the Mervlet
interface, which consists of methods that can be used to help replicate the
Mervlet and synchronize states among replicas. A developer of each Mervlet
can also provide implementations of *adaptation helpers* that can help the
AOE runtime to make adaptation decisions. An application developer can
optionally create user interface widgets used specifically for a particular
device. These widgets can be dynamically bound to the application. For
each application there is an application preference file that is specific for
the particular deployment and will be combined with client and server
preferences to form the runtime preference of the application instance.

The AOE runtime is the environment for the execution of AOE applications. The client and the server each run an instance of the runtime
(Figure 42.9). Hypertext Transfer Protocol (HTTP) requests sent from the
browser are intercepted by the AOE runtime on the client device (termed
client AOE). The client AOE then does one of the following:

- Passes the request to the AOE runtime on the server side (*server
  AOE*) using any transport, one option being a reconfigurable
  messaging system (RMS), in which case the server AOE receives

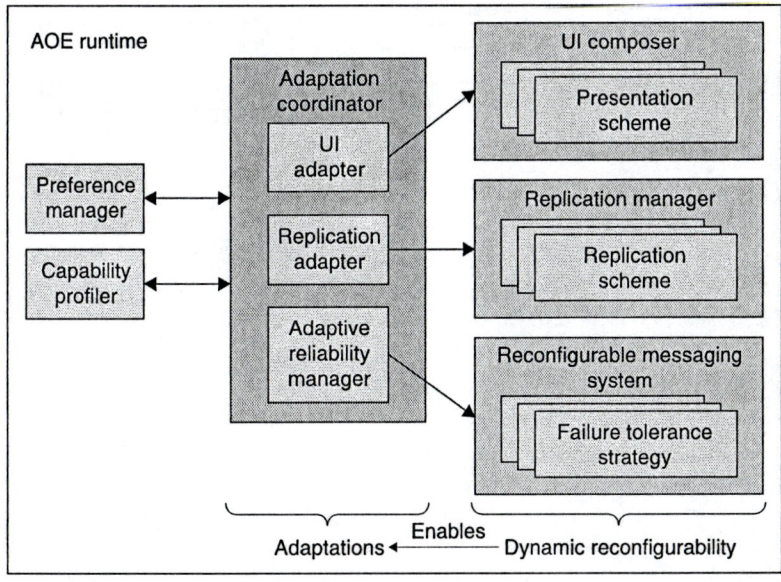

Figure 42.10   Layered components of AOE runtime.

and serves the request and sends response to the client AOE using AOE messages

■ Serves the request locally, when the requested page is locally available or when the client AOE decides to execute the requested Mervlet locally

The response, which is usually the presentation of the application described in languages such as Hypertext Markup Language (HTML), is then optionally rebound with the user interface library deployed on the client device and finally returned to the browser in HTTP format. During the process, when required, the client and server AOEs cooperatively ensure that when the client AOE receives an HTTP request from the browser it will serve according to a reliability guarantee such as processing the request once and only once. Such reliability assurances are provided by the *Adaptive Reliability Manager* inside the AOE runtime, and the ability to decide where to serve the request and the ability to dynamically rebind the user interface of the application are supported by the *Replication Manager* and *UI Composer*, respectively (Figure 42.10). The details of the Replication Manager, UI Composer, and Reliability Manager are described in later sections.

A key feature of our system is that basic system facilities are adaptable. The three key facilities (UI Composer, Replication Manager, and Reconfigurable Messaging System) are made adaptable by using three adaptation managers (UI Adapter, Replication Adapter, and Adaptive Reliability Manager, respectively) that make their adaptation decisions based on input from the Preference Manager and the Capability Profiler. An Adaptation Coordinator coordinates the individual adaptation managers of each facility. Specifically:

- The Adaptation Coordinator is responsible for resolving conflicts between adaptation decisions made by the three adaptation managers (i.e., UI Adapter, Replication Adapter, and Adaptive Reliability Manager). The adaptation managers consult the Adaptation Coordinator about whether they should proceed before making any adaptation decisions.
- The Preference Manager provides a means for devices, servers, and applications to jointly configure and reconfigure an AOE runtime, to allow devices and servers to control the behaviors of applications, and to allow the exchange of preferences between two AOE runtimes.
- The Capability Profiler of an AOE runtime is a repository for maintaining device and network capabilities, as well as some application characteristic data. Some of the capability or characteristic data (such as installed memory) is static, and other data (such as current CPU load) is dynamic.
- The Replication Manager, the UI Composer, and the RMS are three fundamental facilities provided by the AOE for client resource utilization. They rely on the Adaptation Coordinator for agility.

The Replication Manager

Mobile users may perceive poorer experiences because of longer and wider variations in response latency, lower throughput, and a greater likelihood of disconnection. Although existing Web caching and prefetching techniques are applicable to mobile Web applications for static and some dynamically generated Web pages, these technologies do not improve user experience for highly dynamic, highly personalized Web content that is generated by server-side code units (such as Servlets) and widely used in mobile Web applications. The AOE handles such highly dynamic and personalized Web content by allowing the caching of server-side code units onto client devices and by dynamically making per-invocation decisions about whether to invoke the replica on the server or the replica on the device. Details about replication support in AOE are described in the later Replet section.

UI Composer

A problem confronting Web applications targeted to mobile devices today is the heterogeneity in the interaction capabilities of these devices. Ideally, an application should be written only once, independent of any specific device; yet, when the application is used on different devices, its presentation should be automatically adjusted to a form optimal to the device. The AOE allows binding between the abstract description of the presentation and its implementation to occur on the client device whenever there is enough resource on the device to do so. In addition, such binding is dynamic so implementation of the presentation can change in accordance with the change in resource availability.

In AOE, we require applications to write their presentations in tags [16] to translate XML for rendering; however, the tags are not statically bound to the application but are constructed as dynamically attachable libraries that may be chained together at runtime in a specific order. At runtime, the UI Adapter attaches the appropriate library to enable presentation of the application. Presentations are implemented by user interface libraries. Each user interface library (UIL) has an interface (`UILInterface`) and an implementation (`UILImplementation`). Each implementation is tagged by a capability specification.

At runtime, when the generated Mervlet makes a call that involves any aspects of presentation the appropriate user interface library is called. The proxy asks the UI Adapter service to provide an implementation of the specified library interface that is suitable for the device capabilities. The UI Adapter uses the characteristics of the device to find the appropriate library to install and link to the application.

Reliable Messaging System

Transient failures in networks and devices are common for mobile handsets. User reliance on mobile devices for important daily activities such as e-commerce, online shopping, and personal information management is increasing, and such applications require some degree of confidence that these activities will be performed reliably [17]. One problem with current approaches is that they are inflexible and do not adapt to changing system conditions or application requirements, but mobile environments are characterized by change so no one fault-tolerance mechanism will work for all instances or all the time. We believe that dynamically adapting fault tolerance addresses this issue. In this chapter, we propose adaptive fault tolerance through the use of a reconfigurable messaging system.

Support for fault-tolerance is realized with the reconfigurable messaging system (RMS) and the recoverable Mervlets. The RMS provides configurable message delivery functionality in the Mervlet environment. The RMS failure-free strategy interface encapsulates the RMS functionality. An application only calls methods on the interface. The actual implementation to be used is set by the Adaptive Reliability Manager (ARM) based on user or application requirements; for example, the RMS can be configured as a point-to-point messaging service or to use a centralized messaging server (such as the Java Message Service [JMS]). At the application level, recoverable Mervlets are used to provide fault tolerance. During failure-free operation, the Mervlet engine invokes the methods on the recoverable Mervlets by retrieving the current failure-free strategy first and then calling the desired method (e.g., doPost, doGet) on the strategy. Recoverable Mervlets allow the same application to have different fault-tolerance mechanisms in different contexts; for example, a Web mail application may be configured to be more reliable for corporate e-mail than personal e-mail.

Dynamic reconfigurability support in fault tolerance is achieved by allowing the two main components (RMS and recoverable Mervlets) to have different failure-free and recovery strategies which can be set dynamically by the ARM, as shown in Figure 42.11. The separation between failure-free and recovery strategies helps in the development of multiple recovery strategies corresponding to a failure-free strategy; for example, in the case of RMS, one recovery strategy may prioritize the order in which messages are recovered but another recovery strategy may not.

In our current implementation, the adaptability in fault-tolerance support is reflected in the ability to dynamically switch on and off server-side logging, depending on current server load. Under high server load, the Adaptive Reliability Manager can reconfigure the RMS to stop logging on the server side. This, in some cases, can result in marked improvement in the client-perceived response time.

Replets

The Replet project evolved from the Mervlet project and is also part of the Agile Operating Environment. A Replet is a Mervlet that can be shared by a group of users and can be accessed by the same user from different devices. Compared with the Mervlet project, the Replet project fulfills additional requirements of next-generation mobile middleware by targeting dynamic Web applications that provide services to a group of mobile clients. It provides a flexible framework for the on-device replication of such services. This framework in particular addresses the issue of maintaining

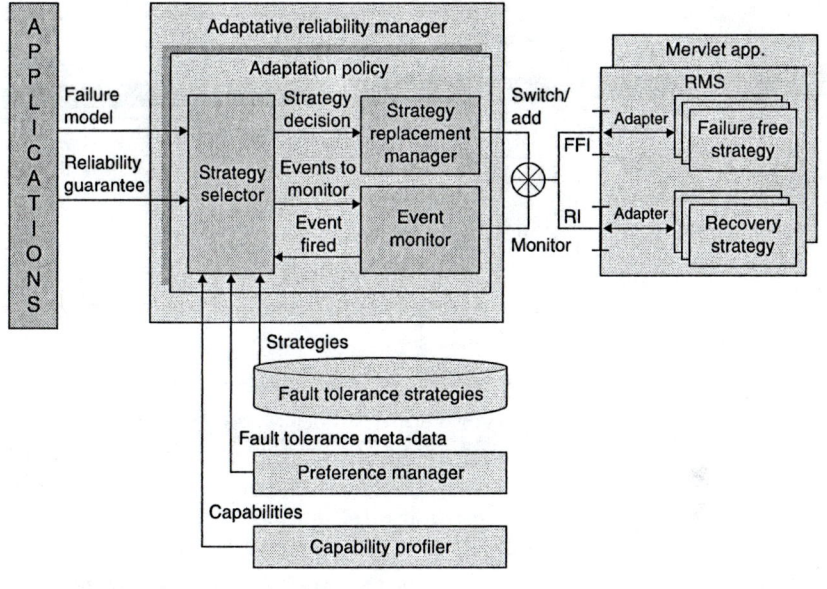

Figure 42.11   Reliability support in AOE.

both state consistency and service availability under unfavorable conditions by providing facilities for exploiting application-, server-, and device-specific consistency and availability requirements [18].

Replet Model

A Replet replica is explicitly divided into code, immutable data, and mutable data. The code part includes the class files that define the Web-related application logic, and it is identical for all replicas of the same Replet; however, the mutable and immutable data (which can be a combination of in-memory objects, files, and mobile database tables with records and attributes tailored to the client's needs) of a specific client replica can be different from that of a server replica or other client replicas. For example, a client replica can have rows of a database table filtered out that are different from another client replica. The mutable data of a replica, which can be modified by clients, is further divided into a public fragment and a private fragment. The public fragment is shared by a number of clients and thus is accessible to the server replica and the client replicas on those clients. The private fragment is specific to a client and is only accessible to the client replica and its server replica.

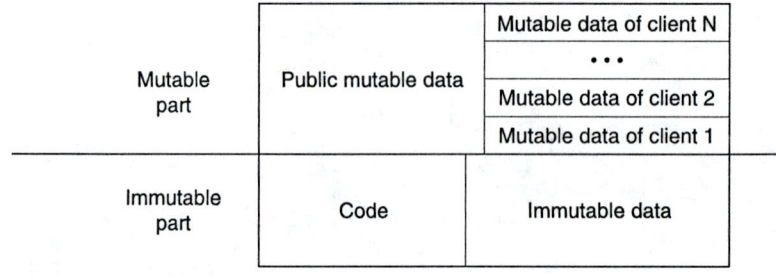

		Mutable data of client N
Mutable part	Public mutable data	•••
		Mutable data of client 2
		Mutable data of client 1
Immutable part	Code	Immutable data

Figure 42.12   Replet server replica.

Figure 42.12 depicts a server replica. The client replica on a client device is slightly different in that it only has the private mutable data for that given client. At a given moment for a given client, a server replica can be in one of following three states (Figure 42.13):

- *App-synchronized* — The server replica has up-to-date public mutable data but does not have up-to-date private mutable data for the client.
- *Client-synchronized* — The server replica has synchronized copies of both public mutable data and the private mutable data for that given client.
- *Invoked* — The server replica was in a client-synchronized state and has been selected to serve a request from the client, and the invocation is in the process.

Note that a server replica can be in the client-synchronized state for one client but in the app-synchronized state for another.

A client replica has an additional *selected* state, meaning that the replica has not yet been populated with code and data, or the code and data have been removed to allow the replication of other applications (see Figure 42.14). Note that, after invocation, a server replica transits into the app-synchronized state, but a client replica transits into the client-synchronized state.

Preference Management

Preference information consists of preferences prespecified by the application server, the user, or the application, such as the required memory and the preferred response time. Preference information helps a Replet to determine which replica to use or whether to download a replica from the server. Three participating entities, also called *roles*, take part in

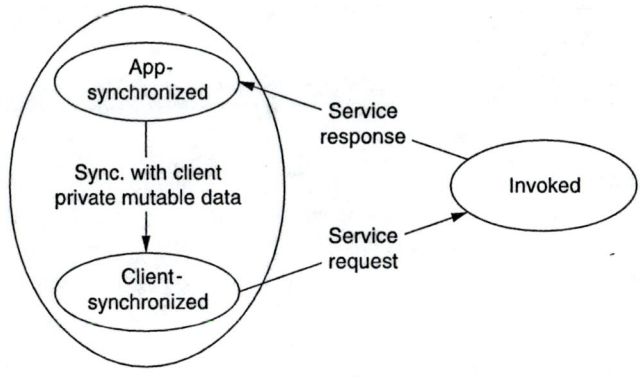

Figure 42.13 State transition of a server replica.

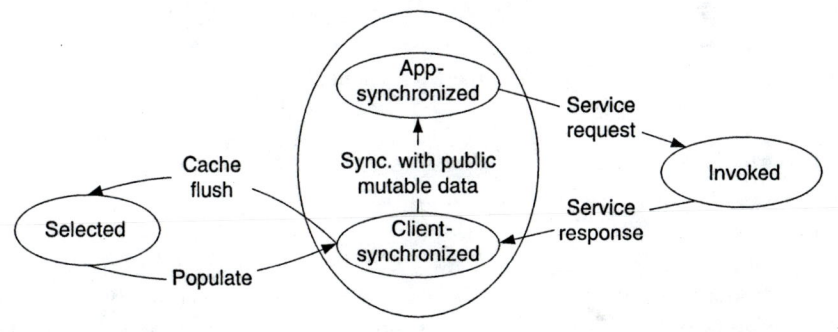

Figure 42.14 State transition of a client replica.

preference derivations: the client, the server, and the application. The entities often have conflicting or overlapping preferences; for example, an impatient user wants to get results in 10 seconds but the application estimates the acceptable response time to be within 15 seconds. Preference Management merges potentially conflicting partial preferences from different roles into a unified global preference.

Figure 42.15 illustrates how Preference Management works. Each entity specifies a partial preference that consists of a set of *partial properties*. A *partial property* is a <property, precedence> pair, where property is a property and precedence is a number denoting priority. Each property is a <name, value> pair, where name is a string and value can be an arbitrary object. For example, as shown in the figure, <<ResponseTime, 10s>, 8> in the Device Preference means that the device preferred response time should be within 10 seconds with a precedence level of 8. To derive a Global Preference, we must first validate

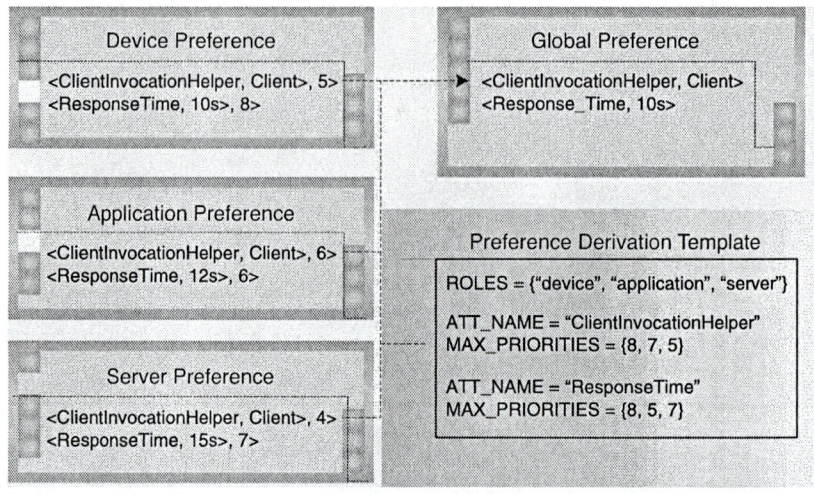

Figure 42.15   Example of Preference Management.

each partial property with a *Preference Derivation Template*. The Preference Derivation Template sets maximum priority levels for each partial property. If a partial property contains a precedence value higher than the maximum level specified in the template, then the partial property will be considered invalid; for example, ResponseTime in the Application Preference will be invalid when its precedence (i.e., 6) exceeds the maximum priority (i.e., 5). Finally, we merge all valid properties into the Global Preference. If there are conflicting properties, we keep only those with higher precedence values. Suppose, for example, that two valid ResponseTime properties are specified in the Client Preference and Server Preference. In this case, we keep the one in the Client Preference because its precedence value (i.e., 8) is higher.

Replication Process

The process of Mervlet replication can be divided into four phases: Selection Phase, Populating Phase, Invocation Phase, and Synchronization Phase. In the Selection Phase, before sending an HTTP request to the server, an AOE-enabled client will check the local Device Preference to see if the device allows Mervlet replication. If it does, the client will insert a field into the header of the HTTP request to indicate its willingness to be considered as a site for replicating the service. Upon receiving such a request, the server will first serve the request and generate a usual response. It then checks whether or not the application allows itself to be replicated, and if it does then it will send application and server preference and capability information to the client, again in the form of HTTP header fields. The profile of the application includes hints on the application's usage of memory, storage, CPU, and the application's consistency requirement and data update frequency. Also included in the header is a private ID created by the server for the client.

In the Populating Phase, the client uses the private ID received from the server earlier to identify proper session states to be downloaded, in addition to classes, immutable data, and shared mutable data. In the Invocation Phase, when a client receives a request from a user or other applications, it will check preference and profiling information, derived both locally and received from the server, to decide whether to serve the request locally or remotely. If the decision is to serve the request remotely, then the service request will be forwarded to the server with no additional header fields. In the Synchronization Phase, state modifications made in the Invocation Phase will be synchronized to maintain a consistency level that satisfies the requirements of the application. Note that only application global states have to be synchronized at all times among replicas. Session states only have to be synchronized when changing from invoking client replicas to server replicas, or *vice versa*.

The selection and populating of a replica are customizable and adaptable in that devices, servers, and applications can define their own triggers for selecting a device as a replication site and populate the site, and the dynamic capabilities provided by the AOE runtime are used to automatically evaluate the prerequisites for the trigger. Adaptation in replica invocation is supported by per-request replica selection for invocation; that is, for each request received by the client, the client- and server-side AOE runtime will collaboratively decide which replica to use for the particular request.

The Synchronization Phase in Replet replication is separated into two stages by the Invocation Phase: a Read Stage before the Invocation Phase and a Write Stage after the Invocation Phase (see Figure 42.16):

- *Read Stage* — On the client side, when the Replication Manager decides to serve a request locally, it consults the client copy of the Synchronization Helper (SH) specified by the preference to see if it is necessary to read the most up-to-date version of the Replet's public data from the server replica. On the server side, when such a read request is received, the Replication Manager consults the server copy of the SH for the client to determine if it is necessary to apply a write-lock to the server replica to prevent concurrent access. The lock expires after a time defined by the preference. Note that, because there is one preference for each client, potentially each client can have a different SH.

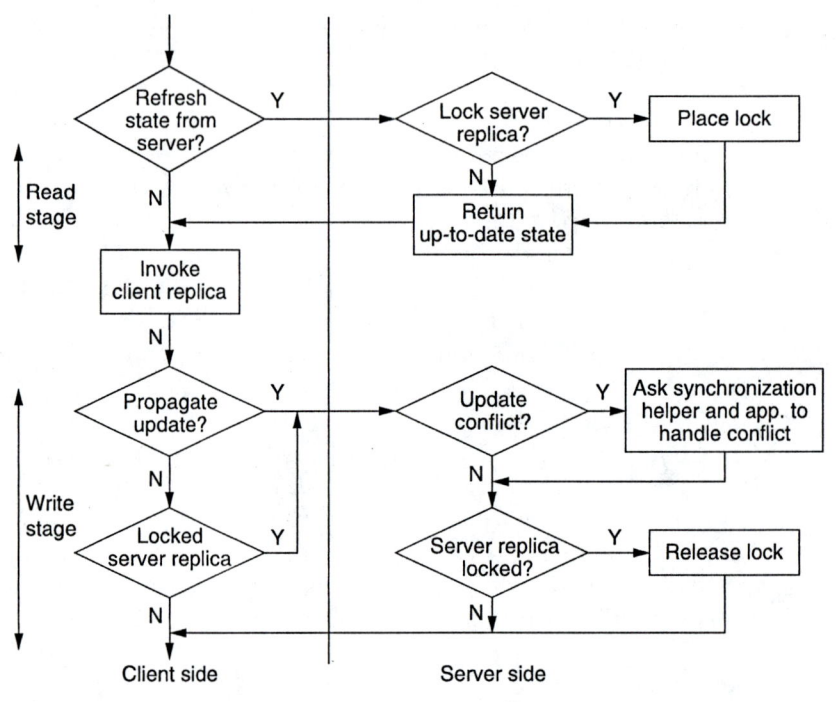

Figure 42.16 Flow chart for data synchronization.

- *Write Stage* — On the client side, after an invocation on the client replica that modifies the Replet's public data, the Replication Manager again consults the SH to see if it is necessary to propagate the modification to the server replica immediately. On the server side, when such a modification propagation message is received, the Replication Manager consults the SH of the client to detect potential update conflicts and resolve such conflicts. The Replication Manager then releases the lock on the server replica if the client acquired one during the Read Stage. Finally, the Replication Manager consults the SHs of other clients to see if the modification has to be pushed to these clients.

Note that the actuation of the Read and Write Stages depends on whether the invoked method reads or writes public data. If the method does not read or write public data, then the Read and Write Stages are bypassed.

The SH is also part of the preference and can potentially be defined by the client, the server, or the application. Some sample consistency maintenance schemes that can be implemented using the SH include:

- *Pessimistic replication*, where one-copy serializability is achieved by demanding refreshment from the server replica before each invocation that reads Replet public data, locking the server replica after the refresh, and contacting the server replica after invocation for update propagation and lock release
- *Optimistic replication with 1/K synchronization*, where the data refreshment and update propagation are carried out once for every K invocations that access public data, and no lock is applied on the server replica
- *Optimistic replication with push approach*, where the client depends on updates pushed from the server to maintain the freshness of the Replet's public data

The pessimistic replication scheme is practical if only a small fraction of the invocations involve access to public data or if the data synchronization overhead is much less costly than the overhead associated with the transport of request and response messages and the processing of the request on the server. Other sample consistency schemes include one that chooses the appropriate update rate depending on client capability and current server load for self-refreshing Web contents and another one enforced by a server under unusually high load to temporarily disable a device from reloading the content until sometime later.

Conclusions

Platform openness, resource availability variation, and service shared access are three main obstacles to constructing middleware for autonomous cellphones that are being increasingly utilized in advanced scenarios. Requirements for such middleware include functionalities such as advanced device management, adaptive application models, and autonomous service infrastructures. Our work in this area has addressed these requirements. The MBB project uses externalization concepts and a state machine approach to enable fine-grained software inspection, diagnosis, and reconfiguration, and it serves as the foundation for advanced device management. The Mervlet and Replet projects together provide mechanisms for dynamically selecting between a standalone version and a client–server version of an application to adaptively respond to variations in resource availability, and the project provides support for state synchronization and consistency for access to shared services. Our future work will further our efforts in this area and target building truly autonomic mobile middleware.

References

[1] http://www.nttdocomo.com/technologies/present/imodetechnology/index.html.
[2] Beaufour, A. and Bonnet, P., Personal servers as digital keys, paper presented at the Second IEEE Int. Conf. on Pervasive Computing and Communications (PerCom 2004), Orlando, FL, March, 2004.
[3] http://www.3gnewsroom.com/3g_news/dec_02/news_2861.shtml.
[4] kUDDI Project, http://kuddi.enhydra.org.
[5] kXML 2 Project, http://kxml.objectweb.org/project/aboutProject/index.html.
[6] Dashofy, E.M., van der Hoek, A., and Taylor, R.N., Towards architecture-based self-healing systems, in *Proc. of the First Workshop on Self-Healing Systems (WOSS 2002)*, Charleston, SC, November 18–19, 2002, pp. 21–26.
[7] Kon, F., Costa, F., Blair, G., and Campbell, R.H., The case for reflective middleware, *Commun. ACM*, 45, 33–38, 2002.
[8] Capra, L., Blair, G., Mascolo, C., and Emmerich, W., Exploiting reflection in mobile computing middleware, *Mobile Comput. Commun. Rev.*, 6, 34–44, 2002.
[9] Maes, P., Concepts and experiments in computational reflection, paper presented at the ACM Conf. on Object-Oriented Programming Systems, Languages, and Applications (OOPSLA'87), Orlando, FL, October 4–8, 1987.
[10] Roman, M., Kon, F., and Campbell, R.H., Design and implementation of runtime reflection in communication middleware: the dynamicTAO case, paper presented at IEEE Int. Conf. on Distributed Computing Systems (ICDCS'99), Austin, TX, May, 1999.
[11] Blair, G., Coulson, G., Andersen, A., Blair, L., Clarke, M. et al., The design and implementation of open ORB v2, *IEEE Distributed Syst. Online*, 2, 2001 (special issue on reflective middleware).
[12] Leymann, F. and Roller, D., *A Quick Overview of BPEL4WS*, IBM, White Plains, NY, 2002.
[13] Islam, N., Zhou, D., Shaoib, S., Ismael, A., and Sajith, K., AOE: a mobile operating environment for Web-based applications, in *Proc. of Int. Symp. on Applications and the Internet (SAINT 2004)*, Tokyo, Japan, January 26–30, 2004.
[14] NTT DoCoMo R&D, Special issue on i-mode services, *DoCoMo Tech. J.*, 1(1), 1999.
[15] van der Heijden, M., Heijden, M., and Taylor, M., *Understanding WAP: Wireless Applications, Devices and Services*, Artech House, Norwood, MA, 2002.
[16] Shannon, B. et al., *Java 2 Platform, Enterprise Edition: Platform and Component Specifications*, Addison-Wesley, Boston, MA, 2000.
[17] Joseph, A. and Kaashoek, F., Building reliable mobile-aware applications using the Rover toolkit, in *Proc. of the 2nd ACM/IEEE Int. Conf. on Mobile Computing and Networking (MOBICOM'96)*, White Plains, NY, November, 1996.
[18] Zhou, D., Islam, N., and Ismael, A., Flexible on-device replication with Replets, in *Proc. of the Thirteenth Int. Conf. on the World Wide Web (WWW 2004)*, New York, May, 2004.

Chapter 43

Middleware for Wearable Computing

Chandra Narayanaswami

CONTENTS

Introduction

Wearable computing has existed as an independent field for more than a dozen years now, and its impact is being felt today by millions of people. So, what is a wearable computer? Broadly speaking, a wearable computer is typically worn directly on the body or attached to clothes worn on the body. Sometimes, the wearable computer is even implanted inside the body. The overall vision is that, by integrating the wearable computer into something the user always wears, it will be available to the user almost all the time. Research in wearable computing [2,5,20,21,25,26,29,39,42,43] attempts to bring the power of computing technologies into our daily lives at a much more pervasive level than the traditional desktop computer. Unlike traditional computing where humans explicitly go to a computing device to use its services, the approach of wearable computing is one where the computer is with the user and continually augments the capabilities of the individual.

Over the years, as shown in Figure 43.1, Figure 43.2, Figure 43.3, and Figure 43.4, wearable computers of several types have been designed, prototyped, and manufactured. They range from general-purpose personal computers that are essentially PCs packaged differently (with a head-mounted display and single-handed keyboard and mouse) to single-function devices such as a one-way pager. Many wristwatches, pagers, hearing aids, pacemakers, and heart-rate monitors are specialized wearable computers. The most popular wearable computer is undoubtedly the cellphone. Though it started out as a device for just voice communication,

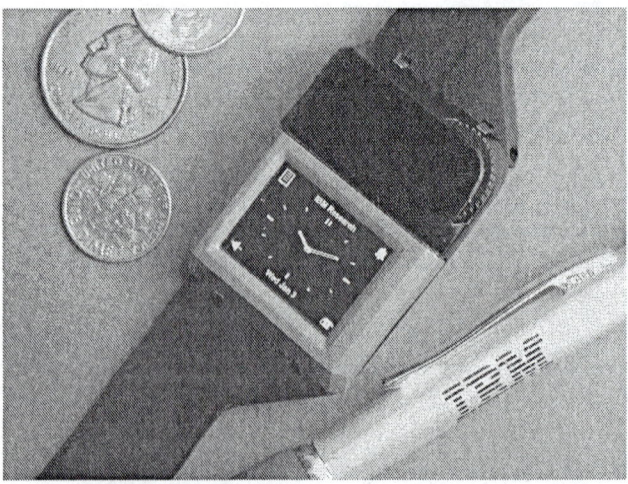

Figure 43.1   IBM's Linux watch prototype.

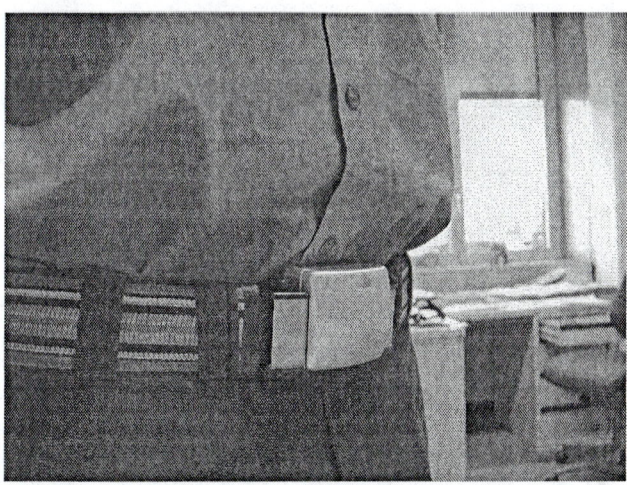

Figure 43.2   Belt computer. (Image courtesy of Wearable Computing Lab, ETH Zurich.)

recent cellphones incorporate powerful processors, disk drives, high-resolution displays, general-purpose operating systems, and a variety of applications. Many cellphones now also include digital cameras and radio-frequency identification (RFID) tags. In the very near future, we expect cellphones to include even RFID readers. Though they fit in a shirt pocket or can be worn as a pendant, the amount of raw computing power in

Figure 43.3   Sony Vaio U70P portable computer.

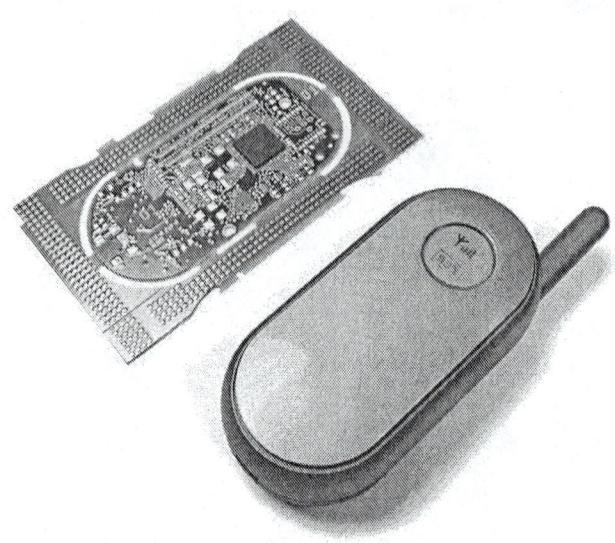

Figure 43.4   IBM Personal Mobile Hub prototype.

today's cellphone far exceeds that of even PCs from a decade ago, and they are regarded as a general-purpose computing platform [48].

Developments in mechanical engineering over the last 200 years have allowed us to reduce the physical strength required to perform many tasks; for example, a person with average physical ability can operate an excavator and succeed in a construction job. Before the advent of such tools, however, jobs in construction were for only the very hardy. Computing technology offers the promise of augmenting the capabilities of individuals in a variety of ways — for example, by keeping track of pieces of information that we find difficult to remember, by performing complex calculations much more rapidly than is humanly possible, and by making information readily available from remote sources. So, just as mechanical engineering advances leveled the playing field in terms of physical ability, wearable computing has the potential to level the playing field in an intellectual sense.

Key aspects of a wearable computer include the following: form factor (size, shape, and where on the body it will be worn), the environments in which it will be used, and whether it is special purpose or general purpose. Clearly, wearable computers have to be smaller and lighter than traditional computers. Wearable computers are designed for users who may be engaged in other tasks, so usability is a critical aspect that has to be addressed. Well-designed wearable computers allow the user to focus on the primary activity of the moment while providing the support the user needs. Users should find interaction with the computer intuitive and easy to understand. Wearable computers cannot afford to place a high degree of cognitive load on users because doing so may result in diverting attention away from their activities.

As the capabilities of wearable computers increase, the complexity of interacting with the wearable computer also increases. User interface paradigms that are popular in traditional desktop systems do not map well to wearable computers. The wearable computer as well as its user interface must be carefully designed, taking into consideration the types of activities the user is engaged in and the kinds of services the wearable device will provide to the user.

The flexibility of wearable computers comes at a price. The size constraints on a truly wearable device permit the inclusion of only a small set of user interface controls. Users also find it difficult to remember which combinations or sequence of button pushes are required to accomplish the functions they are interested in. Other input modes, such as voice, may not be appropriate in many situations for wearable computers, as the user is likely to be in a social setting. The user may also be on the move, and the user's environment may be too noisy for voice input. Size limitations on wearable computers also translate to a limited amount of display area. Any information that is displayed on the device must take into account the user's visual acuity. Although users may have the option of positioning the device closer or farther away from their eyes to make reading the display easier, such positioning may not always be desirable or even possible depending on the task the user is engaged in. The appearance of information presented on the display of a wearable computer must be designed to change depending on whether the user seeks to obtain that information at a glance or is willing to devote more attention and consciously study the display for a longer duration. The amount of battery energy that can be packed into wearable computers is also fairly limited [17–40]. The energy density of batteries in these computers is already at dangerous levels, so one cannot expect significant improvements in battery capacity. Wearable computers also typically communicate through wireless means. Wireless communication consumes more energy, especially at higher data rates and longer ranges; thus, energy efficiency must be considered at all levels, including hardware, base software, middleware, and applications.

The above limitations in input and output capabilities and limited energy capacity can be compensated for to some extent if the user could transfer sessions and running applications seamlessly from the wearable device to more powerful stationary devices when necessary. In this way, the flexibility of wearable devices could be combined with the inherent advantages of stationary devices [33].

Finally, wearable computers are often used in environments that are harsher compared to desktop computers. Being rugged is often a mandatory feature of wearable computers. Another item of concern in domains where the wearable computers may be shared among users is the issue of hygiene. Replaceable covers or similar techniques will be necessary to overcome that barrier. Heat dissipation from wearable computers may have to be addressed, as well [41].

Applications of Wearable Computing

We first consider a few applications of wearable computing before we discuss how suitable middleware can be designed to help address the concomitant challenges. Advances in wearable computing technology and sensors are beginning to impact the administration of health care in both inpatient and outpatient settings. The increasing costs of inpatient hospital stays combined with the growing evidence that faster recovery and immunity from infections are possible when the patient is back at home and ambulating are also motivating the deployment of patient monitoring applications.

Pacemakers have been used for several decades to adjust heart rhythms and help patients live a more satisfying life. Several wearable devices that can measure and track, for example, pulse rates are also available now at modest costs for casual use. Some of them can be connected to fitness equipment and to personal computers. More recently, efforts are being addressed not only at measuring health signs but also on analyzing the measured data and then relaying the measured data to healthcare personnel for immediate consultation or action. Typical areas include glucose monitoring, blood pressure monitoring, coronary event prediction, and pill regimen compliance. Another area where wearable computing technology is being used is in drug trials. By being able to monitor the effects of the drug on the patient in real time and in less controlled settings, researchers are able to improve the process and recruit more subjects for trials while reducing the cost of the trials. In addition, wearable monitors embedded in clothes are allowing more direct and accurate measurement of some parameters that had not been possible earlier. Researchers are also exploring brain–computer interfaces for the disabled [11,28,45]. Signals received from electrodes or sensors attached to the brain can be captured and used to move cursors, select items, etc. Such work is in the early stages but looks promising.

The nature of medical applications dictates that they must be reliable. Rebooting a wearable device in the middle of a monitoring application may not be a viable option. This means that applications must be resilient against loss of connectivity, failure to deliver messages, loss of battery, etc. Fallback options have to be designed from the beginning.

The relaying and sharing of medical information brings some unique challenges in terms of security and privacy. In many parts of the world, people want to keep their health issues private for a variety of reasons and want tight control over who has access to that data. In fact, several regulations such as the Health Insurance Portability and Accountability Act (HIPAA) specify how medical information can be shared and disseminated; thus, wireless security, theft of wearable devices, and unauthorized access to medical databases are some of the issues that have to be handled.

Armed forces personnel were early adopters of wearable computers. Wearable computers provide quick access to necessary information in battlefield situations. The computer is typically in a backpack or worn around the belt, and displays are provided as helmet-mounted units. One can imagine that the latest information, such as enemy area maps and enemy positions, can be sent to soldiers just in time. Information about fallen soldiers can be sent precisely to nearby personnel for possible recovery. The soldier's condition can be monitored regularly and relayed to command centers. Police officers in several civilian communities have also started to adopt wearable computing technology; for example, a wearable computer saves a trip back to the police car to verify a license or compare images or fingerprints.

Repair technicians in certain domains have been early adopters of wearable computing. Bell Canada, for example, reported that its utility workers were able to save about 50 minutes per day on average by not having to get down from poles to go to their trucks to use their computers. Typically, workers are able to bring up repair manuals while they are still in front of complex equipment and thus are better able to correlate data from the manuals with the real equipment. Wearable computers have been used in the aircraft maintenance industry, as well [3,27]. Workers at warehouses typically wear barcode scanners to help manage inventory more efficiently because the fingers are left free and the scanner is exceptionally well suited for package-handling-intensive activities such as order picking, stock pulling, and restocking.

Drishti [35], a prototype wearable system that includes a global position system (GPS) and audio output, helps blind people navigate streets by correlating the GPS data with spatial databases that include walkways, benches, poles, and other obstacles and providing just-in-time information to the user. The remembrance agent [37,38] is another application of wearable computing. The user's wearable computer provides information that is relevant to what the user may be reading or writing, such as e-mails from reviewers or relevant papers to cite. An often-quoted extension is the following: The agent helps the user with the name of the person the user is talking to, the date and time of the last conversation with that person, etc.

Other common wearable computers are found in the entertainment field. Music players such as the iPod® are an example. Echo-canceling headsets and Bluetooth® headsets are other examples.

In summary, wearable computing is transforming several business processes by either increasing the efficiency or expanding the conditions under which work can be done. In addition, wearable computing has improved the quality of life for several individuals and is poised to have a more significant impact in the years to come.

Middleware Components

Given that we now have a reasonable understanding of some of the challenges faced by wearable computers and some of the popular applications, we are in a position to look at middleware components that can help accelerate their reach.

Context

Because a wearable computer may be used in many locations other than the office, the *context* in which it is used is rather varied. Context could include the user's location, physical activity, intellectual activity, emotional state, future activity, and information about his environment, such as the names and pictures of people who are with him and what those people are doing. The context information can then be used to adapt user interfaces, route messages, present pending tasks, etc.

Location is a well-studied topic, and several methods are available to obtain location to a varying degree of accuracy [6]. Context middleware does not particularly care about which physical method is used as long as the location data is reported. Location data can specify the location in some absolute or abstract way, the accuracy of the information, the method that was used, etc. Sometimes location data is stale. In these cases, the middleware function could specify additional location data that specify the last reliable location of the user. The user's estimated trajectory can also be provided.

The user's physical activity can be derived directly from wearable sensors or by more indirect means such as looking up the user's calendar. Several researchers have used accelerometers to reliably determine whether the user is stationary, walking, or running. The user's physical activity can also be determined from studying images from cameras near the user in the environment. The middleware function that reports the user's physical activity should specify the confidence level with which it is reported.

The user's intellectual activity can be determined by analyzing what the user is working on — namely, reading, writing, or speaking. In the future, it may be possible to even determine what the user is thinking by attaching sensors to the brain. In some cases the user's calendar may be accurate enough to determine the activity. If the intellectual activity needs to be deduced it can be done by analyzing the text that is being composed by the user or the words being spoken, looking at the most active and recent processes running on the user's computer. The user's future activity can sometimes be deduced from the above analysis and by looking at the user's calendar. The middleware function that reports the user's mental activity should specify how the activity was determined.

The user's environment can be deduced in many ways; for example, the user's location may give an indication of whether the user is outdoors or indoors or in a warm area or cold, or it can provide the user's time zone. The user's wearable computer can record voices of people around the user and arrive at how many people are in the user's immediate vicinity at a given time. Images from the user's camera may also give an indication of the number of people around the user.

We believe that eXtensible Markup Language (XML) representations and Web services interfaces for the above types of context will be very useful. For one, they will allow a machine to parse the user's context and take appropriate action. Web services also remove the programming language dependencies and allow heterogeneous environments to mix. Standardized toolkit concepts such as Context Toolkit [10] can be adapted to the field of wearable computing.

Another area where middleware can help is in context prediction. Although knowing the user's current context is useful, it can be even more helpful if the user's future context can be deduced. A variety of means can be used for this purpose; for example, if the user is on a bus going from place A to B, the future location of the user is known assuming that information about the bus route and traffic information is available. The user's calendar is another rich source for this information and works if the calendar is accurate. Typically, calendars for more than 50 percent of the users are inaccurate because of the difficulty of keeping calendars up to date. If the user's wearable device knows that the user is going to run into person X in the corridor a minute from now, the system can quickly check to see if any messages need to be communicated in person when person X actually encounters the user.

Sensor Interfaces

Wearable devices must provide low-power hardware interfaces and data formats for a variety of sensors that can help get some of the context information described above. The IBM Personal Mobile Hub [16] (see Figure 43.4) is an example that uses this philosophy. By standardizing the hardware and software interfaces, the sensor manufacturers can focus on designing the best sensor instead of worrying about the complete software and hardware stack. Clearly, this technique has worked well in the PC industry; for example, a GPS sensor designer can make his design more compact by not having to build long-range wireless communication capabilities into the GPS unit and instead relying on short-range wireless communication that consumes less power.

Wireless means are clearly preferred for connecting the sensors to the wearable unit; for example, a sensor in the shoe may transmit a signal indicating whether the user is walking, running, or stationary. Running a wire from the shoe to the user's wearable unit, such as a cellphone, is rather inconvenient. Researchers [18,19] have placed accelerometers in various locations on the human body to determine the user's physical activity more accurately. Some clothes now have moisture, temperature, and ambient light sensors. The user's wearable unit can periodically communicate with all of the body worn sensors to determine the user's context.

Data from the sensors can be used to enhance user experience; for example, imagine that a music player with a disk drive can be told ahead of time that the user is going to jog. The player may be able to avoid disk accesses during the jogging period by copying to RAM or flash memory the songs in the playlist that the user is going to use while jogging, as such memory is less susceptible to read errors while in motion. Ambient light sensors can be used to modify the colors and brightness levels of the user interface or to switch to another modality, such as voice, if appropriate. If the user is expected to travel into a zone with poor wireless coverage based on past history, the user's mail replication schedule could be advanced.

In summary, sensors that we can expect to communicate with the user's wearable unit include temperature, ambient light, moisture, location, and physical activity. A common event model that delivers a piece of information consistently and reliably would significantly help in streamlining data transfer and delivery.

Data Logging and Analysis

Wearable devices are capable of holding several tens of gigabytes of storage today. An iPod with a 60-GB disk drive is already available. Cellphones have begun to incorporate 4-GB drives. We can only expect this number to go up over time. One of the consequences of this is that the user can carry a lot of information on the wearable device itself. In fact, the SoulPad model [7,34] allows users to carry their entire suspended computer state on the disk drive and resume the suspended state on any PC.

The information stored on the wearable unit could be captured from attached sensors or from other traditional sources such as databases and

Web servers. Middleware components that will be useful are embedded versions of databases that can be synchronized with larger versions on a server; for example, a jogger can log details about a particular run, such as the time of day, length of run, period of run, number of paces taken, and instantaneous heart rate. Such data can be stored in databases for further analysis, either locally or in a more global fashion.

Data analysis on captured data is an important area. In our jogger example, the events can be examined to look for patterns. Middleware to specify rules for patterns becomes essential [44]. Statistical packages, event correlation engines, neural networks, etc. can be used to launch software agents that take further action; for example, a simple rule might say that, if the instantaneous heart beat exceeds 165 beats per minute for 45 minutes and the temperature is above 85°F, then the user should be notified and asked to slow down. The rule engine should allow specification of simple rules and include a rule composition language that allows several rules to be compounded to derive a complex rule. More complex rules could look for patterns over longer periods of time and at multiple streams of data; for example, a user attending a conference might specify a rule requesting that his wearable notify him whenever a talk is taking place that cites one of the user's technical papers and the user's schedule is free for that time. Because rules can change over time, the middleware must allow the addition of new rules and refinement of existing rules. It is also important to verify that the compound rules do not contain cyclic dependencies, etc. Formal verification of rules can be performed in the background. Stream data processing is an emerging field [8,9,22] with particular applicability to wearable computing because multiple sensors can stream data to the user.

When local data analysis on the wearable unit will not suffice, the collected data can be sent to remote locations for further analysis, perhaps with larger groups of users; for example, if a rule is triggered and detects a suspicious medical event, the data supporting the triggering of the rule can be sent to an expert for further analysis. The expert may have a larger collection of rules or data samples and may be able to better analyze the data than the user's wearable unit. Standard methods to synchronize different data types are evolving. Technologies such as SyncML® [15] may prove to be useful.

Software, Service, and Device Management

One of the big challenges for wearable computers is the availability of software when it is needed dynamically. Users cannot expect to have all of the necessary software they will need on their wearable computers due to the number and frequency of novel situations they are likely to encounter; for example, if the user wishes to utilize services available in an active space around him, suitable mechanisms have to be available for service discovery and dynamic software provisioning. Fortunately, standards such as the Open Services Gateway Initiative (OSGi) [0] provide methods to publish and dynamically provide code bundles to the wearable device from a bundle server. Mechanisms are also available to update older versions of code and keep the device software up to date.

Several service discovery mechanisms are also available. They include Domain Name System (DNS)-based service discovery (DNS-SD); the Simple Service Discovery Protocol (SSDP); Universal Description, Discovery, and Integration (UDDI) for Web services discovery; the Bluetooth® Service Discovery protocol; etc. Some of these mechanisms such as DNS-SD and Bluetooth Service Discovery are closely coupled with lower levels of the networking stack. Others operate at higher levels and are programming language independent and typically use newer Web-based standards including XML for service descriptions. Protocols such as Universal Plug and Play (UPnP™) leverage SSDP to help discover devices. Near Field Communication (NFC) [49] technology is beginning to gain momentum as an aid to device and service discovery within short ranges. The basic idea is to use RFID tags and readers to discover services that are available in the immediate vicinity.

The next challenge is to create suitable middleware that allows active spaces to be populated with services to make it easy to create new active spaces — say, at bus stops, coffee shops, or campus sign boards. Middleware to test whether the services available in the active spaces are functioning correctly also becomes crucial.

The World Wide Web Consortium (W3C) has several activities directed toward mobile devices. These activities include the Dynamic Properties Framework (DPF) [47] and the Device Independence Activity [47]. DPF defines platform- and language-neutral interfaces that provide Web applications with access to a hierarchy of dynamic properties representing device capabilities, configurations, user preferences, and environmental conditions. The W3C Device Independence Activity is working to ensure seamless Web access with all kinds of devices and the development of worldwide standards for the benefit of Web users and content providers alike. The Bluetooth Human Interface Devices (BT-HID) profile defines a set of services that can be used between a host capable of supporting HID devices and a BT-HID device. More specifically, the HID profile defines a mechanism through which Bluetooth-enabled devices such as phones, PDAs, etc. can communicate with another device. Configuration files on the device define its behavior by mapping particular keys or joystick movements to particular events or actions on the remote device. All of the above activities will help address device heterogeneity in the device space that arises due to the personal preferences of users and intense competition between device vendors to provide well-designed appliances.

Privacy and Security

It is clear that wearable computers can collect and analyze a significant amount of data. A natural question that people often raise is one of privacy. Who can be given access to information about the user's current physical activity? How can we ensure that data from a heart-rate monitor does not fall into the wrong hands? How should privacy permissions be propagated? If someone has agreed to release location information to a colleague, is that colleague at liberty to reveal that person's location to others?

Let us look at data security before we suggest possible approaches to handle the privacy issues above. Because data may be received from body-worn sensors it is necessary to protect such data during transfer. The sensors could have a shared secret with the wearable computer and use it to encrypt data transfers. Recently, it has been shown that Secure Sockets Layer (SSL) can be implemented even on resource-constrained devices [13]. Because body-worn sensors typically have short ranges, the security issues can generally be met without requiring significant innovations. Because wearable devices could potentially be lost, sensitive data on the devices should be protected using standard security mechanisms. Periodic data backups, similar to those for PCs today, are also essential.

Privacy management middleware could help define the level of privacy for each data element; for example, a user could specify that his calendar entries will be viewable by person X between 9 a.m. and noon on Tuesday the 7th. Another example that is coupled to rules is that if the user's heart rate is above 165 beats per minute then person X can access the user's location. By being able to place conditions under which some information is revealed, the user retains the privacy he desires most of the time, while still allowing for divulging data in case of emergencies.

A big challenge to specifying the privacy properties for any data is the usability of the system. If the method for entering the rules is complex, most users will take the default privacy levels that the system offers. The privacy middleware must have good support for cloning rules defined earlier and for editing them. A system than can learn the user's predilections for privacy will be of immense value. When such systems become available, privacy tags can be assigned automatically by the system.

Multimodal and Multiform User Interfaces

Because the user of the wearable computer can be in many different types of environments, the user interface that is most suitable for communicating with the user may change; for example, although pen-based text input may be fine when users are at their desks, the application must seamlessly allow the users to switch to a voice-based text input system when they are walking. In essence, middleware to switch the modality of the user interface becomes critical. The application designer must take into account that the modality of the user interface could change during the course of the application and provide suitable transition points where the modality can be changed.

Wearable computers may have lower resolution direct-view displays and may allow the attachment of high-resolution, eye-mounted displays. Applications have to be cognizant of this fact and adjust to varying display size and also exploit the two displays whenever practical and useful. The eye-mounted display may be more suitable for private information, and the direct view display may be easier to access in a variety of day-to-day settings. Projection-based and flexible displays [23] can also change the size of the display dynamically. Middleware that estimates the size of the active display or displays will be necessary to adjust to viewing conditions. Similarly, the wearable device may be operated with a limited number of dedicated buttons on the device itself or through auxiliary input devices such as full-function keyboards. User interfaces have to reconfigure themselves as different input mechanisms are added.

Tactile- and vibration-based user interfaces are useful in a mobile scenario [12,31]; for example, patterns of vibration can distinguish between calls from family versus non-family. Gestures made by moving the wearable device [14,36] can be sensed by accelerometers housed in the device and can be used to control the device. Middleware for gestures and for vibration-based output have to be developed.

Voice input is generally not used widely in a desktop environment. It appears that, in spite of deficiencies in the keyboard and mouse interface, people still find it superior to other interfaces on the desktop; however, when the user is mobile and the task is more complex, voice-based technologies can play a role. As an example, voice-based dialing on cellphones is popular. Using voice and using natural language processing to specify rules for event correlation seems simpler than doing it with a keyboard and mouse.

The context of the user can be used to modulate the volume of an audio interface. Clearly, the volume can be increased in noisy environments. Middleware that connects the user's context to the output interface is therefore essential. In addition, instead of having the same fixed volume for all applications on the device, volume could be customized for each application depending on the current context. Some applications may warrant a higher volume and others may be muted. These parameters have to be captured by suitable functions in the middleware layer.

In summary, knowing the user's precise context can help tailor the user interface. Evolution of suitable middleware to support multimodal input and output and seamless switching between modalities will improve the usability of wearable computing.

Energy Management

Energy is an important consideration in wearable computing. The simplest form of middleware allows the applications to get the status of the battery powering the device. This middleware will take the user's battery recharging strategy into account, as well; for example, if the user typically charges his wearable computer in the car when returning home, the system will not go into the lowest power mode if it detects a relatively low battery level toward the end of the day. Energy management middleware will help determine the amount of energy that will be consumed for certain tasks. The user may specify in some high-level terms the nature of the task, and the middleware will determine an energy consumption profile for the task depending on other tasks that are running on the system. Such middleware should also measure and learn the energy consumption of installed applications as a background activity. Energy management middleware can help tune the user interface; for example, if the battery level is low and the user is not expected to recharge the battery any time soon, the user interface can be modified to perhaps be a bit less friendly but more frugal from an energy consumption angle. Energy management middleware to perform clock and frequency scaling to reduce energy consumption can be integrated with other contextual information about the user; for example, if a user wants to finish viewing a video before getting off the bus but does not see any benefit to finishing viewing the video any sooner, then that information can be provided to the clock and frequency scaling middleware as a constraint.

Suspend/Resume

Although wearable computers provide the flexibility of being available all the time, they come with several disadvantages compared to stationary counterparts; for example, a wearable computer may have a Twiddler™ keyboard, and the user may have a slower data input rate on such a keyboard. Wearable computers also have smaller displays and may have weaker processors and less memory than stationary PCs. Wearable computers also have limited energy sources. For these reasons, users of wearable computers may want to switch their computing sessions from the wearable to a stationary computer. Essentially, users should be given the ability to suspend the computing state on their wearable computers and then transfer it to a more convenient machine and resume the computation with out losing the context (e.g., the tasks that were running, the windows that were open).

Such functions are indeed provided by the SoulPad approach [34]. The SoulPad technique allows users to carry just a small portable device with sufficient storage so it can be attached to a public PC to give the users access to their complete personal computing environment (data and applications) and allow them to resume prior suspended sessions without requiring network connectivity. The SoulPad software stack has three layers: an autoconfiguring operating system (OS), a virtual machine (VM) monitor, and a suspended VM image (includes guest OS and applications). The SoulPad stack can be carried on a portable disk drive or on devices that include a drive such as an iPod. Because all of the software that runs on the PC comes from SoulPad, it does not rely on any installed software on the PC and does not even access the internal disk on the PC. SoulPad can work with diskless PCs, as well. One area that requires further investigation is the seamless handoff of the user interface when computation is transferred among devices with differing I/O interfaces.

Another useful technology is the Session Initiation Protocol (SIP) [51]. Acharya et al. [1] used SIP to control media sessions to wearable computers. SIP separates the data path for a piece of communication from the control path. In a sense, it is analogous to using pointers to refer to a data object. This technology allows the user, for example, to receive a call on his wearable device that he can subsequently transfer to a stationary device nearby. The wearable device can either remain in the control path or drop off. A user may seamlessly transfer an ongoing conversation from his office phone at the end of his day to his cellphone while on his way to the car and then to his car phone and finally to his home phone without dropping the call.

Seamless mobility between devices will be important in some domains. The application and user interface must both be considered. This may change the way applications are written, because an application can be started on one platform and resumed on another. Applications will have to define suitable points at which the application may be suspended for later resumption.

Rapid Prototyping

Wearable devices may have short lifecycles due to rapid advances in technology. This means that rapid prototyping and development are crucial. Tools that facilitate rapid prototyping of the function of a wearable device will be of significant value; for example, as discussed in Naraya-naswami and Raghunath [24], the design of a new form factor can be separated into a wirelessly connected physical piece that models the input and output devices of the form factor and several software components running on standard PCs that perform the actual function of the device. This separation allows faster debugging of the system and parallel development of system software and applications. Typically, this approach is an improvement over software device emulators because the usability and human interface aspects can be tested more accurately. Middleware that allows users to specify standard input output devices and connect them to software emulators will help this process.

Conclusions

Wearable computing is an emerging field with great potential to change the way business is conducted and the way people entertain themselves. Over the last several years, the wearable computing community has worked toward addressing some of the fundamental challenges that face this emerging field. The broad functionality provided by modern cell-phones demonstrates that the basic challenges have largely been addressed in some areas. We are now at an exciting point where increasingly larger segments of the general population can benefit from wearable computing technologies. Well-designed middleware that makes it less expensive, faster, and easier to develop personalized applications is critical for expanding the possible impact from wearable computing. The development and eventual availability of middleware components discussed in this chapter will also make it easier for users to build their own personal constellation of devices. The user's personal constellation of devices can interact with devices in the environment to make users more productive in a more diverse set of situations than was possible earlier without well-defined middleware components.

References

[1] Acharya, A., Berger, S., and Narayanaswami, C., Unleashing the power of wearable devices in a SIP infrastructure, in *Proc. of the 3rd Int. Conf. on Pervasive Computing and Communications (PerCOM'05)*, Kauai Island, Hawaii, March 8–12, 2005, pp. 159–168.

[2] Bass, L.J., Kasabach, C., Martin, R., Siewiorek, D.P., Smailagic, A., and Stivoric, J., The design of a wearable computer, in *Proc. of ACM SIGCHI Conf. on Human Factors in Computing Systems (CHI'97)*, Atlanta, GA, April 18–23, 1997, pp. 139–146.

[3] Bass, L., Siewiorek, D., Bauer, M., Casciola, R., Kasabach, C. et al., Constructing wearable computers for maintenance applications, in *Fundamentals of Wearable Computers and Augmented Reality*, Barfield, W. and Caudell, T., Eds., Lawrence Earlbaum Associates, Mahwah, NJ, 2001.

[4] Bigus, J.P., Schlosnagle, D.A., Pilgrim, J.R., Mills, III, W.N., and Diao, Y., ABLE: a toolkit for building multiagent autonomic systems, *IBM Syst. J.*, 41(3), 350–371, 2002.

[5] Billinghurst, M. and Starner, T., Wearable devices: new ways to manage information, *IEEE Comput.*, 32(1), 57–64, 1999.

[6] Borriello, G., Chalmers, M., LaMarca, A., and Nixon, P., Delivering real-world ubiquitous location systems, *Commun. ACM*, 48(3), 36–41, 2005.

[7] Caceres, R., Carter, C., Narayanaswami, C., and Raghunath, M.T., SoulPad: reincarnating PCs using portable devices, in *Proc. of the Third Int. Conf. on Mobile Systems, Applications, and Services (MobiSys'05)*, Seattle, WA, June, 2005, pp. 65–78 (www.research.ibm.com/WearableComputing/Soul-Pad/soulpad.html).

[8] Carney, D., Cetintemel, U., Cherniack, M., Convey, C., Lee, S. et al., Monitoring streams, a new class of data management applications, in *Proc. 28th Int. Conf. on Very Large Data Bases (VLDB'02)*, Hong Kong, August, 2002.

[9] Chandrasekaran, S. and Franklin, M.J., Streaming queries over streaming data, in *Proc. of the 28th Int. Conf. on Very Large Data Bases (VLDB'02)*, Hong Kong, August, 2002.

[10] Dey, A.K., Providing Architectural Support for Building Context-Aware Applications, Ph.D. thesis, College of Computing, Georgia Institute of Technology, Atlanta, 2000.

[11] Farwell, L.A. and Donchin, E., Talking off the top of your head: toward a mental prosthesis utilizing event-related brain potentials, *Electroencephalogr. Clin. Neurophysiol.*, 70(6), 510–523, 2002.

[12] Gemperle, F., Ota, N., and Siewiorek, D., The design of a wearable tactile display, in *Proc. of the Fifth IEEE Int. Symp. on Wearable Computers (ISWC'01)*, Zurich, Switzerland, October 7–9, 2001.

[13] Gupta, V., Millard, M., Fung, S., Zhu, Y., Gura, N. et al., Sizzle: a standards-based end-to-end security architecture for the embedded Internet, in *Proc. of the 3rd IEEE Int. Conf. on Pervasive Computing and Communications (PerCom'05)*, Kauai Island, Hawaii, March 8–12, 2005, pp. 247–256.

[14] Hamburgen, W.R., Wallach, D.A., Viredaz, M.A., Brakmo, L.S., Waldspurger, C.A. et al., Itsy: stretching the bounds of mobile computing, *IEEE Comput.*, 34(4), 28–36, 2001.

[15] Hansmann, U., Thompson, P., Mettala, R.M., and Purakayastha, A., *SYNCML: Synchronizing Your Mobile Data*, Prentice Hall, Upper Saddle River, NJ, 2002.

[16] Husemann, D., Narayanaswami, D., and Nidd, M., Personal mobile hub, in *Proc. of the Eighth IEEE Int. Symp. on Wearable Computers (ISWC'04)*, Arlington, VA, October 31–November 3, 2004, pp. 85–91.

[17] Kamijoh, N., Inoue, T., Olsen, C.M., Raghunath, M., and Narayanaswami, C., Energy trade-offs in the IBM wristwatch computer, in *Proc. of the Fifth IEEE Int. Symp. on Wearable Computers (ISWC'01)*, Zurich, Switzerland, October 7–9, 2001, pp. 133–140.

[18] Kern, N., Schiele, B., and Schmidt, A., Multi-sensor activity context detection for wearable computing, in *Proc. of European Symp. on Ambient Intelligence (EUSAI'03)*, Eindhoven, The Netherlands, November 3–4, 2003, pp. 220–232.

[19] Kern, N., Antifakos, S., Schiele, B., and Schwaninger, A., A model for human interruptability: experimental evaluation and automatic estimation from wearable sensors, in *Proc. of the Eighth IEEE Int. Symp. on Wearable Computers (ISWC'04)*, Arlington, VA, October 31–November 3, 2004, pp. 158–165.

[20] Kortuem, G., Bauer, M., and Segall, Z., NETMAN: the design of a collaborative wearable computer system, *Mobile Networks Appl.*, 4(1), 49–58, 1999.

[21] Kortuem, G., Segall, Z., and Bauer, M., Context-aware, adaptive wearable computers as remote interfaces to "intelligent" environments, in *Proc. of the 2nd IEEE Int. Symp. on Wearable Computers (ISWC'98)*, Pittsburgh, PA, October 19–20, 1998, pp. 58–65.

[22] Madden, S., Shah, M.A., Hellerstein, J.M., and Raman, V., Continuously adaptive continuous queries over streams, in *Proc. of ACM SIGMOD Int. Conf. on Management of Data*, Madison, WI, June 3–6, 2002, pp. 49–60.

[23] Narayanaswami, C. and Raghunath, M.T., *Unraveling Flexible OLED Displays for Wearable Computing*, IBM Research Report RC23622, IBM Corp., Armonk, NY, 2005.

[24] Narayanaswami, C. and Raghunath, M.T., Designing a new form factor for wearable computing, *IEEE Pervasive Comput.*, 1(4), 42–48, 2002.

[25] Narayanaswami, C. et al., IBM's Linux watch: the challenge of miniaturization, *IEEE Comput.*, 35(1), 33–41, 2002.

[26] Narayanaswami, C., Raghunath, M., Kamijoh, N., and Inoue, T., *What Would You Do with a Hundred MIPS on Your Wrist?*, IBM Research Report RC22057, IBM Corp., Armonk, NY, 2001.

[27] Nicolai, T., Sindt, T., Kenn, H., and Witt, H., *Case Study of Wearable Computing for Aircraft Maintenance*, Universität Bremen, Germany, Germany Center for Computing Technologies (TZI), 2005.

[28] Nykopp, T., Laitinen, L., Heikkonen, J., and Sams, M., Statistical methods for MEG based finger movement classification, *IEEE Trans. Neural Syst. Rehabilitation Eng.*, 2006 (in press).

[29] Pentland, A., Human design: wearable computers for human networking, in *Proc. of IEEE Int. Conf. on Distributed Computing Systems (ICDCS'03)*, Providence, RI, May, 2003, pp. 264–265.

[30] Pinhanez, C., The everywhere displays projector: a device to create ubiquitous graphical interfaces, in *Proc. of Int. Symp. on Ubiquitous Computing (Ubicomp'01)*, Atlanta, GA, September 30–October 2, 200, pp. 315–331.

[31] Poupyrev, I., Maruyama, S., and Rekimoto, J., Ambient touch: designing tactile interfaces for handheld devices, in *Proc. of the 15th Ann. ACM Symp. on User Interface Software and Technology*, Paris, France, October 27–30, 2002.

[32] Raghunath, M.T. and Narayanaswami, C., User interfaces for applications on a wrist watch, *J. Pers. Ubiquit. Comput.*, 6, 17–30, 2002.

[33] Raghunath, M.T., Narayanaswami, C., and Pinhanez, C., Fostering a symbiotic handheld environment, *IEEE Comput.*, 36(9), 55–65, 2003.

[34] Raghunath, M.T., Narayanaswami, C., Carter, C., and Caceres, R., *Soulpad: Reincarnating PCs Using Portable Devices*, IBM Research Report 23418, IBM Corp. Armonk, NY, 2005.

[35] Ran, L., Helal, S., and Moore, S., Drishti: an integrated indoor/outdoor blind navigation system and service, in *Proc. of the Second IEEE Int. Conf. on Pervasive Computing and Communications (PerCom 2004)*, Orlando, FL, March, 2004, pp. 23–32.

[36] Rekimoto, J., GestureWrist and GesturePad: unobtrusive wearable interaction devices, in *Proc. of the Fifth IEEE Int. Symp. on Wearable Computers (ISWC'01)*, Zurich, Switzerland, October 7–9, 2001, pp. 21–28.

[37] Rhodes, B.J., Using physical context for just-in-time information retrieval, *IEEE Trans. Comput.*,52(8), 1011–1014, 2003.

[38] Rhodes, B.J., The wearable remembrance agent: a system for augmented memory, in *Proc. of the First IEEE Int. Symp. on Wearable Computers (ISWC'97)*, Cambridge, MA, October 13–14, 1997, p. 123.

[39] Smailagic, A., Siewiorek, D., and Luo, L., A system design and rapid prototyping of wearable computers, course 2003, in *Proc. of IEEE Int. Conf. on Microelectronics Systems Education (MSE'03)*, Anaheim, CA, June 1–2, 2003.

[40] Starner, T., Human-powered wearable computing, *IBM Syst. J.*, 35(3/4), 618–629, 1996.

[41] Starner, T. and Maguire, Y., A heat dissipation tutorial for wearable computers, in *Proc. of the 2nd IEEE Int. Symp. on Wearable Computers (ISWC'98)*, Pittsburgh, PA, October 19–20, 1998, pp. 140–148.

[42] Starner, T., The challenges of wearable computing, part 1, *IEEE Micro*, 21(4), 44–52, 2001.

[43] Starner, T., The challenges of wearable computing, part 2, *IEEE Micro*, 21(4), 54–67, 2001.

[44] Yemini, S.A., Kliger, S., Mozes, E., Yemini, Y., and Ohsie, D., High speed and robust event correlation, *IEEE Commun. Mag.*, 34(5), 82–90, 1996.

[45] Wolpaw, J.R., McFarland, D.J., Neat, G.W., and Forneris, C.A., An EEG-based brain-computer interface for cursor control, *Electroencephalogr. Clin. Neurophysiol.*, 78, 252–259, 1991.

[46] Bluetooth Human Interface Devices (HID), Bluetooth SIG, http://www.bluetooth.com/.

[47] W3C Device Independence Activity Statement, http://www.w3.org/2001/di/Activity.

[48] *IEEE Pervasive Comput.*, 4(2), 2005.

[49] Near Field Communication (NFC) Forum, www.nfc-forum.org/.

[50] Open Services Gateway Initiative, http://www.osgi.org/.

[51] Session Initiation Protocol (SIP), http://www.sipforum.org/.

Chapter 44

Middleware for
Mobile Entertainment
Computing

Vittorio Ghini, Fabio Panzieri, and Marco Roccetti

CONTENTS

Introduction

Entertainment is a cornerstone of communication. The shared enjoyment of music, stories, games, and other entertainment experiences allows people to communicate and socialize. Conventional forms of entertainment include reading a book, listening to music, watching a video, playing a game, chatting with friends, and sharing fun. The deployment of Internet-based technology enables the dissemination of these forms of entertainment on a large (possibly planetary) scale, as well as the development of additional, more sophisticated forms of entertainment (e.g., multiparty network games, media streaming, online news) [1].

The architecture of an online entertainment application can be thought of as consisting of the five principal macrocomponents illustrated in Figure 44.1 and summarized below:

- The *content* represents the data the application requires for its execution.
- The *pipe* includes the network infrastructure and the protocol stack (from the physical layer to the application layer) that transports and delivers the content to the user. This component may include wireless and wired links and may require specialized hardware in the user device.
- The *platform* represents the hardware and the operating system on both the server and client end systems. End systems can be general-purpose end systems (e.g., desktop PCs, laptops, PDAs) or specialized devices such as the PSP™, Xbox®, and N-Gage™ game consoles.
- The *user interface* (UI) is the hardware and software required for human interaction (e.g., graphic interface, touch-screen display).
- The *connecting component* represents the hardware and software involved in the service provision (i.e., service discovery, accounting, and billing software).

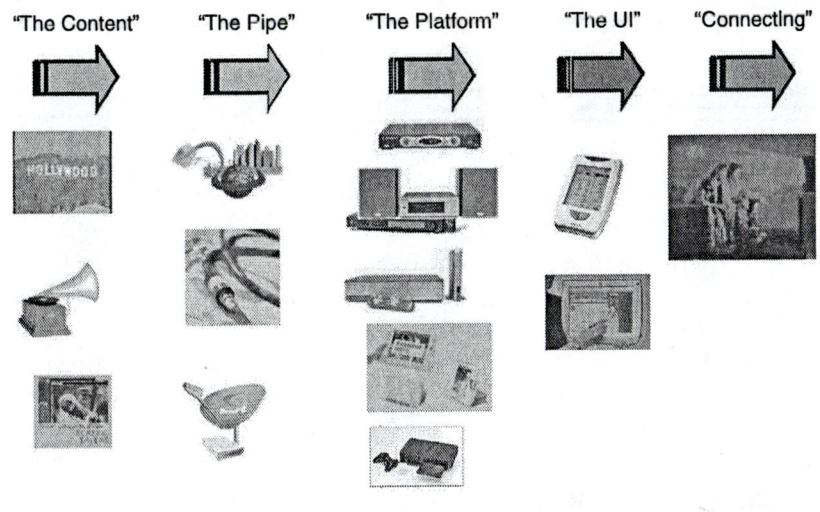

Figure 44.1 Online entertainment components.

Mobile communication technologies are encouraging the development of new entertainment models that meet such emerging user requirements as mobile and nomadic user behavior, multimodality, and context awareness. Thus, new communication protocols must be devised to offer universal access to online entertainment services from anywhere, at any time, and from any device [2]. Examples of online entertainment services that may benefit from universal access include: (1) network games that can be played via complex, possibly dedicated, interfaces, as well as simple keyboards; (2) music downloading services that can be carried out while the user moves across heterogeneous, both wired and wireless, communication technologies; and (3) applications that use such context information as the user location or the current time and temperature to provide their services (e.g., an online city guide application that provides its users with direction information, relative to their current geographical position).

A number of design issues must be addressed in the development of an online mobile entertainment architecture. These issues include the provision of support for communications, scalability, mobility, and content and format runtime adaptation to varying operational conditions. Moreover, the architecture should accommodate applications developed using different programming languages for different application modules; thus, support can be required for general-purpose programming languages, as well as scripting languages for high-level control logic and artificial intelligence (e.g., Quake™ 3, Unreal®, Spades), and dedicated languages for graphic three-dimensional (3D) tools (e.g., MEL™ for Maya®, MAXScript™ for 3D Studio MAX®, VEX for Houdini®). In addition, the architecture should enable the development of a platform that allows the application designer to incorporate third-party modules in its application so as to simplify the application implementation. Finally, that platform should support application code portability.

We argue that, to master and control the complexity inherent in the development of an online entertainment application, the majority of the design issues introduced above can be dealt with in a middleware layer designed to support both application development and execution. This middleware layer should effectively meet the nonfunctional requirements of the application (such as those for scalability, mobility, code portability, etc., mentioned earlier) to allow the application designer to concentrate only on the design and implementation of the application logic.

Entertainment Computing

Online entertainment computing includes a large set of different applications, devices, and technologies. A basic taxonomy of these technologies is illustrated in Figure 44.2. This taxonomy distinguishes online entertainment in two basic categories, based on *passive* and *interactive applications*. The passive category includes those applications that meet the user demand for stored or live information and collect and reproduce that information using different media formats. These applications can be further classified as *on-demand* and *live streaming* applications and are characterized as described below.

A Taxonomy: On-Demand Applications

Downloading and playing music and video represents a typical prestored form of entertainment (also referred to as on-demand). The user may use a wide set of devices, ranging from a so-called *home theater* system (i.e., an enhanced PC-based system located in the home and served by a broadband connection) to a dedicated portable device with a wireless interface that allows communications with a distribution infrastructure. A notable example of home theater systems is the Microsoft® Media Center (MC), illustrated in Figure 44.3. This is a PC-based system, with high-capacity memory and hard disk and a CD-ROM/DVD unit, powered by Windows® XP Media Center Edition 2005. The MC allows the user to store, share, and render photos, music, home video, and even recorded television programs from a large set of devices and interfaces.

In addition to PC-based systems, the class of on-demand applications includes a variety of small portable devices that can store and play music files but are not directly connected to the Internet. Thus, downloading

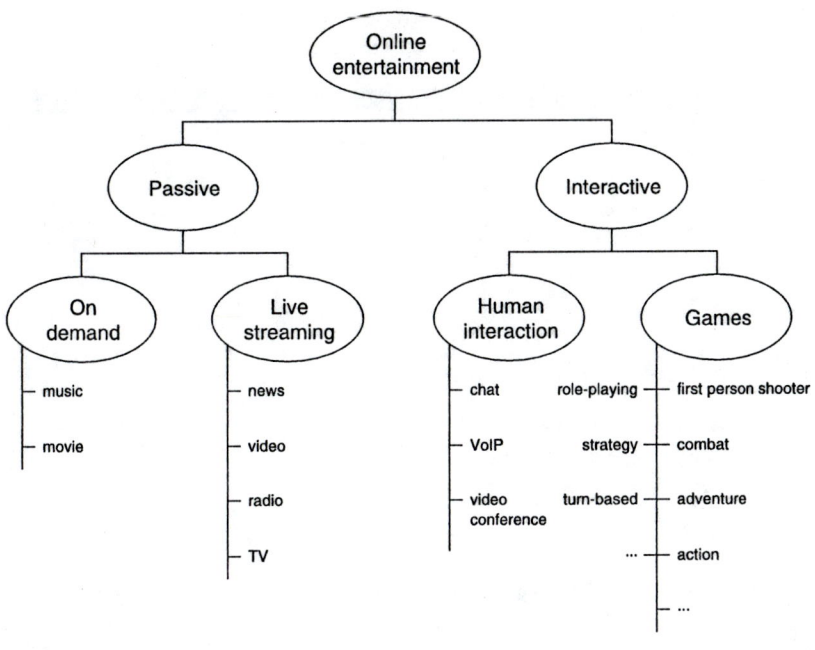

Figure 44.2   A taxonomy for online entertainment.

music via the Internet using these devices requires a PC intermediate system, as depicted in Figure 44.4. A representative of this class of devices is the Apple® iPod® family. The iPod devices can store the music files in an internal hard disk with a capacity ranging from 512 MB to 60 GB. A song library available via the Internet (iTunes®) allows the user to purchase and download songs to a Mac or Windows-based PC connected to the

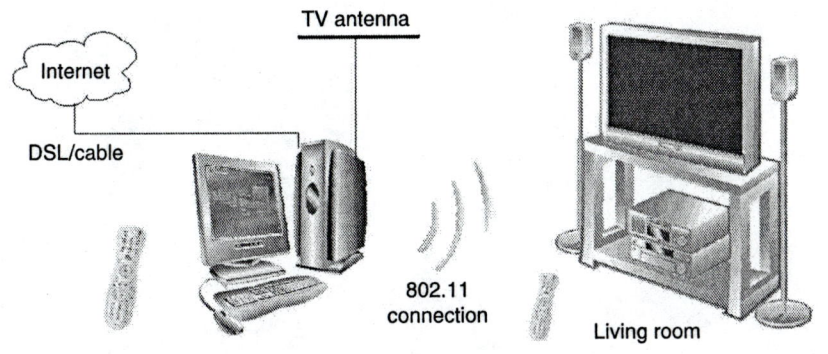

Figure 44.3   Microsoft Media Center.

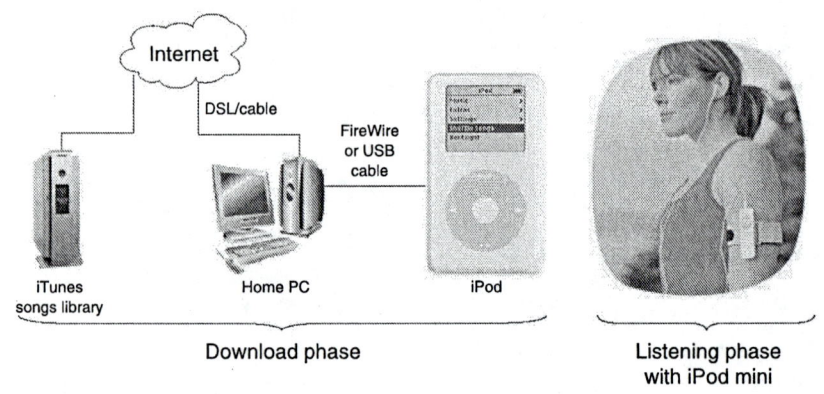

Figure 44.4   Apple iPod/iTunes system.

Internet. The user connects the iPod to the PC through a FireWire™ or USB cable and stores the songs in the iPod for final listening on the move.

The peer-to-peer (P2P) paradigm has enabled the implementation of very popular on-demand applications for online entertainment. These applications are typically based on PCs connected to the Internet. As depicted in Figure 44.5, one or more overlay networks without centralized servers, such as WinMX or Overnet, store and distribute media contents. Each network's node may discover and download contents depending on the availability of other peers and on the connectivity provided by the underlying network subsystem. After the download phase, the user enjoys the contents by using a common application, such as Windows Media Player or VNC Player.

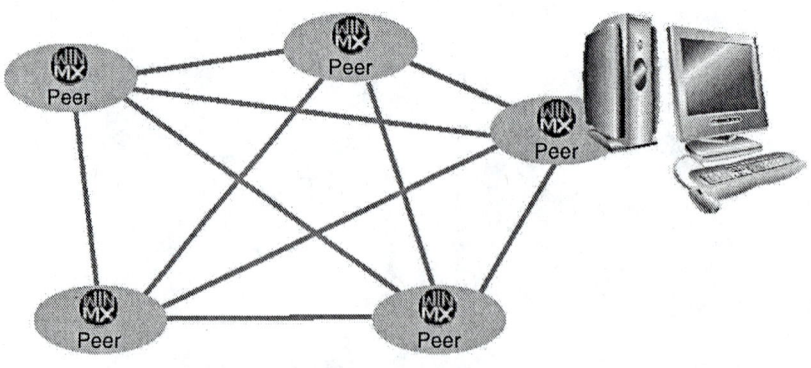

Figure 44.5   A peer-to-peer system.

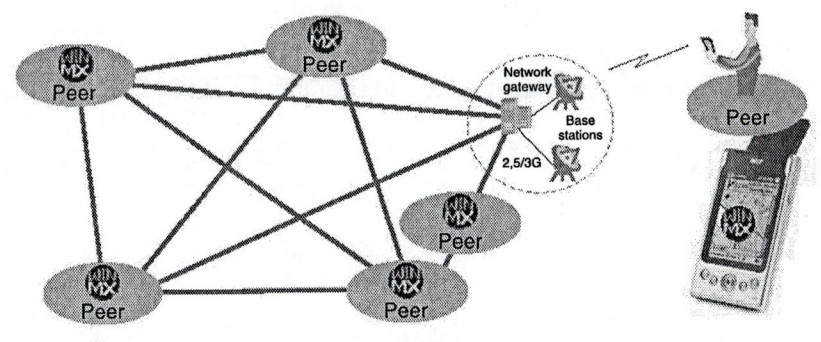

Figure 44.6   A peer-to-peer system over a wireless network.

Peer-to-peer applications can be also applied to a general-purpose laptop or palmtop device connected to the Internet by means of a wireless network (see Figure 44.6); however, the success of a wireless P2P approach depends on the availability of a peer application compiled for the specific target device. Moreover, if this approach is adopted, the problem emerges of providing continuity of content downloads on a wireless link.

In general, issues of scalability and platform independence characterize the design and development of applications for mobile entertainment. Open problems concerning mobility when downloading include: (1) discovering the location or locations from which the contents can be retrieved; (2) guaranteeing that the download can be resumed after temporary unavailability of the wireless link; (3) selecting and using the *best* connection when the user device holds two or more wireless interfaces of possibly different technologies; and, in this latter case; and (4) using concurrently all the connections to augment the aggregated bandwidth available. Finally, an additional desirable functionality is to enable the user to switch the download of a music file among different devices while the download is in progress (e.g., to have the music file downloaded to the best available output device).

A Taxonomy: Live Streaming Applications

Live streaming is a method for transferring digital data with real-time characteristics in such a way that the recipient can enjoy the content while the data is being delivered. The advantage of streaming compared to downloading is that it makes it possible for the recipient to begin using the content as it is delivered. Thus, an entire file does not have to be downloaded and stored on the client device before rendering; however, the quality of a presentation is constrained by the underlying network. Moreover, the user may use the content only once for each download phase.

A stream is a flow of data packets containing media contents. The packets are normally generated by a streaming media server from an arbitrary data source, which can be media content stored or captured from a live source (e.g., camera, microphone, television broadcast). The content data is usually packed using a codec targeted for compressing that particular content. The bit rate of the stream specifies how much compressed payload data is sent in a time unit. The stream bit rate can be either constant or variable. The used bit rate and codec highly affect the quality of the encoded content. The generated data packets are continuously sent to the recipient over a packet-switched network, using some streaming protocol. The recipient runs a streaming media player software, which receives the packets, decodes the contents data with an appropriate codec, and finally renders the presentation to the user.

The live-based streaming applications introduce more issues from a technical standpoint than from a user perspective. In fact, live streaming entertainment, such as watching the Olympic games, imposes strong time constraints that severely tax the network and raise the need for a dedicated set of protocols, data formats, and adaptive services. Streaming is sensitive to errors and delays in the transmission, as a continuous flow of data is required for an uninterrupted presentation. If some data packets are lost or delayed during the transmission, the media player may not be able to decode the data correctly, and some errors, or interruptions, may occur. Hence, properties of the underlying network, such as error rate, latency, delay, and throughput, may have a significant effect on the streaming quality. To compensate for possible delays, streaming media players usually receive some amount of packets before beginning to play the content. This is referred to as *buffering*.

The term *mobile streaming* is used if the content is streamed to a terminal over a mobile network. A terminal is usually a mobile phone or a personal digital assistant (PDA) plus a streaming media player software. A typical mobile streaming architecture is depicted in Figure 44.7. In mobile streaming, interoperability between different streaming components (servers, encoders, and players) is an emergent issue (see Figure 44.8). The most important organizations that define the standards for mobile audio/video streaming are the Internet Streaming Media Alliance (ISMA) and the 3rd Generation Partnership Project (3GPP). ISMA is a consortium focused on the dissemination of an open standard for streaming on IP networks that has defined five functional areas (formats, storage, transport, description, and control) and adopted the MPEG-4 standard for data coding. 3GPP is aimed toward the specification of standards for wireless networks. For the purposes of our discussion, we are interested in the part of these specifications that describes a framework for guaranteeing interoperability among streaming services on mobile networks. That

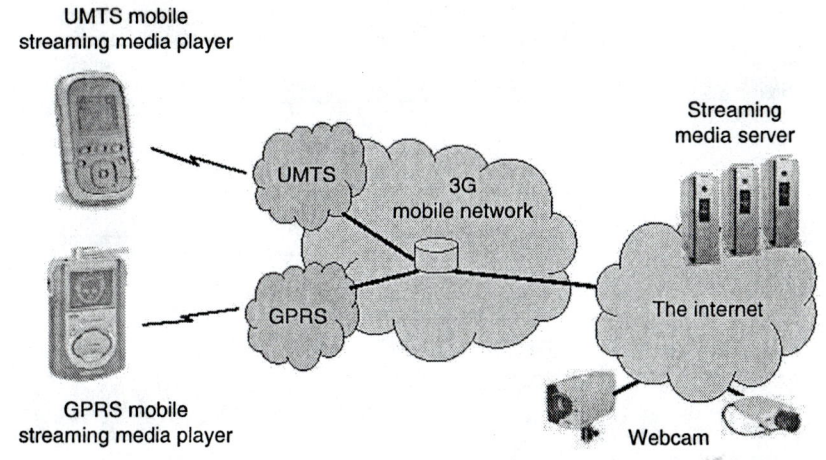

Figure 44.7   A mobile streaming architecture.

framework is named the Packet-Switched Streaming Service (PSS) and reuses the work made available by other organizations (e.g., IETF, W3C, MPEG, ISO, ITU).

The protocol architecture defined by the 3GPP PSS is illustrated in Figure 44.9. The transport service is provided by the Real-Time Transport Protocol (RTP) on the User Datagram Protocol (UDP); the transport control relies on the Real-Time Control Protocol (RTCP); the Real-Time Streaming Protocol (RTSP) performs the session control; and, finally, the multimedia

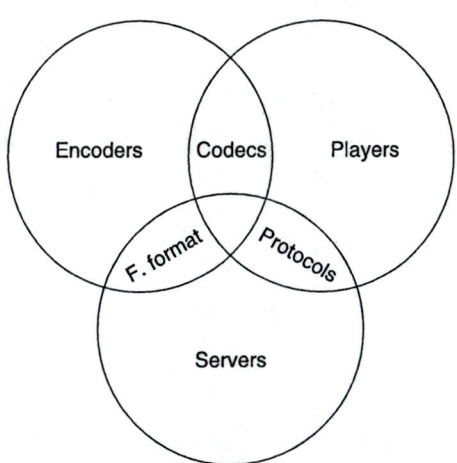

Figure 44.8   Relationships among standards for streaming.

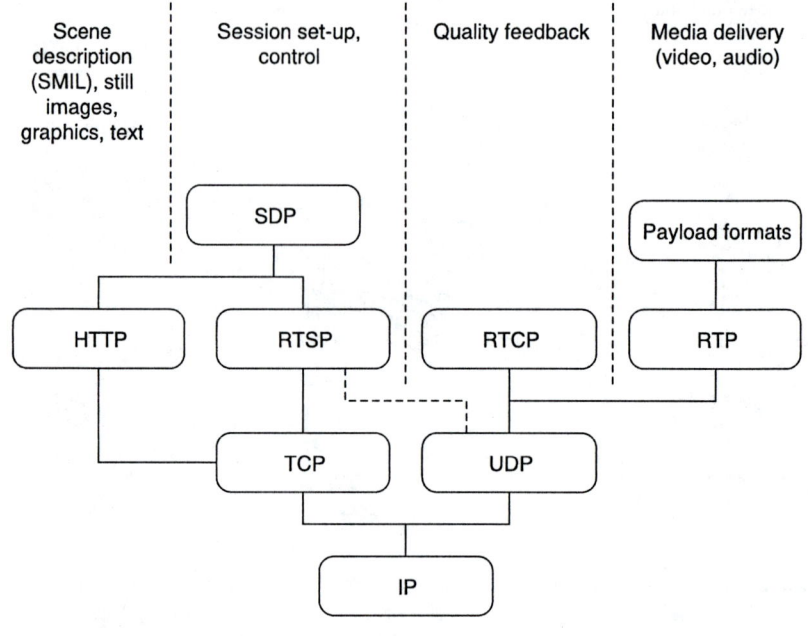

Figure 44.9   3GPP Packet-Switched Streaming Service protocol architecture.

content description is based on the Session Description Protocol (SDP). Synchronized Multimedia Integration Language (SMIL) provides scene description and content adaptation to the devices. In addition, the Session Initiation Protocol (SIP) can be used for commencing the session.

The live streaming applications deliver data using UDP and do not assume reliable communications; thus, these applications may tolerate the loss of a small percentage of data packets without degradation of the quality perceived by the users. With this approach, streaming audio/video from a mobile device can be possible even if the wireless network introduces packet losses. The only requirement is bandwidth availability. In a context in which the network bottleneck is close to the client, while the server holds a broadband connection (as depicted in Figure 44.10a), the streaming server may adapt at runtime the content to the available bandwidth at the client side by changing the codec. In a server-based scenario, the server bandwidth is sufficient to serve many clients.

The 3GPP PSS architecture is rather complex but still leaves some problems unsolved. One of these problems is related to the provision of live streaming services in a peer-based context, in which a single user, rather than a large organization, publishes its contents for free (see Figure 44.10b). In this context, it can be difficult to provide streaming services,

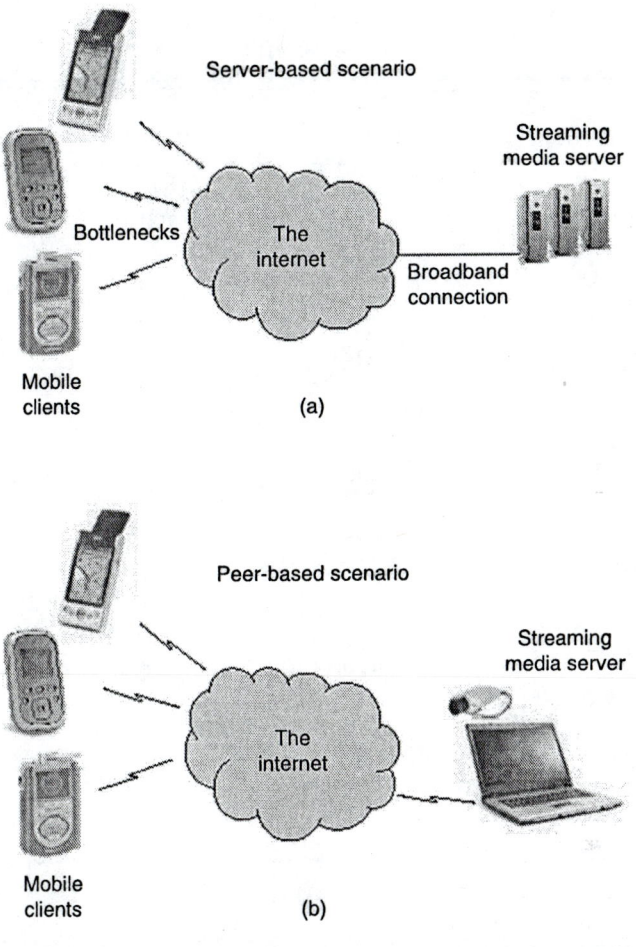

Figure 44.10 Mobile streaming scenario.

as: (1) the bandwidth of the data producer can be very limited, (2) the number of users that enjoy the streaming service is unpredictable and possibly too large to be served effectively with the limited bandwidth available to the data producer, and (3) the communications may be unreliable. Recent works propose the adoption of a multipath approach to be incorporated at a middleware layer, as discussed later in this chapter.

The interactive category in our proposed taxonomy consists of two principal classes of applications — namely, the so-called *human interaction* applications and *games* applications. The former class includes conventional, interactive applications, such as chatting, Voice-over-Internet Protocol (VoIP), and video conference applications, generally used for

Table 44.1   2004 Top 10 Video Game Titles, Ranked By Units Sold

Rank	Title	Platform	Publisher	Release Date	Price
1	Grand Theft Auto: San Andreas	PlayStation® 2	Take-Two Interactive Software	10/2004	$49
2	Halo 2	Xbox®	Microsoft	11/2004	$52
3	Madden NFL 2005	PlayStation® 2	Electronic Arts	8/2004	$49
4	ESPN NFL 2K5	PlayStation® 2	Take-Two Interactive Software	7/2004	$19
5	Need for Speed: Underground 2	PlayStation® 2	Electronic Arts	11/2004	$48
6	Pokemon Fire Red W/Adapter	Game Boy® Advanced	Nintendo	9/2004	$32
7	NBA Live 2005	PlayStation® 2	Electronic Arts	9/2004	$33
8	Spider-Man: The Movie 2	PlayStation® 2	Activision	6/2004	$43
9	Halo	Xbox®	Microsoft	11/2001	$29
10	ESPN NFL 2K5	Xbox®	Take-Two Interactive Software	7/2004	$19

Source: NPD Group/NPD Funworld® 2004 Annual Report on U.S. Video Game Industry Retail Sales, http://www.npdfunworld.com/funServlet?nextpage=pr_body.html&content_id-2076.

enabling interactions among humans. We assume that the reader is familiar with this class of applications and will not discuss them further in this chapter. Instead, the latter class of applications is discussed below.

A Taxonomy: Games

The size and economical relevance of the game market can be perceived by observing the rankings of the top-selling console games in 2004, shown in Table 44.1, where the top game sold over 5.1 million copies in 2004, and the number two game in the classification sold 4.2 million copies. Games can be classified as action, adventure, combat, music, puzzle, racing, role-playing, first-person shooter, sports, strategy, or turn-based, among others. Regardless of the game classification, however, the principal issues

Fixed
general purpose

Desktop PC

Mobile
general purpose

Handheld   Laptop

Mobile
phones

Game consoles

Sony Playstaion 2   Microsoft Xbox

Nintendo GameCube   Sega Dreamcast

Mobile game consoles

Nokia N-Gage

Sony PSP

Nintendo DS

Figure 44.11   Devices for games.

a game designer has to address include: (1) targeting the design and implementation of a game application to the devices that will host that application; (2) evaluating the expected degree of interactivity between the players; (3) anticipating the maximum number of players that are to be allowed to play the game concurrently; (4) examining the characteristics, or properties, of the available network infrastructures; (5) understanding the application requirements for graphic effects, the need for artificial intelligence (AI) and physic simulation modules; and (6) developing the high-level control logic. Some of these issues are introduced below.

A first crucial issue the game designer must address in the design of a game for mobile online entertainment is to identify the set of target devices that will run the game application. This set can include various classes of diverse devices (Figure 44.11). The first class consists of general-purpose devices. Among these devices, we can identify the networked fixed devices, such as desktop PCs, and mobile devices, such as laptop and handheld devices equipped with wireless interfaces. A second class consists of mobile phones equipped with small displays and low-power CPUs that can run simple games only. A third class includes those networked fixed devices explicitly designed for games — that is, the so-called consoles such as the Microsoft® Xbox®, Sony PlayStation® 2, Sega® Dreamcast™, and Nintendo™ GameCube™ (see Table 44.2). These devices are composed of a central box that runs the games and is equipped with cabled or wireless sockets for the controllers and a cabled Ethernet socket for Internet connections. Each user connects the box and plays using one of these controllers. Finally, the fourth class is that of portable consoles, such as the Sony PlayStation Portable (PSP), Nintendo DS, and Nokia N-Gage. In spite of their small size, these devices are equipped with high-performance CPUs and graphics hardware and hold a wireless interface (such as General Packet Radio Service, or GPRS) that allows low-bandwidth communications while the user moves around.

Typically, these devices may differ from each other in such components as the user interface, CPU power, operating systems, library availability, monitor resolutions, graphic cards, accelerators, and so on; however, to meet the demand of a wide market area, game applications should be developed so as to work on different devices, regardless of the heterogeneities among these devices. This problem can be dealt with according to the following two alternative approaches. Using the first approach, different versions of the same game can be implemented, with each version targeted to a given device. In contrast, using the second approach, a middleware layer should first be implemented to resolve issues of application portability across different platforms and to shield the applications from possible platform heterogeneities. Second, the game application can be developed on top of that middleware layer. Provided that this layer can be ported on different platforms, the game application can be ported as well, with no modifications.

A second critical issue the game designers must address is interactivity among users and between the user and the server. In practicality, a turn-based game, such as online chess, does not require strict interactivity. In contrast, a first-person shooter game, in which players must react as quickly as possible to eliminate adversaries, requires high performance from both the user device graphic module and the network layer. Note that the higher the number of players, the heavier the workload for the network subsystem and the CPU (particularly at the server side); thus, the game application architectural support must be designed accurately so as to provide sufficient computing power; different approaches to this design issue include the use of hierarchical server mirroring techniques and P2P architectures. Moreover, the player's device may access the network using different network technologies. In particular, the presence of wireless access technologies may strongly influence the design of the game, as specific software layers and protocols may have to be introduced to provide sufficient network guarantees.

Artificial intelligence techniques can be applied to a variety of tasks in modern electronic games [3]. A game may use probabilistic networks

Table 44.2 Technical Characteristics of Portable Game Consoles

	Nintendo DS	Sony PlayStation Portable (PSP)	Nokia N-Gage QD
Communication type	IEEE 802.11	IEEE 802.11	GPRS (40.2 KB/sec)
Processor type	ARM 9	—	—
Processor speed	67 MHz	333 MHz	—
Audio support	16-channel, ADPCM/PCM stereo speakers	Stereo speakers	—
Battery	10 hours	2.5 hours	6 hours
Display type	LCD TFT (active matrix)	LCD TFT (active matrix)	—
Display size	3 in.	4.3 in.	—
Display resolution	256 × 192	480 × 272	176 × 208
Touch screen?	Yes	No	No
Graphics card memory	656 KB	—	—
Fill rate	30 million pixels/sec	—	—
Polygons per second	120,000	—	—
Maximum number of colors	260,000	16.76 million	—
RAM	4 MB	36 MB	—
Size	3.36 × 1.14 × 5.85 in.	—	13.4 × 7 × 2 cm
Weight	61 lb	—	137 g
Release date	2004	2005	—

to predict the player's next move to precompute graphics. The increasing complexity of AI technology makes necessary the use of third-party middleware to enable code reuse and to simplify the game implementation. The same requirements apply to the design of the game control logic. As of today, middleware-layer scripting languages and interpreters are available to implement high-level control logic; these languages and interpreters (e.g., Quake 3, Unreal, and Spades) can simplify the game design and make it less expensive.

A third critical issue the game designers have to deal with is related to the mathematical knowledge required, in general, in the design of game applications. Typically, notions of both calculus and physics may be necessary in the design and implementation of any kind of simulation, rendering, and signal-processing task used for application development purposes; however, this knowledge may not fall within the field of expertise of the game designer [4]. Thus, supplying that expertise as part of the middleware layer can be a valuable approach to addressing this third issue.

System Architecture for Online Entertainment

In spite of their different scopes, requirements, and behavior, applications for online entertainment present the same set of basic problems and share a common structure consisting of the six principal modules, introduced below.

Discovery Service

Online entertainment applications can be thought of as being constructed out of *entities* (e.g., clients and servers, peer systems) interacting with each other. The number of entities involved in an application may vary at runtime; for example, a game application may enable entities to join or leave the game at runtime, or entities may be forced to leave the game as a consequence of a failure. Thus, any online entertainment application requires a discovery service to locate the set of entities participating in the application itself and maintain this set of entities at runtime. Note that some applications may require a discovery service that simply maps a host name into an Internet Protocol (IP) address and whose implementation can be based on the Domain Name System (DNS). This is the case of a simple video-on-demand (VoD) application; a client can reach a VoD server by means of the server host name, which the DNS resolves in that VoD server host IP address. More complex is the case in which the client requests a download operation involving more than one server; in this case, the discovery service module may have to maintain state information about the locations of all the servers involved in the download and their contents. A further level of complexity can be introduced in the discovery service of an online entertainment application by requiring that entities be allowed to join or leave that application or to change their geographical location at application runtime. In this latter case, the discovery service may have to maintain the state of the application entities, and reconfigure the application at runtime, if the circumstances require it to do so.

Accounting Service

As the discovery service discovers that an entity (a client or a server) is part of an application, that application needs to enter an accounting phase in which the identity and privileges of that entity are controlled. An accounting service can implement this functionality. In addition, that service may well be made responsible for both billing the application entities and providing information on (or even negotiating) the quality of service the application can provide.

User Interface and Graphics

The friendliness and elegance of the user interface play a key role in the success of modern online entertainment applications; thus, a number of tools, such as Maya and 3D Studio MAX, are currently available for building three-dimensional scenes, animation, and sound effects that can characterize the user interface of one such an application. The user interface and graphics module is particularly complex as it must take into account the characteristics of a very large set of different devices on which the applications may be required to run. Several techniques (e.g., simulation-based prerendering) are currently adopted to solve issues of rendering on different devices. Moreover, to support the development of the user interface, languages are available that simplify the use of 3D graphic tools. A relevant example of these languages is MEL, which is interpreted by Maya, MAXScript for 3D Studio Max, and VEX for Houdini.

Event Notification and Management

Online entertainment applications may consist of a number of entities that exchange data, such as audio/video frames (as in VoD and streaming applications) and character movements and actions (as in game applications). In particular, for live streaming applications and games, the data must reach its destination within a given time interval. If a video frame reaches its destination too late to be played out, the video frame results are useless; moreover, the bandwidth used to transmit that frame will have been wasted. Hence, event notification systems are deployed for bandwidth usage optimization and message management [5]. To this end, the event notification and management system is typically strictly integrated with the network subsystem and the event scheduler and makes use of logical clocks and network latency monitoring for event synchronization purposes.

Communication Subsystem

The network plays a key role in online entertainment applications; it may consist of a large set of different communication infrastructures that can operate under varying runtime conditions. Depending on the application requirements, different protocol stacks can be adopted to provide the applications with sufficient interactivity and reliability across the network. For mobile applications, in particular, the communication subsystem must be designed so as to enable the use of heterogeneous, both wired and wireless, networks. Moreover, issues of scalability of live streaming applications can be met at the communication subsystem level by adopting a multipath networking approach that makes this subsystem responsible for setting up and monitoring an overlay network.

Application Logic

Finally, the module implementing the application logic is constructed on top of the five modules introduced above. Specifically, for game applications, this module implements the behavior of the game characters and provides support for rigid body/physics systems. In addition, this module supports the implementation of rag-doll character animations, complex vehicles, separable objects, and the interactions among them. Moreover, the application logic module implements AI techniques that can be deployed to learn the behavior of the users and create virtual clones of those users; thus, if a user is temporarily unreachable while a game is in progress, that user can be replaced by his or her virtual clone, and the game can carry on uninterrupted. Typically, the application logic module is implemented using middleware services and tools based on scripting languages, such as Quake 3, Unreal, and Spades. An example of use of one of these tools is shown in Figure 44.12.

To conclude this section, we wish to observe that the majority of the modules described above can be conveniently instantiated in a middleware layer to provide the online entertainment application developer with adequate services that effectively meet such nonfunctional application requirements as scalability and mobility.

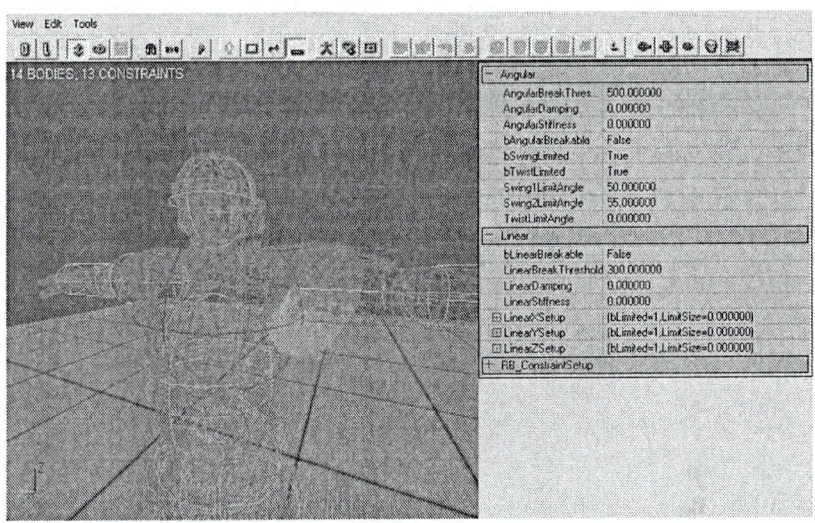

Figure 44.12 Using the development tool provided by Unreal Engine 3.

Mobility

The principal problem in supporting mobility in online entertainment applications originates from the requirement of reaching a large number of potential users using a variety of possibly different and mobile devices. Thus, online entertainment application designers need to develop their applications so they can run on devices such as PCs, PDAs, mobile game consoles, and mobile phones. Moreover, as these devices can typically be managed by different operating systems, a requirement for multiplatform support arises to enable application code portability across different platforms and to reduce the complexity of the application development task. Moreover, issues of process migration arise in mobile entertainment applications such as the so-called *participatory simulation* applications [6]. These applications implement role-playing cooperative activities that simulate the evolution of complex dynamic systems in the time domain [7,8].

Additional design issues originate from the use of both mobile devices and communication networks. These issues derive from the limited computing power a mobile device may offer, its energy consumption, the limited availability of network bandwidth, the possibility of intermittent operation of the device, and the unreliability of the communications. In particular, modern entertainment applications require notable CPU power to run 3D graphic libraries and complex compression algorithms for encoding and decoding the application content. Adding more powerful processors to mobile devices can add complexity to these devices, making

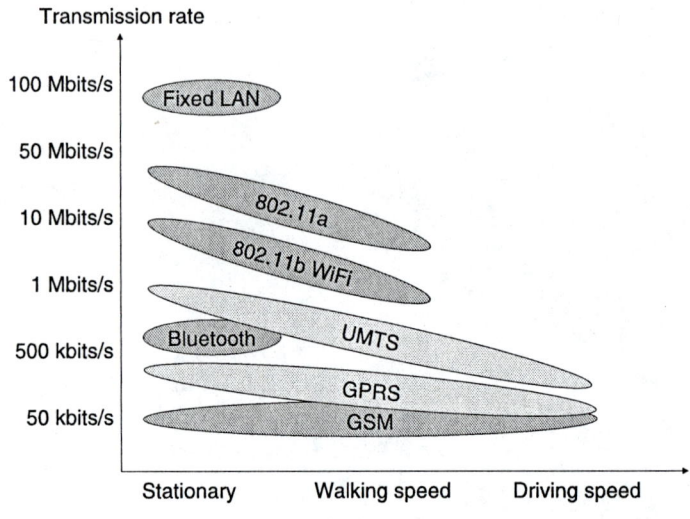

Figure 44.13 Wireless access technology transmission rate versus user movements.

them more expensive and, above all, increasing their power consumption, possibly causing heating problems. Thus, currently, research is underway on the development of hardware technologies that can augment the mobile device capabilities, as well as the implementation of software solutions that can reduce CPU usage and its power consumption. As to these latter solutions, it has been pointed out in Mohapatra et al. [9] that issues of power saving can be addressed at the middleware level by implementing middleware services that, for example, select an appropriate video quality and carry out the fine tuning of the device parameters to meet specific power-saving requirements.

Also, the bandwidth that the current wireless network technologies can make available to mobile applications has notably increased in recent years; however, almost at the same time, advanced multimedia capabilities have been introduced in mobile devices that have increased the bandwidth requirements of these devices. These requirements are not yet fully met by the current wireless network technologies. Because a lack of sufficient network bandwidth can turn out to be a severe limitation for some online entertainment applications (e.g., full-motion content applications), content adaptation coding techniques are necessary to reduce the bandwidth requirements of those applications.

Relatively low network bandwidth is not the only limitation for mobile entertainment applications. Usually, problems are very likely to occur in dense urban areas during peak hours; for example, fast movements may decrease both the communication rate (see Figure 44.13) and reliability perceived by the mobile device. As handovers may occur more frequently when a device moves, the probability increases that the device will enter a cell with insufficient resources; thus, a handover may cause short transmission breaks, resulting in delayed or possibly lost packets. Longer delays and breaks may cause interruptions in the application synchronization and playback or, in the worst case, may result in breaking an entire TCP connection between a mobile device and a server.

Finally, the simultaneous presence of different network technologies (and providers) introduces such new challenges as intersystem handovers. Techniques for providing always-best-connected (ABC) services aim at selecting and using the best connection available when the user's device holds two or more wireless interfaces of possibly different technologies [10].

The issues introduced here may have different consequences on different entertainment applications, depending on the application requirements. A discussion on the effects of the mobility requirement on the diverse application categories we consider is provided later in this chapter, along with a number of research topics relevant in the design of middleware platforms for mobile entertainment.

A Middleware-Based Approach to Mobile Entertainment

This section aims at describing the currently most popular research guidelines in the development of middleware for mobile entertainment applications. As discussed earlier, each application class in our earlier taxonomy presents specific open problems; hence, the solutions proposed for each of these problems are discussed in the following, in isolation.

On-Demand Applications

In essence, media downloading and playing require reliable data transfer from a server to a client. In general, this operation can be implemented using the most popular reliable transport protocol; namely the Transmission Control Protocol (TCP).

Nomadic Behavior and Handovers

Unfortunately, when TCP is deployed in a mobile environment, five conditions negatively affect its behavior and limit its reliability and download continuity. The first condition occurs when the client device is equipped with a single wireless interface and changes the provider providing it with

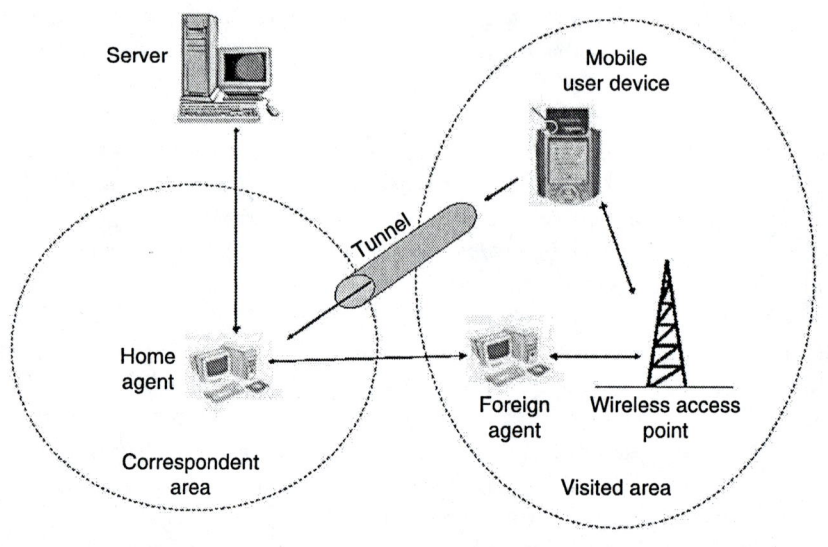

Figure 44.14   Mobile IP scenario.

the network connectivity services at runtime. The new provider assigns a
different IP address to the wireless network interface of the device, thus
changing the identifier of one of the endpoints of the TCP connection that
links the device to the server. This is the only condition that the Mobile
IP approach [11] can deal with, as illustrated in Figure 44.14.

The second condition occurs when the mobile device moves outside
the wireless coverage. For example, when the device user utilizes a lift,
the device wireless interface cannot detect the wireless carrier signal; thus,
the wireless interface becomes unavailable. Typically, signal losses can
interfere with the TCP communications in progress at the time those losses
occur. Some operating systems (e.g., Linux™ and Linux-based operating
systems for handheld devices, such as Familiar [12] and Intimate [13])
tolerate carrier signal losses by waiting until the carrier returns available.
Other operating systems (e.g., Windows CE), when a carrier signal loss
occurs, disable the wireless interface and notify the applications of the
signal loss problem. When the wireless interface is disabled by the
operating system, any application-level TCP socket becomes unusable,
even if the carrier signal returns available (thus, with this operating system
class, the Mobile IP approach mentioned above cannot be deployed to
solve signal loss problems).

The third condition affects the TCP connections maintained by the
operating systems that do not disable the unavailable mobile device
interfaces (e.g., the Linux-based operating systems mentioned above).

When a device embodies two or more wireless network interfaces, a TCP connection can be set up between that device and a server so the data flows through one of the available device interfaces (e.g., the first interface that correctly detects its wireless network carrier signal). It is possible that this interface could move outside the coverage of its wireless network, losing the carrier signal. As an additional wireless interface is available in the device, one can observe that, to maintain connectivity, the application might resume all its currently active communications through that additional interface. Unfortunately, as the operating system masks the loss of the carrier signal to the application, the application is forced to wait for communications occurring through the unavailable interface.

The fourth condition occurs when the mobile device incorporates two or more wireless interfaces, the operating system disables a network interface as it becomes unavailable, and another interface remains available for communications. As in the previous case, communications in progress cannot be resumed through the interface that is still available, as the use of a different network interface requires that each TCP connection through that interface must be identified by a new and unique IP address that identifies the endpoint of the connection at the mobile device end.

Finally, a fifth condition exists in which the application cannot perform as effectively as it might. When a mobile device incorporates more than one wireless interface and more than one of them is operational, it is preferred that the device concurrently use all the available interfaces so as to maximize the overall bandwidth usage. In other words, the application should perform bandwidth load balancing among all the available wireless interfaces.

Seamless and Always-Best-Connected

The solution to the five problems above is to provide the application with ABC service. This means selecting and using the best network connection from those available when the user's device holds two or more wireless interfaces of possibly different technologies. ABC protocols do not have a defined position in the classic Internet protocol stack and are shifted from the data link layer to the application layer, depending on the application context. Typically, ABC protocols at the data link level may exercise control over the wireless link, only; hence, these protocols cannot be designed to deal with issues of performance of the complete path between the mobile device and the server. To overcome this problem, ABC management can be implemented either at the network level [14] or at the transport level [15]; however, neither Fodor et al. [14] nor Eddy [15] addresses issues of nomadic computing, and both require modifications of the TCP/IP infrastructure.

Another ABC-type proposal is the Stream Control Transmission Protocol (SCTP) [16,17]. This protocol provides applications with an advanced and reliable transport-level service supporting multi-homing at the end systems, as well as ensuring congestion avoidance. As illustrated in Figure 44.15, SCTP must assume the role of TCP, which may cause the following four main shortcomings:

- *Packet dropping* — Firewalls present in the path between a mobile device and a server may drop IP packets that carry no TCP data units.
- *Application transparency* — The applications typically rely on TCP; if SCTP is used instead, the application developers need to replace the TCP system calls with SCTP function calls.
- *Implementation availability* — SCTP implementations are not available for all the architectures.
- *No load balancing* — SCTP uses only one of the available network interfaces; the other interfaces are used only in case of failure of the primary connection.

An alternative approach to the provision of ABC services can be obtained by implementing the ABC management protocols in a middleware layer constructed on top of the standard TCP transport layer. The responsibilities of that middleware layer, instantiated in a mobile device, should include the monitoring of:

- The occurrence of possible IP configuration changes in the network interfaces hosted by that device
- The availability of the communications along the different paths between each device network interface and the server

The monitoring middleware can detect interruptions in the TCP connection and hide them to the application. In particular, in a media download scenario, the middleware can resume the download from the point at which it has been interrupted by using a new connection established through the *best available* device interface. Frequently, the word *best* just implies larger bandwidth or lower delay; however, it can be more generally related to a complex set of application requirements that must be met at the middleware level. Figure 44.16 shows the protocol stack that implements the solution proposed above. That protocol stack has been developed, implemented, and tested in the context of a music distribution application [18].

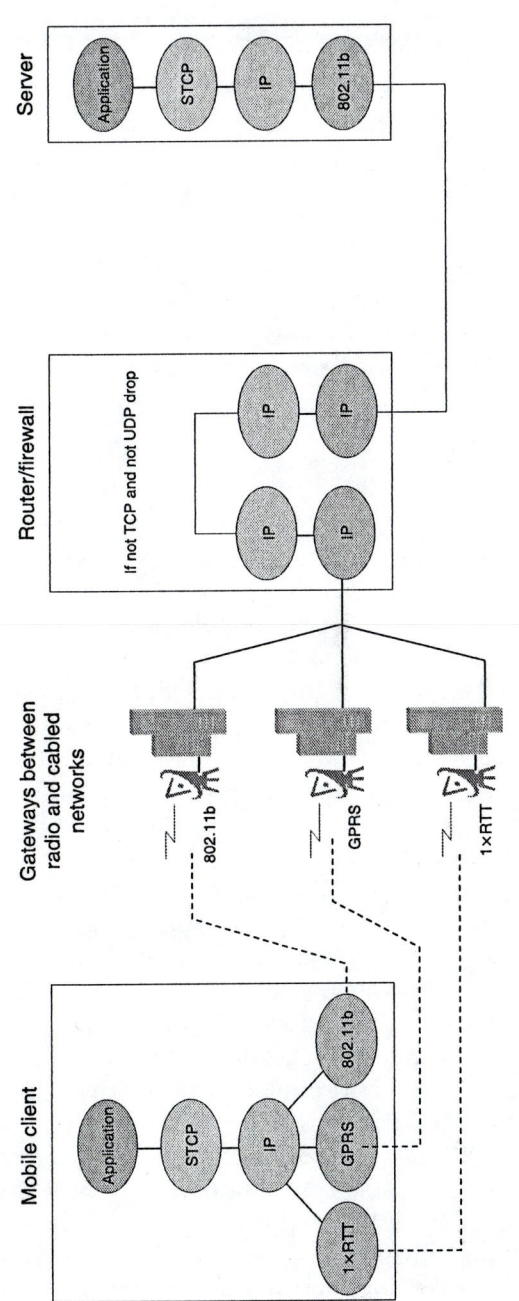

Figure 44.15 SCTP protocol stack.

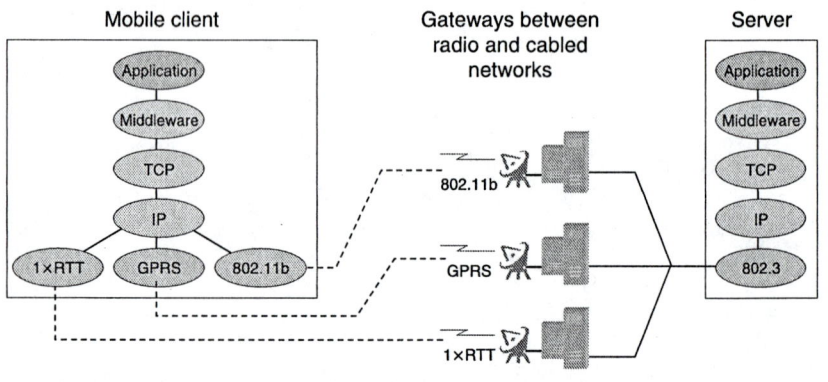

Figure 44.16 A middleware-based protocol stack.

Transparency

Any middleware solution can be implemented by resorting to libraries. Following this approach, the introduction of an ABC-type communication service entails replacing each call to a given TCP function with a call to the corresponding middleware-based module implementing that service. This raises a transparency problem for mobile entertainment applications where ABC services should be introduced at the lowest implementation cost. The challenge is designing a sort of *hidden* middleware able to implement an ABC service for mobile entertainment applications that do not have to be aware of this. The basic idea is to split the middleware core over two separate communicating software modules running on both the server and the mobile client.

As illustrated in Figure 44.17, an appropriate module, termed *Filter*, would be in charge of (1) intercepting any TCP segment generated at the application level, and (2) redirecting it to the companion module, namely the connection manager. The connection manager, in turn, has the responsibility of managing a session protocol implementing the ABC service. That service may implement complex ABC strategies, including, for example, monitoring the availability of multiple network accesses, controlling the link availability, managing the use of different IP interfaces, and balancing the communication load over different network interfaces. Moving the responsibility of implementing an ABC service to a separate software module offers the following great advantage: A separate module (i.e., the configuration manager) deals with problems deriving, for example, from a TCP disconnection, and the application may continue computing without any interruption.

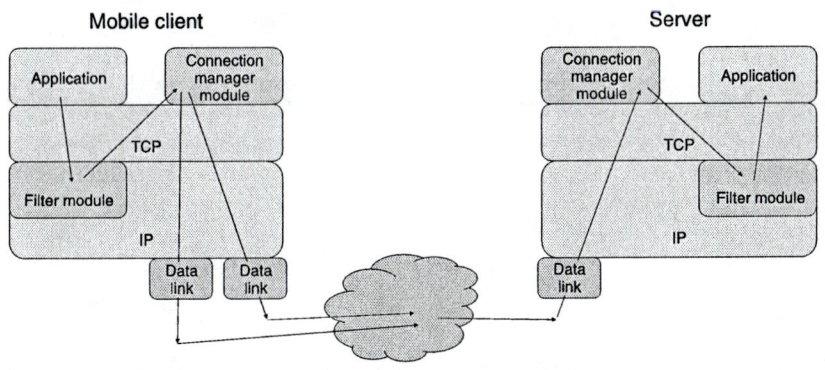

Figure 44.17   A hidden middleware architecture.

A prominent example of that approach is provided in Zandy and Miller [19], where the filter and the communication manager modules have been implemented based on a Linux kernel. The filter module was implemented by exploiting the functions provided by the Netfilter routine and the Linux-based IP tables [20]. These tools enable several actions, including packet filtering, network address (and port) translation, and other packet manipulation activities. Unfortunately, technical problems still remain with this approach; for example, the decision to bind each TCP connection to a given local IP interface can be detrimental when a reconfiguration of that IP interface is needed, as in this case when the operating system typically annuls all the TCP sockets associated with that interface.

A possible solution to this specific problem has been presented by Schulist [21], where the use of a virtual interface (based on the Linux-based Ethertap module) is exploited to keep the IP interface separated by the TCP connection. Another typical problem with this proposal is that of portability, as only specific Linux-based system calls can be exploited. We claim that a solution to this problem may rest upon the use of the LibPcap libraries [22] for packet capturing and of the LibNet libraries [23] for packet injection.

Live Streaming Applications

A live streaming application typically delivers data using UDP. As the communication is assumed to be unreliable, a small percentage of packets can be lost, still maintaining a satisfactory transmission quality. Obviously, an important requirement to be met is that of bandwidth availability.

Bandwidth Availability and Distribution Trees

To cope with the problem of bandwidth availability, the 3GPP Packet-Switched Streaming Service proposes to adapt media contents to the context where the device operates. Unfortunately, this cannot always be the one, unique solution. Take, for example, the case when the number of simultaneous consumers wishing to enjoy a live event is very large. In this case, the scarcity of available bandwidth at the producer side may represent a serious problem even if the media has been reduced in size. Here, the problem is that sending a different copy of each data packet to each different client is not a viable strategy. A possible solution amounts to organizing the set of all clients based on an IP multicast distribution tree, with the root at the producer side. Unfortunately, the IP multicast technology is often disregarded by network providers as impractical due to the amount of traffic it produces. To surpass this problem, an application-level multicast solution can be developed, and an overlay distribution tree may be established at the application layer to interconnect participants. The main advantage of this solution is that the overlay multicast tree can be constructed on the top of any network, providing a simple, unicast transport service, like UDP. An overlay distribution tree offers the further advantage of a greater flexibility in terms of network resource utilization, as simple end hosts can be employed rather than routers.

Hybrid solutions can also be devised where an overlay distribution tree is implemented mixed with the use of IP multicast. The implementation of hybrid approaches rests upon the use of RTP [24]. Two different RTP entities are usually exploited. A *mixer* can encode audio/video data, adapting the size to the available bandwidth. A *translator* can set up IP tunnels with the aim of delivering IP multicast packets passing through networks that do not allow IP multicast. Following this strategy [25,26], all data relays happen via unicast, and a variety of end-to-end streaming adaptation techniques (e.g., frame dropping, transcoding) can be exploited to control in a much finer granularity the streaming rate directed to each client. Second, the overlay multicast tree can be rearranged on the fly, on a session basis, so as to better accommodate requirements of diverse and heterogeneous networks.

Many application-level multicast systems have been proposed for different target applications. Each of them creates a distribution tree in a particular way, yet they generally fall into a specific category, depending on which of the following two metrics is adopted to construct the tree: (1) delay reduction [26–29] or (2) throughput increase [30–32]. They all share the common characteristic of using unreliable protocols to distribute data from a single producer to a variety of different consumers. Although this approach can be effective in cases where only fixed links are exploited, this does not appear to be so for environments characterized by a high

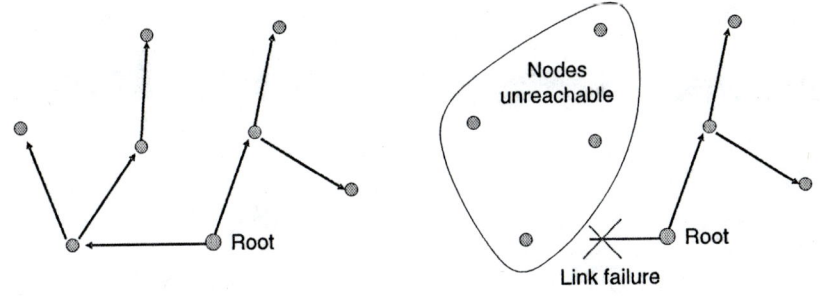

Figure 44.18 An overlay multicast tree (left) may leave many nodes unreachable (right).

mobility. The main drawback here is that when an application-level connection fails, a large number of end systems may become unreachable, thus resulting in disconnection until a point in time when a new distribution tree is recreated (see Figure 44.18). This addresses the further problem of fault tolerance.

Fault Tolerance

A very general solution for the problem mentioned above amounts to the use of a *multiple description coding* (MDC) technique for producing data to be delivered to final consumers through a set of nonoverlapping paths (multiple path transport, or MPT). The MDC technique entails manipulating media streams to obtain multiple substreams based on the following requirements [33]: (1) Each substream must be smaller than the primary stream; (2) each substream can be played out separately at the receiver side, at a lower quality than the primary stream; (3) two or more substreams can be reassembled together at the receiver side for play out; and (4) different substreams can follow different paths (on different distribution trees) to reach their final destinations. As a result, network failures do not significantly affect media transmissions, thus allowing each final consumer to receive a number of substreams that are enough to play the media at a satisfactory quality.

Combining the MDC technique with an MPT service aids traffic dispersion and load balancing and can effectively relieve congestion, increase network utilization, and enable better integration of both wired and wireless links. The benefits of MDC come at a price, however, as developing a multiple description coder is usually more complex than developing a traditional one. An MD coder typically uses more bits. This excess of redundancy is inserted intentionally to make the bitstream more robust to transmission errors; hence, the primary objective in designing an MD coder is to maintain control over this redundancy while meeting the requirement of minimizing end-to-end distortion. In recent years, a variety of practical MD compression algorithms have been proposed that use subsampling strategies in the spatial, temporal, or frequency domains, mixed with improved quantization techniques [34]. MPT is a convincing approach also for wireless entertainment, because: (1) a path can break down frequently because of movements of the device, (2) wireless links can cause frequent packet losses, and (3) wireless links may not have an adequate capacity to support high-bandwidth services.

Several researchers explored the problem of an efficient integration of MDC with MPT, especially in the field of wireless video [35–37], yet a number of issues remain unsolved. The most important one is how to manage, at the middleware level, the construction of a multipath distribution tree. This problem is further exacerbated by the need for a quick tree reconfiguration as soon as a given node becomes unavailable. We feel that suitable tools that might be good candidates for managing a multipath distribution tree at the middleware layer include Pastry [38] and all its companions (Chord [39], FreePastry [40], and Tapestry [41]), as well as Scribe [42] and SplitStream [43]. Pastry is an example of a generic, scalable, self-organizing substrate for peer-to-peer applications, and Scribe and SplitStream provide group communication and event notification capabilities, respectively, for constructing separate multicast trees with disjoint interior nodes.

Games

Besides the typical problems of available *bandwidth* and *delay*, a number of specific issues should be addressed at the middleware layer to make the development of mobile games easier. Some of them refer to the ability to run the game on devices that exhibit very different characteristics, in terms of monitor resolution, graphic cards capability, CPU power, and graphical libraries availability. Other complementary issues are related to the need to simplify the game development efforts when complex mathematical and physics models have to be implemented with the aim of making the game as realistic as possible. These latest issue are typically addressed by exploiting scripting languages and interpreters for high-level control of the implementation of the application logic.

Gaming Architectures

Multiplayer games form a very challenging class of entertainment applications. Existing commercial products are usually based on a centralized client–server architecture where a server maintains the game state. The game clients run on the devices of the individual players. They receive information about the game state from the server and may influence it by sending their local actions to the server. Recently, prototypes have been proposed for developing networked games based on replicated architectures without the need of a centralized server [44]. Along this line, distributed gaming architectures may be devised that are based on the use of mirrored servers. Each mirror maintains a redundant state of the game. The client connects to a given server communicating in a traditional centralized model. Mirrored servers, on the other hand, communicate with each other in a P2P fashion to keep the game state consistent throughout the entire system [45]. Mirrored server architectures are clearly scalable and do not present any single point of failure. When a server crashes, the system has implicit mechanisms to redirect the connected clients to other mirrors, thus ensuring service continuity.

Whether based on a centralized architecture or deployed over a distributed infrastructure, games should also guarantee a high degree of playability independent of the user's location, type of connection, utilized device, and number of players. Unfortunately, especially in the wireless case, network characteristics can have a negative impact on the gaming organization. Obviously, a scarce bandwidth limits the quantity of information that clients and servers may exchange. The definition of an area of interest within a game for a given player may coincide with selecting what information this player needs to receive, thus reducing the need for bandwidth. To determine those players who are interested in a given game event, several group selection and maintenance strategies have been proposed [5,46].

In all gaming architectures, the user's actions are first propagated to the server and then to the other clients; therefore, game events can arrive late at other clients depending on the delay experienced over the network. The problem of managing delay and jitter makes even more crucial the need to maintain a shared and consistent view of the game state for all the players. A middleware approach has been recently introduced aimed at alleviating delay and jitter problems. A proxy-based architecture was proposed where game events are managed in terms of different levels of urgency and relevance [47–49]. Highly urgent events are sent with higher precedence than all the other events, thus reducing their presence in the network. Highly relevant events, on the other hand, are sent via a loss-prevention technique to ensure guaranteed delivery. Other interesting schemes have been proposed to cope with problems deriving from network latency and bandwidth limitations. Among them, it is worth mentioning the *dead reckoning* strategy, which was developed to solve the problem of how to predict the game evolution for situations where the actual information updates do not arrive at the destination on time.

Multiple-Platform Support

To cover a wider market, games should be developed so they can run on different devices and operating systems. A middleware layer can be interposed between the application and the operating system to resolve issues of application portability across different platforms. This layer would be responsible for driving different hardware devices, as well as for guaranteeing better use of operating system capabilities. A critical example is represented by those gaming applications that greatly utilize thread programming. Given the variability of behavior and performance obtained when threads are run on different operating systems, a middleware layer should be able to provide a harmonized solution. Recently, an open-source, object-oriented development platform was proposed that features a portable application programming interface (API) offering a unique threading model with the ability to manage typical problems, including thread starvation and deadlocks [50].

Graphical Libraries

Displaying game events on different wireless devices can be a serious problem. A usual solution amounts to the use of portable graphical libraries, such as OpenRM [51] and OpenGL [52]. Using the OpenRM library, complex graphical scenes can be incrementally built by means of objects to be rendered, along with rendering parameters, such as the position of the viewer, the lights illuminating the scene, textures of the objects, and so forth. OpenRM works on top of OpenGL and is essentially a three-dimensional triangle-based rendering engine. Support is provided for complex images, such as sprites, bitmaps, and depth pixels. Semiprocedural primitives are included that manage both two- and three-dimensional representations, such as spheres, cones, cylinders, boxes, glyph-based markers, and text. OpenRM is extensible as it supports application-supplied raw drawing code fragments, as well as a rich array of callbacks that can be used to implement advanced features, including view-dependent operations. It is worth mentioning that other open-source, platform-independent, game-oriented graphical libraries exist, such as, for example, OpenSceneGraph [53] (see Figure 44.19).

CPU Workload

Often, mobile devices are equipped with a low-power CPU with the intent of limiting energy consumption. This constraint must be taken into consideration when games are developed, in particular for those operations that overload the CPU. This is typical for graphic rendering. Using OpenRM

Figure 44.19 An OpenSceneGraph-based flight simulator.

and OpenGL, a new approach has been recently developed that tries to select the objects to be rendered on each device, based on its relative position within the scene [54]. The idea is to make it possible to visualize only a part of a scene when an entire renderable model cannot reside on a given device due to its hardware and software limitations. This strategy performs rendering actions by issuing OpenGL commands. Level-of-detail techniques are a good example of how that approach works. Portions of the scene that should look far away from the viewer can be rendered using representations at a lower resolution than those apparently closer to the viewer, thereby reducing the rendering load. Another approach aiming to limit the CPU overload for wireless gaming is described in Cacciaguerra et al. [8]. The novelty is that the rendering engine on the wireless device is able to display state updates at a variable frame rate while scaling down the graphical quality when the CPU load increases. Moreover, the rendering engine may drop scene updates to cope with large transmission delays.

Conclusions

Analysts indicate that consumer demand for entertainment will constantly increase. This demand will be characterized by a growing need to enjoy entertainment content on mobile devices. Following this trend, industrial investments are expected for the development of services that can effectively meet this demand. Also, designers of online entertainment applications are following the evolution of user expectations by shifting their attention from the fixed to the mobile environment. We have argued that the majority of the design issues emerging from a mobile entertainment scenario can be dealt with in a middleware layer designed to support both the application development and its execution.

References

[1] Kozbe, B., Roccetti, M., and Ulema, M., Entertainment everywhere: system and networking issues in emerging network-centric entertainment systems, part I and II [guest editorial], *IEEE Commun. Mag.*, 43(6), 67–68, 73–74, 2005.

[2] Bellavista, P., Corradi, A., and Magistretti, E., Lightweight autonomic dissemination of entertainment services in wide-scale wireless environments, *IEEE Commun. Mag.*, 43(6), 94–101, 2005.

[3] Nareyek, A., AI in computer games, *ACM Queue*, 1(10), 58–65, 2004.

[4] Blow, J., Game development: harder than you think, *ACM Queue*, 1(10), 29–37, 2004.

[5] Ferretti, S. and Roccetti, M., A novel obsolescence-based approach to event delivery synchronization in multiplayer games, *Int. J. Intelligent Games Simul.*, 3(1), 7–19, 2004.

[6] Kanter, T.G., Attaching context-aware services to moving locations, *IEEE Internet Comput.*, 7(2), 43–51, 2003.

[7] Wilensky, U. and Stroup, W., Networked gridlock: students enacting complex dynamic phenomena with the HubNet architecture, in *Proc. 4th Ann. Int. Conf. on the Learning Sciences*, Ann Arbor, MI, June, 2000.

[8] Cacciaguerra, S., Lomi, A., Roccetti, M., and Roffilli, M., A wireless software architecture for fast 3D rendering of agent-based multimedia simulations on portable devices, in *Proc. IEEE Consumer Communications and Networking Conference (CCNC'04)*, Las Vegas, NV, January 5–8, 2004.

[9] Mohapatra, S., Venkatasubramanian, N., Dutt, N., Periera, C., and Gupta, R., Energy-aware adaptations for end-to-end video streaming to mobile handheld devices, in *Ultra Low Power Electronics and Design*, Macii, E., Ed., Kluwer Academic, Boston, MA, 2004.

[10] Gustafsson, E. and Jonsson, A., Always best connected, *IEEE Commun. Mag.*, 10(1), 49–55, 2003.

[11] Perkins, C., *IP Mobility Support for IPv4*, Request for Comments 3344, Internet Engineering Task Force (IETF), 2002 (http://www.ietf.org/rfc/rfc3344.txt).

[12] The Familiar Project, http://familiar.handhelds.org/.

[13] The Intimate Project, http://intimate.handhelds.org/index.html.

[14] Fodor, G., Eriksson, A., and Tuoriniemi, A., Providing quality of service in always best connected networks, *IEEE Commun. Mag.*, 41(7), 154–163, 2003.

[15] Eddy, W.M., At what layer does mobility belong?, *IEEE Commun. Mag.*, 42(10), 155–159, 2004.

[16] Xie, Q. et al., *Stream Control Transmission Protocol*, Request for Comments 2960, Internet Engineering Task Force (IETF), 2000 (http://www.ietf.org/rfc/rfc2960.txt).

[17] Stewart, R., Ramalho, M., Xie, Q., Tuexen, M., and Conrad, P., *Stream Control Transmission Protocol (SCTP) Partial Reliability Extension*, Request for Comments 3758, Internet Engineering Task Force (IETF), 2004 (http://www.ietf.org/rfc/rfc3758.txt).

[18] Ghini, V., Pau, G., Roccetti, M., Salomoni, P., and Gerla, M., For here or to go? Downloading music on the move with an ultra reliable wireless internet application, *Comput. Networks*, 49(1), 4–26, 2005.

[19] Zandy, V.C. and Miller, B.P., Reliable network connections, in *Proc. of the 8th ACM/IEEE Int. Conf. on Mobile Computing and Networking (MOBICOM'02)*, Atlanta, GA, September, 2002.

[20] Netfilter framework, http://www.netfilter.org/.

[21] Schulist, J., Ethertap Programming mini-HOWTO, Linux Kernel Documentation, http://www.hu.kernel.org/pub/linux/kernel/v2.6/patch-2.6.12.

[22] LibPcap, http://www.tcpdump.org/pcap3 man.

[23] Libnet, http://www.packetfactory.net/Projects/Libnet/.

[24] Schulzrinne, H., Casner, S., Frederick, R., and Jacobson, V., *RTP: A Transport Protocol for Real-Time Applications*, Request for Comments 3550, Internet Engineering Task Force (IETF), 2003 (http://www.ietf.org/rfc/rfc3550.txt).

[25] Amir, E., McCanne, S., and Katz, R., An active service framework and its application to real-time multimedia transcoding, in *Proc. of ACM SIGCOMM'98*, Vancouver, Canada, October, 1998.

[26] Banerjee, S., Bhattacharjee, B., and Kommareddy, C., Scalable application layer multicast, in *Proc. of ACM SIGCOMM'02*, Pittsburgh, PA, August, 2002.

[27] Chawathe, Y. and Seshadri, M., Broadcast Federation: an application-layer broadcast Internetwork, in *Proc. of the 12th Int. Workshop on Network and Operating Systems Support for Digital Audio and Video (NOSSDAV 2002)*, Miami Beach, FL, May 12–14, 2002.

[28] Chu, Y., Rao, S.G., Seshan, S., and Zhang, H., Enabling conferencing applications on the Internet using an overlay multicast architecture, in *Proc. of ACM SIGCOMM'01*, San Diego, CA, August, 2001.

[29] Pendarakis, D., Shi, S., Verma, D., and Waldvogel, M., ALMI: an application level multicast infrastructure, in *Proc. of the 3rd USENIX Symp. on Internet Technologies and Systems (USITS'01)*, San Francisco, CA, March 26–28, 2001.

[30] Jannotti, J., Gifford, D.K., Johnson, K.L., Kaashoek, Jr., M.F., and O'Toole, J.W., Overcast: reliable multicasting with an overlay network, in *Proc. of the 4th Symp. on Operating Systems Design and Implementation*, San Diego, CA, October, 2000.

[31] Cui, Y., Li, B., and Nahrstedt, K., oStream: asynchronous streaming multicast in application-layer overlay networks, *IEEE J. Selected Areas Commun.*, 22(1), 91–106, 2004.

[32] Kim, M.S., Lam, S.S., and Lee, D., Optimal distribution tree for Internet streaming media, in *Proc. of IEEE Int. Conf. on Distributed Computing Systems (ICDCS'03)*, Providence, RI, May, 2003.

[33] Wang, Y., Reibman, A.R., and Lin, S., Multiple description coding for video delivery, *Proc. IEEE*, 93(1), 57–70, 2005.

[34] Goyal, V.K., Multiple description coding: compression meets the network, *IEEE Signal Process. Mag.*, 18(5), 74–93, 2001.

[35] Gogate, N., Doo-Man, C., Panwar, S.S., and Yao, W., Supporting image and video applications in a mobile multihop radio environment using route diversity and multiple description coding, *IEEE Trans. Circuits Syst. Video Technol.*, 12(9), 777–792, 2002.

[36] Padmanabhan, V.N., Wang, H.J., and Chou, P.A., Resilient peer-to-peer streaming, in *Proc. of Int. Conf. on Network Protocols (ICNP'03)*, Atlanta, GA, November 4–7, 2003.

[37] Mao, S., Lin, S., Panwar, S.S., Wang, Y., and Celebi, E., Video transport over *ad hoc* networks: multistream coding with multipath transport, *IEEE J. Selected Areas in Commun.*, 21(10), 1721–1737, 2003.

[38] Rowstron, A. and Druschel, P., Pastry: scalable, distributed object location and routing for large-scale peer-to-peer systems, in *Proc. of IFIP/ACM Int. Conf. on Distributed Systems Platforms (Middleware)*, Heidelberg, Germany, November, 2001.

[39] Stoica, I., Morris, R., Liben-Nowell, D., Karger, D.R., Kaashoek, M.F., Dabek, F., and Balakrishnan, H., Chord: a scalable peer-to-peer lookup protocol for Internet applications, *IEEE/ACM Trans. Networking*, 11(1), 17–32, 2003.

[40] FreePastry, http://freepastry.rice.edu/.

[41] Zhao, B.Y., Huang, L., Stribling, J., Rhea, S.C., Joseph, A.D., and Kubiatow-icz, J., Tapestry: a resilient global-scale overlay for service deployment, *IEEE J. Selected Areas in Commun.*, 21(1), 41–53, 2004.

[42] Rowstron, A., Kermarrec, A.M., Castro, M., and Druschel, P., SCRIBE: the design of a large-scale event notification infrastructure, in *Proc. of the Third Int. Workshop on Networked Group Communication (NGC'01)*, London, November, 2001.

[43] Castro, M., Druschel, P., Kermarrec, A.M., Nandi, A., Rowstron, A., and Singh, A., SplitStream: high-bandwidth multicast in a cooperative environment, in *Proc. of the 19th ACM Symp. on Operating Systems Principles (SOSP'03)*, Lake George, NY, October 19–22, 2003.

[44] Diot, C. and Gautier, L., A distributed architecture for multiplayer interactive applications on the Internet, *IEEE Networks Mag.*, 13(4), 6–15, 1999.

[45] Cronin, E., Filstrup, B., Jamin, S., and Kurk, A.R., An efficient synchronization mechanism for mirrored game architectures, *Multimedia Tools Appl.*, 23(1), 7–30, 2004.

[46] Zou, L., Ammar, M., and Diot, C., *An Evaluation of Grouping Techniques for State Dissemination in Networked Multi-User Games*, Technical Report, Georgia Institute of Technology, Atlanta, GA, 1999.

[47] Griwodz, C., State replication for multiplayer games, in *Proc. of the First ACM Workshop on Network and System Support for Games (NetGames'02)*, Braunschweig, Germany, April 16–17, 2002.

[48] Griwodz, C., Halvorsen, P., and Munthe-Kaas, E., *MiSMoSS: Middleware Services for Management of Shared State in Large-Scale Distributed Interactive Applications*, http://www.ifi.uio.no/dmms/mismoss/missmoss.html.

[49] Mauve, M., Fischer, S., and Widmer, J., A generic proxy system for networked computer games, in *Proc. of the First ACM Workshop on Network and System Support for Games (NetGames'02)*, Braunschweig, Germany, April 16–17, 2002.

[50] Internet Communications Engine (ICE), http://www.zeroc.com/ice.html.

[51] Bethel, W., *RM Scene Graph*, White Paper, R3vis Corp., Novato, CA, 1999 (http://www.r3vis.com/RM/index.html).

[52] Segal, M. and Akeley, K., *The OpenGL Graphics Systems: A Specification (Version 1.4)*, http://www.opengl.org.

[53] OpenSceneGraph, http://www.openscenegraph.org/.

[54] Bethel, E.W., Humphreys, G., Paul, B., and Brederson, J.D., *Sort-First Distributed Memory Parallel Visualization and Rendering*, White Paper, R3vis Corp., Novato, CA, 2003 (http://www.r3vis.com/Downloads/PVG2003-OpenRMChromium.pdfRM).

Chapter 45

Software Support for Application Development in Wireless Sensor Networks

Chien-Liang Fok, Gruia-Catalin Roman, and Chenyang Lu

CONTENTS

Introduction

Middleware for wireless sensor networks (WSNs) may seem misplaced in a book on mobile middleware. After all, most existing WSNs are installed on static objects such as, for example, the nests of the Leach's Storm Petrels on the Great Duck Island [61], three feet underground at the Pickberry Vineyard in California, and on the housing of semiconductor machinery at Intel. Yet, upon closer inspection, it becomes readily apparent why WSNs are relevant. First, WSNs need not be static; they can be just as easily installed on moving objects; for example, in ZebraNet [48], a WSN was installed on a herd of zebras to track their movement. As the zebras move, the nodes opportunistically create wireless *ad hoc* links for distributing code and data updates. Mobile middleware in the form of Impala [54] was used to ensure flexible and adaptable applications in such a dynamic context. Second, there are many applications that involve a WSN interacting with a mobile user [29,44,47,48,57] or tracking a mobile entity [18,32,38]. Third, as mobile applications mature, they must exhibit context awareness; that is, they must not only understand what their users are doing but also perceive properties about their environment. This collection of context information must be done in real time and at resolutions only WSNs are capable of cost-effectively providing. The combination of user, entity, node, data, and code mobility, along with the reliance of traditional mobile devices on WSNs for information about the surrounding physical environment, render WSN middleware relevant to this book.

Wireless sensor networks are relatively new, having only recently been made possible by advances in microelectromechanical systems (MEMS), battery technology, wireless technology, and system-on-chip (SoC) designs. They consist of tiny autonomous nodes each containing a variety of sensors, microprocessor, memory, battery, and wireless communication interface. They are less expensive to deploy than wired networks and can potentially offer greater reliability and agility by routing data through a wireless mesh rather than fixed hard-wired links. Because each node is nominal, WSNs can scale to thousands of nodes offering higher sensing resolutions than was previously possible. WSNs are currently used for habitat monitoring on the Great Duck Island [61] and James Reserve [27] and for microclimate research around redwood trees [20]. Emerging uses include surveillance [38], emergency medical care [55], and structural integrity monitoring [77]. In the future, WSNs may help automate highways [43] and coordinate military operations [53,72]. As WSNs mature, they will revolutionize the way humans and computers interact.

Programming WSNs is challenging. To ensure seamless integration with the environment, each node must be physically small. Many nodes are the size of a matchbox, if not smaller, severely limiting the capabilities of a node; for example, the Tmote Sky [1] runs on two AA batteries and measures 5.45 cm^2 without the battery pack. It contains a slow 16-bit, 8-MHz Texas Instruments MSP430 processor with a mere 10 KB of data memory, 48 KB of instruction memory, and 1 MB of external flash memory. It is programmable using an onboard USB port, communicates using an IEEE 802.14.4-compliant Chipcon CC2420 radio, and has built-in light, temperature, and humidity sensors, as well as SPI and I^{2}C interfaces for attaching peripherals. The Tmote Sky is relatively new; older nodes such as the MICA2 [2], MICAz [3], NMRC [21], M2030 [4], M1030 [5], Smart Dust [6], ESB platform [7], and Intel iMote [8] share similar, if not weaker, specifications.

Many WSN applications such as habitat and infrastructure monitoring require long deployment intervals, often without human intervention. Throughout this time, the nodes are continuously subjected to a potentially harsh environment, resulting in some nodes dying early and new nodes being deployed to replenish those that have failed. This variability further complicates the already unreliable wireless links. In all, programming applications for WSNs is difficult because they must achieve extraordinarily high levels of efficiency, reliability, and autonomy in an environment that is highly dynamic and resource deprived. Such challenges are further amplified when a WSN must handle all of the aforementioned forms of mobility.

Middleware is often relied upon to address the difficulties associated with programming WSNs. It offers more sophisticated abstractions and often includes very high-level programming languages. This chapter presents an overview of the key principles behind various types of WSN middleware.

Operating System Support

Prior to discussing WSN middleware, we must first understand the basic services provided by the underlying operating system. Although many operating systems work in embedded systems, such as LynxOS [9], ChorusOS [10], Contiki [30], VxWorks [11], NetBSD [12], OSE [13], QNX, OS-9 [14], FreeDOS [15], and eCos [16], not all of them work in the highly resource-constrained setting of WSNs or are flexible enough to work in a dynamic WSN environment. Instead, we focus on TinyOS [42], which is a representative example and arguably the most popular WSN operating system used today. TinyOS is a highly efficient minimalist operating system originally developed for the Atmel Atmega series of microprocessors but ported to the TI MP430. It is event based, with a two-level thread hierarchy consisting of tasks and events. Tasks contain long-running computations that may be preempted by time-critical events. To minimize overhead, a single task queue forces tasks to run sequentially. When an event occurs, other events are disabled until the current event completes. Blocking is avoided using split-phase operations; for example, when the application makes a system call, TinyOS immediately passes control back to the application while it processes the command and later signals completion of the command using an event.

TinyOS provides a component-based programming language called *NesC*. NesC divides an application into components that may be modules or configurations. Program behavior is encapsulated within modules, which are wired together using configurations. Configurations can wire other configurations together, allowing the formation of component trees. An application is implemented as a tree of components, where the root is the TinyOS kernel, branches are configurations, and leaves are modules or hardware components. For simplicity, TinyOS does not provide dynamic memory management. Instead, memory is statically allocated to each module and parameterized interfaces are used in place of multiple component instantiations. To maximize efficiency, TinyOS uses active messages for network communication [25]. Active messages contain an identifier specifying how the receiver should process the message. This, along with the event-based execution model of TinyOS, avoids the need for components to block or poll for messages.

Although programming WSNs using TinyOS is simpler than using assembly or C, it still is not easy. As pointed out in Levis and Culler [52], TinyOS has a high learning curve, particularly for nonprogrammers. It requires one to become familiar with tasks, events, commands, split-phase operations, and component hierarchy in addition to subtle race conditions involving asynchrony and atomicity between tasks and events. Many features are common to other programming languages, such as component instantiations and private state, that are not provided by NesC, mostly to ensure that applications can attain maximum efficiency. Furthermore, TinyOS components tend to encapsulate low-level services; for example, many of the components of TinyOS are part of the hardware presentation layer (HPL) that interfaces directly with the hardware. Other components encapsulate the network stack (GenericComm), flash memory (ByteEEPROM), timers (TimerC), and kernel (Main). In addition, the hardwiring of TinyOS components makes it difficult to develop flexible applications that can adapt to a changing context. To change the behavior of a program, either the new behavior must be precoded into the program or the nodes must be retrieved and reprogrammed, neither of which is scalable. Middleware is clearly needed to provide higher level programming abstractions that hide the complexities of TinyOS and allow programmers to quickly implement, test, and deploy their WSN applications.

Basic Building Blocks

We now discuss middleware packages that provide basic building blocks. Two middleware packages we focus on are the Sensor Network Application Construction Kit (SNACK) [36] and Hood [76]. SNACK provides a high-level language and a library of components that offer application-level services. Hood provides a neighbor list abstraction commonly used by many applications.

Sensor Network Application Construction Kit

Object-oriented programming languages are sometimes not used to write software for embedded systems due to concerns about efficiency. In an object-oriented program, a component may have multiple instantiations, each with private state. This results in a redundancy of state that could otherwise be shared, increasing memory utilization. To avoid this, many WSN applications are written using non-object-oriented languages such as NesC and C. These languages do not allow multiple instantiations of a component; instead, each component and all of its variables are statically defined. When programming applications using these languages, developers need to maintain the state of each virtual instance manually through complex programming constructs such as state arrays and parameterized interfaces. A typical example is the timer within TinyOS. The timer is independently used by many components, but because only one timer exists it relies on a parameterized interface and an internal state array for remembering when the timer of each component should fire. The programmer is given fine-grain control over which variables are shared and can thus maximize memory reuse. Doing this manually, however, is tedious and results in complex code.

SNACK is a middleware that provides a high-level language and a library of application-level services. Like object-oriented languages, SNACK allows multiple instantiations of a component. It is implemented on top of TinyOS but can be ported to other operating systems. To reduce overhead, the SNACK compiler aggressively detects variables that can be shared and

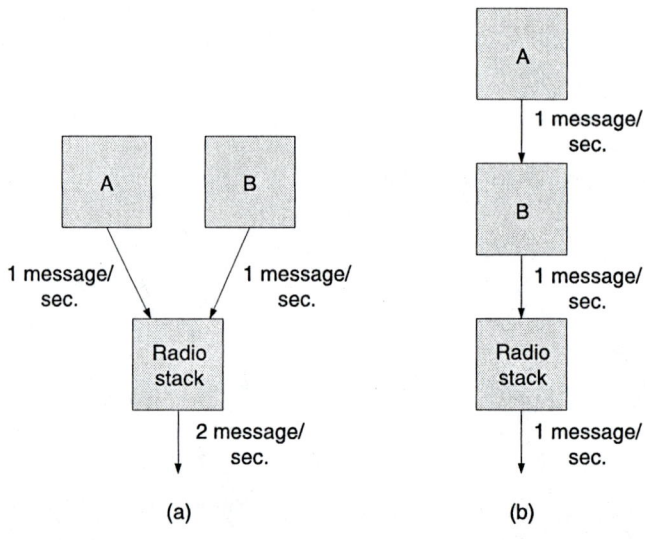

Figure 45.1   Alternative message control flows: (a) Two components, A and B, are independently sending messages, resulting in two messages per second; (b) In a more efficient flow, the two components share the same message, halving the number of messages broadcast over the radio.

reorganizes the program to maximize efficiency. SNACK uses a fairly simple but aggressive mechanism for determining whether state can be shared. When a component is instantiated, all of its state is shared unless the instance is declared using the keyword "my." If "my" is present, all variables inside the component are private.

The SNACK programming language gives the compiler greater flexibility in rearranging components for higher efficiency. Many WSN applications consist of numerous independent components. Changing the control flow between them can significantly reduce overhead; for example, consider the application shown in Figure 45.1a. The figure shows two components that generate messages at a rate of 1 message per second each. Both components send their messages to the radio stack, which forwards them at a rate of two messages per second. If each message contains little data, this will result in high overhead. Rearranging the message flow to that shown in Figure 45.1b can halve the number of transmissions; however, programming components to adhere to this arrangement is tedious and may not be possible if the two components were developed independently. Instead, SNACK provides a messaging service that works with the compiler to achieve the control flow shown in Figure 45.1b.

Allowing the compiler to reorganize the control flow of an application is not trivial. The programming language of SNACK enables this by providing a richer syntax for specifying parameter values and using transitive links. Instead of specifying a specific value for a parameter, the SNACK programming language allows a range to be specified (e.g., "at most," "at least," or "in between"). By relaxing the constraints on the parameter, the compiler is more likely to find values that intersect, allowing it to rearrange the control flow. Transitive links are used to specify links that can be arranged by the compiler. A transitive link simply indicates that one component should be wired to another; it allows the compiler to insert any number of components in between.

SNACK provides a library of components consisting of a messaging service, timer service, and a hash table data structure. The messaging service provides support for the control flow shown in Figure 45.1b. It includes `MsgSrc`, `Network`, and `AttrM` components. The `MsgSrc` generates empty messages and is transitively linked to the `Network`. As the message travels from the `MsgSrc` to the `Network`, other components can use `AttrM` to add their data to the message, merging multiple messages into one. The compiler decides how the final program is structured. The timer service reduces the number of independent timers used by an application. By relying on transitive links, a single timer may pass through multiple components. Finally, SNACK provides a hash table data structure keyed by node ID. The SNACK implementation is much simpler as it can be instantiated using the keyword "my," ensuring all state within it is unshared.

Hood: A Neighbor List

The lack of fixed infrastructure in WSNs coupled with uncertainties associated with individual sensor readings result in algorithms that exhibit a high degree of locality. Nodes need to collaborate with their neighbors to route and aggregate data. They also have to compare their sensor readings to verify and improve accuracy. The localized algorithms that perform these tasks require neighborhood information — namely, a list containing the addresses of nearby nodes whose attributes match certain application-specific criteria. The neighborhood discovery and maintenance protocol is frequently used across a broad range of applications and is, thus, provided as middleware in the form of Hood.

Hood is a middleware for WSNs that provides a neighbor list abstraction. It handles all of the underlying details of neighbor discovery, message passing, and data caching. By using Hood, an application developer can treat the neighbor list as a programming primitive and operate on data shared by the neighbors, thus simplifying programming. Until now, the messaging protocols and filtering algorithms required to create and maintain this list were done from scratch and customized for each application. This was necessary because each application required different algorithms and data structures. Hood avoids this problem by allowing the developer to plug in application-specific code for distributing a node's profile, called the *push policy*, and for processing discovered profiles, called a *filter*.

Hood allows a node to specify a list of attributes to share and a push policy that determines how these attributes are broadcast. The attributes may include anything, such as sensor readings, application state, or remaining power. The push policy may broadcast a message whenever an attribute is set, perform periodic broadcasts, or perform reliable broadcasts (e.g., broadcast an attribute several times whenever it is set). If attributes are rarely broadcast, a bootstrap mechanism may be implemented that broadcasts notifications whenever a node joins a neighborhood. Whenever a bootstrap message is received, the node broadcasts its attributes. To ensure maximum flexibility, the push policy is implemented in native NesC and is passed as a parameter to Hood.

To create a neighborhood, a node passes a filter to Hood. The filter receives all of the attributes broadcast by neighbors and determines which neighbors should be included in the neighbor list. Note that each node independently creates and maintains a neighborhood, meaning neighborhoods are not symmetric. It is possible for node A to consider node B to be a neighbor, but not *vice versa*. A node does not know who considers it to be a neighbor. This decouples the owner of an attribute from the observers that consider the owner a neighbor. Achieving symmetric neighborhoods would require more complex protocols with transactional semantics which may prove infeasible in an environment as dynamic and unreliable as a WSN. The unreliability of WSNs also means Hood cannot provide any consistency guarantees; that is, it cannot guarantee that every node within a certain distance will receive every broadcast.

Hood allows applications to append scribbles to each neighbor. A scribble is a locally derived value like an estimate link quality based on the neighbor's attributes. The filters may include scribbles in their analysis of determining when to include a node in the neighbor list. Complications arise when the filter requires multiple attributes and scribbles because some may be defined while others remain undefined. To address this, a hierarchy of neighborhoods may be created where the members of one neighborhood are dependent on the neighbors in another neighborhood.

Hood is a specific middleware that has been shown to be useful in object-tracking applications. It represents a general class of middleware that provides an abstraction for facilitating neighbor discovery. The key idea of this middleware is that it allows the application developer to specify application-specific details such as the push policy and filtering algorithm, but it provides a generic infrastructure for putting these policies to work. The middleware with a neighbor list abstraction can be easily generalized to provide multiple neighbor lists, each with different attributes, or to provide higher level operations that operate over all members of a particular list.

Data-Centered Abstractions

The primary purpose of a WSN is to gather data about an environment and relay that information to the consumers. Two key challenges to distributing sensor information are (1) mobility and (2) limited resources. The data consumers may include external users, internal nodes, or a combination of both. Many applications such as safe-route navigation and a museum tour guide involve multiple external users that transiently connect to the network and physically move through the sensor field. To complicate matters, WSNs operate under severe resource limitations in terms of power, memory, and computational ability, all of which should be sparingly used. This prevents simply porting existing *ad hoc* routing protocols developed for laptops and PDAs into a WSN. The underlying operating system, in minimizing overhead, provides little support; for example, TinyOS only offers one-hop unicast and broadcast. Clearly, new, more efficient and agile algorithms for distributing sensor data and higher-level programming models to use these algorithms are required. Many types of middleware provide this. We consider two algorithms, geographic hash tables [68] and directed diffusion [39,46], and two models, WSN as a database [35,59,69,79] and abstract regions [75], all of which are available as middleware.

In a WSN, both the sender and the receiver may be transient and mobile. Traditional routing protocols for wireless *ad hoc* networks, such as *Ad Hoc* On-Demand Distance Vector (AODV) [66] and Dynamic Source Routing (DSR) [23], are not ideal because they have high over-head and latency, especially when the network is dynamic. This has resulted in the development of many novel routing algorithms that use data caching and forward pointers to efficiently deliver a message to a moving consumer [51]. Forward pointers, however, leave control state scattered through the network and may still fail if the receiver is moving too rapidly. A geographic hash table does not suffer from this problem. It takes advantage of the spatial property of a WSN. Whenever a sender wishes to send a message to a particular destination, the destination's address is hashed to a particular location. The data is sent to that location, and any node within one hop stores the message, serving as a mailbox. The destination periodically checks its mailbox to receive the message. A geographic hash table guarantees that the message will be delivered as long as no message loss occurs. By choosing a proper hash function, data can be distributed evenly across the entire network, providing load balancing. Drawbacks include the need for each consumer to periodically query its mailbox, the possibility that the mailbox and consumer may be located on opposite sides of the network, and the additional overhead of dealing with node mobility.

Wireless sensor networks consist of potentially thousands of nodes. Many of them will fail mid-deployment, and new nodes may be added. This highly dense and dynamic environment calls for applications that place less emphasis on the data collected by each individual node and instead treat the nodes as an aggregate. Applications for WSNs tend not to require data from a specific sensor but rather from sensors that have certain properties such as, for example, being located in a particular region or having sensor readings above a certain threshold. This observation led to the development of content-based routing where data is not sent to a particular address but is routed based on its attributes. One system that provides this is directed diffusion [46]. Directed diffusion introduces the idea of routing data down interest gradients. When a consumer is interested in a particular type of data, it broadcasts a description of its interest, which is propagated throughout the network. By propagating the interest, a downhill gradient is produced where the consumer is at the bottom. Any data matching the interest is forwarded down the gradient to the consumer. The middleware handles all of the distribution of interests and configuring of "flows" for delivering data down interest gradients. To save power, it also performs in-network data aggregation to reduce the number of packets sent back to the consumer; for example, instead of receiving all of the raw data, it only receives the average over a certain epoch.

Wireless sensor network middleware may also change the fundamental model presented to the application programmer. One example is a database. Using this model, a programmer need not program individual sensors or debug complex data aggregation and routing algorithms. Instead, a WSN is treated like a single table in a database consisting of the aggregate data collected across all WSN nodes. Many middleware packages provide this abstraction. They include TinyDB [59], SINA [69], Cougar [79], and IrisNet [35]. All of these middleware packages implement a distributed query processor that spans the entire network. They provide a Structure Query Language (SQL)-like language tailored to WSNs. A primary difference between implementing a virtual WSN database versus a real database is the fact that the data is continuously generated, often only after a query is issued. Also, in-network data aggregation is necessary in WSNs to reduce message transmissions and save power. To account for this, the database query language is augmented with EPOCH, SAMPLE_PERIOD, and function modifiers that allow a programmer to specify how aggregate data should be generated within the network. To account for the spatial properties of a network, a WHERE modifier is also provided. Recent work in this area has produced more powerful aggregation operations [41] and techniques for compressing aggregate data [71].

Another type of middleware is the notion of abstract regions [75], which provides a new model in place of individual nodes in a WSN. It allows programmers to define regions in the network and to treat each region as if it were an individual sensor. A region may in reality contain many sensors, but all of them aggregate their data, and the application is only given the aggregate. Abstract regions still give application programmers control over resource consumption and provide feedback on how successfully requested aggregate operations were carried out. It uses a tuple-space-like model for distributing data and provides a thread-like model where applications can perform blocking operations. Together, this greatly simplifies the programming model, allowing even non-computer scientists to program a network.

Mobility-Centered Abstractions

A common application for WSNs is to track a mobile entity as it travels through a sensor field. The entity can be, among many other things, a soldier in a field or a car in a parking lot. In the simplest implementation, any node that detects an entity would report to a central base station. This implementation is less accurate as the sensors do not collaborate to verify their readings. Some entities are difficult to track and can only be accurately sensed by multiple sensors concurrently [38]. The simple implementation is also not energy efficient as redundant notifications may be routed through the network when multiple sensors detect the same entity. Other challenges arise when the nodes operate on a sleep schedule. If a node is asleep, it will not be able to detect the entity; thus, complex motion prediction and forewarning algorithms are required to wake up nodes ahead of the entity so they can participate in the event detection process. Tracking mobile entities is a complex process used by many applications. Several middleware packages have been developed to reduce this complexity. One such package is EnviroTrack [18]. Another middleware package that allows mobile users to access the WSN is MobileQuery [57].

EnviroTrack

EnviroTrack provides a context-label abstraction for each entity. A context label is distinguished by type (e.g., "car") and contains a user-defined aggregate state about the entity and application-specific code that operates on this state. The context label is dynamically created when the entity is first detected and logically follows the entity as it moves through the sensor field. EnviroTrack simplifies programming by hiding the details associated with internode communication and group maintenance necessary to detect and track an entity. Programmers interact with the static context label rather than with a continuously changing set of nodes that sense the entity. This is done using a directory service based on a geographic hash table. The type of context label is hashed to a specific location, and nodes within one hop of this location serve as directory objects. Each context label periodically updates its state with its directory objects, and programmers interact with the directory objects.

To use EnviroTrack, the programmer must specify the possible types of context labels. For each type, the programmer must provide a `sense` function, a `state` function, and, optionally, tracking objects. The `sense` function takes the raw sensor readings and determines whether the entity is present. It includes a freshness constraint that determines the maximum age of a sensor reading for it to be considered and a critical mass constraint that determines the minimum number of sensors that sense the entity for the detection to be considered valid. The `state` function produces the aggregate data to maintain within the context label. The tracking objects perform local processing on the nodes that comprise the context label, allowing them to perform context-sensitive tasks that may directly affect the entity (e.g., forming a barrier if the entity is an intruder or activating a sprinkler system if the entity is a fire).

EnviroTrack provides a group maintenance algorithm for organizing nodes that can sense the entity. Ideally, the algorithm will produce a single context label per entity, each with a single leader who is responsible for aggregating the data and updating the directory service of the context label. A context label is formed whenever a node detects an entity using the sense function and is not already part of, or aware of, a group for the entity. Whenever this occurs, it creates a context label and declares itself leader. As the leader, it periodically broadcasts a heartbeat that is propagated a certain number of hops beyond the group's boundary. The heartbeat serves the dual purpose of telling members the leader is still alive and providing a forewarning system. If a member node can sense an entity but does not hear a heartbeat within a certain time, it creates a new group. Whenever a node receives a heartbeat, it remembers the last time it received it. If the entity is sensed within a certain time of receiving the last heartbeat, the node joins the group by periodically sending its sensor readings to the leader. Multiple leaders may occur when the entity is initially detected by multiple nodes or when the heartbeat fails to propagate. To account for this, each leader maintains a weight corresponding to the number of updates it has received from group members. The leader with the lower weight concedes to the one with the higher weight. As the entity moves, the leader may lose the ability to sense the entity. When this occurs, it hands off its leadership to a node that inherits the former leader's weight, thereby reducing the probability that spurious leaders will remain active.

EnviroTrack has been implemented on top of TinyOS. To further simplify programming, it provides a preprocessor that translate EnviroTrack code into NesC. Through simulations and actual implementations on MICA motes, EnviroTrack has been shown to simplify WSN programming.

MobileQuery

MobileQuery [57] is a middleware supporting user mobility. MobileQuery is best motivated using an example. Imagine a user carrying a PDA and traveling through a sensor field. As the user travels, what is the simplest task the user may want to do that is useful to many applications? One such task is to simply query the sensor readings of nodes within a certain vicinity; for example, a possible query could be "What is the average of all temperature readings within a one-mile area?" This requires the dissemination of a query to a certain geographic region and routing and aggregation algorithms for delivering the results back to the user. Because the user continuously moves, this query process should be periodic. MobileQuery provides this service.

Implementing MobileQuery may at first appear simple: Just broadcast a query, and wait for a reply. In networks that do not operate on a sleep schedule and applications that do not require real-time operation, this is probably sufficient. However, many WSN applications such as hazard detection and safe-route navigation software require real-time context information. Also, WSNs often operate on a sleep schedule to prolong network lifetime where the nodes remain asleep for the majority of the time, only briefly waking up to perform application-specific tasks. If the user issues a query when the nodes just went to sleep, the latency will be high. Furthermore, as the query area expands, a naive flooding solution will result in greater energy consumption due to duplicate broadcasts and more message being lost due to contention. To avoid these problems, MobileQuery adapts a prefetching and tree-building scheme where prefetching messages are sent ahead of the user to predefined pickup points (i.e., the location where the user expects to receive query results), and a routing tree rooted at the pickup point is created to collect and aggregate the results. MobileQuery assumes that a user's movements are predictable and that a prefetch message can travel faster than the user. A user's movements may be predictable based on history or in certain scenarios such as driving on the highway or hiking on a trail, when the motions may follow a predefined map. To ensure that prefetch messages can travel faster than the user, a backbone overlay network [28,74,78] is used.

The two types of prefetching are all-at-once and just-in-time. All-at-once prefetching sends the prefetch messages as far as the user will travel. It assumes that the user travels a fixed distance along a known path. By sending the prefetch messages immediately, it can guarantee that all of the nodes in future query areas will be ready to supply sensor readings by the time the user arrives.

Dynamic Reprogramming

The aforementioned middleware systems simplify programming by providing commonly used services. They are integrated into the application at compile time and do little to increase network runtime flexibility. Flexibility is important in WSNs because of their highly dynamic topology and lengthy deployment intervals throughout which the user requirements, or the users themselves, may change. Knowing all possible uses of a WSN prior to deployment is not possible. A network initially deployed for habitat monitoring may later be used for wildfire detection. Large WSNs covering a wide geographic area may be deployed for a single use initially but later may be divided into regions, each running software specialized to features within each environment. Anticipating and incorporating all possible behaviors into an application prior to deployment is not feasible. Many aspects such as algorithms and data structures are difficult to parameterize. Memory constraints prevent including extraneous behavior. To address this, middleware packages have been created to facilitate the dynamic reprogramming of a predeployed network. By enabling dynamic reprogramming, the network can assume any behavior, and can serve transient users as they arrive and move through the sensor field.

Embedded systems often adhere to a Harvard architecture with separates data and instruction memories. This naturally leads to the ability to reprogram a node by reflashing the instruction memory and can be provided at the operating system level [45]. The problem is that: (1) some nodes do not employ rewritable instruction memory due to cost, (2) reflashing memory consumes a lot of power and is unreliable when the batteries are not fully charged, and (3) it requires the transmission of the entire operating system and application over a lossy low-bandwidth radio. Some middleware packages provide a more efficient way to reprogram a WSN. They either allow the instruction memory to be partially reprogrammed or, more often,

Table 45.1   Middleware Supporting Network Reprogramming

Name	Execution Unit	Execution Model	Network Reprogramming	Communication Model
Agilla	Agent	Agent thread	Strong migration	Tuple space
Impala	Modules	Event	Flood code	Message passing
Maté	Capsules	Event	Floor code	Message passing
SensorWare	Script	Event	Weak migration	Message passing
Smart Messages	Smart message	Single thread	Strong migration	Remote: migration Local: tag space

provide a virtual machine that interprets lightweight mobile control scripts. Middleware packages of this sort include Agilla [33], Impala [54], Maté [52], SensorWare [22], and Smart Messages [50]. A summary of their features is shown in Table 45.1.

Among them, Agilla [33] provides a virtual machine that supports mobile agents and highly flexible capabilities for in-network reprogramming. Its middleware architecture is shown in Figure 45.2. Unlike traditional programs that are statically installed on a specific node, mobile agents can move and clone themselves across the network performing

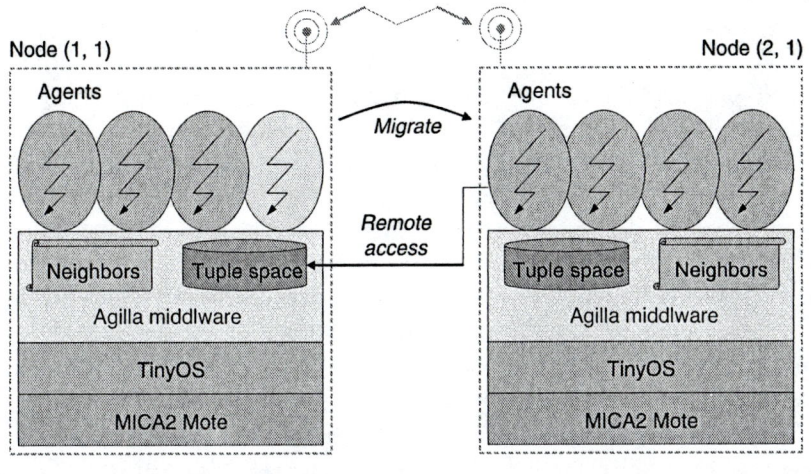

Figure 45.2   The Agilla model.

application-specific tasks. Two types of migration operations are provided: strong and weak. Strong migration captures the entire state of an agent including its program counter and stack. The agent resumes execution at the destination uninterrupted. A weak migration, on the other hand, only captures data state, all execution state is lost, and the agent resumes running at the beginning when it arrives at the destination. Although strong migration may simplify programming, it entails higher overhead.

An application often consists of multiple agents; for example, in a fire detection application, multiple mapping agents may be used to form a perimeter around the fire [32]. When multiple agents are used to carry out an application's task, they must coordinate with each other. In Agilla, coordination is achieved through local Linda-like tuple spaces [34]. A tuple space is a special type of shared memory where the data is addressed by content using templates. One agent can insert a tuple, and another can later read or remove it using pattern matching via a template. To prevent polling, Agilla augments the tuple space with reactions [26,31,49,64] that notify an agent of a tuple matching a particular template when it appears in the tuple space. Tuple spaces decouple agents ensuring that they remain autonomous. The autonomy of an agent is vital in a dynamic environment such as a WSN because inter-agent interactions tend to be highly transient.

The tuple space also serves as a convenient mechanism for agents to discover their context; for example, in Agilla, the tuple space stores the types of sensors available and the identities of the other colocated agents. Tradeoffs regarding what to store in the tuple space versus through a dedicated data structure must be expected; for example, because the address of neighboring nodes is frequently accessed by many applications, Agilla provides an acquaintance list abstraction, but a list of available sensors is accessed less frequently, so it is stored in the tuple space. Providing a dedicated data structure reduces latency but increases memory overhead.

Unlike other mobile middleware [64], Agilla does not support a global tuple space that spans across multiple nodes primarily due to bandwidth and energy constraints. Instead, it supports local tuple spaces where each node maintains a distinct and separate tuple space. If agents were restricted to operate only on the local tuple space, they would only be able to coordinate with colocated agents, but in many applications, agents need to communicate with agents residing on different hosts. Although this may be accomplished by having the agent migrate to the other agent's host, agent migration is relatively expensive; thus, Agilla provides special instructions that allow an agent to perform operations on a remote host's tuple space. These instructions rely on simple mutihop unicast communication and are thus scalable. Sequentially accessing each neighbor's tuple space, however, entails significant overhead. To address this, Agilla provides a group instruction that uses single-hop multicast to query the tuple spaces of all one-hop neighbors. Scalability is ensured because this operation operates only over one hop.

Wireless sensor networks are unique in that spatial properties are important. Sensor nodes detect certain properties of the environment. The location at which they take the measurement is necessary when, for example, determining where the intruder is; in other words, many WSN applications must know their spatial placement to make sense of the sensor data they collect. Agilla embraces this reliance on spatial information by addressing nodes by their geographic location. Sensors can obtain their location through global position system (GPS) or any number of other localization schemes [24,62,67]. As a side benefit, addressing nodes by location enables Agilla primitives to be easily extended to operate over geographic regions and to use geographic routing for multihop interactions.

Agilla is implemented on top of TinyOS and tested on MICA2 motes. It has been shown that careful engineering of the middleware makes programming flexible applications consisting of mobile agents not only feasible but easier. In its current implementation, standard coordination mechanisms such as tuple spaces and acquaintance lists are used. In the future, agents may be able to communicate directly with each other, thereby further reducing overhead, or they may be able to mutate their code, taking on additional capabilities as they gain experience within the network. The possibilities are endless.

Impala [54] and Maté [52] are two similar middleware systems that divide the code of an application into capsules that are then distributed throughout the network. The main difference is that Impala uses native code, whereas Maté uses a virtual machine. Unlike Agilla agents, Impala and Maté capsules have no control over where they execute. When an updated capsule is issued, it is flooded throughout a network. This prevents multiple applications from running concurrently and different areas of the network from running different code. Both Impala and Maté use an event-based programming model where each capsule remains inactive unless an event to which it is sensitive occurs; for example, one capsule within Maté is a timer capsule, and the code within this capsule is executed whenever the timer fires. Another capsule is sensitive to the arrival of a message. By using an event-based model, these middlewares achieve high efficiency by avoiding polling and allowing the execution unit to remain dormant during periods of no events.

Smart Messages [50] and SensorWare [22] are similar to Agilla in that they allow their execution units to control where they are located. Smart Messages allow its execution units, known as a smart message, to perform strong migrations. SensorWare uses mobile scripts as its execution unit, but only supports weak migration. Both systems are implemented as virtual machines on relatively powerful PDAs. Unlike Agilla, Smart Messages only provides a single thread per node and separates local communication (via a tag space) from remote communication (via migration). By having a single thread, multiple applications cannot run concurrently on a node, and the need for a smart message to migrate to communicate with a remote node incurs higher overhead than simply remotely accessing the node. SensorWare, on the other hand, uses an event-based execution model like Impala and Maté. Also like Impala and Maté, SensorWare uses direct message passing for communication between its mobile scripts.

Emerging Strategies

Wireless sensor networks are continuously evolving as technology improves and new applications become feasible. In recent years, WSNs have evolved from rigid application-specific deployments to flexible embedded computing platforms. As WSN nodes improve, they will run more sophisticated applications that demand better middleware support. Middleware designers are currently investigating several emerging strategies, which include providing quality of service (QoS), macroprogramming, and connecting WSNs with traditional networks.

Quality of Service Management

As WSNs mature, they will run more sophisticated applications and multiple applications at a time. Existing middleware such as Agilla already allows multiple applications to be dynamically loaded into a network. Little attention, however, has been paid to QoS, specifically as related to message delivery latency, sensing accuracy, energy consumption, and data throughput. Most existing middleware packages provide services on a best-effort basis and do not consider application-specific semantics when making tradeoffs between QoS and resource consumption. Many applications share the same types of tradeoffs (e.g., decrease sensing accuracy or increase message latency for additional power savings). Programming these tradeoffs within each application is tedious and error prone. Furthermore, when multiple applications share the same network, interactions across applications must be accounted for; for example, if one application is tracking a raging wildfire, it should be given better QoS than an application monitoring the migration patterns of monarch butterflies. Middleware provides a convenient mechanism for adding these QoS provisions.

Two middleware projects that provide QoS are MiLan [40,63] and AutoSec [37]. MiLan takes a specification on the minimum QoS an application requires and adapts the network to achieve this QoS while minimizing resource consumption. Instead of simply observing network parameters and adapting the application, MiLan attempts to control the network; for example, consider a habitat monitoring application running on a WSN deployed throughout a forest. The majority of the time nothing interesting is going on, so the middleware selects a sparse subset of the nodes to monitor the environment at a low resolution to save power; however, when an interesting event is detected, the middleware increases the QoS by activating additional nodes near the phenomenon of interest. This proactive approach allows MiLan to provide high QoS while still consuming low resources. AutoSec differs from MiLan in that it focuses on resource allocation to ensure maximum system throughput. It relies on a directory service that stores information about the current state of the network and, based on this information, chooses a resource allocation policy that divvies up the resources such that each application achieves its desired QoS. Part of the challenge with AutoSec lies in determining how to maintain the directory service and what resource allocation policies should be provided.

Real-time behavior is a specific type of QoS that many applications require. In a real-time application (e.g., surveillance, fire monitoring, and intruder detection), messages and actions must be precisely choreographed for the application to function correctly. Messages must be delivered on time at the right place carrying data of a certain freshness. Two middleware projects that provide real-time functionality are DSWare [58] and RAP [56]. DSWare is a publish–subscribe middleware that relies on standard real-time packet scheduling mechanisms, such as earliest deadline first (EDF). It also provides group management, event detection, data caching, and data storage services, reducing the burden of application developers. RAP is a real-time query service for WSNs. It introduces velocity monotonic scheduling, which takes advantage of the spatial properties of the network to provide real-time message delivery. RAP allows a user to issue a query with certain period, deadline, and data freshness requirements. As data is delivered, its velocity, as measured by how far it has traveled over how long, is used as a local indicator of how urgently the packet must be forwarded. A message that will barely make or miss its deadline traveling at its current velocity will have a higher priority than a message that will easily make its deadline. Both middleware projects are still in the prototype phase, having only been evaluated in simulators.

Macroprogramming

Another emerging strategy being embraced by developers of WSN middleware is the idea of macroprogramming. Macroprogramming relies on new programming languages that allow a programmer to describe, at a high level, what the sensor network should do. The middleware and compiler would then determine the low-level code that executes on each individual node. By hiding the distributed nature of the network, programming it is greatly simplified. Agilla can be viewed as a form of macroprogramming where developers create agents without worrying about precisely where they are installed. Other middleware projects based on macroprogramming include abstract regions [75], virtual markets [60], Regiment [65], and MagnetOS [19].

Abstract regions were discussed earlier. They allow programmers to reason about abstract regions that map to collections of nodes. Virtual markets take a unique approach to achieving new behavior in a WSN. Instead of introducing new code into the network, a virtual market simply changes the value of performing certain tasks. In a virtual market, intelligent agents are distributed throughout the network that can perform certain actions (e.g., take a sensor reading, aggregate data, and forward data). Each action has a value associated with it. By programming the agents to seek maximum profits, the overall system behavior can be controlled by simply changing the value of each action. Regiment and MagnetOS are both high-level programming languages that allow a developer to program a WSN application as if it ran on a single node. The underlying middleware and compiler take the program and determine how it can be divided and distributed across multiple nodes within the WSN. They differ in that Regiment provides a functional language whereas MagnetOS is written in Java.

Integration with Traditional Networks

Another emerging area of WSN middleware research involves developing platforms that allow the seamless integration of WSNs with traditional networks. For WSNs to gain widespread use, they must be easily integrated with the existing computing infrastructure. Currently, the code that bridges WSNs with traditional networks is proprietary relying on custom protocols tailored to each application. Furthermore, as WSNs gain widespread use, more sophisticated applications will want to harness the power of multiple potentially heterogeneous WSNs. These applications are often distributed across multiple administrative domains; for example, a company's inventory management system may reside on servers belonging to the company, supplier, and shipping company. Applications running on traditional networks are mature. They operate across administrative domains by adhering to common protocols and languages such as those proposed by the Open Grid Services Architecture (OGSA) [17]. OGSA, however, introduces too much overhead for use in WSNs. Developing and maintaining custom protocols that facilitate the interactions between WSNs and the fixed infrastructure is a formidable task. It is a service that WSN middleware is just beginning to provide.

One middleware package that links WSNs with traditional networks is Agilla. In Agilla, mobile agents can easily migrate between WSNs and traditional networks. Another project is Hourglass [70]. Hourglass operates over the Internet and handles the delivery of data between consumers and sources located across multiple WSNs. In creates an overlay network over the fixed Internet infrastructure and provides a circuit abstraction that connects the sources with the destinations. Circuits are tailored to handling WSN data by allowing various services to operate over the data flowing through them; for example, some of the services provided by Hourglass include filtering, aggregating, compressing, and buffering. Because Hourglass nodes are more powerful than WSN nodes, they are capable of performing more complex operations on the data. An application developer simply tells Hourglass its data needs, and the middleware takes care of discovering the networks that provide the raw data and assembling the necessary services to produce the required data. Hourglass is still in the prototype stage, having only been simulated in ModelNet [73].

Conclusion

Wireless sensor networks promise to revolutionize the way humans interact with their physical environment. They will soon gain widespread use because they provide many benefits and are relatively inexpensive to deploy; however, to gain widespread use, new middleware solutions are required. Programming WSNs is difficult because they have extremely limited resources and exhibit many forms of mobility involving the users, entities being sensed, sensor nodes, data, and code. To help simplify application development, many middleware packages have been created. Initial WSN middleware provided basic building blocks such as high-level programming languages, neighbor lists, and libraries of components that provide application-level services. As applications matured and gained complexity, new middleware for handling the various forms of mobility and increasing network flexibility through in-network reprogramming were developed. WSNs are relatively new and are rapidly evolving, forcing middleware designers to embrace new strategies. These emerging strategies include providing QoS, macroprogramming, and providing a foundation for connecting WSNs to traditional networks.

Acknowledgments

This research was supported in part by the Office of Naval Research under MURI research contract N00014-02-1-0715 and by the NSF under ITR contract CCR-0325529. Any opinions, findings, and conclusions expressed in this paper are those of the authors and do not necessarily represent the views of the research sponsors.

References

[1] http://www.moteiv.com.
[2] http://www.xbow.com/Products/productsdetails.aspx?sid=72.
[3] http://www.xbow.com/Products/productsdetails.aspx?sid=101.
[4] http://www.dustnetworks.com/docs/M2030.pdf.
[5] http://www.dustnetworks.com/docs/M1030.pdf.
[6] http://robotics.eecs.berkeley.edu/~pister/SmartDust/.
[7] http://www.scatterweb.com/.
[8] http://www.intel.com/research/exploratory/motes.htm.
[9] http://www.lynuxworks.com/.
[10] http://www.experimentalstuff.com/Technologies/ChorusOS/.
[11] http://www.windriver.com/.
[12] http://www.netbsd.org/.
[13] http://www.ose.com/.
[14] http://www.microware.com/.
[15] http://www.freedos.org/.
[16] http://sourceware.org/ecos/.
[17] http://www.globus.org/ogsa/.
[18] Abdelzaher, T., Blum, B., Cao, Q., Chen, Y., Evans, D. et al., EnviroTrack: towards an environmental computing paradigm for distributed sensor networks, in *Proc. of IEEE Int. Conf. on Distributed Computing Systems (ICDCS'04)*, Tokyo, Japan, March, 2003, pp. 582–589.
[19] Barr, R., Bicket, J.C., Dantas, D.S., Du, B., Kim, T.W.D. et al., On the need for system-level support for *ad hoc* and sensor networks, *SIGOPS Oper. Syst. Rev.*, 36(2), 1–5, 2002.
[20] Batalin, M.A., Rahimi, M., Yu, Y., Liu, D., Kansal, A. et al., *Towards Event-Aware Adaptive Sampling Using Static and Mobile Nodes*, Technical Report 38, Center for Embedded Networked Sensing, University of California, Los Angeles, 2004.
[21] Bellis, S., Delaney, K., O'Flynn, B., Barton, J., Razeeb, K., and O'Mathuna, C., Development of field programmable modular wireless sensor network nodes for ambient systems, *Computer Commun.*, 2005 (special issue on wireless sensor networks).
[22] Boulis, A., Han, C.-C., and Srivastava, M., Design and implementation of a framework for efficient and programmable sensor networks, in *Proc. of the First Int. Conf. on Mobile Systems, Applications, and Services (MobiSys'03)*, San Francisco, CA, May, 2003.

[23] Broch, J., Johnson, D.B., and Maltz, D.A., *The Dynamic Source Routing Protocol for Mobile Ad Hoc Networks*, Internet Draft, Internet Engineering Task Force (IETF) Mobile *Ad Hoc* Networking Working Group, 1998.

[24] Bulusu, N., Heidemann, J., and Estrin, D., *GPS-Less Low-Cost Outdoor Localization for Very Small Devices*, Technical Report 00-729, University of Southern California, Los Angeles, 2000.

[25] Buonadonna, P., Hill, J., and Culler, D., *Active Message Communication for Tiny Networked Sensors*, http://www.tinyos.net/papers/ammote.pdf.

[26] Cabri, G., Leonardi, L., and Zambonelli, F., *Reactive Tuple Spaces for Mobile Agent Coordination*, Vol. 1477, Lecture Notes in Computer Science, Springer-Verlag, Heidelberg, 1998, pp. 237–252.

[27] Cerpa, A., Elson, J., Estrin, D., Girod, L., Hamilton, M., and Zhao, J., Habitat monitoring: application driver for wireless communications technology, *SIGCOMM Comput. Commun. Rev.*, 31(2, Suppl.), 20–41, 2001.

[28] Chen, B., Jamieson, K., Balakrishnan, H., and Morris, R., Span: an energy-efficient coordination algorithm for topology maintenance in *ad hoc* wireless networks, in *Proc. of the 7th ACM/IEEE Int. Conf. on Mobile Computing and Networking (MOBICOM'01)*, Seattle, WA, August, 2001, pp. 85–96.

[29] Curino, C., Giani, M., Giorgetta, M., Giusti, A., Murphy, A.L., and Picco, G.P., TinyLime: bridging mobile and sensor networks through middleware, in *Proc. of the 3rd IEEE Int. Conf. on Pervasive Computing and Communications (PerCom'05)*, Kauai Island, Hawaii, March 8–12, 2005, pp. 61–72.

[30] Dunkels, A., Grnvall, B., and Voigt, T., Contiki: a lightweight and flexible operating system for tiny networked sensors, in *Proc. of the First IEEE Workshop on Embedded Networked Sensors (IEEE EmNetS-I)*, Tampa, FL, November, 2004.

[31] Fok, C.-L., Roman, G.-C., and Hackmann, G., A lightweight coordination middleware for mobile computing, in *Proc. of the 6th Int. Conf. on Coordination Models and Languages (Coordination 2004)*, DeNicola, R., Ferrari, G., and Meredith, G., Eds., Vol. 2949, Lecture Notes in Computer Science, Springer-Verlag, Heidelberg, 2004, pp. 135–151.

[32] Fok, C.-L., Roman, G.-C., and Lu, C., Mobile agent middleware for sensor networks: an application case study, in *Proc. of the Fourth Int. Conf. on Information Processing in Sensor Networks (IPSN'05)*, Los Angeles, CA, April 25–27, 2005, pp. 382–387.

[33] Fok, C.-L., Roman, G.-C., and Lu, C., Rapid development and flexible deployment of adaptive wireless sensor network applications, in *Proc. of the 25th Int. Conf. on Distributed Computing Systems (ICDCS'05)*, Columbus, OH, June, 2005, pp. 653–662.

[34] Gelernter, D., Generative communication in Linda, *ACM Trans. Program. Lang. Syst.*, 7(1), 80–112, 1985.

[35] Gibbons, P., Carp, B., Ke, Y., Nath, S., and Seshan, S., IrisNet: an architecture for a worldwide sensor Web, *IEEE Pervasive Comput.*, 2(4), 22–33, 2003.

[36] Greenstein, B., Kohler, E., and Estrin, D., A sensor network application construction kit (SNACK), in *Proc. of the 2nd Int. Conf. on Embedded Networked Sensor Systems (SenSys'04)*, Baltimore, MD, November 3–5, 2004, pp. 69–80.

[37] Han, Q. and Venkatasubramanian, N., AutoSec: an integrated middleware framework for dynamic service brokering, *IEEE Distributed Syst. Online*, 2(7), 2001.

[38] He, T., Krishnamurthy, S., Stankovic, J.A., Abdelzaher, T., Luo, L. et al., Energy-efficient surveillance system using wireless sensor networks, in *Proc. of the Second Int. Conf. on Mobile Systems, Applications, and Services (MobiSys'04)*, Boston, MA, June, 2004, pp. 270–283.

[39] Heidemann, J., Silva, F., and Estrin, D., Matching data dissemination algorithms to application requirements, in *Proc. of the First Int. Conf. on Embedded Networked Sensor Systems (SenSys'03)*, Los Angeles, CA, November 5–7, 2003, pp. 218–229.

[40] Heinzelman, W., Murphy, A., Carvalho, H., and Perillo, M., Middleware to support sensor network applications, *IEEE Network Mag.*, 18, 6–14, Jan. 2004.

[41] Hellerstein, J., Hong, W., Madden, S., and Stanek, K., Beyond average: towards sophisticated sensing with queries, in *Proc. of the Second Int. Conf. on Information Processing in Sensor Networks (IPSN'03)*, Palo Alto, CA, April 21–23, 2003.

[42] Hill, J., Szewczyk, R., Woo, A., Hollar, S., Culler, D., and Pister, K., System architecture directions for networked sensors, in *Proc. of Architectural Support for Programming Languages and Operating Systems (ASPLOS)*, Cambridge, MA, November 12–15, 2000, pp. 93–104.

[43] Hsieh, T.T., Using sensor networks for highway and traffic applications, *IEEE Potentials*, 23(2), 13–16, 2004.

[44] Huang, Q., Lu, C., and Roman, G.-C., Spatiotemporal multicast in sensor networks, in *Proc. of the First Int. Conf. on Embedded Networked Sensor Systems (SenSys'03)*, Los Angeles, CA, November 5–7, 2003, pp. 205–217.

[45] Hui, J. and Culler, D., The dynamic behavior of a data dissemination protocol for network programming at scale, in *Proc. of the 2nd Int. Conf. on Embedded Networked Sensor Systems*, Tampa, FL, November, 2004, pp. 81–94.

[46] Intanagonwiwat, C., Govindan, R., and Estrin, D., Directed diffusion: a scalable and robust communication paradigm for sensor networks, in *Proc. of the 6th ACM/IEEE Int. Conf. on Mobile Computing and Networking (MOBICOM'00)*, Boston, MA, August, 2000, pp. 56–67.

[47] Jea, D., Somasundara, A., and Srivastava, M., Multiple controlled mobile elements (data mules) for data collection in sensor networks, in *Proc. of the Int. Conf. on Distributed Computing in Sensor Systems (DCOSS'05)*, Marina del Rey, CA, June 30–July 1, 2005.

[48] Juang, P., Oki, H., Wang, Y., Martonosi, M., Peh, L.S., and Rubenstein, D., Energy-efficient computing for wildlife tracking: design tradeoffs and early experiences with ZebraNet, *ACM SIGPLAN Notices*, 37(10), 96–107, 2002.

[49] Julien, C. and Roman, G.-C., EgoSpaces: facilitating rapid development of context-aware mobile applications, *IEEE Trans. Software Eng.*, 32(5), 281–298, 2006.

[50] Kang, P., Borcea, C., Xu, G., Saxena, A., Kremer, U., and Iftode, L., Smart messages: a distributed computing platform for networks of embedded systems, *Comput. J.*, 47, 475–494, 2004 (special issue on mobile and pervasive computing).

[51] Kim, H.S., Abdelzaher, T.F., and Kwon, W.H., Minimum-energy asynchronous dissemination to mobile sinks in wireless sensor networks, in *Proc. of the First Int. Conf. on Embedded Networked Sensor Systems (SenSys'03)*, Los Angeles, CA, November 5–7, 2003, pp. 193–204.

[52] Levis, P. and Culler, D., Maté: a tiny virtual machine for sensor networks, in *Proc. of the Tenth Int. Conf. on Architectural Support for Programming Languages and Operating Systems (ASPLOS'02)*, San Jose, CA, November, 2002, pp. 85–95.

[53] Lin, T.-H., Sanchez, H., Kaiser, W.J., and Marcy, H., Wireless integrated network sensors (WINS) for tactical information systems, in *Proc. of the 1998 Government Microcircuit Applications Conf. (GOMAC'98)*, Arlington, VA, March 16–19, 1998.

[54] Liu, T. and Martonosi, M., Impala: a middleware system for managing autonomic, parallel sensor systems, in *Proc. of ACM SIGPLAN Symp. on Principles and Practice of Parallel Programming (PPoPP'03)*, San Diego, CA, June, 2003.

[55] Lorincz, K., Malan, D., Fulford-Jones, T.R.F., Nawoj, A., Clavel, A. et al., Sensor networks for emergency response: challenges and opportunities, *IEEE Pervasive Comput.*, 3(4), 16–23, 2004 (special issue on pervasive computing for first response).

[56] Lu, C., Blum, B.M., Abdelzaher, T.F., Stankovic, J.A., and He, T., Rap: a real-time communication architecture for large-scale wireless sensor networks, in *Proc. of the 8th IEEE Real-Time and Embedded Technology and Applications Symp. (RTAS'02)*, San Jose, CA, September 24–27, 2002.

[57] Lu, C., Xing, G., Chipara, O.L., Fok, C.-L., and Bhattacharya, S., A spatiotemporal query service for mobile users in sensor networks, in *Proc. of IEEE Int. Conf. on Distributed Computing Systems (ICDCS'05)*, Columbus, OH, June, 2005, pp. 381–390.

[58] Lu, S., Son, S., and Stankovic, J., Event detection services using data service middleware in distributed sensor network, in *Proc. of the Second Int. Conf. on Information Processing in Sensor Networks (IPSN'03)*, Palo Alto, CA, April 21–23, 2003.

[59] Madden, S., Franklin, M., Hellerstein, J., and Hong, W., The design of an acquisitional query processor for sensor networks, in *Proc. of ACM SIGMOD Int. Conf. on Management of Data*, San Diego, CA, June, 2003, pp. 491–502.

[60] Mainland, G., Kang, L., Lahaie, S., Parkes, D., and Welsh, M., Using virtual markets to program global behavior in sensor networks, in *Proc. of the 11th ACM SIGOPS European Workshop*, Leuven, Belgium, September, 2004.

[61] Mainwaring, A., Polastre, J., Szewczyk, R., Culler, D., and Anderson, J., Wireless sensor networks for habitat monitoring, in *Proc. of the 1st ACM Workshop on Wireless Sensor Networks and Applications(WSNA'02)*, Atlanta, GA, September, 2002.

[62] Moore, D., Leonard, J., Rus, D., and Teller, S., Robust distributed network localization with noisy range measurements, in *Proc. of the 2nd Int. Conf. on Embedded Networked Sensor Systems (SenSys'04)*, Baltimore, MD, November 3–5, 2004.

[63] Murphy, A. and Heinzelman, W., *MiLAN: Middleware Linking Applications and Networks*, Technical Report TR-795, University of Rochester, 2002.

[64] Murphy, A.L., Picco, G.P., and Roman, G.-C., Lime: a middleware for physical and logical mobility, in *Proc. of IEEE Int. Conf. on Distributed Computing Systems (ICDCS'01)*, Phoenix, AZ, April, 2001, pp. 524–533.

[65] Newton, R. and Welsh, M., Region streams: functional macroprogramming for sensor networks, in *Proc. of the 1st Int. Workshop on Data Management for Sensor Networks (DMSN)*, Toronto, Canada, August 30, 2004.

[66] Perkins, C. and Royer, E., *Ad hoc* on-demand distance vector routing, in *Proc. of the Second IEEE Workshop on Mobile Computing Systems and Applications (WMCSA'99)*, New Orleans, LA, February 25–26, 1999, pp. 90–100.

[67] Priyantha, N., Chakraborty, A., and Balakrishnan, H., The cricket location-support system, in *Proc. of the 6th ACM/IEEE Int. Conf. on Mobile Computing and Networking (MOBICOM'00)*, Boston, MA, August, 2000, pp. 32–43.

[68] Ratnasamy, S., Karp, B., Yin, L., Yu, F., Estrin, D. et al., GHT: a geographic hash table for data-centric storage in sensornets, in *Proc. of the 1st ACM Workshop on Wireless Sensor Networks and Applications (WSNA'02)*, Atlanta, GA, September, 2002.

[69] Shen, C.-C., Srisathapornphat, C., and Jaikaeo, C., Sensor information networking architecture and applications, *IEEE Pers. Commun. Mag.*, 8(4), 52–59, 2001.

[70] Shneidman, J., Pietzuch, P., Ledlie, J., Roussopoulos, M., Seltzer, M., and Welsh, M., *Hourglass: An Infrastructure for Connecting Sensor Networks and Applications*, Technical Report TR-21-04, Harvard University, Boston, MA, 2004.

[71] Shrivastava, N., Buragohain, C., Agrawal, D., and Suri, S., Medians and beyond: new aggregation techniques for sensor networks, in *Proc. of the 2nd Int. Conf. on Embedded Networked Sensor Systems (SenSys'04)*, Baltimore, MD, November 3–5, 2004, pp. 239–249.

[72] Simon, G., Maroti, M., and Ledeczi, A., Sensor network-based countersniper system, in *Proc. of the 2nd Int. Conf. on Embedded Networked Sensor Systems (SenSys'04)*, Baltimore, MD, November 3–5, 2004.

[73] Vahdat, A., Yocum, K., Walsh, K., Mahadevan, P., Kostic, D., Chase, J., and Becker, D., Scalability and accuracy in a large-scale network emulator, in *Proc. of 5th Symp. on Operating Systems Design and Implementation (OSDI'02)*, Boston, MA, December 9–11, 2002.

[74] Wang, X., Xing, G., Zhang, Y., Lu, C., Pless, R., and Gill, C., Integrated coverage and connectivity configuration in wireless sensor networks, in *Proc. of the First Int. Conf. on Embedded Networked Sensor Systems (SenSys'03)*, Los Angeles, CA, November 5–7, 2003, pp. 28–39.

[75] Welsh, M. and Mainland, G., Programming sensor networks using abstract regions, in *Proc. of the First USENIX/ACM Symp. on Networked Systems Design and Implementation (NSDI 2004)*, San Francisco, CA, March 29–31, 2004.

[76] Whitehouse, K., Sharp, C., Brewer, E., and Culler, D., Hood: a neighborhood abstraction for sensor networks, in *Proc. of Second Int. Conf. on Mobile Systems, Applications, and Services (MobiSys'04)*, Boston, MA, June, 2004, pp. 99–110.

[77] Xu, N., Rangwala, S., Chintalapudi, K.K., Ganesan, D., Broad, A. et al., A wireless sensor network for structural monitoring, in *Proc. of the 2nd Int. Conf. on Embedded Networked Sensor Systems (SenSys'04)*, Baltimore, MD, November 3–5, 2004, pp. 13–24.

[78] Xu, Y., Heidemann, J., and Estrin, D., Geography-informed energy conservation for *ad hoc* routing, in *Proc. of the 7th ACM/IEEE Int. Conf. on Mobile Computing and Networking (MOBICOM'01)*, Seattle, WA, August, 2001, pp. 70–84.

[79] Yao, Y. and Gehrke, J., The cougar approach to in-network query processing in sensor networks, *SIGMOD Rec.*, 31(2), 9–18, 2002.

Chapter 46

Mobile Middleware for Automotive Applications

Francesco Lilli

CONTENTS

Introduction

Automotive telematics will be a challenge over the next 10 years. In the 1980s, automobile manufacturers and system suppliers initiated research activities aimed at introducing information and communications systems within vehicles. They began with autonomous systems such as car radios and navigation systems. The continuous growth and availability of innovative technologies such as communication with short and long coverage capabilities, localization systems, and human–machine interface devices allowed improvement of the telematic platform. The integration of ICT required a revision of the in-vehicle architecture to consolidate the technologies in a single open platform. The various systems must communicate and exchange data with each other, and this is possible thanks to common hardware and software platforms. The development of an in-vehicle platform for components belonging to the car as well as for those installed on the external infrastructure represents the killer topic that will propel the penetration of telematics into the automotive market. This chapter is focused on the contribution of mobile middleware to this goal.

The technological challenge requires consolidating the entire telematics platform and services, as well as wide utilization of the platform, at least at a continental level. Many potential services are based on having access to external information, and the application of telematics technologies within automotive and roadside infrastructures will guarantee the availability of such information. In this chapter, we focus on middleware applicable to the automotive telematics platform, but particular attention has to be focused on the relevance of the roadside infrastructure as well as the service and content provider.

Such integration requires compatibility among the different components of the platform but must also address the quality of services and security of the exchanged data. This chapter discusses an appropriate middleware for the in-vehicle architecture, as well as specific issues relevant to the standards for each telematics technology (e.g., 802.11x standards).

It is not sufficient to address only the technologies, quality of services, and security with regard to inclusion of the telematics concept within the automobiles. Legal aspects should also be addressed, which means certification of software platforms and applications, especially as they affect the safety of the driver and other vehicle occupants.

Today's cars are obviously different from personal computers, workstations, or smart mobile phones. All of these devices have clearly defined roles, local computational power, and a set of well-defined peripherals. The car and its contents have radically changed in the last 20 years [1]; if we think about telematics devices in vehicles in the 1970s, we can recall the car radio. At first, they were introduced as aftermarket (AM) products, but then they quickly became original equipment manufacturing (OEM) products. The high penetration of radios in the market and the consequent reduction of costs allowed car manufacturers to include radios as a component of their vehicles. Since then, newer telematics technologies have contributed to shifting the attention of the engineers toward innovative audio functions such as digital audio broadcasting (DAB) receivers or multimedia media players (MP3). The same transition has occurred with the introduction of car navigation systems, following the market model adopted by the Japanese of anticipating OEM products with AM ones.

Today, several computational devices are installed in vehicles, but they have different functional aims, technologies, and system architectures. Some of them (independent systems) perform their tasks with minimal interaction with other systems, such as, for example, the automatic air-conditioning system. Others operate in a strong cooperative mode but within a closed group of devices; these vehicle-related systems include the powertrain controller and the braking system. Yet others perform functions not directly related to vehicle functionalities; these non-vehicle-related systems include an integrated mobile phone. Finally, some systems (dependent systems), such as the dashboard controller, are quite dependent on all the other systems to perform their tasks.

To further illustrate the degree of automotive complexity, a standard rule for integrating devices within the car does not exist, due to the fact that the introduction of devices will depend on the vehicle model (low versus high price), customer choice (standard or optional components), and style (sport versus luxury cars). Sometimes some of the functions are grouped within the same controller, but sometimes they are split among more than one, so the scenario is constantly evolving with the introduction of new functionalities and electronic units. We therefore decided to include a section in this chapter that describes the architecture of a vehicle and will provide the basis for our discussion in the rest of this chapter.

Description of an Automobile Architecture

What we really need here is a simplified vehicle, different from any existing (or future) one but useful for defining the common terminology for in-vehicle devices. Figure 46.1 shows the architecture of our simple vehicle. Our car will have a certain number of vehicle-specific controllers that are interconnected in some way with one or more vehicle networks, which will be considered as a single entity. One of the devices in the car will have the capability to remotely connect with the external world through both a cellular phone system and some kind of short-range communications technology. Such functions can be performed by different devices in a vehicle, but here we will imagine them grouped into a single device we will call the *InfoTelematics Unit* (ITU). For the sake of simplicity, we will assume this unit has a direct connection with the entertainment and the car navigation systems. Also, all interactions with the driver or passenger (except those involving drive-related information such as vehicle

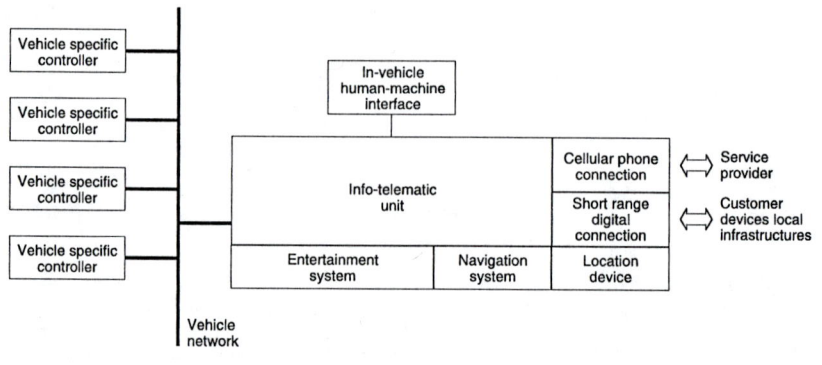

Figure 46.1 In-vehicle architecture.

speed or light lamp status) are provided by a central *human–machine interface* (HMI) controller directly connected and controlled by the ITU.

Now we can examine the internal software architecture of our vehicle. The ITU is the only place in which will reside the telematics applications designed for mobile communications. This device and its peripherals are integrated in a way that is transparent to the user. (Note that detailed information about the software architecture for the next-generation telematics platform of FIAT vehicles will be not explained here for obvious reasons related to intellectual properties rights.) We can imagine that a certain number of telematics building blocks, for the management of internal and external telematics modules, are realized above a software framework; these building block are enablers of the telematics services and application.

In a real implementation, this kind of application can actually be hosted in more than one device, integrated in the car and external to it. The link between these devices can have different characteristics, such as being wired or wireless and having specific peer-to-peer connections or standard buses. Furthermore, as we will describe later on, localization and communication systems or the HMI could also be characterized by external devices [2]. As an example, the driver might wish to use a Bluetooth® global position system (GPS) device, a mobile phone, and a personal digital assistant (PDA) plugged into the dashboard. Adding together all these variables only increases the complexity of the system (and adds to the confusion of the reader).

For such an ideal system, a similarly ideal (albeit simplified and hypothetical) software architecture could be described by the diagram shown in Figure 46.2. As the reader can see, the choice here was to hide most of the complexity of the system; we have omitted all of the lower levels of the protocols and all parts of the operating system. Throughout

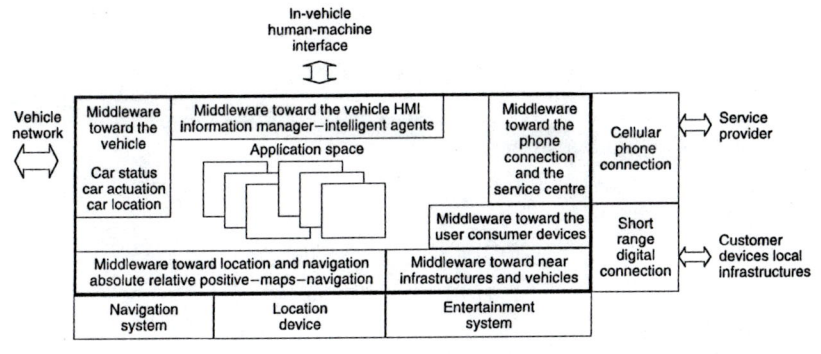

Figure 46.2   Middleware in the software in-vehicle architecture.

this chapter, we focus primarily on the middleware needs of applications, digging into the lower levels only when a specific topic demands it. Refer to Figure 46.2 throughout the rest of the chapter.

We begin our discussion about automotive middleware by addressing aspects related to accessing vehicle data and operating some of the vehicle devices. We then focus on access to the vehicle HMI, investigating the management of providing information to the driver and its impacts on safety. We also investigate how the intelligent agent technology can have a positive impact on the driver's cognitive load.

We then examine middleware for a service center by illustrating ongoing activities directed toward telematics service access, taking into account general-purpose services, location-based services, and traffic information services. The management of telematics applications through the service center helps to provide the appropriate level of quality of service and security. Applications are downloaded from authorized service centers, and the services are accessible by registered users. The quality of information and the timing of their dispatching are controlled, especially for audio and video multimedia applications, to minimize the allocation of in-vehicle resources that are limited. A certain level of trust within the telematics network is necessary to ensure the security of data exchanged between the different architecture components and to authenticate components within the network.

The automotive world is a dynamic network in which vehicles represent dynamic nodes. All aspects related to positioning and navigation, such as location information and digital road map formatting and accessing, are covered on the section devoted to the middleware for location and navigation. Finally, we provide a broad analysis of topics related to short-range communication, the road infrastructure, and surrounding vehicles, with an emphasis on driver assistance.

Middleware Between On-Board Applications and the Vehicle

The vehicle has evolved from primarily a mechanical device, with some electrical support device such as a lamp or windshield wiper, to a complex system integrating its mechanical parts with electronic devices able to manage or to control them [3]. The evolution did not followed a straight path. Electronics have been added whenever they allowed a task to perform better, such as for electronic engine control, or could add new functionalities, such as the antilock brake system (ABS). In some cases, the addition of new electronics has been feasible only when the cost of the device was lower than the equivalent mechanical or electrical solution, as in the case of the dashboard controller. Initially, the electronics were usually applied as single components, but it became apparent that better results and new functions would be obtained by connecting together different systems, such as the engine, gearbox, and brake controllers.

This led to the development of vehicle serial connections and networks, based on proprietary solutions. Then, some standards began to be used [4]. Today, the most widely used vehicle serial connection is the *local interconnect network* (LIN), and the most frequently used network is the *controller area network* (CAN), even when several different solutions (for historical, performance, or economical reasons) are in use. CAN has gained such wide application in cars that people consider it to be the *de facto* standard, but CAN specifies only the two lower layers of the ISO/OSI stack: physical and link. We can assume that new car models have a significant portion of their electronic devices connected to one of the vehicle networks, which are interconnected by one or more gateways.

Several steps must be taken to allow our middleware to access the vehicle data. If, in the future, car makers utilize a standard network (in terms of a single physical network or a common communications protocol) for connecting vehicle electronic devices, then mobile middleware could be enabled by the car's computer, which would manage the vehicle devices and allow access to vehicle data. Various applications will be enabled by knowledge of the vehicle data, such as, for example, preventive diagnoses of the vehicle that could help the driver perform necessary maintenance and avoid unexpected breakdowns.

Middleware Between On-Board Applications and the Vehicle HMI

The growth of in-vehicle system complexity requires accurate studies for the design and the development of a human–machine interface. A suitable HMI, in fact, impacts the operation of different applications and can reduce

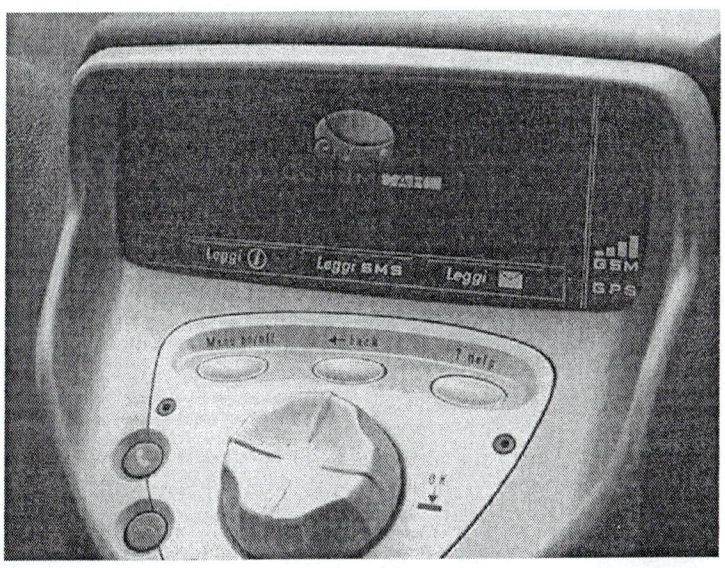

Figure 46.3 COMUNICAR European Project, an innovative HMI concept for telematics services [6].

the workload of drivers, thus improving road safety [5]. The operation of various applications could be made available to users through graphical and acoustic devices, could be activated by traditional input devices, or could be provided by the emerging voice recognition technologies. Moreover, dedicated mechanisms are required to manage modalities and priorities for dispatching information; these strategies help the driver to use the applications correctly and at the same time avoid confusion that could jeopardize road safety.

The mobile middleware concept might be quite suitable for HMI devices that are used as portable means for dispatching application functionalities. Car makers are developing devices that are fully integrated in the dashboard but which are also portable devices such as PDAs or smart mobile phones (see Figure 46.3). The connection of these devices with the in-vehicle platform requires the appropriate middleware to set up and handle the communication link to enable the collection of user data or to show graphical information. Some preliminary products have been based on a wired connection, but some exploitation of short-range communication technologies, such us WiFi or Bluetooth®, is occurring. Broadband wireless technologies could allow the transmission of a huge amount of data between the telematics platform and multimedia devices. The possibility of using flexible solutions for the HMI system architecture via portable devices and suitable mobile middleware offers users the ability to download multimedia information about in-vehicle applications as well as general-purpose functions (when not in the vehicle).

HMI mobile middleware is not simply focused on the management of input and output devices (e.g., hardware and software connections and information priorities); in fact, an emerging concept is the provision of information based on the context and location awareness of users. Behind this concept is intelligent agent technology that, with regard to their application in the automotive sector, is able to run different telematics modules (e.g., localization system) and functions (e.g., traffic information) on the basis of user preferences. The user could select application preferences and at the same time choose the modalities for suitable output; for example, the driver could ask an application to get information about points of interest (POIs) during the drive and show them on the digital road map used by the car navigation system. The context and location are used by the in-vehicle system for automatically selecting user preferences in terms of application content and modalities.

Middleware Between On-Board Applications and a Service Center

An innovative concept for vehicles that has been developed over the last several years by vehicle manufacturers is telematics. Within the automotive sector, car manufacturers began to develop the telematics concept during the 1990s in response to the emerging wireless communication technologies and satellite localization systems. The idea was to integrate within the vehicle a telematics platform able to connect the car with the external world. Early work was concerned with communication between the vehicle and a roadside infrastructure through radiofrequency (RF) technology (Figure 46.4); this approach represented a first tentative step toward the so-called vehicle-to-infrastructure cooperative system.

With the introduction of the Global System for Mobile Communications (GSM) and the growing penetration of the U.S. military satellite GPS in the civilian world, car manufacturers and their research and development centers began to address carrying the telematics concept a step further by connecting vehicles with operations centers to exchange voice and data. Such services (e.g., traffic information, emergency calls, anti-theft alarms) have been studied by specific consortia (e.g., GATS, WAP) and launched on the market by car manufacturers or service providers (e.g., Webraska, TrafficMaster, Tegaron) in the last several years.

Most in-vehicle autonomous applications have benefited from the telematics platform; for example, in-car navigation systems were launched

Figure 46.4   Communication technologies for telematics services.

in the 1980s as stand-alone systems offering suggestions for reaching destinations based on static optimized routes. Since then, the availability of traffic information via a telematics platform (e.g., radio tuners that receive RDS–TMC broadcast information from radio operators) has improved the navigation function, because routes can be dynamically updated according to reported traffic congestion.

In-vehicle applications, then, have been improved thanks to their connection with a service center, and several telematics services have been established based on the navigation platform, where information coming from the service center is matched with the vehicle location or directly filtered in the service center to establish so-called location-based service (LBS). In addition to this beneficial concept, other services have been studied and developed for the mobility sector. The availability of information coming from a private or public fleet equipped with a telematics platform has been leveraged, for example, to identify traffic congestion in urban areas (floating car data concept) and to monitor the quality of freight transport service (freight fleet management).

From a technological point of view, middleware providing the connection to the service center has been designed to manage a single communication device used in different modalities; for example, telematics services based on the GSM technology take advantage of its short message service (SMS), data, and voice capabilities. Over the past several years, the trend for telematics platforms is to be open and expandable for integrating different wireless communication technologies, so middleware designed to access the service center must be designed to manage different communications devices, as well as the different modalities for using each of those.

Middleware Between On-Board Applications and Location and Navigation

On-board navigation was a killer application for the automotive sector in the 1990s. Navigation applications [8] have been able to penetrate the automotive market because of the availability of a worldwide technology that provides the absolute position of a vehicle everywhere and any time, such as the U.S. GPS system and the future European Global Navigation Satellite System (GNSS) (EGNOS and GALILEO) [7], as well as the fact that most of the major digital map companies (Navteq and TeleAtlas) have invested in the production of a road database.

Although telematics services have contributed to development of the navigation application itself, the navigation system has become a building block of the in-vehicle telematics platform. The navigation application has become an off-board navigation service that decentralizes some components of the system in the service center. The fundamental blocks are still in the vehicle, such as the vehicle positioning and route guidance tools, but the digital maps database and the route calculation tools have been moved from the service center. The result is an in-vehicle navigation unit that allows the utilization of updated digital maps and dynamic routes, including traffic information. With regard to road transport and safety, the navigation system has become integral to the in-vehicle platform for location-based services as well as for advanced driver assistance systems (ADAS) that provide location data, as well as information about the surrounding area.

One of the first prototypes of such a platform was developed as part of the GALLANT European project in which a traditional ADAS application — cruise control — was integrated with the EGNOS system [9] (see Figure 46.5). The middleware for the on-board application, which provides functions to the driver and information to the service center, accommodates the navigation system and location receiver. Behind the navigation system, relevant component hardware and software characterize the platform; in particular, from the hardware point of view, the most relevant of these are the digital road map database, supported by the physical storage format (PSF), and the human–machine interface (HMI). Digital mapping is not foreseen as being a feature of off-board

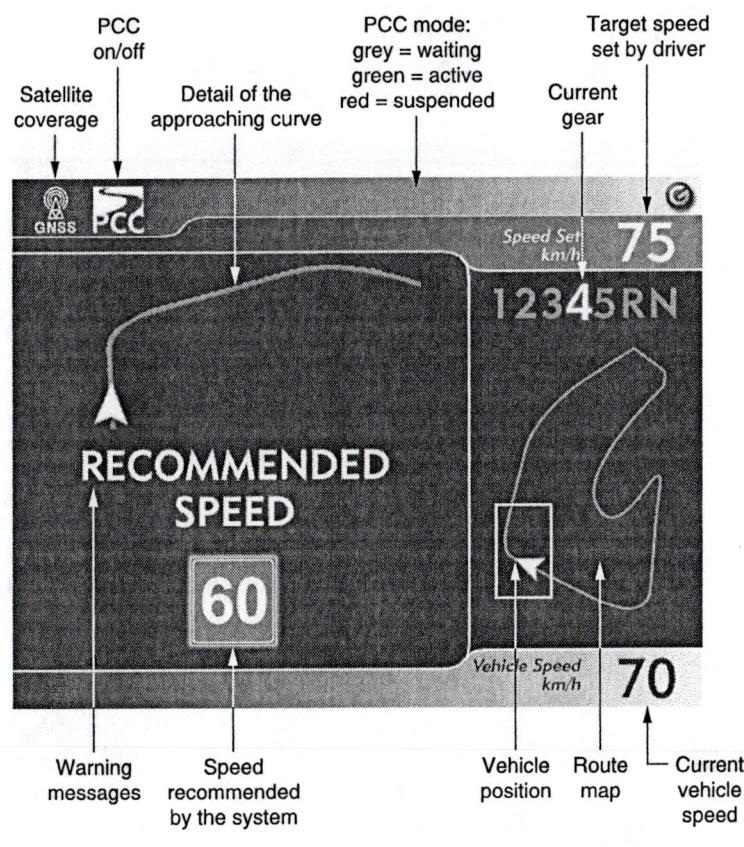

Figure 46.5 GALLANT European Project: integration of ADAS and GNSS systems.

navigation, as the HMI can show only the textual and graphical information downloaded by the service center. At the software level, the building blocks characterizing the navigation application are the vehicle position, map matching, route calculation, and route guidance. For off-board navigation, map matching is performed on the route that is downloaded from the service center, and route calculation is completely handled by the service center. In this case, the route could be updated dynamically taking into account traffic information that might be available. The location receiver could be a stand-alone device designed to work with those LBSs that do not any information other than the position itself (e.g., a point of interest on a digital road map). Obviously, the navigation system is an in-vehicle application that uses the location receiver to determine the vehicle position.

Middleware Between On-Board Applications and Surrounding Infrastructure and Vehicles

We will now take a look at the possibilities, as well as the issues and concerns, raised by the availability of short-range communication links with surrounding infrastructures and vehicles [10]. The range of applications is indeed quite wide, covering such aspects as local-area communication links between vehicle occupants and virtual Tamagotchis living in the car multimedia system and exchanged with (or stolen by) nearby vehicles. Rather than trying to cover all the possibilities, we focus here on a particularly challenging and interesting one: the functions that support ADAS.

Advanced driver assistance systems commonly include several kinds of functions that support the driver while driving. The most well-known ADAS function is adaptive cruise control (ACC). The ACC function is viewed by the driver as a normal cruise control that maintains the car at a certain speed without user intervention. A big advantage offered by the ACC is that, by utilizing a frontal radar (or laser), the system is able to vary the speed of the vehicle according to the speed of the vehicles in front of it.

Several other functions have been developed or investigated to support the driver. Some examples are the lane departure warning system, which alerts drivers when they deviate from their lane; a curve and speed limit information system to alert drivers to the maximum allowable speed for the road; a lane-keeping assistant that helps the driver, through an active intervention on the steering wheel, to follow the correct path in the lane; an obstacle and collision warning system, which warns drivers of sudden, unseen obstacles; and various kinds of intersection support to aid in the perception of infrastructures (signs and lamp) and the behavior of surrounding vehicles. A complete analysis and description of the ADAS functions was developed by the ADASE project. The interested reader will find a longer list of ADAS functions with descriptions and further details in Reference 11.

From a functional viewpoint, ADAS applications can be classified into two main types: one that uses visual, acoustic, and haptic feedback to help drivers make correct choices and another that actively substitutes for the driver. The active function can act on longitudinal control, lateral control, or both of them. Another important classification of ADAS functions depends on the application; most of the ADAS functions simply support drivers and reduce their workload, but others have been developed to take over in emergency situations to the extent necessary. The communications channel requirements for ADAS systems are extremely challenging, resulting in much research into middleware devoted to this kind of service.

Figure 46.6 Communication technologies for ADAS application.

Not all of the ADAS functions have to be based on some kind of wireless short-range connection, but in the vast majority of them the information coming from the other vehicles or from the roadside infrastructure and vulnerable users could represent an added value; the wireless link is precious for bringing the knowledge available from these other sensors [12]. For some applications, such as revealing the presence of a vehicle just beyond a curve in the road or vulnerable pedestrians crossing the road, the communication link will represent the core technology of the system (Figure 46.6).

Acknowledgments

I particularly thank my colleague Walter Savio for his valuable contribution to the production of this chapter. Acknowledgments are also addressed to the ElectroTelematics Systems team of Centro Ricerche FIAT, which over the years has developed the telematics mentioned in this chapter.

References

[1] RocSearch, London (telematics and automotive communications), http://roc-search.ecnext.com/.
[2] SRI Consulting Business Intelligence, *Portable Intelligence: 2004 a Year of Growth for Location Technology Applications in PI Devices* (http://www.sric-bi.com/Explorer/PI-archive.shtml).
[3] AUTOSAR, http://www.autosar.org.
[4] Society of Automotive Engineers (SAE), www.sae.org.
[5] PTV Traffic Mobility Logistics, *The Technology and Trends in Automotive HMI Design* (http://www.english.ptv.de/cgi-bin/mobility/mob_report.pl).

[6] COMUNICAR European Project, http://www.comunicar-eu.org.
[7] European Space Agency (ESA), www.esa.int/esaNA/index.html.
[8] Research and Markets, *European Telematics: Market Trends and Analysis of Embedded and Portable Navigation Systems*, 2005 (http://www.research-andmarkets.com/reports/c18210).
[9] GALLANT (GALileo for safety of Life Application of driver assistaNce in road Transport) European Project, http://ec.europa.eu/dgs/energy_transport/galileo/applications/pilotprojects_en.htm.
[10] Car-to-Car Communication Consortium (C2C CC), http://www.car-to-car.org.
[11] ADASE European Project, www.adase2.net.
[12] Networking the world with sensors, *MIT Technology Insider*, January, 2004 (http://burgaz.mit.edu/OURPRESS/ARTICLES/Jan2004-ILPInsider.pdf).

Chapter 47

A QoS Framework for Multimedia Communication for Wireless Mobile *Ad Hoc* Defense Networks

Raymond Paul, Waseem Sheikh,
Basit Shafiq, and Arif Ghafoor

CONTENTS

Introduction

According to U.S. Army's Future Combat System (FCS) initiative, the future battle force will consist of a large number of mobile sensing systems, unmanned air vehicles (UAVs), unmanned ground vehicles (UGVs), manned helicopters, manned combat vehicles, and several dismounted infantry units, all communicating through a wireless communication network [2]. A typical battlefield scenario and the different components of a battlefield telecommunication system are depicted in Figure 47.1. These components of a battlefield telecommunication system should be able to support command, control, communications, computer, intelligence, surveillance, and reconnaissance (C4ISR) capabilities.

Future defense communications networks will be composed of myriad heterogeneous networks and require the deployment of forces over a wide geographic region where a fixed infrastructure is not available. The support of mission-critical applications in such a scenario requires the defense network to rely heavily on a wireless mobile *ad hoc* network; a typical network of this kind is depicted in Figure 47.1. These networks provide a challenging environment for the transmission of quality of service (QoS)-guaranteed, mission-critical, multimedia, tactical data. In a wireless *ad hoc* network, all of the nodes are mobile and the network topology changes with time [3]. No fixed infrastructure exists, unlike a cellular network. Multiple access techniques such as code-division multiple access (CDMA), frequency-division multiple access (FDMA), or time-division multiple access (TDMA) may be used to provide multiple wireless channels. The internode communication occurs over wireless links and is multihop. The bandwidth of a particular link varies over time, and error rates are very high. The network should be able to support datagram as well as multimedia traffic. Under such an environment, delivery of QoS-guaranteed, multimedia data becomes an extremely difficult task. In this chapter, we address the problem of transmitting QoS-guaranteed, pre-orchestrated, multimedia data in a wireless mobile *ad hoc* network. The multimedia data that we consider may consist of videos or images captured by UAVs, images and videos captured by UGVs, or target-tracking data accumulated by various sensors over the course of time.

Consider a battlefield scenario in which the soldiers, armored vehicles, and aircraft maintain contact with one another via radio links using *ad*

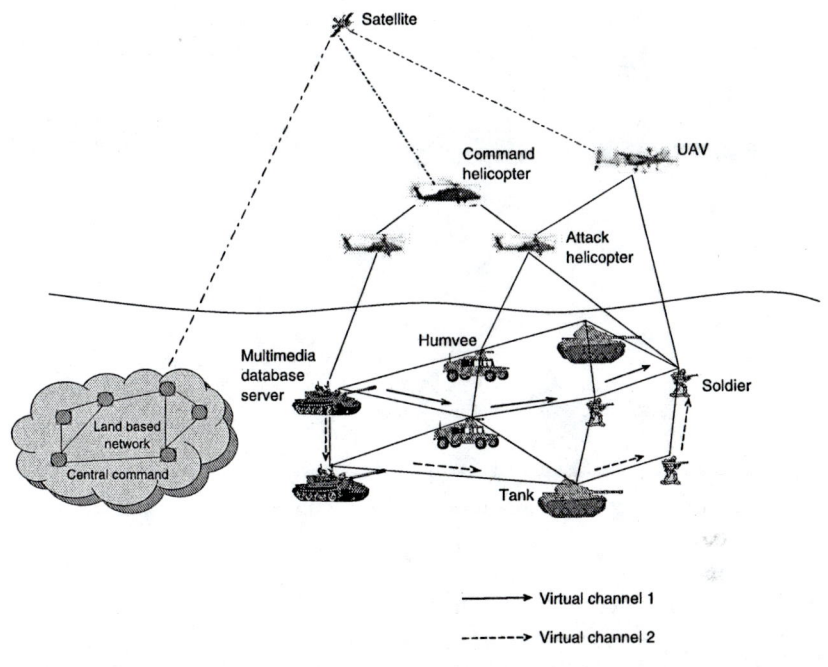

Figure 47.1   Components of the future battlefield communication network.

hoc networks. Some of the personnel may be able to maintain contact with mobile nodes that act as mobile database servers, but others may have to move out of the range of the servers when they enter the battlefield. Transmission of video and audio among the different units of command and control is critical for the success of any military operation, as is the delivery of stored aerial images and video of enemy positions from mobile database servers, which may be installed on UAVs or armored vehicles, to forward-deployed forces. An example of such a network is the Joint Network Node (JNN) built by General Dynamics [1]. Another example of such a network is the Joint Battlespace Infosphere (JBI), which is a system that integrates, aggregates, and distributes information to users at all echelons, from the command center to the battlefield. One of the goals of JBI is to deliver the right information to the right user at the right time in the right format.

Figure 47.1 shows a wireless mobile *ad hoc* network in a battlefield scenario that has several multimedia database servers mounted on tanks and armored vehicles, a population of mobile soldiers and armored vehicles, aircraft, and a satellite connection to a command-and-control center located in a land-based network. The goal of such a network is to provide C4ISR capabilities.

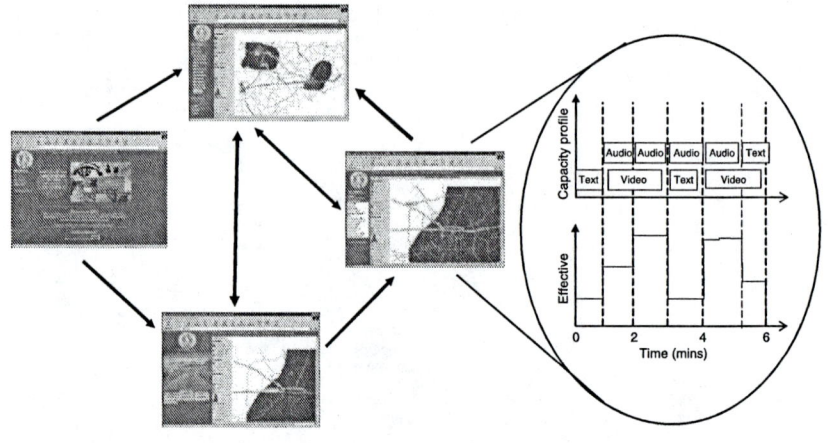

Figure 47.2 Browsing graph of a multimedia document of geographical information system (GIS) map.

Multimedia applications in general are not monolithic in nature and can consist of several objects integrated together, such as video clips, images, text, and audio segments. A multimedia application consisting of multiple objects can be represented as a multimedia document maintained by database servers [4]. An example of a multimedia document is an MPEG-4-based application. MPEG-4 encompasses all types of media and uses scene and object descriptors (SDs and ODs) to define the spatiotemporal features of the component media object. Figure 47.2 shows a collection of multimedia documents of a geographical information system (GIS) map organized for a Web-based browsing environment. Typically, such an environment consists of a collection of documents or SDs for MPEG4-based applications, integrated to allow random browsing by mobile users.

To deliver multimedia data and knowledge for C4ISR-based, mission-critical information with the desired presentation quality, virtual channels are established between the multimedia server and end terminal. In a wireless *ad hoc* defense network, a single virtual channel may not have sufficient bandwidth capacity to satisfy the desired bandwidth requirement of a given multimedia session. In such a case, multiple virtual channels can be used for multimedia data transmission. These virtual channels may differ in their bandwidth capacity, end-to-end delays, and path lifetime. The path lifetime of a virtual channel is the duration for which the channel can provide a communication link between the multimedia server and the mobile user. Due to the diverse characteristics of virtual channels, delivery of QoS-guaranteed, multimedia data over multiple heterogeneous virtual channels (MHVCs) in a mobile *ad hoc* defense network poses additional networking challenges.

In this chapter, we propose a QoS framework for supporting multimedia traffic in a mobile *ad hoc* defense network. Such a framework can be deployed for battlefield management and supporting mission-critical applications over a large geographical area. The QoS parameters such as bandwidth and end-to-end delay can be derived from multimedia document schema stored at multimedia database servers, as shown in Figure 47.1. Our proposed QoS Routing (QoSR) protocol, with the help of mobility management, establishes and maintains multiple heterogeneous virtual channel in an *ad hoc* defense network infrastructure similar to that shown in Figure 47.1. In addition, we present a scheduling scheme for transmitting multiple data streams over established MHVCs in an efficient and timely manner.

Challenges Involved in Supporting Multimedia Applications in Mobile *Ad Hoc* Defense Networks

Transmission of QoS-guaranteed, multimedia data in a wireless *ad hoc* defense network involves numerous challenges; in this chapter, we consider and propose solutions using our QoSR protocol for mobility management and the scheduling of multimedia data over MHVCs. A flat network architecture as shown in Figure 47.1 is well suited for our QoSR scheme, as MHVCs may exist between source and destination nodes. This architecture does not subdivide the overall network. Routing in a flat architecture may be easier to accomplish but such an architecture has certain limitations with regard to its scalability. In a hierarchical architecture, nodes are partitioned into different clusters, and a hierarchical network has the advantages of spatial reuse of shared channels, minimal amount of control information exchanged to maintain routing information, power control, and mobility management [5]. Routing, however, is suboptimal in a hierarchical network because of the limited number of routes between source and destination nodes.

A major challenge for multimedia services in a mobile *ad hoc* defense network is the characterization of the bandwidth requirements of multimedia documents. Different objects such as video clips, text, and images in a multimedia document can have different bandwidth requirements. Depending on the concurrency level of objects, the quality parameters associated with individual multimedia objects, and the presentation duration of these objects, the overall bandwidth profile of a multimedia document may change considerably over a period of time. Accordingly, the network resource requirements also vary. To provide quality-based multimedia services, the underlying network must accommodate such changes by allocating bandwidth resources in an efficient and timely manner. Prestored data provides considerable flexibility in allocating resources for managing multimedia traffic. By identifying the temporal characteristics and the required presentation quality of the multimedia information being accessed by users, the overall resources required can be determined in advance.

Another important issue to be considered in ensuring seamless transmission of multimedia data is the breakdown of already established MHVCs as a result of user's mobility. Typically, the lifetime of a multimedia session can be greater than the lifetime of MHVCs. In this case, the mobility management scheme should be able to predict the lifetime of MHVCs, and the QoSR protocol should find new MHVCs before the expiry of old ones to ensure QoS guaranteed transmission.

In addition to addressing the above-mentioned challenges, we also present a solution for the scheduling of multiple data streams belonging to a multimedia document over MHVCs [9]. During the transmission of video and audio data streams, network delays must be bounded to maintain inter-stream and intra-stream temporal synchronization. Inter-stream synchronization deals with the synchronized playback of related streams, whereas intra-stream synchronization is required for the continuous jitter-free presentation of each stream. The scheduling scheme discussed in this article helps to achieve these objectives.

Past research in *ad hoc* networks has focused on efficient routing for data traffic, particularly on finding the minimum hop path from source to destination; however, not much work has been done on the transmission of multimedia data in mobile *ad hoc* networks. Alwan et al. [6] have considered the problem of delivery of real-time multimedia traffic in an *ad hoc* network, but they have addressed this problem in a clustered architecture. Resource management for multimedia data in a fixed cellular wireless network is considered in Shafiq et al. [7], and the problem of network resource management for land-based networks is addressed in Baqai et al. [8]. Scheduling the transmission of multimedia streams over MHVCs is presented in Woo et al. [10], but their work deals only with land-based networks. Not much research has been done on the problem of scheduling multimedia data streams over MHVCs in an *ad hoc* network environment.

Multimedia Document Model

In this section, we present a multimedia document model for documents stored in database servers, including Internet-based Web servers. This model helps reserve network resources in advance and provides better control for synchronization purpose. Traffic patterns generated by pre-orchestrated multimedia data are different from the traffic variations exhibited by a monolithic multimedia object such as a variable bit rate (VBR) video stream. In the latter case, the bandwidth variations are due to interframe compression. Data streams generated by a multimedia document server, on the other hand, can have drastic bandwidth variations due to the presence of multiple objects as well as the random retrieval of documents by a user, as illustrated in Figure 47.2. Document-level variations result from the changing levels of concurrency and characteristics of the component objects within the multimedia document. Figure 47.3b illustrates the bandwidth profile of a multimedia document.

Effective capacity approximation can be used to estimate the bandwidth profile at the object level. Because objects within a multimedia document can generate diverse traffic patterns, such an approximation cannot fully characterize document-level variations. In addition, bandwidth variations at the browsing level are the most difficult to characterize, as these variations are dependent on the random browsing activity of the user. In summary, the random nature of the browsing process, the changing levels of concurrency, and the quality and time attributes of component objects within a multimedia document result in a statistically varying workload.

Various document specification models, such as eXtensible Markup Language (XML) and graphical models [4], have been proposed in the literature for specifying temporal, synchronization, and quality parameters for all objects in a document. One such emerging model for multimedia documents is the Synchronized Multimedia Integration Language (SMIL). SMIL is a collection of XML elements and attributes that can be used to describe the temporal and spatial coordinates of multimedia objects in a multimedia document. SMIL belongs to the family of XML-related standards and is used to create multimedia documents. The various features of SMIL consist of media content, layout, timing, linking, and adaptivity. The most important among these is the timing feature that is used to describe the temporal behavior of a multimedia document. A SMIL presentation uses three basic timing containers to model temporal constraints among various media objects: sequential, parallel, and exclusive. SMIL provides a logical timing framework in which the structured relationship of media objects can be used to define the timing relationships among objects. As a result, in a SMIL presentation the structured composition of media objects determines the timeline. In addition, SMIL provides a set of attributes to control the timing of media objects. These attributes are timing control, extended activation, object persistence, repeating control, synchronization, and XML timing integration [20].

Another multimedia document specification model is the Object Composition Petri-Net (OCPN) [4]. Figure 47.3a shows the OCPN specification for a multimedia document consisting of multiple objects and their temporal

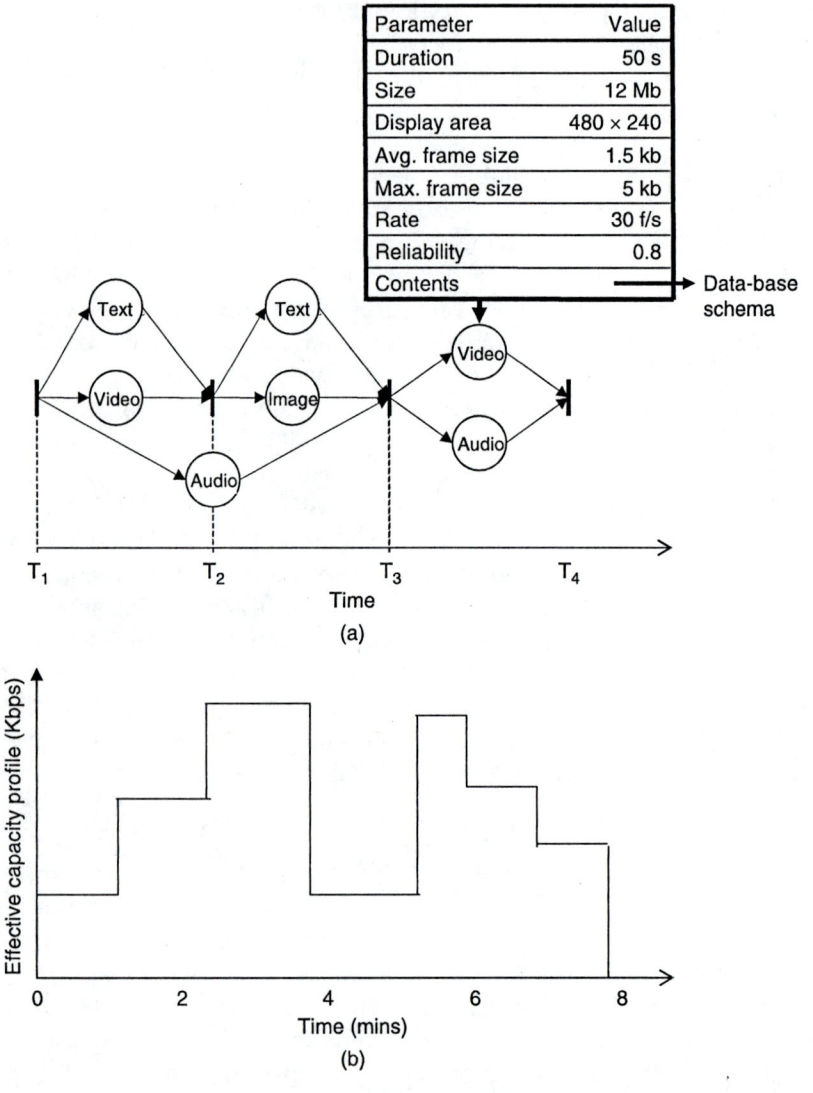

Parameter	Value
Duration	50 s
Size	12 Mb
Display area	480 × 240
Avg. frame size	1.5 kb
Max. frame size	5 kb
Rate	30 f/s
Reliability	0.8
Contents	→ Data-base schema

Figure 47.3 (a) OCPN of a multimedia document, and (b) bandwidth profile of a multimedia document.

synchronization and presentation requirements. In addition, the model allows specification of quality-of-presentation (QoP) attributes specific to a particular object, as shown in Figure 47.3a. The QoP attributes may include resolution, frame rate, reliability, and synchronization requirements. The reliability requirement specifies the maximum percentage of multimedia document that can be dropped if the allocated link capacity is limited.

Synchronization is required for isochronous objects, such as audio and video, to have meaningful presentation.

Given a document specification model, the bandwidth requirement of the component objects can be extracted by either using some effective bandwidth approximation method or specifying the peak or average bandwidth requirements [8]. The bandwidth profile shown in Figure 47.3b can be stored with the document when it is created [4] and can be provided to each node in the wireless network at the time of connection establishment. Because the presentation schedule is available *a priori*, the wireless network can efficiently allocate resources in advance by evaluating the bandwidth profile in a manner that maintains the desired QoP within acceptable bounds.

The multimedia objects must be further decomposed into smaller units for synchronization purposes. This decomposition into finer granularity results in better control of the transmission and playout of isochronous data such as video and audio. These fine-grained data units are called *synchronization interval units* (SIUs) [9]. The transmission of an object is basically the transmission of a stream of SIUs. The playout duration of an SIU is referred to as the *synchronization interval*. This is the atomic unit for the presentation process, and it depends on the type of object to which such an SIU belongs; for example, an audio object can be decomposed into SIUs where each SIU may consist of an audio sample.

The media type and QoS requirements of an SIU are the same as those of the multimedia object from which it is derived. The QoS requirements may include bandwidth and bounds on end-to-end delays and jitter for the object. From the QoS parameters, a bound on end-to-end transit delay that includes both propagation delay and jitter can be derived for each SIU. Similarly, the size of SIUs for uncompressed data can be found from the overall size of the object, its duration, and the length of the synchronization interval. For compressed data, we assume that information regarding the size of SIUs is stored with the data.

User Mobility Management

In a wireless *ad hoc* defense network, the path between source and destination nodes changes with the movement of the nodes. In such an environment, traditional QoS routing protocols cannot function properly because the established virtual channels may not be available for the entire duration of the multimedia session. To ensure continuous delivery of multimedia data to mobile users regardless of their mobility patterns, the expiry time of a particular route has to be predicted so new routes can be established before the old routes expire. Several schemes have been proposed to predict the mobility pattern of a user [13].

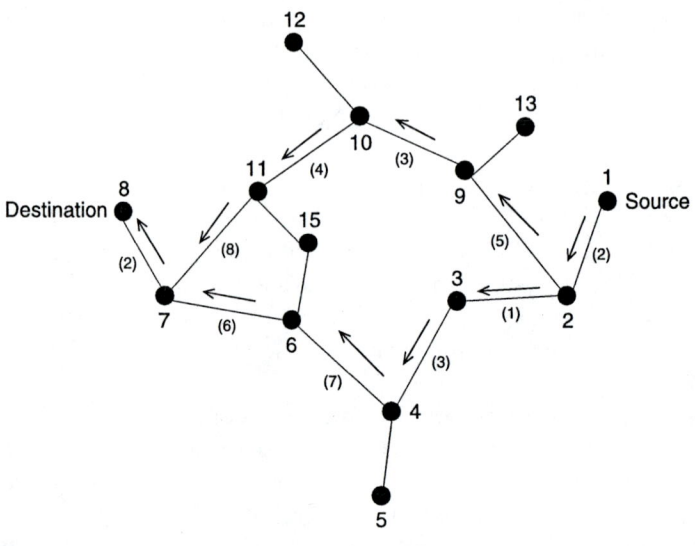

Figure 47.4 Route selection based on the values of LLT.

Due to the rapidly changing network topology in a mobile *ad hoc* defense network, link information has to be updated regularly to guarantee uninterrupted data delivery. By predicting the lifetime (the time duration for which the two nodes remain connected) of a link, alternative routes can be established well in advance between source and destination nodes to avoid any disruption in the connection. If the motion parameters (e.g., speed, direction, radio propagation range) of two neighboring nodes are known, we can determine the lifetime of the link between the two nodes [13]. We assume that all nodes are global position system (GPS) equipped and, hence, these parameters are known. Suppose two nodes i and j are within transmission range r of each other, with their coordinates given as (x_i, y_i) and (x_j, y_j). Let v_i and v_j be the speed and θ_i and θ_j ($0 \leq \theta_i, \theta_j \leq 2\pi$) be the moving directions of nodes i and j, respectively. Then, the *link lifetime* (LLT) can be predicted as follows [13].

$$\text{LLT} = \frac{-(ab+cd) + \sqrt{\left(a^2 + c^2\right)r^2 - (ad - bc)^2}}{a^2 + c^2} \qquad (47.1)$$

where $a = v_i\cos\theta_i - v_j\cos\theta_j$; $b = x_i - x_j$; $c = v_i\sin\theta_i - v_j\sin\theta_j$; and $d = v_i - v_j$. Note that when $v_i = v_j$ and $\theta_i = \theta_j$, LLT becomes ∞.

The lifetime of a route (RLT) is the minimum LLT over all the links along the route. A route selection example is shown in Figure 47.4. Two routes are available from the source node (node 1) to the destination

Table 47.1 Route Lifetime (RLT) Values of Routes *A* and *B*

	A	*B*
Path	1–2–9–10–11–7–8	1–2–3–4–6–7–8
Link lifetime	2	1

node (node 8). Route *A* consists of nodes 1–2–9–10–11–7–8, and route *B* goes along the nodes 1–2–3–4–6–7–8. The LLT values are shown in Table 47.1. The lifetime of route *A* is 2(min(2,5,3,4,8,2)), and that of route *B* is 1(min(2,1,3,7,6,2)).

Proposed Approaches for QoS Routing and Synchronization of Multimedia Data over MHVCs

In the following, we describe the QoS routing protocol and the scheduling scheme for transmitting multimedia data streams over MHVCs.

QoS Routing Protocol

Various routing protocols have been proposed for mobile *ad hoc* networks [14]. Most of these protocols minimize only the hop distance metric and do not support QoS. For multimedia traffic, the underlying network protocol has to take into consideration the QoS characteristics of different paths between source and destination nodes [12]. In this section, we propose a QoSR protocol based on mobility prediction.

Most of the existing *ad hoc* network routing protocols can be broadly classified into two categories: *proactive* and *reactive* protocols. The former are table-driven routing protocols that maintain consistent up-to-date routing information from each mobile node to every other mobile node in the network. Examples of proactive protocols include Destination Sequenced Distance Vector (DSDV) routing, Clusterhead Gateway Switch Routing (CGSR), and Wireless Routing Protocol (WRP) [14]. Reactive protocols are source-initiated, on-demand protocols that create a route only when desired by the source mobile node. Examples of reactive protocols include *Ad Hoc* On-Demand Distance Vector (AODV) routing, Dynamic Source Routing (DSR), Temporally Ordered Routing Algorithm (TORA), Associativity-Based Routing (ABR), and Signal Stability Routing (SSR) [14].

The choice between proactive and reactive protocols involves a tradeoff between latency of route discovery and route discovery/maintenance overhead. Proactive protocols have a lower latency of route discovery because routes are maintained at all times; however, these protocols have high routing overhead. In a mobile *ad hoc* defense network, resources such as bandwidth, battery power, and buffer size are limited; therefore, proactive protocols are not well suited for this scenario [15]. Reactive protocols may have a higher latency of route discovery because a route from source to destination will be found only when a source node attempts to send data to the destination node. These protocols have generally lower routing overhead than proactive protocols. Which approach achieves a better tradeoff depends on the traffic characteristics, mobility patterns, and applications. Because, in this chapter, we are considering the transmission of pre-orchestrated multimedia data, the initial session setup delay could be tolerated by delaying the playout time at the destination; however, in the case of applications such as video conferencing involving VBR video stream, the initial setup delay becomes a critical issue.

QoSR Protocol Description

As mentioned earlier, QoS requirements (bandwidth and delay) are specified by the OCPN parameters of the multimedia document. These requirements must be satisfied during data transmission to ensure QoS-guaranteed transmission. The proposed QoSR protocol takes this information into consideration and proceeds in two phases: route discovery and establishment and route maintenance. Figure 47.5 is a flow diagram of the QoSR protocol for the establishment of MHVCs in a multimedia *ad hoc* defense network. This is a distributed protocol, as its various blocks are executed by different nodes.

Route Discovery and Establishment

When a mobile user requests a document, the database server initiates a route discovery process to determine potential paths to the requesting node. It broadcasts a route request packet (RREQ) to its neighbors, which then forward the request to their neighbors, and so on, until the RREQ packets reach the destination node. Each RREQ packet carries detailed information, including the location of server and client nodes, the QoS information of the document to be transmitted, and some mobility estimates. In particular, the packet has the following fields:

- SEQ_ID, the sequence number of the RREQ packet
- Route_record, the IDs of all the nodes over which the packet has traversed

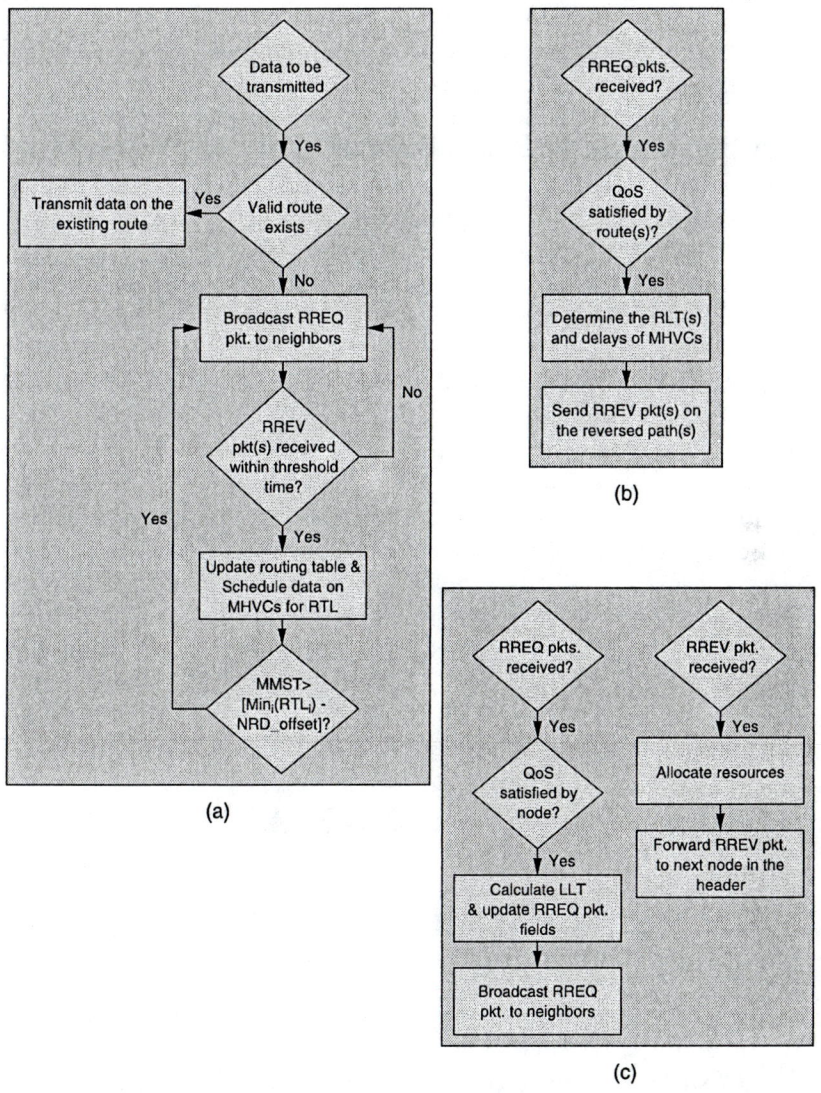

Figure 47.5 Flow diagram of QoSR protocol for (a) source node, (b) destination node, and (c) intermediate node.

- (x_i, y_i), the current coordinates of the node sending the RREQ packet
- θ_i, the direction of motion of the node with respect to some reference direction
- v_i, the velocity of the node
- SRC_ID, the source node's ID

- DEST_ID, the destination node's ID
- OCPN, which contains the QoS requirements of the multimedia document which include bandwidth and delay information
- node_info, which includes related information for the node (e.g., link delay, available link bandwidth)

Each node receiving the RREQ packet adds its own ID to the Route_record of the packet and then forwards the packet along its outgoing links. To limit the number of RREQs propagated on the outgoing links of a node, a mobile node forwards the RREQ packet only if the request has not yet been seen by the node and if the mobile's address does not already appear in the route record. When an intermediate node (a node that is neither source nor destination node) receives the RREQ packet, it performs the following procedure. If the link between the transmitting and receiving nodes cannot satisfy the QoS requirements specified by OCPN parameters, then the RREQ packet is discarded; however, if the link can satisfy these requirements and the receiving node is not the destination node, then the RREQ packet is forwarded to all of its neighbors. This forwarded packet will have its Route_record, x_i, y_i, v_i, θ_i, OCPN, and node_info fields updated. The RREQ packet is forwarded in the above manner until it reaches the destination node. Every node along the path from source to destination calculates the *link lifetime* (LLT) and the link delay and appends it to the RREQ packet. The destination node, on receiving the RREQ packet, finds the *route lifetime* (RLT), which is the minimum of all of the LLTs along the path, and the total delay of the route.

Each RREQ packet that reaches the destination has a route record consisting of the sequence of hops taken by it to reach the destination. A destination node may receive multiple copies of a RREQ packet having different Route_record entries corresponding to multiple routes between source and destination. The destination node selects MHVCs that satisfy the required bandwidth and delay information from the set of all discovered MHVCs. Destination sequence numbers may be employed to ensure all routes are loop free and contain the most recent route information. Figure 47.6a shows a typical scenario of the propagation of RREQ packets through a network.

When the destination node receives the RREQ packet, it reverses the Route_record field contained in the RREQ packet and places it into the *route reservation* (RREV) packet. Because we assume that all links are bidirectional, then by reversing the Route_record field we can obtain the path from the destination to the source node. Figure 47.6b shows the transmission of the RREV packet. The RREV packet reserves resources on each node along the path. When the source node receives the RREV

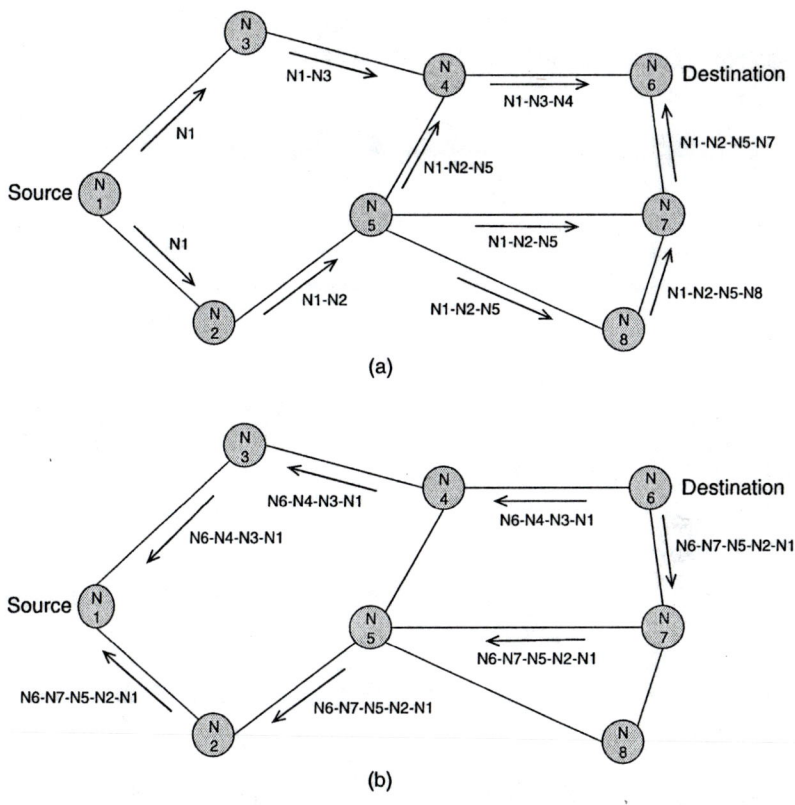

Figure 47.6 **(a) Propagation of RREQ packet along with the `Route_record` during the route discovery phase, and (b) propagation of a RREV packet with the `Route_record` along two MVCs.**

packet, then resources are reserved along the established paths that satisfy the QoS requirements. MHVCs may be established as a result of the above-mentioned procedure. When the RLTs of all of the available paths are known, the scheduler schedules multimedia streams over the established MHVCs.

We assume that every node along the path established maintains a dynamic routing table that contains information for the next hop node for the destination. When the RREV packet backtracks on the established path, it sets the next hop node for every node along its path accordingly. This way we avoid the excessive overhead of appending the route to every data packet that we send. The LLT and RLTs can also be updated periodically during the flow of data packets, and new route discovery can be initiated if a route is near expiration.

The aforementioned scheme does have some drawbacks. First, the packet header size grows with route length due to source routing; however, we only use packet header during the route discovery and establishment phase. Second, the flood of RREQs may reach all nodes in the network, but the scope of the route request flood can be reduced by using schemes such as location-aided routing (LAR) [16] and query localization [17]. Third, collisions may occur between route requests propagated between neighboring nodes, but this problem can be resolved by inserting random delays between the forwarding of RREQ packets.

When the route has been established and resources have been reserved, data packets are forwarded using routing tables at each node along the path established using the next hop information for a particular destination. The second phase of the QoSR protocol consists of route maintenance.

Route Maintenance Phase

When the source node receives the RREV packet, it checks if the route has expired. In this case, a new RREQ packet is transmitted again; otherwise, the document is transmitted over the established routes for the period of the MHVC's lifetime. If the multimedia document session time (MMST) exceeds the route lifetime of any one of the established MHVCs, then the source node reinitiates the route discovery phase. We define the new route discovery offset (NRD_offset) as the time period before the expiry of a route that is dedicated to finding new routes so data can be redirected to the new routes before the expiry of the current existing route to ensure seamless transmission of multimedia data.

Scheduling of Multimedia Documents Over MHVCs

When MHVCs have been established between a multimedia database server and mobile users, the server-based scheduler transmits multiple data streams over MHVCs in some optimal manner. In particular, the problem of scheduling transmission of multimedia data can be viewed as the problem of optimally scheduling the transmission of synchronization interval units (SIUs) while maintaining the temporal relations between SIUs to ensure a synchronized playout of the multimedia data at the destination.

As mentioned earlier, synchronization of a multimedia document has to be achieved at two levels — that is, at the inter-stream and intra-stream levels [9]. This can be achieved by delivering the SIUs to the destination before their playout deadline; however, the deadline of each SIU can only be met when the network provides a set of channels with enough bandwidth and bounded delay. Because a wireless *ad hoc* defense network is a resource-constrained system in which bandwidth and buffer capacity are severely limited, it is essential to use some efficient scheduling scheme to present multimedia documents synchronously without extensive buffering at the destination node.

In a mobile *ad hoc* defense network, a channel may not have sufficient bandwidth and bounded delay to deliver the document in a synchronized fashion. In some cases, the required number and type of MHVCs may not be available, so the document cannot be delivered in a synchronous manner. One way to address this problem is to prefetch data at the destination node and play it out later. This scheme has the drawback of initial session setup delay incurred due to buffering. This delay has to be long enough to ensure both inter-stream and intra-stream synchronization after the destination node begins presentation. Minimizing the initial startup delay becomes an important consideration in an *ad hoc* defense network environment in which a mobile user may have limited buffering capability.

We assume that the multimedia document requested by the mobile user consists of multimedia objects having a total of n SIUs. Let the set of n SIUs be denoted by $S = \{SIU_1, SIU_2, ..., SIU_n\}$. Each SIU_i has two scheduling-related parameters: its size s_i and its playout deadline d_i. Let $C = \{C_1, C_2, C_3, ..., C_m\}$ be a set of m MHVCs available during a time interval Δt, where:

$$\Delta t = \min_{1 \leq i \leq m} RLT_i$$

In our scheme, the time interval Δt defines the scheduling horizon (i.e., the set of SIUs is rescheduled for transmission after Δt time units). During this time interval, the number of channels available as well as their characteristics remain the same. It can be assumed that each channel C_j provides a guaranteed effective bandwidth rate c_j and bound on the transit delay δ_j for a time interval Δt.

Suppose SIU_i is scheduled for transmission on channel C_j at some time α_j according to some scheduling policy. Let A_i denote the arrival time of SIU_i at the destination node. Then, this arrival time is given as follows:

$$A_i = \alpha_j + \frac{s_i}{c_j} + \delta_j \qquad (47.2)$$

The transit delay δ_j of channel C_j may not be the same as the required transit delay of SIU_i. The *tardiness* of SIU_i with respect to its playout deadline is defined as $T_i = \max[0, A_i - d_i]$. If $T_i > 0$, then SIU_i misses its playout deadline, resulting in intra-stream as well as inter-stream asynchrony. One way to avoid this is to delay the start of presentation until the tardy SIUs become available at the destination node. In this case, the playout of each SIU has to be delayed by the maximum tardiness $T_{max} = \max_{1 \le i \le n}\{T_i\}$. In other words, the earliest feasible playout start time of a multimedia object is equal to $d_i + T_{max}$. The quantity T_{max} in the induced playout deadlines is the initial delay in the presentation process. This delay is the end-to-end delay in the network that is incurred if the network does not provide sufficient MHVCs with the required bandwidth.

The scheduling problem is defined as follows. Given a set of n SIUs, $S = \{SIU_1, SIU_2, \ldots, SIU_n\}$ along with their sizes and playout deadlines, and a set of m channels $C = \{C_1, C_2, C_3, \ldots, C_m\}$ with their bandwidth and transit delays and the scheduling horizon Δt, find a schedule for the n SIUs on the m channels that would result in minimizing maximum tardiness T_{max}.

The above scheduling problem can be formulated as a parallel processor scheduling problem in which m heterogeneous channels can be modeled as m uniform processors with different speeds. A set of SIUs is equivalent to a set of independent jobs. The processing time of a job i on channel j is given as (s_i/c_j). Because the server can access the desired multimedia data at any time, it can be assumed that all SIUs have the same release time, although they have different playout deadlines. Because the playout deadlines of SIUs are known *a priori*, this is a deterministic scheduling problem. We assume that an SIU is transmitted on a channel without interruption, corresponding to a non-preemptive environment. The transit delay δt associated with each channel distinguishes this problem from other processor scheduling problems. This transit delay is independent of the SIU being transmitted; hence, using the notation of Lageweg et al. [19], the scheduling problem may be represented as $Qm/\delta_j/T_{max}$. Here, Qm represents m heterogeneous channels, δ_j represents the presence of transit delays, and T_{max} is the performance measure of max tardiness.

A special case of $Qm/\delta_j/T_{max}$ is scheduling n jobs on m parallel identical processors to minimize max tardiness ($Pm//T_{max}$), which is an NP-hard problem. Therefore, the scheduling problem $Qm/\delta_j/T_{max}$ is also NP hard [10].

Different heuristic algorithms have been proposed for the above-mentioned scheduling problem [9]. These heuristics involve a tradeoff between SIU deadline misses and buffer overflow. One of the heuristic algorithms in Baqai et al. [9] results in reduced buffer usage while minimizing the maximum tardiness. This heuristic algorithm given in Figure 47.7 is more suited for a mobile *ad hoc* defense network environment because of the limited buffer capacity at mobile nodes. This algorithm first orders SIUs in increasing order of their playout deadlines, then a

1. Sort SIUs in increasing order of their playout deadlines. If two SIUs have same deadline then the SIU with larger size precedes the smaller one.

2. Initialize $L_j = \phi$ and $\alpha_j = 0$ for all channels, $1 \le j \le m$; L_j represent the schedule for channel C_j.

3. Start from the head of the list and schedule each SIU $\in$ S closest to its deadline on the channel that results in ECT.
$$L_j = L_j \cup \{SIU_i\}$$
where $j = \arg\min_{1 \le k \le m} \left\{ \max[\alpha_k, d_i] + \frac{s_i}{c_k} + \delta_k \right\}$

Initial scheduling time for SIU_i, $S_i' = \max\{\alpha_j, d_i\}$, and the expected arrival time
$$A_i' = \max\{\alpha_j, d_i\} + \frac{s_i}{c_j} + \delta_j$$

4. Update α_j as $\alpha_j = \max\{\alpha_j, d_i\} + \frac{s_i}{c_j}$.

5. Repeat steps 1-4 until all SIUs in S are processed.

6. T_{max} is then calculated as $\max_{1 \le k \le n}[0, A_i' - d_i]$.

7. Set the scheduling time for each SIU_i as $S_i = S_i' - T_{max}$. The arrival time
$$A_i' = S_i' - T_{max} + \frac{s_i}{c_j} + \delta_j$$

Figure 47.7   Heuristic scheduling algorithm for minimizing maximum tardiness.

---

tentative schedule is constructed by scheduling an SIU closest to its deadlines on a channel that results in the earliest completion time (ECT); that is, the one that minimizes its arrival time at the destination. The maximum tardiness T_{max} among all SIUs is calculated and is subtracted from the tentative schedule to obtain the final schedule.

The overall QoS framework presented in this chapter is depicted in Figure 47.8. This framework consists of various QoS-related functional components that interact with one another to ensure QoS-guaranteed transmission of multimedia data in a wireless mobile *ad hoc* defense network.

Conclusion

In this chapter, we presented a QoS framework for the transmission of multimedia data in a battlefield wireless mobile *ad hoc* defense network for supporting mission-critical applications over large geographical areas. This framework consists of a QoS routing component that establishes multiple virtual channels between a multimedia database server and mobile user that satisfy user-specified QoS requirements of a given session. Heterogeneity may exist in these virtual channels as a result of differences

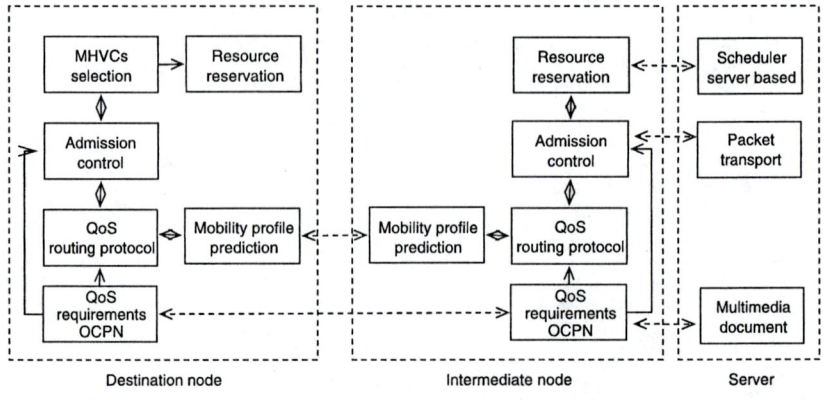

Figure 47.8 Functional components and information flow in QoS framework

in the available bandwidth and end-to-end delay of virtual channels due to a high degree of mobility. We have formulated the multimedia data scheduling problem over multiple heterogeneous virtual channels as a uniform processor scheduling problem. We also discussed a heuristic algorithm to minimize the maximum tardiness of the multimedia document. The proposed technique can be used for developing viable mobile multimedia applications for defense networks.

References

[1] Tiboni, F., Battlefield communications, *Federal Commun. Week (FCW)*, March 14, 2005 (http://www.fcw.com/article88262).

[2] Jameson, S.M., Architectures for distributed information fusion to support situation awareness on the digital battlefield, in *Proc. of the Fourth Int. Conf. on Data Fusion*, Montreal, Canada, August 7–10, 2001 (http://www.atl.external.lmco.com/overview/papers/1030.pdf).

[3] Perkins, C.E., *Ad Hoc Networking*, Addison-Wesley, Boston, MA, 2000.

[4] Little, T.D.C. and Ghafoor, A., Multimedia synchronization protocols for broadband integrated services, *IEEE J. Selected Areas Commun. (JSAC)*, 9, 1368–1382, 1991.

[5] Lin, C.R. and Gerla, M., Adaptive clustering for mobile wireless networks, *IEEE J. Selected Areas Commun. (JSAC)*, 15(7), 1265–1275, 1997.

[6] Alwan, A. et al., Adaptive mobile multimedia networks, *IEEE Pers. Commun. Mag.*, 3(2), 34–51, 1996.

[7] Shafiq, B., Ghafoor, A., Baqai, S., Fahmi, H., and Khokhar, A., Wireless network resource management for Web-based multimedia document services, *IEEE Commun. Mag.*, 41(3), 138–145, 2003.

[8]   Baqai, S., Woo, M., and Ghafoor, A., Network resource management for enterprise-wide multimedia services, *IEEE Commun. Mag.*, 34, 78–85, 1996.

[9]   Baqai, S., Khan, M.F., Woo, M., Shinkai, S., Khokhar, A.A., and Ghafoor, A., Quality-based evaluation of multimedia synchronization protocols for distributed multimedia information systems, *IEEE J. Selected Areas Commun. (JSAC)*, 14(7), 1388–1403, 1996.

[10] Woo, M., Uzsoy, R., and Ghafoor, A., Media streams scheduling for synchronization in distributed multimedia systems, *J. Parallel Distributed Comput.*, 53(3), 272–295, 1999.

[11] Woo, M., Qazi, N., and Ghafoor, A., A synchronization framework for communication of pre-orchestrated multimedia information, *IEEE Network*, 8, 52–61, 1994.

[12] Shah, S.H. and Nahrstedt, K., Predictive location-based QoS routing in mobile *ad hoc* networks, in *Proc. of IEEE Int. Conf. on Communications (ICC'02)*, New York City, April 28–May 2, 2002, pp. 1022–1027

[13] Su, W., Lee, S.-J., and Gerla, M., Mobility prediction in wireless networks, in *Proc. of Military Communications Conf. (MILCOM 2000)*, Los Angeles, October, 2000.

[14] Royer, E. and Toh, C.-K., A review of current routing protocols for *ad hoc* mobile networks, *IEEE Pers. Commun. Mag.*, 6(2), 46–55, 1999.

[15] Kim, Y.-W., Ryu, J.-H., and Cho, D.-H., A novel adaptive routing scheme for the QoS-based multimedia services in mobile *ad hoc* networks, in *Proc. IEEE Vehicular Technology Conf. (VTC'99)*, Houston, TX, May, 1999, pp. 396–400.

[16] Ko, Y. and Vaidya, N., Location-aided routing in mobile *ad hoc* networks, in *Proc. of the 4th ACM/IEEE Int. Conf. on Mobile Computing and Networking (MOBICOM'98)*, Dallas, TX, October, 1998.

[17] Castaneda, R. and Das, S.R., Query localization techniques for on-demand routing protocols in *ad hoc* networks, in *Proc. of the 5th ACM/IEEE Int. Conf. on Mobile Computing and Networking (MOBICOM'99)*, Seattle, WA, August, 1999

[18] Little, T.D.C. and Ghafoor, A., Synchronization and storage models for multimedia objects, *IEEE J. Selected Areas Commun. (JSAC)*, 8, 413–427, 1990.

[19] Lagewag, B.J., Lenstra, J.K., and Rinnooy Kan, A.H.G., *Computer-Aided Complexity Classification of Deterministic Scheduling Problems*, Research Report No. BW 138/81, Mathematisch Centrum, Amsterdam, 1981.

[20] Bulterman, D.C.A., SMIL 2.0, part 1: overview, concepts, and structure, *IEEE Multimedia*, 8, 82–88, 2001.

Chapter 48

Mobile Middleware for Rescue and Emergency Scenarios

Ellen Munthe-Kaas, Ovidiu Drugan,
Vera Goebel, Thomas Plagemann, Matija Puzar,
Norun Sanderson, and Katrine S. Skjelsvik

CONTENTS

Introduction

Efficient collaboration among rescue personnel from various organizations is a mission-critical key element for a successful operation in emergency and rescue situations. The two central requirements for efficient collaboration are the incentive to collaborate, which is a given situation for rescue personnel, and the ability to efficiently communicate and share information. Mobile *ad hoc* networks (MANETs) could provide the technical platform for efficient information sharing in such scenarios, assuming that all rescue personnel are carrying and using mobile computing devices with wireless network interfaces. Appropriate applications are needed to turn a working infrastructure of a MANET into a useful system; however, application development for MANETs is far from easy. MANETs are typically very dynamic networks in terms of available communication partners, available network resources, connectivity, etc. Furthermore, the end-user devices are very heterogeneous, ranging from high-end laptops to low-end PDAs and mobile phones. CPU storage space, bandwidth, and battery power represent important resources. The diversity of organizational affiliations of end users and their roles within their respective organizations also introduces heterogeneity to the system. Finally, many application scenarios, such as coordination of rescue teams, also have quite hard nonfunctional requirements, such as availability (including reliability, fault tolerance, and survivability), dependability, efficient resource utilization, integrity, security, and privacy. Thus, providing sufficient quality in information access and sharing in such an environment faces many obstacles. Obviously, solving these issues in every new MANET application from scratch is not meaningful; instead, a set of middleware services that support the development of applications for MANETs is needed.

Application Scenario and Requirements

Rescue Scenarios

Rescue scenarios typically involve rescue personnel from various organizations, such as policemen, firemen, physicians, and paramedics, who must collaborate across organizational boundaries. This forces a variety of heterogeneity upon the network, and the middleware must therefore find a way to present the information so all organizations can understand it. This implies supporting functionality akin to high-level distributed database system functionality, keeping track of what information is available in the network and supporting querying of available information. The participating rescue organizations may use different domain ontologies, standards, etc., so a major challenge for knowledge management is to support such information sharing across organizations.

Each organization has its own portfolio of procedures, tailored to fit the nature and size of the rescue operation. The rescue procedures, among other things, cover the command structure in force. Cross-organizational procedures comprise governmental and other authorities and include a cross-organizational command structure. Thus, the middleware should support both intra- and interorganizational structures. An obvious benefit would be to have contextual support, in that contexts can be used for reflecting specific rescue procedures in force; for supporting user role and device profiling and personalization; for providing temporal and spatial information, movement patterns, etc.

Another concern is that of security, integrity, and privacy. The need for security services that span across the organizations involved is obvious. Depending on the nature of the incident, the rescue leaders can also involve voluntary personnel from idealistic and private organizations. Such parties might depend on being able to partake in some of the information exchange, but it is unlikely that an *a priori* understanding would have been reached with the authorities and core organizations with regard to security credentials, data formats, and the like. Even for non-volunteer personnel, the security precautions should still be fairly flexible. Different organizations may have conflicting security policies; in such cases, privileged users could supply input on how to resolve the conflict. Also, if security slows things down or conflicts with issues of higher priority, it must be possible to relax the security precautions.

In such a scenario, a multitude of devices are brought into the area by rescue personnel. There may also be elements of a fixed infrastructure (e.g., sensors or devices) already present on the rescue site. Some of these devices might serve as a gateway between the MANET and the Internet, provided the necessary security requirements are fulfilled; however, we cannot assume or rely on a fixed backbone of devices running the middleware services in a MANET. When a train accident takes place inside a tunnel, one cannot expect radio, mobile phone, or global position system (GPS) coverage. In such cases, the services will probably have to run on mobile devices that might lose contact due to network partitioning resulting from, for example, topological obstacles. Because the middleware should provide basic services to support a mission-critical task (i.e., information sharing), it has to be designed in such a way that it is of high availability. In the case of network partitioning, the services should support information sharing in the different network partitions, and it should work as well as possible if arbitrary devices are switched off, including nodes running the middleware services; therefore, a centralized solution is not possible, and a distributed solution has to provide a sufficient degree of redundancy through replication. The dynamic nature of MANETs means that middleware services based on synchronous communication are not a good choice, because they are too vulnerable with respect to communication disruptions. The alternative is systems based on message passing or event-based systems; thus, middleware support for a distributed event notification system is required.

The number of devices expected to be part of a MANET in emergency and rescue applications probably would not exceed several hundred, but small sensors might increase the number of devices up to tens of thousands. Even though this covers a spectrum of several magnitudes, it is still significantly smaller than Internet scale. These devices are typically of a heterogeneous nature also with respect to available memory and disk space. It is necessary, therefore, to design the middleware services to be configurable such that small, resource-weak devices run only a minimal set of protocols or subsets or simplified, lightweight versions of the services contained in the protocols; devices with sufficient resources might implement full-scale services. Furthermore, it is necessary to keep track of which resources are available so other devices can be used as proxies or to make meaningful decisions on where to place replicas to increase the system availability.

Performance and efficient resource utilization are also important, but typically a tradeoff exists between these two requirements and availability. There is no general solution for this tradeoff, and its resolution often depends on the particular application and even on the particular emergency situation; therefore, it is necessary to allow the application to define policies on how to handle these tradeoffs.

Even though the collection of possible rescue scenarios is diverse, with respect to the size and type of incident, the personnel skills required, the rescue procedures to be followed, etc., it is still possible to extract out some commonalities. We have identified six different phases in such a scenario:

- *Phase 1. A priori* — Before the incident, the different organizations in cooperation with the authorities will exchange information on data format and make agreements on procedures and working methods.
- *Phase 2. Briefing* — The first step after an incident has occurred involves gathering information about the disaster. Some preliminary decisions about rescue procedures and working methods are made, according to the nature of the incident.

- *Phase 3. Bootstrapping the network* — This phase takes place at the rescue site. Events such as registrations of nodes and appointing rescue leaders take place.
- *Phase 4. Running the network* — Different events may happen that will affect the middleware services: A node may join or leave the network, the network may be partitioned, and network partitions may be merged again. Information is collected, exchanged, and distributed. In the earlier stages of this phase, the role as rescue leader may be transferred from one person to another. New organizations or personnel groups may appear at the site. Interorganizational *ad hoc* personnel groups may form.
- *Phase 5. Closing the network* — At the end of the rescue operation, all services must be terminated.
- *Phase 6. Postprocessing* — After the rescue operation, it might be useful to analyze resource use, user movements, and how and what type of information was shared to gain knowledge for future situations.

Notice that, although one cannot rely on fixed networks or Internet access during a rescue operation (phases 3 to 5), this is not the case during the initial phases (phases 1 to 2). This means that a lack of networks and fixed resources during the rescue operation can be compensated for by careful preparation in the initial phases.

Requirements

The challenges presented are reflected in a number of requirements for the middleware: We need support for intra- and interorganizational information flow and knowledge exchange, as well as the means to announce and discover information sources. Contextual support enables applications to adapt better to particular scenarios and allows them to fine-tune according to spatial and temporal data. Profiling and personalization can assist in filtering and presenting information in accordance with the needs of the users and devices, as well as displaying their capabilities. The middleware should provide support for an organizational structure and for the creation of groups on the fly. Security must be dynamic, enabling privileged users to grant group memberships at the rescue site, as well as to influence changes to the security regime when circumstances demand it. Communication must be available, reliable, and efficient even in the presence of frequent network partitioning. Extensive support for resource sharing between devices is necessary, including ways to register and discover available resources of different types. To allow graceful degradation, the middleware must monitor and prepare for it.

This leaves us with nine articulated requirements and goals for the middleware: (1) intra- and interorganizational information flow, (2) service availability, (3) context management, (4) profiling and personalization, (5) group and organizational support, (6) dynamic security, (7) communication, (8) resource sharing, and (9) graceful degradation. In addition, a tenth requirement is the ever-present need for (10) data sharing and storage.

Applications

Applications range from being special-purpose, organization-internal to cross-organization and general-purpose, from ones that push data to ones that collect data to be pulled at need, from low-priority and low-resource applications to applications that multicast high-priority data to a selected group. Some concrete examples are monitoring sensors (e.g., tracking heartbeats, temperature, oxygen flow, or position and movement of casualties); dispatching and coordinating rescue personnel and equipment (e.g., by pushing information from a team leader to the rest of his group or from the rescue leader to all team leaders across organizations); providing access to stored data, documents, and multimedia material (e.g., rescue procedures, passenger lists, freight lists, explosives handling procedures, maps, building plans); publishing multimedia material (e.g., sharing of medical information, surveillance camera output, VoIP); and collecting evidence (e.g., logging facts for forensics reports). From the point of view of an application, the phases contain the following activities:

- *Phase 1. A priori* — Required certificates are installed. The application is installed and run, allowing it to complete an initial self-configuration phase. The middleware prepares an application-tailored communication and knowledge environment. Contexts reflecting different scenarios are prepared. Based on user profiles, group memberships are set up and data replication strategies are chosen.
- *Phase 2. Briefing* — It is possible to configure and prepare the application further, based on information collected during the briefing phase. The relevant rescue contexts and profiles are put in force. Security levels are chosen.
- *Phase 3. Bootstrapping the network* — The middleware enriches the application's working environment by preparing communication, taking care of security restrictions in force.
- *Phase 4. Running the network* — The application can now communicate, using whatever knowledge is provided by the middleware about available resources and capabilities of the nodes in the

network. It can update to changes to the resource landscape as the network is evolving, query for more data or information as it becomes available, and adjust its configuration and behavior accordingly. Computing resources, processing environments, and applications of neighbors can be utilized by using resource information provided by the middleware and obeying accepted policies for resource sharing. Replicas and proxies can be placed at strategic nodes in the network. Interorganizational groups can be formed on an at-need basis. The nodes in the network will receive event notifications based on relevance and priority. As nodes join and leave the network, the middleware keeps track of the available resources and adjusts its communication and knowledge environment accordingly.

- *Phase 5. Closing the network* — The application adapts to closing of the network by acting on received information about degradation of the capabilities and resources of the network.
- *Phase 6. Postprocessing* — Depending on the nature of the application, it may have gathered statistical or other information for later scenario analysis or future use.

Middleware Framework and State of the Art

We address our ten requirements by identifying six middleware concerns that together constitute a foundation for a middleware framework covering the required services for MANETs in rescue scenarios (see Figure 48.1):

- *Knowledge management* — to handle ontologies and support metadata integration and interpretation
- *Context management* — to manage context models, context sharing, profiling, and personalization
- *Data management* — to provide distributed database-like capabilities
- *Communication infrastructure* — to support distributed event notification, publish and subscribe services, and message mediation
- *Resource management* — to register and discover information sources and Web services as well as the resources available to handle neighbor awareness, computation and application sharing, mobile agents, proxy and replica placement, and movement prediction
- *Security management* — to provide access control, signing and encryption of messages, support of group and organizational structures, assignment of group keys, and dynamic security services

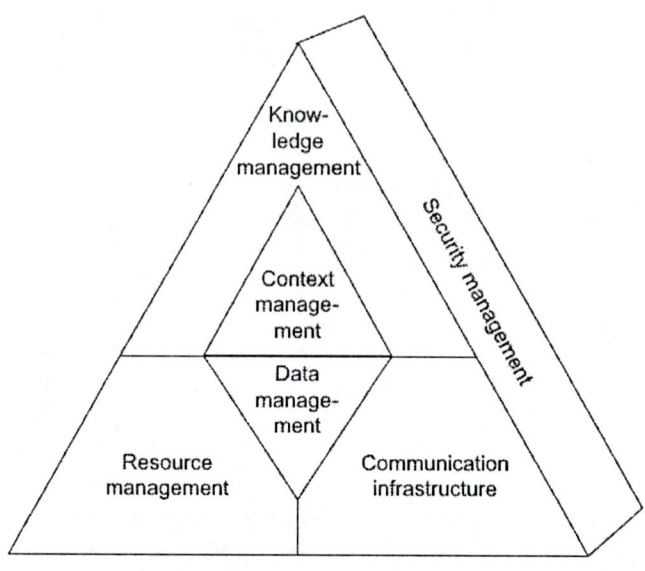

Figure 48.1 The six middleware concerns.

An in-depth detailing of all aspects of the six concerns is outside the scope of this chapter, so data management is primarily elaborated upon when necessary to explain its boundaries to the other concerns. For each of the other concerns, we provide a discussion on the main requirements that any middleware framework for the concern should address, a selection of issues that must be faced, problems that must be solved, the state of the art, and approaches to solving some of the problems. As will become evident, the middleware frameworks all rely heavily on each other's services.

Knowledge Management and Context Management

Knowledge management covers support for the dissemination, sharing, and interpretation of ontologies, as well as browsing and querying of ontologies and ontology contents. Thus, a distributed knowledge base functionality and a global view of what knowledge is available in the network are necessary. The requirements indicate a set of issues that must be addressed: *understanding* across domains and organizations through the use of knowledge management techniques; avoiding *information overflow* through content filtering and personalization; managing the *availability* of information, metadata, and ontologies; offering information *query and retrieval* services; and supporting *information exchange*. The issues translate into the following set of subconcerns: a *semantic metadata and*

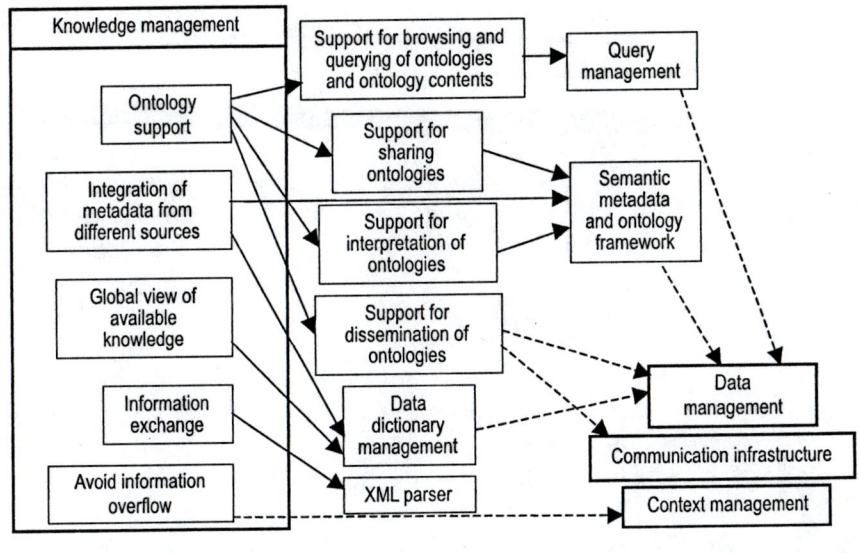

Figure 48.2 Knowledge management issues.

ontology framework for handling the sharing and interpretation of ontologies, *data dictionary management* for the management of metadata in local and global data dictionaries, *query management* for querying ontology and metadata as well as for retrieval of relevant information, and *XML parsing* for information exchange. In addition, the *Context Management Framework* supports filtering and personalization. An overview of subconcerns and issues for the *Knowledge Management Framework* is provided in Figure 48.2.

Knowledge Management and Context Management Subconcerns

To enable sharing of information and knowledge in a heterogeneous environment, the different systems and domains have to understand each other; that is, a mapping or translation of the different conceptual models, schemas, and languages that are used is necessary. This is a well-known issue in the distributed databases domain. In general, the three kinds of heterogeneity [1] are *syntactic*, which involves the data formats and languages used and standardization of these (e.g., XML); *structural*, which concerns data structures and schemas; and *semantic*, which is the meaning of the information. It can be said that ontologies provide a metaperspective on data and information by considering a higher level of abstraction. In addition to providing domain-specific descriptions of information, ontologies may be used to solve problems of semantic heterogeneity [2]. The semantic metadata and ontology framework should give certain recommendations as to what kind of ontology and modeling languages are most appropriate to use, but it may have to support more than one language. Requirements for ontology languages address issues such as expressiveness, completeness, correctness, and efficiency, as well as interoperability with relevant standards [3]. For knowledge exchange, protocols such as the Knowledge Query and Manipulation Language (KQML) and the Key Encryption Protocol (KEP) [4] are necessary.

With regard to the dissemination of ontologies, the availability of information (here, metadata and ontologies) implies a need for the management of data dictionaries that can provide a global view of knowledge available in the network at any one time. Three kinds of metadata originate from different concerns: *semantic metadata*, which describes the meaning of information (which includes to some extent ontologies); *context and profile metadata*, which describes current contexts and profiles of users and devices; and *structure and content describing metadata*, which has to do with how information items are structured and descriptions of the intellectual content of these. Both context management and knowledge management are producers and consumers of the metadata stored and thus build on the services provided by data management. The main challenges for data dictionary management are related to the availability of partial information (e.g., for retrieval and traversal of knowledge in the network, update propagation, and fault tolerance). Solutions will depend largely on issues of data distribution and replication.

The structural heterogeneity inherent in cross-organizational information sharing may require support for different approaches to querying and retrieval (e.g., by structure, content, context, and naming) and possibly support for both structured and unstructured query languages. Filtering and ranking retrieved results are necessary. Due to the need for profiles and context for filtering and personalization, query management concerns border very closely to context management concerns.

With the advent of XML as a *de facto* standard for information exchange in and among rescue organizations [5–8], parsing of XML documents is a necessity, which implies a need for lightweight XML parsers that can function well on mobile devices. Depending on the ontology languages used, Resource Description Framework (RDF) parsing may also be needed to facilitate understanding. XML parsing is typically expensive with regard to processing and memory which presents a challenge given the scarcity of resources for mobile devices.

The term *context* is used in many different ways in the literature [9–12]. Profiles viewed as contexts give information regarding the "what" and "who" of an entity or person and are fairly static. Spatiotemporal and

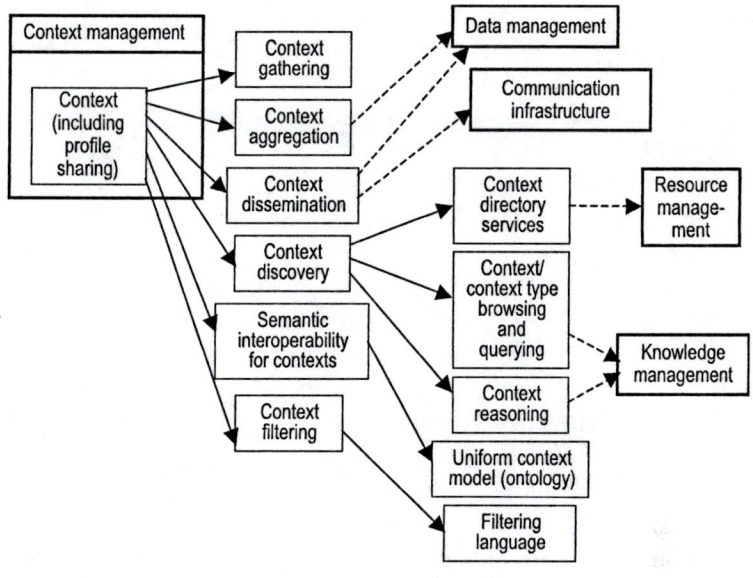

Figure 48.3   Context management issues.

situation context information is dynamic and concerns the "where," "when," and "why" of an entity. Such contexts may have a short lifespan and may not be subject to any persistency actions. Another frequent use of the term is to denote the domain knowledge and background situation where an information object gets semantic meaning. The Context Management Framework must be flexible enough to allow a wide range of interpretations. It should offer a uniform context model that facilitates context sharing and semantic interoperability and is flexible enough to accept user-defined models. The context manager should provide the means by which to collect context data and transform it to fit the context model, thus providing a sharable, semantics-based format, as well as filtering facilities. An issue is whether and how this can be achieved in light of possible resource restrictions. To support context aggregation, discovery, query, reasoning, and dissemination, the context manager can build on solutions from the other frameworks. See Figure 48.3 for a representation of Context Management Framework concerns.

State of the Art

Approaches for solving syntactic and structural heterogeneity have been addressed in standardization work [5,6,13] and in the distributed database domain. For semantic heterogeneity, approaches for information and ontology integration can be found in Stuckenschmidt and van Harmelen [1] and Wache et al. [2]. Ontology-based solutions for information sharing and Semantic Web are described in Davies et al. [14] and Fensel et al. [15]. Various XML parsing paradigms are presented in References 16 and 17.

In the Shark approach [18,19], topic-based knowledge ports are used to handle knowledge management tasks. The knowledge ports are defined as topic types and declare topics for knowledge exchange. The (mobile) users form groups, and knowledge is shared both within a group and across group boundaries (i.e., both intra- and intergroup knowledge exchange). A drawback is that the architecture relies on stationary server nodes for knowledge and synchronization management. MoGATU [10] is relevant for knowledge and context management with regard to the use of profiles and ontologies for filtering and prioritization of data. For handling metadata, information managers functioning as local metadata repositories are used, allowing for semantic-based caching. One common language, DARPA Agent Markup Language (DAML), is used for metadata representation. This solution does not offer a global view of knowledge available in the network.

Relevant for data management and context management is DBGlobe [11,20], which is a service-oriented and data-centric approach. It relies on fixed network servers to keep track of the movement of mobile units and store profile and metadata describing each mobile device, including context and which resources are offered. The devices form data-sharing communities that together make up an *ad hoc* database of the collection of data on devices that exist around a specific context (e.g., location or user). For service location and query routing, distributed indexes based on Bloom filters are used.

The reliance on a fixed network is a shortcoming with regard to emergency and rescue scenarios. AmbientDB [21,22] adds high-level data management functionalities to the distributed middleware layer by providing a global database abstraction over a MANET. This is a noncentralized, *ad hoc*/dynamic approach using distributed hash tables (DHTs) for indexing and structured queries for querying. It constitutes a full-fledged distributed database system but does not support the use of ontologies or methods from knowledge management.

A survey of different context models and context-aware systems as well as a reference architecture for context-aware systems can be found in Anagnostopoulos et al. [23]. The Cabot project [24] advocates the importance of supporting context processing in the middleware. Both Cabot and WASP [25] are architectures that include context models, ontologies, and subscription languages. None of these addresses the challenges of MANETs.

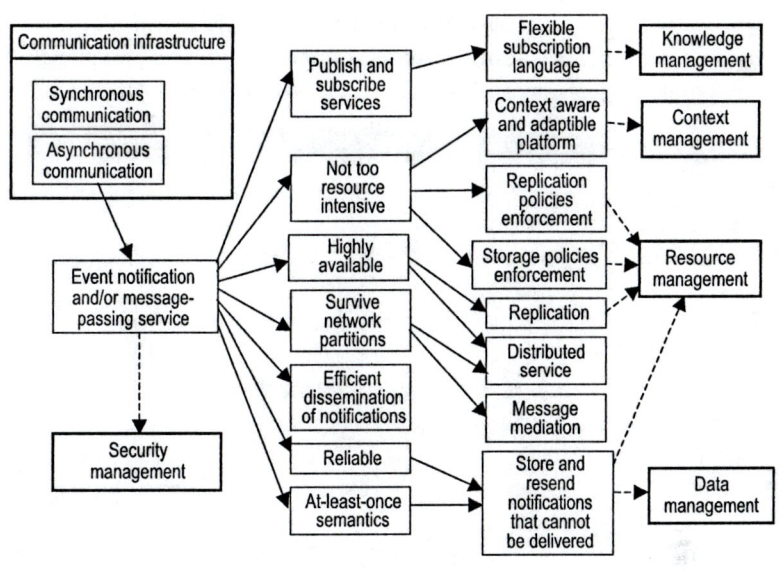

Figure 48.4   Communication infrastructure issues.

Communication Infrastructure

Although synchronous communication has its uses in the application domain, communication cannot be based on synchronism alone due to the dynamic nature of MANETs. Synchronous communication such as remote procedure calls, Remote Method Invocation, etc., blocks the caller until a response is returned, which creates problems when communication disruptions are frequent. The alternative is message-passing or event-based systems. The availability of new important information, new resources, network partitioning, and merging are all events; thus, middleware support for a distributed event notification system is necessary. In mediation-based publish–subscribe communication, the sender (*publisher*) and receiver (*subscriber*) are decoupled in time, space, and synchronization; thus, they do not have to be active at the same time, they do not need to know each other's addresses or identities, and senders are not blocked during communication [26]. This loosely coupled cooperation fits the nature of MANETs. Consult Figure 48.4 for an overview of communication infrastructure issues.

Communication Infrastructure Concerns

Event-based communication and publish–subscribe services for our scenario must be designed to handle situations where resources are scarce. Still, the services provided should be highly available and work in the presence of frequent network partitioning and merging. This leads us to solutions providing redundancy through replication and where communication services are run on a dynamic, distributed set of nodes acting as mediators, collecting subscriptions, and efficiently disseminating notifications. When electing suitable mediator nodes, the Resource Management Framework might be consulted, both for evaluating the resource assets of a node and for predicting its continuous availability in the near future.

In emergency and rescue applications, information could be a life saver and must therefore not be lost; however, in MANETs, it is not possible to guarantee an *at-least-once* message delivery semantics because of network partitionings that become permanent. Because disconnection is the rule rather than the exception, handling network partitioning and merging is vital. A distinction can be made between notifications having information that is valid only around the time of an event and information that could also be useful in the future. A delivery semantics that comes as close as possible to at-least-once should be implemented for notifications of the latter category. One way of doing this is to store notifications that cannot be delivered (e.g., because a subscriber resides in a network partition other than the publisher when an event occurs) and to resend notifications when networks merge. Also, the publisher or a mediator carrying the message may be subject to network partitioning; thus, the storage management regime is crucial. Because of limited storage space, a storage management policy is required that determines which notifications should be stored, where, for how long, and, if buffer overflow occurs, which ones to delete. To have a flexible solution and to use the possibly scarce resources wisely, the service should be adaptable to the situation at hand. This can be achieved through context awareness; for example, contexts can be used to decide the degree of replication or choice of storage policy. Applications can influence policies through the provisioning of context information.

To tailor message delivery to the needs of the subscriber, publish–subscribe systems offer subscription languages for filtering messages. The most common way to group subscription languages is by *channel, subject,* or *content.* In content-based systems, the subscriber may filter notifications based on the content of the notification and not just the subject; thus, such languages are probably the most suitable for our scenario. To reduce traffic, the framework might support subscriber-initiated filtering on or close to the publisher node. Security issues must also be tackled, such as making sure that a node cannot subscribe to information for which it is not authenticated.

State of the Art

Event services are used for both small-scale centralized applications and large-scale distributed systems. They differ in the structure of the events that can be dispatched, the way events are observed, the mechanisms for event subscription, and their overall runtime architecture [27]. Publish–subscribe systems are often categorized by the type of network; subscription language; where filtering of notifications are done; whether mediators are used and, if so, how they are organized; how and where subscriptions are stored; and how notifications are disseminated. Many of the first publish–subscribe systems were tailored for static, non-mobile environments with stationary publishers and subscribers and fixed communication paths. Some of these systems have later been adapted to manage mobile clients, but they usually assume a fixed infrastructure for the nodes acting as mediators for the publishers and subscribers (i.e., the nodes running the event service).

Siena [28] is a content-based event notification system that is tailored for a wide area network. Support for mobile clients is provided, but it must be handled explicitly by the clients. In JEDI [29], the clients choose one event dispatcher to connect to, and if a client moves or disconnects it may invoke moveOut and moveIn operations at dispatchers. Cugola and Nitto [30] describe a way to adapt the routing of notifications when a client reconnects to another dispatcher. REBECA [31] uses virtual clients and presubscriptions to manage mobility. Each client has several virtual clients, but it is only connected to one at a time, the other virtual clients will buffer received notifications. Virtual clients are started on every mediator to which the client may connect in the near future, using mobility patterns to predict movements [32]. The REBECA system also uses histories to provide access to past notifications. TIB/Rendezvous [33] retransmits a notification after 60 seconds to implement reliable message delivery, but, in cases of a device being disconnected, 60 seconds will often not be enough. Elvin [34] uses a caching proxy to manage disconnected clients; the proxies act as normal clients to the server but as a proxy server to the clients. A permanent connection to the Elvin server is maintained, so a disconnected client will receive its notifications on reconnection. Pronto [35] is a middle-ware system for mobile applications using messaging in infrastructure-based mobile networks based on the Java Message Service (JMS). JMS provides asynchronous communication between distributed components. The service is topic based and uses the Java Naming and Directory Interface (JNDI) for topic directory service.

An *ad hoc* setting where even the mediators running the service may get disconnected introduces challenges on how to deliver notifications. Huang and Garcia-Molina [36] provide an overview of different architectures, both centralized and distributed. STEAM [37] is an event-based middleware service tailored for *ad hoc* networks, where the middleware is fully distributed over all machines. A publisher will send notifications directly to its subscribers in the proximity using a multicast protocol; when a subscriber enters the area, it may join the group and receive and filter notifications. The work is based on the assumption that the closer subscribers are located to a publisher, the more likely they are to be interested in its events. This assumption may not be valid in a rescue operation where rescue personnel in charge may be interested in events happening in the entire area.

The Epidemic Messaging Middleware for *Ad Hoc* Networks (EMMA) [38] is based on JMS adapted for MANETs. If a subscriber is not reachable and the subscription is durable, they propose using an asynchronous epidemic routing protocol. Acknowledgment messages are sent to the senders to inform them about successful delivery and to delete possible replicated messages still in the network. In Vollset et al. [39], notifications are disseminated using a multicast routing protocol that maps JMS topics to multicast addresses, thus being subject based. Cugola et al. [40] presented an algorithm for managing changes in the topology of mobile event dispatchers for JEDI. It was developed to manage link breaks in a tree of dispatching servers; however, it does not consider network partitions, only partitioned subtrees where a new route merging the trees may be found. Skjelsvik et al. [41] described the design of a content-based distributed event notification service tailored for MANETs in a rescue operation. All nodes may be mobile, even the nodes running the service and acting as mediators. Monitoring agents are installed on the publisher to perform subscription-based filtering of events. If a notification is not delivered due to network partitioning, it will be replicated among the mediator nodes and stored for later delivery.

Resource Management

The main task of the Resource Management Framework is to promote sharing of all kinds of resources among the devices involved in the network, which means that it must gather and disseminate information about available resources and facilitate resource access and sharing. Utilizing remote access to resources can improve the availability of information and services and promote graceful degradation behavior. Figure 48.5 provides an overview.

Subconcerns for Resource Management

For distributed applications running in a network of many devices with limited resources, it is imperative to make good use of all available resources. Given the strong incitement for cooperation across organizations,

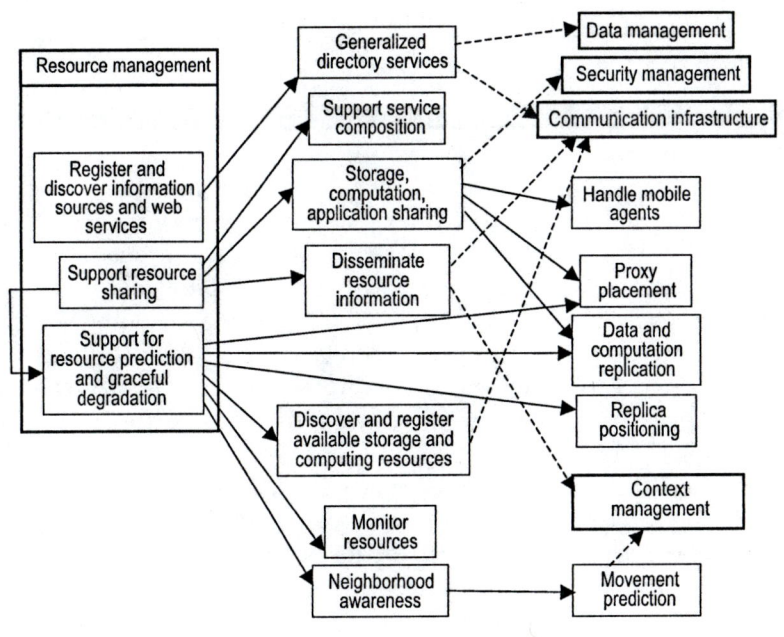

Figure 48.5 Resource management issues.

MANETs used in rescue and emergency scenarios form collaborative environments where nodes are inclined to share resources. To keep track of available resources and services and enable resource sharing, a distributed solution is necessary.

The resource manager should provide the means for the applications and frameworks to register and discover services and information sources and make such information available throughout the network. For this, information about locally available services and information sources, as well as existing directory services, might be stored in a resource sharing profile at each node. This kind of resource information might be disseminated using the capabilities of the Data Management Framework. Information about possible shareable physical resources can also be stored in the resource sharing profile, but physical resources and running services require frequent monitoring (e.g., using mechanisms provided by the operating system). A means to achieve the dissemination of frequently changing resource information is to share (accumulated) information about available resources and running services with neighboring nodes and have some strategy on how to perform the information sharing. Notice, however, that certain security concerns exist relative to information accumulation, with respect to both the trustworthiness (quality) of the information and possible misuse (access rights). Such concerns must be addressed in close cooperation with the Security Management Framework. The Resource Management Framework should facilitate remote resource access by providing mechanisms to request, negotiate, and administer remote resource access. Allowing access to remote resources involves diverse security-related measures such as access control, user traceability, and data integrity, which are all in the realm of the Security Management Framework.

Due to the dynamic nature of MANETs, sharing distributed resources based on traditional resource reservation is not a good choice because such a system is too vulnerable with respect to communication disruptions. Instead, resource reservation might be treated as a soft state that is only valid for a specified time, either for the time period the resource is needed exclusively by a process or for the time period the resources are (with high probability) accessible.

A resource management middleware for MANETs can benefit from predicting the future availability of resources, not only to establish meaningful time-outs for soft-state reservations but also to increase the availability of information and services through replication and graceful degradation, if necessary. Several approaches may be used to address the prediction of future connectivity and, through this, future access to resources and services. One approach is to analyze the location and movement of nodes with GPS information; however, GPS devices might not always work (e.g., in buildings and tunnels). An alternative is the use of history information on neighborhood relations to predict the future adjacency or non-adjacency of nodes and, thus, network partitions, as well. Such neighborhood awareness information can be used together with the resource sharing profile of the node to create a *resource-oriented context*. This context can be exploited by the applications and frameworks to achieve pervasive data storage and computation and increase service availability and quality.

To further increase availability and allow for graceful degradation, services should be able to detect and react to the imminent loss of resources. Receiving a warning regarding the absence of resources, a service might choose to downscale and eventually terminate in a controlled manner; however, if it knows that suitable alternative resources are available, a service might instead attempt to survive by exporting applications for execution in another node's environment or by using mobile agents, proxies, or the like. Service composition facilities can help applications make better use of available services by supporting the composition of complex services from more simple ones. They can also enhance service availability by reacting to a missing service by replacing it with a similar service provided by another node.

State of the Art

Most of the existing work on resource sharing in MANETs is oriented toward studies of quality of service (QoS) [42–44], bandwidth management [45], and mobility management [46]. Some of the existing works suggest the use of node mobility information to improve information accessibility in MANETs. For example, Chen et al. [47] propose a framework for a distributed data accessibility service to access multimedia data within a heterogeneous cooperative group. It is assisted by a predictive, location-based routing protocol that tries to maintain a specific set of QoS parameters. For this, they assume that nodes move in groups and follow predictable movement patterns. Each node constructs movement patterns of its neighboring nodes, relying on information such as the geographical location of nodes, movement direction and velocity, transmission range of the node, and received periodic positions broadcast from the nodes. Using movement patterns, each node participating in a transmission is capable of predicting the future location of the intermediate nodes and destination. Under similar assumptions, NonStop [48] constructs the movement patterns for a set of mobile nodes that exhibit similar mobility patterns in their movements. They are used to guarantee the continuous availability of multimedia streaming. NonStop estimates the occurrence of network partitioning to replicate data to a streaming server that has a low probability of being disconnected from a requesting client during a streaming session.

The network Media Access Control (MAC) layer is an important source of information. Hu and Johnson [49] propose a solution based on the use of congestion information to avoid network hotspots by locally monitoring the network interface transmission queue length and MAC layer behavior at each node. For optimization, MARE [50] tries to reduce bandwidth requirements by moving operations rather than data across a network; information on available resources is shared by periodically announcing their availability through distributed tuple spaces. Allia [51] uses peer-to-peer caching and policy-driven agents to facilitate cross-platform service discovery. Relevant work on replication strategies for MANETs includes strategies proposed by Hara [52]; the emphasis is on access frequency and network topology, and the proposed strategies also consider the periodic updating of replicated data.

The above approaches cannot be used directly in our scenario. For example, Chen et al. [47] and Li and Wang [48] assume that every node has a means of determining its position, which does not apply in our case. Additionally, the nodes possess highly heterogeneous capabilities, and it is reasonable to believe that not all of them have the ability to predict partitions or to participate in replications.

Security Management

Wireless communication is by nature more susceptible to eavesdropping than wired media. In most cases the data involved should not be available to third parties. At the same time, some data should not even be shared between all of the rescue personnel. Examples of sensitive data to which access should be controlled include medical and police records and confidential voice communication. Some services are intra-application or even intra-organizational within an application. In these cases, a group concept is of immense value [53,54]. Support for an organizational structure and for *ad hoc* groups across organizations is also desirable. Much of this can be achieved through the use of group keys and accessibility strategies and so is in the natural realm of security services. Applications must be allowed to require different levels or types of security policies. This may lead to incompatible policies (e.g., across organizations). Security policies may even change during the cause of an operation (e.g., due to a lack of resources). It should be ensured that such situations do not present a problem for the operation. A problem that always emerges when bringing up the subject of security is user friendliness. Security should be automated and transparent to users as much as possible, especially rescue operations where human lives are involved and there is no time to think about synchronizing network keys, for example. Security management subconcerns and issues are illustrated in Figure 48.6.

Security and Middleware

To achieve highly efficient and truly integrated and appropriate security services, security must be considered from an early stage in the development of applications, middleware services, and all their components. In our case, this means integrating security with all the other concerns. In addition to basic security requirements, such as the authentication of nodes and message integrity, the Security Management Framework provides direct services to the aforementioned components. For example, the resource manager might need to evaluate the trustworthiness of data received, the communication infrastructure might have to implement restricted subscription to certain events, etc. It is of great importance to provide simple, yet efficient, protocols for such purposes.

The main characteristics of the traditional layered architecture [55] is its strict separation of layers, preventing the unauthorized leaking of data between different layers. Although the layered approach provides a high level of security, it lacks flexibility and thus places many limitations on both development and usage — for example, how to establish an IPsec security association dynamically between a fireman and a policeman and how to relate that to trust as defined in the system at the application layer.

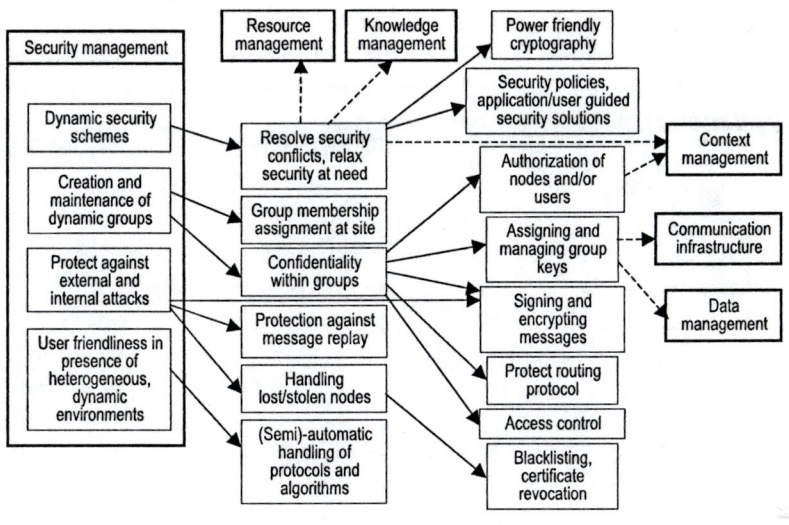

Figure 48.6 Security management issues.

The limitations of the layered approach are most notable in a heterogeneous, dynamic environment, making an adaptable and cross-layer middleware solution a logical alternative. Extension techniques, such as reflection [56,57], allow middleware services to adapt to the environment and target some of the challenges faced. Some solutions, such as Open ORB 2 [58] and ReMMoC [59], are lightweight examples of such architectures, suitable for mobile devices. The programming language OBOL [60] is a step toward solving security issues in middleware-based architectures. Aspect-oriented programming [61] allows the incorporation of cross-concerns in the development phase and might therefore address some of the challenges faced. The mentioned approaches by themselves do not present complete solutions for the integration of security with the other concerns but might provide a good starting point.

The Security Management Framework makes use of services from the other frameworks. Some examples are use of the communication infrastructure for the distribution of keys, data management for the storage of keys and certificates, context management and resource management to get information on changes in the environment, and knowledge management to possibly analyze security conflicts.

State of the Art

Possible security and privacy attacks can be roughly divided into external and internal [62]. Most of the problems related to external attacks in MANETs can be solved relatively easily by means of standard encryption and digital signature techniques; however, denial-of-service attacks on lower layers that cause battery drain or network congestion, such as jamming or flooding, cannot be handled at the middleware layer. Internal attacks are more difficult to detect and a much bigger threat because such attacks come from nodes that have previously been authenticated but later either lost or stolen.

Authorized nodes have to be distinguished from foreign nodes, so some form of authentication has to be ensured [63,64]. This can be achieved through use of public key infrastructures that allow nodes to authenticate each other without the need for an online server. Capkun et al. [65] proposed a fully self-organized public key management system that does not rely on trusted authorities; however, in contrast to pure *ad hoc* networks, in our scenario the *a priori* phase can be used to preinstall certificates with a common certificate authority on the devices to ensure the privacy of sensitive data. The advantages are twofold: The data in the network is rendered more secure, and establishing trust becomes much more efficient.

Whether the certificates on the nodes will identify devices or actual users handling them (who would then present the certificate to the device by means of a token such as a smart card) is another issue. The decision made impacts the way lost and stolen nodes are handled (e.g., revoking certificates and blacklisting such nodes). One can cryptographically bind a node's IP address to its certificate [66] to prevent nodes from impersonating others. Because no central authority exists, a way to decide which node, person, or role/rank can perform the task of revoking certificates is necessary. It might be safe to assume that the devices of the leaders will be physically better protected than the others, making such devices a possible candidate.

In MANETs, protecting the routing protocol is of particular importance, as incorrect routing messages could cause the network to function improperly [62]. To sign and verify messages, a shared network key can be used. It is highly probable, though, especially during the bootstrap procedure, that several networks with different keys will form, causing network key inconsistency. Solutions based on the Diffie–Hellman key exchange [67] do not have this problem due to the very nature of the algorithm. In such protocols [68–71], all nodes contribute to the final shared key, causing rekeying every time a node joins or leaves the group. Most of these protocols rely on some kind of hierarchy and a group manager to deploy and maintain shared keys. In the presence of frequent network partitioning and merging, however, the use of contributory protocols would cause significant computational and bandwidth costs which cannot be afforded, and centralized key management is inappropriate. Also, in such protocols a fully working routing infrastructure has to be established prior to key exchanges. Because the routing protocol is one of the main things to be protected, this requirement is a major drawback; however, these protocols might be very useful for the creation of dynamic groups or teams.

A characteristic of key predistribution protocols is that a group of nodes can compute a shared key out of predistributed sets of keys present on each node. These sets of keys are either given by a trusted entity before the nodes come to the scene [72,73] or chosen and managed by the nodes themselves, as is done in the Distributed Key Predistribution Scheme (DKPS) [74]. Designed for emergency and rescue operations, the SKiMPy protocol [75] makes use of preinstalled certificates to perform authentication of nodes and to produce shared keys while aiming at keeping the computational and bandwidth costs of the protocol low. Nevertheless, having signed messages does not necessarily mean they are valid. A malicious node could listen to all traffic, save certain messages locally, and decide to replay them at a later point, easily causing disruption of the network; therefore, protection from replayed messages (such as time-stamp checking in the OLSR security plug-in [76]) is also of extreme importance.

Including the devices of bystanders in the network might be a desired feature, but a challenging issue to solve with regard to security. This would introduce an even wider heterogeneity of devices and operating systems, making the compatibility problem a major issue. In addition, such devices would not have proper application support nor would they carry valid credentials to be able to join the network. The latter problem could be solved by means of leaders having the ability to install necessary software packages and issue certificates valid only for a short period of time and giving such devices restricted access.

With regard to bandwidth, messages carrying certificates could become large if several levels in the certificate chain are present, and as such their presence in the air should be reduced to a minimum. Authentication processes using asymmetric cryptography are computationally very expensive, which in our case becomes a problem with regard to battery power. Special care has to be taken when choosing algorithms for both symmetric and asymmetric cryptography [77]; for example, elliptic curve cryptography [78,79] is a good alternative to RSA [80] with regard to power consumption, offering equivalent security with smaller key sizes and thus faster computation, lower memory consumption, and lower bandwidth usage [81].

Conclusions and Open Issues

By analyzing the current state of research and applications for MANETs, we have identified a strong need for middleware services to facilitate efficient development of applications over MANETs. For MANETs used in rescue and emergency scenarios, we have performed a thorough analysis, including studies of today's approach of rescue teams to collaborate in operations. Through this analysis we have distinguished ten requirements and goals for middleware services for MANETs, and we have identified six middleware *concerns* that together constitute a foundation for a middleware framework covering the required services for MANETs in rescue scenarios. The analysis has resulted in the following insights: Information sharing is a key element for successful collaboration, knowledge management must address distribution and the various data representations and models being used, and information and resources have to be protected with security mechanisms. Furthermore, a decentralized solution is needed that builds on an asynchronous event-notification/message-passing system and that uses sufficient redundancy to reach high availability.

Valuable sources of requirements can be found within the areas of mobile distributed systems and ubiquitous and mobile computing. Mascolo et al. [82] provide a categorization of *ad hoc* systems and review various middleware systems designed to support mobility. Modahl et al. [83] introduce a five-class infrastructure taxonomy based on orthogonal functionalities of the most commonly occurring ubiquitous computing subsystems. The taxonomy is comparable to our six concerns but places registration and discovery (service availability) and computation sharing (resource sharing) in different taxonomy classes, whereas we (although definitely recognizing them as different requirements) place both within the Resource Management Framework. Grace et al. [59] and Sørensen et al. [84] advocate the importance of providing *ad hoc* networking middleware that is flexible and adaptable and propose reflection as an appropriate technology to accomplish adaptability and self-configuration. Grace et al. [59] and Mascolo et al. [82] provide a number of references to related work, although none of them is concerned with rescue scenarios. Plagemann et al. [85] discuss application requirements for rescue scenarios and propose an architecture that consists of five main building blocks: knowledge management, a local and a distributed event notification service, resource management, and security and privacy management.

Little work can be found that covers the entire problem area, and (with the exception of Plagemann et al. [85]) they barely (if at all) touch upon security issues; at most, the need for security is noted but no details are provided. We believe that security should be addressed in particular, even if it belongs to any standard set of general critical-system engineering requirements, because of the particular challenges posed by security in MANETs. To the extent that knowledge management issues are recognized in work on middleware platforms for MANETs, they are related to the presence of context awareness, through context ontologies and reasoning.

On the other hand, state-of-the-art information and knowledge management solutions provide for sharing and integration of information and content but do not consider challenges posed by MANETs. Our claim is that dynamic environments will benefit from combining the middleware infrastructure provided by MANETs with solutions provided by information and knowledge management for information sharing.

An alternative approach to realizing some of the concerns can be found in the Limone coordination model [86]. It is based on Linda tuple spaces [87] but is tailored toward MANETs. It supports context awareness by being reactive to tuple space changes and asynchronous communication by using timeouts, and it can also accomplish message passing and data sharing through the appropriate use of tuple spaces. We believe, however, that the notion of tuple spaces, although a middleware issue, is better implemented on top of a framework as sketched in this chapter, thus providing application developers with a particular application programming model and environment that can coexist with other computational models, such as publish–subscribe models and object-oriented models (e.g., the sentient object programming model [84]).

Acknowledgments

This work was funded by the Norwegian Research Council in the IKT-2010 Program, Project No. 152929/431, and by a sabbatical year grant from the Faculty of Mathematics and Natural Sciences, University of Oslo.

References

[1] Stuckenschmidt, H. and van Harmelen, F., *Information Sharing on the Semantic Web*, Springer-Verlag, Heidelberg, 2005.

[2] Wache, H. et al., Ontology-based integration of information: a survey of existing approaches, in *Proc. of Int. Joint Conf. on Artificial Intelligence (IJCAI'01), Workshop on Ontologies and Information Sharing*, Seattle, WA, August 4–10, 2001, pp. 108–117.

[3] Fensel, D. et al., OIL: an ontology infrastructure for the Semantic Web, *IEEE Intelligent Syst.*, 16(2), 38–45, 2001.

[4] Schulz, S. et al., Towards trust-based knowledge management in mobile communities, in *Agent-Mediated Knowledge Management*, Technical Report SS-03-01, Stanford University, Stanford, CA, 2003.

[5] *Informasjonsteknologi for helse og velferd* [in Norwegian], http://www.kith.no/.

[6] *European Standardization of Health Informatics*, http://www.centc251.org/.

[7] KITH rapport nr R25/02 [in Norwegian], http://www.kith.no/arkiv/rapporter/rammeverk-v09-testing-2.pdf.

[8] OASIS, http://www.oasis-open.org/.

[9] Schwotzer, T., Context driven spontaneous knowledge exchange, in *Proc. of the First German Workshop on Experience Management (GWEM'02)*, Berlin, Germany, 2002, pp. 131–138.

[10] Perich, F. et al., Profile driven data management for pervasive environments, in *Proc. of the 13th Int. Workshop on Database and Expert Systems Applications (DEXA'02)*, Aix-en-Provence, France, September 2–6, 2002, pp. 361–370.

[11] Pfoser, D., Pitoura, E., and Tryfona, N., Metadata modeling in a global computing environment, in *Proc. of the 10th ACM Int. Symp. on Advances in Geographic Information Systems*, McLean, VA, 2002, pp. 68–73.

[12] Chen, H., Finin, T., and Joshi, A., An ontology for context-aware pervasive computing environments, *Knowledge Eng. Rev.*, 18(3), 197–207, 2003.

[13] OASIS news, http://www.oasis-open.org/news/oasis_news_03_29_04.php.

[14] Davies, J., Fensel, D., and Van Harmelen, F., Eds., *Towards the Semantic Web: Ontology-Driven Knowledge Management*, John Wiley & Sons, New York, 2003.

[15] Fensel, D. et al., On-to-knowledge: ontology-based knowledge management, *IEEE Comput.*, 35(11), 56–59, 2002.

[16] XML and Perl: Now Let's Start Digging, http://www.informit.com/articles/article.asp?p=30010.

[17] *A Survey of APIs and Techniques for Processing XML*, http://www.xml.com/pub/a/2003/07/09/xmlapis.html.

[18] Schwotzer, T. and Geihs, K., Shark: a system for management, synchronization and exchange of knowledge in mobile user groups, in *Proc. of the 2nd Int. Conf. on Knowledge Management* (I-KNOW'02); *J. Universal Comput. Sci.*, 8(6), 644–651, 2002.

[19] Schwotzer, T. and Geihs, K., Mobiles verteiltes Wissen: Modellierung, Speicherung und Austausch [in German], *Datenbank-Spektrum* 5, 30–39, 2003.

[20] Pitoura, E. et al., DBGlobe: a service-oriented P2P system for global computing, *ACM SIGMOD Rec.*, 32(3), 77–82, 2003.

[21] Fontijn, W. and Boncz, P., AmbientDB: P2P data management middleware for ambient intelligence, in *Proc. of the Second IEEE Int. Conf. on Pervasive Computing and Communications (PerCom'04), Workshop on Middleware Support for Pervasive Computing (PerWare'04)*, Orlando, FL, March 14, 2004, pp. 203–207.

[22] Boncz, P.A. and Treijtel, C., *AmbientDB: Relational Query Processing in a P2P Network*, Technical Report INS-R0306, CWI, Amsterdam, The Netherlands, 2003.

[23] Anagnostopoulos, C., Tsounis, A., and Hadjiefthymiades, S., Context awareness in mobile computing environments: a survey, in *Proc. of Mobile eConference*, August, 2004.

[24] Xu, C. et al., Cabot: on the ontology for the middleware support of context-aware pervasive applications, in *Proc. of IFIP NPC Workshop on Building Intelligent Sensor Networks* (BISON 2004), Vol. 3222, Lecture Notes in Computer Science, Springer-Verlag, Heidelberg, 2004, pp. 568–575.

[25] Rios, D. et al., Using ontologies for modeling context-aware services platforms, in *Proc. of ACM Conf. on Object-Oriented Programming Systems, Languages, and Applications (OOPSLA 2003), Workshop on Ontologies to Complement Software Architectures*, Anaheim, CA, October, 2003.

[26] Eugster, P. et al., The many faces of publish/subscribe, *ACM Comp. Surv.*, 35(2), 114–131, 2003.

[27] Rosenblum, D.S. and Wolf, A.L., A design framework for Internet-scale event observation and notification, in *Proc. of the 6th European Software Engineering Conf.*, Zurich, Switzerland, September 22–25, 1997, pp. 344–360.

[28] Carzaniga, A., Rosenblum, D.S., and Wolf, A.L., Design and evaluation of a wide-area event notification service, *ACM Trans. Comp. Sys.*, 19(3), 332–383, 2001.

[29] Cugola, G., Nitto, E.D., and Fuggetta, A., The JEDI event-based infrastructure and its applications to the development of the OPSS WFMS, *IEEE Trans. Software Eng.*, 27, 827–850, 2001.

[30] Cugola, G. and Nitto, E.D., Using a publish/subscribe middleware to support mobile computing, in *Proc. of the Workshop on Middleware for Mobile Computing*, Heidelberg, Germany, November 16, 2001.

[31] Fiege, L. et al., Dealing with uncertainty in mobile publish/subscribe middleware, in *Proc. of the First Int. Workshop on Middleware for Pervasive and Ad Hoc Computing (MPAC'03)*, Rio de Janeiro, Brazil, June, 2003.

[32] Cilia, M. et al., Looking into the past: enhancing mobile publish/subscribe middleware, in *Proc. of Int. Workshop on Distributed Event-Based Systems (DEBS'03)*, San Diego, CA, June, 2003, pp. 1–8.

[33] TIBCO, TIB/Rendezvous Concepts, http://www.tibco.com.

[34] Segall, B. et al., Content based routing with Elvin4, in *Proc. AUUG2K*, Canberra, Australia, June 28–30, 2000.

[35] Yoneki, E., Pronto: mobile gateway with publish-subscribe paradigm over wireless networks, Middleware'03 Work in Progress Session, *IEEE Distributed Syst. Online*, 4(5), 2003.

[36] Huang, Y. and Garcia-Molina, H., Publish/subscribe in a mobile environment, in *Proc. of the 2nd ACM Int. Workshop on Data Engineering for Wireless and Mobile Access (MobiDe'01)*, Santa Barbara, CA, May, 2001, pp. 27–34.

[37] Meier, R. and Cahill, V., Steam: event-based middleware for wireless *ad hoc* networks, in *Proc. of Int. Workshop on Distributed Event-Based Systems (DEBS'02)*, Vienna, Austria, July 2–3, 2002, pp. 639–644.

[38] Musolesi, M., Mascolo C., and Hailes, S., EMMA: epidemic messaging middleware for *ad hoc* networks, *Pers. Ubiquitous Comput. J.*, 10(1), 28–36, 2006.

[39] Vollset, E., Ingham, D., and Ezhilchelvan, P., JMS on mobile *ad hoc* networks, in *Proc. of Personal Wireless Communications (PWC'03)*, Venice, Italy, September, 2003, pp. 40–52.

[40] Cugola, G., Picco, G., and Murphy, A., Towards dynamic reconfiguration of distributed publish-subscribe middleware, in *Proc. of the 3rd Int. Workshop on Software Engineering and Middleware* (SEM'02), Orlando, FL, May 20–21, 2002, pp. 187–202.

[41] Skjelsvik, K.S., Goebel, V., and Plagemann, T., Distributed event notification service for mobile *ad hoc* networks, *IEEE Distributed Syst. Online*, 5(8), 2004.

[42] Phanse, K.S., DaSilva, L.A., and Midkiff, S.F., Design and demonstration of policy-based management in a multi-hop *ad hoc* network, *Ad Hoc Networks*, 3(3), 389–401, 2005.

[43] Cardei, I. et al., Resource management for *ad hoc* wireless networks with cluster organization, *J. Cluster Comput. Internet*, 7(1), 91–103, 2004.

[44] Lee, S.-B. et al., INSIGNIA: an IP-based quality of service framework for mobile *ad hoc* networks, *J. Parallel Distributed Comput.*, 60(4), 374–406, 2000.

[45] Ahn, K.-M. and Kim, S., Optimal bandwidth allocation for bandwidth adaptation in wireless multimedia networks, *Comput. Operations Res.*, 30(13), 1917–1929, 2003.

[46] Pei, G. and Gerla, M., Mobility management for hierarchical wireless networks, *Mobile Networks Appl. Arch.*, 6(4), 331–337, 2001.

[47] Chen, K., Shah, S.H., and Nahrstedt, K., Cross-layer design for data accessibility in mobile *ad hoc* networks, *J. Wireless Personal Commun.*, 21(1), 49–76, 2002.

[48] Li, B. and Wang, K.H., NonStop: continuous multimedia streaming in wireless *ad hoc* networks with node mobility, *IEEE J. Selected Areas Commun.*, 21(10), 1627–1641, 2003.

[49] Hu, Y.C. and Johnson, D.B., Exploiting MAC layer information in higher layer protocols in multihop wireless *ad hoc* networks, in *Proc. of the 24th Int. Conf. on Distributed Computing Systems (ICDCS'04)*, Tokyo, Japan, March, 2004, pp. 301–310.

[50] Storey, M., Blair, G., and Friday, A., MARE: resource discovery and configuration in *ad hoc* networks, *Mobile Networks Appl.*, Kluwer Academic Publishers, 7(5), 377–387, 2002.

[51] Ratsimor, O. et al., Allia: alliance-based service discovery for *ad hoc* environments, in *Proc. of the 2nd ACM Mobicom Int. Workshop on Mobile Commerce* (WMC'02), Atlanta, GA, September, 2002, pp. 1–9.

[52] Hara, T., Replica allocation methods in *ad hoc* networks with data update, *J. Mobile Networks Appl. (MONET)*, 8(4), 343–354, 2003.

[53] Jung, E., Liu, X.-Y.A., and Gouda, M.G., Key bundles and parcels: secure communication in many groups, in *Proc. of the 5th Int. Workshop on Networked Group Communications (NGC'03)*, Vol. 2816, Lecture Notes in Computer Science, Springer-Verlag, Heidelberg, 2003, pp. 119–130.

[54] Zhou, D. and Wu, J., Survivable multi-level *ad hoc* group operations, in *Proc. IEEE Int. Conf. on Distributed Computing Systems (ICDCS'03)*, Providence, RI, May, 2003, pp. 70–75.

[55] Gutmann, P., The design of a cryptographic security architecture, in *Proc. of the 8th USENIX Security Symp.*, Washington D.C., August, 1999.

[56] Maes, P., Concepts and experiments in computational reflection, in *Proc. of ACM Conf. on Object-Oriented Programming Systems, Languages, and Applications (OOPSLA 1987)*, Orlando, FL, October 4–8, 1987, pp. 147–155.

[57] Blair, G. S. et al., An architecture for next generation middleware, in *Proc. of IFIP Int. Conf. on Distributed Systems Platforms and Open Distributed Processing (Middleware'98)*, The Lake District, England, September, 1998.

[58] Blair, G. S. et al., The design and implementation of Open ORB 2, *IEEE Distributed Syst. Online*, 2(6), 2001.

[59] Grace, P., Blair, G.S., and Samuel, S., *Interoperating with Services in a Mobile Environment*, Technical Report (MPG-03-01), Lancaster University, Lancaster, U.K., 2003.

[60] Andersen, A. et al., Reflective middleware and security: OOPP meets Obol, in *Proc. of the 2nd Workshop on Reflective and Adaptive Middleware*, Rio de Janeiro, Brazil, June 17, 2003, pp. 100–104.

[61] Kiczales, G. et al., Aspect-oriented programming, in *Proc. of the 11th European Conf. on Object-Oriented Programming (ECOOP'97)*, Jyväskylä, Finland, June, 1997, pp. 220–242.

[62] Kärpijoki, V., Security in *ad hoc* networks, in *Seminars on Network Security*, Telecommunications Software and Multimedia Laboratory, Helsinki University of Technology, Helsinki, Finland, 2000.

[63] Balfanz, D. et al., Talking to strangers: authentication in *ad hoc* wireless networks, in *Proc. of the 9th Annual Network and Distributed System Security Symp. (NDSS'02)*, San Diego, CA, February 6–8, 2002.

[64] Stajano, F. and Anderson, R., The resurrecting duckling: security issues for *ad hoc* wireless networks, in *Proc 7th Int. Workshop on Security Protocols*, Vol. 1796, Lecture Notes in Computer Science, Springer-Verlag, Heidelberg, 1999, pp. 172–194.

[65] Capkun, S., Buttyán, L., and Hubaux, J.-P., Self-organized public-key management for mobile *ad hoc* networks, *IEEE Trans. Mobile Comput.*, 2(1), 52–64, 2003.

[66] Montenegro, G. and Castelluccia, C., Statistically unique and cryptographically verifiable (SUCV) identifiers and addresses, in *Proc. of the 9th Annual Network and Distributed System Security Symp. (NDSS'02)*, San Diego, CA, February 6–8, 2002.

[67] Diffie, W. and Hellman, M.E., New directions in cryptography, *IEEE Trans. Inform. Theory*, 22(6), 644–654, 1976.

[68] Becker, K. and Wille, U., Communication complexity of group key distribution, in *Proc. of the 5th ACM Conf. on Computer and Communications Security (CCS'98)*, San Francisco, CA, November 3–5, 1998, pp. 1–6.

[69] Bresson, E., Chevassut, O., and Pointcheval, D., Provably authenticated group Diffie–Hellman key exchange: the dynamic case, in *Proc. of the 7th Int. Conf. on the Theory and Application of Cryptology and Information Security (ASIACRYPT '01)*, Vol. 2248, Lecture Notes in Computer Science, Springer-Verlag, Heidelberg, 2001, pp. 290–309.

[70] Di Pietro, R., Mancini, L.V., and Jajodia, S., Efficient and secure keys management for wireless mobile communications, in *Proc. of the 2nd ACM Int. Workshop on Principles of Mobile Computing (POMC'02)*, Toulouse, France, October, 2002, pp. 66–73.

[71] Steiner, M., Tsudik, G., and Waidner, M., CLIQUES: a new approach to group key agreement, in *Proc. IEEE Int. Conf. on Distributed Computing Systems (ICDCS'98)*, Amsterdam, The Netherlands, May, 1998, pp. 380–387.

[72] Blom, R., An optimal class of symmetric key generation system, in *Advances in Cryptology (Eurocrypt'84)*, Vol. 209, Lecture Notes in Computer Science, Springer-Verlag, Heidelberg, 1985, pp. 335–338.

[73] Matsumoto, T. and Imai, H., On the key predistribution systems: a practical solution to the key distribution problem, in *Advances in Cryptology (CRYPTO'87)*, Vol. 293, Lecture Notes in Computer Science, Springer-Verlag, Heidelberg, 1988, pp. 185–193.

[74] Chan, A.C.-F., Distributed symmetric key management for mobile *ad hoc* networks, in *Proc. of IEEE INFOCOM'04*, Hong Kong, March, 2004.

[75] Puzar, M. et al., SKiMPy: a simple key management protocol for MANETs in emergency and rescue operations, in *Proc. of the 2nd European Workshop on Security and Privacy in Ad Hoc and Sensor Networks (ESAS 2005)*, Vol. 3813, Lecture Notes in Computer Science, Springer-Verlag, Heidelberg, 2005, pp. 14–26.

[76] Hafslund A. et al., Secure extension to the OLSR protocol, in *Proc. of OLSR Interop & Workshop*, San Diego, CA, August 6–7, 2004.

[77] Potlapally, N.R. et al., Analyzing the energy consumption of security protocols, in *Proc. of IEEE Int. Symp. on Low Power Electronics and Design*, Long Beach, CA, August, 2003, pp. 30–35.

[78] Aydos, M., Sunar, B., and Koç, Ç.K., An elliptic curve cryptography based authentication and key agreement protocol for wireless communication, in *Proc. of the 2nd Int. Workshop on Discrete Algorithms and Methods for Mobile Computing and Communications*, Dallas, TX, October 30, 1998.

[79] Aydos, M., Sunar, B., and Koç, Ç.K., Implementing network security protocols based on elliptic curve cryptography, in *Proc. of the 4th Symp. on Computer Networks*, Instanbul, Turkey, May, 1999, p. 130–139.

[80] Rivest, R.L., Shamir, A., and Adelman, L.M., A method for obtaining digital signatures and public-key cryptosystems, *Commun. ACM*, 21(2), 120–126, 1978.

[81] Gupta, V. et al., Performance analysis of elliptic curve cryptography for SSL, in *Proc. of the 8th ACM/IEEE Int. Conf. on Mobile Computing and Networking (MOBICOM'02)*, ACM Workshop on Wireless Security (WiSe), Atlanta, GA, September, 2002, pp. 87–94.

[82] Mascolo, C., Capra, L., and Emmerich, W., Mobile computing middleware, in *Advanced Lectures on Networking*, Vol. 2497, Lecture Notes in Computer Science, Springer-Verlag, Heidelberg, 2002, pp. 20–58.

[83] Modahl, M. et al., Toward a standard ubiquitous computing framework, in *Proc. of the ACM 2nd Workshop on Middleware for Pervasive and Ad Hoc Computing (MPAC'04)*, Toronto, Canada, October 18–22, 2004, pp. 135–139.

[84] Sørensen, C.-F. et al., A context-aware middleware for applications in mobile *ad hoc* environments, in *Proc. of the ACM 2nd Workshop on Middleware for Pervasive and Ad Hoc Computing (MPAC'04)*, Toronto, Canada, October 18–22, 2004, pp. 107–110.

[85] Plagemann, T. et al., Middleware services for information sharing in mobile ad-hoc networks: challenges and approach, in *Proc. of IFIP World Computer Conf., Workshop on Challenges of Mobility*, Vol. 169, International Federation for Information Processing (IFIP), Laxenburg, Austria, 2005, pp. 225–236

[86] Fok, C.-L., Roman, G.-C., and Hackmann, G., A lightweight coordination middleware for mobile computing, in *Proc. of the Sixth Int. Conf. on Coordination Models and Languages*, Vol. 2949, Lecture Notes in Computer Science, Springer-Verlag, Heidelberg, 2004, pp. 135–151.

[87] Gelernter, D., Generative communication in Linda, *ACM Trans. Programming Languages Syst.*, 7(1), 80–112, 1985.

Index